The Mechatronics Handbook
Second Edition

Edited by
Robert H. Bishop

Mechatronic Systems, Sensors, and Actuators
Fundamentals and Modeling

Mechatronic System Control, Logic, and Data Acquisition

The Electrical Engineering Handbook Series

Series Editor

Richard C. Dorf

University of California, Davis

Titles Included in the Series

The Handbook of Ad Hoc Wireless Networks, Mohammad Ilyas
The Avionics Handbook, Second Edition, Cary R. Spitzer
The Biomedical Engineering Handbook, Third Edition, Joseph D. Bronzino
The Circuits and Filters Handbook, Second Edition, Wai-Kai Chen
The Communications Handbook, Second Edition, Jerry Gibson
The Computer Engineering Handbook, Second Edition, Vojin G. Oklobdzija
The Control Handbook, William S. Levine
The CRC Handbook of Engineering Tables, Richard C. Dorf
The Digital Avionics Handbook, Second Edition Cary R. Spitzer
The Digital Signal Processing Handbook, Vijay K. Madisetti and Douglas Williams
The Electrical Engineering Handbook, Second Edition, Richard C. Dorf
The Electric Power Engineering Handbook, Second Edition, Leonard L. Grigsby
The Electronics Handbook, Second Edition, Jerry C. Whitaker
The Engineering Handbook, Third Edition, Richard C. Dorf
The Handbook of Formulas and Tables for Signal Processing, Alexander D. Poularikas
The Handbook of Nanoscience, Engineering, and Technology, Second Edition William A. Goddard, III, Donald W. Brenner, Sergey E. Lyshevski, and Gerald J. Iafrate
The Handbook of Optical Communication Networks, Mohammad Ilyas and Hussein T. Mouftah
The Industrial Electronics Handbook, J. David Irwin
The Measurement, Instrumentation, and Sensors Handbook, John G. Webster
The Mechanical Systems Design Handbook, Osita D.I. Nwokah and Yidirim Hurmuzlu
The Mechatronics Handbook, Second Edition, Robert H. Bishop
The Mobile Communications Handbook, Second Edition, Jerry D. Gibson
The Ocean Engineering Handbook, Ferial El-Hawary
The RF and Microwave Handbook, Second Edition, Mike Golio
The Technology Management Handbook, Richard C. Dorf
The Transforms and Applications Handbook, Second Edition, Alexander D. Poularikas
The VLSI Handbook, Second Edition, Wai-Kai Chen

MECHATRONIC SYSTEM CONTROL, LOGIC, AND DATA ACQUISITION

Edited by

Robert H. Bishop
The University of Texas at Austin
U.S.A.

CRC Press
Taylor & Francis Group
Boca Raton London New York

CRC Press is an imprint of the
Taylor & Francis Group, an **informa** business

CRC Press
Taylor & Francis Group
6000 Broken Sound Parkway NW, Suite 300
Boca Raton, FL 33487-2742

Printed in the United States of America on acid-free paper
10 9 8 7 6 5 4 3 2 1

International Standard Book Number-13: 978-0-8493-9260-3 (Hardcover)

Library of Congress Cataloging-in-Publication Data

Mechatronic system control, logic, and data acquisition / editor, Robert H. Bishop.
p. cm.
"A CRC title."
Includes bibliographical references and index.
ISBN 978-0-8493-9260-3 (alk. paper)
1. Programmable controllers. 2. Mechatronics. I. Bishop, Robert H., 1957-

TJ223.P76M43 2008
621--dc22 2007025442

Visit the Taylor & Francis Web site at
http://www.taylorandfrancis.com

and the CRC Press Web site at
http://www.crcpress.com

Preface

According to the original definition of mechatronics proposed by the Yasakawa Electric Company and the definitions that have appeared since, many of the engineering products designed and manufactured in the last 30 years integrating mechanical and electrical systems can be classified as *mechatronic systems.* Yet many of the engineers and researchers responsible for those products were never formally trained in mechatronics per se. The *Mechatronics Handbook, 2nd Edition* can serve as a reference resource for those very same design engineers to help connect their everyday experience in design with the vibrant field of mechatronics.

The *Handbook of Mechatronics* was originally a single-volume reference book offering a thorough coverage of the field of mechatronics. With the need to present new material covering the rapid changes in technology, especially in the area of computers and software, the single-volume reference book quickly became unwieldy. There is too much material to cover in a single book. The topical coverage in the *Mechatronics Handbook, 2nd Edition* is presented here in two books covering *Mechatronic Systems, Sensors, and Actuators: Fundamentals and Modeling* and *Mechatronic System Control, Logic, and Data Acquisition.* These two books are intended for use in research and development departments in academia, government, and industry, and as a reference source in university libraries. They can also be used as a resource for scholars interested in understanding and explaining the engineering design process.

As the historical divisions between the various branches of engineering and computer science become less clearly defined, we may well find that the mechatronics specialty provides a roadmap for nontraditional engineering students studying within the traditional structure of most engineering colleges. It is evident that there is an expansion of mechatronics laboratories and classes in the university environment worldwide. This fact is reflected in the list of contributors to these books, including an international group of academicians and engineers representing 13 countries. It is hoped that the books comprising the *Mechatronics Handbook, 2nd Edition* can serve the world community as the definitive reference source in mechatronics.

Organization

The *Mechatronics Handbook, 2nd Edition* is a collection of 56 chapters covering the key elements of mechatronics:

a. Physical Systems Modeling
b. Sensors and Actuators
c. Signals and Systems
d. Computers and Logic Systems
e. Software and Data Acquisition

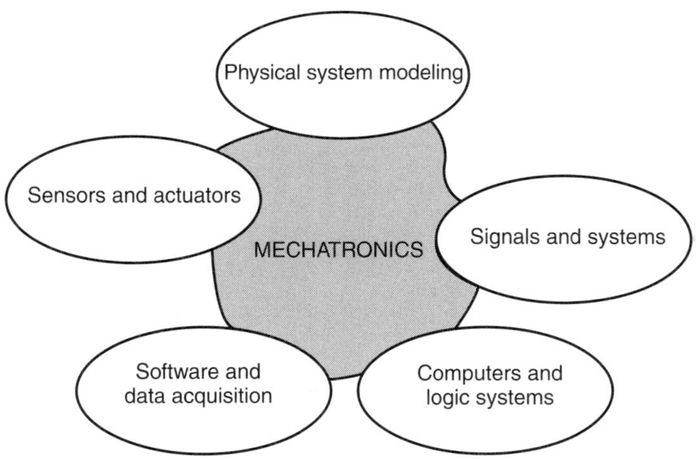

Key Elements of Mechatronics

Mechatronic System Control, Logic, and Data Acquisition

An overview of signals and system control, computers, logic systems, software, and data acquisition is presented in this book. These are most rapidly changing areas of mechatronics.

Section I—Mechatronic System Control

Since there is a significant body of readily available material to the reader on the general subject of signals and systems, there is no overriding need to repeat that material here. Instead, the goal of this book is to present the relevant aspects of signals and systems of special importance to the study of mechatronics. The book begins with chapters on the role of control in mechatronics and on the role of modeling in mechatronic design. These chapters set the stage for the more fundamental discussions on signals and systems comprising the bulk of the material in this section. Modern aspects of control design using optimization techniques from H_2 theory, adaptive and nonlinear control, neural networks, and fuzzy systems are also included as they play an important role in modern engineering system design. The book includes chapters on design optimization for mechatronic systems, and real-time monitoring and control. The chapters, listed in order of appearance, are

1. The Role of Controls in Mechatronics
2. The Role of Modeling in Mechatronics Design
3. Signals and Systems
    - 3.1 Continuous- and Discrete-Time Signals
    - 3.2 *z* Transforms and Digital Systems
    - 3.3 Continuous- and Discrete-Time State-Space Models
    - 3.4 Transfer Functions and Laplace Transforms
4. State Space Analysis and System Properties
5. Response of Dynamic Systems
6. The Root Locus Method
7. Frequency Response Methods
8. Kalman Filters as Dynamic System State Observers
9. Digital Signal Processing for Mechatronic Applications
10. Control System Design via H^2 Optimization
11. Adaptive and Nonlinear Control Design
12. Neural Networks and Fuzzy Systems
13. Advanced Control of an Electrohydraulic Axis
14. Design Optimization of Mechatronic Systems
15. Motion Control
16. Real-Time Monitoring and Control
17. Micromechatronics and Microelectromechanical Motion Devices

Section II—Computers and Logic Systems

The development of the computer, and then the microcomputer, embedded computers, and associated information technologies and software advances, has impacted the world in a profound manner. This is especially true in mechatronics where the integration of computers with electromechanical systems has led to a new generation of smart products. The future is filled with promise of better and more intelligent products resulting from continued improvements in computer technology and software engineering. In this section, the focus is on computer hardware and associated issues of logic, communication, networking, architecture, fault analysis, embedded computers, and programmable logic controllers. The chapters, listed in order of appearance, are

18. Introduction to Computers and Logic Systems
19. Digital Logic Concepts and Combinational Logic Design
20. System Interfaces
21. Communications and Computer Networks
22. Fault Analysis in Mechatronic Systems
23. Logic System Design

24. Architecture
25. Control with Embedded Computers and Programmable Logic Controllers
26. Graphical System Design for Embedded Systems
27. Field-Programmable Gate Arrays
28. Graphical Programming for Field-Programmable Gate Arrays: Applications in Control and Mechatronics

Section III—Software and Data Acquisition

Given that computers play a central role in modern mechatronics products, it is very important to understand how data is acquired and how it makes its way into the computer for processing and logging. The final section of this book is devoted to the issues surrounding computer software and data acquisition. The chapters, listed in order of appearance, are

29. Introduction to Data Acquisition
30. Measurement Techniques: Sensors and Transducers
31. A/D and D/A Conversion
32. Signal Conditioning
33. Virtual Instrumentation Systems
34. Software Design and Development
35. Data Recording and Logging

Acknowledgments

I wish to express my heartfelt thanks to all the contributing authors. Taking time in otherwise busy and hectic schedules to author the excellent chapters appearing in this book is much appreciated.

This handbook is a result of a collaborative effort expertly managed by CRC Press. My thanks to the editorial and production staff:

Nora Konopka	Acquisitions Editor
Theresa Delforn	Project Coordinator
Joette Lynch	Project Editor

Thanks to my friend and collaborator Professor Richard C. Dorf for his continued support and guidance. And finally, a special thanks to Lynda Bishop for managing the incoming and outgoing draft manuscripts. Her organizational skills were invaluable to this project.

Editor

Robert H. Bishop is a professor of aerospace engineering and engineering mechanics at The University of Texas at Austin and holds the Joe J. King Professorship. He received his BS and MS from Texas A&M University in aerospace engineering, and his PhD from Rice University in electrical and computer engineering. Prior to coming to The University of Texas at Austin, he was a member of the technical staff at the MIT Charles Stark Draper Laboratory. Dr. Bishop is a specialist in the area of planetary exploration with emphasis on spacecraft guidance, navigation and control. He is a fellow of the American Institute of Aeronautics and Astronautics. Currently, Dr. Bishop is currently working with the NASA Johnson Space Center on techniques for achieving precision landing on the moon and Mars. He is an active researcher authoring and co-authoring over 100 journal and conference papers. He was twice selected a faculty fellow at the NASA Jet Propulsion Laboratory and as a Welliver faculty fellow by The Boeing Company. Dr. Bishop co-authors *Modern Control Systems* with Professor R. C. Dorf, and he has authored two other books entitled *Learning with LabView* and *Modern Control System Design and Analysis Using Matlab and Simulink.* He received the John Leland Atwood Award by the American Society of Engineering Educators and the American Institute of Aeronautics and Astronautics that is given periodically to "a leader who has made lasting and significant contributions to aerospace engineering education." Dr. Bishop is a member of the Academy of Distinguished Teachers at The University of Texas at Austin.

List of Contributors

Maruthi R. Akella
Department of Aerospace Engineering and Engineering Mechanics
The University of Texas at Austin
Austin, Texas

Craig Anderson
National Instruments, Inc.
Austin, Texas

Dragos Arotaritei
Department of Computer Science and Engineering
Aalborg University
Esbjerg, Denmark

Brian Betts
Data Acquisition and Signal Conditioning
National Instruments, Inc.
Austin, Texas

Tomas Brezina
Technical University of Brno
Brno, Czech Republic

George I. Cohn
California State University
Los Angeles, California

Daniel A. Connors
Department of Electrical and Computer Engineering
University of Colorado at Boulder
Boulder, Colorado

Kevin C. Craig
Department of Mechanical, Aerospace, and Nuclear Engineering
Rensselaer Polytechnic Institute
Troy, New York

Timothy P. Crain II
Department of Aerospace Engineering and Engineering Mechanics
NASA Johnson Space Center
Houston, Texas

Raymond A. de Callafon
Department of Mechanical and Aerospace Engineering
University of California, San Diego
La Jolla, California

Darcy Dement
National Instruments, Inc.
Austin, Texas

Santosh Devasia
Department of Mechanical Engineering
University of Washington
Seattle, Washington

C. Nelson Dorny
Moore School of Electrical Engineeering
University of Pennsylvania
Philadelphia, Pennsylvania

Stephen A. Dyer
Electrical and Computer Engineering
Kansas State University
Manhattan, Kansas

Jeannie Sullivan Falcon
National Instruments, Inc.
Austin, Texas

Daniel R. Fay
University of Colorado at Boulder
Boulder, Colorado

Gerardo Garcia
National Instruments, Inc.
Austin, Texas

Shelley Gretlein
National Instruments, Inc.
Austin, Texas

Margaret H. Hamilton
Hamilton Technologies, Inc.
Tucson, Arizona

Cecil Harrison
University of Southern Mississippi
Hattiesburg, Mississippi

Bonnie S. Heck
School of Electrical and Computer Engineering
Georgia Institute of Technology
Atlanta, Georgia

Wen-Mei W. Hwu
University of Illinois
Urbana, Illinois

Mohammad Ilyas
Florida Atlantic University
Boca Raton, Florida

Florin Ionescu
University of Applied Sciences
Konstanz, Germany

Hugh Jack
Product Design and Manufacturing Engineering
Grand Valley State University
Grand Rapids, Michigan

Jeffrey A. Jalkio
University of St. Thomas
St. Paul, Minnesota

Rolf Johansson
Department of Automatic Control
Lund Institute of Technology
Lund, Sweden

Jayantha Katupitiya
The University of New South Wales
Sydney, New South Wales, Australia

Ctirad Kratochvil
Technical University of Brno
Brno, Czech Republic

Rahul Kulkarni
Industrial Data Acquisition and Control
National Instruments, Inc.
Austin, Texas

Thomas R. Kurfess
Department of Mechanical Engineering
Clemson University
Clemson, South Carolina

Kam K. Leang
Department of Mechanical Engineering
University of Washington
Seattle, Washington

Sergey Edward Lyshevski
Department of Electrical Engineering
Rochester Institute of Technology
Rochester, New York

Thomas N. Moore
Department of Mechanical Engineering
Queen's University
Kingston, Ontario, Canada

Leila Notash
Department of Mechanical Engineering
Queen's University
Kingston, Ontario, Canada

Cestmir Ondrusek
Technical University of Brno
Brno, Czech Republic

Hitay Özbay
Department of Electrical Engineering
The Ohio State University
Columbus, Ohio

M. K. Ramasubramanian
Department of Mechanical and Aerospace Engineering
North Carolina State University
Raleigh, North Carolina

Armando A. Rodriguez
Department of Electrical Engineering
Arizona State University
Tempe, Arizona

Momoh-Jimoh Eyiomika Salami
International Islamic University of Malaysia
Kuala Lumpur, Malaysia

Mario E. Salgado
Departamento de Electrónica
Universidad Técnica Federico Santa Maria
Valparaíso, Chile

Jyh-Jong Sheen
Department of Mechanical Engineering and Marine Engineering
National Taiwan Ocean University
Keelung, Taiwan

Andrew Sterian
Padnos College of Engineering and Computing
Grand Valley State University
Grand Rapids, Michigan

Fred Stolfi
Xerox Mechanical Engineering Sciences Laboratory
Rensselaer Polytechnic Institute,
Troy, New York

Michael J. Tordon
The University of New South Wales
Sydney, New South Wales, Australia

Michael Trimborn
National Instruments, Inc.
Austin, Texas

Job van Amerongen
University of Twente
Enschede, The Netherlands

Crina Vlad
Politehnica University of Bucharest
Bucharest, Romania

Bogdan M. Wilamowski
Department of Electrical and Computer Engineering
Auburn University
Auburn, Alabama

Juan I. Yuz
Universidad Tećnica Federico Santa Maria
Valparaíso, Chile

Qingze Zou
University of Washington
Seattle, Washington

Contents

SECTION I Mechatronic System Control

SECTION II Computers and Logic Systems

SECTION III Software and Data Acquisition

I

Mechatronic System Control

1

The Role of Controls in Mechatronics

Job van Amerongen
University of Twente

1.1 Introduction

"Mechatronic design deals with the integrated and optimal design of a mechanical system and its embedded control system." This definition implies that the mechanical system is enhanced with electronic components in order to achieve a better performance, a more flexible system, or just reduce the cost of the system. In many cases the electronics are present in the form of a computer-based embedded (control) system. This does not imply that every controlled mechanical system is a mechatronic system because in many cases the control is just an add-on to the mechanical system in a sequential design procedure. A real mechatronics approach requires that an optimal choice be made with respect to the realization of the design specifications in the different domains. In control engineering the design of an optimal control system is well understood and for linear systems standard methods exist. The optimization problem is formulated as: given a process to be controlled, and given a performance index (cost function), find optimal controller parameters such that the cost function is minimized. With a state feedback controller and a quadratic cost function, solutions for the optimal controller gains can be found with standard controller design software, such as MATLAB[1] (Figure 1.1).

Mechatronic design on the contrary requires that not only the controller be optimized. It requires optimization of the system as a whole. In the ideal case all the components in the system: the process itself, the controller, as well as the sensors and actuators, should be optimized simultaneously (Figure 1.2).

In general this is not feasible. The problem is ill defined and has to be split into smaller problems that can be optimized separately. Later on the partial solutions have to be combined and the performance of the complete system has to be evaluated. After eventually readjusting some parts of the system this leads to a sub-optimal solution.

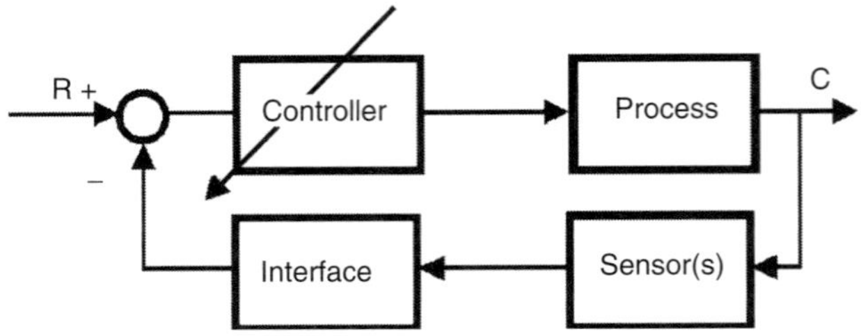

FIGURE 1.1 Optimization of the controller.

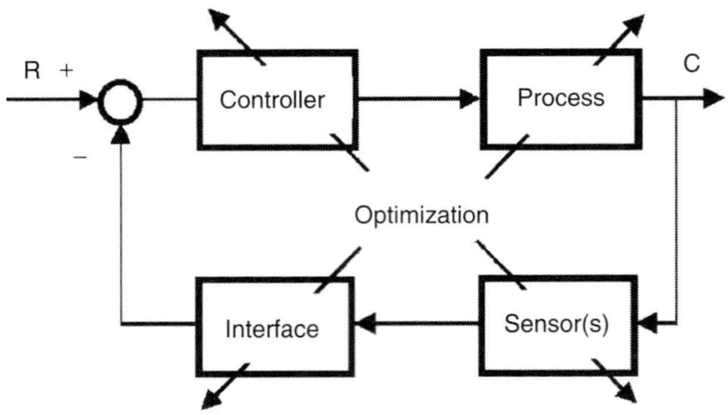

FIGURE 1.2 Optimization of the all system components simultaneously.

In the initial conceptual design phase it has to be decided which problems should be solved mechanically and which problems electronically. In this stage decisions about the dominant mechanical properties have to be made, yielding a simple model that can be used for controller design. Also a rough idea about the necessary sensors, actuators, and interfaces has to be available in this stage. When the different partial designs are worked out in some detail, information about these designs can be used for evaluation of the complete system and be exchanged for a more realistic and detailed design of the different parts.

Although the word mechatronics is new, mechatronic products have been available for some time. In fact, all electronically controlled mechanical systems are based on the idea of improving the product by adding features realized in another domain. Good mechatronic designs are based on a *real systems approach*. But mostly, control engineers are confronted with a design in which major parameters are already fixed, often based on static or economic considerations. This prohibits optimization of the system as a whole, even when optimal control is applied.

In the last days of gramophones, the more sophisticated designs used tacho feedback in combination with a light turntable to achieve a constant number of revolutions. But a really new design was the compact disc player. Instead of keeping the number of revolutions of the disc constant, it aims for a constant speed of the head along the tracks of the disc. This means that the disc rotates slower when tracks with a greater diameter are read. The bits read from the CD are buffered electronically in a buffer that sends its information to the DA converter, controlled by a quartz crystal. This enables the realization of a very constant bit rate and eliminates all audible speed fluctuations. Such a performance could never be obtained from a pure mechanical device only, even if it were equipped with a good speed control system. In fact, the control loop for the disc speed does not need to have very strict specifications. It should only prevent overflow or underflow of the buffer. The high accuracy is obtained in an open loop mode, steered by a quartz crystal (Figure 1.3).

The flexibility introduced by the combination of precision mechanics and electronic control has allowed the development of CD-ROM players, running at speeds more than 50 times faster than the original audio CDs. A new way of thinking was necessary to come to such a new solution. On the other

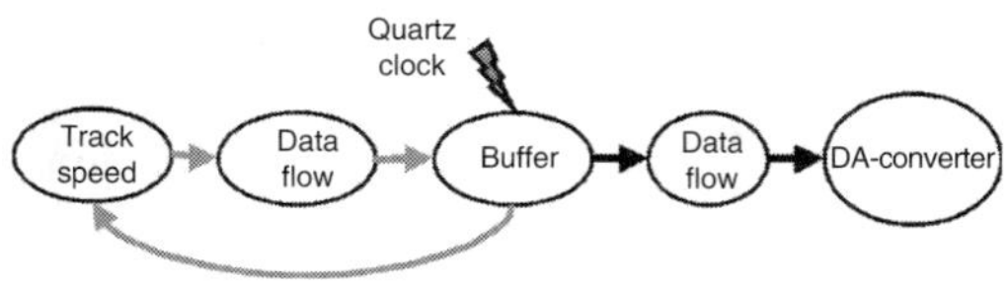

FIGURE 1.3 Combination of closed-loop and open-loop control in a CD player.

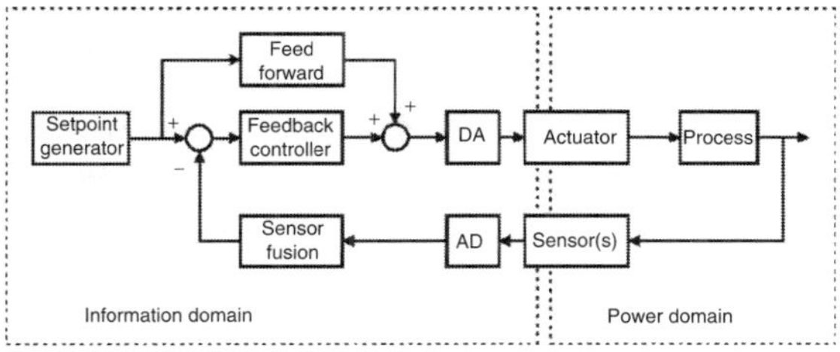

FIGURE 1.4 Mechatronic system.

hand, the CD player is still a sophisticated piece of precision mechanics. No solid-state electronic memory device can compete yet economically with the opto-mechanical storage capabilities of the CD and its successor the DVD. But this may change rapidly.

1.2 Key Elements of Controlled Mechatronic Systems

A mechatronic system consists by definition of a mechanical part that has to perform certain motions and an electronic part (in many cases an embedded computer system) that adds intelligence to the system. In the mechanical part of the system power plays a major role. This in contrast to the electronic part of the system where information processing is the main issue. Sensors convert the mechanical motions into electrical signals where only the information content is important or even into pure information in the form of numbers (if necessary, through an AD converter). Power amplifiers convert signals into modulated power. In most cases the power supply is electrical, but other sources such as hydraulic and pneumatic power supplies are possible as well. A controlled mechanical motion system thus typically consists of a mechanical construction, one or more actuators to generate the desired motions, and a controller that steers the actuators based on feed-forward and sensor-based feedback control (Figure 1.4).

1.3 Integrated Modeling, Design and Control Implementation

1.3.1 Modeling

During the design of mechatronic systems it is important that changes in the construction and the controller be evaluated simultaneously. Although a proper controller enables building a cheaper construction, a badly designed mechanical system will never be able to give a good performance by adding a sophisticated controller. Therefore, it is important that during an early stage of the design a proper choice can be made with respect to the mechanical properties needed to achieve a good performance of the controlled system. On the other hand, knowledge about the abilities of the controller to compensate for mechanic imperfections may enable that a cheaper mechanical construction be built. This requires that in an early stage of the design a simple model is available that reveals the performance limiting factors of the system. Still there is a gap between modeling and simulation software used for evaluation of mechanical constructions and software used for controller design. Mechanical engineers are used to finite element packages to examine the dynamic properties of mechanical constructions. It is only after reduction to low-order models (modal analysis) that these models can be used for controller design. On the other hand, typical control-engineering software does not directly support the mechatronic design process either; in the modeling process the commonly used transfer functions and state space descriptions often have lost the relation with the physical parameters of the mechanical construction. Tools are required that allow modeling of mechanical systems in a way that the dominant physical parameters (like mass and dominant stiffness) are preserved in the model and simultaneously provide an interface to the controller design and simulation tools control engineers are used to (Coelingh;[2] Coelingh et al.[3]).

Simulation is an important tool to evaluate the design of mechatronic systems. Most simulation programs like Simulink[1] use block diagram representations and do not support physical modeling in a way that direct tuning of the physical parameters of the mechanical construction and those of the controller is possible as required in the design of mechatronic systems. Recently, programs that allow physical modeling in *various physical domains* became available. They use an object-oriented approach that allows hierarchical modeling and reuse of models. The order of computation is only fixed after combining the subsystems. Examples of these programs are 20-sim,[4] described by Broenink[5] as CAMAS and Dymola.[6]

In this section the modeling and simulation program 20-sim (pronounced Twente Sim) will be used to illustrate the simultaneous design of construction and controller in a mechatronic system. 20-sim supports object-oriented modeling. Power and signal ports to and from the outside world determine each object[7]. Inside the object there can be other objects or, on the lowest level, equations. Various *realizations* of an object can contain different or more detailed descriptions as long as the interface (number and type of ports) is identical. Modeling can start by a simple interconnection of (empty) submodels. Later they can be filled with realistic descriptions with various degrees of complexity. De Vries[8] refers to this as *polymorphic* modeling. Submodels can be constructed from other submodels in hierarchical structures. Proper physical modeling is achieved by coupling the submodels by means of the *flow of energy*, rather than by *signals* such as voltage, current, force, and speed. This way of modeling is well suited for mechatronics system design. It will be illustrated with an example. We want to consider the design of a simple servo system, considering the use of a voltage source, a DC motor, and a mechanical load driven through a transmission (Figure 1.5).

The transmission is disregarded for the time being. The belt is considered as infinitely stiff and the transformation ratio is taken care of by changing the motor constant. If a power amplifier driven by a signal generator describes the voltage source, we can draw the iconic diagram of Figure 1.6. At this stage the different *components* in this model are still empty. But all components have electrical and/or mechanical "ports." With the proper interfaces (ports) defined, the components can be connected to each other.

FIGURE 1.5 Simple DC-servo system.

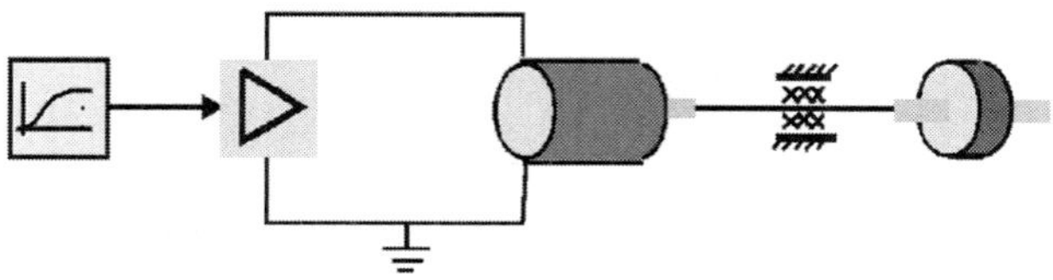

FIGURE 1.6 Iconic diagram of the simple servo system.

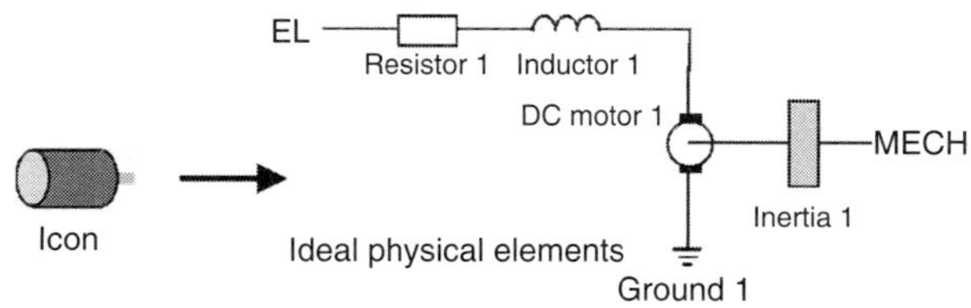

FIGURE 1.7 Icon of the motor expanded to ideal physical elements.

In the next step we can detail the description of the DC motor. One solution could be the description given in Figure 1.7. The motor is now described by a number of *ideal physical elements*, each representing a basic physical relation. The motor has an electrical (EL) as well as a mechanical port (MECH).

Each of the *elements* in this figure can be described as an element with an electrical and/or mechanical port. The idea of ports is made more explicit in so-called bond graphs.[9–12] For the electrical elements these are the voltage difference over the element and the current through the element. For the mechanical elements these are the torque and the (angular) velocity. The products of these conjugated variables ($P = ui$ or $P = T\omega$) represent power.

If we go down a step further into the hierarchy, we arrive at the level of equations. For instance, an electrical resistor can be described by the equation:

$$p \cdot u = R * p \cdot i \tag{1.1}$$

where the variables $p \cdot u$ and $p \cdot i$ indicate the conjugated variables u and i of the electrical port p. Note that this is an equation and not an assignment statement. It could have been written equally well in the form:

$$p \cdot i = 1/R * p \cdot u \tag{1.2}$$

In a similar way the inductance can be described by the equations:

$$p \cdot u = L * \mathrm{ddt}(p \cdot i) \quad \text{or} \quad (p \cdot i) = 1/L * \mathrm{int}(p \cdot u) \tag{1.3}$$

where $\mathrm{ddt}(p \cdot i)$ denotes di/dt and $\mathrm{int}(p \cdot u)$ denotes $\int u\, dt$. In case of an R-element there is no preference for one of the two forms. For the I-element the integral form is preferred in the simulations. 20-sim determines the preferred causal form and derives the equations automatically.

The energy flow or *power P* is the product of *two conjugated signals*, called effort (e) and flow (f):

$$P = ef \tag{1.4}$$

Examples of this expression in the mechanical and electrical domain are

$$P = Fv \quad \text{or} \quad P = T\omega \tag{1.5}$$

$$P = ui \tag{1.6}$$

where F is force, v is velocity, T is torque, ω is angular velocity, u is voltage, and i is current.

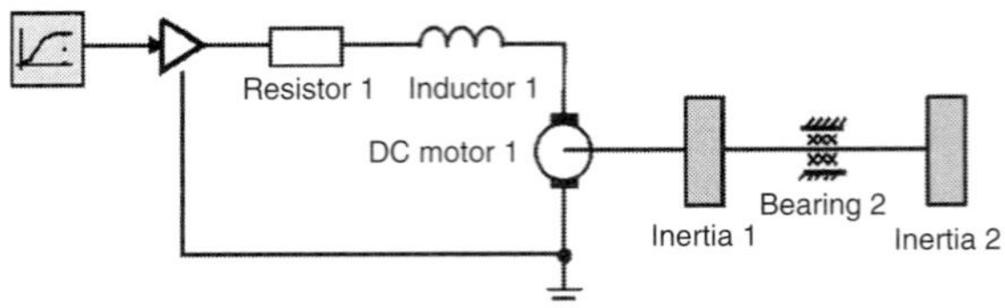

FIGURE 1.8 Complete model in the form of ideal physical elements.

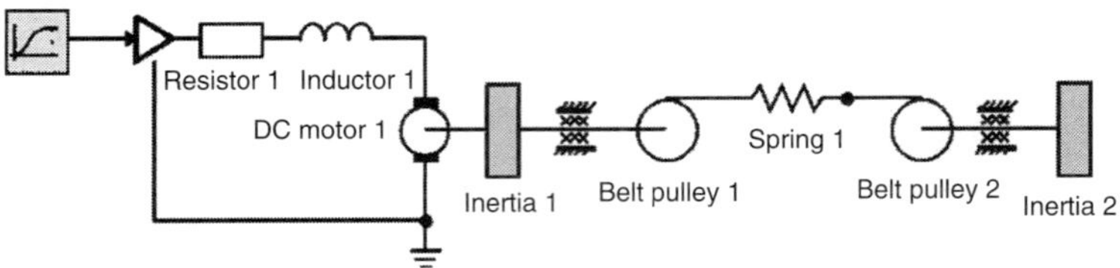

FIGURE 1.9 Model extended with transmission.

When we expand the complete Figure 1.6 we obtain Figure 1.8. When this model is processed a message pops up that indicates that inertia 2 has a dependent state. The two inertias in this model always have the same speed, and therefore, they are dependent. They cannot have independent initial conditions. The message indicates that this element can only be written in derivative form:

$$T = J\, d\omega/dt \tag{1.7}$$

There are several ways to deal with this problem.

1. The two inertias can be combined into one inertia (the program will do this automatically). A message pops up that the dependency of the two inertias has been solved symbolically.
2. Dealing with the derivative causality by means of an implicit integration algorithm.
3. The transmission can be added, including some flexibility in the belt.

If the flexibility is negligible, solution 1 leads to the simplest model. On the other hand, the warning raises the question whether the flexibility of the belt can be disregarded indeed. If not, the model has to be extended with a spring element. It should be noted that this should not be done for numerical reasons only. If the transmission were very stiff, this would result in high-frequency dynamics and lead to unnecessary slow simulations. On the other hand, if the flexibility is important, as it is in this system, the warning draws the designer's attention to the fact that the model may be oversimplified. In Figure 1.9 the transmission, including a spring element, has been added. Processing of this model does not produce any warnings.

This example illustrates how modern software can help to come up with a model that has the complexity that is needed for a particular problem. Physical models, in the form of an iconic diagram, based on connecting elements by means of power ports, may help in this modeling process. The user can select the preferred view, whether this is a bond graph, an iconic diagram with ideal physical element, or a view using higher lever submodels, like in Figure 1.6. In the next section it will be shown how to use this model for the design of controllers.

1.3.2 Control System Design Methodologies

Many processes can be reasonably well controlled by means of PID controllers. This is due to the fact that these processes can be more or less accurately described by means of a second-order model. Tuning rules, like those of Ziegler Nichols, enable less experienced people to tune such controllers. Relatively simple models can also describe many mechatronic systems. A mechatronic system mostly consists of an actuator, some form of transmission, and a load. A fourth-order model can properly describe such a system. The performance-limiting factor in these systems is the resonance frequency. A combination of position and tacho feedback (basically a PD controller) can be applied here as well. But due to the resonant poles proper selection of the signals to be used in the feedback is essential. Efforts have been made[2,3,13] to derive recipes for tuning such systems, in addition to selecting the proper feedback signals. Computer support tools are essential to enable less experienced designers to use these recipes (Van Amerongen, Coelingh, and De Vries[14]). Coelingh[2] and Coelingh et al.[3] describe a structural design method for mechatronic systems. The method starts with reducing the conceptual design to a fourth-order model that represents the dominant properties of the system in terms of the total mass to be moved and the dominant stiffness. This model still has physical meaningful parameters. In this model appropriate sensors are chosen, as well as a path generator. In the conceptual design phase a simple controller is developed and mechanical properties are changed, if necessary. Then a more detailed design phase follows where also parameter uncertainties are taken into account.

1.3.3 Servo System Design

Here we will consider some simple aspects of the design of a servo system in order to illustrate the advantage of the use of physical models and to illustrate the need for an integrated design approach. We consider the model discussed before, a load driven by an electric motor, through a flexible transmission. The iconic diagram of this model was given in Figure 1.9. In this example a current amplifier has replaced the voltage amplifier allowing the removal of the electrical resistor and the inductance. In the step responses of Figure 1.10 the resonance due to the flexible transmission is clearly visible.

From the equations used for the simulation, 20-sim can automatically derive a model in a form suitable for controller design, such as a state-space description, a transfer function, or poles and zeros. An interface is provided to MATLAB[1] enabling, for instance, to use MATLAB algorithms to compute the gains of advanced controllers like an LQR (optimal state feedback) or LQG controller (with a Kalman filter for state estimation and optimal state feedback). The diagram of the process together with an LQG controller is given in Figure 1.11 and some responses in Figure 1.12.

A properly designed P(I)D controller is able to perform almost similarly, especially when the amount of noise is small. A first attempt could be to use only measurements of the load angle and load speed.

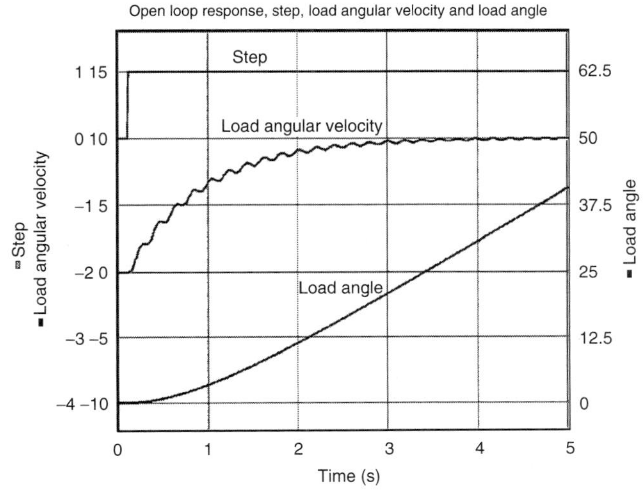

FIGURE 1.10 Open loop responses.

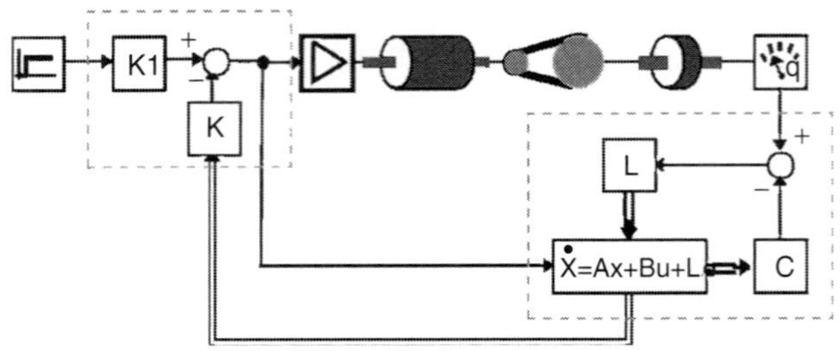

FIGURE 1.11 Process with Kalman filter and state feedback.

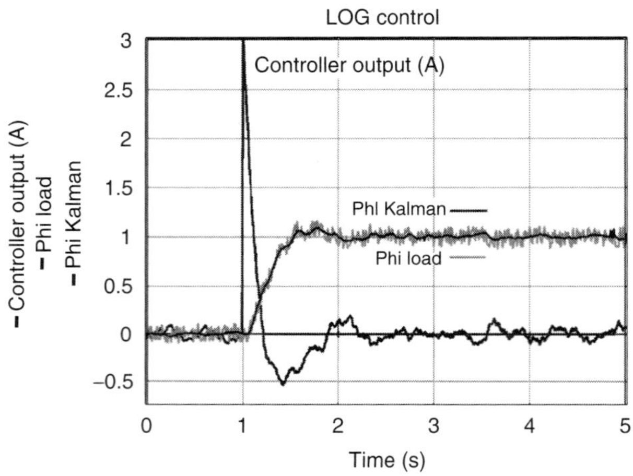

FIGURE 1.12 Response of the LQG-controlled system.

This attempt fails, because feedback of the load speed leads almost immediately to an unstable system as can be seen from the root locus for variations in the gain of the velocity feedback. From the responses of Figure 1.10, 20-sim can easily determine the transfer function between the motor current and the load speed and plot the root locus (Figure 1.13).

Figure 1.14 clearly shows that even a small amount of velocity feedback will lead to an unstable system. It is well known that feedback of the motor speed is a better solution. Using again the model of Figures 1.9 and 1.10 to determine the transfer from input current to motor speed yields the root locus of Figure 1.15.

Complex zeros now accompany the complex poles and because they are close together their influence on the response will be almost negligible. The branch of the root locus on the real axis now shows the desired behavior: moving the dominant pole to the left in the *s*-plane. Combining the feedback of the motor speed with feedback of the load angle yields the PD-controller structure of Figure 1.15 and the responses of Figure 1.16. Except for the noise there is not much difference in the responses of the system with the Kalman filter, although the PD-controlled system is simpler. The observations made here are generally applicable. A system with two resonant (complex) poles and no zeros, such as in Figure 1.13, is difficult to control by means of a simple controller. If complex zeros accompany the resonant poles with an imaginary part smaller than that of the poles, stable control is easily achieved. In the frequency domain this is seen as an anti-resonance, followed by a resonance (type *AR*). On the contrary a type *RA* system, where the resonance frequency is lower than the anti-resonance frequency (the imaginary part of the poles is smaller than that of the zeros), is just as difficult to control as in the case of only resonant poles. The existence and location of resonant zeros is completely determined by the (geometrical) location of

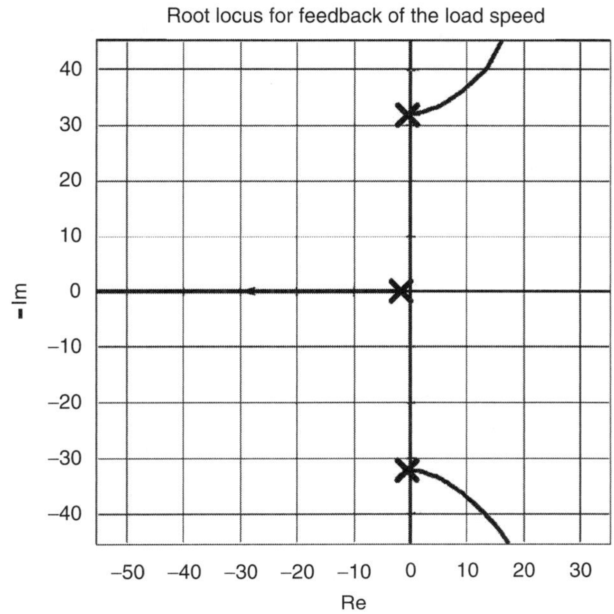

FIGURE 1.13 Root locus for velocity feedback of load axis.

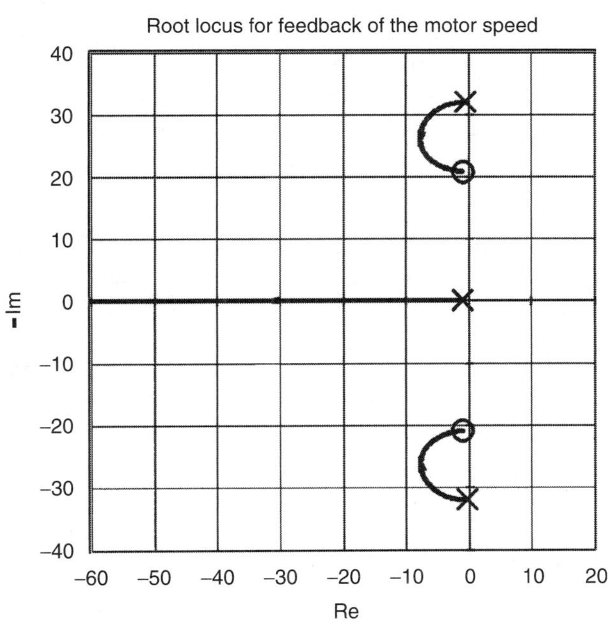

FIGURE 1.14 Root locus for velocity feedback of motor axis.

the sensors in the mechanical system. A careful choice of these sensor locations is therefore crucial for the successful application of a controller. It should be noted that using a properly designed set-point generator could prevent resonance, as seen in Figure 1.10. The set point generator should not excite the resonance frequencies, for instance, by using a low pass filter with bandwidth lower than the resonance frequencies. However, such a set-point generator does not solve the above-mentioned stability problems.

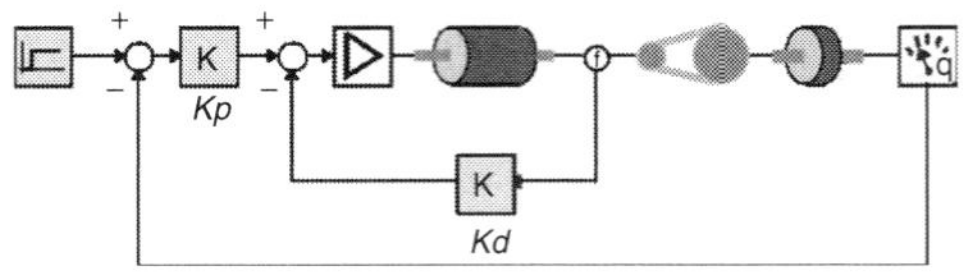

FIGURE 1.15 Servo system with PD-controller.

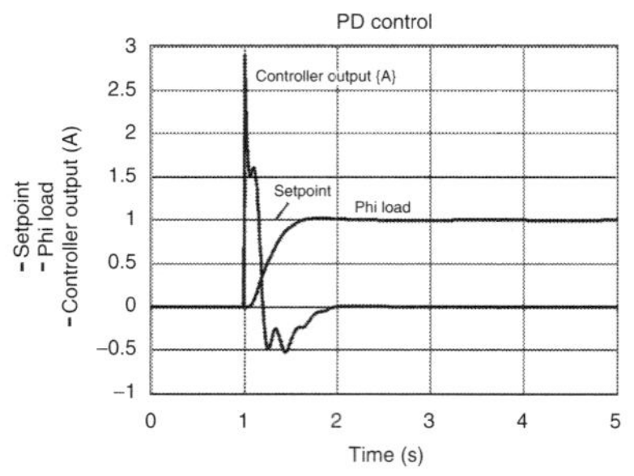

FIGURE 1.16 Responses of the system of Figure 1.16.

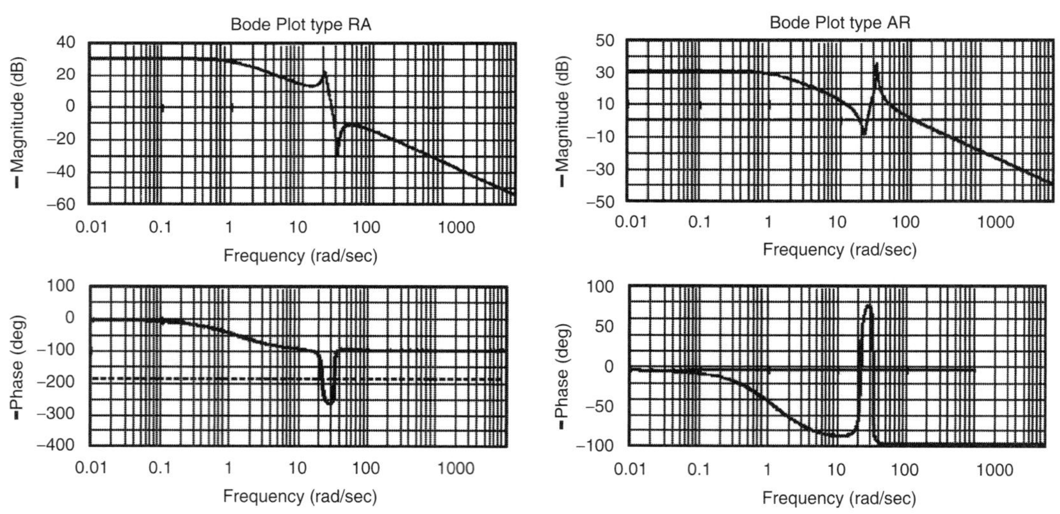

FIGURE 1.17 Bode plots of type *RA* and *AR* systems.

1.3.4 Design of a Mobile Robot

A typical example of the early design procedure is the conceptual design of a mobile assembly robot. Already in a very early stage of the design conflicting demands have to be resolved. Such a robot should be able to collect parts all around a production facility and do the assembly while driving. Because a high accuracy is required between the gripper of the robot and the surface where the parts are located, it is important that floor irregularities and vibration modes of the structure do not prevent proper assembly. On the other hand the path controller, partly based on dead reckoning (i.e., measuring of the wheel speed and orientation), requires that the wheels be very stiff. Damping of disturbances has to be realized by another means of suspension. This has led to the concept of an upper frame and a lower frame, connected by means of springs (Figure 1.18).

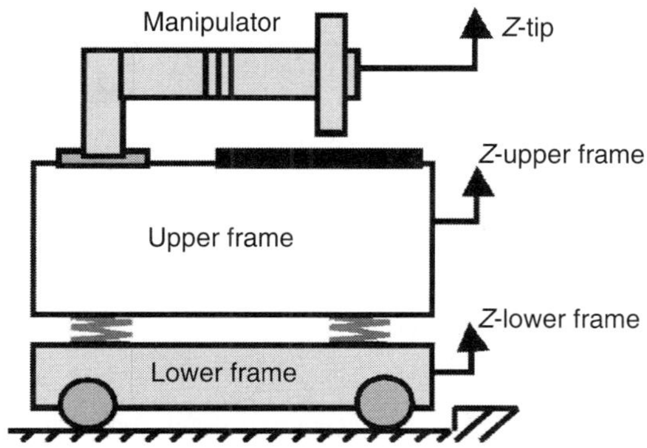

FIGURE 1.18 Conceptual design of the mobile robot.

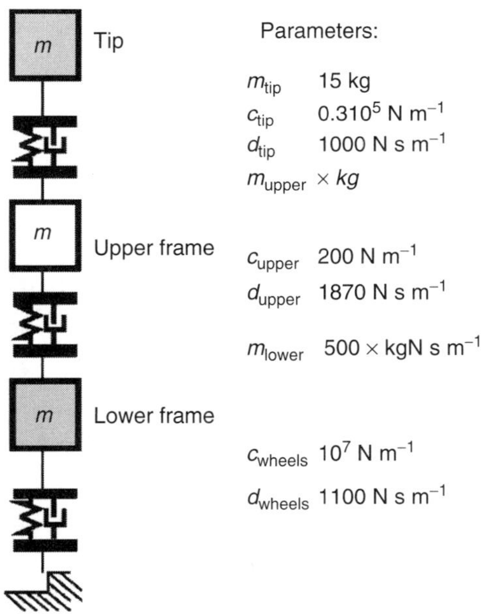

FIGURE 1.19 Simple model with ideal physical elements to compute the error e_{tip}.

The robot can be mounted at the upper frame and should have sufficient bandwidth such that the position error ($e_{tip} = z_{tip} - z_{upper\ frame}$) between the tip of the robot (z_{tip}) and the upper frame ($z_{upper\ frame}$) is small enough.

The next step is to derive a simple model, in order to have some parameters for the weight distribution and the stiffness and damping of the springs. In the model of Figure 1.22 the robot is confronted with a bump in the floor at a speed of 1 m/s.

Based upon the payload—mainly the weight of the batteries—the total mass of the vehicle was estimated to be 500 kg. Stiffness and damping of the wheels follow from the demands for the accuracy of the position estimation. The mass and bandwidth of the controlled manipulator were already known from other studies, yielding the effective stiffness and damping for the robot tip. When also initial estimates of the stiffness and damping of the springs between the upper and lower frame are made, the only parameter to be varied is the weight distribution between the upper frame and lower frame. By using the optimization feature of 20-sim, the optimal weight distribution can easily be found. In order to minimize the error between the tip of the robot and the upper frame (Figure 1.19), the weight has to be placed as much as possible in the

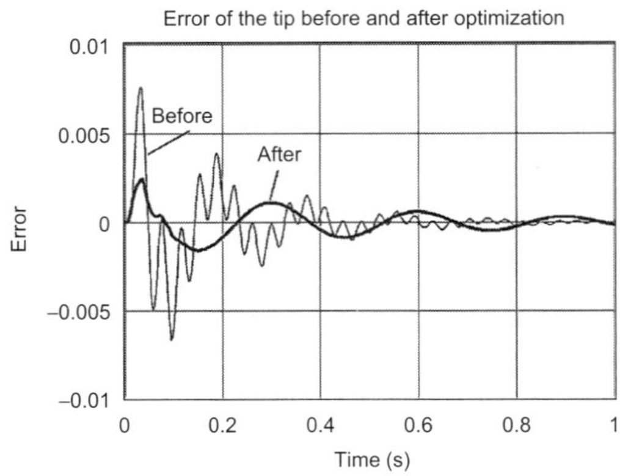

FIGURE 1.20 Error of the tip before and after optimization of the weight distribution between upper and lower frame.

FIGURE 1.21 The mobile robot (MART) after completion.

upper frame (Figure 1.20). This example illustrates how the mechanical configuration of the system is determined by the requirements for good path control and accurate control of the assembly task.

A next step could be to optimize the properties of the suspension between upper and lower frame. This will further improve the error. This decision made in a very early stage of the design directed other design decisions. After completion of the project it appeared that the different parameters of the final construction were close to these early estimates (Figure 1.21).

1.4 Modern Examples of Mechatronic Systems in Action

A few examples have already been treated in the previous sections. In this section two more examples will be given.

1.4.1 Rudder Roll Stabilization of Ships

Nowadays most ships use an autopilot to control the heading of the ship. A rudder is the most commonly used actuator. Some ships, like ferries and naval ships, need also roll stabilization. This can be achieved passively by means of two connected tanks filled with water that generate stabilizing forces that should be in counter phase with the forces of the waves. In order to make the system effective for varying frequencies of the waves, the water flow between the two tanks should be controlled. For fast ships mostly stabilizing fins are used. These are a kind of actively controlled "wings" that generate the moments needed to counteract the moments of the waves. The fins not only influence the roll motions but also have influence on the heading. On the other hand, the rudder not only influences the heading but also induces roll. In control engineering terms this leads to a multivariable system that requires a multivariable controller design for optimum performance. In practice such a multivariable system is seldom seen and two separate control systems are used.

Another approach is to use only one of the actuators (rudder or fins) to achieve course control and roll reduction. Because the frequencies of the roll motions are outside the bandwidth of the course-control system this is possible. The rudder is most suited as actuator. An additional advantage for naval ships is that removing the fins will reduce the underwater noise of the vessel.

Redesigning the course controller in order to stabilize the roll as well, demonstrates the feasibility of this approach, but also makes clear that the "process"—the ship—should be modified. The most important modification is needed for the steering machine. The maximum speed of the steering machine appears to be the limiting factor for such a system (it should increase from the commonly used values of 3–7°/s to 20–25°/s). By means of dynamic simulations the demands for the steering machine can be found in terms of the maximum speed of the steering machine and the maximum time constant that is allowed for reaching this speed. This requires reengineering of the hydraulic steering machine. A step further would be to consider also changes in the shape of the ship, in order to optimize the parameters that determine the effectiveness of the rudder roll stabilization system.

In order to decide whether this new solution is better, it should be evaluated whether the redesigned steering machine is less expensive than the original rudder and fin actuators. These design issues have to be solved in a very early stage of the design. Rudder roll stabilization has been successfully applied on naval as well as merchant marine ships.[15]

1.4.2 Compensation of Nonlinear Effects in a Linear Motor

Many mechanical systems suffer from nonlinear effects that limit the accuracy that can be achieved. Friction and cogging are two examples. A (linear) feedback controller can diminish the influence of nonlinearities, but complete compensation may be difficult. For systems that perform repetitive motions, an Iterative Learning Controller can help to further improve the performance.[16,17] The basic idea is explained in Figure 1.22.

When only the feedback loop is present and under the assumption that there are no disturbances, the error signal and thus the controller signal U_C will be the same for each repetitive motion. It is obvious that the accuracy can be improved when in the next motion the controller signal from the former cycle is used as a feed-forward signal, U_F. The feedback will generate a signal that further compensates for the remaining error by updating the feed-forward signal U_F with the formula

$$U_F^{k+1} = U_F^k + LE^k \tag{1.8}$$

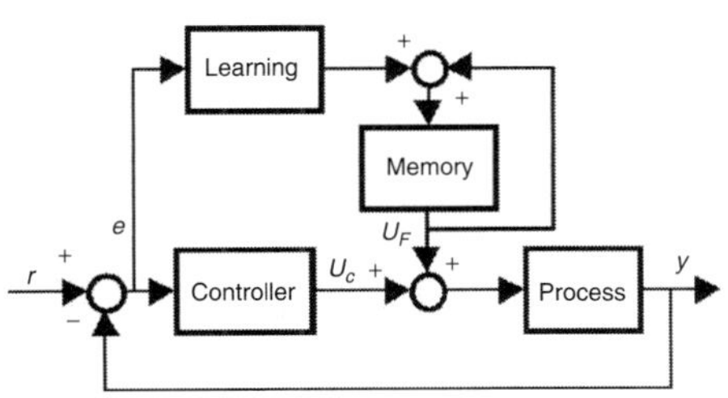

FIGURE 1.22 Principle of iterative learning control.

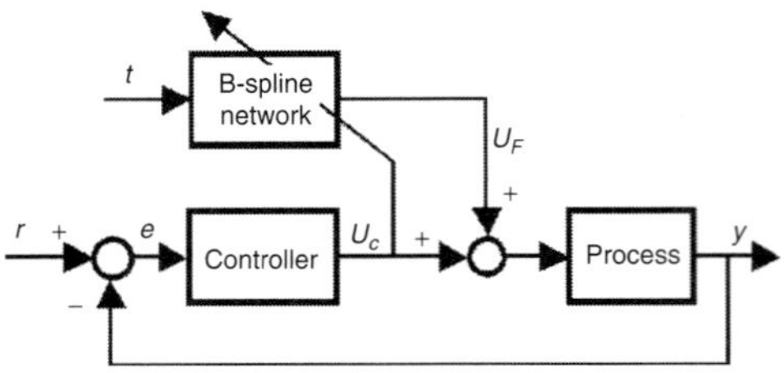

FIGURE 1.23 Learning feed-forward controller for repetitive motions.

where L is the transfer function of the learning filter. The superscript k denotes the kth repetitive motion. The signal U_F should converge to a feed-forward signal that compensates for all repetitive errors. An example of a situation where such errors are present is, for instance, a CD player that has to compensate for the eccentricity of the disk.

A variation on this idea and even more straightforward is the learning feed-forward controller (LFFC) setup of Figure 1.23. When the feed-forward signal would be perfect, the output of the controller would be zero. This implies that this output can be used as a training signal for a neural network. An adaptive B-spline network enables learning of complex nonlinear characteristics. Also support vector machines have been used to implement the learning feed forward.[18] The input of the B-spline network is the time t. It is reset each time a new motion starts. This is called a time-indexed LFFC. Instead of the time, also the reference signal and its derivatives—obtained from a path generator—could be used as index for the network (path-indexed LFFC). The advantage of this structure is that after proper training the LFFC can successfully be used for nonrepetitive motions as well. Velthuis has given a stability analysis for time-indexed as well as path-indexed LFFC.[19] The stability analysis is relatively easy for the time-indexed case. For the path-indexed case it is more complex and some heuristics are required to guarantee a stable system. The main issue is that the number of B-splines should not be too large. On the other hand a sufficiently dense B-spline distribution is desired for an accurate approximation of the nonlinear process. LFFC has successfully been applied to compensate for cogging in an industrial Linear Motor [20] and for compensation of (Coulomb) friction of a linear motor used in a flight simulator.[19] It has also been applied to the tracking control of the mobile robot described in the section Design of a Mobile Robot.[21]

The application to cogging compensation of a linear motor will be described in a little bit more detail. Such a motor is a commonly used element in assembly machines. Even with the best magnets and accurate assembly the error could not be made smaller than 100 μ, with a PID controller in combination with nonlearning feed-forward control. The design goal was to improve the maximally achievable accuracy from 100 μ to less than 10 μ. Figure 1.24 shows a picture of a linear motor.

According to the structure of Figure 1.23 the linear motor is controlled by means of a PID controller, while a B-spline neural network is present to learn the inverse motor model, including the nonlinearity due to cogging. Cogging occurs in DC motors with permanent magnets. It causes more or less sinusoidal shaped forces that depend on the position of the translator with respect to the stator. If these forces really had a sinusoidal shape, they would be easy to compensate for by means of a feed-forward compensator. However, this would require magnets with completely similar magnetic properties and very accurate spacing of the magnets. An alternative is to design a controller that learns the disturbance pattern and compensates it by means of a learning feed-forward compensator. An additional advantage is that such a system can also be used to compensate for other nonlinear effects, such as friction. This has also been demonstrated in a part of a flight simulator (a control stick) where friction forces spoil the feeling of a realistic simulation especially at almost zero speed. Figure 1.25 shows that learning is almost completed after six training cycles.

Learning feed-forward control is an attractive method to compensate for nonlinearities that are present in mechatronic systems, such as cogging and friction. The use of B-spline neural networks results in fast convergence, relatively low computational effort, and a good generalizing ability. Because of recently obtained results with respect to the stability of such systems, robust control systems can be designed.

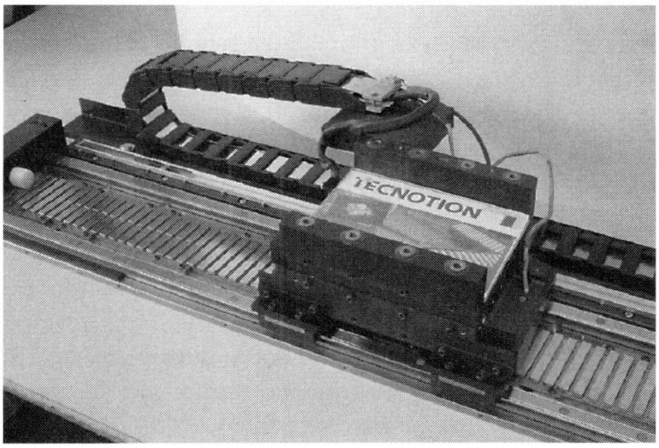

FIGURE 1.24 Linear motor. The magnets that cause the cogging are clearly visible.

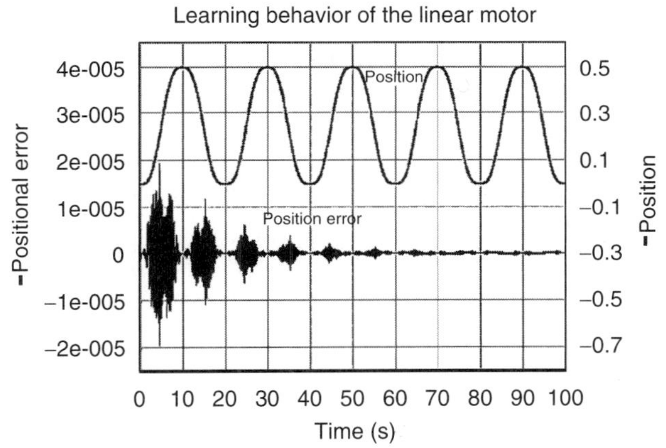

FIGURE 1.25 Position and error signal during learning of the LFFC.

A mechatronic view on this design problem raises the question whether it is possible to use the same techniques to build a less expensive linear motor, when maximum accuracy is not the main goal of the design. It has been demonstrated that a motor constructed with less expensive components and less demanding assembly specifications but with LFFC can compete well with the more expensive construction. The accuracy can typically be improved by a factor 10.

1.5 Special Requirements of Mechatronics that Differentiate from "Classic" Systems and Control Design

The main difference between "ordinary controller design" and mechatronic system design is that the latter deals with the design of the system as a whole. This approach can be considered as optimization of all components of the system simultaneously, although there are no algorithms to do this automatically. In practice the problem is often split into smaller problems that can be optimized. After integration of all the partial solutions a suboptimal system is achieved that can be further optimized by retuning the different parts, taking into account the already available intermediate design of the overall system. In order to achieve optimization of the system as a whole, it is desired that the mechanical part, where power plays a role, and the information processing part (the controller) can be simulated and adjusted simultaneously. This requires that mechanical parameters like masses and compliances be available in simulations of the controlled system. Examples have been given of modeling and simulation with 20-sim that allows for such an approach.

Mechatronic designers should constantly be aware of the fact that solutions can be found in different domains. Not every mechanical deficiency can easily be solved by control. A good mechanical design may be easier and cheaper to achieve. On the other hand, a good controller may be able to achieve the desired performance much easier and cheaper than a complex mechanical construction. In some cases the combination can even achieve performances that would never have been possible without a mechatronic design.

The same holds for the design of sensors. Each sensor could be fitted with a filter to remove noise from the measurements. But if several sensors are being combined, sensor fusion in a Kalman filter algorithm will benefit from the availability of the raw data.

Communication between all the designers involved and transparency of the design decisions in the various domains are essential for the success of a true mechatronic design.

References

1. Matworks, *The Mathworks: Developers of MATLAB and Simulink,* 2000, www.mathworks.com
2. Coelingh, H.J., *Design Support for Motion Control Systems,* Ph.D. thesis, University of Twente, 2000, also www.rt.el.utwente.nl/clh/
3. Coelingh, H.J., de Vries, T.J.A., van Amerongen, J., Design support for motion control systems—application to the Philips fast component mounter, in *Mechatronics Forum 7th Int. Conf., Mechatronics 2000,* Atlanta, Ga, USA.
4. Controllab Products, *20-sim,* www.20sim.com
5. Broenink, J.F., *Computer-Aided Physical-Systems Modeling and Simulation: a Bond-Graph Approach,* Ph.D. thesis, University of Twente, 1990.
6. Dynasim, *Dymola,* www.dynasim.se/
7. Weustink, P.B.T., de Vries, T.J.A., Breedveld, P.C., *Object Oriented Modeling and Simulation of Mechatronic Systems with 20-sim 3.0,* in *Mechatronics 98,* J. Adolfson and J. Karlsén (Eds.), Elsevier Science, 1998.
8. De Vries, T.J.A., *Conceptual Design of Controlled Electro-Mechanical Systems,* Ph.D. thesis, University of Twente, 1994.
9. Breedveld, P.C., Fundamentals of bond graphs, in *IMACS Annals of Computing and Applied Mathematics, Vol. 3: Modelling and Simulation of Systems,* Basel, 1989, pp. 7–14.
10. Cellier, F.E., Elmqvist, H., Otter, M., Modeling from physical principles, in *The Control Handbook,* W.S. Levine (Ed.), CRC Press, 1996, pp. 99–108.
11. Gawthrop, P., Lorcan Smith, L., *Metamodelling: Bond Graphs and Dynamic Systems,* Prentice-Hall, NJ, 1996.
12. Van Amerongen, J., Modelling, simulation and controller design for mechatronic systems with 20-sim 3.0, in *Proc. 1st IFAC Conf. on Mechatronic Systems,* Darmstadt, Germany, September 2000, pp. 831–836.
13. Groenhuis, H., *A Design Tool for Electromechanical Servo Systems,* Ph.D. thesis, University of Twente, 1991.
14. Van Amerongen, J., Coelingh, H.J., de Vries, T.J.A., "Computer support for mechatronic control system design," *Robotics and Autonomous Systems,* vol. 30, no. 3, pp. 249–260, PII: SO921-8890 (99)00090-1, 2000.
15. Van Amerongen, J., van der Klugt, P.G.M., van Nauta Lemke, H.R., Rudder roll stabilization for ships, *Automatica,* vol. 26, no. 4, pp. 679–690.
16. Arimoto, S., A brief history of iterative learning control, in *Iterative Learning Control: Analysis, Design, Integration and Applications,* Kluwer Academic Publishers, 1988, pp. 3–7.
17. De Vries, T.J.A., Velthuis, W.J.R., van Amerongen, J., Learning feed-forward control: a survey and historical note, in *1st IFAC Conf. on Mechatronic Systems,* Darmstadt, Germany, September 2000, pp. 949–954.

18. De Kruif, Bas J., de Vries, T.J.A., On using a support vector machine in learning feed-forward control, in *Proc. 2001 IEEE/ASME Int. Conf. on Advanced Intelligent Mechatronics,* Como, Italy, 8–12 July, 2001.
19. Velthuis, W.J.R., *Learning Feed-Forward Control—Theory, Design and Applications,* Ph.D. thesis, University of Twente, 2000, also http://www.rt.el.utwente.nl/vts/
20. Otten, G., de Vries, T.J.A., van Amerongen, J., Rankers, A.M., Gaal, E., Linear motor motion control using a learning feedforward controller, *IEEE/ASME Transactions on Mechatronics,* vol. 2, no. 3, ISSN 1083-4435, 1997, pp. 179–187.
21. Starrenburg, J.G., van Luenen, W.T.C., Oelen, W., van Amerongen, J., Learning feed-forward controller for a mobile robot vehicle, *Control Engineering Practice,* vol. 4, no. 9, 1996, pp. 1221–1230.

2

The Role of Modeling in Mechatronics Design

Jeffrey A. Jalkio

University of St. Thomas

If mechatronics design is more than just the combination of electronic, software, and mechanical design, the additional feature must lie in the ability of the mechatronic designer to optimize a design solution across these disparate fields. This requires a sufficient understanding of each of these fields to determine which portions of an engineering problem are best solved in each of these domains given the current state of technology. In turn, this requires the ability to model the problem and potential solutions using techniques that are domain independent or at least permit easy comparison of solutions and tools from different domains.

For example, the optical inspection system shown in Figure 2.1 depends on optical components in precise alignment, mechanical elements capable of precise motion, transducers for sensing and providing mechanical power, electrical systems to control motion and filter sensor signals, and software for image analysis and motion control. Only by dividing these tasks appropriately among electronics, mechanical components, and software can the system be optimized. This requires an understanding of all the system requirements and limitations as well as the capabilities of each component in the various domains. Modeling of requirements and systems is crucial in determining whether a proposed solution is acceptable as well as in documenting these determinations for future use. In this article we shall examine the varieties of models used at different points in the design process, the diverse roles of these models and their relative strengths and weaknesses in each of these roles, and finally the specific tradeoffs involved in choosing dynamic models for signals and systems analysis.

2.1 Modeling as Part of the Design Process

Models serve different purposes at different points in the design process; so to decide which modeling tools are most effectively employed in different phases we must examine the design process itself. Many descriptions of the design process are available that have been developed by researchers around the world.[1–3] Typically these descriptions serve to systematize the process to improve the productivity of

FIGURE 2.1 An optical inspection system for printed circuit boards. (Used by permission © CyberOptics Corporation 2001, all rights reserved.)

designers or to describe techniques that provide improved product quality, lower cost, or other benefits. However, since our purpose is to examine the modeling needs of the design process, we can consider a simple model that distinguishes phases of the design process in terms of types of design activity rather than a more complex model that may be preferable for other purposes. For this purpose, we can consider a four-phase process consisting of requirements analysis, concept generation, analysis and selection, and detailed design. In the first phase of this process the designer focuses on analysis of the problem without considering possible solutions. In the second phase, conceptual solutions are generated with the hope that an acceptable solution can be found from these initial concepts via combination or modification of concepts or by variation of parameters present in one of the conceptual solutions. In the third phase, these concepts are evaluated and a design is chosen for implementation. The fourth phase consists of identifying design problems that need to be solved to implement the chosen concept and applying the design process to those smaller problems. We shall consider the activities of each of these phases in detail.

2.1.1 Phase 1

The requirements analysis phase consists in obtaining a sufficient understanding of the problem to be solved. The difficulty of this process varies with the scale of the problem, the designers' familiarity with the problem domain, the variability of market needs, and the presence of hidden requirements that are poorly articulated in the initial problem statement. Depending on the nature of the design problem, the requirements identified in this phase may be the needs of a single customer, the common needs of a group of potential customers identified via a market survey, or societal needs identified by government regulations. Most design problems include some combination of these as well as internal requirements such as design guidelines and company policies. The key objective of this phase is to obtain enough detail to know when a design has solved the problem satisfactorily. Models in this phase serve primarily as communication and documentation tools since the primary problem in this phase is the clear communication and documentation of the criteria for design success. Examples of models that aid in this process include specification listings, use case diagrams, sequence diagrams, and context diagrams. Many of these modeling tools have now been standardized as parts of the Unified Modeling Language (UML), which is becoming increasingly important as an analysis tool.[4]

Use case diagrams model the interactions between a system and its users at a very high level of abstraction in terms of purpose of the interaction. Figure 2.2 gives an example of a use case diagram used to document the various operations required of a network printer. The use case diagram helps us avoid overlooking important but rare use cases such as maintenance. It is important to note that use case diagrams do not

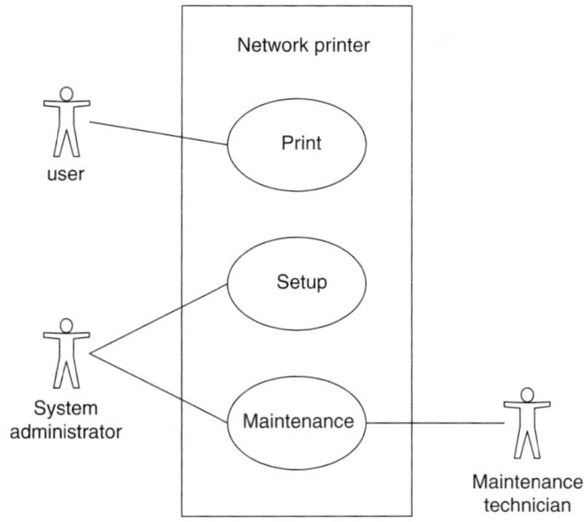

FIGURE 2.2 Use case diagram for a network printer.

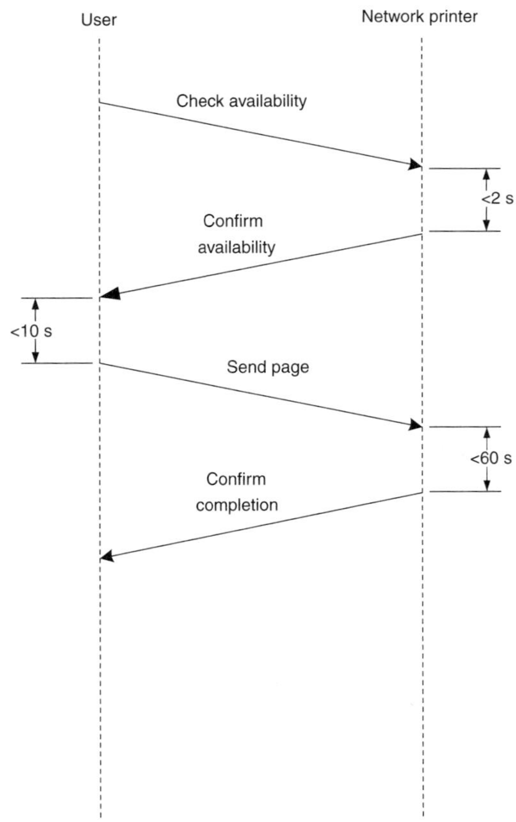

FIGURE 2.3 A sequence diagram for network printer use.

provide information about the nature of the transaction, but rather documents its existence. It serves as a top level model of a system only.

Sequence diagrams can be drawn to describe the details of individual use cases in order to clarify the interactions that must occur, timing constraints, and interactions between the system and multiple elements of the environment. Figure 2.3 shows an example of a sequence diagram for the "print" use case of Figure 2.2.

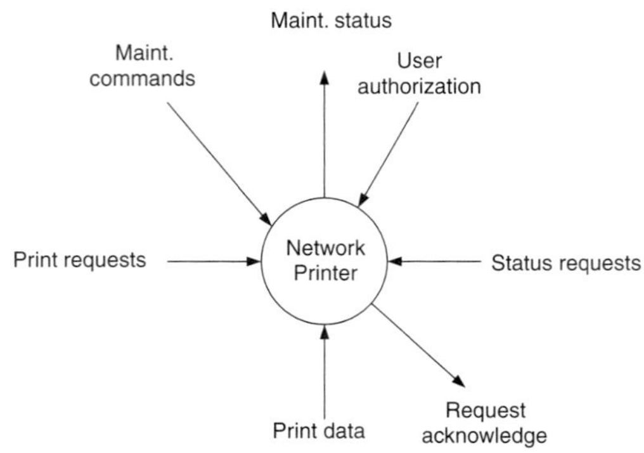

FIGURE 2.4 Context diagram for network printer.

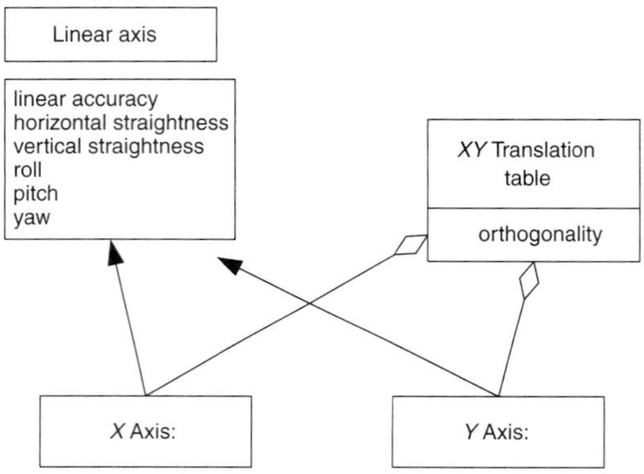

FIGURE 2.5 Class diagram for linear translation stages.

Note that the sequence diagram documents a particular instance of a particular use case. The sequence diagram can show both the direction of information flow and timing requirements. In this regard the sequence diagram and the traditional timing diagram serve similar purposes.

Context diagrams do not capture timing elements but capture the types of information that flow between the system and its environment.[5] While use case diagrams focus attention on the scenarios of system use, context diagrams focus attention on information flows that must exist to enable those scenarios. Both serve useful and complementary purposes. Figure 2.4 shows a context diagram for an inspection system like the example mentioned above. Notice that the context diagram summarizes information flow shown in the various interaction diagrams for all use cases. Regulatory and safety requirements are often in the form of limits on the interactions the system may have with aspects of its environment. These requirements must all be captured for use in later design phases and communicated back to the source of the requirement for verification of accuracy. The context diagram is also used as the top level of a data flow diagram for a system if structured analysis is used.[6]

Just as the context diagram ties to structured analysis techniques, object oriented analysis techniques provides the class relationship diagram as a means of documenting the relationships between a system and its environment and between components of a system.[7] Class relationship diagrams show how a system is composed of subsystems, how components are similar to one another, and how they differ. For example, Figure 2.5 shows a class relationship diagram for a two-axis translation system consisting of two single axis subsystems. This diagram documents the existence of 13 error components that must be specified in the requirements phase. It does this by showing that the *X* and *Y* axis components are instances of a class of single axis translation stages, that each have six error components, and that they are components of a system that adds a single additional error. By documenting relationships such as composition and inheritance, requirements and the interdependence of requirements can be represented in a compact and comprehensible way.

One key aspect of the requirements definition phase is the importance of defining requirements without specifying a preferred solution embodiment. Hence, modeling methods should be chosen to document these requirements and their intrinsic relationships without implying a particular solution.

2.1.2 Phase 2

In the concept generation phase, our objective is to generate multiple design concepts that might satisfy the requirements identified in phase 1. Here we need modeling techniques that allow us to describe possible solutions with varying levels of detail dependent on the degree of detail needed to document the key elements of the concept. Since individual concepts generated in this phase may only satisfy some portions of the design requirements, it is critical that modeling at this point allow for partial descriptions of embodiments and for the easy combination of design concepts. For this reason, our models must clearly document the portions of the requirements satisfied as well as any unspecified parameters or additional requirements introduced in the concept. For some problems, block diagrams showing interconnections between components solving portions of the problem are a useful modeling tool at this stage. Figure 2.6 shows two possible block diagrams describing a given design concept The first specifies a particular control algorithm, sensor, and actuator, while the second leaves these particulars unspecified and simply describes a closed loop controller. Depending on the situation either of these may be appropriate descriptions. The first provides details and is closer to a complete design while the second, being more generic, is easier to combine with other concepts to generate hybrid solutions.

Block diagrams are not the only modeling tool appropriate at this stage. For other problems, schematics showing arrangements of components or equations or psuedocode of proposed algorithms may be employed. As this phase continues, design concepts are often combined to form potential solutions to the overall design problem; therefore, it is useful if the model of each concept can contain descriptions of preconditions for its use, results, and other parameters that help a design team determine how to combine concepts. Similarly, once potential solutions are formed by concatenating concepts, it is often useful to clean up the final concept by combining features from different component concepts or eliminating overlapping features that are no longer needed.[8] This process is facilitated by modeling techniques that allow simultaneous modeling of mechanical, thermal, electrical, and software components and concepts. These techniques include the familiar linear graph and bond graph models.[9,10]

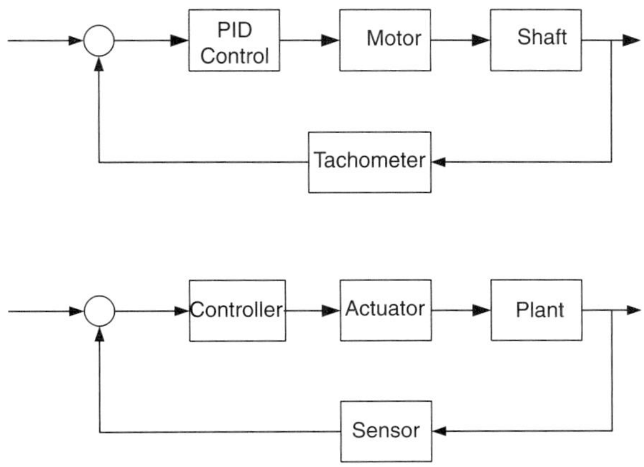

FIGURE 2.6 Block diagrams at two levels of detail.

2.1.3 Phase 3

In the third phase we evaluate potential solutions in terms of the problem requirements. This phase is eased if we have a model that allows us to compare problem requirements with design features. A particular design methodology may include various quality criteria in addition to those set by customer requirements and the regulatory environment, for example, the design axioms of Suh prefer designs with minimum information content and which satisfy each requirement by independent features of the design.[11]

One approach to evaluation is to attempt to find numerical criteria for all requirements and evaluate solutions via minimization of an overall cost function (or equivalently, the maximization of an overall value function). This approach has proven effective in certain types of problems where requirements are amenable to quantification (e.g., the optimal solution being the one that meets requirements with minimal weight or economic cost). Even in these cases, it can be difficult to determine the relative importance of diverse requirements in the formulation of the value function. For example, it may be desirable for a design to have minimal parts cost and minimal weight. An appropriate cost function might be

$$\Psi = \sqrt[n]{\left(\frac{c}{c_0}\right)^p + \left(\frac{w}{w_0}\right)^q}$$

where c and w are the cost and weight, respectively, and c_0 and w_0 are scaling factors indicating when the two factors are of equal importance, while the exponents p, q, and n express the relative importance of minimizing the two factors. Clearly, there are many possible choices for these parameters and many other models that can be used. This highlights the difficulties of this method. An example of this difficulty in a relatively simple case is determining the cost of a component being out of tolerance. Models for this problem range from a step function of cost = 0 within tolerance and cost = C outside tolerance to Taguchi's quadratic cost function.[12] Needless to say, the "optimal" solution found is typically dependent on the cost function chosen. On the other hand, in some cases, the optimum found is not strongly dependent on precise choice of cost function and a satisfactory evaluation can be made without expending excessive effort to obtain an exact cost function. In fact, if there are relatively few design options from which to choose, it is possible to invert this problem. Rather than finding the design that minimizes a particular cost function, one can instead find the range of cost function parameters that result in any particular solution being optimal. It is often easy to determine which range of parameters is most realistic.

2.1.4 Phase 4

The fourth design phase is detailed design, in which the entire process is repeated to resolve open design details for the individual components of the resulting design. This is in keeping with the design heuristic "allocate resources as long as the cost of not knowing exceeds the cost of finding out."[13] The process is inherently recursive, with each high level design decision producing a simpler design problem at a lower level of abstraction with simpler requirements. To accommodate this, we need to be able to model the design at multiple levels of abstraction and to allow the specification of interfaces between components. These models must allow us to define static as well as dynamic (behavioral) components of the interface.

In addition to this recursive aspect of design, there is of course iteration also possible, as dead ends are discovered in an otherwise promising chain of design decisions and as changes in the marketplace demand revisions of a product. To accommodate this iteration, design models should document not only decisions made, but also reasons for those decisions. Otherwise, the reasoning must be rediscovered in each iteration, or worse, may not be rediscovered until it is too late.

2.2 The Goals of Modeling

We have seen that models serve many purposes in mechatronics design. First, models serve to document concepts, assumptions, and requirements and allow the communication of these concepts to others in the design group and the various stakeholders of the design process. Second, models provide a hierarchical framework which allows division of labor in the design process and permits concurrent work on separate components.

Third, models can provide us with insights into the behavior of a system which might be obscured when we attempt to see the system in its full complexity. Fourth, they allow us to grasp similarities with other systems and use prior experience to help solve current problems. Fifth, models provide us a way of identifying and testing our ignorance of the actual behavior of the system by identifying unknown parameters and providing hypotheses worthy of testing. In this section, we shall consider these diverse purposes of modeling and examine the characteristics that cause a particular modeling methodology to be better or more poorly suited for each of these purposes.

2.2.1 Documentation and Communication

The first purpose of a model that we shall consider is documentation. We often forget that aside from the various purposes of modeling in the design process, our models are often our best tools for communicating with our colleagues, our customers, and our successors. Every document generated in the design process is a model of some part of the system. To the extent that we can avoid double documentation of the same concept, we not only reduce workload and decrease time to market, but we also reduce the likelihood of inconsistencies in our documentation. How do our models serve as documentation? Consider a circuit diagram. It is not the circuit, but documentation of the circuit's design It is also a model of the behavior we are seeking from the circuit. Nonidealities and component variations will cause the actual circuit to behave differently from this model, but the model still serves its purpose. Earlier in the design process, requirements documents also document the product, but in terms of requirements rather than the features that satisfy them.

Design documentation is important for at least four distinct groups. The engineering team itself needs communication among its members to ensure that effort is not wasted on elements that do not interoperate or that do not contribute to the overall product meeting requirements. Communication is required between the design team and the customer to ensure that requirements are correctly understood and implemented in the design. Management needs to understand the status of the project in terms of outstanding risks, ongoing expenses, and tradeoffs that would affect market size. Finally, future design teams can benefit from documentation of earlier design concepts, analyses, and decisions. This role of documentation as a communication with future workers in the field has been recognized as a key part of all scientific disciplines[14] but is often overlooked by engineers focused on meeting short-term deadlines.

What characteristics of a model lead to good documentation and clear communication? Clearly, the model must be understood by others. This requirement encourages the use of standardized modeling techniques whenever possible. It also encourages the use of modeling techniques that are interdisciplinary whenever feasible. When choosing models for documentation purposes, one must remember that each model highlights certain features of the design while hiding others. For example, the sequence diagram clearly documents communication between two objects while hiding details of how the objects generate outgoing messages or interpret incoming ones. If multiple models of the system are used to document the various aspects of the system, there should be some way to cross-reference items between models (e.g., determine that timing requirements documented in a sequence diagram model are met by the dynamics of a system described in a block diagram). Also, if the documentation is to be useful to future designers, it should be possible to cross reference design models based on a variety of criteria including system requirements and included components, so that future designs can benefit from current ones. Clearly, any modeling technique that serves some other purpose in design will also serve as documentation, but the model must be chosen so as to match the features one wishes to document.

2.2.2 Hierarchical Framework

Real mechatronic systems include a large number of components that interact in many ways. Our models simplify these complex interfaces and describe products and processes in terms of simply interacting components with well-defined interfaces and behaviors. This simplification allows us to subdivide complex problems into a set of simpler problems whose solutions can be easily integrated. In complex mechatronic systems where we must consider components that interact in several energy domains, this simplification is essential. When we draw a block diagram of a system we are dividing the system's behavior into a set of defined elements with modeled behavior and a known set of interactions between those elements, which is itself amenable to analysis. This role of modeling would lead us to select modeling techniques that allow us to divide the model into portions that can be independently tackled by independent engineers. For example, systems of linear differential equations may give us great insight into the fundamental modes of a system, but give us little insight on how to divide a design problem among members of a team. However, a block diagram, a class diagram, or a data flow diagram provides clear interfaces between elements that allow for subdivision.

2.2.3 Insights

Regarding insights into the behavior of the system, different models provide different types of insights. Signal flow graphs of electromechanical systems show the presence of feedback loops that stabilize or destabilize the system. Differential equations provide insights regarding the time scales of various behaviors and the relative importance of various factors as well as providing estimates of the validity of simplifying assumptions. Similarly, sequence and timing diagrams can provide insight regarding the processing power required to meet timing requirements for various use cases. In each case the model provides insights into the problem or the characteristics of a proposed solution by bringing into focus certain aspects of the modeled system while hiding others from view.

2.2.4 Analogies

Related to the insights a model can give us regarding the system under consideration are the insights that a model can provide in allowing us to see similarities to other systems that we or others have seen before. In this regard, modeling techniques that use analogies between various domains can be useful. However, analogies must be chosen with care. Consider the common analogy between electrical and mechanical quantities shown in the left half of Figure 2.7. In this analogy, velocity is analogous to voltage and force is analogous to current. However, as seen in the right half of Figure 2.7, the equations for the electrical system are unchanged in the dual system, in which current and voltage are exchanged and the roles of inductance and resistance are exchanged with capacitance and conductance, respectively. This dual system results in a different mechanical analogy, in which velocity is analogous to current and force is analogous to voltage. Each of these analogies can prove useful in moving design parameters from one energy domain to another, as the duality between the two electrical models can prove useful in circuit design. With any of these analogies, it must be remembered that real components deviate substantially from their idealized models and that the analogies do not strictly hold. This can be both a bane and a blessing, since a design that suffers from component nonidealities might be replaced by an analogous design from a domain with different but less detrimental nonidealities.

Element	Electrical	Mechanical	Electrical	Mechanical
Capacitance	$i = C\frac{dv}{dt}$	$f = M\frac{dv}{dt}$	$v = \frac{1}{C}\int i\,dt$	$f = k\int v\,dt$
Inductance	$i = \frac{1}{L}\int v\,dt$	$f = k\int v\,dt$	$v = L\frac{di}{dt}$	$f = M\frac{dv}{dt}$
Resistance	$i = \frac{1}{R}v$		$v = Ri$	$f = Bv$
Conductance	$i = Gv$	$f = Bv$	$v = \frac{1}{G}i$	

FIGURE 2.7 Electrical–mechanical analogies.

2.2.5 Identification of Ignorance

The final role of modeling is to help us identify our ignorance. This takes two forms. First, the construction of the model often requires us to name parameters whose numerical value is unknown. This leads us to estimate the possible range of such parameters or to design experiments to determine the parameter's value. Identifying missing details is the first step in determining them. In this way models help us identify our ignorance of details of the problem or our proposed solution. Secondly, the predictions we make with the help of our models are testable. The adequacy of a model to describe system behavior can be determined by comparing actual system behavior with predictions based on that model. Differences between actual system performance and a model's predictions point to errors in the model that may be significant. By testing the validity of our models we identify aspects of the real system that are inadequately reproduced in our model or artifacts of the model that do not exist in the real system. These discrepancies may point to an inappropriate assumption or to a fundamental misunderstanding of the physical system. In either case, the source of these discrepancies should be understood in order to determine if our model is adequate for its intended purpose.

2.3 Modeling of Systems and Signals

In the area of systems and signals we typically deal with the modeling of system dynamics and information flows through the system and its environment. As can be seen from the discussion above, this is but a small part of the modeling needed for system design. However, there are a number of decisions that must be made in selecting a model for the behavior of a dynamic system. These include the choice between analytic and numerical modeling, the use of distributed or lumped parameters, the approach used to model random factors, and the choice between linear and nonlinear models. These issues are considered in the paragraphs below.

2.3.1 Analytical versus Numerical Models

In this area both analytical and numerical tools are available to support our efforts, so we have a number of decisions to make regarding the level of complexity of the models we choose to use for various tasks. Analytical tools provide insights into overall system behaviors and can allow us to make comparisons between dissimilar systems based on a similarity in their models. These analogous systems often prove useful in the concept generation phase of design because they allow us to bring to bear solutions to diverse problems to the solution of the problem at hand. However, as described above, these analogies usually break down upon closer inspection. In the case of analogies in dynamics, they are typically based on systems described by equivalent differential equations (e.g., electronic and mechanical oscillators). However, these linear differential equations are but simplifications that ignore nonlinearities and time-dependent behavior that may greatly affect system behavior. Therefore, when applying analogies one must always be careful to explicitly state assumptions made in the modeling process.

Numerical tools, on the other hand, do not yield the same insights of analytical tools, but provide other and more detailed insights into real system behavior and performance. Typically these tools become most useful later in the design process, both for tradeoff analysis and for detailed design work. The availability of these tools opens many doors to modeling options that have in the past been closed and allows us to consider options that have historically been considered unwieldy. For example, numerical solution of nonlinear differential equations allows modeling of systems that had previously been approached only with linear approximations.

2.3.2 Partial versus Ordinary Differential Equations

One modeling choice that must be made is whether we must model the dynamics of the system using ordinary or partial differential equations. Partial differential equations must be used whenever values of interest vary spatially as well as temporally. For example, if we are considering the design of a robot arm, we are free to use ordinary differential equations if we consider the arm to be rigid, partial differential equations if we wish to describe the bending of a flexible arm in detail. However, what if we recognize a flexibility in the arm but do not need to know the details of its motion but only the effect of its flexure on the end effector? In this case we can make a lumped parameter model of the arm that summarizes its dynamics in terms of end effector motion and dynamic forces on the driver and end effector. When can this lumped parameter model be used? There are two restrictions. First, the details of the variation of behavior over space must be uninteresting to us. If we need to know these details (e.g., to determine stress concentration, or temperature distribution) we cannot apply the simplified model. Secondly, the partial differential equation may not be amenable to forming a lumped form due to its complexity.

2.3.3 Stochastic versus Deterministic Models

One choice that must be made in the modeling of system behavior is how uncontrolled variability will be represented in the model.[15] The approach taken depends a great deal on the source and characteristics of the variability. For example, the values of some system parameters may not be precisely known, but may be constant over time. Others may vary slowly over time in unknown ways. In the first case, the sensitivity of the system to parameter value can be determined, using a variety of both analytical and numerical techniques. In the latter case, one must examine the speed of parameter variation to determine if the system can be analyzed in quasi-steady state or if the dynamics of the parameter variations are coupled with the dynamics of the system. For this purpose, it is often useful to write equations in dimensionless form where time and scale constants indicate critical speeds for coupling.

When the variation of parameters over time must be considered or when some inputs to a system have uncontrolled variability, we must choose whether to model the input as an arbitrary signal or a signal with known statistical parameters. In the first case, we have just another input and can analyze the system for sensitivity to this input. In the second case, we can characterize the system in terms of autocorrelation or equivalent spectral measures and use optimal filtering techniques to reduce uncertainty.[16] A key point in the statistical modeling of a system is to base the model on the known variability of the process rather than assuming additive, white, Gaussian noise at every turn.[17]

2.3.4 Linear versus Nonlinear

A wealth of techniques are available for the analysis of linear time invariant systems. Unfortunately, the world gives us nonlinear components. One advantage of modeling techniques that allow us to divide a complex system into simpler sub-components is that we can often isolate nonlinearities as individual elements and develop linear models for them. For small variations about a differentiable point on a nonlinear curve, we can always find a linear model, but the value of this model might be extremely limited depending on its accuracy over a reasonable range of input values. In the case of a discontinuous or nondifferentiable nonlinearity, there is a greater problem. A linear model can be found only when we are far from the discontinuity or when the discontinuity is small compared to the linear portion of the model.

However, nonlinearities also add essentially new behaviors to systems that are not possible in purely linear systems. For this reason alone, nonlinear models are needed in certain circumstances. As an example, for a linear system, stable oscillations are only theoretically possible for differential equations with purely imaginary eigenvalues. Positive real parts lead to growing oscillations, while negative real parts lead to damped oscillations. Clearly, the behavior of such a system is highly sensitive to system parameters. Small variations will either kill the oscillation or drive the system into nonlinear regions of operation. Furthermore, even for perfect oscillators, the amplitude of oscillations is unconstrained. However, a nonlinear system such as the Van der Pol equation:

$$\ddot{y} = -\omega^2 y + \alpha\left[1 - \left(\frac{y}{y_0}\right)^2\right]\dot{y}$$

includes a damping term that increases with the magnitude of the oscillation. For oscillations small compared to y_0 the oscillations grow, while larger oscillations are damped. This and other essentially nonlinear behavior cannot be modeled adequately with linear models. Although once this behavior is recognized through the use of a nonlinear model, the resulting steady-state system (a stable oscillator) can be modeled for many purposes using its linear (although physically unrealizable) equivalent.

When using differential equations to model a system, we must be particularly careful in our choice of units of measure and in how we group parameters. For example, in the Van der Pol equation above, we can choose an alternative scale for y such that y_0 is 1 and a time scale such that ω is 1. In this case we have

$$z'' = -z + \varepsilon(1 - z^2)z'$$

where $z = y/y_0$, $\tau = \omega t$, and $\varepsilon = \alpha/\omega$. This scaling provides a dimensionless equation as well as providing an indication of the system's natural time scale, the scale of y, and an indication that the size of α relative to ω is important. In general, we find that writing equations in dimensionless form provides three advantages. First, as mentioned above, the scaling factors provide a sense of time scale and magnitude of various features of the system's dynamic behavior. Second, the dimensionless parameters that result from combining physical system parameters provide a way of characterizing the behavior of a wide range of systems in terms of parameters that are easily mapped. Third, by putting the equation into dimensionless form, we are able to categorize it and recognize similar equations in different domains.

References

1. Hubka, V. and Eder, W.E., *Engineering Design—General Procedural Model of Engineering Design,* Heuista, Zurich, 1992.
2. Pahl, G. and Beitz, W., *Engineering Design: A Systematic Approach,* Springer-Verlag, Berlin, 1988.
3. Dym C.L., *Engineering Design—A Synthesis of Views,* Cambridge University Press, New York, 1994.
4. Booch, G., Jacobson, I., Rumbaugh, J., and Rumbaugh, J., *The Unified Modeling Language User Guide,* Addison-Wesley, New York, 1998.
5. Douglass, B., *Doing Hard Time: Developing Real-Time Systems with UML, Objects, Frameworks and Patterns,* Addison-Wesley, New York, 1999.
6. Demarco, T., *Structured Analysis and System Specification,* Yourdon Press, NJ, 1978.
7. Coad, P. and Yourdon, E., *Object-Oriented Analysis,* Yourdon Press, NJ, 1990.
8. Glegg, G.L., *The Design of Design,* Cambridge University Press, London, 1969.
9. Shearer, Murphy, and Richardson, *Introduction to Dynamic Systems,* Addison-Wesley, Reading, MA, 1967.
10. Karnopp, D. and Rosenberg, R., *System Dynamics: A Unified Approach,* Wiley, New York, 1975.
11. Suh, N.P., *Principles of Design,* Oxford University Press, New York, 1990.
12. Taguchi, G. and Yokoyama, Y., *Taguchi Methods: Design of Experiments,* American Supplier Institute, Dearborn, MI, 1994.
13. Koen, B.V., *Definition of the Engineering Method,* American Society for Engineering Education, Washington, 1985.
14. Lonergan, B., *Method in Theology,* University of Toronto, Toronto, 1990.
15. Zhou, K., *Essentials of Robust Control,* Prentice-Hall, NJ, 1998.
16. Papoulis, A., *Probability, Random Variables, and Stochastic Processes,* McGraw-Hill, New York, 1991.
17. Law, A. and Kelton, W., *Simulation, Modeling and Analysis,* McGraw-Hill, New York, 1991.

3

Signals and Systems

Momoh-Jimoh Eyiomika Salami
International Islamic University of Malaysia

Rolf Johansson
Lund Institute of Technology

Kam K. Leang
Qingze Zou
Santosh Devasia
University of Washington

C. Nelson Dorny
University of Pennsylvania

3.1 Continuous- and Discrete-Time Signals

Momoh-Jimoh Eyiomika Salami

Signals are physical variables or quantities measured at various parts of a system, which when processed yield the desired information. A wide variety of signals are often encountered in describing many practical systems. Electrical signal, in form of current and voltage, is the most easily measured quantity, hence the need to use sensors and transducers to transform other non-electrical quantity into electrical signals. These signals must be processed by appropriate techniques if desirable results are to be obtained. Several methods of signal representation, suitable for effective signal processing in both time and frequency domains, are discussed in this section.

3.1.1 Signal Classification [1–4]

Signals are broadly classified as either continuous-time (CT) or discrete-time (DT) signals, and each of these may in turn be categorized as deterministic or random signals. A deterministic signal can always

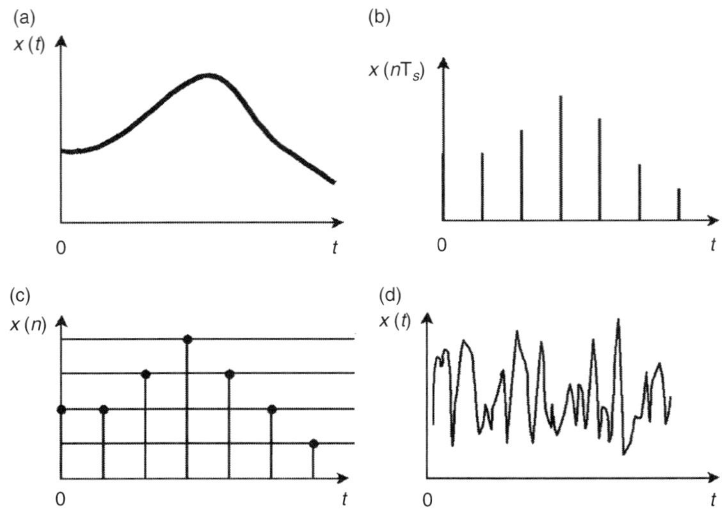

FIGURE 3.1 Description of some classes of signals: (a) continuous-time analog, (b) sampled-data, (c) digital signal, (d) random signal.

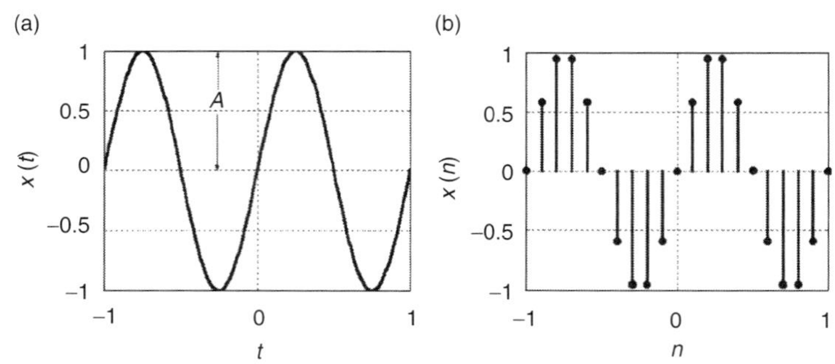

FIGURE 3.2 Description of periodic signals: (a) CT, (b) DT.

be expressed mathematically, whereas the time of occurrence or value of a random signal cannot be predicted with certainty. A CT signal, $x(t)$, has a specified value for every value of time, t, while a DT signal, $x(n)$, has specified a value only at discrete points, that is, for integer values of n. Closely related to CT and DT signals are analog and digital signals, respectively. If the amplitude of a signal can take on any value in a continuous range, then it is an analog signal. On the other hand, the amplitude of a digital signal can have only finite number of values at discrete points. Examples of continuous-time, discrete-time, digital, and random signals are shown in Figure 3.1.

Deterministic signals fall into two main categories, namely periodic and aperiodic signals. A periodic signal has the same values at times separated by one period, T, that is, $x(t)$ satisfies the relation, $x(t) = x(t + T)$, $-\infty < t < \infty$. An example of a CT periodic signal is a sinusoidal waveform of the form $x(t) = A\sin(\Omega t + \theta)$, where θ is the phase in radians, $\Omega = 2\pi F$ is the frequency in radians per second, and F is the frequency in hertz. It should be noted that the frequency range for the analog sinusoids is $-\infty < F < \infty$. A periodic DT signal is described by $x(n) = x(n + N)$, $-\infty, < n < \infty$, where N represents the period. An example of this is the sinusoidal waveform $x(n) = A\sin(2\pi rn + \theta)$, $-\infty < n < \infty$, where r is the signal frequency per sample frequency and has values in the range $-\frac{1}{2} \leq r \leq \frac{1}{2}$. Samples of sinusoids having frequencies within this range are unique and distinct. However, DT sinusoids whose frequencies are separated by an integer multiple of 2π are identical. Examples of analog and DT sinusoids are depicted in Figure 3.2.

Any deterministic signal that is not periodic is referred to as aperiodic or nonperiodic. Damped sinusoids and exponential decaying signals are common examples of aperiodic signals. For some applications, it may be quite useful to classify the signals according to their energy or power content. The total energy over the range $t \in (-\infty, \infty)$ for a CT signal is given by

$$E = \lim_{T\to\infty} \int_{-T/2}^{T/2} |x(t)|^2 dt \tag{3.1}$$

and the average power is defined as

$$P = \lim_{T\to\infty} \left[\frac{1}{T} \int_{-T/2}^{T/2} |x(t)|^2 dt \right] \tag{3.2}$$

Consequently, $x(t)$ is an energy signal if and only if $0 < E < \infty$, which implies that $P = 0$. Similarly, $x(t)$ is a power signal if and only if $0 < P < \infty$, indicating $E = \infty$. A signal that fails to satisfy either definition is, therefore, neither energy nor power signal. These definitions are also applicable to DT signals except that the integral in Equations 3.1 and 3.2 is replaced by summation. In general, periodic signals exist for all the time and as such have infinite energy. However, they have finite average power, hence they are power signals. On the other hand, bounded finite-duration signals are energy signals. The classification of a signal to finite energy, finite power, or neither is important so that appropriate and effective procedures can be selected for its analysis.

3.1.2 Singularity Functions

Singularity functions are useful for signal modeling, that is, they serve as basis for representing complex signals to simplify their analysis.

3.1.2.1 The Unit Impulse Function

The impulse or delta function is a mathematical model for representing physical phenomena that occurs within very small time duration; this time duration can be assumed to be equal to zero. The unit delta function is not a mathematical function in the usual sense; rather it is a distribution or a generalized function. Thus, the impulse function can be described by its effect on the *test function* $\phi(t)$, that is,

$$\int_{-\infty}^{+\infty} \phi(t)\delta(t)\,dt = \phi(0) \tag{3.3}$$

provided $\phi(t)$ is continuous at $t = 0$. This equation shows the shifting property of the delta function. A graphical plot of the delta function is shown in Figure 3.3a. Table 3.1 shows the operating properties of the delta function.

3.1.2.2 The Unit Step Function

The unit step function is particularly useful for the mathematical analysis of CT signals. This is depicted in Figure 3.3b, and is defined as

$$u(t) = \begin{cases} 1, & t > 0 \\ 0, & t < 0 \end{cases} \tag{3.4}$$

TABLE 3.1 Properties of the Delta Function

Property	Mathematical Expression
Sampling	$\int_{-\infty}^{+\infty} x(t)\,\delta(t-a)\,dt = x(a)$
Shifting	$x(t)\,\delta(t-a) = x(a)\,\delta(t-a)$
Scaling	$\delta(at \pm b) \equiv \frac{1}{\lvert a\rvert}\delta\left(t \pm \frac{b}{a}\right)$
Convolution	$x(t) * \delta(t-a) = \int_{-\infty}^{+\infty} x(\tau)\,\delta(t-\tau-a)\,d\tau = x(t-a)$

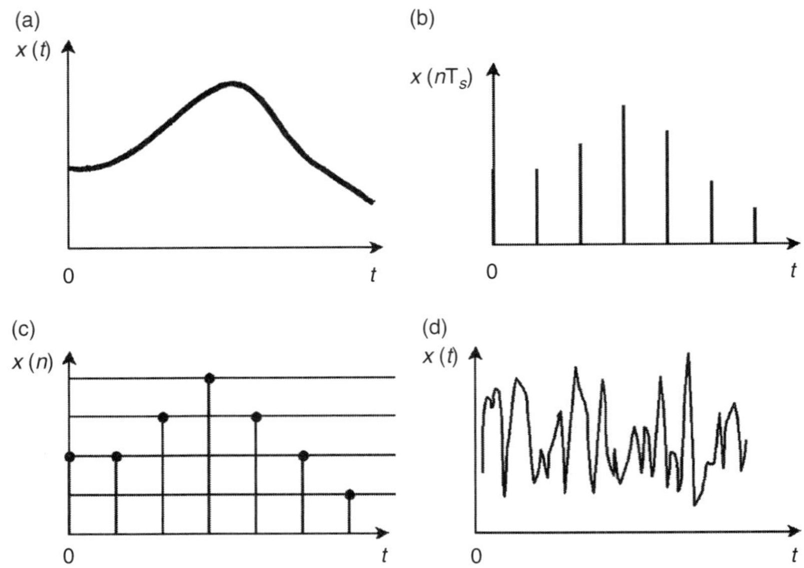

FIGURE 3.3 Description of singularity function: (a) impulse, (b) step, (c) ramp, (d) parabolic.

The signum function is derived from the step function according to

$$\text{sgn}(t) = 2u(t) - 1$$

3.1.2.3 The Ramp Function

Integrating (3.4) yields the ramp function shown in Figure 3.3c. The unit ramp function $r(t)$ is expressed as

$$r(t) = \begin{cases} t, & t \geq 0 \\ 0, & t < 0 \end{cases} \tag{3.5}$$

Furthermore, integrating $r(t)$ yields a unit parabolic signal of the form

$$a(t) = \begin{cases} \frac{t^2}{2}, & t \geq 0 \\ 0, & t < 0 \end{cases} \tag{3.6}$$

This is depicted in Figure 3.3d.

3.1.3 Basic Continuous-Time Signals

Figure 3.4 shows some of the elementary signals that are often encountered in signal analysis. Some of these signals can be derived directly from the singular functions discussed above. For example, the unit rectangular pulse signal that extends from $-\tau/2$ to $\tau/2$ can be expressed as

$$\prod(t) = u\left(t + \frac{\tau}{2}\right) - u\left(t - \frac{\tau}{2}\right) \tag{3.7}$$

and this is depicted in Figure 3.4a.

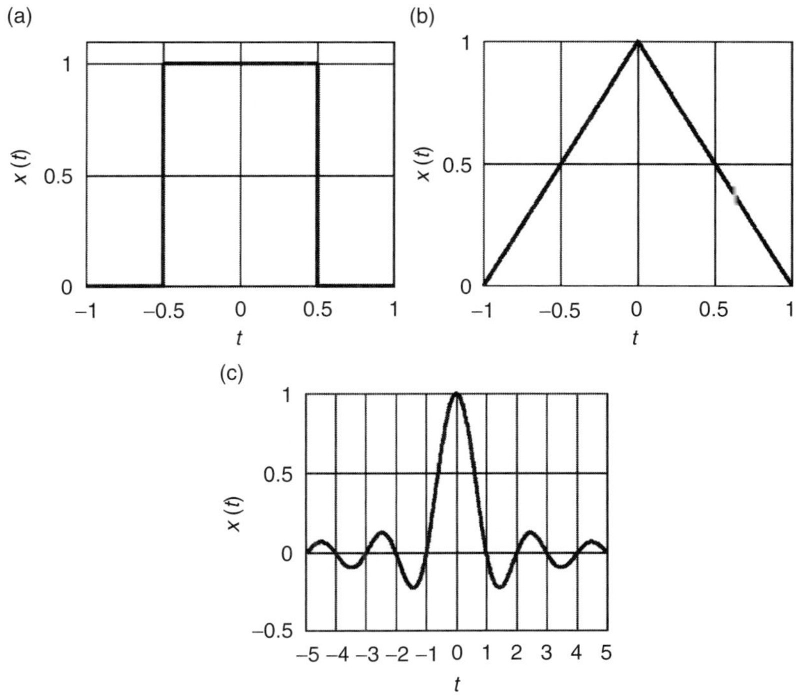

FIGURE 3.4 Description of basic CT signal: (a) rectangular pulse, (b) triangular pulse, and (c) sin *c* function.

The triangular function, denoted as $\Lambda(t)$, is defined as

$$\Lambda(t) = \begin{cases} 1 - |t|, & |t| < 1 \\ 0, & \text{otherwise} \end{cases} \tag{3.8}$$

Using the convolution theorem, to be discussed in Section 3.1.5, it can be shown that

$$\Lambda(t) = \prod(t) * \prod(t)$$

The sin*c* signal is defined as

$$\operatorname{sin}c(t) = \begin{cases} \dfrac{\sin(\pi t)}{\pi t}, & t \neq 0 \\ 1, & t = 0 \end{cases} \tag{3.9}$$

Both the triangular and sin*c* signals are shown in Figures 3.4b and 3.4c, respectively.

3.1.4 Basic Discrete-Time Signals

The unit sample sequence, denoted as $\delta(n)$, is defined as

$$\delta(n) = \begin{cases} 1, & n = 0 \\ 0, & \text{otherwise} \end{cases}$$

It is also referred to as the unit impulse sequence or Kronecker delta function. The working properties of the unit sample sequence are analogous to that of $\delta(t)$ and these are shown here:

$$\sum_{n=-\infty}^{\infty} x(n)\delta(n-m) = x(m)$$

$$x(n)\delta(n-m) = x(m)\delta(n-m)$$

$$\delta(an \pm b) = \delta\left(n \pm \frac{b}{a}\right)$$

$$x(n) * \delta(n-m) = \sum_{n=-\infty}^{\infty} x(r)\delta(n-r-m) = x(n-m)$$

Note that the scaling property is only applicable when both a and b/a are integers. Two other basic signals that are useful for analysis are the unit step and unit ramp signals. The unit step sequence, $u(n)$, is defined as

$$u(n) = \begin{cases} 1, & n \geq 0 \\ 0, & n < 0 \end{cases} \tag{3.10}$$

whereas the unit ramp signal, denoted as $r(n)$, is given by

$$r(n) = \begin{cases} n, & n \geq 0 \\ 0, & n < 0 \end{cases}$$

The above three sequences are related as follows:

$$\delta(n-k) = u(n-k) - u(n-k-1)$$

$$u(n) = \sum_{m=-\infty}^{n} \delta(n-m)$$

$$r(n) = u(n) * u(n-1)$$

Figure 3.5 illustrates the above DT sequences.

3.1.5 Analysis of Continuous-Time Signals

3.1.5.1 Basic Operations on Signals

There are some important operations that are often performed on signals so as to understand either their characteristics or the physical phenomena generating them. The three most common operations are shifting, time scaling, and reflection. Examples of these operations are illustrated in Figure 3.6, where $x(t)$ is expressed as

$$x(t) = \begin{cases} t+1, & -1 \leq t \leq 3 \\ 3, & 3 < t \leq 6 \\ 0, & \text{otherwise} \end{cases}$$

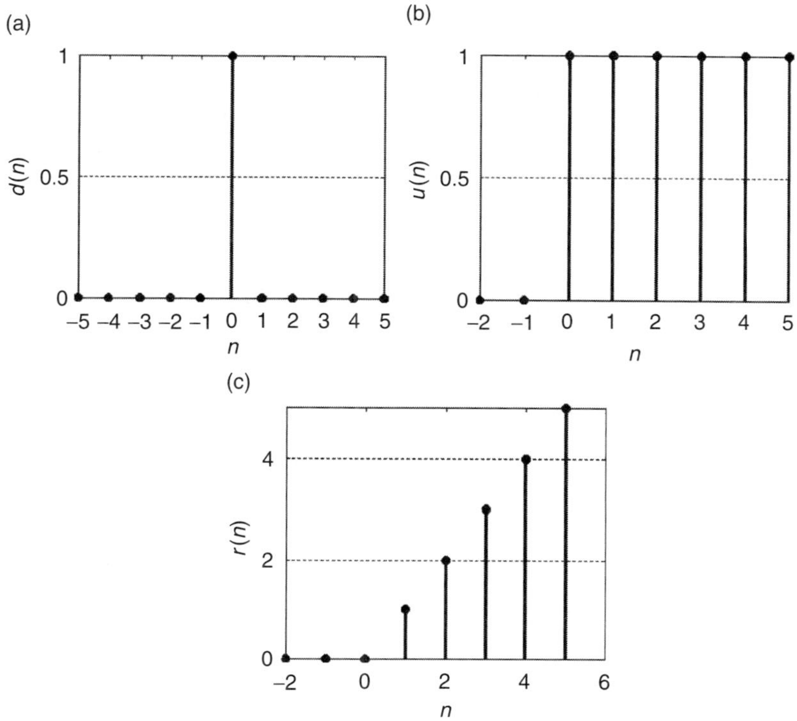

FIGURE 3.5 Description of basic DT signal: (a) unit pulse sequence, (b) unit step sequence, (c) ramp sequence.

3.1.5.2 The Convolution and Correlation Integrals[2]

Though the convolution operation is often associated with systems studies, occasionally this operation may be needed in analyzing signals obtained from a physical system. The convolution of two CT signals $x(t)$ and $y(t)$ yields $z(t)$, where

$$z(t) = x(t) * y(t) = \int_{-\infty}^{\infty} x(\tau)y(t-\tau)d\tau \tag{3.11}$$

Convolution is not limited to time-domain since it is also used to determine the frequency-domain spectrum associated with the product of two time-domain signals. The cross-correlation function between $x(t)$ and $y(t)$, denoted as $R_{xy}(t)$, is defined as

$$R_{xy}(t) = x(t) \oplus y(t) = \int_{-\infty}^{\infty} x(\tau)y^{*}(\tau-t)d\tau \tag{3.12}$$

Unlike the convolution there is no reflection operation here. Furthermore, the variable lag, t, is the scanning parameter that measures the degree of similarity between these two signals. If $x(t) = y(t)$, then (3.12) describes the autocorrelation function. Some properties of the correlation functions are given in Table 3.2.

Both convolution and correlation integrals are applicable to the energy as well as power signals. In the case of power signals, the integral is taken over the period T and the result is scaled by $1/T$. Correlation analysis is important as it leads to the computation of the energy spectral density for the transient signals, and power spectral density for both periodic and random signals.

TABLE 3.2 Properties of the Correlation Function

Property	Autocorrelation	Crosscorrelation
Even/Reorder	$R_{xx}(t) = R_{xx}(-t)$	$R_{xy}(t) = R_{yx}(-t)$
Upper Bound	$R_{xx}(0) \geq R_{xx}(t)$, for any t	$\lvert R_{xy}(t)\rvert \leq \sqrt{R_{xx}(0)R_{yy}(0)}$

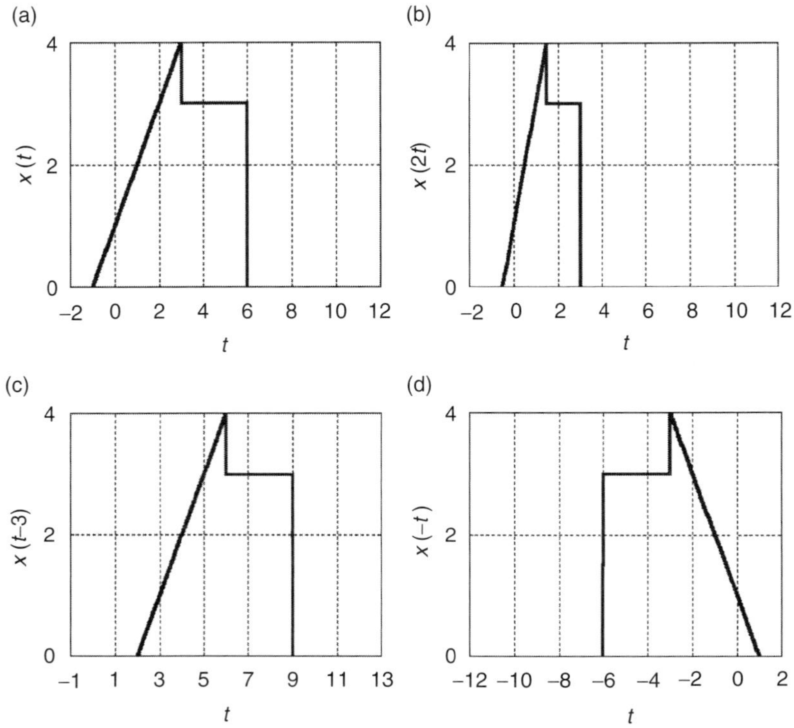

FIGURE 3.6 Basic operations on signal: (a) original signal, (b) scaling, (c) shifting, and (d) reflection.

3.1.6 Fourier Analysis of CT Signals

So far we have discussed only the time-domain methods of analyzing CT signals. The convolution integral is of particular interest since this can be used to study how a signal is modified as it passes through a system. There is need to consider the frequency-domain methods of analysis since the convolution analysis can be laborious. Furthermore, the formulation of convolution integral is based on representing the signals by shifted δ-functions. In many applications, it is more appropriate and desirable to choose a set of orthogonal functions as the basic signals since this approach leads to a reduction in computational complexity as well as providing a graphical representation of the frequency components in a given signal.

3.1.6.1 Orthogonal Basis Functions[2,3]

It is mathematically convenient to represent arbitrary signals as a weighted sum of orthogonal waveforms as this leads to a very much simplified signal analysis as well as showing the fundamental similarity between signals and vectors. Consider a set of basis function $\phi(t)$, $i = 0, \pm 1, \pm 2, \ldots$. This is said to be orthogonal over an interval (t_1, t_2) if

$$\int_{t_1}^{t_2} \phi_m(t)\phi_k^*(t)dt = E_k\delta(m-k) \tag{3.13}$$

where $\phi_k^*(t)$ stands for the complex conjugate of the signal. If E_k is equal to unity for all values of k, then the $\phi(t)$ is an orthonormal set. It is relatively easy to approximate a given signal by an appropriate set of orthonormal functions as this leads to minimum error between the actual signal and its approximation. Thus, a given signal $x(t)$ with finite energy over the interval $t_1 < t < t_2$ can be expressed as

$$x(t) = \sum_{k=-\infty}^{\infty} c_k \phi_k(t) \tag{3.14}$$

where

$$c_k = \int_{t_1}^{t_2} x(t)\phi_k^*(t)\,dt, \quad k = 0, \pm 1, \pm 2, \ldots$$

This equation is referred to as the generalized Fourier series of $x(t)$, and the constants c_k, $k = 0, \pm 1, \pm 2, \ldots$, are called the Fourier series coefficients with respect to the orthogonal set $\{\phi(t)\}$.

Denoting the first M terms in Equation 3.14 as $\hat{x}(t)$, the resulting error function is

$$e_M(t) = x(t) - \sum_{k=0}^{M} c_k \phi_k^*(t) \tag{3.15}$$

By computing the average power of this error function and setting its derivatives with respect to c_k to zero, yields an optimal set of $\{c_k\}$ that minimizes the error energy. Also, if $\lim_{M\to\infty} \hat{x}(t) = x(t)$, then the basis functions are said to be complete, that is, the error energy is equal to zero. When dealing with periodic signals, the time interval (t_1, t_2) is equal to the period, T, of the signal. In addition, $\phi_n(t) = \exp\{jn\omega_0\}$ is often selected as the set of basis functions, for $n = 0, \pm 1, \pm 2, \ldots$, and $\omega_0 = 2\pi/T$. The methods of computing the Fourier series coefficients are subsequently discussed.

3.1.6.2 The Complex Exponential Fourier Series

Let the signal $x(t)$ be such that $x(t) = x(t + T)$ and that it satisfies the Dirichlet conditions:[3]

1. $x(t)$ is absolutely integrable over its period, that is,

$$\int_{t_1}^{t_1+T} |x(t)|\,dt < \infty$$

2. the number of maxima and minima of $x(t)$ in each period is finite,
3. the number of discontinuities of $x(t)$ in each period is finite,

then $x(t)$ can be expanded as

$$x(t) = \sum_{k=-\infty}^{\infty} c_k e^{jk\omega_0 t}, \quad \omega_0 = 2\pi f_0 \tag{3.16}$$

where

$$c_k = \frac{1}{T}\int_{t_1}^{t_1+T} x(t)e^{-jk\omega_0 t}\,dt \tag{3.17}$$

for any arbitrary value of t_1.

The coefficients c_k are called the complex Fourier series coefficients for the signal $x(t)$, which, in general, may be complex numbers.

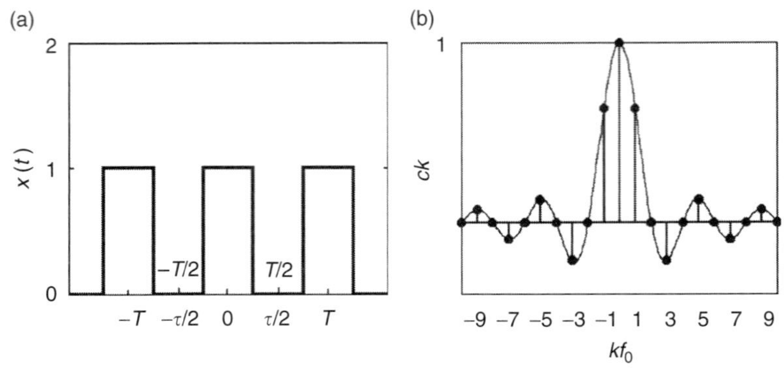

FIGURE 3.7 Rectangular periodic signal and its Fourier series coefficients.

The Dirichlet conditions are only sufficient conditions for the existence of the Fourier series expansion. The Fourier series expansion of signals, which does not satisfy these conditions, can still be obtained.

The complex Fourier series coefficients can be determined by either directly evaluating the integral in (3.17) or using the method of differentiation. The latter method relies on differentiating $x(t)$, for a certain number of times, to produce a train of impulses. These two methods of determining c_k will now be illustrated.

Consider the periodic signal shown in Figure 3.7a, which can be expressed as

$$x(t) = \frac{1}{T}\sum_{n=-\infty}^{\infty}\prod\left(\frac{t-nT}{\tau}\right)$$

Substituting this in Equation 3.17 we have

$$\begin{aligned} c_k &= \frac{1}{T}\int_{-T/2}^{T/2} x(t)e^{-j2\pi k f_0 t}\,dt \\ &= \frac{1}{T}\int_{-\tau/2}^{\tau/2} e^{-j2\pi k f_0 t}\,dt \\ &= \frac{1}{\pi k}\sin(k\pi f_0\tau) \\ &= \tau f_0 \operatorname{sinc}(k f_0 \tau) \end{aligned}$$

Thus,

$$x(t) = \sum_{n=-\infty}^{\infty} \tau f_0 \operatorname{sinc}(k f_0\tau)e^{j2\pi k f_0 t}$$

A graph of the magnitude and phase spectra of the complex Fourier series coefficients are shown in Figure 3.7b for varied τ/T.

Consider the computation of the complex Fourier series coefficients of the signal shown in Figure 3.8a using the method of differentiation. This signal is expanded according to Equation 3.16. Differentiating this equation twice with respect to t yields

$$x''(t) = \sum_{k=-\infty}^{\infty} (j2\pi k f_0)^2 c_k e^{j2\pi k f_0 t}$$

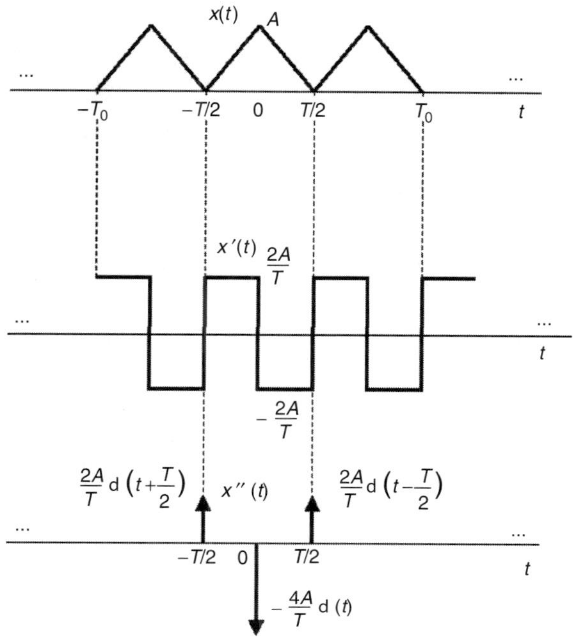

FIGURE 3.8 Illustration of the differentiation technique for Fourier series coefficient computation.

which can be written as

$$x''(t) = \sum_{k=-\infty}^{\infty} \beta_k e^{j2\pi k f_0 t}$$

Figure 3.8c shows the result of differentiating $x(t)$. It is noted that if a signal is periodic, its derivatives will also be periodic. This implies that β_k is the complex Fourier series coefficient of $x''(t)$ and can be computed from

$$\beta_k = \frac{1}{T}\int_{-T/2}^{T/2} x''(t) e^{-j2\pi k f_0 t} dt$$

where

$$x''(t) = 2\frac{A}{T}\left[\delta\left(t+\frac{T}{2}\right) - 2\delta(t) + \delta\left(t-\frac{T}{2}\right)\right]$$

Thus,

$$\beta_k = -8\frac{A}{T^2}\sin^2\left(\frac{\pi k T f_0}{2}\right) = (j2\pi f_0 k)^2 c_k$$

That is,

$$c_k = \begin{cases} \dfrac{2A}{\pi^2 k^2}, & k \text{ odd} \\ 0, & \text{otherwise} \end{cases}$$

and

$$c_0 = \frac{1}{T}\int_{-T/2}^{T/2} x(t)\,dt = \frac{A}{2}.$$

This method of determining c_k, where applicable, is less laborious than the direct method.

3.1.6.3 Fourier Series for Real Signals[3]

A real signal $x(t)$ that satisfies the Dirichlet conditions can be expanded by the following procedure. Recall Equation 3.17 and replace k by $-k$, we obtain

$$c_{-k} = \frac{1}{T}\int_{t_1}^{t_1+T} x(t)e^{j2\pi k f_0 t}\,dt = \left[\frac{1}{T}\int_{t_1}^{t_1+T} x(t)e^{-j2\pi k f_0 t}\,dt\right]^* = c_k^* \tag{3.18}$$

That is, the positive and negative coefficients are complex conjugates of each other for real signals. For such signals, $|c_k|$ has even symmetry and $\angle c_k$ has odd symmetry with respect to $k = 0$. Denote $c_k = (a_k - jb_k)/2$, then $c_{-k} = (a_k + jb_k)/2$, so that

$$c_k e^{j2\pi k f_0 t} + c_{-k} e^{-j2\pi k f_0 t} = \frac{a_k - jb_k}{2} e^{j2\pi k f_0 t} + \frac{a_k + jb_k}{2} e^{-j2\pi k f_0 t}$$

Since c_0 is real and defined as $c_0 = a_0/2$, then we can write

$$x(t) = \frac{a_0}{2} + \sum_{k=1}^{\infty} [a_k \cos(2\pi k f_0 t) + b_k \sin(2\pi k f_0 t)] \tag{3.19}$$

This relationship, which only holds for a real periodic signal $x(t)$, is called the trigonometric Fourier series expansion. Since

$$\begin{aligned} c_k = \frac{a_k - jb_k}{2} &= \frac{1}{T}\int_{t_1}^{t_1+t} x(t)e^{-j2\pi k f_0 t}\,dt \\ &= \frac{1}{T}\int_{t_1}^{t_1+T} x(t)[\cos(2\pi k f_0 t) - j\sin(2\pi k f_0 t)]\,dt \end{aligned} \tag{3.20}$$

Hence

$$a_k = \frac{2}{T}\int_{t_1}^{t_1+T} x(t)\cos(2\pi k f_0 t)\,dt \tag{3.21a}$$

$$b_k = \frac{2}{T}\int_{t_1}^{t_1+T} x(t)\sin(2\pi k f_0 t)\,dt \tag{3.21b}$$

and

$$a_0 = \frac{2}{T}\int_{t_1}^{t_1+T} x(t)\,dt$$

The third method of Fourier series expansion of real signals is given by

$$x(t) = c_0 + \sum_{k=1} c_k \cos(2\pi k f_0 t + \theta_k) \tag{3.22}$$

TABLE 3.3 Fourier Series Symmetry Conditions

Type of Symmetry	Real Fourier Series Coefficient	Complex Fourier Series Coefficients	Comments
Even periodic $x(t) = x(-t)$	$a_k = \frac{4}{T}\int_0^{T/2} x(t)\cos(2\pi k f_0 t)\,dt$ $b_k = 0$	$c_k = \frac{1}{2}a_k$ c_k has real value	Phase of c_k is either zero or π
Odd periodic $x(t) = -x(-t)$	$a_k = 0$ $b_k = \frac{4}{T}\int_0^{T/2} x(t)\sin(2\pi k f_0 t)\,dt$	$c_k = -j\frac{1}{2}b_k$ c_k has imaginary values	Phase of c_k is either $\pi/2$ or $-\pi/2$
Half-wave even symmetry $x(t) = x\left(t+\frac{T}{2}\right)$	a_{2k} and b_{2k} may have nonzero values but $a_{2k+1} = 0$, $b_{2k+1} = 0$	$c_{2k} \neq 0$ $c_{2k+1} = 0$	
Half-wave odd symmetry $x(t) = -x\left(t+\frac{T}{2}\right)$	a_{2k+1} and b_{2k+1} may have nonzero values but $a_{2k} = 0$, $b_{2k} = 0$	$c_{2k} = 0$ $c_{2k+1} \neq 0$	

TABLE 3.4 Properties of the Fourier Series

Property	Signal Description	Fourier Series Coefficients, c_k
Linearity	$ax(t) + by(t)$; a, b constants	$a\alpha_k + b\beta_k$
Multiplication	$x(t)y(t)$	$\alpha_k * \beta_k$
Convolution	$x(t) * y(t)$	$\alpha_k \beta_k$
Parseval's theorem	$\frac{1}{T}\int_{t_1}^{t_1+T} \lvert x(t)^2 \rvert\,dt$	$\sum_{k=-\infty}^{\infty} \lvert \alpha_k \rvert^2$
Time shift	$x(t \pm \tau)$	$\alpha_k e^{\pm j2\pi f_0 \tau}$
Differentiation	$\frac{d^n}{dt^n}x(t)$	$(j2\pi k f_0)^n \alpha_k$
Integration	$\int_T x(\tau)\,d\tau$	$(j2\pi k f_0)^{-1}\alpha_k$, $\alpha_0 = 0$

where c_0 is the dc component, c_k and θ_k represent the amplitude and phase angle of the kth harmonic, respectively. Equation 3.22 is called the harmonic form of Fourier series expansion of $x(t)$. The parameters c_k and θ_k are related to a_k and b_k according to

$$c_0 = \frac{a_0}{2}, \qquad c_k = \sqrt{a_k^2 + b_k^2}, \qquad \theta_k = \tan^{-1}\frac{b_k}{a_k}$$

3.1.6.4 Properties of the Fourier Series[1,4]

Knowledge of signal symmetry can simplify its complex Fourier series coefficients computation. While many forms of symmetry can be established, the following important types of symmetry are more often encountered in signal analysis:

- Even symmetry, $x(t) = x(-t)$
- Odd symmetry, $x(t) = -x(t)$
- Half-wave odd symmetry, $x(t) = -x(t + T/2)$

The effects of symmetry on the Fourier series computations are shown in Table 3.3. The other properties on Fourier series are summarized in Table 3.4, where α_k and β_k are the complex Fourier series coefficients of $x(t)$ and $y(t)$, respectively.

3.1.7 Fourier Transform

Frequency domain method of analyzing periodic CT signals has been presented in the previous section. A different technique, termed the Fourier transform, is used for the analysis of aperiodic signals. The development of the Fourier transform is based on Equations 3.16 and 3.17. Substituting Equations 3.17 into 3.16 gives

$$x(t) = \sum_{k=-\infty}^{\infty}\left(\frac{1}{T}\int_{-T/2}^{T/2} x(\lambda)e^{-j2\pi kf_0\lambda}d\lambda\right)e^{j2\pi kf_0 t} \tag{3.23}$$

Allowing $T \to \infty$, then $1/T \to df$, $kf_0 \to f$, and Equation 3.23 becomes

$$X(f) = \int_{-\infty}^{\infty} x(t)e^{-j2\pi kft}dt \tag{3.24a}$$

and

$$x(t) = \int_{-\infty}^{\infty} X(f)e^{j2\pi ft}df \tag{3.24b}$$

3.1.8 Properties of the Fourier Transform[5,6]

Some of the basic properties of the Fourier transform that are often used in analysis are given in Table 3.5.

The Fourier transform pairs of the following basic signals are also useful for analysis:

$$K \leftrightarrow K\delta(f)$$

$$\text{sgn}(t) \leftrightarrow \frac{1}{j\pi f}$$

$$u(t) \leftrightarrow \frac{1}{2}\delta(f) + \frac{1}{j2\pi f}$$

$$\cos(2\pi f_0 t) \leftrightarrow \frac{1}{2}\delta(f-f_0) + \frac{1}{2}\delta(f+f_0)$$

A two-step procedure is required in order to determine the Fourier transform of a periodic signal. If $x(t)$ is a periodic signal with period T, then $x(t)$ can be Fourier series expanded as

$$x(t) = \sum_{k=-\infty}^{\infty} c_k e^{j2\pi kf_0 t}, \quad f_0 = \frac{1}{T} \tag{3.25}$$

Applying the linearity property to this equation gives

$$X(f) = \sum_{k=-\infty}^{\infty} c_k\delta(f-kf_0) \tag{3.26}$$

as the Fourier transform of any periodic signal $x(t)$. Equation 3.26 explains the difference between the frequency spectrum produced by the Fourier series analysis and the amplitude spectral density produced by the Fourier transformation. Thus, the frequency spectrum is a (discrete) display of c_k versus kf_0, whereas

TABLE 3.5 Properties of the Fourier Transform

Property	Signal Description	Fourier Transform
Linearity	$ax(t) + by(t)$; a, b constants	$aX(f) + bY(f)$
Evenness and oddness	$x(-t) = x(t)$ $x(-t) = -x(t)$	$X(f) = 2\int_0^{\infty} x(t)\cos(2\pi ft)\,dt$ $X(f) = -2\int_0^{\infty} x(t)\sin(2\pi ft)\,dt$
Time shift	$x(t-\tau)$	$e^{-j2\pi f\tau}X(f)$
Time scale	$x(at)$	$\frac{1}{\lvert a\rvert}X\left(\frac{f}{a}\right)$
Time reversal	$x(-t)$	$X^{*}(f)$
Duality	$X(t)$	$x(-f)$
Time convolution	$x(t) * y(t)$	$X(f)\,Y(f)$
Frequency convolution	$x(t)\,y(t)$	$X(f) * Y(f)$
Modulation	$x(t)e^{j2\pi f_0 t}$	$X(f-f_0)$
Time differentiation	$\frac{d^n}{dt^n}x(t)$	$(j2\pi f)^n X(f)$
Frequency differentiation	$t^n x(t)$	$\left(\frac{j}{2\pi}\right)^n \frac{d^n}{df^n}X(f)$
Integration	$\int_{-\infty}^{\infty} x(\tau)\,d\tau$	$\frac{1}{j2\pi f}X(f) + \frac{1}{2}X(0)\delta(f)$
Correlation	$R_{xy}(\tau) = \int_{-\infty}^{\infty} y(t)x(t-\tau)\,dt$	$Y(-f)\,X(f)$
Parseval's theorem	$\int_{-\infty}^{\infty} \lvert x(t)\rvert^2 dt$	$\int_{-\infty}^{\infty} \lvert X(f)\rvert^2 df$

the amplitude spectral density gives a continuous display of the amplitude density spectrum, which in this case is in the form of impulses, rather than just a number.

Similarly the Fourier transform of a train of impulses of the form

$$p(t) = \sum_{n=-\infty}^{\infty} \delta(t-nT) \tag{3.27}$$

is given by

$$P(f) = \frac{1}{T}\sum_{k=-\infty}^{\infty} \delta(f-kF_s), \quad F_s = \frac{1}{T} \tag{3.28}$$

3.1.8.1 Energy and Power Spectral Density[6]

Suppose $x(t)$ is an aperiodic signal with a Fourier transform $X(f)$, then its energy is given by

$$E = R_{xx}(0) = \int_{-\infty}^{\infty} \lvert x(t)\rvert^2 dt = \int_{-\infty}^{\infty} \lvert X(f)\rvert^2 df \tag{3.29}$$

This is Parseval's theorem and it shows that the principle of conservation of energy in the time and frequency domains holds. The amplitude spectrum $X(f)$ can be expressed as

$$X(f) = \lvert X(f)\rvert \angle X(f)$$

and denoting

$$S_{xx}(f) = |X(f)|^2$$

then the total energy of the signal is given by

$$E = \int_{-\infty}^{\infty} S_{xx}(f)df \tag{3.30}$$

where $S_{xx}(f)$ represents the distribution of the signal energy as a function of frequency. $S_{xx}(f)$ is termed the energy spectral density for the finite energy signal $x(t)$.

Consider a periodic signal $x(t)$ with an autocorrelation function

$$R_{xx}(\tau) = \lim_{T\to\infty} \frac{1}{T}\int_{-T/2}^{T/2} x(t)x^*(t-\tau)dt$$

Then,

$$R_{xx}(0) = \lim_{T\to\infty} \frac{1}{T}\int_{-T/2}^{T/2} |x(t)|^2 dt \tag{3.31}$$

which equals the signal power. Following the same procedure for energy signals, we define $S_{xx}(f)$ as the Fourier transform of $R_{xx}(\tau)$ so that

$$P = R_{xx}(0) = \int_{-\infty}^{\infty} S_{xx}(f)df \tag{3.32}$$

and $S_{xx}(f)$ is termed the power spectral density of the periodic signal $x(t)$.

The need to analyze stationary random signals also arises in many practical situations. The properties of such signals can be inferred from their correlation functions. For example the autocorrelation function, $\phi_{xx}(\tau)$ of a stationary random signal decreases and goes to zero as τ increases since the events become uncorrelated for a large separation of time. Hence, $\phi_{xx}(\tau) = \phi_{xx}(-\tau)$ and its Fourier transform exists. Consequently we can write

$$\phi_{xx}(0) = \int_{-\infty}^{\infty} \Gamma_{xx}(f)df \tag{3.33}$$

where $\Gamma_{xx}(f)$ and $\phi_{xx}(0)$ represent, respectively, the power spectral density and the average power of a random process.

3.1.9 Sampled Continuous-Time Signals

Discrete-time (DT) signals arise either naturally or by sampling continuous-time (CT) signals; however, the latter form is more often encountered in practice. In this case, a digital signal is formed from a CT signal through the process of analog-to-digital conversion. The first part of this process is the sampling of the analog signal, that is, the conversion of $x(t)$ into $x(nT_s)$, where T_s is the sampling period and its reciprocal, $F_s = 1/T_s$, is the sampling frequency in samples per second. The sampling frequency must be appropriately selected to avoid spectral distortion (aliasing), thereby ensuring that $x(t)$ can be reconstructed from its samples. To gain a good understanding of this procedure the sampling process is examined in the frequency domain.

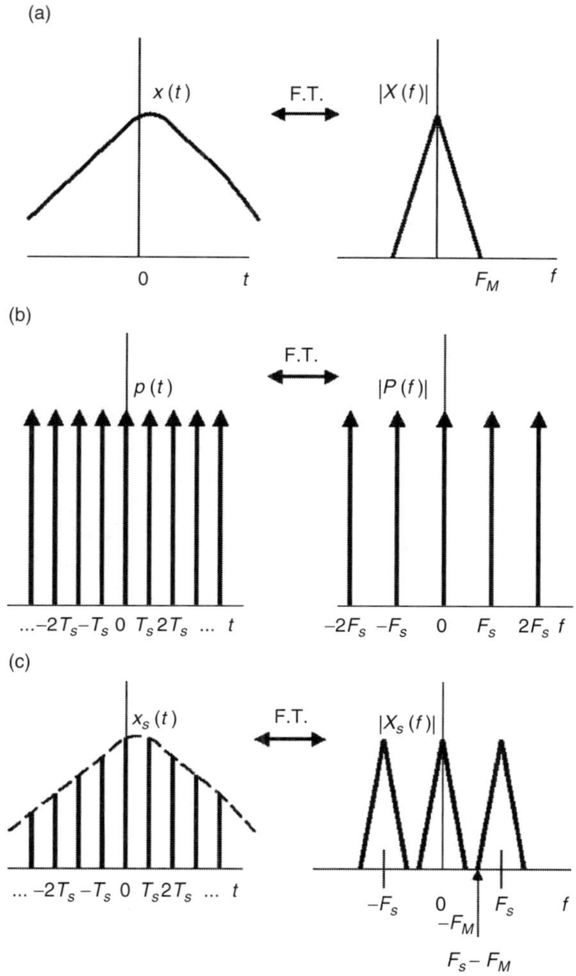

FIGURE 3.9 Ideal impulse sampling process: (a) bandlimited continuous-time signal and its Fourier transform, (b) train of impulses and its spectra, (c) sampled signal and its spectra.

3.1.9.1 Impulse Sampling[6–9]

Consider an idealized impulse-sampling process shown in Figure 3.9, where $x(t)$ is to be sampled using the train of impulses $p(t)$, so that

$$x_s(t) = x(t)p(t) = x(t)\sum_{n=-\infty}^{\infty}\delta(t-nT_s) = \sum_{n=-\infty}^{\infty}x(nT_s)\delta(t-nT_s) \tag{3.34}$$

The main difference between $x_s(t)$ and $x(nT_s)$ is that the former is essentially a CT signal with zero values except at the integer values of T_s, while the latter is a perfectly selected sample of $x(t)$ as a result of impulse sampling.

The Fourier transform of Equation 3.34 yields

$$X_s(f) = X(f) * P(f) = F_s\sum_{k=-\infty}^{\infty}X(f-F_s) \tag{3.35}$$

where $X(f)$ is the spectrum of $x(t)$.

It is observed that $X_s(f)$ consists of a scaled and periodically repeated version of $X(f)$ with period F_s. As shown in Figure 3.9c, it is obvious that when $F_s - F_M \geq F_M$, there is no overlapping of the spectra and the signal $x(t)$ can be recovered completely from $x_s(t)$. However, if $F_s - F_M < F_M$ the replicas of $X(f)$ will overlap, resulting in a distorted spectrum and as such, $x(t)$ can no longer be recovered from its sampled version. Consequently, in order to recover $x(t)$ from its samples, the sampling frequency should be such that

$$F_s - F_M \geq F_M$$

that is,

$$F_s \geq 2F_M$$

This is called the Nyquist sampling theorem. The minimum sampling frequency $F_s = 2F_M$ is called the Nyquist frequency. Sampling a signal at less than the Nyquist frequency results in a spectral distortion termed aliasing. Furthermore, sampling a signal at a frequency of at least Nyquist frequency implies that an ideal low-pass filter (LPF) with a gain of $1/F_s$ and cutoff frequency F_c can be used to recover its original spectrum, where $F_M \leq F_c \leq F_s - F_M$.

Suppose we want to reconstruct $x(t)$ from its samples. Assume that $X(f)$ is the spectrum of $x(nT_s)$, with no aliasing, as shown in Figure 3.9c. Thus,

$$X_a(f) = \begin{cases} \frac{1}{F_s} X(f), & |f| \leq \frac{F_s}{2} \\ 0, & |f| > \frac{F_s}{2} \end{cases} \tag{3.36}$$

Note that

$$X_a(f) = \sum_{n=-\infty}^{\infty} x(nT_s) e^{-j2\pi nfT_s} \tag{3.37}$$

so that its inverse Fourier transform is given by

$$\begin{aligned} x_a(t) &= \frac{1}{F_s} \sum_{n=-\infty}^{\infty} x(nT_s) \int_{-F_s/2}^{F_s/2} e^{j2\pi f(t-nT_s)} df \\ &= \sum_{n=-\infty}^{\infty} x(nT_s) \frac{\sin[\pi(t-nT_s)/T_s]}{\pi(t-nT_s)/T_s} \end{aligned} \tag{3.38}$$

This is the formula for the reconstruction of $x(t)$ from its samples. That is, $x(t)$ is generated by multiplying the appropriately shifted function $g(t) = \mathrm{sinc}(tF_s)$ by the corresponding samples of $x(nT_s)$.

3.1.9.2 Practical Sampling[8–10]

The above discussion on sampling is based on the idealized models of periodic impulse sampling and bandlimited interpolation. In practice, CT signals are not precisely bandlimited just as impulse signals and ideal low-pass filters do not exist physically. Figure 3.10 represents the block diagram for the conversion of continuous signals into their discrete forms. The continuous-time signal is prefiltered, sampled, quantized, and finally encoded into finite-length words of, say, b bits. The prefilter, which is also called anti-aliasing filter (AAF), is a low-pass filter that is needed to limit the input signal bandwidth to $F_s/2$ prior to sampling to avoid aliasing. In practice this filter will possess non-ideal characteristics, hence

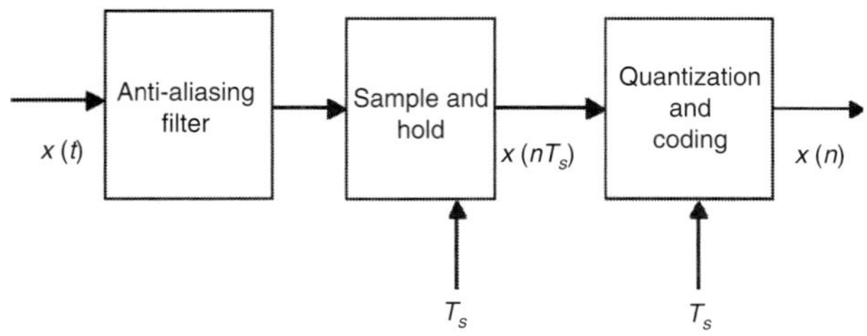

FIGURE 3.10 Practical sampling of continuous-time signals.

it should be designed to provide sufficient attenuation, usually to a level undetectable by the analog-to-digital converter (ADC) at frequencies above the Nyquist frequency.

The prefiltered signal is fed into the ADC where it will be converted to a DT signal. The ADC has an in-built sample-and-hold circuit and it operates at the sampling rate of F_s; however, the sampling function has a finite width as opposed to the impulse sampling discussed above. The sampling operation can be modeled by the finite-width pulse sampler shown in Figure 3.11b, in which the sampling gate is open for τ-out of T_s seconds and shorted to ground the remainder of the sampling interval. Here $p(t)$ is expressed as

$$p(t) = \sum_{n=-\infty}^{\infty} \Pi\left(\frac{t-nT_s}{\tau}\right)$$

which can be Fourier series expanded as

$$p(t) = \sum_{n=-\infty}^{\infty} c_k e^{j2\pi k f_s t}$$

where

$$c_k = \frac{\tau}{T_s}\sin c\left(k\frac{\tau}{T_s}\right)$$

The Fourier transform of the sampled signal can be written as

$$X_s(f) = \sum_{n=-\infty}^{\infty} c_k X(f-kF_s)$$

Note that c_k is not constant in this expression (as opposed to the impulse sampling) since its value depends on the harmonic number (k) as well as the duty cycle τ/T_s. The discrete-time signal, $x(nT_s)$, is fed into the quantizer where each sample is transformed into one of the nearest finite sets of prescribed values, that is, $\hat{x}(n) = Q(x(nT_s))$, where $\hat{x}(n)$ is the quantized sample. The quantization process is shown in Figure 3.11d for zero-order sample hold ADC, where L_i denotes the quantization level and Δ is the quantization step. The quantization error (or noise) incurred in this process is

$$L_i - \frac{\Delta}{2} < x(nT_s) < L_i + \frac{\Delta}{2}$$

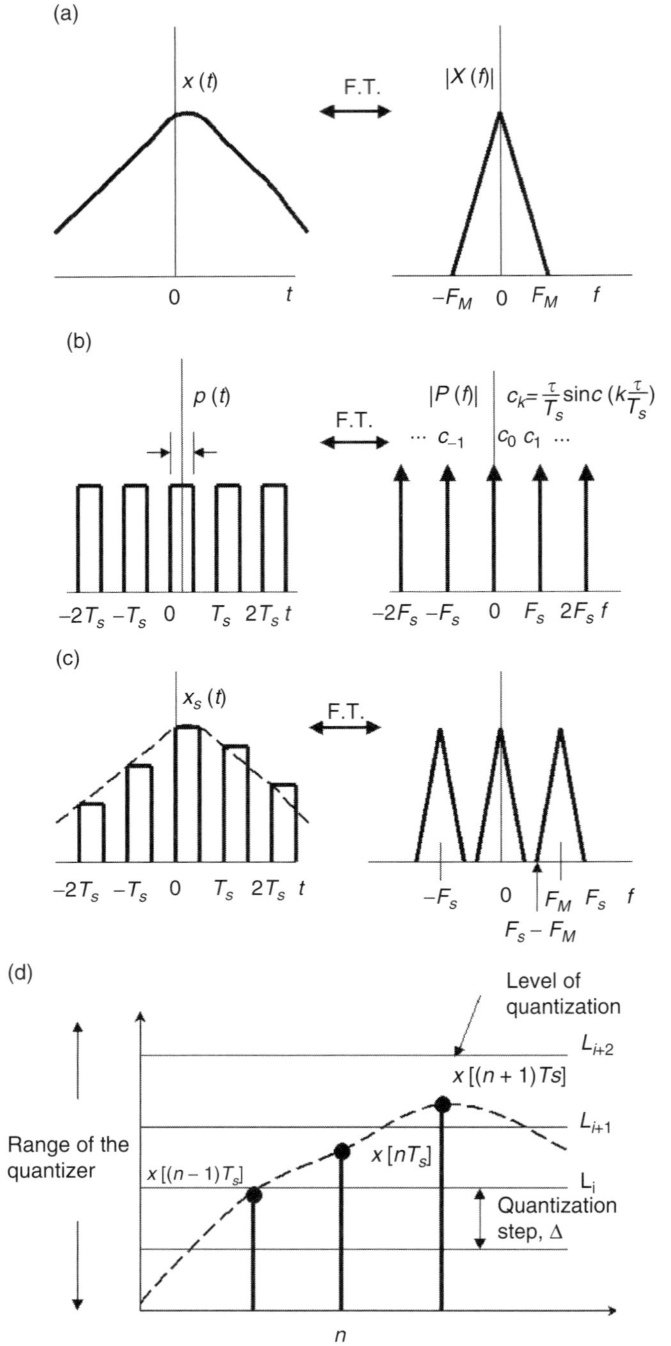

FIGURE 3.11 Finite-width pulse sampling signals and spectra: (a) bandlimited signal and spectrum, (b) finite-width train of pulses and its transform, (c) sampled signal and its spectra, (d) quantization process.

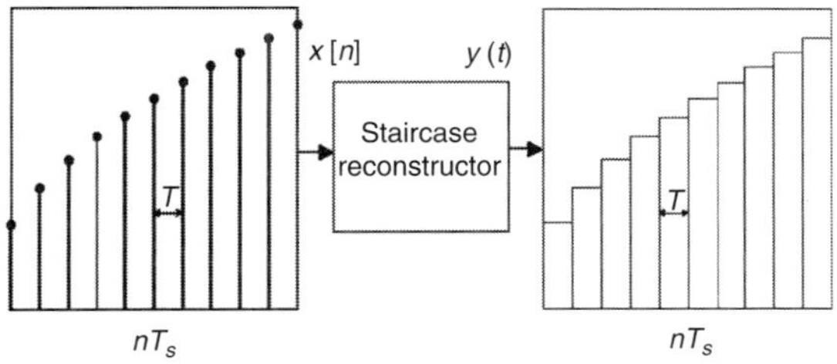

FIGURE 3.12 Digital-to-analog conversion process.

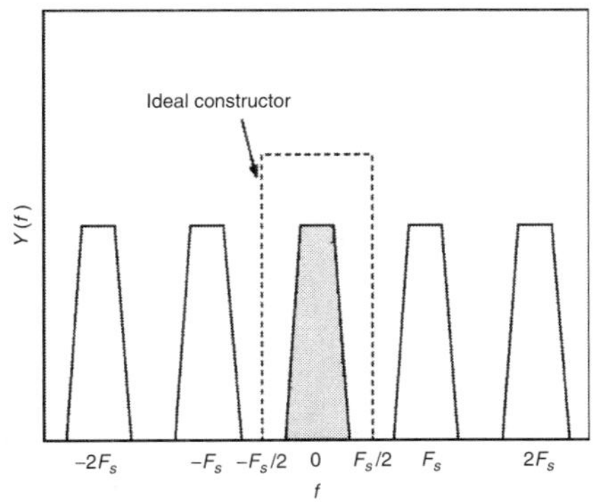

FIGURE 3.13 Frequency response of the ideal reconstruction filter.

From statistical considerations, the noise power is found to be $\Delta^2/12$ watts. A common measure of the performance of the ADC is the ratio of the signal power to noise power, and this is called the signal-to-quantization noise power, which expressed in decibels (dB) is

$$\text{SQNR(dB)} = 1.76 + 6.02b$$

3.1.9.3 Digital-to-Analog Conversion[8–12]

Reconstruction of the analog signal from its sampled form is closely akin to lowpass filtering of the sampled signal. Figure 3.12 shows how an analog signal can be reconstructed by filling the gaps between samples and holding the current value constant till the next sample is received. Consider an ideal reconstruction filter with an impulse response function $h(t)$, then its response is given by

$$y(t) = \sum_{n=-\infty}^{\infty} x(nT_s)h(t - nT_s)$$

Taking the Fourier transform of this equation gives

$$Y(f) = H(f)X(f)$$

where $X(f)$ is the periodic spectrum of $x(nT_s)$ as shown in Figure 3.13 and $h(t) = \text{sinc}(F_s t)$. Note that $h(t)$ is noncausal; hence it cannot be used for real-time applications. Furthermore, since $h(t)$ is not time-limited, an infinite number of impulse responses must be used for interpolating between values in order to obtain exact results. Consequently, alternative reconstruction filters such as zero-order hold (staircase), first-order, or fractional-order holds are used in practice. However, the staircase reconstruction filter is, by and large, the simplest and most widely used in practice. The impulse response of this filter is given

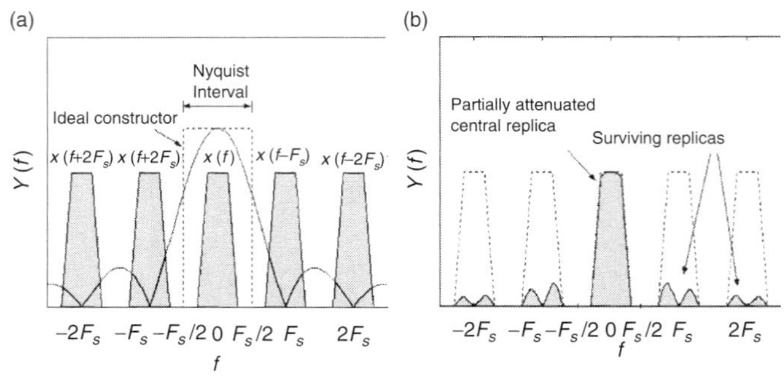

FIGURE 3.14 (a) Staircase reconstruction in frequency domain, (b) Result of staircase reconstruction in frequency domain.

as $h_o(t) = u(t) - u(t - T_s)$ since it must have a duration of T_s seconds to fill the entire gap between the samples. Thus, in effect this filter, as its name implies, generates a staircase approximation to the original analog signal. The frequency response of the filter is

$$H_o(f) = T\frac{\sin(\pi f T_s)}{\pi f T_s}e^{-j\pi f T_s}$$

Though the output of the staircase reconstruction filter is smoother than its sampled form, see Figure 3.14a, it contains spurious high-frequency components due to the sudden jumps in the staircase levels as different sampled values are considered. In addition, holding each of $x(nT_s)$ by T_s seconds introduces a time shift of $T_s/2$ to the output signal. However, this time delay has virtually no effect on the quality of the output signal. Figure 3.14b compares the signal spectra before and after the staircase reconstruction filter. It is noted that the output spectrum is slightly distorted due to non-ideal characteristics of $H_o(f)$ and that distorted and attenuated versions of $X(f)$ remain centered at nonzero multiples of F_s. These remaining spectral replicas may be removed by using an anti-imaging filter.[8,11,12] In essence, the anti-imaging filter smoothens out the discontinuities produced by the staircase reconstruction filter as illustrated in Figure 3.15.

3.1.10 Frequency Analysis of Discrete-Time Signals

The analysis of discrete-time (DT) signals in the frequency domain is very much similar to that of continuous-time (CT) signals. As in the CT analysis, the techniques for the analysis depend on the type of signal. Analysis of aperiodic DT signals will be considered first.

3.1.10.1 Discrete-Time Fourier Transform[6–8]

The decomposition of an aperiodic DT signal into its frequency components is carried out using discrete-time Fourier transformation (DTFT). Thus, the DTFT of $x(n)$ is given by

$$X(f) = \sum_{n=-\infty}^{\infty} x(n)e^{-j2\pi fn} \tag{3.39}$$

Unlike the Fourier transform of analog signal, $X(f)$ is periodic with period F_s; hence, the frequency range for a DT signal is unique over the frequency interval $(-F_s/2, F_s/2)$ or, equivalently $(0, F_s)$. Note that Equation 3.39 must be absolutely summable in order for $X(f)$ to exist, that is,

$$\sum_{n=-\infty}^{\infty} |x(n)| < \infty$$

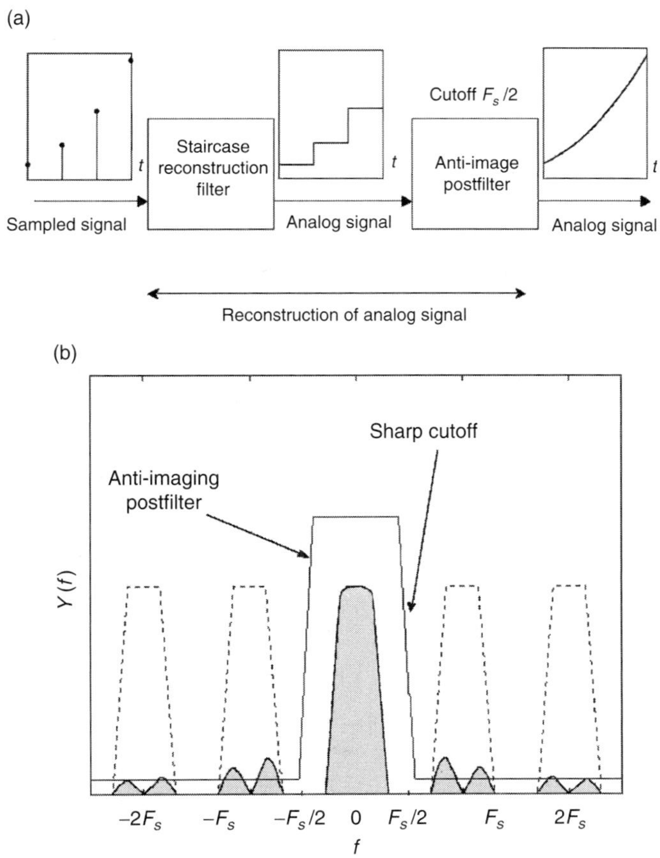

FIGURE 3.15 (a) Reconstruction of analog signal. (b) effect of anti-image postfiltering.

If the spectrum of a signal exists, then we can find the signal from its spectrum through the inverse DTFT. The inverse DTFT of $X(f)$ is given as

$$x(n) = \frac{1}{F_s}\int_{-F_s/2}^{F_s/2} X(f)e^{j2\pi fn}df$$

The properties of the DTFT of a DT signal are shown in Table 3.6.

For DT signals, the convolution of two sequences $x(n)$ and $y(n)$ is expressed as

$$x(n) * y(n) = \sum_{m=-\infty}^{\infty} x(m)y(n-m) \tag{3.40}$$

while the cross-correlation function of $x(n)$ and $y(n)$ is given by

$$R_{xy}(n) = x(n) \oplus y(n) = \sum_{m=-\infty}^{\infty} x(m)y^*(m-n)$$

When $x(n) = y(n)$, then the autocorrelation of $x(n)$ is

$$R_{xx}(n) = \sum_{m=-\infty}^{\infty} x(m)x^*(m-n)$$

TABLE 3.6 Properties of DTFT

Property	Signal Description	Frequency Domain
Even symmetry (real signal)	$x_e(n) = \frac{1}{2}\{x(n)+x(-n)\}$	$X_e(f) = \sum_{n=-\infty}^{\infty} x_e(n)\cos(2\pi nf)$
Odd symmetry (real signal)	$x_o(n) = \frac{1}{2}\{x(n)-x(-n)\}$	$X_o(f) = -\sum_{n=-\infty}^{\infty} x_o(n)\sin(2\pi nf)$
Linearity	$ax(n) + by(n)$	$aX(f) + bY(f)$
Time shifting	$x(n-m)$	$e^{-j2\pi fm}X(f)$
Time reversal	$x(-n)$	$X(-f)$
Convolution	$x(n) * y(n)$	$X(f)Y(f)$
Correlation	$R_{xy}(n) = x(n) \oplus y(n)$	$S_{xy}(f) = X(f)Y(-f)$
Wiener–Khintchine theorem	$R_{xx}(n)$	$S_{xx}(f)$
Frequency shifting	$e^{j2\pi f_0 n}x(n)$	$X(f-f_0)$
Modulation	$x(n)\cos(2\pi nf_0)$	$\frac{1}{2}\{X(f+f_0)+X(f-f_0)\}$
Multiplication	$x(n)y(n)$	$\frac{1}{F_s}\int_{F_s} X(\lambda)Y(f-\lambda)d\lambda$
Differentiation in the frequency domain	$nx(n)$	$\frac{j}{2\pi}\frac{dX(f)}{df}$
Time differencing	$x(n) - x(n-1)$	$(1-e^{-j2\pi f})X(f)$
Summation	$\sum_{m=-\infty}^{n} x(m)$	$\frac{X(f)}{(1-e^{-j2\pi f})} + \frac{F_sX(0)}{2}\sum_{m=-\infty}^{\infty}\delta(f-mF_s)$
Conjugation	$x^*(n)$	$X^*(-f)$
Parseval's theorem	$\sum_{n=-\infty}^{\infty} x(n)y^*(n)$	$\frac{1}{F_s}\int_{F_s} X(f)Y^*(f)df$

With the exception of the reflection operation (for the convolution), the procedures for computing the convolution and correlation are the same. Hence, it is more computationally efficient to use the same algorithm for evaluating both functions. To achieve this, one of the sequences is reflected (only for the correlation analysis), followed by convolution operation, that is,

$$R_{xy}(n) = x(n) * y^*(-n)$$

and

$$R_{xx}(n) = x(n) * x^*(-n)$$

The energy of an aperiodic signal is computed from

$$E = \sum_{n=-\infty}^{\infty} \left|x(n)^2\right| = \sum_{n=-\infty}^{\infty} x(n)x^*(n)$$

Substituting the conjugated form of 3.40 into this equation gives

$$E = \frac{1}{F_s}\int_{F_s} |X(f)|^2 df = \frac{1}{F_s}\int_{F_s} S_{xx}(f)df \tag{3.41}$$

This expression relates the distribution of the energy of an aperiodic signal to frequency. The quantity $|X(f)|^2$ is called the energy spectral density of $x(n)$. The DTFT of some commonly encountered signals are shown in Table 3.7.

TABLE 3.7 DTFT of Common DT Signals

$x(n)$	Frequency-Domain Representation, $X(f)$
$\delta(n)$	1
$A, -\infty < n < \infty$	$AF_s \sum_{k=-\infty}^{\infty} \delta(f - mF_s),\ F_s = \frac{1}{T}$
$u(n)$	$\frac{1}{1-e^{-j2\pi f}} + \frac{F_s}{2} \sum_{k=-\infty}^{\infty} \delta(f - kF_s)$
$\Pi\left(\frac{n}{2q+1}\right)$	$\frac{\sin[(2q+1)\pi f]}{\sin(\pi f)}$
$\Lambda\left(\frac{n}{q}\right)$	$\frac{\sin^2(\pi f q)}{q \sin^2(\pi f)}$
$\mathrm{sgn}(n)$	$\frac{2}{1-e^{-j2\pi f}}$
$\alpha^n u(n)$	$\frac{1}{1-\alpha e^{-j2\pi f}},\ \lvert\alpha\rvert < 1$
$\alpha^{\lvert n\rvert}$	$\frac{1-\alpha^2}{1-2\alpha\cos(2\pi f)+\alpha^2},\ \lvert\alpha\rvert < 1$
$n\alpha^n u(n)$	$\frac{\alpha e^{-j2\pi f}}{(1-\alpha e^{-j2\pi f})^2},\ \lvert\alpha\rvert < 1$
$e^{-\alpha n} u(n)$	$\frac{1}{1-e^{-(\alpha + j2\pi f)}},\ \lvert\alpha\rvert > 0$
$e^{-\alpha\lvert n\rvert}$	$\frac{1-e^{-2\alpha}}{1-2e^{-\alpha}\cos(2\pi f)+e^{-2\alpha}},\ \lvert\alpha\rvert > 0$
$e^{j2\pi f_0 n}$	$F_s \sum_{k=-\infty}^{\infty} \delta(f - f_0 + kF_s)$
$\cos(2\pi f_0 n + \theta)$	$\frac{F_s}{2} \sum_{k=-\infty}^{\infty} \{e^{j\theta}\delta(f - f_0 + kF_s) + e^{-j\theta}\delta(f + f_0 + kF_s)\}$
$\frac{\sin(2\pi f_c n)}{\pi n}$	$\Pi\left(\frac{f}{2f_c}\right)$

3.1.10.2 Discrete Fourier Series[6–8]

Suppose $x(n)$ is a periodic DT signal with period N, then it is possible to obtain its discrete Fourier series (DFS) expansion in a manner analogous to the computation of the complex Fourier series for the CT signals. The orthogonal basis function for the DFS is $W_N^{-n} = e^{j2\pi n/N}$ so that the decomposition of $x(n)$ into a sum of N harmonically related complex exponentials is expressed as

$$x(n) = \sum_{k=0}^{N-1} c_k e^{j2\pi kn/N} = \sum_{k=0}^{N-1} c_k W_N^{-kn} \tag{3.42}$$

where c_k represents the discrete Fourier series coefficients. Multiplying both sides of 3.42 by W_N^{mn}, summing over one period, and using the fact that

$$\sum_{k=0}^{N-1} W_N^{n(k-m)} = \begin{cases} N, & k = m \\ 0, & k \neq m \end{cases}$$

TABLE 3.8 Properties of the Discrete Fourier Series

Property	Signal Description	Discrete Fourier Series Coefficients
Linearity	$\sum_{m=0}^{Q} \beta_m x_m(n)$	$\sum_{m=0}^{Q} \beta_m c_{m,k}$
Time shift	$x(n-m)$	$c_k W_N^{km}$
Time-reversal	$x(-n)$	c_{-k}
Modulation	$x(n)\, W_N^{-mn}$	c_{k-m}
Real $x(n)$	$x(n)$	$c_k = c_k^*$
$x_e(n)$ is an even DT signal	$x_e(n)$	$\mathrm{Re}(c_k)$
$x_0(n)$ is an odd DT signal	$x_o(n)$	$j\{\mathrm{Im}(c_k)\}$

results in

$$c_k = \frac{1}{N}\sum_{n=0}^{N-1} x(n)\, W_N^{kn}, \quad k = 0, 1, 2, \ldots, N-1 \tag{3.43}$$

The DFS coefficients $\{c_k\}$ provide the description of $x(n)$ in the frequency domain since c_k represents the amplitude and phase spectra associated with the kth harmonic component. Note that $\{c_k\}$ is a periodic sequence with fundamental period N, that is, $c_k = c_{k+N}$. Hence, any N consecutive samples of the signal or its DFS coefficients provide a complete description of the signal in the time or frequency domains. As shown in Table 3.8, the properties of the DFS follow closely those given for the DTFT. One of the main advantages of DFS over the DTFT is the replacement of the infinite summation in the DTFT by a finite sum in the DFS, thus allowing the computations of DFS and inverse DFS by digital computers.

Consider a periodic DT signal with period N, then its average power is

$$P = \frac{1}{N}\sum_{n=0}^{N-1} |x(n)|^2 \tag{3.44}$$

Using Equation 3.42 in 3.44 gives

$$P = \frac{1}{N}\sum_{n=0}^{N-1} x(n)x^*(n) = \frac{1}{N}\sum_{n=0}^{N-1} x(n)\left(\sum_{k=0}^{N-1} c_k^* W_N^{kn}\right) = \sum_{k=0}^{N-1} |c_k|^2 \tag{3.45}$$

This is a Parseval's relation for the DT periodic signals, which indicates that the average power in $x(n)$ is the sum of the harmonics power. Consequently, the sequence $|c_k|^2$ is the power spectral density as this represents the distribution of power as a function of frequency.

3.1.10.3 The Discrete Fourier Transform[6,8,13]

The discrete Fourier transform (DFT) is an important and extremely powerful technique for analyzing DT signals. Contrary to the DTFT, the DFT is applicable to finite-length sequences and it produces finite-length discrete spectra. Consequently, this transformation is amenable to digital computations and it is suitable for use in digital hardware implementations. As a result of its practicability, DFT has become a very useful tool for analyzing various waveforms or data that arise in many disciplines.

The DFT is a mapping of an N sample sequence, $x(n)$, into another N sequence $X(k)$ in the frequency domain, that is,

$$X(k) = \sum_{n=0}^{N-1} x(n)\, W_N^{kn}, \quad k = 0, 1, \ldots, N-1 \tag{3.46}$$

TABLE 3.9 Properties of DFT

Property	Signal Description	Discrete Fourier Transform
Linearity	$\sum_{m=0}^{Q} a_m x_m(n)$	$\sum_{m=0}^{Q} a_m X_m(k)$
Circular shift	$x[(n-m)\text{mod } N]$	$W_N^{km} X(k)$
Modulation	$W_N^{-qn} x(n)$	$X[(k-q)\text{mod } N]$
Time reversal	$x^*[-n \text{ mod } N]$	$x^*(k)$
Complex conjugation	$x^*(n)$	$X^*[-k \text{ mod } N]$
Circular convolution	$\sum_{m=0}^{N-1} x(m)y[(n-m)\text{mod } N]$	$X(k)Y(k)$
Multiplication	$x(n)y(n)$	$\frac{1}{N}\sum_{m=0}^{N-1} X(m)\, Y[(k-m)\text{mod } N]$
Parseval's theorem	$\sum_{k=0}^{N-1} x(n)y^*(n)$	$\frac{1}{N}\sum_{k=0}^{N-1} X(k)\,Y^*(k)$
Real part of signal	$\text{Re}\{x(n)\}$	$\frac{1}{2}\{X(k)+X^*(N-k)\}$
Imaginary part of signal	$j\,\text{Im}\{x(n)\}$	$\frac{1}{2}\{X(k)-X^*(N-k)\}$
Complex even	$x_{ce}(n) = \frac{1}{2}\{x(n)+x^*(N-n)\}$	$X_R(k)$
Complex odd	$x_{co}(n) = \frac{1}{2}\{x(n)-x^*(N-n)\}$	$jX_I(k)$
Any real signal	$x(n)$	$X(k) = X^*(N-k)$ $X_R(k) = X_R(N-k)$ $X_I(k) = -X_I(N-k)$ $\lvert X(k)\rvert = \lvert X(N-k)\rvert$ $\angle X(k) = -\angle X(N-k)$

$X(k)$ is called the kth harmonic and this exists provided all the samples of $x(n)$ are bounded. In order to recover $x(n)$ from $X(k)$ we need to perform inverse transformation. Thus, the inverse discrete Fourier transform (IDFT) of $X(k)$ is defined as

$$x(n) = \frac{1}{N}\sum_{k=0}^{N-1} X(k)\,W_N^{-kn}, \quad n = 0,1,\ldots,N-1 \tag{3.47}$$

The DFT and DFS are conceptually different in the sense that the DFS is only applicable to periodic signals whereas DFT assumes that the signal is periodic, that is, there is an inherent windowing of nonperiodic signal if analyzed by the DFT technique. Also the scaling factor is applicable to the synthesis equation for the DFT operation, whereas it is used in the analysis equation in the DFS analysis.

The DFT properties bear strong resemblance to those of the DFS and DTFT as shown in Table 3.9.

The following should be noted in the DFT computation: Circular shift operation can be considered as wrapping the part of the sequence that falls outside of 0 to $N-1$ to the front of the sequence, that is, $x[(-n) \text{ mod } N]$ is equivalent to $x(N-n)$.

As a result of the circular shift, linear convolution as given by Equation 3.40 is different from circular convolution given in Table 3.9. Thus,

$$x(n)\otimes_N y(n) = \sum_{m=0}^{N-1} x(m)y[(n-m)\text{mod } N] = \sum_{m=0}^{N-1} x[(n-m)\text{mod } N]y(m)$$

where $x[(n-m) \bmod N]$ is the reflected and circularly translated version of $x(n)$. However, by appropriate selecting the value of N, both the circular and linear convolution can be the same. Thus, if signals $x(n)$ and $y(n)$ are of length N_1 and N_2, respectively, then circular and linear convolution are the same provided $N \geq N_1 + N_2 - 1$.

Circular correlation can be implemented by circular convolution since the cross correlation of the two sequences $x(n)$ and $y(n)$ can be expressed as

$$R_{xy}(n) = x(n) \otimes_N y(n) = x(n) \otimes_N y^*[(-n) \bmod N]$$

From Equations 3.46 and 3.47, it is noted that to compute the DFT coefficients would require N^2 complex multiplications and $N(N-1)$ complex additions. This can be a computational load if N is very large. Fast Fourier transforms (FFTs) are different types of algorithms that have been developed to speed up the computation of the DFT coefficients. Readers can refer to many textbooks where the development of the various forms of the FFT algorithms has been discussed.[7,8] It can be shown that the number of computations required for the FFT algorithms can be expressed as a constant times $N * \log_2 N$. Consequently, there is a reduction in computation when an FFT algorithm is used for the DFT computation.

3.1.10.4 Remarks on the DFT Processing of Signals

Zero Padding[6,8]

The use of FFT for DFT coefficients computation imposes some constraints on the value of N, for example N has to be a power of 2 for radix-2 FFT algorithm. Also, circular convolution is an undesirable solution as it manifests from the IDFT of the product of the transform of two sequences. Zero-padding is a technique for remedying the above situations, that is, the zero-padding is used either to augment the sequence length so that either a radix-2 FFT algorithm can be used or to ensure that both the linear and circular convolution are the same. In addition, this procedure is also used to provide a better-looking display of the signal spectrum since the frequency spacing of the FFT samples decreases as N increases.

Error Sources[8,10]

The resultant spectrum of a sampled signal comprises the analog spectrum repeated at the integer multiples of the sampling frequency. The overlapping of the analog signal spectrum with its shifted versions causes *aliasing.* In practice, excessive errors due to aliasing are minimized by either increasing the sampling rate or prefiltering the signal to remove the high-frequency spectral components. Another source of error in the DFT processing of signals is the *spectral leakage.* This is caused by using a window to truncate an infinitely long signal to obtain a finite-length data for DFT processing. It is known that windowing the samples of a signal in the time domain transforms to convolution of sampled signal spectrum and the window spectrum in the frequency domain. Suppose the window width does not correspond to an integer multiple of periods of all the frequency components of a discrete-time signal, then a single frequency component will spread (leak) into other frequency locations in the DFT of the truncated data. This phenomenon causes spectral distortion and makes it difficult to determine whether or not two closely adjacent frequencies are present in a signal. Spectral resolution becomes better if the window width is increased, or by choosing a window function with low sidelobes. The *picket-fence effect* arises because only a finite number of frequency points of a continuous-frequency spectrum are produced by the DFT. It is therefore possible to miss the peak of a particular frequency component in a signal because it is located between two adjacent frequency points in the spectrum. Since the frequency spacing $\Delta f = F_s/N$, this problem can be alleviated by increasing N the number of DFT points while maintaining the same sampling rate, implying having more samples in the DFT or a employing zero-padding technique.

DFT Parameter Selection[6]

The sampling period, T_s, frequency resolution (spacing), Δf, and the DFT length, N, are the three parameters that must be specified in performing DFT signal processing. Since $F_s = N\Delta f = 1/T_s$, these parameters are related according to $\Delta f = 1/NT_s$. If T_x denotes the length of the sampled data, then $T_x = MT_s$. Thus, there is no spectral distortion if $N \geq M = T_x/T_s$. Equivalently, there will be no spectral distortion if

$$\Delta f = \frac{1}{T} = \frac{1}{NT_s} \leq \frac{1}{MT_s} = \frac{1}{T_x}$$

Parameters for the DFT processing of a sampled continuous signal must be carefully selected to avoid spectral distortion due to aliasing or data truncation. Assuming a window width of T_x seconds and that the signal has a maximum bandwidth of B hertz, then based on the sampling theorem we would have negligible aliasing, provided $F_s = 1/T_s \geq 2B$. Spectral leakage due to sharp data truncation is avoided provided the frequency resolution is selected to satisfy $1/\Delta f = T \geq T_x$. Consequently, spectral distortion due to aliasing and spectral leakage can be avoided if the length of the DFT is selected to satisfy $N = F_s/\Delta f \geq 2BT_x$.

References

1. Soliman, S.S., and Srinath, M.D., *Continuous and Discrete Signals and Systems,* Prentice-Hall, 1998.
2. O'Flynn, M., and Moriarty, E., *Linear Systems: Time Domain and Transform Analysis,* Wiley, 1987.
3. Lathi, B.P., *Modern Digital and Analog Communication Systems,* Oxford University Press, Third Edition, 1998.
4. Proakis, J.G., and Salehi, M., *Communication Systems Engineering,* Prentice-Hall, 1994.
5. Taylor, F.J., *Principles of Signals and Systems,* McGraw-Hill, 1994.
6. Carlson, G.E., *Signal and Linear System Analysis,* Wiley, Second Edition, 1998.
7. Proakis, J.G., and Manolakis, D.G., *Digital Signal Processing: Principles, Algorithms and Applications,* Prentice-Hall, 1996.
8. Oppenheim, A.V., and Schafer, R.W., with Buck, J.R., *Discrete-Time Signal Processing,* Prentice-Hall, Second Edition, 1999.
9. Houts, R.C., *Signal Analysis in Linear Systems,* Saunders College Publishing, 1991.
10. Ziemer, R.E., Tranter, W.H., and Fannin, D.R., *Signals and Systems: Continuous and Discrete,* MacMillan Publishing Company, Third Edition, 1993.
11. Orfanidis, S.J., *Introduction to Signal Processing,* Prentice-Hall, 1996.
12. Haykin, S., and Veen, B.V., *Signals and Systems,* Wiley, 1999.
13. Taylor, F., and Mellot, J., *Hands-On Digital Signal Processing,* McGraw-Hill, 1998.

3.2 z Transforms and Digital Systems

Rolf Johansson

A digital system (or discrete-time system or sampled-data system) is a device such as a digital controller or a digital filter or, more generally, a system intended for digital computer implementation and usually with some periodic interaction with the environment and with a supporting methodology for analysis and design. Of particular importance for modeling and analysis are recurrent algorithms—for example, difference equations in input–output data—and the z transform is important for the solution of such problems.

The z **transform** is being used in the analysis of linear time-invariant systems and discrete time signals—for example, for digital control or filtering—and may be compared to the Laplace transform as used in the analysis of continuous-time signals and systems, a useful property being that the convolution of two time-domain signals is equivalent to multiplication of their corresponding z transforms. The z transform is important as a means to characterize a linear time-invariant system in terms of its pole–zero locations, its transfer function and Bode diagram, and its response to a large variety of signals. In addition, it provides important relationships between temporal and spectral properties of signals. The z transform generally appears in the analysis of difference equations as used in many branches of engineering and applied mathematics.

3.2.1 The *z* Transform

The *z* transform of the sequence $\{x_k\}_{-\infty}^{+\infty}$ is defined as the generating function

$$X(z) =]\{x\} = \sum_{k=-\infty}^{\infty} x_k z^{-k} \tag{3.48}$$

where the variable *z* has the essential interpretation of a forward shift operator so that

$$]\{x_{k+1}\} = z]\{x_k\} = zX(z) \tag{3.49}$$

The *z* transform is an infinite power series in the complex variable z^{-1} where $\{x_k\}$ constitutes a sequence of coefficients. As the *z* transform is an infinite power series, it exists only for those values of *z* for which this series converges and the *region of convergence* of $X(z)$ is the set of *z* for which $X(z)$ takes on a finite value. A sufficient condition for existence of the *z* transform is convergence of the power series

$$\sum_{k=-\infty}^{\infty} |x_k| \cdot |z^{-k}| < \infty \tag{3.50}$$

The region of convergence for a finite-duration signal is the entire *z* plane except $z = 0$ and $z = \infty$. For a one-sided infinite-duration sequence $\{x_k\}_{k=0}^{\infty}$, a number *r* can usually be found so that the power series converges for $|z| > r$. Then, the *inverse z transform* can be derived as

$$x_k = \frac{1}{2\pi i}\oint X(z) z^{k-1}\, dz \tag{3.51}$$

where the contour of integration encloses all singularities of $X(z)$. In practice, it is standard procedure to use tabulated results; some standard *z* transform pairs are to be found in Table 3.10.

3.2.2 Digital Systems and Discretized Data

Periodic sampling of signals and subsequent computation or storing of the results requires the computer to schedule sampling and to handle the resulting sequences of numbers. A measured variable $x(t)$ may be available only as periodic observations of $x(t)$ as sampled with a time interval T (the sampling period). The sample sequence can be represented as

$$\{x_k\}_{-\infty}^{\infty}, \quad x_k = x(kT) \quad \text{for } k = \ldots, -1, 0, 1, 2, \ldots \tag{3.52}$$

and it is important to ascertain that the sample sequence adequately represents the original variable $x(t)$; see Figure 3.16. For ideal sampling it is required that the duration of each sampling be very short and the sampled function may be represented by a sequence of infinitely short impulses $\delta(t)$ (the Dirac impulse). Let the sampled function of time be expressed thus:

$$x_\Delta(t) = x(t) \cdot T \sum_{k=-\infty}^{\infty} \delta(t - kT) = x(t) \cdot \sqcup\!\sqcup_T(t) \tag{3.53}$$

where

$$\sqcup\!\sqcup_T(t) \overset{\Delta}{=} T \sum_{k=-\infty}^{\infty} \delta(t - kT) \tag{3.54}$$

TABLE 3.10 Properties of the z Transform

z transform	$(\{f_k\}) = F(z)$	
Convolution	$(\{f_k * g_k\}) = (\{f_k\}) \cdot (\{g_k\})$	
	$(\{f_k \cdot g_k\}) = (\{f_k\}) * (\{g_k\})$	
Forward shift	$(\{f_{k+1}\}) = z(\{f_k\}) = zF(z)$	
Backward shift	$(\{f_{k-1}\}) = (\{f_k\}) = z^{-1}F(z)$	
Linearity	$(\{af_k + bg_k\}) = a(\{f_k\}) + b(\{g_k\})$	
Multiplication	$(\{a^k f_k\}) = F(a^{-1}z)$	
Final value	$\lim_{k\to\infty} f_k = \lim_{k\to 1}(1 - z^{-1})F(z)$	
Initial value	$f_0 = \lim_{z\to\infty} F(z)$	
	Time Domain	z Transform
Impulse	$\delta_k = \begin{cases} 1, & k = 0 \\ 0, & k \neq 0 \end{cases}$ $\leftrightharpoons$	$\{\delta_k\} = 1, \quad z \in C$
Step function	$\sigma_k = \begin{cases} 0, & k < 0 \\ 1, & k \geq 0 \end{cases}$ $\leftrightharpoons$	$](\sigma_k) = \frac{z}{z-1}, \quad \lvert z\rvert > 1$
Ramp function	$x_k = k \cdot \sigma_k$ $\leftrightharpoons$	$X(z) = \frac{z}{(z-1)^2}, \quad \lvert z\rvert > 1$
Exponential	$x_k = a^k \cdot \sigma_k$ $\leftrightharpoons$	$X(z) = \frac{z}{z-a}, \quad \lvert z\rvert > \lvert a\rvert$
Sinusoid	$x_k = \sin \omega k \cdot \sigma_k$ $\leftrightharpoons$	$X(z) = \frac{z \sin \omega}{z^2 - 2z\cos\omega + 1}, \quad \lvert z\rvert > 1$

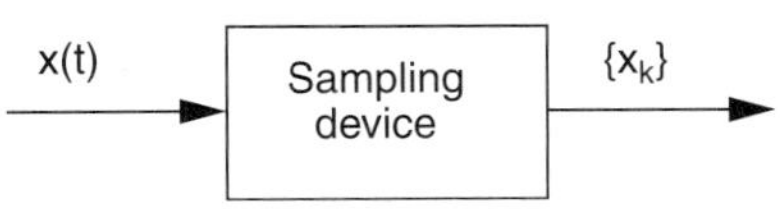

FIGURE 3.16 A continuous-time signal $x(t)$ and a sampling device that produces a sample sequence $\{x_k\}$.

and where the sampling period T is multiplied to ensure that the averages over a sampling period of the original variable x and the sampled signal x_Δ, respectively, are of the same magnitude. A direct application of the discretized variable $x_\Delta(t)$ in Equation 3.53 verifies that the spectrum of x_Δ is related to the z transform $X(z)$ as

$$X_\Delta(i\omega) = \hat{}\{x(t) \cdot ⧢_T(t)\} = T\sum_{k=-\infty}^{\infty} x_k \exp(-i\omega kT) = TX(e^{i\omega T}) \tag{3.55}$$

Obviously, the original variable $x(t)$ and the sampled data are not identical, and thus it is necessary to consider the distortive effects of discretization. Consider the spectrum of the sampled signal $x_\Delta(t)$ obtained as the Fourier transform

$$X_\Delta(i\omega) = \hat{}\{x_\Delta(t)\} = \hat{}\{x(t)\} * \hat{}\{⧢_T(t)\} \tag{3.56}$$

where

$$\hat{}\{⧢_T(t)\} = \sum_{k=-\infty}^{\infty} \delta\left(\omega - \frac{2\pi}{T}k\right) = \frac{T}{2\pi} ⧢_{2\pi/T}(\omega) \tag{3.57}$$

so that

$$X_\Delta(i\omega) = \hat{}\{x(t)\} * \hat{}\{ \sqcup\!\sqcup_T(t)\} = \sum_{k=-\infty}^{\infty} X\left[i\left(\omega - \frac{2\pi}{T}k\right)\right] \tag{3.58}$$

Thus, the Fourier transform X_Δ of the sampled variable has a periodic extension of the original spectrum $X(i\omega)$ along the frequency axis with a period equal to the sampling frequency $\omega_s = 2\pi/T$. There is an important result based on this observation known as the *Shannon sampling theorem*, which states that the continuous-time variable $x(t)$ may be reconstructed from the samples $\{x_k\}_{-\infty}^{+\infty}$ if and only if the sampling frequency is at least twice that of the highest frequency for which $X(i\omega)$ is nonzero. The original variable $x(t)$ may thus be recovered as

$$x(t) = \sum_{k=-\infty}^{\infty} x_k \frac{\sin\frac{\pi}{T}(t-kT)}{\frac{\pi}{T}(t-kT)} \tag{3.59}$$

The formula given in Equation 3.59 is called *Shannon interpolation*, which is often quoted though it is valid only for infinitely long data sequences and would require a noncausal filter to reconstruct the continuous-time signal $x(t)$ in real-time operation. The frequency $\omega_n = \omega_s/2 = \pi/T$ is called the *Nyquist frequency* and indicates the upper limit of distortion-free sampling. A nonzero spectrum beyond this limit leads to interference between the sampling frequency and the sampled signal (*aliasing*); see Figure 3.17.

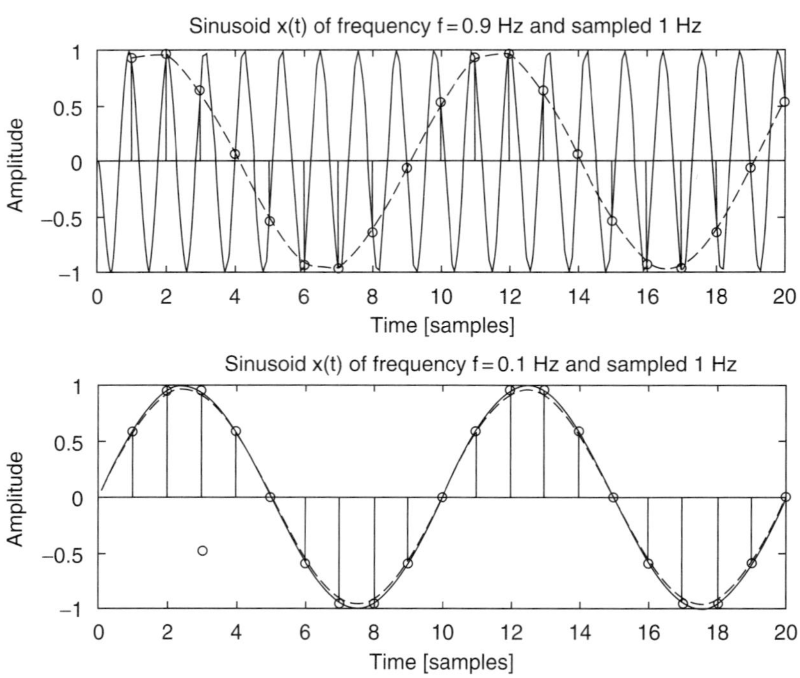

FIGURE 3.17 Illustration of aliasing appearing during sampling of a sinusoid $x(t) = \sin 2\pi \cdot 0.9t$ at the insufficient sampling frequency 1 Hz (sampling period $T = 1$) (*upper graph*). The sampled signal exhibits aliasing with its major component similar to a signal $x(t) = \sin 2\pi \cdot 0.1t$ sampled with the same rate (*lower graph*).

3.2.3 The Discrete Fourier Transform

Consider a finite length sequence $\{x_k\}_{k=0}^{N-1}$ that is zero outside the interval $0 \le k \le N-1$. Evaluation of the z transform $X(z)$ at N equally spaced points on the unit circle $z = \exp(i\omega_k T) = \exp[i(2\pi/NT)kT]$ for $k = 0, 1, \ldots, N-1$ defines the *discrete Fourier transform* (DFT) of a signal x with a sampling period h and N measurements:

$$X_k = \mathrm{DFT}\{x(kT)\} = \sum_{l=0}^{N-1} x_l \exp(-i\omega_k lT) = X(e^{i\omega_k T}) \tag{3.60}$$

Notice that the discrete Fourier transform $\{X_k\}_{k=0}^{N-1}$ is only defined at the discrete frequency points

$$\omega_k = \frac{2\pi}{NT}k, \quad \text{for } k = 0, 1, \ldots, N-1 \tag{3.61}$$

In fact, the discrete Fourier transform adapts the Fourier transform and the z transform to the practical requirements of finite measurements. Similar properties hold for the discrete Laplace transform with $z = \exp(sT)$, where s is the Laplace transform variable.

3.2.4 The Transfer Function

Consider the following discrete-time linear system with input sequence $\{u_k\}$ (stimulus) and output sequence $\{y_k\}$ (response). The dependency of the output of a linear system is characterized by the convolution-type equation and its z transform,

$$\begin{aligned} y_k &= \sum_{m=0}^{\infty} h_m u_{k-m} + v_k = \sum_{m=-\infty}^{k} h_{k-m} u_m + v_k, \quad k = \ldots, -1, 0, 1, 2, \ldots \\ Y(z) &= H(z)U(z) + V(z) \end{aligned} \tag{3.62}$$

where the sequence $\{v_k\}$ represents some external input of errors and disturbances and with $Y(z) = \{y\}$, $U(z) = \{u\}$, $V(z) = \{v\}$ as output and inputs. The *weighting function* $h(kT) = \{h_k\}_{k=0}^{\infty}$, which is zero for negative k and for reasons of causality is sometimes called *pulse response* of the digital system (compare *impulse response* of continuous-time systems). The pulse response and its z transform, the *pulse transfer function*,

$$H(z) = \{h(kT)\} = \sum_{k=0}^{\infty} h_k z^{-k} \tag{3.63}$$

determine the system's response to an input $U(z)$; see Figure 3.18. The pulse transfer function $H(z)$ is obtained as the ratio

$$H(z) = \frac{X(z)}{U(z)} \tag{3.64}$$

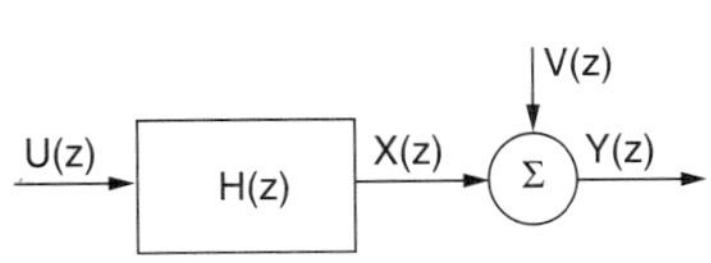

FIGURE 3.18 Block diagram with an assumed transfer function relationship $H(z)$ between input $U(z)$, disturbance $V(z)$, intermediate $X(z)$, and output $Y(z)$.

and provides the frequency domain input–output relation of the system. In particular, the Bode diagram is evaluated as $|H(z)|$ and arg $H(z)$ for $z = \exp(i\omega_k T)$ and for $|\omega_k| < \omega_n = \pi/T$, that is, when $H(z)$ is evaluated for frequency points up to the Nyquist frequency ω_n along the unit circle.

3.2.5 State-Space Systems

Alternatives to the input–output representations by means of transfer functions are the state-space representations. Consider the following finite dimensional discrete state-space equation with a state vector $x_k \in R^n$, input $u_k \in R^p$, and observations $y_n \in R^m$.

$$\begin{cases} x_{k+1} = \Phi x_k + \Gamma u_k \\ \qquad\qquad\qquad k = 0, 1, \ldots \\ y_k = Cx_k + Du_k \end{cases} \tag{3.65}$$

with the pulse transfer function

$$H(z) = C(zI - \Phi)^{-1}\Gamma + D \tag{3.66}$$

and the output variable

$$Y(z) = C\sum_{k=0}^{\infty} \Phi^k z^{-k} x_0 + H(z)U(z) \tag{3.67}$$

where possible effects of initial conditions x_0 appear as the first term. Notice that the initial conditions x_0 can be viewed as the net effects of the input in the time interval $(-\infty, 0)$.

3.2.6 Digital Systems Described by Difference Equations (ARMAX Models)

An important class of nonstationary stochastic processes is one in which some deterministic response to an external input and a stationary stochastic process are superimposed. This is relevant, for instance, when the external input cannot be effectively described by some probabilistic distribution. A discrete-time model can be formulated in the form of a difference equation with an external input $\{u_k\}$ that is usually considered to be known:

$$y_k = -a_1 y_{k-1} - \cdots - a_n y_{k-n} + b_1 u_{k-1} + \cdots + b_n u_{k-n} + w_k + c_1 w_{k-1} + \cdots + c_n w_{k-n} \tag{3.68}$$

Application of the z transform permits formulation of Equation 3.68 as

$$A(z^{-1})Y(z) = B(z^{-1})U(z) + C(z^{-1})W(z) \tag{3.69}$$

where

$$\begin{aligned} A(z^{-1}) &= 1 + a_1 z^{-1} + \cdots + a_n z^{-n} \\ B(z^{-1}) &= 1 + b_1 z^{-1} + \cdots + b_n z^{-n} \\ C(z^{-1}) &= 1 + c_1 z^{-1} + \cdots + c_n z^{-n} \end{aligned} \tag{3.70}$$

Stochastic models including the A polynomial, according to Equations 3.69 and 3.70, are known as **autoregressive (AR) models** and models including the C polynomial are known as **moving-average (MA) models**, whereas the B polynomial determines the effects of the external input (X). Notice that the term *moving average* is here somewhat misleading, as there is no restriction that the coefficients should add to 1 or that the coefficients are nonnegative. An alternative description is *finite impulse response* or *all-zero filter.*

Thus, the full model of Equation 3.69 is an **autoregressive moving average model** with external input (ARMAX) and its pulse transfer function $H(z) = B(z^{-1})/A(z^{-1})$ is stable if and only if the *poles*—that is, the complex numbers $z_1,\ldots, z_n$ solving the equation $A(z^{-1}) = 0$—are strictly inside the unit circle, that is, $|z_i| < 1$. The *zeros* of the system—that is, the complex numbers $z_1,\ldots, z_n$ solving the equation $B(z^{-1}) = 0$—may take on any value without any instability arising, although it is preferable to obtain zeros located strictly inside the unit circle, that is, $|z_i| < 1$ (*minimum-phase zeros*). By linearity $\{y_k\}$ can be separated into one purely deterministic process $\{x_k\}$ and one purely stochastic process $\{v_k\}$:

$$\begin{cases} A(z^{-1})X(z) = B(z^{-1})U(z) \\ A(z^{-1})V(z) = C(z^{-1})W(z) \end{cases} \quad \text{and} \quad \begin{cases} y_k = x_k + v_k \\ Y(z) = X(z) + V(z) \end{cases} \tag{3.71}$$

The type of decomposition (Equation 3.71) that separates the deterministic and stochastic processes is known as the *Wold decomposition.*

3.2.7 Prediction and Reconstruction

Consider the problem of predicting the output d steps ahead when the output $\{y_k\}$ is generated by the ARMA model,

$$A(z^{-1})Y(z) = C(z^{-1})W(z) \tag{3.72}$$

which is driven by a zero-mean white noise $\{w_k\}$ with covariance $\mathcal{E}\{w_i w_j\} = \sigma_w^2 \delta_{ij}$. In other words, assuming that observations $\{y_k\}$ are available up to the present time, how should the output d steps ahead be predicted optimally? Assume that the polynomials $A(z^{-1})$ and $C(z^{-1})$ are mutually prime with no zeros for $|z| \geq 1$. Let the C polynomial be expanded according to the *Diophantine equation,*

$$C(z^{-1}) = A(z^{-1})F(z^{-1}) + z^{-d}G(z^{-1}) \tag{3.73}$$

which is solved by the two polynomials

$$\begin{aligned} F(z^{-1}) &= 1 + f_1 z^{-1} + \cdots + f_{nF} z^{-nF}, & n_F &= d-1 \\ G(z^{-1}) &= g_0 + g_1 z^{-1} + \cdots + g_{nG} z^{-nG}, & n_G &= \max(n_A - 1, n_C - d) \end{aligned} \tag{3.74}$$

Interpretation of z^{-1} as a *backward shift operator* and application of Equations 3.72 and 3.73 permit the formulation

$$y_{k+d} = F(z^{-1})w_{k+d} + \frac{G(z^{-1})}{C(z^{-1})} y_k \tag{3.75}$$

Let us, by $\hat{y}_{k+d|k}$, denote linear d-step predictors of y_{k+d} based upon the measured information available at time k. As the zero-mean term $F(z^{-1})w_{k+d}$ of Equation 3.75 is unpredictable at time k, it is natural to suggest the following d-step predictor:

$$\hat{y}_{k+d|k} = \frac{G(z^{-1})}{C(z^{-1})} y_k \tag{3.76}$$

The prediction error satisfies

$$\begin{aligned} \varepsilon_{k+d} &= (\hat{y}_{k+d|k} - y_{k+d}) \\ &= \frac{G(z^{-1})}{C(z^{-1})} y_k - \frac{A(z^{-1})F(z^{-1}) + z^{-d}G(z^{-1})}{C(z^{-1})} y_{k+d} \\ &= -F(z^{-1})w_{k+d} \end{aligned} \tag{3.77}$$

Let $\{\cdot|\ _k\}$ denote the *conditional mathematical expectation* relative to the measured information available at time k. The conditional mathematical expectation and the covariance of the d-step prediction relative to available information at time k is

$$\begin{aligned} \%\{\hat{y}_{k+d|k} - y_{k+d}|\wedge_k\} &= \%\{-F(z^{-1})w_{k+d}|\wedge_k\} = 0 \\ \%\{(\hat{y}_{k+d|k} - y_{k+d})^2|\wedge_k\} &= \%\{[F(z^{-1})w_{k+d}]^2|\wedge_k\} \\ &= \%\{(w_{k+d} + f_1 w_{k+d-1} + \cdots + f_{d-1}w_{k+1})^2|\wedge_k\} \\ &= (1 + f_1^2 + \cdots + f_{nF}^2)\sigma_w^2 = 0 \end{aligned} \tag{3.78}$$

It follows that the predictor of Equation 3.76 is unbiased and that the prediction error only depends on future, unpredictable noise components. It is straightforward to show that the predictor of Equation 3.76 achieves the lower bound of Equation 3.78 and that the predictor of Equation 3.76 is optimal in the sense that the prediction error variance is minimized.

Example 3.1—An Optimal Predictor for a First-Order Model

Consider for the first-order ARMA model

$$y_{k+1} = -a_1 y_k + w_{k+1} + c_1 w_k \tag{3.79}$$

The variance of a one-step-ahead predictor $\hat{y}_{k+1|k}$ is

$$\begin{aligned} \%\{(\hat{y}_{k+1|k} - y_{k+1})^2|\wedge_k\} &= \%\{(\hat{y}_{k+1|k} + a_1 y_k - c_1 w_k)^2|\wedge_k\} + \%\{w_{k+1}^2|\wedge_k\} \\ &= \%\{(\hat{y}_{k+1|k} + a_1 y_k - c_1 w_k)^2|\wedge_k\} + \sigma_w^2 \geq \sigma_w^2 \end{aligned} \tag{3.80}$$

The optimal predictor satisfying the lower bound in Equation 3.80 is obtained from Equation 3.80 as

$$\hat{y}_{k+1|k}^{o} = -a_1 y_k + c_1 w_k \tag{3.81}$$

which, unfortunately, is not realizable as it stands because w_k is not available to measurement. Therefore, the noise sequence $\{w_k\}$ has to be substituted by some function of the observed variable $\{y_k\}$. A linear predictor chosen according to Equation 3.76 is

$$\hat{y}_{k+1|k} = \frac{G(z^{-1})}{C(z^{-1})}y_k = \frac{c_1 - a_1}{1 + c_1 z^{-1}}y_k \tag{3.82}$$

3.2.8 The Kalman Filter

Consider the linear state-space model

$$\begin{aligned} x_{k+1} &= \Phi x_k + v_k, \quad x_k \in \mathrm{R}^n \\ y_k &= Cx_k + w_k, \quad y_k \in \mathrm{R}^m \end{aligned} \tag{3.83}$$

where $\{v_k\}$ and $\{w_k\}$ are assumed to be independent zero-mean white-noise processes with covariances Σ_v and Σ_w, respectively. It is assumed that $\{y_k\}$, but not $\{x_k\}$, is available to measurement and that it is desirable to predict $\{x_k\}$ from measurements of $\{y_k\}$.

Introduce the state predictor,

$$\begin{aligned} \hat{x}_{k+1|k} &= \Phi\hat{x}_{k|k-1} - K_k(\hat{y}_k - y_k), \quad \hat{x}_{k|k-1} \in \mathrm{R}^n \\ \hat{y}_k &= C\hat{x}_{k|k-1}, \quad y_k \in \mathrm{R}^m \end{aligned} \tag{3.84}$$

The predictor of Equation 3.84 has the same dynamics matrix Φ as the state-space model of Equation 101.83 and, in addition, there is a correction term $K_k(\hat{y}_k - y_k)$ with a factor K_k to be chosen. The prediction error is

$$\tilde{x}_{k+1|k} = \hat{x}_{k+1|k} - x_{k+1} \tag{3.85}$$

The prediction-error dynamics is

$$\tilde{x}_{k+1} = (\Phi - K_kC)\tilde{x}_k + v_k - K_k w_k \tag{3.86}$$

The mean prediction error is governed by the recursive equation

$$\%\{\tilde{x}_{k+1}\} = (\Phi - K_kC)\%\{\tilde{x}_k\} \tag{3.87}$$

and the mean square error of the prediction error is governed by

$$\begin{aligned} \%\{\tilde{x}_{k+1}\tilde{x}_{k+1}^T\} &= \%\{[(\Phi - K_kC)\tilde{x}_k + v_k - K_kw_k][(\Phi - K_kC)\tilde{x}_k + v_k - K_kw_k]^T\} \\ &= (\Phi - K_kC)\%\{\tilde{x}_k\tilde{x}_k^T\}(\Phi - K_kC)^T + \Sigma_v + K_k\Sigma_wK_k \end{aligned} \tag{3.88}$$

If we denote

$$P_k = \%\{\tilde{x}_k\tilde{x}_k^T\}, \qquad Q_k = \Sigma_w + CP_kC^T \tag{3.89}$$

then Equation 3.88 is simplified to

$$P_{k+1} = \Phi P_k\Phi^T - K_kCP_k\Phi - \Phi^TP_kC^TK_k^T + \Sigma_v + K_kQ_kK_k^T \tag{3.90}$$

By completing squares of terms containing K_k we find

$$P_{k+1} = \Phi P_k \Phi^T + \Sigma_v - \Phi P_k C^T Q_k^{-1} C P_k \Phi^T + (K_k - \Phi P_k C^T Q_k^{-1}) Q_k (K_k - \Phi P_k C^T Q_k^{-1})^T \quad (3.91)$$

where only the last term depends on K_k. Minimization of P_{k+1} can be done by choosing K_k such that the positive semidefinite K_k-dependent term in Equation 3.91 disappears. Thus P_{k+1} achieves its lower bound for

$$K_k = \Phi P_k C^T (\Sigma_w + C P_k C^T)^{-1} \quad (3.92)$$

and the *Kalman filter* (or *Kalman-Bucy filter*) takes the form

$$\begin{aligned} \hat{x}_{k+1|k} &= \Phi \hat{x}_{k|k-1} - K_k(\hat{y}_k - y_k) \\ \hat{y}_k &= C\hat{x}_{k|k-1}, \qquad K_k = \Phi P_k C^T (\Sigma_w + C P_k C^T)^{-1} \\ P_{k+1} &= \Phi P_k \Phi^T + \Sigma_v - \Phi P_k C^T (\Sigma_w + C P_k C^T)^{-1} C P_k \Phi^T \end{aligned} \quad (3.93)$$

which is the optimal predictor in the sense that the mean square error (Equation 3.88) is minimized in each step.

Example 3.2—Kalman Filter for a First-Order System

Consider the state-space model

$$x_{k+1} = 0.95 x_k + v_k, \qquad y_k = x_k + w_k \quad (3.94)$$

where $\{v_k\}$ and $\{w_k\}$ are zero-mean white-noise processes with covariances $\%\{v_k^2 = 1$ and $\%\{w_k^2\} = 1$, respectively.

The Kalman filter takes on the form

$$\begin{aligned} \hat{x}_{k+1|k} &= 0.95\hat{x}_{k|k-1} - K_k(\hat{x}_{k|k-1} - y_k) \\ K_k &= \frac{0.95 P_k}{1 + P_k} \\ P_{k+1} &= 0.95^2 P_k + 1 - \frac{0.95^2 P_k^2}{1 + P_k} \end{aligned} \quad (3.95)$$

The result of one such realization is shown in Figure 3.19.

Defining Terms

Autoregressive (AR) model: An autoregressive time series of order n is defined via $y_k = -\Sigma_{m=1}^{n} a_m y_{k-m} + w_k$. The sequence $\{w_k\}$ is usually assumed to consist of zero-mean identically distributed stochastic variables w_k.

Autoregressive moving average (ARMA) model: An autoregressive moving average time series of order n is defined via $y_k = -\Sigma_{m=1}^{n} a_m y_{k-m} + \Sigma_{m=0}^{n} c_m w_{k-m}$. The sequence $\{w_k\}$ is usually assumed to consist of zero-mean identically distributed stochastic variables w_k.

Discrete Laplace transform: The discrete Laplace transform is a counterpart to the Laplace transform with application to discrete signals and systems. The discrete Laplace transform is obtained from the z transform by means of the substitution $z = \exp(sT)$, where T is the sampling period.

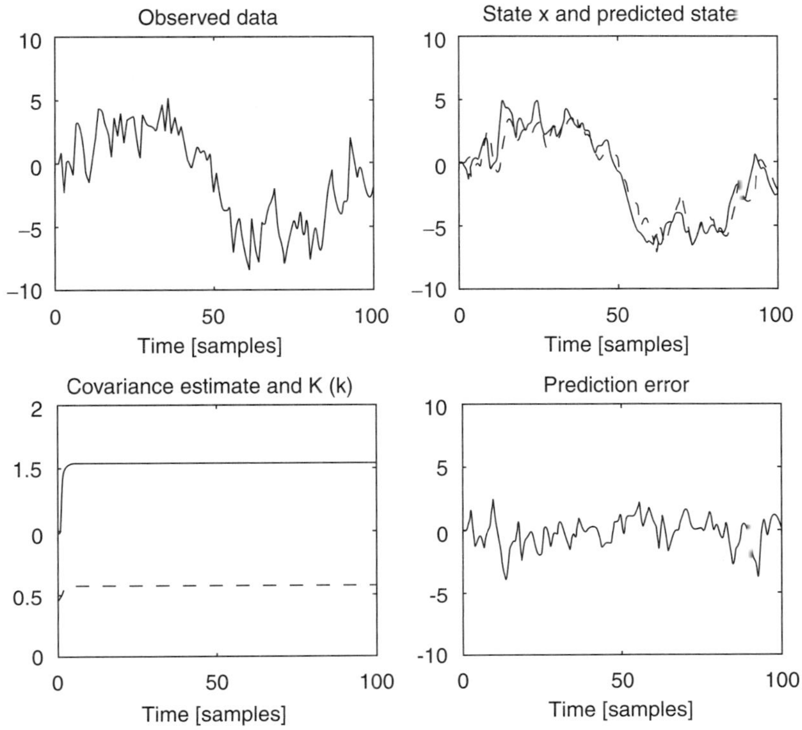

FIGURE 3.19 Kalman filter applied to one-step-ahead prediction of x_{k+1} in Equation 3.94. The observed variable $\{y_k\}$, the state $\{x_k\}$, and the predicted state $\{\hat{x}_k\}$, the estimated variance $\{P_k\}$ and $\{K_k\}$, and the prediction error $\{\tilde{x}_k\}$ are shown in a 100-step realization of the stochastic process. (From Johansson, R. 1993. *System Modeling and Identification.* Prentice-Hall, Englewood Cliffs, NJ.)

Moving average (MA) process: A moving average time series of order n is defined via $y_k = \sum_{m=0}^{n} c_m w_{k-m}$. The sequence $\{w_k\}$ is usually assumed to consist of zero-mean identically distributed stochastic variables w_k.

Rational model: AR, MA, ARMA, and ARMAX are commonly referred to as rational models.

Time Series: A sequence of random variable $\{y_k\}$, where k belongs to the set of positive and negative integers.

***z* transform:** A generating function applied to sequences of data and evaluated as a function of the complex variable z with interpretation of frequency.

References

Box, G. E. P. and Jenkins, G. M. 1970. *Time Series Analysis: Forecasting and Control.* Holden-Day, San Francisco, CA.

Hurewicz, W. 1947. Filters and servo systems with pulsed data. In *Theory of Servomechanisms,* H. M. James, N. B. Nichols, and R. S. Philips, eds., McGraw-Hill, New York.

Jenkins, G. M. and Watts, D. G. 1968. *Spectral Analysis and Its Applications.* Holden-Day, San Francisco, CA.

Johansson, R. 1993. *System Modeling and Identification.* Prentice-Hall, Englewood Cliffs, NJ.

Jury, E. I. 1956. Synthesis and critical study of sampled-data control systems. *AIEE Trans.* 75: 141–151.

Kalman, R. E. and Bertram, J. E. 1958. General synthesis procedure for computer control of single and multi-loop linear systems. *Trans. AIEE.* 77: 602–609.

Kolmogorov, A. N. 1939. Sur l'interpolation et extrapolation des suites stationnaires. *C. R. Acad. Sci.* 208: 2043–2045.
Ragazzini, J. R. and Zadeh, L. A. 1952. The analysis of sampled-data systems. *AIEE Trans.* 71:225–234.
Tsypkin, Y. Z. 1950. Theory of discontinuous control. *Avtomatika i Telemekhanika.* Vol. 5.
Wiener, N. 1949. *Extrapolation, Interpolation and Smoothing of Stationary Time Series with Engineering Applications.* John Wiley & Sons, New York.

Further Information

Early theoretical efforts developed in connection with servomechanisms and radar applications (Hurewicz, 1947). Tsypkin (1950) introduced the discrete Laplace transform and the formal z transform definition was introduced by Ragazzini and Zadeh (1952) with further developments by Jury (1956). Much of prediction theory was originally developed by Kolmogorov (1939) and Wiener (1949) whereas state-space methods were forwarded by Kalman and Bertram (1958). Pioneering textbooks on time-series analysis and spectrum analysis are provided by Box and Jenkins (1970) and Jenkins and Watts (1968).

Detailed accounts of time-series analysis and the z transform and their application to signal processing are to be found in

- Oppenheim, A. V. and Schafer, R. W. 1989. *Discrete-Time Signal Processing.* Prentice-Hall, Englewood Cliffs, NJ.
- Proakis, J. G. and Manolakis, D. G. 1989. *Introduction to Digital Signal Processing.* Maxwell MacMillan Int. Ed., New York.

Theory of time-series analysis and its application to discrete-time control is to be found in

- Åström, K. J. and Wittenmark, B. 1990. *Computer-Controlled Systems,* 2nd ed., Prentice-Hall, Englewood Cliffs, NJ.

Theory of time-series analysis and methodology for determination and validation of discrete-time models and other aspects of system identification are to be found in

- Johansson, R. 1993. *System Modeling and Identification.* Prentice-Hall, Englewood Cliffs, NJ.

Good sources to monitor current research are

- *IEEE Transactions on Automatic Control*
- *IEEE Transactions on Signal Processing*

Examples of easy-to-read survey articles for signal processing applications are

- Cadzow, J. A. 1990. Signal processing via least-squares error modeling. *IEEE ASSP Magazine.* 7:12–31, October.
- Schroeder, M. R. 1984. Linear prediction, entropy, and signal analysis. *IEEE ASSP Magazine.* 1:3–11, July.

3.3 Continuous- and Discrete-Time State-Space Models

Kam K. Leang, Qingze Zou, and Santosh Devasia

3.3.1 Introduction

In this section we introduce the modeling of continuous- and discrete-time systems using the state-space approach. The state-space approach is a technique that uses a set of first order differential equations to represent the behavior of a system in the time-domain. The state-space approach has an advantage over frequency-domain approaches such as the transfer-function approach: it can be used to model linear, nonlinear, time-varying, and multivariable systems, whereas the transfer-function approach is suited to linear time-invariant (LTI) systems [1, Chapter 3]. In addition, models expressed in first order state-space form in the time-domain can be readily solved by a digital computer or microprocessor, which makes this approach quite useful for the design and control of modern mechatronic systems. Furthermore, there is a wide variety of available computer software, such as MATLAB [2], that take advantage of the state-space form for analyzing and solving design problems. Therefore, the state-space approach can be used to investigate the behavior and facilitate in the design of both continuous- and discrete-time systems, the fundamentals of which will be the focus of this section.

In the following, we begin with an example: the modeling of a piezoceramic actuator and use the example throughout the section. The concept of a system state is introduced and we explain the state-space equation for linear systems and present its solution. The topic of linearization of nonlinear systems is briefly mentioned. The relationships between time- and frequency-domain models are discussed and a procedure for obtaining a state-space model using experimental frequency-domain (frequency-response) data is presented. This section closes with a discussion of discrete-time state-space modeling and concluding remarks. Useful MATLAB commands are also included as footnotes.

3.3.2 States and the State-Space

3.3.2.1 An Example Piezoceramic Actuator

We begin by modeling a piezoceramic actuator, which is an example mechatronic (electromechanical) system. When a voltage is applied to a piezoceramic material, its dimension changes. This change in dimension can be used to precisely position an object or tool (such as a sensor), therefore making piezoceramics suitable actuators for a wide variety of applications. For example, due to their ability to achieve positioning with sub-nanometer level precision, piezoceramic actuators have become ideal for emerging nanotechnologies. In particular, a piezo-tube actuator is used in scanning probe microscopes (SPMs, see Figure 3.20) to precisely position a probe tip for high-precision nanofabrication, surface modification, and the acquisition of images of atoms [3]. The probe tip can be positioned in the three coordinate axes (x, y, and z), with each motion controlled by an independent voltage source ($V_x(t)$, $V_y(t)$, and $V_z(t)$). Scanning of the probe is performed parallel to the sample surface along the x- and y-axis; the z-axis movement allows motion of the probe perpendicular to the sample surface. An accurate mathematical model of the dynamics of a piezo-tube actuator is required for the analysis and design of SPM systems. A designer can exploit the known information of the system from its model to improve or optimize a design for building faster and more reliable SPMs. For example, an approach that has been

FIGURE 3.20 The main components of a scanning probe microscope (SPM) used for surface analysis, which includes the piezo-tube actuator, the probe tip, and the sample. The configuration of the probe tip and sample with respect to the coordinate axes (x, y, and z) are shown in the magnified view.

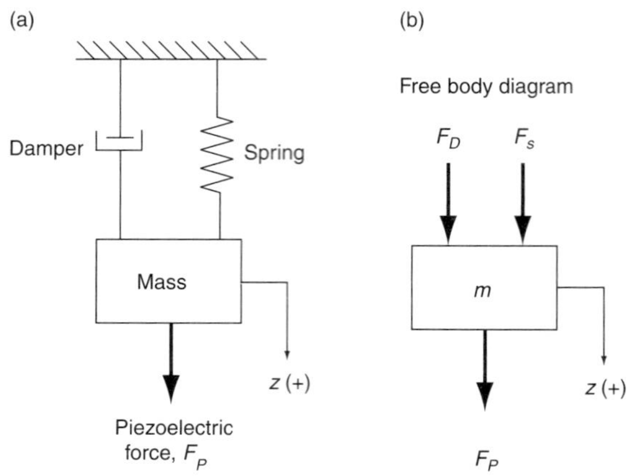

FIGURE 3.21 (a) A simple *lumped model* of the piezo-tube actuator modeled along the *z*-axis consisting of a mass, a spring, and a damper [4]. The positive *z*-direction is indicated by the arrow and the "+" sign. (b) The forces acting on the mass (free body diagram).

successfully implemented is the inversion-based control method, which finds the inputs required to achieve exact tracking by inverting the system model [3]. This technique works best when the dynamics of the system are well characterized and understood. In general, the analysis and design of control systems also requires a system model. Thus for analysis and design, it is crucial to obtain an accurate mathematical model that describes the behavior of a system. Modeling of the example piezo-tube system is considered in the following.

Simple Model of a Piezo-Tube Actuator

We will model the dynamics of the piezo-tube actuator along the *z*-axis where the input is the applied voltage $V_z(t)$ and the output of the system is the displacement of the probe tip $z(t)$. We begin the modeling by simplifying the system as an isolated mass, an ideal spring, and a damper as shown in Figure 3.21a. The entire mass of the piezo-tube is lumped into one mass element *m*, the internal elastic behavior of the piezo-tube is modeled as a spring, and the structural damping in the piezo-tube is modeled as a damper or a viscous friction element (such models are referred to as *lumped models* [4]). A mathematical relationship between the applied voltage $V_z(t)$ and the displacement of the probe tip $z(t)$ can be obtained using physical laws. Applying Newton's second law (the sum of all external forces F_i acting on a body is equal the product of its mass *m* and acceleration $\ddot{z}(t)$) we can write the equation of motion as

$$\sum_i F_i(t) = m\ddot{z}(t) \tag{3.96}$$

As shown in Figure 3.21b (the free body diagram), there are three external forces acting on the piezo-tube. First, the force exerted by the spring is assumed to be proportional to the displacement of the probe tip, that is,

$$F_s(t) = -kz(t) \tag{3.97}$$

where k is the spring constant with SI units [N/m]. Second, the damping force is considered to be proportional to the velocity of the probe tip $\dot{z}(t)$, that is,

$$F_D(t) = -c\dot{z}(t) \tag{3.98}$$

where c is the viscous friction or damping coefficient with SI units [N · s/m]. Third, induced strain ε in the piezoceramic material is proportional to the applied voltage $V_z(t)$ [5], and by Hooke's Law, the induced stress σ is proportional to the induced strain ε. Hence, the induced force $F_p(t)$ (stress σ times the cross-sectional area) is proportional to the applied voltage $V_z(t)$, that is,

$$F_p(t) = bV_z(t) \tag{3.99}$$

where b is a constant with SI units [N/V]. Rewriting Equation 3.96 in terms of the three external forces, the equation of motion becomes

$$\sum_{i=1}^{3} F_i(t) = F_S(t) + F_D(t) + F_p(t) = -kz(t) - c\dot{z}(t) + bV_z(t) = m\ddot{z}(t) \tag{3.100}$$

which is called the *mass-spring-damper* model. Note that the relationship between the input voltage $V_z(t)$ and the displacement $z(t)$ of the probe tip (i.e., the model of the dynamics) is a second order differential equation. The response of the probe tip (displacement of mass m) to an applied voltage $V_z(t)$ can be obtained in the frequency-domain by using the Laplace transform technique [6, Chapter 2, Section 5]; however, the state-space approach can be used to obtain the solution directly in the time-domain. In the remaining sections, the state-space approach to modeling is presented and the *mass-spring-damper* model of the piezo-tube actuator will be used as an example.

3.3.3.2 States of a System

We begin by introducing the concept of a state, which is the basis for the state-space approach. In general, a state can be defined as the following:

> The state $x(t_0)$ of a dynamic system at time t_0 is a set of variables that, together with the input $u(t)$, for $t \geq t_0$, determines the behavior of the system for all $t \geq t_0$ [7, Chapter 2, Section 1.1].

Fundamental to this definition is the notion that the state summarizes the current configuration of a system. Therefore, the memory of a dynamical system is preserved in the state variables at the current time t_0 (called initial condition), and the future behavior of the system is determined by the initial condition $x(t_0)$ and the applied input $u(t)$, for $t \geq t_0$. The state of a system can be written as the set

$$x(t) = \begin{bmatrix} x_1(t) \\ x_2(t) \\ \vdots \\ x_n(t) \end{bmatrix} \tag{3.101}$$

where n is the number of states.* Any set of variables that satisfy the above definition can be a valid state, hence the state is not unique [8, Chapter 2, Section 2].

Example

The state variables required to describe the mass-spring-damper system can be chosen as the position $z(t)$ and velocity $\dot{z}(t)$ of the mass. We can write the state vector as

$$x(t) = \begin{bmatrix} x_1(t) \\ x_2(t) \end{bmatrix} = \begin{bmatrix} z(t) \\ \dot{z}(t) \end{bmatrix} \tag{3.102}$$

*For a discussion on the minimal set of states required to describe a system (minimal realization), see [7, Chapter 7].

where the number of states is two ($n = 2$). If the position $z(t)$ and velocity $\dot{z}(t)$ of the mass are known at time t_0, along with the applied voltage $V_z(t)$ defined for $t \geq t_0$, then the future behavior of the system (i.e., the state $x(t)$) can be determined by solving the differential Equation 3.100.

3.3.3.3 The Linear State-Space Equation and Its Solution

For a linear system, the evolution of the states of a system over time can be described by a set of linear first order differential equations of the form:

$$
\begin{aligned}
x_1(t) &= \frac{dx_1(t)}{dt} = a_{11}(t)x_1(t) + \cdots + a_{1n}(t)x_n(t) + b_{11}(t)u_1(t) + \cdots b_{1p}(t)u_p(t) \\
x_2(t) &= \frac{dx_2(t)}{dt} = a_{21}(t)x_1(t) + \cdots + a_{2n}(t)x_n(t) + b_{21}(t)u_1(t) + \cdots b_{2p}(t)u_p(t) \\
&\vdots \\
x_n(t) &= \frac{dx_n(t)}{dt} = a_{n1}(t)x_1(t) + \cdots + a_{nn}(t)x_n(t) + b_{n1}(t)u_1(t) + \cdots b_{np}(t)u_p(t)
\end{aligned}
\tag{3.103}
$$

where n is the number of states (or the order of the system) and p is the number of inputs.* Defining the input vector as

$$
u(t) = \begin{bmatrix} u_1(t) \\ u_2(t) \\ \vdots \\ u_p(t) \end{bmatrix}
\tag{3.104}
$$

and the state vector $x(t)$ as defined in Equation 3.101, the set of first order differential equations given by Equation 3.103 can be rewritten in compact matrix form as [8, Chapter 2, Section 2]

$$
\begin{aligned}
x(t) &= \begin{bmatrix} a_{11}(t) & a_{12}(t) & \cdots & a_{1n}(t) \\ a_{21}(t) & a_{22}(t) & \cdots & a_{2n}(t) \\ \vdots & \vdots & \ddots & \vdots \\ a_{n1}(t) & a_{n2}(t) & \cdots & a_{nn}(t) \end{bmatrix} x(t) + \begin{bmatrix} b_{11}(t) & b_{12}(t) & \cdots & b_{1p}(t) \\ b_{21}(t) & b_{22}(t) & \cdots & b_{2p}(t) \\ \vdots & \vdots & \ddots & \vdots \\ b_{n1}(t) & b_{n2}(t) & \cdots & b_{np}(t) \end{bmatrix} u(t) \\
&= A(t)x(t) + B(t)u(t)
\end{aligned}
\tag{3.105}
$$

where $A(t)$ is an $n \times n$ matrix and $B(t)$ is an $n \times p$ matrix. For a system defined with q outputs $y(t)$, which are assumed to be a linear combination of the state $x(t)$ and input $u(t)$, we can write the output equation as

$$
\begin{aligned}
y(t) &= \begin{bmatrix} c_{11}(t) & c_{12}(t) & \cdots & c_{1n}(t) \\ c_{21}(t) & c_{22}(t) & \cdots & c_{2n}(t) \\ \vdots & \vdots & \ddots & \vdots \\ c_{q1}(t) & c_{q2}(t) & \cdots & c_{qn}(t) \end{bmatrix} x(t) + \begin{bmatrix} d_{11}(t) & d_{12}(t) & \cdots & d_{1p}(t) \\ d_{21}(t) & d_{22}(t) & \cdots & d_{2p}(t) \\ \vdots & \vdots & \ddots & \vdots \\ d_{q1}(t) & d_{q2}(t) & \cdots & d_{qp}(t) \end{bmatrix} u(t) \\
&= C(t)x(t) + D(t)u(t)
\end{aligned}
\tag{3.106}
$$

---

*Given a higher order differential equation, a set of first order differential equations can be obtained by a procedure known as reduction to first order as presented in [9].

where $C(t)$ is a $q \times n$ matrix and $D(t)$ is a $q \times p$ matrix. In general, the matrices $A(t)$, $B(t)$, $C(t)$, and $D(t)$ are time varying; however, in this chapter we will only consider the time-invariant case where A, B, C, and D are constant matrices, then Equations 3.107 and 3.108 are called the linear time-invariant (LTI) state and output equations, respectively.*

$$x(t) = Ax(t) + Bu(t) \tag{3.107}$$

$$y(t) = Cx(t) + Du(t) \tag{3.108}$$

The response of the system to an applied input can be quantified by the evolution of the system state $x(t)$ and the output $y(t)$. The state-space Equation 3.107 is a set of first order differential equations in matrix form, which can be solved in time for a given initial condition $x(t_0)$ as [8, Chapter 3]

$$x(t) = e^{A(t-t_0)}x(t_0) + \int_{t_0}^{t} e^{A(t-\tau)}Bu(\tau)\,d\tau \tag{3.109}$$

Note that solution (3.109) is the sum of two terms: the first term is the effect of initial condition $x(t_0)$ and the second is the effect of the applied input $u(t)$ between $t_0 \le \tau \le t$.† Using the output Equation 3.108 and the state solution given by (3.109), the output $y(t)$ becomes

$$y(t) = Ce^{A(t-t_0)}x(t_0) + \int_{t_0}^{t} Ce^{A(t-\tau)}Bu(\tau)\,d\tau + Du(t) \tag{3.110}$$

The system response $y(t)$ to an applied input $u(t)$ is characterized by the system matrices (A, B, C, D). For example, the output $y(t)$ will be bounded for any bounded input if the system is stable and the system is stable if the real parts of all the eigenvalues of A are less than zero (strictly negative) [8, Chapter 4, Section 4].‡

Example

For the mass-spring-damper example system, the state-space equation can be found by differentiating the states $x(t)$ defined in Equation 3.102 and using the equation of motion 3.100 to obtain

$$\begin{aligned} x_1(t) = z(t) &= x_2(t) \\ x_1(t) = z(t) &= -\left(\frac{k}{m}\right)z(t) - \left(\frac{c}{m}\right)z(t) + \left(\frac{b}{m}\right)V_z(t) = -\left(\frac{k}{m}\right)x_1(t) - \left(\frac{c}{m}\right)x_2(t) + \left(\frac{b}{m}\right)u(t) \end{aligned} \tag{3.111}$$

We choose the position of the mass $z(t)$ to be the output of the system, and write the state-space and output equation in the form given by Equations 3.107 and 3.108 as

$$x(t) = \begin{bmatrix} 0 & 1 \\ -(k/m) & -(c/m) \end{bmatrix} x(t) + \begin{bmatrix} 0 \\ b/m \end{bmatrix} u(t) \tag{3.112}$$

$$y(t) = \begin{bmatrix} 1 & 0 \end{bmatrix} x(t) \tag{3.113}$$

*For a detailed discussion of the solution of linear time-varying equations, see [7, Chapter 4, Section 5].

†The MATLAB command `lsim` simulates the time response of LTI models to arbitrary inputs.

‡The MATLAB command `eig (A)` returns the eigenvalues of the system matrix `A`.

3.3.3.4 Linearization of Nonlinear Systems

A general form of the state-space equation (for nonlinear systems) is

$$x(t) = g(x, u) \tag{3.114}$$

$$y(t) = h(x, u) \tag{3.115}$$

where g and h can be nonlinear functions.* The behavior of nonlinear systems is beyond the scope of this section; however, a detailed discussion can be found in [10]. The behavior of a nonlinear system can be approximated by a linear model in a neighborhood of an equilibrium point. Such linearizations can simplify the analysis and design of nonlinear systems because the tools developed for linear systems can be applied under certain conditions [10]. Let x_0 and u_0 be the equilibrium point and equilibrium input, respectively, such that [10, Chapter 1]

$$g(x_0, u_0) = 0 \tag{3.116}$$

$$h(x_0, u_0) = y_0 \tag{3.117}$$

Consider small perturbations in the equilibrium point $x(t) = x_0 + \bar{x}(t)$, the input $u(t) = u_0(t) + \bar{u}(t)$, and the output $y(t) = y_0 + \bar{y}(t)$. If the perturbation $\bar{x}(t)$ is small for all t, we obtain the following by expanding Equation 3.114 in Taylor series (neglecting higher order terms of $\bar{x}(t)$ and $\bar{u}(t)$):

$$\begin{aligned} x_0 + \dot{x}(t) &= g(x_0 + x(t), u_0 + \bar{u}(t)) \\ \dot{x}(t) &= g(x_0, u_0) + \left.\frac{\partial g}{\partial x}\right|_{\substack{x=x_0\\u=u_0}} x(t) + \left.\frac{\partial g}{\partial u}\right|_{\substack{x=x_0\\u=u_0}} \bar{u}(t) \end{aligned} \tag{3.118}$$

Recognizing that $g(x_0, x_0) = 0$ we obtain

$$\dot{x}(t) = \bar{A}\bar{x}(t) + \bar{B}\bar{u}(t) \tag{3.119}$$

where

$$\bar{A} = \left.\frac{\partial g}{\partial x}\right|_{\substack{x=x_0\\u=u_0}} \quad \text{and} \quad \bar{B} = \left.\frac{\partial g}{\partial u}\right|_{\substack{x=x_0\\u=u_0}} \tag{3.120}$$

The matrices $\bar{A}$ and $\bar{B}$ are the Jacobians evaluated at x_0 and u_0. Equation 3.119 is a linear state equation and is valid for small perturbations about x_0 and u_0. A similar result can be obtained for the change $\bar{y}(t)$ in the output from the equilibrium value y_0 as

$$y(t) = \bar{C}x(t) + \bar{D}\bar{u}(t) \tag{3.121}$$

where

$$\bar{C} = \left.\frac{\partial h}{\partial x}\right|_{\substack{x=x_0\\u=u_0}} \quad \text{and} \quad \bar{D} = \left.\frac{\partial h}{\partial u}\right|_{\substack{x=x_0\\u=u_0}} \tag{3.122}$$

*The MATLAB command `ode45` can be used to obtain the numeric solution to the general nonlinear state space equation.

3.3.4 Relationship between State Equations and Transfer-Functions

3.3.4.1 State-Space to Transfer-Function

The input-to-output relationship of a dynamic system in the frequency-domain is represented by a transfer-function, which can be obtained by taking the Laplace transform of (3.107) and (3.108) with zero initial conditions as follows [8, Chapter 3, Section 5]:

$$sX(s) = AX(s) + BU(s), \tag{3.123}$$

$$Y(s) = CX(s) + DU(s), \tag{3.124}$$

where s is the Laplace variable. Solving (3.123) for $X(s)$ and substituting into (3.124), the ratio of the output $Y(s)$ to input $U(s)$ for a single-input single-output system (SISO) can be found as

$$\begin{aligned} G(s) &= \frac{Y(s)}{U(s)} = C(sI-A)^{-1}B + D \\ &= \frac{N(s)}{D(s)} \end{aligned} \tag{3.125}$$

where I is an $n \times n$ identity matrix. In Equation 3.125, $N(s)$ and $D(s)$ are referred to as the numerator and denominator polynomial of $G(s)$, respectively.*

Analogous to the state-space equation, the boundedness of the output response $y(t)$ to a bounded input $u(t)$ is characterized by the roots of the denominator polynomial $D(s)$, that is, the values of s for which $D(s) = 0$. In particular, the output $y(t)$ will be bounded for any bounded input, that is, system is stable if the real parts of all the roots of $D(s)$ are less than zero (strictly negative).† Alternatively, a convenient method to determine stability without having to find the roots of $D(s)$ explicitly is the Routh–Hurwitz stability criterion [6, Chapter 6].

Example

With the state-space description of the mass-spring-damper system defined in Equations 3.112 and 3.113, the transfer-function realization using Equation 3.125 becomes

$$\begin{aligned} G(s) &= \frac{Y(s)}{U(s)} = \begin{bmatrix} 1 & 0 \end{bmatrix} \left[s\begin{bmatrix} 1 & 0 \\ 0 & 1 \end{bmatrix} - \begin{bmatrix} 0 & 1 \\ -(k/m) & -(c/m) \end{bmatrix} \right]^{-1} \begin{bmatrix} 0 \\ b/m \end{bmatrix} + [0] \\ &= \frac{b/m}{s^2 + (c/m)s + k/m} \end{aligned} \tag{3.126}$$

The input to the system is the applied voltage $V_z(t)$ and the output is the displacement of the mass $z(t)$.

3.3.4.2 Frequency-Response Using Transfer-Functions

Consider a linear single-input single-output (SISO) stable system with transfer-function description $G(s)$. When the system $G(s)$ is excited by a sinusoidal input of the form

$$u(t) = P\sin(\omega t) \tag{3.127}$$

with amplitude P and frequency ω, the output response (after the transients decay) will also be a sinusoid of the form

$$y(t) = MP\sin(\omega t + \phi) \tag{3.128}$$

---

*The MATLAB command `ss2tf` can be used to convert a state-space realization to a transfer-function.

†The MATLAB command `roots (den)` can be used to find the roots of `den`, where `den` is the coefficients of $D(s)$.

with the same frequency ω and a phase shift ϕ [6, Chapter 8]. The output amplitude is the input amplitude scaled by M, the magnitude gain. The magnitude gain M is found by taking the magnitude of $G(s)$ evaluated at $s = j\omega$, that is,

$$M = |G(s)|_{s=j\omega} \tag{3.129}$$

Usually, the magnitude gain M is expressed in units of decibels (dB), i.e., M [dB] = 20log M. The phase shift ϕ is the angle of $G(s)$ evaluated at $s = j\omega$, that is,

$$\phi = \angle G(s)|_{s=j\omega} \tag{3.130}$$

with units of degrees. The plot of the magnitude gain M and the phase shift ϕ versus the frequency ω gives a graphical representation of the frequency-response (Bode plots) of a system.* These plots can be obtained experimentally by measuring the magnitude gain and phase shift between the input and output response of a system over a range of frequencies. Additionally, a system's transfer-function can be obtained from an experimental frequency-response data by using curve-fitting software. In Section 3.3.5, we present this approach to determine a model for a system using experimental frequency-response data.

3.3.4.3 Transfer-Function to State-Space

In Section 3.3.4, a transfer-function model was obtained for a system in state-space form. In the following, an approach for realizing a state-space model from a transfer-function $G(s)$ is presented. For a realizable transfer function $G(s)$ of a SISO system of the form

$$G(s) = \frac{b_0 s^n + b_1 s^{n-1} + \cdots + b_n}{s^n + a_1 s^{n-1} + \cdots + a_n} \tag{3.131}$$

the controllable canonical state-space form written in terms of the coefficients of $G(s)$ is

$$x(t) = \begin{bmatrix} -a_1 & -a_2 & \cdots & -a_{n-1} & -a_n \\ 1 & 0 & \cdots & 0 & 0 \\ 0 & 1 & \cdots & 0 & 0 \\ \vdots & \vdots & \ddots & \vdots & \vdots \\ 0 & 0 & \cdots & 1 & 0 \end{bmatrix} x(t) + \begin{bmatrix} 1 \\ 0 \\ 0 \\ \vdots \\ 0 \end{bmatrix} u(t) \tag{3.132}$$

$$y(t) = [(b_1 - a_1 b_0)\ (b_2 - a_2 b_0)\ \cdots\ (b_n - a_n b_0)]x(t) + [b_0]u(t) \tag{3.133}$$

where the number of states n is equal to the highest power of the denominator of $G(s)$.† The smallest possible dimension for realizing a system, referred to as the minimum realization, is an important factor to consider in analysis and design.‡ Models of minimum order require less computational power in simulation and implementation compared to higher order models. For information about other equivalent canonical state-space forms, refer to [7, Chapter 4, Sections 3 and 4].

*The MATLAB command `bode` plots the magnitude gain and phase shift for a linear system.

†The MATLAB command `tf2ss` generates the controllable canonical form realization of a transfer function $G(s)$. Other useful commands include `ss2tf`, `zp2tf`, and `tf2zp`.

‡For a detailed discussion of minimal realizations for multi-input multi-output systems, see [7, Chapter 7].

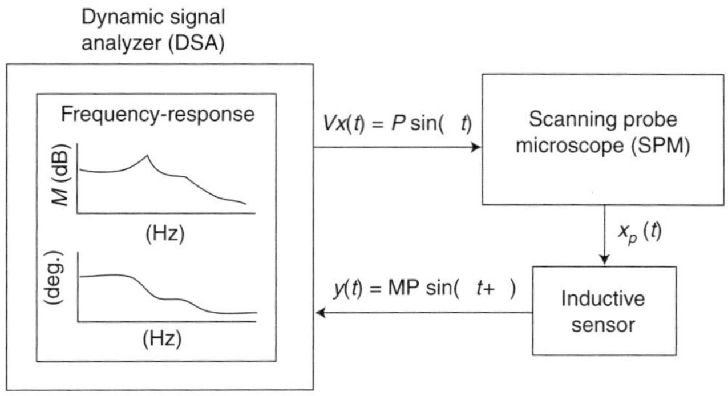

FIGURE 3.22 A schematic of the experimental setup used to determine the frequency-response of the piezo-tube actuator. An inductive sensor measured the displacement of the actuator along the x-axis, and the frequency-response data from the DSA were used to estimate the system model.

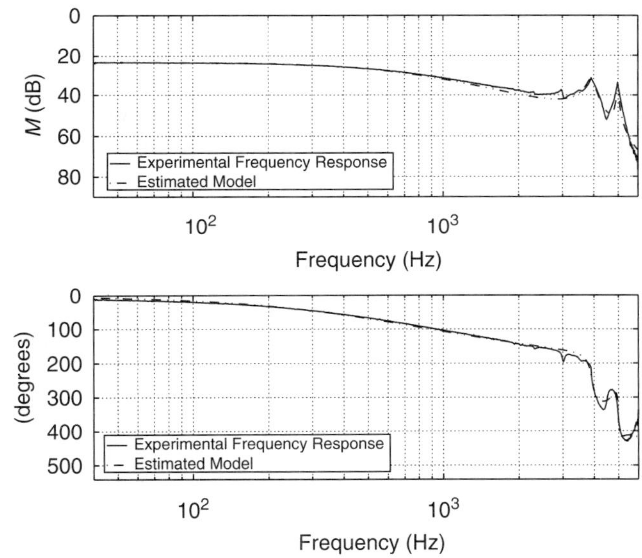

FIGURE 3.23 The experimental magnitude gain and phase versus frequency plots for the piezo-tube actuator measured along the x-axis with superimposed model frequency-response. Solid line represents experimental data; dashed line represents results from estimated model.

3.3.5 Experimental Modeling Using Frequency-Response

An approach to modeling using experimental frequency-response data is presented in this section. Using a dynamic signal analyzer (DSA), the frequency-response of the dynamics along the x-axis for the piezo-tube actuator was measured.* A sinusoidal input voltage $V_x(t)$ with frequency varying between 10 Hz and 6 kHz was generated by a DSA and applied to the scanning probe microscope (SPM) system as shown in Figure 3.22. Using an inductive sensor, the displacement $x_p(t)$ of the piezo-tube along the x-axis was measured and fed back to the DSA to compute the frequency-response (M and ϕ versus frequency ω plots). Figure 3.23 shows the Bode plots obtained by the DSA between the applied voltage $V_x(t)$ and the output of the inductive sensor $y(t)$. An estimate of the system model from the frequency-response

*Stanford Research Systems, model SRS785.

data was then found with the MATLAB software.* The transfer-function between the applied input voltage $V_x(t)$ and the output of the inductive sensor $y(t)$ was found to be

$$G_1(s) = \frac{Y(s)}{V_x(s)}$$
$$= \frac{5.544\times 10^5 s^4 - 7.528\times 10^9 s^3 + 1.476\times 10^{15} s^2 - 4.571\times 10^{18} s + 9.415\times 10^{23}}{s^6 + 1.255\times 10^4 s^5 + 1.632\times 10^9 s^4 + 1.855\times 10^{13} s^3 + 6.5\times 10^{17} s^2 + 6.25\times 10^{21} s + 1.378\times 10^{25}} \quad (3.134)$$

with units of V/V. Equation 3.135 was scaled by the inductive sensor gain (30 Å/V) and the transfer-function between the applied voltage $V_x(t)$ and the actual displacement of the piezo-tube $x_p(t)$ is given by

$$G_2(s) = \frac{X_p(s)}{V_x(s)}$$
$$= \frac{1.663\times 10^7 s^4 - 2.258\times 10^{11} s^3 + 4.427\times 10^{16} s^2 - 1.371\times 10^{20} s + 2.825\times 10^{25}}{s^6 + 1.255\times 10^4 s^5 + 1.632\times 10^9 s^4 + 1.855\times 10^{13} s^3 + 6.5\times 10^{17} s^2 + 6.25\times 10^{21} s + 1.378\times 10^{25}} \quad (3.135)$$

with units of Å/V.

3.3.5.1 Time Scaling of a Transfer-Function Model

We present below an approach for rescaling time for $G_2(s)$ from seconds [s] to milliseconds [ms]. We briefly recall the time scaling property of the Laplace transform presented in [1, Chapter 3, Section 1.4]. Let $F(s)$ be the Laplace transform of $f(t)$, that is,

$$f(t) \xrightarrow{L} F(s) \quad (3.136)$$

where L denotes the Laplace transform operator. Now, consider a new time scale defined as $\hat{t} = at$, where a is a constant. The Laplace transform of $f(\hat{t}) = f(at)$ is given by

$$f(\hat{t}) = f(at) \xrightarrow{L} \frac{1}{|a|}F\left(\frac{s}{a}\right) = \hat{F}(s) \quad (3.137)$$

Using relation (3.137), we can reduce the coefficients of $G_2(s)$ by changing the time units of both the input signal $u(t)$ and output signal $y(t)$ as follows:

$$\hat{G}(s) = \frac{\hat{Y}(s)}{\hat{U}(s)} = \frac{Y(s/a)/|a|}{U(s/a)/|a|} = \frac{Y(s/a)}{U(s/a)} = G\left(\frac{s}{a}\right) \quad (3.138)$$

Therefore, to rescale time for $G_2(s)$ from seconds [s] to millisecond [ms], we choose $\hat{t} = at = 0.001t$ and the new rescaled transfer $G_2(s)$ becomes

$$\hat{G}_2(s) = G_2\left(\frac{s}{a}\right)\bigg|_{a=0.001}$$
$$= G_2(1000s)$$
$$\hat{G}_2(s) = \frac{16.63s^4 - 225.8s^3 + 4.427\times 10^4 s^2 - 1.371\times 10^5 s + 2.825\times 10^7}{s^6 + 12.55s^5 + 1.632\times 10^3 s^4 + 1.855\times 10^4 s^3 + 6.5\times 10^5 s^2 + 6.25\times 10^6 s + 1.378\times 10^7} \quad (3.139)$$

*The MATLAB command `invfreqs` gives real numerator and denominator coefficients of experimentally determined frequency response data.

Note that the time unit of the input and output signal of $\hat{G}_2(s)$ are now in milliseconds, not seconds! The coefficients of the numerator and denominator polynomials are smaller and this form ($\hat{G}_2(s)$) is less prone to computational errors due to round off than the form $G_2(s)$, Equation 3.135.

3.3.5.2 The State-Space Model

The state-space realization for $\hat{G}_2(s)$ expressed in controllable canonical form (Equations 3.132 and 3.133) is given by the following:

$$\dot{x}(t) = \begin{bmatrix} -12.55 & -1.632\times 10^3 & -1.855\times 10^4 & -6.50\times 10^5 & -6.25\times 10^6 & -1.378\times 10^7 \\ 1 & 0 & 0 & 0 & 0 & 0 \\ 0 & 1 & 0 & 0 & 0 & 0 \\ 0 & 0 & 1 & 0 & 0 & 0 \\ 0 & 0 & 0 & 1 & 0 & 0 \\ 0 & 0 & 0 & 0 & 1 & 0 \end{bmatrix} x(t) + \begin{bmatrix} 1 \\ 0 \\ 0 \\ 0 \\ 0 \\ 0 \end{bmatrix} u(t) \tag{3.140}$$

$$y(t) = [0 \;\; 16.63 \;\; -225.8 \;\; 4.427\times 10^4 \;\; -1.371\times 10^5 \;\; 2.825\times 10^7]\,x(t) \tag{3.141}$$

The time unit for Equations 3.140 and 3.141 are milliseconds [ms]. If the initial state at t_0 is known, along with the applied voltage $V_x(t)$ defined for $t \geq t_0$, the future behavior of the system, that is, the state $x(t)$ and output $y(t)$, can be determined from Equations 3.140 and 3.141, respectively.

3.3.6 Discrete-Time State-Space Modeling

3.3.6.1 Introduction

The study of discrete-time systems is important to the analysis and the design of modern mechatronics systems where digital computers or small microprocessors are predominantly used to control systems. Digital computers and microprocessors output or acquire information at discrete time instants. For example, the input applied by a digital computer to actuate the piezo-tube changes at discrete time instants. Similarly, the displacment of the piezo-tube can only be measured at specified time instants using digital computers; therefore, in comparison to a continuous-time control system where the input signals change continuously over time, the input of a discrete-time system changes once in a while. Such discrete-time systems are studied next.

Consider a continuous-time system with continuous input $u(t)$ and output $y(t)$ as described by Equations 3.107 and 3.108. Let a digital computer or microprocessor be used to provide the input $u[k]$ and measure the output $y[k]$ as depicted in Figure 3.24 (such systems with continuous and discrete signals are

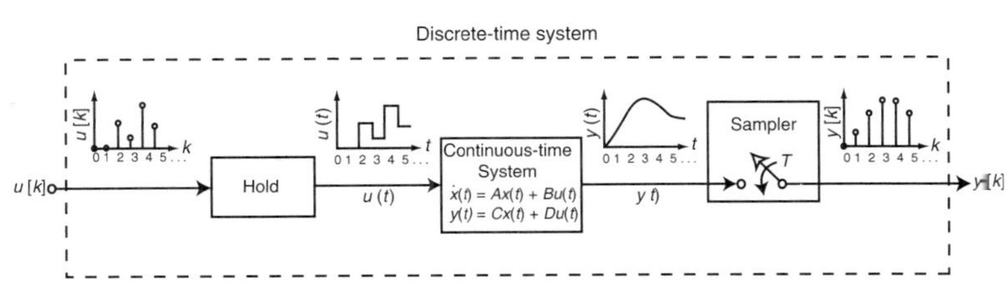

FIGURE 3.24 A block diagram of a discrete-time system showing signals in graphic form. Note that $u[k] = u(k \cdot T)$ and $y[k] = y(k \cdot T)$, for $k = 0,1,2,\ldots$, and the sampling period T is assumed to be constant.

called sampled-data systems). The input $u[k]$ and output $y[k]$ of this system are discrete with $u[k] = u(k \cdot T)$ and $y[k] = y(k \cdot T)$ for $k = 0,1,2,\ldots$, where T is the constant sampling period. The discrete input $u[k]$ is applied to the continuous system from a digital computer or microprocessor and is held constant during the time interval T (zero-order hold). A sampler acquires the output of the continuous system at each time instant T yielding the discrete output $y[k]$. The discrete system is between the input $u[k]$ and the output $y[k]$ [11, Chapter 1].* The equivalent discrete-time state-space representation of the continuous-time state-space model given by Equations 3.107 and 3.108 is given by (the details of the formulation can be found in [11, Chapter 5, Section 5])

$$x[k+1] = A_D x[k] + B_D u[k] \tag{3.142}$$

$$y[k] = C_D x[k] + D_D u[k] \tag{3.143}$$

where

$$A_D = e^{AT}, \quad B_D = \left(\int_0^T e^{A\lambda} d\lambda\right) B, \quad C_D = C, \text{ and } D_D = D \tag{3.144}$$

and matrices C_D and D_D are not changed by the sampling.[†] This discrete model (Equations 3.142 and 3.143) is the representation of the sampled-data system shown in Figure 3.24.

3.3.6.2 Solutions to the Discrete-Time State-Space Equations

The solution to the discrete model (Equations 3.142 and 3.143) is given by

$$x[k] = A_D^k x[0] + \sum_{j=0}^{k-1} A_D^{k-j-1} B_D u[j] \tag{3.145}$$

$$y[k] = CA_D^k x[0] + C\sum_{j=0}^{k-1} A_D^{k-j-1} B_D u[j] + Du[k] \tag{3.146}$$

for each sampling step k. Details of the formulation can be found in [11, Chapter 5, Section 3]. The state response $x[k]$ to an applied input $u[k]$ is characterized by the system matrices (A_D, B_D, C_D, D_D). In particular, the output $y[k]$ will be bounded for any bounded input $u[k]$ if the system is stable. A system in the form given by Equation 3.142 is stable if the magnitude of all the eigenvalues of A_D are less than unity, that is, lie within the unit circle center at the origin of the z-plane [11, Chapter 5, Section 6].

3.3.6.3 The z Transform and Relationship with the State-Space

The input-to-output relationship in the frequency-domain for a discrete-time system is represented by a discrete transfer-function called the z transform, written in terms of the variable z [12, Chapter 4]. Analogous to the continuous-time case, the model of a dynamic system in discrete transfer-function form can be useful in the design and control of systems [12, Chapter 7]. If the system model is available in discrete transfer-function form, then a state-space realization can be found as follows. Given a discrete system described by the following z-transform $G(z)$:

$$G(z) = \frac{d_0 + d_1 z^{-1} + \cdots + d_n z^{-n}}{1 + c_1 z^{-1} + \cdots + c_n z^{-n}} \tag{3.147}$$

*We do not discuss quantizing and quantization error. See [11, Chapter 1, Section 3] for details.

†Given a continuous-time state-space model (A, B, C, D), the MATLAB command `c2d`, gives the discrete time equivalent for a specified sampling period T.

the controllable canonical discrete state-space realization for $G(z)$ is

$$x[k+1] = \begin{bmatrix} -c_1 & -c_2 & \cdots & -c_{n-1} & -c_n \\ 1 & 0 & \cdots & 0 & 0 \\ 0 & 1 & \cdots & 0 & 0 \\ \vdots & \vdots & & \vdots & \vdots \\ 0 & 0 & \cdots & 1 & 0 \end{bmatrix} x[k] + \begin{bmatrix} 1 \\ 0 \\ 0 \\ \vdots \\ 0 \end{bmatrix} u[t] \tag{3.148}$$

$$y[k] = [(d_1 - c_1 d_0) \ (d_2 - c_2 d_0) \ \ldots \ (d_n - c_n d_0)]x[k] + [d_0]u[k] \tag{3.149}$$

The number of states n is equivalent to the highest power of the denominator of $G(z)$. For information about other equivalent canonical state-space forms, refer to [11, Chapter 5, Section 2].

Example

Consider the continuous-time state-space model of the piezo-tube system described by Equations 3.140 and 3.141. A digital computer with the sampling rate of 10 kHz ($T = 1.0 \times 10^{-4}$) is used to provide the control input $u[k]$ and measure its displacement along the x-axis (output $y[k]$). The discrete-time state-space model with (A_D, B_D, C_D, and D_D) given by Equation 3.144 is

$$x[k+1] = \begin{bmatrix} 0.999 & -0.163 & -1.85 & -65.0 & -624.5 & -1377.1 \\ 9.99\times10^{-5} & 0.999 & -9.26\times10^{-5} & -3.25\times10^{-3} & -3.12\times10^{-2} & -6.69\times10^{-2} \\ 5.00\times10^{-9} & 1.00\times10^{-4} & 1 & -1.08\times10^{-7} & -1.04\times10^{-6} & -2.30\times10^{-6} \\ 1.67\times10^{-13} & 5.00\times10^{-9} & 1.00\times10^{-4} & 1 & -2.60\times10^{-11} & -5.74\times10^{-11} \\ 4.17\times10^{-18} & 1.67\times10^{-13} & 5.00\times10^{-9} & 1.00\times10^{-4} & 1 & -1.15\times10^{-15} \\ 8.33\times10^{-23} & 4.17\times10^{-18} & 1.67\times10^{-13} & 5.00\times10^{-9} & 1.00\times10^{-4} & 1 \end{bmatrix} x[k]$$

$$+ \begin{bmatrix} 9.99\times10^{-5} \\ 4.99\times10^{-9} \\ 1.67\times10^{-13} \\ 4.17\times10^{-18} \\ 8.33\times10^{-23} \\ 1.39\times10^{-27} \end{bmatrix} u[k] \tag{3.150}$$

$$y[k] = \lfloor 0 \ \ 16.63 \ \ -225.8 \ \ 4.427\times10^{4} \ \ -1.371\times10^{5} \ \ 2.825\times10^{7} \rfloor x[k] \tag{3.151}$$

The realization given by Equations 3.150 and 3.151 was found using the MATLAB command 'c2d'.

3.3.7 Summary

We presented tools for modeling continuous- and discrete-time systems using the state-space approach in this section. The state-space approach to modeling is a powerful technique for the analysis and design of mechatronic and dynamic systems, and can take advantage of tools available in modern digital computers and microprocessors. The discussion of the system states and the state-space was motivated by an example piezo-tube actuator system. We considered the modeling of linear systems and a technique for linearizing nonlinear systems was briefly introduced. The frequency-response of a system and an approach to modeling using experimental frequency-response data was presented. Relationships between models expressed in the frequency- and time-domain for both continuous- and discrete-time systems was discussed. For additional details about the concepts mentioned in this section and those not covered, it is recommended that the reader consider the attached references for further reading.

References

1. Franklin, G. F., et al., *Feedback Control of Dynamic Systems,* 3rd ed., Addison-Wesley, New York, 1994.
2. Hanselman, D., and Littlefield, B., *The Student Edition of Matlab, Version 5, User's Guide,* Prentice-Hall, Upper Saddle River, NJ, 1997.
3. Croft, D., et al., Creep, hysteresis, and vibration compensation for piezoactuators: atomic force microscopy application, *ASME J. Dyn. Syst., Meas., Control,* 123, 35, 2001.
4. Dorny, C. N., *Understanding Dynamic Systems-Approaches to Modeling, Analysis, and Design,* Prentice-Hall, Englewood Cliffs, NJ, 1993.
5. Locatelli, M., et al., Easy method to characterize a piezoelectric ceramic tube as a displacer, *Rev. Sci. Instrum.,* 59, 4, 1988.
6. Dorf, R. C., and Bishop, R. H., *Modern Control Systems,* 9th ed., Prentice-Hall, Upper Saddle River, NJ, 2001.
7. Chen, T. C., *Linear System Theory and Design,* Oxford University Press, New York, 1999.
8. Friedland, B., *Control System Design: An Introduction to State-Space Methods,* McGraw-Hill, New York, 1986.
9. Gillis, J. T., State space, in *The Control Handbook,* Levine, W. S., CRC Press, Salem, MA, 1996, Chap. 5.
10. Khalil, H. K., *Nonlinear Systems,* 2nd ed., Prentice-Hall, Upper Saddle River, NJ, 1996.
11. Ogata, K., *Discrete-Time Control Systems,* 2nd ed., Prentice-Hall, Englewood Cliffs, NJ, 1995.
12. Franklin, G. F., et al., *Digital Control of Dynamic Systems,* 3rd ed., Addison-Wesley, Menlo Park, 1998.

3.4 Transfer Functions and Laplace Transforms

C. Nelson Dorny

We perceive a system primarily through its behavior. Therefore, our mental image of a system usually includes representative response **signals**. The *step response,* the behavior when we suddenly turn on the system, is such a system-characterizing signal. We should view the step response as a description of the system. The *impulse response* is another description of the system. For a system represented by linear differential equations, the unit-step response is the integral of the unit-impulse response.

Let us represent time differentiation (d/dt) by the *time-derivative operator, p.* Then we can denote the time derivative of a signal y by py, its second derivative by p^2y, its integral with respect to time by $(1/p)y$, and so on. This *operator notation* simplifies the expressions for differential equations. We shall use the expression *system equations* to mean a set of differential equations that determines fully the behaviors of the dependent variables that appear in those equations. We can reduce a *linear* set of system equations to a single **input-output system equation** by eliminating all but one dependent variable from the set. The *transfer function* associated with that dependent variable is a mathematical expression that contains all the essential information embodied in the system differential equation.

The Laplace transformation converts signals (functions of time) to functions of a *complex-frequency variable,* $s = \sigma + j\omega$. There is a one-to-one correspondence between a signal and its Laplace transform. We can retrieve the time function by inverse transformation. Laplace transformation produces images that have some properties that are more convenient than those of the original signals. In particular, time differentiating a signal corresponds to multiplying its Laplace transform by the complex-frequency variable s. Hence, the transformation converts linear constant-coefficient differential equations to linear algebraic equations. Such simplifications of time-domain operations make Laplace transformation useful.

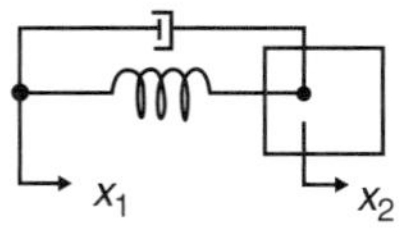

FIGURE 3.25 The lumped model of a mechanical system.

The Laplace transformation also converts the impulse response of a system variable to the transfer function for that variable. As a consequence, we can view the differential equation that represents a linear system as an expression of the response of that system to an impulsive input.

3.4.1 Transfer Functions

The node displacements x_1 and x_2 and the compressive forces f_1 and f_2 within the branches of the lumped model of Figure 3.25 are related to each other by the spring equation, the damper equation, and the balance of forces at node 2. The spring equation is $f_2 = k(x_1 - x_2)$. The equation for the damper is $f_1 = b(px_1 - px_2)$. The balance of forces requires that $f_1 + f_2 = mp^2x_2$. These equations describe fully the behavior of the system if the spring and mass are unenergized. (If the mass were moving and/or the spring were compressed, we would have to express separately their initial energy states to describe fully the future relations among the variables.)

Eliminate f_1 and f_2 from the equations to obtain the operational equation:

$$(mp^2 + bp + k)x_2 = (bp + k)x_1 \tag{3.152}$$

This differential equation describes fully the **zero-state** relation between x_1 and x_2. Rearrange Equation 3.152 to form the ratio

$$\frac{x_2}{x_1} = \frac{bp + k}{mp^2 + bp + k} \tag{3.153}$$

We call Equation 3.152 the *transfer function* from x_1 to x_2. The transfer function focuses attention on the mathematical operations that characterize the behavioral relationships rather than on the particular natures of the variables. (Note that the transfer function from v_1 to v_2, where $v_1 = px_1$ and $v_2 = px_2$, is the same as the transfer function given by Equation 3.153.)

In general, suppose that y_1 and y_2 are two variables related (in operator notation) by the linear differential equation

$$y_2 = G(p)y_1 \tag{3.154}$$

We formally define the *transfer function* from y_1 to y_2 by

$$G(p) = \left.\frac{y_2}{y_1}\right|_{ZS} \tag{3.155}$$

where the notation ZS means **zero state.** If y_1 is an independent variable, then $G(p)$ is the *input-output transfer function* for the variable y_2 and accounts fully for its behavior owing to the **input signal** y_1. We can determine from that transfer function the behavior of the system for any source waveform and any initial state.

3.4.2 The Laplace Transformation

The *one-sided Laplace transformation*, $\mathcal{L}$, is an integral operator that converts a signal $f(t)$ to a complex-valued function $F(s)$ in the following fashion:

$$+[f(t)] \equiv F(s) \triangleq \int_{0^-}^{\infty} f(t)e^{-st}\,dt \tag{3.156}$$

We refer to the transformed function $F(s)$ as the *Laplace transform* of the signal $f(t)$. Picture the lower limit 0^- of the integral as a *specific* instant prior to but infinitesimally close to $t = 0$. It is customary to use a lowercase symbol (f) to represent a signal waveform and an uppercase symbol (F) to represent its Laplace transform. (Although we speak here of time signals, there is nothing in Equation 3.156 that requires $f(t)$ to be a function of time. The transformation can be applied to functions of any quantity t.)

We shall use the Laplace transformation to transform the signals of **time-invariant** linear systems. The behavior of such a system for t Š 0 depends only on the input signal for $t \geq 0$ and on the prior **state** of the output variable (at $t = 0^-$). Hence, it does not matter that the Laplace transformation ignores $f(t)$ for $t < 0^-$.

The process of finding the time function $f(t)$ that corresponds to a particular Laplace transform $F(s)$ is called *inverse Laplace transformation*, and is denoted by $\mathcal{L}^{-1}$. We also call $f(t)$ the *inverse Laplace transform* of $F(s)$. Since the one-sided Laplace transformation ignores $t < 0^-$, $F(s)$ contains no information about $f(t)$ for $t < 0^-$. Therefore, inverse Laplace transformation cannot reconstruct $f(t)$ for $t < 0^-$. We shall treat all signals as if they are defined only for $t \geq 0^-$. Then there is a one-to-one relation between $f(t)$ and $F(s)$.

To illustrate the Laplace transformation, we find the Laplace transform of the decaying exponential, $f(t) = e^{-\alpha t}$, $t \geq 0^-$. The transform is

$$\begin{aligned} F(s) &= \int_{0^-}^{\infty} e^{-\alpha t} e^{-st}\,dt = \left.\frac{e^{-(s+\alpha)t}}{-(s+\alpha)}\right|_{0^-}^{\infty} \\ &= \left.\frac{e^{-(\sigma+\alpha)t}e^{-j\omega t}}{-(s+\alpha)}\right|_{0^-}^{\infty} = \frac{1}{s+\alpha} \quad \text{for } \mathrm{Re}[s] > -\alpha \end{aligned} \tag{3.157}$$

We must require $\sigma > -\alpha$, where σ is the real part of s, in order that the real-exponent factor converge to zero at the upper limit. (The magnitude of the complex-exponent factor remains 1 for all t.) Therefore, the Laplace transform of the decaying exponential is defined only for $\mathrm{Re}[s] > -\alpha$. This restriction on the domain of F in the complex s plane is comparable to the restriction $t \geq 0^-$ on the domain of f.

The significant features of the complex-frequency function $1/(s + \alpha)$ are the existence of a single pole and the location of that pole, $s = -\alpha$ [rad/s]. (The pole defines the left boundary of that region of the complex s plane over which the transform $1/(s + \alpha)$ is defined.) The significant features of the corresponding time function are the fact of decay and the rate of decay, with the exponent $-\alpha$ [rad/s]. There are clear parallels between the features of $f(t)$ and $F(s)$. We should think of the whole complex-valued function F as representing the whole time waveform f.

As a second transformation example, let $f(t) = \delta(t)$, the unit impulse, essentially a unit-area pulse of very short duration. It acts at $t = 0$, barely within the lower limit of the Laplace integral. It has value zero at $t = 0^-$. (Because we use 0^- as the lower limit of the defining integral, it does not matter whether the impulse straddles $t = 0$ or begins to rise at $t = 0$.) The impulse is nonzero only for $t \approx 0$, where $e^{-st} \approx 1$. Therefore, the Laplace transform is

$$\Delta(s) = \int_{0^-}^{\infty} \delta(t)e^{-st}\,dt \approx \int_{0^-}^{\infty} \delta(t)(1)\,dt = 1 \tag{3.158}$$

Table 3.11 Laplace Transform Pairs

$f(t) = \mathcal{L}^{-1}[F(s)], t \geq 0^-$		$F(s) = \mathcal{L}[f(t)]$
1. Unit impulse $\delta(t)$		1
2. Unit step $u_s(t)$	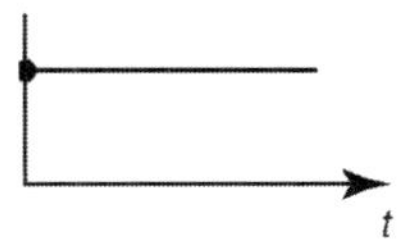	$\frac{1}{s}$
3. t^n, $n = 1, 2, \ldots$	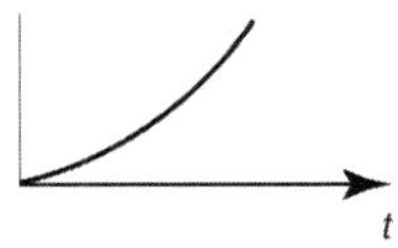	$\frac{n!}{s^{n+1}}$
4. $e^{-\alpha t}$	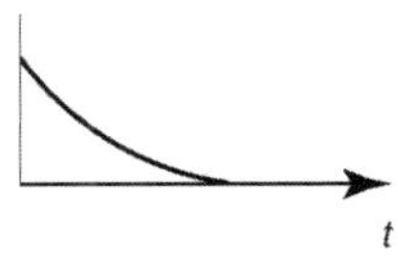	$\frac{1}{s+\alpha}$
5. $t^n e^{-\alpha t}$, $n = 1, 2, \ldots$	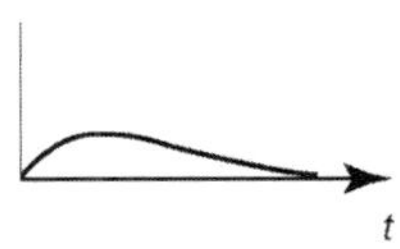	$\frac{n!}{(s+\alpha)^{n+1}}$
6. $\sin(\omega_0 t)$	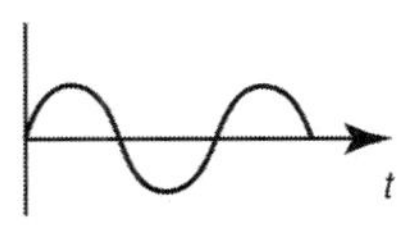	$\frac{\omega_0}{s^2+\omega_0^2}$
7. $\cos(\omega_0 t)$	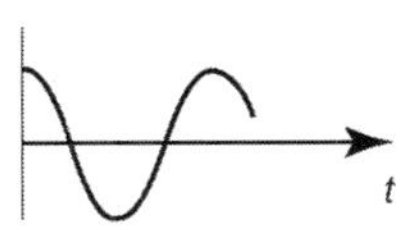	$\frac{s}{s^2+\omega_0^2}$
8. $e^{-\alpha t}\sin(\omega_d t)$	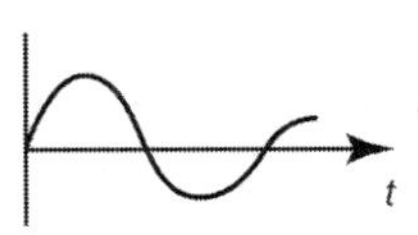	$\frac{\omega_d}{(s+\alpha)^2+\omega_d^2}$
9. $e^{-\alpha t}\cos(\omega_d t)$	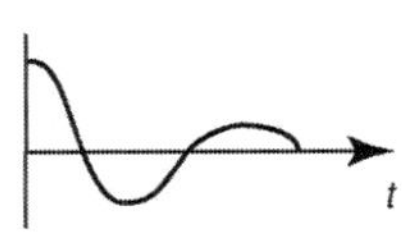	$\frac{s+\alpha}{(s+\alpha)^2+\omega_d^2}$

Source: Dorny, C. N. 1993. *Understanding Dynamic Systems*, p. 412. Prentice-Hall, Englewood Cliffs, NJ. With permission.

It is not necessary to derive the Laplace transform for each signal that we use in the study of systems. Table 3.11 gives the transforms for some signal waveforms that are common in dynamic systems.

3.4.3 Transform Properties

A number of useful properties of the Laplace transformation $\mathcal{L}$ are summarized in Table 3.12. According to the derivative property, the multiplier s acts precisely like the time-derivative operator, but in the domain of Laplace-transformed signals. When we Laplace transform the equation for an energy-storage element such as a mass or a spring, the derivative property automatically incorporates the prior energy

Table 3.12 Properties of the Laplace Transformation, □

1. Magnification	$+[af(t)] = a\mathrm{F}(s)$
2. Addition	$+[f_1(t)+f_2(t)] = \mathrm{F}_1(s)+\mathrm{F}_2(s)$
3. Derivative	$+[\dot{f}(t)] = s\mathrm{F}(s)-f(0^-)$
4. Derivatives	$+[\ddot{f}(t)] = s^2\mathrm{F}(s)-sf(0^-)-\dot{f}(0^-)$
5. Integral	$+\left[\int_{0^-}^{t} f(t)\,dt\right] = \frac{\mathrm{F}(s)}{s}$
6. Convolution	$+\left[\int_{0^-}^{t} f_1(\lambda)f_2(t-\lambda)\,d\lambda\right] = \mathrm{F}_1(s)\mathrm{F}_2(s)$
7. Initial value	$f(0^+) = \lim_{t\to 0^+} f(t) = \lim_{s\to\infty} s\mathrm{F}(s)$
8. Final value	$f(\infty) = \lim_{t\to\infty} f(t) = \lim_{s\to 0} s\mathrm{F}(s)$ if finite
9. Definite integral	$\int_0^{\infty} f(t)\,dt = \lim_{s\to 0} s\mathrm{F}(s)$ if finite
10. Exponential decay	$+[e^{-\alpha t}f(t)] = \mathrm{F}(s+\alpha)$
11. Delay	$+[f(t-t_0)u_s(t-t_0)] = e^{-t_0 s}\mathrm{F}(s)$ for $t_0 \geq 0$
12. Time multiplication	$+[tf(t)] = -\frac{d\mathrm{F}(s)}{ds}$
13. Time division	$+\left[\frac{f(t)}{t}\right] = \int_s^{\infty} \mathrm{F}(s)\,ds$
14. Time scaling	$+[f(at)] = \frac{\mathrm{F}(s/a)}{a}$

Source: Dorny, C. N. 1993. *Understanding Dynamic Systems*, p. 413. Prentice-Hall, Englewood Cliffs, NJ. With permission.

state of the element—essentially the value of the variable at $t = 0^-$. When we Laplace transform the input–output system equation for a particular system variable, the derivative property automatically incorporates the whole prior system state. As a consequence, we can find the solution to the system equation without having to determine the initial conditions (at $t = 0^+$)—a considerable simplification of the solution process.

Since $F(s)$ contains all information about $f(t)$ for $t \geq 0^-$, it is possible to find some features of the signal $f(t)$ from the transform $F(s)$ without performing an inverse Laplace transformation. Properties 7–9 of Table 3.12 provide three of these features, namely the initial value ($t \to 0^+$), the final value ($t \to \infty$), and the area under the waveform. The remaining properties in the table show the effect on the transform of various changes in the signal waveform.

The usual approach to finding inverse transforms is to use a table of transform pairs. That table might be stored in a software package such as CC, MATLAB, MAPLE, and so on. Table 3.11 demonstrates that transforms of typical system signals are ratios of polynomials in s. A ratio of polynomials can be decomposed into a sum of *simple* polynomial fractions—a process referred to as *partial fraction expansion.* Hence, the inversion process can be accomplished by a computer program that incorporates a brief table of transforms.

3.4.4 Transformation and Solution of a System Equation

Suppose that an independent external source applies a specific velocity pattern $v_1(t)$ to node 1 of Figure 3.25. To obtain the input–output system equation that relates the velocity v_2 of node 2 to the input signal v_1, multiply Equation 3.153 by p and substitute v_1 for px_1 and v_2 for px_2. The result is

$$(mp^2 + bp + k)v_2 = (bp + k)v_1 \tag{3.159}$$

The two sides of Equation 3.159 are identical functions of time. Therefore, the Laplace transforms of the two sides of Equation 3.159 are equal. Since the Laplace transformation is linear (properties 1 and 2 of Table 3.12), and since the coefficients of the differential equation are constants, the Laplace transform can be applied separately to the individual terms of each side. The result is

$$m[s^2 V_2(s) - sv_2(0^-) - \dot{v}_2(0^-)] + b[sV_2(s) - v_2(0^-)] + kV_2(s) = b[sV_1(s) - v_1(0^-)] + kV_1(s) \tag{3.160}$$

where the derivative properties of the Laplace transformation (properties 3 and 4 of Table 3.12) introduce the prior values $v_1(0^-)$, $v_2(0^-)$, and $\dot{v}_2(0^-)$ into the equation. According to Equation 3.160 to fully determine the transform $V_2(s)$ of the behavior $v_2(t)$, we must specify these prior values and also $V_1(s)$. It can be shown that specifying the three prior values is equivalent to specifying the energy states of the spring and mass.

Let us assume that the independent source applies the constant velocity $v_1(t) = v_c$ beginning at $t = 0$. The corresponding transform, by item 2 of Table 3.11 and property 1 of Table 3.12, is $V_1(s) = v_c/s$. Substitute the transform $V_1(s)$ into Equation 3.160 and solve for

$$V_2(s) = \frac{(bs + k)v_c + ms\dot{v}_2(0^-) + bs[v_2(0^-) - v_1(0^-)] + ms^2 v_2(0^-)}{s(ms^2 + bs + k)} \tag{3.161}$$

We could find the output signal waveform $v_2(t)$ as a function of the model parameters m, k, b, the source-signal parameter v_c, and the prior state information $v_1(0^-)$, $v_2(0^-)$, and $\dot{v}_2(0^-)$, but the expression for the solution would be messy. Instead, we complete the solution process for specific numbers: $m = 2$ kg, $b = 4$ N $\cdot$ s/m, $k = 10$ N/m, $\dot{v}_2(0^-) = 0$ m/s^2, $v_1(0^-) = 0$ m/s, $v_2(0^-) = -1$ m/s, and $v_c = 1$ m/s. The partial-fraction expansion of the transform and the inverse transform, both obtained by a commercial computer program, are

$$V_2(s) = \frac{1}{s} - \frac{2s + 2}{(s + 1)^2 + 2^2} \tag{3.162}$$

$$v_2(t) = 1 - 2e^{-t}\cos(2t), \quad \text{for } t \geq 0 \tag{3.163}$$

We can take Laplace transforms of the system equations at any stage in their development. We can even write the equations directly in terms of transformed variables if we wish. The process of eliminating variables can be carried out as well in one notation as in another. For example, the operator $G(p)$ in Equation 3.154 represents a ratio of polynomials in the time-derivative operator p. Therefore, Laplace transforming the differential equation, Equation 3.154, introduces the prior values of various derivatives of y_1 and y_2. If the prior values of all these derivatives are zero, then the Laplace-transformed equation is

$$Y_2(s) = G(s)Y_1(s) \tag{3.164}$$

where the operator p in Equation 3.154 is replaced by the complex-frequency variable s in Equation 3.164. It is appropriate, therefore, to define the transfer function directly in terms of Laplace-transformed signals:

$$G(s) = \left.\frac{Y_2(s)}{Y_1(s)}\right|_{\text{PV}=0} \tag{3.165}$$

where $Y_1(s)$ and $Y_2(s)$ are the Laplace transforms of the signals $y_1(t)$ and $y_2(t)$, and the notation PV = 0 means that the prior values (at $t = 0^-$) of $y_1(t)$ and $y_2(t)$ and the various derivatives mentioned above in connection with Equation 3.164 are set to zero. The *frequency domain* definition, Equation 3.165, is equivalent to the *time domain* definition, Equation 3.155.

Suppose that the input signal $y_1(t)$ is the unit impulse $\delta(t)$. Then the response signal $y_2(t)$ is the unit-impulse response of the system. Since the Laplace transform of the unit impulse is $Y_1(s) = \Delta(s) = 1$ by entry 1 of Table 3.11, Equation 3.164 shows that the Laplace transform $Y_2(s)$ of the unit-impulse response is identical to the zero-state transfer function (expressed in the transform domain).

The transfer function for a linear system has two interpretations. Both interpretations characterize the system. In the frequency domain, the transfer function $G(s)$ is the multiplier that produces the response—by multiplying the source-signal transform, as in Equation 3.164. In the time domain, we use a representative response signal—the impulse response—to characterize the system. The transfer function $G(s)$ is the Laplace transform of that characteristic response.

Defining Terms

Input: An independent variable.

Input-output system equation : A differential equation that describes the behavior of a single dependent variable as a function of time. The dependent variable is viewed as the system output. The independent variable(s) are the inputs.

Output: A dependent variable.

Signal: An observable variable; a quantity that reveals the behavior of a system.

State: The state of an nth-order linear system corresponds to the values of a dependent variable and its first $n - 1$ time derivatives.

Time invariant: A system that can be represented by differential equations with constant coefficients.

Zero state: A condition in which no energy is stored or in which all variables have the value zero.

References

Dorny, C. N. 1993. *Understanding Dynamic*, p. 413, Prentice-Hall, Englewood Cliffs, NJ.

Franklin, G. F., Powell, J. D., and Emami-Naeini, A. 1994. *Feedback Control of Dynamic Systems*, 3rd ed., Addison Wesley, Reading, MA.

Kuo, B. C. 1991. *Automatic Control Systems*, 6th ed., Prentice-Hall, Englewood Cliffs, NJ.

Nise, N. S. 1992. *Control Systems Engineering*, Benjamin Cumming, Redwood City, CA.

Further Information

A thorough mathematical treatment of Laplace transforms is presented in *Advanced Engineering Mathematics*, by C. Ray Wylie and Louis C. Barrett. *Understanding Dynamic Systems*, by C. Nelson Dorny, applies transfer functions and related concepts in a variety of contexts. The following journals publish papers that use transfer functions and Laplace transforms:

IEEE Transactions on Automatic Control. Published monthly by the Institute of Electrical and Electronics Engineers.

IEEE Transactions on Systems, Man, and Cybernetics. Published bimonthly.

Journal of Dynamic Systems, Measurement, and Control. Published quarterly by the American Society of Mechanical Engineers.

4

State Space Analysis and System Properties

Mario E. Salgado
Juan I. Yuz
Universidad Técnica Federico Santa María

4.1 Models: Fundamental Concepts

An essential connection between an engineer/scientist and a system relies on his/her ability to describe the system in a way which is useful to understand and to quantify its behavior.

Any description supporting that connection is a **model**. In system theory, models play a fundamental role, since they are needed to analyze, to synthesize, and to design systems of all imaginable sorts.

There is not a unique model for a given system. Firstly, the need for a model may obey different purposes. For instance, when dealing with an electric motor, we might be interested in the electromechanical energy conversion process, alternatively, we might be interested in modelling the motor either as a thermal system, or as a mechanical system to study vibrations, the strength of the materials, and so on.

A second source of that nonuniqueness is the fact that models are always inaccurate, since real systems are usually infinitely complex. One of the key decisions for an engineer when facing the task of **modelling** a system is to decide which are the essential features that the model should capture, and that decision is also closely related to the purpose of the model.

The theory supporting modelling is by itself a vast field, where first principles, signal theory, mathematics and numerical tools combine in different ways to generate rich methodologies. A model is rarely built in one go, the model building process is usually iterative, and it progresses according to the quality of the results obtained when using the model in a particular application. Iterations may also include changes in modelling methodology.

In this chapter we will deal with a special class of models to describe *dynamic systems*. Dynamic systems are those where the system variables are interdependent not only algebraically, but also in a way where we observe the intervention of accumulated effects and rate of change. Models for dynamic systems can be built in the continuous time domain, in the discrete time domain, or in a continuous-discrete time framework (for hybrid systems, involving sampled systems). We will cover the three situations.

In this quest we will put the emphasis on concepts, fundamental properties, physical interpretations, and examples. We will include neither proofs nor intricate theoretical developments. Sometimes we will sacrifice rigor for the sake of an easier understanding. To cover in depth the theory supporting our presentation we refer the interested reader to the specialized literature such as [6,8,10–14].

4.2 State Variables: Basic Concepts

4.2.1 Introduction

One of the most frequently used class of models is that defined by a set of equations on a set of system inner variables. These inner variables are known as **state variables.** The values they have at a specific time instant form a set known as the **system state**, although we will often use the expressions *state variables* and *system state* as synonyms.

The above definition is too vague since it would fit to any set of system variables. What is distinctive in the set of state variables is clarified in the following definition.

A set of state variables for the given system is a set of system inner variables such that any system variable can be computed as a function of the present state and the present and future system inputs.

In this definition we have preferred to stress the physical meaning of state variables. However, a more abstract definition is also possible. The definition also implies that if we know the state at time t we can then compute the energy stored in the system at that instant. The energy stored in a system depends on some system variables (speed, voltage, current, position, temperature, pressure, etc.) and all of them, by definition, can be computed from the system state.

The above definition suggests that one can think of the state in a more general way: the state variables can be chosen as a **function** (e.g., a linear combination) of inner system variables. This generalization builds some distance between the state and its physical interpretation. However, it has the advantage of making the framework more general. It also makes more evident an interesting feature: the **choice of state variables is not unique.**

Another important observation is that the time evolution of the state, the state trajectory itself, can be computed from the present value of the state and the present and future inputs. Thus, the models involved are first order differential (continuous time) or one-step recursive (discrete time) equations.

4.2.2 Basic State Space Models

If we denote by x the vector corresponding to a particular choice of state variables, the general form of a state variable model is as follows:

For continuous-time systems

$$\frac{d\mathbf{x}}{dt} = \mathbf{F}(\mathbf{x}(t), \mathbf{u}(t), t) \tag{4.1}$$

$$\mathbf{y}(t) = \mathbf{G}(\mathbf{x}(t), \mathbf{u}(t), t) \tag{4.2}$$

where $\mathbf{u}(t)$ is the system **input vector** and $\mathbf{y}(t)$ is an **output vector.**

For discrete-time systems

$$\mathbf{x}[t+1] = \mathbf{F}_d(\mathbf{x}[t], \mathbf{u}[t], t) \tag{4.3}$$

$$y[t] = \mathbf{G}_d(\mathbf{x}[t], \mathbf{u}[t], t) \tag{4.4}$$

Similarly to the continuous-time case, $\mathbf{u}[t]$ is the system **input vector** and $\mathbf{y}[t]$ is an **output vector.**

Note that throughout this chapter we will use the symbol t to denote continuous and discrete time, but the difference will be made on using [and] to enclose the argument in the discrete-time case, when $t \in \mathbb{Z}$.

To obtain a first glimpse at the concepts underlying the state space approach, we consider the following example.

Example 4.1

In Figure 4.1, an external force $f(t)$ is applied to a mass-spring system. The position $d(t)$ is measured with respect to the mass position when the spring is relaxed and no external force is applied. The mass movement is damped by a viscous friction force proportional to the mass velocity, $v(t)$.

From first principles we know that to be able to compute the mass position and the mass velocity we must know the initial mass velocity, and the initial spring stretching. Thus, the state vector must have two components, that is, $\mathbf{x}(t) = [x_1(t) \quad x_2(t)]^T$, and a natural state choice is

$$x_1(t) = d(t) \tag{4.5}$$

$$x_2(t) = v(t) = \dot{x}_1(t) \tag{4.6}$$

With this choice, one can apply Newton laws to obtain

$$f(t) = M\frac{dv(t)}{dt} + Kd(t) + Dv(t) = M\dot{x}_2(t) + Kx_1(t) + Dx_2(t) \tag{4.7}$$

where D is the viscous friction proportional constant. We are now in position to write the state equations as

$$\dot{x}_1(t) = x_2(t) \tag{4.8}$$

$$\dot{x}_2(t) = -\frac{K}{M}x_1(t) - \frac{D}{M}x_2(t) + \frac{1}{M}f(t) \tag{4.9}$$

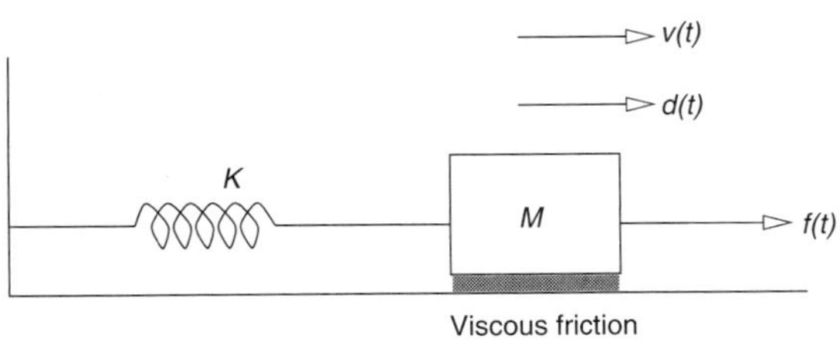

FIGURE 4.1 Mechanical system.

We also observe that the energy, $w(t)$, stored in the system is given by

$$w(t) = \frac{1}{2}K(d(t))^2 + \frac{1}{2}M(v(t))^2 = \mathbf{x}(t)^T \Lambda \mathbf{x}(t) \tag{4.10}$$

where Λ is a diagonal matrix: $\Lambda = \text{diag}\left\{\frac{K}{2}, \frac{M}{2}\right\}$.

Finally, the nonuniqueness of the state vector can be appreciated if, instead of the choices made in 4.8, we choose a new state $\bar{\mathbf{x}}(t)$ related to $\mathbf{x}(t)$ by a nonsingular matrix $\mathrm{T} \in \mathbb{R}^{2\times 2}$, that is,

$$\bar{\mathbf{x}}(t) = \mathrm{T}\mathbf{x}(t) \tag{4.11}$$

More on this will be said in Section 4.3.3.

4.2.3 Signals and State Space Description

The state space framework can also be used to describe a wide variety of signals using a model of the form

$$\frac{d\mathbf{x}(t)}{dt} = \mathrm{A}\mathbf{x}(t), \qquad y(t) = \mathrm{C}\mathbf{x}(t) \quad \text{for continuous-time signals} \tag{4.12}$$

$$\mathbf{x}[t+1] = \mathrm{A_q}\mathbf{x}[t], \qquad y[t] = \mathrm{C_q}\mathbf{x}[t] \quad \text{for discrete-time signals} \tag{4.13}$$

To illustrate the idea we consider a continuous-time signal given by

$$f(t) = 2 + 4\cos(5t) - \sin(5t) \tag{4.14}$$

This signal can be interpreted as the solution for the homogeneous differential equation

$$\frac{d^3 f(t)}{dt^3} + 25\frac{df(t)}{dt} = 0, \quad \text{subject to } f(0) = 6,\ \dot{f}(0) = -5 \text{ and } \ddot{f}(0) = -100 \tag{4.15}$$

If we now choose, as state variables, $x_1(t) = f(t)$, $x_2(t) = \dot{f}(t)$, and $x_3(t) = \ddot{f}(t)$, then the state space model for this signal is

$$\frac{d\mathbf{x}(t)}{dt} = \begin{bmatrix} 0 & 1 & 0 \\ 0 & 0 & 1 \\ 0 & -25 & 0 \end{bmatrix}\mathbf{x}(t), \qquad y(t) = [1 \quad 0 \quad 0]\,\mathbf{x}(t) \tag{4.16}$$

In this usage of state space models, the state variables have no particular physical meaning. However, this description is particularly useful in signal reconstruction theory and when dealing with disturbances in control system synthesis.

4.3 State Space Description for Continuous-Time Systems

In this section the state space description for continuous-time systems is presented. The analysis is focused on the class of **linear and time invariant** systems; to do that, we first show how to build a linear model from the nonlinear Equations 4.1 and 4.2.

An additional restriction is that, at this stage, the systems under study have no pure time delays. This feature generates an infinite dimensional state vector. However, we will see in section 4.4 that this class of systems can be successfully dealt with using sampled data models.

4.3.1 Linearization

Since we will concentrate on time invariant systems, (4.1) and (4.2) can be rewritten as

$$\frac{d\mathbf{x}}{dt} = \mathbf{F}(\mathbf{x}(t), \mathbf{u}(t)) \tag{4.17}$$

$$\mathbf{y}(t) = \mathbf{G}(\mathbf{x}(t), \mathbf{u}(t)) \tag{4.18}$$

We assume that the model (4.17) and (4.18) has at least one *equilibrium point* given by $\{\mathbf{x}_Q, \mathbf{u}_Q, \mathbf{y}_Q\}$. This is a triad conformed by three constant vectors satisfying

$$0 = \mathbf{F}(\mathbf{x}_Q, \mathbf{u}_Q) \tag{4.19}$$

$$\mathbf{y}_Q = \mathbf{G}(\mathbf{x}_Q, \mathbf{u}_Q) \tag{4.20}$$

Note that the equilibrium point is defined by the state derivatives equal to zero.

If we now consider a neighborhood around the equilibrium point, then we can approximate the model (4.17) and (4.18) by a truncated Taylor's series having the form

$$\dot{\mathbf{x}}(t) \approx \mathbf{F}(\mathbf{x}_Q, \mathbf{u}_Q) + \left.\frac{\partial \mathbf{F}}{\partial \mathbf{x}}\right|_{\substack{\mathbf{x}=\mathbf{x}_Q\\ \mathbf{u}=\mathbf{u}_Q}} (\mathbf{x}(t) - \mathbf{x}_Q) + \left.\frac{\partial \mathbf{F}}{\partial \mathbf{u}}\right|_{\substack{\mathbf{x}=\mathbf{x}_Q\\ \mathbf{u}=\mathbf{u}_Q}} (\mathbf{u}(t) - \mathbf{u}_Q) \tag{4.21}$$

$$\mathbf{y}(t) \approx \mathbf{G}(\mathbf{x}_Q, \mathbf{u}_Q) + \left.\frac{\partial \mathbf{G}}{\partial \mathbf{x}}\right|_{\substack{\mathbf{x}=\mathbf{x}_Q\\ \mathbf{u}=\mathbf{u}_Q}} (\mathbf{x}(t) - \mathbf{x}_Q) + \left.\frac{\partial \mathbf{G}}{\partial \mathbf{u}}\right|_{\substack{\mathbf{x}=\mathbf{x}_Q\\ \mathbf{u}=\mathbf{u}_Q}} (\mathbf{u}(t) - \mathbf{u}_Q) \tag{4.22}$$

Equations 4.21 and 4.22 can then be written as

$$\frac{d\Delta\mathbf{x}(t)}{dt} = \mathbf{A}\Delta\mathbf{x}(t) + \mathbf{B}\Delta\mathbf{u}(t) \tag{4.23}$$

$$\Delta\mathbf{y}(t) = \mathbf{C}\Delta\mathbf{x}(t) + \mathbf{D}\Delta\mathbf{u}(t) \tag{4.24}$$

where

$$\Delta\mathbf{x}(t) = \mathbf{x}(t) - \mathbf{x}_Q, \qquad \Delta\mathbf{u}(t) = \mathbf{u}(t) - \mathbf{u}_Q, \qquad \Delta\mathbf{y}(t) = \mathbf{y}(t) - \mathbf{y}_Q \tag{4.25}$$

and

$$\mathbf{A} = \left.\frac{\partial \mathbf{F}}{\partial \mathbf{x}}\right|_{\substack{\mathbf{x}=\mathbf{x}_Q\\ \mathbf{u}=\mathbf{u}_Q}}, \qquad \mathbf{B} = \left.\frac{\partial \mathbf{F}}{\partial \mathbf{u}}\right|_{\substack{\mathbf{x}=\mathbf{x}_Q\\ \mathbf{u}=\mathbf{u}_Q}}, \qquad \mathbf{C} = \left.\frac{\partial \mathbf{G}}{\partial \mathbf{x}}\right|_{\substack{\mathbf{x}=\mathbf{x}_Q\\ \mathbf{u}=\mathbf{u}_Q}}, \qquad \mathbf{D} = \left.\frac{\partial \mathbf{G}}{\partial \mathbf{u}}\right|_{\substack{\mathbf{x}=\mathbf{x}_Q\\ \mathbf{u}=\mathbf{u}_Q}} \tag{4.26}$$

The linearization ideas presented above are illustrated in the following example.

Example 4.2

Consider the magnetic levitation system shown in Figure 4.2.

The metallic sphere is subject to two forces: its own weight, mg, and the attraction force generated by the electromagnet, $f(t)$. The electromagnet is commanded through a voltage source, $e(t) > 0$, $\forall t$.

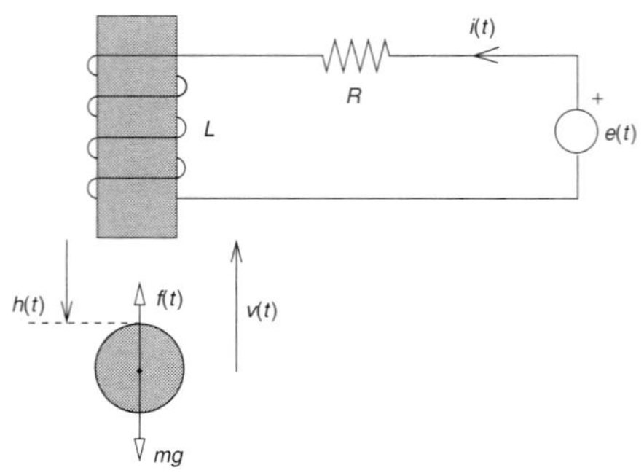

FIGURE 4.2 Magnetic levitation system.

The attraction force on the sphere, $f(t)$, depends on the distance $h(t)$ and the current, $i(t)$. This relation can be approximately described by

$$f(t) = \frac{K_1}{h(t) + K_2} i(t) \tag{4.27}$$

where K_1 and K_2 are positive constants.

Using first principles we can write

$$e(t) = Ri(t) + L\frac{di(t)}{dt} \tag{4.28}$$

$$v(t) = -\frac{dh(t)}{dt} \tag{4.29}$$

$$f(t) = \frac{K_1}{h(t) + K_2} i(t) = mg + m\frac{dv(t)}{dt} \tag{4.30}$$

We next choose as state variables: the current $i(t)$, the sphere position $h(t)$, and the sphere speed $v(t)$, that is,

$$x(t) = [x_1(t) \quad x_2(t) \quad x_3(t)]^T = [i(t) \quad h(t) \quad v(t)]^T \tag{4.31}$$

Then, from 4.28–4.30 we can set the system description as in 4.1 yielding

$$\frac{di(t)}{dt} = \frac{dx_1(t)}{dt} = -\frac{R}{L}x_1(t) + \frac{1}{L}e(t) \tag{4.32}$$

$$\frac{dh(t)}{dt} = \frac{dx_2(t)}{dt} = -x_3(t) \tag{4.33}$$

$$\frac{dv(t)}{dt} = \frac{dx_3(t)}{dt} = \frac{K_1}{m(x_2(t) + K_2)}x_1(t) - g \tag{4.34}$$

Before one can build the linearized model, an equilibrium point has to be computed. The driving input in this system is the source voltage $e(t)$. Say that the equilibrium point is obtained with $e(t) = E_Q$.

Hence, the state in equilibrium can be computed from (4.32) to (4.34), setting all the derivatives equal to zero, that is,

$$-\frac{R}{L}x_{1Q}+\frac{1}{L}E_Q = 0 \Rightarrow x_{1Q} = \frac{E_Q}{R} \tag{4.35}$$

$$-x_{3Q} = 0 \Rightarrow x_{3Q} = 0 \tag{4.36}$$

$$\frac{K_1}{m(x_{2Q}+K_2)}x_{1Q}-g = 0 \Rightarrow x_{2Q} = \frac{K_1}{mg}x_{1Q}-K_2 = \frac{K_1E_Q}{mgR}-K_2 \tag{4.37}$$

The setting now is adequate to build the linearized model in the incremental input $\Delta e(t)$ and the incremental state $\Delta \mathbf{x}(t) = [\Delta x_1(t) \quad \Delta x_2(t) \quad \Delta x_3(t)]^T$. The result is

$$\frac{d\Delta x_1(t)}{dt} = -\frac{R}{L}\Delta x_1(t)+\frac{1}{L}\Delta e(t) \tag{4.38}$$

$$\frac{d\Delta x_2(t)}{dt} = -\Delta x_3(t) \tag{4.39}$$

$$\frac{d\Delta x_3(t)}{dt} = \frac{Rg}{E_Q}\Delta x_1(t)-\frac{Rmg^2}{K_1E_Q}\Delta x_2(t) \tag{4.40}$$

If we define as the system output, the sphere position $h(t)$, we can then compare the above equations with (4.23) and (4.24) to obtain

$$\mathbf{A} = \begin{bmatrix} -\frac{R}{L} & 0 & 0 \\ 0 & 0 & -1 \\ \frac{Rg}{E_Q} & -\frac{Rmg^2}{K_1E_Q} & 0 \end{bmatrix}, \quad \mathbf{B} = \begin{bmatrix} \frac{1}{L} \\ 0 \\ 0 \end{bmatrix}, \quad \mathbf{C} = \begin{bmatrix} 0 \\ 1 \\ 0 \end{bmatrix}, \quad \mathbf{D} = 0 \tag{4.41}$$

In the sequel we will drop the prefix Δ, but the reader should bear in mind that the model above is linear in the **incremental** components of the state, the inputs and the outputs around a chosen equilibrium point.

4.3.2 Linear State Space Models

Our starting point is now the linear time invariant state space model

$$\frac{d\mathbf{x}(t)}{dt} = \mathbf{A}\mathbf{x}(t)+\mathbf{B}\mathbf{u}(t) \tag{4.42}$$

$$\mathbf{y}(t) = \mathbf{C}\mathbf{x}(t)+\mathbf{D}\mathbf{u}(t) \tag{4.43}$$

The solution to Equation 4.42, subject to $\mathbf{x}(t_o) = \mathbf{x}_o$, is given by

$$\mathbf{x}(t) = e^{\mathbf{A}(t-t_o)}\mathbf{x}_o+\int_{t_o}^{t} e^{\mathbf{A}(t-\tau)}\mathbf{B}\mathbf{u}(\tau)\,d\tau \quad \forall t \geq t_o \tag{4.44}$$

where the **transition matrix** e^{At} satisfies

$$e^{At} = I + \sum_{k=1}^{\infty} \frac{1}{k!} A^k t^k \tag{4.45}$$

The interested reader can check that (4.44) satisfies (4.43). To do that he/she should use the Leibnitz's rule for the derivative of an integral.

With the above result, the solution for (4.43) is given by

$$y(t) = Ce^{A(t-t_o)}x_o + C\int_{t_o}^{t} e^{A(t-\tau)} Bu(\tau)\, d\tau + Du(t) \tag{4.46}$$

4.3.2.1 System Dynamics

The state of the system has two components: the **unforced** component, $x_u(t)$, and the **forced** component, $x_f(t)$, where

$$x_u(t) = e^{A(t-t_o)} x_o \tag{4.47}$$

$$x_f(t) = \int_{t_o}^{t} e^{A(t-\tau)} Bu(\tau)\, d\tau \tag{4.48}$$

To gain insight into the state space model and its solution, consider the case when $t_o = 0$ and $u(t) = 0$ $\forall t \geq 0$, that is, the state has only the **unforced part**. Then

$$x(t) = e^{At} x_o \tag{4.49}$$

Further assume that $A \in \mathbb{R}^n$ and that, for simplicity, it has distinct eigenvalues $\lambda_1, \lambda_2, \ldots, \lambda_n$ with n (linearly independent) eigenvectors $v_1, v_2, \ldots, v_n$. Then there always exists a set of constants $\alpha_1, \alpha_2, \ldots, \alpha_n$ such that

$$x_o = \sum_{\ell=1}^{n} \alpha_\ell v_\ell, \quad \alpha_\ell \in \mathbb{C} \tag{4.50}$$

A well-known result from linear algebra tells us that the eigenvalues of A^k are $\lambda_1^k, \lambda_2^k, \ldots, \lambda_n^k$ with corresponding eigenvectors $v_1, v_2, \ldots, v_n$. The application of this result yields

$$x(t) = e^{At} x_o = I + \sum_{\ell=1}^{n} \alpha_\ell \sum_{k=1}^{\infty} \frac{1}{k!} \underbrace{A^k v_\ell}_{\lambda_\ell^k v_\ell} t^k = \sum_{\ell=1}^{n} \alpha_\ell e^{\lambda_\ell t} v_\ell \tag{4.51}$$

This equation shows that the unforced component of the state is a linear combination of **natural modes**, $\{e^{\lambda_\ell t}\}$, each of which is associated with an eigenvalue of **A**. Hence the matrix **A** determines:

- Structure of the unforced response
- Stability (or otherwise) of the system
- Speed of response

When the matrix **A** does not have a set of n independent eigenvectors, Jordan forms can be used (see, e.g., [9,10]).

4.3.2.2 Structure of the Unforced Response

In the absence of input, the state evolves as a combination of natural modes which belong to a defined class of functions: all those generated by exponentials with either real or complex exponents. Hence these modes include constants, real exponentials, pure sine waves, exponentially modulated sine waves, and some other specials functions arising from repeated eigenvalues.

To illustrate these ideas and their physical interpretation consider the system in Example 4.1. For that system

$$\mathbf{A} = \begin{bmatrix} 0 & 1 \\ -\frac{K}{M} & -\frac{D}{M} \end{bmatrix} \tag{4.52}$$

Hence, the system eigenvalues are solutions to the equation

$$\det(\lambda \mathbf{I} - \mathbf{A}) = \lambda^2 + \frac{D}{M}\lambda + \frac{K}{M} = 0 \tag{4.53}$$

that is,

$$\lambda_{1,2} = -\frac{D}{2M} \pm \sqrt{\frac{D^2}{4M^2} - \frac{K}{M}} \tag{4.54}$$

Hence, when the damping is zero ($D = 0$), the system eigenvalues are a couple of conjugate imaginary numbers, and the two natural (complex) modes combine to yield a sustained oscillation with angular frequency $\omega_o = \sqrt{K/M}$. This is in agreement with our physical intuition, since we expect a sustained oscillation to appear when the system has nonzero initial conditions even if the external force, $f(t)$, is zero.

When the system is slightly damped ($D^2 < 4KM$), the matrix eigenvalues are conjugate complex numbers, and the associated complex natural modes combine to yield an exponentially damped sine wave. This also agrees with intuition, since the energy initially stored in the mass and the spring will periodically go from the mass to the spring and vice versa but, at the end, it will completely dissipate, as heat, in the viscous friction.

Finally if the damping is high ($D^2 > 4KM$), the matrix eigenvalues are a couple of negative real numbers, and the natural modes are two decaying exponentials. The heavy damping will preclude oscillations and the initial energy will dissipate quickly.

The three different situation are illustrated in Figure 4.3. For this simulation we have used three different values of the viscous friction constant D and

$$M = 2 \text{ kg}, \quad K = 0.1 \text{ N/m}, \quad d(0) = 0.3 \text{ m}, \quad v(0) = 0.05 \text{ m/s} \tag{4.55}$$

Note that, except when there is no friction ($D = 0$), the mass comes to rest asymptotically.

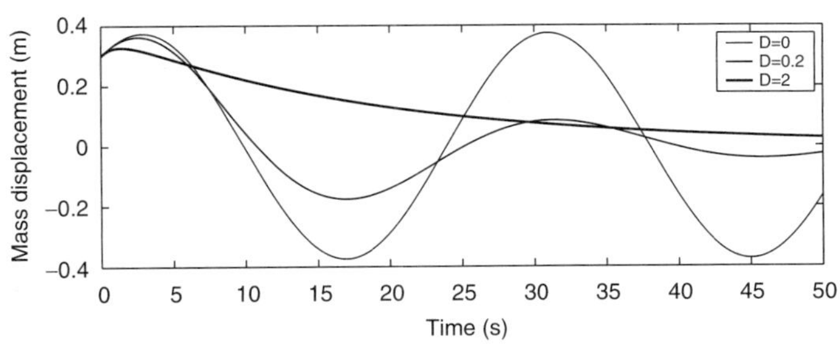

FIGURE 4.3 Unforced response of a mass-spring system.

4.3.2.3 Structure of the Forced Response

When the initial state is zero, the state will exhibit only the forced component. The forced component of the state will include natural modes and some additional **forced** or **particular** modes, which depend on the nature of the system input $\mathbf{u}(t)$. In general the forcing modes in the input will also appear in the state. However, some special cases arise when some of the forcing modes in $\mathbf{u}(t)$ coincide with some system natural modes.

4.3.2.4 System Stability

Stability in linear, time-invariant systems can also be analyzed using the state matrix **A**.

All systems variables can be expressed as linear functions of the state and the system input. When the system input $\mathbf{u}(t)$ is a vector of bounded time functions, then the boundeness of the system variables depends on the state to be bounded.

We then have the following result:

Theorem 4.1 *Consider a system with the state description* (4.42) *and* (4.43) *where* **B**, **C**, *and* **D** *have bounded elements. Then the system state (and hence the system output) is bounded for all bounded inputs if and only if the eigenvalues of* **A** *have negative real parts.*

To illustrate this theorem we again consider the magnetic levitation system from Example 4.2. For that system the matrix **A** (in the linearized model) is given by

$$\mathbf{A} = \begin{bmatrix} -\frac{R}{L} & 0 & 0 \\ 0 & 0 & -1 \\ \frac{Rg}{E_Q} & -\frac{Rmg^2}{K_1 E_Q} & 0 \end{bmatrix} \tag{4.56}$$

and its eigenvalues are the roots of $\det(\lambda\mathbf{I} - \mathbf{A}) = 0$, where

$$\det(\lambda\mathbf{I} - \mathbf{A}) = \left(\lambda + \frac{R}{L}\right)\left(\lambda - \sqrt{\frac{Rmg^2}{K_1 E_Q}}\right)\left(\lambda + \sqrt{\frac{Rmg^2}{K_1 E_Q}}\right) \tag{4.57}$$

One can then see that the set of matrix eigenvalues includes one which is **real and greater than zero.** This implies that the system is **unstable**. This is in agreement with physical reasoning. Indeed, at least theoretically, we can position the sphere in equilibrium (this is described by x_{2Q} in (4.37)). However, this is an unstable equilibrium point, since as soon as we slightly perturb the sphere, it accelerates either towards the ground or towards the magnet.

4.3.2.5 Speed of Response and Resonances

Even if the system is stable there are still some questions regarding other fundamental properties.

To start with, in stable systems the real part of the eigenvalues determines the speed at which the associated mode converges to zero. The slowest modes, the *dominant* modes, determine the speed at which the system output settles at its steady state value, that is, determine the system speed of response. For example, if the system dominant eigenvalues are $\lambda_{1,2} = -\sigma \pm j\omega_o$, $\sigma > 0$, the combined natural modes generate an exponentially damped sine wave $y(t) = Ae^{-\sigma t}\sin(\omega_o t + \alpha)$. We then observe that this signal decays faster for a larger σ.

A second issue, of special importance for flexible structures, is the presence of resonances, which have associated complex eigenvalues. In physical systems, the existence of complex eigenvalues is intimately connected to the presence of two forms of energy. The resonance describes the (poorly damped) oscillation between those two forms of energy. In electric circuits those energies are the electrostatic energy in capacitors and the electromagnetic energy in inductors. In mechanical systems we have the kinetic energy of moving masses and the potential energy in springs. Flexible structures may have many resonant modes. One of the main problems with resonances occurs when the input contains energy at a frequency close to the resonant frequency. For example, if a system has eigenvalues $\lambda_{1,2} = -0.05 \pm j$, that is, the resonant frequency is 1 rad/s and, additionally, one of the **input** components is a sine wave of frequency 0.9 rad/s, then the system output exhibits a very large (forced) oscillation with amplitude initially growing almost linearly and later, stabilizing to a constant value. In real situations this phenomenon may destroy the system (recall the Tacoma bridge case).

4.3.3 State Similarity Transformation

We have already said that the choice of state variables is nonunique. Say that we have a system with input $\mathbf{u}(t)$, output $\mathbf{y}(t)$, and two different choices of state vectors: $\mathbf{x}(t) \in \mathbb{R}^n$ with an associated 4-tuple (**A**, **B**, **C**, **D**), and $\bar{\mathbf{x}}(t) \in \mathbb{R}^n$ with an associated 4-tuple $(\bar{\mathbf{A}}, \bar{\mathbf{B}}, \bar{\mathbf{C}}, \bar{\mathbf{D}})$. Then there exists a nonsingular matrix $\mathbf{T} \in \mathbb{R}^{n\times n}$ such that

$$\bar{\mathbf{x}}(t) = \mathbf{T}\mathbf{x}(t) \Leftrightarrow \mathbf{x}(t) = \mathbf{T}^{-1}\bar{\mathbf{x}}(t) \tag{4.58}$$

This leads to the following equivalences:

$$\bar{\mathbf{A}} = \mathbf{T}\mathbf{A}\mathbf{T}^{-1}, \quad \bar{\mathbf{B}} = \mathbf{T}\mathbf{B}, \quad \bar{\mathbf{C}} = \mathbf{C}\mathbf{T}^{-1} \tag{4.59}$$

Different choices of state variables may or may not respond to different phenomenological approaches to the system analysis. Sometimes it is just a question of mathematical simplicity, as we shall see in Section 4.6. In other occasions, the decision is made considering relative facility to measure certain system variables. However, what is important is that, no matter which state description is chosen, certain fundamental system characteristics do not change. They are related to the fact that the system eigenvalues are invariant with respect to similarity transformations, since

$$\det(\lambda\mathbf{I} - \bar{\mathbf{A}}) = \det(\lambda\mathbf{T}\mathbf{T}^{-1} - \mathbf{T}\mathbf{A}\mathbf{T}^{-1}) = \det(\mathbf{T})\det(\lambda\mathbf{I} - \mathbf{A})\det(\mathbf{T}^{-1}) \tag{4.60}$$

$$= \det(\lambda\mathbf{I} - \mathbf{A}) \tag{4.61}$$

Hence, stability, nature of the unforced response, and speed of response are invariants with respect to similarity transformations.

Example 4.3

Consider the electric network shown in Figure 4.4

We choose the state vector $\mathbf{x}(t) = [x_1(t) \quad x_2(t)]^T = [i_L(t) \quad v_c(t)]^T$. Also $u(t) = v_f(t)$. Using first principles we have that

$$\frac{d\mathbf{x}(t)}{dt} = \begin{bmatrix} 0 & \frac{1}{L} \\ -\frac{1}{C} & -\frac{R_1 + R_2}{R_1 R_2 C} \end{bmatrix}\mathbf{x}(t) + \begin{bmatrix} 0 \\ \frac{1}{R_1 C} \end{bmatrix}u(t) \tag{4.62}$$

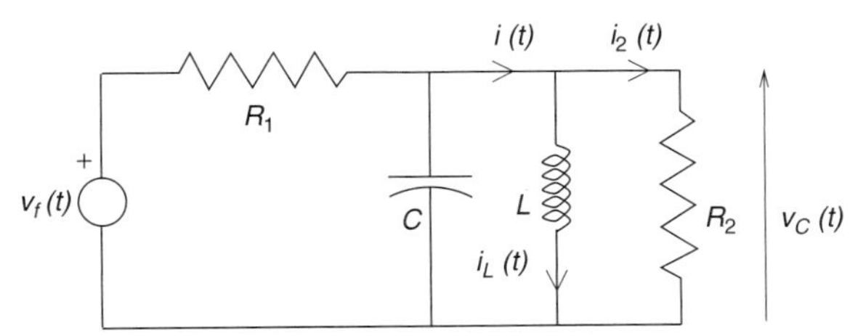

FIGURE 4.4 Electric network.

An alternative state vector is $\bar{\mathbf{x}}(t) = [x_1(t) \quad x_2(t)]^T = [i(t) \quad i_2(t)]^T$. It is straightforward to show that

$$\bar{\mathbf{x}}(t) = \underbrace{\frac{1}{R_2}\begin{bmatrix} R_2 & 1 \\ 0 & 1 \end{bmatrix}}_{\mathbf{T}}\mathbf{x}(t) \tag{4.63}$$

4.3.4 State Space and Transfer Functions

The state space description of linear time invariant systems is an alternative description to that provided by transfer functions. Strictly speaking, the state space description has a wider scope, as we shall see in this subsection.

For a linear time invariant system with input $\mathbf{u}(t) \in \mathbb{R}^m$ and output $\mathbf{y}(t) \in \mathbb{R}^p$, the transfer function, $\mathbf{H}(s) \in \mathbf{C}^{p \times m}$, is defined by the equation

$$\mathbf{Y}(s) = \mathbf{H}(s)\mathbf{U}(s), \quad \text{where} \quad [\mathbf{H}(s)]_{ij} = \frac{Y_i(s)}{U_j(s)} \tag{4.64}$$

that is, the (i, j) element in matrix $\mathbf{H}(s)$ is the Laplace transformation of the response in the ith output when a unit impulse is applied at the jth input, with zero initial conditions and with the remaining inputs equal to zero for all $t \geq 0$.

On the other hand, if we Laplace-transform (4.42) and (4.43) with zero initial conditions, we obtain

$$\mathbf{X}(s) = (s\mathbf{I} - \mathbf{A})^{-1}\mathbf{B}\mathbf{U}(s) \tag{4.65}$$

$$\mathbf{Y}(s) = \mathbf{C}\mathbf{X}(s) + \mathbf{D}\mathbf{U}(s) = \underbrace{(\mathbf{C}(s\mathbf{I} - \mathbf{A})^{-1}\mathbf{B} + \mathbf{D})}_{\mathbf{H}(s)}\mathbf{U}(s) \tag{4.66}$$

For simplicity, and to be able to go deeper into the analysis, **in the remaining part of this section we will focus our attention on the class of scalar systems**, that is, systems with a single input and a single output (SISO systems). This means that $m = p = 1$, $\mathbf{B}$ becomes a column vector, $\mathbf{C}$ is a row vector, and $D = H(\infty)$ (in real systems it usually holds that $D = H(\infty) = 0$). For SISO systems, $H(s)$ is a quotient of polynomials in s, that is,

$$H(s) = \frac{\mathbf{C}\,\mathrm{Adj}(s\mathbf{I} - \mathbf{A})\mathbf{B} + \mathbf{D}\,\det(s\mathbf{I} - \mathbf{A})}{\det(s\mathbf{I} - \mathbf{A})} \tag{4.67}$$

where Adj(o) denotes the adjoint matrix of (o).

A key issue is that the transfer function poles are eigenvalues of matrix $\mathbf{A}$. However, it is not true, in general, that the set of transfer function poles is identical to the set of matrix $\mathbf{A}$ eigenvalues. This can be appreciated through the following example.

Example 4.4

Let

$$\mathbf{A} = \begin{bmatrix} -2 & 1 \\ 0 & -3 \end{bmatrix}, \quad \mathbf{B} = \begin{bmatrix} 1 \\ 0.5 \end{bmatrix}, \quad \mathbf{C} = \begin{bmatrix} 0 & 1 \end{bmatrix}, \quad D = 0 \tag{4.68}$$

Then

$$H(s) = \mathbf{C}(s\mathbf{I}-\mathbf{A})^{-1}\mathbf{B} = \frac{1}{(s+2)(s+3)}\begin{bmatrix}0 & 1\end{bmatrix}\begin{bmatrix}s+3 & 1\\ 0 & s+2\end{bmatrix}\begin{bmatrix}1\\ 0.5\end{bmatrix} \tag{4.69}$$

$$= \frac{0.5(s+2)}{(s+2)(s+3)} = \frac{0.5}{(s+3)} \tag{4.70}$$

Therefore, the transfer function has only one pole, although matrix **A** has two eigenvalues. We observe that there is a pole–zero cancellation in $H(s)$. This phenomenon is closely connected to the question of system properties, which is the central topic in Section 4.6.

To acquire a phenomenological feeling on this issue, consider again the magnetic levitation system in Example 4.2. If we define the current $i(t)$ as the system output we can immediately see that the transfer function from the input $e(t)$ to this output has only one pole. This contrasts with the fact that the dimension of the state is equal to three. The explanation for this is that, in our simplified physical model, the current $i(t)$ is unaffected by the position and the speed of the metallic sphere (note that we have neglected the changes in the inductance due to changes in the sphere position).

The key result is that **the transfer function may not provide the same amount of information than the state space model** for the same system.

An interesting problem is to obtain a state space description from a given transfer function. The reader must be aware that the resulting state space model does not reveal pole-zero cancellations; for that reason, the obtained description is known as a **minimal realization.**

There are many methods to go from the transfer function to a state space model. We present below one of those methods.

Consider a transfer function given by

$$H_T(s) = \frac{B_o(s)}{A_o(s)} + H_T(\infty) = \frac{b_{n-1}s^{n-1} + b_{n-2}s^{n-2} + \cdots + b_1 s + b_0}{s^n + a_{n-1}s^{n-1} + \cdots + a_1 s + a_0} + H_T(\infty) \tag{4.71}$$

We first recall that $D = H_T(\infty)$. We can thus concentrate on the transfer function $H(s) = H_T(s) - H_T(\infty)$, which is a strictly proper transfer function.

Consider next a variable $v_\ell(t) \in \mathbb{R}$ whose Laplace transform, $V(s)$, satisfies

$$V_\ell(s) = \frac{s^{\ell-1}}{A_o(s)}U(s), \quad \ell \in \{1, 2, \ldots, n\} \tag{4.72}$$

This implies that

$$v_\ell(t) = \frac{dv_{\ell-1}(t)}{dt}, \quad \ell \in \{2, \ldots, n\} \tag{4.73}$$

$$Y(s) = \sum_{\ell=1}^{n} b_{\ell-1} V_\ell(s) \tag{4.74}$$

$$U(s) = \frac{A_o(s)}{A_o(s)}U(s) = \underbrace{\frac{s^n}{A_o(s)}U(s)}_{sV_n(s)} + \sum_{\ell=1}^{n} a_{\ell-1} \underbrace{\frac{s^{\ell-1}}{A_0(s)}U(s)}_{V_\ell(s)} \tag{4.75}$$

Now choose as state variables,

$$x(t) = v(t) \tag{4.76}$$

The above equations yield

$$A = \begin{bmatrix} 0 & 1 & 0 & \cdots & 0 & 0 \\ 0 & 0 & 1 & \cdots & 0 & 0 \\ A & A & A & & A & A \\ -a_0 & -a_1 & -a_2 & \cdots & -a_{n-2} & -a_{n-1} \end{bmatrix}, \quad B = \begin{bmatrix} 0 \\ 0 \\ A \\ 0 \\ 1 \end{bmatrix} \tag{4.77}$$

$$C = [b_0 \quad b_1 \quad b_2 \quad \cdots \quad b_{n-1}], \qquad D = H_T(\infty) \tag{4.78}$$

Example 4.5

The transfer function of a system is given by

$$H(s) = \frac{4s - 10}{(s+2)^2(s-1)} = \frac{4s - 10}{s^3 + 3s^2 - 4} \tag{4.79}$$

Then a minimal realization for this system is

$$A = \begin{bmatrix} 0 & 1 & 0 \\ 0 & 0 & 1 \\ 4 & 0 & -3 \end{bmatrix}, \qquad B = \begin{bmatrix} 0 \\ 0 \\ 1 \end{bmatrix} \tag{4.80}$$

$$C = \begin{bmatrix} -10 & 4 & 0 \end{bmatrix}, \qquad D = 0 \tag{4.81}$$

A key result is that a **system transfer function is invariant with respect to state similarity transformations.**

4.4 State Space Description for Discrete-Time and Sampled Data Systems

In this section we will present an overview of the state space description for discrete time systems, mainly based on the results presented for the continuous time case.

Discrete time models may arise from two different sources:

- From a *pure* discrete-time system, usually nonlinear, whose variables are defined only at specific time instants t_k. Systems like that can be found in economic systems, stochastic process theory, etc.
- From a discretization of a continuous-time system. In this case, we are only concerned with the value of some system variables at specific time instants. These models are useful when digital systems, such as microcontrollers, computers, PLCs, or others, interact with continuous-time *real systems* such as mechanical structures, valves, tanks, analog circuits or a whole industrial process*. These are called **sampled data systems.**

In both cases our analysis will be focused on the class of **linear and time invariant** models.

*Through *digital-to-analog* and *analog-to-digital* converters (DAC and ADC, respectively).

4.4.1 Linearization of Discrete Time-Systems

The discrete time equivalents to (4.3) and (4.4) are given by the nonlinear equations

$$\mathbf{x}[t+1] = \mathbf{F}_d(\mathbf{x}[t], \mathbf{u}[t]) \tag{4.82}$$

$$\mathbf{y}[t] = \mathbf{G}_d(\mathbf{x}[t], \mathbf{u}[t]) \tag{4.83}$$

The linearization of models for discrete time systems follows along the same lines to that for continuous ones. Consider firstly an equilibrium point given by $\{\mathbf{x}_Q, \mathbf{u}_Q, \mathbf{y}_Q\}$:

$$\mathbf{x}_Q = \mathbf{F}_d(\mathbf{x}_Q, \mathbf{u}_Q) \tag{4.84}$$

$$\mathbf{y}_Q = \mathbf{G}_d(\mathbf{x}_Q, \mathbf{u}_Q) \tag{4.85}$$

Note that an equilibrium point is defined by a set of **constant** values of the state and **constant** values of the input which satisfy (4.82) and (4.83). This yields a constant system output. The discrete model can then be linearized around this equilibrium point. Defining

$$\Delta\mathbf{x}[t] = \mathbf{x}[t] - \mathbf{x}_Q, \qquad \Delta\mathbf{u}[t] = \mathbf{u}[t] - \mathbf{u}_Q, \qquad \Delta\mathbf{y}[t] = \mathbf{y}[t] - \mathbf{y}_Q \tag{4.86}$$

we have the state space model

$$\Delta\mathbf{x}[t+1] = \mathbf{A}_d\Delta\mathbf{x}[t] + \mathbf{B}_d\Delta\mathbf{u}[t] \tag{4.87}$$

$$\Delta\mathbf{y}[t] = \mathbf{C}_d\Delta\mathbf{x}[t] + \mathbf{D}_d\Delta\mathbf{u}[t] \tag{4.88}$$

where

$$\mathbf{A}_d = \left.\frac{\partial \mathbf{F}_d}{\partial \mathbf{x}}\right|_{\substack{\mathbf{x}=\mathbf{x}_Q \\ \mathbf{u}=\mathbf{u}_Q}}, \qquad \mathbf{B}_d = \left.\frac{\partial \mathbf{F}_d}{\partial \mathbf{u}}\right|_{\substack{\mathbf{x}=\mathbf{x}_Q \\ \mathbf{u}=\mathbf{u}_Q}}, \qquad \mathbf{C}_d = \left.\frac{\partial \mathbf{G}_d}{\partial \mathbf{x}}\right|_{\substack{\mathbf{x}=\mathbf{x}_Q \\ \mathbf{u}=\mathbf{u}_Q}}, \qquad \mathbf{D}_d = \left.\frac{\partial \mathbf{G}_d}{\partial \mathbf{u}}\right|_{\substack{\mathbf{x}=\mathbf{x}_Q \\ \mathbf{u}=\mathbf{u}_Q}} \tag{4.89}$$

4.4.2 Sampled Data Systems

As we have already said, discrete time models are frequently obtained by sampling inputs and outputs in continuous-time systems. When a digital device is to be used to act upon a continuous-time system, the command signals need only to be defined at specific instants, and not at *all* time. However, to be able to act upon the continuous-time system, we need a continuous-time signal. This is usually built with a *zero order hold*, which generates a staircase signal. Also, when we want to digitally measure a system variable this is done at some specific time instants. This means that we must **sample** the output signals. Figure 4.5 illustrates these concepts. If we assume a periodic sampling, with period Δ, we are only interested in the signals at time $k\Delta$. In the sequel we will drop Δ from the arguments, using $\mathbf{u}(k\Delta) = \mathbf{u}[t]$ for the input, $\mathbf{y}(k\Delta) = \mathbf{y}[t]$ for the output, and $\mathbf{x}(k\Delta) = \mathbf{x}[t]$ for the system state.

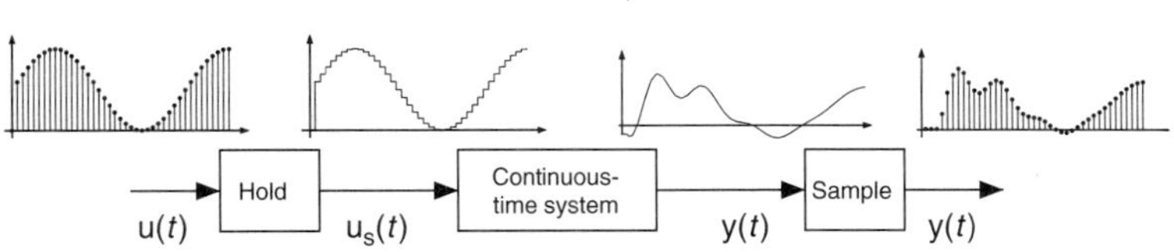

FIGURE 4.5 Schematic representation of a sampled data system.

If we consider the continuous, time-invariant, and linear state space model defined by Equations 4.42 and 4.43, with initial state $\mathbf{x}(k_0\Delta) = \mathbf{x}_0$, we can use Equation 4.44 to calculate the next value of the state:

$$\mathbf{x}(k_0\Delta + \Delta) = e^{\mathbf{A}(k_0\Delta+\Delta-k_0\Delta)}\mathbf{x}(k_0\Delta) + \int_{k_0\Delta}^{k_0\Delta+\Delta} e^{\mathbf{A}(k_0\Delta+\Delta-\tau)}\mathbf{B}\,\mathbf{u}(\tau)\,d\tau \tag{4.90}$$

Furthermore, if a zero order hold is used, that is, $\mathbf{u}(\tau) = \mathbf{u}(k_0\Delta)$ for $k_0\Delta \le \tau < k_0\Delta + \Delta$, we obtain

$$\mathbf{x}(k_0\Delta + \Delta) = e^{\mathbf{A}\Delta}\mathbf{x}(k_0\Delta) + \int_0^{\Delta} e^{\mathbf{A}\eta}d\eta\mathbf{B}\,\mathbf{u}(k_0\Delta) \tag{4.91}$$

And, if we know the state and the input at time $k_0\Delta$, the output is defined by Equation 4.43:

$$\mathbf{y}(k_0\Delta) = \mathbf{C}\mathbf{x}(k_0\Delta) + \mathbf{D}\,\mathbf{u}(k_0\Delta) \tag{4.92}$$

We can now conclude that given a continuous-time model with state space matrices {**A, B, C, D**}, and we sample inputs and outputs every Δ seconds then, the equivalent sampled data systems will be described by the discrete-time state space model:

$$\mathbf{x}(k\Delta + \Delta) = \mathbf{A}_d\mathbf{x}(k\Delta) + \mathbf{B}_d\mathbf{u}(k\Delta) \tag{4.93}$$

$$\mathbf{y}(k\Delta) = \mathbf{C}_d\mathbf{x}(k\Delta) + \mathbf{D}_d\mathbf{u}(k\Delta) \tag{4.94}$$

where

$$\mathbf{A}_d = e^{\mathbf{A}\Delta}, \quad \mathbf{B}_d = \int_0^{\Delta} e^{\mathbf{A}\eta}d\eta\mathbf{B}, \quad \mathbf{C}_d = \mathbf{C}, \quad \mathbf{D}_d = \mathbf{D} \tag{4.95}$$

There are different methods to obtain $\mathbf{A}_d$ defined in (4.95), but a simple way to calculate this matrix is to use Laplace transformation. This yields

$$\mathbf{A}_d = e^{\mathbf{A}\Delta} = \mathcal{L}^{-1}\{(s\mathbf{I} - \mathbf{A})^{-1}\}\big|_{t=\Delta} \tag{4.96}$$

Example 4.6

Consider the mechanical system of Example 4.1 on p. 4, that was described by the state space model:

$$\begin{bmatrix} x_1(t) \\ x_2(t) \end{bmatrix} = \begin{bmatrix} 0 & 1 \\ -\frac{K}{M} & -\frac{D}{M} \end{bmatrix} \begin{bmatrix} x_1(t) \\ x_2(t) \end{bmatrix} + \begin{bmatrix} 0 \\ -\frac{1}{M} \end{bmatrix} f(t) \tag{4.97}$$

where $f(t)$ is the external force, and where we can choose either the mass position, $x_1(t)$, or the mass velocity, $x_2(t)$, of the mass, as the system output.

For the purpose of a numerical illustration, we set $M = 1$ kg, $D = 1.2$ N s/m, and $K = 0.32$ N/m.

The matrix $\mathbf{A}_d$ is obtained from (4.96), applying inverse Laplace transformation

$$\mathbf{A}_d = \mathcal{L}^{-1}\left\{\begin{bmatrix} s & -1 \\ 0.32 & s+1.2 \end{bmatrix}^{-1}\right\}\Bigg|_{t=\Delta} = \begin{bmatrix} 2e^{-0.4\Delta} - e^{-0.8\Delta} & 2.5(e^{-0.4\Delta} - e^{-0.8\Delta}) \\ 0.8(e^{-0.4\Delta} - e^{-0.8\Delta}) & -e^{-0.4\Delta} + 2e^{-0.8\Delta} \end{bmatrix} \tag{4.98}$$

and the B_d matrix is obtained from (4.95):

$$B_d = \int_0^{\Delta} \begin{bmatrix} 2e^{-0.4\eta} - e^{-0.8\eta} & 2.5(e^{-0.4\eta} - e^{-0.8\eta}) \\ 0.8(e^{-0.4\eta} - e^{-0.8\eta}) & -e^{-0.4\eta} + 2e^{-0.8\eta} \end{bmatrix} d\eta \begin{bmatrix} 0 \\ 1 \end{bmatrix}$$
$$\Rightarrow B_d = \begin{bmatrix} -6.25e^{-0.4\Delta} + 3.125e^{-0.8\Delta} + 3.125 \\ 2.5(e^{-0.4\Delta} - e^{-0.8\Delta}) \end{bmatrix} \tag{4.99}$$

Note that both, A_d and B_d are functions of Δ. Thus, the sampling period, Δ, has a strong presence in the dynamic behavior of the sampled system, as we shall observe in the following subsections.

4.4.3 Linear State Space Models

We will analyze the linear time invariant state space model

$$x[t+1] = A_d x[t] + B_d u[t] \tag{4.100}$$

$$y[t] = C_d x[t] + D_d u[t] \tag{4.101}$$

This can be a linearized discrete time model like (4.87) and (4.88), or a sampled data system like (4.93) and (4.94) where Δ has been dropped from the time argument.

The solution to Equations 4.100 and 4.101, subject to $x[t_o] = x_o$, is given by

$$x[t] = A_d^{(t-t_o)} x_o + \sum_{i=0}^{(t-t_o)-1} A_d^{(t-t_o)-i-1} B_d u[i+t_o] \quad \forall t \geq t_o \tag{4.102}$$

where $A_d^{(t-t_o)}$ is the **transition matrix**.

The reader can check easily that (4.102) satisfies (4.100). With the above result, the solution for (4.101) is given by

$$y[t] = C_d A_d^{(t-t_o)} x_o + C_d \sum_{i=0}^{(t-t_o)-1} (A_d^{(t-t_o)-i-1} B_d u[i+t_o]) + D_d u[t] \tag{4.103}$$

4.4.3.1 System Dynamics

The state of the system has two components: the **unforced** component, $x_u[t]$, and the **forced** component, $x_f[t]$, where

$$x_u[t] = A_d^{(t-t_o)} x_o \tag{4.104}$$

$$x_f[t] = \sum_{\tau=0}^{(t-t_o)-1} A_d^{(t-t_o)-i-1} B_d u[i+t_o] \tag{4.105}$$

To gain insight into the state space model and its solution consider the case when $t_o = 0$ and $u[t] = 0$, $\forall t \geq 0$, that is, the state has only the **unforced part**. Then

$$x[t] = A_d^t x_o \tag{4.106}$$

Further assume that $A_d \in R^{n\times n}$ and that, for simplicity, it has n distinct eigenvalues η, with n linearly independent eigenvectors $\mathbf{v}$. Then there always exists a set of n constants α such that

$$\mathbf{x_o} = \sum_{\ell=1}^{n} \alpha_\ell \mathbf{v}_\ell, \quad \alpha_\ell \in \mathbb{C} \tag{4.107}$$

A well known result from linear algebra tells us that the eigenvalues of A_d^k are $\eta_,^k$, for $k \in N$, with corresponding eigenvectors $\mathbf{v}$. The application of this result yields

$$\mathbf{x}[t] = \mathbf{A}_d^t\mathbf{x_o} = \mathbf{A}_d^t \sum_{\ell=1}^{n} \alpha_\ell \mathbf{v}_\ell = \sum_{\ell=1}^{n} \alpha_\ell \underbrace{\mathbf{A}_d^t \mathbf{V}_\ell}_{\eta_\ell^t \mathbf{v}_\ell} \tag{4.108}$$

$$\mathbf{x}[t] = \sum_{-=1}^{n} \alpha_- \eta_-^t \mathrm{v}_- \tag{4.109}$$

This equation shows that the unforced component of the state is a linear combination of **natural modes**, $\{\eta_-^t\}$, and each one is associated with an eigenvalue of A_d, which are also known as **natural frequencies** of the model. Thus, we again have that the matrix A_d determines:

- Structure of the unforced response
- Stability (or otherwise) of the system
- Speed of response

4.4.3.2 Structure of the Unforced Response

In the absence of input, the state evolves as a combination of natural modes which belong to a defined class of functions: the powers of the model eigenvalues, either real or complex. These modes are discrete functions related to constants, real exponentials, pure sine waves, exponentially modulated sine waves, and some other specials functions arising from repeated eigenvalues.

To illustrate these ideas and their physical interpretation consider the sampled system in Example 4.6. If $\Delta = 1$, the state space matrices are

$$A_d = \begin{bmatrix} 0.8913 & 0.5525 \\ -0.1768 & 0.2283 \end{bmatrix}, \quad B_d = \begin{bmatrix} 0.3397 \\ 0.5525 \end{bmatrix} \tag{4.110}$$

Hence, the system eigenvalues are solutions to the equation

$$\det(\eta I - A_d) = \det\left(\begin{bmatrix} \eta - 0.8913 & -0.5525 \\ 0.1768 & \eta - 0.2283 \end{bmatrix}\right) \tag{4.111}$$

$$= (\eta - 0.6703)(\eta - 0.4493) = 0 \tag{4.112}$$

that is, $\eta_1 = 0.6703$, $\eta_2 = 0.4493$, and the unforced response is

$$\mathbf{x}_u[t] = C_1(0.6702)^t + C_2(0.4493)^t \tag{4.113}$$

where $C_1$and C_2 depend on the initial conditions only. We can observe that, when t tends to infinity, $\mathbf{x}_u[t]$ decays to zero, because $|\eta_{1,2}| < 1$. Also these eigenvalues are positive real numbers, so there is no oscillation in the natural modes. This last observation is consistent with the parameter choice in Example 1.6, which made the mass-spring system to be overdamped.

4.4.3.3 Structure of the Forced Response

Consider the Equation 4.102. Then, when the initial state is zero, the state will only exhibit the forced component. However, the forced component will still include natural modes plus some additional **forced** or **particular** modes, which depend on the nature of the system input $\mathbf{u}[t]$. In general, the forcing modes in the input will also appear in the state. However, special cases arise when a forcing mode in $\mathbf{u}[t]$ coincides with a system natural mode.

4.4.3.4 System Stability

Stability in linear time-invariant systems can also be analyzed using the state matrix $\mathbf{A}_d$. As we said, all systems variables can be expressed as linear functions of the state and the system input. When the system input $\mathbf{u}[t]$ is a vector of bounded time functions, then the boundedness of the system variables depends on the state to be bounded. We then have the following result:

Theorem 4.2 *Consider a system with the state description* (4.100) *and* (4.101) *where* $\mathbf{B}_d$, $\mathbf{C}_d$, *and* $\mathbf{D}_d$ *have bounded elements. Then the system state is bounded for all bounded inputs if and only if the eigenvalues of* $\mathbf{A}_d$ *lies inside the unit disc, that is,* $|\eta| < 1, \forall,$.

4.4.3.5 Speed of Response and Resonances

We recall that the natural modes of discrete-time systems are the powers of the eigenvalues η. Since those eigenvalues can always be described as complex quantities, we can then write the natural modes as

$$(\eta_\ell)^t = (|\eta_\ell|e^{j\theta_\ell})^t = |\eta_\ell|^t e^{j\theta_\ell t}, \quad \text{where } \theta_\ell = \angle\eta_\ell \tag{4.114}$$

Therefore, we have that

- $0 < |\eta| < \infty$ determines the *speed* at which the mode decays to zero for stable systems ($|\eta| < 1$), or grows to infinity for unstable systems ($|\eta| > 1$)
- $-\pi < \theta \leq \pi$ determines the *frequency* of the natural mode, measured in radians

Although the natural modes of stable systems decay to zero, their nature determines the system transient response.

To illustrate these issues the **step response**, with zero initial conditions, is frequently used.

Example 4.7

Consider the first order, single-input single-output discrete-time system

$$\mathbf{x}[t+1] = \eta\cdot\mathbf{x}[t] + \mathbf{u}[t] \tag{4.115}$$

$$\mathbf{y}[t] = (1-\eta\cdot)\mathbf{x}[t] \tag{4.116}$$

To obtain the step response, we can use the Equation 4.103, where $\mathbf{x}_o = 0$, $\mathbf{u}[t] = 1, \forall t \geq 0$.

$$\mathbf{y}[t] = \mathbf{C}_d\left(\sum_{i=0}^{t-1}\mathbf{A}_d^{t-i-1}\right)\mathbf{B}_d \tag{4.117}$$

$$= (1-\eta_\ell)\left(\sum_{i=0}^{t-1}\eta_\ell^{t-i-1}\right) = (1-\eta_\ell)\eta_\ell^{t-1}\frac{1-\eta_\ell^{-t}}{1-\eta_\ell^{-1}} \tag{4.118}$$

$$= 1-\eta^t \tag{4.119}$$

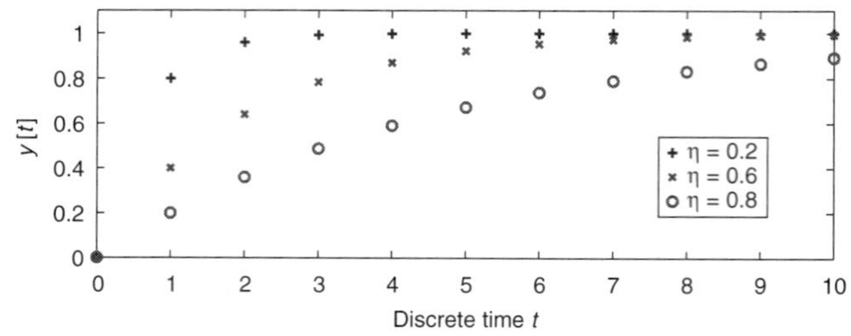

FIGURE 4.6 Step response of the system for different eigenvalues.

The output signal, $y[t] = y_h[t] + y_p[t]$, is shown in Figure 4.6, for different values of the eigenvalue η. The transient is given by $y_h[t] = -\eta^t$, and the steady state response by $y_p[t] = 1$.

We observed in Equation 4.114 that the system eigenvalues define the damping of its transient response, but also determine its frequency of oscillation (when the eigenvalues have a nonzero imaginary part). The potential problem when resonant modes exist is the same problem we found in the context of continuous-time systems, that is, the system input contains a sine wave or another kind of signal, with energy at a frequency close to one of the natural frequencies of the system. The system output still remains bounded, although it grows to undesirable amplitudes.

Example 4.8

Consider the discrete-time system described by the state space model

$$\mathbf{x}[t+1] = \begin{bmatrix} 1.2796 & -0.81873 \\ 1 & 0 \end{bmatrix} \mathbf{x}[t] + \begin{bmatrix} 1 \\ 0 \end{bmatrix} u[t] \tag{4.120}$$

$$y[t] = \begin{bmatrix} 0 & 0.5391 \end{bmatrix} \mathbf{x}[t] \tag{4.121}$$

The eigenvalues of the system are obtained from $\mathbf{A}_d$:

$$\eta_{1,2} = 0.6398 \pm j0.6398 = 0.9048\ (e^{j\pi/4}) \tag{4.122}$$

And the associated natural modes, present in the transient response, are

$$\eta_{1,2}^t = 0.9048^t e^{j\frac{\pi}{4}t} = 0.9048^t \left[\cos\left(\frac{\pi}{4}t\right) \pm j\sin\left(\frac{\pi}{4}t\right)\right] \tag{4.123}$$

The natural modes are slightly damped, because $|\eta_{1,2}|$ is close to 1, and they show an oscillation of frequency $\pi/4$.

In the plots shown in Figure 4.7 we appreciate a strongly resonant output. The upper plot corresponds to an input $u[t] = \sin(\frac{\pi}{4}t)$, that is, the input frequency coincides with the frequency of the natural modes. In the lower plot the input is a square wave of frequency input signal $\pi/12$. In this case, the input **third** harmonic has a frequency equal to the frequency of the natural modes.

4.4.3.6 Effect of Different Sampling Periods

We observe in Equation 4.95 that $\mathbf{A}_d$ and $\mathbf{B}_d$ depend on the choice of the sampling period Δ. This choice determines the position of the eigenvalues of the system too. If we look at the Equation 4.96, assuming that $\mathbf{A}$ has been diagonalized, we have that

$$\mathbf{A}_d = e^{\text{diag}\{\lambda_1,\ldots,\lambda_n\}\Delta} = \text{diag}\{e^{\lambda_1\Delta},\ldots, e^{\lambda_n\Delta}\} \tag{4.124}$$

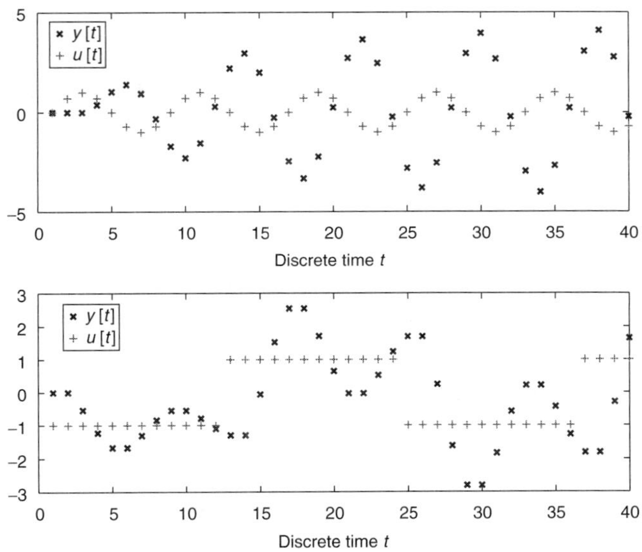

FIGURE 4.7 Resonant effect in the system output.

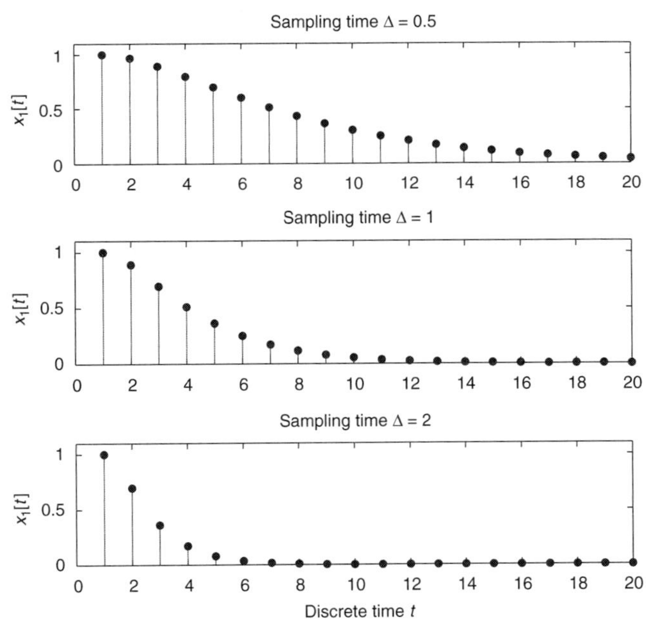

FIGURE 4.8 Effect of sampling in natural modes.

where $\{\lambda_1, \ldots, \lambda_n\}$are the eigenvalues of the underlying continuous-time systems. Then, these eigenvalues are mapped to the eigenvalues of the sampled-data system by equation:

$$\eta_\ell = e^{\lambda_\ell \Delta} \tag{4.125}$$

In Figure 4.8 we observe the response of the sampled system of Example 4.6, choosing $x_1[t]$ as the system output, when the initial condition is $\mathbf{x}_o = [1 \quad 0]^T$, for different values of Δ. Observe that the horizontal axis corresponds to t, so the *real* instants times are $t\Delta$.

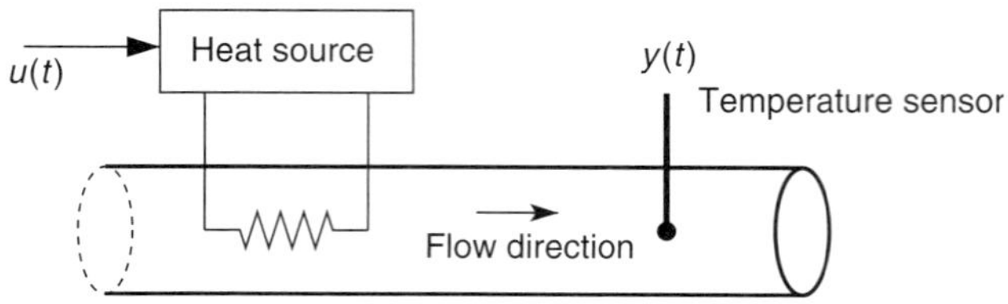

FIGURE 4.9 Heating system with time delay.

A fundamental issue regarding sampling of continuous-time signals is that the sampling period has to be chosen small enough to capture the essential nature of the signal to be sampled. To exemplify an ill-chosen Δ, assume that the signal $f(t) = A\sin(\omega_o t)$ is sampled every Δ seconds, with $\Delta = 2 \;\; \pi/\omega_o, \quad \in \mathbb{N}$. Then the resulting discrete time signal is $f[t] = 0, \; \forall t \in Z$.

4.4.3.7 Sampled Data Systems and Time Delays

We said in Section 4.3 that one cannot use continuous-time state space models to describe systems with time delays, because they are infinite dimensional systems. It was also said there that we would be able to tackle this problem using sampled signals. This is done using the following example.

Example 4.9

Consider the heating system sketched in Figure 4.9.

The measured temperature, $y(t)$, of the flow depends on the power injected by the heat source. This source is commanded by a control signal $u(t)$. Changes in $u(t)$ yield changes in the temperature $y(t)$, but with a significant time delay. The linearized system can thus be represented by the transfer function:

$$\frac{Y(s)}{U(s)} = H(s) = \frac{e^{-\tau s}K}{s+\lambda} \tag{4.126}$$

where $U(s)$ and $Y(s)$ are the Laplace transforms of $u(t)$ and $y(t)$, respectively.

We next assume that the input and output signals are sampled every $\Delta[s]$. The time delay τ, in seconds, is a function of the flow velocity and we can assume, for simplicity, that τ it is a multiple of the sampling interval Δ, that is, $\tau = m\Delta, \; m \in Z^+$. These delays translate in a factor z^m in the denominator of the Z-transform transfer function. In other words, the delay gives rise to a set of m poles at the origin. Furthermore, the continuous-time system eigenvalue at $s = -\lambda$ becomes a discrete-time system eigenvalue at $z = e^{-\lambda\Delta}$ (see Equation 4.125). The resulting transfer function is

$$\frac{Y[z]}{U[z]} = H[z] = \frac{K}{\lambda}\frac{1-e^{-\lambda\Delta}}{z^m(z-e^{-\lambda\Delta})} \tag{4.127}$$

And this transfer function can be expressed as the discrete state space model

$$x_1[t+1] = x_2[t] \tag{4.128}$$

$$x_2[t+1] = x_3[t] \tag{4.129}$$

A A

$$x_m[t+1] = x_{m+1}[t] \tag{4.130}$$

$$x_{m+1}[t+1] = e^{-\lambda\Delta}x_{m+1}[t] + \frac{K}{\lambda}(1-e^{-\lambda\Delta})u[t] \tag{4.131}$$

$$y[t] = x_1[t] \tag{4.132}$$

We can then think of the states variables $x_{m+1}[t], \ldots, x_1[t]$ as the temperature at equally spaced points, between the heat source and the temperature sensor.

When the time delay τ is not a multiple of the sampling period Δ, an additional pole at the origin and an additional zero appear in the discrete transfer function. The details can be found elsewhere, for example, in [7].

4.4.4 State Similarity Transformation

The idea of transforming the state via a similarity transformation equally applies to discrete-time systems. The system properties also remain unchanged.

4.4.5 State Space and Transfer Functions

For discrete-time systems the relation between state space and transfer function models is basically the same as in the continuous-time case. As we said then, the state space description of linear time invariant systems is an alternative description to that provided by transfer functions, although in some situations it provides more information on the system.

For a linear discrete-time invariant system with input $\mathbf{u}[t] \in \mathbb{R}^m$ and output $\mathbf{y}[t] \in \mathbb{R}^p$, the transfer function, $\mathbf{H}[z] \in \mathbb{C}^{p \times m}$, is defined by the equation

$$\mathbf{Y}[z] = \mathbf{H}[z]\mathbf{U}[z], \quad \text{where} \quad [\mathbf{H}[z]]_{ij} = \frac{Y_i[z]}{U_j[z]} \tag{4.133}$$

that is, the (i, j) element in matrix $\mathbf{H}[z]$ is the Zeta transformation of the response in the ith output when a unit Kronecker's delta is applied at the jth input, with zero initial conditions and with the remaining inputs equal to zero for all $t \geq 0$.

On the other hand, if we apply Zeta transform to the discrete time state space model (4.100) and (4.101), with zero initial conditions, we have

$$\mathbf{X}[z] = (z\mathbf{I} - \mathbf{A}_d)^{-1}\mathbf{B}_d\mathbf{U}[z] \tag{4.134}$$

$$\mathbf{Y}[z] = \mathbf{C}_d\mathbf{X}[z] + \mathbf{D}_d\mathbf{U}[z] \tag{4.135}$$

leading to

$$\mathbf{C}_d(z\mathbf{I} - \mathbf{A}_d)^{-1}\mathbf{B}_d + \mathbf{D}_d = \mathbf{H}[z] \tag{4.136}$$

In the following analysis, we will focus on the class of scalar systems, that is, $m = p = 1$, $\mathbf{B}_d$, $\mathbf{C}_d^T$ are column vectors, and $\mathbf{D}_d = H[\infty]$. We can then see that $H[z]$ is a quotient of polynomials in z, that is,

$$H[z] = \frac{\mathbf{C}_d \operatorname{Adj}(z\mathbf{I} - \mathbf{A}_d)\mathbf{B}_d + \mathbf{D}_d \det(z\mathbf{I} - \mathbf{A}_d)}{\det(z\mathbf{I} - \mathbf{A}_d)} \tag{4.137}$$

where Adj(o) denotes the adjoint matrix of (o).

We have again, paralleling the continuous-time case, that the transfer function poles are eigenvalues of $\mathbf{A}_d$. However, it is not true in general that the set of transfer function poles is identical to the set of eigenvalues of the matrix. It is important to realize that transfer function models can hide cancellations between poles and zeros, with the consequences described in Section 4.6.1 and Section 4.6.2.

A key result for discrete-time system is the same for continuous-time systems: **the transfer function may not provide the same amount of information than the state space model** for the same system.

One way to obtain the state space model is to use the same method proposed in Section 4.3.4 "State Space and Transfer Functions," applying Zeta transformation instead of Laplace transformation, and using the fact that

$$F[z] = \mathcal{Z}\{f[t]\} \Leftrightarrow zF[z] = \mathcal{Z}\{f[t+1]\} \tag{4.138}$$

Example 4.10

The transfer function of a system is given by

$$H[z] = \frac{2z^2 - z + 1}{(z-0.8)(z-0.6)} = \frac{1.8z + 0.04}{z^2 - 1.4z + 0.48} + 2 \tag{4.139}$$

Then a minimal realization for this system is

$$\mathbf{A}_d = \begin{bmatrix} 0 & 1 \\ -0.48 & -1.4 \end{bmatrix}, \quad \mathbf{B}_d = \begin{bmatrix} 0 \\ 1 \end{bmatrix} \tag{4.140}$$

$$\mathbf{C}_d = [0.04 \quad 1.8], \qquad \mathbf{D}_d = 2 \tag{4.141}$$

In discrete-time models it also happens that the **system transfer function is invariant with respect to state similarity transformations.**

4.5 State Space Models for Interconnected Systems

To build state space models for complex systems it is sometimes useful (and possible) to describe them as the interconnection of simpler systems. That interconnection is usually a combination of three basic interconnection structures: series, parallel, and feedback. In those three basic cases our aim is to obtain a state space model for the composite system.

In the following analysis we will use two systems, which are defined by

System 1:
$$\frac{d\mathbf{x}_1(t)}{dt} = \mathbf{A}_1\mathbf{x}_1(t) + \mathbf{B}_1\mathbf{u}_1(t) \tag{4.142}$$

$$\mathbf{y}_1(t) = \mathbf{C}_1\mathbf{x}_1(t) + \mathbf{D}_1\mathbf{u}_1(t) \tag{4.143}$$

System 2:
$$\frac{d\mathbf{x}_2(t)}{dt} = \mathbf{A}_2\mathbf{x}_2(t) + \mathbf{B}_2\mathbf{u}_2(t) \tag{4.144}$$

$$\mathbf{y}_2(t) = \mathbf{C}_2\mathbf{x}_2(t) + \mathbf{D}_2\mathbf{u}_2(t) \tag{4.145}$$

4.5.1 Series Connection

The system interconnection shown in Figure 4.10 is known as a series or cascade connection. To build the desired state space model, we first observe that $\mathbf{y}_2(t) = \mathbf{u}_1(t)$. Also, the composite system input is $\mathbf{u}(t) = \mathbf{u}_2(t)$,

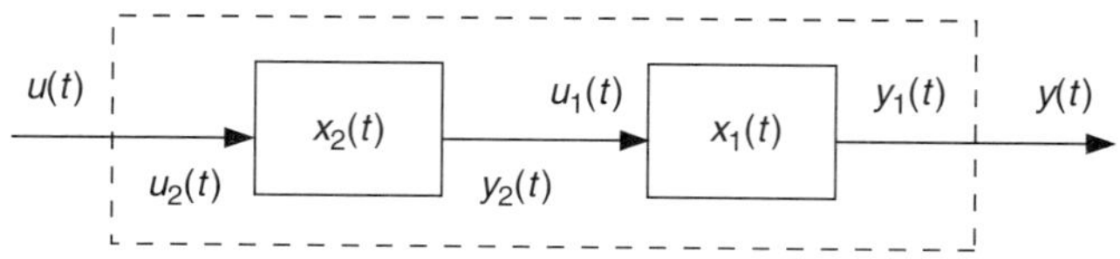

FIGURE 4.10 Series connection.

and the composite system output is $\mathbf{y}(t) = \mathbf{y}_1(t)$. We thus obtain

$$\begin{bmatrix} \dot{\mathbf{x}}_1(t) \\ \dot{\mathbf{x}}_2(t) \end{bmatrix} = \begin{bmatrix} \mathbf{A}_1 & \mathbf{B}_1\mathbf{C}_2 \\ 0 & \mathbf{A}_2 \end{bmatrix} \begin{bmatrix} \mathbf{x}_1(t) \\ \mathbf{x}_2(t) \end{bmatrix} + \begin{bmatrix} \mathbf{B}_1\mathbf{D}_2 \\ \mathbf{B}_2 \end{bmatrix} \mathbf{u}(t) \tag{4.146}$$

$$\mathbf{y}(t) = [\mathbf{C}_1 \quad \mathbf{D}_1\mathbf{C}_2] \begin{bmatrix} \mathbf{x}_1(t) \\ \mathbf{x}_2(t) \end{bmatrix} + [\mathbf{D}_1\mathbf{D}_2]\, \mathbf{u}(t) \tag{4.147}$$

4.5.2 Parallel Connection

The system interconnection shown in Figure 4.11 is known as a parallel connection. To obtain the desired state space model we observe that the input is $\mathbf{u}(t) = \mathbf{u}_1(t) = \mathbf{u}_2(t)$ and the output for the whole system is $\mathbf{y}(t) = \mathbf{y}_1(t) + \mathbf{y}_2(t)$. We obtain

$$\begin{bmatrix} \dot{\mathbf{x}}_1(t) \\ \dot{\mathbf{x}}_2(t) \end{bmatrix} = \begin{bmatrix} \mathbf{A}_1 & 0 \\ 0 & \mathbf{A}_2 \end{bmatrix} \begin{bmatrix} \mathbf{x}_1(t) \\ \mathbf{x}_2(t) \end{bmatrix} + \begin{bmatrix} \mathbf{B}_1 \\ \mathbf{B}_2 \end{bmatrix} \mathbf{u}(t) \tag{4.148}$$

$$\mathbf{y}(t) = [\mathbf{C}_1 \quad \mathbf{C}_2] \begin{bmatrix} \mathbf{x}_1(t) \\ \mathbf{x}_2(t) \end{bmatrix} + [\mathbf{D}_1 + \mathbf{D}_2]\, \mathbf{u}(t) \tag{4.149}$$

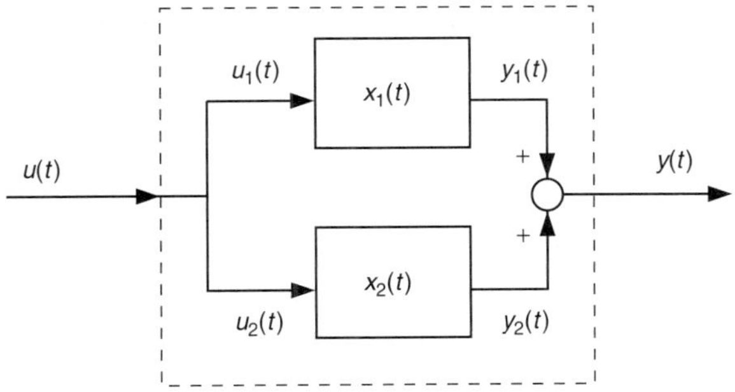

FIGURE 4.11 Parallel connection.

4.5.3 Feedback Connection

The system interconnection shown in Figure 4.12 is known as feedback connection (with unit negative feedback), and it corresponds to the basic structure of a control loop, where S_1 is the *plant* and S_2 is the *controller*. To build the composite state space model we observe that the overall system input satisfies the equation $\mathbf{u}(t) = \mathbf{u}_2(t) + \mathbf{y}_1(t)$, and the overall system output is $\mathbf{y}(t) = \mathbf{y}_1(t)$. Furthermore, we assume

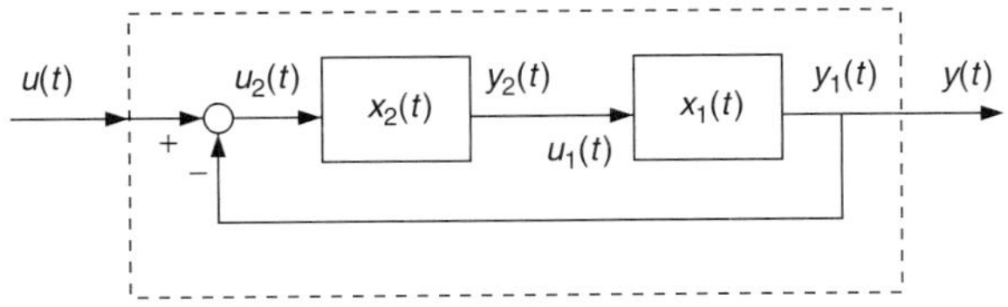

FIGURE 4.12 Feedback connection.

that the system S_1 (the plant) is strictly proper, that is, $D_1 = 0$. We then obtain

$$\begin{bmatrix} \dot{x}_1(t) \\ \dot{x}_2(t) \end{bmatrix} = \begin{bmatrix} A_1 - B_1D_2C_1 & B_1C_2 \\ -B_2C_1 & A_2 \end{bmatrix} \begin{bmatrix} x_1(t) \\ x_2(t) \end{bmatrix} + \begin{bmatrix} B_1D_2 \\ B_2 \end{bmatrix} u(t) \quad (4.150)$$

$$y(t) = [C_1 \quad 0] \begin{bmatrix} x_1(t) \\ x_2(t) \end{bmatrix} \quad (4.151)$$

The same results apply, mutatis mutandis, to discrete-time interconnected systems. More details can be found elsewhere, for example, in [15].

4.6 System Properties

4.6.1 Controllability, Reachability, and Stabilizability

A very important question that we must be interested in regarding control systems using state space models is whether or not we can steer the state via the control input to certain locations in the state space. We must remember that the states of a system frequently are internal variables like temperature, pressure, level of tanks, or others. These are sometimes critical variables that we want to keep between specific values.

4.6.1.1 Controllability

The issue of controllability is concerned with whether or not a given initial state $\mathbf{x}_0$ can be steered to the origin in finite time using the input $\mathbf{u}(t)$.

Example 4.11

If we examine the model defined in (4.152), we note that the input $u(t)$ has no effect over the state $x_2(t)$.

$$\begin{bmatrix} x_1(t) \\ x_2(t) \end{bmatrix} = \begin{bmatrix} 0 & 1 \\ 0 & 0 \end{bmatrix} \begin{bmatrix} x_1(t) \\ x_2(t) \end{bmatrix} + \begin{bmatrix} 1 \\ 0 \end{bmatrix} u(t) \quad (4.152)$$

Given an initial state $[x_1(0), x_2(0)]^T$, the input $u(t)$ can be chosen to steer $x_1(t)$ to zero, while $x_2(t)$ remains unchanged.

Formally, we have the following definition:

Definition 4.1 A state x_o is said to **controllable** if there exists a finite interval $[0, T]$ and an input $\{u(t), t \in [0, T]\}$ such that $x(T) = 0$. If all states are controllable, then the system is said to be **completely controllable.**

4.6.1.2 Reachability

A related concept is that of **reachability**, used sometimes in discrete-time systems. It is formally defined as follows:

Definition 4.2 A state $\bar{\mathbf{x}} \neq 0$ is said to be **reachable**, from the origin, if given $\mathbf{x}(0) = 0$, there exists a finite time interval $[0, T]$ and an input $\{\mathbf{u}(t), t \in [0, T]\}$ such that $\mathbf{x}(T) = \bar{\mathbf{x}}$. If all states are reachable the system is said to be **completely reachable.**

For continuous, time-invariant, linear systems, there is no distinction between complete controllability and reachability. However, the following example illustrates that there is a subtle difference in the discrete-time case. Consider the system and the output

$$x[t+1] = \underbrace{\begin{bmatrix} 0.5 & 1 \\ -0.25 & -0.5 \end{bmatrix}}_{A_d} x[t] \Rightarrow x[t] = \begin{bmatrix} 0.5 & 1 \\ -0.25 & -0.5 \end{bmatrix}^t x[0] \tag{4.153}$$

We can see that this system is completely controllable since $x[t] = 0$, $\forall t \geq 2$ and $\forall x[0] \in R^2$. This implies that every initial state is controllable. However, no nonzero state is reachable.

In view of the distinction between controllability and reachability in discrete time, we will use the term *controllability* in the sequel to cover the stronger of the two concepts.

Usually, in the context of linear time invariant systems, controllability and reachability are used interchangeably.

4.6.1.3 Controllability Test

We now present a systematic way to determine the complete controllability of a system.

Theorem 4.3 *Consider the linear, time-invariant, state space model where* $A \in R^{n\times n}$*:*

$$\dot{x}(t) = Ax(t) + Bu(t) \tag{4.154}$$

$$y(t) = Cx(t) + Du(t) \tag{4.155}$$

i) *The set of all controllable states is the* ***range space*** *of the controllability matrix* $\Gamma_c[A, B]$ *where*

$$\Gamma_c[A, B] \triangleq [B \quad AB \quad A^2B \quad \cdots \quad A^{n-1}B] \tag{4.156}$$

ii) *The model is completely controllable if and only if* $\Gamma_c[A, B]$ *has* ***full row rank.***

Example 4.12

Consider the state space model given in (4.152), with state space matrices

$$A = \begin{bmatrix} 0 & 1 \\ 0 & 0 \end{bmatrix}, \quad B = \begin{bmatrix} 1 \\ 0 \end{bmatrix} \tag{4.157}$$

The controllability matrix for this system, is given by

$$\Gamma_c[A, B] = [B \quad AB] = \begin{bmatrix} 1 & 0 \\ 0 & 0 \end{bmatrix} \tag{4.158}$$

Clearly, rank $\Gamma_c[A, B] = 1$, thus the system is **not** completely controllable.

The result above applies to continuous-time models, and it holds equally well for reachability of discrete-time models.

Also we can see that the controllability of a system is a property that does not depend on the choice of state variables. To see that, consider the similarity transformation defined in Section 4.4.4. Then, observing that $\bar{A}^i = T^{-1}A^iT$, we have

$$\Gamma_c[\bar{A}, \bar{B}] = T^{-1}\Gamma_c[A, B] \tag{4.159}$$

which implies that $\Gamma_c[\bar{A}, \bar{B}]$ and $\Gamma_c[A, B]$ have the same rank.

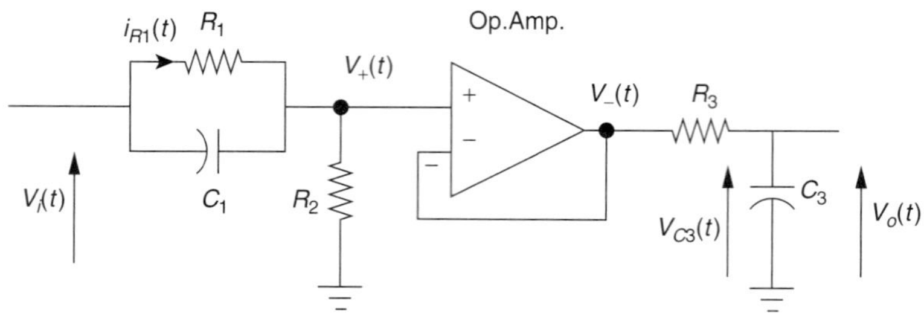

FIGURE 4.13 Electronic circuit.

The reader may wish to check that the state space models used to describe signals in Section 4.2.3 are uncontrollable. Indeed, it is always true **that any state space model where B = 0 is completely uncontrollable.**

4.6.1.4 Loss of Controllability

Lack of controllability is sometimes a structural feature. However, in some other cases, it depends on the numerical value of certain parameters. We illustrate this in the following example.

Example 4.13

Consider the electronic circuit shown in Figure 4.13.

We first build a state space model for the circuit. We choose, as state variables, $x_1(t) = i_{R1}(t)$ and $x_2(t) = v_{C3}(t)$. Using first principles on the left half of the circuit we have that

$$i_{C1} = C_1\frac{d}{dt}(v_i - v_+), \qquad i_{R1} = \frac{v_i - v_+}{R_1}, \qquad i_{R2} = \frac{v_+}{R_2}, \qquad i_{C1} = i_{R2} - i_{R1} \tag{4.160}$$

This yields

$$\frac{di_{R1}(t)}{dt} = -\frac{(R_1 + R_2)}{C_1R_1R_2}i_{R1}(t) + \frac{1}{C_1R_1R_2}v_i(t) \tag{4.161}$$

$$v_+(t) = -R_1i_{R1}(t) + v_i(t) \tag{4.162}$$

And, similarly, from the right half of the circuit we obtain

$$\frac{dv_{C3}(t)}{dt} = -\frac{1}{R_3C_3}v_{C3}(t) + \frac{1}{R_3C_3}v_-(t) \tag{4.163}$$

$$v_o(t) = v_{C3}(t) \tag{4.164}$$

The (ideal) operational amplifier ensures that $v_+(t) = v_-(t)$, so we can combine the state space models given in Equations 4.161–4.164 to obtain

$$\begin{bmatrix} \dfrac{di_{R1}(t)}{dt} \\ \dfrac{dv_{C3}(t)}{dt} \end{bmatrix} = \begin{bmatrix} -\dfrac{(R_1+R_2)}{C_1R_1R_2} & 0 \\ -\dfrac{R_1}{R_3C_3} & -\dfrac{1}{R_3C_3} \end{bmatrix} \begin{bmatrix} i_{R1}(t) \\ v_{C3}(t) \end{bmatrix} + \begin{bmatrix} \dfrac{1}{C_1R_1R_2} \\ \dfrac{1}{C_3R_3} \end{bmatrix} v_i(t) \tag{4.165}$$

$$v_o(t) = [0 \quad 1]\begin{bmatrix} i_{R1}(t) \\ v_{C3}(t) \end{bmatrix} \tag{4.166}$$

The controllability matrix is then given by

$$\Gamma_c[\mathbf{A},\mathbf{B}] = [\mathbf{B} \quad \mathbf{AB}] = \begin{bmatrix} \frac{1}{R_1R_2C_1} & \frac{-(R_1+R_1)}{(R_1R_2C_1)^2} \\ \frac{1}{R_3C_3} & \frac{-(R_2C_1+R_3C_3)}{(R_3C_3)^2R_2C_1} \end{bmatrix} \tag{4.167}$$

and

$$\det(\Gamma_c[\mathbf{A},\mathbf{B}]) = \frac{R_2}{(R_1R_2R_3C_1C_2)^2}(-R_1C_1+R_3C_3) \tag{4.168}$$

where we can observe that the system is completely controllable if, and only if, $R_1C_1 \neq R_3C_3$.

This issue has a very important interpretation if we analyze it from the transfer function point of view. Applying Laplace transform to Equations 4.161–4.164, the transfer function from $v_i(t)$ to $v_o(t)$ (recall that $V_+(s) = V_-(s)$) is given by

$$\frac{V_o(s)}{V_i(s)} = \frac{V_o(s)}{V_-(s)}\frac{V_+(s)}{V_i(s)} = \frac{\frac{1}{R_3C_3}}{\left(s+\frac{1}{R_3C_3}\right)} \cdot \frac{\left(s+\frac{1}{R_1C_1}\right)}{\left(s+\frac{R_1+R_2}{R_1R_2C_1}\right)} \tag{4.169}$$

where we can observe that the loss of complete controllability, when $R_1C_1 = R_3C_3$ obtained from (4.168), means that there is a **zero-pole cancellation** in the transfer function, that is, the zero from the left half of the circuit in Figure 4.13 is cancelled by the pole from the other part of the circuit. This issue will be discussed in more detail in Section 4.6.3.

4.6.1.5 Controllability Gramian

The test of controllability gives us a *yes or no* answer about the controllability of a system model. However, to conclude that a system is completely controllable says nothing about the *degree* of controllability. For **stable** systems, we can quantify the effort to control the system state through the energy involved in the input signal $\mathbf{u}(t)$ applied from $t = -\infty$ to reach the state $\mathbf{x}(0) = \mathbf{x}_0$ at $t = 0$:

$$J(\mathbf{u}) = \int_{-\infty}^{0} \|\mathbf{u}(t)\|^2 dt = \int_{-\infty}^{0} \mathbf{u}(t)^T\mathbf{u}(t)\,dt \tag{4.170}$$

It can be shown that the minimal *control energy* is

$$J(\mathbf{u}_{\text{opt}}) = \mathbf{x}_o^T\mathbf{P}^{-1}\mathbf{x}_o \tag{4.171}$$

where

$$\mathbf{P} = \int_0^{\infty} e^{\mathbf{A}t}\mathbf{B}\mathbf{B}^Te^{\mathbf{A}^Tt}dt \tag{4.172}$$

The matrix $\mathbf{P}$ is called the **controllability gramian**, and it measures the controllability of the state vector $\mathbf{x}(0)$. If this matrix is *small*, it means that we need a lot of energy in the control input $\mathbf{u}(t)$ to steer the state vector to $\mathbf{x}_0$. Indeed, we can appreciate the necessary effort for each one of the state variables, making, for example $\mathbf{x}_0 = [0,\ldots, 0, 1, 0,\ldots, 0]^T$.

It is important to emphasize that the existence of the integral defined in (4.172) is guaranteed only if the eigenvalues of A have negative real part, that is, the system must be stable.

Also, the controllability gramian P defined in (4.172) satisfies the Lyapunov equation

$$\mathbf{AP} + \mathbf{PA}^T + \mathbf{BB}^T = 0 \tag{4.173}$$

For discrete-time systems we have the following equations for the controllability gramian:

$$\mathbf{P}_d = \sum_{k=0}^{\infty} \mathbf{A}_d^k \mathbf{B}_d \mathbf{B}_d^T (\mathbf{A}_d^T)^k \tag{4.174}$$

which satisfies

$$\mathbf{A}_d \mathbf{P}_d \mathbf{A}_d^T - \mathbf{P}_d + \mathbf{B}_d \mathbf{B}_d^T = 0 \tag{4.175}$$

The sum defined in (4.174) is bounded if and only if the discrete-time system is stable, that is, its eigenvalues lie inside the unit disc.

Example 4.14

We can analyze the model of the Example 4.13, where the electronic circuit was described by the state space models (4.165) and (4.166). If we want to appreciate the information that we can obtain from the controllability gramian, defined in (4.172), when the model is close to losing complete controllability, we can choose suitable values of the parameters that ensure $R_1C_1 \approx R_3C_3$.

If we choose

$$R_1 = R_2 = R_3 = 10^3\ \Omega, \qquad C_1 = 0.9 \times 10^3\ \mu\text{F}, \qquad C_3 = 10^3\ \mu\text{F} \tag{4.176}$$

the model will be described by

$$\begin{bmatrix} \dot{i}_{R1}(t) \\ \dot{v}_{C3}(t) \end{bmatrix} = \begin{bmatrix} -\frac{20}{9} & 0 \\ -10^3 & -1 \end{bmatrix} \begin{bmatrix} i_{R1}(t) \\ v_{C3}(t) \end{bmatrix} + \begin{bmatrix} \frac{0.01}{9} \\ 1 \end{bmatrix} v_i(t) \tag{4.177}$$

$$v_o(t) = [0 \quad 1] \begin{bmatrix} i_{R1}(t) \\ v_{C3}(t) \end{bmatrix} \tag{4.178}$$

If we look at the relative magnitude of the elements of $\mathbf{B}$, we can *a priori* say that the effect of the input $u(t)$ upon the state $i_{R1}(t)$ will be much weaker than its effect upon the state $v_{C3}(t)$. To verify this we can compute the controllability gramian defined in (4.172), solving

$$0 = \mathbf{AP} + \mathbf{PA}^T + \mathbf{BB}^T \tag{4.179}$$

$$0 = \begin{bmatrix} -\frac{20}{9} & 0 \\ -10^3 & -1 \end{bmatrix} \begin{bmatrix} p_{11} & p_{12} \\ p_{21} & p_{22} \end{bmatrix} + \begin{bmatrix} p_{11} & p_{12} \\ p_{21} & p_{22} \end{bmatrix} \begin{bmatrix} -\frac{20}{9} & -10^3 \\ 0 & -1 \end{bmatrix} + \begin{bmatrix} \frac{0.01}{9} \\ 1 \end{bmatrix} \begin{bmatrix} \frac{0.01}{9} & 1 \end{bmatrix} \tag{4.180}$$

We have

$$\mathbf{P} = \begin{bmatrix} 0.28\times10^{-6} & 0.000258620 \\ 0.000258620 & 0.99999948 \end{bmatrix}, \qquad \mathbf{P}^{-1} = \begin{bmatrix} 4736624.0 & -1224.9 \\ -1224.9 & 1.3 \end{bmatrix} \tag{4.181}$$

So we can obtain the minimal control energy to steer the state $\mathbf{x}(t)$, from 0 in $t = -\infty$ to $\mathbf{x}_0$ in $t = 0$, from Equation 4.171.

$$\mathbf{x}_0 = [1, 0]^T \Rightarrow J(u_{\text{opt}}) = 4736624.0 \tag{4.182}$$

$$\mathbf{x}_0 = [0, 1]^T \Rightarrow J(u_{\text{opt}}) = 1.3 \tag{4.183}$$

We can thus verify that the control energy to attain $i_{R1}(0) = 1$ is six orders of magnitude greater than the necessary energy to attain $v_{C3}(0) = 1$.

Also, if we substitute the parameter values in Equation 4.169, we have that the transfer function is given by

$$\frac{V_o(s)}{V_i(s)} = \frac{1}{s+1} \cdot \frac{s+1+\frac{1}{9}}{s+\frac{20}{95}} \tag{4.184}$$

from where we observe a **zero-pole quasi cancellation.**

The idea of gramian has been extended to include the unstable case; see [16].

4.6.1.6 Canonical Decomposition and Stabilizability

If we have a system which is not completely controllable, it can be **decomposed** into a controllable subsystem and a completely uncontrollable subsystem in the following way.

Lemma 4.1 *Consider a system having rank* $\{\Gamma_c[\mathbf{A}, \mathbf{B}]\} = k < n$. *Then there exists a similarity transformation* T *such that* $\bar{x} = \mathbf{T}^{-1}x$,

$$\bar{\mathbf{A}} = \mathbf{T}^{-1}\mathbf{A}\mathbf{T}, \qquad \bar{\mathbf{B}} = \mathbf{T}^{-1}\mathbf{B} \tag{4.185}$$

and $\bar{\mathbf{A}}$, $\bar{\mathbf{B}}$ *have the form*

$$\bar{\mathbf{A}} = \begin{bmatrix} \bar{\mathbf{A}}_c & \bar{\mathbf{A}}_{12} \\ 0 & \bar{\mathbf{A}}_{nc} \end{bmatrix}, \qquad \bar{\mathbf{B}} = \begin{bmatrix} \bar{\mathbf{B}}_c \\ 0 \end{bmatrix} \tag{4.186}$$

where $\bar{\mathbf{A}}_c$ *has dimension k and* $(\bar{\mathbf{A}}_c, \bar{\mathbf{B}}_c)$ *is completely controllable.*

The above result tells us what states we can and what states we cannot steer to zero. To appreciate this, we express the state and output equations in the form

$$\begin{bmatrix} \dot{\bar{\mathbf{x}}}_c \\ \dot{\bar{\mathbf{x}}}_{nc} \end{bmatrix} = \begin{bmatrix} \bar{\mathbf{A}}_c & \bar{\mathbf{A}}_{12} \\ 0 & \bar{\mathbf{A}}_{nc} \end{bmatrix} \begin{bmatrix} \bar{\mathbf{x}}_c \\ \bar{\mathbf{x}}_{nc} \end{bmatrix} + \begin{bmatrix} \bar{\mathbf{B}}_c \\ 0 \end{bmatrix} \mathbf{u} \tag{4.187}$$

$$\mathbf{y} = \begin{bmatrix} \bar{\mathbf{C}}_c & \bar{\mathbf{C}}_{nc} \end{bmatrix} \begin{bmatrix} \bar{\mathbf{x}}_c \\ \bar{\mathbf{x}}_{nc} \end{bmatrix} + \mathbf{D}\mathbf{u} \tag{4.188}$$

The **controllable subspace** of a state space model is composed of all states generated through every possible linear combination of the states in $\bar{\mathbf{x}}_c$. The stability of this subspace is determined by the location of the eigenvalues of $\bar{\mathbf{A}}_c$.

On the other hand, the **uncontrollable subspace** is composed of all states generated through every possible linear combination of the states in $\bar{\mathbf{x}}_{nc}$. The stability of this subspace is determined by the location of the eigenvalues of $\bar{\mathbf{A}}_{nc}$.

Hence, the input will have no effect over the uncontrollable subspace, so the best we can hope is that this uncontrollable subspace is stable, since then the state in this subspace will go to the origin. In this case the state space model is said to be **stabilizable.**

A key feature of the descriptions (4.187) and (4.188) arises from the fact that the transfer function is given by

$$\mathrm{H}(s) = \bar{\mathrm{C}}_c(s\mathrm{I} - \bar{\mathrm{A}}_c)^{-1}\bar{\mathrm{B}}_c + \mathrm{D} \tag{4.189}$$

Equation 4.189 says that the eigenvalues of the uncontrollable subspace do not belong to the set of poles of the system transfer function. This implies that there is a cancellation of all poles corresponding to the roots of $(s\mathbf{I} - \bar{\mathbf{A}}_{nc})$.

4.6.1.7 Controllability Canonical Form

Lemma 4.2 *Consider a completely reachable state space model for a SISO system. Then, there exists a similarity transformation which converts the state space model into the following* ***controllability canonical form:***

$$\mathrm{A}' = \begin{bmatrix} 0 & 0 & \cdots & 0 & -\alpha_0 \\ 1 & 0 & \cdots & 0 & -\alpha_1 \\ 0 & 1 & \cdots & 0 & -\alpha_2 \\ \mathrm{A} & \mathrm{A} & \ddots & \mathrm{A} & \mathrm{A} \\ 0 & 0 & \cdots & 1 & -\alpha_{n-1} \end{bmatrix}, \quad \mathrm{B}' = \begin{bmatrix} 1 \\ 0 \\ 0 \\ \mathrm{A} \\ 0 \end{bmatrix} \tag{4.190}$$

where $\lambda^n + \alpha_{n-1}\lambda^{n-1} + \cdots + \alpha_1\lambda + \alpha_0 = \det(\lambda\mathrm{I} - \mathrm{A})$ *is the characteristic polynomial of* A.

Lemma 4.3 *Consider a completely controllable state space model for a SISO system. Then, there exists a similarity transformation which converts the state space model into the following* ***controller canonical form:***

$$\mathrm{A}'' = \begin{bmatrix} -\alpha_{n-1} & -\alpha_{n-2} & \cdots & -\alpha_1 & -\alpha_0 \\ 1 & 0 & \cdots & 0 & 0 \\ 0 & 1 & \cdots & 0 & 0 \\ \mathrm{A} & \mathrm{A} & \ddots & \mathrm{A} & \mathrm{A} \\ 0 & 0 & \cdots & 1 & 0 \end{bmatrix}, \quad \mathrm{B}'' = \begin{bmatrix} 1 \\ 0 \\ 0 \\ \mathrm{A} \\ 0 \end{bmatrix} \tag{4.191}$$

where $\lambda^n + \alpha_{n-1}\lambda^{n-1} + \cdots + \alpha_1\lambda + \alpha_0 = \det(\lambda\mathrm{I} - \mathrm{A})$ *is the characteristic polynomial of* A.

4.6.2 Observability, Reconstructibility, and Detectability

If we consider the state space model of a system, one might conjecture that if one observes the output over some time interval then this might tell us some information about the state. The associated model property is called observability (or reconstructibility).

4.6.2.1 Observability

Observability is concerned with the issue of what can be said on the state if we measure the plant output.

Example 4.15

If we look at the system defined by state space model

$$\begin{bmatrix} \dot{x}_1(t) \\ \dot{x}_2(t) \end{bmatrix} = \begin{bmatrix} -1 & 0 \\ 1 & -1 \end{bmatrix} \begin{bmatrix} x_1(t) \\ x_2(t) \end{bmatrix}, \qquad y(t) = \begin{bmatrix} 1 & 0 \end{bmatrix} \begin{bmatrix} x_1(t) \\ x_2(t) \end{bmatrix} \tag{4.192}$$

we can see that the output $y(t)$ only is determined by $x_1(t)$, and the other state variable $x_2(t)$ has no influence on the output. So the system is not completely observable.

A formal definition is as follows:

Definition 4.3 The state $\mathbf{x}_o \neq \mathbf{0}$ is said to be **unobservable** if given $\mathbf{x}(0) = \mathbf{x}_o$, and $\mathbf{u}(t) = \mathbf{0}$ for $t \geq 0$, then $\mathbf{y}(t) = \mathbf{0}$ for $t \geq 0$, that is, we cannot see any effect of $\mathbf{x}_o$ on the system output.

The system is said to be **completely observable** if there exists no nonzero initial state that it is unobservable.

4.6.2.2 Reconstructibility

There is another concept, closely related to observability, called **reconstructibility**. Reconstructibility is concerned with what can be said about $\mathbf{x}(T)$, having observed the **past** values of the output, y, for $0 \leq t \leq T$. For linear time invariant, continuous-time systems, the distinction between observability and reconstructibility is unnecessary. However, the following example illustrates that in discrete time, the two concepts are different. Consider

$$\mathbf{x}[t+1] = 0, \qquad \mathbf{x}[0] = \mathbf{x}_o \tag{4.193}$$

$$\mathbf{y}[t] = 0 \tag{4.194}$$

This system is clearly reconstructible for all $T \geq 1$, since we know for certain that $\mathbf{x}[T] = 0$ for $T \geq 1$. However, it is completely unobservable since $\mathbf{y}[t] = 0, \forall k$ irrespective of $\mathbf{x}_o$.

In view of the subtle difference between observability and reconstructibility, we will use the term *observability* in the sequel to cover the stronger of the two concepts.

4.6.2.3 Observability Test

A test for observability of a system is established in the following theorem.

Theorem 4.4 *Consider the linear, continuous, time-invariant, state space model where* $\mathbf{A} \in \mathbb{R}^{n \times n}$

$$\dot{\mathbf{x}}(t) = \mathbf{A}\mathbf{x}(t) + \mathbf{B}\mathbf{u}(t) \tag{4.195}$$

$$\mathbf{y}(t) = \mathbf{C}\mathbf{x}(t) + \mathbf{D}\mathbf{u}(t) \tag{4.196}$$

i) *The set of all unobservable states is equal to the null space of the observability matrix* $\Gamma_o[\mathbf{A}, \mathbf{C}]$ where

$$\Gamma_o[\mathbf{A}, \mathbf{C}] \triangleq \begin{bmatrix} \mathbf{C} \\ \mathbf{CA} \\ \vdots \\ \mathbf{CA}^{n-1} \end{bmatrix} \tag{4.197}$$

ii) *The system is completely observable if and only if* $\Gamma_o[\mathbf{A}, \mathbf{C}]$ *has* ***full column rank*** n.

Example 4.16

Consider the following state space model:

$$A = \begin{bmatrix} -3 & -2 \\ 1 & 0 \end{bmatrix}, \quad B = \begin{bmatrix} 1 \\ 0 \end{bmatrix}, \quad C = [1 \quad -1] \tag{4.198}$$

The observability matrix is given by

$$\Gamma_o[A, C] = \begin{bmatrix} C \\ CA \end{bmatrix} = \begin{bmatrix} 1 & -1 \\ -4 & -2 \end{bmatrix} \tag{4.199}$$

Hence rank $\Gamma_o[A, C] = 2$, which says that the system is completely observable.

Example 4.17

If we look at the model defined in (4.192), we have

$$A = \begin{bmatrix} -1 & 0 \\ 1 & -1 \end{bmatrix}, \quad C = [1 \quad 0] \tag{4.200}$$

The observability matrix is

$$\Gamma_o[A, C] = \begin{bmatrix} 1 & 0 \\ -1 & 0 \end{bmatrix} \tag{4.201}$$

Hence rank $\Gamma_o[A, C] = 1 < 2$ and the system is not completely observable.

The above result also applies to discrete-time models.

The observability is a system property that does not depend on the choice of state variables. It can be proved that the rank of the matrix defined in Equation 4.197 does not change when a similarity transformation T is used (see Section 4.3.3).

4.6.2.4 Loss of Observability

Lack of observability may arise from structural system features. However, it is also possible that lack of observability occurs when certain system parameters take some specific numerical values. This is the same phenomenon, for controllability, we analyzed in Section 4.6.1. We expect that those parameters will affect the complete observability of the model in a similar way. Let us look at the following example.*

Example 4.18

Consider the electronic circuit in Figure 4.14. We can see this is the same as that in Figure 4.13 where the left and right halves were swaped, so we can use similar equations to obtain a state space model. The state variables have been chosen to be $x_1(t) = v_{C3}(t)$ and $x_2(t) = i_{R1}(t)$.

For the left half of the circuit, we have

$$\frac{dv_{C3}(t)}{dt} = -\frac{1}{R_3C_3}v_{C3}(t) + \frac{1}{R_3C_3}v_i(t) \tag{4.202}$$

$$v_+(t) = v_{C3}(t) \tag{4.203}$$

*Which is the *dual* of Example 4.13.

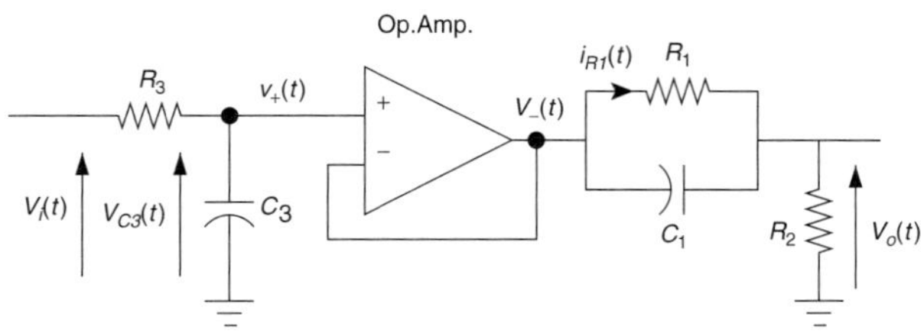

FIGURE 4.14 Electronic circuit.

And for the right half, we have

$$\frac{di_{R1}(t)}{dt} = -\left(\frac{R_1 + R_2}{C_1 R_1 R_2}\right) i_{R1}(t) + \frac{1}{C_1 R_1 R_2} v_-(t) \tag{4.204}$$

$$v_o(t) = -R_1 i_{R1}(t) + v_-(t) \tag{4.205}$$

The operational amplifier, in voltage follower connection, ensures that $v_+(t) = v_-(t)$, so we can combine the state space models given in Equations 4.202–4.205:

$$\begin{bmatrix} \frac{dv_{C3}(t)}{dt} \\ \frac{di_{R1}(t)}{dt} \end{bmatrix} = \begin{bmatrix} -\frac{1}{R_3 C_3} & 0 \\ \frac{1}{C_1 R_1 R_2} & -\frac{R_1 + R_2}{C_1 R_1 R_2} \end{bmatrix} \begin{bmatrix} v_{C3}(t) \\ i_{R1}(t) \end{bmatrix} + \begin{bmatrix} \frac{1}{R_3 C_3} \\ 0 \end{bmatrix} v_i(t) \tag{4.206}$$

$$v_o(t) = [1 \ -R_1] \begin{bmatrix} v_{C3}(t) \\ i_{R1}(t) \end{bmatrix} \tag{4.207}$$

The observability matrix is given by

$$\Gamma_c[\mathbf{C}, \mathbf{A}] = \begin{bmatrix} \mathbf{C} \\ \mathbf{CA} \end{bmatrix} = \begin{bmatrix} 1 & -R_1 \\ -\frac{1}{R_3 C_3} - \frac{1}{R_2 C_1} & \frac{R_1 + R_2}{R_2 C_1} \end{bmatrix} \tag{4.208}$$

To determine the complete observability, or otherwise, we need to compute the matrix determinant

$$\det(\Gamma_c[\mathbf{C}, \mathbf{A}]) = \frac{1}{R_3 C_3 C_1}(-R_1 C_1 + R_3 C_3) \tag{4.209}$$

from where we conclude that the model system is completely observable if and only if, $R_1 C_1 \neq R_3 C_3$, which is the same condition we obtained in Example 4.13.

Applying Laplace transform to Equations 4.204–4.203 we obtain the transfer function from $V_i(s)$ to $V_o(s)$:

$$\frac{V_o(s)}{V_i(s)} = \frac{V_+(s)}{V_i(s)} \frac{V_o(s)}{V_-(s)} = \frac{s + \frac{1}{R_1 C_1}}{s + \frac{R_1 + R_2}{R_1 R_2 C_1}} \cdot \frac{\frac{1}{R_3 C_3}}{s + \frac{1}{R_3 C_3}} \tag{4.210}$$

The condition $R_1C_1 = R_3C_3$ produces the loss of complete observability, leading to a **pole-zero cancellation** in the model transfer function, i.e., the pole from the left half of the circuit in Figure 4.14 is cancelled by the zero from the right half. There is subtle difference between the transfer functions in (4.210) and (4.169). The final result is the same, but the order the cancellation is different in each case. The **zero-pole cancellation** is connected to the loss of complete observability and the **pole-zero cancellation** is connected to the loss of complete controllability. These issues will be discussed in more detail in Section 4.6.3.

4.6.2.5 Observability Gramian

The observability test in Theorem 4.4 answers *yes* or *no* to the question about completely observability of a model. However, sometimes we are interested in the *degree* of observability for a particular model. So we can quantify the energy of the output signal $\mathbf{y}(t)$, when there is no input ($\mathbf{u}(t) = \mathbf{0}$) and the state is $\mathbf{x}(0) = \mathbf{x}_0$ at $t = 0$

$$E(\mathbf{x}_0) = \int_0^\infty \mathrm{i}\mathbf{y}(t)\mathrm{i}^2 dt = \int_0^\infty \mathbf{y}(t)^T\mathbf{y}(t)dt \tag{4.211}$$

It can be proved that the *ouput energy* is

$$E(\mathbf{x}_0) = \int_0^\infty \mathrm{i}\mathbf{y}(t)\mathrm{i}^2 dt = \mathbf{x}_0^T\mathbf{Q}\mathbf{x}_0 \tag{4.212}$$

where

$$\mathbf{Q} = \int_0^\infty e^{\mathbf{A}^T t}\mathbf{C}^T\mathbf{C}\, e^{\mathbf{A}t} dt \tag{4.213}$$

The matrix $\mathbf{Q}$ is called **observability gramian**, and it measures the observability of the state vector $\mathbf{x}(0)$. If this matrix is *small*, it means that we have a weak contribution of the initial state $\mathbf{x}_0$ in the energy of the output $\mathbf{y}(t)$. Indeed, we can appreciate the effect of each one of the state variables taking, for example, $\mathbf{x}_0 = [0,\ldots,0,1,0,\ldots,0]^T$.

Note that the existence of the integral defined in (4.213) is guaranteed if and only if the system is stable, i.e., if and only if the eigenvalues of $\mathbf{A}$ have negative real part.

Also, the observability gramian $\mathbf{Q}$ defined in (4.213) satisfies the Lyapunov equation

$$\mathbf{A}^T\mathbf{Q} + \mathbf{Q}\mathbf{A} + \mathbf{C}^T\mathbf{C} = 0 \tag{4.214}$$

For stable discrete-time systems, the controllability gramian is defined by

$$\mathbf{Q}_\mathrm{d} = \sum_{k=0}^{\infty}(\mathbf{A}_\mathrm{d}^T)^k\mathbf{C}_\mathrm{d}^T\mathbf{C}_\mathrm{d}\mathbf{A}_\mathrm{d}^k \tag{4.215}$$

which satisfies

$$\mathbf{A}_\mathrm{d}^T\mathbf{Q}_\mathrm{d}\mathbf{A}_\mathrm{d} - \mathbf{Q}_\mathrm{d} + \mathbf{C}_\mathrm{d}^T\mathbf{C}_\mathrm{d} = 0 \tag{4.216}$$

Example 4.19

We will use the model of Example 4.18, described by the state space models (4.206) and (4.207), to appreciate the utility of the observability gramian (4.213), especially when the model is close to losing complete observability, i.e., when $R_1C_1 \approx R_3C_3$.

Assuming the same component values as in Example 4.14 for R_1, R_2, R_3, C_1, and C_3 we have

$$\begin{bmatrix} \dot{v}_{C3}(t) \\ \dot{i}_{R1}(t) \end{bmatrix} = \begin{bmatrix} -1 & 0 \\ 10^{-3} & -\frac{20}{9} \end{bmatrix} \begin{bmatrix} v_{C3}(t) \\ i_{R1}(t) \end{bmatrix} + \begin{bmatrix} 1 \\ 0 \end{bmatrix} v_i(t) \quad (4.217)$$

$$v_o(t) = [1 \quad -10^3] \begin{bmatrix} v_{C3}(t) \\ i_{R1}(t) \end{bmatrix} \quad (4.218)$$

If we look at the relative magnitude of the components of **C** matrix, we can foretell *a priori* that the output $v_o(t)$ will be mainly determined by state $i_{R1}(t)$. To verify this we compute the observability gramian defined in (4.172), solving

$$0 = \mathbf{A}^T\mathbf{Q} + \mathbf{Q}\mathbf{A} + \mathbf{C}^T\mathbf{C} \quad (4.219)$$

$$0 = \begin{bmatrix} -1 & 10^{-3} \\ 0 & -\frac{20}{9} \end{bmatrix} \begin{bmatrix} q_{11} & q_{12} \\ q_{21} & q_{22} \end{bmatrix} + \begin{bmatrix} q_{11} & q_{12} \\ q_{21} & q_{22} \end{bmatrix} \begin{bmatrix} -1 & 0 \\ 10^{-3} & -\frac{20}{9} \end{bmatrix} + \begin{bmatrix} 1 \\ -10^3 \end{bmatrix} [1 \quad 10^{-3}] \quad (4.220)$$

We have

$$\mathbf{Q} = \begin{bmatrix} 0.57 & 69.83 \\ 69.83 & 225000 \end{bmatrix} \quad (4.221)$$

From there we can compute the contribution of each state to the total energy in the output. Doing this, we verify that the state variable $i_{R1}(t)$ has an effect over the output greater than the effect of $v_{C3}(t)$, as defined in Equation 4.212:

$$\mathbf{x}_0 = [1, 0]^T \Rightarrow E(\mathbf{x}_0) = 0.57 \quad (4.222)$$

$$\mathbf{x}_0 = [0, 1]^T \Rightarrow E(\mathbf{x}_0) = 225000 \quad (4.223)$$

The transfer function is

$$\frac{V_o(s)}{V_i(s)} = \frac{s + 1 + \frac{1}{9}}{s + \frac{20}{9}} \cdot \frac{1}{s+1} \quad (4.224)$$

We observe that there is a pole-zero **quasi-cancellation.**

4.6.2.6 Duality Principle

We observe a remarkable similarity between the results in Theorem 4.3 and in Theorem 4.4, and also for the definitions of the gramians (4.172) and (4.213). This is known as the **duality principle**, and it can be formalized as follows:

Theorem 4.5 (Duality) *Consider a state space model described by the 4-tuple* (**A**, **B**, **C**, **D**). *Then the system is completely controllable if and only if the dual system* ($\mathbf{A}^T$, $\mathbf{C}^T$, $\mathbf{B}^T$, $\mathbf{D}^T$) *is completely observable.*

4.6.2.7 Canonical Decomposition and Detectability

The above theorem can often be used to go from a result on controllability to one on observability and vice versa. The *dual* of Lemma 4.1 is:

Lemma 4.4 *If rank* $\{\Gamma_o[A, C]\} = k < n$, *there exists a similarity transformation* T *such that with* $\bar{x} = T^{-1}x$, $\bar{A} = T^{-1}AT$, $\bar{C} = CT$, *then* $\bar{C}$ *and* $\bar{A}$ *take the form*

$$\bar{A} = \begin{bmatrix} \bar{A}_o & 0 \\ \bar{A}_{21} & \bar{A}_{no} \end{bmatrix}, \qquad \bar{C} = [\bar{C}_o \quad 0] \tag{4.225}$$

where $\bar{A}_o$ *has dimension k and the pair* $(\bar{C}_o, \bar{A}_o)$ *is completely observable.*

This result has a relevance similar to that of the controllability property and the associated decomposition. To appreciate this, we apply the dual of Lemma 4.1 to express the (transformed) state and output equations in partitioned form as

$$\begin{bmatrix} \dot{\bar{x}}_o(t) \\ \dot{\bar{x}}_{no}(t) \end{bmatrix} = \begin{bmatrix} \bar{A}_o & 0 \\ \bar{A}_{21} & \bar{A}_{no} \end{bmatrix} \begin{bmatrix} \bar{x}_o(t) \\ \bar{x}_{no}(t) \end{bmatrix} + \begin{bmatrix} \bar{B}_o \\ \bar{B}_{no} \end{bmatrix} u(t) \tag{4.226}$$

$$y(t) = \begin{bmatrix} \bar{C}_o & 0 \end{bmatrix} \begin{bmatrix} \bar{x}_o(t) \\ \bar{x}_{no}(t) \end{bmatrix} + Du(t) \tag{4.227}$$

The above description reveals why one can be in trouble when trying to control a system using only the system output. The output has no information on the state $\bar{x}_{no}$.

The **observable subspace** of a model is the space composed of all states generated through every possible linear combination of the states in $\bar{x}_o$. The stability of this subspace is determined by the location of the eigenvalues of $\bar{A}_o$.

The **unobservable subspace** of a model is the space composed of all states generated through every possible linear combination of the states in $\bar{x}_{no}$. The stability of this subspace is determined by the location of the eigenvalues of $\bar{A}_{no}$.

If the unobservable subspace is stable we say that the system is **detectable.**

A key feature of the descriptions (4.226) and (4.227) arises from the fact that the transfer function is given by

$$H(s) = \bar{C}_0(sI - \bar{A}_o)^{-1}\bar{B}_o + D \tag{4.228}$$

Equation (4.228) says that the eigenvalues of the unobservable subspace do not belong to the set of poles of the system transfer function. This implies that there is a cancellation of all poles corresponding to the roots of $(sI - \bar{A}_{no})$.

4.6.2.8 Observability Canonical Form

There are also duals of the canonical forms given in Lemmas 4.2 and 4.3. For example, the dual of Lemma 4.3 is:

Lemma 4.5 *Consider a completely observable SISO system. Then there exists a similarity transformation that converts the model to the* ***observer canonical form:***

$$x(t) = \begin{bmatrix} -\alpha_{n-1} & 1 & & \\ A & & \ddots & \\ A & & & 1 \\ -\alpha_0 & 0 & & 0 \end{bmatrix} x(t) + \begin{bmatrix} b_{n-1} \\ A \\ A \\ b_0 \end{bmatrix} u(t) \tag{4.229}$$

$$y(t) = [1 \quad 0 \quad \cdots \quad 0]x(t) + Du(t) \tag{4.230}$$

4.6.3 Canonical Decomposition

Further insight into the structure of linear dynamical systems is obtained by considering those systems which are only partially observable or controllable. These systems can be separated into completely observable and completely controllable systems.

The two results of Lemmas 4.1 and 4.4 can be combined for those systems, which are neither completely observable nor completely controllable. We can see it as follows.

Theorem 4.6 (Canonical Decomposition Theorem) *Consider a system described in state space form. Then, there always exists a similarity transformation* T *such that the transformed model for* $\boldsymbol{x} = \mathrm{T}^{-1}\boldsymbol{x}$ *takes the form*

$$\bar{\mathrm{A}} = \begin{bmatrix} \bar{\mathrm{A}}_{co} & 0 & \bar{\mathrm{A}}_{13} & 0 \\ \bar{\mathrm{A}}_{21} & \bar{\mathrm{A}}_{22} & \bar{\mathrm{A}}_{23} & \bar{\mathrm{A}}_{24} \\ 0 & 0 & \bar{\mathrm{A}}_{33} & 0 \\ 0 & 0 & \bar{\mathrm{A}}_{34} & \bar{\mathrm{A}}_{44} \end{bmatrix}, \qquad \bar{\mathrm{B}} = \begin{bmatrix} \bar{\mathrm{B}}_1 \\ \bar{\mathrm{B}}_2 \\ 0 \\ 0 \end{bmatrix}, \qquad \bar{\mathrm{C}} = [\bar{\mathrm{C}}_1 \;\; 0 \;\; \bar{\mathrm{C}}_2 \;\; 0] \tag{4.231}$$

where

i) *The subsystem* $[\bar{\mathrm{A}}_{co}, \bar{\mathrm{B}}_1, \bar{\mathrm{C}}_1]$ *is both completely controllable and completely observable and has the same transfer function as the original system (see Lemma 4.6).*

ii) *The subsystem*

$$\begin{bmatrix} \bar{\mathrm{A}}_{co} & 0 \\ \bar{\mathrm{A}}_{21} & \bar{\mathrm{A}}_{22} \end{bmatrix}, \quad \begin{bmatrix} \bar{\mathrm{B}}_1 \\ \bar{\mathrm{B}}_2 \end{bmatrix}, \quad [\bar{\mathrm{C}}_1 \;\; 0] \tag{4.232}$$

is completely controllable.

iii) *The subsystem*

$$\begin{bmatrix} \bar{\mathrm{A}}_{co} & \bar{\mathrm{A}}_{13} \\ 0 & \bar{\mathrm{A}}_{33} \end{bmatrix}, \quad \begin{bmatrix} \bar{\mathrm{B}}_1 \\ 0 \end{bmatrix}, \quad [\bar{\mathrm{C}}_1 \;\; \bar{\mathrm{C}}_2] \tag{4.233}$$

is completely observable.

The canonical decomposition described in Theorem 4.6 leads to an important consequence for the transfer function of the model, which will take only the completely observable and completely controllable subspace.

Lemma 4.6 *Consider the transfer function matrix* H(*s*) *given by*

$$Y(s) = \mathrm{H}(s)U(s) \tag{4.234}$$

Then

$$\mathrm{H} = \mathrm{C}(s\mathrm{I} - \mathrm{A})^{-1}\mathrm{B} + \mathrm{D} = \bar{\mathrm{C}}_1(s\mathrm{I} - \bar{\mathrm{A}}_{co})^{-1}\bar{\mathrm{B}}_1 + \mathrm{D} \tag{4.235}$$

where $\bar{\mathrm{C}}_1$, $\bar{\mathrm{A}}_{co}$, *and* $\bar{\mathrm{B}}_1$ *are as in Equation 4.231. This state description is a minimal realization of the transfer function.*

If $\mathbf{M}$ is any square matrix and we denote by $\Lambda\{\mathbf{M}\}$ the set of eigenvalues of $\mathbf{M}$, then

$$\Lambda\{\bar{\mathbf{A}}\} = \Lambda\{\bar{\mathbf{A}}_{co}\} \cup \Lambda\{\bar{\mathbf{A}}_{22}\} \cup \Lambda\{\bar{\mathbf{A}}_{33}\} \cup \Lambda\{\bar{\mathbf{A}}_{44}\} \tag{4.236}$$

where

$\Lambda\{\bar{\mathbf{A}}\}$ = eigenvalues of the system,
$\Lambda\{\bar{\mathbf{A}}_{co}\}$ = eigenvalues of the controllable and observable subsystem,
$\Lambda\{\bar{\mathbf{A}}_{22}\}$ = eigenvalues of the controllable but unobservable subsystem,
$\Lambda\{\bar{\mathbf{A}}_{33}\}$ = eigenvalues of the uncontrollable but observable subsystem,
$\Lambda\{\bar{\mathbf{A}}_{44}\}$ = eigenvalues of the uncontrollable and unobservable subsystem.

We observe that controllability for a given system depends on the structure of the input ports, that is, where, in the system, the manipulable inputs are applied. Thus, the states of a given subsystem may be uncontrollable for a given input, but completely controllable for another. This distinction is of fundamental importance in control system design since not all plant inputs can be manipulated (consider, for example, disturbances) and, therefore, cannot be used to steer the plant to reach certain states.

Similarly, the observability property depends on which outputs are being considered. Certain states may be unobservable from a given output, but they may be completely observable from some other output. This also has a significant impact on output feedback control systems, since some states may not appear in the plant output being measured and feeded back. However, they may appear in crucial internal variables and thus be important to the control problem.

4.6.4 PBH Test

An alternative test for controllability and observability is provided by the following lemma known as PBH test.

Lemma 4.7 *Consider a state space model* $(\mathbf{A}, \mathbf{B}, \mathbf{C})$. *Then*

(i) *The system is not completely observable if and only if there exists a nonzero vector* $\mathbf{x} \in \mathbb{C}^n$ *and a scalar* $\lambda \in \mathbb{C}$ *such that*

$$\mathbf{A}\mathbf{x} = \lambda\mathbf{x}, \quad \mathbf{C}\mathbf{x} = 0 \tag{4.237}$$

(ii) *The system is not completely controllable if and only if there exists a nonzero vector* $\mathbf{x} \in \mathbb{C}^n$ *and a scalar* $\lambda \in \mathbb{C}$ *such that*

$$\mathbf{x}^T\mathbf{A} = \lambda\mathbf{x}^T, \quad \mathbf{x}^T\mathbf{B} = 0 \tag{4.238}$$

4.7 State Observers

4.7.1 Basic Concepts

When the state variables have to be measured for monitoring, implementing control systems, or other purposes, there are hard technical and economical issues to face. Observers are a way to estimate the state variables based upon a system model, measurements of the plant output $\mathbf{y}(t)$, and measurements of the plant input $\mathbf{u}(t)$. This problem is a generalization of that of indirectly measuring a system variable using a system model and the measurement of some other easier-to-measure variable.

4.7.2 Observer Dynamics

Assume that the system has a state space model given by (4.42) and (4.43) with $\mathbf{D} = \mathbf{0}$ (a strictly proper system has been assumed). Then, the general structure of a classic observer for the system state is as shown in Figure 4.15, where the matrix $\mathbf{J}$ is the **observer gain**.

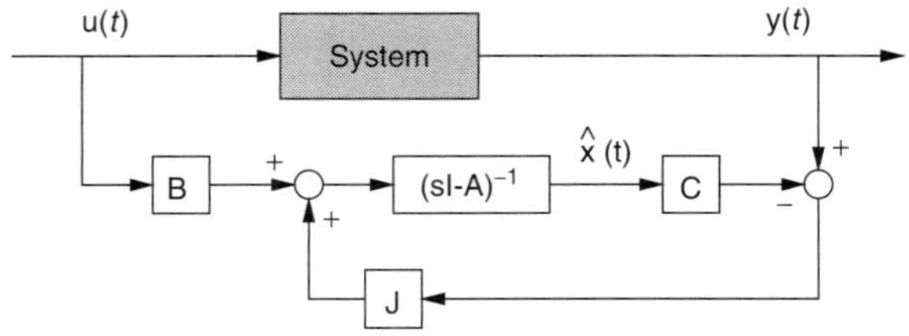

FIGURE 4.15 Classic state observer.

Therefore, the observer equation is

$$\frac{d\hat{\mathbf{x}}(t)}{dt} = \mathbf{A}\hat{\mathbf{x}}(t) + \mathbf{B}\mathbf{u}(t) + \mathbf{J}(\mathbf{y}(t) - \mathbf{C}\hat{\mathbf{x}}(t)) \tag{4.239}$$

An obvious question is: if we know an exact system model and the system input, why do we need to feed the system output? The answer is that **we need the output measurement since we do not know the system initial state.** This can be appreciated from the equation for the state estimation error, $\tilde{\mathbf{x}}(t) = \mathbf{x}(t) - \hat{\mathbf{x}}(t)$. That equation can be obtained subtracting (4.239) from (4.42). This leads to

$$\frac{d\tilde{\mathbf{x}}(t)}{dt} = (\mathbf{A} - \mathbf{J}\mathbf{C})\tilde{\mathbf{x}}(t) \tag{4.240}$$

From (4.240) we observe that the estimation error will converge to zero for a nonzero initial error if and only if all the eigenvalues of the matrix $\mathbf{A} - \mathbf{J}\mathbf{C}$ have negative real parts, that is, if the **observer polynomial** $E(s) = \det(sI - \mathbf{A} + \mathbf{J}\mathbf{C})$ is strictly Hurwitz.

Discussion

- Equation (4.240) is valid only if the model is a perfect representation of the system under study. Modelling errors will impact the observer. This will normally lead to nonzero state estimation errors.
- If the pair $(\mathbf{A}, \mathbf{C})$ is completely observable, then the eigenvalues of $\mathbf{A} - \mathbf{J}\mathbf{C}$ can be arbitrarily located (in the stability region). Thus, the speed of the estimation convergence is a designer's choice. Those eigenvalues are known as the **observer poles.**
- If the pair $(\mathbf{A}, \mathbf{C})$ is detectable, then the observer will yield zero steady state error asymptotically, although not all the eigenvalues of $\mathbf{A} - \mathbf{J}\mathbf{C}$ can be placed at will.
- If the system is not completely observable, and the unobservable subspace contains unstable modes, then the observer will never converge.

To illustrate the observer techniques we refer to Example 4.5.

Example 4.20

Assume that we want the observer poles for the state model in Example 4.5 to be located at $s = -4$, $s = -6$, and $s = -8$. We can then compute the observer gain, **J**, using a software such as MATLAB. This yields

$$J = [-4.5247 \quad -7.5617 \quad -4.1543]^T \tag{4.241}$$

To appreciate the observer dynamics, assume that the initial system state is $\mathbf{x}(0) = [-1 \quad 2 \quad 1]^T$ and that the system input is a square wave of amplitude 1, and frequency equal to 1 rad/s. The observer is initialized with $\hat{\mathbf{x}}(0) = 0$. Then the norm of the estimation error, $||\hat{\mathbf{x}}(t)||$, evolves as shown in Figure 4.16. It is

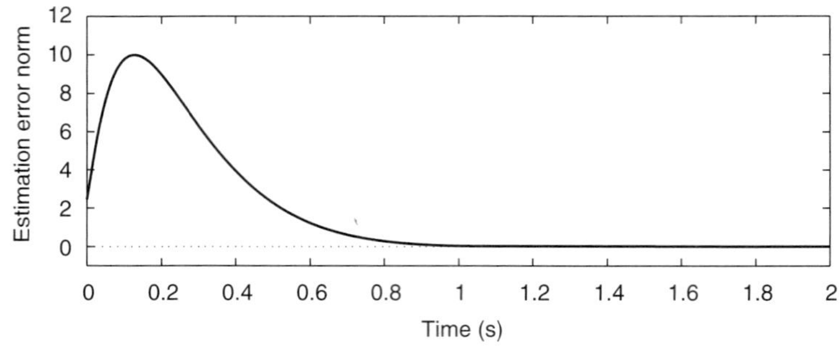

FIGURE 4.16 State estimation error.

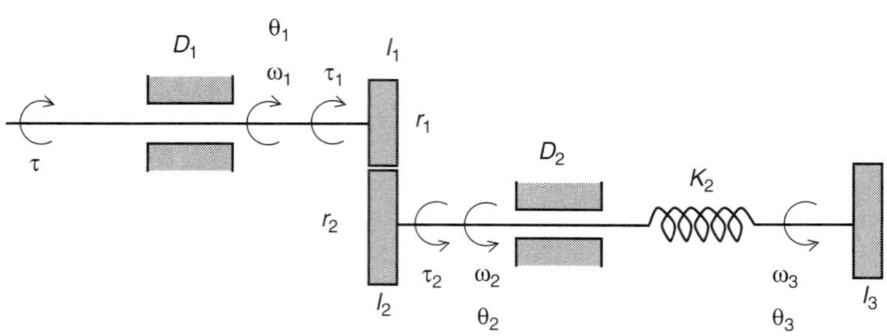

FIGURE 4.17 Rotational system.

important to point out that, in this example, the plant is unstable. This means that the state and the state estimation grow unbounded. However, under the assumption of perfect modelling, the estimation error converges to zero.

To gain physical insight into the observer philosophy, we consider the following application.

Example 4.21

Figure 4.17 shows the schematics of a rotational system driven by a torque $\tau(t)$. The system power is transmitted through a gear system built with two wheels with radii r_1 and r_2 and inertias I_1 and I_2, respectively. The rotation of both shafts is damped by viscous friction with coefficients D_1 and D_2, and a significant torsional spring in shaft 2 has also been modelled. The system load is modelled as an inertia I_3. We want to estimate the load speed ω_3 based on the measurement of the speed in shaft 1, ω_1.

We first need to build a state space model. To do that we choose a minimum set of system variables, which quantify the energy stored in the system. The system has four components able to store energy: three inertias and a spring. Nevertheless, the energy stored in I_1 and I_2 can be computed either from ω_1 or from ω_2, that is, we need only one of these speeds, since they satisfy

$$\frac{\omega_1(t)}{\omega_2(t)} = \frac{r_2}{r_1} \quad \text{and} \quad \tau_1(t)\omega_1(t) = \tau_2(t)\omega_2(t) \tag{4.242}$$

Thus, a physically oriented choice of state variables is

$$x_1(t) = \omega_1(t) \tag{4.243}$$

$$x_2(t) = \theta_2(t) - \theta_3(t) \tag{4.244}$$

$$x_3(t) = \omega_3(t) \tag{4.245}$$

From first principles we have

$$\tau(t) = D_1\omega_1(t) + I_1\frac{d\omega_1(t)}{dt} + \tau_1(t) \quad (4.246)$$

$$\tau(t) = \frac{r_2}{r_1}\tau_1(t) = D_2\omega_2(t) + I_2\frac{d\omega_2(t)}{dt} + K_2(\theta_2(t) - \theta_3(t)) \quad (4.247)$$

$$0 = K_2(\theta_3(t) - \theta_2(t)) + I_3\frac{d\omega_3(t)}{dt} \quad (4.248)$$

Since we have chosen $\omega_1(t)$ as the measurable system variable, we finally obtain

$$\frac{d\mathbf{x}(t)}{dt} = \underbrace{\begin{bmatrix} -\frac{r_1^2D_2 + r_2^2D_1}{r_1^2I_2 + r_2^2I_1} & -\frac{r_1r_2K_2}{r_1^2I_2 + r_2^2I_1} & 0 \\ \frac{r_1}{r_2} & 0 & -1 \\ 0 & \frac{K_2}{I_3} & 0 \end{bmatrix}}_{\mathbf{A}} \mathbf{x}(t) + \underbrace{\begin{bmatrix} \frac{r_2^2}{r_1^2I_2 + r_2^2I_1} \\ 0 \\ 0 \end{bmatrix}}_{\mathbf{B}} \tau(t) \quad (4.249)$$

$$\omega_1(t) = \underbrace{[1 \quad 0 \quad 0]}_{\mathbf{C}}\mathbf{x}(t) \quad (4.250)$$

To evaluate the observability properties of this system, numerical values for the parameters are chosen as follows:

$$r_1 = 0.25 \text{ m}, \qquad r_2 = r_3 = 0.50 \text{ m}, \qquad D_1 = D_2 = 10 \text{ Nms/rad} \quad (4.251)$$

$$K_2 = 30 \text{ Nm/rad}, \qquad I_1 = 2.39 \text{ Nms}^2\text{/rad}, \qquad I_2 = I_3 = 38.29 \text{ Nms}^2\text{/rad} \quad (4.252)$$

With these values we have that

$$\mathbf{A} = \begin{bmatrix} -1.045 & -1.254 & 0 \\ 0.5 & 0 & -1 \\ 0 & 0.784 & 0 \end{bmatrix}, \qquad \mathbf{B} = \begin{bmatrix} 0.084 \\ 0 \\ 0 \end{bmatrix} \quad (4.253)$$

We next use the test presented in Section 4.6.2. This yields

$$\Gamma_o = \begin{bmatrix} \mathbf{C} \\ \mathbf{CA} \\ \mathbf{CA}^2 \end{bmatrix} = \begin{bmatrix} 1.0000 & 0 & 0 \\ -1.0450 & -1.2540 & 0 \\ 0.4650 & 1.3104 & 1.2540 \end{bmatrix} \quad (4.254)$$

From this expression we see that Γ_o is a full rank matrix. Thus, the system state is completely observable from $\omega_1(t)$.

Once we have a state estimate, $\hat{\mathbf{x}}(t)$, an estimate, $\hat{\omega}_3(t)$ for ω_3, is obtained from

$$\hat{\omega}_3(t) = \underbrace{[0 \quad 0 \quad 1]}_{\mathbf{K}_3^T}\hat{x}(t) \tag{4.255}$$

where $\hat{\omega}_3(t)$ can be obtained from (4.239). This yields

$$\frac{d\hat{\omega}_3(t)}{dt} = \mathbf{K}_3^T\frac{d\hat{\mathbf{x}}(t)}{dt} = \mathbf{K}_3^T(\mathbf{A}-\mathbf{JC})\hat{\mathbf{x}}(t) + \underbrace{\mathbf{K}_3^T\mathbf{B}}_{0}\,\tau(t) + \mathbf{K}_3^T\mathbf{J}\omega_1(t) \tag{4.256}$$

4.7.3 Observers and Measurement Noise

In the theory above we have assumed that both the system input, $\mathbf{u}(t)$, and the system output, $\mathbf{y}(t)$, are available with no errors. This assumption is usually correct with regard to $\mathbf{u}(t)$, since the same equipment generating $\mathbf{u}(t)$ is normally used to estimate the state. However, that assumption is not usually valid with respect to $\mathbf{y}(t)$, since the measurement of this variable is normally corrupted with noise. To analyze the effect of this error, let us denote by $\mathbf{y}_m(t)$ the noisy measurement, that is, $\mathbf{y}_m(t) = \mathbf{y}(t) + \mathbf{v}(t)$, where $\mathbf{v}(t)$ is the additive measurement noise. Therefore, the state estimation error satisfies

$$\frac{d\tilde{\mathbf{x}}(t)}{dt} = (\mathbf{A}-\mathbf{JC})\tilde{\mathbf{x}}(t) + \mathbf{J}\mathbf{v}(t) \tag{4.257}$$

We then have that

$$\tilde{X}(s) = (s\mathbf{I}-\mathbf{A}+\mathbf{JC})^{-1}\tilde{\mathbf{x}}(0) + (s\mathbf{I}-\mathbf{A}+\mathbf{JC})^{-1}\mathbf{J}\,V(s) \tag{4.258}$$

Hence, the error is small if the transfer function $(s\mathbf{I} - \mathbf{A} + \mathbf{JC})^{-1}\mathbf{J}$ filters out the noise. Consider the following example.

Example 4.22

A system has a state space model given by

$$\mathbf{A} = \begin{bmatrix} -2 & 1 \\ 1 & -3 \end{bmatrix}, \quad \mathbf{B} = \begin{bmatrix} 1 \\ 0.5 \end{bmatrix}, \quad \mathbf{C} = [1 \quad -1], \quad D = 0 \tag{4.259}$$

Assume that we want to estimate a system variable $z(t) = \gamma^T\mathbf{x}(t)$, where $\gamma^T = [1 \quad 1]$. Then, a suitable observer-based estimate is $\hat{z}(t)$, which is given by

$$\hat{z}(t) = \gamma^T\hat{\mathbf{x}}(t) \tag{4.260}$$

Then, the noise term in the estimation of $z(t)$ is $z_v(t)$, whose Laplace transform satisfies

$$Z_v(s) = H_v(s)V(s), \quad \text{where } H_v(s) = \gamma^T(s\mathbf{I}-\mathbf{A}+\mathbf{JC})^{-1}\mathbf{J} \tag{4.261}$$

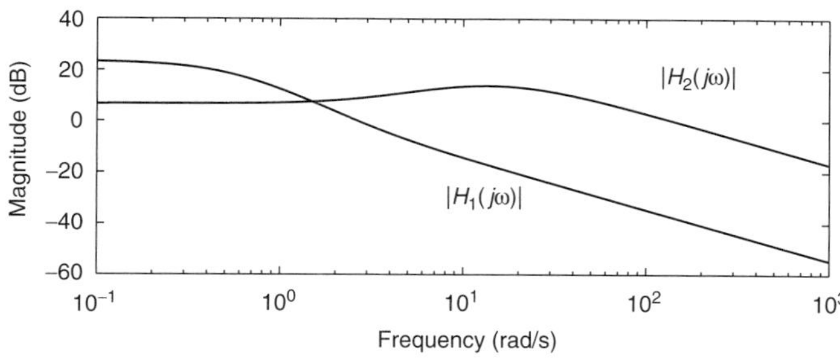

FIGURE 4.18 Observer filtering characteristics.

We next consider two different choices for the observer polynomial $E(s)$. They are

$$E_1(s) = (s+0.5)(s+0.75) \quad \text{and} \quad E_2(s) = (s+10)(s+20) \tag{4.262}$$

The reader can appreciate that the resulting observers will have very different speeds the first observer being much slower than the second one.

With those choices we compute the observer gains, J_1 and J_2, and the corresponding filter functions

$$H_1(s) = \gamma^T(s\mathbf{I} - \mathbf{A} + \mathbf{J}_1\mathbf{C})^{-1}\mathbf{J}_1 = \frac{1.875s + 5.625}{s^2 + 1.25s + 0.375} \tag{4.263}$$

$$H_2(s) = \gamma^T(s\mathbf{I} - \mathbf{A} + \mathbf{J}_2\mathbf{C})^{-1}\mathbf{J}_2 = \frac{144s + 432}{s^2 + 30s + 200} \tag{4.264}$$

To compare both cases we compute and plot the frequency response of each filter. The result is shown in Figure 4.18.

From Figure 4.18 we observe that for a high frequency noise, the slowest filter is more immune to noise than the fast filter.

The above case exemplifies the trade-off between observer speed and noise immunity. A systematic way to face this dilemma is to use an optimal filter theory, such as Kalman–Bucy filtering. The interested reader is referred to [2].

4.8 State Feedback

4.8.1 Basic Concepts

When all the system states can be measured, and the system is completely reachable (in the sense explained in Section 4.6.1), we can control the system using state feedback to achieve full command of the loop dynamics. This idea is captured in Figure 4.19.

Figure 4.19 shows the most basic form of state feedback: the plant input has a component that is proportional to the state (the other component is an external signal $\bar{r}(t)$).

State feedback is a very simple, almost naive idea. A careful analysis shows that this idea has some shortcomings and potentially dangerous features, such as

- It requires as many sensors as state variables. This is not only very expensive but also, in some cases, its implementation may become impossible.
- Each state measurement is a source of error because of its limited accuracy.
- Each measurement introduces noise, which has deleterious effect on the control system performance.

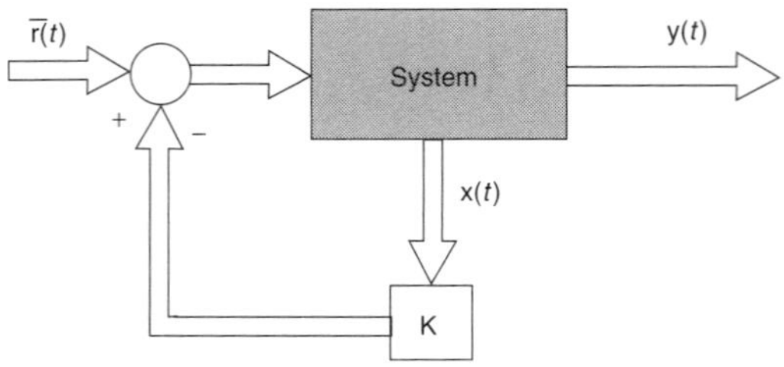

FIGURE 4.19 State feedback.

- The correct overall performance relies on the correct functioning of a complex set of equipments. This poses several questions regarding performance degradation and system integrity.

In spite of these weak points, state feedback is by itself a powerful concept, since it works as a basis for more sophisticated and robust control schemes. The key reason for this is that any linear controller can be explained as the combination of a state observer and state feedback.

4.8.2 Feedback Dynamics

Assume that the system to be controlled has a transfer function $\mathbf{H}(s)$ and a state space representation given by (4.42) and (4.43), with $\mathbf{D} = \mathbf{0}$. If the plant input is generated according to

$$\mathbf{u}(t) = -\mathbf{K}\mathbf{x}(t) + \bar{\mathbf{r}}(t) \tag{4.265}$$

then the state space representation for the complete control loop is given by

$$\frac{d\mathbf{x}(t)}{dt} = \mathbf{A}\mathbf{x}(t) + \mathbf{B}(-\mathbf{K}\mathbf{x}(t) + \bar{\mathbf{r}}(t)) \tag{4.266}$$

$$\mathbf{y}(t) = \mathbf{C}\mathbf{x}(t) \tag{4.267}$$

It can be shown that the relationship between $\bar{\mathbf{R}}(s)$ and $\mathbf{Y}(s)$ is given by

$$\mathbf{Y}(s) = \underbrace{\mathbf{C}(s\mathbf{I} - \mathbf{A})^{-1}\mathbf{B}}_{\mathbf{H}(s)}(\mathbf{I} + \mathbf{K}(s\mathbf{I} - \mathbf{A})^{-1}\mathbf{B})^{-1}\bar{\mathbf{R}}(s) \tag{4.268}$$

This shows that the state feedback loop preserves the system zeros and shifts the poles to the roots of det $(s\mathbf{I} - \mathbf{A} + \mathbf{B}\mathbf{K})$.

4.8.3 Optimal State Feedback—The Optimal Regulator

Consider a linear time invariant system having a state space representation given by (4.42) and (4.43), with $\mathbf{D} = \mathbf{0}$, subject to the initial state $\mathbf{x}(0) = \mathbf{x}_o$.

Assume that the control objective is to steer the plant from the the initial state, $\mathbf{x}_o$, to the smallest possible value as soon as possible in the interval $[0, t_f]$. We additionally require that the steering process does not demand too much control effort. Then, the optimal regulator problem is defined as the problem of finding an optimal control $\mathbf{u}(t)$ over the interval $[0, t_f]$ such that a quadratic cost function is minimized. This cost function is chosen as

$$J_u(\mathbf{x}_o) = \int_0^{t_f} [\mathbf{x}(t)^T\mathbf{Q}x(t) + \mathbf{u}(t)^T\mathbf{R}\mathbf{u}(t)]\,dt + \mathbf{x}(t_f)^T\mathbf{Q}_f\mathbf{x}(t_f) \tag{4.269}$$

where $\mathbf{Q} \in \mathbb{R}^{n \times n}$, $\mathbf{Q}_f \in \mathbb{R}^{n \times n}$ are symmetric nonnegative definite matrices and $\mathbf{R} \in \mathbb{R}^{m \times m}$ is a symmetric positive definite matrix. The requirements on the weighting matrices are set so that the cost function makes sense. For instance, if **Q** is allowed to be negative, then the *optimal* cost could even be negative while the state could grow unbounded in magnitude. Also, if we allow **R** to have eigenvalues at the origin (that is, **R** is allowed to be a nonnegative definite matrix, instead of requiring it to be a strictly positive definite matrix) then the control $\mathbf{u}(t)$ could also grow unbounded (in the directions of the associated eigenvectors) without that situation being revealed by the cost function.

A time invariant linear control law is asymptotically obtained when $t_f \rightarrow \infty$. Under this condition, the optimal control law is given by

$$\mathbf{u}^{o}(t) = -\mathbf{K}^{o}x(t) \tag{4.270}$$

with

$$\mathbf{K}^{o} = -\mathbf{R}^{-1}\mathbf{B}^{T}\mathbf{P}_{\infty} \tag{4.271}$$

and where $\mathbf{P}_{\infty}$ is the only nonnegative solution of the algebraic Riccati equation

$$0 = \mathbf{Q} - \mathbf{P}_{\infty}\mathbf{B}\mathbf{R}^{-1}\mathbf{B}^{T}\mathbf{P}_{\infty} + \mathbf{P}_{\infty}\mathbf{A} + \mathbf{A}^{T}\mathbf{P}_{\infty} \tag{4.272}$$

For this solution to exist, it is necessary that certain technical conditions are satisfied (for a detailed discussion of these issues see, for instance, [5]).

Discussion

- The solution for the LQR problem minimizes the cost function (4.269) and, when $t_f \rightarrow \infty$, always stabilizes the plant.
- A key issue is how to choose the weighting matrices Q and R. A frequent choice for Q is $\mathbf{Q} = \mathbf{C}^T\mathbf{C}$. With this choice, the magnitude of the plant output is directly introduced into the cost function.
- For a given Q, the *size* of R strongly influences the location of the closed loop poles. The larger R is, the slower is the control loop.

Further reading on optimal quadratic regulators can be found in the literature. See, for example, [1,3,4,8,9].

4.9 Observed State Feedback

4.9.1 Separation Strategy

Due to the drawbacks inherent in the measuring of the state, feedback of the estimated state can be used instead. The resulting control system integrates an observer and a feedback mechanism for the observed states.

The combination of a state observer and the feedback of the estimated state conform the structure shown in Figure 4.20.

In Figure 4.20, the (matrix) transfer functions $\mathbf{T}_1(s)$ and $\mathbf{T}_2(s)$ can be obtained from Figure 4.15. This yields

$$\mathbf{T}_1(s) = (s\mathbf{I} - \mathbf{A}_o + \mathbf{J}\mathbf{C}_o)^{-1}\mathbf{B}_o \tag{4.273}$$

$$\mathbf{T}_2(s) = (s\mathbf{I} - \mathbf{A}_o + \mathbf{J}\mathbf{C}_o)^{-1}\mathbf{J} \tag{4.274}$$

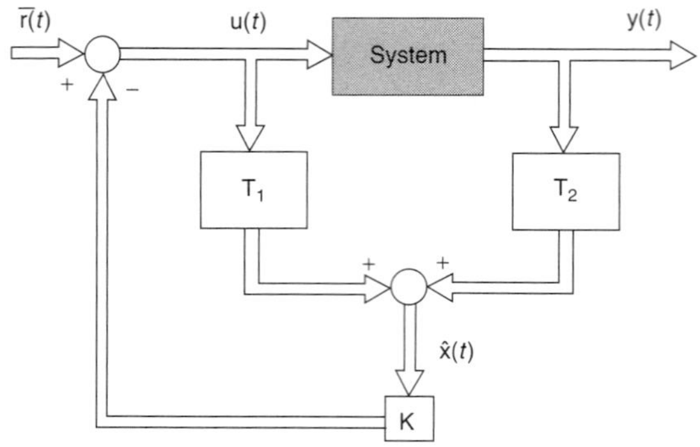

FIGURE 4.20 Estimated state feedback.

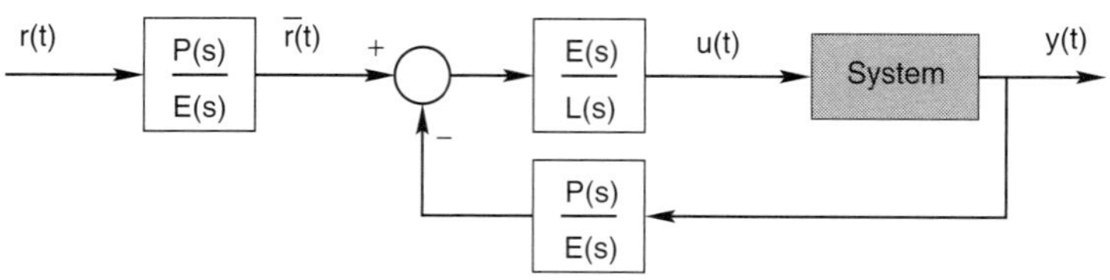

FIGURE 4.21 Equivalent control loop.

4.9.2 Transfer Function Interpretation for the Single-Input Single-Output Case

Consider a SISO plant having transfer function

$$G_o(s) = \mathbf{C}(s\mathbf{I} - \mathbf{A}_o)^{-1}\mathbf{B} = \frac{N_o(s)}{M_o(s)} \tag{4.275}$$

where $M_o(s)$ and $N_o(s)$ are polynomials in s.

First, a state feedback gain, **K**, is chosen to obtain a closed loop polynomial $F(s)$, where $F(s) = \det(s\mathbf{I} - \mathbf{A}_o + \mathbf{B}_o\mathbf{K})$. Next, an observer gain, **J**, is computed to obtain an observer polynomial $E(s) = \det(s\mathbf{I} - \mathbf{A}_o + \mathbf{J}\mathbf{C}_o)$.

If the observer and the observed state feedback are combined, the resulting control loop can be made equivalent (by a suitable choice of $\bar{r}(t)$) to the classical control loop shown in Figure 4.21.

In Figure 4.21 the polynomials $P(s)$ and $L(s)$ satisfy the Diophantine equation

$$M_o(s)L(s) + N_o(s)P(s) = E(s)F(s) \tag{4.276}$$

This result says that the set of closed loop poles is the union of the set of observer poles and the set of state feedback poles.

References

1. Anderson, B.D.O. and Moore, J., *Linear optimal Control.* Prentice-Hall, Englewood Cliffs, NJ, 1971.
2. Anderson, B.D.O. and Moore, J., *Optimal Filtering.* Prentice-Hall, Englewood Cliffs, NJ, 1979.
3. Athans, M. and Falb, P., *Optimal Control.* McGraw HIll, 1966.
4. Dennis Bernstein and Wassim Haddad. LQG control with an H_∞ performance bound: A Riccati equation approach. *IEEE Transactions on Automatic Control,* 34(3): L293–305, 1989.
5. Bittanti, S., Laub, A.J., and Willems, J.C., *The Riccati Equation.* Springer Verlag, Berlin, 1996.

6. Dorf, R.C. and Bishop, R., *Modern Control Systems.* Prentice-Hall, Englewood Cliffs, NJ, 1997.
7. Franklin, G.F. and Powel, J.D., *Digital Control of Dynamics Systems.* 2nd ed., Addison-Wesley, 1990.
8. Goodwin, G.C., Graebe, S., and Salgado, M.E., *Control System Design.* Prentice-Hall, NJ, 2001.
9. Kwakernaak, H. and Sivan, R., *Linear Optimal Control System.* Wiley–Interscience, New York, 1972.
10. Ogata, K., *State Space Analysis of Control Systems.* Prentice-Hall, Englewood Cliffis, NJ, 1967.
11. Rosenbrock, H.H., *State Space and Multivariable Theory.* John Wiley and Sons, New York, 1970.
12. Schultz, D.G. and Melsa, J.L., *State Function and Linear Control Systems.* McGraw Hill, New York, 1967.
13. Wiberg, D.W., *Theory and Problems of State Space and Linear Systems.* McGraw Hill, New York, 1971.
14. Zadeh, L.A. and Desoer, C.A., *Linear System Theory: The State Space Approach.* McGraw Hill, New York, 1963.
15. Zhou, K., *Essentials of Robust Control.* Prentice-Hall, Englewood Cliffs, NJ, 1998.
16. Zhou, K., Salomon, G., and Wu, E., Balanced realization and model reduction for unstable systems. *International Journal of Robust and Nonlinear Control,* 9:183–198, 1999.

5

Response of Dynamic Systems

Raymond A. de Callafon
University of California

5.1 System and Signal Analysis

In dynamic system design and analysis it is important to predict and understand the dynamic behavior of the system. Examining the dynamic behavior can be done by using a mathematical model that describes the relevant dynamic behavior of the system in which we are interested. Typically, a model is formulated to describe either continuous or discrete time behavior of a system. The corresponding equations that govern the model are used to predict and understand the dynamic behavior of the system.

A rigorous analysis can be done for relatively simple models of a dynamic system by actually computing solutions to the equations of the model. Usually, this analysis is limited to linear first and second order models. Although limited to small order models, the solutions tend to give insight in the typical responses of a dynamic system. For more complicated, higher order and possibly nonlinear models, numerical simulation tools provide an alternative for the dynamic system analysis.

In the following we review the analysis of linear models of discrete and continuous time dynamic systems. The equations that describe and relate continuous and discrete time behavior are presented. For the analysis of continuous time systems extensive use is made of the Laplace transform that converts linear differential equations into algebraic expressions. For similar purposes, a z-transform is used for discrete time systems.

5.1.1 Continuous Time Systems

Models that describe the linear continuous time dynamical behavior of a system are usually given in the form of differential equations that relate an input signal $u(t)$ to an output signal $y(t)$. The differential equation of a time invariant linear continuous time model has the general format

$$\sum_{j=0}^{n_a} a_j \frac{d^j}{dt^j} y(t) = \sum_{j=0}^{n_b} b_j \frac{d^k}{dt^j} u(t) \tag{5.1}$$

in which a linear combination is taken using the jth order time derivatives d^j/dt^j of a single output $y(t)$ and a single input $u(t)$. In (5.1), the scalar real valued numbers a_j for $j = 0,\ldots n_a$, $a_{n_a} \neq 0$ and b_j for $j = 0,\ldots, n_b$, $b_{n_b} \neq 0$, respectively, are called the denominator and numerator coefficients. The input $u(t)$ is distinguished from the output $y(t)$ in (5.1) by requiring $n_a \geq n_b$. As a result, the n_ath derivative is the highest derivative of the output $y(t)$ and n_a is used to indicate the order of the differential equation.

An alternative representation of a model of a continuous time system can be obtained by rewriting the n_ath order differential equation in (5.1) into a set of (coupled) first order differential equations. This can be done by introducing a state variable $x(t)$ and rewriting the higher order differential equation into

$$\begin{aligned} \frac{d}{dt}x(t) &= Ax(t) + Bu(t) \\ y(t) &= Cx(t) + Du(t) \end{aligned} \tag{5.2}$$

where A, B, C, and D are real valued matrices. The set of first order differential equations given in (5.2) is referred to as a state space representation. The state variable $x(t)$ is a column vector and contains n_a variables, where n_a is the order of the differential equation.

The size of the matrices in (5.2) corresponds to the order of differential equation from which the state space realization is derived. For generalization purposes, consider multiple inputs and outputs rearranged in $m \times 1$ input column vector $u(t)$ and a $p \times 1$ output column vector $y(t)$. Given the $n_a \times 1$ size of the state vector, the state matrix A has size $n_a \times n_a$, the input matrix has size $n_a \times m$, the output matrix C has size $p \times n_a$, and the feedthrough matrix D has size $m \times p$. From these size considerations it can be observed that the state space realization in (5.2) easily generalizes the model description of multi-input multi-output systems.

To illustrate the concepts, consider the differential equation

$$m\frac{d^2}{dt^2}y(t) + c\frac{d}{dt}y(t) + ky(t) = u(t) \tag{5.3}$$

that describes the dynamical behavior of the one cart system given in Figure 5.1. The differential Equation 5.3 is found by writing Newton's second law for the cart mass m with position output $y(t)$, spring force $ky(t)$, damper force $c(d/dt)y(t)$, and force input $u(t)$. Comparing with (5.1) it can be seen that $n_a = 2 \geq n_b = 0$, making (5.3) a second order differential equation. The differential equation can be rewritten into a state space representation (5.2) by defining the state variable

$$x(t) := \begin{bmatrix} y(t) \\ \frac{d}{dt}y(t) \end{bmatrix}$$

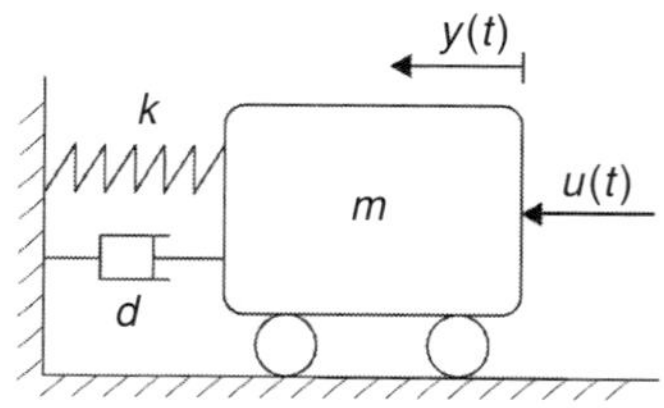

FIGURE 5.1 One cart system representing a single mass dynamical system with cart mass m, spring constant k, and damping constant c.

that consists the position and velocity of the mass. With this state variable (5.3) can be rewritten into

$$\frac{d}{dt}x(t) = \begin{bmatrix} 0 & 1 \\ -\frac{k}{m} & -\frac{d}{m} \end{bmatrix} x(t) + \begin{bmatrix} 0 \\ \frac{1}{m} \end{bmatrix} u(t)$$

$$y(t) = \begin{bmatrix} 1 & 0 \end{bmatrix} x(t) + 0u(t)$$

which yields a state space model similar to (5.2). In this case, the size of the state matrix A is 2×2, the input matrix B is 2×1, the output matrix C is 1×2, and the feedthrough matrix $D = 0$ is scalar.

5.1.2 Discrete Time Systems

Discrete time models approximate and describe the sampled data behavior of a continuous time dynamical system. In some applications, such as digital control, the dynamical control system is inherently discrete time. In these situations, analysis with discrete time equivalent models in necessary.

For analysis purposes, both input $u(t)$ and output $y(t)$ are assumed to be sampled on a regular discrete time interval

$$t = k\Delta T, \qquad k = 0,1,2,\ldots$$

where ΔT indicates the sampling time. To maintain uniform notation throughout the analysis, the sampling time ΔT is normalized to $\Delta T = 1$ and the time dependency t is assumed to be discrete with $t = k = 0,1,2,\ldots$.

Given sampled or discrete time input/output data, a linear discrete time model can be formulated in the form of a difference equation

$$\sum_{i=0}^{n_c} c_k y(k+j) = \sum_{i=0}^{n_d} d_k u(k+j) \tag{5.4}$$

in which a linear combination is taken of positive time shifted inputs $u(k)$ and outputs $y(k)$. To distinguish the differential equation from the differential Equation 5.1, different scalar real valued numbers c_j for $j = 0,\ldots,n_c$, $c_{n_c} \neq 0$ and d_j for $j = 0,\ldots,n_d$, $d_{n_d} \neq 0$ are used. The input $u(k)$ is distinguished from the output $y(k)$ in (5.1) by requiring $n_c \geq n_d$ for causality purposes. As a result, the n_c is the largest time shift of the output $y(k)$ and n_c is used to indicate the order of the difference equation.

The simplicity with which the difference equation can be represented also allows an algebraic representation of (5.4). Introducing the time shift operator

$$qu(k) := u(k+1) \tag{5.5}$$

allows (5.4) to be rewritten into the algebraic expression

$$y(k)\sum_{i=0}^{n_c} c_j q^j = u(k)\sum_{i=0}^{n_d} d_j q^j$$

Following this analysis, the discrete time output $y(k)$ can be represented by the difference model

$$y(k) = G(q)u(k), \qquad \text{with } G(q) = \frac{\sum_{j=0}^{n_d} d_j q^j}{\sum_{j=0}^{n_c} c_j q^j} \tag{5.6}$$

where the scalar real valued numbers c_j for $j = 0,\ldots,n_c$, $c_{n_c} \neq 0$ and d_j for $j = 0,\ldots,n_d$, $d_{n_d} \neq 0$, respectively, indicate the denominator and numerator coefficients.

Similar to the continuous time system representation, the higher order difference Equation 5.4 can also be rewritten into a set of (coupled) first order difference equations for analysis purposes. This can be done by introducing a state variable $x(k)$ and rewriting the higher order difference equation into

$$\begin{aligned} qx(k) &= Fx(k) + Gu(k) \\ y(k) &= Hx(k) + Ju(k) \end{aligned} \tag{5.7}$$

where $qx(k) = x(k + 1)$, according to (5.5). The state variable $x(k)$ is a column vector and contains n_c variables, where n_c is the order of the difference equation. The state space matrices in (5.7) are labeled differently to distinguish them from the continuous time state space model.

5.1.3 Laplace and *z* Transform

An important mathematical concept for the analysis of models described by linear differential equations such as (5.1) and (5.2) is the Laplace transform. As indicated before, the Laplace transform converts linear differential equations into algebraic expressions. With this conversion, proper algebraic manipulation can be used to recover solutions of the differential equation. In a similar manner, the z transform is used for discrete time models described by difference equations. Although it was shown in (5.6) that a difference equation can be written as an algebraic expression, the z transform allows complex analysis of the discrete time models.

The Laplace transform of a signal $u(t)$ is defined to be

$$L\{u(t)\} := u(s) = \int_{t=0}^{\infty} u(t)e^{-st}\,dt \tag{5.8}$$

where the integration over t eliminates the time dependency and the transform $u(s)$ is a function of the Laplace variable only. This is indicated in the transform $u(s)$ where the dependency of t has been dropped, and $u(s)$ is a function of the (complex valued) Laplace variable s only.

The integral (5.8) exists for most commonly used signals $u(t)$, provided certain conditions on s are imposed. To illustrate the transform, consider a (unity) step signal

$$u(t) := \begin{cases} 0, & t < 0 \\ 1, & t \geq 0 \end{cases}$$

where the shape of $u(t)$ resembles a stepwise change of an input signal. With the definition of the Laplace transform in (5.8) the transform of the step signal becomes

$$u(s) = \int_{t=0}^{\infty} u(t)e^{-st}\,dt = \int_{t=0}^{\infty} e^{-st}\,dt = -\frac{e^{-st}}{s}\bigg|_0^{\infty} = \frac{1}{s} \tag{5.9}$$

where it is assumed that the real part of s is greater than zero so that $\lim_{t\to\infty} e^{-st} = 0$.

If a signal $u(k)$ is given at discrete time samples $k = 0, 1, 2\ldots$, the integral expression of (5.8) cannot be applied. Instead, a transform similar to the Laplace transform can be used and denoted by the z transform. The z-transform of a discrete time signal $u(k)$ is defined as

$$L\{u(k)\} := u(z) = \sum_{k=0}^{\infty} u(k)z^{-k} \tag{5.10}$$

The series (5.10) converges if it is assumed that there exist values r_l and r_u with $r_l < |z < r_u$ as bounds on the magnitude of the complex variable z.

The z transform has the same role in discrete time systems that the Laplace transform has in continuous time systems. In case of sampling, the complex variable z of the z transform is related to the complex variables s in the Laplace transform via

$$z = e^{s\Delta T} \tag{5.11}$$

where ΔT is the sampling time used for sampling. Both the Laplace and z-transform are linear operators and satisfy

$$L\{\alpha u(t) + \beta y(t)\} = \alpha L\{u(t)\} + \beta\{y(t)\} \tag{5.12}$$

Using the definition in (5.8) and the linearity property in (5.12), the transform of most commonly used functions has been precalculated and tabulated.

Of particular interest for the analysis of linear differential equations such as (5.1) and (5.2) is the Laplace transform of a derivative:

$$\begin{aligned} L\left\{\frac{d}{dt}u(t)\right\} &= \int_{t=0}^{\infty}\frac{d}{dt}u(t)e^{-st}\,dt \\ &= u(t)e^{-st}\Big|_0^{\infty} + s\int_{t=0}^{\infty}u(t)e^{-st}\,dt \\ &= su(s) - u(0) \end{aligned}$$

With $u(0) = 0$ it can be seen that the Laplace transform of the derivative of $u(t)$ is simply s times the Laplace transform of $u(s)$. This result can be extended to higher order derivatives and the result for the nth derivative is given by

$$L\left\{\frac{d^n}{dt^n}u(t)\right\} = s^n u(s) - \sum_{j=1}^{n} s^{n-j}\frac{d^{j-1}}{dt^{j-1}}u(t)\Big|_{t=0}$$

In case the signal $u(t)$ satisfies the initial zero conditions

$$\frac{d^{j-1}}{dt^{j-1}}u(t)\Big|_{t=0} = 0 \quad \text{for } j = 1,\ldots,n$$

the formula reduces to

$$L\left\{\frac{d^n}{dt^n}u(t)\right\} = s^n u(s)$$

and the Laplace transform of an nth order derivative is simply s^n times the transform $u(s)$.

For discrete time systems the interest lies in the z-transform of a time-shifted signal. Similar to the Laplace transform, the z transform of an n time-shifted signal can be computed and is given by

$$L\{q^n u(k)\} = z^n u(z) - \sum_{j=0}^{n-1} z^{n-j} u(j)$$

In case the discrete time signal $u(k)$ satisfies the initial zero conditions $u(j) = 0$ for $j = 0,\ldots,n-1$, the formula reduces to

$$L\{q^n u(k)\} = z^n u(z)$$

and the z transform of an n time-shifted discrete time signal is simply z^n times the transform $u(z)$.

5.1.4 Transfer Function Models

The results of the Laplace and z-transform can be used to reduce linear differential Equations 5.1 and difference Equation 5.4 to the algebraic expressions. Starting with the differential equations for continuous time models and assuming zero initial conditions for both the input $u(t)$ and output signal $y(t)$, the Laplace transform of (5.1) yields

$$y(s)\sum_{j=0}^{n_a} a_j s^j = u(s)\sum_{j=0}^{n_b} b_j s^j$$

which can be written in transfer function format

$$y(s) = G(s)u(s), \quad \text{with } G(s) = \frac{\sum_{j=0}^{n_b} b_j s^j}{\sum_{j=0}^{n_a} a_j s^j} \tag{5.13}$$

In (5.13), the transfer function $G(s)$ is the ratio of the numerator polynomial $\sum_{j=0}^{n_b} b_j s^j$ and the denominator polynomial $\sum_{j=0}^{n_a} a_j s^j$. As indicated before, the scalar real valued numbers a_j for $j = 0,\ldots,n_a$, $a_{n_a} \neq 0$ and b_j for $j = 0, \ldots, n_b$, $b_{n_b} \neq 0$, respectively, are called the denominator and numerator coefficients.

Similarly for the discrete time model, assuming zero initial conditions for both the input $u(k)$ and output signal $y(k)$, the z-transform of (5.4) yields

$$y(z)\sum_{j=0}^{n_c} c_j z^j = u(z)\sum_{j=0}^{n_d} b_j z^j$$

which can be written in transfer function format

$$y(z) = G(z)u(z), \quad \text{with } G(z) = \frac{\sum_{j=0}^{n_c} c_j z^j}{\sum_{j=0}^{n_a} a_j z^j} \tag{5.14}$$

From the transfer function representations, poles and zeros of the dynamic system can be computed for dynamic system analysis. The poles of the system are defined as the roots of the denominator polynomial. The zeros of the system are defined as the roots of the numerator polynomial.

The Laplace and z-transform can also be used to reduce the state space representation to a set of algebraic expressions that consists of (coupled) first order polynomials. Assuming zero initial conditions for the state vector $x(t)$, application of the Laplace transform to (5.2) yields

$$\begin{aligned} sx(s) &= Ax(s) + Bu(s) \\ y(s) &= Cx(s) + Du(s) \end{aligned}$$

in which the state vector $x(s)$ can be eliminated. Solving for $x(s)$ gives $x(s) = (sI - A)^{-1} Bu(s)$ and the above transform can be rewritten into a transfer function representation

$$y(s) = G(s)u(s), \quad \text{with } G(s) = D + C(sI - A)^{-1}B \tag{5.15}$$

Under mild technical conditions involving controllability and observability of the state space model, the transfer function representations in (5.13) and (5.15) are similar in case the state space model in (5.2) is derived from the differential Equation 5.1 and vice versa.

5.2 Dynamic Response

The Laplace and z transform offer the possibility to compute the dynamic response of a dynamic system by means of algebraic manipulations. The analysis of the dynamic response gives insight into the dynamic behavior of the system by addressing the response to typical test signals such as impulse, step, and sinusoid excitation of the system.

The response can be computed for relatively simple continuous or discrete dynamical systems given by low order differential or difference equations. Both the state space model and the transfer function descriptions provide helpful representations in the analysis of a dynamic system. The result are presented in the following.

5.2.1 Pulse and Step Response

A possible way to evaluate the response of a dynamic system is by means of pulse and step based test signals. For continuous time systems an input impulse signal is defined as a δ function

$$u_{\text{imp}}(t) := \delta(t) = \begin{cases} \infty, & t = 0 \\ 0, & t \neq 0 \end{cases}$$

with the property

$$\int_{t=-\infty}^{\infty} f(t)\delta(t) = f(0)$$

where $f(t)$ is an integrable function over $(-\infty, \infty)$. Although an impulse signal is not practical from an experiment point of view, the computation or simulation of the impulse response gives insight into the transient behavior of the dynamical system.

With the properties of the impulse function $\delta(t)$ mentioned above, the Laplace transform of the impulse function is given by

$$L\{\delta(t)\} = \delta(s) = \int_{t=0}^{\infty} \delta(t)e^{-st}\,dt = e^{-s0} = 1$$

Hence the output $y(s)$ due to an impulse input is given by $y_{\text{imp}}(s) = G(s)u_{\text{imp}}(s) = G(s)\delta(s) = G(s)$. As a result, an immediate inverse Laplace transform of the continuous time transfer function $G(s)$,

$$y_{\text{imp}}(t) = L^{-1}\{G(s)\}$$

gives the dynamic response $y_{\text{imp}}(t)$ of the system to an impulse input response.

The computation of the step response is done in a similar way. In (5.9), the Laplace transform of the step signal

$$u_{\text{step}}(t) := \begin{cases} 0, & t < 0 \\ 1, & t \geq 0 \end{cases}$$

is given as $u_{\text{step}}(s) = 1/s$. Consequently, with $y_{\text{step}}(s) = G(s)u_{\text{step}}(s) = G(s)/s$, the inverse Laplace transform of $G(s)/s$

$$y_{\text{step}}(t) = L^{-1}\left\{\frac{G(s)}{s}\right\}$$

will yield the dynamic response $y_{\text{step}}(t)$ of the system to a step input response.

From a practical point of view, the computation of an inverse Laplace transform is limited to low order models of first or second order. However, the results give insight into the dominant behavior of most dynamic systems. This is illustrated in the following examples.

- Consider a first order continuous model given by the transfer function

  $$G(s) = \frac{K}{\tau s + 1}$$

  where K and τ indicate, respectively, the static gain and the time constant of the system. Such a transfer function may arise from a simple RC network with $\tau = \text{RC}$. In order to compute the step response of the system, the inverse Laplace transform of $G(s)/s$ needs to be computed. This inverse Laplace transform is given by

  $$y_{\text{step}}(t) = L^{-1}\left\{\frac{G(s)}{s}\right\} = \frac{K}{\tau}(1 - e^{-t/\tau})$$

  and it can be seen that the step response is an exponential function. For stability the time constant τ needs to satisfy $\tau > 0$. It can also be observed that the smaller the time constant, the faster the response.
- Consider a second order continuous time model given by the transfer function

  $$G(s) = \frac{\omega_n^2}{s^2 + 2\beta\omega_n s + \omega_n^2} \tag{5.16}$$

  where ω_n and β, respectively, indicate the undamped resonance frequency and the damping coefficient of the system. This model can be derived from the dynamical behavior of the one cart system depicted in Figure 5.1 and given in (5.3). For $\beta < 1$ (underdamped), the inverse Laplace transform of $G(s)$ is given by

  $$y_{\text{imp}}(t) = \frac{\omega_m}{\sqrt{1-\beta^2}} e^{-\beta\omega_n t} \sin\left(\omega_n\sqrt{1-\beta^2}\, t\right)$$

  From this expression it can be observed that the response is a decaying sinusoid with a resonance frequency of $\omega_n\sqrt{1-\beta^2}$. For stability, both $\omega_n > 0$ and $\beta > 0$ and the larger ω_n, the faster the decay of the sinusoid and the higher is the frequency of the response $y_{\text{imp}}(t)$. Illustration of the impulse response of this second order system have been depicted in Figures 5.2 and 5.3 where variations in the undamped resonance frequency ω_n and the damping coefficient β illustrate the dynamic behavior of the system.

For discrete systems, the analysis of the pulse response is based on the discrete time pulse function

$$u_{\text{imp}}(k) := \delta(k) = \begin{cases} 1, & k = 0 \\ 0, & k \neq 0 \end{cases}$$

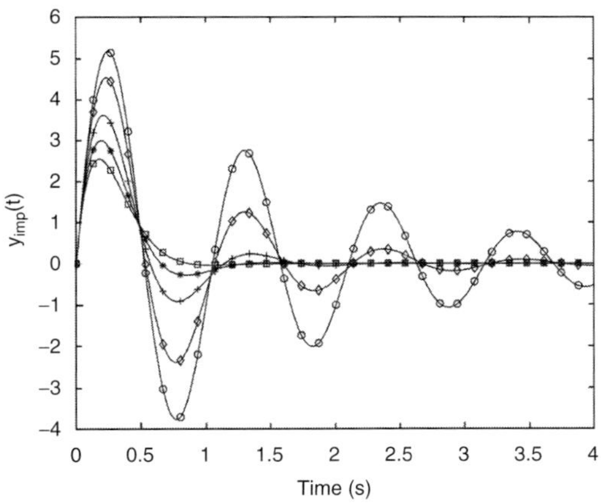

FIGURE 5.2 Variations in impulse response $y_{\mathrm{imp}}(t)$ of second order system with $\omega_n = 6$ and $\beta = 0.1(\bigcirc)$, $0.2(\diamond)$, $0.4(+)$, $0.6\ (*)$, $0.8(\square)$.

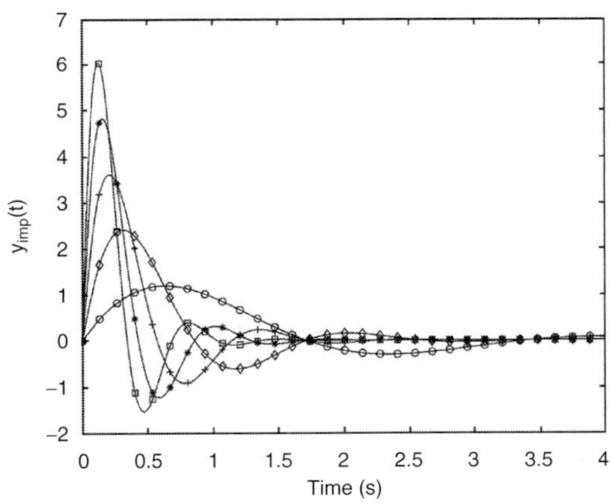

FIGURE 5.3 Variations in impulse response $y_{\mathrm{imp}}(t)$ of second order system with $\beta = 0.4$ and $\omega_n = 2(\bigcirc)$, $4(\diamond)$, $6(+)$, $8(*)$, $10(\square)$.

which has a value of 1 at $k = 0$ and zero anywhere else. The step signal is similar to the continuous time signal and is given by

$$u_{\mathrm{step}}(k) := \begin{cases} 0, & k < 0 \\ 1, & k \geq 1 \end{cases}$$

In order to characterize the discrete time pulse and step response a similar procedure as for the continuous time model can be followed by using the z transform. It is easy to show that the z-transform $u_{\mathrm{imp}}(z) = 1$ and the z transform of the step signal equals $u_{\mathrm{step}} z = z/(z-1)$. Hence, the response of the discrete time system to a pulse or step signal can be computed with

$$y_{\mathrm{imp}}(k) = L^{-1}\{G(z)\}, \qquad y_{\mathrm{step}}(k) = L^{-1}\left\{\frac{G(z)z}{z-1}\right\}$$

In addition to the approach using a z-transform, the ratio of the polynomials in the difference model (5.6) can be written in a series expansion:

$$G(q) = \frac{\sum_{j=0}^{n_d} d_j q^j}{\sum_{j=0}^{n_c} c_j q^j} = \sum_{j=0}^{\infty} g_k q^{-k}$$

With the discrete time pulse function $u_{\text{imp}}(k)$ as an input, it can be observed that

$$y_{\text{imp}}(k) = \sum_{j=0}^{\infty} g_k q^{-k} \delta(k) = g_k$$

and it can be concluded that the pulse response $y_{\text{imp}}(k)$ equals the coefficients in the series expansion of the difference equation. Similarly, with the discrete time step function $u_{\text{step}}(k)$ as an input, it can be observed that

$$y_{\text{imp}}(k) = \sum_{j=0}^{\infty} g_k q^{-k} u_{\text{step}}(k) = \sum_{j=0}^{k} g_k$$

and it can be concluded that the step response $y_{\text{step}}(k)$ values are computed as a finite sum of the coefficients in the series expansion of the difference equation. The computation of a discrete time pulse response for a first order discrete time model is given in the following example.

- Consider a first order discrete model given by the difference model

  $$G(q) = \frac{1}{q+d}$$

  where d indicates the discrete time constant of the system. The series expansion of the difference model can be computed as follows:

  $$G(q) = \frac{1}{q-d} = \sum_{j=0}^{\infty} d^j$$

  and it can be seen that the discrete time pulse response

  $$y_{\text{imp}}(k) = d^k$$

  is an exponential function. For stability the discrete constant d needs to satisfy $|d| < 1$. Similar as in the continuous time model it can be observed that the smaller the time constant, the faster the response. Additionally, the first order discrete time model may exhibit an oscillation in case $-1 < d < 0$.

5.2.2 Sinusoid and Frequency Response

So far we have considered transient effects caused by step, pulse, and impulse inputs to investigate the dynamic properties of a dynamical system. However, periodic inputs occur frequently in practical situations and the analysis of a dynamic system to periodic inputs and especially sinusoidal inputs can yield more insight into the behavior of the system.

The response of a linear system to a sinusoidal input is referred to as the frequency response of the system. An input signal, $u(t) = U\sin\omega t$, that is, a sine wave with amplitude U and frequency ω_j, has a Laplace transform

$$u(s) = \frac{U\omega}{s^2 + \omega^2}.$$

Consequently, the response of the system is given by

$$y(s) = G(s)\frac{U\omega}{s^2 + \omega^2}$$

and a partial fraction expansion of $y(s)$ will result in terms that represent the (stable) transient behavior of $y(s)$ and the term associated to the sinusoidal input $u(s)$. Elimination of the transient effects and performing an inverse Laplace transform will yield a periodic time response $y(t)$ of the same frequency ω_j given by

$$y(t) = AU\sin(\omega t + \phi)$$

where the amplitude magnification A and the phase shift ϕ are given by

$$A = G(s)|_{s=i\omega}, \qquad \phi = \angle G(s)|_{s=i\omega} \tag{5.17}$$

By evaluating the transfer function $G(s)$ along the imaginary axis $s = i\omega$, $\omega \geq 0$, the magnitude $|G(i\omega)|$ gives information on the relative amplification of the sinusoidal input, whereas the phase $\angle G(i\omega)$ gives information on the relative phase shift between input and output.

This analysis can be easily extended to discrete time systems by employing the relation between the Laplace variable s and the z-transform variable in (5.11) to obtain the discrete time sinusoidal response

$$y(k) = AU\sin(\omega k + \phi)$$

where the amplitude magnification A and the phase shift ϕ are given by

$$A = G(z)|_{z=e^{i\Delta T\omega}}, \qquad \phi = \angle G(z)|_{z=e^{i\Delta T\omega}} \tag{5.18}$$

Due to the sampling nature of the discrete time system, the transfer function $G(z)$ is now evaluated on the unit circle

$$e^{i\Delta T\omega}, \quad 0 \leq \omega < \frac{\pi}{\Delta T}$$

to attain information of the magnitude and phase shift of the sinusoidal response.

Plotting the frequency response of a dynamical system gives insight in the pole locations (resonance modes) and zero locations of the dynamical system. As an example, the frequency response of the second order system given in (5.16) has been depicted in Figure 5.4. It can be seen from the figure that, as expected, the second order system is less damped for smaller damping coefficients β and this results in a larger amplitude response of the second order system at the resonance frequency $\omega_n = 6$ rad/s. It can also be observed that the phase change at the resonance frequency becomes more abrupt for smaller damping coefficients.

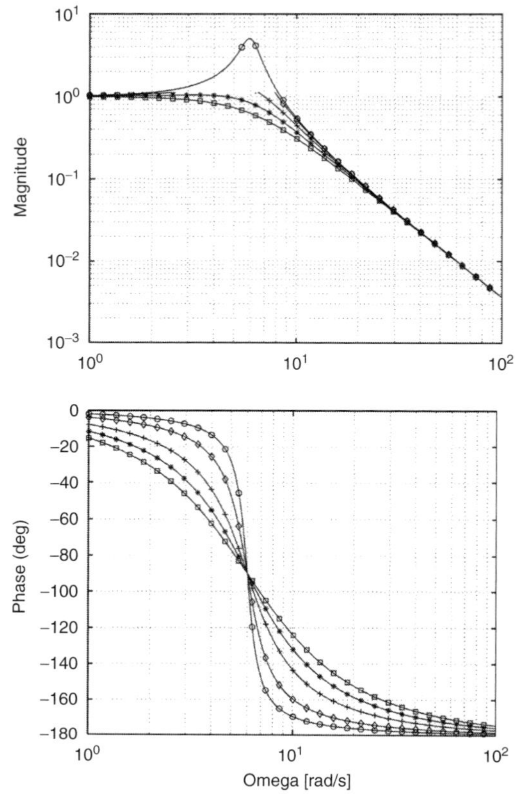

FIGURE 5.4 Variations in frequency response of second order system $G(s)$ with $\omega_n = 6$ and $\beta = 0.1(\bigcirc), 0.2(\diamond)$, $0.4(+), 0.6(*), 0.8(\square)$.

5.3 Performance Indicators for Dynamic Systems

5.3.1 Step Response Parameters

Specifications for dynamic systems often involve requirements on the transient behavior of the system. Transient behavior requirements can be formulated on the basis of a step response and the most significant parameters have been summarized below and illustrated in Figure 5.5.

- Steady state or DC value y_s of step response output.
- The steady state error y_{se} is the error between steady state value y_s and desired DC value of step response output.
- The maximum overshoot A_m is the maximum deviation of the step response output above its steady state value y_s.
- The peak time t_p is the time at which the maximum overshoot occurs.
- Settling time t_s is the time at which the step response input stays within some small percentage rage of the steady state value y_s. Typically, a percentage of 2% or 5% is chosen to determine the settling time.
- The rise time t_r is usually defined as the time required for the step response output to rise from 10% to 90% of the steady state value y_s.
- The delay time t_d is defined as the time required to reach 50% of the steady state value y_s.

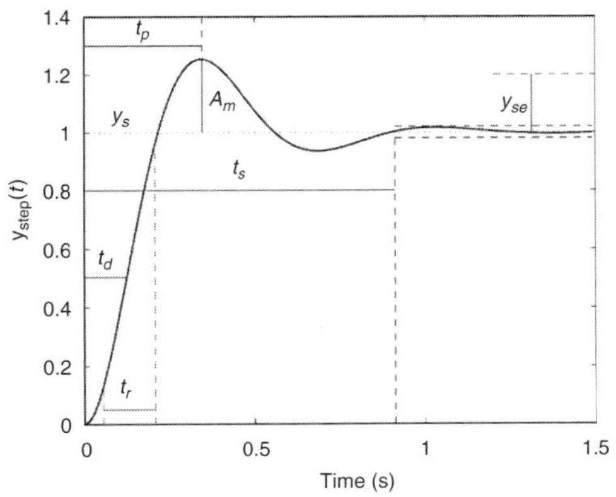

FIGURE 5.5 Parameters for step-response behavior: steady state value y_s, steady state error y_{se}, maximum overshoot A_m, peak time t_p, settling time t_s, rise time t_r, and delay time t_d.

Most of the above value can be obtained from an experimentally determined step response. In general, they cannot be obtained in an analytical form, except for low order models. For the second order model of the one mass system given in (5.3), some analytical results can be obtained. For a second order model of (5.3), the maximum overshoot A_m is determined by

$$A_m = 100e^{-\pi\beta}/\sqrt{1-\xi^2}, \quad \text{where } \xi = \frac{A}{\sqrt{\pi^2 + A^2}}, \quad A = \ln\left(\frac{100}{A_m}\right)$$

The peak time t_p can be computed by

$$t_p = \frac{\pi}{\omega_n\sqrt{1-\xi^2}}$$

whereas the delay time t_d can be approximated by

$$t_d \approx \frac{1 + 0.7\xi}{\omega_n}$$

As the maximum overshoot increases with a smaller damping coefficient β in the system, the maximum overshoot is often used to indicate the relative stability of the system.

5.3.2 Frequency Domain Parameters

With the frequency domain analysis of dynamic systems, specifications for the dynamic properties of a system can also be stated in the frequency domain. Frequency domain specifications in filter design often address ripple, bandwidth, roll-off, and phase lag parameters. Similar characteristics can also be specified for dynamic systems in case the model of the system is analyzed in the frequency domain. The most significant parameters have been summarized below and illustrated in Figure 5.6.

- The bandwidth ω_b is a notion for the maximum frequency at which the output will track a sinusoidal input in a satisfactory manner. By convention, the bandwidth is defined as the frequency at which the output is attenuated −3 dB (0.707).
- The resonant frequency ω_r is the first frequency at which a significant resonance mode with low damping occurs. The resonance mode can, if uncontrolled, negatively influence the settling time of the dynamic system and plays an important role in characterization of performance

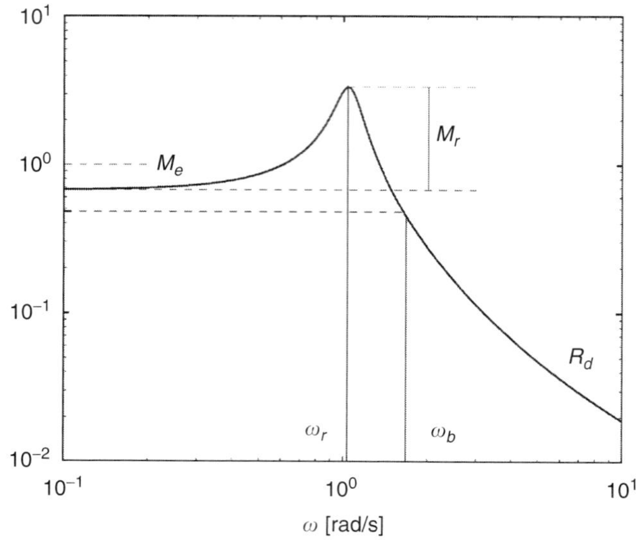

FIGURE 5.6 Parameters for frequency response behavior: bandwidth ω_b, resonance frequency ω_r, resonant peak M_r, steady state error M_e, and roll-off R_d.

- The resonant peak M_r is the height of a resonance mode. The resonant peak is a measure for the damping. As illustrated in Figure 5.2 for a second order model, the resonance mode increases at lower damping coefficients.
- Steady state errors M_e can also be analyzed in the frequency response of a system. Using the final value theorem for continuous time systems

$$\lim_{t\to\infty} y(t) = y_s = \lim_{s\to 0} sy(s)$$

the presence of steady state errors can be inspected in the frequency domain by evaluation $|G(s)|$ at $s = i\omega = 0$ or for small values of the frequency vector ω. This can be seen as follows. As the Laplace transform $u_{\text{step}}(s)$ of a step input signal $u_{\text{step}}(t)$ is $u_{\text{step}}(s) = 1/s$,

$$\lim_{t\to\infty} y_{\text{step}}(t) = \lim_{s\to 0} sy_{\text{step}}(s) = \lim_{s\to 0} sG(s)\frac{1}{s} = \lim_{s\to 0} G(s)$$

By evaluating $|G(i\omega)|$ for small frequencies ω, the steady state behavior of $G(s)$ can be studied.

A similar result exist for discrete time systems, where the final value theorem reads as follows. If $u(z)$ converges for $|z| > 1$ and all poles of $(z-1)u(z)$ are inside the unit circle, then

$$\lim_{k\to\infty} u(k) = \lim_{z\to 1} (z-1)u(z)$$

Hence, for discrete time systems the steady state behavior of a transfer function $G(z)$ can be studied by evaluating $|G(e^{i\omega\Delta T})|$ for small frequencies ω.
- Roll-off R_d at high frequencies is defined as the negative slope of the frequency response at higher frequencies. The roll-off determines the performance of the dynamic system as high frequent disturbances can be amplified if a dynamic system does not have enough high frequent roll-off.

6

The Root Locus Method

Hitay Özbay
The Ohio State University

6.1 Introduction

The root locus technique is a graphical tool used in feedback control system analysis and design. It has been formally introduced to the engineering community by W. R. Evans [3,4], who received the Richard E. Bellman Control Heritage Award from the American Automatic Control Council in 1988 for this major contribution.

In order to discuss the root locus method, we must first review the basic definition of bounded input bounded output (BIBO) stability of the standard linear time invariant feedback system shown in Figure 6.1, where the plant, and the controller, are represented by their transfer functions $P(s)$ and $C(s)$, respectively.* The plant, $P(s)$, includes the physical process to be controlled, as well as the actuator and the sensor dynamics.

The feedback system is said to be stable if none of the closed-loop transfer functions, from external inputs r and v to internal signals e and u, have any poles in the closed right half plane, $\overline{\mathbb{C}}_+ := \{s \in \mathbb{C} : \mathrm{Re}(s) \geq 0\}$. A necessary condition for feedback system stability is that the closed right half plane zeros of $P(s)$ (respectively $C(s)$) are distinct from the poles of $C(s)$ (respectively $P(s)$). When this condition holds, we say that there is no unstable pole–zero cancellation in taking the product $P(s)C(s) =: G(s)$, and then checking feedback system stability becomes equivalent to checking whether all the roots of

$$1 + G(s) = 0 \tag{6.1}$$

are in the open left half plane, $\mathbb{C}_- := \{s \in \mathbb{C} : \mathrm{Re}(s) < 0\}$. The roots of (6.1) are the closed-loop system poles. We would like to understand how the closed-loop system pole locations vary as functions of a real parameter of $G(s)$. More precisely, assume that $G(s)$ contains a parameter K, so that we use the notation

*Here we consider the continuous time case; there is essentially no difference between the continuous time case and the discrete time case, as far as the root locus construction is concerned. In the discrete time case the desired closed-loop pole locations are defined relative to the unit circle, whereas in the continuous time case desired pole locations are defined relative to the imaginary axis.

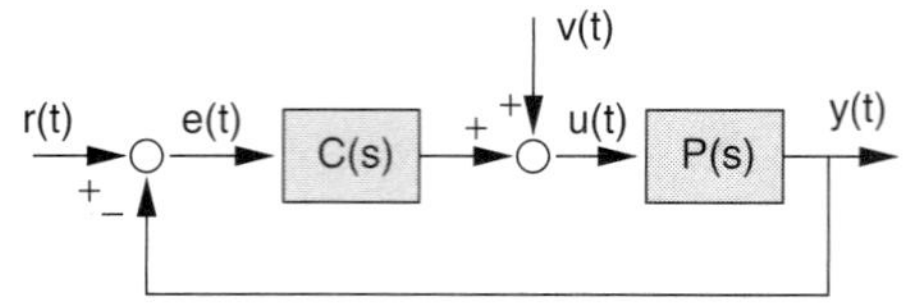

FIGURE 6.1 Standard unity feedback system.

$G(s) = G_K(s)$ to emphasize the dependence on K. The *root locus* is the plot of the roots of (6.1) on the complex plane, as the parameter K varies within a specified interval.

The most common example of the root locus problem deals with the uncertain (or adjustable) gain as the varying parameter: when $P(s)$ and $C(s)$ are fixed rational functions, except for a gain factor, $G(s)$ can be written as $G(s) = G_K(s) = KF(s)$, where K is the uncertain/adjustable gain, and

$$F(s) = \frac{N(s)}{D(s)} \quad \text{where} \quad \begin{aligned} N(s) &= \prod_{j=1}(s - z_j) \\ D(s) &= \prod^{n}(s - p_i), \end{aligned} \qquad n \geq m \tag{6.2}$$

with $z_1,\ldots,z_m$, and $p_1,\ldots,p_n$ being the open-loop system zeros and poles. In this case, the closed-loop system poles are the roots of the characteristic equation

$$\chi(s) := D(s) + KN(s) = 0 \tag{6.3}$$

The *usual root locus* is obtained by plotting the roots $r_1(K),\ldots,r_n(K)$ of the characteristic polynomial $\chi(s)$ on the complex plane, as K varies from 0 to $+\infty$. The same plot for the negative values of K gives the *complementary root locus.* With the help of the root locus plot the designer identifies the admissible values of the parameter K leading to a set of closed-loop system poles that are in the desired region of the complex plane. There are several factors to be considered in defining the "desired region" of the complex plane in which all the roots $r_1(K),\ldots,r_n(K)$ should lie. Those are discussed briefly in the next section. Section 6.3 contains the root locus construction procedure, and design examples are presented in Section 6.4.

The root locus can also be drawn with respect to a system parameter other than the gain. For example, the characteristic equation for the system $G(s) = G_\lambda(s)$, defined by

$$G_\lambda(s) = P(s)C(s), \quad P(s) = \frac{(1 - \lambda s)}{s(1 + \lambda s)}, \quad C(s) = K_c\left(1 + \frac{1}{T_I s}\right)$$

can also be transformed into the form given in (6.3). Here K_c and T_I are given fixed PI (Proportional plus Integral) controller parameters, and $\lambda > 0$ is an uncertain plant parameter. Note that the phase of the plant is

$$\angle P(j\omega) = -\frac{\pi}{2} - 2\tan^{-1}(\lambda\omega)$$

so the parameter λ can be seen as the uncertain phase lag factor (e.g., a small uncertain time delay in the plant can be modeled in this manner, see [9]). It is easy to see that the characteristic equation is

$$s^2(\lambda s + 1) + K_c(1 - \lambda s)\left(s + \frac{1}{T_I}\right) = 0$$

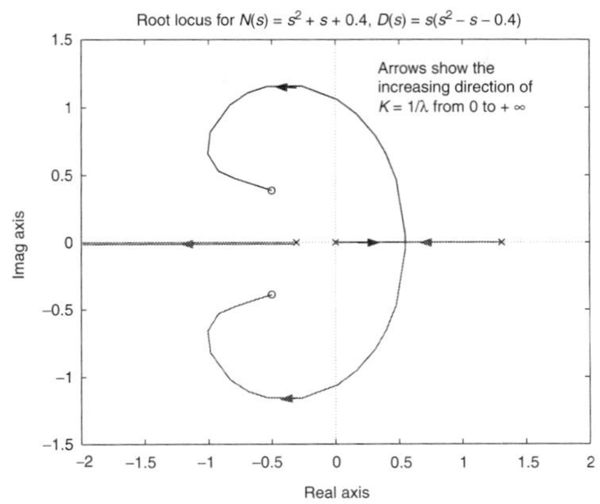

FIGURE 6.2 The root locus with respect to $K = 1/\lambda$.

and by rearranging the terms multiplying λ this equation can be transformed to

$$1 + \frac{1}{\lambda}\frac{(s^2 + K_c s + K_c/T_I)}{s(s^2 - K_c s - K_c/T_I)} = 0$$

By defining $K = \lambda^{-1}$, $N(s) = (s^2 + K_c s + K_c/T_I)$, and $D(s) = s(s^2 - K_c s - K_c/T_I)$, we see that the characteristic equation can be put in the form of (6.3). The root locus plot can now be obtained from the data $N(s)$ and $D(s)$ defined above; that shows how closed-loop system poles move as λ^{-1} varies from 0 to $+\infty$, for a given fixed set of controller parameters K_c and T_I. For the numerical example $K_c = 1$ and $T_I = 2.5$, the root locus is illustrated in Figure 6.2.

The root locus construction procedure will be given in Section 6.3. Most of the computations involved in each step of this procedure can be performed by hand calculations. Hence, an approximate graph representing the root locus can be drawn easily. There are also several software packages to generate the root locus automatically from the problem data $z_1,\ldots,z_m$, and $p_1,\ldots,p_n$.

If a numerical computation program is available for calculating the roots of a polynomial, we can also obtain the root locus with respect to a parameter which enters into the characteristic equation nonlinearly. To illustrate this point let us consider the following example: $G(s) = G_{\omega_o}(s)$ where

$$G_{\omega_o}(s) = P(s)C(s), \quad P(s) = \frac{(s-0.1)}{(s^2 + 1.2\omega_o s + \omega_o^2)(s+0.1)}, \quad C(s) = \frac{(s-0.2)}{(s+2)}$$

Here $\omega_o \geq 0$ is the uncertain plant parameter. Note that the characteristic equation

$$1 + \frac{\omega_o(1.2s + \omega_o)(s+0.1)(s+2)}{s^2(s+0.1)(s+2) + (s-0.2)(s-0.1)} = 0 \tag{6.4}$$

cannot be expressed in the form of $D(s) + KN(s) = 0$ with a single parameter K. Nevertheless, for each ω_o we can numerically calculate the roots of (6.4) and plot them on the complex plane as ω_o varies within a range of interest. Figure 6.3 illustrates all the four branches, $r_1(K),\ldots,r_4(K)$, of the root locus for this system as ω_o increases from zero to infinity. The figure is obtained by computing the roots of (6.4) for a set of values of ω_o by using MATLAB.

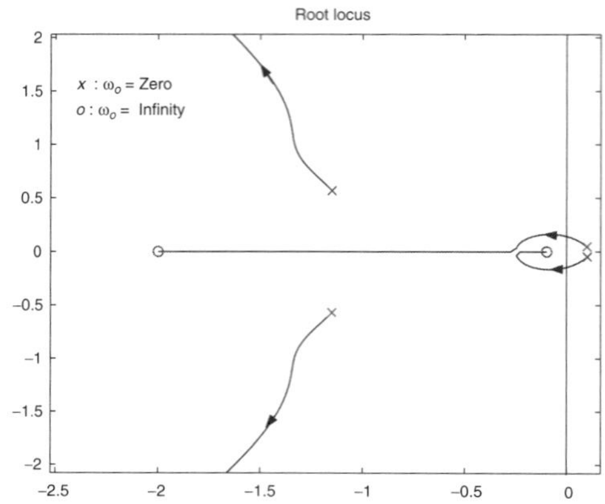

FIGURE 6.3 The root locus with respect to ω_o.

6.2 Desired Pole Locations

The performance of a feedback system depends heavily on the location of the closed-loop system poles $r_i(K) = 1,\ldots,n$. First of all, for stability we want $r_i(K) \in \mathbb{C}_-$ for all $i = 1,\ldots,n$. Clearly, having a pole "close" to the imaginary axis poses a danger, that is, "small" perturbations in the plant might lead to an unstable feedback system. So the desired pole locations must be such that stability is preserved under such perturbations (or in the presence of uncertainties) in the plant. For second-order systems, we can define certain stability robustness measures in terms of the pole locations, which can be tied to the characteristics of the step response. For higher order systems, similar guidelines can be used by considering the dominant poles only.

In the standard feedback control system shown in Figure 6.1, assume that the closed-loop transfer function from $r(t)$ to $y(t)$ is in the form

$$T(s) = \frac{\omega_o^2}{s^2 + 2\zeta\omega_o s + \omega_o^2}, \quad 0 < \zeta < 1, \quad \omega_o \in \mathrm{R}$$

and $r(t)$ is the unit step function. Then, the output is

$$y(t) = 1 - \frac{e^{-\zeta\omega_o t}}{\sqrt{1-\zeta^2}} \sin(\omega_d t + \theta), \quad t \geq 0$$

where $\omega_d := \omega_o\sqrt{1-\zeta^2}$ and $\theta := \cos^{-1}(\zeta)$. For some typical values of ζ, the step response $y(t)$ is as shown in Figure 6.4. The maximum *percent overshoot* is defined to be the quantity

$$\mathrm{PO} := \frac{y_p - y_{ss}}{y_{ss}} \times 100\%$$

where y_p is the peak value. By simple calculations it can be seen that the peak value of $y(t)$ occurs at the time instant $t_p = \pi/\omega_d$, and

$$\mathrm{PO} = e^{-\pi\zeta/\sqrt{1-\zeta^2}} \times 100\%$$

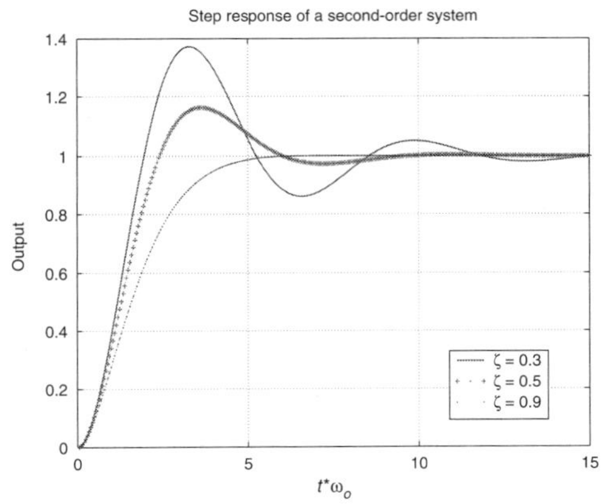

FIGURE 6.4 Step response of a second-order system.

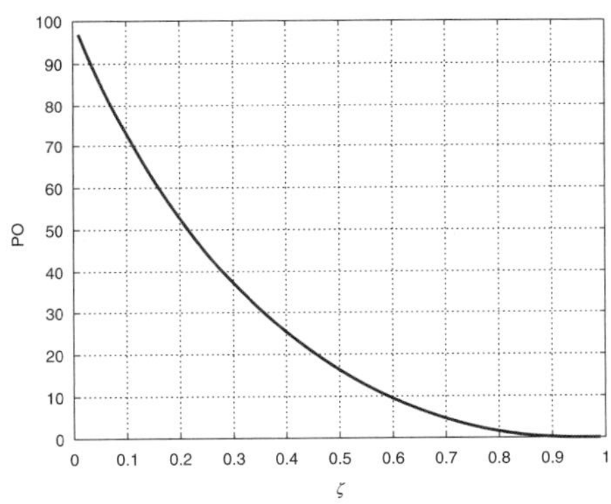

FIGURE 6.5 PO versus ζ.

Figure 6.5 shows PO versus ζ. The *settling time* is defined to be the smallest time instant t_s, after which the response $y(t)$ remains within 2% of its final value, that is,

$$t_s := \min\{t': |y(t) - y_{ss}| \leq 0.02 y_{ss} \ \forall t \geq t'\}$$

Sometimes 1% or 5% is used in the definition of settling time instead of 2%; conceptually, there is no difference. For the second-order system response, we have

$$t_s \approx \frac{4}{\zeta\omega_o}$$

So, in order to have a fast settling response, the product $\zeta\omega_o$ should be large.

The closed-loop system poles are

$$r_{1,2} = -\zeta\omega_o \pm j\omega_o\sqrt{1-\zeta^2}$$

Therefore, once the maximum allowable settling time and PO are specified, we can define the region of desired pole locations by determining the minimum allowable ζ and $\zeta\omega_o$. For example, let the desired PO and t_s be bounded by

$$\text{PO} \leq 10\% \quad \text{and} \quad t_s \leq 8\,\text{s}$$

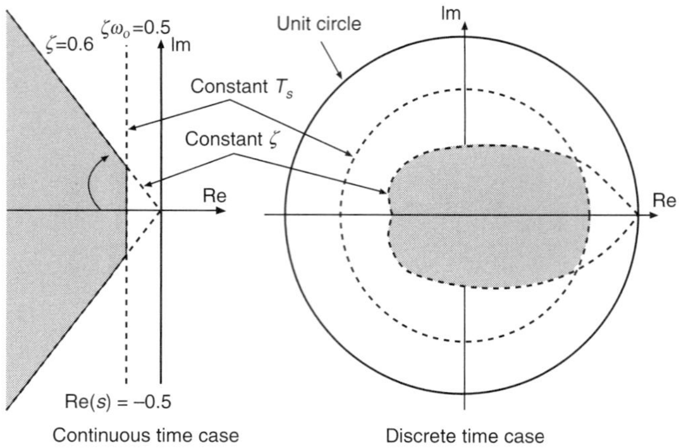

FIGURE 6.6 Region of the desired closed-loop poles.

The PO requirement implies that $\zeta \geq 0.6$, equivalently $\theta \leq 53°$ (recall that $\cos(\theta) = \zeta$). The settling time requirement is satisfied if and only if $\text{Re}(r_{1,2}) \leq -0.5$. Then, the region of desired closed-loop poles is the shaded area shown in Figure 6.6. The same figure also illustrates the region of desired closed-loop poles for similar design requirements in the discrete time case.

If the order of the closed-loop transfer function $T(s)$ is higher than two, then, depending on the location of its poles and zeros, it may be possible to approximate the closed-loop step response by the response of a second-order system. For example, consider the third-order system

$$T(s) = \frac{\omega_o^2}{(s^2 + 2\zeta\omega_o s + \omega_o^2)(1 + s/r)} \quad \text{where } r \gg \zeta\omega_o$$

The transient response contains a term e^{-rt}. Compared with the envelope $e^{-\zeta\omega_o t}$ of the sinusoidal term, e^{-rt} decays very fast, and the overall response is similar to the response of a second-order system. Hence, the effect of the third pole $r_3 = -r$ is negligible.

Consider another example,

$$T(s) = \frac{\omega_o^2[1 + s/(r+e)]}{(s^2 + 2\zeta\omega_o s + \omega_o^2)(1 + s/r)} \quad \text{where } 0 < e \ll r$$

In this case, although r does not need to be much larger than $\zeta\omega_o$, the zero at $-(r + e)$ cancels the effect of the pole at $-r$. To see this, consider the partial fraction expansion of $Y(s) = T(s)R(s)$ with $R(s) = 1/s$:

$$Y(s) = \frac{A_0}{s} + \frac{A_1}{s - r_1} + \frac{A_2}{s - r_2} + \frac{A_3}{s + r}$$

where $A_0 = 1$ and

$$A_3 = \lim_{s \to -r} (s + r)Y(s) = \frac{\omega_o^2}{2\zeta\omega_o r - (\omega_o^2 + r^2)}\left(\frac{e}{r + e}\right)$$

Since $|A_3| \to 0$ as $e \to 0$, the term $A_3 e^{-rt}$ is negligible in $y(t)$.

In summary, if there is an approximate pole–zero cancellation in the left half plane, then this pole–zero pair can be taken out of the transfer function $T(s)$ to determine PO and t_s. Also, the poles closest to the imaginary axis dominate the transient response of $y(t)$. To generalize this observation, let $r_1, \ldots, r_n$ be the poles of $T(s)$, such that $\mathrm{Re}(r_k) \ll \mathrm{Re}(r_2) = \mathrm{Re}(r_1) < 0$, for all $k \geq 3$. Then, the pair of complex conjugate poles $r_{1,2}$ are called the *dominant poles*. We have seen that the desired transient response properties, for example, PO and t_s, can be translated into requirements on the location of the dominant poles.

6.3 Root Locus Construction

As mentioned above, the root locus primarily deals with finding the roots of a characteristic polynomial that is an affine function of a single parameter, K,

$$\chi(s) = D(s) + KN(s) \tag{6.5}$$

where $D(s)$ and $N(s)$ are fixed monic polynomials (i.e., coefficient of the highest power is normalized to 1). If N and/or D are not monic, the highest coefficient(s) can be absorbed into K.

6.3.1 Root Locus Rules

Recall that the usual root locus shows the locations of the closed-loop system poles as K varies from 0 to $+\infty$. The roots of $D(s)$, $p_1, \ldots, p_n$, are the poles, and the roots of $N(s)$, $z_1, \ldots, z_m$, are the zeros, of the open-loop system, $G(s) = KF(s)$. Since $P(s)$ and $C(s)$ are proper, $G(s)$ is proper, and hence $n \geq m$. So the degree of the polynomial $\chi(s)$ is n and it has exactly n roots.

Let the closed-loop system poles, that is, roots of $\chi(s)$, be denoted by $r_1(K), \ldots, r_n(K)$. Note that these are functions of K; whenever the dependence on K is clear, they are simply written as $r_1, \ldots, r_n$. The points in C that satisfy (6.5) for some $K > 0$ are on the root locus. Clearly, a point $r \in \mathbb{C}$ is on the root locus if and only if

$$K = -\frac{1}{F(r)} \tag{6.6}$$

The condition (6.6) can be separated into two parts:

$$|K| = -\frac{1}{|F(r)|} \tag{6.7}$$

$$\angle K = 0^\circ = -(2, +1) \times 180^\circ - \angle F(r), \quad , = 0, \pm 1, \pm 2, \ldots \tag{6.8}$$

The phase rule (6.8) determines the points in C that are on the root locus. The magnitude rule (6.7) determines the gain $K > 0$ for which the root locus is at a given point r. By using the definition of $F(s)$, (6.8) can be rewritten as

$$(2, +1) \times 180^\circ = \sum_{i=1}^{n} \angle(r - p_i) - \sum_{i=1}^{m} \angle(r - z_j) \tag{6.9}$$

Similarly, (6.7) is equivalent to

$$K = \frac{\prod_{i=1}^{n} |r - p_i|}{\prod_{j=1}^{m} |r - z_j|} \tag{6.10}$$

6.3.2 Root Locus Construction

There are several software packages available for generating the root locus automatically for a given $F = N/D$. In particular, the related MATLAB commands are `rlocus` and `rlocfind`. In many cases, approximate root locus can be drawn by hand using the rules given below. These rules are determined from the basic definitions (6.5), (6.7), and (6.8).

1. The root locus has n branches: $r_1(K),\ldots,r_n(K)$.
2. Each branch starts $(K \cong 0)$ at a pole p_i and ends (as $K \to \infty$) at a zero z_j, or converges to an asymptote, $Me^{j\alpha_,}$, where $M \to \infty$ and

$$\alpha_, = \frac{2, + 1}{n - m} \times 180°, \quad , = 0,\ldots,(n - m - 1)$$

3. There are $(n - m)$ asymptotes with angles α. The center of the asymptotes (i.e., their intersection point on the real axis) is

$$\sigma_a = \frac{\sum_{i=1}^{n} p_i - \sum_{j=1}^{m} z_j}{n - m}$$

4. A point $x \in \mathrm{R}$ is on the root locus if and only if the total number of poles p_i's and zeros z_j's to the right of x (i.e., total number of p_i's with $\mathrm{Re}(p_i) > x$ plus total number of z_j's with $\mathrm{Re}(z_j) > x$) is odd. Since $F(s)$ is a rational function with real coefficients, poles and zeros appear in complex conjugates, so when counting the number of poles and zeros to the right of a point $x \in \mathrm{R}$ we just need to consider the poles and zeros on the real axis.
5. The values of K for which the root locus crosses the imaginary axis can be determined from the Routh–Hurwitz stability test. Alternatively, we can set $s = j\omega$ in (6.5) and solve for real ω and K satisfying

$$D(j\omega) + KN(j\omega) = 0$$

   Note that there are two equations here, one for the real part and one for the imaginary part.
6. The break points (intersection of two branches on the real axis) are feasible solutions (satisfying rule 4) of

$$\frac{d}{ds}F(s) = 0 \tag{6.11}$$

7. Angles of departure $(K \cong 0)$ from a complex pole, or arrival $(K \to +\infty)$ to a complex zero, can be determined from the phase rule. See example below.
   Let us now follow the above rules step by step to construct the root locus for

$$F(s) = \frac{(s+3)}{(s-1)(s+5)(s+4+j2)(s+4-j2)}$$

First, enumerate the poles and zeros as $p_1 = -4 + j2$, $p_2 = -4 - j2$, $p_3 = -5$, $p_4 = 1$, $z_1 = -3$. So, $n = 4$ and $m = 1$.

1. The root locus has four branches.
2. Three branches converge to the asymptotes whose angles are 60°, 180°, and –60°, and one branch converges to $z_1 = -3$.

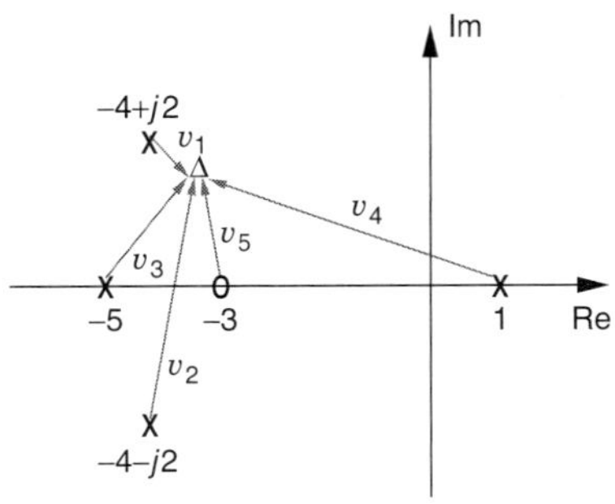

FIGURE 6.7 Angle of departure from $-4 + j2$.

3. The center of the asymptotes is $\sigma = (-12 + 3)/3 = -3$.
4. The intervals $(-\infty, -5]$ and $[-3, 1]$ are on the root locus.
5. The imaginary axis crossings are the feasible roots of

$$(\omega^4 - j12\omega^3 - 47\omega^2 + j40\omega - 100) + K(j\omega + 3) = 0 \tag{6.12}$$

for real ω and K. Real and imaginary parts of (6.12) are

$$\omega^4 - 47\omega^2 - 100 + 3K = 0$$

$$j\omega(-12\omega^2 + 40 + K) = 0$$

They lead to two feasible pairs of solutions ($K = 100/3$, $\omega = 0$) and ($K = 215.83$, $\omega = \pm 4.62$).
6. Break points are the feasible solutions of

$$3s^4 + 36s^3 + 155s^2 + 282s + 220 = 0$$

Since the roots of this equation are $-4.55 \pm j1.11$ and $-1.45 \pm j1.11$, there is no solution on the real axis, hence no break points.
7. To determine the angle of departure from the complex pole $p_1 = -4 + j2$, let Δ represent a point on the root locus near the complex pole p_1, and define v_i, $i = 1,\ldots,5$, to be the vectors drawn from p_i, for $i = 1,\ldots,4$, and from z_1 for $i = 5$, as shown in Figure 6.7. Let $\theta_1,\ldots,\theta_5$ be the angles of $v_1,\ldots,v_5$. The phase rule implies

$$(\theta_1 + \theta_2 + \theta_3 + \theta_4) - \theta_5 = \pm 180° \tag{6.13}$$

As Δ approaches p_1, θ_1 becomes the angle of departure and the other θ_i's can be approximated by the angles of the vectors drawn from the other poles, and from the zero, to the pole p_1. Thus θ_1 can be solved from (6.13), where $\theta_2 \approx 90°$, $\theta_3 \approx \tan^{-1}(2)$, $\theta_4 \approx 180° - \tan^{-1}\left(\frac{2}{5}\right)$, and $\theta_5 \approx 90° + \tan^{-1}\left(\frac{1}{2}\right)$. That yields $\theta_1 \approx -15°$.

The exact root locus for this example is shown in Figure 6.8. From the results of item 5 above, and the shape of the root locus, it is concluded that the feedback system is stable if

$$33.33 < K < 215.83$$

that is, by simply adjusting the gain of the controller, the system can be made stable. In some situations we need to use a dynamic controller to satisfy all the design requirements.

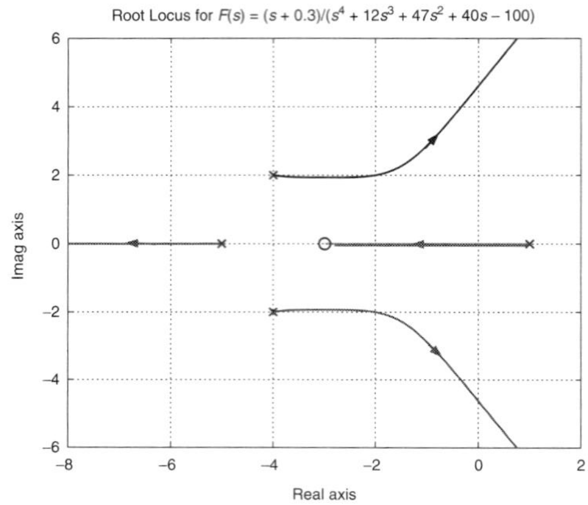

FIGURE 6.8 Root locus for $F(s) = \dfrac{(s+3)}{(s-1)(s+5)(s+4+j2)(s+4-j2)}$.

6.3.3 Design Examples

Example 1

Consider the standard feedback system with a plant

$$P(s) = \frac{1}{0.72}\frac{1}{(s+1)(s+2)}$$

and design a controller such that

- Feedback system is stable.
- $\text{PO} \le 10\%$, $t_s \le 4$ s, and steady state error is zero when $r(t)$ is unit step.
- Steady state error is as small as possible when $r(t)$ is unit ramp.

It is clear that the second design goal cannot be achieved by a simple proportional controller. To satisfy this condition, the controller must have a pole at $s = 0$, that is, it must have integral action. If we try an integral control of the form $C(s) = K_c/s$, with $K_c > 0$, then the root locus has three branches, the interval $[-1, 0]$ is on the root locus; three asymptotes have angles $\{60°, 180°, -60°\}$ with a center at $\sigma_a = -1$; and there is only one break point at $-1 + \frac{1}{\sqrt{3}}$, see Figure 6.9. From the location of the break point, center, and angles of the asymptotes, it can be deduced that two branches (one starting at $p_1 = -1$, and the other one starting at $p_3 = 0$) always remain to the right of p_1. On the other hand, the settling time condition implies that the real parts of the dominant closed-loop system poles must be less than or equal to -1. So, a simple integral control does not do the job. Now try a PI controller of the form

$$C(s) = K_c\left(\frac{s - z_c}{s}\right), \quad K_c > 0$$

In this case, we can select $z_c = -1$ to cancel the pole at $p_1 = -1$ and the system effectively becomes a second-order system. The root locus for $F(s) = 1/s(s + 2)$ has two branches and two asymptotes, with center $\sigma_a = -1$ and angles $\{90°, -90°\}$; the break point is also at -1. The branches leave -2 and 0, and go toward each other, meet at -1, and tend to infinity along the line $\text{Re}(s) = -1$. Indeed, the closed-loop system poles are

$$r_{1,2} = -1 \pm \sqrt{1 - K}, \quad \text{where } K = K_c/0.72$$

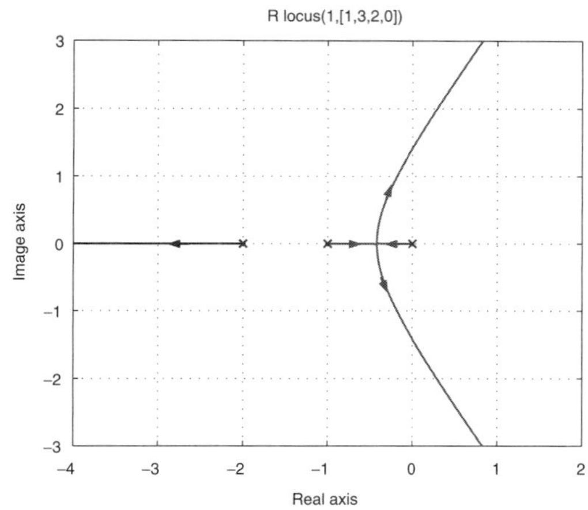

FIGURE 6.9 Root locus for Example 1.

The steady state error, when $r(t)$ is unit ramp, is $2/K$. So K needs to be as large as possible to meet the third design condition. Clearly, $\text{Re}(r_{1,2}) = -1$ for all $K \geq 1$, which satisfies the settling time requirement. The percent overshoot is less than 10% if ζ of the roots $r_{1,2}$ is greater than 0.6. A simple algebra shows that $\zeta = 1/\sqrt{K}$, hence the design conditions are met if $K = 1/0.36$, that is $K_c = 2$. Thus a PI controller that solves the design problem is

$$C(s) = 2\left(\frac{s+1}{s}\right)$$

The controller cancels a stable pole (at $s = -1$) of the plant. If there is a slight uncertainty in this pole location, perfect cancellation will not occur and the system will be third-order with the third pole at $r_3 \cong -1$. Since the zero at $z_o = -1$ will approximately cancel the effect of this pole, the response of this system will be close to the response of a second-order system. However, we must be careful if the pole–zero cancellations are near the imaginary axis because in this case small perturbations in the pole location might lead to large variations in the feedback system response, as illustrated with the next example.

Example 2

A flexible structure with lightly damped poles has transfer function in the form

$$P(s) = \frac{\omega_1^2}{s^2(s^2 + 2\zeta\omega_1 s + \omega_1^2)}$$

By using the root locus, we can see that the controller

$$C(s) = K_c\frac{(s^2 + 2\zeta\omega_1 s + \omega_1^2)(s + 0.4)}{(s + r)^2(s + 4)}$$

stabilizes the feedback system for sufficiently large r and an appropriate choice of K_c. For example, let $\omega_1 = 2$, $\zeta = 0.1$, and $r = 10$. Then the root locus of $F(s) = P(s)C(s)/K$, where $K = K_c\omega_1^2$, is as shown in Figure 6.10. For $K = 600$, the closed-loop system poles are

$$\{-10.78 \pm j2.57, -0.94 \pm j1.61, -0.2 \pm j1.99, -0.56\}$$

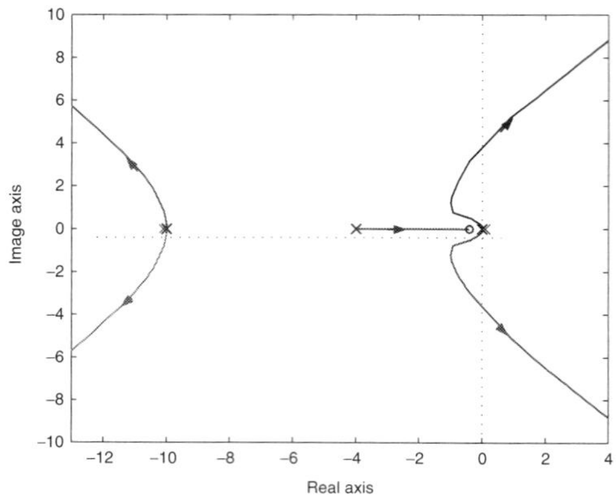

FIGURE 6.10 Root locus for Example 2(a).

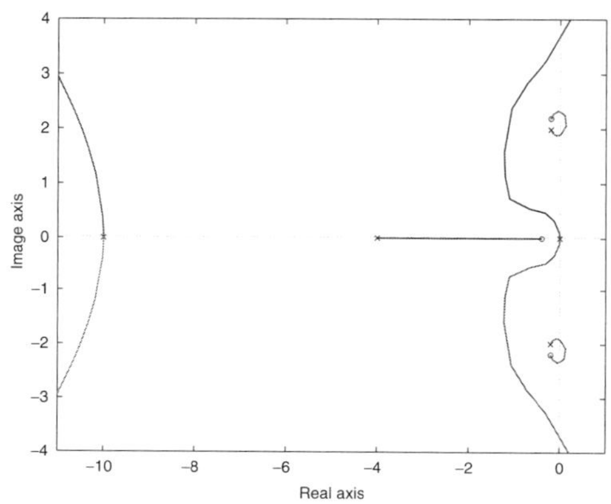

FIGURE 6.11 Root locus for Example 2(b).

Since the poles $-0.2 \pm j1.99$ are canceled by a pair of zeros at the same point in the closed-loop system transfer function $T = G(1 + G)^{-1}$, the dominant poles are at -0.56 and $-0.94 \pm j1.61$ (they have relatively large negative real parts and the damping ratio is about 0.5).

Now, suppose that this controller is fixed and the complex poles of the plant are slightly modified by taking $\zeta = 0.09$ and $\omega_1 = 2.2$. The root locus corresponding to this system is as shown in Figure 6.11. Since lightly damped complex poles are not perfectly canceled, there are two more branches near the imaginary axis. Moreover, for the same value of $K = 600$, the closed-loop system poles are

$$\{-10.78 \pm j2.57, -1.21 \pm j1.86, 0.05 \pm j1.93, -0.51\}$$

In this case, the feedback system is unstable.

Example 3

One of the most important examples of mechatronic systems is the DC motor. An approximate transfer function of a DC motor [8, pp. 141–143] is in the form

$$P_m(s) = \frac{K_m}{s(s + 1/\tau_m)}, \quad \tau_m > 0$$

Also note that if τ_m is large, then $P_m(s) \approx P_b(s)$, where $P_b(s) = K_b/s^2$ is the transfer function of a rigid beam. In this example, the general class of plants $P_m(s)$ will be considered. Assuming that $p_m = -1/\tau_m$ and K_m are given, a first-order controller

$$C(s) = K_c\left(\frac{s - z_c}{s - p_c}\right) \tag{6.14}$$

will be designed. The aim is to place the closed-loop system poles far from the Im-axis. Since the order of $F(s) = P_m(s)C(s)/K_mK_c$ is three, the root locus has three branches. Suppose the desired closed-loop poles are given as p_1, p_2, and p_3. Then, the pole placement problem amounts to finding $\{K_c, z_c, p_c\}$ such that the characteristic equation is

$$\begin{aligned}\chi(s) &= (s - p_1)(s - p_2)(s - p_3)\\ &= s^3 - (p_1 + p_2 + p_3)s^2 + (p_1p_2 + p_1p_3 + p_2p_3)s - p_1p_2p_3\end{aligned}$$

But the actual characteristic equation, in terms of the unknown controller parameters, is

$$\begin{aligned}\chi(s) &= s(s - p_m)(s - p_c) + k(s - z_c)\\ &= s^3 - (p_m + p_c)s^2 + (p_mp_c + K)s - Kz_c\end{aligned}$$

where $K := K_mK_c$. Equating the coefficients of the desired $\chi(s)$ to the coefficients of the actual $\chi(s)$, three equations in three unknowns are obtained:

$$\begin{aligned}p_m + p_c &= p_1 + p_2 + p_3\\ p_mp_c + K &= p_1p_2 + p_1p_3 + p_2p_3\\ Kz_c &= p_1p_2p_3\end{aligned}$$

From the first equation p_c is determined, then K is obtained from the second equation and finally z_c is computed from the third equation.

For different numerical values of p_m, p_1, p_2, and p_3 the shape of the root locus is different. Below are some examples, with the corresponding root loci shown in Figures 6.12–6.14.

(a) $p_m = -0.05$, $p_1 = p_2 = p_3 = -2 \Rightarrow$

$$K = 11.70, \quad p_c = -5.95, \quad z_c = -0.68$$

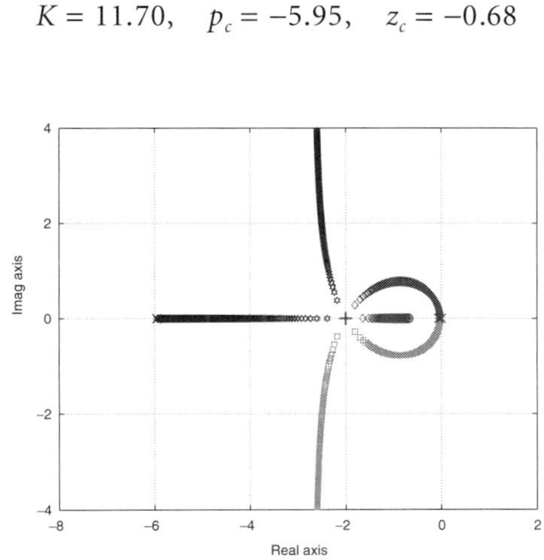

FIGURE 6.12 Root locus for Example 3(a).

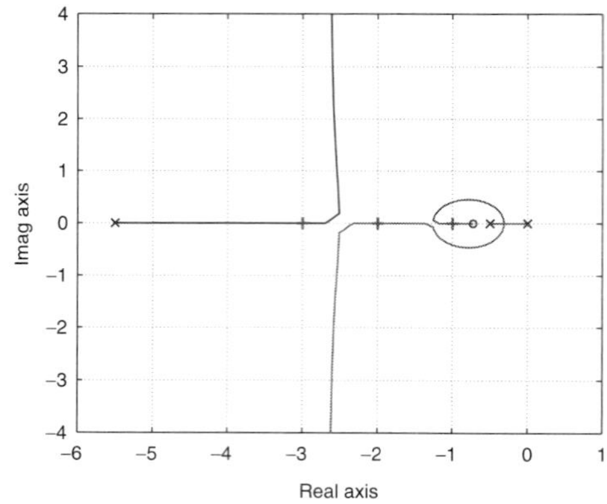

FIGURE 6.13 Root locus for Example 3(b).

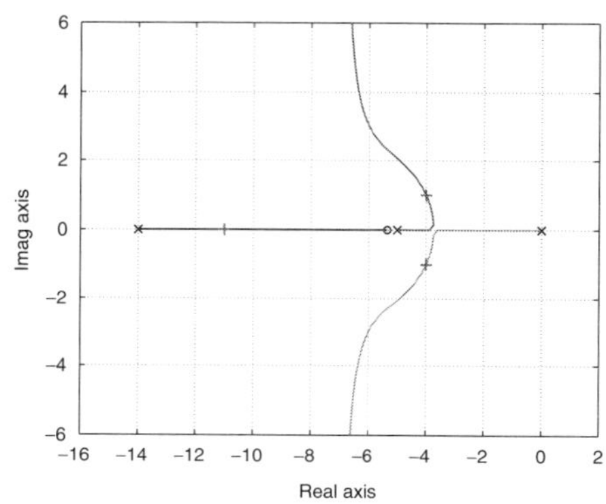

FIGURE 6.14 Root locus for Example 3(c).

(b) $p_m = -0.5,\ p_1 = -1,\ p_2 = -2,\ p_3 = -3 \Rightarrow$

$$K = 8.25, \quad p_c = -5.50, \quad z_c = -0.73$$

(c) $p_m = -5,\ p_1 = -11,\ p_2 = -4 + j1,\ p_3 = -4 - j1 \Rightarrow$

$$K = 35, \quad p_c = -14, \quad z_c = -5.343$$

Example 4

Consider the open-loop transfer function

$$P(s)C(s) = K_c \frac{(s^2 - 3s + 3)(s - z_c)}{s(s^2 + 3s + 3)(s - p_c)}$$

where K_c is the controller gain to be adjusted, and z_c and p_c are the controller zero and pole, respectively. Observe that the root locus has four branches except for the non-generic case $z_c = p_c$. Let the desired dominant closed-loop poles be $r_{1,2} = -0.4$. The steady state error for unit ramp reference input is

$$e_{ss} = \frac{p_c}{K_c z_c}$$

Accordingly, we want to make the ratio $K_c z_c / p_c$ as large as possible.

The characteristic equation is

$$\chi(s) = s(s^2 + 3s + 3)(s - p_c) + K_c(s^2 - 3s + 3)(s - z_c)$$

and it is desired to be in the form

$$\chi(s) = (s + 0.4)^2 (s - r_3)(s - r_4)$$

for some $r_{3,4}$ with $\text{Re}(r_{3,4}) < 0$, which implies that

$$\chi(s)|_{s=-0.4} = 0, \quad \frac{d}{ds}\chi(s)|_{s=-0.4} = 0 \qquad (6.15)$$

Conditions (6.15) give two equations:

$$0.784(0.4 + p_c) - 4.36 K_c(0.4 + z_c) = 0$$
$$4.36 K_c - 0.784 - 1.08(0.4 + p_c) + 3.8 K_c(0.4 + z_c) = 0$$

from which z_c and p_c can be solved in terms of K_c. Then, by simple substitutions, the ratio to be maximized, $K_c z_c / p_c$, can be reduced to

$$\frac{K_c z_z}{p_c} = \frac{3.4776 K_c - 0.784}{24.2469 K_c - 3.4776}$$

The maximizing value of K_c is 0.1297; it leads to $p_c = -0.9508$ and $z_c = -1.1637$. For this controller, the feedback system poles are

$$\{-1.64 + j0.37, -1.64 - j0.37, -0.40, -0.40\}$$

The root locus is shown in Figure 6.15.

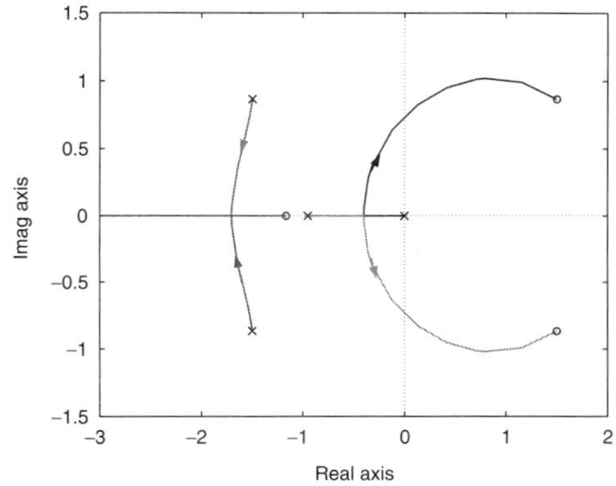

FIGURE 6.15 Root locus for Example 4.

6.4 Complementary Root Locus

In the previous section, the root locus parameter K was assumed to be positive and the phase and magnitude rules were established based on this assumption. There are some situations in which controller gain can be negative as well. Therefore, the complete picture is obtained by drawing the usual root locus (for $K > 0$) and the complementary root locus (for $K < 0$). The complementary root locus rules are

$$\times 360^\circ = \sum_{i=1}^{n} \angle(r - p_i) - \sum_{j=1}^{m} \angle(r - z_j), \quad , = 0, \pm 1, \pm 2, \ldots \tag{6.16}$$

$$|K| = \frac{\prod_{i=1}^{n} |r - p_i|}{\prod_{j=1}^{m} |r - z_j|} \tag{6.17}$$

Since the phase rule (6.16) is the 180° shifted version of (6.9), the complementary root locus is obtained by simple modifications in the root locus construction rules. In particular, the number of asymptotes and their center are the same, but their angles α's are given by

$$\alpha_{,} = \frac{2,}{(n - m)} \times 180^\circ, \quad , = 0, \ldots, (n - m - 1)$$

Also, an interval on the real axis is on the complementary root locus if and only if it is not on the usual root locus.

Example 3 (revisited)

In the Example 3 given above, if the problem data is modified to $p_m = -5$, $p_1 = -20$, and $p_{2,3} = -2 \pm j$, then the controller parameters become

$$K = -10, \quad p_c = -19, \quad z_c = 10$$

Note that the gain is negative. The roots of the characteristic equation as K varies between 0 and $-\infty$ form the complementary root locus; see Figure 6.16.

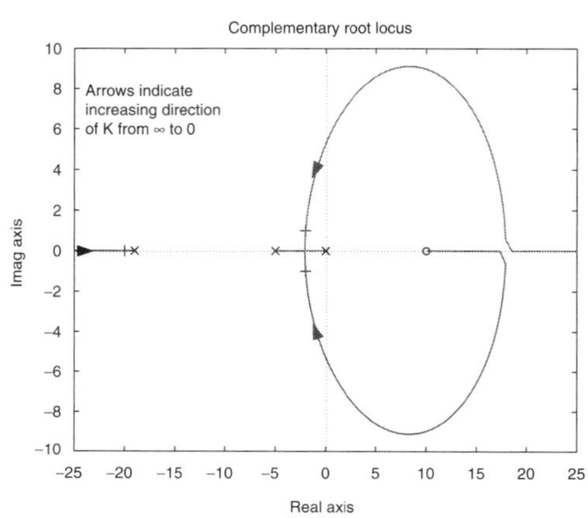

FIGURE 6.16 Complementary root locus for Example 3.

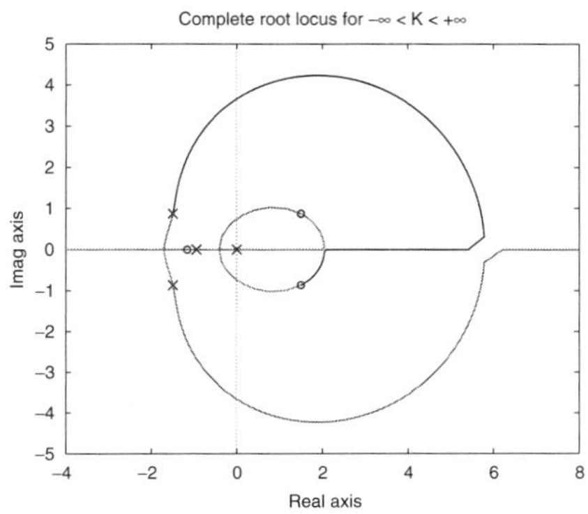

FIGURE 6.17 Complementary and usual root loci for Example 4.

Example 4 (revisited)

In this example, if K increases from $-\infty$ to $+\infty$, the closed-loop system poles move along the complementary root locus, and then the usual root locus, as illustrated in Figure 6.17.

6.5 Root Locus for Systems with Time Delays

The standard feedback control system considered in this section is shown in Figure 6.18, where the controller C and plant P are in the form

$$C(s) = \frac{N_c(s)}{D_c(s)}$$

and

$$P(s) = e^{-hs}P_0(s) \quad \text{where } P_0(s) = \frac{N_p(s)}{D_p(s)}$$

with (N_c, D_c) and (N_p, D_p) being coprime pairs of polynomials with real coefficients.* The term e^{-hs} is the transfer function of a pure delay element (in Figure 6.18 the plant input is delayed by h seconds). In general, time delays enter into the plant model when there is

- Sensor (or actuator) processing delay
- Software delay in the controller
- Transport delay in the process

In this case the open-loop transfer function is

$$G(s) = G_h(s) = e^{-hs}G_0(s)$$

where $G_0(s) = P_0(s)C(s)$ corresponds to the no delay case, $h = 0$.

Note that magnitude and phase of $G(j\omega)$ are determined from the identities

$$|G(j\omega)| = |G_0(j\omega)| \tag{6.18}$$

$$\angle G(j\omega) = -h\omega + \angle G_0(j\omega) \tag{6.19}$$

*A pair of polynomials is said to be coprime pair if they do not have common roots.

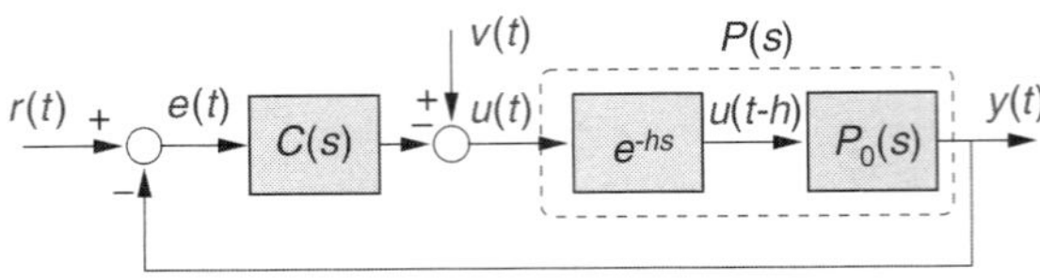

FIGURE 6.18 Feedback system a with time delay.

6.5.1 Stability of Delay Systems

Stability of the feedback system shown in Figure 6.18 is equivalent to having all the roots of

$$\chi(s) = D(s) + e^{-hs}N(s) \tag{6.20}$$

in the open left half plane, C_-, where $D(s) = D_c(s)D_p(s)$ and $N(s) = N_c(s)N_p(s)$. We assume that there is no unstable pole–zero cancellation in taking the product $P_0(s)C(s)$, and that $deg(D) > deg(N)$ (here N and D need not be monic polynomials). Strictly speaking, $\chi(s)$ is not a polynomial because it is a transcendental function of s. The functions of the form (6.20) belong to a special class of functions called *quasi-polynomials*. The closed-loop system poles are the roots of (6.20).

Following are known facts (see [1,10]):

(i) If r_k is a root of (20), then so is $\bar{r}_k$ (i.e., roots appear in complex conjugate pairs as usual).
(ii) There are infinitely many poles $r_k \in C$, $k = 1, 2,\ldots,$satisfying $\chi(r_k) = 0$.
(iii) And r_k's can be enumerated in such a way that $\text{Re}(r_k + 1) \leq \text{Re}(r_k)$; moreover, $\text{Re}(r_k) \to -\infty$ as $k \to \infty$.

Example

If $G_h(s) = e^{-hs}/s$, then the closed-loop system poles r_k, for $k = 1, 2,\ldots,$ are the roots of

$$1 + \frac{e^{-h\sigma_k} e^{-jh\omega_k}}{\sigma_k + j\omega_k} e^{\pm j2k\pi} = 0 \tag{6.21}$$

where $r_k = \sigma_k + j\omega_k$ for some $\sigma_k, \omega_k \in$ R. Note that $e^{\pm j2k\pi} = 1$ for all $k = 1, 2,\ldots$. Equation (6.1) is equivalent to the following set of equations:

$$e^{-h\sigma_k} = |\sigma_k + j\omega_k| \tag{6.22}$$

$$\pm(2k-1)\pi = h\omega_k + \angle(\sigma_k + j\omega_k), \quad k = 1,2,\ldots \tag{6.23}$$

It is quite interesting that for $h = 0$ there is only one root $r = -1$, but even for infinitesimally small $h > 0$ there are infinitely many roots. From the magnitude condition (6.22), it can be shown that

$$\sigma_k \geq 0 \Rightarrow |\omega_k| \leq 1 \tag{6.24}$$

Also, for $\sigma_k \geq 0$, the phase $/(\sigma_k + j\omega_k)$ is between $-\pi/2$ and $+\pi/2$, therefore (6.23) leads to

$$\sigma_k \geq 0 \Rightarrow h|\omega_k| \geq \frac{\pi}{2} \tag{6.25}$$

By combining (6.24) and (6.25), it can be proven that the feedback system has no roots in the closed right half plane when $h < \pi/2$. Furthermore, the system is unstable if $h \geq \pi/2$. In particular, for $h = \pi/2$ there are two roots on the imaginary axis, at $\pm j1$. It is also easy to show that, for any $h > 0$ as $k \to \infty$, the roots converge to

$$r_k \to \frac{1}{h}\left[-\ln\left(\frac{2k\pi}{h}\right) \pm j2k\pi\right]$$

As $h \to 0$, the magnitude of the roots converge to ∞.

As illustrated by the above example, property (iii) implies that for any given real number σ there are only finitely many r_k's in the region of the complex plane

$$\mathrm{C}_\sigma := \{s \in \mathrm{C}: \mathrm{Re}(s) \geq \sigma\}$$

In particular, with $\sigma = 0$, this means that the quasi-polynomial $\chi(s)$ can have only finitely many roots in the right half plane. Since the effect of the closed-loop system poles that have very large negative real parts is negligible (as far as closed-loop systems' input–output behavior is concerned), only finitely many "dominant" roots r_k, for $k = 1,\ldots,m$, should be computed for all practical purposes.

6.5.2 Dominant Roots of a Quasi-Polynomial

Now we discuss the following problem: given $N(s)$, $D(s)$, and $h \geq 0$, find the dominant roots of the quasi-polynomial

$$\chi(s) = D(s) + e^{-hs}N(s)$$

For each fixed $h > 0$, it can be shown that there exists $\sigma_{\max}$ such that $\chi(s)$ has no roots in the region $\mathrm{C}_{\sigma_{\max}}$, see [11] for a simple algorithm to estimate $\sigma_{\max}$, based on Nyquist criterion. Given $h > 0$ and a region of the complex plane defined by $\sigma_{\min} \leq \mathrm{Re}(s) \leq \sigma_{\max}$, the problem is to find the roots of $\chi(s)$ in this region.

Clearly, a point $r = \sigma + j\omega$ in $\mathbb{C}$ is a root of $\chi(s)$ if and only if

$$D(\sigma + j\omega) = -e^{-h\sigma}e^{-jh\omega}N(\sigma + j\omega)$$

Taking the magnitude square of both sides of the above equation, $\chi(r) = 0$ implies

$$A_\sigma(x) := D(\sigma + x)D(\sigma - x) - e^{-2h\sigma}N(\sigma + x)N(\sigma - x) = 0$$

where $x = j\omega$. The term $D(\sigma + x)$ stands for the function $D(s)$ evaluated at $\sigma + x$. The other terms of $A_\sigma(x)$ are calculated similarly. For each fixed σ, the function $A_\sigma(x)$ is a polynomial in the variable x. By symmetry, if x is a zero of $A_\sigma(\cdot)$, then $(-x)$ is also a zero.

If $A_\sigma(x)$ has a root x whose real part is zero, set $r = \sigma + x$. Next, evaluate the magnitude of $\chi(r)$; if it is zero, then n is a root of $\chi(s)$. Conversely, if $A_\sigma(x)$ has no root on the imaginary axis, then $\chi(s)$ cannot have a root whose real part is the fixed value of σ from which $A_\sigma(\cdot)$ is constructed.

6.5.2.1 Algorithm

Given $N(s)$, $D(s)$, h, $\sigma_{\min}$, and $\sigma_{\max}$:

Step 1. Pick σ values $\sigma_1,\ldots,\sigma_M$ between $\sigma_{\min}$ and $\sigma_{\max}$ such that $\sigma_{\min} = \sigma_1$, $\sigma_i < \sigma_{i+1}$, and $\sigma_M = \sigma_{\max}$. For each σ_i perform the following.

Step 2. Construct the polynomial $A_i(x)$ according to

$$A_i(x) := D(\sigma_i + x)D(\sigma_i - x) - e^{-2h\sigma_i}N(\sigma_i + x)N(\sigma_i - x)$$

Step 3. For each imaginary axis roots x of A_i, perform the following test:

Check if $|\chi(\sigma_i + x)| = 0$; if yes, then $r = \sigma_i + x$ is a root of $\chi(s)$; if not, discard x.

Step 4. If $i = M$, stop; else increase i by 1 and go to Step 2.

Example

We will find the dominant roots of

$$1 + \frac{e^{-hs}}{s} = 0 \tag{6.26}$$

for a set of critical values of h. Recall that (6.26) has a pair of roots $\pm j1$ when $h = \pi/2 = 1.57$. Moreover, dominant roots of (6.26) are in the right half plane if $h > 1.57$, and they are in the left half plane if $h < 1.57$. So, it is expected that for $h \in (1.2, 2.0)$ the dominant roots are near the imaginary axis. Take $\sigma_{min} = -0.5$ and $\sigma_{max} = 0.5$, with $M = 400$ linearly spaced σ_i's between them. In this case

$$A_i(x) = \sigma_i^2 - e^{-2h\sigma_i} - x^2$$

Whenever $e^{-2h\sigma_i} \geq \sigma_i^2$, $A_i(x)$ has two roots:

$$x_{<} = \pm j\sqrt{e^{-2h\sigma_i} - \sigma_i^2}, \quad < = 1, 2$$

For each fixed σ_i satisfying this condition, let $r = \sigma_i + x$ (note that x is a function of σ_i, so r is a function of σ_i) and evaluate

$$f(\sigma_i) := \left|1 + \frac{e^{-hr_{<}}}{r_{<}}\right|$$

If $f(\sigma_i) = 0$, then r is a root of (6.26). For 10 different values of $h \in (1.2, 2.0)$, the function $f(\sigma)$ is plotted in Figure 6.19. This figure shows the feasible values of σ_i for which r (defined from σ_i) is a root of (6.26).

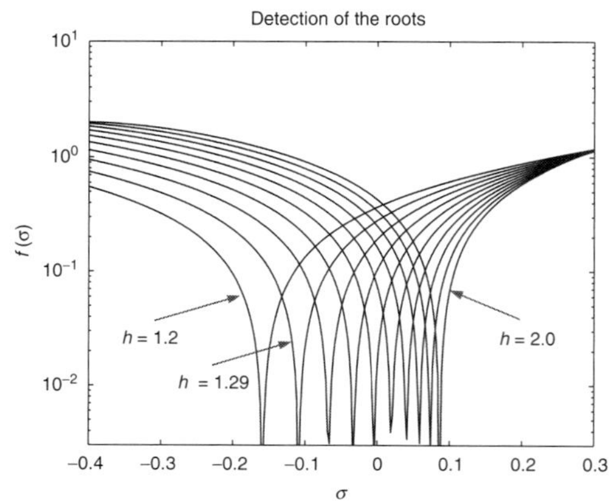

FIGURE 6.19 Detection of the dominant roots.

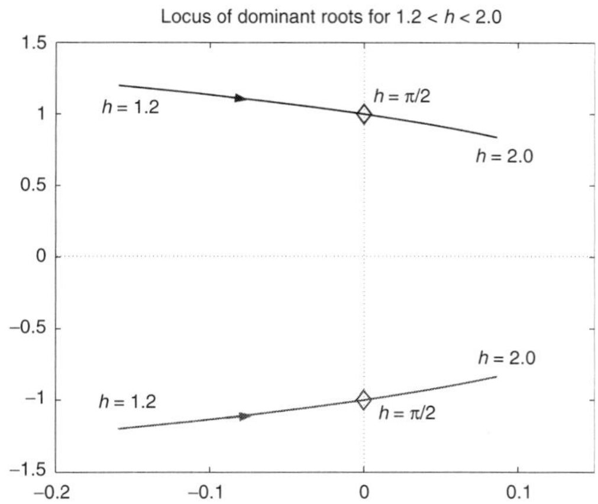

FIGURE 6.20 Dominant roots as h varies from 1.2 to 2.0.

The dominant roots of (6.26), as h varies from 1.2 to 2.0, are shown in Figure 6.20. For $h < 1.57$, all the roots are in $\mathbb{C}_-$. For $h > 1.57$, the dominant roots are in $\mathbb{C}_+$, and for $h = 1.57$, they are at $\pm j1$.

6.5.3 Root Locus Using Padé Approximations

In this section we assume that $h > 0$ is fixed and we try to obtain the root locus, with respect to uncertain/adjustable gain K, corresponding to the dominant poles. The problem can be solved by numerically calculating the dominant roots of the quasi-polynomial

$$\chi(s) = D(s) + KN(s)e^{-hs} \tag{6.27}$$

for varying K, by using the methods presented in the previous section. In this section an alternative method is given that uses Padé approximation of the time delay term e^{-hs}. More precisely, the idea is to find polynomials $N_h(s)$ and $D_h(s)$ satisfying

$$e^{-hs} \approx \frac{N_h(s)}{D_h(s)} \tag{6.28}$$

so that the dominant roots

$$D(s)D_h(s) + KN(s)N_h(s) = 0 \tag{6.29}$$

closely match the dominant roots of $\chi(s)$, (6.27). How should we do the approximation (6.28) for this match?

By using the stability robustness measures determined from the Nyquist stability criterion, we can show that for our purpose we may consider the following cost function in order to define a meaningful measure for the approximation error:

$$\Delta_h =: \sup_{\omega} \left|\frac{K_{\max}N(j\omega)}{D(j\omega)}\right|\left|e^{-jh\omega} - \frac{N_h(j\omega)}{D_h(j\omega)}\right|$$

where $K_{\max}$ is the maximum value of interest for the uncertain/adjustable parameter K.

The th order Padé approximation is defined as follows:

$$N_h(s) = \sum_{k=0} (-1)^k c_k h^k s^k$$

$$D_h(s) = \sum c_k h^k s^k$$

where coefficients c_k's are computed from

$$c_k = \frac{(2, - k)!,!}{2,!k!(, - k)!}, \quad k = 0, 1, \ldots,,$$

First- and second-order approximations are in the form

$$\frac{N_h(s)}{D_h(s)} = \begin{cases} \dfrac{1 - hs/2}{1 + hs/2}, & , = 1 \\ \dfrac{1 - hs/2 + (hs)^2/12}{1 + hs/2 + (hs)^2/12}, & , = 2 \end{cases}$$

Given the problem data $\{h, K_{\max}, N(s), D(s)\}$, how do we find the smallest degree, , of the Padé approximation, so that $\Delta_h \leq \delta$ (or $\Delta_h / K_{\max} \leq \delta'$) for a specified error δ, or a specified relative error δ'? The answer lies in the following result [7]: for a given degree of approximation we have

$$\left| e^{-jh\omega} - \frac{N_h(j\omega)}{D_h(j\omega)} \right| \leq \begin{cases} 2\left(\dfrac{eh\omega}{4,}\right)^{2,+1}, & \omega \leq \dfrac{4,}{eh} \\ 2, & \omega \geq \dfrac{4,}{eh} \end{cases}$$

In light of this result, we can solve the approximation order selection problem by using the following procedure:

1. Determine the frequency ω_x such that

$$\left| \frac{K_{\max} N(j\omega)}{D(j\omega)} \right| \leq \frac{\delta}{2}, \quad \text{for all } \omega \geq \omega_x$$

and initialize = 1.

2. For each ≥ 1 define

$$\omega_< = \max\left\{ \omega_x, \frac{4<}{eh} \right\}$$

and plot the function

$$\Phi_<(\omega) := \begin{cases} 2\left| \dfrac{K_{\max} N(j\omega)}{D(j\omega)} \right| \left(\dfrac{eh\omega}{4<}\right)^{2<+1}, & \text{for } \omega \leq \dfrac{4<}{eh} \\ 2\left| \dfrac{K_{\max} N(j\omega)}{D(j\omega)} \right|, & \text{for } \omega_< \geq \omega \geq \dfrac{4<}{eh} \end{cases}$$

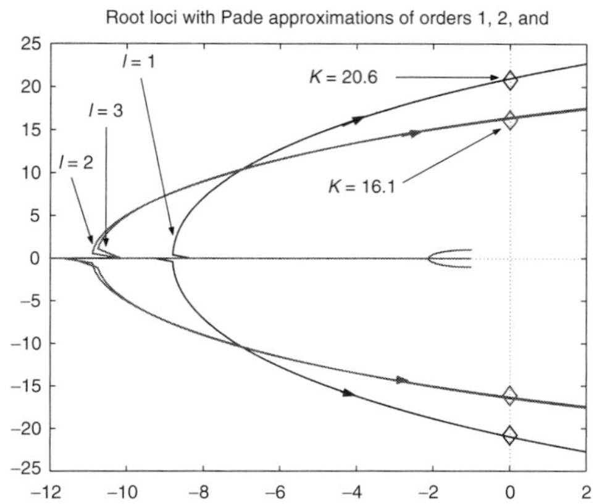

FIGURE 6.21 Dominant root for = 1.

3. Check if

$$\max_{\omega \in [0,\omega_x]} \Phi_{<}(\omega) \leq \delta \tag{6.30}$$

   If yes, stop, this value of satisfies the desired error bound: $\Delta_h \leq \delta$. Otherwise, increase by 1, and go to Step 2. Note that the left-hand side of the inequality (6.30) is an upper bound of Δ_h.

Since we assumed $deg(D) > deg(N)$, the algorithm will pass Step 3 eventually for some finite $\ell \geq 1$. At each iteration, we have to draw the error function $\Phi_\ell(\omega)$ and check whether its peak value is less than δ. Typically, as δ decreases, ω_x increases, and that forces ℓ to increase. On the other hand, for very large values of ℓ, the relative magnitude c_0/c_ℓ of the coefficients becomes very large, in which case numerical difficulties arise in analysis and simulations. Also, as time delay h increases, ℓ should be increased to keep the level of the approximation error δ fixed. This is a fundamental difficulty associated with time delay systems.

Example

Let $N(s) = s + 1$, $D(s) = s^2 + 2s + 2$ and $h = 0.1$, and $K_{max} = 20$. Then, for $\delta = 0.05$, applying the above procedure we calculate $\ell = 2$ as the smallest approximation degree satisfying $\Delta_h/K_{max} < \delta$. Therefore, a second-order approximation of the time delay should be sufficient for predicting the dominant poles for $K \in [0, 20]$. Figure 6.21 shows the approximate root loci obtained from Padé approximations of degrees $\ell = 1, 2, 3$. There is a significant difference between the root loci for $\ell = 1$ and $\ell = 2$. In the region $\text{Re}(s) \geq -12$, the predicted dominant roots are approximately the same for $\ell = 2, 3$, for $K \in [0, 20]$. So, we can safely say that using higher order approximations will not make any significant difference as far as predicting the behavior of the dominant poles for the given range of K.

6.6 Notes

This chapter in the handbook is an edited version of related parts of the author's book [9]. More detailed discussions of the root locus method can be found in all the classical control books, such as [2, 5, 6, 8]. As mentioned earlier, extension of this method to discrete time systems is rather trivial: the method to find the roots of a polynomial as a function of a varying real parameter is independent of the variable s (in the continuous time case) or z (in the discrete time case). The only difference between these two cases is the definition of the desired region of the complex plane: for the continuous time systems, this is defined relative to the imaginary axis, whereas for the discrete time systems the region is defined with respect to the unit circle, as illustrated in Figure 6.6.

References

1. Bellman, R. E., and Cooke, K. L., *Differential Difference Equations,* Academic Press, New York, 1963.
2. Dorf, R. C., and Bishop, R. H., *Modern Control Systems,* 9th ed., Prentice-Hall, Upper Saddle River, NJ, 2001.
3. Evans, W. R., "Graphical analysis of control systems," *Transac. Amer. Inst. Electrical Engineers,* vol. 67 (1948), pp. 547–551.
4. Evans, W. R., "Control system synthesis by root locus method," *Transac. Amer. Inst. Electrical Engineers,* vol. 69 (1950), pp. 66–69.
5. Franklin, G. F., Powell, J. D., and Emami-Naeini, A., *Feedback Control of Dynamic Systems,* 3rd ed., Addison Wesley, Reading, MA, 1994.
6. Kuo, B. C., *Automatic Control Systems,* 7th ed., Prentice-Hall, Upper Saddle River, NJ, 1995.
7. Lam, J., "Convergence of a class of Padé approximations for delay systems," *Int. J. Control,* vol. 52 (1990), pp. 989–1008.
8. Ogata, K., *Modern Control Engineering,* 3rd ed., Prentice-Hall, Upper Saddle River, NJ, 1997.
9. Özbay, H., *Introduction to Feedback Control Theory,* CRC Press LLC, Boca Raton, FL, 2000.
10. Stepan, G., *Retarded Dynamical Systems: Stability and Characteristic Functions,* Longman Scientific & Technical, New York, 1989.
11. Ulus, C., "Numerical computation of inner-outer factors for a class of retarded delay systems," *Int. J. Systems Sci.,* vol. 28 (1997), pp. 897–904.

7

Frequency Response Methods

Jyh-Jong Sheen

National Taiwan Ocean University

7.1 Introduction

The analysis and design of industrial control systems are often accomplished utilizing frequency response methods. By the term frequency response, we mean the steady-state response of a linear constant coefficient system to a sinusoidal input test signal. We will see that the response of the system to a sinusoidal input signal is also a sinusoidal output signal at the same frequency as the input. However, the magnitude and phase of the output signal differ from those of the input signal, and the amount of difference is a function of the input frequency. Thus, we will be investigating the relationship between the transfer function and the frequency response of linear stable systems.

Consider a stable linear constant coefficient system shown in Figure 7.1. Using Euler's formula, $e^{j\omega t} = \cos \omega t + j \sin \omega t$, let us assume that the input sinusoidal signal is given by

$$u(t) = U_0 e^{j\omega t} = U_0 \cos \omega t + jU_0 \sin \omega t \tag{7.1}$$

Taking the Laplace transform of $u(t)$ gives

$$U(s) = \frac{U_0}{s - j\omega} = U_0 \frac{s + j\omega}{s^2 + \omega^2} = \frac{U_0 s}{s^2 + \omega^2} + j\frac{U_0 \omega}{s^2 + \omega^2} \tag{7.2}$$

The first term in Equation 7.2 is the Laplace transform of $U_0 \cos \omega t$, while the second term, without the imaginary number j, is the Laplace transform of $U_0 \sin \omega t$.

Suppose that the transfer function $G(s)$ can be written as

$$G(s) = \frac{n(s)}{d(s)} = \frac{n(s)}{(s + p_1)(s + p_2)\cdots(s + p_n)} \tag{7.3}$$

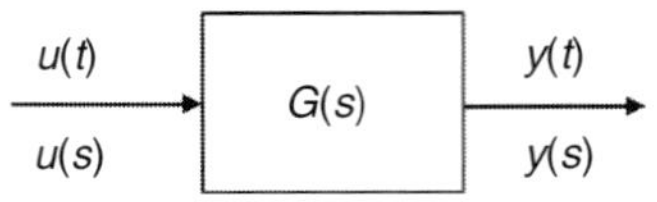

FIGURE 7.1 A stable linear constant coefficient system.

where p_i, $i = 1, 2,\ldots, n$, are assumed to be distinct poles. The Laplace transform of the output $Y(s)$ is then

$$Y(s) = G(s)U(s) = G(s)\frac{U_0}{s-j\omega} \tag{7.4}$$

Taking the partial fraction expansion of $Y(s)$ gives

$$Y(s) = \frac{k_1}{s+p_1} + \cdots + \frac{k_n}{s+p_n} + \frac{\alpha}{s-j\omega} \tag{7.5}$$

The coefficient α can be determined by

$$\alpha = [(s-j\omega)Y(s)]\big|_{s=j\omega} = [U_0 G(s)]\big|_{s=j\omega} = U_0 G(j\omega)$$

Therefore, the inverse Laplace transform of $Y(s)$ yields

$$y(t) = k_1 e^{-p_1 t} + \cdots + k_n e^{-p_n t} + U_0 G(j\omega)e^{j\omega t}, \quad t \geq 0 \tag{7.6}$$

For a stable system, all $-p_i$ have negative nonzero real parts and, therefore, all the terms $k_i e^{-p_i t}$, $i = 1, 2,\ldots, n$, approach zero as t approaches infinity. Thus, at steady state, the output $y(t)$ becomes

$$y_{ss}(t) = \lim_{t\to\infty} y(t) = U_0 G(j\omega)e^{j\omega t} = U_0|G(j\omega)|e^{j(\omega t+\phi)} \tag{7.7}$$

The sinusoidal transfer function, $G(j\omega)$, is written in exponential form

$$G(j\omega) = |G(j\omega)|e^{j\phi}$$

where

$$|G(j\omega)| = \sqrt{\{\mathrm{Re}[G(j\omega)]\}^2 + \{\mathrm{Im}[G(j\omega)]\}^2} \tag{7.8a}$$

and

$$\phi = \angle G(j\omega) = \tan^{-1}\frac{\mathrm{Im}[G(j\omega)]}{\mathrm{Re}[G(j\omega)]} \tag{7.8b}$$

Equation 7.7 shows that for a stable system subject to a sinusoidal input, the steady-state response is a sinusoidal output of the same frequency as the input. The amplitude of the output is that of the input times $|G(j\omega)|$, and the phase angle differs from that of the input by the amount $\phi = \angle G(j\omega)$.

Example 1

A first-order low-pass filter is shown in Figure 7.2. The transfer function of this filter is

$$G(s) = \frac{V_o(s)}{V_i(s)} = \frac{1}{RCs+1}$$

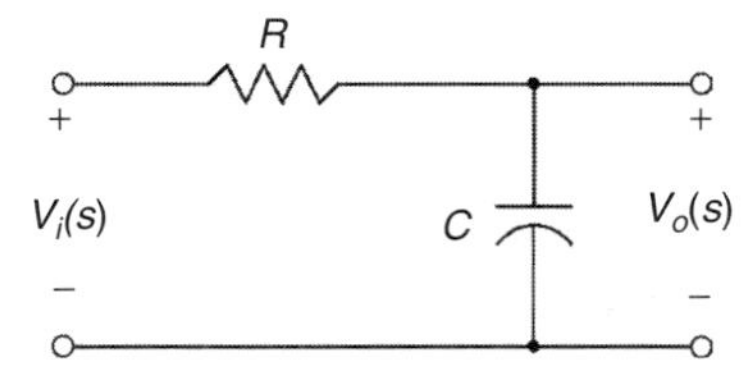

FIGURE 7.2 A first-order low-pass filter.

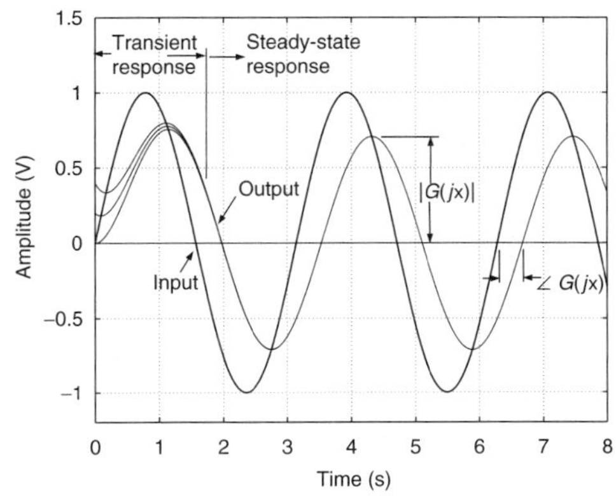

FIGURE 7.3 Frequency response of $G(s) = 1/(0.5s + 1)$ to $u(t) = \sin 2t$.

The sinusoidal transfer function is given by

$$G(j\omega) = \frac{1}{j\omega(RC)+1} = \frac{1}{j(\omega/\omega_1)+1}$$

where $\omega_1 = 1/RC$. The magnitude and phase angle of the frequency response are

$$|G(j\omega)| = \frac{1}{\sqrt{1+(\omega/\omega_1)^2}} \quad \text{and} \quad \phi(\omega) = -\tan^{-1}\frac{\omega}{\omega_1}$$

Figure 7.3 shows the response of the system with $RC = 0.5$ to the input $u = \sin 2t$. It can be seen that the steady-state response is irrelevant to the initial conditions, and the steady-state amplitude of the output is $1/\sqrt{2}$ and the phase angle is $-45°$.

7.2 Bode Plots

There are three commonly used displays of frequency response of a system. They are:

1. Bode diagram or logarithmic plot
2. Polar plot
3. Log-magnitude versus phase plot or Nichols chart

We will present Bode diagrams of sinusoidal transfer functions in this section, followed by the sections on polar plots and log-magnitude versus phase plots.

The main advantages in using the logarithmic plot are the capability of plotting low and high frequency characteristics of the transfer function in one diagram, and the relative ease of adding the separate terms of a high-order transfer function graphically. The basic types of factors that may occur in a transfer function are as follows:

1. Constant gain K
2. Poles (or zeros) at the origin $(j\omega)^{\pm n}$
3. Poles (or zeros) on the real axis $(j\omega\tau + 1)^{\pm 1}$
4. Complex conjugate poles (or zeros) $[(j\omega/\omega_n)^2 + 2\varsigma(j\omega/\omega_n) + 1]^{\pm 1}$

The curves of logarithmic magnitude and phase angle for these four factors can easily be drawn and then added together graphically to obtain the curves for the complete transfer function. The process of drawing the logarithmic plot can be further simplified by using asymptotic approximations to these curves and obtaining the actual curves at specific important frequencies.

7.2.1 Constant Gain K

The logarithmic gain for the constant gain K is

$$20\log|K| = \text{constant in decibel}, \quad \angle K = \begin{cases} 0°, & \text{if } K > 0 \\ -180°, & \text{if } K < 0 \end{cases}$$

The gain and phase curves are simply horizontal lines on the Bode diagram.

7.2.2 Poles (or Zeros) at the Origin $(j\omega)^{\pm n}$

Since

$$20\log|j\omega|^{\pm n} = \pm 20 n \log\omega, \quad \angle(j\omega)^{\pm n} = \pm n \times 90°$$

the slopes of the magnitude curves are $\pm 20n$ dB/decade for the factor $(j\omega)^{\pm n}$ and the phase angles are constants equal to $\pm n \times 90°$.

7.2.3 Poles (or Zeros) on the Real Axis $(j\omega\tau + 1)^{\pm 1}$

For a pole factor $(j\omega\tau + 1)^{-1}$,

$$\left|\frac{1}{j\omega\tau + 1}\right| = \frac{1}{\sqrt{\omega^2\tau^2 + 1}}$$

The magnitude of the pole factor is 1 when $\omega \ll 1/\tau$, and $1/(\omega\tau)$ when $\omega \gg 1/\tau$. Thus, there are two asymptotic curves for the pole factor,

$$20\log\left|\frac{1}{j\omega\tau + 1}\right| \approx \begin{cases} 0\ \text{dB}, & \text{when } \omega \ll \dfrac{1}{\tau} \\ -20\log\omega\tau = -20\left(\log\omega - \log\dfrac{1}{\tau}\right), & \text{when } \omega \gg \dfrac{1}{\tau} \end{cases}$$

The slope of the asymptotic curve when $\omega \gg 1/\tau$ is -20 dB/decade for the pole factor. The two asymptotes intersect at $\omega = 1/\tau$, the break frequency or the corner frequency. The actual logarithmic gain at $\omega = 1/\tau$ is -3 dB. The phase angle is $\phi(\omega) = -\tan^{-1}\omega\tau$.

The Bode diagram of a zero factor $(j\omega\tau + 1)$ is obtained in the same manner. However, the slope of the magnitude asymptotic curve when $\omega \gg 1/\tau$ is $+20$ dB/decade, and the phase angle is $\phi(\omega) = +\tan^{-1}\omega\tau$. The Bode diagrams of first-order factors are shown in Figure 7.4. Linear approximations to the phase angle curves are also presented.

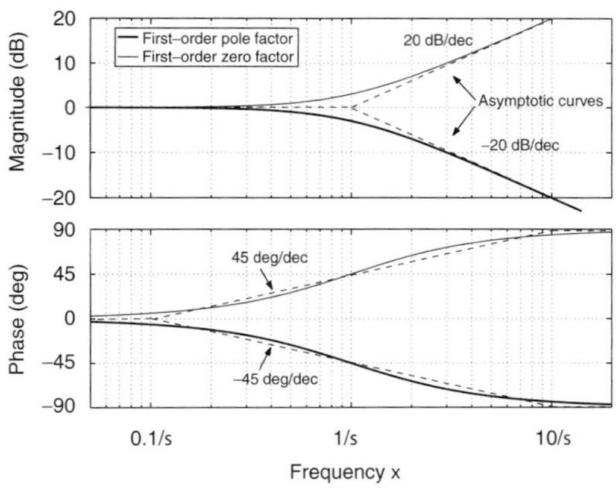

FIGURE 7.4 Bode diagrams for the first-order factors $(j\omega\tau + 1)^{\pm 1}$.

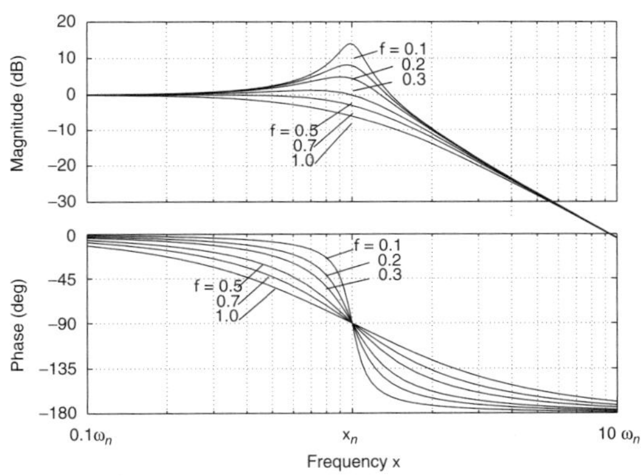

FIGURE 7.5 Bode diagram for the quadratic pole factor $[(j\omega/\omega_n)^2 + 2\zeta(j\omega/\omega_n) + 1)]^{-1}$.

7.2.4 Complex Conjugate Poles (or Zeros) $[(j\omega/\omega_n)^2 + 2\varsigma(j\omega/\omega_n) + 1]^{\pm 1}$

The magnitude and phase angle of the complex conjugate poles $[(j\omega/\omega_n)^2 + 2\zeta(j\omega/\omega_n + 1]^{-1}$ are

$$\left|\left(j\frac{\omega}{\omega_n}\right)^2 + 2\varsigma\left(j\frac{\omega}{\omega_n}\right) + 1\right|^{-1} = \left[\left(1-\frac{\omega^2}{\omega_n^2}\right)^2 + \left(2\varsigma\frac{\omega}{\omega_n}\right)^2\right]^{-1/2}$$

$$\angle\left[\left(j\frac{\omega}{\omega_n}\right)^2 + 2\varsigma\left(j\frac{\omega}{\omega_n}\right) + 1\right]^{-1} = -\tan^{-1}\frac{2\varsigma\omega/\omega_n}{1-\omega^2/\omega_n^2}$$

The magnitude of the complex conjugate pole factor is 1 when $\omega \ll \omega_n$, and $(\omega/\omega_n)^{-2}$ when $\omega \gg \omega_n$. Therefore, the two asymptotic curves for the complex conjugate pole factor are

$$20\log\left|\left(j\frac{\omega}{\omega_n}\right)^2 + 2\varsigma\left(j\frac{\omega}{\omega_n}\right) + 1\right|^{-1} \approx \begin{cases} 0 \text{ dB}, & \text{when } \omega \ll \omega_n \\ -40(\log\omega - \log\omega_n), & \text{when } \omega \gg \omega_n \end{cases}$$

The slope of the asymptotic curve when $\omega \gg \omega_n$ is –40 dB/decade for the complex conjugate pole factor. The magnitude asymptotes intersect at $\omega = \omega_n$, the natural frequency. The actual gain at $\omega = \omega_n$ is $G(j\omega_n) = 1/2\zeta$. The Bode diagram of a complex conjugate pole factor is shown in Figure 7.5. It is seen from Figure 7.5 that the difference between the actual magnitude curve and the asymptotic approximation is a function of damping ratio. The resonant frequency ω_r is defined as the frequency where the peak value of the frequency response M_r occurs. When the damping ratio approaches zero, ω_r approaches ω_n. The resonant frequency can be determined by taking the derivative of the magnitude with respect to the frequency, and setting it equal to zero. The resonant frequency and the peak value of the magnitude are represented by

$$\omega_r = \omega_n\sqrt{1-2\varsigma^2}, \quad \varsigma < 0.707 \tag{7.9a}$$

and

$$M_r = \frac{1}{2\varsigma\sqrt{1-\varsigma^2}}, \quad \varsigma < 0.707 \tag{7.9b}$$

Example 2

Let us consider the transfer function

$$G(s) = \frac{10(s/5+1)}{s(s+1)[(s/10)^2+(s/10)+1]}$$

We first list the basic factors of $G(s)$ in Table 7.1 in the order of increasing corner or natural frequencies.

The complete asymptotic magnitude curve for $G(j\omega)$ is produced by adding together the asymptotic logarithmic magnitudes of each factor, as shown by the solid line in Figure 7.6. Since the dc gain of each factor is 1, these factors have no effect on the asymptotic magnitude until the frequency approaches their corner or natural frequencies. Thus, the asymptotic magnitude can be quickly obtained by plotting each asymptote in order as frequency increases. The asymptotic curve intersects 20 dB at $\omega = 1$ with the slope −20 dB/decade due to the pole at the origin and the constant gain $K = 10$. At $\omega = 1$ the slope further decreases to −40 dB/decade due to the pole at $\omega = 1$. Then at $\omega = 5$ the slope increases to −20 dB/decade

TABLE 7.1 The Basic Factors of $G(j\omega)$

Type of Factors	Constant Gain	Pole	Pole	Zero	Complex Poles
Corner frequency	$K = 10$	0	1	5	10
Order	0	−1	−1	+1	−2

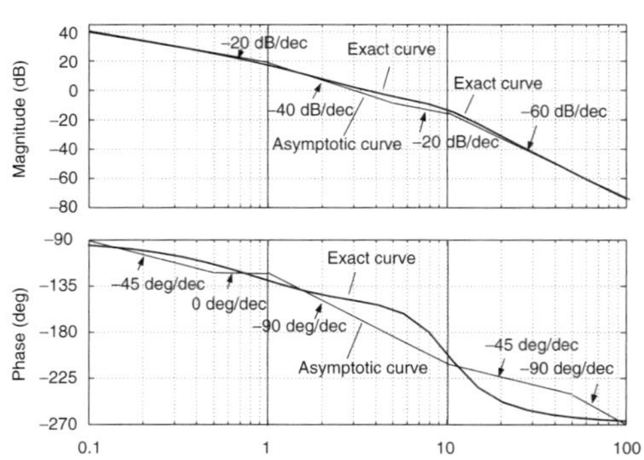

FIGURE 7.6 The Bode plot of the transfer function in Example 2.

due to the zero at $\omega = 5$. Finally at $\omega = 10$ the slope becomes −60 dB/decade due to the complex conjugate poles at $\omega_n = 10$.

The exact magnitude is obtained by calculating the actual magnitude at important frequencies such as the corner or natural frequencies of each factor. The phase curve can be obtained by adding the phase due to each factor. Although the linear approximation of the phase characteristic for a single pole or zero is suitable for initial analysis, the error between the exact phase curve and the linear approximation of complex conjugate poles can be large, as seen in Figure 7.6. Hence, if the accurate phase angle curve is required, a computer program such as Matlab or Ctrl-C can be utilized to generate the actual phase curve.

7.3 Polar Plots

The polar plot of a sinusoidal transfer function $G(j\omega)$ is a plot of both the magnitude and the phase of the frequency response in polar coordinates as the frequency ω varies from zero to infinity. Since the sinusoidal transfer function $G(j\omega)$ can be expressed as

$$G(j\omega) = \mathrm{Re}[G(j\omega)] + j\mathrm{Im}[G(j\omega)] = |G(j\omega)|e^{j\phi}$$

the polar plot of $G(j\omega)$ is a plot of $\mathrm{Re}[G(j\omega)]$ on the horizontal axis versus $\mathrm{Im}[G(j\omega)]$ on the vertical axis in the complex $G(s)$-plane as ω varies from zero to infinity. Hence, for each value of ω, a polar plot of $G(j\omega)$ is defined by a vector of length $|G(j\omega)|$ and a phase angle $\phi = \angle G(j\omega)$, as in Equation 7.8.

We can investigate the general shapes of polar plots according to the system types and relative degrees of transfer functions. Relative degree of a transfer function is defined as the difference between the degree of the denominator polynomial and that of the numerator. Consider a transfer function of the form

$$\begin{aligned} G(j\omega) &= \frac{K(1 + j\omega\tau_a)(1 + j\omega\tau_b)\cdots}{(j\omega)^N(1 + j\omega\tau_1)(1 + j\omega\tau_2)\cdots} \\ &= \frac{b_0(j\omega)^m + b_1(j\omega)^{m-1} + \cdots}{a_0(j\omega)^n + b_1(j\omega)^{n-1} + \cdots} \end{aligned}$$

where $K > 0$ and the relative degree $n - m \geq 0$. The magnitudes and phase angles of $G(j\omega)$ as ω approaches zero and infinity are presented in Table 7.2. The general shapes of the polar plots of various system types in the low-frequency portion are shown in Figure 7.7. The high-frequency portions of the polar plots of various relative degrees are shown in Figure 7.8. It can be seen that the $G(j\omega)$ loci are parallel to either the horizontal or the vertical axes with infinite magnitude as $\omega \to 0^+$ for system types greater than zero. If the relative degree is greater than zero, the $G(j\omega)$ loci converge to the origin clockwise and are tangent to one or the other axes. Note that the polar plot curves can be very complicated due to the numerator and denominator dynamics over the intermediate frequency range. Therefore, the polar plot of $G(j\omega)$ in the frequency range of interest must be accurately determined.

TABLE 7.2 $G(j\omega)$ versus System Type and Relative Degree as $\omega \to 0^+$ and ∞

System Type N	$\omega \to 0^+$	Relative Degree $n - m$	$\omega \to \infty$
0	$K \angle 0°$	0	$b_0/a_0 \angle 0°$
1	$\infty \angle -90°$	1	$0 \angle -90°$
2	$\infty \angle -180°$	2	$0 \angle -180°$
3	$\infty \angle -270°$	3	$0 \angle -270°$

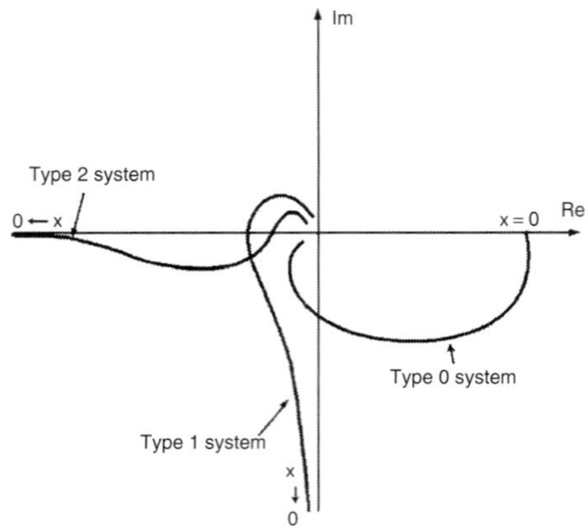

FIGURE 7.7 Polar plots of system with various system types as $\omega \rightarrow 0$.

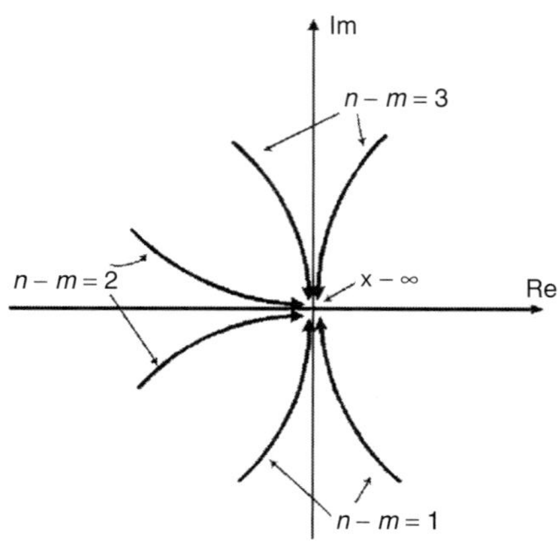

FIGURE 7.8 Polar plots of system with various relative degrees as $\omega \rightarrow \infty$.

We will see that for a closed-loop system, the polar plot of the loop transfer function is useful in determining the stability of the system. The polar plots of some simple systems are shown in Figure 7.9.

7.4 Log-Magnitude versus Phase Plots

Another approach to presenting the frequency response of a system by a single graph is to plot its logarithmic magnitude versus the phase angle over a frequency range of interest. The resulting curve is a function of the frequency ω. Such log-magnitude versus phase plots are called Nichols charts.

Advantages of the Nichols chart are that the relative stability of the closed-loop system can be determined quickly and that the process of closed-loop compensation can be carried out easily. The Nichols charts of the systems in Figure 7.9 are depicted in Figure 7.10 for comparison. Figure 7.11 displays three different frequency-response curves of the second-order system

$$G(s) = \frac{\omega_n^2}{s^2 + 2\varsigma\omega_n s + \omega_n^2}$$

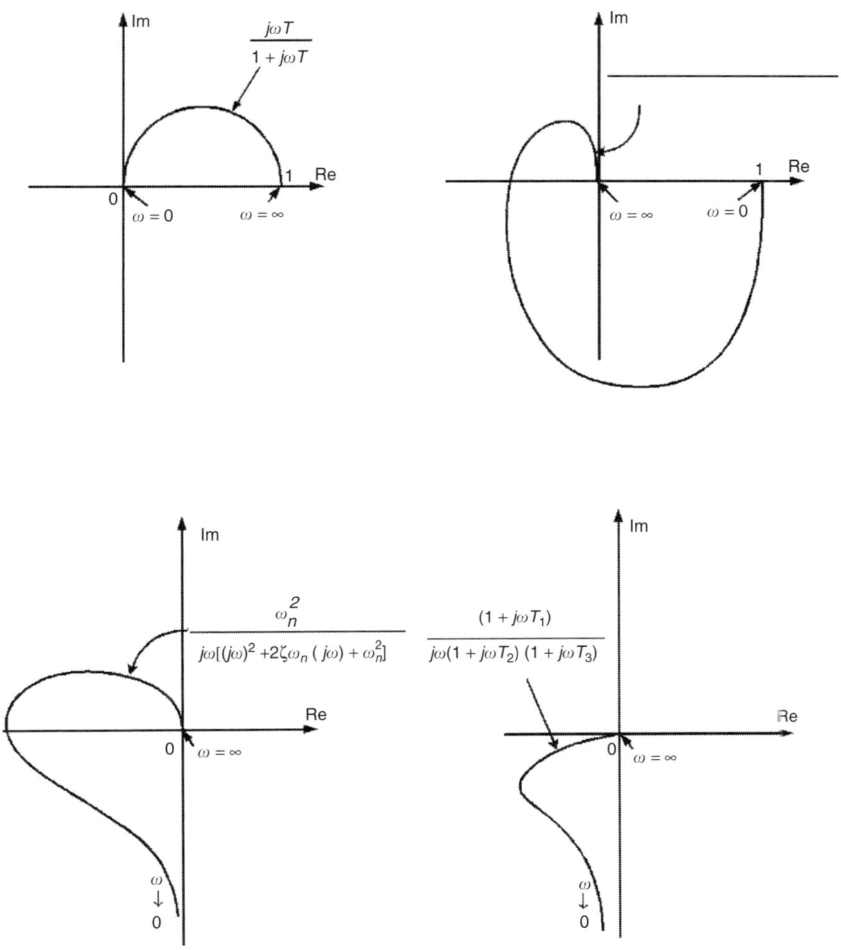

FIGURE 7.9 Polar plots of simple transfer functions.

7.5 Experimental Determination of Transfer Functions

We can obtain a transfer function model from frequency-response measurements of a stable system. First, the Bode diagram of the frequency response is plotted from the measurements. Then the open-loop transfer function can be deduced from the magnitude and phase plots based on the relationships of the basic pole and zero factors.

A wave analyzer is a device to measure the amplitudes and phases of the steady-state response as the frequency of the input sinusoidal wave is altered. A transfer function analyzer can be used to measure the open-loop and closed-loop transfer functions.

We will use a computer program combined with an analog-to-digital and digital-to-analog (AD and DA) card to generate the sinusoidal input signal and to measure the frequency response of a system. Consider the second-order Sallen-Key low-pass filter in Figure 7.12. The transfer function of the filter is given by

$$G(s) = \frac{V_o(s)}{V_i(s)} = \frac{K}{s^2/\omega_n^2 + 2\varsigma(s/\omega_n) + 1} \tag{7.10}$$

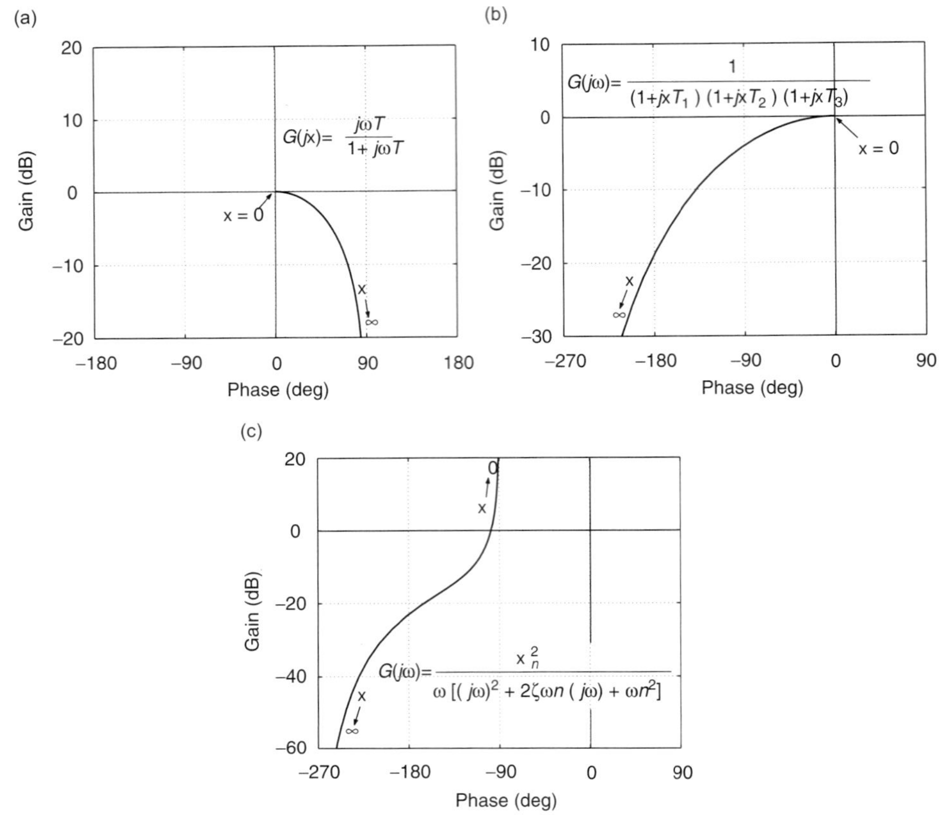

FIGURE 7.10 Nichols charts of simple transfer functions.

where

$$K = \frac{R_1 + R_2}{R_2}, \quad \omega_n = \frac{1}{\sqrt{R_A R_B C_A C_B}}$$

and

$$\varsigma = \frac{1}{2}\left[\sqrt{\frac{C_B}{C_A}}\frac{R_A + R_B}{\sqrt{R_A R_B}} + (1 - K)\sqrt{\frac{R_A C_A}{R_B C_B}}\right]$$

The Real-Time Windows Target in MATLAB is used with an Advantech PCL-818L AD and DA card. The sampling time is 0.001 s. The measured magnitudes and phase angles are shown in Figure 7.13. From the Bode plot, we can find that the dc gain is equal to 1.995 and the natural frequency $\omega_n = 17.90$ rad/s. From Equation 7.9b and $M_r = 1.993$, we have $\zeta = 0.26$.

An alternative to estimating the transfer function is to use an excitation signal that is sufficiently rich in the frequency contents of interest and to measure the corresponding output. System identification technique is then applied to find the order and parameters of the transfer function. Suitable excitation signals are the impulse signal, sweep sine signal, random sequence, and so forth. Figure 7.14 presents the sweep sine input and the corresponding output. The MATLAB System Identification Toolbox is then

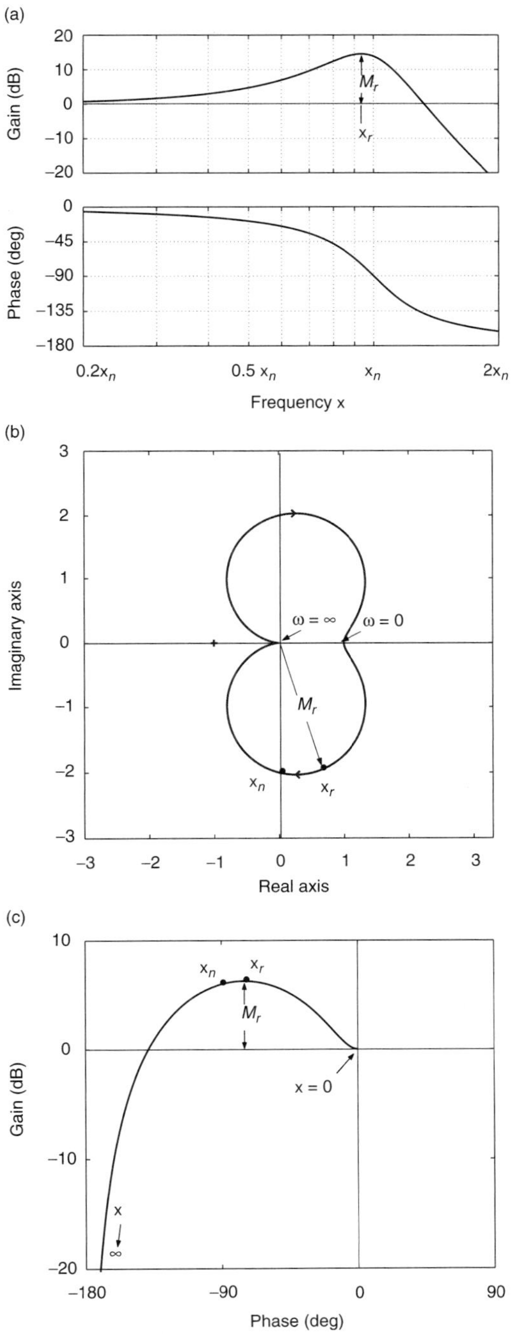

FIGURE 7.11 Three frequency response representations of $G(s) = \omega_n^2/(s^2 + 2\zeta\omega_n s + \omega_n^2)$: (a) Bode diagram, (b) polar plot, and (c) Nichols chart.

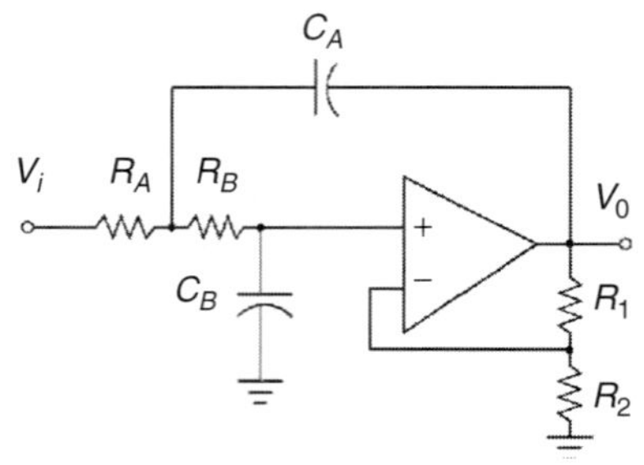

FIGURE 7.12 Sallen-Key low-pass filter.

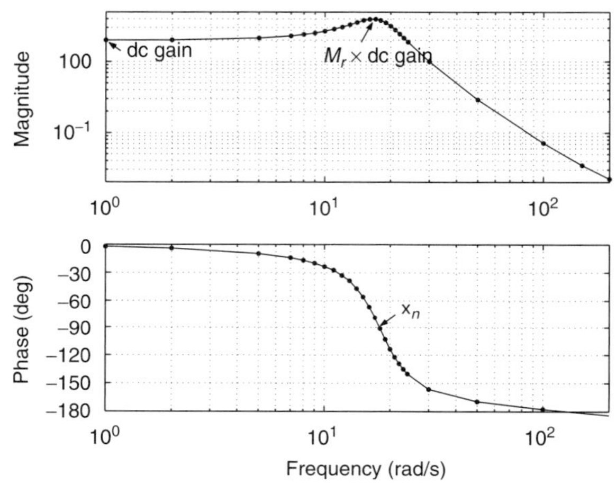

FIGURE 7.13 Frequency response of the Sallen-Key filter from experimental data.

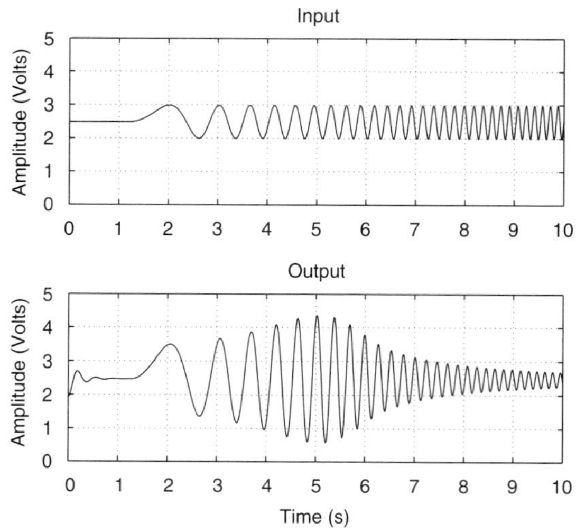

FIGURE 7.14 Sweep sine response of the Sallen-Key filter.

TABLE 7.3 Estimated Transfer Functions for the Second-Order Low-Pass Filter

1. Ideal op-amp circuit: $G(s) = \dfrac{1.997}{(s/18.09)^2 + 2 \times 0.271(s/18.09) + 1}$

   where the measured values of resistors and capacitors are substituted in Equation 7.10 with $R_1 = 98.4$ kΩ, $R_2 = 98.7$ kΩ, $R_A = 51.3$ kΩ, $R_B = 98.5$ kΩ, $C_A = 1.083$ μF, and $C_B = 0.564$ μF.

2. From the Bode plot: $G(s) = \dfrac{1.995}{(s/17.90)^2 + 2 \times 0.259(s/17.90) + 1}$

3. System identification: $G(s) = \dfrac{1.997}{(s/17.78)^2 + 2 \times 0.255(s/17.78) + 1}$

utilized to estimate the transfer function. The resulting transfer functions from the ideal op-amp circuit in Equation 7.10, the Bode plot, and system identification are shown in Table 7.3 for comparison. It is seen that the differences among the three transfer functions are very small. However, the task of determining transfer functions from Bode plots can be very difficult as various pole or zero factors of close corner frequencies can complicate the magnitude and phase plots for high-order systems. Thus, it is recommended that system identification technique be used for determination of high-order transfer functions.

7.6 The Nyquist Stability Criterion

The Nyquist stability criterion provides a graphical procedure for determining the closed-loop stability from the open-loop frequency-response curves. The criterion is based on a result from complex variables theory known as the argument principle, due to Cauchy.

Suppose $F(s)$ is a rational function of s with real coefficients that are analytic everywhere in the s-plane except at its poles. Let Γ_s be a closed, clockwise contour in the s-plane that does not pass through any zeros or poles of $F(s)$. The contour map Γ_F is defined by substituting the values of s on the contour Γ_s for s in $F(s)$. The resulting map is also a closed continuous contour in the $F(s)$-plane. The principle of the argument can be stated as follows:

> A contour map Γ_F of a complex function $F(s)$ defined on Γ_s in the s-plane will only encircle the origin of the $F(s)$-plane if the contour contains a pole or zero of the function. The net number that Γ_F encircles the origin in the clockwise direction is
>
> $$N = Z - P \tag{7.11}$$
>
> where Z and P are, respectively, the numbers of zeros and poles of $F(s)$ enclosed by a closed clockwise contour Γ_s in the s-plane.

Example 3

To illustrate the argument principle, consider a rational function

$$F(s) = \frac{(s+3)(s+4)}{(s+1)(s+2)}$$

which has zeros at $s = -3, -4$ and poles at $s = -1, -2$. The various contour maps of $F(s)$ are shown in Figure 7.15, where Γ_r denotes the contour map of a clockwise circular contour of radius r in the s-plane. We have the following observations from Figure 7.15:

1. The contour map $\Gamma_{0.5}$ does not encircle the origin of the $F(s)$-plane as the contour in the s-plane does not encircle any pole or zero.
2. $\Gamma_{1.99}$ encircles the origin once in the counterclockwise direction as the contour encircles the pole at $s = -1$ in the clockwise direction in the s-plane, and from Equation 7.11, $N = Z - P = 0 - 1 = -1$. Note that $\Gamma_{1.99}$ is a closed contour with two loops and only the one encircling the origin is shown in Figure 7.15.

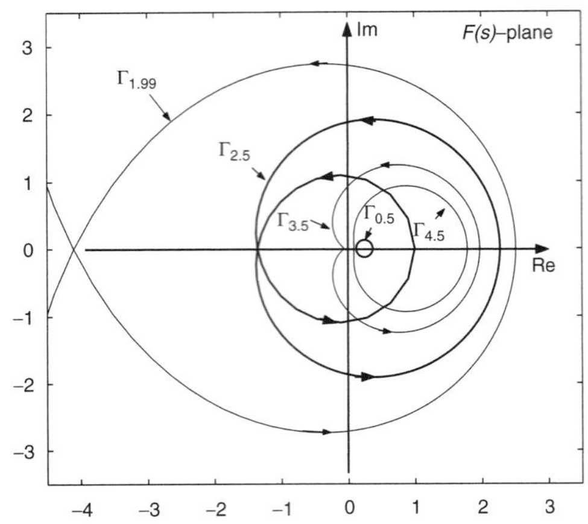

FIGURE 7.15 The contour maps of $F(s)$ in Example 3.

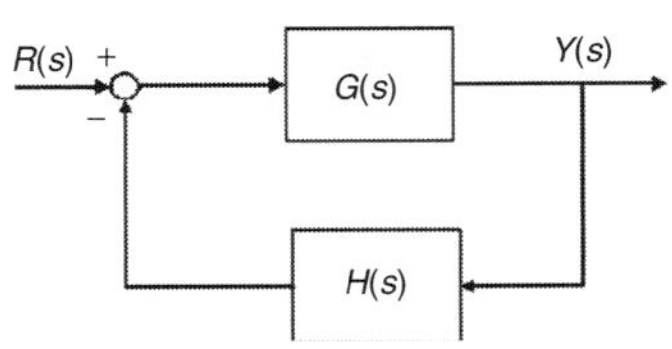

FIGURE 7.16 Closed-loop system.

3. $\Gamma_{2.5}$ encircles the origin twice in the counterclockwise direction as the contour contains two poles at $s = -1, -2$ and $N = Z - P = 0 - 2 = -2$.
4. When the radius of the contour is increased to contain the poles at $s = -1, -2$ and the zero at $s = -3$, then $N = Z - P = 1 - 2 = -1$ and a contour map like $\Gamma_{3.5}$ encircles the origin once in the counterclockwise direction.
5. When the radius of the contour is further increased to encircle the two poles and two zeros, then $N = 2 - 2 = 0$ and the contour map like $\Gamma_{4.5}$ does not encircle the origin.

We now apply Cauchy's principle of argument to develop the Nyquist stability criterion. Suppose that the characteristic equation of the closed-loop system in Figure 7.16 is

$$F(s) = 1 + G(s)H(s) = 0$$

Let $L(s) = G(s)H(s)$, the loop transfer function. Using the argument principle, let us assume that none of the poles or zeros of $F(s)$ lie on the imaginary axis in the s-plane. We now define the Nyquist path, Γ_s, that is composed of the imaginary axis and a semicircle of infinite radius. This contour completely encloses the entire complex right-half plane as depicted in Figure 7.17a. The corresponding contour map Γ_F is shown in Figure 7.17b. It follows from the argument principle that N corresponds to the net number of clockwise encirclements of the origin of the $1 + L(s)$-plane by Γ_F. P is the number of poles of $F(s)$ in the right-half s-plane and thus is the number of poles of the loop transfer function $L(s)$ in the right-half s-plane. Z is the number of zeros of the characteristic equation $F(s)$ of the closed-loop system in the right-half s-plane. Therefore, Z must be zero for the closed-loop system to be stable.

In practice, a modification is made to simplify the application of the Nyquist criterion. Instead of plotting Γ_F in the $1 + L(s)$-plane, we plot just $L(s)$ evaluated along the contour Γ_s. The resulting contour map Γ_L is in the $L(s)$-plane and has the same shape as Γ_F but is shifted 1 unit to the left, as shown in Figure 7.17c. It thus follows that N is the net number of encirclements of the -1 point in the $L(s)$-plane.

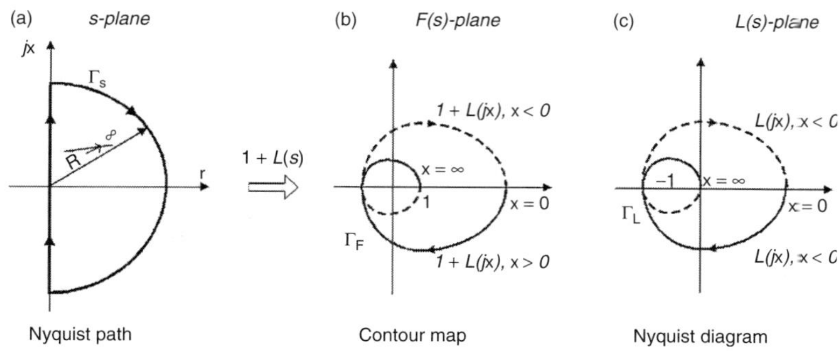

FIGURE 7.17 Nyquist diagram.

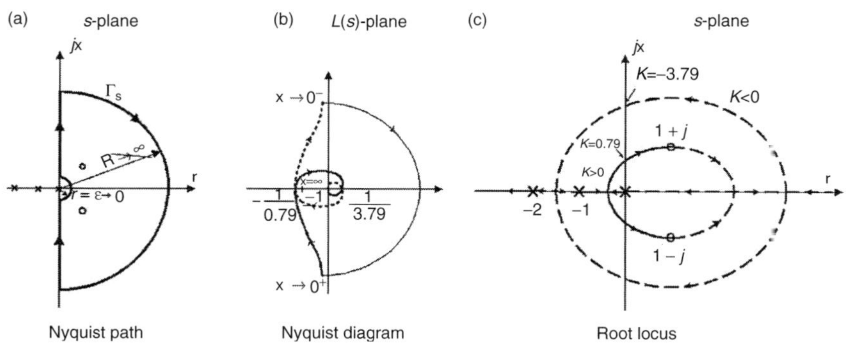

FIGURE 7.18 Nyquist diagram and root locus of Example 4.

The Nyquist stability criterion can now be stated as follows:

A necessary and sufficient condition for the closed-loop stability of a system defined by the loop transfer function $L(s)$ is that

$$Z = N + P \tag{7.12}$$

be equal to zero, where N is the net number of encirclements of the -1 point in the $L(s)$-plane, and P is the number of unstable poles of the loop transfer function $L(s)$.

Example 4

Consider the system with the loop transfer function

$$KL(s) = KG(s)H(s) = K\frac{s^2 - 2s + 2}{s(s+1)(s+2)} \tag{7.13}$$

Let us determine the range of the gain K such that the closed-loop system is stable. Since there is a pole at $s = 0$, we need to modify the Nyquist path to detour around the origin. The contour is shown in Figure 7.18a, where the detour is chosen to be a semicircle of radius approaching zero in the limit. We use the following procedure to sketch the Nyquist plot in Figure 7.18b:

1. Determine $L(j\omega)$ as $\omega \to 0^+$: $L(s)$ is of system type 1 and thus

$$L(j\omega) \approx \frac{1}{j\omega} = \infty\angle{-90^\circ}$$

according to Table 7.2.

2. Determine $L(j\omega)$ as $\omega \to \infty$: $L(s)$ has a relative degree of 1 and

$$L(j\omega) \approx \frac{1}{j\omega} = 0\angle -90°$$

according to Table 7.2.

3. From the Bode plot, draw the polar plot of $L(j\omega)$ as ω varies from 0^+ to ∞. Although the magnitude curve of the factor $(s^2 - 2s + 2)$ is the same as the factor $(s^2 + 2s + 2)$, the phase of the factor $(s^2 - 2s + 2)$ changes from 0° to −180°. Thus, a sketch of the Bode diagram shows that the magnitude curve varies from infinity to zero and the phase changes from −90° to −450°. Since there are two points at which the phases are −180° and −360°, there will be two intersections of the $L(j\omega)$ locus with the real axis in the $L(s)$-plane.
4. Draw the polar plot of $L(j\omega)$ as ω varies from 0^- to $-\infty$ by reflecting the curve of $L(j\omega)$ in procedure 3 with respect to the real axis in the $L(s)$-plane.
5. Determine the contour map of the small detour around the origin of the s-plane to complete the plot. On the detour,

$$s = \lim_{\varepsilon \to 0} \varepsilon e^{j\theta}, \quad -90° \le \theta \le 90°$$

The contour map of the detour can then be determined by

$$\lim_{\varepsilon \to 0} L(\varepsilon e^{j\theta}) = \lim_{\varepsilon \to 0} \frac{(\varepsilon e^{j\theta})^2 - 2\varepsilon e^{j\theta} + 2}{\varepsilon e^{j\theta}(\varepsilon e^{j\theta} + 1)(\varepsilon e^{j\theta} + 2)} = \lim_{\varepsilon \to 0} \frac{1}{\varepsilon e^{j\theta}} = \lim_{\varepsilon \to 0} \frac{1}{\varepsilon} \angle -\theta$$

The resulting map is a large semicircle of radius approaching infinity. This semicircle starts at the point $L(j0^-)$ and swings 180° in the counterclockwise direction to connect the point $L(j0^+)$ in the $L(s)$-plane.

6. Calculate the intersections of the $L(j\omega)$ locus with the real-axis, for these points are related to the relative stability of the system. Suppose that the $L(j\omega)$ locus intersects the real axis for some critical frequency ω_{cr}. Then

$$L(j\omega_{cr}) = \begin{cases} 180° + k360°, & \text{for } K > 0 \\ 0° + k360°, & \text{for } K < 0 \end{cases}$$

where $k = 0, \pm 1, \pm 2, \pm 3, \ldots$. This phase condition at the critical frequency is directly related to the angle condition of the root locus when the root locus crosses the imaginary axis. Therefore, we can utilize the Routh–Hurwitz criterion to determine the points where the $L(j\omega)$ locus crosses the real axis. The characteristic equation of the system (7.13) can be written as

$$s^3 + (K + 3)s^2 + (2 - 2K)s + 2K = 0$$

Thus, the Routh array is

$$\begin{array}{c|cc} s^3 & 1 & 2 - 2K \\ s^2 & K + 3 & 2K \\ s^1 & c & \\ s^0 & 2K & \end{array}$$

where

$$c = \frac{(K + 3)(2 - 2K) - 2K}{K + 3}$$

Let $c = 0$, and solving for K, we get the critical gains

$$K_{cr} = \frac{-3 \pm \sqrt{21}}{2} = 0.79, -3.79$$

Substituting the values of K_{cr} in the auxiliary equation

$$(K_{cr} + 3)s^2 + 2K_{cr} = 0$$

we obtain the critical frequencies

$$\omega_{cr} = \sqrt{\frac{2K_{cr}}{K_{cr} + 3}} = \begin{cases} 0.65, & \text{when } K_{cr} = 0.79 \\ 3.10, & \text{when } K_{cr} = -3.79 \end{cases}$$

At the critical frequency, we have the characteristic equation

$$1 + K_{cr} L(j\omega_{cr}) = 0$$

Hence the points of the $L(j\omega)$ locus that cross the real-axis are

$$L(j\omega_{cr}) = -\frac{1}{K_{cr}} = -\frac{1}{0.79}, \frac{1}{3.79}$$

The complete Nyquist plot is shown not to scale in Figure 7.18b. The range of the gain K for which the system is stable can be determined using Nyquist criterion. For different values of K, the Nyquist diagram needs to be redrawn in order to count the number of encirclement of the -1 point. We can avoid this by counting the number of encirclement of $-1/K$ point instead. From the Nyquist criterion, $Z = N + P$, where $P = 0$. It can be seen from Figure 7.18b that there are four cases of the encirclements of the $-1/K$ point.

1. $K > 0$ and $-1/K < -1/0.79 \Rightarrow 0 < K < 0.79$, and $N = 0$. We have $Z = 0$ and the system is stable.
2. $K > 0$ and $-1/K > -1/0.79 \Rightarrow K > 0.79$, and $N = 2$. We have $Z = 2$ and the system has two unstable poles.
3. $K < 0$ and $-1/K < 1/3.79 \Rightarrow K < -3.79$, and $N = 3$. We have $Z = 3$ and the system has three unstable poles.
4. $K < 0$ and $-1/K > 1/3.79 \Rightarrow -3.79 < K < 0$ and $N = 1$. We have $Z = 1$ and the system has one unstable pole.

The root locus of system (7.13) is also shown in Figure 7.18c for comparison.

7.7 Relative Stability

In designing a control system, it is required that the system be stable. In addition to stability, there are important concerns such as acceptable transient response and capability to deal with model uncertainty. Since the model used in the design and analysis of a control system is never exact, it may suggest a stable system; but the physical system turns out to be unstable. Therefore, it is required that the system not only be stable but also have some stability margin or adequate relative stability.

Suppose that the sinusoidal loop transfer function $L(j\omega)$ locus intersects the -1 point for some critical frequency ω_{cr}. Then

$$L(j\omega_{cr}) = G(j\omega_{cr})H(j\omega_{cr}) = -1$$

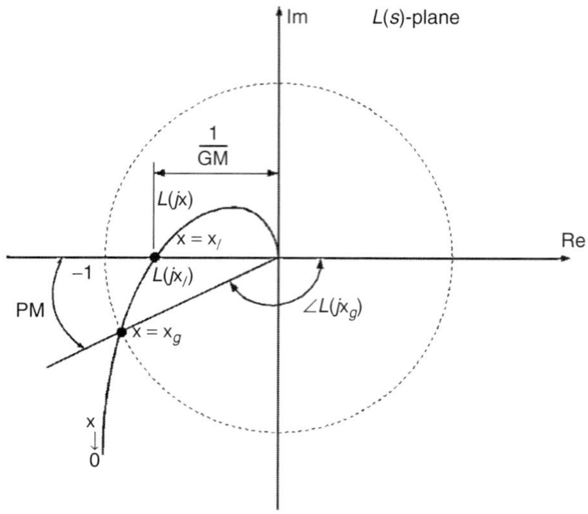

FIGURE 7.19 Gain and phase margins.

or

$$1 + L(j\omega_{cr}) = 1 + G(j\omega_{cr})H(j\omega_{cr}) = 0$$

This indicates that the closed-loop system has a pair of complex poles at $s = \pm j\omega_{cr}$. Hence, the system is marginally stable and oscillates with the frequency ω_{cr}, provided that all other closed-loop system poles are in the left half s-plane. In general, the closer the $L(j\omega)$ locus comes to the $-1 + j0$ point in the Nyquist plot, the more oscillatory is the system response. For this reason, the closeness of the $L(j\omega)$ locus to the -1 point can be used as a measure of the stability margin. Two traditional measures of the stability margin are gain margin and phase margin.

Gain margin and phase margin are usually defined for stable closed-loop systems that are characterized by a minimum phase, loop transfer function $L(s)$. The gain margin is the factor by which the open-loop gain of a stable closed-loop system can be increased before the system goes unstable. The phase margin is the amount of additional phase lag at the gain crossover frequency required to make the stable closed-loop system marginally stable. Thus we have the following definitions:

Gain margin (GM): The gain margin is the reciprocal of the magnitude $|L(j\omega)|$ at the phase crossover frequency ω_ϕ, where the phase of $L(j\omega_\phi)$ reaches $-180°$. The gain margin is given by

$$\text{GM} = \frac{1}{|L(j\omega_\phi)|} \quad \text{or} \quad \text{GM(dB)} = -20\log|L(j\omega_\phi)|$$

Phase margin (PM): The phase margin is defined as the angle between the phase of the loop transfer at the gain crossover frequency ω_g where $|L(j\omega_g)| = 1$ and the angle $-180°$, or $\text{PM} = \angle L(j\omega_g) + 180°$.

The gain and phase margins are shown in Figure 7.19. Gain and phase margins are stability margins for single-input single-output systems. They cannot apply for multi-input multi-output systems. In addition, they can be a poor indication of stability margin in the face of combined gain and phase variations, as shown in Figure 7.20. This is due to the fact that gain and phase margins are measures of stability margin in terms of only pure gain and phase variations, but not a combination of both. As a consequence, a system may have good gain and phase margins, but it is close to instability, as indicated in Figure 7.20. To make up for the insufficiencies of gain and phase margins, a third stability margin, return difference, is used in modern control theory. We will only consider single-input single-output systems.

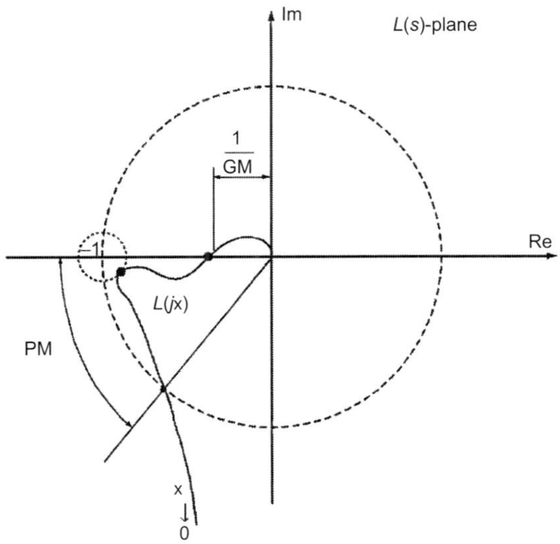

FIGURE 7.20 Insufficiency of gain and phase margins.

Minimum return difference: The minimum return difference is the minimum value of $|1 + L(j\omega)|$, for $0 < \omega < \infty$. It can be seen from Figure 7.20 that the minimum return difference is the minimum distance from the Nyquist plot to the −1 point. Therefore, the gain and phase margins are special cases of the minimum return difference. The gain margin is directly related to the case when the minimum return difference occurs at the phase crossover frequency, and the phase margin is corresponding to the case that the minimum return difference occurs at the gain crossover frequency.

Although the minimum return difference is a better measure of stability margin than the gain and phase margins, it is seldom used in the classical control theory. This is because the classical control analysis and design is usually carried out using the Bode diagram or the Nichols chart instead of the Nyquist plot. The gain and phase margins are more easily determined from the Bode diagram or the Nichols chart than the Nyquist plot. Despite the fact that the minimum return difference can be easily evaluated from the Nyquist plot, it is difficult to determine the minimum return difference from the Bode plot or the Nichols chart.

We now correlate the phase margin and the damping ratio ς of an underdamped second-order system. Consider the standard second-order system

$$T(s) = \frac{\omega_n^2}{s^2 + 2\varsigma\omega_n s + \omega_n^2} \tag{7.14}$$

We assume that the transfer function $T(s)$ comes from a unity feedback configuration and can be rewritten as

$$T(s) = \frac{G(s)}{1 + G(s)}$$

where the open-loop transfer function $G(s)$ is given by

$$G(s) = \frac{\omega_n^2}{s(s + 2\varsigma\omega_n)} \tag{7.15}$$

The phase margin occurs at the gain crossover frequency ω_c when $|G(j\omega_c)| = 1$, or

$$\frac{\omega_n^2}{\omega_c(\omega_c^2 + 4\varsigma^2\omega_n^2)^{1/2}} = 1$$

This equation can be rewritten as

$$(\omega_c^2)^2 + 4\varsigma^2\omega_n^2(\omega_c^2) - \omega_n^4 = 0$$

Solving for positive ω_c, we obtain

$$\frac{\omega_c^2}{\omega_n^2} = (4\varsigma^4 + 1)^{1/2} - 2\varsigma^2$$

Substituting $s = j\omega_c$ into Equation 7.15, the phase margin for the system is

$$\begin{aligned} \text{PM} &= 180° + \angle G(j\omega_c) \\ &= 180° - 90° - \tan^{-1}\left(\frac{\omega_c}{2\varsigma\omega_n}\right) \\ &= 90° - \tan^{-1}\left(\frac{1}{2\varsigma}[(4\varsigma^4 + 1)^{1/2} - 2\varsigma^2]^{1/2}\right) \\ &= \tan^{-1}\left(2\varsigma\left[\frac{1}{(4\varsigma^4 + 1)^{1/2} - 2\varsigma^2}\right]^{1/2}\right) \end{aligned} \tag{7.16}$$

Equation 7.16 relates the damping ratio of the standard second-order system (7.14) to the phase margin of its corresponding open-loop system (7.15) in a unity feedback configuration. This equation provides a correlation between the frequency response and the time response. A plot of ζ versus PM is shown in Figure 7.21. The curve of ζ versus PM can be approximated by a dashed line in Figure 7.21. The linear approximation can be expressed as

$$\varsigma = 0.01\ \text{PM} \tag{7.17}$$

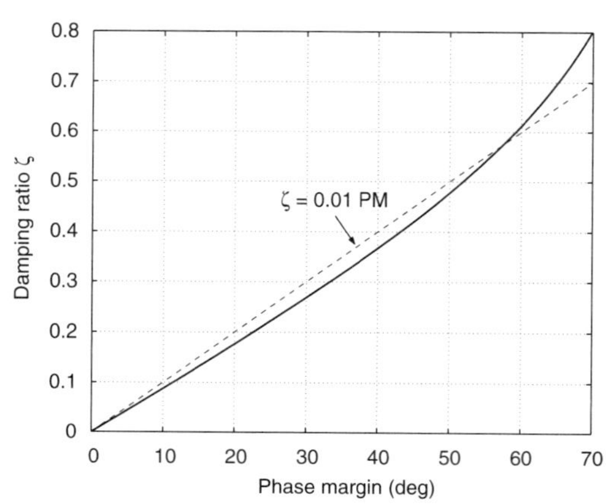

FIGURE 7.21 Damping ratio versus phase margin for a second-order system.

This approximation is reasonably accurate for $\zeta \leq 0.7$ and is useful in relating the frequency response to the transient performance of a second-order system. Equation 7.17 may also be used for higher-order systems if the system can be assumed to have a pair of dominant underdamped complex poles.

References

1. Dorf, R.C., and Bishop, R.H., *Modern Control Systems,* 9th ed., Prentice-Hall, 2001.
2. Ogata, K., *Modern Control Engineering,* 2nd ed., Prentice-Hall, 1990.
3. Kuo, B.C., *Control Systems,* 7th ed., Prentice-Hall, 1995.
4. Franklin, G.F., Powell, J.D., and Emami-Naeini, A., *Feedback Control of Dynamic Systems,* 3rd ed., Addison-Wesley, 1994.
5. Phillips, C.L., and Harbor, R.D., *Feedback Control Systems,* 4th ed., Prentice-Hall, 2000.
6. Wolovich, W.A., *Automatic Control Systems: Basic Analysis and Design,* Harcourt Brace College Publishing, 1994.

8

Kalman Filters as Dynamic System State Observers

Timothy P. Crain II
NASA Johnson Space Center

8.1 The Discrete-Time Linear Kalman Filter

Distilled to its most fundamental elements, the Kalman filter [1] is a predictor–corrector estimation algorithm that uses a dynamic system model to predict state values and a measurement model to correct this prediction. However, the Kalman filter is capable of a great deal more than just state observation in such a manner. By making certain stochastic assumptions, the Kalman filter carries along an internal metric of the statistical confidence of the estimate of individual state elements in the form of a covariance matrix. The essential properties of the Kalman filter are derived from the requirements that the state estimate be

- Linear combination of the previous state estimate and current measurement information
- Unbiased with respect to the true state
- Optimal in terms of having minimum variance with respect to the true state

Starting with these basic requirements an elegant and efficient formulation for the implementation of the Kalman filter may be derived.

The Kalman filter processes a time series of measurements to update the estimate of the system state and utilizes a dynamic model to propagate the state estimate between measurements. The observed measurement is assumed to be a function of the system state and can be represented via

$$Y(t) = h(X(t), \beta, t) + v(t) \quad (8.1)$$

where $Y(t)$ is an m dimensional observable, h is the nonlinear measurement model $X(t)$ is the n-dimensional system state, β is a vector of modeling parameters, and $v(t)$ is a random process accounting for measurement noise.

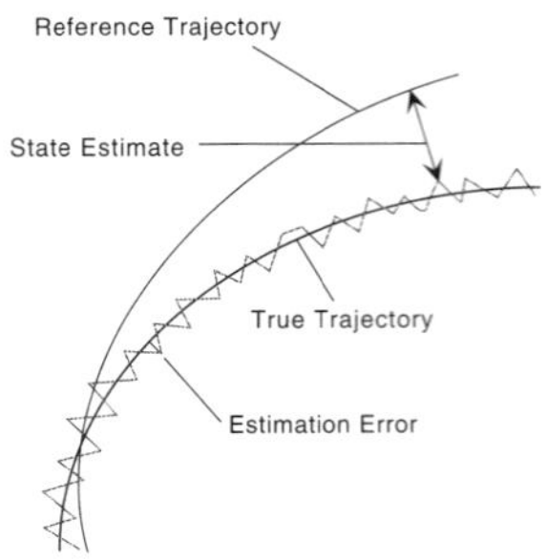

FIGURE 8.1 LKF tracking of a two-dimensional trajectory.

The true dynamic system is described by a general first-order, ordinary differential equation

$$\dot{\mathrm{X}}(t) = f(\mathrm{X}(t), \alpha, t) + \mathbf{w}(t) \tag{8.2}$$

where f is the nonlinear dynamics function that incorporates all significant deterministic effects of the environment, α is a vector of parameters used in the model, and $\mathbf{w}(t)$ is a random process that accounts for the noise present from mismodeling in f or from the quantum uncertainty of the universe, depending on the accuracy of the deterministic model in use.

With these general models available, a linear Kalman filter (LKF) may be derived in a discrete-time formulation. The dynamics and measurement functions are linearized about a known reference state, $\tilde{\mathrm{X}}(t)$, which is related to the true environment state, $\mathrm{X}(t)$, via

$$\tilde{\mathrm{X}}(t) + \mathbf{x}(t) = \mathrm{X}(t) \tag{8.3}$$

The LKF state estimate is related to the true difference by

$$\hat{\mathbf{x}}_k^{(\pm)} = \mathbf{x}_k + \delta\mathbf{x}_k^{(\pm)} \tag{8.4}$$

where the "$\hat{}$" denotes the state estimate (or filter state), $\delta\mathbf{x}_k^{(\pm)}$ is the estimation error, and "±" indicates whether the estimate and error are evaluated instantaneously before (−) or after (+) measurement update at discrete time t_k.

It is important to emphasize that the LKF filter state is the estimate of the difference between the environment and the reference state. The LKF mode of operation will therefore carry along a reference state and the filter state between measurement updates. Only the filter state is at the time of measurement update. Figure 8.1 illustrates the generalized relationship between the true, reference, and filter states in an LKF estimating a two-dimensional trajectory.

8.1.1 Linearization of Dynamic and Measurement System Models

The dynamics and measurement functions may be linearized about the known reference state, $\tilde{\mathrm{X}}(t)$, according to

$$f(\mathrm{X}, \alpha, t)\ .\ f(\tilde{\mathrm{X}}(t), \alpha, t) + \mathrm{F}(\tilde{\mathrm{X}}(t), \alpha, t)\mathbf{x}(t) + \mathbf{w}(t) \tag{8.5}$$

$$h(\mathrm{X}, \alpha, t)\ .\ h(\tilde{\mathrm{X}}(t), \beta, t) + \mathrm{H}(\tilde{\mathrm{X}}(t), \beta, t)\mathbf{x}(t) + \mathbf{v}(t) \tag{8.6}$$

where **F** is the dynamics partial derivative matrix and **H** is the measurement partial derivative matrix defined by

$$\mathbf{F}(\tilde{\mathbf{X}}(t), \alpha, t) = \left.\frac{\partial f}{\partial \mathbf{X}}\right|_{\mathbf{X}=\tilde{\mathbf{X}}} \tag{8.7}$$

$$\mathbf{H}(\tilde{\mathbf{X}}(t), \beta, t) = \left.\frac{\partial h}{\partial \mathbf{X}}\right|_{\mathbf{X}=\tilde{\mathbf{X}}} \tag{8.8}$$

and $\mathbf{x}(t)$ is the true state to be estimated representing the difference between the environment and reference states

$$\mathbf{x}(t) = \mathbf{X}(t) - \tilde{\mathbf{X}}(t) \tag{8.9}$$

After linearizing the dynamic and measurement models, the effect of neglecting the higher order terms is assumed to be included in the random processes $\mathbf{w}(t)$ and $\mathbf{v}(t)$. The linearization is an acceptable approximation if $\mathbf{x}(t)$ is sufficiently small.

The reference and filter states are propagated according to the discrete-time linear relationship

$$\tilde{\mathbf{X}}_{k+1} = \Phi(t_{k+1}, t_k)\tilde{\mathbf{X}}_k \tag{8.10}$$

$$\hat{\mathbf{x}}_{k+1}^{(-)} = \Phi(t_{k+1}, t_k)\hat{\mathbf{x}}_k^{(-)} \tag{8.11}$$

where $\Phi(t_{k+1}, t_k)$ is the state transition matrix from time t_k to time t_{k+1} and has the following properties:

$$\begin{aligned} \Phi(t_k, t_k) &= \mathbf{I} \\ \dot{\Phi}(t_{k+1}, t_k) &= \mathbf{F}(\tilde{\mathbf{X}}(t), \alpha, t)\Phi(t_{k+1}, t_k) \\ \Phi(t_{k+2}, t_k) &= \Phi(t_{k+2}, t_{k+1})\Phi(t_{k+1}, t_k) \end{aligned} \tag{8.12}$$

Note that the system dynamics are now incorporated into the propagation of the reference and filter states by the integration of the dynamics partial derivative in Equation 8.13.

Mathematically, the true difference state is propagated in a similar fashion with the addition of a process noise random value

$$\mathbf{x}_{k+1} = \Phi(t_{k+1}, t_k)\mathbf{x}_k + \mathbf{w}_k \tag{8.13}$$

In general, it is not required that the reference dynamic model be exactly the same as the truth dynamics or that the modeling parameter α be equivalent to the true modeling vector. This notation is left in place to simplify the derivation of the Kalman filter formulation. A number of innovative approaches have been developed for adapting reference model parameters to improve fidelity with the unknown real-world system model [2–6] and can be used to enhance filter operation.

The LKF also requires a linearized measurement, $\mathbf{y}_k = \mathbf{Y}_k - \mathbf{Y}_k$, modeled by

$$\mathbf{y}_k = \mathbf{H}_k\mathbf{x}_k + \mathbf{v}_k \tag{8.14}$$

For the development of the Kalman filters presented here, the random contributions $\mathbf{v}_k$ and $\mathbf{w}_k$ are assumed to be discrete realizations of the continuous zero mean Gaussian process in Equations 8.1 and 8.2 and are defined by

$$E\left[\mathbf{v}_k\mathbf{v}_i^T\right] = \mathbf{R}_k\delta_{ki} \tag{8.15}$$

$$E\left[\mathbf{w}_k\mathbf{w}_i^T\right] = \mathbf{Q}_k\delta_{ki} \tag{8.16}$$

Generally, it is assumed that the process noise and measurement noise sequences are uncorrelated so that

$$E\left[\mathbf{w}_k \mathbf{v}_i^T\right] = 0 \quad \forall k, \forall i \tag{8.17}$$

However, the Kalman filter can be configured to operate in systems where this assumption does not apply [7].

8.1.2 Linear Kalman Filter Error Covariance Propagation

The propagation of the filter and reference states in the LKF were outlined in the previous section in Equations 8.11 and 8.13. However, all Kalman filter formulations must also propagate a confidence metric of the state estimate in the form of a state error covariance matrix. The state error covariance, **P**, is defined as the expectation of the outer product of the estimation error vector

$$\mathbf{P}_k^{(\pm)} = E\left[\delta\mathbf{x}_k^{\pm}\delta\mathbf{x}_k^{\pm T}\right] \tag{8.18}$$

The state error covariance matrix is $n \times n$ and symmetric, and must remain positive definite to retain filter stability. The mechanism for propagating the covariance can be derived by taking the covariance just before measurement update at time t_{k+1}

$$\mathbf{P}_{k+1}^{(-)} = E\left[\delta\mathbf{x}_{k+1}^{(-)}\delta\mathbf{x}_{k+1}^{(-)T}\right] \tag{8.19}$$

and substituting the estimation error and propagation definitions in Equations 8.4, 8.11, and 8.13 to yield

$$\mathbf{P}_{k+1}^{(-)} = E\left[\Phi(t_{k+1}, t_k)\left(\hat{\mathbf{x}}_k^{(+)} - \mathbf{x}_k\right)\left(\hat{\mathbf{x}}_k^{(+)} - \mathbf{x}_k\right)^T \Phi(t_{k+1}, t_k)^T + \mathbf{w}_k\mathbf{w}_k^T\right] \tag{8.20}$$

Utilizing the definitions of process noise covariance in Equation 8.16 and state error covariance in Equation 8.18 the propagation equation reduces to

$$\mathbf{P}_{k+1}^{(-)} = \Phi(t_{k+1}, t_k)\mathbf{P}_k^{(+)}\Phi(t_{k+1}, t_k)^T + \mathbf{Q}_k \tag{8.21}$$

The propagation equation can be interpreted as the sum of the mapping of the previous post-update error covariance through the system dynamics and the system process noise induced uncertainty. Thus, process noise acts to increase the state error covariance between measurement updates.

8.1.3 Linear Kalman Filter Update

The LKF seeks an unbiased, minimum variance solution for the difference state, $\mathbf{x}_k$, by combining previous state information with available measurements. The state estimate after measurement update is therefore assumed to be a linear combination of the pre-update state and the linearized measurement information

$$\hat{\mathbf{x}}_k^{(+)} = \mathbf{K}_k^*\hat{\mathbf{x}}_k^{(-)} + \mathbf{K}_k\mathbf{z}_k \tag{8.22}$$

Substituting Equations 8.4 and 8.14 into Equation 8.22 and solving for the estimation error yields

$$\delta\mathbf{x}_k^{(+)} = (\mathbf{K}_k^* + \mathbf{K}_k\mathbf{H}_k - \mathbf{I})\mathbf{x}_k + \mathbf{K}_k^*\delta\mathbf{x}_k^{(-)} + \mathbf{K}_k\mathbf{v}_k \tag{8.23}$$

By definition $E[\mathbf{v}_k] = \mathbf{0}$ and $E[\delta\mathbf{x}_k^{(-)}] = \mathbf{0}$ by assumption of unbiased estimation. Therefore, the updated state estimation error is unbiased

$$E\left[\delta\mathbf{x}_k^{(+)}\right] = 0 \tag{8.24}$$

only if

$$\mathbf{K}_k^* + \mathbf{K}_k\mathbf{H}_k - \mathbf{I} = 0 \tag{8.25}$$

Substitution of Equation 8.25 into Equation 8.22 results in an expression for the updated state estimate

$$\hat{\mathbf{x}}_k^{(+)} = \hat{\mathbf{x}}_k^{(-)} + \mathbf{K}_k\left(\mathbf{z}_k - \mathbf{H}_k\hat{\mathbf{x}}_k^{(-)}\right) \tag{8.26}$$

with estimation error

$$\delta\mathbf{x}_k^{(+)} = (\mathbf{I} - \mathbf{K}_k\mathbf{H}_k)\,\delta\mathbf{x}_k^{(-)} + \mathbf{K}_k\mathbf{v}_k \tag{8.27}$$

The post-measurement error covariance in Equation 8.18 may be expanded to

$$\mathbf{P}_k^{(+)} = (\mathbf{I} - \mathbf{K}_k\mathbf{H}_k)\mathbf{P}_k^{(-)}(\mathbf{I} - \mathbf{K}_k\mathbf{H}_k)^T + \mathbf{K}_k\mathbf{R}_k\mathbf{K}_k^T \tag{8.28}$$

by substitution of Equation 8.27 and applying the conditions of uncorrelated process and measurement noise, zero mean measurement noise, and the definition of the pre-measurement state estimation error covariance. At this point, only the requirement that the Kalman filter be an unbiased estimator has been satisfied, so now we will select the Kalman gain $\mathbf{K}_k$ that delivers the minimum summed variance on the post-measurement state estimation error. In other words, we seek the gain that will minimize

$$J_k = trace\left[\mathbf{P}_k^{(+)}\right] \tag{8.29}$$

The necessary condition for minimality of J_k is that its partial derivative with respect to the Kalman gain is zero. By employing the following relationship:

$$\frac{\partial}{\partial\mathbf{A}}[trace(\mathbf{ABA}^T)] = 2\mathbf{AB} \tag{8.30}$$

where $\mathbf{B}$ is a symmetric matrix, on the components of Equation 8.28 with respect to $\mathbf{K}_k$ results in

$$(\mathbf{I} - \mathbf{K}_k\mathbf{H}_k)\mathbf{P}_k^{(-)}\mathbf{H}_k^T + \mathbf{K}_k\mathbf{R}_k = 0 \tag{8.31}$$

The optimal gain (the Kalman gain) is therefore

$$\mathbf{K}_k = \mathbf{P}_k^{(-)}\mathbf{H}_k^T\left[\mathbf{H}_k\mathbf{P}_k^{(-)}\mathbf{H}_k^T + \mathbf{R}_k\right]^{-1} \tag{8.32}$$

which is sometimes written as

$$\mathbf{K}_k = \mathbf{P}_k^{(-)}\mathbf{H}_k^T\mathbf{W}_k^{-1} \tag{8.33}$$

where the term $\mathbf{W}_k$ is referred to as the innovations covariance

$$\mathbf{H}_k\mathbf{P}_k^{(-)}\mathbf{H}_k^T + \mathbf{R}_k \tag{8.34}$$

8.2 Other Kalman Filter Formulations

In addition to the LKF, there are several other formulations of the Kalman filter that may be employed to more closely follow the characteristics of specific state observation scenarios. The LKF may be varied according to the temporal nature of the dynamic and measurement systems to be continuous in dynamics and measurements or continuous in dynamics and discrete in measurements [12]. Also, there are applications when the dynamic system is energetic or the measurement quality is poor that may cause the reference state in the LKF to quickly leave the region of linearity about the environment state. In such systems, the reference state can be updated through addition of the filter state into an implementation known as the extended Kalman filter (EKF). The EKF is highly suited to real-time applications but is nonlinear in the sense that the reference state is essentially reinitialized at the time of each measurement update. Both the continuous–discrete LKF and EKF will be developed in the following sections.

8.2.1 The Continuous–Discrete Linear Kalman Filter

There may quite naturally arise an application where the reference state, filter state, and state error covariance are more suitably propagated in a continuous fashion than through the linear application of the state transition matrix. Also, it is common for the measurement system to deliver discrete-time observations even when the dynamics are best modeled continuously. In such a situation the update mechanization is unchanged from the previous LKF derivation while the propagation between updates is carried out through continuous integration. Without loss of generality, it may be stated that the reference dynamics of a continuous Kalman filter may be represented by

$$\dot{\tilde{\mathbf{X}}}(t) = f(\tilde{\mathbf{X}}(t), \alpha, t) \tag{8.35}$$

Furthermore, by taking time derivatives of the filter state and covariance propagation (Equations 8.11 and 8.21) and substituting in Equation 8.13 for the derivative of the state transition matrix, the continuous-time filter state and covariance relations are found to be

$$\dot{\hat{\mathbf{x}}}^{(-)}(t) = f(\tilde{\mathbf{X}}(t), \alpha, t) + \mathbf{F}(\tilde{\mathbf{X}}(t), \alpha, t)[\hat{\mathbf{x}}^{(-)}(t) - \tilde{\mathbf{X}}(t)] \tag{8.36}$$

$$\dot{\mathbf{P}}(t) = \mathbf{F}(t)\mathbf{P}(t) + \mathbf{P}(t)\mathbf{F}^T(t) + \bar{\mathbf{Q}}(t) \tag{8.37}$$

where $\bar{\mathbf{Q}}(t)$ is the spectral density of the dynamic process noise at time t and the explicit functional dependency of F was dropped for notational convenience. In this mechanization of the LKF, the state transition matrix need not be calculated as the dynamics are included directly via the partial derivative matrix and the reference state, filter state, and error covariance are propagated continuously.

The process and measurement noise representations in this formulation are continuous and discrete for the respective models, and are again assumed to be zero mean processes governed by the continuous dynamic process noise covariance

$$E[\mathbf{w}(t)\mathbf{w}^T(\tau)] = \mathbf{Q}(t)\delta(t-\tau) \tag{8.38}$$

and the discrete measurement noise covariance

$$E\left[\mathbf{v}_k\mathbf{v}_j^T\right] = \mathbf{R}_k\delta_{kj} \tag{8.39}$$

It is also assumed here that the process and measurement noises are uncorrelated so that

$$E\left[\mathbf{w}(t)\mathbf{v}_k^T\right] = 0 \tag{8.40}$$

although the formulation can be modified to accommodate process and measurement noise correlations if necessary [7].

8.2.2 The Continuous–Discrete Extended Kalman Filter

In applications where the reference state may quickly deviate beyond the linear region of the environment state, the reference may be directly updated at the time of measurement update by adding the LKF filter state to the reference in an EKF. The EKF is similar to the LKF, in that measurements are processed to provide an estimate of the difference between the true state and reference state of the spacecraft. Also, the EKF evaluates dynamics and measurement partials with respect to the reference state in a manner similar to the LKF. However, the reference state about which these partials are evaluated is modified through the addition of measurement information

$$\tilde{X}(t_k)^{(+)} = \tilde{X}(t_k)^{(-)} + \hat{x}(t_k) \tag{8.41}$$

The reference state dynamics model used in the EKF formulation is given by Equation 8.35, but the measurement model is the discrete form given by Equation 8.1. The filter state representing the estimated difference between the true state and the reference state is only calculated at the time of measurement update via dropping the previous estimate information term from Equation 8.26:

$$\hat{x}_k = K_k Z_k \tag{8.42}$$

where the innovation is now the actual measurement residual

$$Z_k = Y_k - h\left(\tilde{X}_k^{(-)}, \beta, t_k\right) \tag{8.43}$$

Therefore, in the EKF there is not a separate filter state that needs to be propagated to the time of the next measurement, as the filter state has been incorporated into the updated reference state.

As before, the error covariance at each measurement is updated by

$$P_k^{(+)} = (I - K_k H_k) P_k^{(-)} (I - K_k H_k)^T + K_k R_k K_k^T, \tag{8.44}$$

and the EKF Kalman gain and innovations covariance are analogous to their LKF counterparts

$$K_k = P_k^{(-)} H_k W_k^{-1} \tag{8.45}$$

$$W_k = H_k P_k^{(-)} H_k + R_k. \tag{8.46}$$

The difference between EKF operation and LKF operation is illustrated by revisiting the two-dimensional trajectory illustration in Figure 8.2. The reference trajectory can now be seen to respond to measurement information availability and tracks the true environment trajectory.

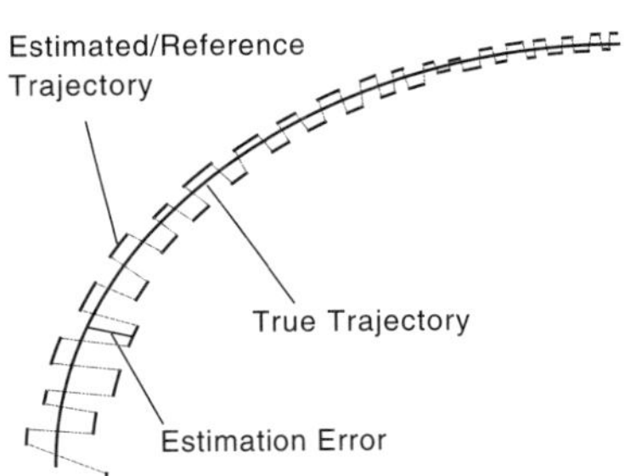

FIGURE 8.2 EKF tracking of a two-dimensional trajectory.

8.3 Formulation Summary and Review

The LKF discrete–discrete formulation was given by the following propagation equations:

$$\begin{aligned}
\tilde{X}_{k+1} &= \Phi(t_{k+1}, t_k)\tilde{X}_k \\
\hat{x}_{k+1}^{(-)} &= \Phi(t_{k+1}, t_k)\hat{x}_k^{(-)} \\
P_{k+1}^{(-)} &= \Phi(t_{k+1}, t_k)P_k^{(-)}\Phi(t_{k+1}, t_k)^T + Q_k
\end{aligned}$$

and update equations

$$\begin{aligned}
\hat{x}_k^{(+)} &= \hat{x}_k^{(-)} + K_k\left[z_k - H_k\hat{x}_k^{(-)}\right] \\
P_k^{(+)} &= (I - K_kH_k)P_k^{(-)}(I - K_kH_k)^T + K_kR_kK_k^T \\
K_k &= P_k^{(-)}H_k^T\left[H_kP_k^{(-)}H_k^T + R_k\right]^{-1}
\end{aligned}$$

In the discrete time LFK mechanization, the reference state is unaffected by the incorporation of measurement information into the filter state.

In a slight variation of this approach, the dynamics of the LKF may be made continuous, and the filter state, reference state, and covariance propagated without the use of a state transition matrix.

$$\begin{aligned}
\dot{\tilde{X}}(t) &= f(\tilde{X}(t), \alpha, t) \\
\dot{\hat{x}}^{(-)}(t) &= F(\tilde{X}(t), \alpha, t)\hat{x}^{(-)}(t) \\
\dot{P}(t) &= F(t)P(t) + P(t)F^T(t) + \bar{Q}(t)
\end{aligned}$$

When the application requires that the reference state be modified to remain in the linear vicinity of the environment state, the EKF continuous–discrete formulation may be appropriate. In the continuous–discrete EKF formulation, the propagation is carried out according to

$$\begin{aligned}
\dot{\tilde{X}}(t) &= f(\tilde{X}(t), \alpha, t) \\
\dot{P}(t) &= F(t)P(t) + P(t)F^T(t) + \bar{Q}(t)
\end{aligned}$$

and the measurement update according to

$$\begin{aligned}
\tilde{X}(t_k)^{(+)} &= \tilde{X}(t_k)^{(-)} + \hat{x}(t_k) \\
\hat{x}_k &= K_k\left[Y_k - h\left(\tilde{X}_k^{(-)}, \beta, t_k\right)\right] \\
P_k^{(+)} &= (I - K_kH_k)P_k^{(-)}(I - K_kH_k)^T + K_kR_kK_k^T \\
K_k &= P_k^{(-)}H_k^T\left[H_kP_k^{(-)}H_k^T + R_k\right]^{-1}
\end{aligned}$$

The reference state will change with the incorporation of measurement information into the EKF and the partials evaluated along this changing reference.

8.4 Implementation Considerations

It is commonly held among designers of Kalman filters that the implementation of the formulas listed above represent only a portion of the effort required to develop an accurate and robust Kalman filter application. Once the dynamics, measurements, and partial derivatives have been coded, the task remains to tune the noise magnitudes represented in the process noise covariance **Q** and the measurement noise covariance **R**. While the measurement noise can be based in realistic hardware performance specifications, the process noise is often used as a tuning parameter to ensure that the filter operates correctly. This process of tuning the filter crosses over into the area of design and is nearly an art form of such myriad approaches that it is beyond the scope of this work to outline. However, a Kalman filter checklist is provided for the newcomer to the field to reduce the time of the implementation and tuning learning curve:

- Because the linear Kalman filter does not change the reference state in the presence of measurement information, the reference state and partial derivatives for an LKF application may be computed prior to operation. This makes the LKF more amenable to computationally restricted applications or hypothesis testing where differing process noise and measurement noise parameters are being evaluated in parallel [8].
- Process noise serves to keep the filter from becoming overconfident in its estimate (i.e., a covariance with near zero diagonal values) and converging prematurely. Examining the propagation equations for the Kalman filters presented previously, it can easily be seen how the addition of process noise increases the magnitude of the state error covariance between measurements.
- The innovations covariance should ideally converge to describe the variance in the filter measurement residuals. Adaptive techniques have been implemented where the filter noise parameters are tuned according to a metric linking residual statistics with the innovations covariance [5]. In an ideal filter, the innovations covariance should approach the measurement noise covariance as the process noise magnitude approaches zero.
- When multiple measurements are available at the same time, they may be processed as a series of scalar observations as long as they are uncorrelated (i.e., **R** is a diagonal matrix). The effect of processing scalar measurements is that the innovations covariance becomes a scalar, and a numerical division rather than a matrix inversion is required to calculate the Kalman gain.
- Measurement editing may be employed to prevent spurious data from causing filter divergence in a number of ways. One of the most common is to reject measurements when the ratio of the measurement residual squared to the scalar innovations covariance

$$\frac{r_k^2}{W_k} \tag{8.47}$$

  is above a user-defined threshold. The threshold value may either be a constant or may be time varying after long propagation periods to allow for a smooth transition to a steady state innovations covariance.
- The covariance should always be positive definite. If filter divergence is a chronic problem in a particular application, the numerical integrity of the covariance may provide insight into the nature of the divergence. There are also several numerical implementations of the covariance update equation that take advantage of its symmetry and positive definiteness to enhance its stability while reducing computational load [9].
- Process noise may be enhanced by including time correlated states such as first-order Gauss–Markov processes to the filter to account for specific dynamic effects. The biases associated with these processes can be included in the filter state for estimation.

As a final note it should be stressed that the Kalman filter is not the state observer algorithm best suited for all applications. Its strengths lie in light computational requirements and real-time availability of a state estimate in the presence of accurate measurement information. However, batch estimation techniques such as least-squares estimation may be more appropriate in applications where the dynamic process is modeled to a high degree of fidelity, measurements are not uniformly accurate, and real-time operation is not an issue. A number of quality texts [10–12] have been written on the subject of stochastic estimation in general and specifically Kalman filtering that the the reader is encouraged to pursue for more detailed information.

References

1. Kalman, R. E., "A new approach to linear filtering and prediction problems," *Transactions of the ASME, Ser. D, Journal of Basic Equations,* March 1960, pp. 35–45.
2. Burkhart, P. and Bishop, R., "Adaptive orbit determination for interplanetary spacecraft," *Journal of Guidance, Control, and Dynamics,* Vol. 19, No. 3, 1997, pp. 693–701.
3. Chaer, W., Bishop, R., and Ghosh, J., "Hierarchical adaptive Kalman filtering for interplanetary orbit determination," *IEEE Transactions on Aerospace and Electronic Systems,* Vol. 34, No. 3, 1998, pp. 1–14.
4. Crain, T. and Bishop, R., "The mixture-of-experts gating network: integration into the ARTSN extended Kalman filter," Technical Memorandum CSR-TM-99-01, Center for Space Research, March 1999.
5. Ely, T., Bishop, R., and Crain, T., "Adaptive interplanetary navigation using genetic algorithms," *The Journal of Astronautical Sciences,* 2000, Accepted for Publication.
6. Crain, T. and Bishop, R., "Unmodeled impulse detection and identification during Mars pathfinder cruise," Technical Memorandum CSR-TM-00-01, Center for Space Research, March 2000.
7. Chaer, W. and Bishop, R., "Adaptive Kalman filtering with genetic algorithms," *Advances in the Astronautical Sciences,* edited by R. Proulx, J. Liu, P. Siedelmann, and S. Alfano, Vol. 89, Univelt, San Diego, CA, 1995, pp. 141–156, Pt. 1.
8. Gholson, N. and Moose, R., "Maneuvering target tracking using adaptive state estimation," *IEEE Transactions on Aerospace and Electronic Systems,* Vol. 13, No. 3, May 1997, pp. 310–317.
9. Bierman, G., *Factorization Methods for Discrete Sequential Estimation,* Academic Press, 1977.
10. Brown, R. G. and Huang, P. Y. C., *Introduction to Random Signals and Applied Kalman Filtering,* John Wiley & Sons, 1992.
11. Lewis, F., *Applied Optimal Control and Estimation,* Prentice-Hall, Englewood Clifis, NJ, 1992.
12. Gelb, A., *Applied Optimal Estimation,* The MIT Press, Cambridge, MA, 1974.

9

Digital Signal Processing for Mechatronic Applications

Bonnie S. Heck
Georgia Institute of Technology

Thomas R. Kurfess
Clemson University

9.1 Introduction

Most engineers work in the world of mechatronics as there are relatively few systems that are purely mechanical or electronic. There are a variety of means by which electrical systems augment mechanical systems and vise versa. For example, most microprocessors found in a computer today have some sort of heat sink and perhaps a fan attached to them to keep them within their operational temperature zone. Electrical systems are widely employed to monitor and control a wide variety of mechanical systems. With the advent of inexpensive digital processing chips, digital filtering and digital control for mechanical systems is becoming commonplace. Examples of this can be seen in every automobile and most household appliances. For example, sensor signals used in monitoring and controlling of mechanical systems require some form of signal processing. This signal processing can range from simply "cleaning-up" the signal using a low pass filter to more advanced analyses such as torque and power monitoring in a DC servo motor. This chapter presents a brief overview of digital signal processing methods suitable for mechanical systems. Since this chapter is limited in space, it does not give any derivation or details of analysis. For a more detailed discussion, see references [1,2].

9.2 Signal Processing Fundamentals

A few fundamental concepts on signal processing must be introduced before a discussion of filtering or control can be undertaken.

9.2.1 Continuous-Time Signals

Laplace transforms are used for system analysis of continuous-time systems, solving for system response, and control design. The single-sided Laplace transform of a continuous-time signal, $x(t)$, is given by

$$X(s) = \int_0^{\infty} x(t)e^{-st}\,dt$$

A transfer function of a linear system, $H(s)$, can be found as the ratio of the Laplace transforms of the output over that of the input (with zero initial conditions).

The Fourier transform is used to determine the frequency content of a signal. The Fourier transform of $x(t)$ is given by

$$X(\omega) = \int_{-\infty}^{\infty} x(t)e^{-j\omega t}\,dt \tag{9.1}$$

where ω is in units of radians per second. Notice that when $x(t) = 0$ for $t \leq 0$, the Laplace transform is equivalent to the Fourier transform by setting $s = j\omega$. (It should be noted that there are some additional convergence considerations for the Fourier transform.) The frequency response of a system is defined as the ratio of the Fourier transforms of the output over that of the input. Equivalently, it can be found from the transfer function as $H(\omega) \equiv H(j\omega) = H(s)|_{s=j\omega}$. For simplicity of notation, the j is usually not shown in the argument list, giving rise to the notation $H(\omega)$ to represent the frequency response. The bandwidth of a system is defined as the frequency at which $H(\omega) = 0.707H(0)$.

9.2.2 Discrete-Time Signals

The z-transform is useful for solving a difference equation and for performing system analysis. The z-transform of a discrete-time signal, $x[n]$, is defined as

$$X(z) = \sum_{n=-\infty}^{\infty} x[n]z^{-n}$$

The discrete-time Fourier transform (DTFT) is used to determine the frequency content of a signal. The DTFT and the inverse DTFT of a signal are defined by

$$X(\Omega) = \sum_{n=-\infty}^{\infty} x[n]e^{-j\Omega n} \tag{9.2}$$

and

$$x[n] = \frac{1}{2\pi}\int_{-\pi}^{\pi} X(\Omega)e^{j\Omega n}\,d\Omega \tag{9.3}$$

Note that the DTFT can be derived from the z-transform by setting $z = e^{j\Omega}$. (Again, there are some assumptions on convergence in this derivation.) Since the DTFT is periodic with period 2π, it is typically displayed over the range $[-\pi, \pi]$ or $[0, 2\pi]$, where the frequencies of general interest are from $\Omega = 0$ (low frequency) to $\Omega = \pi$ (high frequency). The frequency response of a discrete-time system can be found as the ratio of the DTFT of the output signal over that of the input signal. Alternatively, it can be found from the transfer function as $H(\Omega) \equiv H(e^{j\Omega}) = H(z)|_{z=e^{j\Omega}}$. The notation $H(\Omega)$ is preferred over $H(e^{j\Omega})$ for its simplicity. As in the continuous-time case, the bandwidth is defined as the frequency at which $H(\Omega) = 0.707H(0)$.

While the DTFT is continuous with respect to the frequency variable Ω, the discrete Fourier transform (DFT) contains points that are discrete with respect to a parameter k. Consider a finite duration sequence $x[n]$, where $x[n] = 0$ for $n < 0$ and for $n \geq N$. The DFT of $x[n]$ and the inverse DFT are defined as

$$X_k = \sum_{n=0}^{N-1} x[n]e^{-j2\pi nk/N}, \quad k = 0, 1, \ldots, N-1 \tag{9.4}$$

and

$$x_n = \frac{1}{N}\sum_{k=0}^{N-1} X_k e^{j2\pi nk/N}, \quad n = 0, 1, \ldots, N-1$$

Note that the DFT is a discretized version of the DTFT where $X_k = X(\Omega)|_{\Omega=2\pi k/N}$ over the range $\Omega = 0$ to $\Omega = 2\pi$. Calculating a closed-form solution for the DTFT can be done only for simple signals such as a square pulse or a triangular pulse. Therefore, the DFT is generally used as a numerical method to calculate the DTFT at discrete points in frequency in the range $0 \leq \Omega \leq 2\pi$. In particular, to obtain a plot of the DTFT, plot X_k versus k where k is scaled by $2\pi/N$. For an arbitrary signal, such as obtained from measurements of a physical device, computing the DFT instead of the DTFT is the preferred method to find the frequency content of the signal. To get more resolution in plotting a DTFT from the points calculated by a DFT, zeros can be added to the end of the sequence so that the value of N is increased.

Suppose a time domain signal is not finite in duration, so that there is no value of N such that $x[n] = 0$ for n Š N. In order to perform the DFT, the signal must be truncated. There are two cases to be considered: the case where $x[n]$ is decaying to zero and the case where $x[n]$ has periodic components. The case when $x[n]$ decays to zero is handled by choosing N to be large enough so that the signal is negligible beyond that value. The resulting DFT is an approximation (not a discretized version) of the DTFT. If the signal is periodic, the DTFT cannot be computed numerically since the resulting DTFT would have impulses in it. However, the frequencies present in the signal could still be determined if the value of N used for the truncation is chosen so that the truncated signal goes through an integer number of cycles. If this not done, the resulting DFT will have leakage in the frequency plot when compared to the DTFT of the true signal. For example, consider a signal $x[n] = \cos(0.4\pi n)$. This is periodic with period $n = 5$ and has DTFT given by $X(\Omega) = \pi[\delta(\Omega + 0.4\pi) + \delta(\Omega - 0.4\pi)]$ for $-\pi \leq \Omega \leq \pi$. All the frequency content is located at $\Omega = 0.4\pi$ and $\Omega = -0.4\pi$. Since the DTFT is periodic with 2π, there is also an impulse at $\Omega = 2\pi - 0.4\pi$. The DFT is computed for two truncations of the signal, one at $N = 20$ (four complete cycles) and the other at $N = 22$. The DFT for $N = 20$ is plotted in Figure 9.1a where the independent variables k are scaled by $2\pi/N$ for the plot. This plot shows zero frequency content except at $\Omega = 0.4\pi$ (=1.2566) and $\Omega = 2\pi - 0.4\pi$ (=5.0265), giving the correct location of the impulses in the DTFT. Similarly, the DFT for $N = 22$ is plotted in Figure 9.1b, notice the resulting leakage in the frequency characteristics.

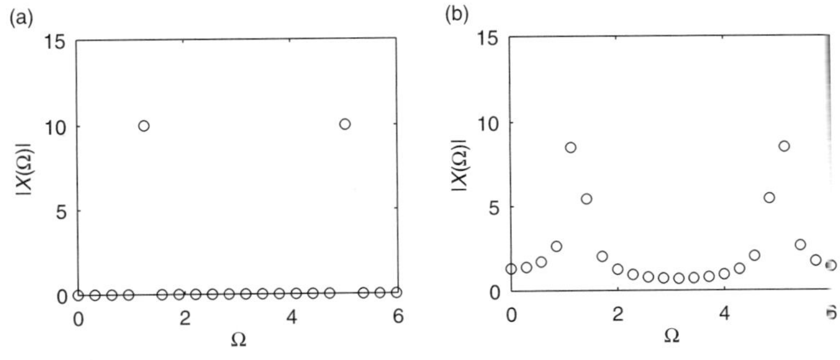

FIGURE 9.1 DFT of periodic signal (a) truncated after 4 complete cycles and (b) truncated after 4.4 cycles.

If a signal has periodic content, but is not periodic, such as $x[n] = \cos(0.5\pi n) + \cos(0.2n)$, then leakage cannot be avoided by a selection of N. An alternate means of reducing leakage is to first taper the signal to zero at the initial and end points of the sequence prior to computing the DFT. This process, known as *windowing* the data, is accomplished by multiplying $x[n]$ by a window function $w[n]$ and then performing the DFT on the product $x[n]w[n]$. Three common windows are the rectangular window, which is a sharp truncation, the Hanning window, and the Hamming window [1].

Rectangular Window:

$$w[n] = 1, \quad 0 \le n \le N-1$$

Hanning Window:

$$w[n] = \frac{1}{2}\left(1 - \cos\left(\frac{2\pi n}{N-1}\right)\right), \quad 0 \le n \le N-1$$

Hamming Window:

$$w[n] = 0.54 - 0.46\cos\left(\frac{2\pi n}{N-1}\right), \quad 0 \le n \le N-1$$

If the value of N in a DFT is a power of 2, there is a fast method to compute the DFT called the fast Fourier transform (FFT). If the value of N is not a power of 2, zeros can be padded to the end of the signal in order to use the FFT. This does not affect the accuracy of the result, but it does improve the resolution of the resulting plot when the DFT (or FFT) is used to compute the DTFT. In many cases, the expression used in Equation 9.4 suffices to compute the DFT since the added computational power of today's processors lessens the need for the numerical efficiency of the FFT. The details of the algorithm for the FFT are beyond the scope of this handbook. See [1] or [2] for details.

9.3 Continuous-Time to Discrete-Time Mappings

While most physical systems operate in continuous-time, computers operate in discrete-time. Therefore, in order to use computers to process measurements taken from continuous-time systems, there must be ways of mapping between the continuous-time world to the discrete-time world.

9.3.1 Discretization

Before an analog signal can be analyzed using digital techniques, it must be discretized (i.e., converted into a discrete-time signal). The ideal method for discretization is *sampling*, where the values of the signal are determined at discrete points in time. Generally, the signal is sampled at a fixed rate known as the *sampling period*. The sampling rate (in hertz) is the inverse of the sampling period. Figure 9.2 depicts a 1 Hz signal that has been sampled at two rates. The dark points are sampled at 15 ms intervals, while the lighter points are sampled at 250 ms intervals. From Figure 9.2, the waveform approximation clearly degrades as the sampling frequency is reduced and approaches the signal frequency. In fact, it can be shown that a signal must be sampled at a frequency that is higher than twice its maximum frequency content. This is known as the Nyquist Sampling Theorem. For example, if the signal in Figure 9.2 is sampled at 0.5 Hz, it is possible for every sample to have a value of 0 as at 0, 500, 1000 ms, etc… the value of the signal is 0. The erroneous interpretation of the signal due to a sample frequency that is too low is known as aliasing. There are two means by which the Nyquist Sampling Theorem can be satisfied. The first is by employing a sample frequency that is more than twice the highest frequency content of the signal being sampled. This value frequency is known as the Nyquist frequency. As one never is sure of the actual frequency content of a real signal, a low pass filter may be used to ensure that a signal does not possess frequencies above a certain cut-off level. Such a filter is commonly

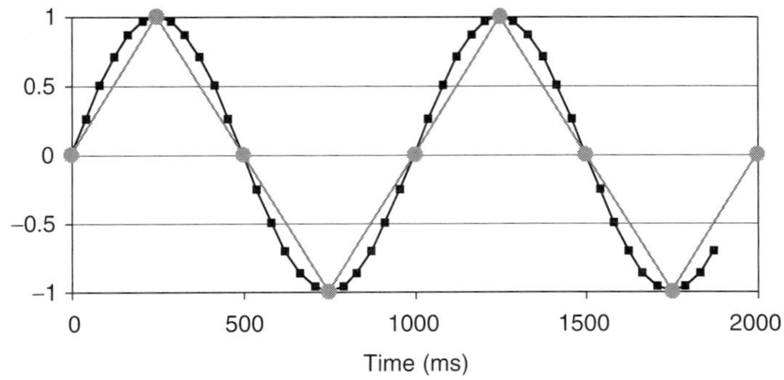

FIGURE 9.2 A 1 Hz signal.

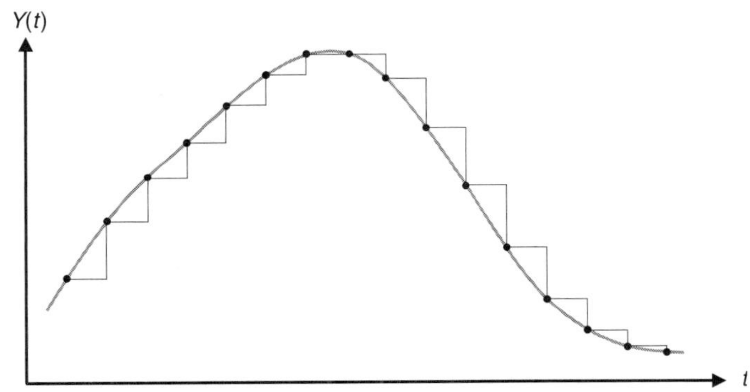

FIGURE 9.3 A signal sampled and reconstructed using a zero order hold.

called an anti-aliasing filter. This is the second and most practical method that is used to satisfy the Nyquist Sampling Theorem. Thus, a combination of a well-designed anti-aliasing filter as well as a sampling frequency that is well above the cut-off frequency of the filter will ensure that the Nyquist Sampling Theorem is satisfied.

There are two important points that should be noted when using an anti-aliasing filter. First, it is important that the anti-aliasing filter be used before the signal is sampled as sampling is what causes aliasing. Basically, this requires that the anti-aliasing filter is implemented using an analog filter prior to the signal being digitized. Once a signal has been aliased during sampling, it cannot be corrected using digital filtering. The second point is that, in practice, the cutoff frequency of the anti-aliasing filter should be a factor of 5–10 below the value of the Nyquist frequency. It should be noted that an anti-aliasing filter adds phase lag to the measurement, which might deteriorate stability and performance in a feedback loop unless the bandwidth of the anti-aliasing filter is much higher than that of the closed loop system. Commercially available devices that perform sampling are analog-to-digital converters (ADCs), and the anti-aliasing filter is used before this device.

The converse of sampling is reconstruction where a discrete-time signal is converted into a continuous-time signal. The Nyquist sampling rate ensures that if a continuous-time signal is sampled at a rate that is at least twice the highest frequency component in the signal, then the continuous-time signal can be reconstructed exactly from the samples. However, this theorem assumes that an ideal reconstruction process is available, which is not practical. The most common practical means to reconstruct a signal is a zero-order hold (ZOH). The ZOH assumes that the value of the signal is constant between samples. This approximation is quite reasonable if the sampled signal does not change substantially between individual samples. Figure 9.3 is an example of a signal and its ZOH representation. The gray, smooth line represents the original analog signal. The black points along the signal indicate sample values of the signal. Each black point is connected to the next via a horizontal then vertical straight line. The horizontal line is representative of the ZOH assumption that the value of the signal remains constant between samples. The vertical line is the reality that the signal does not remain constant over the sample period. As the time between sample points is increased, the accuracy of the ZOH decreases. Conversely, as the sample period is decreased, the accuracy of the ZOH is improved. Commercially available devices that perform reconstruction are digital-to-analog converters (DACs), which generally use the ZOH method.

9.3.2 *s*-Plane to *z*-Plane Mappings

One method to relate the *s*-plane to the *z*-plane is to derive a continuous-time mathematical representation of the sampled signal for $x(t)$ and compute its Laplace transform. The resulting Laplace transform of the sampled signal can be related to the *z*-transform of $x[n]$ by setting $z = e^{sT}$ where T is the sampling period. The relationship $z = e^{sT}$ is commonly termed the *exact mapping* between the *z*-plane and the *s*-plane. (For details of this derivation, see [1]). For example, a digital representation, $H_d(z)$, of a continuous-time system, $H(s)$, can be obtained using this mapping $H_d(z) = H(s)|_{z=e^{sT}}$. However, this mapping results in a nonrational function for $H_d(z)$.

Approximate mappings between the *s*-plane and the *z*-plane are commonly used that do result in a rational function for $H_d(z)$. Three such mappings are the bilinear transformation, forward transformation, and backward transformation.

Bilinear transformation:

$$s = \frac{2(z-1)}{T(z+1)}$$

Forward transformation:

$$s = \frac{1}{T}(z-1)$$

Backward transformation:

$$s = \frac{1}{Tz}(z-1)$$

The bilinear transformation (also known as Tustin's rule or the trapezoidal rule) is the most accurate of these mappings. It maps the entire left-hand side of the *s*-plane into the unit circle of the *z*-plane, so that it preserves stability. Consider a first-order example of $H(s) = 1/(s + 2)$. The discrete-time representation of this transfer function is

$$H_d(z) = H(s)|_{s=\frac{2(z-1)}{T(z+1)}} = \frac{T(z+1)}{2z-2+2T}$$

Note that the resulting transfer function is rational in *z*.

An alternate method of mapping transfer functions between the continuous-time and the discrete-time domains is the response-matching mapping.

Response-matching: Suppose $x(t)$ is the input to a system $H(s)$ with the resulting output $y(t)$. Let $x[n]$ and $y[n]$ be the sampled versions of $x(t)$ and $y(t)$. Then, $H_d(z)$ is found from the ratio of *z*-transforms of $x[n]$ and $y[n]$. The most common response matching is step-response matching where $x(t)$ is a step function, $x(t) = 1$ for $t \geq 0$ and $x(t) = 0$ for $t < 0$. An expression for $H_d(z)$ is found from the following operation:

$$H_d(z) = (1 - z^{-1})Z\left[\frac{H(s)}{s}\right]$$

where $Z[H(s)/s]$ represents the z-transform of the sampled version of the step response of the continuous-time system. The form for a generic first-order system is given below:

$$H(s) = \frac{a}{s+a} \Leftrightarrow H_d(z) = \frac{(1-e^{-aT})z^{-1}}{1-e^{-aT}z^{-1}}$$

The response matching method (especially with step inputs) is commonly used to map a continuous-time plant to discrete-time when designing a digital controller in the discrete-domain. Since most digital controllers are implemented using a ZOH on the output of the digital controller, the plant sees a stepped signal, one that looks like a sum of delayed step signals. Therefore, the step-response matching method is the most accurate way to map a plant that has a ZOH on its input.

9.3.3 Frequency Domain Mappings

The continuous-time Fourier transform can be related to the DTFT through the expression:

$$X(\omega) = TX(\Omega)|_{\Omega=\omega T} \quad \text{for } -\pi \le \Omega \le \pi$$

where $X(\omega)$ is defined in Equation 9.1 and represents the continuous-time Fourier transform of $x(t)$, while $X(\Omega)$ is defined in Equation 9.2 and represents the DTFT of the sampled signal $x[n]$ This mapping is very useful for computing the Fourier transform of measured data. In particular, suppose a continuous-time signal is measured by sampling it through an ADC and storing it as a discrete-time sequence. If the signal $x(t)$ is finite in duration, the DTFT of $x[n]$ can be computed at discrete points in frequency by using the DFT, X_k, as given in Equation 9.4. Using the relationships $\omega = \Omega/T$, $\Omega = 2\pi k/N$, and $X_k = X(\Omega)|_{\Omega=2\pi k/N}$, where N is the length of the sequence for $x[n]$ and T is the sampling period, gives the relationship:

$$X(\omega)|_{\omega=2\pi k/NT} = TX_k \quad \text{for } 0 \le \omega \le \omega_s/2, \; 0 \le k \le (N-1)/2$$

where $\omega_s = 2\pi/T$ is the sampling frequency in radian per second. Accuracy can be improved by decreasing the sampling period T, and the resolution in the plot can be increased by increasing NT.

If the signal $x(t)$ is not finite in duration, it must be truncated in order to use this numerical method to calculate the continuous-time Fourier transform. As discussed in Section 9.2.2, if the sampled signal $x[n]$ decays to zero, choose the number of sampled points N to be large enough so that $x(t)$ is negligible beyond that value. If the sampled signal $x[n]$ is periodic, choose the sampling period T and the number of points N such that the sampled signal $x[n]$ goes through an integer number of cycles. For example, consider the signal $x(t) = \cos(\pi t)$. If the sampling period is chosen as $T = 0.4$ s, the discretized signal would be $x[n] = x(nT) = \cos(0.4\pi n)$, which is the same signal analyzed in Section 9.2.2. Choosing $N = 5$, 10, 15, etc. would yield correct results in the DFT while any other value would result in leakage. If the signal $x(t)$ has periodic content, but does not appear to be periodic, then use a windowing function as discussed in Section 9.2.2 to reduce leakage when computing the DFT.

Note that the DFT can also be used to determine the Fourier coefficients of periodic signals. Consider a Fourier series in the form

$$x(t) = \sum_{k=-\infty}^{\infty} c_k e^{j\omega kt}$$

Sample the signal $x(t)$ by first making sure that the sampled signal goes through an integer number of cycles. The coefficients c_k for $k = 0,\ldots,(N-1)/2$ can be found as $c_k = X_k/N$ where X_k is found from the DFT. The rest of the coefficients are obtained from $c_{-k} = c_k^*$.

Since the frequency response of a system is the ratio of the Fourier transforms of the output over the input, a mapping between a continuous-time system, $H(\omega)$, and a corresponding discrete-time system, $H_d(\Omega)$, can be derived from the previous mapping as

$$H(\omega) = H_d(\Omega)|_{\Omega=\omega T} \quad \text{for } -\pi \leq \Omega \leq \pi$$

This mapping is useful for the design of both digital filters and digital controllers.

9.4 Digital Filter Design

The frequency response function of a discrete-time system describes how the system processes input signals of different frequencies. Consider an input signal $x[n] = A\cos(\Omega_0 n)$ to a system with frequency response $H(\Omega)$ where 0 ð Ω_0 ð 2π. The corresponding output is given by

$$y[n] = |H(\Omega_0)|\cos(\Omega_0 n + \angle H(\Omega_0))$$

For aperiodic signals, the filtering property of Fourier transforms gives the relationship:

$$Y(\Omega) = H(\Omega)X(\Omega)$$

Thus, if $|H(\Omega)|$ is small over a certain range of frequencies, then input signals with frequency content in that range are attenuated as they pass through the system.

It is often convenient to filter continuous-time signals through a digital filter as shown in Figure 9.4. The analog-to-digital converter (ADC) samples the continuous-time signal creating a sequence of discrete-time signals for processing by the computer or digital signal processing board. The filtered signal can be stored digitally for further study or it can be sent through a digital-to-analog converter (DAC). The digital filter can be implemented in software by a recursive equation obtained from the difference equation. Consider a digital filter with transfer function:

$$H(z) = \frac{b_1 z^N + b_2 z^{N-1} + \cdots + b_{N+1}}{a_1 z^N + a_2 z^{N-1} + \cdots + a_{N+1}} = \frac{b_1 + b_2 z^{-1} + \cdots + b_{N+1} z^{-N}}{a_1 + a_2 z^{-1} + \cdots + a_{N+1} z^{-N}}$$

The recursion used to calculate the current value of the output $y[n]$ is given by the difference equation:

$$y[n] = \frac{1}{a_1}(b_1 x[n] + b_2 x[n-1] + \cdots + b_{N+1} x[n-N] - a_2 y[n-1] - \cdots - a_{N+1} y[n-N]) \quad (9.5)$$

Notice that the past values of y and x must be stored for use in the recursion.

Now consider the impulse response of a digital filter, where $y[n]$ is calculated for an input $x[n]$ equal to an impulse (that is, $\delta[n] = 1$ when $n = 0$ and $\delta[n] = 0$ otherwise). The recursion shown above results in a response for $y[n]$ that has infinite duration (that is, there is no value of M so that $y[n] = 0$ for all $n > M$). This type of filter is called an *infinite impulse response (IIR) filter.*

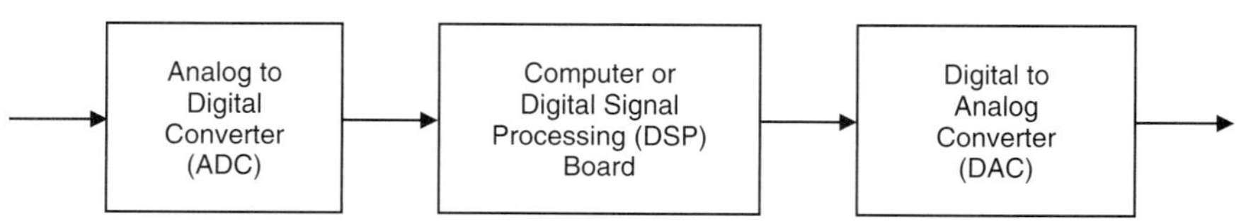

FIGURE 9.4 Configuration for standard digital signal processing hardware.

Now consider the case where the coefficients of the filter $a_m = 0$ for $m > 1$. The resulting expression for $y[n]$ from Equation 9.5 would no longer be recursive since it would depend only on present and past values of x, and not on past values of y. As a result, the impulse response would have duration N. This type of filter is called *a finite impulse response (FIR) filter.* FIR filters are sometimes preferred over IIR filters since they have linear phase in the frequency response. Linear phase means that the angle of the frequency response is given by $-\theta\Omega$, where θ is a constant. This corresponds to a delay in the time domain. Design methods for both types of filters are described in the next two sections.

9.4.1 IIR Filter Design

The two methods for designing IIR filters are termed *analog emulation* (or indirect design) and *direct design.* Analog emulation involves designing an analog filter first and then using one of the mapping techniques described in Section 9.3.2 to convert it to a digital filter. This method has advantages in that there is a wealth of design techniques for analog filters that can be used in digital filter design. Direct design methods generally involve numerical techniques, and they are often preferred over analog emulation when the sampling period is not very small. Direct design is beyond the scope of this handbook; consult reference [2] for more information on the topic.

Analog filter design begins by selecting a bandwidth, a prototype of filter, and an order of the filter. Additional specifications may be set on the amount of ripple that is allowed in the passband or stopband. Two common analog prototypes are the Butterworth filters and the Chebyshev filters.

Butterworth filter: The Butterworth filter is characterized by having no zeros and having poles that are situated on a semicircle in the left-half of the s-plane. The distance of the poles to the origin is the bandwidth frequency and is denoted as ω_b. The angle of the poles can be determined by equally spacing out twice the number of poles around a full circle with radius ω_b and then keeping only the poles in the left-half plane, as shown in Figure 9.5. An Nth order Butterworth filter is given by

$$H(s) = \frac{\omega_b^N}{\prod_k (s - \omega_b p_k)}$$

where

$$p_k = \begin{cases} e^{jk\pi/N}, & k = \dfrac{N+1}{2} \text{ to } \dfrac{3N-1}{2} \quad \text{for } N \text{ odd} \\[2ex] e^{j(k+0.5)\pi/N}, & k = \dfrac{N}{2} \text{ to } \dfrac{3N-2}{2} \quad \text{for } N \text{ even} \end{cases}$$

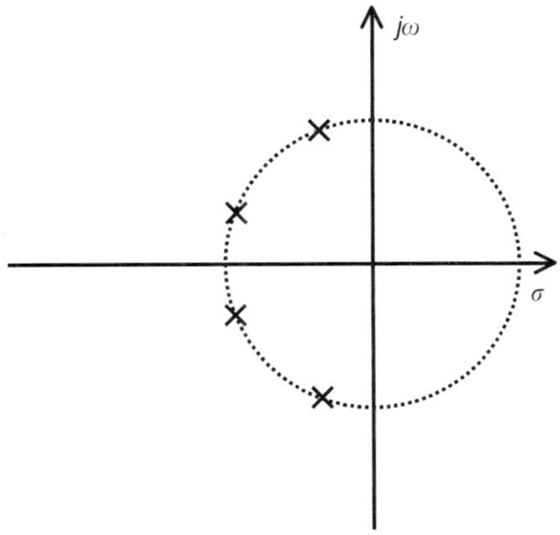

FIGURE 9.5 Pole distribution of a fourth-order Butterworth filter.

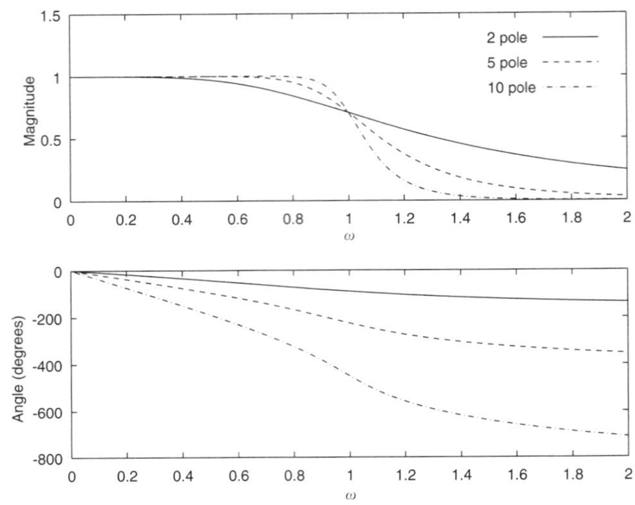

FIGURE 9.6 Comparison of analog Butterworth filters.

This filter is lowpass in that the magnitude of the frequency response is reasonably flat and close to a value of 1 for $\omega < \omega_b$ and drops off sharply beyond the bandwidth frequency. The larger the order of the filter, the sharper the drop off. Three filters are compared in Figure 9.6. Notice that the sharp transitions offered by the larger order filters do come with a price: the phase is also decreased dramatically. The phasing becomes important in real-time measurement systems such as that required by feedback controllers.

Chebyshev filter: Unlike the monotonic behavior of the Butterworth filter, the Chebyshev filter allows some ripple in the magnitude plot for either the passband or the stopband. The Type 1 Chebyshev filter allows for ripple in the passband while the Type 2 Chebyshev filter allows for ripple in the stopband. Allowing for a ripple results in the Chebyshev filters having sharper transitions near the bandwidth than are achievable by a Butterworth filter of the same order. In Chebyshev design, the cutoff frequency ω_c is usually specified as opposed to the bandwidth. The cutoff frequency is the frequency at which the magnitude of the filter decays to a preset ratio of the DC value. When this ratio is 0.707, the cutoff frequency is the bandwidth. Often, in Chebyshev design, this ratio is chosen to correspond to the amount of ripple allowed in the passband. A Type 1 Chebyshev lowpass filter is defined by the relationships:

$$|H(\omega)| = \frac{1}{\sqrt{1 + \varepsilon^2 C_N^2(\omega/\omega_c)}}$$

and

$$C_N(x) = 2xC_{N-1}(x) - C_{N-2}(x)$$

The $C_N(x)$ expression is called the *N*th order Chebyshev polynomial, and it is calculated recursively starting with $C_0(x) = 1$ and $C_1(x) = x$. The value of $\varepsilon > 0$ determines the amount of ripple allowed in the passband; in particular, the ripple exists between the values of 1 and $1/\sqrt{1+\varepsilon^2}$. Consider, for example, the Type 1 Chebyshev filters shown in Figure 9.7; these filters were designed to have 1 dB of ripple in the passband ($\varepsilon = 0.51$). Note that $\varepsilon = 1$ for 3 dB of ripple.

As mentioned above, these prototype filters are lowpass. To design another type of filter, first the lowpass filter is designed, $H(s)$, with a cutoff frequency ω_c (typically chosen to be 1). Then a frequency transformation is used to convert the filter to the desired type. The standard frequency transformations are given below.

Lowpass to lowpass: To obtain a lowpass filter with cutoff frequency ω_1, replace s in the original $H(s)$ by $s\omega_c/\omega_1$.

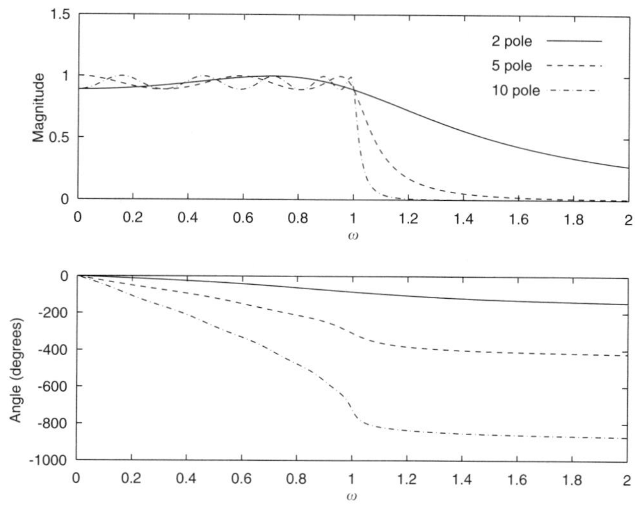

FIGURE 9.7 Comparison of analog Type I Chebyshev filters.

Lowpass to highpass: To obtain a highpass filter with a passband running from ω_1 to ∞, replace s in the original $H(s)$ by $\omega_1\omega_c/s$.

Lowpass to bandpass: To obtain a bandpass filter with a passband running from ω_1 to ω_2, replace s in the original $H(s)$ by

$$\frac{s^2 + \omega_2\omega_1}{s(\omega_2 - \omega_1)}$$

Lowpass to bandstop: To obtain a bandstop filter with stopband running from ω_1 to ω_2, replace s in the original $H(s)$ by

$$\frac{s(\omega_2 - \omega_1)}{s^2 + \omega_2\omega_1}$$

9.4.2 FIR Filter Design

One way to obtain an FIR filter is to truncate the impulse response of an ideal IIR filter. For example, an ideal IIR lowpass filter has the frequency response:

$$H(\Omega) = \begin{cases} A, & -\Omega_c \le \Omega \le \Omega_c \\ 0, & \text{otherwise} \end{cases}$$

where A is a constant and Ω_c is the cutoff frequency. The impulse response of this filter is found from taking the inverse DTFT using Equation 9.3:

$$h[n] = \frac{A\Omega_c}{\pi} e^{jn\Omega_c/2} \operatorname{sinc}\left(\frac{\Omega_c n}{\pi}\right)$$

Notice that this has infinite duration for both $n < 0$ and $n > 0$. Creating an FIR filter would entail truncating the impulse response for $n < -N$ and for $n > N$. However, the original IIR filter and the resulting truncated FIR filter are both noncausal; that is, the impulse response is nonzero for $n < 0$.

Noncausal filters need future values of the input in order to calculate the present value of the output; hence, they cannot be implemented in real-time. For this reason, typical IIR design uses nonideal filters (often based on analog prototypes) that approximate the ideal frequency response. When filtering a signal off-line that has been stored, causality is no longer required since all of the values of the signal are available (including "future" values).

In order to perform real-time implementation of an FIR filter that was generated by truncating an ideal IIR filter, the filter must be delayed so that all of the significant information of the impulse response occurs for n Š 0. This delay in the time domain is equivalent to a linear phase lag in the frequency domain.

Thus, an FIR filter can be designed by first selecting an ideal IIR filter (lowpass, highpass, etc.), then taking the inverse DTFT to find the impulse response, and then truncating the impulse response, and finally, delaying it in time. An equivalent and more preferred method is to rearrange the steps described above. First, add a phase lag in the frequency response of the ideal IIR filter. This is done by multiplying the frequency response by $e^{j(N-1)/2}$. Then, take the inverse DTFT and truncate it for $n < 0$ and $n > N - 1$. The result is a causal FIR filter with order N.

The following are generic FIR filters of order N that have been generated using the method described above. Let $m = (N - 1)/2$.

Lowpass FIR filter with cutoff frequency Ω_c:

$$h[n] = \begin{cases} \dfrac{\Omega_c}{\pi}, & n = 0 \\ \dfrac{\Omega_c}{\pi}\text{sinc}\left[\dfrac{\Omega_c(n-m)}{\pi}\right], & \text{for } 0 < n \leq N-1 \\ 0, & \text{otherwise} \end{cases}$$

where $\text{sinc}(x) = \sin(\pi x)/\pi x$.

Highpass FIR filter with passband from Ω_1:

$$h[n] = \begin{cases} 1 - \dfrac{\Omega_1}{\pi}, & \text{for } n = 0 \\ -\dfrac{\Omega_1}{\pi}\text{sinc}\left[\dfrac{\Omega_1(n-m)}{\pi}\right]0, & \text{for } 0 < n \leq N-1 \\ 0, & \text{otherwise} \end{cases}$$

Bandpass FIR filter with passband from Ω_1 *to* Ω_2:

$$h[n] = \begin{cases} \dfrac{\Omega_2 - \Omega_1}{\pi}, & \text{for } N = 0 \\ \dfrac{\Omega_2}{\pi}\text{sinc}[\Omega_2(n-m)/\pi] - \dfrac{\Omega_1}{\pi}\sin[\Omega_1(n-m)/\pi], & \text{for } 0 < n \leq N-1 \\ 0, & \text{otherwise} \end{cases}$$

To implement these filters, the coefficients in Equation 9.5 are set to $b_m = h[m - 1]$, $a_1 = 1$, and $a_m = 0$ for $m > 1$.

The FIR filters designed using this method have frequency responses that have rather sharp transitions between the passband and the stopband (the larger the order, the sharper the transition), but they tend to give rise to a ripple in the passband and stopband. This ripple results from the sharp truncation of the IIR filter's impulse response. A more gradual truncation using a window can be performed that smooths the ripple in the frequency response. The windows discussed in Section 9.3.2 that are employed in data collection are also used in FIR filter design, where the modified filter is given as $h[r]w[n]$. FIR design using different windows is discussed in further detail in [1,2].

9.4.3 Computer-Aided Design of Digital Filters

MATLAB™ is a common software package for signal processing analysis and design. The signal processing toolbox contains several commands for designing and simulating digital filters. For example, the commands butter and cheby1 automatically design a prototype analog filter for an IIR and then use the bilinear transformation to map the filter to the discrete-time domain. Lowpass, highpass, bandstop, and bandpass filters can be designed using these commands as long as the digital cutoff frequencies normalized by π, are specified. To design a digital lowpass filter based on the analog Butterworth filter with cutoff frequency $w1$, use the command $[b, a]$ = butter(N, $w1 * T$/pi) where N is the number of poles, T is the sampling period, and $w1 * T$ is digital cutoff frequency. This command puts the coefficients of the filter, defined in Equation 9.5, in vectors b and a in ascending order. To design a digital highpass filter with analog cutoff frequency $w1$, use the commands $[b, a]$ = butter(N, $w1 * T$/pi, 'high'). To design a digital bandpass filter with analog passband from $w1$ to $w2$, define $w = [w1, w2]$ and use the command $[b, a]$ = butter (N, $w * T$/pi). To design a digital bandstop filter with stopband from $w1$ to $w2$, define $w = [w1, w2]$ and use the command $[b, a]$ = butter(N, $w * T$/pi, 'stop'). The design for an Nth order Type I Chebyshev filter is accomplished using the same methods as for butter except that "butter" is replaced by "cheby1."

The signal processing toolbox also provides commands for designing FIR filters. To obtain a lowpass FIR filter with length N and analog cutoff frequency $w1$, use the command h = fir1($N - 1$, $w1 * T$/pi). The resulting vector h contains the impulse response of the FIR where $h(1)$ is the value of $h[0]$. The values in the vector h also equal the coefficients of b in Equation 9.5 in ascending order. (Recall, that $a_1 = 1$ and $a_m = 0$ for $m > 1$.) A length N highpass FIR filter with analog cutoff frequency $w1$ is designed by using the command h = fir1($N - 1$, $w1 * T$/pi, 'high'). A bandpass FIR filter with passband from $w1$ to $w2$ is obtained by typing h = fir1($N - 1$, $w * T$/pi) where $w = [w1, w2]$. A bandstop FIR filter with stopband from $w1$ to $w2$ is obtained by typing h = fir1($N - 1$, $w * T$/pi, 'stop') where $w = [w1, w2]$. The fir1 command uses the Hamming window by default. Other windows are obtained by adding an option of "hanning" or "boxcar" (which is the rectangular window) to the arguments; for example, h = fir1($N - 1$, $w1 * T$/ pi, 'high,' boxcar(N)) creates a highpass FIR filter with analog cutoff frequency $w1$ using a rectangular window.

The filter command in MATLAB is used to compute an output of a digital filter given its input sequence. An example of its use is y = filter(b, a, x) where b and a are the coefficients of the filter and x is the input sequence.

9.4.4 Filtering Examples

Quite often, 60 Hz noise is encountered in measurements of electromechanical systems due to standard line voltage. (Note, in Europe noise at a 50-Hz frequency is typically encountered.) For demonstration purposes, a 60-Hz signal is superimposed on a lower frequency signal shown in Figure 9.8. To alleviate the detrimental effects of the 60-Hz noise, a bandstop filter may be employed. Typically, most systems

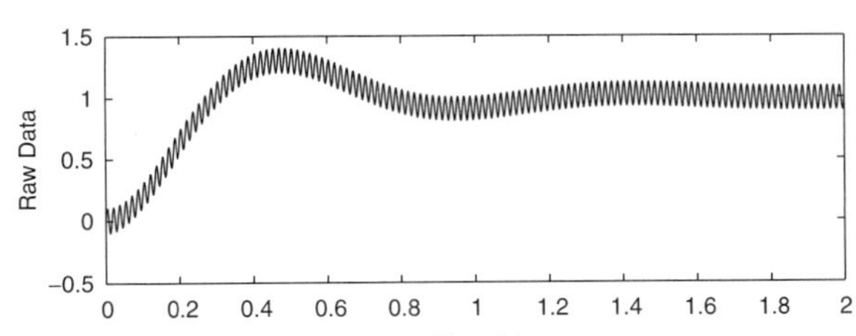

FIGURE 9.8 Measurement corrupted with 60-Hz noise.

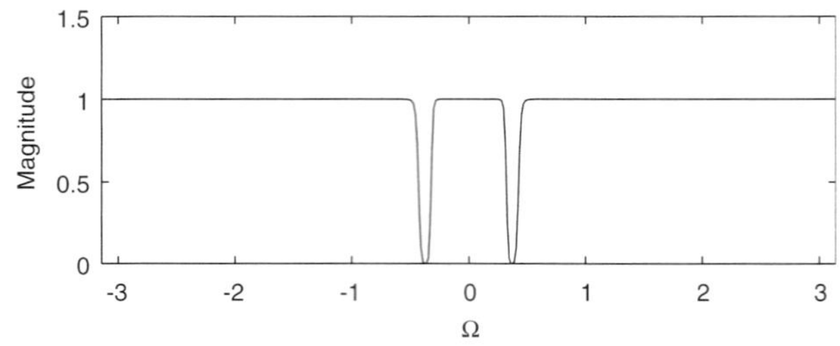

FIGURE 9.9 Bandstop filter.

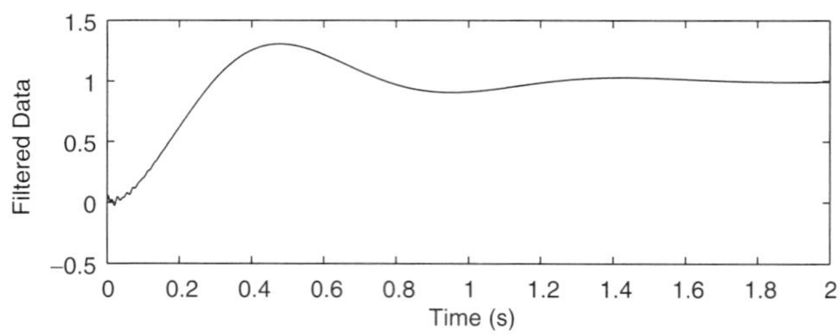

FIGURE 9.10 Filtered measurement.

have a bandstop filter designed around 60 Hz to avoid the type of response seen in Figure 9.8. The following MATLAB commands can be employed to design an eighth-order digital Butterworth bandstop filter whose break frequencies are 50 and 70 Hz. Thus, the filter should reject the 60-Hz noise.

```
T = 0.001; %Sample period
n = 4; % half the order of filter
low_freq = 50 * (2*pi); %Stop signals between 50 and 70 Hz
high_freq = 70 * (2*pi);

w1 = low_freq*(T/pi); % normalized digital break frequencies
w2 = high_freq*(T/pi);
w = [w1 w2];

[b,a] = butter(n,w,'stop'); % filter coefficients

W = -pi:pi/200:pi; % define a digital frequency vector
H = freqz(b,a,W); % computes the frequency response for plotting
```

Figure 9.9 shows the magnitude of the frequency response for the resulting IIR filter. Note that the frequency variable is plotted for the range $[-\pi, \pi]$ where DC frequency corresponds to $\Omega = 0$ and the highest frequency allowable is $\Omega = \pi$. In this example, the digital break frequencies correspond to $\Omega_1 = 50(2\pi)T = 0.314$ and $\Omega_2 = 70(2\pi)T = 0.44$. Figure 9.10 shows the result of applying this filter to the noisy signal. For all practical purposes, the 60-Hz noise is completely attenuated. As can be seen in Figure 9.10, there are some initial system transients during the first 100 ms of the step response. This is a combination of the fourth-order Butterworth filter and the initial system transients to the 60-Hz signal. It should be noted that the sample frequency of 1000 kHz is fast enough to accurately capture the 60-Hz signal. If a sample frequency of less than 120 Hz is used, the 60-Hz signal will be aliased, and no amount of digital filtering would be able to eliminate the effects of the 60-Hz disturbance.

Another application of digital filtering in mechatronics is used when estimating displacement from an acceleration measurement. A simplistic approach to calculating the displacement is to integrate the acceleration twice. In the *s*-domain, this double integration is equivalent to multiplying by $1/s^2$. Using the

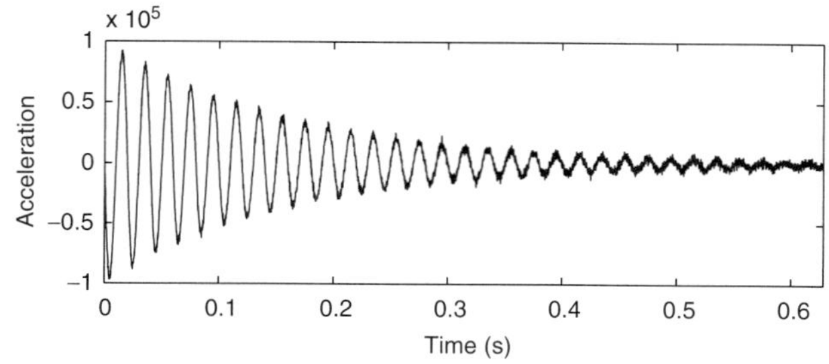

FIGURE 9.11 Acceleration measurement.

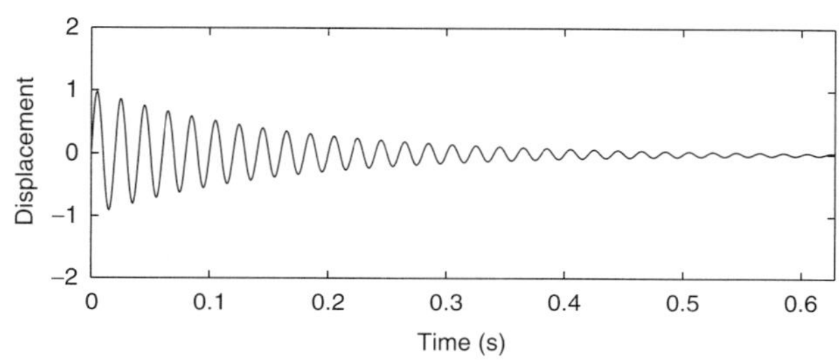

FIGURE 9.12 Actual displacement.

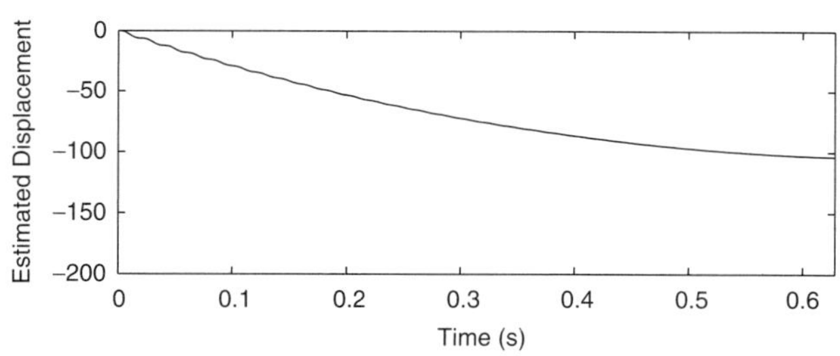

FIGURE 9.13 Estimated displacement without use of a prefilter.

bilinear transformation to convert $1/s^2$ to the z-domain yields the following transfer function,

$$H(z) = H(s)\Big|_{s=\frac{2(z-1)}{T(z+1)}} = \frac{1}{s^2}\Big|_{s=\frac{2(z-1)}{T(z+1)}} = \left(\frac{T^2}{4}\right)\left(\frac{z^2+2z+1}{z^2-2z+1}\right) = \left(\frac{T^2}{4}\right)\left(\frac{1+2z^{-1}+z^{-2}}{1-2z^{-1}+z^{-2}}\right)$$

The corresponding difference equation used to calculate the displacement $y[\bullet]$ from the acceleration $y_{dd}[\bullet]$ is $4(y[n] - 2y[n-1] + y[n-2]) = T^2(y_{dd}[n] + 2y_{dd}[n-1] + y_{dd}[n-2])$. However, accelerometers generally do not have good response at low frequencies; in fact, they often insert a bias in the data yielding a drift in the calculated displacement. They also are very sensitive to random vibrations. An alternate approach is to process the acceleration data through a bandpass filter before using the difference equation to integrate it numerically. The bandpass range must contain the natural frequencies in the system.

Consider, for example, the acceleration data shown in Figure 9.11, which has some random noise. This signal is sampled at a rate of 6400 Hz, where the natural frequency of the system is 50 Hz. Figure 9.12 shows the actual displacement, while Figure 9.13 shows the estimated displacement calculated by numerically integrating the acceleration data, using the difference equation given above. This estimation is very poor. Alternatively, an analog eighth-order Chebyshev Type I bandpass filter with passband 25–500 Hz

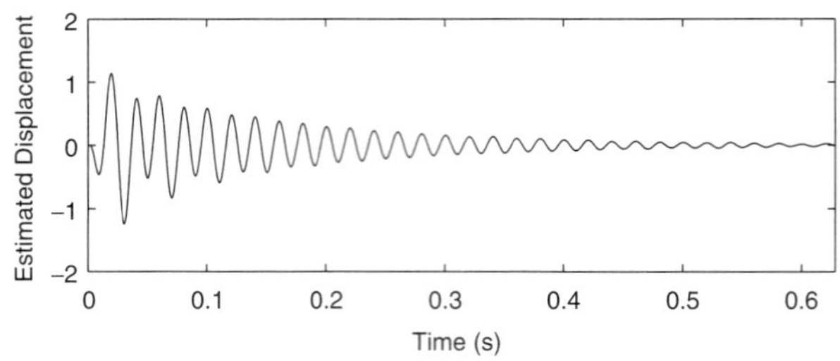

FIGURE 9.14 Estimated displacement with IIR prefilter.

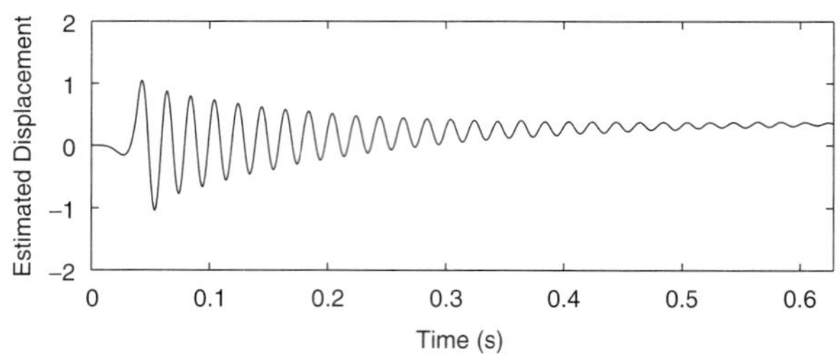

FIGURE 9.15 Estimated displacement with FIR prefilter.

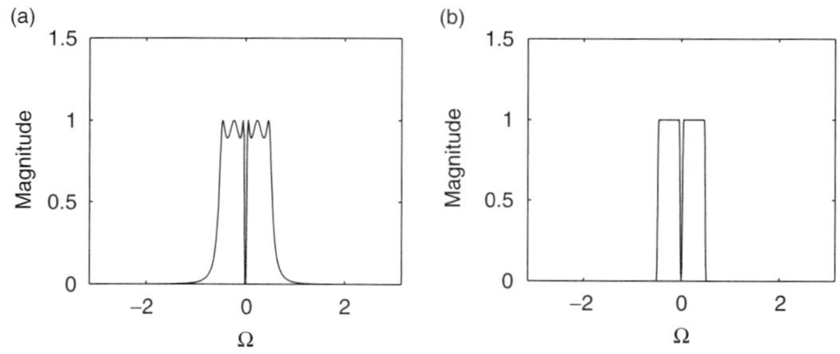

FIGURE 9.16 Digital bandpass filters: (a) Chebyshev IIR and (b) FIR filter.

is designed and then discretized using the bilinear transformation. The acceleration data is processed through this filter first, and then the filtered data are numerically integrated with the result shown in Figure 9.14. Notice that the estimate is much better than that obtained without the bandpass filter. A 500th order FIR bandpass is also designed for this example with passband 25–500 Hz. After passing the data through the FIR filter, it is then numerically integrated resulting in the estimated displacement shown in Figure 9.15. Due to the linear phase characteristic of the FIR filter, it has less transient distortion than the IIR filter, but it adds a larger lag. The larger the order, the more accurate the result, since less significant information is lost in truncating the impulse response of the ideal IIR bandpass filter, but the lag is larger. The magnitudes of the IIR filter and the FIR bandpass filters are shown in Figure 9.16.

Two observations on this example should be mentioned:

1. The unfiltered calculation was extremely sensitive to bias in the data (as expected from the double integration). Therefore, the bias was removed from the acceleration before processing. Both filters effectively removed bias, so the results were virtually unchanged if the bias was present.
2. The FIR filter shows some drift. Presumably, the cause of the drift is that the filter seems to have some difficulty with the small stopband region near the origin. Increasing the stopband region

does reduce the drift. This can be done by decreasing the sample frequency or by increasing the passband frequency. Both of these remedies decrease the drift but increase other errors in the signal. Increasing the length of the filter decreases the drift error without introducing other errors.

Some of the MATLAB commands used to design the filters and generate the results are:

```
[num,den] = c2dm(1,[1 0 0],T,'tustin'); %digitize 1/s^2
y1 = filter(num,den,ydd); % double integration of ydd

Wbreak = [2*pi*25*T, 2*pi*500*T}; % digital break frequencies
[b,a] = cheby1(4,1,Wbreak); % design IIR filter with 1dB ripple

W = -pi:pi/200:pi; % define digital frequency range for plot
H = freqz(b,a,W); % get frequency response
plot(W,abs(H)); % plot magnitude of frequency response

yddfilt = filter(b,a,ydd); % calculate output of IIR filter
y2 = filter(num,den,yddfilt); % double integration of yddfilt

hfir = fir1(500,Wbreak); % design FIR filter of order 500
yddfilt = filter(hfir,1,ydd); % calculate output of FIR filter
y3 = filter(num,den,yddfilt); % double integration of yddfilt
```

9.5 Digital Control Design

As in the digital filter design case, there are two general methods for designing a digital controller: an *indirect method* that is based on discretizing an analog design, and a *direct method* that is based on discretizing a plant (usually using the step-response matching method) and then designing the controller directly in the discrete domain. Most engineers learn classical continuous-time controls, and it is common for them to have more training in continuous-time control design than in discrete-time or digital control design. Fortunately, continuous-time control tools can often be used when designing digital control systems. To make use of controllers designed in the continuous-time domain, an *s*-plane to *z*-plane mapping is used. Any of the mappings discussed in this chapter can be used for a variety of controllers. It is always best to determine the mapping that is most efficient for a particular control or filter. Even though the bilinear approximation is more complex than the forward or backward approximations, it is used for most mechatronic systems. This is due to the fact that most modern controllers have enough computational power to manage the increased complexity at the required bandwidth of the mechatronic system.

As an example of the indirect design method, consider a PD (proportional derivative) controller that may be used to enhance the performance of a system. The derivative and proportional gains for the controller are K_d and K_p, respectively. The PD controller, $K(s)$, is given by

$$K(s) = K_d s + K_p. \tag{9.6}$$

Equation 9.6 can be implemented digitally using any of the *s*-plane to *z*-plane mappings discussed earlier in this chapter. As an example, the bilinear transformation is used generating the digital controller, $K(z)$.

$$K(z) = K(s)|_{s=\frac{2(z-1)}{T(z+1)}} = \frac{(2K_d + TK_p)z + (TK_p - 2K_d)}{Tz + T} \tag{9.7}$$

Besides the control gains, the only factor that is needed for Equation 9.7 is T, the sample time. As previously stated, the sample time should be at least a factor of 5–10 times the fastest system time constant.

However, sampling times are often chosen to be several hundred times faster than the fastest system time constant. An alternative strategy for a feedback system is to choose the sample rate to be at least 20 times the desired closed loop bandwidth. Having sampling times that are substantially faster than the actual system mitigates any differences between the controller as it is designed in the continuous domain and the implementation in the discrete domain. It should be noted that as the sampling frequency becomes higher, the control gains become smaller. For example, in Equation 9.7, as the sampling time becomes smaller, T becomes smaller requiring better numerical resolution for the controller gains. If T becomes smaller than the controller's numerical gain resolution, it may be erroneously implemented at a value of 0 (zero) yielding an incorrect control law.

9.5.1 Digital Control Example

Consider a high speed position motor with motor dynamics governed by the first-order equation

$$G(s) = \frac{\omega(s)}{V_{\text{in}}(s)} = \frac{K_m}{T_m s + 1}$$

where K_m is the motor gain constant, T_m is the motor time constant, $\omega(s)$ is the Laplace transform of the motor velocity, and $V_{\text{in}}(s)$ is the Laplace transform of the motor input voltage. To determine the values of T_m and K_m, the velocity step response of the motor is used. Figure 9.17 is the response of the motor to a 1-V step input. The motor gain, K_m, is the steady-state value of the final motor speed and is 5. This result can also be determined using the Final Value Theorem as

$$\begin{aligned}\lim_{t\to\infty}\omega(t) &= \lim_{s\to 0} s\omega(s) = \lim_{s\to 0} sG(s)V_{\text{in}}(s)\\ &= \lim_{s\to 0} sG(s)\frac{1}{s} = \lim_{s\to 0} s\frac{K_m}{T_m s+1}\frac{1}{s} = K_m\end{aligned}$$

The motor time constant, T_m, can be computed by determining the motor velocity for the step response at time $t = T_m$ as follows:

$$\omega(t = T_m) = K_m(1 - e^{-t/T_m}) = K_m(1 - e^{-1}) = 0.632K_m$$

So the time required for the motor to reach 63.2% of its steady-state step response is its time constant. From Figure 9.18, the time constant of this motor is 0.05 s. Thus, the transfer function for the motor is given by

$$G(s) = \frac{\omega(s)}{V_{\text{in}}(s)} = \frac{K_m}{T_m s+1} = \frac{5}{0.05s+1} \tag{9.8}$$

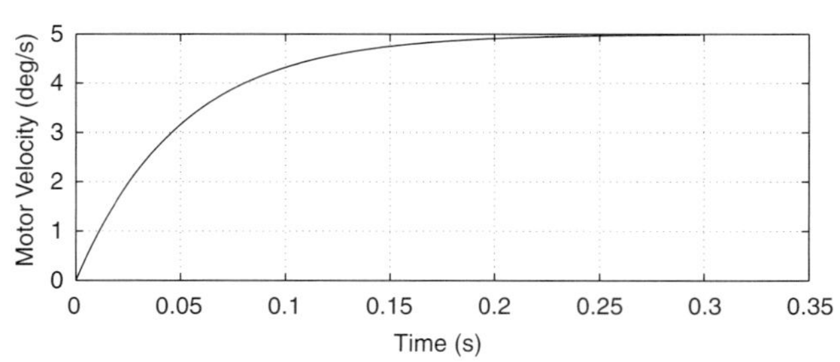

FIGURE 9.17 Motor velocity step response.

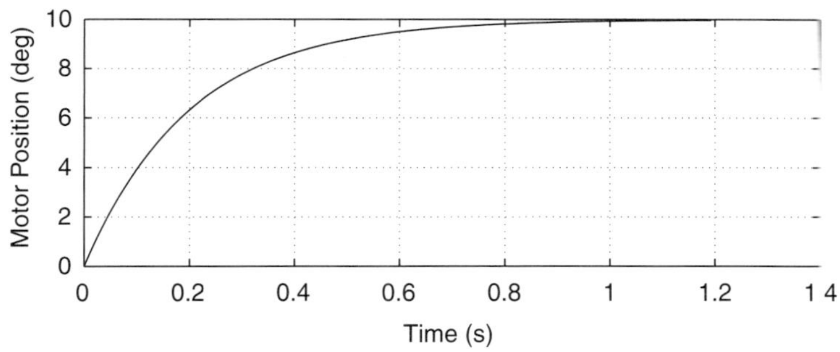

FIGURE 9.18 Closed-loop position response.

For this example, the motor is used in position control mode. Since the motor position is the integral of its velocity, Equation 9.8 can be augmented with an integrator to generate the transfer function of the motor relating the input voltage to the output position, $\theta(s)$:

$$G_p(s) = \frac{\theta(s)}{V_{\text{in}}(s)} = \frac{K_m}{s(T_m s+1)} \tag{9.9}$$

A PD controller is chosen for use in this example in order to enhance the system performance. To achieve a fast response with no overshoot, the derivative gain, K_d, and the proportional gain, K_p are chosen to be 0.05 and 1, respectively, yielding the following control law:

$$K(s) = K_d s + K_p = 0.05s + 1 \tag{9.10}$$

Nominally, this design cancels the high frequency pole of the motor dynamics given in Equation 9.9.

A sample period of 1 ms is chosen for this example as it is significantly faster than the system's time constants, and it is not an unreasonable value given modern digital controllers. As previously discussed, using a 1-kHz (1 ms) sample frequency mitigates any differences between the controller as it is designed in the continuous domain and its implementation is in the discrete domain. Using the bilinear transformation given in Section 9.3.2 results in a digital controller of the form:

$$K_D(z) = \frac{101z-99}{z+1}$$

In fact, the closed-loop response of the system using the digital controller cannot be easily distinguished from that of the system using the analog controller given by Equation 9.10. The closed-loop position response of the motor for a 10° command input is shown in Figure 9.18.

As mentioned in Section 9.4.4, 60 Hz noise is often present in measurements of electro-mechanical systems, so a bandstop filter is often used to attenuate the noise. In the closed-loop operation, the digital bandpass filter is cascaded with the digital PD controller.

References

1. Kamen, E.W., and Heck, B.S., *Signals and Systems Using the Web and Matlab,* 2nd ed., Prentice-Hall, Englewood Cliffs, NJ, 2000.
2. Britton Rorabaugh, C., *Digital Filter Designer's Handbook: with C++ Algorithms,* 2nd ed., McGraw-Hill, New York, 1997.

10

Control System Design Via $\mathcal{H}^2$ Optimization

Armando A. Rodriguez

Arizona State University

10.1 Introduction

This chapter addresses control system design via $\mathcal{H}^2$ (quadratic) optimization. A unifying framework based on the concept of a generalized plant and weighted optimization permits designers to address state feedback, state estimation, dynamic output feedback, and more general structures in a similar fashion. The framework permits one to easily incorporate design parameters and/or weighting functions that may be used to influence the outcome of the optimization, satisfy desired design specifications, and systematize the design process. Optimal solutions are obtained via well-known Riccati equations; for example, Control Algebraic Riccati Equation (CARE) and Filter Algebraic Riccati Equation (FARE). While dynamic weighting functions increase the dimension of the Riccati equations being solved, solutions are readily obtained using today's computer-aided design software (e.g., MATLAB, robust control toolbox, μ-synthesis toolbox, etc.).

In short, $\mathcal{H}^2$ optimization generalizes all of the well-known quadratic control and filter design methodologies:

- Linear Quadratic Regulator (LQR) design methodology [7,11]
- Kalman–Bucy Filter (KBF) design methodology [5,6]
- Linear Quadratic Gaussian (LQG) design methodology [4,10,11]

$\mathcal{H}^2$ optimization may be used to systematically design constant gain state feedback control laws, state estimators, dynamic output controllers, and much more.

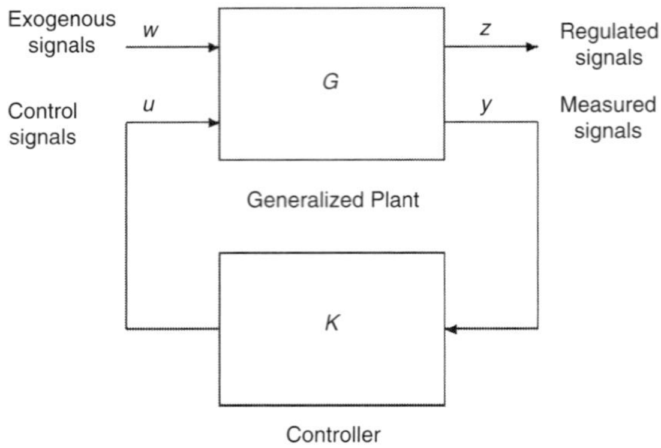

FIGURE 10.1 Generalized feedback system.

10.2 General Control System Design Framework

In this section, we present a general framework for control system (and estimator) design. Toward this end, we consider the generalized feedback system in Figure 10.1. In this figure, G represents a *generalized plant.* G contains a model for the actual plant P (physical system) to be controlled. It may also contain additional (frequency dependent) weighting functions that are used to address closed loop design objectives. K represents a controller or compensator to be designed.

10.2.1 Central Idea: Design Via Optimization

The central idea here is that many important problems that arise in controls, estimation, filtering, and other areas of engineering may be cast in terms of a generalized plant G and a controller K to be designed by minimizing some norm (e.g., $\mathcal{H}^2$) on the closed loop transfer function matrix T_{wz} from the signals w to the signals z.

10.2.2 The Signals

To appreciate the flexibility of our generalized feedback system structure, it suffices to consider the nature of the signals z, u, w, and y in the figure. These signals may be described as follows:

- *Regulated Signals.* The signals $z \in \mathrm{R}^{n_z}$ represent *regulated signals* or signals that we would like to keep "small" in some sense, which depends on the application and desired performance objectives. Such signals might include tracking errors, actuator or control inputs, signal estimation errors, etc.
- *Control Signals.* The signals $u \in \mathrm{R}^{n_u}$ represent *control signals* or *manipulated variables* that are generated by the controller K. Control signals might include fuel flow to an engine, voltage applied to a dc motor, etc. They might also include state estimates provided by K. The idea is for K to manipulate and coordinate control signals u in a manner which keeps the regulated signals z "small." In practice, we typically have more signals that require "regulation" than controls (i.e., $n_z \geq n_u$). It should be noted, however, that generally if we want to independently control m quantities, then we need at least m independent controls. This basic tenet must be adhered to in practice. The more independent controls u that are available, the easier (in principle) it is to influence the signals z to be regulated.
- *Exogenous Signals.* The signals $w \in \mathrm{R}^{n_w}$ represent *exogenous (or external) signals* that act upon the system. Exogenous signals may include reference commands issued to the control system, disturbances acting on the system, sensor noise, etc.

- *Measurement Signals.* The signals $y \in \mathrm{R}^{n_y}$ represent measurements or signals that are directly available to the controller K. Measurements may include a portion of or all of the plant state variables, measurable plant "outputs," measurable control signals, measurable exogenous signals, etc. In practice, we typically have more exogenous signals than measurements (i.e., $n_w \geq n_y$). Generally, the more independent measurements we have the better—since, in theory, more useful information can be extracted.

Comment 10.1 (Toward a Separation Principle)

It is natural to associate the controls u with the regulated signals z. One might argue that the pair implicitly defines a regulation or control problem. This is analogous to the situation addressed in classical LQR problems. In such problems, one trades off control action (size) versus speed of regulation.

Similarly, it is natural to associate the exogenous signals w with the measurements y. One might argue that the pair implicitly defines an information extraction or estimation problem. This is analogous to the situation addressed in classical KBF problems. In such problems, one trades off sensor cost (or immunity to noise) versus speed of estimate construction.

Such associations suggest that just as in classical LQG problems, our surprisingly general structure may give rise to a natural separation principle. Indeed, this will be the case for the so-called $\mathcal{H}^2$ output feedback problem that we consider. ■

10.2.3 General $\mathcal{H}^2$ Optimization Problem

The so-called general $\mathcal{H}^2$ optimization problem may be stated as follows:

- 'Find a proper real-rational (finite dimensional) controller K that internally stabilizes G such that the $\mathcal{H}^2$ norm of the closed loop system transfer function matrix $T_{wz}(K)$ is minimized:

$$\min_K \|T_{wz}(K)\|_{\mathrm{H}^2} \tag{10.1}$$

where

$$\|F\|_{\mathrm{H}^2} \stackrel{\text{def}}{=} \sqrt{\frac{1}{2\pi}\int_{-\infty}^{\infty} trace\{F^H(j\omega)F(j\omega)\}\,d\omega} \tag{10.2}$$

$$= \sqrt{\int_0^{\infty} trace\{f^H(t)f(t)\}\,dt} \tag{10.3}$$

$$= \|f\|_{\mathrm{L}^2(\mathrm{R}_+)} \tag{10.4}$$

and f is the impulse response matrix associated with the transfer function matrix F.

Comment 10.2 (Use of Two Norm: Wide Band Exogenous Signals)

Noting that the two norm measures the energy of the response to an impulse and noting that the transform of unit Dirac delta function δ is unity, it follows that the two norm is appropriate when the exogenous signals w are wide band in nature. This can always be justified by introducing appropriate (low pass) filters within G. It should be noted that these ideas have stochastic interpretations as well. Instead of unit delta functions, one instead deals with white noise with unit intensity. ■

Comment 10.3 (Control and Estimation Problems)

Although we are seeking an $\mathcal{H}^2$ optimal controller, it must be noted that the generalized plant framework will enable the design of state estimators as well as dynamic and constant gain control laws. ■

Given the above problem statement, it is appropriate to recall the following elementary result:

Lemma 10.1 (Two Norm of a Stable System)

Consider a causal stable LTI strictly proper system $F = [A, B, C]$. It follows that

$$\|F\|_{H^2} = \|f\|_{L^2(R^+)} = \sqrt{CL_cC^H} = \sqrt{B^HL_oB} \tag{10.5}$$

where L_c is the system controllability gramian and L_o is the system observability gramian. The controllability gramian

$$L_c \stackrel{\text{def}}{=} \int_0^\infty e^{At}BB^He^{A^Ht}dt \tag{10.6}$$

is the unique symmetric (at least) positive semi-definite solution of the algebraic Lyapunov equation

$$AL_c + L_cA^H + BB^H = 0 \tag{10.7}$$

L_c is positive definite if and only if (A, B) is controllable. The observability gramian

$$L_o \stackrel{\text{def}}{=} \int_0^\infty e^{A^Ht}C^HCe^{At}dt \tag{10.8}$$

is the unique symmetric (at least) positive semi-definite solution of the algebraic Lyapunov equation

$$A^HL_o + L_oA + C^HC = 0 \tag{10.9}$$

L_o is positive definite if and only if (A, C) is observable. ■

Comment 10.4 ($\mathcal{H}^2$ Norm May Mislead—L^∞ Norm Is Important)

It is important to note that the $\mathcal{H}^2/\mathcal{L}^2$ norm (or energy) of a function may be very small, while the function itself may be very large in amplitude. Consider a tall thin pulse, for example. This observation is critical because there are many important cases in which we are very concerned with the height of a function—more so than its energy. A good example of this comes from classical Nyquist stability theory [2,8]. Nyquist taught us that the peak magnitude of the sensitivity function $S = 1/(1 + L)$ associated with a standard negative feedback loop is very important in terms of the feedback loop's stability robustness. A large sensitivity means that the Nyquist plot comes close to the critical −1 point—implying that a small perturbation (or unanticipated modeling error) may cause the closed loop system to go unstable. To assist us with this fundamental issue we may use frequency dependent weighting functions, but what we really need is a norm that directly addresses such concerns. This motivates the so-called $\mathcal{H}^\infty$ and $\mathcal{L}^\infty$ norms as well as $\mathcal{H}^\infty/\mathcal{L}^\infty$ control theory [4,11]. ■

Comment 10.5 (Computation of $\mathcal{H}^2$ Norm in MATLAB)

The $\mathcal{H}^2$ norm of a system $F = [A, B, C, D]$ may be computed using the following MATLAB command sequence:

```
lc = lyap (a, b*b')
twonorm = sqrt (trace(c*lc*c'))
```

or

```
lo = lyap (a', c*c')
twonorm = sqrt (trace(b'*lo*b)).
```

■

10.2.4 Generalized Plant

The generalized plant G is assumed to possess the following two-port state space structure:

$$G = \left[\begin{array}{c|c} G_{11} & G_{12} \\ \hline G_{21} & G_{22} \end{array}\right] = \left[\begin{array}{c|cc} A & B_1 & B_2 \\ \hline C_1 & 0_{n_z \times n_w} & D_{12} \\ C_2 & D_{21} & 0_{n_y \times n_u} \end{array}\right] = \left[\begin{array}{c|c} A & B \\ \hline C & D \end{array}\right] \tag{10.10}$$

where $G_{ij}(s) = C_i(sI - A)^{-1}B_j$, $A \in \mathbb{R}^{n \times n}$, $B_1 \in \mathbb{R}^{n \times n_w}$, $B_2 \in \mathbb{R}^{n \times n_u}$, $C_1 \in \mathbb{R}^{n_z \times n}$, $C_2 \in \mathbb{R}^{n_y \times n}$, $D_{12} \in \mathbb{R}^{n_z \times n_u}$, $D_{21} \in \mathbb{R}^{n_u \times n_z}$.

Comment 10.6 (Weighting Functions: Satisfying Closed Loop Design Specifications)

As stated earlier, the generalized plant G may contain frequency dependent weighting functions as well as a model for the physical system P (plant) being controlled. Typically $P = G_{22} = [A, B_2, C_2]$. Weighting functions within G may be viewed as design parameters (mathematical knobs) that may be manipulated by a designer to influence the $\mathcal{H}^3$ problem in a manner which results in a controller that is not just optimal—a notion that is often irrelevant in practical applications—but which satisfies desired closed loop design specifications. Weighting functions may be used to weight (penalize) tracking errors, actuator and other signal levels, state estimation errors, etc. By making the weight on a signal large in a specific frequency range, we are indirectly telling the optimization problem to find a controller that makes the signal small in that frequency range. By making the weight on a signal small in a specific frequency range, we are indirectly conveying our willingness to tolerate a signal which is large in that frequency range. This idea can be illustrated via example. ■

Comment 10.7 ($D_{11} = 0$ Necessary, $D_{22} = 0$ Not Necessary)

$D_{11} = 0$ *Necessary.* Note that we have assumed that $D_{11} = 0$; that is, there is no direct path from the exogenous signals w to the regulated signals z. This assumption is essential for the $\mathcal{H}^2$ norm of the closed loop transfer function T_{wz} to be finite. If $D_{11} \neq 0$, then $||T_{wz}||_{\mathrm{H}^2}$ will be infinite and the $\mathcal{H}^2$ problem will be ill-posed; that is, make no sense. If we have a nonzero D_{11}, adding strictly proper filters on either w or z (e.g., $[1000/(s+1000)]I$) will result in $D_{11} = 0$. In this sense, the assumption is not restrictive.

$D_{22} = 0$ *Not Necessary.* It has also been assumed that $D_{22} = 0$; that is, the transfer function matrix D_{22} from controls u to measurements y is strictly proper. This assumption is very realistic since G_{22} (our plant P) is typically strictly proper in practice. If not, high frequency dynamics (e.g., actuator dynamics, flexible modes, parasitics, etc.) may be included to make it strictly proper. One might even include a simple high bandwidth low pass filter (e.g., $1000/(s+1000)$) to make G_{22} strictly proper. If this is not desirable because of the increased dimension, there is an alternative that does not increase the dimension of G.

- One can always remove D_{22} from G_{22} to obtain a new generalized plant $\hat{G}$ with $\hat{D}_{22} = 0$. The term D_{22} is then absorbed into an augmented controller K by noting that u is related to y as follows:

$$u = K[y + D_{22}u] = [I - KD_{22}]^{-1}Ky \tag{10.11}$$

Noting this, it follows that the augmented controller, denoted $\hat{K}$, is given by

$$\hat{K} = [I - KD_{22}]^{-1}K \tag{10.12}$$

The $\mathcal{H}^2$ problem can then be carried out for $\hat{G}$ and $\hat{K}$ (without regard to D_{22}) When the optimal controller $\hat{K}$ for $\hat{G}$ is obtained, one can compute the optimal controller K for G using the relationship

$$K = \hat{K}[I + D_{22}\hat{K}]^{-1} \tag{10.13}$$

With this stated, the assumption $D_{22} = 0$ is made without any loss of generality. ■

10.2.5 Closed Loop Transfer Function Matrices

Given the structure for the generalized plant G, we have the following closed loop relationships:

$$u = Ky \tag{10.14}$$

$$= K(G_{21}w + G_{22}u) \tag{10.15}$$

$$= [I - KG_{22}]^{-1}KG_{21}w \tag{10.16}$$

$$= K[I - G_{22}K]^{-1}G_{21}w \tag{10.17}$$

$$y = [I - G_{22}K]^{-1}G_{21}w \tag{10.18}$$

$$z = G_{11}w + G_{12}u \tag{10.19}$$

$$= G_{11}w + G_{12}Ky \tag{10.20}$$

$$= [G_{11} + G_{12}K[I - G_{22}K]^{-1}G_{21}]w \tag{10.21}$$

From this, we have the following closed loop transfer function matrices:

$$T_{wu} = K[I - G_{22}K]^{-1}G_{21} \tag{10.22}$$

$$T_{wy} = [I - G_{22}K]^{-1}G_{21} \tag{10.23}$$

$$T_{wz} = G_{11} + G_{12}K[I - G_{22}K]^{-1}G_{21} \tag{10.24}$$

We say that each of these is a *linear fractional transformation* (*LFT*) involving K.

Comment 10.8 (Well Posedness of Closed Loop System)

In the above manipulation, it has been assumed that the inverse $[I - G_{22}K]^{-1}$ is well defined. This well posedness condition is guaranteed by our assumption that $D_{22} = 0$. This assumption implies that $G_{22}(j\infty) = D_{22} = 0$ and hence that the inverse is well defined. ■

The following example shows how to formulate a so-called Weighted $\mathcal{H}^3$ Mixed Sensitivity Problem to address feedback control system design issues.

Example 10.1 (Weighted $\mathcal{H}^2$ Mixed Sensitivity Problem: Design Philosophy)

This example considers the design of a controller K for a plant $P = [A_p, B_p, C_p, D_p]$ as shown in Figure 10.2. To obtain K, we will formulate an $\mathcal{H}^2$ optimization that considers (directly or indirectly) various issues that are of importance in the design of a good feedback loop.

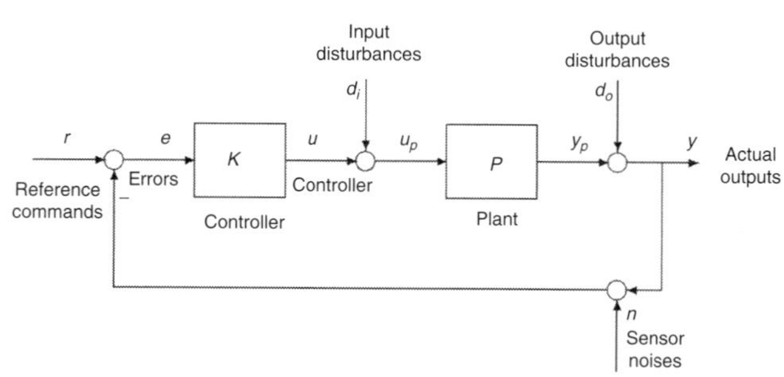

FIGURE 10.2 Standard negative feedback loop.

Feedback System Performance Issues. Generally, in designing a feedback controller K as shown in Figure 10.2, a designer must consider each of the following closed loop performance issues:

1. *Closed Loop Stability.* The closed loop system should be stable. This involves all closed loop transfer function matrices since we generally want all of them to be stable. A stricly proper closed loop transfer function matrix whose $\mathcal{H}^2$ norm is infinite, for example, implies that the transfer function matrix is unstable (or marginally stable). Stable strictly proper transfer function matrices necessarily have a finite $\mathcal{H}^3$ norm.

2. *Command Following.* The closed loop system should exhibit good low frequency reference command following; i.e., the output y (not to be confused with generalized plant measurements) should track low frequency reference commands r that are issued to the feedback system. This typically requires that the sensitivity transfer function matrix

$$S \overset{\text{def}}{=} [I + PK]^{-1} \tag{10.25}$$

be small at low frequencies.

3. *Disturbance Attenuation.* The closed loop system should exhibit good low frequency disturbance attenuation. For disturbances d_o modeled at the plant output, this requires that the sensitivity transfer function matrix be small at low frequencies. For disturbances d_i modeled at the plant input, this requires that

$$T_{d_i y} \overset{\text{def}}{=} SP \tag{10.26}$$

be small at low frequencies.

4. *Sensor Noise Attenuation.* The closed loop system should exhibit good high frequency noise n attenuation. This typically requires that the complementary sensitivity transfer function matrix

$$T \overset{\text{def}}{=} I - S \tag{10.27}$$

be small at high frequencies.

5. *Stability Robustness.* The closed loop system should exhibit robustness with respect to high frequency unmodeled dynamics (e.g., flexible modes, parasitic dynamics, time delays, etc.); This typically requires that the "peak" of some closed loop transfer function matrix be small at high frequencies.
   - *Multiplicative Modeling Error.* For a plant modeled as

$$P_{\text{act}} = [I + \Delta]P \tag{10.28}$$

   where P_{act} represents the actual plant, P represents a nominal model, and Δ represents a stable multiplicative perturbation at the plant output, the relevant closed loop transfer function matrix (that seen by Δ) is T.
   - *Additive Modeling Error.* For a plant modeled as

$$P_{\text{act}} = P + \Delta \tag{10.29}$$

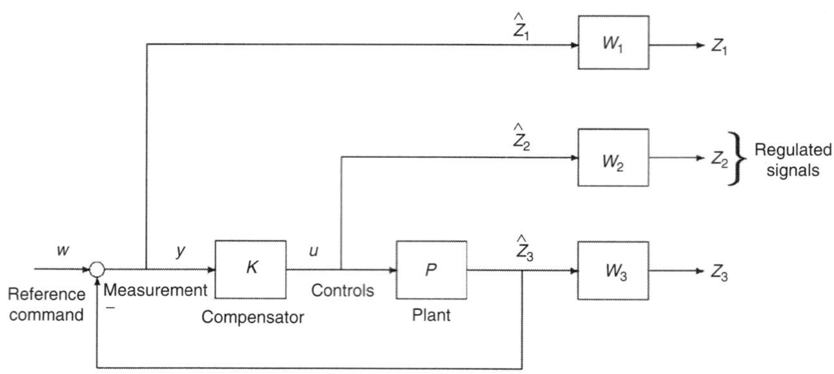

FIGURE 10.3 Negative feedback system for weighted mixed sensitivity problem.

where P_{act} represents the actual plant, P represents a nominal model, and Δ represents a stable additive perturbation, the relevant closed loop transfer function matrix (that seen by Δ) is KS.

- *Reasonable Control Action.* The closed loop system should exhibit reasonably sized control action for typical reference commands and sensor noise. This typically requires that the "size" of KS be controlled. Too much lead (i.e., derivative action) in K may help in terms of stabilization, achieving a high bandwidth and phase margin, but it may result in controls that are unnecessarily large in the presence of typical reference commands r and sensor noise n.

The above list suggests that there are many important issues that impact the control system design process. Part of a designer's job, however, is to prioritize and select issues that are most important. Toward this end, we turn our attention away from Figure 10.2 and consider instead the "fictitious" (mathematical) system depicted in Figure 10.3.

Weighting Functions and Closed Loop Transfer Function Matrix

Figure 10.3 includes specific weighting functions that will help us formulate an $\mathcal{H}^2$ optimization that (directly or indirectly) addresses some of the issues mentioned above. The figure shows a weighting W_1 on the signal y (the tracking error), a weighting W_2 on the controls u, and a weighting W_3 on the plant outputs $\hat{z}_3$. From the figure, it follows that regulated signals

$$z = \begin{bmatrix} z_1 \\ z_2 \\ z_3 \end{bmatrix}$$

are related to the exogenous signals w as follows:

$$z_1 = W_1\hat{z}_1 = W_1 S w \tag{10.30}$$

$$z_2 = W_2\hat{z}_2 = W_2 K S w \tag{10.31}$$

$$z_3 = W_3\hat{z}_3 = W_3 T w \tag{10.32}$$

From this, it follows that the closed loop transfer function matrix from w to z is given by

$$T_{wz} = \begin{bmatrix} W_1 S \\ W_2 K S \\ W_3 T \end{bmatrix} \tag{10.33}$$

Since T_{wz} involves various "sensitivity" transfer function matrices, we say that we have a weighted mixed sensitivity problem.

Selection of Weighting Functions

Typically, the weighting functions W_1, W_2, W_3 are selected to be stable transfer function matrices that are (at least initially) diagonally structured.

- *Sensitivity Weighting.* One might select the sensitivity weighting W_1 on the sensitivity S as follows:

$$W_1 = \left[\frac{k_1}{s+\varepsilon}\right] I_{n_y \times n_y} \tag{10.34}$$

  where k_1, $\varepsilon > 0$. The parameter k_1 is typically selected to be large. The parameter ε is typically selected to be small. Such selections are made so that S is heavily penalized at low frequencies—precisely where we want K to make S small.

- *Control Weighting.* One might select the control weighting W_2 on KS as follows:

$$W_2 = k_2 I_{n_u \times n_u} \tag{10.35}$$

  where $k_2 > 0$ provides a nonsingular penalty on the controls u (i.e., on KS).

- *Output Weighting.* One might select the output weighting W_3 on T as follows:

$$W_3 = \left[\frac{k_3(s+z_3)}{s+p_3}\right] I_{n_y \times n_y} \tag{10.36}$$

  with $k_3 > 0$ and $z_3 < p_3$. Such a weighting would penalize T more heavily at higher frequencies.

In general, care must be taken in selecting the structure of weighting functions. Inappropriate selections may result in an ill-posed problem and a very arduous design process. For example, W_1 must be strictly proper for $\mathcal{H}^2$ problems—otherwise the $\mathcal{H}^2$ norm of $W_1 S$ makes no sense (since S approaches the identity at high frequencies). While there exists no precise systematic method for the selection of weighting functions, the above structures seem to work well (as starting points) in many applications.

Input–Output Representation for Generalized Plant G

To obtain an input–output (transfer function matrix) description for our generalized plant, we must express the regulated signals z_1, z_2, z_3 and the measurements y in terms of the exogenous signals w and the controls u. Doing so yields

$$z_1 = W_1 \hat{z}_1 = W_1(w - Pu) = W_1 w - W_1 P u \tag{10.37}$$

$$z_2 = W_2 \hat{z}_2 = W_2 u \tag{10.38}$$

$$z_3 = W_3 \hat{z}_3 = W_3 P u \tag{10.39}$$

$$y = w - \hat{z}_3 = w - Pu \tag{10.40}$$

From this, we obtain the following input–output (transfer function matrix) description for our generalized plant G:

$$\left[\begin{array}{c} z \\ \hline y \end{array}\right] = \left[\begin{array}{c|c} G_{11} & G_{12} \\ G_{21} & G_{22} \end{array}\right] \left[\begin{array}{c} w \\ \hline u \end{array}\right] \tag{10.41}$$

or

$$\begin{bmatrix} z_1 \\ z_2 \\ z_3 \\ \hline y \end{bmatrix} = \begin{bmatrix} W_1 & -W_1 P \\ 0 & W_2 \\ 0 & W_3 P \\ \hline I & -P \end{bmatrix} \begin{bmatrix} w \\ u \end{bmatrix} \tag{10.42}$$

State Space Representation for Generalized Plant G

Next we obtain a two-port state space representation for G. To do so, we assume the following state space representations:

$$P = [A_p,\ B_p,\ C_p,\ D_p] \quad \text{with state } x_p \tag{10.43}$$

$$W_1 = [A_1,\ B_1,\ C_1,\ D_1] \quad \text{with state } x_1 \tag{10.44}$$

$$W_2 = [A_2,\ B_2,\ C_2,\ D_2] \quad \text{with state } x_2 \tag{10.45}$$

$$W_3 = [A_3,\ B_3,\ C_3,\ D_3] \quad \text{with state } x_3 \tag{10.46}$$

To obtain the desired state space representation for G, we need to express the signals $(\{x_i\}_{i=1}^3, x_p, \{z_i\}_{i=1}^3, y)$ in terms of the signals $(\{x_i\}_{i=1}^3, x_p, w, u)$. This is just a matter of simple bookkeeping. Doing so yields the following:

$$x_1 = A_1x_1 + B_1y = A_1x_1 + B_1(w - C_px_p - D_pu) = A_1x_1 - B_1C_px_p - B_1D_pu \tag{10.47}$$

$$x_2 = A_2x_2 + B_2u \tag{10.48}$$

$$x_3 = A_3x_3 + B_3\hat{z}_3 = A_3x_3 + B_3(C_px_p + D_pu) = A_3x_3 + B_3C_px_p + B_3D_pu \tag{10.49}$$

$$x_p = A_px_p + B_pu \tag{10.50}$$

$$z_1 = C_1x_1 + D_1y = C_1x_1 + D_1(w - C_px_p - D_pu) = C_1x_1 - D_1C_px_p + D_1w - D_1D_pu \tag{10.51}$$

$$z_2 = C_2x_2 + D_2u \tag{10.52}$$

$$z_3 = C_3x_3 + D_3\hat{z}_3 = C_3x_3 + D_3(C_px_p + D_pu) = C_3x_1 + D_3C_px_p + D_3D_pu \tag{10.53}$$

$$y = w - C_px_p - D_pu = -C_px_p + w - D_pu \tag{10.54}$$

The above equations may be written in standard two-port form:

$$\begin{bmatrix} x \\ \hline z \\ \hline y \end{bmatrix} = \begin{bmatrix} A & B_{11} & B_{12} \\ \hline C_{11} & D_{11} & D_{12} \\ \hline C_{21} & D_{21} & D_{22} \end{bmatrix} \begin{bmatrix} x \\ \hline w \\ \hline u \end{bmatrix} \tag{10.55}$$

as follows:

$$\begin{bmatrix} \dot{x}_1 \\ \dot{x}_2 \\ \dot{x}_3 \\ \dot{x}_p \\ \hline z_1 \\ z_2 \\ z_3 \\ \hline y \end{bmatrix} = \left[\begin{array}{cccc|c|c} A_1 & & & -B_1C_p & & -B_1D_p \\ & A_2 & & & & B_2 \\ & & A_3 & B_3C_p & & B_3D_p \\ & & & A_p & & B_p \\ \hline C_1 & & & -D_1C_p & D_1 & -D_1D_p \\ & C_2 & & & & D_2 \\ & & C_3 & D_3C_p & & D_3D_p \\ \hline & & & -C_p & I & -D_p \end{array}\right] \begin{bmatrix} x_1 \\ x_2 \\ x_3 \\ x_p \\ \hline w \\ \hline u \end{bmatrix} \tag{10.56}$$

Checking Assumptions

In selecting the weights, W_1, W_2, W_3, one must make sure that none of the "standard" $\mathcal{H}^2$ problem assumptions are violated. Thus far, we require that $D_{11} = 0$ and $D_{22} = 0$. To ensure that $D_{11} = 0$, we need

$$D_1 = 0 \tag{10.57}$$

To ensure that $D_{22} = 0$, we need

$$D_p = 0. \tag{10.58}$$

This results in

$$\begin{bmatrix} \dot{x}_1 \\ \dot{x}_2 \\ \dot{x}_3 \\ \dot{x}_p \\ \hline z_1 \\ z_2 \\ z_3 \\ \hline y \end{bmatrix} = \left[\begin{array}{cccc|c|c} A_1 & & & -B_1C_p & & B_2 \\ & A_2 & & & & \\ & & A_3 & B_3C_p & & \\ & & & A_p & & B_p \\ \hline C_1 & & & & & \\ & C_2 & & & & D_2 \\ & & C_3 & \Delta_3C_p & & \\ \hline & & & -C_p & I & \end{array}\right] \begin{bmatrix} x_1 \\ x_2 \\ x_3 \\ x_p \\ \hline w \\ \hline u \end{bmatrix} \tag{10.59}$$

In subsequent sections, additional assumptions will be imposed on the two-port state space representation for the generalized plant G. The additional assumptions imposed will depend upon the specific $\mathcal{H}^2$ problem being considered.

Weighted $\mathcal{H}^2$ Optimal Mixed Sensitivity Problem

Given the above, the weighted $\mathcal{H}^2$ optimal mixed sensitivity problem is then to find a real-rational (finite-dimensional) proper internally stabilizing controller K that minimizes $\|T_{wz}\|_{H^2}$; that is

$$\min_K \|T_{wz}\|_{H^2} = \min_K \left\| \begin{bmatrix} W_1S \\ W_2KS \\ W_3T \end{bmatrix} \right\|_{H^2} \tag{10.60}$$

We will show how this optimal control problem—and problems like it—can be readily solved using computer-aided design soflware (e.g., MATLAB, robust control toolbox, μ-tools). ■

Comment 10.9 (Construction of Generalized Plant G)

Generalized plants G are very easy to construct within SIMULINK. Input port blocks may be used to specify exogenous signals w and controls u. Output port blocks may be used to specify regulated signals z and measurements y. The "linmod" command may be applied to the constructed block diagram (SIMULINK file) to obtain a two-port state space (A, $B = [B_1\ B_2]$, $C = [C_1; C_2]$, $D = [D_{11}\ D_{12}; D_{21}\ D_{22}]$) representation for G. The syntax for the command is as follows:

```
[ a, b, c, d ] = linmod ('filename')
```

This method enables one to create generalized plant models quickly. ■

10.2.6 Overview of $\mathcal{H}^2$ Optimization Problems to Be Considered

Three fundamental problems are considered in this chapter:

1. ***$\mathcal{H}^2$ Output Feedback Problem.*** The solution to this problem is an optimal model-based dynamic compensator possessing the structure

$$K_{\text{opt}} = \left[\begin{array}{c|c} A - B_2G_c - H_fC_2 & H_f \\ \hline -G_c & O_{n_u \times n} \end{array}\right] \tag{10.61}$$

   where G_c, is a control gain (state feedback) matrix and H_f is a filter gain (observer) matrix. G_c is found by using the solution of a Control Algebraic Riccati Equation (CARE)—similar to that found in Linear Quadratic Regulator (LQR) problems. H_f is found by using the solution of a Filter Algebraic Riccati Equation (FARE)—similar to that found in Kalman–Bucy Filtering (KBF) problems. The structure of K_{opt}, G_c, and H_f can be thought of as the solution to a classical Linear Quadratic Gaussian (LQG) control problem which gives rise to the well known separation principle: closed loop poles are the eigenvalues of $A - B_2G_c$, and the eigenvalues of $A - H_fC_2$.
2. ***$\mathcal{H}^2$ State Feedback Problem.*** The solution to this problem is an optimal constant gain (state feedback) compensator possessing the structure

$$K_{\text{opt}} = -G_c \tag{10.62}$$

   where G_c is a control gain (state feedback) matrix found by using the solution to a CARE—similar to that found in LQR problems. The poles of the resulting closed loop system are the eigenvalues of $A- B_2G_c$. In short, this problem should be viewed as a mechanism for computing control gain matrices G_c that may be used in a state feedback application or in a model-based compensator application.
3. ***$\mathcal{H}^2$ Output Injection Problem.*** The solution to this problem is an optimal constant gain (static) compensator possessing the structure

$$K_{\text{opt}} = -H_f \tag{10.63}$$

   where H_f is a filter gain (observer) matrix found by using the solution to a FARE—similar to that found in KBF problems. The poles of the resulting closed loop system are the eigenvalues of $A - H_fC_2$. In short, this problem should be viewed as a mechanism for computing filter gain matrices H_f that may be used in a state estimation application or in a model-based compensator application.

10.3 $\mathcal{H}^2$ Output Feedback Problem

In this section, we consider the $\mathcal{H}^2$ output feedback problem. This problem results in model-based (dynamic) compensators involving a control gain (state feedback) matrix G_c and a filter gain (observer) matrix H_f. As such, the problem generalizes the ideas presented in classical LQG theory.

The following "standard" $\mathcal{H}^2$ output feedback problem assumption is now made.

Assumption 10.1 ($\mathcal{H}^2$ Output Feedback Problem)

Throughout this section, it will be assumed that

1. *Plant G_{22} Assumption.* (A, B_2, C_2) stabilizable and detectable.
   - This assumption is necessary and suficient for the existance of a proper internally stabilizing controller K. With this assumption, the following model based (observer based) controller stabilizes the feedback loop in Figure 10.1:

$$K = \left[\begin{array}{c|c} A - B_2 G_c - H_f(C_2 - D_{22} G_c) & H_f \\ \hline -G_c & O_{n_u \times n} \end{array}\right] \tag{10.64}$$

   provided that $(A - B_2 G_c)$ and $(A - H_f C_2)$ are stable, as suggested by the classical separation principle from the theory of linear systems.

   This assumption mandates that *all of the "bad" open loop poles (right half plane and imaginary) must be controllable through the controls u and observable through the measurements y.*

   Suppose that G satisfies the assumption. Consider the augmentation of an integrator I/s (weighting function). Such an augmentation can result in the assumption being violated. Absorbing an integrator I/s (weighting function) on the exogenous signals w into G, for example, would violate the stabilizability assumption since it would introduce an open loop pole on the imaginary axis that is not controllable through the controls u. Absorbing an integrator on the regulated signals z into G would violate the detectability assumption since it would introduce an open loop pole on the imaginary axis that is not observable through the measurements y. Using $I/(s + \epsilon)(\epsilon > 0)$ instead of I/s—in either case—would result in a G that does not violate the assumption.
2. *Nonsingular Control Weighting Assumption.* $R = D_{12}^T D_{12} > 0$.
   - This assumption implies that $D_{12} \in \mathbb{R}^{n_z \times n_u}$ has full column rank (i.e., rank $D_{12} = n_u$) and hence that every control (direction) u influences the regulated signals z through D_{12} (i.e., D_{12} has no right null space). The matrix D_{12} must therefore be "tall" and "thin;" that is,

$$\text{(number of regulated signals) } n_z \geq n_u \text{ (number of control signals)} \tag{10.65}$$

   The matrix $R = D_{21}^T D_{12} \in \mathbb{R}^{n_u \times n_u}$ may be interpreted as a weighting on the controls u—just like the control weighting matrix "R" in LQR problems. As in the LQR problem, we might say that the control weighting R on u is nonsingular. The larger R, the smaller we want our controls to be—sacrificing speed of regulation. A large R results in a low "regulation" bandwidth. The smaller R, the larger we will permit our controls u to be—in order to speed up regulation. A small R results in a high "regulation" bandwidth.
3. *Regulator Assumption.* $\begin{bmatrix} j\omega I - A & -B_2 \\ C_1 & D_{12} \end{bmatrix}$ has full column rank $(n + n_u)$ for all ω.
   - This assumption implies that transfer function matrix from control signals u to regulated signals z has no (right) zero on the imaginary axis. Together with (1) and (2), it will guarantee that the Hamiltonian $\mathcal{H}_{con}$ involving (A, B_2, C_1, D_{12}, R)—that is, associated with the controls u and regulated signals z—will belong to *dom(Ric)*. This, in turn, guarantees that the solution of the associated CARE results in a control gain matrix $G_c \in \mathbb{R}^{n_u \times n}$ such that $A - B_2 G_c$ is stable.

The assumption implies that G has no imaginary modes that are unobservable through the regulated signals z; that is, all open loop poles on the imaginary axis must be observable through the regulated signals z. (A, C_1), therefore, cannot possess unobservable imaginary modes. This is a necessary condition. It is not suficient. An integrator hanging on the measurements y, for example, would violate this.

Since D_{12} has full column rank, the assumption is equivalent to the pair

$$(A - B_2 R^{-1} D_{12}^T C_1, \; (I - D_{12} R^{-1} D_{12}^T) C_1) \tag{10.66}$$

having no unobservable imaginary modes:

- If D_{12} is square, then it is invertible and the assumption is equivalent to $A - B_2 R^{-1} D_{12}^T C_1$ having no imaginary modes.
- If $D_{12}^T C_1 = 0$ (no cross penalty between controls and states), then the assumption is equivalent to (A, C_1) having no unobservable imaginary modes.

4. *Nonsingular Measurement Weighting Assumption.* $\Theta = D_{21} D_{12}^T > 0$.
   - This assumption implies that $D_{21} \in \mathbb{R}^{n_y \times n_w}$ has full row rank (i.e., rank $D_{21} = n_y$) and hence that the measurements y are linearly independent through D_{21} (i.e., D_{21} has no left null space). The matrix D_{21} must therefore be "short" and "fat;" that is,

$$\text{(number of measurements) } n_y \leq n \text{ (number of exogenous signals)} \tag{10.67}$$

The matrix $\Theta = D_{21} D_{21}^T \in \mathbb{R}^{n_y \times n_y}$ may be interpreted as the intensity of sensor noise impacting the measurements y—just like the sensor noise intensity matrix "Θ" found in KBF problems. As in the KBF problem, we say that the intensity matrix Θ associated with the measurements y is nonsingular. The larger Θ, the more we want to low pass filter the measurements y—sacrificing speed of estimation. A large Θ results in a low bandwidth for the associated estimator (observer). The smaller Θ, the less we want to low pass filter the measurements y—trading off our immunity to noise for speed of estimation. A small Θ results in a high bandwidth for the associated estimator (observer).

5. *Filter Assumption.* $\begin{bmatrix} j\omega I - A & -B_1 \\ C_2 & D_{21} \end{bmatrix}$ has full row rank $(n + n_y)$ for all ω.
   - This assumption implies that transfer function matrix from exogenous signals w to measurements y has no (left) zero on the imaginary axis. Together with (1) and (3), it will guarantee that the Hamiltonian $\mathcal{H}_{\text{fil}}$ involving $(A, B_1, C_2, D_{21}, \Theta)$—that is, involving exogenous signals z and measurements y—will belong to *dom*(*Ric*). This, in turn, guarantees that the solution of the associated FARE results in a filter gain matrix $H_f \in \mathbb{R}^{n \times n_y}$ such that $A - H_f C_2$ is stable.

The assumption implies that G has no imaginary modes that are uncontrollable through the exogenous signals w; that is, all open loop poles on the imaginary axis must be controllable through the exogenous signals w. (A, B_1), therefore, cannot possess uncontrollable imaginary modes. This is a necessary condition. It is not suficient. An integrator hanging on the controls u, for example, would violate this.

Since D_{21} has full row rank, the assumption is equivalent to the pair

$$(A - B_1 D_{21}^T \Theta^{-1} C_2, \; B_1 (I - D_{21}^T \Theta^{-1} D_{21})) \tag{10.68}$$

having no uncontrollable imaginary modes.

- If D_{21} is square, then it is invertible and the assumption is equivalent to $A - B_1 D_{21}^T \Theta^{-1} C_2$ having no imaginary modes.
- If $B_1 D_{21}^T = 0$ (uncorrelated process and sensor noise), then the assumption is equivalent to (A, B_1) having no uncontrollable imaginary modes. ■

Comment 10.10 (Duality Relationships)

In the above discussion, we note the following dual relationships:

$$A \longleftrightarrow A^T \tag{10.69}$$

$$B_2 \longleftrightarrow C_2^T \tag{10.70}$$

$$C_1 \longleftrightarrow B_1^T \tag{10.71}$$

$$D_{12} \longleftrightarrow D_{21}^T \tag{10.72}$$

$$R = D_{12}^T D_{12} \longleftrightarrow \Theta = D_{21} D_{21}^T \tag{10.73}$$

These imply that

- Controls u are dual to measurements y.
- Regulated signals z are dual to exogenous signals w. ■

10.3.1 Hamiltonian Matrices

Associated with our $\mathcal{H}^2$ optimal control problem are the following two Hamiltonian matrices:

$$H_{\text{con}} = \begin{bmatrix} A & 0 \\ -C_1^T C_1 & -A^T \end{bmatrix} - \begin{bmatrix} B_2 \\ -C_1^T D_{12}^T \end{bmatrix} R^{-1} [D_{12}^T C_1 \quad B_2^T] \tag{10.74}$$

$$= \begin{bmatrix} A - B_2 R^{-1} D_{12}^T C_1 & -B_2 R^{-1} B_2^T \\ -C_1^T (I - D_{12}^T R^{-1} D_{12}^T) C_1 & -(A - B_2 R^{-1} D_{12}^T C_1)^T \end{bmatrix} \tag{10.75}$$

$$H_{\text{fil}} = \begin{bmatrix} A^T & 0 \\ -B_1 B_1^T & -A \end{bmatrix} - \begin{bmatrix} C_2^T \\ -B_1 D_{21}^T \end{bmatrix} \Theta^{-1} \begin{bmatrix} D_{21} B_1^T & C_2 \end{bmatrix} \tag{10.76}$$

$$= \begin{bmatrix} (A - B_1 D_{21}^T \Theta^{-1} C_2)^T & -C_2^T \Theta^{-1} C_2 \\ -B_1 (I - D_{21}^T \Theta^{-1} D_{21}) B_1^T & -(A - B_1 D_{21}^T \Theta^{-1} C_2) \end{bmatrix} \tag{10.77}$$

The first Hamiltonian is associated with an optimal state feedback control or regulator problem. The second is associated with an optimal filtering or estimation problem.

The solution to the $\mathcal{H}^2$ output feedback problem is now given [11, pp. 261–262].

Theorem 10.1 (Solution to $\mathcal{H}^2$ Output Feedback Problem Subject to Standard Assumptions)

Suppose that G satisfies the assumptions given in Assumption 10.1—the so-called standard $\mathcal{H}^2$ output feedback problem assumptions. Given this, we have the following.

The unique minimizing $\mathcal{H}^2$ optimal controller is n dimensional (like generalized plant G) and is given by

$$K_{\text{opt}} = \left[\begin{array}{c|c} A - B_2 G_c - H_f C_2 & H_f \\ \hline -G_c & O_{n_u} \times n_y \end{array}\right] \tag{10.78}$$

where the control gain matrix $G_c \in \mathbb{R}^{n_u \times n}$ *is given by*

$$G_c = R^{-1}[B_2^T X + D_{12}^T C_1] \tag{10.79}$$

$X = Ric(\mathcal{H}_{con}) \geq 0$ *is the unique (at least) positive semi-definite solution of the CARE:*

$$(A - B_2 R^{-1} D_{12}^T C_1)^T X + X(A - B_2 R^{-1} D_{12}^T C_1) + C_1^T (1 - D_{12}^T R^{-1} D_{12}^T) C_1 - X B_2 R^{-1} B_2^T X = 0 \tag{10.80}$$

and the filter gain matrix $H_f \in \mathbb{R}^{n \times n_y}$ *is given by*

$$H_f = [Y C_2^T + B_1 D_{21}^T] \Theta^{-1} \tag{10.81}$$

$Y = Ric(\mathcal{H}_{fil}) \geq 0$ *is the unique (at least) positive semi-definite solution of the FARE:*

$$(A - B_1 D_{12}^T \Theta^{-1} C_2) Y + Y(A - B_1 D_{21}^T \Theta^{-1} C_2)^T + B_1 (I - D_{21}^T \Theta^{-1} D_{21}) B_1^T - Y C_2^T \Theta^{-1} C_2 Y = 0 \tag{10.82}$$

Moreover, the minimum norm is given by

$$\| T_{wz}(K_{\text{opt}}) \|_{L^2} = \sqrt{\| M_c B_1 \|_{L^2}^2 + \| R^{1/2} G_c M_f \|_{L^2}^2} \tag{10.83}$$

$$= \sqrt{trace(B_1^T X B_1) + trace(R G_c Y G_c^T)} \tag{10.84}$$

where

$$M_c = [A - B_2 G_c, I_{n \times n}, C_1 - D_{12} G_c] \tag{10.85}$$

$$M_f = [A - H_f C_2, I_{n \times n}, B_1 - H_f D_{21}] \tag{10.86}$$

Finally, the closed loop poles are the eigenvalues of $A - B_2 G_c$ *and* $A - H_f C_2$. ■

Comment 10.11 (Computing Optimal $\mathcal{H}^2$ Controller in MATLAB)

The following MATLAB command sequence may be used to compute the optimal $\mathcal{H}^2$ controller K_{opt} and the resulting closed loop transfer function matrix T_{wz}:

```
tss_g = mksys(a, [b1 b2], [c1; c2], [0*ones(nz, nw) d12; d21
0*ones(ny, nu),'tss')
[ss_k ss_twz] = h21qg(tss_g, 'schur')
[a_k, b_k, d_k] = branch(ss_k, 'a,b,c,d')
```

The "mksys" command packs the two-port state space data for the generalized plant G into a column vector data structure (called a tree) possessing the "tss" (two-port state space) variable designation. All dimension information is encoded into the column vector. The "h21qg" command computes the optimal $\mathcal{H}^2$ controller K_{opt} and the associated closed loop system from the exogenous signals w to the regulated signals z. An eigenvalue-eigenvector method is the default method used to solve the two relevant algebraic Riccati equations. A Schur method—based on Schur's unitary transformation of a matrix to upper triangular form—may be used by including the "schur" option. The results are stored in the tree vectors ss_k and ss_twz, respectively. The 'branch' command is then used to retrieve the state space representation for K_{opt} from the tree vector ss_k. ■

Comment 10.12 (Relationship to LQG, Stability Robustness Margins)

Theorem 10.1 shows that the optimal $\mathcal{H}^2$ output feedback controller is identical in structure to that found in classical LQG problems. While certain LQR, KBF, and LQG/LTR problem formulations do result in feedback loops possessing stability robustness margins, LQG controllers need not possess margins [3].

The same is true for $\mathcal{H}^2$ output feedback designs. We will show how the $\mathcal{H}^2$ framework presented can be manipulated to solve LQG/LTR problems which yield model-based controllers with desirable stability robustness margins—comparable to those found in feedback designs resulting from suitably formulated LQR and KBF problems (e.g., infinite upward gain margin, at least 6 dB downward gain margin, at least ±60° phase margin). ■

The following example shows how weighted $\mathcal{H}^2$ mixed sensitivity optimization may be used to design a controller for an unstable system with a time delay.

Example 10.2 (Weighted H ^{2} Mixed Sensitivity Design for Unstable System with Time Delay)

In this example, we consider an unstable system with a time delay $\Delta = 0.05$ s (50 ms). The system is modeled (approximately) as follows:

$$P \approx \frac{1}{s-1}\left[\frac{2/\Delta - s}{2/\Delta + s}\right] = \frac{1}{s-1}\left[\frac{40-s}{40+s}\right] \tag{10.87}$$

Design Specifications. The objective is to design a controller K that satisfies the following closed loop specifications: (1) closed loop stability, (2) sensitivity below −60 dB for all frequencies below 0.1 rad/s, (3) sensitivity gain crossover between 2 and 3 rad/s, (4) peak sensitivity below 5 dB, (5) peak complementary sensitivity below 10 dB.

Weighted $\mathcal{H}^2$ Mixed Sensitivity Problem. To achieve the above specifications, we formulated a weighted $\mathcal{H}^2$ mixed sensitivity problem—with a weighting W_1 on the sensitivity S and a weighting W_2 on KS; i.e.,

$$\min_K \|T_{wz}\|_{H^2} = \min_K \left\| \begin{bmatrix} W_1 S \\ W_2 KS \end{bmatrix} \right\|_{H^2} \tag{10.88}$$

The weighting functions used were as follows:

$$W_1 = \frac{k_1}{s+p_1} = \frac{10}{s+0.01} \tag{10.89}$$

$$W_2 = \frac{k_2(s+z_2)}{s+p_2} = \frac{0.1(s+40)}{s+2} \tag{10.90}$$

W_1 penalizes the sensitivity S heavily at low frequencies (e.g., below 0.001 rad/s). Above 0.1 rad/s, W_1 is small and W_2 penalizes KS (with magnitude greater than unity) until about 4 rad/s. Since the solution of our $\mathcal{H}^2$ optimization depends in a very complex manner on the parameters that define W_1 and W_2, it should be no surprise that it took a while to determine suitable parameters.

Construction of Generalized Plant. The generalized plant G was assembled using SIMULINK and the "linmod" command. The resulting two-port state space representation is as follows:

$$G = \left[\begin{array}{c|c} 0 & -W_1 P \\ 0 & W_2 \\ \hline 1 & -P \end{array}\right] = \left[\begin{array}{c|c|c} A & B_1 & B_2 \\ \hline C_1 & D_{11} & D_{12} \\ \hline C_2 & D_{21} & D_{22} \end{array}\right] = \left[\begin{array}{cccccc} -0.01 & 0 & -40 & 1 & 1 & 0 \\ 0 & -2 & 0 & 0 & 0 & 1 \\ 0 & 0 & 0 & 1 & 0 & 0 \\ 0 & 0 & 40 & -39 & 0 & 1 \\ \hline 10 & 0 & 0 & 0 & 0 & 0 \\ 0 & 3 & 0 & 0 & 0 & 0.1 \\ \hline 0 & 0 & -40 & 1 & 1 & 0 \end{array}\right] \tag{10.91}$$

Computation of $\mathcal{H}^2$ Optimal Controller. The "mksys" command was used to pack the above two-port state space into a tree vector data structure. The "h2lqg" command was then used to obtain the optimal controller. Note that the generalized plant is 4th order (two for plant P, one for sensitivity weighting W_1, one for control weighting W_2). The optimal controller:

$$K_{\text{opt}} = \frac{191.0813(s+40)(s+2)(s+0.526)}{(s+1.915)(s+0.01)(s^2+84.15s+2133)} \tag{10.92}$$

is also 4th order—the order of the generalized plant G. The pole at $s = -0.01$ is an approximate integrator—a consequence of the heavy weighting that W_1 places on the sensitivity at low frequencies.

Closed Loop Analysis. The resulting closed loop poles (two plant $P = G_{22}$, four from controller K_{opt}) are as follows:

$$s = -1, -2.0786 \pm j0.8302, -40, -40, -39.9216 \tag{10.93}$$

The resulting sensitivity, KS, and complementary sensitivity frequency responses are shown in Figures 10.4–10.6, respectively. The figures show that all of the design specifications are met (or nearly met). The peak sensitivity is about 4.855 dB. The peak complementary sensitivity is about 8.71 dB. The KS response shows the impact of the compensators' lead between 0.1 and 10 rad/s.

Computation of Minimum $\mathcal{H}^2$ Norm. The minimum two-norm was computed using the following MATLAB command sequence:

```
lc = lyap(acl, bcl*bcl)
minnorm = sqrt( trace( ccl*c*c'))
```

The minimum two-norm was found to be 9.0648. ■

The following simple example illustrates how the $\mathcal{H}^2$ output feedback problem solution can be used to solve classical LQG problems.

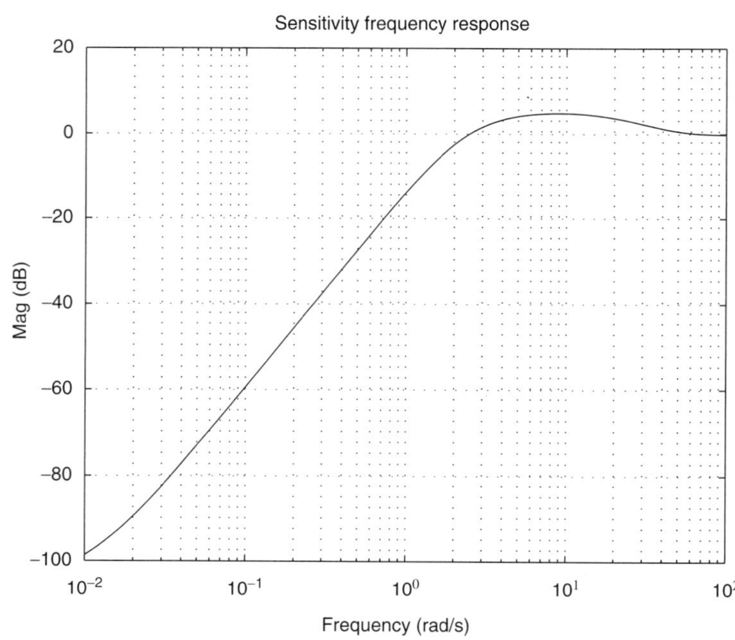

FIGURE 10.4 $\mathcal{H}^2$ Design sensitivity frequency response.

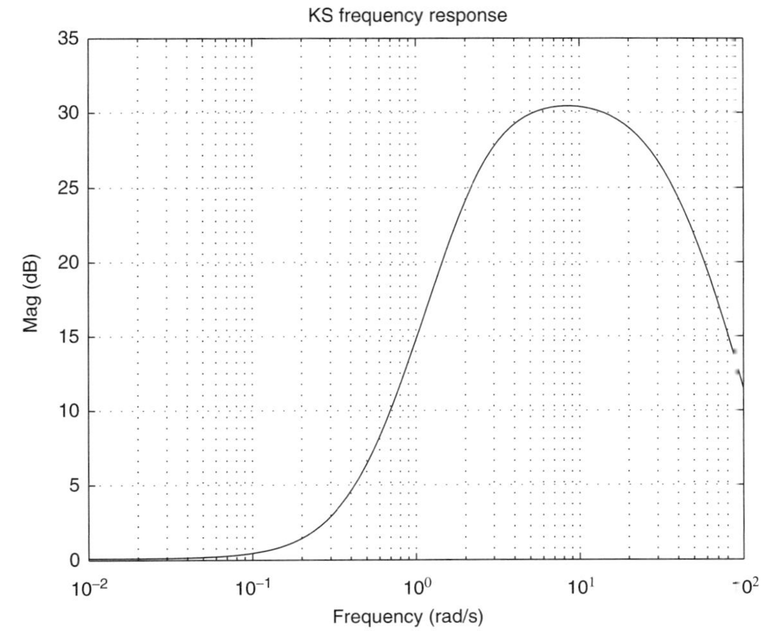

FIGURE 10.5 $\mathcal{H}^2$ Design *KS* frequency response.

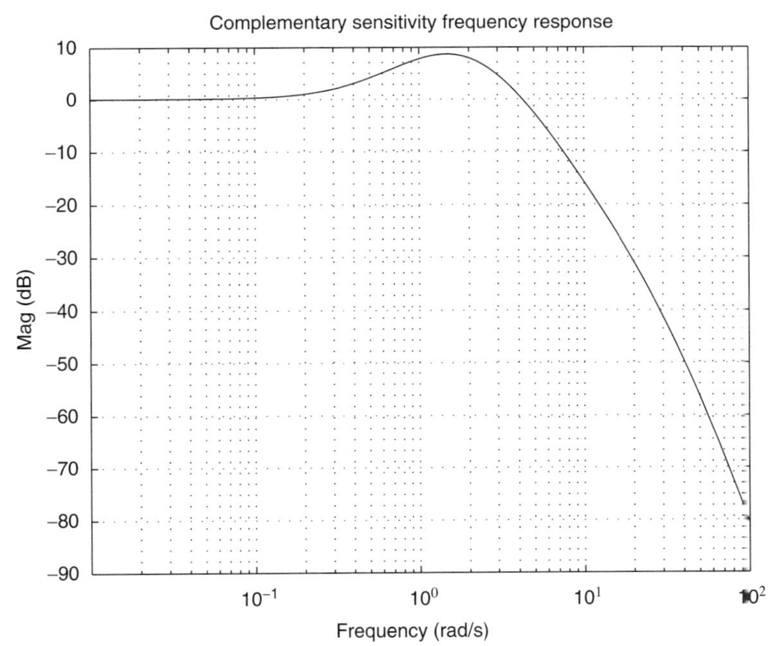

FIGURE 10.6 $\mathcal{H}^2$ Design complementary sensitivity frequency response.

Example 10.3 (LQG/LTR Design for First Order Unstable Missile Model)

We consider an unstable missile described by a simple first order model with state x (pitch attitude), control input u (fin elevator deflection), process noise $w_1 = \xi$ (angular wind gust), and sensor noise $w_2 = \theta$.

It is assumed that the missile's center of gravity (c.g.) is aft of its center of pressure (c.p.)—where lift is concentrated. This assumption results in a missile pitch instability. It is also assumed that the missile's moment of inertia about its c.g. is very small. This assumption leads to a simple first order model. The missile's angular velocity x is assumed to be proportional to its attitude x and the process noise $w_1 = \xi$. Regulated signals $z = [z_1 \ z_2]^T$ include the vehicle's pitch attitude $z_1 = x$ and a weighted control input $z_2 = \sqrt{\rho}u$. Here, $\rho > 0$ is a design parameter to be selected below. The vehicle's pitch attitude is measured. The pitch attitude measurement y includes additive sensor noise $w_2 = \theta$.

Missile Model. The (generalized) missile model is given as follows:

$$x = x + \xi + u \tag{10.94}$$

$$z = \begin{bmatrix} x \\ \sqrt{\rho}u \end{bmatrix} \tag{10.95}$$

$$y = x + \sqrt{\mu}\theta \tag{10.96}$$

where $\mu > 0$ is design parameter to be selected below.

Design Specifications. The goal is to design a real-rational proper model-based $\mathcal{H}^2$ optimal compensator (i.e., minimizes $\|T_{wz}\|_{H^2}$) which results in a stable closed loop system with a dominant closed loop pole at $s = -5$ (settling time $t_s \approx 1$ s).

Construction of Generalized Plant. The above model may be rewritten as follows:

$$x = x + \begin{bmatrix} 1 & 0 \end{bmatrix} \begin{bmatrix} \xi \\ \theta \end{bmatrix} + u \tag{10.97}$$

$$z = \begin{bmatrix} 1 \\ 0 \end{bmatrix} x + 0_{2\times 2} \begin{bmatrix} \xi \\ \theta \end{bmatrix} + \begin{bmatrix} 0 \\ \sqrt{\sigma} \end{bmatrix} u \tag{10.98}$$

$$y = x + \begin{bmatrix} 0 & \sqrt{\mu} \end{bmatrix} \begin{bmatrix} \xi \\ \theta \end{bmatrix} + 0_{1\times 1} u \tag{10.99}$$

From this, it follows that

$$A = 1, \qquad B_1 = \begin{bmatrix} 1 & 0 \end{bmatrix}, \qquad B_2 = 1 \tag{10.100}$$

$$C_1 = \begin{bmatrix} 1 \\ 0 \end{bmatrix}, \qquad D_{11} = 0_{2\times 2}, \qquad D_{12} = \begin{bmatrix} 0 \\ \sqrt{\rho} \end{bmatrix} \tag{10.101}$$

$$C_2 = 1, \qquad D_{21} = \begin{bmatrix} 0 & \sqrt{\mu} \end{bmatrix}, \qquad D_{22} = 0_{1\times 1} \tag{10.102}$$

$\mathcal{H}^2$ Problem Assumptions. We now check each of the $\mathcal{H}^2$ output feedback problem assumptions, as stated in Assumption 10.2. From the above data, it follows that $D_{11} = 0_{2\times 2}$, $D_{22} = 0_{1\times 1}$, and (A, B_2, C_2) is stabilizable and detectable, and

$$R = D_{12}^T D_{12} = \rho > 0 \tag{10.103}$$

$$\Theta = D_{21}^T D_{21} = \mu > 0 \tag{10.104}$$

Since

$$D_{12}^T C_1 = 0 \tag{10.105}$$

$$B_1 D_{21}^T = 0 \tag{10.106}$$

the imaginary axis rank conditions involving (A, B_2, C_1, D_{12}) and (A, B_1, C_2, D_{21}) in Assumption 10.2 become equivalent to (A, C_1) having no imaginary unobservable modes and (A, B_1) having no imaginary uncontrollable modes. These are clearly satisfied since $A = 1$ has no imaginary modes. Given this, it follows that all of the $\mathcal{H}^2$ output feedback problem assumptions in Assumption 10.2 are satisfied.

Plant. Finally, we note that the so-called plant (or missile) transfer function $P = G_{22}$ is given by

$$P = G_{22} = C_2(sI - A)^{-1} B_2 \tag{10.107}$$

$$= \frac{1}{s-1} \tag{10.108}$$

G_{22} is unstable with a right half plane pole at s = 1. G_{22} is also minimum phase (i.e., no zeros in Res > 0).

Filter Gain Matrix H_f. Since $B_1 D_{21}^T = 0$, the associated FARE is given by

$$AY + YA^T + B_1 B_1^T - YC_2^T \Theta^{-1} C_2 Y = Y + Y + 1 - \frac{1}{\mu} Y^2 = 0 \tag{10.109}$$

or

$$Y^2 - 2\mu Y - \mu = 0 \tag{10.110}$$

Application of the quadratic formula and selecting the positive (stabilizing) root yields:

$$Y = \mu + \sqrt{\mu^2 + \mu} \tag{10.111}$$

This yields the following filter gain matrix:

$$H_f = YC_2^T \Theta^{-1} = 1 + \sqrt{1 + \frac{1}{\mu}} \tag{10.112}$$

We now select μ to achieve the given dominant pole specification:

$$A - H_f C_2 = 1 - 1 - \sqrt{1 + \frac{1}{\mu}} = -5 \tag{10.113}$$

This yields

$$\mu = \frac{1}{24} \tag{10.114}$$

The associated KBF open loop transfer function is given by

$$G_{KF} = -C_2(sI - A)^{-1} H_f \tag{10.115}$$

$$= \frac{-6}{s-1} \tag{10.116}$$

We will see below that this will be the approximate open loop transfer function PK_{opt} for the final design. In this sense, G_{KF} represents our target open loop transfer function.

Control Gain Matrix G_c. Since $D_{12}^T C_1 = 0$, the associated CARE is given by

$$A^T X + XA + C_1^T C_1 - XB_2 R^{-1} B_2 X = X + X + 1 - \frac{1}{\rho} X^2 = 0 \tag{10.117}$$

or

$$X^2 - 2\rho X - \rho = 0 \tag{10.118}$$

Application of the quadratic formula and selecting the positive (stabilizing) root yields:

$$X = \rho + \sqrt{\rho^2 + \rho} \tag{10.119}$$

This yields the following control gain matrix:

$$G_c = R^{-1} B_2^T X = 1 + \sqrt{1 + \frac{1}{\rho}} \tag{10.120}$$

This results in a closed loop (regulator) pole at

$$A - B_2 G_c = 1 - 1 - \sqrt{1 + \frac{1}{\rho}} = -\sqrt{1 + \frac{1}{\rho}} \tag{10.121}$$

Note that for large ρ (referred to as expensive control in LQR problems) we have a closed loop pole at $s = -1$, at the left half plane reflection of the plant pole at $s = 1$. We will select the design parameter ρ to be small (referred to as cheap control in LQR problems) so that this closed loop (regulator) pole $s \approx -1/\sqrt{\rho}$ is fast and the closed loop filter pole at $s = -5$ at is the dominant closed loop pole.

$\mathcal{H}^2$ Optimal Output Feedback Model-Based Compensator. The resulting $\mathcal{H}^2$ optimal output feedback model-based compensator is given by

$$K_{opt} = \left[\begin{array}{c|c} A - B_2 G_c - H_f C_2 & H_f \\ \hline -G_c & 0_{n_u \times n_y} \end{array}\right] \tag{10.122}$$

where

$$A - B_2 G_c - H_f C_2 = 1 - 1 - \sqrt{1 + \frac{1}{\rho}} - 1 - \sqrt{1 + \frac{1}{\mu}} \tag{10.123}$$

$$= 1 - 1 - \sqrt{1 + \frac{1}{\rho}} - 1 - \sqrt{1 + 24} \tag{10.124}$$

$$= -6 - \sqrt{1 + \frac{1}{\rho}} \tag{10.125}$$

$$H_f = 1 + \sqrt{1 + \frac{1}{\mu}} \tag{10.126}$$

$$= 1 + \sqrt{1 + 24} \tag{10.127}$$

$$= 6 \tag{10.128}$$

$$G_c = 1 + \sqrt{1 + \frac{1}{\rho}} \tag{10.129}$$

Given this, the compensator transfer function is given by

$$K_{\text{opt}} = -G_c(sI - A + B_2G_c + H_fC_2)^{-1}H_f \tag{10.130}$$

$$= \frac{-6(1+\sqrt{1+1/\rho})}{s+6+\sqrt{1+1/\rho}} \tag{10.131}$$

For small ρ (cheap control), this yields

$$K_{\text{opt}} \approx \frac{-6(1/\sqrt{\rho})}{s+1/\sqrt{\rho}} \tag{10.132}$$

Open Loop Transfer Function. The associated open loop transfer function is given by

$$PK_{\text{opt}} = -C_2(sI-A)^{-1}B_2 \quad G_c(sI - A + B_2G_c + H_fC_2)^{-1}H_f \tag{10.133}$$

$$= \frac{1}{s-1}\left[\frac{-6(1+\sqrt{1+1/\rho})}{s+6+\sqrt{1+1/\rho}}\right] \tag{10.134}$$

For small ρ (cheap control), this becomes

$$PK_{\text{opt}} \approx \frac{1}{s-1}\left[\frac{-6(1/\sqrt{\rho})}{s+1/\sqrt{\rho}}\right] \tag{10.135}$$

Loop Transfer Recovery (LTR). From this, we see that as control weighting parameter ρ approaches zero (cheap control), the open loop transfer function approaches the KBF open loop transfer function G_{KF}; that is,

$$\lim_{\rho\to 0^+} G_{22}K_{\text{opt}} = \frac{-6}{s-1} \tag{10.136}$$

$$= G_{\text{KF}} \tag{10.137}$$

This shows that as ρ approaches zero (cheap control), the actual open loop transfer function PK_{opt} approaches the target open loop transfer function G_{KF}. The above procedure of recovering a target open loop transfer function (with desirable closed loop properties) using an LQG controller is called *LQG with loop transfer recovery or LQG/LTR.*

Selection of Far Away Closed Loop Regulator Pole. For small ρ, the closed loop system is stable with closed loop poles at $s = -5$ and $s \approx -1/\sqrt{\rho}$. A good selection for ρ might be $\rho = 1/2500$. This results in a fast closed loop pole at $s \approx -50$ and makes the closed loop filter pole at $s = -5$ the dominant closed loop pole, as required.

Stability Robustness Margins. It is well known that $\mathcal{H}^2$ and LQG designs need not possess good stability robustness margins. In fact, they can be arbitrarily bad [3]. LQG/LTR designs for minimum phase plants (such as ours: P = 1/(s − 1) have guaranteed stability robustness margins. LQG/LTR designs provide margins that approach those associated with LQR and KBF designs; i.e., infinite upward gain margin, at least 6 dB downward gain margin, and at least ±60° phase margin. Our final LQG/LTR design

$$PK_{\text{opt}} = \frac{-6}{s-1}\left[\frac{50}{s+50}\right] \tag{10.138}$$

offers an infinite upward gain margin and a downward gain margin of 1/6 (−15.56 dB). The resulting unity gain crossover frequency is $\omega_g = \sqrt{35} = 5.92$ rad/s and the associated phase margin is about 99.59°. Not bad.

The following example extends the LQG/LTR ideas presented in Example 10.3 to the general MIMO setting—enabling the design of feedback loops (with nominal robustness margins) via $\mathcal{H}^2$ optimization. ■

Example 10.4 (MIMO LQG and LQG/LTR Control Design Via $\mathcal{H}^2$ Optimization)

We consider a MIMO plant *P* defined by the state space representation

$$x = Ax + Bu \tag{10.139}$$

$$y = Cx \tag{10.140}$$

It is assumed that the plant $P = [A, B, C]$ is stabilizable and detectable.

The goal is to demonstrate how the $\mathcal{H}^2$ optimal output feedback solution that has been presented may be used to solve MIMO LQG control problems. We specifically would like to present a method which lends itself to the concept of LTR—whereby we use a model-based LQG controller to recover a target loop transfer function matrix with desirable closed loop properties. Our motivation is not optimal stochastic LQG control problems; it is the design of control laws with desirable closed loop properties.

Construction of Generalized Plant G. With our final objective being a model-based compensator defined by a control gain matrix G_c and a filter gain matrix H_f, we consider the following generalized plant:

$$x = Ax + L\xi + Bu \tag{10.141}$$

$$z = \left[\frac{Mx}{\sqrt{\rho}I_{n_u \times n_u}}\right] \tag{10.142}$$

$$y = Cx + \sqrt{\mu}\theta \tag{10.143}$$

where u is the control, x is the (generalized) plant state, $w_1 = \xi$ represents process noise in the state equation, $w_2 = \theta$ represents sensor noise in the measurement equation, $A \in \mathcal{R}^{n\times n}$, $L \in \mathrm{R}^{n\times n_u}$, $B \in \mathrm{R}^{n\times n_u}$, $M \in \mathrm{R}^{n_y\times n}$, $C \in \mathrm{R}^{n_y\times n}$, $n_y = n_u$, $\rho > 0$, $\mu > 0$.

Design Parameter Assumptions. It is assumed that either:

(*A*, *L*) has no imaginary uncontrollable modes and (*A*, *M*) is detectable, or
(*A*, *L*) is stabilizable and (*A*, *M*) has no imaginary unobservable modes.

Here, L, M, μ, and ρ should be viewed as "design parameters" that are selected in order to obtain control and filter gain matrices G, and H_f such that the resulting model-based compensator exhibits desirable closed loop properties.

Two-Port State Space Representation for Generalized Plant G. The above model may be rewritten in two-port state space form as follows:

$$\left[\begin{array}{c} x \\ \hline z \\ \hline y \end{array}\right] = \left[\begin{array}{c|c|c} A & \begin{bmatrix} L & 0_{n\times n_y} \end{bmatrix} & B \\ \hline \begin{bmatrix} M \\ 0_{n_u\times n} \end{bmatrix} & \begin{bmatrix} 0_{n_y\times n_u} & 0_{n_y\times n_y} \\ 0_{n_u\times n_u} & 0_{n_u\times n_y} \end{bmatrix} & \begin{bmatrix} 0_{n_y\times n_u} \\ \sqrt{\rho}I_{n_u\times n_u} \end{bmatrix} \\ \hline C & \begin{bmatrix} 0_{n_y\times n_u} & \sqrt{\mu}I_{n_y\times n_y} \end{bmatrix} & \end{array}\right] \left[\begin{array}{c} x \\ \hline \xi \\ \theta \\ \hline u \end{array}\right] \tag{10.144}$$

Check on $\mathcal{H}^2$ Output Feedback Assumptions. We now make sure that all of the $\mathcal{H}^2$ output feedback problem assumptions in Assumption 10.2 are satisfied:

- *Plant $P = G_{22}$ Assumptions.* Since the plant $P = G_{22} = [\ A, B, C\ ]$ is stabilizable and detectable, it follows that $(A, B_2 = B, C_2 = C)$ is stabilizable and detectable.
- *Regulator Assumptions.* Since

$$D_{12} = \begin{bmatrix} 0_{n_y \times n_u} \\ \sqrt{\rho} I_{n_u \times n_u} \end{bmatrix}$$

has full column rank, it follows that the control weighting matrix $R = D_{12}^T D_{12} = \rho I_{n_u \times n_u} > 0$ is nonsingular.

Since $D_{12}^T C_1 = 0$, it follows that the imaginary axis (column) rank condition involving (A, B_2, C_1, D_{12}) in Assumption 10.2 is equivalent to $(A - B_2 R^{-1} D_{12}^T C_1, (I - D_{12} R^{-1} D_{12}^T) C_1) = (A, C_1)$ having no unobservable imaginary modes. Since (A, M) is either detectable or has no imaginary unobservable modes, it follows that

$$A, C_1 = \begin{bmatrix} M \\ 0_{n_u \times n} \end{bmatrix}$$

has no unobservable imaginary modes. The associated Hamiltonian H_{con} will, therefore, yield a Riccati solution and control gain matrix G_c such that $A - BG_c$ is stable.
- *Filter Assumptions.* Since $D_{21} = [0_{n_y \times n_u} \ \ \sqrt{\mu} I_{n_y \times n_y}]$ has full row rank, it follows that the measurement weighting matrix $\Theta = D_{21} D_{21}^T = \mu I_{n_y \times n_y} > 0$ is nonsingular.

Since $B_1 D_{21}^T = 0$, it follows that the imaginary axis (row) rank condition involving (A, B_1, C_2, D_{21}) in Assumption 10.2 is equivalent to $(A - B_1 D_{21}^T \Theta^{-1} C_2, B_1(I - D_{21}\Theta^{-1} D_{21})) = (A, B_1)$ having no uncontrollable imaginary modes. Since (A, L) is either stabilizable or has no uncontrollable imaginary modes, it follows that $(A, B_1 = [L \ \ 0_{n \times n_y}])$ has no uncontrollable imaginary modes. The associated Hamiltonian H_{fil} will therefore yield a Riccati solution and filter gain matrix H_f such that $A - H_f C$ is stable.

Given the above, it follows that all of the $\mathcal{H}^2$ output feedback problem assumptions in Assumption 10.2 are satisfied.

Control Gain Matrix. It follows that the control gain matrix G_c is given by

$$G_c = R^{-1} B^T X \tag{10.145}$$

where $X \geq 0$ is the unique (at least) positive semi-definite solution of the CARE:

$$A^T X + XA + C^T C - XBR^{-1} BX = 0 \tag{10.146}$$

Moreover, $A - BG$ is stable.

Filter Gain Matrix. It follows that the filter gain matrix H_f is given by

$$H_f = YC^T \Theta^{-1} \tag{10.147}$$

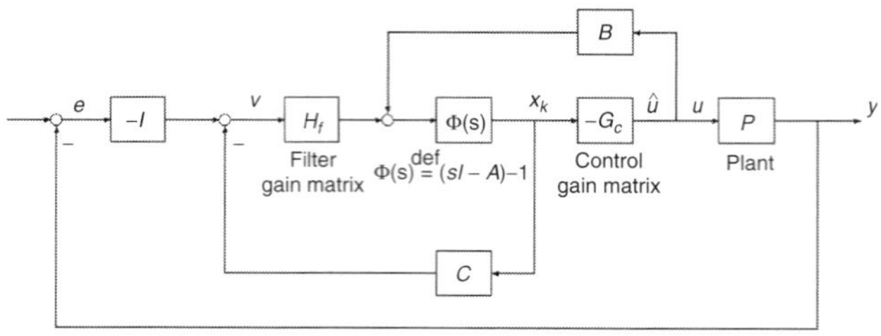

FIGURE 10.7 Negative feedback loop with LQG model-based compensator and plant.

where $Y > 0$ is the unique (at least) positive semi-definite solution of the FARE:

$$AY + YA^T + LL^T - YC^T\Theta^{-1}CY = 0 \tag{10.148}$$

Moreover, $A - HC$ is stable.

$\mathcal{H}^2$ Optimal (LQG) Compensator. The $\mathcal{H}^2$ optimal compensator that minimizes the $\mathcal{H}^2$ norm of the transfer function matrix from the exogenous signals $w = \begin{bmatrix} \xi \\ \theta \end{bmatrix}$ to the regulated signals $z = \begin{bmatrix} C_x \\ \sqrt{\rho} I_{n_u \times n_u} \end{bmatrix}$ is then given by

$$K_{\text{opt}} = \left[\begin{array}{c|c} A - BG_c - H_f C & H_f \\ \hline G_c & 0_{n_u \times n} \end{array}\right] \tag{10.149}$$

Note that the minus sign on G_c (lower left hand entry of K_{opt}) has been removed in anticipation of the negative feedback system implementation shown in Figure 10.7. By the separation principle, the closed loop poles are the eigenvalues of $A - BG_c$ and $A - H_f C$.

Stability Robustness Margins. It should be emphasized that the resulting controller K_{opt}, although stabilizing, may possess arbitrarily bad stability robustness margins [3]. This is despite the fact that the associated regulator loop

$$G_{LQ} = G_c(sI - A)^{-1}B \tag{10.150}$$

and filter loop

$$G_{KF} = C(sI - A)^{-1}H_f \tag{10.151}$$

when viewed as MIMO open loop transfer function matrices within their own negative feedback loops, possess the following well-known stability robustness margins: infinite upward gain margin, at least 6 dB downward gain margin, and at least ±60° phase margin. This gives rise to the following natural question:

Is there a way that we can select the control gain matrix G_c and the filter gain matrix H_f so that the resulting model-based compensator K_{opt} results in a feedback loop which possesses the above nice margins?

Fortunately, the answer to this is a definitive yes! Two methods which result in comparable stability margins at the plant input or at the plant output (but not both simultaneously) are now presented.

Loop Transfer Recovery (LTR) Methods. The approach we take to achieve a feedback design with good stability margins is as follows. The process involves two steps.

1. *Target Loop Design.* The first step is to design a target open loop transfer function matrix that possesses desired closed loop properties. The target loop may be associated with the plant output. If so, we denote it L_o. In such a case, L_o represents our desired PK_{opt}. If associated with the plant input, we denote it L_i. In such a case, L_i represents our desired K_{opt}. (In general, $PK_{\text{opt}}P \neq K_{\text{opt}}P$.)

2. *Target Loop Recovery Via Model-Based Compensator.* The second step is to use a model-based compensator $K_{opt} = [A - BG_c - H_f C, H, G]$ to recover the target loop (either L_o or L_i).

   If we want to recover L_o (i.e., good properties at that plant output), then we want $PK_{opt} \approx L_o$. This is called loop transfer recovery at the plant output (LTRO).

   If we want to recover L_i (i.e., good properties at that plant input), then we want $K_{opt}P \approx L_i$. This is called loop transfer recovery at the plant input (LTRI).

   Note: In general, the properties associated with breaking the loop at the plant output (properties of PK_{opt}) are different (perhaps very different) from those associated with breaking the loop at the plant input (properties of $K_{opt}P$). It is usually very difficult for PK_{opt} and $K_{opt}P$ to both possess great properties (e.g., margins, etc.). Typically, a designer must trade off nice properties at the plant output for nice properties at the plant input, or vice versa.

$\mathcal{H}^2$-based methods for LTRO and LTRI are now presented.

a. *Loop Transfer Recovery at Plant Output (LTRO)*

1. *Design of Target Loop L_o.* The first step is to design a target loop $L_o = C(sI - A)^{-1}H_f$ with desirable closed loop properties (e.g., stability, sensitivity, complementary sensitivity, stability robustness margins, etc). This may be done using any method! (Any method you feel comfortable enough with.)

   One procedure that results in good properties at the plant output is based on *KBF* methods. The idea is to select the design (shaping) matrix L so that the singular values $G_{FOL} = C(sI - A)^{-1}L$ look nice; for example, large minimum singular value at low frequencies, small maximum singular value at high frequencies, singular values cross 0 dB with slopes of −20 dB/dec, etc.

   We then solve the FARE with A, L, C, $\Theta = \mu I_{n_y \times n_y}$—using $\mu > 0$ to adjust the bandwidth of our target loop $L_o = G_{KF} = C(sI - A)^{-1}H_f$. A smaller (larger) μ results in a larger (smaller) bandwidth.

   Guidelines for Shaping of Target Loop $L_o = G_{KF}$.

- The so-called Kalman Frequency Domain Equality (KFDE) guides our loop shaping:

$$[I + G_{KF}(j\omega)][I + G_{KF}(j\omega)]^H = I + \left[\frac{1}{\sqrt{\mu}}G_{FOL}(j\omega)\right]\left[\frac{1}{\sqrt{\mu}}G_{FOL}(j\omega)\right]^H \tag{10.152}$$

  From this, it follows that

$$\sigma_i[I + G_{KF}(j\omega)] = \sqrt{1 + \frac{1}{\mu}\sigma_i^2[G_{FOL}(j\omega)]} \tag{10.153}$$

  This suggests that by shaping G_{FOL}, we can shape the target loop $L_o = G_{KF}$. Specifically, if G_{FOL} is large at low frequencies, then we expect

$$G_{KF}(j\omega) \approx \frac{1}{\sqrt{\mu}}G_{FOL}(j\omega) \tag{10.154}$$

  at low frequencies. This shows that the matrix L should be used for shaping the target loop $L_o = G_{KF}$ while $\mu > 0$ is used to adjust the target loop bandwidth—decreasing/increasing μ to raise/lower the target loop bandwidth. The resulting loop $L_o = G_{KF}$ is guaranteed to possess nice closed loop properties as described below.

- The above singular value relation implies that

$$\sigma_{min}[I + G_{KF}(j\omega)] \geq 1 \tag{10.155}$$

for all ω. This, in turn, implies that the associated sensitivity singular values satisfy

$$\sigma_{\max}[S_{\mathrm{KF}}(j\omega)] = \frac{1}{\sigma_{\min}[S_{\mathrm{KF}}(j\omega)]^{-1}} \leq 1 \quad (0\,\mathrm{dB}) \tag{10.156}$$

for all ω, where

$$S_{\mathrm{KF}}(j\omega) = [I + G_{\mathrm{KF}}(j\omega)]^{-1} \tag{10.157}$$

- From the above sensitivity singular value relationship, we obtain the follow celebrated KBF loop margins:

  infinite upward gain margin
  at least $\frac{1}{2}$ (6 dB) downward gain margin
  at least ±60° phase margin

  The above gain margins apply to simultaneous and independent gain perturbations when the loop is broken at the output. The same holds for the above phase margins. The above margins are NOT guaranteed for simultaneous gain and phase perturbations. It should be noted that these margins can be easily motivated using elementary SISO Nyquist stability arguments [2,8].
- From the above sensitivity singular value relations, we obtain the following complementary sensitivity singular value relationship:

$$\sigma_{\max}[T_{\mathrm{KF}}(j\omega)] = \sigma_{\max}[I - S_{\mathrm{KF}}(j\omega)] \leq 1 + \sigma_{\max}[S_{\mathrm{KF}}(j\omega)] \leq 2 \quad (6\mathrm{dB}) \tag{10.158}$$

for all ω, where

$$T_{\mathrm{KF}} = I - S_{\mathrm{KF}} = G_{\mathrm{KF}}[1 + G_{\mathrm{KF}}]^{-1} \tag{10.159}$$

2. *Recovery of Target Loop L_o Using Model-Based Compensator.* The second step is to use a model-based compensator $K_{\mathrm{opt}} = [\ A - BG_c - H_fC,\ H_f,\ G_c\ ]$ where G_c is found by solving the CARE with A, B, $M = C$, $R = \rho I_{n_u \times n_u}$ with ρ a small positive scalar. Since ρ is small, we call this a cheap control problem.
   - If the plant $P = [A, B, C]$ is minimum phase, then it can be shown that

$$\lim_{\rho \to 0^+} X = 0 \tag{10.160}$$

$$\lim_{\rho \to 0^+} \sqrt{\rho} G_c = WC \tag{10.161}$$

   for some orthonormal W (i.e., $W^TW = WW^T = I$)

$$\lim_{\rho \to 0^+} PK_{\mathrm{opt}} = L_o \tag{10.162}$$

   In such a case, $PK_{\mathrm{opt}} \approx L_o$ for small ρ and hence PK_{opt} will possess stability margins that are close to those of L_o (at the plant output)—whatever method was used to design L_o. It must be noted that the minimum phase condition on the plant P is a suficient condition. It is not necessary. Moreover, G_c need not be computed using a CARE. In fact, any G_c which (1) satisfies a limiting condition $\lim_{\rho \to 0^+} \sqrt{\rho} G_c = WC$ for some invertible matrix W and which (2) ensures that $A - BG_c$ is stable (for small ρ), will result in LTR at the plant output. This result is a consequence of the structure of model-based compensators and has nothing to do with optimal control and filtering problems.

- Assuming that a limiting condition $\lim_{\rho \to 0^+} \sqrt{\rho} G_c = WC$ holds for some invertible matrix W, loop transfer recovery of the target loop $L_o = C(sI - A)^{-1}H_f$ may be proven as follows: For small ρ we have

$$G_c \approx \frac{WC}{\sqrt{\rho}} \tag{10.163}$$

which gives yields the following:

$$PK_{\text{opt}} = PG_c(sI - A + BG_c + H_fC)^{-1}H_f \tag{10.164}$$

$$\approx P\frac{WC}{\sqrt{\rho}}\left(sI - A + B\frac{WC}{\sqrt{\rho}}\right)^{-1}H_f \tag{10.165}$$

$$\approx P\frac{WC}{\sqrt{\rho}}(sI - A)^{-1}\left[I + B\frac{WC}{\sqrt{\rho}}(sI - A)^{-1}\right]^{-1}H_f \tag{10.166}$$

$$\approx P\frac{WC(sI - A)^{-1}}{\sqrt{\rho}}\left[I + B\frac{WC(sI - A)^{-1}}{\sqrt{\rho}}\right]^{-1}H_f \tag{10.167}$$

$$\approx P\left[I + \frac{WC(sI - A)^{-1}}{\sqrt{\rho}}B\right]^{-1}\frac{WC(sI - A)^{-1}}{\sqrt{\rho}}H_f \tag{10.168}$$

$$\approx P\left[\frac{WP}{\sqrt{\rho}}\right]^{-1}W\frac{C(sI - A)^{-1}H_f}{\sqrt{\rho}} \tag{10.169}$$

$$\approx C(sI - A)^{-1}H_f = L_o \tag{10.170}$$

The central idea (underneath the algebra) is that as ρ goes to zero, the C feedback path within the compensator $K_{\text{opt}} = [A - BG_c - H_fC, G_c, H_f]$ is broken (see Figure10.7) and the nice properties that hold at the so-called innovations v (e.g., open loop transfer function matrix at v is $L_o = C(sI - A)^{-1}H_f$) in Figure 10.7 get transferred to the error signal e (compensator input, or plant output) within the feedback loop.

b. *Loop Transfer Recovery at Plant Input (LTRI).*

1. *Design of Target Loop L_i.* The first step is to design a target loop $L_i = G_C(sI - A)^{-1}B$ with desirable closed loop properties (e.g., stability, sensitivity, complementary sensitivity, stability robustness margins, etc). This may be done using any method! (Any method you feel comfortable enough with.)

   One procedure that results in good properties at the plant input is based on LQR methods. The idea is to select the design (shaping) matrix M so that the singular values of $G_{\text{OL}} = M(sI - A)^{-1}B$ look nice; for example, large minimum singular value at low frequencies, small maximum singular value at high frequencies, singular values cross 0 dB with slopes of −20 dB/dec, etc.

   We then solve the CARE with A, B, M, $R = \rho I_{n_u \times n_u}$, using $\rho > 0$ to adjust the bandwidth of our target loop $L_i = G_{\text{LQR}} G_c(sI - A)^{-1}B$. A smaller (larger) ρ results in a larger (smaller) bandwidth.

   Guidelines for Shaping of Target Loop $L_i = G_{\text{LQ}}$.

- The so-called LQ frequency domain equality (LQFDE) guides our loop shaping:

$$[I + G_{\text{LQ}}(j\omega)]^H[I + G_{\text{LQ}}(j\omega)] = I + \left[\frac{1}{\sqrt{\rho}}G_{\text{OL}}(j\omega)\right]^H\left[\frac{1}{\sqrt{\rho}}G_{\text{OL}}(j\omega)\right] \tag{10.171}$$

From this, it follows that

$$\sigma_i[I + G_{LQ}(j\omega)] = \sqrt{I + \frac{1}{\rho}\sigma_i^2[G_{OL}(j\omega)]} \tag{10.172}$$

This suggests that by shaping G_{OL}, we can shape the target loop $L_i = G_{LQ}$. Specifically, if G_{OL} is large at low frequencies, then we expect

$$G_{LQ}(j\omega) \approx \frac{1}{\sqrt{\rho}} G_{OL}(j\omega) \tag{10.173}$$

at low frequencies. This shows that the matrix M should be used for shaping the target loop $L_i = G_{LQ}$ while $\rho > 0$ is used to adjust the target loop bandwidth—decreasing/increasing ρ to raise/lower the target loop bandwidth. The resulting loop $L_i = G_{LQ}$ is guaranteed to possess nice closed loop properties as described below.

- The above singular value relation implies that

$$\sigma_{\min}[I + G_{LQ}(j\omega)] \geq 1 \tag{10.174}$$

for all ω. This, in turn, implies that the associated sensitivity singular values satisfy

$$\sigma_{\max}[S_{LQ}(j\omega)] = \frac{1}{\sigma_{\min}[S_{LQ}(j\omega)^{-1}]} \leq 1 \quad (0\text{dB}) \tag{10.175}$$

for all ω, where

$$S_{LQ} = [I + G_{LQ}]^{-1} \tag{10.176}$$

- From the above sensitivity singular value relationship, we obtain the follow celebrated LQR loop margins:

  infinite upward gain margin,
  at least $\frac{1}{2}$(6 dB) downward gain margin,
  at least ±60° phase margin.

  The above gain margins apply to simultaneous and independent gain perturbations when the loop is broken at the input. The same holds for the above phase margins. The above margins are NOT guaranteed for simultaneous gain and phase perturbations. It should be noted that these margins can be easily motivated using elementary SISO Nyquist stability arguments [2,8].
- From the above sensitivity singular value relations, we obtain the following complementary sensitivity singular value relationship:

$$\sigma_{\max}[T_{LQ}(j\omega)] = \sigma_{\max}[I - S_{LQ}(j\omega)] \leq 1 + \sigma_{\max}[S_{LQ}(j\omega)] \leq 2 \quad (6\text{dB}) \tag{10.177}$$

for all ω, where

$$T_{LQ} = I - S_{LQ} = G_{LQ}[1 + G_{LQ}]^{-1} \tag{10.178}$$

2. Recovery of Target Loop L_i Using Model-Based Compensator. The second step is to use $K_{opt} = [A - BG_c, -H_fC, H_f, G_c]$ where H_f is found by solving the FARE with A, $L = B$, C, $\Theta = \mu I_{n_y \times n_y}$ with μ a small positive scalar. Since μ is small, we call this an expensive sensor problem.

- If the plant $P = [A, B, C]$ is minimum phase, then it can be shown that

$$\lim_{\mu\to 0^+} Y = 0 \tag{10.179}$$

$$\lim_{\mu\to 0^+} \sqrt{\mu}H_f = BV \tag{10.180}$$

for some orthonormal V (i.e., $V^TV = VV^T = I$)

$$\lim_{\mu\to 0+} K_{\text{opt}}P = L_i \tag{10.181}$$

In such a case, $K_{\text{opt}}\, P \approx L_i$ for small μ and hence $K_{\text{opt}}\, P$ will possess stability margins that are close to those of L_i (at the plant input)—whatever method was used to design L_i.

It must be noted that the minimum phase condition on the plant P is a sufficient condition. It is not necessary. Moreover, H_f need not be computed using a FARE. In fact, any H_f which (1) satisfies a limiting condition $\lim_{\mu\to 0^+} \sqrt{\mu}H_f = BV$ for some invertible matrix V and which (2) ensures that $A - H_fC$ is stable (for small μ), will result in LTR at the plant input; that is, $\lim_{\mu\to 0^+} K_{\text{opt}}P = L_i$. This result is a consequence of the structure of model-based compensators and has nothing to do with optimal control and filtering problems.

- Assuming that a limiting condition $\lim_{\mu\to 0^+} \sqrt{\mu}H_f = BV$ holds for some invertible matrix V, loop transfer recovery of the target loop $L_i = G_c(sI - A)^{-1}$ B may be proven as follows. For small μ we have

$$H_f \approx \frac{BV}{\sqrt{\mu}} \tag{10.182}$$

which gives the following:

$$K_{\text{opt}}P = G_c(sI - A + BG_c + H_fC)^{-1}H_fP \tag{10.183}$$

$$\approx G_c\left(sI - A + BG_c + \frac{BV}{\sqrt{\mu}}C\right)^{-1}\frac{BV}{\sqrt{\mu}}P \tag{10.184}$$

$$\approx G_c\left(sI - A + \frac{BV}{\sqrt{\mu}}C\right)^{-1}\frac{BV}{\sqrt{\mu}}P \tag{10.185}$$

$$\approx G_c(sI - A)^{-1} + \left[I + \frac{BV}{\sqrt{\mu}}C(sI - A)^{-1}\right]^{-1}\frac{BV}{\sqrt{\mu}}P \tag{10.186}$$

$$\approx G_c(sI - A)^{-1}\frac{BV}{\sqrt{\mu}}\left[I + C(sI - A)^{-1}\frac{BV}{\sqrt{\mu}}\right]^{-1} P \tag{10.187}$$

$$\approx G_c(sI - A)^{-1}\frac{BV}{\sqrt{\mu}}\left[C(sI - A)^{-1}\frac{BV}{\sqrt{\mu}}\right]^{-1} P \tag{10.188}$$

$$\approx G_c(sI - A)^{-1}B\frac{V}{\sqrt{\mu}}\left[P\frac{V}{\sqrt{\mu}}\right]^{-1} P \tag{10.189}$$

$$\approx G_c(sI - A)^{-1}B = L_i \tag{10.190}$$

The central idea (underneath the algebra) is that as μ goes to zero, the B feedback path within the compensator $K_{\text{opt}} = [A - BG_c - H_fC, G_c, H_f]$ is broken (see Figure 10.7) and the nice properties that hold at $\hat{u}$ (e.g., open loop transfer function matrix at $\hat{u}$ is $L_i = G_c(sI - A)^{-1}B$) in Figure 10.7 get transferred to the plant input u (compensator output) within the feedback loop. ■

Comment 10.13 (Stability Margins and Peak Sensitivity)

The peak on the sensitivity plot is very important in the design of a feedback system. A large peak, for example, may be due to a closed loop pole near the imaginary axis. This certainly is undesirable. We thus want the peak to be "small." It can be shown that the peak necessarily establishes gain and phase margin bounds.

Suppose that the peak sensitivity is bounded above by $\alpha \geq 1$; that is, $\sigma_{\max} S(j\omega) < \alpha$ for all ω. It can be shown that the feedback loop then enjoys the following nominal multivariable stability robustness (gain and phase) margin bounds:

$$\uparrow GM > \frac{\alpha}{\alpha - 1} \tag{10.191}$$

$$\downarrow GM < \frac{\alpha}{\alpha + 1} \tag{10.192}$$

$$|PM| > 2\sin^{-1}\left(\frac{1}{2\alpha}\right) \tag{10.193}$$

These bounds may be easily motivated using SISO Nyquist [2,8] ideas as follows:

If

$$|S(j\omega)| < \alpha \tag{10.194}$$

for all ω, then it follows that

$$\frac{1}{\alpha} < |1 + L(j\omega)| \tag{10.195}$$

for all ω. This, however, implies that the Nyquist plot associated with L cannot penetrate a circle centered at -1 with radius $1/\alpha$, left most end point at $-[(\alpha+1)/\alpha]$, and right most end point at $-[(\alpha-1)/\alpha]$. The upward gain margin bound follows from the right most point of the circle. The downward gain margin bound follows from the left most point of the circle. The phase margin bound can be obtained with a little geometry. ■

The following example considers the application of $\mathcal{H}^2$ theory to a robotic manipulator.

Example 10.5 ($\mathcal{H}^2$-LQG/LTR Design for PUMA 560 Robotic Manipulator)

In this example, we show how $\mathcal{H}^2$ optimization may be used to design an LQG/LTR controller for a PUMA 560 robotic manipulator. The manipulator is shown in Figure 10.8.

A two degree-of-freedom (dof) linear model $P = [A_p, B_p, C_p]$ was used to initiate the design process. Linearizing the PUMA's nonlinear model [9] about the equilibrium point $\theta_1 = 90°$ $\theta_2 = 0°$ (both links vertical), results in the following linear model:

$$x_p = A_p x_p + B_p \mu_p \tag{10.196}$$

$$y_p = C_p x_p \tag{10.197}$$

$$u_p = [\tau_1 \quad \tau_2]^T \tag{10.198}$$

$$x_p = [\theta_1 \quad \theta_2 \quad \dot{\theta}_1 \quad \dot{\theta}_2]^T \tag{10.199}$$

$$y_p = [\theta_1 \quad \theta_2]^T \tag{10.200}$$

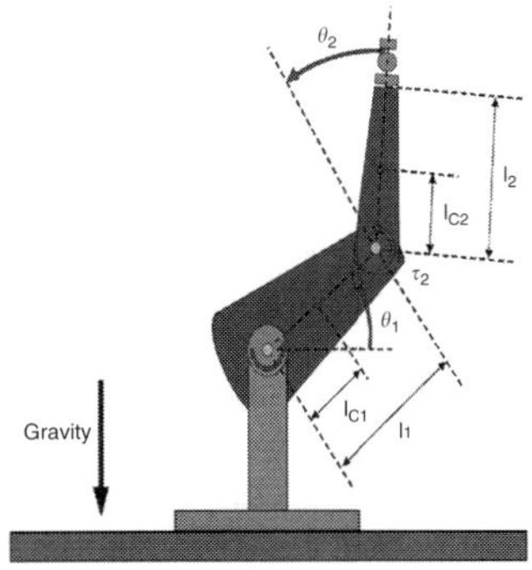

FIGURE 10.8 Two degree-of-freedom PUMA 560 robotic manipulator.

where

$$A_p = \begin{bmatrix} 0.0000 & 0.0000 & 1.0000 & 0.0000 \\ 0.0000 & 0.0000 & 0.0000 & 1.0000 \\ 31.7613 & -33.0086 & 0.0000 & 0.0000 \\ -56.9381 & 187.7089 & 0.0000 & 0.0000 \end{bmatrix} \tag{10.201}$$

$$B_p = \begin{bmatrix} 0.0000 & 0.0000 \\ 0.0000 & 0.0000 \\ 1037.7259 & -3919.6674 \\ -3919.6674 & 2030.8306 \end{bmatrix} \tag{10.202}$$

$$C_p = [I_{2\times 2} \quad 0_{2\times 2}] \tag{10.203}$$

The system poles are $s = \pm 14.1050$, $s = \pm 4.5299$. Eigenvector analysis shows that the fast instability at $s = 14.1050$ is primarily associated with the upper (shorter) link, while the slower instability at $s = 4.5299$ is primarily associated with the lower (longer) link. The system does not possess any natural integrators (i.e., no zero eigenvalues) and, as expected, the singular values $\sigma_i[P(j\omega)]$ are flat at low frequencies (see Figure 10.9).

Closed Loop Objectives

A controller to be implemented within a negative feedback loop is sought. The closed loop system should exhibit the following properties: (1) closed loop stability, (2) zero steady state error to step reference commands, (3) good low frequency reference command following (step commands followed with little overshoot within 3 s), (4) good low frequency disturbance attenuation, (5) good high frequency noise attenuation, (6) good stability robustness margins at the plant output.

Each step of the control system design process is now described. A central idea is the formation of a so-called design plant P_d from the original plant P. The design plant P_d is what is submitted to our $\mathcal{H}^2$-LQG/LTR design machinery.

Step 1: Augment Plant P with Integrators to Get Design Plant Pd = [A, B, C]

In order to guarantee zero steady-state error to step reference commands, we begin by augmenting the plant $P = [A_p, B_p, C_p]$ with integrators—one in each control channel—to form the design plant $P_d = [A, B, C]$; that is, $P_d = P(I_{2\times 2}/s)$. This is done as follows:

$$A = \begin{bmatrix} 0_{2\times 2} & 0_{2\times 4} \\ B_p & A_p \end{bmatrix} \tag{10.204}$$

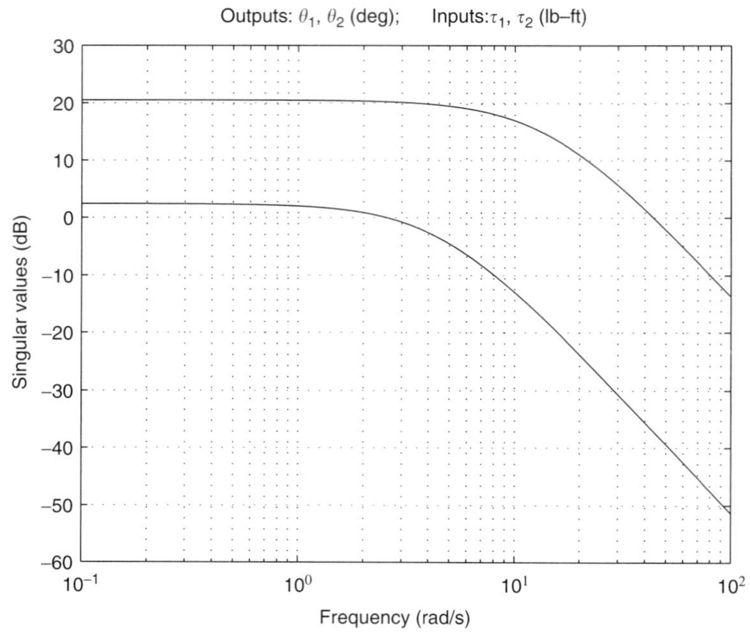

FIGURE 10.9 PUMA 560 robotic manipulator singular values.

$$B = \begin{bmatrix} 1_{2\times 2} \\ 0_{4\times 2} \end{bmatrix} \tag{10.205}$$

$$C = [0_{2\times 2} \quad C_p] \tag{10.206}$$

The state of this system is $x = \begin{bmatrix} x_i \\ x_p \end{bmatrix}$ where x_i is the integrator state and x_p is the plant state. The singular values for the augmented system P_d exhibit a slope of –20 dB/dec at low frequencies as expected (see Figure 10.10). The minimum singular value crosses zero dB just above 1 rad/s. The maximum singular value crosses zero dB at about 8 rad/s.

Step 2: Design Target Open Loop Transfer Function Matrix $L_o = G_{KF} = C(sI - A)^{-1}H_f$

Next we design a target open loop transfer function matrix $L_o = G_{KF} = C(sI - A)^{-1}H_f$ that has desirable closed loop properties (e.g., sensitivity singular values, pole locations, stability margins, etc.) at the output. To do this, we use Kalman Filtering ideas. Like LQR loops designed without a cross-state-control-coupling penalty, Kalman Filter loops designed in similar fashion exhibit desirable stability robustness margins (e.g., infinite upward gain margin, at least 6 dB downward gain margin, at least ±60° phase margin). This target loop design is carried out as follows:

- Consider the augmented system shown in Figure 10.11.

  It will be used to design a target loop transfer function matrix $L_o = G_{KF}$ with desirable closed loop properties at the output. To do so, we begin by forming an augmented system $G_{FOL} = C(sI - A)^{-1}L$ with

$$L = \begin{bmatrix} L_L \\ L_H \end{bmatrix} \tag{10.207}$$

$$L_L = [C_p(-A_p)^{-1}B_p]^{-1} \tag{10.208}$$

$$L_H = (-A_p)^{-1}B_pL_L \tag{10.209}$$

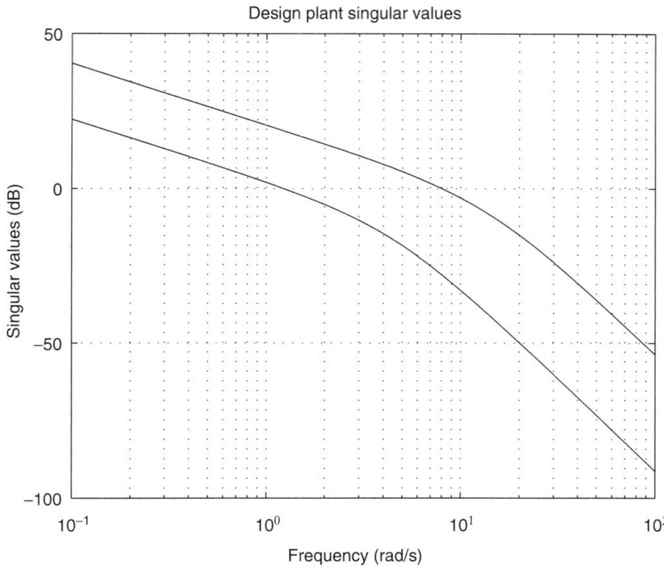

FIGURE 10.10 PUMA 560 robotic manipulator design plant singular values.

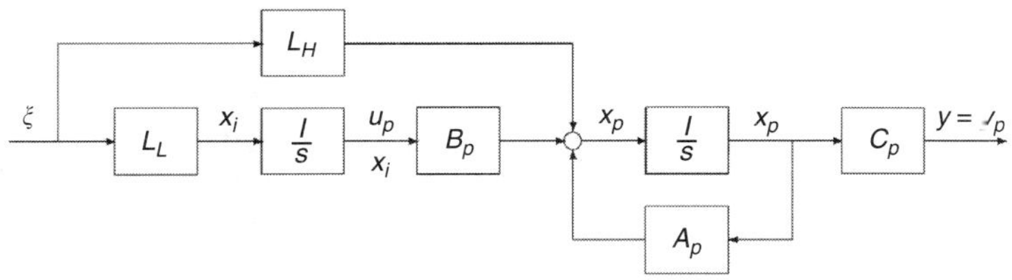

FIGURE 10.11 Augmented system used for designing target loop.

The matrix L_L matches the singular values of $G_{\text{FOL}} = C(sI - A)^{-1}L$ at low frequencies. The matrix L_H matches the singular values at high frequencies. Together, L_L and L_H match the singular values of $G_{\text{FOL}} = C(sI - A)^{-1}L$ at all frequencies (see Figure 10.12). Why is this?

This selection for L_L and L_H results in

$$G_{\text{FOL}} = C_p(sI - A_p)^{-1}L_H + C_p(sI - A_p)^{-1}B_p\left(\frac{I}{s}\right)L_L \tag{10.210}$$

$$= C_p(sI - A_p)^{-1}\left[L_H + B_p\left(\frac{I}{s}\right)L_L\right] \tag{10.211}$$

$$= C_p(sI - A_p)^{-1}\left[(-A_p)^{-1}B_pL_L + B_p\left(\frac{I}{s}\right)L_L\right] \tag{10.212}$$

$$= C_p(sI - A_p)^{-1}[sI - A_p](-A_p)^{-1}B_pL_L\left(\frac{I}{s}\right) \tag{10.213}$$

$$= C_p(-A_p)^{-1}B_pL_L\left(\frac{I}{s}\right) \tag{10.214}$$

$$= \frac{I}{s} \tag{10.215}$$

The resulting gain crossover frequency in Figure 10.12 is 1 rad/s, as expected.

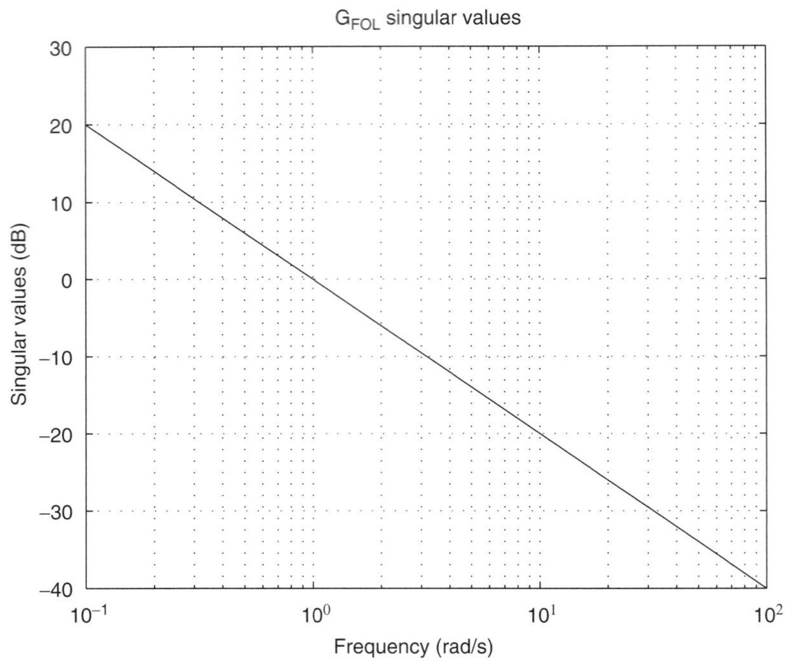

FIGURE 10.12 PUMA 560 robotic manipulator G_{FOL} singular values.

Why match the singular values of G_{FOL} in this manner? From the so-called Kalman Frequency Domain Equality (KFDE), it follows that

$$\sigma_i[I + G_{\text{KF}}(j\omega)] = \sqrt{1 + \frac{1}{\mu}\sigma_i^2[G_{\text{FOL}}(j\omega)]} \tag{10.216}$$

This suggests that by shaping G_{FOL}, we can shape the target $L_o = G_{\text{KF}}$. Specifically, if G_{FOL} is large at low frequencies, then we expect (from KFDE)

$$L_o(j\omega) = G_{\text{KF}}(j\omega) \approx \frac{1}{\sqrt{\mu}} G_{\text{FOL}}(j\omega) \tag{10.217}$$

at low frequencies. This shows that the matrix L should be used for shaping the target loop $L_o = G_{\text{KF}}$ while $\mu > 0$ is used to adjust the target loop bandwidth—decreasing/increasing μ to raise/lower the target loop bandwidth.

Note that through our selection of L, we have made all of the plant's unstable modes uncontrollable through L. Hence, (A, L) is NOT stabilizable! While this might appear to be troublesome, it is not. What matters is that the associated Hamiltonian belongs to *dom*(*Ric*) so that a stabilizing H_f exists. A necessary and suficient condition for this, however, is that (A, C) be detectable and (A, L) has no unobservable modes on the imaginary axis. Since each of these conditions are indeed satisfied, we can use the "are" command to find a stabilizing solution to the FARE.

- Next we solved the FARE with $\Theta = \mu I_{2\times 2}$ $(\mu = 0.1)$:

$$AY + YA^T + LL^T - YC^T\Theta^{-1}CY = 0 \tag{10.218}$$

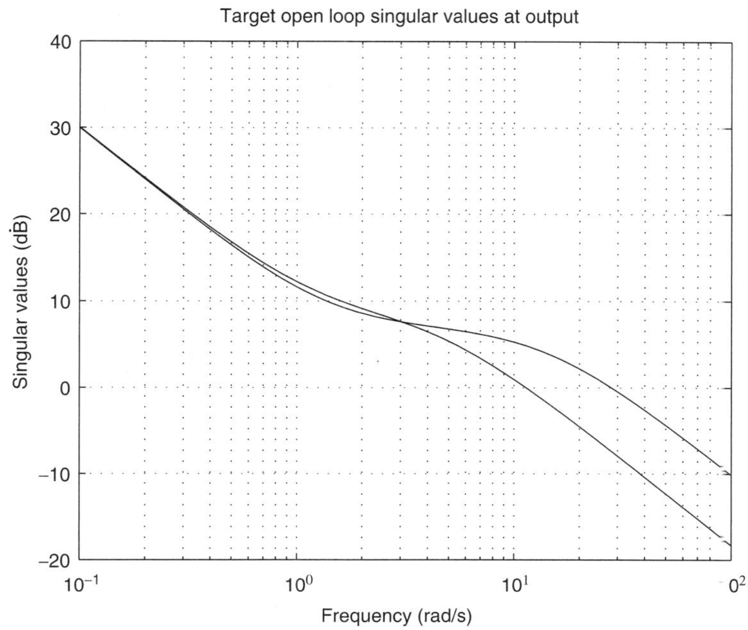

FIGURE 10.13 PUMA 560 robotic manipulator target loop G_{KF} singular values.

for $Y \geq 0$. The "are" command was used to do this, as it returns a stabilizing solution (provided that one exists). We then formed the filter gain matrix

$$H_f = YC^T\Theta^{-1} \tag{10.219}$$

$$= \begin{bmatrix} 2.3635 & 0.0384 \\ 0.4085 & 0.3091 \\ 13.1371 & -4.2300 \\ -4.2300 & 30.4572 \\ 90.2377 & -83.4384 \\ -100.9668 & 467.7679 \end{bmatrix} \tag{10.220}$$

Doing so results in the following target closed loop poles ($\lambda_i(A - H_fC)$):

$$s = -3.1623, -3.1623, -4.5299, -4.5299, -14.1050, -14.1050 \tag{10.221}$$

The singular values for the resulting target open loop transfer function matrix $L_o = G_{\mathrm{KF}} = C(sI - A)^{-1}H_f$ are shown in Figure 10.13.

The target open loop singular values—as expected from the KFDE—are matched at low frequencies with a slope of −20 dB/dec. They remain matched til about 1 rad/s, then they separate. This is expected since $G_{\mathrm{FOL}} = I/s$ is not an achievable loop. (Not if closed loop stability matters!) The resulting filter gain matrix provides the necessary bandwidth to stabilize the unstable robotic manipulator, with open loop instabilities at $s = 14.1050, 4.5299$. One singular value crosses 0 dB just above 10 rad/s, the other just below 30 rad/s. μ was used to adjust the bandwidth.

The corresponding target sensitivity $S_{\mathrm{KF}} = [I + G_{\mathrm{KF}}]^{-1}$ singular values and complementary sensitivity $T_{\mathrm{KF}} = G_{\mathrm{KF}}[I + G_{\mathrm{KF}}]^{-1}$ singular values are shown in Figures 10.14 and 10.15, respectively. The associated

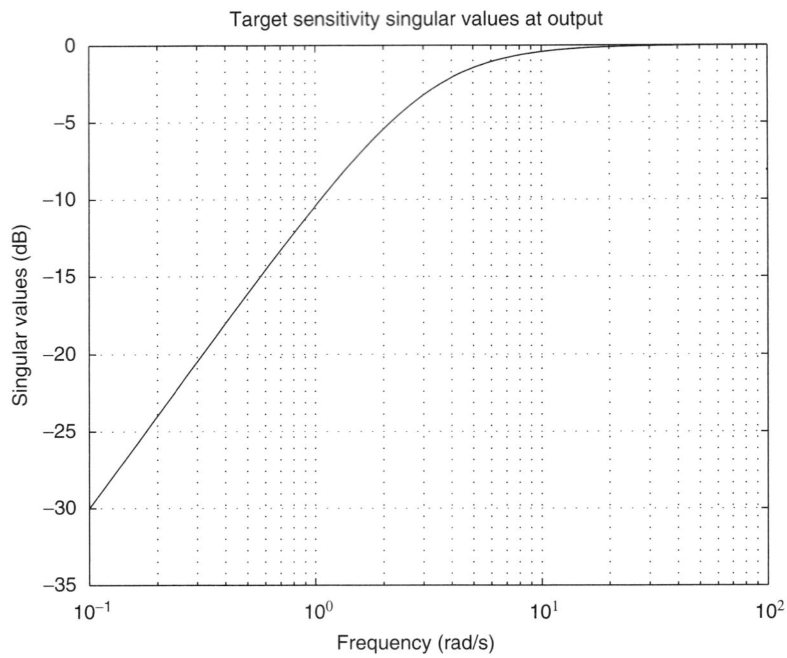

FIGURE 10.14 PUMA 560 robotic manipulator target sensitivity $S_{KF} = [I + G_{KF}]^{-1}$ singular values.

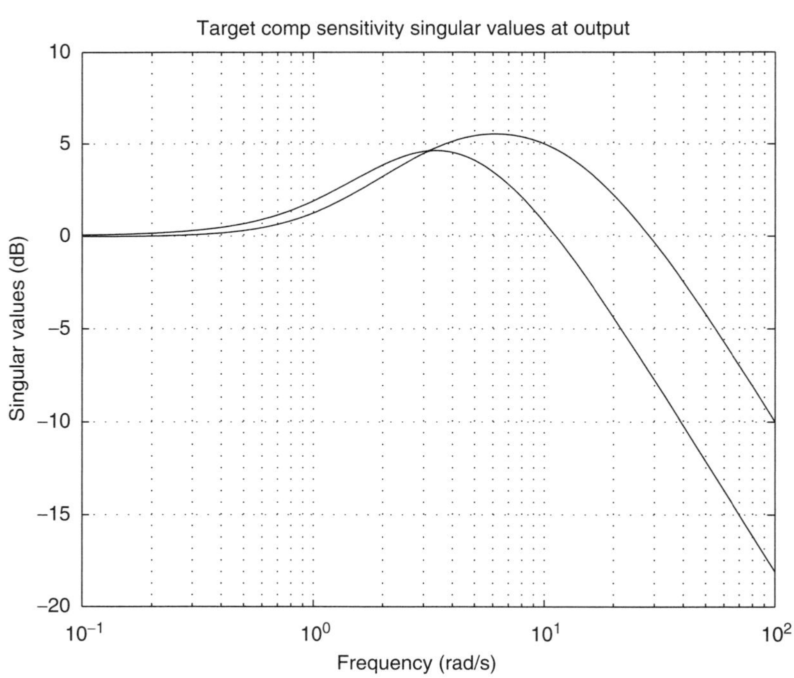

FIGURE 10.15 PUMA 560 robotic manipulator target complementary sensitivity $T_{KF} = G_{KF}[I + G_{KF}]^{-1}$ singular values.

sensitivity and complementary sensitivity singular values are desirable in that they suggest that the target loop will possess:

- Good low frequency command following properties
- Good low frequency disturbance attenuation properties
- Good high frequency sensor noise attenuation properties
- Good MIMO stability margins (nearly infinite upward gain margin, at least 6 dB downward gain margin, and at least ±60° phase margin) at the output

The complementary sensitivity singular values suggest that a reference command prefilter W would reduce overshoot due to step reference commands. The design of such a filter will be considered below.

Step 3: Solve Cheap Control Problem to Recover Target Loop at Plant Output

Next we solve an appropriately formulated "cheap LQR control problem" that would produce a control gain matrix G_c such that the $\mathcal{H}^2$ optimal model-based compensator $K_d = [A - BG_c - H_fC, H_f, G_c]$ with $P_d = [A, B, C]$ approximates ("recovers") the target loop transfer function matrix $L_o = G_{KF}$; that is,

$$P_d K_d \approx L_o = G_{KF} \tag{10.222}$$

This was done by solving the following CARE (using the "lqr" command) with $R = \rho I_{2\times 2}(\rho = 10^{-13})$:

$$XA + A^T X + C^T C - XBR^{-1}B^T X = 0 \tag{10.223}$$

for $X \geq 0$ and forming the control gain matrix

$$G_c = R^{-1}B^T X \tag{10.224}$$

$$= \begin{bmatrix} 987.9832 & -543.0034 & 3162945.2928 & 56.9921 & 13941.9005 & 2069.3324 \\ -543.0034 & 3657.5891 & 11.7867 & 3162634.3919 & 2069.7987 & 3765.7978 \end{bmatrix} \tag{10.225}$$

Doing so yields the following closed loop regulator poles ($\lambda_i(A - BG_c)$):

$$s = -440.8808, -220.4404 \pm j381.7871, -1881.9053, -940.9527 \pm j1629\ 7168 \tag{10.226}$$

All have damping factors greater than or equal to $\zeta = 0.5$. As a practical note to facilitate real-time implementation of the resulting controller, one might use model reduction techniques [10] to remove some of the very high frequency poles in the compensator. Doing so would permit using a larger integration step size in any real-time embedded system or microprocessor implementation.

Step 4: Construct Final Controller K

Next we form the final controller as follows:

$$K = \frac{K_d}{s} \tag{10.227}$$

$$= \frac{[A - BG_c - H_fC, H_f, G_c]}{s} \tag{10.228}$$

$$= [A_k, B_k, C_k] \tag{10.229}$$

A state space representation for this controller is given by

$$A_K = \begin{bmatrix} 0_{2\times 2} & G_c \\ 0_{6\times 2} & A - BG_c - H_fC \end{bmatrix}, \quad B_K = \begin{bmatrix} 0_{2\times 2} \\ H_f \end{bmatrix} \tag{10.230}$$

$$C_K = \begin{bmatrix} I_{2\times 2} & 0_{2\times 6} \end{bmatrix} \tag{10.231}$$

With this selection, we have

$$PK = P\frac{K_d}{s} \tag{10.232}$$

$$= P\frac{I_{2\times 2}}{s}K_d \tag{10.233}$$

$$= P_d K_d \tag{10.234}$$

$$\approx L_o = G_{\mathrm{KF}} \tag{10.235}$$

Through this selection of K, we have recovered the target loop transfer function matrix $L_o = G_{\mathrm{KF}}$. That is, K has approximately inverted P (from the right) in order to achieve $PK \approx L_o = G_{\mathrm{KF}}$. An examination of the singular values for the actual loop PK shows that the actual singular values agree with the target singular values up to and beyond 100 rad/s.

Loop Transfer Recovery. Why were we able to recover the target loop? The recovery was permitted by the model-based structure of the compensator K, the Riccati equations used to obtain the gain matrices G_c and H_f, and the fact that the plant $P = [A_p, B_p, C_p]$ (and hence the design plant $P_d = [A, B, C] = P(I/s)$ is minimum phase. The minimum phase condition, specifically, is a suficient condition which guarantees that there exists an orthonormal matrix $U(U^T U = UU^T = I)$ such that

$$\lim_{\rho \to 0^+} \sqrt{\rho} G_c = UC \tag{10.236}$$

This limiting behavior relating the control gain matrix and the design plant's C matrix, however, can be used to prove that loop transfer recovery takes place; that is,

$$\lim_{\rho \to 0^+} P_d K_d = \lim_{\rho \to 0^+} PK = L_o = G_{\mathrm{KF}} \tag{10.237}$$

Step 5: Design Command Pre-filter W

The MATLAB command

$$t_0(a - b * g - h * c, h, g) \tag{10.238}$$

can be used to find the compensator's transmission zeros. These are also zeros of the closed loop transfer function matrix from r to y. The final compensator (as well as the target loop G_{KF}) has zeros near $s \approx -1.2$. Given this, a reference command prefilter

$$W = \frac{1.2}{s + 1.2} I_{2\times 2} \tag{10.239}$$

was added outside the loop to filter reference commands. By so doing, we ensure that step reference commands for θ_1 and θ_2 are followed in the steady state (due to integrators in controller) without excessive overshoot during the transient.

Sensitivity Frequency Response

The resulting sensitivity singular values are plotted in Figure 10.16. The plot suggests that low frequency reference commands r will be followed and low frequency output disturbances d_o will be attenuated. More precisely, reference commands r with frequency content below 0.3 rad/s should be followed to within about 20 dB; that is, with a steady-state error of about 10%. Similarly, output disturbances d_o with frequency content below 0.3 rad/s should be attenuated by approximately 20 dB.

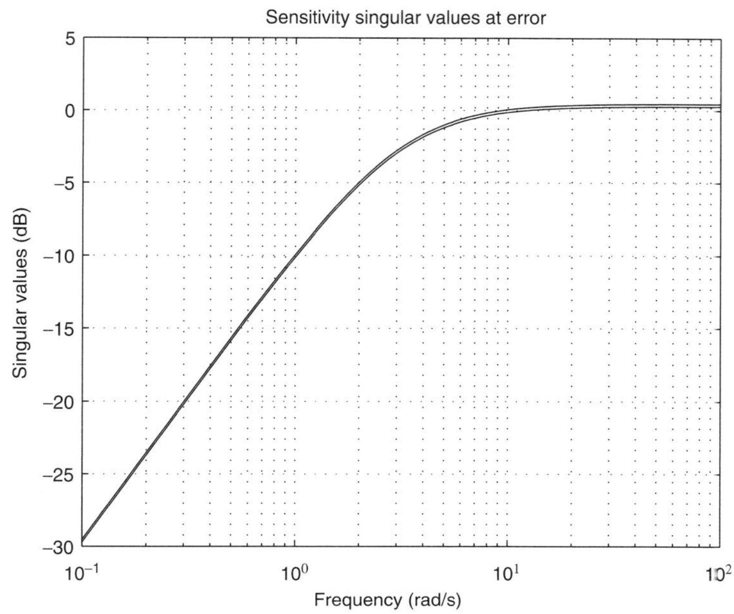

FIGURE 10.16 PUMA 560 sensitivity frequency response at error.

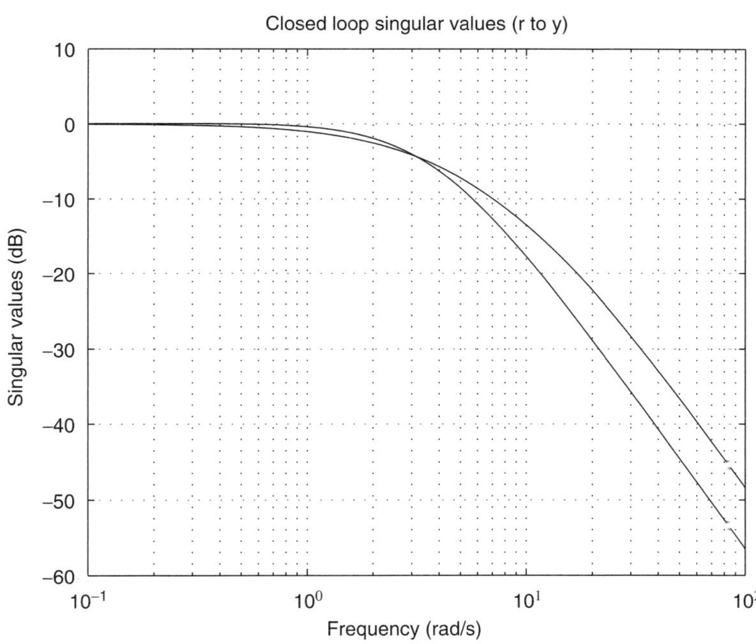

FIGURE 10.17 PUMA 560 reference to output frequency response.

Reference to Output Frequency Response

The transfer function matrix from reference commands r to link angles y is

$$T_{ry} = [I + PK]^{-1}PKW \tag{10.240}$$

Its singular values are plotted in Figure 10.17. The plot suggests that low frequency reference commands will be followed in the steady state and that little overshoot will result during the transient.

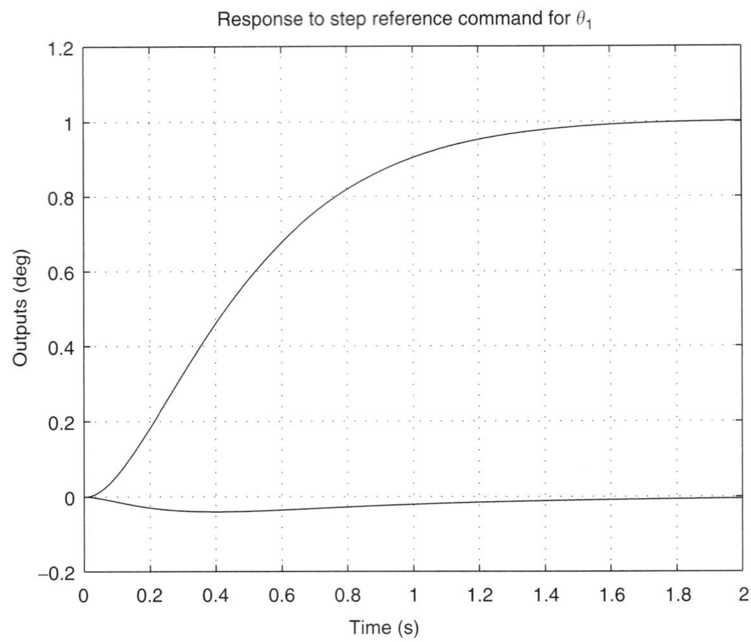

FIGURE 10.18 PUMA 560 outputs: response to θ_1 reference command.

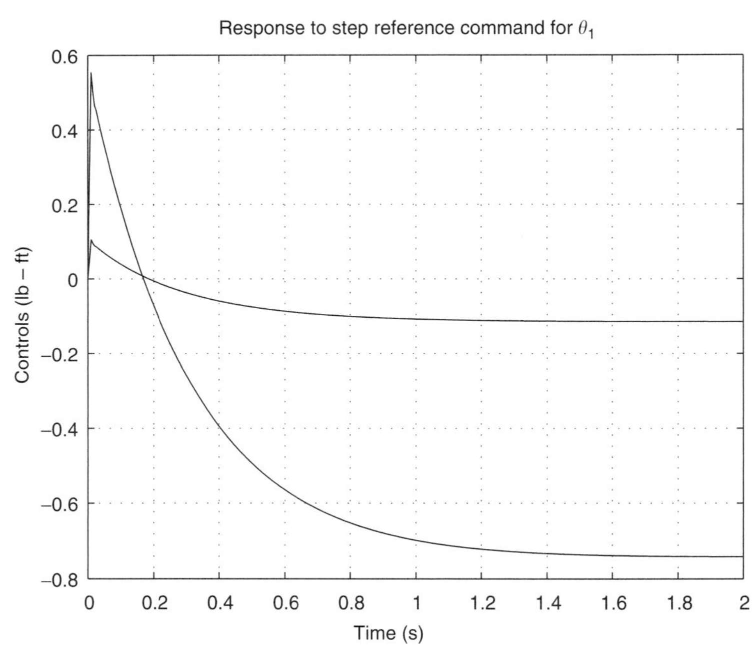

FIGURE 10.19 PUMA 560 controls: response to θ_1 reference command.

Response to Θ_1 Step Reference Command

The response to a unit step θ_1 command is plotted in Figure 10.18. As expected, θ_1 follows the step command well, with no overshoot and settling in about 1.6 s. The associated θ_2 response is small, indicating little cross coupling in the final closed loop system. The corresponding controls are plotted in Figure 10.19. They are acceptable in size.

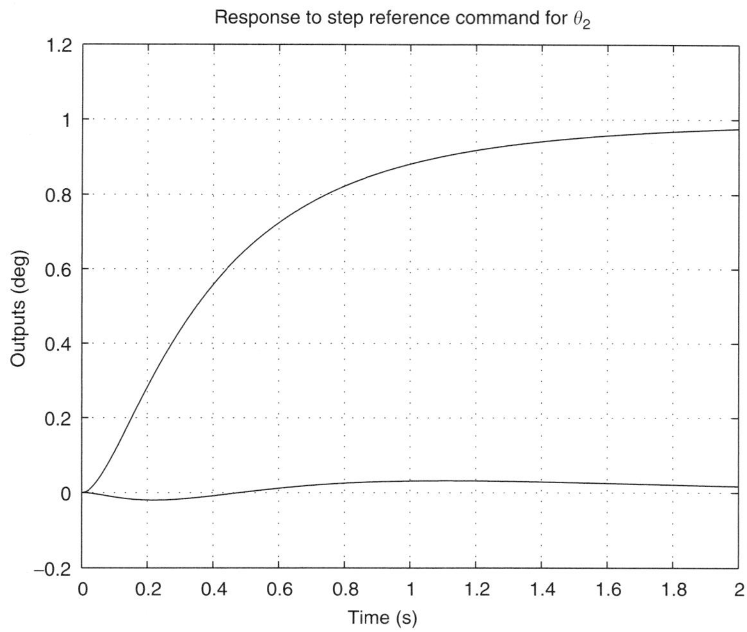

FIGURE 10.20 PUMA 560 outputs: response to θ_2 reference command.

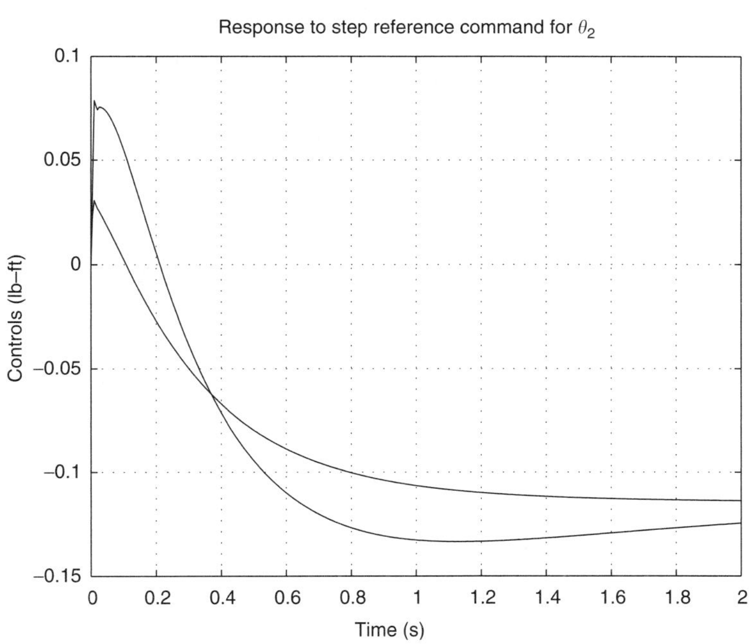

FIGURE 10.21 PUMA 560 controls: response to θ_2 reference command.

Response to Θ_2 Step Reference Command

The response to a unit step θ_2 command is plotted in Figure 10.20. As expected, θ_2 follows the step command well, with no overshoot and settling in about 3 s. The associated θ_1 response is small, indicating little cross coupling in the final closed loop system. The corresponding controls are plotted in Figure 10.21. They are acceptable in size. ■

10.4 $\mathcal{H}^2$ State Feedback Problem

This section shows that the methods presented for output feedback may be readily adopted to permit the design of $\mathcal{H}^2$ optimal constant gain state feedback control laws (control gain matrices G_c) as well.

10.4.1 Generalized Plant Structure for State Feedback

For this case, the generalized plant G (including plant P and weighting functions) takes the following form:

$$G = \left[\begin{array}{c|c} G_{11} & G_{12} \\ \hline G_{21} & G_{22} \end{array}\right] = \left[\begin{array}{c|cc} A & B_1 & B_2 \\ \hline C_1 & 0_{n_z \times n_w} & D_{12} \\ I_{n \times n} & 0_{n_y \times n_w} & 0_{n_y \times n_u} \end{array}\right] = \left[\begin{array}{c|c} A & B \\ \hline C & D \end{array}\right] \tag{10.241}$$

This implies that the measured signals y are the states x of the generalized plant G. As such, all of the modes of A are observable through $C_2 = I_{n\times n}$.

10.4.2 State Feedback Assumptions

The standard state feedback assumptions are a subset of those required for the output feedback problem formulation. The state feedback assumptions are as follows.

Assumption 10.2 ($\mathcal{H}^2$ State Feedback Problem)

Throughout this section, it will be assumed that

1. *Plant G_{22} Assumption.* (A, B_2) stabilizable.
2. *Nonsingular Control Weighting Assumption.* $R = D_{12}^T D_{12} > 0$ (D_{12} full column rank).
3. *Regulator Assumption.* $\begin{bmatrix} j\omega I - A & -B_2 \\ C_1 & D_{12} \end{bmatrix}$ has full column rank for all ω. ■

It should be noted that if $D_{12}^T C_1 = 0$, then (3) is equivalent to (A, C_1) having no unobservable imaginary modes. If (A, C_1) is detectable, then this is satisfied.

$\mathcal{H}^2$ Optimal State Feedback Control Law

The $\mathcal{H}^2$ optimal controller is given by

$$K_{\text{opt}} = -G_c \tag{10.242}$$

where the control gain matrix $G_c \in \mathrm{R}^{n_u \times n}$ is given by

$$G_c = R^{-1}[B_2^T X + D_{12}^T C_1] \tag{10.243}$$

where $X \geq 0$ is the unique (at least) positive semi-definite solution of the CARE:

$$(A - B_2 R^{-1} D_{12}^T C_1)^T X + X(A - B_2 R^{-1} D_{12}^T C_1) + C_1^T (I - D_{12}^T R^{-1} D_{12}^T) C_1 - X B_2 R^{-1} B_2^T X = 0 \tag{10.244}$$

The closed loop poles that result from the above constant gain state feedback control law are the eigenvalues of $A - B_2 G_c$. The minimum closed loop norm is given by

$$\min_K \|T_{wz}\|_{H^2} = \sqrt{trace(B_1^T X B_1)} \tag{10.245}$$

State Feedback Loop Shaping

If one selects

$$B_2 = B \tag{10.246}$$

$$C_1 = \begin{bmatrix} M \\ 0_{n_u \times n} \end{bmatrix} \tag{10.247}$$

$$D_{12} = \begin{bmatrix} 0_{n_y \times n_u} \\ \sqrt{\rho} I_{n_u \times n_u} \end{bmatrix} \tag{10.248}$$

$$R = \rho I_{n_u \times n_u} \tag{10.249}$$

then $D_{12}^T C_1 = 0$ and we have

$$G_c = \frac{1}{\rho} B_2^T X \tag{10.250}$$

where $X \geq 0$ is the unique (at least) positive semi-definite solution of the CARE:

$$A^T X + XA + M^T M - XB\frac{1}{\rho}B^T X = 0 \tag{10.251}$$

The following LQFDE may be derived from the CARE:

$$[I + G_{\text{LQ}}(j\omega)]^H [I + G_{\text{LQ}}(j\omega)] = I + \left[\frac{1}{\sqrt{\rho}} G_{\text{OL}}(j\omega)\right]^H \left[\frac{1}{\sqrt{\rho}} G_{\text{OL}}(j\omega)\right] \tag{10.252}$$

where

$$G_{\text{OL}} = M(sI - A)^{-1} B \tag{10.253}$$

$$G_{\text{LQ}} = G_c(sI - A)^{-1} B \tag{10.254}$$

Given this, the loop shaping ideas discussed earlier are applicable. A designer may use the matrix M and the scalar $\rho > 0$ to shape G_{OL} in an effort to get a desirable loop G_{LQ}. The matrix M, specifically, may be used to match singular values at low frequencies, high frequencies, all frequencies, etc. Assuming that (A, B) is stabilizable and (A, M) has no imaginary modes that are unobservable, a stabilizing solution is guaranteed to exist. Moreover, the resulting G_{LQ} loop will possess nominal sensitivity and stability robustness properties—a consequence of the LQFDE. The resulting control gain matrix G_c may be used within a state feedback loop, a modified state feedback loop, or within a model-based compensator.

10.5 $\mathcal{H}^2$ Output Injection Problem

This section shows how the methods presented for output feedback may be readily adopted to permit the design of $\mathcal{H}^2$ optimal state estimators (filter gain matrices H_f) as well.

10.5.1 Generalized Plant Structure for Output Injection

For this case (dual to the state feedback case), the generalized plant G (including plant P and weighting functions) takes the following form:

$$G = \left[\begin{array}{c|c} G_{11} & G_{12} \\ \hline G_{21} & G_{22} \end{array}\right] = \left[\begin{array}{c|cc} A & B_1 & I_{n\times n} \\ \hline C_1 & 0_{n_z\times n_w} & 0_{n_z\times n_u} \\ C_2 & D_2 & 0_{n_y\times n_u} \end{array}\right] = \left[\begin{array}{c|c} A & B \\ \hline C & D \end{array}\right] \tag{10.255}$$

This implies that the control signals u directly impact all of the generalized plant states x. As such, all of the modes of A are controllable through $B_2 = I_{n\times n}$.

10.5.2 Output Injection Assumptions

The standard output injection assumptions are a subset of those required for the output feedback problem formulation. The output injection assumptions are as follows.

Assumption 10.3 ($\mathcal{H}^2$ Output Injection Problem)

Throughout this section, it will be assumed that

1. *Plant G_{22} Assumption.* (A, C_2) detectable.
2. *Nonsingular Measurement Weighting Assumption.* $\Theta = D_{21}D_{21}^T > 0$ (D_{21} full row rank).
3. *Filter Assumption.* $\begin{bmatrix} j\omega I - A & -B \\ C_1 & D_{21} \end{bmatrix}$ has full row rank for all ω. ■

It should be noted that if $B_1D_{21}^T = 0$, then (3) is equivalent to (A, B_1) having no uncontrollable imaginary modes. If (A, B_1) is stabilizable, then this is satisfied.

$\mathcal{H}^2$ Optimal Output Injection Law

The $\mathcal{H}^2$ optimal controller is then given by

$$K_{\text{opt}} = -H_f \tag{10.256}$$

where the filter gain matrix $H_f \in \mathrm{R}^{n\times n_y}$ is given by

$$H_f = [YC_2^T + B_1D_{21}^T]\Theta^{-1} \tag{10.257}$$

where $Y \geq 0$ is the unique (at least) positive semi-definite solution of the FARE:

$$(A - B_1D_{21}^T\Theta^{-1}C_2)Y + Y(A - B_1D_{21}^T\Theta^{-1}C_2)^T + B_1(I - D_{21}^T\Theta^{-1}D_{21})B_1^T - YC_2^T\Theta^{-1}C_2Y = 0 \tag{10.258}$$

The closed loop poles that result from the above output injection law are the eigenvalues of $A - H_fC_2$. The minimum closed loop norm is given by

$$\min_K \|T_{wz}\|_{H^2} = \sqrt{trace\left(C_1YC_1^T\right)} \tag{10.259}$$

where Y is the solution to the FARE.

Estimator (Filter) Loop Shaping
If one selects

$$B_1 = [L \quad 0_{n \times n_y}] \tag{10.260}$$

$$D_{21} = [0_{n_y \times n_u} \quad \sqrt{\mu} I_{n_y \times n_y}] \tag{10.261}$$

$$C_2 = C \tag{10.262}$$

$$\Theta = \mu I_{n_y \times n_y} \tag{10.263}$$

then $B_1 D_{21}^T = 0$ and we have

$$H_f = Y C_2^T \frac{1}{\mu} \tag{10.264}$$

where $Y \geq 0$ is the unique (at least) positive semi-definite solution of the FARE:

$$YA^T + AY + LL^T - YC^T \Theta^{-1} CY = 0 \tag{10.265}$$

Given this, the following KFDE may be derived from the FARE:

$$[I + G_{\text{KF}}(j\omega)][I + G_{\text{KF}}(j\omega)]^H = I + \left[\frac{1}{\sqrt{\mu}} G_{\text{FOL}}(j\omega)\right]\left[\frac{1}{\sqrt{\mu}} G_{\text{FOL}}(j\omega)\right]^{H} \tag{10.266}$$

where

$$G_{\text{FOL}} = C(sI - A)^{-1} L \tag{10.267}$$

$$G_{\text{KF}} = C(sI - A)^{-1} H_f \tag{10.268}$$

Given this, the loop shaping ideas discussed earlier are applicable. A designer may use the matrix L and the scalar $\mu > 0$ to shape G_{FOL} in an effort to get a desirable loop G_{KF}. The matrix L, specifically, may be used to match singular values at low frequencies, high frequencies, all frequencies, etc. Assuming that (A, C) is detectable and (A, L) has no imaginary modes that are uncontrollable, then a stabilizing solution is guaranteed to exist. Moreover, the resulting G_{KF} loop will possess nominal sensitivity and stability robustness properties—a consequence of the KFDE. The resulting filter (output injection) gain matrix H_f may be used within an estimator (feedback) loop, a modified estimator (feedback) loop, or within a model-based compensator.

10.6 Summary

This chapter has presented a general framework for control system design via $\mathcal{H}^2$ optimization. While the focus has been on continuous time LTI systems, the methods are very flexible and have wide application. They may be used to design constant gain state feedback control laws, constant gain state estimators, dynamic output feedback controllers, and much more. Weighting functions are easily accommodated within the generalized plant framework presented. Such functions may be used to achieve closed loop design objectives. All of the ideas presented may be extended with subtle (all be it very important) modifications, to accommodate control system design via $\mathcal{H}^\infty$ optimization. Additional details may be found in.

The methods presented in this chapter may be extended to discrete time linear shift invariant (LSI) systems. Extensions to sampled data systems are also possible [1].

References

1. Chen, T. and Francis, B., *Optimal Sampled-Data Control Systems,* Springer, London, 1995.
2. Dorf, R.C. and Bishop, R.H., *Modern Control Systems,* 8th ed, Addison Wesley, CA, 1998.
3. Doyle, J.C., "Guaranteed margins for LQG regulators," *IEEE Transactions on Automatic Control,* Vol. AC-23, No. 4, August 1978, pp. 756–757.
4. Doyle, J.C., Glover, K., Khargonekar, P.P., and Francis, B.A., "State-space solutions to standard $\mathcal{H}^2$ and $\mathcal{H}^\infty$ control problems," *IEEE Transactions on Automatic Control,* Vol. AC-34, No. 8, 1989, pp. 831–847. Also see *Proceedings of the 1988 American Control Conference,* Atlanta, Georgia, June, 1988.
5. Kalman, R.E., "A new approach to linear filtering and prediction problems," *ASME Journal of Basic Engineering,* Vol. 85, 1960, pp. 34–45.
6. Kalman, R.E. and Bucy, R.S., "New results in linear filtering and prediction problems," *ASME Journal of Basic Engineering,* 1960, pp. 95–108.
7. Kwakernaak, H. and Sivan, R., *Linear Optimal Control Systems,* Wiley-Interscience, New York, 1972.
8. Rodriguez, A.A., *A Practical Neo-Classical Approach to Feedback Control System Analysis and Design,* Control3D, 2000.
9. Spong, M.W. and Vidyasagar, M., *Robot Dynamics and Control,* John Wiley & Sons, New York, 1989.
10. Zhou, K., Doyle, J.C., and Glover, K., *Robust and Optimal Control,* Prentice-Hall, NJ, 1996.
11. Zhou, K. and Doyle, J.C., *Essentials of Robust Control,* Prentice-Hall, NJ, 1998.

11

Adaptive and Nonlinear Control Design

Maruthi R. Akella
The University of Texas at Austin

11.1 Introduction

The most important challenge for modern control theory is that it should deliver acceptable performance while dealing with poor models, high nonlinearities, and low-cost sensors under a large number of operating conditions. The difficulties encountered are not peculiar to any single class of systems and they appear in virtually every industrial application. Invariably, these systems contain such a large amount of model and parameter uncertainty that "fixed"controllers can no longer meet the stability and performance requirements. Any reasonable solution for such problems must be a suitable amalgamation between nonlinear control theory, adaptive elements, and information processing. Such are the factors behind the birth and evolution of the field of adaptive control theory, strongly motivated by several practical applications such as chemical process control and design of autopilots for high-performance aircraft, which operate with proven stability over a wide variety of speeds and altitudes.

A commonly accepted definition for an adaptive system is that it is any physical system that is designed from an adaptive standpoint![1] All existing stability and convergence results, in the field of adaptive control theory, hinge on the crucial assumption that the unknown parameters must occur linearly within the plant containing known nonlinearities. Conceptually, the overall process makes the parameter estimates themselves as state variables, thus enlarging the dimension of the state space for the original system. By nature, adaptive control solutions for both linear and nonlinear dynamical systems lead to nonlinear time-varying formulations wherein the estimates of the unknown parameters are updated using input–output data. A parameter adaptation mechanism (typically nonlinear) is used to update the parameters within the control law. Given the nonlinearity due to adaptive feedback, there is the need to ensure that the closed-loop stability is preserved. It is thus an unmistakable fact that the fields of adaptive control and nonlinear system stability are intrinsically related to one another and any new insights gained in one field would be of potential benefit to the other. Many formalisms in nonlinear stability theory can be employed such as the Lyapunov direct method and passivity-based methods. We will first present some important mathematical and analytical tools for studying the stability of nonlinear dynamical systems.

11.2 Lyapunov Theory for Time-Invariant Systems

The Lyapunov direct method is a commonly adopted and arguably one of the most popular methods for proving closed-loop stability in the adaptive control area. It is not restricted to local system behavior and determines the stability properties of the nonlinear system by considering the time evolution of the system solutions with respect to an "energy-like" scalar function, often known as the Lyapunov function.

Consider any dynamical system represented by the following nonlinear autonomous differential equation

$$x = f(x), \qquad f(0) = 0 \tag{11.1}$$

Obviously $x(t) = 0$ is a solution. A sufficient condition for the existence and uniqueness of solutions for Equation 11.1 is that $f(x)$ be locally Lipschitz, that is,

$$\| f(x) - f(y)\| \le L\|x - y\| \tag{11.2}$$

for all x and y in a finite neighborhood of the origin. We are interested in the stability of the solutions of Equation 11.1 in the presence of perturbations. Before discussing the main Lyapunov stability theorems, we present some important definitions.

Definition: **Lyapunov stability**
The solution $x(t) = 0$ of Equation 11.1 is called *stable* in the sense of Lyapunov if for all $\epsilon > 0$, there exists a $\delta(\epsilon) > 0$ such that for all initial conditions satisfying $\|x(0)\| < \delta$, we have $\|x(0)\| < \epsilon$ for $t \in [0, \infty)$. The solution is *unstable* if it is not stable. The solution is *asymptotically stable* if it is stable and there exists a $\delta > 0$ such that every initial condition that satisfies $\|x(0)\| < \delta$ has the property

$$\lim_{t\to\infty} \|x(t)\| = 0$$

The solution is *globally asymptotically stable* if it is asymptotically stable for all initial conditions. These definitions refer to stability of particular solutions of Equation 11.1 with respect to initial conditions and not to the stability of differential equations.

Definition: **Positive definite and semidefinite functions**
Any continuously differentiable function $V: \Re^n \to \Re$ is called *positive definite* if (i) $V(0) = 0$ and (ii) $V(x) > 0$ for all $x \neq 0$. A function is *positive semidefinite* if condition (ii) is replaced by $V(x) \ge 0$ for all $x \neq 0$.

Theorem: **Lyapunov's stability theorem for time-invariant systems**
If there exists a positive definite function $V: \Re^n \to \Re$ such that the time derivative of V along the solution of $\dot{x} = f(x)$ given by

$$\frac{d}{dt}V = \left[\frac{\partial V}{\partial x}\right]^T \dot{x} = \left[\frac{\partial V}{\partial x}\right]^T f(x)$$

is negative semidefinite, then the solution $x(t) = 0$ of Equation 11.1 is stable. In this case, the solution converges to the set $\{x \in \Re^n : \dot{V}(x) = 0\}$. If $\dot{V}$ is negative definite, then the solution is asymptotically stable. Furthermore, if $\dot{V}$ is negative definite and $V(x) \to \infty$ as $\|x\| \to \infty$, then the solution is globally asymptotically stable. The function $V(x)$ is called a Lyapunov function for the system described by Equation 11.1.

Remark: Lyapunov's theorem, though simple to state, has powerful applications in the stability analysis of nonlinear systems. However, since the theorem provides only a sufficient condition in terms of the Lyapunov function, we are often encountered with the difficult problem of finding a suitable Lyapunov function. In the special case when Equation 11.1 is a stable linear system,

$$\dot{x} = A_m x$$

a quadratic Lyapunov function $V = x^T P x$ exists where P is a symmetric positive definite matrix satisfying the so-called Lyapunov equation

$$A_m^T P + P A_m = -Q \tag{11.3}$$

for any symmetric positive definite Q matrix. On the other hand, there is no general recipe for construction of Lyapunov functions for nonlinear systems. As a rule of thumb, in the case of mechanical systems, "energy-like" quantities are good candidates for a first attempt.

11.3 Lyapunov Theory for Time-Varying Systems

We are now ready to consider the stability of solutions for a time-varying (nonautonomous) differential equation

$$\dot{x} = g(x,t), \quad g(0,t) = 0 \ \ \forall t \leq 0 \tag{11.4}$$

The function g is assumed to be piecewise continuous with respect to t and locally Lipschitz in x about a neighborhood of the solution $x(t) = 0$. This would guarantee that the origin is an equilibrium for Equation 11.4. In order to investigate the stability of equilibrium for this nonautonomous system, it is important to recognize that any solution of Equation 11.4 depends not only on time t but also on the initial time t_0. Thus, we need to revisit our previous definitions of stability.

Definition: **Uniform Lyapunov stability**
The solution $x(t) = 0$ for Equation 11.4 is *uniformly stable* if for every $\epsilon > 0$, there exists a $\delta(\epsilon) > 0$ that is independent of the initial time t_0 such that

$$\|x(t_0)\| < \delta \quad \text{implies} \quad \|x(t)\| < e \quad \forall t \geq t_0 \geq 0$$

The solution is *uniformly asymptotically stable* if it is uniformly stable and there is a positive constant ρ independent of t_0 such that $\|x(t)\| \to 0$ as $t \to \infty$ for all $\|x(t_0)\| < \rho$. The solution is *globally uniformly asymptotically stable* if it is uniformly asymptotically stable for all initial conditions.

The main stability theorem for nonautonomous systems requires the definition of certain *class K functions.*

Definition: **Class K functions**
A continuous function α: $[0, a) \to [0, \infty)$ is said to belong to *class K* if it is strictly increasing and $\alpha(0) = 0$. It is said to belong to class K_∞, or radially unbounded, if $a = \infty$ in such a way that $\alpha(r) \to \infty$ as $r \to \infty$.

Theorem: **Lyapunov's stability theorem for time-varying systems**
Consider a set $D = \{x \in \mathbb{R}^n: \|x\| \leq R\}$ about the equilibrium $x(t) = 0$ for Equation 11.4. If there exists a scalar function V: $\mathbb{R}^n \times \mathbb{R}^+ \to \mathbb{R}$ with continuous partial derivatives such that

(i) $\alpha_1(\|x\|) \leq V(x,t) \leq \alpha_2(\|x\|)$ *positive definite* and *decrescent*

(ii) $\dot{V} = \dfrac{\partial V}{\partial t} + \left[\dfrac{\partial V}{\partial x}\right]^T g(x,t) \leq \alpha_3(\|x\|)$

for all $t \geq 0$, where α_1, α_2, and α_3 are class K functions, then the equilibrium point $x = 0$ is uniformly asymptotically stable.

Remark: Note that in order to show stability for nonautonomous systems, it is necessary to bound the function $V(x, t)$ by the class K functions that do not depend upon time t. A detailed treatment of all the definitions and proof for this theorem can be found in Slotine and Li[2] and Khalil.[3]

Remark: In the recent years, several interesting converse Lyapunov results have been obtained. In particular, for every uniformly stable (or uniformly asymptotic stable) system, there exists a positive definite Lyapunov function with a negative semidefinite time derivative (see Sastry and Bodson[4]). These results are particularly useful from a closed-loop performance point of view because they allow us to explicitly estimate the convergence rates in some cases of nonlinear adaptive control systems.

The application of Lyapunov's stability theorem for nonautonomous systems arising out of adaptive control often leads us to negative semidefinite time derivatives of the Lyapunov function. Therefore, asymptotic stability analysis is a much harder problem and the following result, known as Barbalat's Lemma, is extremely useful in such situations.

Lemma: **Barbalat.**
Consider a uniformly continuous function ϕ: – → – defined at all real values of $t \geq 0$. If

$$\lim_{t\to\infty}\int_0^t \phi(s)\, ds$$

exists and is finite, then $\phi(t) \to 0$ as $t \to \infty$.

Remark: A consequence of this result is that if $\phi \in \mathcal{L}_2$ and $\dot{\phi} \in \mathcal{L}_\infty$, then $\phi(t) \to 0$ as $t \to \infty$ (see Slotine and Li[2] and Tao[5] for discussion and proof).

11.4 Adaptive Control Theory

In contrast to a fixed or ordinary controller, an adaptive controller is one with adjustable parameters and an adjustment mechanism. The following are some basic concepts that are necessary for any discussion of adaptive control theory.

11.4.1 Regulation and Tracking Problems

The desired objective for any control problem is to maintain the plant output either at its desired value or within specified/acceptable bounds of the desired value. If these desired values are constant with respect to time, we have a regulation problem, otherwise it is a tracking problem.

11.4.2 Certainty Equivalence Principle

This principle has been the bedrock of most adaptive control design methods and has received considerable attention during the past two decades.[4,6,7] Adaptive controllers based on this approach are obtained by independently designing a control law that meets the control objective assuming complete knowledge of all the unknown plant parameters (deterministic case), along with a parameter update law, which is usually a differential equation that generates online parameter estimates that are used to replace the unknown parameters within the control law. Such a controller would have perfect output tracking capability in the case when the plant parameters are exactly known. In the presence of parameter uncertainty, the adaptation mechanism will adjust the controller parameters so that the tracking objective is asymptotically achieved. The main issue, thus, in adaptive controller design is to synthesize the adaptation mechanism (parameter update law) that will guarantee that the control system remains stable and the output tracking error converges to zero as the parameter values are updated.

11.4.3 Direct and Indirect Adaptive Control

There exist two philosophically distinct approaches within adaptive control for plants containing unknown or uncertain parameters. The first of those is the so-called *direct approach* where the controller parameters are directly adjusted by the adaptation mechanism in such a way to optimize some pre-specified performance index based on the output. The second approach is the *indirect approach* wherein plant parameters are directly estimated and updated by the adaptation law and these estimated values are then used to compute the controller parameters. Direct adaptive control eliminates the need for this additional computation. Consequently, indirect adaptive control is plant parameter adaptive, whereas direct adaptive control is output performance adaptive. The plant parameter identification process is explicit within the direct approach while implicit in the direct approach. Hence, they have also been referred to as explicit and implicit approaches. In both of these cases, the controller structure remains the same and is determined from the certainty equivalence principle.

11.4.4 Model Reference Adaptive Control (MRAC)

The model reference adaptive control (MRAC) framework consists of four parts: (i) the plant containing the unknown parameters, (ii) a suitable reference model for specifying the desired output characteristics, (iii) a feedback control law that contains adjustable parameters, and (iv) an adaptation mechanism that updates the adjustable parameters within the control law. A schematic sketch for this framework is shown in Figure 11.1.

The plant is assumed to have a known structure with unknown parameters. For the case of linear systems, this means that the number of poles and zeros are assumed to be known, but the exact locations of poles and zeros are unknown. In the case of nonlinear systems, the structure of the governing equations of motion is assumed to be known, but some of the parameters appearing linearly within those equations can be unknown. The reference model specifies the desired output behavior expected from the plant as a result of the external reference input. It provides the ideal plant response which the adaptation mechanism should seek to track while updating the parameter estimates. The choice of the reference plant lies at the heart of any MRAC design and any acceptable selection must essentially satisfy two crucial requirements. The first of these requirements is that the reference model must accurately reflect the closed-loop performance specifications such as rise time, settling time, overshoot, and other transient performance characteristics. The other requirement is that given the assumed structure of the plant dynamics, the reference model's output behavior should be asymptotically achievable by the adaptive control system implying certain extra conditions on the relative degree of the reference model and persistent excitation conditions on the reference input. The controller structure is dictated by the certainty equivalence

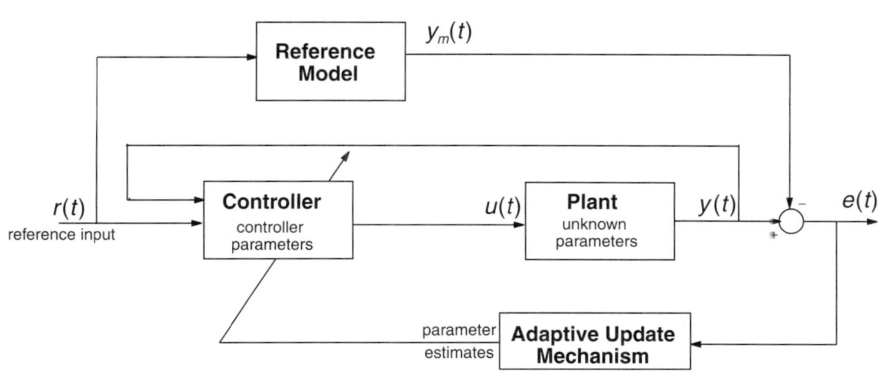

FIGURE 11.1 The model reference adaptive control framework.

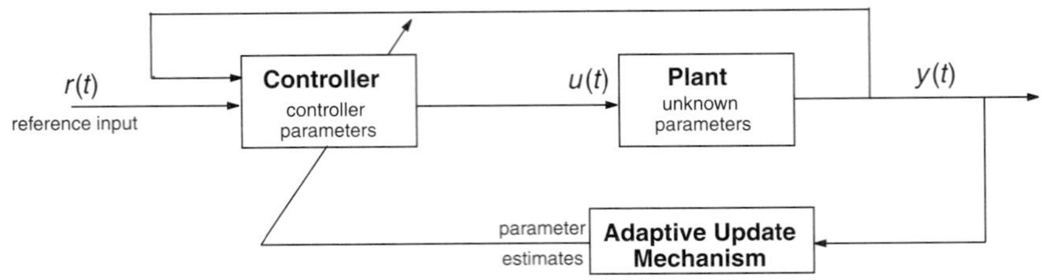

FIGURE 11.2 The self-tuning control architecture.

principle and both direct and indirect parameter update procedures can be adopted within the MRAC framework. Much of the work in this area deals with continuous time systems.

11.4.5 Self-Tuning Controller (STC)

In contrast to MRAC, there is no reference model in the self-tuning controller (STC) design. A schematic sketch is shown in Figures 11.2. In this formulation, the controller parameters of the plant parameters are estimated in real time, depending on whether it is a direct or indirect approach. These estimates are then used as if they are equal to the true parameters (certainty equivalence design). Parameter estimation involves finding the *best-fit* set of parameters based on the plant input–output data. This is different from the MRAC parameter adaptation scheme, where the parameter estimates are updated in such a way to achieve asymptotic tracking between the tracking error between the plant and the reference model. In several STC estimation schemes, it is also possible to quantify a measure of the quality of the parameter estimates, which can be used in the design of the controller. Many different combinations of the estimation methods can be adopted and can be applied to both continuous time and discrete time plants. Due to the "separation" between parameter estimation and control in STC, there is greater fiexibility in design. However, stability and convergence are difficult to prove and stronger conditions on input signals are required (persistent excitation) to guarantee parameter convergence. Historically speaking, STC designs arose in the study of the stochastic regulation problem and much of the literature is devoted to discrete time plants using an indirect approach. In spite of the seeming difference between MRAC and STC, a direct correspondence exists between problems from both the areas.[8]

11.5 Nonlinear Adaptive Control Systems

For the most general case of nonlinear systems, there exists very limited theory in the field of adaptive control. Even though there is great interest in this area due to potential applications in a wide variety of complex machanical systems, theoretical diffculties exist because of the lack of general analysis tools. However, some important special cases are well understood by now, and we summarize the conditions that these classes of systems satisfy:

1. The unknown parameters within the nonlinear plant are linearly parameterized.
2. The complete state vector is measured.
3. When the unknown parameters are assumed known, the control input can cancel all the nonlinearities in a feedback-linearization sense and any remaining internal dynamics should be stable. The adaptive design is then accomplished by certainty equivalence.

We now show a typical nonlinear MRAC methodology to deal with a situation in which the nonlinear plant model has unknown parameters. Consider the nonlinear system

$$\dot{x} = \theta f(x) + u \tag{11.5}$$

where θ is a constant and unknown matrix parameter, and f is a known and differentiable nonlinear vector function. In analogy with the MRAC methodology, we assume that it is desired to have the state x asymptotically track the state x_m of a reference system that satisfies

$$\dot{x}_m = A_m x_m + r \tag{11.6}$$

where $r(t)$ is any piecewise continuous and bounded reference input and A_m is a Hurwitz matrix. Introduce an error vector $e = x - x_m$ so that the error dynamics can be established by taking the difference between Equations 11.5 and 11.6 as follows:

$$\dot{e} = \theta f(x) - A_m x_m - r + u \tag{11.7}$$

If the parameter θ is assumed to be known, selecting the control input $u = A_m x + r - \theta f(x)$ would render the following structure for the error dynamics:

$$\dot{e} = A_m e$$

which would achieve the control objective. However, such a choice of control law is not impossible because θ is unknown. Hence we retain the same structure for the control law except for replacing θ by its time-varying estimate $\hat{\theta}$ so that the certainty-equivalence-based adaptive control law is given by

$$u = A_m x + r - \hat{\theta} f(x) \tag{11.8}$$

Application of the control law in Equation 11.7 leads us to the following closed-loop error dynamics:

$$\dot{e} = A_m e - \tilde{\theta} f(x) \tag{11.9}$$

where we have introduced the variable $\tilde{\theta}(t)$ to represent the parameter estimation error $\hat{\theta}(t) - \theta$. There are two things remaining to be done: (i) to show the stability and asymptotic convergence of $e(t)$ to zero as $t \to \infty$, (ii) to provide an appropriate parameter adaptation mechanism for $\hat{\theta}(t)$. We accomplish both these tasks by adopting the Lyapunov method. Given that A_m is Hurwitz, for any choice of symmetric and positive definite matrix Q, there exists a symmetric, positive definite matrix P that satisfies the Lyapunov equation given in Equation 11.3. Choosing a Lyapunov function in terms of such a P matrix,

$$V = e^T P e + \mathrm{tr}[\tilde{\theta}^T \Gamma^{-1} \tilde{\theta}] \tag{11.10}$$

where Γ is a symmetric positive definite learning rate matrix. Taking the time derivative of V along the solutions of Equation 11.9 we find that

$$\dot{V} = e^T (P A_m + A_m^T P) e - 2 e^T P \tilde{\theta} f(x) + 2\,\mathrm{tr}[\tilde{\theta}^T \Gamma^{-1} \dot{\hat{\theta}}] \tag{11.11}$$

Using several matrix trace identities,[9] it is possible to show that

$$e^T P \tilde{\theta} f(x) = \mathrm{tr}[P \tilde{\theta} f(x) e^T] = \mathrm{tr}[\tilde{\theta} f(x) e^T P] = \mathrm{tr}[\tilde{\theta}^T P e f^T(x)]$$

so that we can combine the last two terms on the right-hand side of Equation 11.11 as follows:

$$\dot{V} = e^T \underbrace{(P A_m + A_m^T P)}_{-Q} e + 2\,\mathrm{tr}\,[\tilde{\theta}^T \{\Gamma^{-1} \dot{\hat{\theta}} - P e f^T(x)\}] \tag{11.12}$$

Since θ is constant, $\dot{\theta} = 0$ and $\dot{\hat{\theta}} = \dot{\tilde{\theta}}$. Thus, if the adaptive law for updating $\hat{\theta}$ is chosen as

$$\dot{\hat{\theta}} = \Gamma P e f^T(x) \tag{11.13}$$

then the derivative of the Lyapunov function in Equation 11.12 becomes

$$\dot{V} = -e^T Q e \tag{11.14}$$

which is negative semidefinite, but not negative definite. This implies that $V(t) \leq V(0)$ for all $t \geq 0$, and thus, e and $\tilde{\theta}$ must be bounded. This further implies that $x = e + x_m$ also is bounded. Also, $V \geq 0$ and $\dot{V} \leq 0$, which means $\lim_{t\to\infty} V(t) \doteq V_\infty$ exists and is finite. Now,

$$\int_0^\infty \dot{V}(\tau)d\tau = -\int_0^\infty e^T(\tau)Qe(\tau)d\tau = V_\infty - V(0)$$

implying that $e \in \mathcal{L}_2 \cap \mathcal{L}_\infty$. From Equation 11.9, it is obvious that $\dot{e} \in \mathcal{L}_\infty$. Thus, we can invoke Barbalat's lemma to claim that $e(t) \to 0$ as $t \to \infty$. Notice, however, that the parameter error $\tilde{\theta} = \hat{\theta} - \theta$ will not necessarily converge to zero. True parameter convergence can occur only when the reference input $r(t)$ satisfies certain uniform observability and persistent excitation conditions.[4]

Remark: Note in the above MRAC design that while stability and tracking error convergence are guaranteed for any value of A_m, Q, and Γ, the performance of the controller will depend critically on the learning rate Γ. "Smaller" learning rates mean that the adaptation will be slow leading to large tracking errors and large transients. Conversely, the upper limit on the learning rate is limited by the presence of unmodeled dynamics, because too large a value for the learning rate will lead to highly oscillatory parameter estimates that can adversely excite the high frequency unmodeled plant dynamics.

Remark: The controller design methodology is based upon three crucial steps: (i) finding the appropriate controller structure in the spirit of feedback linearization, (ii) derivation of the tracking error dynamics that depend upon the parameter error terms, and (iii) finding a suitable Lyapunov function that can be used to derive the parameter update law such that the tracking error will go to zero. Determining the controller structure for the known parameter case is probably the most crucial step within any adaptive design because it turns out that adaptive feedback linearization cannot be alway applied to systems that are linearizable by feedback in the known parameter case. This happens because higher derivatives of the parameter estimates appear in the control law for systems of higher order, making difficult the application of the certainty equivalence principle. Other relatively new approaches deviate from the conventional certainty equivalence principles by adopting integrator backstepping, nonlinear damping, and tuning functions.[10] In these methods, the adaptive law estimates the unknown plant parameters directly, thereby permitting full utilization of any prior knowledge and therefore eliminating the possibility for overparameterization introduced by traditional direct MRAC methods. The design methodology and stability proof are obtained through a recursive process,[10,11] an overview for which can be obtained from Kokotovic[12] and subsequent results by his research group.

Remark: Given the fact that there always exist model errors and other unknown disturbance effects in addition to the unknown parameters, adaptive control solutions would have to address the robustness question. Since the parameter error is always unknown, the Lyapunov function time derivative is always negative semidefinite. This implies that the closed-loop equations are not exponentially stable, nor even uniformly asymptotically stable. Any external or unmodeled disturbance would immediately make $\dot{V}$ indefinite, and most methods that modify the stability proof for robustness are fixed to ensure $\dot{V}$ to be negative outside a compact neighborhood of the equilibrium state. By introducing an additional term in the adaptive law Equation 11.13 (referred to as σ-modification), Ioannou[7] accomplishes robust stability. This method, though very popular, suffered from the drawback that when the disturbance is absent, the tracking error would not converge to zero. To overcome this problem, other schemes such as the $\square$-modification[6] have been suggested to ensure robustness within the adaptive designs.

11.6 Spacecraft Adaptive Attitude Regulation Example

Consider the problem of a rigid spacecraft with an initial nonzero attitude and body angular velocity vector that has to be brought to rest at a zero attitude vector. This rigid body adaptive attitude regulation problem based on the feedback linearization approach has been derived by Schaub, Akella, and Junkins.[13] The governing equations are described by Euler's rotational equations of motion and the desired linear closed-loop dynamics (LCLD) can be of either PD or PID form.[13, 14] Only a crude estimate of the moment of inertia matrix is assumed to be known. An adaptive control law is presented, which includes an integral feedback term in the desired closed-loop dynamics and achieves asymptotic stability even in the presence of unmodeled external disturbances.

The resulting simulation is illustrated in Figure 11.3. The attitude vector is specified in terms of the modified Rodrigues parameter (MRP) whose components σ_i are shown in Figure 11.3a. Without any adaptation, the open-loop control is still asymptotically stable. However, the transient attitude errors don't match those of the desired LCLD well at all. With adaptation turned on, the performance matches that of the ideal LCLD very closely.

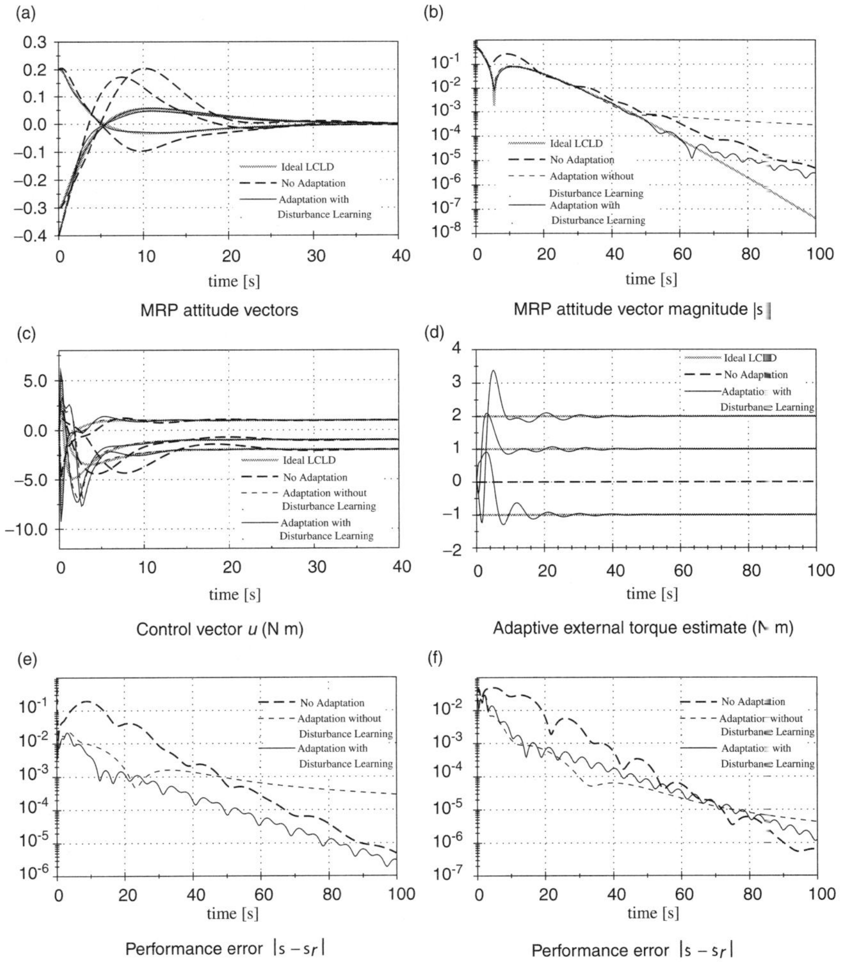

FIGURE 11.3 Rigid body stabilization while enforcing LCLD in the presence of large inertia and external disturbance ignorance.

Figure 11.3b shows the magnitude of the MRP attitude error vector σ on a logarithmic scale. Again the large transient errors of the open-loop, adaptation-free control law are visible during the first 20 s of the maneuver along with the good final convergence characteristics. The ideal LCLD performance is indicated again through the dotted line. Two versions of the adaptive control law are compared here, which differ only by whether or not the external disturbance is adaptively estimated too. On this figure both adaptive laws appear to enforce the desired LCLD very well for the first 40 s of the maneuver. After this the adaptive law without disturbance learning starts to decay at a slower rate, slower even than the open-loop (nonadaptive) solution. Including the external disturbance, adaptation clearly improves the final convergence rate. Note, however, that neither adaptive case starts to deviate from the ideal LCLD case until the MRP attitude error magnitude has decayed to roughly 10^{-3}. This corresponds to having a principal rotation error of roughly 0.23°. With external disturbance adaptation, the tracking error at which the LCLD deviations appear is about two orders of magnitude smaller.

The performance of the adaptive control law can be greatly varied by choosing different learning rates. However, since *large* initial inertia matrix and external disturbance model errors are present, the adaptive learning rates were reduced to avoid radical transient torques. The control torque vector components u_i for various cases are shown in Figure 11.3c. The open-loop torques don't approach the ideal LCLD torque during the transient part of the maneuver. The torques required by either adaptive case are very similar. The difference is that the case with external disturbance learning is causing some extra oscillation of the control about the LCLD case. However, note that with the chosen adaptive learning rates neither control law exhibits any radical transient torques about the ideal LCLD torque profile. Figure 11.3d illustrates that the adaptive external disturbance estimate F_e indeed asymptotically approaches the true external disturbance F_e^*. By reducing the external disturbance adaptive learning rate γ_{F_e} the transient adaptive estimate errors are kept within a reasonable range.

Figure 11.3f shows the absolute performance error in attitude rates. Both cases with adaptation added show large reductions in attitude rate errors compared to the nonadaptive case.

The purpose of the adaptive control discussed in this example is to enforce the desired LCLD. The previous figures illustrate that the resulting overall system remains asymptotically stable. Figure 11.3e illustrates the absolute performance error between the actual motion $\sigma(t)$ and the desired linear reference motion $\sigma_r(t)$. This figure demonstrates again the large performance error that results from using the open-loop control law with the incorrect system model. Adding adaptation improves the transient performance tracking by up to two orders of magnitude. Without including the external disturbance learning, the final performance error decay rate flattens out. This error will decay to zero. However, with the given learning gains, it does so at a slower rate than if no adaptation is taking place. Adding the external disturbance learning greatly improves the final performance error decay since the system is obtaining an accurate model of the actual constant disturbance. If the initial model estimates were more accurate, more aggressive adaptive learning rates could be used, resulting in even better LCLD performance tracking. This simulation illustrates though that even in the presence of large system uncertainty it is possible to track the desired LCLD very well.

11.7 Output Feedback Adaptive Control

In contrast to the state-space approaches, the input–output approach treats the plant as a black box that transforms the applied inputs into the corresponding output space. Stability theory for nonlinear systems from an input–output viewpoint is important in the context of adaptive output feedback control design. Solution to the problem of adaptive observer design involving state estimation of systems with unknown parameters is often the stepping stone towards resolving the output feedback control problem. There has been fairly recent breakthroughs in this area where the nonlinear adaptive observer design procedure has been extended to a slightly more general case of systems where the coefficients of the unknown parameters can depend on the entire state, and not just on the measured part.[15]

To a large extent, some powerful results have been made possible by exploiting certain "passivity-like"conditions coupled with the usual persistent excitation conditions. Crucial to this discussion is the concept of passivity, which is really an abstract representation of the idea of energy dissipation in both linear and nonlinear systems. Passive systems are most common in mechanical and electrical engineering applications. A mechanical system consisting of masses, springs, and viscous dashpots is a common example for a passive system. We now give the following definitions.

Definition: **Truncation of a signal**
Let Y be the space of real-valued functions defined on $[0, \infty)$. Let x be an element of Y. Then the *truncation* of x at some $T > 0$ is defined by

$$x_T(t) = \begin{cases} x(t) & \text{for } 0 \le t \le T \\ 0 & \text{for } t > T \end{cases}$$

Definition: **Extended space**
If X is a normed linear subspace of Y, then the *extended space* X_e is defined by the set

$$\{x \in Y : x_T \in X \text{ for some fixed } T \ge 0\}$$

The extended $\mathcal{L}_2$ space is denoted by $\mathcal{L}_{2e}$.

Definition: **Scalar product between two signals**
The scalar product between two real-valued time signals $x, y \in \mathcal{L}_{2e}$ is defined as

$$\langle x \mid y \rangle = \int_0^\infty x^T(\tau)y(\tau)\, d\tau = \int_0^T x^T(\tau)y(\tau)\, d\tau$$

Definition: **Passive systems**
A system with input $u(t)$ and output $y(t)$ is *passive* if

$$\langle y | u \rangle \ge 0$$

The system is *input strictly passive* if $\exists\ \square > 0$ such that

$$\langle y | u \rangle \ge e \| u \|^2$$

The system is said to be *output strictly passive* if $\exists\ \square > 0$ such that

$$\langle y | u \rangle \ge e \| y \|^2$$

11.8 Adaptive Observers and Output Feedback Control

We now state the nonlinear adaptive observer problem formulated by Besancon:[15]

$$\begin{aligned} \dot{x} &= f(x, u, t) + g(x, u, t)\theta \\ y &= h(x) \end{aligned} \tag{11.15}$$

where functions f and g are C^∞ with respect to all their arguments and θ is a constant and unknown parameter. Variables x, u, and y respectively denote the state, input, and output vectors. The input signals may be assumed to belong to some set of measurable and bounded functions. By the phrase adaptive observer, we imply the problem of reconstructing a state estimate $\hat{x}(t)$ using the input u and output y

in the presence of the unknown parameter θ such that $\lim_{t\to\infty}\|\hat{x}(t) - x(t)\| = 0$. The conditions for the existence of such an observer are now available,[15] which can be stated as follows. If a corresponding observer exists in the case when θ is known, and if this deterministic case observer is such that when a parameter error $\tilde{\theta} \doteq \hat{\theta} - \theta$ is made and the state estimation error system is passive between the "input" $\tilde{\theta}$ and the output error $h(\hat{x}) - y$, then an asymptotic state observer can be designed even when θ is unknown. In addition to this passivity requirement, parameter error convergence, as usual, would further need persistence of excitation with respect to u.

This powerful result finds immediate applications within the problem of spacecraft attitude tracking in the absence of angular velocity measurements.[16] It is now well known that the governing equations of the rigid-body attitude control problem in terms of the MRP vector satisfy certain passivity conditions[17, 18] between the angular velocity vector and the MRP vector. A very important consequence of passivity in this context is the fact that feedback control laws for attitude control can be implemented in a Lyapunov-based construction without requiring angular velocity measurements. In such a case, the only signal needed for feedback purposes would be the attitude vector. The resulting control laws provide *almost* global asymptotic stability in the sense of Tsiotras.[18]

11.9 Concluding Remarks

Historically speaking, the development and application of modern adaptive control theory for generic nonlinear systems adopted the philosophical approach of extending existing linear system methodologies. In some limited cases such as regulator theory, this approach of paralleling linear system methods has been highly successful. However, obtaining the same degree of success has been elusive in other research areas such as trajectory tracking, controller synthesis, and state reconstruction.

It is not difficult to fathom the reason for this bottleneck. Nonlinear systems occur in a vast variety of ways, and not all of them can be handled by simple extensions to existing linear adaptive control methodologies. One promising approach for the purpose of future research would be to specialize the study to mechanical systems, thereby restricting the class of nonlinear systems considered, and thus enabling the introduction of "structure" and additional constraints. Whereas in the case of output feedback control for general nonlinear systems, separate designs of stable observers and controllers do not necessarily guarantee stability for their combination (no separation principle), some structured approaches utilizing state transformations have already been shown to help recover the separation properties in some cases.[15] As a result, these so-called structured approaches also enabled the formulation of global and semi-global tracking controllers based on output (partial state) feedback. It is quite possible that a focused pursuit of the same approach has the potential for providing a key to solving several other problems arising out of electromechanical systems that are otherwise intractable.

References

1. Narendra, K. S., "Parameter adaptive control—The End ... or The Beginning?," *Proceedings of the 33rd Conference on Decision and Control.* Lake Buena Vista, Florida, December 1994.
2. Slotine, J. E. and Li, W., *Applied Nonlinear Control.* Prentice-Hall, Englewood Cliffs, NJ, 1991.
3. Khalil, H. K., *Nonlinear Systems.* Macmillan, New York, NY, 1992.
4. Sastry, S. and Bodson, M., *Adaptive Control: Stability, Convergence and Robustness.* Prentice-Hall, 1989.
5. Tao, G., "A simple alternative proof to the Barbalat Lemma," *IEEE Transactions on Automatic Control,* Vol. 42, No. 5, May 1997, p. 698.
6. Narendra, K. S. and Annaswamy, A. M., *Stable Adaptive Systems.* Prentice-Hall, 1989.
7. Ioannou, P. A. and Sun, J., *Stable and Robust Adaptive Control.* Prentice-Hall, Upper Saddle River, NJ, 1995, pp. 85–134.
8. Astrom, K. J. and Wittenmark, B., *Adaptive Control.* Addison-Wesley, Reading, MA, 1995.
9. Gantmacher. *The Theory of Matrices,* Vol. I. Chelsea Publishing Company, NY, 1977, pp. 353–354.

10. Krstić, M., Kanellakopoulos, I., and Kokotović, P. V., "Transient performance improvement with a new class of adaptive controllers," *Systems & Control Letters,* Vol. 21, 1993, pp. 451–461.
11. Krtić, M., Kanellakopoulos, I., and Kokotović, P. V., "Nonlinear design of adaptive controllers for linear systems," *IEEE Transactions on Automatic Control,* Vol. 39, 1994, pp. 738–752.
12. Kokotovic, P. V., "The joy of feedback: nonlinear and adaptive control," *IEEE Control Systems Magazine,* Vol. 12, No. 3, 1992, pp. 7–17.
13. Schaub, H., Akella, M. R., and Junkins, J. L., "Adaptive control of nonlinear attitude motions realizing linear closed loop dynamics," *Journal of Guidance, Control and Dynamics,* Vol. 24, No. 1, Jan.–Feb. 2001.
14. Akella, M. R., Schaub, H., and Junkins, J. L., "Adaptive realization of linear closed loop tracking dynamics in the presence of large system model errors," *Journal of Astronautical Sciences,* Vol. 48, No. 4, 2000.
15. Besançon, G., "Global output feedback tracking control for a class of Lagrangian systems," *Automatica,* Vol. 36, 2000, pp. 1915–1921.
16. Akella, M. R., "Rigid body attitude tracking without angular velocity feedback," *Systems & Control Letters,* Vol. 42, No. 4, 2001.
17. Lizarralde, F. and Wen, J. T., "Attitude control without angular velocity measurement: a passivity approach," *IEEE Transactions on Automatic Control,* Vol. 41, No. 3, 1996, pp. 468–472.
18. Tsiotras, P., "Further passivity results for the attitude control problem," *IEEE Transactions on Automatic Control,* Vol. 43, No. 11, 1998, pp. 1597–1600.

12

Neural Networks and Fuzzy Systems

Bogdan M. Wilamowski
Auburn University

12.1 Neural Networks and Fuzzy Systems

New and better electronic devices have inspired researchers to build intelligent machines operating in a fashion similar to the human nervous system. Fascination with this goal started when McCulloch and Pitts (1943) developed their model of an elementary computing neuron and when Hebb (1949) introduced his *learning rules.* A decade later Rosenblatt (1958) introduced the **perceptron** concept. In the early 1960s Widrow and Holf (1960, 1962) developed intelligent systems such as ADALINE and MADALINE. Nillson (1965) in his book *Learning Machines* summarized many developments of that time. The publication of the Mynsky and Paper (1969) book, with some discouraging results, stopped for some time the fascination with artificial neural networks, and achievements in the mathematical foundation of the **backpropagation** algorithm by Werbos (1974) went unnoticed. The current rapid growth in the area of neural networks started with the Hopfield (1982, 1984) recurrent network, Kohonen (1982) unsupervised training algorithms, and a description of the backpropagation algorithm by Rumelhart et al. (1986).

12.2 Neuron Cell

A biological neuron is a complicated structure, which receives trains of pulses on hundreds of *excitatory* and *inhibitory* inputs. Those incoming pulses are summed with different weights (averaged) during the time period of *latent summation.* If the summed value is higher than a threshold, then the neuron itself is generating a pulse, which is sent to neighboring neurons. Because incoming pulses are summed with time, the neuron generates a pulse train with a higher frequency for higher positive excitation. In other words, if the value of the summed weighted inputs is higher, the neuron generates pulses more frequently. At the same time, each neuron is characterized by the nonexcitability for a certain time after the firing pulse. This so-called *refractory period* can be more accurately described as a phenomenon where after excitation the threshold value increases to a very high value and then decreases gradually with a certain time constant. The refractory period sets soft upper limits on the frequency of the output pulse train. In the biological neuron, information is sent in the form of frequency modulated pulse trains.

This description of neuron action leads to a very complex neuron model, which is not practical. McCulloch and Pitts (1943) show that even with a very simple neuron model, it is possible to build logic and memory circuits. Furthermore, these simple neurons with thresholds are usually more powerful than typical logic gates used in computers. The McCulloch–Pitts neuron model assumes that incoming and outgoing signals may have only binary values 0 and 1. If incoming signals summed through positive or negative weights have a value larger than threshold, then the neuron output is set to 1. Otherwise, it is set to 0.

$$T = \begin{cases} 1, & \text{if } net \geq T \\ 0, & \text{if } net < T \end{cases} \tag{12.1}$$

where T is the threshold and *net* value is the weighted sum of all incoming signals:

$$net = \sum_{i=1}^{n} w_i x_i \tag{12.2}$$

Examples of McCulloch–Pitts neurons realizing OR, AND, NOT, and MEMORY operations are shown in Figure 12.1. Note that the structure of OR and AND gates can be identical. With the same structure, other logic functions can be realized, as Figure 12.2 shows.

The perceptron model has a similar structure. Its input signals, the weights, and the thresholds could have any positive or negative values. Usually, instead of using variable threshold, one additional constant input with a negative or positive weight can added to each neuron, as Figure 12.3 shows. In this case,

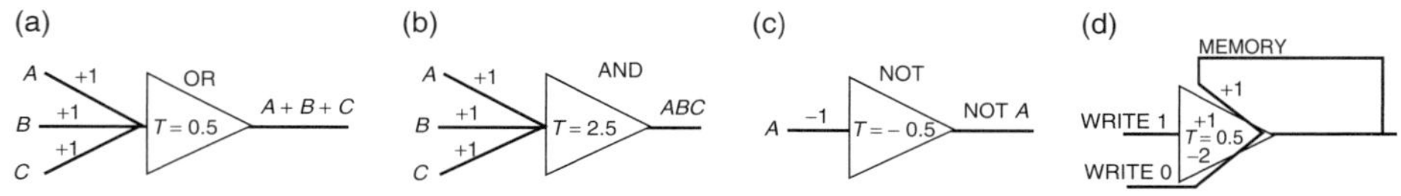

FIGURE 12.1 OR, AND, NOT, and MEMORY operations using networks with McCulloch–Pitts neuron model.

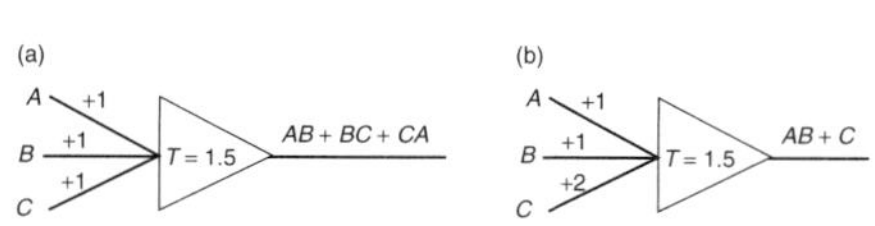

FIGURE 12.2 Other logic function realized with McCulloch–Pitts neuron model.

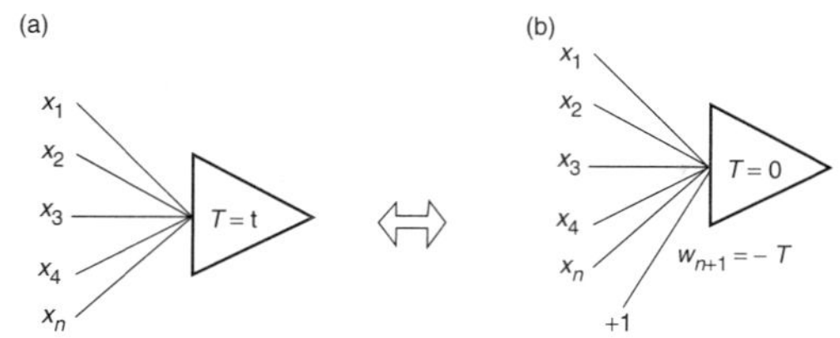

FIGURE 12.3 Threshold implementation with an additional weight and constant input with +1 value: (a) neuron with threshold T, (b) modified neuron with threshold $T = 0$ and additional weight equal to $-T$.

the threshold is always set to be zero and the net value is calculated as

$$net = \sum_{i=1}^{n} w_i x_i + w_{n+1} \tag{12.3}$$

where w_{n+1} has the same value as the required threshold and the opposite sign. Single-layer perceptrons were successfully used to solve many pattern classification problems. The hard threshold activation functions are given by

$$o = f(net) = \frac{\text{sgn}(net) + 1}{2} = \begin{cases} 1, & \text{if } net \geq 0 \\ 0, & \text{if } net < 0 \end{cases} \tag{12.4}$$

for **unipolar** neurons and

$$o = f(net) = \text{sgn}(net) = \begin{cases} 1, & \text{if } net \geq 0 \\ -1, & \text{if } net < 0 \end{cases} \tag{12.5}$$

for **bipolar** neurons. For these types of neurons, most of the known training algorithms are able to adjust weights only in single-layer networks.

Multilayer neural networks usually use continuous activation functions, either unipolar

$$o = f(net) = \frac{1}{1 + \exp(-\lambda net)} \tag{12.6}$$

or bipolar

$$o = f(net) = \tanh(0.5\lambda net) = \frac{2}{1 + \exp(-\lambda net)} - 1 \tag{12.7}$$

These continuous activation functions allow for the gradient-based training of multilayer networks. Typical activation functions are shown in Figure 12.4. In the case when neurons with additional threshold input are used (Figure 12.3b), the λ parameter can be eliminated from Equations 12.6 and 12.7 and the steepness of the neuron response can be controlled by the weight scaling only. Therefore, there is no real need to use neurons with variable gains.

Note, that even neuron models with continuous activation functions are far from an actual biological neuron, which operates with frequency modulated pulse trains.

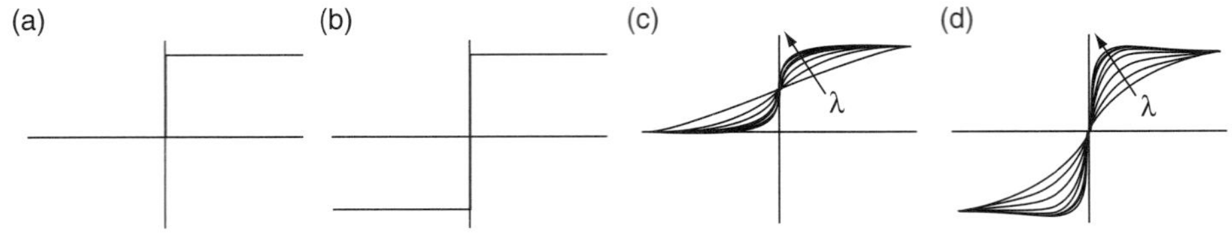

FIGURE 12.4 Typical activation functions: (a) hard threshold unipolar, (b) hard threshold bipolar, (c) continuous unipolar, (d) continuous bipolar.

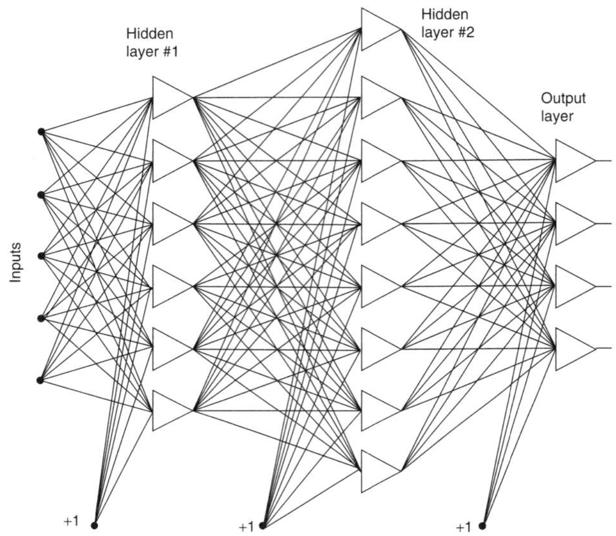

FIGURE 12.5 An example of the three-layer feedforward neural network, which is sometimes known also as the backpropagation network.

12.3 Feedforward Neural Networks

Feedforward neural networks allow only one-directional signal flow. Furthermore, most feedforward neural networks are organized in layers. An example of the three-layer feedforward neural network is shown in Figure 12.5. This network consists of input nodes, two *hidden layers*, and an output layer.

A single neuron is capable of separating input patterns into two categories, and this separation is linear. For example, for the patterns shown in Figure 12.6, the separation line is crossing x_1 and x_2 axes at points x_{10} and x_{20}. This separation can be achieved with a neuron having the following weights: $w_1 = 1/x_{10}$, $w_2 = 1/x_{20}$, and $w_3 = -1$. In general for n dimensions, the weights are

$$w_i = \frac{1}{x_{i0}} \quad \text{for } w_{n+1} = -1 \tag{12.8}$$

One neuron can divide only linearly separated patterns. To select just one region in n-dimensional input space, more than $n + 1$ neurons should be used. If more input clusters are to be selected, then the number of neurons in the input (hidden) layer should be properly multiplied. If the number of neurons in the input (hidden) layer is not limited, then all classification problems can be solved using the three-layer network. An example of such a neural network, classifying three clusters in the two-dimensional space, is shown in Figure 12.7. Neurons in the first hidden layer create the separation lines between input clusters.

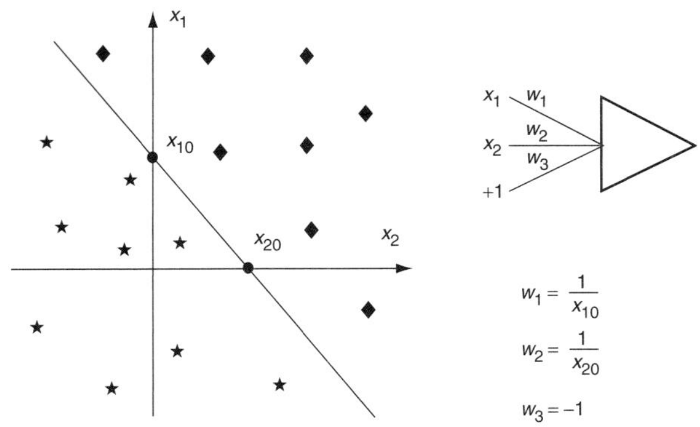

FIGURE 12.6 Illustration of the property of linear separation of patterns in the two-dimensional space by a single neuron.

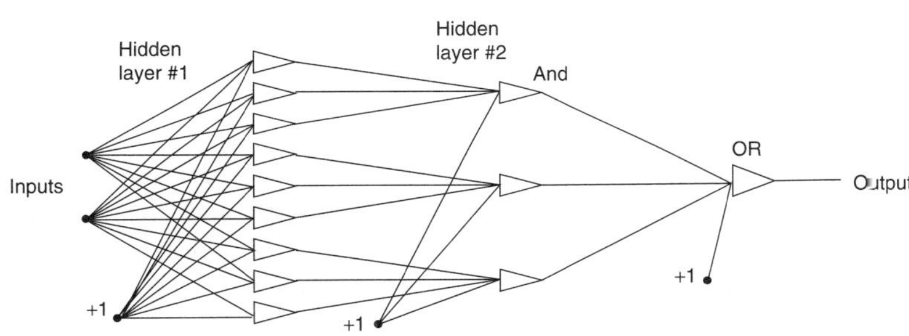

FIGURE 12.7 An example of the three-layer neural network with two inputs for classification of three different clusters into one category. This network can be generalized and can be used for solution of all classification problems.

Neurons in the second hidden layer perform the AND operation, as shown in Figure 12.1b. Output neurons perform the OR operation, as shown in Figure 12.1a, for each category. The linear separation property of neurons makes some problems especially difficult for neural networks, such as exclusive OR, parity computation for several bits, or to separate patterns laying on two neighboring spirals.

The feedforward neural network is also used for nonlinear transformation (mapping) of a multidimensional input variable into another multidimensional variable in the output. In theory, any input–output mapping should be possible if the neural network has enough neurons in hidden layers. (size of output layer is set by the number of outputs required). In practice, this is not an easy task. Presently, there is no satisfactory method to define how many neurons should be used in hidden layers. Usually, this is found by the trial-and-error method. In general, it is known that if more neurons are used, more complicated shapes can be mapped. On the other hand, networks with large numbers of neurons lose their ability for generalization, and it is more likely that such networks will also try to map noise supplied to the input.

12.4 Learning Algorithms for Neural Networks

Similarly to the biological neurons, the weights in artificial neurons are adjusted during a training procedure. Various learning algorithms were developed, and only a few are suitable for multilayer neuron networks. Some use only local signals in the neurons, others require information from outputs; some require a supervisor who knows what outputs should be for the given patterns, and other unsupervised algorithms need no such information. Common learning rules are described in the following sections.

12.4.1 Hebbian Learning Rule

The Hebb (1949) learning rule is based on the assumption that if two neighbor neurons must be activated and deactivated at the same time, then the weight connecting these neurons should increase. For neurons operating in the opposite phase, the weight between them should decrease. If there is no signal correlation, the weight should remain unchanged. This assumption can be described by the formula

$$\Delta w_{ij} = c x_i o_j \tag{12.9}$$

where

w_{ij} = weight from *i*th to *j*th neuron,
c = learning constant,
x_i = signal on the *i*th input,
o_j = output signal.

The training process starts usually with values of all weights set to zero. This learning rule can be used for both soft and hard threshold neurons. Since desired responses of neurons are not used in the learning procedure, this is the **unsupervised learning** rule. The absolute values of the weights are usually proportional to the learning time, which is undesired.

12.4.2 Correlation Learning Rule

The correlation learning rule is based on a similar principle as the Hebbian learning rule. It assumes that weights between simultaneously responding neurons should be largely positive, and weights between neurons with opposite reaction should be largely negative. Contrary to the Hebbian rule, the correlation rule is the **supervised learning**. Instead of actual response, o_j, the desired response, d_j, is used for the weight change calculation

$$\Delta w_{ij} = c x_i d_j \tag{12.10}$$

This training algorithm usually starts with initialization of weights to zero values.

12.4.3 Instar Learning Rule

If input vectors and weights are normalized, or they have only binary bipolar values (-1 or $+1$), then the *net* value will have the largest positive value when the weights and the input signals are the same. Therefore, weights should be changed only if they are different from the signals

$$\Delta w_i = c(x_i - w_i) \tag{12.11}$$

Note, that the information required for the weight is taken only from the input signals. This is a very local and unsupervised learning algorithm.

12.4.4 Winner Takes All (WTA)

The winner takes all (WTA) is a modification of the instar algorithm where weights are modified only for the neuron with the highest *net* value. Weights of remaining neurons are left unchanged. Sometimes this algorithm is modified in such a way that a few neurons with the highest net values are modified at the same time. Although this is an unsupervised algorithm because we do not know what are desired outputs, there is a need for a "judge" or "supervisor" to find a winner with a largest net value. The WTA algorithm, developed by Kohonen (1982), is often used for automatic clustering and for extracting statistical properties of input data.

12.4.5 Outstar Learning Rule

In the outstar learning rule, it is required that weights connected to a certain node should be equal to the desired outputs for the neurons connected through those weights

$$\Delta w_{ij} = c(d_j - w_{ij}) \tag{12.12}$$

where d_j is the desired neuron output and c is the small learning constant, which further decreases during the learning procedure. This is the supervised training procedure because desired outputs must be known. Both instar and outstar learning rules were developed by Grossberg (1969).

12.4.6 Widrow–Hoff LMS Learning Rule

Widrow and Hoff (1960, 1962) developed a supervised training algorithm, which allows training a neuron for the desired response. This rule was derived so the square of the difference between the net and output value is minimized.

$$Error_j = \sum_{p=1}^{P} (net_{jp} - d_{jp})^2 \tag{12.13}$$

where
$Error_j$ = error for jth neuron,
P = number of applied patterns,
d_{jp} = desired output for jth neuron when pth pattern is applied,
net = given by Equation 12.2.

This rule is also known as the least mean square (LMS) rule. By calculating a derivative of Equation 12.13 with respect to w_{ij}, a formula for the weight change can be found:

$$\Delta w_{ij} = cx_i \sum_{p=1}^{P} (d_{jp} - net_{jp}) \tag{12.14}$$

Note that weight change Δw_{ij} is a sum of the changes from each of the individual applied patterns. Therefore, it is possible to correct the weight after each individual pattern was applied. This process is known as *incremental updating; cumulative updating* is when weights are changed after all patterns have been applied. Incremental updating usually leads to a solution faster, but it is sensitive to the order in which patterns are applied. If the learning constant c is chosen to be small, then both methods give the same result. The LMS rule works well for all types of activation functions. This rule tries to enforce the *net* value to be equal to desired value. Sometimes this is not what the oberver is looking for. It is usually not important what the *net* value is, but it is important if the *net* value is positive or negative. For example, a very large *net* value with a proper sign will result in correct output and in large error as defined by Equation 12.13 and this may be the preferred solution.

12.4.7 Linear Regression

The LMS learning rule requires hundreds or thousands of iterations, using formula 12.14, before it converges to the proper solution. Using the linear regression rule, the same result can be obtained in only one step.

Considering one neuron and using vector notation for a set of the input patterns **X** applied through weight vector **w**, the vector of *net* values **net** is calculated using

$$\mathbf{Xw} = \mathbf{net} \tag{12.15}$$

where

$\mathbf{X}$ = rectangular array $(n + 1) \times p$,
n = number of inputs,
p = number of patterns.

Note that the size of the input patterns is always augmented by one, and this additional weight is responsible for the threshold (see Figure 12.3b). This method, similar to the LMS rule, assumes a linear activation function, and so the *net* values **net** should be equal to desired output values **d**

$$\mathbf{X}\mathbf{w} = \mathbf{d} \tag{12.16}$$

Usually $p > n + 1$, and the preceding equation can be solved only in the least mean square error sense. Using the vector arithmetic, the solution is given by

$$\mathbf{w} = (\mathbf{X}^T\mathbf{X})^{-1}\mathbf{X}^T\mathbf{d} \tag{12.17}$$

When traditional method is used, the set of p equations with $n + 1$ unknowns, Equation 12.16, has to be converted to the set of $n + 1$ equations with $n + 1$ unknowns

$$\mathbf{Y}\mathbf{w} = \mathbf{z} \tag{12.18}$$

where elements of the $\mathbf{Y}$ matrix and the $\mathbf{z}$ vector are given by

$$y_{ij} = \sum_{p=1}^{P} x_{ip}x_{jp}, \qquad z_i = \sum_{p=1}^{P} x_{ip}d_p \tag{12.19}$$

Weights are given by Equation 12.17 or they can be obtained by a solution of Equation 12.18.

12.4.8 Delta Learning Rule

The LMS method assumes linear activation function **net** = **0**, and the obtained solution is sometimes far from optimum, as is shown in Figure 12.8 for a simple two-dimensional case, with four patterns belonging to two categories. In the solution obtained using the LMS algorithm, one pattern is misclassified. If error

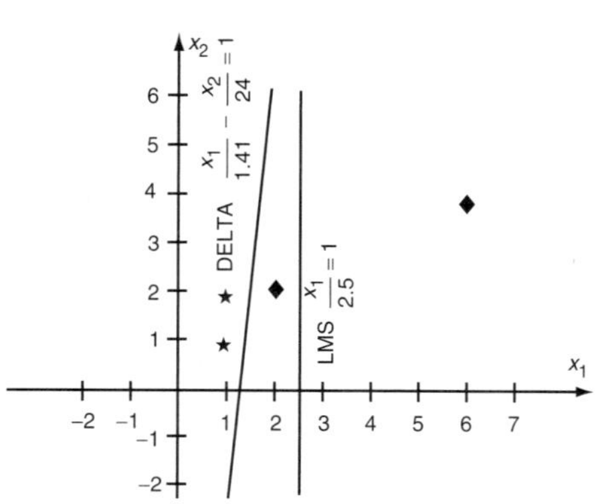

FIGURE 12.8 An example with a comparison of results obtained using LMS and delta training algorithms. Note that LMS is not able to find the proper solution.

is defined as

$$Error_j = \sum_{p=1}^{P} (o_{jp} - d_{jp})^2 \tag{12.20}$$

then the derivative of the error with respect to the weight w_{ij} is

$$\frac{d\,Error_j}{dw_{ij}} = 2\sum_{p=1}^{P} (o_{jp} - d_{jp}) \frac{df(net_{jp})}{d\,net_{jp}} x_i \tag{12.21}$$

since $o = f(net)$ and the *net* is given by Equation 12.2. Note that this derivative is proportional to the derivative of the activation function $f'(net)$. Thus, this type of approach is possible only for continuous activation functions and this method cannot be used with hard activation functions (12.4) and (12 5). In this respect the LMS method is more general. The derivatives' most common continuous activation functions are

$$f' = o(1 - o) \tag{12.22}$$

for the unipolar, Equation 12.6, and

$$f' = 0.5(1 - o^2) \tag{12.23}$$

for the bipolar, Equation 12.7.

Using the cumulative approach, the neuron weight w_{ij} should be changed with a direction of gradient

$$\Delta w_{ij} = cx_i \sum_{p=1}^{P} (d_{jp} - o_{jp}) f'_{jp} \tag{12.24}$$

In case of the incremental training for each applied pattern

$$\Delta w_{ij} = cx_i f'_j (d_j - o_j) \tag{12.25}$$

the weight change should be proportional to input signal x_i, to the difference between desired and actual outputs $d_{jp} - o_{jp}$, and to the derivative of the activation function f'_{jp}. Similar to the LMS rule, weights can be updated in both the incremental and the cumulative methods. In comparison to the LMS rule, the delta rule always leads to a solution close to the optimum. As it is illustrated in Figure 12.8, when the delta rule is used, all four patterns are classified correctly.

12.4.9 Error Backpropagation Learning

The delta learning rule can be generalized for multilayer networks. Using an approach similiar to the delta rule, the gradient of the global error can be computed with respect to each weight in the network. Interestingly,

$$\Delta w_{ij} = cx_i f'_j E_j \tag{12.26}$$

where

c = learning constant,
x_i = signal on the *i*th neuron input,
f'_j = derivative of activation function.

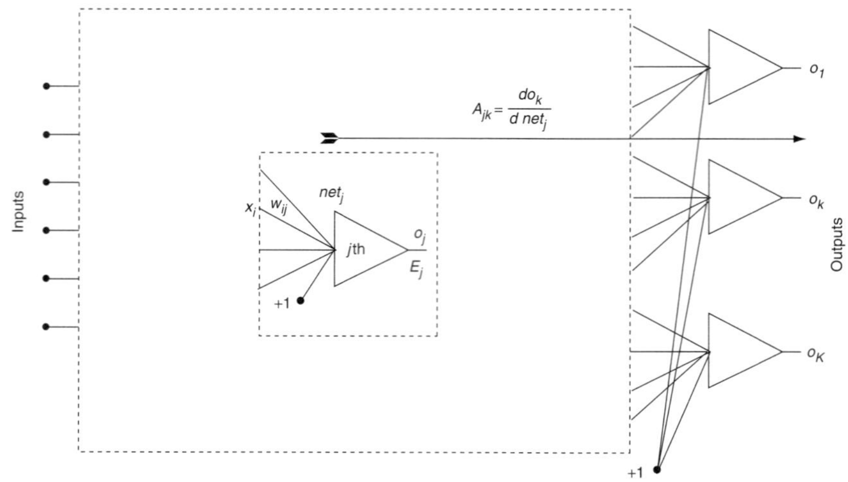

FIGURE 12.9 Illustration of the concept of gain computation in neural networks.

The cumulative error E_j on neuron output is given by

$$E_j = \frac{1}{f'_j}\sum_{k=1}^{K}(o_k - d_k)A_{jk} \tag{12.27}$$

where K is the number of network outputs and A_{jk} is the small signal gain from the input of the jth neuron to the kth network output, as Figure 12.9 shows. The calculation of the backpropagating error starts at the output layer and cumulative errors are calculated layer by layer to the input layer. This approach is not practical from the point of view of hardware realization. Instead, it is simpler to find signal gains from the input of the jth neuron to each of the network outputs (Figure 12.9). In this case, weights are corrected using

$$\Delta w_{ij} = cx_i\sum_{k=1}^{K}(o_k - d_k)A_{jk} \tag{12.28}$$

Note that this formula is general, regardless of whether the neurons are arranged in layers or not. One way to find gains A_{jk} is to introduce an incremental change on the input of the jth neuron and observe the change in the kth network output. This procedure requires only forward signal propagation, and it is easy to implement in a hardware realization. Another possible way is to calculate gains through each layer and then find the total gains as products of layer gains. This procedure is equally or less computationally intensive than a calculation of cumulative errors in the error backpropagation algorithm.

The backpropagation algorithm has a tendency for oscillation. To smooth the process, the weights increment Δw_{ij} can be modified according to Rumelhart et al. (1986):

$$w_{ij}(n+1) = w_{ij}(n) + \Delta w_{ij}(n) + \alpha\Delta w_{ij}(n-1) \tag{12.29}$$

or according to Sejnowski and Rosenberg (1987),

$$w_{ij}(n+1) = w_{ij}(n) + (1-\alpha)\Delta w_{ij}(n) + \alpha\Delta w_{ij}(n-1) \tag{12.30}$$

where α is the momentum term.

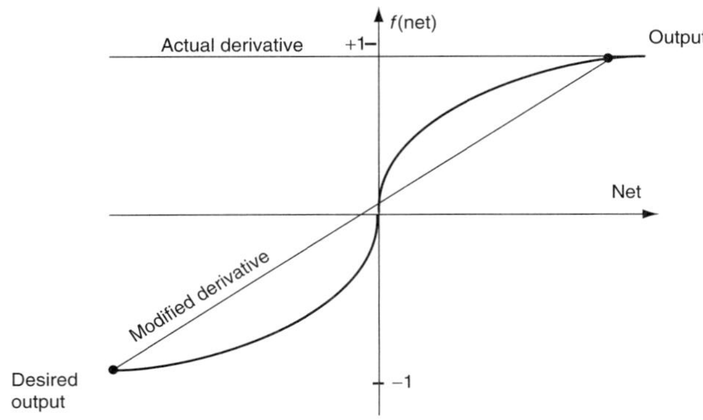

FIGURE 12.10 Illustration of the modified derivative calculation for faster convergency of the error backpropagation algorithm.

The backpropagation algorithm can be significantly sped up, when, after finding components of the gradient, weights are modified along the gradient direction until a minimum is reached. This process can be carried on without the necessity of a computationally intensive gradient calculation at each step. The new gradient components are calculated once a minimum is obtained in the direction of the previous gradient. This process is only possible for cumulative weight adjustment. One method of finding a minimum along the gradient direction is the *tree step process* of finding error for three points along gradient direction and then, using a parabola approximation, jump directly to the minimum. The fast learning algorithm using the described approach was proposed by Fahlman (1988) and is known as the *quickprop*.

The backpropagation algorithm has many disadvantages, which lead to very slow convergency. One of the most painful is that in the backpropagation algorithm, the learning process almost perishes for neurons responding with the maximally wrong answer. For example, if the value on the neuron output is close to +1 and desired output should be close to −1, then the neuron gain $f'(net) \approx 0$ and the error signal cannot backpropagate, and so the learning procedure is not effective. To overcome this difficulty, a modified method for derivative calculation was introduced by Wilamowski and Torvik (1993). The derivative is calculated as the slope of a line connecting the point of the output value with the point of the desired value, as shown in Figure 12.10.

$$f_{modif} = \frac{o_{desired} - o_{actual}}{net_{desired} - net_{actual}} \qquad (12.31)$$

Note that for small errors, Equation 12.31 converges to the derivative of activation function at the point of the output value. With an increase of system dimensionality, the chances for local minima decrease. It is believed that the described phenomenon, rather than a trapping in local minima is responsible for convergency problems in the error backpropagation algorithm.

12.5 Special Feedforward Networks

The multilayer backpropagation network, as shown in Figure 12.5, is a commonly used feedforward network. This network consists of neurons with the sigmoid type continuous activation function presented in Figures 12.4c and 12.4d. In most cases, only the one hidden layer is required, and the number of neurons in the hidden layer are chosen to be proportional to the problem complexity. The number of neurons in the hidden layer is usually found by a trial-and-error process. The training process starts with all weights randomized to small values, and the error backpropagation algorithm is used to find a solution. When the learning process does not converge, the training is repeated with a new set of randomly chosen weights.

Nguyen and Widrow (1990) proposed an experimental approach for the two-layer network weight initialization. In the second layer, weights are randomly chosen in the range from −0.5 to +0.5. In the first layer, initial weights are calculated from

$$w_{ij} = \frac{\beta z_{ij}}{\|z_j\|}, \qquad w_{(n+1)j} = \text{random}(-\beta, +\beta) \tag{12.32}$$

where z_{ij} is the random number from −0.5 to +0.5 and the scaling factor β is given by

$$\beta = 0.7P^{1/N} \tag{12.33}$$

where n is the number of inputs and N is the number of hidden neurons in the first layer. This type of weight initialization usually leads to faster solutions.

For adequate solutions with backpropagation networks, typically many tries are required with different network structures and different initial random weights. It is important that the trained network gains a generalization property. This means that the trained network also should be able to handle correctly patterns that were not used for training. Therefore, in the training procedure, often some data are removed from the training patterns and then these patterns are used for verification. The results with backpropagation networks often depend on luck. This encouraged researchers to develop feedforward networks, which can be more reliable. Some of those networks are described in the following sections.

12.5.1 Functional Link Network

One-layer neural networks are relatively easy to train, but these networks can solve only linearly separated problems. One possible solution for nonlinear problems was presented by Nilsson (1965) and was then elaborated by Pao (1989) using the functional link network shown in Figure 12.11. Using nonlinear terms with initially determined functions, the actual number of inputs supplied to the one-layer neural network is increased. In the simplest case, nonlinear elements are higher order terms of input patterns. Note that the functional link network can be treated as a one-layer network, where additional input data are generated off-line using nonlinear transformations. The learning procedure for one-layer is easy and fast. Figure 12.12 shows an XOR problem solved using functional link networks. Note that when the functional link approach is used, this difficult problem becomes a trivial one. The problem with the functional link network is that proper selection of nonlinear elements is not an easy task. In many practical cases, however, it is not difficult to predict what kind of transformation of input data may linearize the problem, and so the functional link approach can be used.

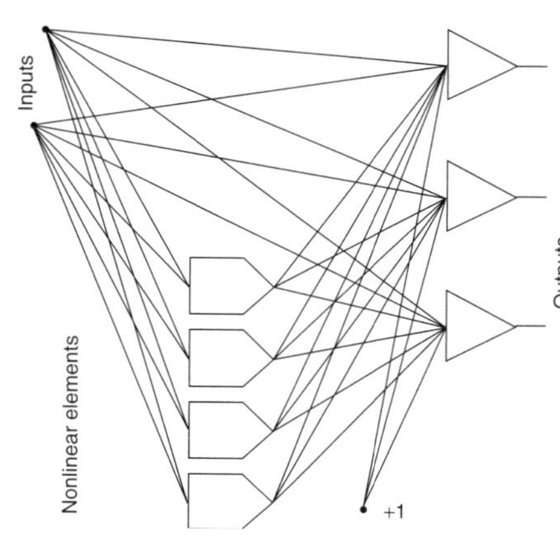

FIGURE 12.11 The functional link network.

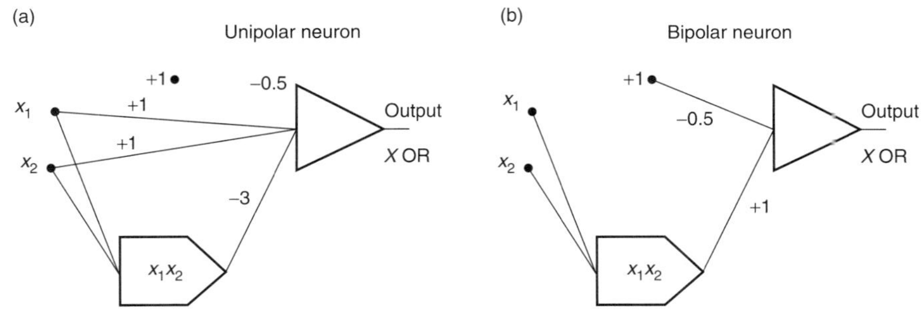

FIGURE 12.12 Functional link networks for solution of the X OR problem: (a) using unipolar signals, (b) using bipolar signals.

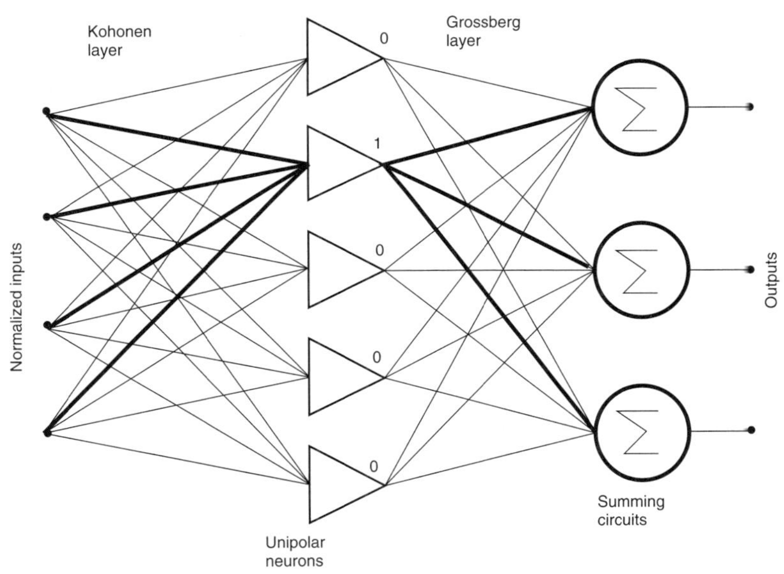

FIGURE 12.13 The counterpropagation network.

12.5.2 Feedforward Version of the Counterpropagation Network

The *counterpropagation network* was originally proposed by Hecht-Nilsen (1987). In this section a modified feedforward version as described by Zurada (1992) is discussed. This network, which is shown in Figure 12.13, requires numbers of hidden neurons equal to the number of input patterns, or more exactly, to the number of input clusters. The first layer is known as the Kohonen layer with unipolar neurons. In this layer only one neuron, the winner, can be active. The second is the Grossberg outstar layer. The Kohonen layer can be trained in the unsupervised mode, but that need not be the case. When binary input patterns are considered, the input weights must be exactly equal to the input patterns. In this case,

$$net = \mathbf{x}^t\mathbf{w} = [n - 2HD(\mathbf{x}, \mathbf{w})] \tag{12.34}$$

where

n = number of inputs,
$\mathbf{w}$ = weights,
$\mathbf{x}$ = input vector,
$HD(\mathbf{w}, \mathbf{x})$ = *Hamming distance* between input pattern and weights.

For a neuron in the input layer to be reacting just for the stored pattern, the threshold value for this neuron should be

$$w_{(n+1)} = -(n-1) \tag{12.35}$$

If it is required that the neuron must also react for similar patterns, then the threshold should be set to $w_{n+1} = -[n-(1 + \mathrm{HD})]$, where HD is the Hamming distance defining the range of similarity. Since for a given input pattern only one neuron in the first layer may have the value of 1 and remaining neurons have 0 values, the weights in the output layer are equal to the required output pattern.

The network, with unipolar activation functions in the first layer, works as a lookup table. When the linear activation function (or no activation function at all) is used in the second layer, then the network also can be considered as an analog memory. For the address applied to the input as a binary vector, the stored set of analog values, as weights in the second layer, can be accurately recovered. The feedforward counterpropagation network may also use analog inputs, but in this case all input data should be normalized,

$$\mathbf{w}_i = \hat{\mathbf{x}}_i = \frac{\mathbf{x}_i}{\|\mathbf{x}_i\|} \tag{12.36}$$

The counterpropagation network is very easy to design. The number of neurons in the hidden layer is equal to the number of patterns (clusters). The weights in the input layer are equal to the input patterns, and the weights in the output layer are equal to the output patterns. This simple network can be used for rapid prototyping. The counterpropagation network usually has more hidden neurons than required. However, such an excessive number of hidden neurons are also used in more sophisticated feedforward networks such as the *probabilistic neural network* (PNN) Specht (1990) or the *general regression neural networks* (GRNN) Specht (1992).

12.5.3 WTA Architecture

The winner take all (WTA) network was proposed by Kohonen (1988). This is basically a one-layer network used in the unsupervised training algorithm to extract a statistical property of the input data, Figure 12.14a. At the first step, all input data are normalized so that the length of each input vector is the same and, usually, equal to unity, Equation 12.36. The activation functions of neurons are unipolar and continuous. The learning process starts with a weight initialization to small random values. During the learning process the weights are changed only for the neuron with the highest value on the output—the winner:

$$\Delta \mathbf{w}_w = c(\mathbf{x} - \mathbf{w}_w) \tag{12.37}$$

where

$\mathbf{w}_w$ = weights of the winning neuron,
$\mathbf{x}$ = input vector,
c = learning constant.

Usually, this single-layer network is arranged into a two-dimensional layer shape, as shown in Figure 12.14b. The hexagonal shape is usually chosen to secure strong interaction between neurons. Also, the algorithm is modified in such a way that not only the winning neuron but also neighboring neurons are allowed for the weight change. At the same time, the learning constant c in Equation 12.37 decreases with the distance from the winning neuron. After such an unsupervised training procedure, the Kohonen layer is able to organize data into clusters. Output of the Kohonen layer is then connected to the one- or two-layer feedforward network with the error backpropagation algorithm. This initial data organization in the WTA layer usually leads to rapid training of the following layer or layers.

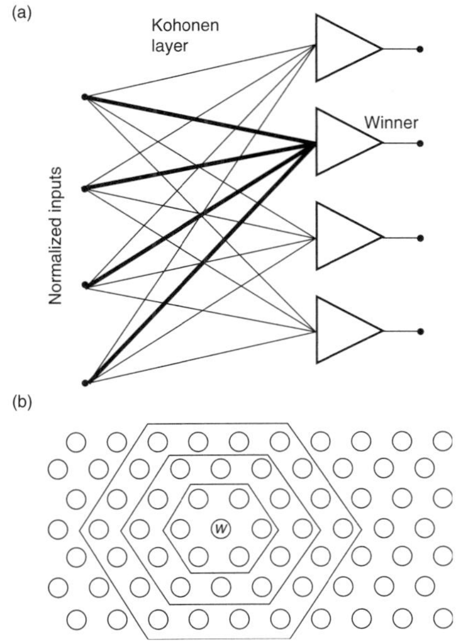

FIGURE 12.14 A winner take all architecture for cluster extracting in the unsupervised training mode: (a) network connections, (b) single-layer network arranged into a hexagonal shape.

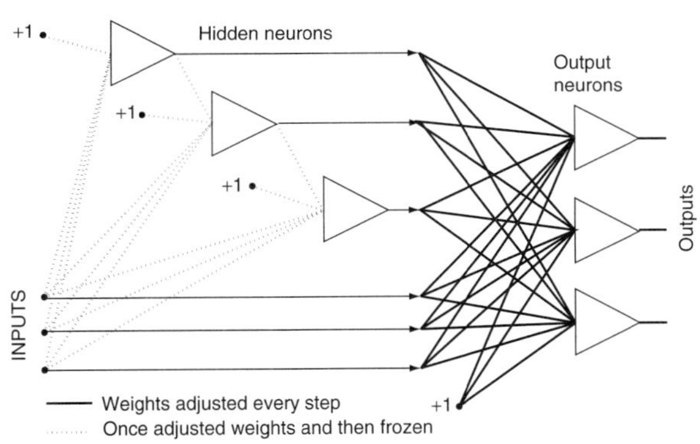

FIGURE 12.15 The cascade correlation architecture.

12.5.4 Cascade Correlation Architecture

The cascade correlation architecture was proposed by Fahlman and Lebiere (1990). The process of network building starts with a one-layer neural network and hidden neurons are added as needed. The network architecture is shown in Figure 12.15. In each training step, a new hidden neuron is added and its weights are adjusted to maximize the magnitude of the correlation between the new hidden neuron output and the residual error signal on the network output to be eliminated. The correlation parameter S must be maximized:

$$S = \sum_{o=1}^{O} \left| \sum_{p=1}^{P} (V_p - \bar{V})(E_{po} - \bar{E}_o) \right| \tag{12.38}$$

where

O = number of network outputs,
P = number of training patterns,
V_p = output on the new hidden neuron,
E_{po} = error on the network output.

$\bar{V}$ and $\bar{E}_o$ are average values of V_p and E_{po}, respectively. By finding the gradient, $\delta S/\delta w_i$, the weight, adjustment for the new neuron can be found as

$$\Delta w_i = \sum_{o=1}^{O}\sum_{p=1}^{P} \sigma_o (E_{po} - \bar{E}_o) f'_p x_{ip} \tag{12.39}$$

where

σ_o = sign of the correlation between the new neuron output value and network output,
f'_p = derivative of activation function for pattern p,
x_{ip} = input signal.

The output neurons are trained using the delta or quickprop algorithms. Each hidden neuron is trained just once and then its weights are frozen. The network learning and building process is completed when satisfactory results are obtained.

12.5.5 Radial Basis Function Networks

The structure of the radial basis network is shown in Figure 12.16. This type of network usually has only one hidden layer with special neurons. Each of these neurons responds only to the inputs signals close to the stored pattern. The output signal h_i of the ith hidden neuron is computed using formula

$$h_i = \exp\left(-\frac{\|\mathbf{x} - \mathbf{s}_i\|^2}{2\sigma^2}\right) \tag{12.40}$$

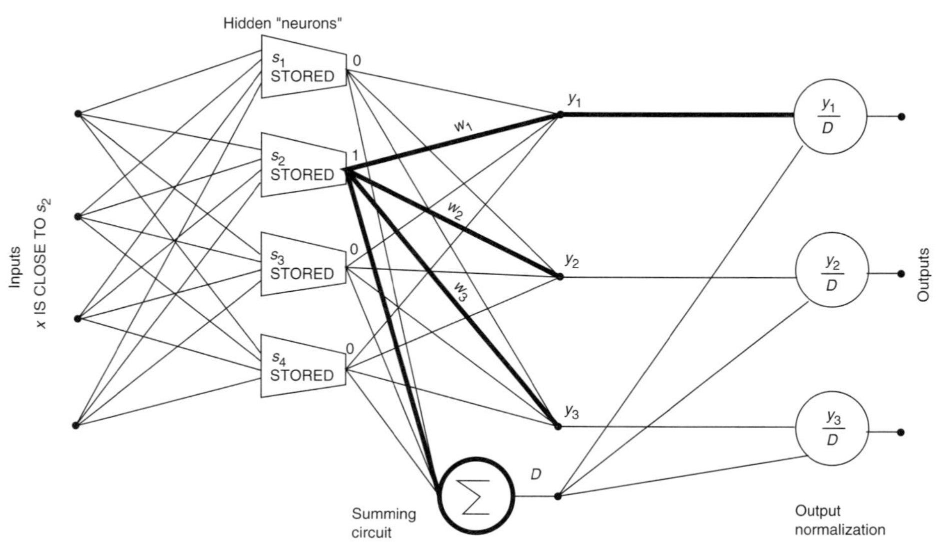

FIGURE 12.16 A typical structure of the radial basis function network.

where

$\mathbf{x}$ = input vector,

$\mathbf{s}_i$ = stored pattern representing the center of the i cluster,

σ_i = radius of the cluster.

Note that the behavior of this "neuron" significantly differs form the biological neuron. In this "neuron," excitation is not a function of the weighted sum of the input signals. Instead, the distance between the input and stored pattern is computed. If this distance is zero, the neuron responds with a maximum output magnitude equal to one. This neuron is capable of recognizing certain patterns and generating output signals that are functions of a similarity. Features of this neuron are much more powerful than a neuron used in the backpropagation networks. As a consequence, a network made of such neurons is also more powerful.

If the input signal is the same as a pattern stored in a neuron, then this neuron responds with 1 and remaining neurons have 0 on the output, as is illustrated in Figure 12.16. Thus, output signals are exactly equal to the weights coming out from the active neuron. This way, if the number of neurons in the hidden layer is large, then any input–output mapping can be obtained. Unfortunately, it may also happen that for some patterns several neurons in the first layer will respond with a nonzero signal. For a proper approximation, the sum of all signals from the hidden layer should be equal to one. To meet this requirement, output signals are often normalized, as shown in Figure 12.16.

The radial-based networks can be designed or trained. Training is usually carried out in two steps. In the first step, the hidden layer is usually trained in the unsupervised mode by choosing the best patterns for cluster representation. An approach similar to that used in the WTA architecture can be used. Also in this step, radii σ_i must be found for a proper overlapping of clusters.

The second step of training is the error backpropagation algorithm carried on only for the output layer. Since this is a supervised algorithm for one layer only, the training is very rapid, 100–1000 times faster than in the backpropagation multilayer network. This makes the radial basis-function network very attractive. Also, this network can be easily modeled using computers; however, its hardware implementation would be difficult.

12.6 Recurrent Neural Networks

In contrast to feedforward neural networks, with **recurrent networks** neuron outputs can be connected with their inputs. Thus, signals in the network can continuously circulate. Until recently, only a limited number of recurrent neural networks were described.

12.6.1 Hopfield Network

The single-layer recurrent network was analyzed by Hopfield (1982). This network, shown in Figure 12.17, has unipolar hard threshold neurons with outputs equal to 0 or 1. Weights are given by a symmetrical square matrix $\mathbf{W}$ with zero elements ($w_{ij} = 0$ for $i = j$) on the main diagonal. The stability of the system is usually analyzed by means of the *energy function*

$$E = -\frac{1}{2}\sum_{i=1}^{N}\sum_{j=1}^{N} W_{ij}v_iv_j \tag{12.41}$$

It has been proved that during signal circulation the energy E of the network decreases and the system converges to the stable points. This is especially true when the values of system outputs are updated in the asynchronous mode. This means that at a given cycle, only one random output can be changed to the required values. Hopfield also proved that those stable points, which the system converges, can be

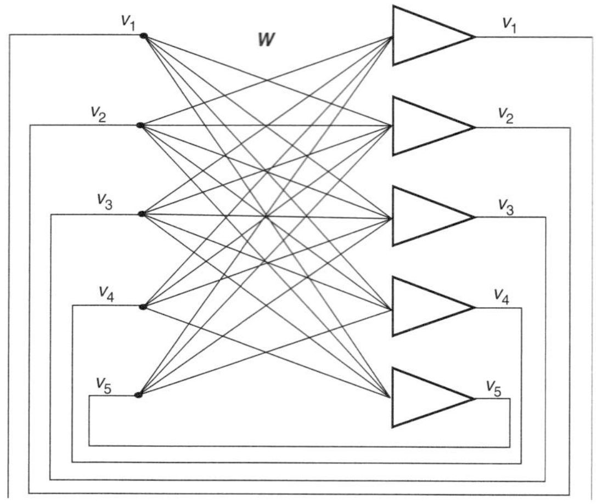

FIGURE 12.17 A Hopfield network or autoassociative memory.

programmed by adjusting the weights using a modified Hebbian rule,

$$\Delta w_{ij} = \Delta w_{ji} = (2v_i - 1)(2v_j - 1)c \tag{12.42}$$

Such memory has limited storage capacity. Based on experiments, Hopfield estimated that the maximum number of stored patterns is $0.15N$, where N is the number of neurons.

Later the concept of energy function was extended by Hopfield (1984) to one-layer recurrent networks having neurons with continuous activation functions. These types of networks were used to solve many optimization and linear programming problems.

12.6.2 Autoassociative Memory

Hopfield (1984) extended the concept of his network to autoassociative memories. In the same network structure as shown in Figure 12.17, the bipolar hard-threshold neurons were used with outputs equal to −1 or +1. In this network, pattern $\mathbf{s}_m$ are stored into the weight matrix **W** using the autocorrelation algorithm

$$W = \sum_{m=1}^{M} \mathbf{s}_m \mathbf{s}_m^T - MI \tag{12.43}$$

where M is the number of stored patterns and **I** is the unity matrix. Note that **W** is the square symmetrical matrix with elements on the main diagonal equal to zero (w_{ji} for $i = j$). Using a modified formula 12.42, new patterns can be added or subtracted from memory. When such memory is exposed to a binary bipolar pattern by enforcing the initial network states, after signal circulation the network will converge to the closest (most similar) stored pattern or to its complement. This stable point will be at the closest minimum of the energy

$$E(\mathbf{v}) = -\frac{1}{2}\mathbf{v}^T \mathbf{W} \mathbf{v} \tag{12.44}$$

Like the Hopfield network, the autoassociative memory has limited storage capacity, which is estimated to be about $M_{max} = 0.15N$. When the number of stored patterns is large and close to the memory capacity, the network has a tendency to converge to spurious states, which were not stored. These spurious states are additional minima of the energy function.

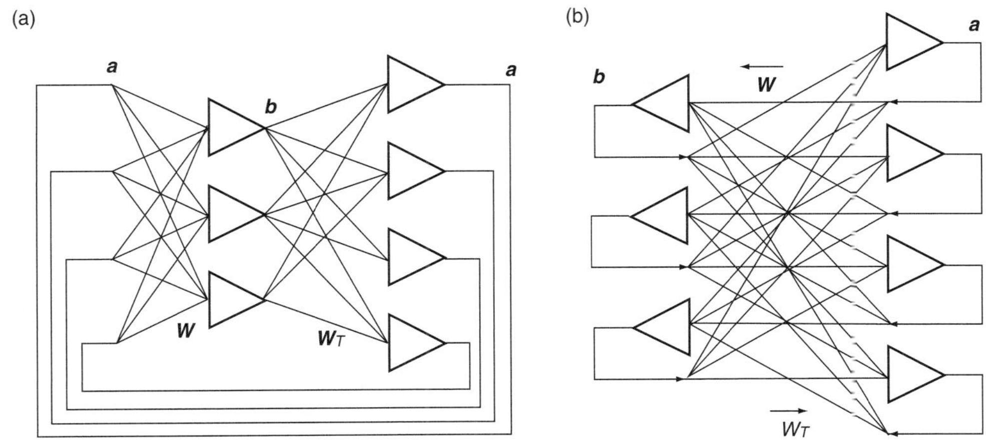

FIGURE 12.18 An example of the bidirectional autoassociative memory: (a) drawn as a two-layer network with circulating signals, (b) drawn as a two-layer network with bidirectional signal flow.

12.6.3 Bidirectional Associative Memories (BAM)

The concept of the autoassociative memory was extended to bidirectional associative memories (BAMs) by Kosko (1987, 1988). This memory, shown in Figure 12.18, is able to associate pairs of the patterns **a** and **b**. This is the two-layer network with the output of the second layer connected directly to the input of the first layer. The weight matrix of the second layer is $\mathbf{W}^T$ and **W** for the first layer. The rectangular weight matrix **W** is obtained as a sum of the cross-correlational matrices

$$\mathbf{W} = \sum_{m=1}^{M} \mathbf{a}_m \mathbf{b}_m \qquad (12.45)$$

where M is the number of stored pairs, and $\mathbf{a}_m$ and $\mathbf{b}_m$ are the stored vector pairs. If the nodes **a** or **b** are initialized with a vector similar to the stored one, then after signal circulations, both stored patterns $\mathbf{a}_m$ and $\mathbf{b}_m$ should be recovered. The BAM has limited memory capacity and memory corruption problems similar to the autoassociative memory. The BAM concept can be extended for association of three or more vectors.

12.7 Fuzzy Systems

The main applications of neural networks are related to the nonlinear mapping of n-dimensional input variables into m-dimensional output variables. Such a function is often required in control systems, where, for specific measured variables, certain control variables must be generated. Another approach for nonlinear mapping of one set of variables into another set of variables is the *fuzzy controller.* The principle of operation of the fuzzy controller significantly differs from neural networks. The block diagram of a fuzzy controller is shown in Figure 12.19. In the first step, analog inputs are converted into a set of fuzzy variables. In this step, for each analog input, 3–9 fuzzy variables typically are generated. Each fuzzy variable has an analog value between zero and one. In the next step, a fuzzy logic is applied to the input fuzzy variables and a resulting set of output variables is generated. In the last step, known as *defuzzification,* from a set of output fuzzy variables, one or more output analog variables are generated, which are used as control variables.

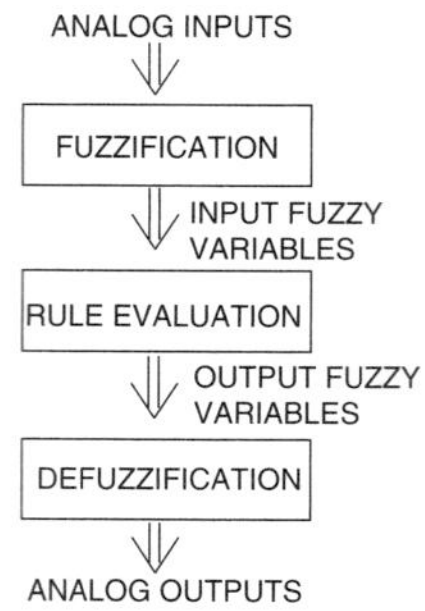

FIGURE 12.19 The block diagram of the fuzzy controller.

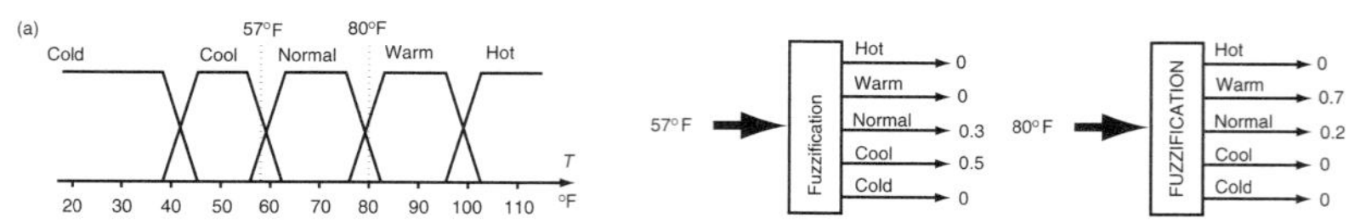

FIGURE 12.20 Fuzzification process: (a) typical membership functions for the fuzzification and the defuzzification processes, (b) example of converting a temperature into fuzzy variables.

12.7.1 Fuzzification

The purpose of fuzzification is to convert an analog variable input into a set of fuzzy variables. For higher accuracy, more fuzzy variables will be chosen. To illustrate the fuzzification process, consider that the input variable is the temperature and is coded into five fuzzy variables: cold, cool, normal, warm, and hot. Each fuzzy variable should obtain a value between zero and one, which describes a *degree of association* of the analog input (temperature) within the given fuzzy variable. Sometimes, instead of the term *degree of association*, the term *degree of membership* is used. The process of fuzzification is illustrated in Figure 12.20. Using Figure 12.20 we can find the degree of association of each fuzzy variable with the given temperature. For example, for a temperature of 57°F, the following set of fuzzy variables is obtained: [0, 0.5, 0.2, 0, 0], and for $T = 80$°F, it is [0, 0, 0.25, 0.7, 0]. Usually only one or two fuzzy variables have a value other than zero. In the example, trapezoidal functions are used for calculation of the degree of association. Various different functions such as triangular or Gaussian can also be used, as long as the computed value is in the range from zero to one. Each membership function is described by only three or four parameters, which have to be stored in memory.

For proper design of the fuzzification stage, certain practical rules should be used:

- Each point of the input analog variable should belong to at least one and no more than two membership functions.
- For overlapping functions, the sum of two membership functions must not be larger than one. This also means that overlaps must not cross the points of maximum values (ones).
- For higher accuracy, more membership functions should be used. However, very dense functions lead to frequent system reaction and sometimes to system instability.

12.7.2 Rule Evaluation

In contrary to boolean logic where variables can have only binary states, in fuzzy logic all variables may have any values between zero and one. The fuzzy logic consists of the same basic: ∧—AND, ∨—OR, and NOT operators:

$$A \wedge B \wedge C \quad \Rightarrow \min\{A, B, C\}\text{—smallest value of } A \text{ or } B \text{ or } C$$
$$A \vee B \vee C \quad \Rightarrow \max\{A, B, C\}\text{—largest value of } A \text{ or } B \text{ or } C$$
$$\bar{A} \quad \Rightarrow 1\ 1{-}A\text{—one minus value of } A$$

(a)

	y_1	y_2	y_3
x_1	z_1	z_1	z_2
x_2	z_1	z_3	z_3
x_3	z_2	z_4	z_4
x_4	z_1	z_2	z_3
x_5	z_1	z_2	z_4

(b)

	y_1	y_2	y_3
x_1	t_{11}	t_{12}	t_{13}
x_2	t_{21}	t_{22}	t_{23}
x_3	t_{31}	t_{32}	t_{33}
x_4	t_{41}	t_{42}	t_{43}
x_5	t_{51}	t_{52}	t_{53}

FIGURE 12.21 Fuzzy tables: (a) table with fuzzy rules, (b) table with the intermediate variables t_{ij}.

FIGURE 12.22 Illustration of the defuzzification process.

For example, $0.1 \wedge 0.7 \wedge 0.3 = 0.1$, $0.1 \vee 0.7 \vee 0.3 = 0.7$, and $\overline{0.3} = 0.7$. These rules are also known as *Zadeh* AND, OR, and NOT operators (Zadeh, 1965). Note that these rules are true also for classical binary logic.

Fuzzy rules are specified in the *fuzzy table* as it is shown for a given system. Consider a simple system with two analog input variables x and y, and one output variable z. The goal is to design a fuzzy system generating z as $f(x, y)$. After fuzzification, the analog variable x is represented by five fuzzy variables: x_1, x_2, x_3, x_4, x_5 and an analog variable y is represented by three fuzzy variables: y_1, y_2, y_3. Assume that an analog output variable is represented by four fuzzy variables: z_1, z_2, z_3, z_4. The key issue of the design process is to set proper output fuzzy variables z_k for all combinations of input fuzzy variables, as shown in the table in Figure 12.21. The designer has to specify many rules such as if inputs are represented by fuzzy variables x_i and y_j, then the output should be represented by fuzzy variable z_k. Once the fuzzy table is specified, the fuzzy logic computation proceeds in two steps. First, each field of the fuzzy table is filled with intermediate fuzzy variables t_{ij}, obtained from AND operator $t_{ij} = \min\{x_i, y_j\}$, as shown in Figure 12.21b. This step is independent of the required rules for a given system. In the second step, the OR (max) operator is used to compute each output fuzzy variable z_k. In the given example in Figure 12.21, $z_1 = \max\{t_{11}, t_{12}, t_{21}, t_{41}, t_{51}\}$, $z_2 = \max\{t_{13}, t_{31}, t_{42}, t_{52}\}$, $z_3 = \max\{t_{22}, t_{23}, t_{43}\}$, $z_4 = \max\{t_{32}, t_{34}, t_{53}\}$. Note that the formulas depend on the specifications given in the fuzzy table shown in Figure 12.21a.

12.7.3 Defuzzification

As a result of fuzzy rule evaluation, each analog output variable is represented by several fuzzy variables. The purpose of defuzzification is to obtain analog outputs. This can be done by using a membership function similar to that shown in Figure 12.20. In the first step, fuzzy variables obtained from rule evaluations are used to modify the membership function employing the formula

$$\mu_k^*(z) = \min\{\mu_k(z), z_k\} \tag{12.46}$$

For example, if the output fuzzy variables are 0, 0.2, 0.7, 0.0, then the modified membership functions have shapes shown by the thick line in Figure 12.22. The analog value of the z variable is found as a *center* *of gravity* of modified membership functions $\mu_k^*(z)$,

$$z_{\text{analog}} = \frac{\sum_{k=1}^{n} \int_{2\infty}^{+\infty} \mu_k^*(z) z \, dz}{\sum_{k=1}^{n} \int_{2\infty}^{+\infty} \mu_k^*(z) \, dz} \tag{12.47}$$

In the case where shapes of the output membership functions $\mu_k(z)$ are the same, the equation can be simplified to

$$z_{\text{analog}} = \frac{\sum_{k=1}^{n} z_k z c_k}{\sum_{k=1}^{n} z_k} \tag{12.48}$$

where

n = number of membership function of z_{analog} output variable,
z_k = fuzzy output variables obtained from rule evaluation,
zc_k = analog values corresponding to the center of kth membership function.

Equation 12.47 is usually too complicated to be used in a simple microcontroller based system; therefore, in practical cases, Equation 12.48 is used more frequently.

12.7.4 Design Example

Consider the design of a simple fuzzy controller for a sprinkler system. The sprinkling time is a function of humidity and temperature. Four membership functions are used for the temperature, three for humidity, and three for the sprinkle time, as shown in Figure 12.23. Using intuition, the fuzzy table can be developed, as shown in Figure 12.24a.

Assume a temperature of 60°F and 70% humidity. Using the membership functions for temperature and humidity, the following fuzzy variables can be obtained for the temperature: [0, 0.2, 0.5, 0], and for the humidity: [0, 0.4, 0.6]. Using the min operator, the fuzzy table can be now filled with temporary fuzzy variables, as shown in Figure 12.24b. Note that only four fields have nonzero values. Using fuzzy rules, as shown in Figure 12.24a, the max operator can be applied in order to obtain fuzzy output variables: short $\rightarrow o_1 = \max\{0, 0, 0.2, 0.5, 0\} = 0.5$, medium $\rightarrow o_2 = \max\{0, 0, 0.2, 0.4, 0\} = 0.4$, long $\rightarrow o_3 = \max\{0, 0\} = 0$.

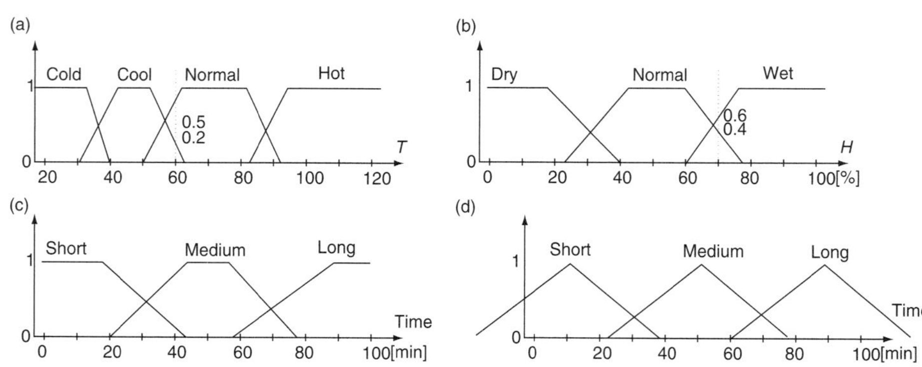

FIGURE 12.23 Membership functions for the presented example: (a) and (b) are membership functions for input variables, (c) and (d) are two possible membership functions for the output variable.

(a)

	Dry	Normal	Wet
Cold	M	S	S
Cool	M	M	S
Warm	L	M	S
Hot	L	M	S

(b)

		Dry	Normal	Wet
		0	0.4	0.6
Cold	0	0	0	0
Cool	0.2	0	0.2	0.2
Warm	0.5	0	0.4	0.5
Hot	0	0	0	0

FIGURE 12.24 Fuzzy tables: (a) fuzzy rules for the design example, (b) fuzzy temporary variables for the design example.

Using Equation 12.47 and Figure 12.23c, a sprinkle time of 28 min is determined. When the simplified approach is used with Equation 12.46 and Figure 12.23d, then sprinkle time is 27 min.

12.8 Genetic Algorithms

The success of the artificial neural networks encouraged researchers to search for other patterns in nature to follow. The power of genetics through evolution was able to create such sophisticated machines as the human being. Genetic algorithms follow the evolution process in nature to find better solutions to some complicated problems. The foundations of genetic algorithms are given by Holland (1975) and Goldberg (1989). After initialization, the steps *selection, reproduction with a crossover*, and *mutation* are repeated for each generation. During this procedure, certain strings of symbols, known as chromosomes, evaluate toward a better solution. The genetic algorithm method begins with coding and an initialization. All significant steps of the genetic algorithm will be explained using a simple example of finding a maximum of the function $(\sin^2(x) - 0.5 * x)^2$ with the range of x from 0 to 1.6. Note that in this range, the function has a global maximum at $x = 1.309$, and a local maximum at $x = 0.262$.

12.8.1 Coding and Initialization

At first, the variable x has to be represented as a string of symbols. With longer strings, the process usually converges faster, so the fewer symbols for one string field that are used, the better. Although this string may be the sequence of any symbols, the binary symbols 0 and 1 are usually used. In our example, 6-bit binary numbers are used for coding, having a decimal value of $40x$. The process starts with a random generation of the initial population given in Table 12.1.

12.8.2 Selection and Reproduction

Selection of the best members of the population is an important step in the genetic algorithm. Many different approaches can be used to rank individuals. In this example the ranking function is given. Highest rank has member number 6, and lowest rank has member number 3. Members with higher rank should have higher chances to reproduce. The probability of reproduction for each member can be obtained as a fraction of the sum of all objective function values. This fraction is shown in the last column

TABLE 12.1 Initial Population

String Number	String	Decimal Value	Variable Value	Function Value	Fraction of Total
1	101101	45	1.125	0.0633	0.2465
2	101000	40	1.000	0.0433	0.1686
3	010100	20	0.500	0.0004	0.0016
4	100101	37	0.925	0.0307	0.1197
5	001010	10	0.250	0.0041	0.0158
6	110001	49	1.225	0.0743	0.2895
7	100111	39	0.975	0.0390	0.1521
8	000100	4	0.100	0.0016	0.0062
Total				0.2568	1.0000

of Table 12.1. Note that to use this approach, our objective function should always be positive. If it is not, the proper normalization should be introduced at first.

12.8.3 Reproduction

The numbers in the last column of Table 12.1 show the probabilities of reproduction. Therefore, most likely members numbers 3 and 8 will not be reproduced, and members 1 and 6 may have two or more copies. Using a random reproduction process, the following population, arranged in pairs, could be generated:

$$\begin{array}{llll} 101101 \rightarrow 45 & 110001 \rightarrow 49 & 100101 \rightarrow 37 & 110001 \rightarrow 49 \\ 100111 \rightarrow 39 & 101101 \rightarrow 45 & 110001 \rightarrow 49 & 101000 \rightarrow 40 \end{array}$$

If the size of the population from one generation to another is the same, two parents should generate two children. By combining two strings, two other strings should be generated. The simplest way to do this is to split in half each of the parent strings and exchange substrings between parents. For example, from parent strings 010100 and 100111, the following child strings will be generated: 010111 and 100100. This process is known as the *crossover.* The resultant children are

$$\begin{array}{llll} 101111 \rightarrow 47 & 110101 \rightarrow 53 & 100001 \rightarrow 33 & 110000 \rightarrow 48 \\ 100101 \rightarrow 37 & 101001 \rightarrow 41 & 110101 \rightarrow 53 & 101001 \rightarrow 41 \end{array}$$

In general, the string need not be split in half. It is usually enough if only selected bits are exchanged between parents. It is only important that bit positions are not changed.

12.8.4 Mutation

In the evolutionary process, reproduction is enhanced with mutation. In addition to the properties inherited from parents, offspring acquire some new random properties. This process is known as mutation. In most cases mutation generates low-ranked children, which are eliminated in the reproduction process. Sometimes, however, the mutation may introduce a better individual with a new property. This prevents the process of reproduction from degeneration. In genetic algorithms, mutation usually plays a secondary role. For very high levels of mutation, the process is similar to random pattern generation, and such a searching algorithm is very inefficient. The mutation rate is usually assumed to be at a level well below 1%. In this example, mutation is equivalent to the random bit change of a given pattern. In this simple case, with short strings and a small population, and with a typical mutation rate of 0.1%, the patterns remain practically unchanged by the mutation process. The second generation for this example is shown in Table 12.2.

TABLE 12.2 Population of Second Generation

String Number	String	Decimal Value	Variable Value	Function Value	Fraction of Total
1	010111	47	1.175	0.0696	0.1587
2	100100	37	0.925	0.0307	0.0701
3	110101	53	1.325	0.0774	0.1766
4	010001	41	1.025	0.0475	0.1084
5	100001	33	0.825	0.0161	0.0368
6	110101	53	1.325	0.0774	0.1766
7	110000	48	1.200	0.0722	0.1646
8	101001	41	1.025	0.0475	0.1084
Total				0.4387	1.0000

Note that two identical highest ranking members of the second generation are very close to the solution $x = 1.309$. The randomly chosen parents for the third generation are:

$$\begin{array}{llll} 010111 \rightarrow 47 & 110101 \rightarrow 53 & 110000 \rightarrow 48 & 101001 \rightarrow 41 \\ 110101 \rightarrow 53 & 110000 \rightarrow 48 & 101001 \rightarrow 41 & 110101 \rightarrow 53 \end{array}$$

which produces the following children:

$$\begin{array}{llll} 010101 \rightarrow 21 & 110000 \rightarrow 48 & 110001 \rightarrow 49 & 101101 \rightarrow 45 \\ 110111 \rightarrow 55 & 110101 \rightarrow 53 & 101000 \rightarrow 40 & 110001 \rightarrow 49 \end{array}$$

The best result in the third population is the same as in the second one. By careful inspection of all strings from the second or third generation, it may be concluded that using crossover, where strings are always split in half, the best solution 110100 → 52 will never be reached, regardless of how many generations are created. This is because none of the population in the second generation has a substring ending with 100. For such crossover, a better result can be only obtained due to the mutation process, which may require many generations. Better results in the future generation also can be obtained when strings are split in random places. Another possible solution is that only randomly chosen bits are exchanged between parents.

The genetic algorithm is very rapid, and it leads to a good solution within a few generations. This solution is usually close to global maximum, but not the best.

Defining Terms

Backpropagation: Training technique for multilayer neural networks.
Bipolar neuron: Neuron with output signal between −1 and +1.
Feedforward network: Network without feedback.
Perceptron: Network with hard threshold neurons.
Recurrent network: Network with feedback.
Supervised learning: Learning procedure when desired outputs are known.
Unipolar neuron: Neuron with output signal between 0 and +1.
Unsupervised learning: Learning procedure when desired outputs are unknown.

References

Fahlman, S.E. 1988. Faster-learning variations on backpropagation: An empirical study. *Proceedings of the Connectionist Models Summer School,* D. Touretzky, G. Hinton, and T. Sejnowski, Eds., Morgan Kaufmann, San Mateo, CA.

Fahlman, S.E. and Lebiere, C. 1990. The cascade correlation learning architecture. *Adv. Ner. Inf. Proc. Syst.*, 2, D.S. Touretzky, ed., pp. 524–532. Morgan Kaufmann, Los Altos, CA.

Goldberg, D.E. 1989. *Genetic Algorithm in Search, Optimization and Machine Learning.* Addison-Wesley, Reading, MA.

Grossberg, S. 1969. Embedding fields: a theory of learning with physiological implications. *Journal of Mathematical Psychology* 6:209–239.

Hebb, D.O. 1949. *The Organization of Behavior, a Neuropsychological Theory.* John Wiley & Sons, New York.

Hecht-Nielsen, R. 1987. Counterpropagation networks. *Appl. Opt.* 26(23):4979–4984.

Hecht-Nielsen, R. 1988. Applications of counterpropagation networks. *Neural Networks* 1:131–139.

Holland, J.H. 1975. *Adaptation in Natural and Artificial Systems.* Univ. of Michigan Press, Ann Arbor, MI.

Hopfield, J.J. 1982. Neural networks and physical systems with emergent collective computation abilities. *Proceedings of the National Academy of Science* 79:2554–2558.

Hopfield, J.J. 1984. Neurons with graded response have collective computational properties like those of two-state neurons. *Proceedings of the National Academy of Science* 81:3088–3092.

Kohonen, T. 1988. The neural phonetic typerater. *IEEE Computer* 27(3):11–22.

Kohonen, T. 1990. The self-organized map. *Proc. IEEE* 78(9):1464–1480.

Kosko, B. 1987. Adaptive bidirectional associative memories. *App. Opt.* 26:4947–4959.

Kosko, B. 1988. Bidirectional associative memories. *IEEE Trans. Sys. Man, Cyb.* 18:49–60.

McCulloch, W.S. and Pitts., W.H., 1943. A logical calculus of the ideas imminent in nervous activity. *Bull. Math. Biophys.* 5:115–133.

Minsky, M. and Papert, S. 1969. *Perceptrons.* MIT Press, Cambridge, MA.

Nilsson, N.J. 1965. *Learning Machines: Foundations of Trainable Pattern Classifiers.* McGraw-Hill, New York.

Nguyen, D. and Widrow, B. 1990. Improving the learning speed of 2-layer neural networks, by choosing initial values of the adaptive weights. *Proceedings of the International Joint Conference on Neural Networks* (San Diego), CA, June.

Pao, Y.H. 1989. *Adaptive Pattern Recognition and Neural Networks.* Addison-Wesley, Reading, MA.

Rosenblatt, F. 1958. The perceptron: a probabilistic model for information storage and organization in the brain. *Psych. Rev.* 65:386–408.

Rumelhart, D.E., Hinton, G.E., and Williams, R.J. 1986. Learning internal representation by error propagation. *Parallel Distributed Processing.*, Vol. 1, pp. 318–362. MIT Press, Cambridge, MA.

Sejnowski, T.J. and Rosenberg, C.R. 1987. Parallel networks that learn to pronounce English text. *Complex Systems* 1:145–168.

Specht, D.F. 1990. Probalistic neural networks. *Neural Networks* 3:109–118.

Specht, D.F. 1992. General regression neural network. *IEEE Trans. Neural Networks* 2:568–576.

Wasserman, P.D. 1989. *Neural Computing Theory and Practice.* Van Nostrand Reinhold, New York.

Werbos, P., 1974. Beyond regression: new tools for prediction and analysis in behavioral sciences. Ph.D. diss., Harvard University.

Widrow, B. and Hoff, M.E. 1960. Adaptive switching circuits. 1960 IRE Western Electric Show and Convention Record, Part 4 (Aug. 23):96–104.

Widrow, B. 1962. Generalization and information storage in networks of adaline Neurons. In *Self-organizing Systems*, M.C. Jovitz, G.T. Jacobi, and G. Goldstein, eds., pp. 435–461. Sparten Books, Washington, D.C.

Wilamowski, M. and Torvik, L., 1993. Modification of gradient computation in the back-propagation algorithm. *ANNIE'93 - Artificial Neural Networks in Engineering.* November 14–17, 1993, St. Louis, Missou.; also in C.H Dagli, ed. 1993. *Intelligent Engineering Systems Through Artificial Neural Networks*, Vol. 3, pp. 175–180. ASME Press, New York.

Zadeh, L.A. 1965. Fuzzy sets. *Information and Control* 8:338–353.

Zurada, J. 1992. *Introduction to Artificial Neural Systems.* West Publ.

13

Advanced Control of an Electrohydraulic Axis

Florin Ionescu
University of Applied Sciences

Crina Vlad
Politehnica University of Bucharest

Dragos Arotaritei
Aalborg University Esbjerg

13.1 Introduction

Due to the development of technology in the last few years, robots are seen as advanced mechatronic systems which require knowledge from mechanics, actuators, and control in order to perform very complex tasks. Different kinds of servo-systems, especially electrohydraulic, could be met at the executive level of the robots. Taking into account the most advanced control approaches, this paper deals with the implementation of advanced controllers besides conventional ones which are used in an electrohydraulic system. The considered electrohydraulic system is one of the axes of a robot. These robots possess three or more electrohydraulic axes, which are identical with the axis studied in this chapter.

An electrohydraulic axis whose mathematical model (MM) is described in this chapter presents a multitude of nonlinearities. Conventional controllers are becoming increasingly inappropriate to control the systems with an imprecise model where many nonlinearities are manifested. Therefore, advanced techniques such as neural networks and fuzzy algorithms are deeply involved in the control of such systems. Neural networks, initially proposed by McCulloch and Pitts, Rosenblatt, Widrow, had several limitations that restricted the domain of applications. An important change took place in the 1980s when Hopfield's theory regarding recurrent neural networks, the model of self-organization developed by Kohonen, and cellular neural networks (Chua) relaunched this research field. The development of some efficient algorithms dedicated specifically to the architecture of neural networks, and the application of these networks in control, represents an interesting area of research in the contemporary world of science.

Fuzzy systems, in conjunction with neural networks, hold an important place in advanced techniques of control. These systems have origins in fuzzy set theory initiated by L. Zadeh. One essential feature of fuzzy systems is the approximate reasoning in which the variables are described in a qualitative manner. Due to the capability of fuzzy systems to deal with imprecise information, they are strongly recommended in order to express knowledge in the form of linguistic rules. In this way, the human operator's knowledge, which is linguistic or numerical, is used to generate the set of fuzzy if-then rules as a basis for a fuzzy controller. A main drawback of fuzzy systems is the difficulty to design them on the basis of a systematic methodology. To overcome this drawback, the learning procedures from neural networks are applied successfully in order to tune the parameters of membership functions.

The merging of these two fields has led to the emergence of neuro-fuzzy systems, which have been applied with promising results in the field of control-engineering. In order to improve dynamic and static performances of the systems characterized by nonlinearities and uncertainties, neuro-fuzzy controllers are used.

The present contribution is organized as follows. An introduction of the electrohydraulic systems with an emphasis on the control of such devices is realized in Section 13.2. Section 13.3 is devoted to the MM of electrohydraulic axes, and the subsequent sections treat the control of electrohydraulic axes through conventional methods (Section 13.4), fuzzy systems (Section 13.5), neural networks (Section 13.6), and neuro-fuzzy techniques (Section 13.7). Conclusions are given in Section 13.8.

13.2 Generalities Concerning ROBI_3, a Cartesian Robot with Three Electrohydraulic Axes

The automated installation, which uses electrohydraulic axes, whose mathematical model is described in Section 13.3, is a Cartesian robot named ROBI_3. ROBI_3 has three identical electrohydraulic axes and is built from aluminium profiles [21]. The slides are actuated by hydraulic servoactuators (Rexroth). Slades move on the linear guideways with balls and two recirculation paths. The hydraulic supply installation is placed under the robot table and has a cooling and controlling installation with air. The mechanical structure of the robot is depicted in Figure 13.1. The three axes of ROBI_3 are identically controlled by the controlling software named TORCH, which runs in Windows. The 32-bit dSPACE controlling hardware endowed with 10 A/D and D/A interfaces is plugged into the PC and serves as the interface between the PC and each of the axes.

An electrohydraulic axis consists of a servovalve and a hydraulic cylinder and has a nonlinear structure. The control system of one axis consists of:

1. The controller represented by a personal computer endowed with a process card
2. The electrohydraulic converter
3. The actuator (a linear hydraulic servomotor (LHM))
4. The mechanical process to be controlled, characterized by the slade position
5. The position transducer

The control system of robot ROBI_3 is illustrated in Figure 13.2, where the presence of three electrohydraulic axes, as well as the structure of one axes, identical to the others, can be observed.

The corresponding mathematical model for one axis, on the basis of which the control of the robot is achieved, is described in Section 13.3. Through numerical simulations of the three axes of the robot, the necessary mechanical structure interface data is obtained.

Preliminary experiments with driven and controlled variable: position and velocity are made in order to achieve experience (Figures 13.2 and 13.3). The diagram of a closed-loop position control with

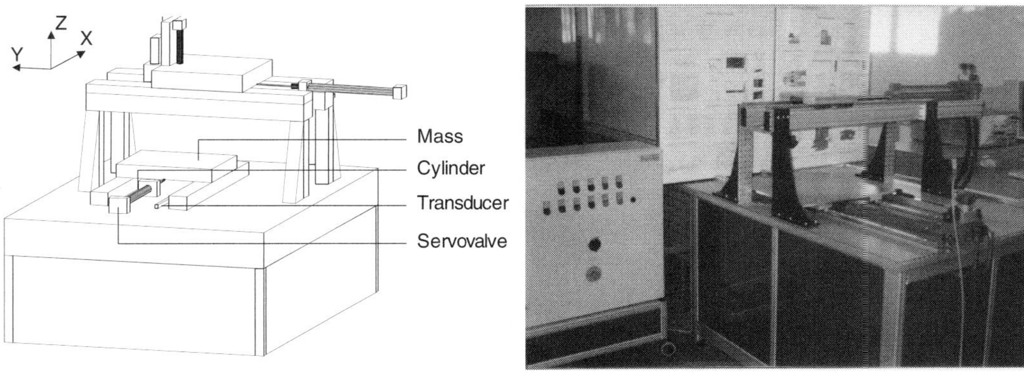

FIGURE 13.1 ROBI_3, a Cartesian robot with three axes [21]. (a) Design. (b) Practical allignment.

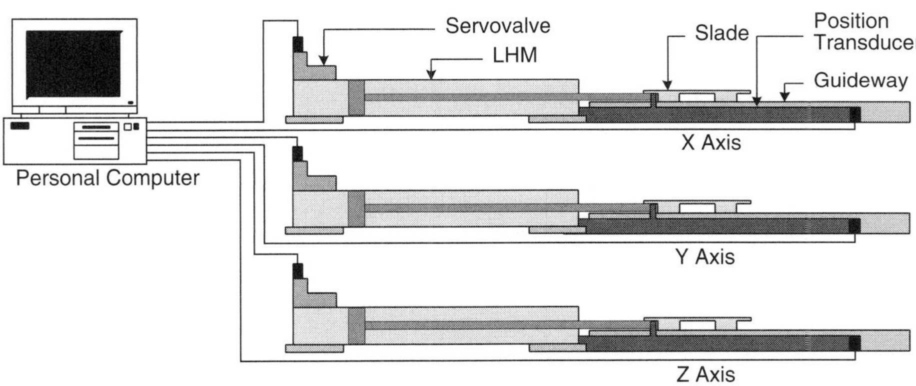

FIGURE 13.2 The control system of the robot [21,24].

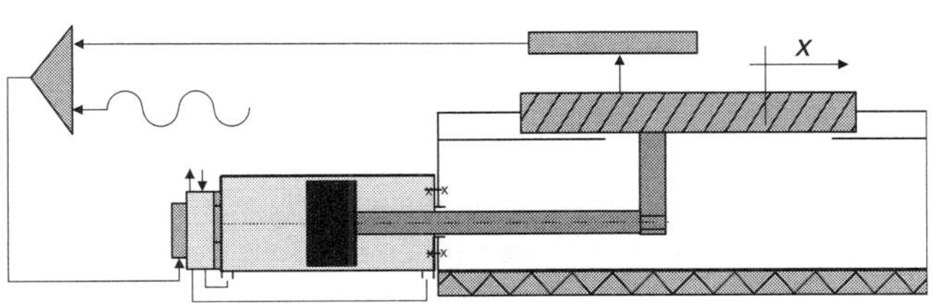

FIGURE 13.3 Diagram of a closed-loop position control with direct measurement.

direct measurement (driven by means of a servovalve) is shown in Figure 13.3, and the closed-loop position control with indirect measurement at spindle (actuated by means of a servovalve) is shown in Figure 13.4.

A volumetric Q-rate regulation with constant pressure ($Q \neq$ const; $p \cong$ const) is shown in Figure 13.5. However, this classic model, useful for application, was used only for preliminary results in simulation. One of the main reasons to use this is because we need a well-known mathematical model with well-studied behavior in order to test the controllers (neural and neuro-fuzzy, namely).

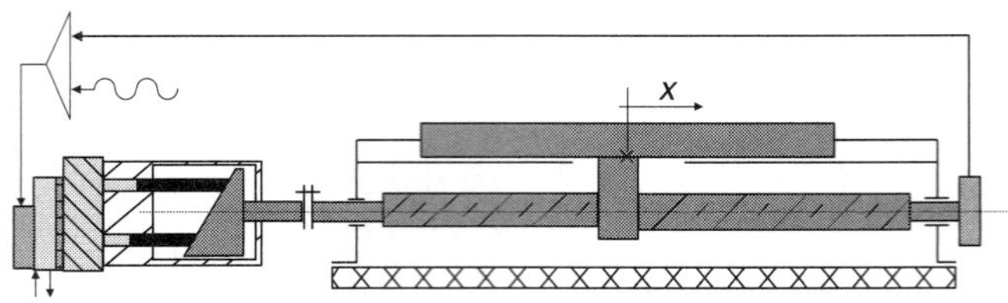

FIGURE 13.4 Closed loop-position control with indirect measurement.

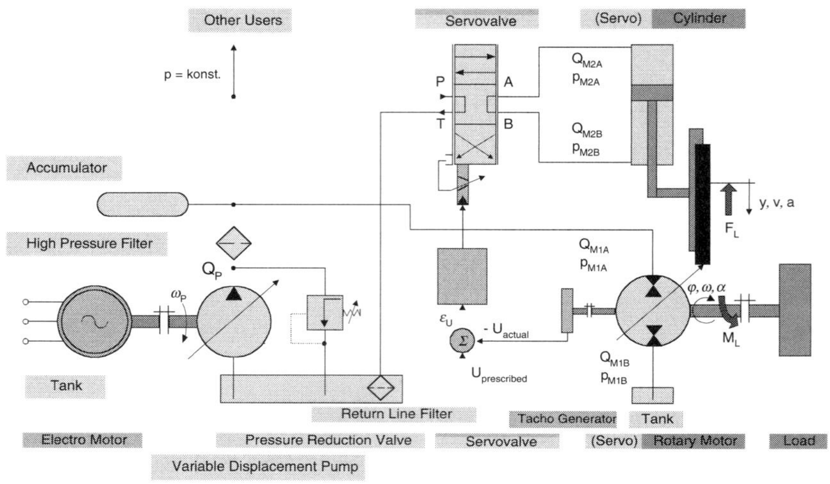

FIGURE 13.5 Volumetric Q-rate regulation with constant pressure.

13.3 Mathematical Model and Simulation of Electrohydraulic Axes

Section 13.3 deals with the analytical findings of the mathematical model (MM) of an electrohydraulic axis, a component part of robot ROBI_3. This method of analysis is advantageous because it offers the possibility to use this MM for other electrohydraulic axes as well, regardless of the different number of stages, and also allows the testing of dynamic performances of the axis at the design level.

In this section, the following were realized:

1. Static models of electrohydraulic system components (servovalve, hydraulic linear motor).
2. Parameters involved in MM, based on constructive and flowing regime characteristics.
3. Nonlinear MM of the proposed electrohydraulic system.
4. Structural scheme of the hydraulic axis in order to simulate its behavior (SIMULINK).
5. Investigations regarding MM, which certify the stability of the system and the fact that the modelled process is a rapid one.

Values of parameters that describe the MM are set based on hydraulic characteristics and on constructive parameters of the considered system.

13.3.1 The Extended Mathematical Model

The studied system consists of a servovalve and an asymmetric motor. In most cases, the control of an electrohydraulic axis is directed towards position control, velocity control, pressure control, or force control. The position control of the axis is studied. The position control loop for the proposed installation is illustrated in Figure 13.6, together with the denomination of some data.

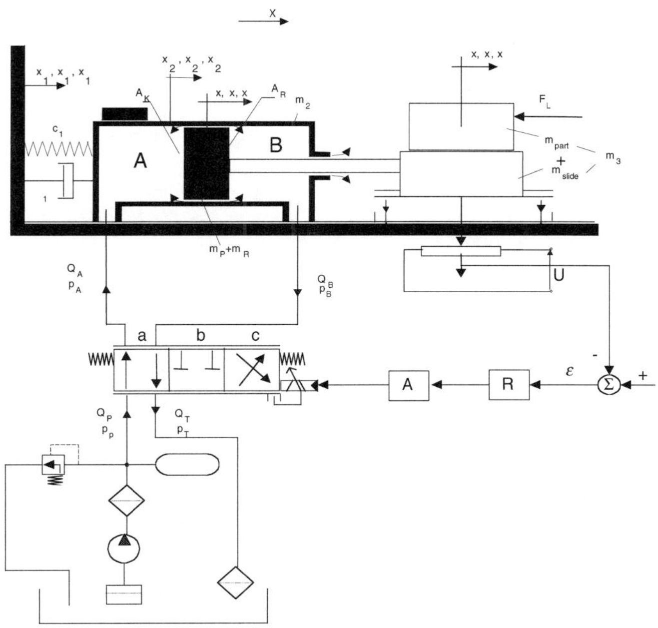

FIGURE 13.6 Control scheme of the servosystem.

13.3.2 Nonlinear Mathematical Model of the Servovalve

As previously mentioned, the servovalve used at the proposed electrohydraulic axis has four paths and three positions. The servovalve has four active control edges at the second hydraulic stage. Three stages could be distinguished by this servovalve: the electromechanical, the mechanohydraulical, and the hydromechanical one. Regarded as a system, a servovalve is complex, with various types of nonlinearities being manifested. Different static and dynamic nonlinearities such as dead zone, jump in origin, saturation, Coulombian and Newtonian frictions with hysteresis, and asymmetry appear in each of these three levels and also in the actuator of the electrohydraulic motor. These were taken into account in the modeling of the servovalve and of the cylinder behavior.

For the studied servovalve, the circulation of the fluid is considered directed from the pump to the admission chamber *A* (Q_A) and from the discharge chamber *B* to the reservoir (Q_B). Figure 13.6 presents the control loop where the transducer is placed on the feedback path and the controller *R* and the amplifier *A* are on the direct path. Electrical signal (±10 V or ±300 mA) is converted into the displacement x_v of the valve, and in this way in a flow *Q*, which is transmitted to the linear hydraulic motor.

From the point of view of control characteristics, the number of active edges serves as a method of classification of slide valves [3]. A servovalve with four active control edges was considered. The lightly simplified mathematical model, which describes the functionality of the servovalve, consists of the following equations:

$$u(t) = L \cdot \frac{di(t)}{dt} + R \cdot i(t) \tag{13.1}$$

where *L* [H], the inductance of electrical level; *R* [W], the resistance of electrical level; $u(t)$ [V], the control voltage; $i(t)$ [A], the control intensity.

$$m \cdot x_v(t) + d \cdot x_v(t) + c \cdot x_v(t) = \sum F \tag{13.2}$$

where *m* [kg], the mass of valve; *d* [N/(m/s)], the linearized gradient of viscous friction for the piston of the valve; *c*[N/m], the coefficient of hydraulic elasticity; x_v [m], the spool displacement; ΣF [N], the

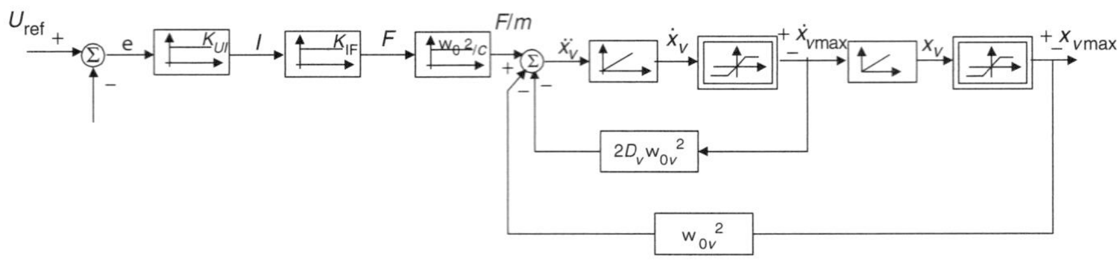

FIGURE 13.7 Nonlinear MM of the first stages of the servovalve.

resulting force, which actuates on the valve spool.

$$x_v(t) + 2 \cdot D_v \cdot \omega_{0v} \cdot x_v(t) + \omega_{0v}^2 \cdot x_v(t) = \frac{\Sigma F}{m}$$

$$x_v(t) = k^* - 2 \cdot D_v \cdot \omega_{0v} \cdot x_v(t) - \omega_{0v}^2 \cdot x_v(t) \quad \text{where } k^* = \frac{\Sigma F}{m} \tag{13.3}$$

The displacement x_v, obtained based upon the mentioned equations, is implemented in SIMULINK using the scheme from Figure 13.7. This module of nonlinear MM includes two stages of electrohydraulic axis: the electrohydraulic and the hydromechanical one.

The corresponding equations for the four flows that go through the servovalve are

$$\begin{aligned}
Q_{PA} &= \alpha_D \cdot \pi \cdot D_v \cdot \sqrt{\frac{2}{\rho}} \cdot (x_0 + x_v(t)) \cdot \sqrt{p_P - p_A(t)}, \quad x_v \in [-x_0, x_{\max}] \\
Q_{AT} &= \alpha_D \cdot \pi \cdot D_v \cdot \sqrt{\frac{2}{\rho}} \cdot (x_0 - x_v(t)) \cdot \sqrt{p_A(t) - p_T}, \quad x_v \in [-x_{\max}, x_0] \\
Q_{PB} &= \alpha_D \cdot \pi \cdot D_v \cdot \sqrt{\frac{2}{\rho}} \cdot (x_0 - x_v(t)) \cdot \sqrt{p_P - p_B(t)}, \quad x_v \in [-x_{\max}, x_0] \\
Q_{BT} &= \alpha_D \cdot \pi \cdot D_v \cdot \sqrt{\frac{2}{\rho}} \cdot (x_0 + x_v(t)) \cdot \sqrt{p_B(t) - p_T}, \quad x_v \in [-x_0, x_{\max}]
\end{aligned} \tag{13.4}$$

where Q_{PA} [m^3/s], the flow to the hydraulic motor, from pump to the chamber *A* of the motor; Q_{AT} [m^3/s], the flow from the chamber *A* to reservoir; Q_{PB} [m^3/s], the flow from pump to the chamber *B* of the motor; Q_{BT} [m^3/s], the flow from the chamber *B* to reservoir; α_D [-], the discharge coefficient; D_v [m], the spool's diameter; x_0 [m], the dimension of the lap of the spool; x_v [m], the spool's displacement; p_A [N/m^2], the fluid pressure in chamber *A*; p_B [N/m^2], the fluid pressure in chamber *B* [kg/m^3] fluids density. The flows, which are transmitted to LHM and evacuated from LHM, are Q_A and Q_B, which are computed as following:

$$Q_A = Q_{PA} - Q_{AT}, \qquad Q_B = Q_{BT} - Q_{PB} \tag{13.5}$$

The lap of the spool is considered to be zero ($x_0 = 0$), and, therefore, the static characteristic is linear around the origin and also in the rest. With $Q_0 = \alpha_D \cdot \pi \cdot D_v \cdot \sqrt{2/\rho}$, the flow equations become

$$\begin{aligned}
Q_{PA} &= Q_0 \cdot x_v \cdot \sqrt{p_P - p_A}, \qquad Q_{AT} = Q_0 \cdot (-x_v) \cdot \sqrt{p_A - p_T} \\
Q_{PB} &= Q_0 \cdot (-x_v) \cdot \sqrt{p_P - p_B}, \qquad Q_{BT} = Q_0 \cdot x_v \cdot \sqrt{p_B - p_T}
\end{aligned} \tag{13.6}$$

13.3.3 Nonlinear Mathematical Model of Linear Hydraulic Motor

The differential equations, based on which the MM of the linear hydraulic motor (LHM) was achieved, are

i. Equation of the dynamic equilibrium of the forces reduced to the motor's rod
ii. Equation of movement and flow continuity

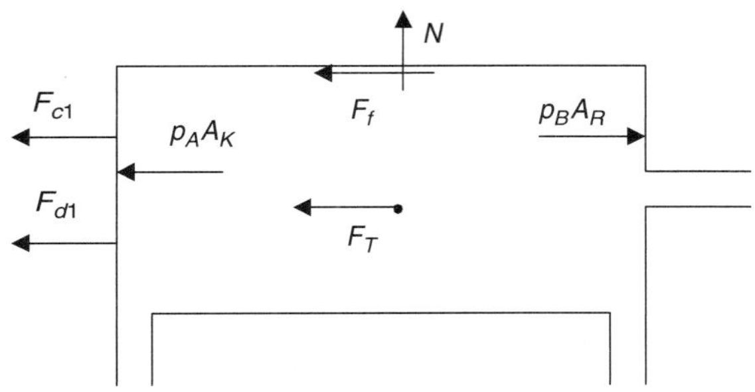

FIGURE 13.8 D'Alembert's principle applied to the cylinder.

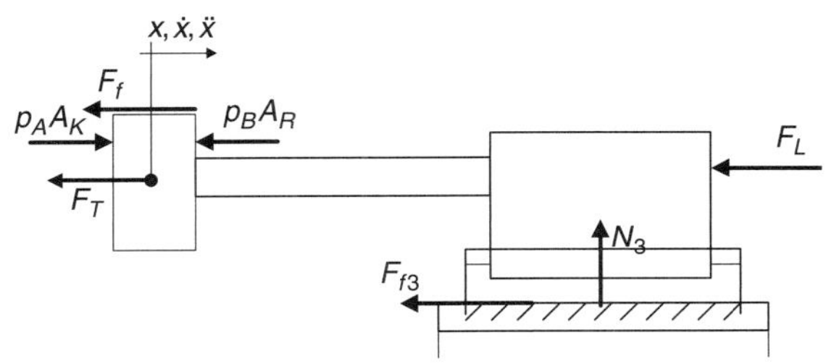

FIGURE 13.9 D'Alembert's principle applied to the rod, piston, and load.

Modeling the LHM, several simplifications (most of them concerning the Coulombian and Newtonian friction) were used. The forces that actuate on the LHM cylinder are depicted in Figure 13.8:

Applying D'Alembert's principle, the equation of dynamic equilibrium of forces for the cylinder of LHM is

$$x_2 = \frac{p_B \cdot A_R - p_A \cdot A_k + c_1 \cdot (x_1 - x_2) + d_1 \cdot (x_1 - x_2) + c_{\text{fu}} \cdot N \cdot \text{sgn}(x - x_2) + d_z \cdot (x - x_2)}{m_2 \cdot s} \tag{13.7}$$

where c_1 [N/m], the elasticity; d_1 [N/(m/s)], the linearized coefficient of the viscous Newtonian friction in the connection actuators' cylinder and wall; c_{fu} [-], the coefficient of the dry Coulombian friction in the cylinder and rod seals; d_z [N/(m/s)], the coefficient of Newtonian friction in the piston and rod seals; m_2 [kg], the cylinder mass; p_A [N/m^2], the fluid pressure in the admission chamber *A* of the actuator; p_B [N/m^2], the fluid pressure in the discharge chamber *B* of the actuator; A_K [m^2], the piston active area in chamber *A*; A_R [m^2], the piston active area in chamber *B*; *N* [N], the normal force, which determines the friction force between piston and cylinder; *x* [m], the piston displacement; x_1 [m], the wall displacement; and x_2 [m], the cylinder displacement.

The forces acting on the rod, piston, and working element are illustrated in Figure 13.9

The velocity corresponding to the rod, piston, and mass m_3 (slade, guideway, and loading are considered stiff fastened) is inferred from the equilibrium equation:

$$x = \frac{p_A \cdot A_K - p_B \cdot A_R - F_L - c_{\text{fu}} \cdot N \cdot \text{sgn}(x - x_2) - d_z \cdot (x - x_2) - c_{\text{fu3}} \cdot N_3 \cdot \text{sgn}(x - x_1) - d_3 \cdot (x - x_1)}{(m_p + m_T + m_3) \cdot s} \tag{13.8}$$

where m_p [kg], the piston mass; m_T [kg], the rod mass; m_3 [kg], load mass (reduced at the rod of piston); c_{fu3} [-], the coefficient of Coulombian friction between guide ways and slade; d_3 [N/(m/s)], the coefficient of Newtonian friction between guideways and slade; N_3 [N], the normal force which appears between loading and table; F_L [N], the loading force.

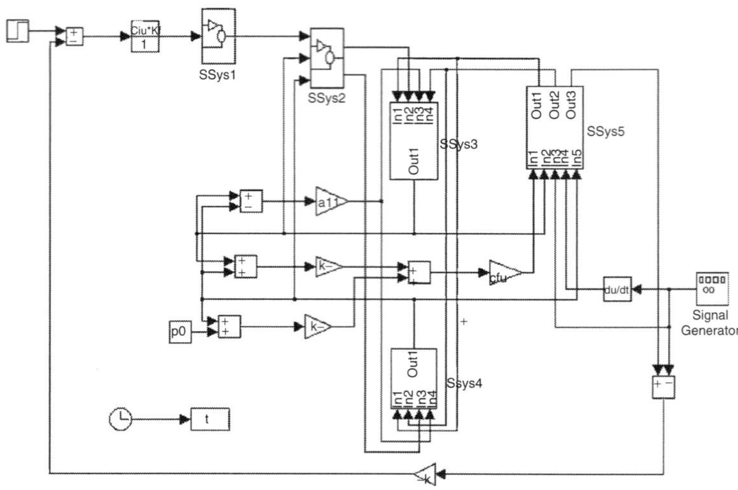

FIGURE 13.10 Electrohydraulic axis implemented in SIMULINK.

From the equations of continuity the pressures p_A and p_B are inferred:

$$p_A = [Q_A - A_K \cdot (x - x_2) - a_{11} \cdot (p_A - p_B)] \cdot \frac{E_{ers}}{V_{0K} + A_K \cdot (x - x_2)} \tag{13.9}$$

$$p_B = [A_R . (x - x_2) + a_{11} \cdot (p_A - p_B) - a_{12} . (p_B - p_0) - Q_B] \cdot \frac{E_{ers}}{(V_{0R} + A_R \cdot h) - A_R \cdot (x - x_2)} \tag{13.10}$$

where a_{11}, a_{12} [(N/m^2)/(m^3/s)], the gradients of leakages; $V_{0K,0R}$ [m^3], the initial average volume of chambers *A* and *B*, respectively; E_{ers} [N/m^2], the equivalent bulk modulus of oil; $p_{A,B}$ [N/m^2], the fluid pressure in chambers *A* and *B*, respectively; and *h* [m], the stroke of the piston-rod.

The LHM operation is based on the equations described above, namely (13.7)–(13.10). The MM proposed in this section is implemented in SIMULINK 2.1/ MATLAB 5.1 and has the structure presented in Figure 13.10.

The signal generator icon from the above figure generates the displacement of the wall x_1, which has a sinusoidal form with the frequency 0.5 Hz and amplitude 0.0001 m.

The subsystems Ssys1 and Ssys2 have as outputs the valve displacement x_v, and the flows Q_A and Q_B, respectively. Ssys3 is the block that implemented Equation 13.9, while the subsystem Ssys4 modelled Equation 13.10. The equations that describe the displacement of the wall cylinder and the LHM piston are modelled by subsystem Ssys5. The reference signal is a step one whose values are in the range 0–10 V.

13.4 Conventional Controllers Used to Control the Electrohydraulic Axis

This section is organized as follows: the first part presents the bibliographic research concerning the traditional directions of the control system, and the second one contains the testing of several classic control structures (PID and control algorithms with Luenberger observer) through simulations of the electrohydraulic axis endowed with these controllers.

The testing of the MM is performed with SIMULINK and has a goal of the achievement of reference experimental results in order to perform a comparative study of classical controllers and advanced control structures applied to the electrohydraulic axis.

13.4.1 PID, PI, PD with Filtering

The conventional control structures used in this chapter are PI (proportional-integral), PID (proportional-integral-derivative), and PD (proportional-derivative) with filtering coefficient.

The transfer function of the PI controller has the following expression:

$$H_{PI} = \frac{U(s)}{\varepsilon(s)} = K_R \cdot \left(1 + \frac{1}{T_i \cdot s}\right) = K_R \cdot \frac{T_i \cdot s + 1}{T_i \cdot s} \tag{13.11}$$

K_R is the proportional factor, and T_i is the time constant of the integrative component.

The transfer function of the PID controller is described by the following equation:

$$H_{PID} = \frac{U(s)}{\varepsilon(s)} = K_R \cdot \left(1 + \frac{1}{T_i \cdot s} + T_d \cdot s\right) = K_R \cdot \frac{T_i \cdot T_d \cdot s^2 + T_i \cdot s + 1}{T_i \cdot s} \tag{13.12}$$

K_R and T_i have the same significance as previously mentioned, and T_d is the time constant of derivative component.

The transfer function corresponding to PD with filtering has the expression:

$$H_{PDF}(s) = \frac{U(s)}{\varepsilon(s)} = K_R \cdot \frac{1 + T_d \cdot s}{1 + \alpha \cdot T_d \cdot s} \tag{13.13}$$

where the coefficient α could have values in the range 0.1–0.125.

Generally speaking, PID controllers are commonly used in industrial control systems and, therefore, are well established. Nevertheless, the results obtained using a PID controller for complex control loops are not very satisfactory because it could be costly and time consuming to retune such regulators. PI controller is enough in situations where derivative action is not frequently used.

13.4.2 Observer

The theory of observers, started with the work of Luenberger and Ackermann, is fairly complete and comprehensive. For the proposed axis, an $(n - m - 1)$ order structure of the observer is adopted, where $n = 5$ represents the order of the system and $m = 1$ is the number of outputs [25]. The model of the servodrive is described by five state variables: two of them for the second-order model of the servovalve and the other three for the third-order servoactuator. The use of a linear observer as a parallel model reconstitutes the state-variables of the installation and delivers them to the controller. Two possibilities could be followed: with partial and with global reconstruction. The solution chosen was with partial reconstruction [12]. The complete system consists of the installation with nonlinearities, a parallel second-order model for the servovalve, a third-order linear servoactuator, a correction matrix for the observer, and a controller with five loops for the five state variables [23,25,26].

The block diagram of an electrohydraulic axis controlled with a third-order observer is shown in Figure 13.11, where **A**, **b**, $\mathbf{c}^T$ are the characteristic matrices of the linear system (electrohydraulic axis), **k** is the correction matrix, and **R** represents the matrix corresponding to the controller.

The simulation diagram of the electrohydraulic axis, controlled by a third-order observer, as it is depicted in SIMULINK, is illustrated in Figure 13.12.

The algorithm used to compute the matrix **k** and **r** consists of the following steps:

i. Achievement of the MM for servovalve and servoactuator
ii. Setting the state-variables of the process
iii. Obtaining the controller upon the dynamics of the closed-loop system
iv. Computing the correction matrix by using the desired poles for the observer [12]

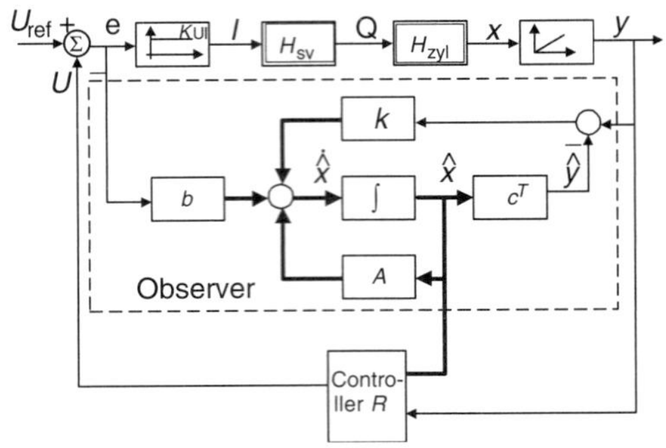

FIGURE 13.11 Block diagram of the control loop using the observer.

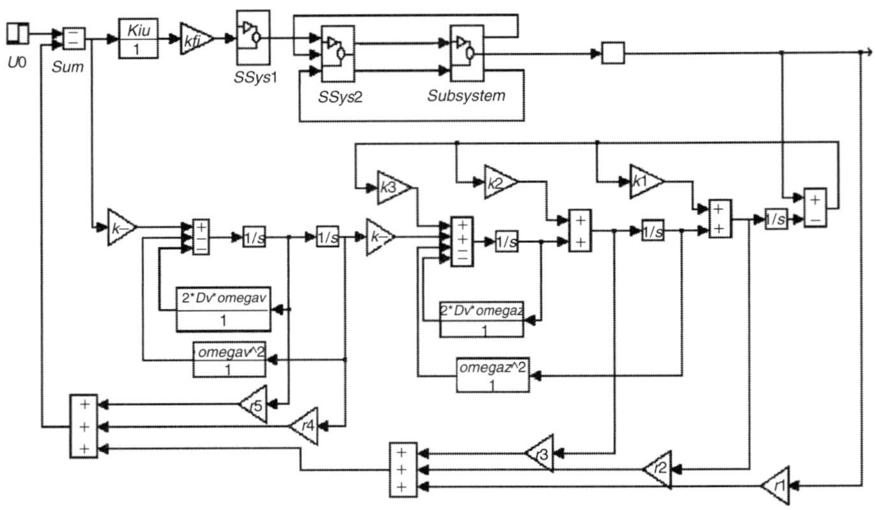

FIGURE 13.12 Control structure with observer for the electrohydraulic axis with SIMULINK.

13.4.2.1 The Linear Mathematical Models

The linear mathematical models (LMM) of the servovalve and of the servoactuator are obtained by using the appropriate transfer functions H_{SV}, H_Z:

$$H_{SV}(s) \hat{=} \frac{Q_V(s)}{I_V(s)} = \frac{k_V \omega_V^2}{s^2 + 2D_V \omega_V s + \omega_V^2} \tag{13.14}$$

$$H_Z(s) \hat{=} \frac{Y_Z(s)}{Q_V(s)} = \frac{k_Z \omega_Z^2}{(s^2 + 2D_Z \omega_Z s + \omega_Z^2)s} \tag{13.15}$$

with the following meanings: Q_V, the servovalve's flow; I_V, the current intensity, Y_Z, the rod's position.

13.4.2.2 LMM in the State of Space

The used variables are: x_1, the rod position; x_2, the rod velocity; x_3, the rod acceleration; x_4, the spool position; and x_5, the spool velocity.

$$\begin{aligned} x_1 &= x_2(t), \quad x_2 = x_3(t), \quad x_4 = x_5(t) \\ x_3 &= -\omega_Z^2 x_2(t) - 2D_Z\omega_Z x_3(t) + k_Z\omega_Z^2 x_4(t) \\ x_5 &= -\omega_V^2 x_4(t) - 2D_V\omega_V x_5(t) + k_V\omega_V^2 u(t) \end{aligned} \tag{13.16}$$

Thus the MM of the axis in state-space form becomes

$$\begin{aligned} \dot{\mathbf{x}}(t) &= \mathbf{A}\mathbf{x}(t) + \mathbf{b}u(t) \\ y(t) &= \mathbf{c}^T\mathbf{x}(t) \end{aligned} \tag{13.17}$$

where

$$\mathbf{A} = \begin{pmatrix} 0 & 1 & 0 & 0 & 0 \\ 0 & 0 & 1 & 0 & 0 \\ 0 & -\omega_Z^2 & -2D_Z\omega_Z & k_Z & 0 \\ 0 & 0 & 0 & 0 & 1 \\ 0 & 0 & 0 & -\omega_V^2 & -2D_V\omega_V \end{pmatrix}, \quad \mathbf{b} = \begin{pmatrix} 0 \\ 0 \\ 0 \\ 0 \\ k_V \end{pmatrix}, \quad \mathbf{c}^T = (1\ 0\ 0\ 0\ 0) \tag{13.18}$$

13.4.2.3 Controller Design

The characteristic polynomial is obtained from $\det[s\mathbf{I} - (\mathbf{A}_C - \mathbf{b}_C\mathbf{r}^T)] = 0$, where $\mathbf{A}_C$ and $\mathbf{b}_C$ are the controllable forms of the matrices $\mathbf{A}$ and $\mathbf{b}$. If $\mathbf{A} \neq \mathbf{A}_C$, the use of transformation matrix $\mathbf{T}$ is advisable, in order to obtain $\mathbf{A}_C$ and $\mathbf{b}_C$. Thus $\mathbf{A}_C = \mathbf{T}\mathbf{A}\mathbf{T}^{-1}$, and $\mathbf{b}_C = \mathbf{T}\mathbf{b}$. The matrix $\mathbf{F} = \mathbf{A}_C - \mathbf{b}_C\mathbf{r}^T$ has the form

$$\mathbf{F} = \begin{pmatrix} 0 & 1 & .. & 0 \\ 0 & 0 & .. & 0 \\ . & . & .. & . \\ 0 & 0 & .. & 1 \\ -a_0 - r_1 & -a_1 - r_2 & .. & -a_{n-1} - r_n \end{pmatrix} \tag{13.19}$$

The characteristic polynomial of the matrix $\mathbf{F}$ is

$$s^n + (a_{n-1} + r_n)s^{n-1} + \cdots + (a_1 + r_2)s + (a_0 + r_1) \tag{13.20}$$

The poles chosen for the closed-loop determine the polynomial

$$s^n + p_{n-1}s^{n-1} + p_{n-2}s^{n-2} + \cdots + p_1 s + p_0 \tag{13.21}$$

The polynomials (13.20) and (13.21) are identical; therefore, the coefficients of matrix $\mathbf{r}_R^T$ are

$$r_\nu = p_{\nu-1} - a_{\nu-1}, \quad \nu = 1,\ldots,n$$

If

$$\mathbf{A} = \mathbf{A}\mathrm{c}, \quad \mathbf{r}^T = \mathbf{r}_R^T$$

otherwise

$$\mathbf{r}^T = \mathbf{r}_R^T \mathbf{T} \tag{13.22}$$

13.4.2.4 Correction Matrix Design

The matrix F is $\mathbf{F} = \mathbf{A}^* - \mathbf{K}\mathbf{c}^T$, where $\mathbf{A}^*$ is the matrix of the observer. For F the chosen poles are $s_1, s_2, \ldots, s_n$, and

$$\det[s\mathbf{I}\text{-}\mathbf{F}] = (s - s_1)(s - s_2)\cdots(s - s_n) \tag{13.23}$$

$$\det[s\mathbf{I}\text{-}\mathbf{F}] = s^n + f_{n-1}s^{n-1} + \cdots + f_1 s + f_0 \tag{13.24}$$

From these two equations, the coefficients $k_1, k_2, \ldots, k_n$ are obtained.

In this case, $\mathbf{c}^T = (1 \ \ 0 \ \ 0)$ and the matrix of the third observer is

$$\mathbf{A}^* = \begin{pmatrix} 0 & 1 & 0 \\ 0 & 0 & 1 \\ 0 & -\omega_Z^2 & -2D_Z\omega_Z \end{pmatrix} \tag{13.25}$$

The correction matrix influences the transient behavior; the further the poles of F from the poles of $\mathbf{A}^*$ the quicker the response.

13.4.3 Simulation Results of Electrohydraulic Axis with Conventional Controllers

Based on the above algorithm in order to determine the correction matrix and the controller matrix, the SIMULINK implementation of the observer involves the following values:

$$r_1 = 19.95854, \quad r_2 = 0.069481, \quad r_3 = -7.06024 \times 10^{-4},$$

$$r_4 = -3.158688 \times 10^2, \quad r_5 = -3.451209 \times 10^{-1}$$

for the controller, and

$$k_1 = 1.67 \times 10^{-2}, \ k_2 = 3.7028 \times 10^4, \quad k_3 = -6.969698 \times 10^6 \text{ for the correction matrix.}$$

When the reference signal is a step signal with $U = 10$ V, the simulation results are shown in Figures 13.13 and 13.14.

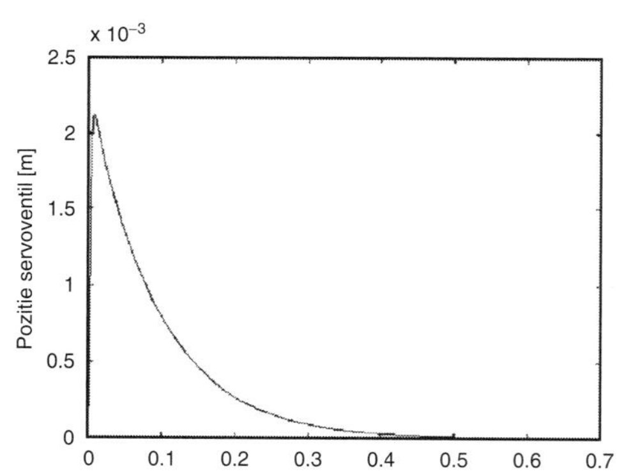

FIGURE 13.13 Position of servovalve for MM with observer.

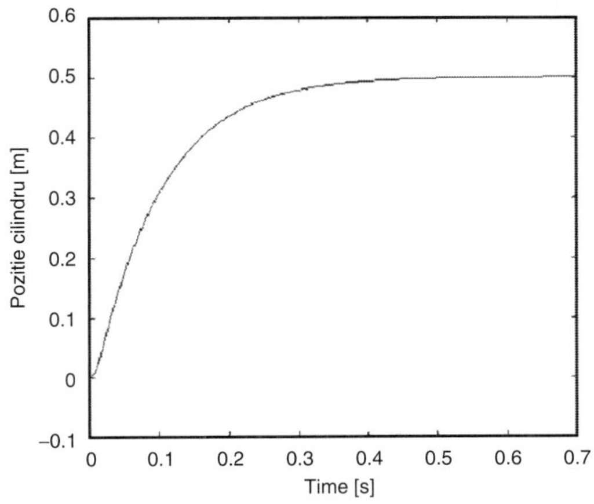

FIGURE 13.14 Cylinder position.

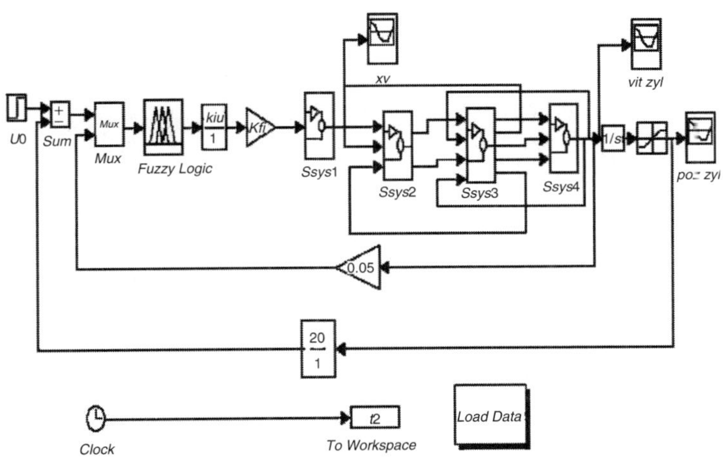

FIGURE 13.15 Electrohydraulic axis control with a fuzzy controller with two inputs.

13.5 Control of Electrohydraulic Axis with Fuzzy Controllers

Section 13.5 of this contribution is devoted to the presentation and the testing of nontraditional controllers based on fuzzy sets, which model the behavior of a human operator in the control process. The simulation results of an electrohydraulic axis with SUGENO and MAMDANI controllers are depicted. For the same number of inference rules extracted from the knowledge base, simulations proved that dynamic performances are improved for a fuzzy controller with two inputs.

The scheme achieved with SIMULINK to control the electrohydraulic axis with two inputs fuzzy controller and a MAMDANI or SUGENO inference is depicted in Figure 13.15.

The results presented concern the simulation of the hydraulic axis endowed with a fuzzy controller, which is based on MAMDANI inference [59]. "Fuzzy Logic" toolbox gives the user the possibility to create MAMDANI or SUGENO fuzzy systems using graphic interfaces. FIS (Fuzzy Inference System) Editor, Membership Function Editor, and Inference Rules Editor are several of the tools available in SIMULINK. For instance, the corresponding FIS editor and the Membership Function Editor of each input for the proposed fuzzy controller with MAMDANI inference and two inputs are illustrated in Figures 13.16 and 13.17.

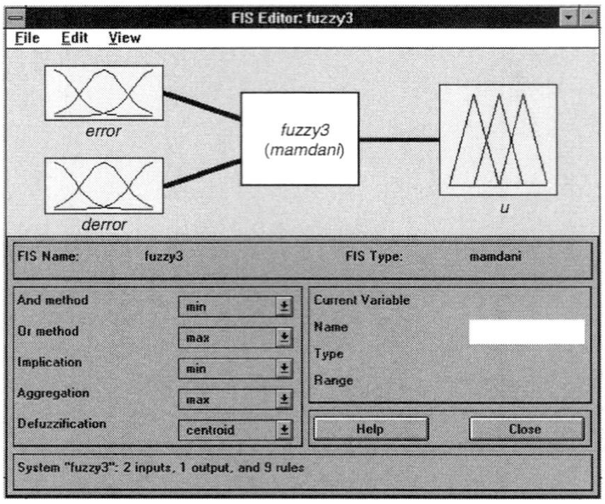

FIGURE 13.16 FIS Editor for fuzzy system based on MAMDANI inference.

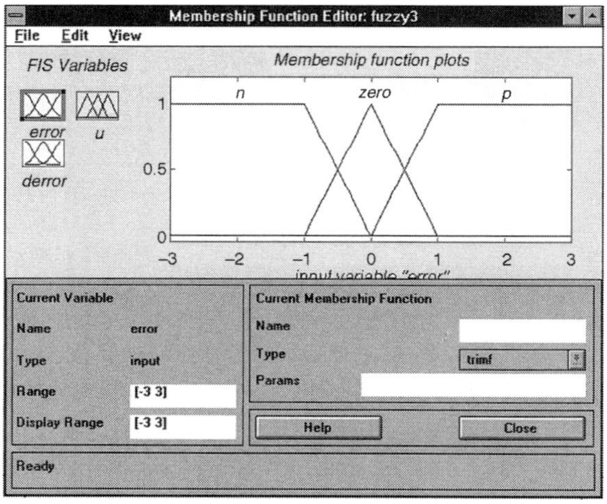

FIGURE 13.17 Membership function associated to the inputs.

For this fuzzy controller there were chosen nine inference rules, which could be visualized using the Inference Rules Editor of SIMULINK. Several simulation results of electrohydraulic axes obtained with the proposed fuzzy controller are shown in Figures 13.18 and 13.19 and depict graphically the position and the velocity of the cylinder.

13.6 Neural Techniques Used to Control the Electrohydraulic Axis

Section 13.6 has as its goals: to emphasize MATLAB's possibilities of using its resources in order to design control systems based on advanced control techniques such as neural networks; to test through simulation these neural algorithms; and to verify performances of the neural control architecture applied to the studied electrohydraulic axis.

There are two main research directions involved in the neural control. One of these implies the developing of one controller going from a neural network, and the other one embeds several controllers

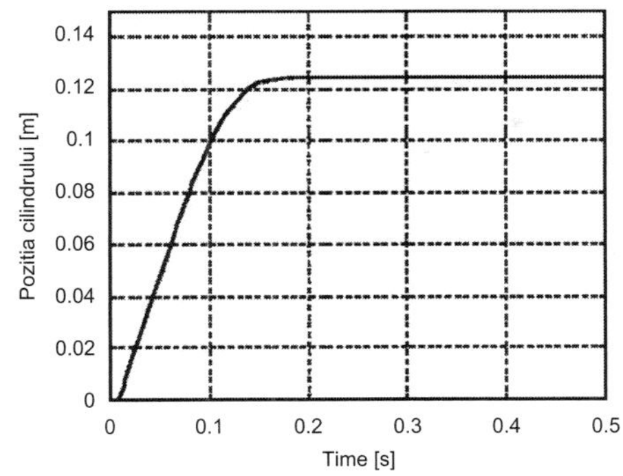

FIGURE 13.18 LHM position.

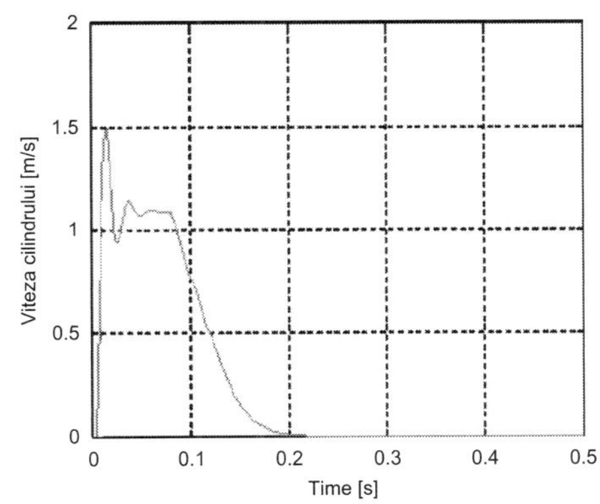

FIGURE 13.19 LHM velocity.

inside a neural network [50]. This section deals with the control of an electrohydraulic axis using a neural controller that has a widely spread structure, namely, multilayer perceptron (MLP).

13.6.1 Neural Control Techniques

13.6.1.1 Learning Based on Mimic

Inspired from biological systems, learning by mimic is applied to control systems. A supervised neural network can mimic the behavior of another system. A first method to develop a neural controller is to replicate a human controller. The neural controller tries to behave like the human operator. Neural training means learning the correspondence between the information received by the human operator and the control input (Figure 13.20).

13.6.1.2 Inverse Learning

The purpose of inverse control is to control a system by using its inverse dynamic. In this case, the neural network receives the output of the system as input and has the input of the system as output. The system works in open loop and has to be in the region where the controller will operate. Inverse learning (Figure 13.21) is an indirect approach to minimize the network output error instead of the overall system error.

13.6.1.3 Specialized Inverse Learning

According to Psaltis, who proposed in 1988 a *specialized inverse learning*, the neural network should be trained online in order to minimize the control error $e_y = r - y$ (see Figure 13.22).

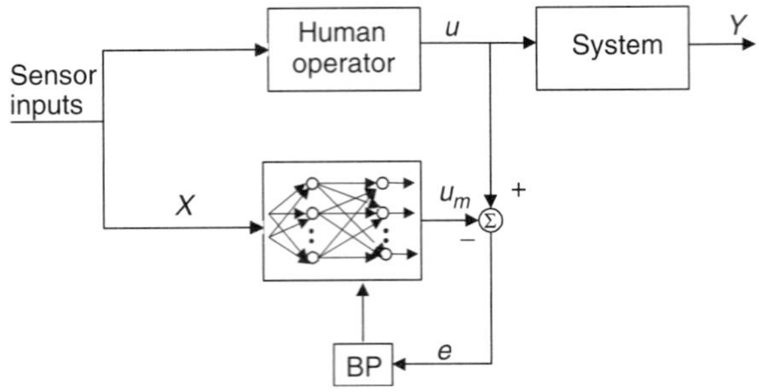

FIGURE 13.20 Diagram for learning based on mimic.

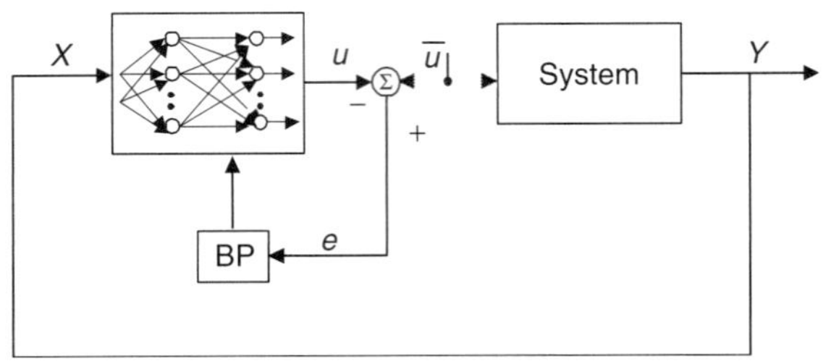

FIGURE 13.21 Training phase at inverse learning.

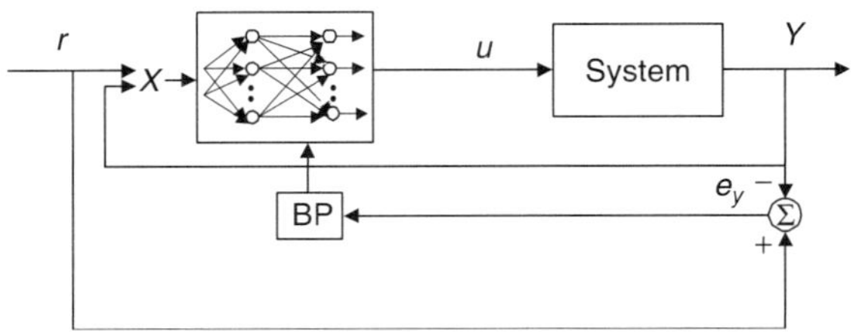

FIGURE 13.22 Specialized inverse control architecture (after [50]).

The neural controller used to control the position of an electrohydraulic axis is a *feed-forward* multilayer neural network, whose learning algorithm is *back-propagation*. In order to adapt the weights which preserve the learned information, two steps are gone through with: a *forward* propagation procedure of the useful signal and a *backward* propagation of the error. The control structure is implemented in SIMULINK as it is shown in Figure 13.23. The neural control of the electrohydraulic axis and the achievement of controller parameters are performed online.

A neural network with four layers, having two neurons on the first layer, a neuron on the last layer, and five neurons on each hidden layer, is proposed. The graphic characteristic corresponding to the axis position and obtained using the neural network described above is illustrated in Figure 13.24.

13.7 Neuro-Fuzzy Techniques Used to Control the Electrohydraulic Axis

This chapter deals with several computer-aided design techniques of hybrid control algorithms. This paper concentrates on these types of algorithms, because the performances achieved through simulation of an electrohydraulic axis with a neuro-fuzzy controller are comparable or superior to those yielded by other control algorithms. Taking into account the novelty of neuro-fuzzy algorithms and the absence in

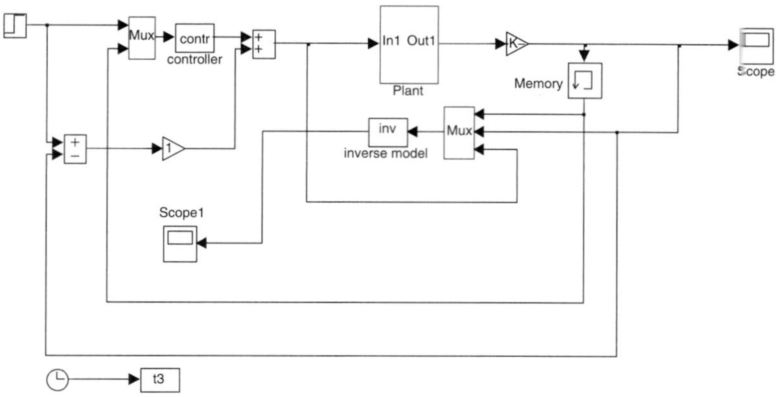

FIGUGE 13.23 The control structure for proposed controllers.

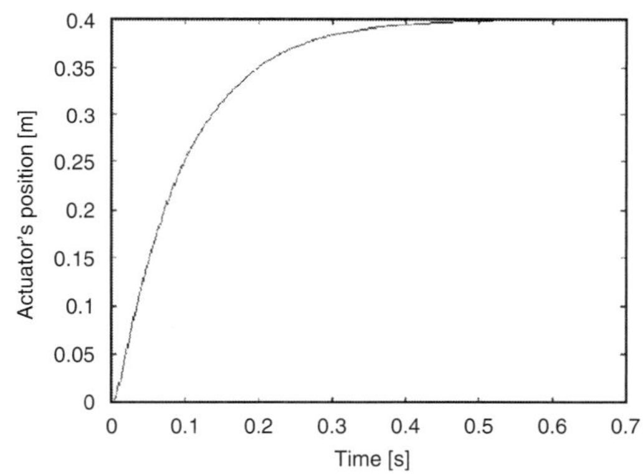

FIGURE 13.24 The axis position for $U = 8$ V input voltage.

SIMULINK of a toolbox devoted to them, the research was oriented to the achievement of a library of C++ programs, which can afford the use of SIMULINK in the design of such controllers. Thus, online adaptation procedures of fuzzy controller parameters are implemented. The comparative study of different classic and advanced algorithms is performed on the basis of integral squared error computed on the transitory horizon.

Because of the capability of fuzzy systems to treat imprecise information, they are strongly recommended in order to express knowledge in the form of linguistic rules. In this way, the human operator's knowledge, which is linguistic or numerical, is used to generate the set of fuzzy if-then rules as a basis for a fuzzy controller. A main drawback of fuzzy systems is the difficulty to design them based on a systematic methodology. To overcome this drawback, the learning procedures from neural networks are successfully applied in order to tune the parameters of membership functions. The merger of neural networks and fuzzy logic has led to the existence of neuro-fuzzy controllers. It can be asserted that neuro-fuzzy controllers embed essential features of both fuzzy systems and neural networks.

The proposed neuro-fuzzy controller has a structure based on the Takagi-Sugeno method and it is depicted in Figure 13.25.

A learning procedure in fact represents a parameter estimation problem. The learning procedure for the proposed neuro-fuzzy controller is gradient-descendent. The method applied to design such a controller is called inverse learning in which an online technique is used to model the inverse dynamics of the plant. The obtained neuro-fuzzy model—the inverse dynamics of the plant—is used to generate control actions.

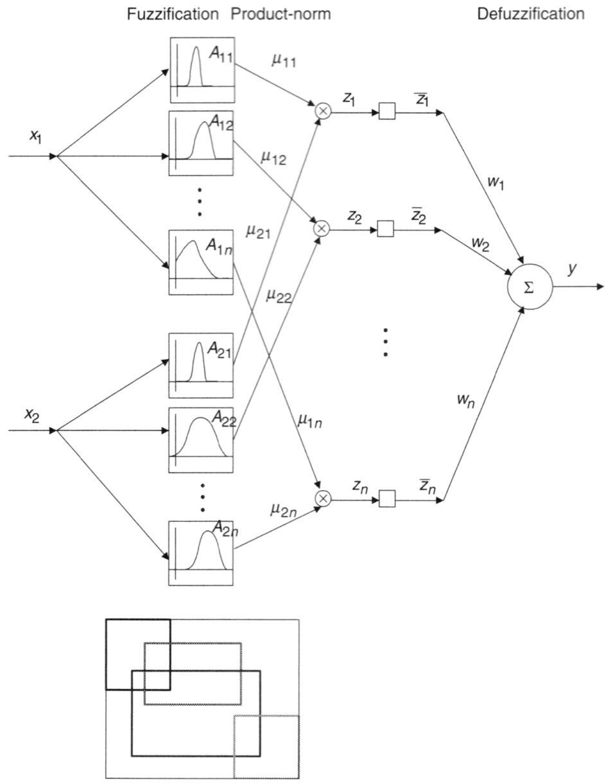

FIGURE 13.25 Structure of the neuro-fuzzy controller.

The neuro-fuzzy controller is a multilayer connectionist system, a multi-input and single-output fuzzy logic system. The network has three layers: one input layer with $n \times m$ units, one hidden layer with n units, and one output layer with one unit [15]. The partition used for this model is a scatter partition [33].

Figure 13.25 presents a particular case where the fuzzy controller has only two inputs and one output. In a general case, the fuzzy controller has m inputs and one output.

The fuzzy rule base contains a set of n linguistic rules in the form:

R_i: If x_1 is A_{1i} and x_2 is A_{2i}
and…
and x_m is A_{mi}
then y is w_i, $i = 1, 2, \ldots, n$

where i is the index of the rule; A_{ji} is a fuzzy set for the jth linguistic variable and the ith rule; and w_i is a number that represents the consequent part.

The membership functions assigned to each input are Gaussian functions. The centers of the membership functions are chosen such that these functions are uniformly distributed over the universe of discourse:

$$\mu_{ji} = e^{-(x_j - a_{ji})^2/2b_{ji}^2} \tag{13.26}$$

The fuzzy inference involved in this neuro-fuzzy controller is the product operator T-norm defined as an *and* conjunction. The firing strength of every rule is

$$z_i = \mu_{1i} \cdot \mu_{2i} \cdot \ldots \cdot \mu_{mi}, \quad i = 1, \ldots, n \tag{13.27}$$

The output is a crisp value obtained as a result of the evaluation of a center of gravity:

$$y = \frac{\sum_{i=1}^{n} z_i \cdot w_i}{\sum_{i=1}^{n} z_i} = \frac{z_1}{\sum_{i=1}^{n} z_i} \cdot w_1 + \frac{z_2}{\sum_{i=1}^{n} z_i} \cdot w_2 + \cdots + \frac{z_n}{\sum_{i=1}^{n} z_i} \cdot w_n = \sum_{i=1}^{n} z_i \cdot w_i \tag{13.28}$$

The parameters to be estimated are obtained by finding the minimum of the following cost function:

$$J(k) = \frac{1}{2} \cdot (y(k) - y_d(k))^2 \tag{13.29}$$

where $y_d(k)$ is the desired output and $y(k)$ is the obtained response at time k.

To minimize this cost function, the stochastic approximation method is used. The learning procedure means the estimation of parameters and is based on the least-mean square algorithm. The parameters to be estimated are

$$p = (a_{11}, \ldots, a_{nm}, b_{11}, \ldots, b_{nm}, w_1, \ldots, w_n) \tag{13.30}$$

The equations to adapt the parameters are the following:

$$\begin{aligned} a_{ji}(t+1) &= a_{ji}(t) - \lambda_a \frac{z_i}{\sum_{l=1}^{n} z_l} \cdot (y - y_d) \cdot (w_i - y) \cdot \frac{x_j - a_{ji}(t)}{b_{ji}^2} \\ b_{ji}(t+1) &= b_{ji}(t) - \lambda_b \frac{z_i}{\sum_{l=1}^{n} z_l} \cdot (y - y_d) \cdot (w_i - y) \cdot \frac{(x_j - a_{ji}(t))^2}{b_{ji}^3} \\ w_i(t+1) &= w_i(t) - \lambda_w \frac{z_i}{\sum_{l=1}^{n} z_l} \cdot (y - y_d) \end{aligned} \tag{13.31}$$

where the learning factors λ_a, λ_b, λ_w are predefined.

In the learning process, parameters that could be modified are (a_{ji}, b_{ji}) which describe Gaussian functions, and w_i, the conclusion values. If the structure of the membership function is established, the only values that could be modified are w_i.

13.7.1 Control Structure

In order to design the neuro-fuzzy controller proposed above, the inverse learning method is applied. The control of an electrohydraulic axis involves the use of an online technique to model the inverse dynamics of the plant. The block diagram for online inverse learning is presented in Figure 13.26.

This scheme is in open loop and it is also found by the controller output error method (COEM) [1] to online tune or adapt the parameters of a fuzzy controller. This method does not require the plant output error to be propagated at the input. There is another constraint, namely the controller has to be capable of stabilizing the plant before the commencement of tuning. To avoid this requirement, a modified COEM (MCOEM) [2] is used. The diagram block in this case is depicted in Figure 13.27.

A proportional feedback controller P is introduced and in this situation the plant input is the sum of $u'(k)$ and $u_p(k)$. The consequent singletons are initialized to zero and the controller P is chosen in such a way that it stabilizes the plant. The structure and the parameters of inverse model and of neuro-fuzzy controller are identical.

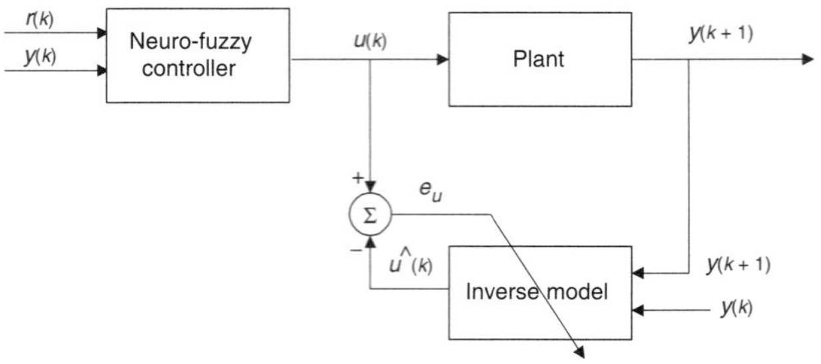

FIGURE 13.26 Diagram of control based on inverse learning.

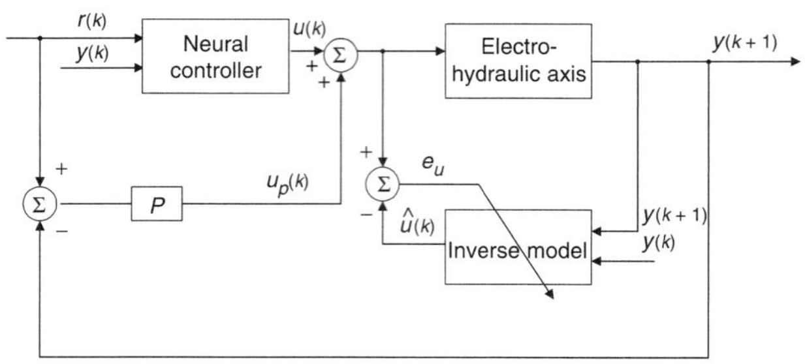

FIGURE 13.27 Block diagram for inverse learning with proportional controller.

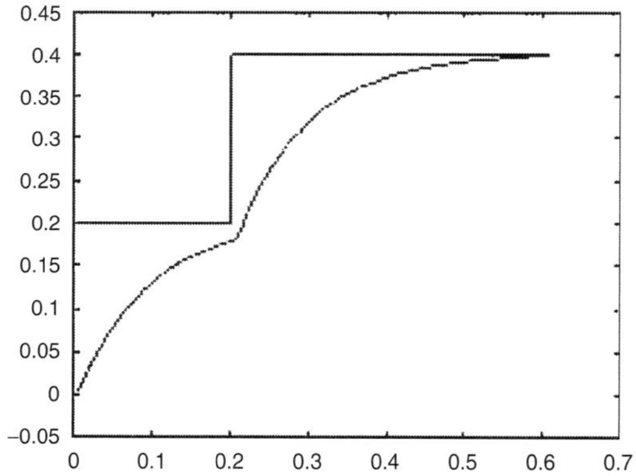

FIGURE 13.28 The position control with neuro-fuzzy controller.

There are two phases in the design of such a controller: the control and the adaptation. In the control phase, the plant output and the reference signal determine a control command $u(k)$. The plant input becomes $u(k)$, the sum of the $u(k)$ and $u_p(k)$. In the adaptation phase, the inverse model, which has as inputs $y(k+1)$ and $y(k)$, produces the signal $\hat{u}\,(k)$ as an output. This signal is used to compute the error $e_u(k)$, which determines the value of the cost function $J(k)$ that has to be minimized.

$$J(k) = \frac{1}{2} \cdot e_u^2(k) = \frac{1}{2} \cdot (u(k) - \hat{u}(k))^2 \tag{13.32}$$

This procedure was used at the control of the electrohydraulic axis position, where the controller parameters are determined online. The actuator position obtained when the reference signal is changed from $U = 4$ V to $U = 8$ V is depicted in Figure 13.28.

TABLE 13.1 Results Obtained by Applying IAE and ISE

Regulator	IAE	ISE
PID (chapter 4)	0.8042	3.4754
PI (chapter 4)	0.8006	3.4618
PD (chapter 4)	0.7928	3.4537
Neural (chapter 6)	0.8027	3.4622
Neuro-fuzzy (chapter 7)	0.7911	3.4501

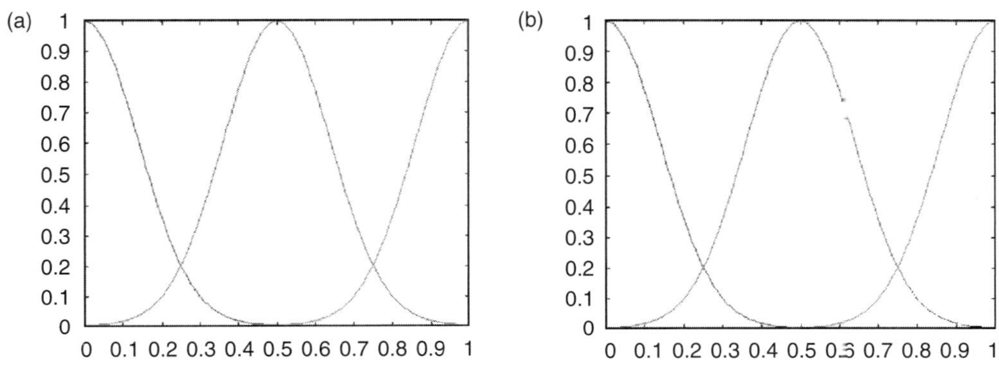

FIGURE 13.29 (a) Membership functions before learning for the variable x_1. (b) Membership functions before learning for the variable x_2.

In order to achieve a comparison of the modern control algorithms (included in this thesis) to the conventional structures, two spread integral criteria, namely, the integral of absolute error (IAE) performance index and the integral of squared error (ISE), are used. The results obtained applying these criteria are included in Table 13.1.

According to previous results, it can be inferred that the described neuro-fuzzy controller exhibits superior performances compared to those obtained with the neural controller based on MLP, or with the classic controllers (PID, PI, PD with filtering) presented in this chapter. The simulation results emphasize the neuro-fuzzy controller, arguing that it represents a very useful tool for practical applications with many nonlinearities.

Optimized results were obtained through variation of data sets and number of iterations. In order to test the performance of the proposed neuro-fuzzy controller, one nonlinear function given by an analytical equation was approximated. The membership functions of input variables x_1 and x_2 before learning are shown in Figure 13.29a,b. The surface obtained after simulation is depicted in Figure 13.30c. One may observe the accuracy of the reconstruction after 300 learning iterations by comparison with the surface to be obtained.

Sets of intermediary results obtained with different simulation data sets are presented below. Different data sets of simulations were used in order to achieve optimized results. Some of them are presented in Figures 13.30–13.34 without comment.

In order to obtain good performances from the model, 10 membership functions are used for each input variable. The learning factors λ_a, λ_b, λ_w were chosen as 0.01. The control algorithm is capable of handling the change in operating range. The results of the electrohydraulic axis simulation with the proposed neuro-fuzzy controller are obtained for various inputs. Those in time domain, results presented in Figure 13.35a,b, correspond to input voltages of 8 and 10 V.

1st set: gw = 0.1; ga = 0.05; gb = 0.05; nepoc = 100; nesant = 100; niter = 200; threshold error = 0.001; Vmax = 0.04.

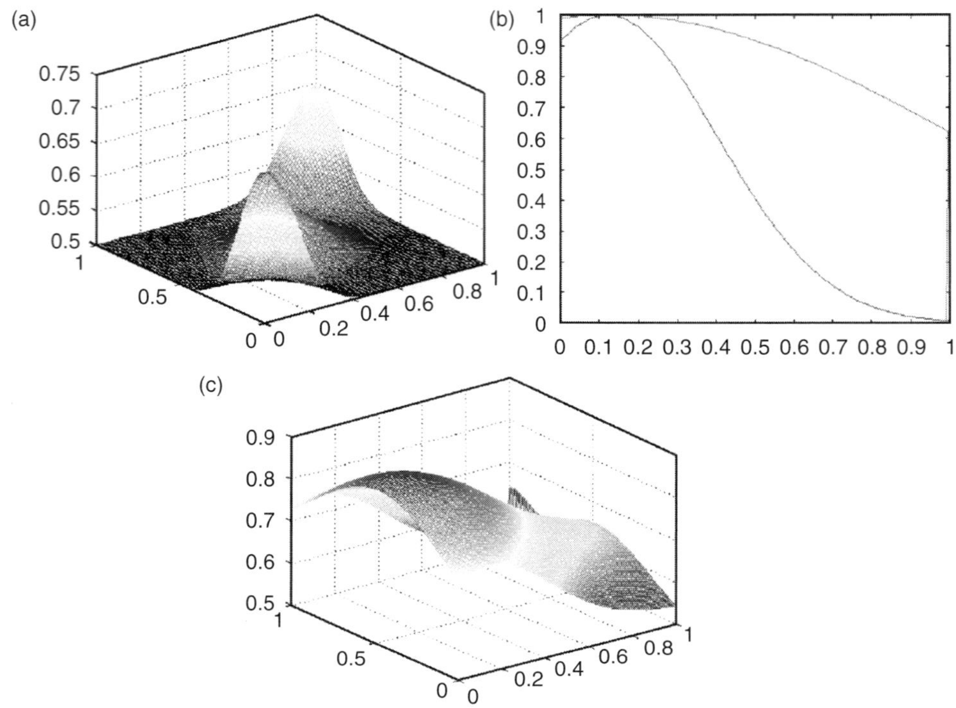

FIGURE 13.30 (a) The surface obtained after the first iteration, (b) membership functions after learning for the variable x_2, (c) the surface obtained after simulation.

2nd set: gw = 0.1; ga = 0.07; gb = 0.05; nepoc = 100; nesant = 21; niter = 200; threshold error = 0.001; V_{max} = 0.0475.

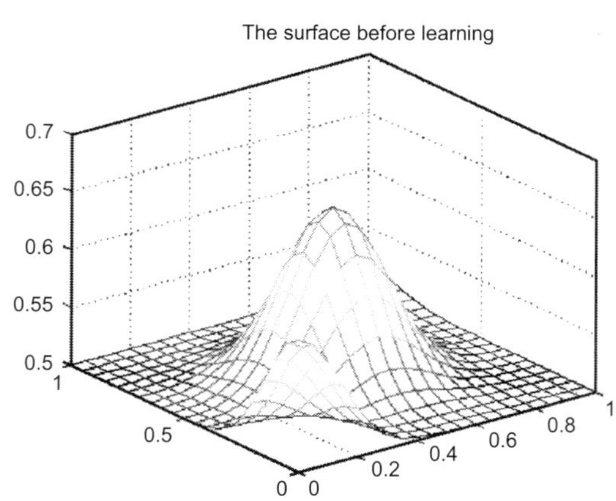

FIGURE 13.31 The surface obtained after the first iteration.

3rd set: gw = 0.5; ga = 0.07; gb = 0.03; nepoc = 200; nesant = 21; niter = 200; threshold error = 0.001; Vmax = 0.047515.

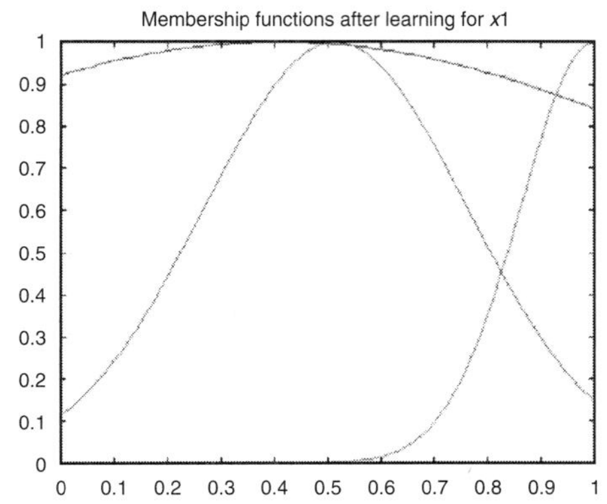

FIGURE 13.32 Membership functions after learning for the variable x_2.

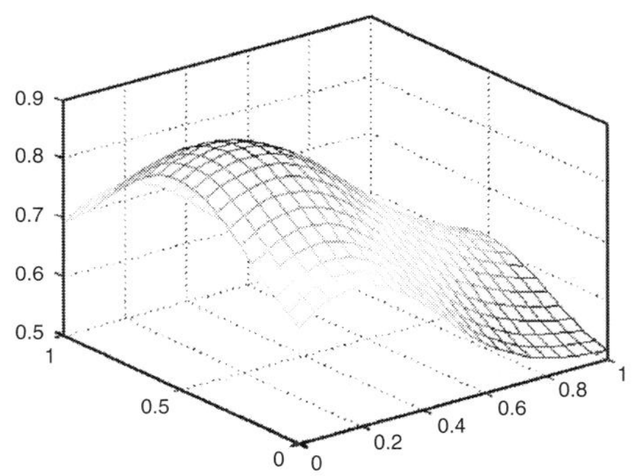

FIGURE 13.33 The surface obtained after learning.

13.8 Software Considerations

The MM of electrohydraulic axis studied in this thesis is supported by a physical installation existing in the mechatronics laboratory of UAS-Konstanz (see Figure 13.1b). Two variants of nonlinear MM are set forth in Section 13.3 and add in static and dynamic nonlinearities that arise in the function of electrohydraulic axis [23,58].

The MM of hydraulic drive presented in the structure of ROBI_3 was implemented in SIMULINK in order to study the dynamic behavior of the axis [26,27]. The extended variant of MM hydraulic axis was done taking into account the relative motion of the constituent parts of this servodrive.

The neural and neuro-fuzzy controller (Takagi-Sugeno) was developed in Borland C++ and implemented in SIMULINK for controlling the electrohydraulic axis. SIMULINK offers the user a FUZZY LOGIC library that allows the designing and modeling of SUGENO or MAMDANI fuzzy inference systems. The lack of a dedicated software to design neuro-fuzzy controllers persuaded the implementation of such a controller in C++ and afterwards the use of it in SIMULINK [26–28].

The support for simulation, SIMULINK 2.1 and MATLAB 5.2 (under Windows), offers solutions to implement our controllers as modules and corresponding icon in a specialized toolbox. In our experiments, we used the facility offered by S-functions and C MEX in conjunction with Borland C++ 5.0 to compile them. We have chosen the C S-function because of the speed necessary to process the information in our block that implement the controller. The block that implements the controller has two inputs (even three inputs are available, though the adaptation process is more complicated) and one output.

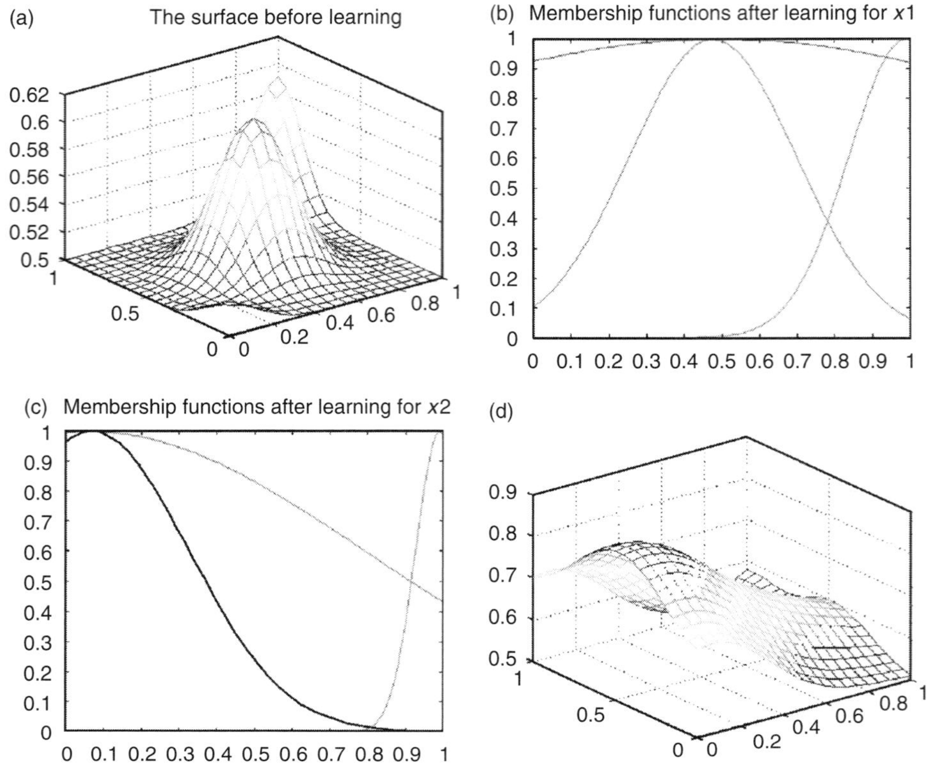

FIGURE 13.34 (a) The surface obtained after first iteration, (b) membership functions after learning for the variable x_1, (c) membership functions after learning for the variable x_2, (d) the surface after learning.

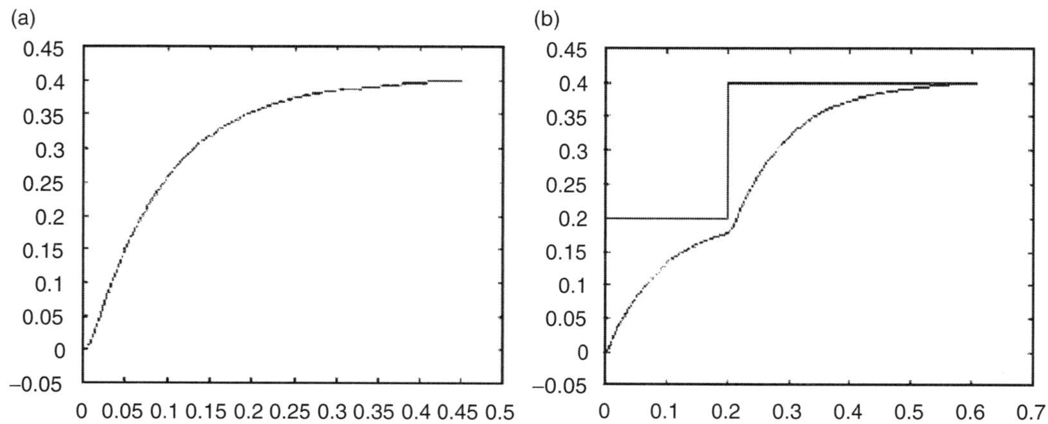

FIGURE 13.35 (a) The position control with neuro-fuzzy controller ($U = 8$ V), (b) the position control with neuro-fuzzy controller ($U = 10$ V).

The adapting parameters (weights, centers, and spread for Gaussian function) must be persistent. Declaring global, static or using the workspace in order to store the are useful techniques to accomplish the task.

The newest version of Simulink offers the possibility to write wrapper S-function, to use the callbacks functions, and as an alternative ADA or Fortran programming language.

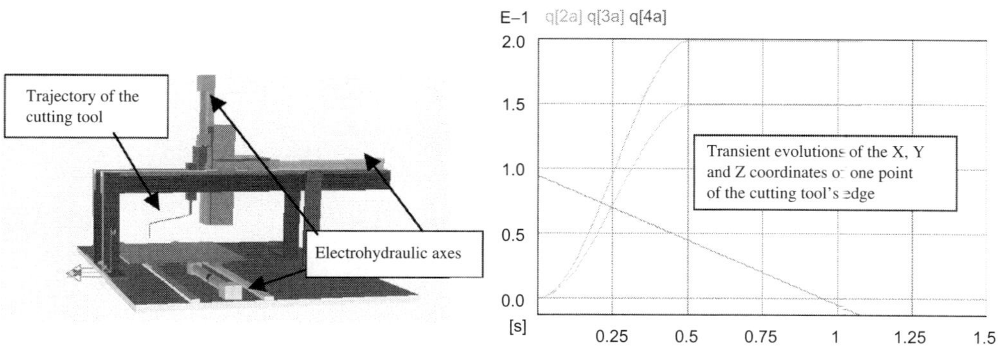

FIGURE 13.36 3-Axes Cartesian robot modeled, controlled, and simulated in SDS [21,24].

13.9 Conclusions

The research achieved as part of this chapter has, as an essential purpose, the development of improved control structure based on advanced techniques (neural and fuzzy), in relation to the conventional one. The studied electrohydraulic axis is a component of the Cartesian robot ROBI_3 implemented in the mechatronics laboratory of UAS-Constance. 3D-simulations with direct and inverse dynamics and implemented controllers by using the SDS-Modelling and Simulation software of the real installation were performed [21,23,25,28]. Model and simulation results are presented in Figure 13.36.

As an overview, Section 13.1 deals with the introduction of this chapter. In Section 13.2, the most important aspects of electrohydraulic system control and of nonlinearities that arise with this type of installation are pointed out. The robot, ROBI_3, is presented from both a component and control perspective.

The mathematical model (MM) of ROBI_3's hydraulic axis is described in Section 13.3. The nonlinear MM is achieved based upon technical data of different components of installation and also taking into account theoretic assessments of electrohydraulic installation functionality. Simulation results of nonlinear MM placed into a position loop are obtained with simulation environment SIMULINK/MATLAB. As a general remark, it should be mentioned that all simulations included in this chapter are achieved by using MATLAB/SIMULINK, while the advanced control algorithms (neural and neuro-fuzzy) are developed in Borland C++ 5.0.

Section 13.4 contains a short overview of the theory devoted to conventional controllers PID, PD, PI, and observer, followed by simulation results of the electrohydraulic axis endowed with the above control structures.

Following the scientific goal of this contribution, Section 13.5 reviews fuzzy system theory and presents simulations of electrohydraulic axis with fuzzy controllers. Fuzzy system theory has contributed greatly to system modeling, and the development of a theoretical frame appropriate to implement the qualitative reasoning specific to human beings. This kind of reasoning is very useful to model complex systems, which are characterized by nonlinearities or imprecise information. Simulation results of hydraulic axis are obtained using SUGENO and MAMDANI fuzzy controllers.

A short introduction of neural networks theory, the most widespread neural structures and also neural control techniques are presented in the beginning of Section 13.6. Neural networks work quantitatively, numerically. If fuzzy logic has an inference based on uncertainty, then neural networks learn by training, at the end of which the network will approximate a desired function. The analysis of trained NN involves many challenges, and as a result, the rules are usually not extracted from trained NN. Simulation results included in Section 13.6 are obtained with a multilayer NN.

Neuro-fuzzy systems preserve the characteristics of NN and also of fuzzy systems, and have been used successfully in control in recent years. Section 13.7 is devoted to the neuro-fuzzy system theory, to the presentation of neuro-fuzzy controller implemented in Borland C++ and applied in SIMULINK to electrohydraulic axis, to the simulation results achieved in this case, and to the comparative study of conventional and modern controllers.

Section 13.8 contains a concise presentation of this chapter, the main contributions to the subject area presented, as well as a listing of perspective areas of interest in order to pursue further research in this direction.

Without intending to confine the parameters of this chapter, following is a listing of possible research directions and development perspectives that may be followed in future research endeavors:

- Applying various controllers implemented in SIMULINK not only to control the electrohydraulic axis discussed, but also for systems with very complex structure which are involved in large hydraulic installations, offering the user a neuro-fuzzy controller's library;
- The hardware implementation of described neuro-fuzzy controller;
- Continued research in the development of an optimal controller, systemically based (through the further study of stability utilizing linear matrix inequalities—LMI);
- The integration of presented controllers in software packages dedicated to hydraulic and pneumatic fields (for instance in HYPAS[23], DSH, etc.);
- The development of controller design in order to promote those controllers, which allow a better symbiosis between classical and advanced methods (neuro-fuzzy, genetic algorithms);
- The extension of preoccupations and extrapolation of research results regarding control of velocity, acceleration, pressure, flow, force, moment, and power.

References

1. Andersen, H.C., Lotfi, A., Tsoi, A.C. A new approach to adaptive fuzzy control: the controller output error method, *IEEE Trans. on Systems, Man, and Cybernetics*, SMC-27-B(4), August 1997.
2. Abonyi, J., Nagy, L., Szeifert, F. Indirect adaptive Sugeno fuzzy control, *Proceedings in Artificial Intelligence*, FNS'98, München, Germany, march 19–20.
3. Backé, W. *Systematik der hydraulischen Widerstandsschaltungen in Ventilen und Regelkreisen*. Krauskopf-Verlag, Mainz, 1974.
4. Costa Branco, P.J., Dente, J.A. *Inverse-Model Compensation Using Fuzzy Modeling and Fuzzy Learning Schemes*. Intelligent Engineering Systems through Artificial Neural Networks, Smart Engineering Systems: Fuzzy Logic and Evolutionary Programming, Ed. C.H. Dagli, M. Akay et al., Vol. 6, ASME Press, New York, pp. 237–242, 1996.
5. Brown, M., Harris, C. *Neuro-Fuzzy Adaptive Modelling and Control*. Prentice-Hall, Englewood Cliffs, NJ, 1994.
6. Catana, I., Vasiliu, D., Vasiliu, N. *Servomecanisme electrohidraulice. Constructie, functionare, modelare, simulare si proiectare asistata de calculator*. U.P.B. Bucuresti, 1995.
7. Cybenko, G. *Mathematical Problems in Neural Computing*. Signal Processing Scattering and Operator Theory and Numerical Processing, M.A. Kashoek, J.H. van Schupper, A.C. Ram, 1989. Vol. 3. pp. 47–64.
8. Driankov, D., Hellendoorn, H., Reinfrank, M. *An Introduction to Fuzzy Control*. Springer-Verlag, Berlin, 1993.
9. Dubois, D., Prade, H., Ughetto, L. Checking the coherence and redundancy of fuzzy knowledge bases, *IEEE Trans. on Fuzzy Systems*, 5(5):398–417, 1997.
10. Dumitrache, I. sa. *Automatizari Electronice*. Editura Didactica si Pedagogica, Bucuresti, 1993.
11. Dumitrache, I., Catana, I., Militaru, A. *Fuzzy Controller for Hydraulic Servosystems*. IFAC International Workshop on Trends in H&P Components & Systems, Chicago, IL, 1994.
12. Föllinger, O. *Regelungstechnick*. Dr. A. Hütig Verlag, Heidelberg, Germany, 1978.
13. Friedrich, A. *Logik und Fuzzy-Logik*. Expert-Verlag, 1997.
14. Ghaoui, L. El. Reduced-order multimodel control using linear matrix inequalities: sufficient conditions, *Proc. Od ACC 1993*, pp. 633–634, 1993.

15. Godjevac, J., Steele, N. Adaptive neuro-fuzzy controller for navigation of mobile robot, *International Symposium on Neuro-Fuzzy Systems AT'96,* Conf. Report, EPFL-Lausanne, 1996.
16. Gupta, M.M. *Fuzzy Logic and Neural Networks,* Proc. of the 2nd International Conference on Fuzzy Logic & Neural networks, Iizuka, Japan, 17–22 July, pp. 187–188, 1992.
17. Healey, M. *Principles of Automatic Control.* The English Universities Press Ltd., 1975.
18. Haykin, S. *Neural Networks,* MacMillan College Publishing Company, New York, 1994.
19. Ionescu, Fl. *Computer aided design of hydraulic and electrohydraulic drive installations,* Proc. of the9th Triennial World IFAC Congress, Budapesta, Ungaria, Pergamon Press, Vol. 1, pp. 569–574, 1984.
20. Ionescu, Fl., Stoffel, B. *Contribution to the Automatic Generation of Mathematical Models for the Computer Assisted Analysis and Synthesis of Hydraulic Drive Systems.* Proc. of the 2nd Intern. Conf. on Fluid Power, Tampere, Finland, 19–21 March, pp. 469–482, 1991.
21. Ionescu, Fl., Haszler, Fl. *TORCH: A Control Software for Electrohydraulic Cartesians Robots.* Proceed. of the 6th Intern. IMEKO Symposium on Measurement and Control in Robotics, ICMR'96, Bruxelles, Belgium, 9–11 May, pp. 484–489, 1996.
22. Ionescu, Fl. *Non-Linear Problems in the Hydraulic Drive Systems.* 2nd World Congress of Nonlinear Analysts, Athena, Greece, 10–17 July 1996, Pergamon Press, Vol. 30, part 3, pp. 1447–1461.
23. Ionescu, Fl., Vlad, C.I. Tools of HYPAS for the control of electrohydraulic drive installations, *Proc. of 7th Symposium on Computer Aided Control Systems Design,* Gent, Belgia, 1997, pp. 311–316.
24. Ionescu, Fl., Borangiu, Th., Vlad, C.I. High integrated CAD strategies for control design of electro-hydraulic systems, *Proc. 3rd IFAC Conference SSC,* Bucharest, pp. 390–395, 1997.
25. Ionescu, Fl., Vlad, C.I. *Hypas tools for the control of electro-hydraulic drive installations Journal* ***a***, Vol. 38, No. 3, Belgium, pp. 38–41, 1997.
26. Ionescu, Fl., Vlad, C.I. Sugeno and hypas fuzzy-control solutions for electro-hydraulic drive installations, *Proceedings EUFIT'97,* Aachen, Germany, 8–11 Sept., Vol. 2, pp. 1238–1242, 1997.
27. Ionescu, Fl., Vlad, C.I., Arotaritei, D. Fuzzy and neuro-fuzzy HYPAS controllers implemented for an electro-hydraulic axis, *International ICSC Symposium on Engineering of Intelligent Systems EIS'98,* Teneriffe, Spain, Feb. 11–13, 1998.
28. Ionescu, Fl., Arotaritei, D., Vlad, C.I. *Modelling of Nonlinearities, Signal Reconstruction and Predictive Solutions Applied in Mechatronics Systems by Using Neuro-Fuzzy Systems,* Internal Report, Department of Mechatronics, FH-University of Applied Sciences-Konstanz, 1998.
29. Ionescu,V., Varga, A. *Teoria Sistemelor. Sinteza robusta. Metode numerice de calcul.* Editura ALL, Bucuresti, 1995.
30. Isermann, R. *Digitale Regelsysteme,* Springer-Verlag, Berlin, 1987.
31. Isermann, R. *Zur Anwendung der Fuzzy-Logik in der Regelungstechnik.* Automatisierungstechnische Praxis (atp) Fuzzy-Control, 38, Oldenbourg Verlag, Germany, 1996.
32. Isermann, R. On fuzzy logic application for automatic control, supervision, and fault diagnosis, *IEEE Transactions on Systems, Man, and Cybernetics—Part A: Systems and Humans,* Vol. 28, No. 2, March 1998, pp. 221–235.
33. Jang, J.-S.R., Sun, C.-T., Mizutani, E. Neuro-fuzzy and soft computing, *A Computational Approach to Learning and Machine Intelligence,* Prentice-Hall, Englewood Cliffs, NJ, 1997.
34. Joh, J., Hong, S.K., Nam, Y., Chung, W.J. On the systematic design of Takagi-Sugeno fuzzy control systems, *International ICSC Symposium on Engineering of Intelligent Systems EIS'98,* Tenerife, Feb. 1998.
35. Kandel, E.R. *Nerves Cell and Behavior,* Principles of Neural Sciences, 3rd ed., pp. 18–36, 1992.
36. Knappe, H. Comparison of conventional and fuzzy-control of non-linear systems, In: Kruse, R., *Fuzzy Systems in Computer Science,* Verlag Vieweg, Wiesbaden, Germany, 1994.
37. Kokotovic, P.V. *Lectures Notes in Control and Information Sciences.* Springer-Verlag, Berlin, 1991.
38. Kosko, B. *Neural Networks and Fuzzy Systems.* Prentice-Hall, Englewood Cliffs, NJ, 1992.
39. Kovacic, Z., Balenovic, M., Bogdan, S. Sensitivity-based self-learning fuzzy logic for a servo-system, *IEEE Control Systems,* June, 1998.

40. Lippmman, R. An introduction to computing with neural nets, *IEEE ASSPMagazine,* April 1987, pp. 4–22.
41. MATLAB 5.2. MathWorks Corp, USA. 1998.
42. Miller, Th., Sutton, R., Werbos, P.J. *Neural Networks for Control.* MIT Press, 1990.
43. Nauck, D., Klawonn, F., Kruse, R. *Neuronale Netze und Fuzzy-Systeme.* Grundlagen des Konnektionismus, Neuronaler Fuzzy-Systeme und der Kopplung mit wissensbasierten Methoden. Vieweg, 1994, Germany.
44. Nesterov, Y., Nemirovski, A. *Interior Point Polynomial Methods in Convex Programming: Theory and Applications,* SIAM, Philadelphia, 1994.
45. Pedrycz, W. *Fuzzy Control and Fuzzy Systems,* Wiley, New York, 2nd ed., 1993.
46. Piechnik, M., Feuser, A. *Simulation mit Komfort–HYVOS 4.0 und MOSIHS 1.0,* Ö & P, 38, 1994.
47. Postlethwaite, B.E. A model-based fuzzy controller, *Trans IChemE,* Vol. 72, Part A, Jan. 1994.
48. Postlethwaite, B.E. Building a model-based fuzzy controller, *Fuzzy Sets and Systems,* 79 (1996), Elsevier.
49. Rehfeldt, K., Shöne, A., Büngener, N. *Einsatz von Fuzzy-Reglern zur Drehzahlregelung einer Hydraulikpumpe, Ölhydraulik und Pneumatic,* 36, Nr. 6, pp. 397–402, 1992.
50. Ronco, E., Gawthrop, P.J. *Neural Networks for Modelling and Control.* Technical Report: csc97008, Centre for System and Control, Dept. of Mechanical Engineering, Univ. of Glasgow, 10 Nov. 1997.
51. Simulink, Dynamic System Simulation for MATLAB, Writing S-functions, The Math Works Inc., 1998.
52. Sontag, E.D. Mathematical control theory, *Deterministic Finite Dimensional Systems.* Springer-Verlag, Berlin, 1990.
53. Takagi, T., Sugeno, M. Fuzzy identification of systems and its applications to modeling and control, *IEEE Trans. Systems, Man, and Cybernetics,* Vol SMC-15, No. 1, pp. 116–132, 1985.
54. Tanaka, K., Sugeno, M. Stability analysis and design of fuzzy control systems, *Fuzzy Sets and Systems,* Vol. 45, pp. 135–156, 1992.
55. Teodorescu, H.N. *Sisteme Fuzzy si Aplicatii.* Institutul Politehnic Iasi, Romania, 1989.
56. Tertisco, M., Penescu, C., Ionescu, G., Ceanga, E. *Identificarea Experimentala a Proceselor Automatizate.* Editura Tehnica, Bucuresti, 1971.
57. Viersma, T. J. *Analysis, Synthesis and Design of Hydraulic Servosystems and Pipelines.* Elsevier, Amsterdam-New York, 1980.
58. Vlad, C.I. Contributions to the Direct Computer Control of Electrohydraulic Axes for Industrial Robots. Technical University "Politehnica," Bucharest, Romania, 1998.
59. Wang, L., Liu, G.P., Harris, C.J., Brown, M. *Advanced Adaptive Control,* Pergamon, 1997.
60. Werbos, B. *Overview* of *Design and Capabilities.* In Neural Networks for Control, pp. 59–65, MIT Press, MA, 1990.
61. Westcott, J.H. The minimum-moment-of-error-squared criterion: a new performance criterion for servo mechanisms, *Proc. of IEE.,* Measurements Section, pp. 471–480, 1954.
62. Yager, R., Zadeh, L. *Fuzzy Sets, Neural Networks and Soft Computing,* 1994.
63. Zadeh, Lotfi. *Fuzzy sets, Information & Control,* No. 8, pp. 338–353, 1965.
64. Zadeh, L., King-Sun Fu, Tanaka, K., Shimura, M. *Fuzzy Sets and their Applications to Cognitive and Decision Processes.* Academic Press, 1975.
65. Zimmermann, H.-J. *Fuzzy Sets Theory—and Its Applications,* Kluwer Academic Publishers, Dordrecht, The Netherlands, 1990.

14

Design Optimization of Mechatronic Systems

Tomas Brezina
Ctirad Kratochvil
Cestmir Ondrusek
Technical University of Brno

14.1 Introduction

Electromechanical systems form an integral part of mechanical and mechatronic systems. Their optimization is a necessary condition for a product to be competitive. In engineering practice, a large number of optimization and identification problems exist that could not be solved without the use of computers [5]. The present level of technological development is characterized by increasing the performance of machines with the production costs kept at a satisfactory level. The demands on the reliability and safety of operation of the designed machines are also considerable.

From practical experience we know that the dynamic properties of electromechanical systems have a considerable influence on their reliability and safety. On the other hand, the tendency to push the price of a machine down often leads to unfavorable dynamic properties that result in increased vibrations and noise during operation. Also, electrical properties dramatically deteriorate as the amount of active materials in a machine is reduced. The increased load leads to, among other things, excessive heat formation, which, in turn, has a negative effect on insulation, shortening the service life of a machine.

14.2 Optimization Methods

14.2.1 Principles of Optimization

The properties of electromechanical systems can be described mathematically using physical quantities. The degree of these properties is then described using mathematically formulated objective (preference) functions. Structural parameters ranging between limit values given as satisfying secondary conditions are the independent variables of these functions. The particular form of the functions depends on the type of machine and its mathematical description. The solutions of a mathematically formulated optimization problem together with optimization methods allow a considerable number of different design variants of a machine to be calculated in a relatively short time. They also make it possible to perform these calculations at production planning stages for a prototype to possess the qualities given by a chosen criteria function. In this way, the design of a machine is not only analyzed but also modified and reconstructed in terms of its electromechanical properties with the aim to improve these properties as much as possible (or optimize them).

From a physical point of view, these are actually problems that, to a certain degree, are inverse to those of calculus. A problem of calculus assumes a fixed, mathematically described model of a real machine to be used for deriving its resulting properties. In problems of calculus, we define properties and try to find out which parameters of a chosen mathematical model possess those properties. In problems of parametric optimization, we look for those parameters that, by a chosen preference function, provide the best properties. It is clear that problems of synthesis and optimization are much more sophisticated than those of calculus.

14.2.2 Parametric Optimization

As the aim is to find the values of certain structural parameters of a machine, we shall deal with this notion in more detail. By a parametric optimization of electromechanical systems, we mean the process of finding those parameters of a mathematical description of the system (arranged in a vector $\bar{p}$) from a set P of admissible parameters at which a suitably selected objective (preference) function $\psi(p)$ of these parameters reaches its extreme.

The objective function $\psi(p)$ quantifies the degree of the properties of an electromagnetic system that has to be made extreme (the parameters with the best degree of this property have to be chosen). When defining an admissible set P, we are guided by the structural possibilities of changes in individual parameters (variables), or we can introduce secondary criteria of the type "the degree of properties may not exceed given critical limits." The possibility of taking into consideration the structural changes of parameters leads to the so-called trivial (natural) constraints of the type $p_i^d \leq p_i \leq p_i^h$, where p_i^d is the lower and p_i^h the upper bound of the ith optimization variable. The introduction of secondary criteria leads to the definition of limiting functions q_i of optimization variables for which we have $q_i^d \leq q_i(p) \leq q_i^h$, where q_i^d is the lower and q_i^h is the upper bound of the relevant function.

Thus, from the mathematical point of view, parametric optimization of electromechanical systems is formulated as the problem of finding a point p in the admissible set P, at which the preference function ψ reaches its global extreme value (maximum or minimum) with regard to P. The admissible set is generally described by m inequalities defined by functions $q_j(p)$, where $j = 1, 2, \ldots, m$. If $P = R^s$, where s is the number of variables to be optimized, we say that the optimization is unconditional. In all other cases we say that the optimization is conditional.

To solve the problem of optimizing the selected properties of a system, the following has to be done:

- A mathematical description has to be formulated.
- It has to be analyzed at the starting point.
- The desired form of the objective function ψ has to be specified.
- The optimization variables have to be selected.
- The desired form of the constraining functions q_j has to be specified.
- A suitable optimization method has to be selected.
- The resulting mathematically formulated optimization problem has to be solved.
- Using the mathematical model, the results have to be transformed back into the dynamic model (for dynamic problems only).

14.2.3 General Aspects of the Optimization Process

If the aim of an optimization process is to optimize several properties that simultaneously affect the system (such as minimizing the size values while respecting the electrical properties), we obtain a multi-criteria objective function. The objective function then takes the form of a weighted sum of single-criterion functions. Each of these functions generally assumes its local minima at different points of the optimization parameter spaces. This is the reason why a multi-criteria function can have a large number of shallow local minima or is insensitive to changes in the optimization parameters. Due to this fact, the selection of an optimization method is of great importance. The result is averaged in the sense that several criteria may participate simultaneously in a reduction of the multi-criteria function, while some other criteria may increase.

A more suitable method may be to select a single-criterion objective function, including all criteria in the constraints. Only the most significant criterion is chosen for the objective function to be specified in the subsequent process. All other criteria included in the constraints are kept within specified limits without being optimized. Thus, the results of an optimization process are dependent on the degree of reduction of the admissible set given by the inequality-type constraints.

Generally, we specify the constraints in a form similar to the objective function

$$q_i(p) = f_i(p) - f_i^h, \quad i = 1, 2, \ldots, m^* \tag{14.1}$$

Here f_i are suitable functions of a vector variable and f_i^h their maximum admissible values.

The selection of optimization variables is given by the sensitivity of the objective function to changes of relevant optimization variables. This sensitivity is described by the gradient vector of the objective function.

$$\text{grad } \psi(p) = \left[\frac{\delta\psi(p)}{\delta p_1}, \ldots, \frac{\delta\psi(p)}{\delta p_s}\right]^T \tag{14.2}$$

14.2.4 Types of Optimization Methods

14.2.4.1 Standard Optimization Methods

Most practical problems lead to nonlinear (transcendental) systems of equations. These may only be solved using numerical optimization methods. According to the order of the derivatives used in the application of a method, numerical methods of finding local minima of functions of several variables may be divided into:

1. Zero-order methods (comparative)
   - Methods of co-ordinate comparison
   - Simplex methods
   - Stochastic methods
2. First-order methods (gradient and quasi-gradient)
   - Methods of associated directions
   - Variable-metric methods
3. Second-order method (Newton method)

Stochastic Methods

These methods consist of calculating the values of the objective function at a large number of selected points. The points are selected by such criteria that each point in the space has an equal probability of being selected. The best points are then determined by comparing the function values. From the outlined strategy, it follows that these methods lead to computing the function values at a large number of points, which may protract the calculation. On the other hand, we can more easily reach the global optimum of the function to be optimized. These methods also comprise the evolution methods since the first solution population is generated completely by random. The difference only consists in the strategy of selecting better solutions.

14.2.4.2 Evolutional Optimization Methods

Since some problems are difficult to solve by standard numeric optimization methods, even if they converge to an acceptable optimum in a reasonable time, new approaches had to be developed. Therefore, a number of new methods have been designed based on the laws of natural genetics copying them in various degrees. These methods employ random generation of input parameters and so can be thought of as stochastic methods. In general, stochastic optimizing algorithms (including virtually all the evolutional algorithms) optimize using multi-parameter function with "wild" behavior, that is, with many minima or with an unknown gradient. Stochastic optimization methods are necessarily slower than heuristic approaches, which take advantage of the fact that they know the type and details of the function to be optimized. Unless the conditions for the global optimum are previously known, we can never be sure whether we have reached the global optimum to be able to terminate the optimization process. However, stochastic optimization methods also bring numerous benefits. They are generally very well specified and thus applicable virtually to any problem, and they can get out of the trap of a local minimum. The evolutional process of searching the space of potential solutions requires an equilibrium of two objectives:

- To find the nearest (mostly local) minimum as quickly as possible
- To search the space of all potential solutions in the optimum manner

The methods differ in their orientation towards these two objectives and they can be roughly ordered in a sequence starting with methods tending to local minima to methods searching a large number of potential solutions:

1. Stochastic "hill climbing" algorithms
2. Tabu search algorithms
3. Simulated annealing algorithms
4. Genetic algorithms

Hill Climbing Algorithm

This is the simplest optimization algorithm being a variant of the gradient method "without gradient" where the direction of the steepest climb is determined by searching the neighborhood. This algorithm also has all the drawbacks of gradient methods, in that it is very likely to end up in a local extreme without reaching the global minimum. Here the starting solution is generated at random. For the currently designed solution, a certain neighborhood is generated using a finite set of transformations and the best minimum is chosen from this neighborhood. The local solution obtained in this way is then used as the center of a new neighborhood in which the optimization is repeated. This process is iterated a specified number of times. In the course of this process the subsequent best solutions are recorded to be finally used as the resulting minimum. The basic drawback of this algorithm is that, after a number of iterations, it may revert to a local minimum that has already been passed in a previous step (the problem of looping). This problem can be avoided by running the algorithm several times with different randomly generated initial values to eventually choose the best result achieved.

Tabu Search Algorithm

At the end of the 1980s, Professor Fred Glover designed a new approach to solving the problem of finding the global minimum, which he called *tabu search*. At present, this method is among those used most frequently to solve combinatorial problems and problems of finding the global minimum. Based on the hill-climbing algorithm, it tries to eliminate the problem of looping. The hill-climbing algorithm is equipped with a so-called short-time memory, which, for a short previous interval of the algorithm history, remembers the inverse transformations to locally optimal solution transformations used to obtain the new centers in iterations. These inverse transformations are prohibited (tabu) when the new neighborhood is created for a given current solution. In this way, the looping caused by falling into the trap of a local minimum may substantially be reduced. A hill-climbing algorithm modified in this way systematically searches the entire area in which the global minimum of a function is to be found.

Simulated Annealing Algorithm

Apart from the stochastic methods and methods based on natural evolution, there is another possibility of simulating the evolution of systems based on the physical evolution of macroscopic systems. The annealing of a solid body in order to remove the internal stress is a simple example of this kind of evolution. For a physical interpretation of this process, consider a body that is heated until it reaches a high temperature. The temperature is then gradually lowered. The atoms of a body heated to a high temperature can easily overcome the local energetic barriers to reach equilibrium states. When the temperature is lowered, atoms are fixed in this state and the cooled off body is without internal stress.

This principle was used to design the method of simulated annealing. First, an initial temperature T_{max} is set, whose value is important for the method to be efficient. The simulated annealing algorithm then searches the space of all potential solutions in a strongly stochastic way, also accepting the states that correspond to solutions worse than the current one. This property of simulated annealing is a characteristic feature of this method and provides a way of escaping from a local minimum trap, thus allowing the search of another area of the entire solution space. However, as the annealing temperature T is lowered, the probability of accepting worse states as well is diminishing. For small temperature values then, only solutions better than the current one are considered.

Genetic Algorithm (GA)

Genetic algorithms (GAs) are most frequently used to optimize the parameters of an unknown system whose mathematical description is either too complicated or unknown [5]. When applying a GA, it is mostly sufficient to know a function assigning a price to each individual in the population. This may be the error of the solution for randomly selected parameters during GA. Since a GA is looking for a maximum, the error, which, on the contrary, is being minimized, must be transformed into looking for a maximum. This may be done in several different ways: by subtracting the error from the maximum error occurring, by calculating the inverted value of the error, or by using another transforming function that approaches zero as the error approaches one. Increased attention should be paid to setting up the program implementing the pricing function since it consumes the most computing time compared with the other GA components.

Apart from general optimization problems, GAs are mostly applied to neural networks. Here the tendency is to employ GAs at two different levels. First, for finding suitable weights for a neural network and second, when optimizing the structure of a neural network, that is, when selecting the algorithm, the number of input neurons in the hidden layers, the number of hidden layers, etc. Using a genetic algorithm to optimize the parameters of another genetic algorithm (the size of the population, the number of crossbreedings, the extent of mutations, the frequency of mutations) is a very revolutionary idea (optimization of the computation time where the computation time is a pricing function of the GA). As far as applications of GAs to problems encountered in research of electric machines are concerned, GAs have been used to identify the parameters of the substitution diagram of an induction motor.

By way of conclusion, it may be added that genetic algorithms perform surprisingly well when all other algorithms fail, such as for incomplete problems where the computation time is an exponential or factorial function of the number of variables. There is no point in using GAs to optimize relatively simple functions or functions for which special algorithms exist for their description. Considering the necessity to calculate the function values for tens or hundreds of genetic chains in a population and the necessity to evaluate hundreds or even thousands of populations during a single run of the program, GAs are rather time-consuming.

Despite the positive results achieved by using GAs, it is clear that nature must use even more intricate and, at the same time, not very sophisticated methods. The GAs described above only correspond to very primitive examples observed in nature, particularly those related to asexual reproduction with a single chromosome. Since nature has taken billions of years to test its algorithms, it is highly efficient to further learn from it. It is interesting that it needs no mathematics to solve complicated problems of optimization. Nevertheless there are other optimization methods suitable to solve the problems of the design [2–4].

14.2.5 Selection of a Suitable Optimization Method

The standard gradient method is still one of the most frequently used methods. Gradient methods or even the standard non-gradient methods (such as the simplex method) are not suitable if the finding of the global minimum is required of a function with many local minima. Mostly, these methods only reach an insignificant minimum close to the starting point (the initial solution) in which they are trapped. This deficiency is mostly removed by repeatedly selecting at random the initial solution of an optimization problem and taking the best result for the solution. The stochastic character of this process can only be seen in the random selection of the initial solution. The subsequent optimization algorithm then proceeds without any randomness. Then the evolutional optimization methods are thought of as stochastic ones despite their employing of a certain strategy when choosing the getter points. The following are the main differences between a genetic algorithm and the more frequently used gradient method:

- GA performs no gradient computation, which might be difficult and time consuming particularly for large systems.
- GA works with randomly generated solutions and may converge more quickly to the global minimum.

To optimize the draft design of an induction motor, an optimization method was employed using a genetic algorithm. This method is described in more detail in the following chapter.

14.3 Optimum Design of Induction Motor

14.3.1 IM Design Introduction

Actual design of an induction motor (IM) usually depends on the requirements of individual customers, who specifically define parameters which a designed machine should accomplish. In this way, with the same machine output, we can obtain different implementations that meet individual conditions more or less. It is possible to require a good quality of one parameter only with the deterioration of other parameters. We are going to deal with a design of motors of (0,6–200) kW outputs. Motors are designed for permanent load and with the project assignment the following input values are required:

Machine output P_n [kW], voltage U_{1n} [V], winding connection Y/D, number of poles 2**p** or rotation speed n [min^{-1}], grid frequency f [Hz], efficiency η[%], power factor cos φ, insulation class, IP implementation, and the shape of the machine.

We consider squirrel-cage motors in closed implementation with framework and cooling ribs. There is a cast aluminum rotor cage. For the design, data such as conductors and slots dimension or magnetic characteristics deducted from tables and graphs that are given by the standard or by the manufacturer's measurement, are needed. The actual design is a compromise between individual design parameters, so that a resulting machine would have the best possible operating characteristics with a perfect heat and material utilization. The actual motor design is described in the following section.

14.3.2 Classical IM Design Evaluation

An induction motor design, when carried out manually, represents hundreds of calculations, which can last tens of hours even with an experienced constructor. As computers made their way into practically all branches of design and analysis, a series of programs which co-operate with a designer in an interactive fashion and speed up a calculation were created.

In spite of indisputable advantages of this design process, we have to realize that there is a remarkable quantity of various design implementations of the given motor which, more or less, achieve the required motor operating characteristics. This approximates the global minimum of an objective function, which evaluates the design quality.

Thus, the idea to create a program for searching the whole state space of all possible solutions and selecting such a variant, which is the most appropriate to an evaluating objective function (the required

TABLE 14.1 Generated Parameters List and Setup of Their Limits

Parameter Name	Symbol	Dimension	High Limit	Low Limit
Stator outside diameter	D_e	mm	User optional	User optional
Stator inside diameter	D	mm	User optional	User optional
Ideal iron length	l_i	mm	User optional	User optional
Air gap induction	$B\delta$	T	0.5	1.0
Stator slot filling	k_{dr1}	—	0.6	0.75 (0.8)
Air gap size	δ	mm	0.2	0.4
Stator current density	σ_1	A mm^{-2}	3.0	15.0
Rotor rod current density	σ_t	A mm^{-2}	2.0	6.0
Rotor ring current density	σ_k	A mm^{-2}	2.0	4.0
Teeth magnetic induction	B_z	T	1.6	2.0
Slot number per pole and stator phase	q_1	—	2.0	5.0

motor characteristics) using some of the optimization methods, was developed. The stochastic evolutionary method genetic algorithm was selected, because it searches the whole state space of all possible solution in a best way.

14.3.3 Description of a Solved Problem

14.3.3.1 Generated Parameters

The values given in Table 14.1 are recommended only, and, for the motors of outputs below 200 kW, they are mostly limiting values. Varying the parameter values or substituting one parameter for another is possible only by intervention in the program source text. Diameter limits D_e and D from the input file are considered only in the case that a motor design without regard to standardized axis height is required. In the case that standardized axis height is entered, these limits are calculated. Limits of an ideal rotor length are appropriate to enter as narrow as possible for faster convergence to a limit. But this is not a required condition. Generally, the lower the range of individual parameters, the faster the convergence to a global minimum, and the number of local minimums is lower.

14.3.3.2 Objective (Criterion) Function

It is not just the form of the objective function, but also the selection of optimized parameters that is important for good optimization results. Selected parameters must sufficiently describe a quality solution to the given problem. In the case of an induction motor design optimization task, the following parameters were selected:

Motor volume	V [dm^3]
Motor temperature rise	ϑ_n [K]
Motor nominal power factor	$\cos\varphi_n$ [–]
Motor nominal efficiency	η_n [–]
Torque overload capacity	m_{pn} [–]

These parameters are most important for the quality of the design and describe the design sufficiently.

The total error is based on the relation given in Equation 14.1, a sum of individual partial errors of each controlled parameter. If we put more emphasis on some parameter, we increase a corresponding weight coefficient, thus achieving its improvement in the final design. At the same time, values of other parameters will decrease. Finding an optimal setup of gain coefficients is one of the most important and difficult problems. The term "optimal setup" means that the designed motor has the highest power factor, efficiency, and torque overload capacity values at the minimal volume and simultaneously does not exceed a permitted temperature rise for a selected insulation class. We have the relationship

$$\begin{aligned}\varepsilon(\mathrm{GR}_i) &= \mathrm{abs}(kV\cdot V) + \mathrm{abs}(k\vartheta(0.89\,\vartheta_d - \vartheta_n)) + \mathrm{abs}(k_{\cos\varphi}(1-\cos\varphi_n)) \\ &\quad + \mathrm{abs}(k\eta(1-\eta_n)) + \mathrm{abs}(k_{mp}(m_p + 1 - m_{pn}))\end{aligned} \tag{14.3}$$

TABLE 14.2 Input Values of 5.5 kW, 380 V Motor

Quantity Name	Symbol	Dimension	Value
Nominal power output	P_n	W	5500
Nominal voltage	U_{1n}	V	380
Required power factor	$\cos\varphi$	—	0.81
Required efficiency	η	—	0.86
Grid frequency	f	Hz	50
Motor axis height	H	mm	132
Number of pole pairs	p	—	3
Temperature class of insulation	TT	—	F
Torque overload capacity	m_p	—	2

where kV is the volume weight coefficient, $k\vartheta$ is the temperature factor weight coefficient, $k_{\cos\varphi}$ is the power factor weight coefficient, $k\eta$ is the efficiency weight coefficient, k_{mp} is the torque overload capacity weight coefficient.

14.3.4 Achieved Results

14.3.4.1 5.5 kW, 380 V Motor Design Description

During program development and tuning, an optimization was performed on the motor described in Table 14.2. In this section we describe the results and problems encountered in the optimization process. The symbols and quantities, which are not explained in detail, were either used in the previous text or are listed at the list of used quantities at the beginning of the document. The motor input parameters are given in Table 14.2.

14.3.4.2 Other Results

It follows from the physical principle that optimized quantities are closely related. Increase of a gain of one quantity results in a disadvantage of the quantities. Based on the performed optimizations, it can be concluded that two kinds of motors exist, depending on the content of iron and copper:

1. Motor with prevailing iron content, high stator current density, good power factor, for a price of worse efficiency of the motor and with slightly worse torque overload capacity than the second motor.
2. Motor with high copper content and conversely low stator current density, good efficiency, and worse power factor value. Torque overload capacity is good.

The type of motor is determined based on the setting of gain coefficients. A sum of temperature and power factor errors on one side impacts the sum of volume and efficiency errors on the other side. The torque overload capacity can be good for both kinds of motors.

Results of individual optimizations are listed in Table 14.3, ordered by volume value from smallest to largest. Different varieties of motors were obtained, depending on values of gain coefficients. It is difficult to determine which solutions are good or bad, because the selections depend on actual customers' requirements. The solution that gives the best value of optimized quantity is marked in bold. Solution numbers 1, 5, 8, 23, and 25 can be considered successful from this perspective. The motor described above (solution no. 2) serves for depicting of the task. Previously-mentioned relations between individual quantities can be observed in Table 14.3, which lists the optimization results without limiting the generated parameter, thus using the requirements in Table 14.1.

Next, a motor optimization was performed with just one optimized parameter, when the gain of the other parameters was set equal to zero.

1. *Volume optimization.* In this case, the algorithm selected as the best solution motors with minimal dimensions, when parameters D_e, D, and l_i were converging to minimum preset limits.
2. *Temperature optimization.* The algorithm reached first reached a local minimum with temperature at the maximum based on the required insulation class, and mostly stayed on this value.

TABLE 14.3 Motor $P = 5.5$ kW, $U = 380$ V Solutions List, without Generated Parameters Limited

Number	V [dm^3]	ϑ [K]	$\cos\varphi$ [–]	η [–]	m_p [–]	Directory
1	**3.96**	88.1	0.798	0.834	1.72	Motor1
2	4.20	86.9	0.818	0.843	1.90	Motor2
3	4.31	74.9	0.787	0.865	1.77	Motor3
4	4.32	88.8	0.836	0.817	1.78	Motor4
5	4.33	75.1	0.690	**0.973**	1.07	Motor5
6	4.50	89.0	0.836	0.834	1.79	Motor6
7	4.51	86.8	0.818	0.818	1.93	Motor7
8	4.54	**90.0**	0.884	0.812	1.98	Motor8
9	4.56	84.6	0.857	0.816	1.74	Motor9
10	4.58	86.5	0.836	0.817	1.77	Motor10
11	4.63	68.2	0.792	0.858	2.10	Motor11
12	4.69	88.4	0.862	0.808	1.80	Motor12
13	4.70	73.4	0.845	0.830	2.25	Motor13
14	4.73	61.0	0.799	0.871	1.90	Motor14
15	4.78	78.1	0.853	0.858	1.67	Motor15
16	4.78	71.0	0.767	0.870	1.80	Motor16
17	4.81	70.6	0.703	0.934	1.28	Motor17
18	4.97	54.5	0.804	0.883	1.90	Motor18
19	5.08	55.5	0.762	0.877	2.20	Motor19
20	5.12	88.2	0.879	0.806	2.05	Motor20
21	5.96	44.0	0.784	0.870	2.55	Motor21
22	6.35	42.4	0.803	0.882	2.69	Motor22
23	6.40	87.5	**0.887**	0.853	2.27	Motor23
24	6.57	59.0	0.747	0.956	1.16	Motor24
25	7.05	42.3	0.793	0.865	**3.00**	Motor25
26	7.39	52.9	0.714	0.986	1.03	Motor26

3. *Power factor optimization.* An effort to achieve the first type of motor (see above discussion) with low copper content, high current density σ_1, worse value of efficiency. Torque overload capacity was good.
4. *Efficiency optimization.* The designed motor corresponded to the second type of motor (see above discussion) with prevailing copper content, low current density σ_1, and good efficiency, however, with worse power factor values. Torque overload capacity was good.
5. *Torque overload capacity optimization.* The motor is designed with high number of slots for pole and phase, resulting in gradual spread of conductors on the perimeter. The motor can have prevailing iron or copper content depending on a local solution, to which it converged. It can have good values of power factor and efficiency for a price of machine volume increase.

14.4 The Use of a Neuron Network for the Identification of the Parameters of a Mechanical Dynamic System

The basic step used to solve the dynamic tasks by means of any type of modeling is to create a set of important quantities that include both the quantities describing structure, conditions, and the interactions of technical objects and the quantities that characterize the consequences (i.e., their demonstration and behavior).

The methods of creating the mathematical models in drive systems, in general an interactive process, utilize

- The applications of well-known physical principles that describe the phenomena in drive systems (e.g., the second Newton's principle, Kirchhoff's laws, etc.)
- The applications of the methods based on artificial intelligence algorithms (e.g., genetic algorithms [1] and artificial neuron networks [6,7])

The theories on which the methods of artificial intelligence are based replace the "standard" analytical and numerical methods when

- These are the only theories which can solve the problem.
- They exhibit better properties from the point of view of the problem solution (e.g., a better conditionality considering the changes of input values).
- They allow the problems to be solved more effectively.

The last case is typical when we want to approach the real operational conditions as close as possible. From various methods of artificial intelligence, the stochastic evolution algorithms and the artificial neuron networks are being increasingly utilized in the field of the modelling of drive interactive systems. In the following section, two methods are shown that are applied to the problems of the analysis of dynamic properties in drive systems.

The solution of dynamics by means of the algorithms of artificial intelligence represents a solution of the following partial problems:

- Specifying the set of important (relevant) quantities
- Selecting the theory which is suitable to solve the problems
- Arranging the relations among relevant quantities so that they allow the selected algorithm of artificial intelligence to be used
- Generating the training data, and the selection of the method of teaching, for example, in neuron networks
- Testing the quality of the results reached and their evaluation

14.4.1 Practical Application

Many identification methods are known and very often verified quite well in practical terms. The limit factors that make these procedures more difficult (e.g., the assumptions about system linearity, stationarity, and normality of the phenomena which occur in the systems, etc.) are also known. Hence, we have used an untraditional approach to the problem *f* identifying the dynamic properties in mechanical systems for which the use of neuron systems seems to be promising, and at the same time available for engineers' thinking.

14.4.1.1 A Practical Application—Gearbox

The first case that was analyzed is a vehicle gearbox. Inputs (engine load), outputs (frequency-amplitude spectra of torsional oscillations), and the gearbox structure were known. The ilete information was related to the selected parameters of stiffness and damping in the drive. Due to variable operational conditions, the magnitude of damping may vary enough to significantly affect output characteristics. If many experimental results are evaluated, some typical failures can be identified and their possible occurrence can be anticipated from output files. We have used the data that were measured in the real system. The frequency-amplitude spectrum of torsional oscillations was measured at the gearbox shaft, which is seated on four bearings with five speed gears (Figure 14.1). Originally, the measurement was carried out to determine the resonance of frequency systems with the goal to reduce noise. The following parameters were set in the system:

Stiffness $K \in \{0.3, 12.0, \infty\}$
Damping $B \in \{-, 0, 0.3, 12.0\}$

The measurements were done with testing frequencies $f = 512$ Hz and $f = 1024$ Hz. To record significant oscillation harmonics, the excitation frequency was flexible in both cases:

1. From 2.5 to 14.0 Hz in steps of 0.5 Hz
2. From 14.0 to 40.0 Hz in steps of 1.0 Hz

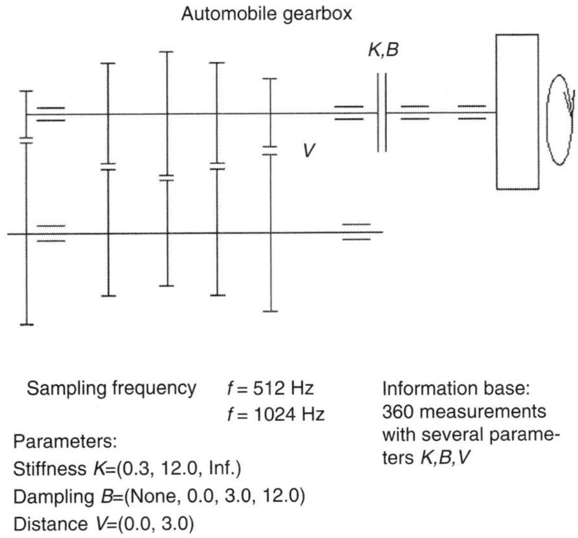

FIGURE 14.1 Five speed gear.

TABLE 14.4 Expected Natural and Excitation Frequencies for the Gearbox

Table of Expected Frequencies		[Hz]
Low frequencies		up to 5.0
Operational frequency (OF) (speed)	OF	14.16
	2 × OF	28.32
	3 × OF	42.48
Interharmonic frequency (IHF)	0.5 × OF	7.08
	1.5 × OF	21.25
	2.5 × OF	35.42
Natural frequency (NF)	I.NF	43.91
	0.5 × I.NF	21.96
	2 × I.NF	87.82
	II.NF	322.1
	0.5 × II.NF	161.1
	2.0 × II.NF	644.2
Combination frequency (CF)	2 × OF + 0.5 × I.NF	50.28
	OF + I.NF	58.07
	2 × OF + I.NF	72.23
	2 × OF + 2 × I.NF	116.1
Tooth frequency (TF) {TF = z.OF}	1st speed gear	184.1
	2nd speed gear	325.7
	3rd speed gear	424.9
	4th speed gear	580.6

The spectrum always included 512 spectral lines. The measurement was repeated 360 times for different variations of the parameters K and B (the system adjustment). The values of natural and excitation frequencies, which are expected for the gearbox to be analyzed, are shown in Table 14.4.

14.4.1.2 Task Definition

The task was originally defined in the following way: to estimate the corresponding magnitudes of parameters K and B (this means, to recognize the adjustment of parameters used in the mechanical dynamic system) by means of the artificial neuron network on the basis of the frequency-amplitude

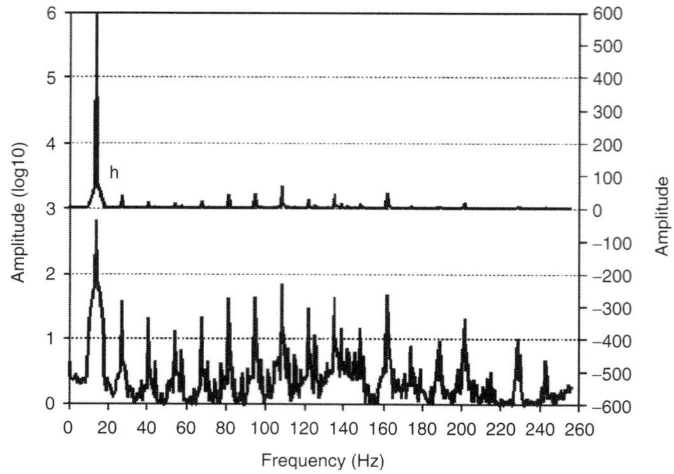

FIGURE 14.2 Stimulus vectors normalization (top, before normalization; bottom, after normalization).

spectrum of torsional shaft oscillations in the given system. A multilayer perceptron with three layers (i.e., input, hidden, and output) is used as the configuration of the neuron network. Select the signal neuron functions as the linear, hidden, and output layers of the logistic function, $f(x) = 1/(1 + e^{-x})$.

According to this definition, this is to identify the system parameters on the basis of the measured frequency-amplitude spectra. However, the parameters are taken from discrete sets (and very low), and the task could be redefined as the "standard" task of the spectrum classification according to seven attributes (each attribute corresponded to one of the possible values of the parameters K and B). The application of neuron networks to solve such a problem is more successful when compared to the solution of the original task.

The amplitudes of spectral lines were expressed in logarithm scale, and a reduction of spectral dynamics with an increase of their informative quality has been achieved. Considering the nonlinear nature of the activation neuron functions used, which extends beyond the saturation range for the input interval $\langle 0.5, 0.95 \rangle$, the network cannot respond well to stimulus vectors with a high range of the values in the individual components. This is illustrated in Figure 14.2. The input network layer was configured to 512 input neurons. The amplitude logarithmic value of one spectral line was entered into each input. The individual neurons in the output layer correspond to the classification attributes. Because there are seven attributes, seven neurons were configured in the output layer.

The only-hidden layer was set as the arithmetic mean of the number of input and output neurons. Two hundred sixty neurons were configured to the only-hidden layer, as illustrated in Figure 14.3. Each item corresponded to one measurement of the frequency spectrum (a stimulus vector) with a corresponding attribute vector (a vector of the required responses). The specific variation of the parameters was expressed by the required network response to two corresponding output neurons equal to 1, and the remaining output neurons equal to 0.

From the original 360 items, 36 items were randomly separated (10% of total) for the future tests. We ensured that the network tests would be carried out with the items that have not passed the training network process (the network was not trained to these situations). This is necessary to verify the generalization model properties. The training set was formed by the remaining 324 items. The sequential strategy of teaching was used, i.e., the items from the training set were used in the teaching process with the fixed sequence (cyclic passages through the training set). Taking into consideration the size of the neuron network to be configured, the method of feedback moment propagation has been selected as the teaching method that exploits only the information up to the first-order inclusively (the values of a special function—a teacher and his gradient), and it has not used Hessian or its estimate, which, in this case, would be very demanding.

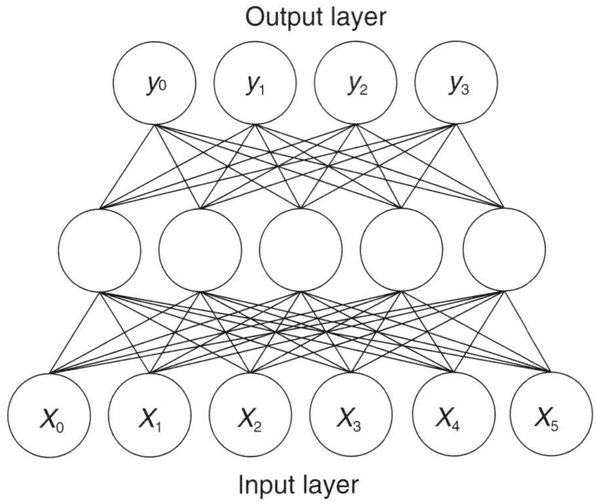

FIGURE 14.3 Network configuration (hidden layer contains neurons without labels).

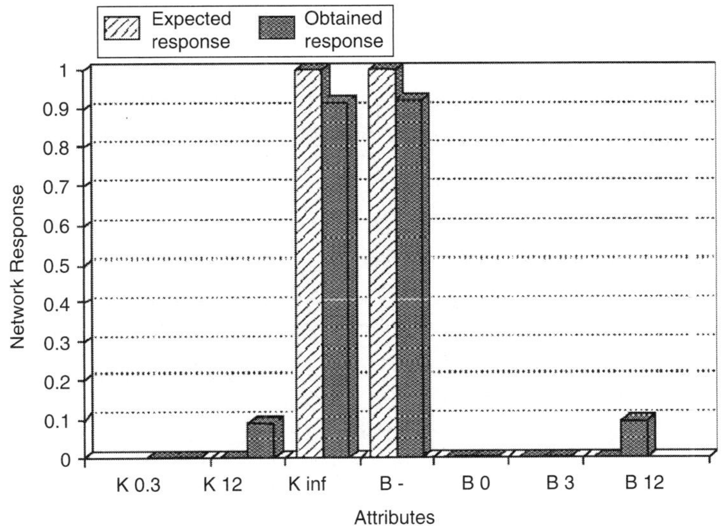

FIGURE 14.4 Successful response of neural net.

14.4.1.3 Results

The neuron model of the mechanical system has manifested a high rate of success during the verification by test sets. The network was taught with random selections of the test items. During testing of the individual models, the responses of the network were successful in 85–95% of all cases (see Figures 14.4 and 14.5). Moreover, the estimate of the values K and B that correspond to the parameters is available within a couple of seconds for the frequency spectrum in the active mode. However, it is possible that a model with higher quality will be achieved if special optimizing techniques are used in the future.

In summary, the neuron model of the mechanical system described above can be assessed as usable in practical terms.

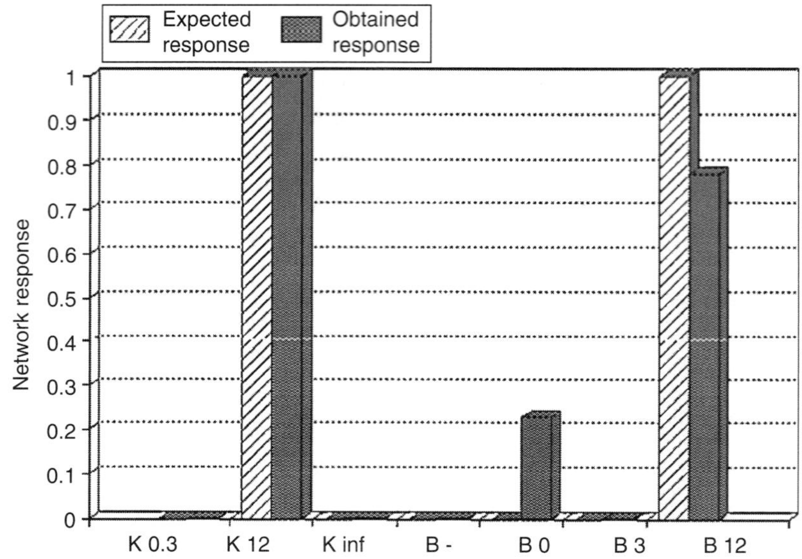

FIGURE 14.5 Failure response of neural net.

References

1. Goldberg, D., *Genetic Algorithms in Searching, Optimisation and Machine Learning,* Reading, MA: Addison-Wesley, 1989.
2. Glover, F., Lagunai, M., Marti, R., Fundamentals of scatter search and path relinking, *Control and Cybernetics,* pp. 653–684, 2000.
3. Glover, F., *Scatter Search and Star-Paths—Beyond the Genetic Metaphor,* New York: Springer-Verlag, pp. 125–137, September 1995.
4. Glover, F., Kelly, J.P., Langunai, M., Genetic algorithm and tabu search—hybrids for optimization, *Computers and Operations Research,* pp. 111–134, January 1995.
5. Lee, J., Hajela, P., Parallel genetic algorithm implementation in multidisciplinary rotor blade design, *Journal of Aircraft,* Vol. 33, No.5, pp. 962–969, September–October 1996.
6. Hagan, M.T., Demuth, H., Beale, M., *Neural Network Design,* Boston: PWS Publishing, 1996.
7. Kosko, B., *Neural Networks and Fuzzy Systems,* Englewood Cliffs, NJ: Prentice-Hall, 1992.
8. Ye, X., Loh, N., Dynamic system identification using recurrent radial basis function network, *Neural Networks Theory, Technology, and Applications,* New York: IEEE Technology Update Series, 1996.

15

Motion Control

Rahul Kulkarni
National Instruments, Inc

15.1 Introduction to Motion Control

15.1.1 Definition

Motion control involves precisely controlling the position, velocity, and torque of a rotational or linear electromechanical device. Examples of such devices include hydraulic pumps, linear actuators, and electric motors.

15.1.2 History

In 1831, Michael Faraday discovered electromagnetic induction—electricity being generated in a wire by means of the electromagnetic effect of a current in another wire, and magneto-electric induction—electricity being generated in a wire by means of the electromagnetic effect of a magnet. These results led to the creation of the modern electric motor.

The industrial revolution put the motor to use in a variety of applications. A single motor was typically used by multiple machines connected to a drive train using belts for transmission and gears for speed reduction. A clutch was used to engage and disengage different machines on the drive train.

Complex machines were designed for mass production using a motor connected to amechanical system based on gears and cams. Hundreds of parts were produced on the same line as long as no parameter in the design changed.

As world moves from mass production to mass customization, machine designers want the same machine to handle a variety of different products at different times in the day, with minimal

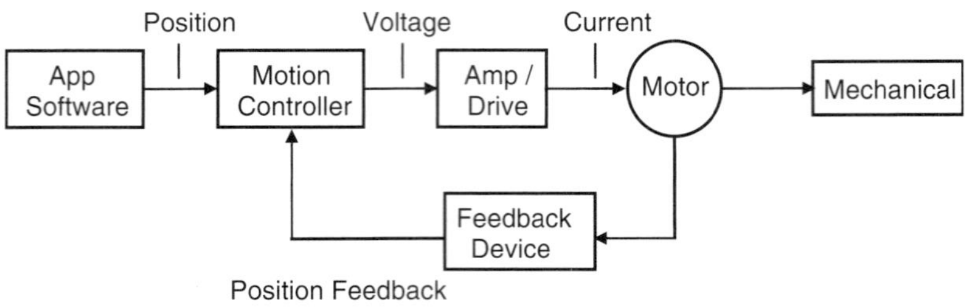

FIGURE 15.1 Components of a motion control system.

changeover time. Servo-based motion control systems enabled tight control over position and move profiles. Thus most machines today are complex electromechanical systems involving servo-based control.

15.2 Components of a Typical Motion Control System

A typical motion control system consists of a variety of components including user interface, motion controller, drive, motor, feedback device, and motion I/O. These components work together seamlessly to control the position, velocity, or torque of an electromechanical device. Figure 15.1 shows the different components of a motion control system.

15.2.1 User Interface

The user interface component accepts commands from the engineer or operator and sends them to the motion controller for execution. These commands include parameters such as target position, and motion control profiles for smooth movement. The user interface typically includes a graphical interface with switches and knobs to resemble the control panel of a machine in operation, and elements such as graphs or XY plots to represent the current position of the object in motion.

15.2.2 Motion Controller

A motion controller is at the center of a typical motion system, which consists of supervisory control, trajectory generation, and a control loop. The controller converts high-level user commands into command signals that drives use to move actuators. The motion controller also monitors the system for error conditions, faults, and asynchronous events that can cause the system to change speed, direction, or start/stop the actuators.

15.2.3 Drive

In a motion control system, the drive is the component that translates the command signals from the controller to current that produces torque or rotation in the motor. Drives can be classified on the basis of the type of motor they actuate—brushed servo, brushless servo, or stepper, or on the basis of the level of computation carried out in the drive—analog or digital. We will learn more about different types of drives in Sections 15.4 and 15.5.

15.2.4 Motor

Motors turn electrical energy into mechanical energy and produce the torque required to move to the desired target position. Motors are designed to provide torque to some mechanical elements including linear slides, robotic arms, and special actuators. There are three main types of motors—brushed DC

TABLE 15.1 Types of Motors

	Pros	Cons	Applications
Stepper Motors	Inexpensive, generally run open loop, good low-end torque, clean rooms	Noisy and resonant, poor high-speed torque, not for hot environments, not for variable loads	Positioning, micro movement
Brushed DC Servo Motors	Inexpensive, moderate speed, good high-end torque, simple drives	Maintenance required, no clean rooms, brush sparking causes EMI and danger in explosive environments	Velocity control, high-speed position control
Brushless Servo Motors	Maintenance-free, long lifetime, no sparking, high speeds, clean rooms, quiet, run cool	Expensive and complicated drives	Robotics, pick-and-place, high-torque applications

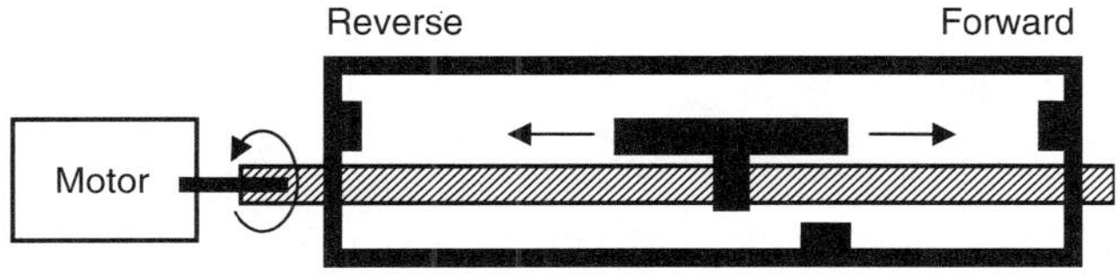

FIGURE 15.2 Limit and home switches in a motion control system.

servo motors, brushless servo motors, and stepper motors. Table 15.1 summarizes the pros, cons, and typical application areas for each type of motor.

15.2.5 Feedback Device

In most motion control systems, amethod for measuring the characteristic being controlled (i.e., position, velocity, etc.) needs to be implemented. This is known as feedback. Some of the different kinds of feedback used in motion control are relative position, absolute position, velocity, and analog voltage feedback. For relative position feedback, the most common device used is the incremental encoder. This device increments as it turns and can be used to keep track of the total distance traveled from some point of reference. The most common absolute position feedback devices are absolute encoders and resolvers. An absolute encoder is similar to an incremental encoder except that it does not need to have a reference position and can know what position the shaft is in on start-up. A resolver is also another absolute position device that is commonly used for both position and velocity feedback. For velocity feedback, a tachometer is commonly used. Some applications require feedback other than the type that has been mentioned above. In these cases, analog feedback describing some characteristic of the system such as temperature or pressure can be useful in controlling motors. A position feedback device is not required for some motion control applications, such as controlling stepper motors, but is vital for servo motors.

15.2.6 Motion I/O

A motion control system also includes digital I/O such as limit switches, home switches, position compare outputs, and position capture inputs.

Limit switches provide information about the end of travel to help you avoid damaging your system. When a motion system hits a limit switch, it typically stops moving. Home switches, on the other hand, indicate the system home position to help you define a reference point. This is important for applications such as pick-and-place (Figure 15.2).

Triggers such as position compare outputs or position capture inputs help when synchronizing moves with other external devices and sensors such as data acquisition and vision.With position compare outputs, you can set up a trigger that fires at a defined position. This type of action is very useful in operations such as scanning, where you might want to trigger a system to take measurements at a series of defined positions. Position capture inputs, on the other hand, cause the motion controller to immediately capture a position on the occurrence of an external event such as a sensor being encountered while moving. This is useful for resynchronizing the position of an axis (registration moves for "cut on the fly" applications) or recording positions for analysis (scanning defects on a printed circuit board (PCB)).

15.3 Functions of a Motion Controller

If we look at the functions of a motion controller, they can be broadly split up into three components—trajectory generation that involves path planning on the basis of the profile desired by the user, supervisory control that includes system initialization, event handling, fault detection, and command response, and the control loop that includes closing the position and velocity loops on the basis of feedback. Figure 15.3 shows the parts and processes of a typical motion controller.

These three components of a motion controller work in synchronization with each other. The supervisory control is the interface that handles user-defined requests and commands the trajectory generator to start calculating set points. The trajectory generator calculates a set point per iteration and writes it to the control loop. The control loop takes this set point and using a control algorithm ensures that the actuator follows this set point accurately. In the case, the control loop can run faster than the trajectory generator; the control loop uses an interpolation algorithm that creates intermediate set points per iteration of the control loop.

15.3.1 Supervisory Control

The supervisory control is the main loop of the motion control system. This loop intercepts commands from the user and signals the trajectory generator to start/stop moves. The supervisory control loop also monitors all I/O needed to perform initialization tasks, such as finding the reference or origin. Supervisory controlmonitors I/O for fault conditions and user-defined events and accordingly commands the trajectory generator to take action.

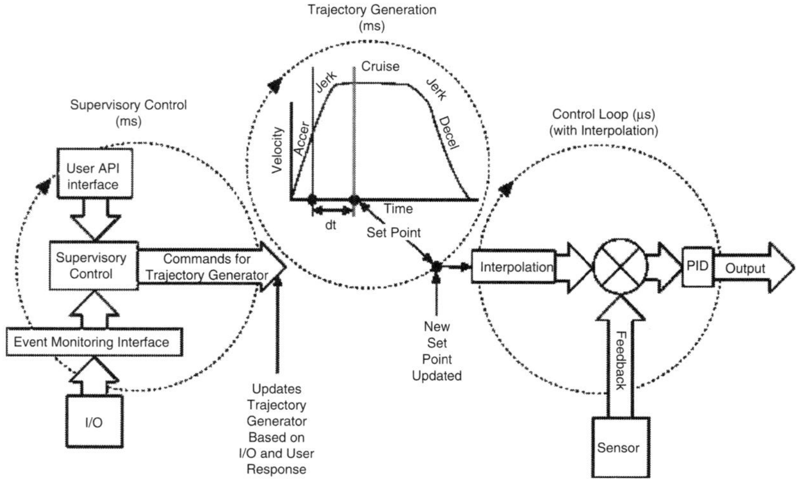

FIGURE 15.3 Motion controller architecture.

15.3.1.1 Axis Initialization (Reference Moves)

A reference move resets the axes of a motion script to a known location or state. Reference moves are generally used at the beginning of a motion script to prepare the motion system for subsequent motion. The following are valid types of reference moves—finding the home switch, finding the encoder index, finding the forward limit switch, finding the reverse limit switch, and finding the center for a system.

Finding the Home Switch
Home reference moves perform a course reference move by traversing the entirety of the motion path in search of a physical home switch. A home switch consists of a logical ON state bordered by a front edge and a back edge. Home reference moves can operate when there is a physical home switch present or if there is a physical end-of-travel present, such as on a stage.

Finding the Encoder Index
Index reference moves perform a find reference move by moving the axis until the index state on the encoder is reached. The index state is a specific position on the 360° encoder that is passed during each revolution of the motor. An index reference move is usually performed after a home reference move for fine reference adjustment. The encoder index signal is accurate to one encoder count and provides amuch more repeatable reference than using just a home switch edge.

Finding the Forward Limit and Reverse Limit Switches
A forward limit reference move initiates a search sequence to find the forward limit. A reverse limit reference move initiates a search sequence to find the reverse limit.When the search operation is initiated, the axis starts moving in the direction of the limit. When the limit in that direction is encountered, the search is completed. A fine-tuning operation is performed when the limit is reached to minimize the effects of motion system windup, backlash, and/or home sensor hysteresis. This type of reference move is used when a home switch is not available on a system.

Finding the Center
Center reference moves initiate a search sequence positioning the axis in the middle of the forward and reverse limits.When the search direction is forward, the axis starts moving in the forward direction.When the limit in that direction is encountered, the direction is reversed to find the reverse limit. After both limits are reached, a center position can be calculated from the recorded positions. The axis then proceeds to the calculated center.

After finding the reference position for an axis the internal position register for that axis is typically updated by:

Resetting the Position Counter Resetting the position counter for an axis causes the axis to start counting its position at a specified starting number. This starting number is typically zero, but resetting to a nonzero number can be useful when you want to begin the moves relative to an absolute zero, rather than the current position. For example, a steering system for a vehicle might come online with the wheels turned 7000 counts to the right of absolute zero. In this example, absolute zero points the wheels straight ahead. To ensure the motion system recognizes that the wheels are not starting off pointed straight ahead, reset the position to the absolute count of 7000.

Offsetting the Axis Position This is achieved by moving the axis by an amount equal to a predetermined offset. Offsetting the position preserves the value of performing the reference moves, but adds another move to create distance from the home switch or index. This is usually done to prevent the axis from sitting on a switch or encoder index.

15.3.1.2 Electronic Gearing

Electronic gearing allows one slave motor to be driven in proportion to a master motor or feedback sensor, such as an encoder or torque (analog) sensor. As the slave follows the master position at a constant ratio, the effect is similar to that of two axes mechanically geared. For example, every turn of the master axis may cause a slave axis to turn twice.

Electronic gearing has several advantages over mechanical gears. The most notable is flexibility because you can change gear ratios on-the-fly. The other major advantage to electronic gearing is that you can superimpose amove over a geared axis. The superimposedmove is added to the geared profile of the slave axis, which allows the slave axis to be synchronized on-the-fly (Figure 15.4).

The gear ratio is used to determine how far the slave axis must move in proportion to the master when gearing is enabled. The gear ratio can be absolute or relative.

$$\text{Slave axis move} = \text{Master axis position} \times \text{Gear ratio}$$

For example, if you have a gearing ratio of 2:1 (slave:master), the slave moves 20 counts when the master device moves 10 counts.

Absolute gearing behaves similarly to relative gearing in that when gearing is enabled, the slave axis follows the master axis movement as it is defined by the gear ratio. The difference between relative and absolute gearing is that the reference position calculated for the master axis is updated only when gearing is enabled. This difference is apparent when the gear ratio is updated on-the-fly (Figure 15.5).

For example, if the gear ratio is 2:1, the current master position is 1010, the current slave position is 3020, and the gear ratio is changed to 3:1; the slave axis jumps from 3020 to 3030 but the master position remains the same (Figure 15.6).

Changing a gear ratio on-the-fly during absolute gearing allows you to quickly synchronize the slave axis with the master axis.

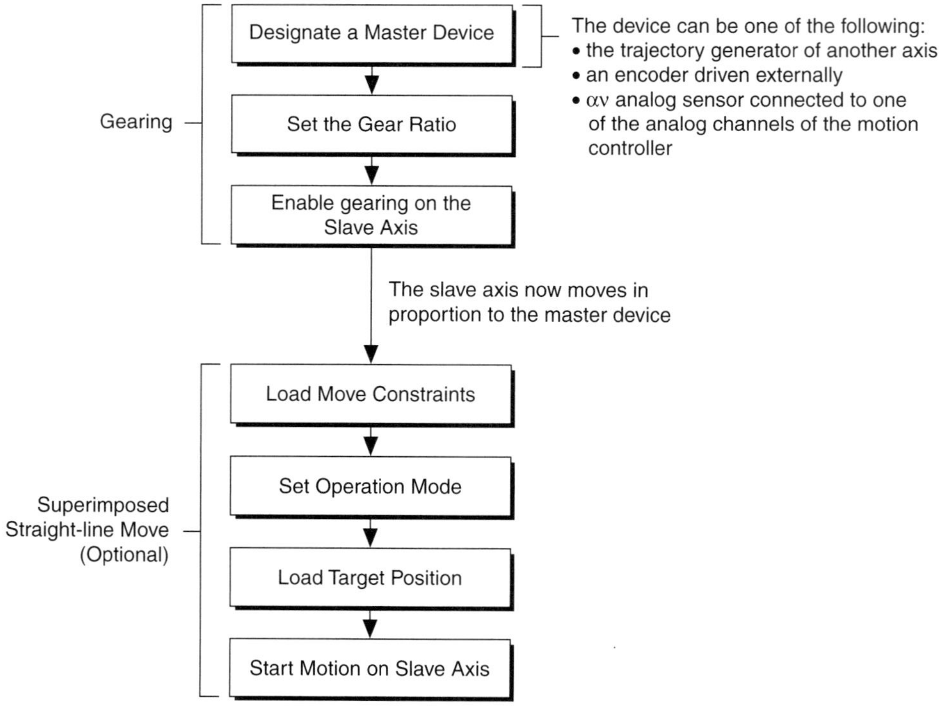

FIGURE 15.4 Electronic gearing algorithm.

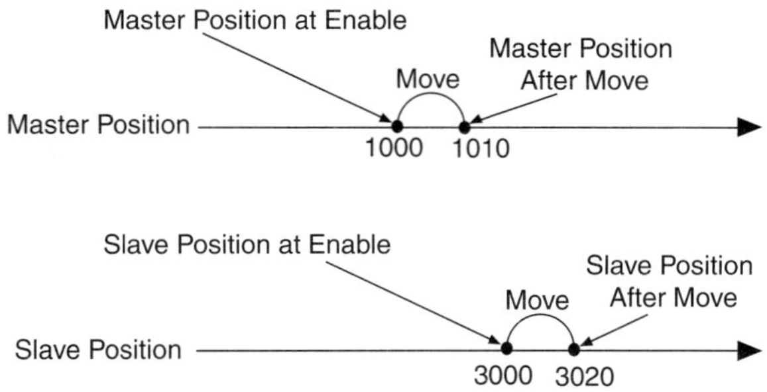

FIGURE 15.5 Relative gearing at enable.

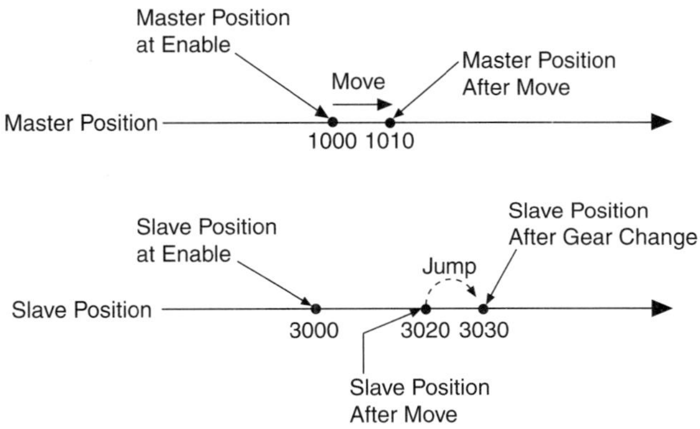

FIGURE 15.6 Absolute gearing at gear ratio change.

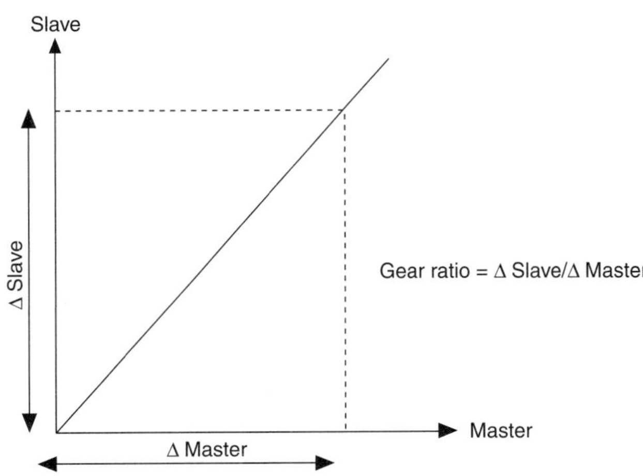

FIGURE 15.7 Master/slave ratio in gearing.

15.3.1.3 Electronic Camming (Includes Registration Moves)

Electronic camming operates similarly to electronic gearing in that the move distance of an axis is proportional to the move distance of its master device. Camming differs from gearing in how the master/slave ratio is handled by the motion controller. Gearing is used in applications where a constant gear value creates a linear slave position profile, as shown in Figure 15.7.

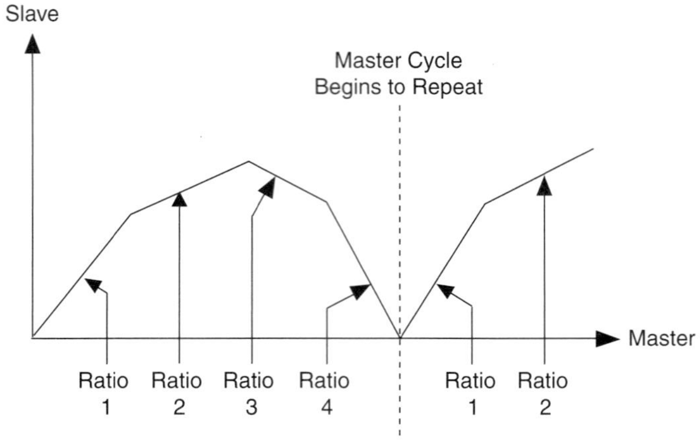

FIGURE 15.8 Multiple camming gear ratios.

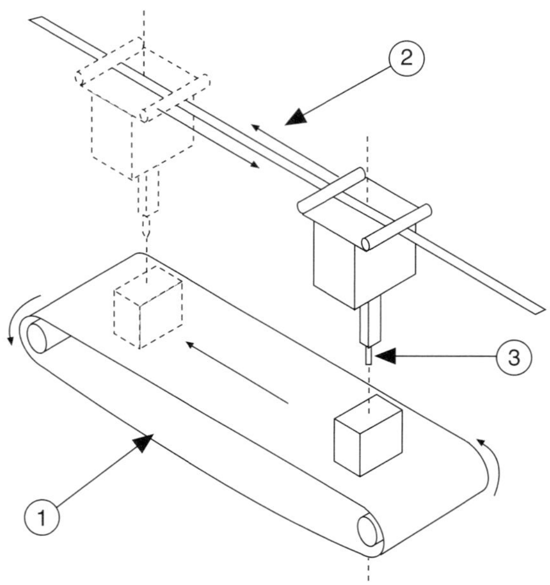

FIGURE 15.9 Welding application showing (1) conveyor belt, (2) movement of the welder as it follows the object and then returns to the initial position, and (3) welding point.

Camming creates a more flexible profile by using more master/slave ratios. These ratios are handled automatically by the motion controller, allowing precise switching of the gear ratios, as shown in Figure 15.8. Camming is used in applications where the slave axis follows a nonlinear profile from a master device.

Figure 15.9 shows that a welding point moves to the first position, and then welds the material for a couple of seconds. Because the conveyor belt keeps moving at a constant rate, the welding point must follow the material at the same speed as the conveyor belt during the weld process. When the welding process is finished for one item, the welding point must quickly return to its initial position, and the process repeats.

In this application, the master device is the position encoder attached to the conveyor belt, and the slave axis is the actuator that moves the welding point. The slave axis repeatedly performs a two-segment movement—first, it follows the material with the same speed as the conveyor belt, and next, it returns to

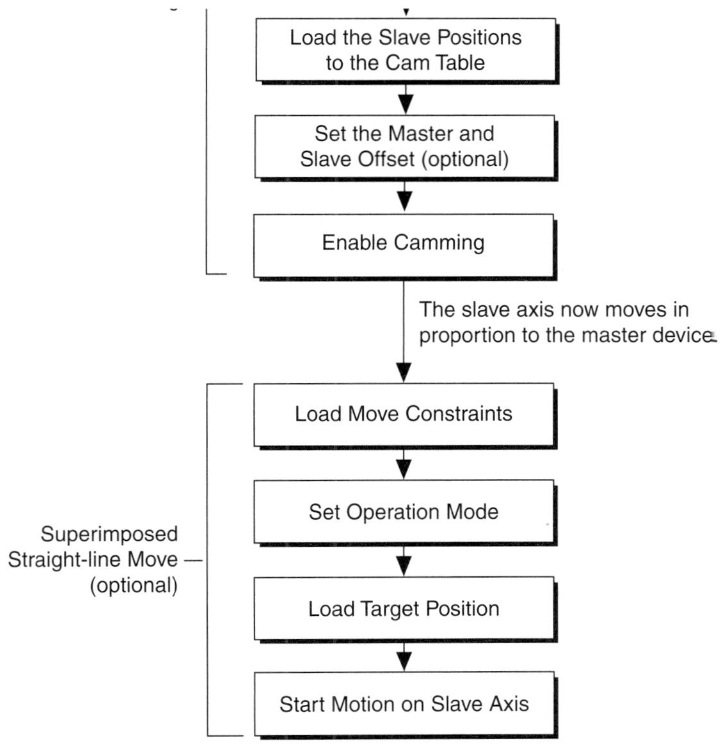

FIGURE 15.10 Camming algorithm.

the initial position as the next material approaches. Each segment of the move is represented with a gear ratio that dictates how fast and which direction the welding point is moving relative to the conveyor belt. This application requires the slave axis to switch from one ratio to the other at the correct master position, otherwise the welding process is not repeatable. If this application used gearing instead of camming, the latency, or delay, of loading a new gear ratio might cause an accumulation of position errors.

Similar to gearing, in a camming application, a slave axis can perform any move when camming is enabled. The move profile is superimposed over the camming profile (Figure 15.10).

15.3.1.4 Exception Handling (E-Stop and Limit Switches)

A motion controller needs to be able to detect fault conditions and handle these exceptions in order of severity. Examples of exceptions include scenarios where the actuator hit a forward or reverse limit switch or if the E-stop switch was pressed.

Limit switches can be either hardware or software limits. The hardware limit inputs are typically connected to end-of-travel limit switches or sensors. An enabled limit input causes a full torque halt stop on the axis when the input becomes active. Active limit inputs also prohibit attempts to start motion that would cause additional travel in the direction of the limit. Similarly, software limits are often used to restrict the range of travel further and avoid ever hitting the hardware limit switches. An enabled software limit causes the axis to smoothly decelerate to a stop when the limit position is reached or exceeded.

15.3.1.5 I/O Synchronization (Captures, Compares, Time-Sampled Data Logging)

You may need to synchronize motion control with data acquisition or image acquisition devices within the same systemusing position compares, position captures and time-sampled data logging. Let us understand first what each of these terms mean.

In scanning applications, you typically inspect an object such as a wafer or an LCD display for defects using a camera. The wafer or LCDdisplay rests on an XY stage that moves in two dimensions. The objective of the scan is to cover as much space on the wafer as possible in the shortest amount of time and trigger the camera to take an image at specific points along the scan. In such applications, you need to send a digital signal (called a breakpoint) to the trigger line of a camera to command it to take an image based on a specific position reached by the XY stage. This functionality is called position compares (breakpoints), which can be classified into absolute breakpoints, relative position breakpoints, modulo breakpoints, periodically occurring breakpoints, and buffered breakpoints.

In some cases, you may need to synchronize position with some measurement occurring externally to the motion controller. You can think of this conceptually as a conveyor belt application running at a constant velocity. Each time a manufactured item moves down the conveyor belt and breaks a light beam across the belt, a stamp is placed on the item, a set distance from the point where the beam is broken. This allows all items to be stamped in the same relative position without necessarily being equally spaced along the conveyor belt. You need to command amove of a specified length relative to the position at which a trigger input is received to perform such a stamping operation. This functionality is called high-speed position captures (triggers), which could be buffered or nonbuffered.

In vision-guided motion applications, you may need to log position and velocity data periodically for off-line analysis and correlation with vision data obtained concurrently. This functionality is known as time-sampled data logging. Data is collected at hardware/firmware timed period and stored in the motion controller's internal memory. This data can be retrieved from the motion controller asynchronously.

Position Compares (Breakpoints)

You can configure a motion controller to generate a trigger signal or toggle the state of an I/O line when the motor reaches a specified position. This specified position is called a breakpoint and is similar to a switch that is triggered after the motor passes a certain position. To use breakpoints, you must have a board with closed-loop encoder feedback, because the controller needs the encoder position to determine where the breakpoint occurs. You can specify breakpoints for both servo motors and stepper motors. You can use breakpoints as a trigger for another piece of hardware because it sends a signal to an I/O line. For example, when a motor reaches a certain position, you can use a breakpoint signal to initiate an image capture or a data acquisition operation. The five ways in which breakpoints can be configured are as follows:

Absolute Mode: With absolute position breakpoints, you can trigger external activities as the motors reach specified positions. For example, if you need to use an image acquisition device to capture an image from a certain position while the device under test is in continuous motion, the motion controller must be able to trigger the image acquisition device as it reaches those positions. The current position is continuously compared against the specified breakpoint position by the encoder circuitry to produce a latency of less than 100 ns. After a breakpoint triggers, you must re-enable it for the breakpoint to work again. In certain cases, such as buffered and periodic breakpoints, the motion controller automatically re-enables the breakpoints.

Relative Mode: Relative position breakpoints trigger events on the basis of a change in position relative to the position at which the breakpoint was enabled. Instead of keeping track of absolute positions and the current position, you can use relative breakpoints to specify the breakpoint relative to the position where the breakpoint is enabled. For example, if you are creating a motion control system to control the 2D movement of a microscope, you might use relative position breakpoints to move the microscope a specific distance in a direction, and then hit a breakpoint that triggers a camera snap. The relative breakpoint is useful in this example because the current position is not important. The application must move the axis a specific number of counts from wherever it is, and then generate a breakpoint.

Periodic Mode: You can program the motion controller to generate multiple breakpoints at fixed and exact intervals, regardless of the direction of travel or velocity. Periodic breakpoints require that you specify an initial breakpoint and an ongoing repeat period.When enabled, the periodic breakpoints begin when the initial breakpoint occurs. From then on, a new breakpoint occurs each time the axis

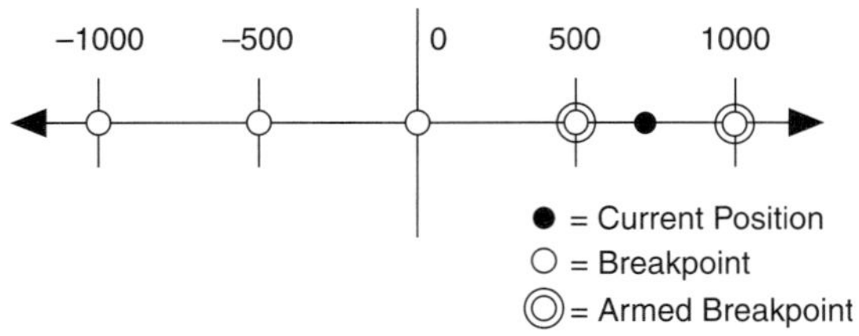

FIGURE 15.11 Periodic breakpoint every 1000 counts/steps.

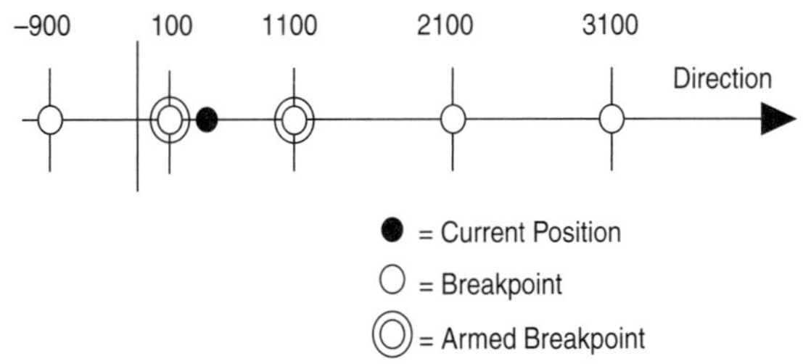

FIGURE 15.12 Breakpoint modulus of 500.

moves a distance equal to the repeat period, with no re-enabling required. For example, if an axis is enabled at position zero, the initial breakpoint is set for position 100, and the breakpoint period is set at 1000, then the axis behaves as shown in Figure 15.11.

Modulo Mode: Modulo breakpoints use a breakpoint window that defines an area around the current position. The two breakpoints around the current position are always enabled. The breakpoint modulus creates a repeat period for the breakpoints, and the breakpoint position is the offset from absolute zero. For example, to create a breakpoint every 500 counts, set the repeat period to 500 and the breakpoint position to 0. If the breakpoint is enabled when the axis is at 710, the breakpoints at 1000 and 500 are both armed, as shown in Figure 15.12.

Buffered Mode: Buffered breakpoints are a way of specifying absolute or relative breakpoint positions in a buffer that can be preloaded into the motion controller memory. The motion controller then automatically arms breakpoint positions from the buffer as and when the breakpoints triggers. The, arming of breakpoints occurs on a firmware-timed basis, which provides higher bandwidth.

Position Captures (Triggers)

Some motion control applications require that you execute amove and record the position where external triggers occur. For example, you might be aligning two fiber-optic cables, in which case, the maximum optical power needs to correspond with the alignment position. To align the fibers, the external device that is recording the optical power must trigger the motion controller so that you can synchronize and analyze positions and optical power measurements. This functionality is known as high-speed capture or trigger inputs. The implementation for high-speed capture is divided into the buffered and nonbuffered high-speed capture methods:

Buffered High-Speed Captures: With buffered high-speed capture, you can create a buffer that holds captured positions that you can read asynchronously from the motion controller. The motion controller automatically arms the next high-speed capture, and writes the captured high-speed data into its onboard buffer. The enabling of high-speed capture occurs on a firmware-timed basis, which provides better frequency than the nonbuffered high-speed capture method.

Nonbuffered High-Speed Captures: With nonbuffered, high-speed capture, you can configure a single high-speed capture event. For multiple high-speed captures, you must re-enable the high-speed capture each time before it triggers.

Time-Sampled Data Logging

Some motion controllers can acquire a hardware-timed buffer of position and velocity data for time-sampled data logging. After you command the motion controller to acquire position and velocity data, a separate acquire data task is created in the motion controller, which reads time-sampled position and velocity data into a FIFO buffer on the motion controller. You can read data in from this buffer number of samples that consist of position data, in counts or steps, and velocity data, in counts/s or asynchronously from the user interface on the host computer. Typically, the FIFO buffer holds a certain steps/s.

15.3.2 Trajectory Generation

The trajectory generator on a motion controller is responsible for the path planning on the basis of geometry specified by the user. Different kinds of geometries are typically supported: straight lines, circular arcs, helical arcs, and spherical arcs. In addition to the predefined geometries, the user can specify a custom geometry via a buffer of points that the motion controller uses to interpolate through using a cubic spline algorithm. This is known as a contoured move.

The user can also specify the velocity profiles that he wants used when traversing desired geometries. This can be specified using move constraints that include maximum velocity, maximum acceleration, and maximum deceleration and jerk (rate of change of acceleration/deceleration). These parameters are used to generate a trapezoidal or s-curve profile. The trajectory generator on the motion controller ensures that these move constraints are never violated while making a move. In addition to this, the user can specify custom profiles by providing position, velocity, and time data (Position Velocity Time, PVT). The trajectory generator also typically supports blending velocity profiles. In this case, the users can concatenate multiple profiles specifying the point of concatenation.

The output of the trajectory generator are set points that are fed to the control loop to act on the basis of the path that needs to be followed.

To be highly reactive and allow for on-the-fly changes, the trajectory generator on the motion controller creates these set points in real time. This allows users to pre-empt the current trajectory by changing end positions, velocity, and acceleration. It also allows for stopping and restarting a move that is currently in progress.

15.3.2.1 Straight-Line Moves

A straight-line move executes the shortest move between two points. Straight-line moves can be position based or velocity based.

Position Based

Position-based straight-line moves use the specified target position to generate the move trajectory. For example, if the motor is currently at position zero, and the target position is 100, a position-based move creates a trajectory that moves 100 counts (steps).

The controller requires the following information to move to another position in a straight line:

- Start position—current position, normally held over from a previous move or initialized to zero
- End position—also known as the target position, or where you want to move to
- Move constraints—maximum velocity, maximum acceleration, maximum deceleration, and maximum jerk

The motion controller uses the given information to create a trajectory that never exceeds the move constraints and that moves an axis or axes to the end position you specify. The controller generates the trajectory in real time, so you can change any of the parameters while the axes are moving. Figure 15.13 illustrates a typical algorithm for a position-based straight-line move.

Velocity Based

Some motion applications require moves that travel in a straight line for a specific amount of time at a given speed. This type of move is known as velocity profiling or jogging. You can use a motion control

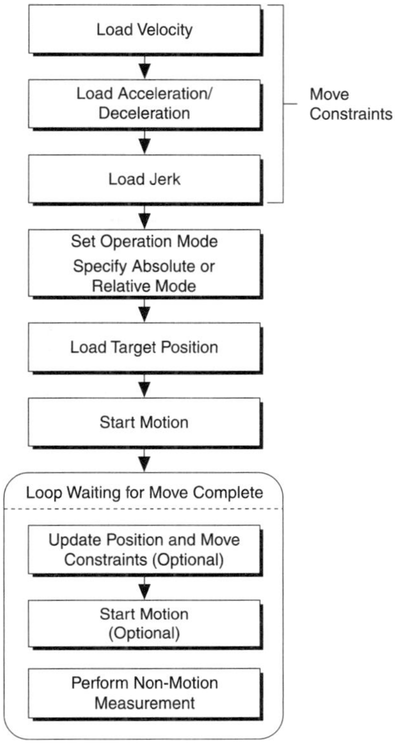

FIGURE 15.13 Position-based straight-line move algorithm.

application to move a motor at a given speed for a specific time, and then change the speed without stopping the axis. The sign of the loaded velocity specifies the direction of motion. Figure 15.14 illustrates an algorithm for a typical velocity-based straight-line move.

15.3.2.2 Arc Moves

An arc move causes a coordinate space of axes to move on a circular, spherical, or helical path. You can move two-dimensional vector spaces in a circle only on a 2D plane. You can move a three-dimensional vector space on a spherical or helical path. Each arc generated by the motion controller passes through a cubic spline algorithm that ensures the smoothest arc. A cubic spline algorithm generates multiple points between every two points of the arc, ensuring smooth motion, minimum jerk, and maximum accuracy at all times.

Circular Arcs

A circular arc defines an arc in the XY plane of a 2D or 3D coordinate space. The arc is specified by a radius, start angle, and travel angle. In addition, like all coordinate space moves, the arc uses the values of move constraints—maximum velocity, maximum acceleration, and maximum deceleration.

To move axes in a circular arc, the motion controller needs the following information:

- Radius—specifies the distance from the center of the arc to its edge.
- Start Angle—orients the arc on its plane using the starting point as an axis to spin around. Because the starting point for a new arc is fixed on the basis of the current position, moving its center around the starting point alters the orientation of a new arc.
- Travel Angle—indicates how far the arc travels in a 360° circle. For example, a travel angle of 90° executes a quarter circle, a travel angle of 360° creates a full circle, and a travel angle of 720° creates two full circles (Figure 15.15).

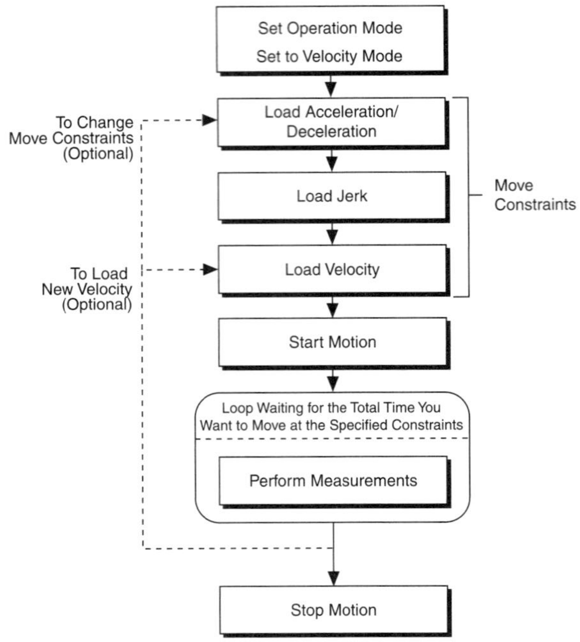

FIGURE 15.14 Velocity-based straight-line move algorithm.

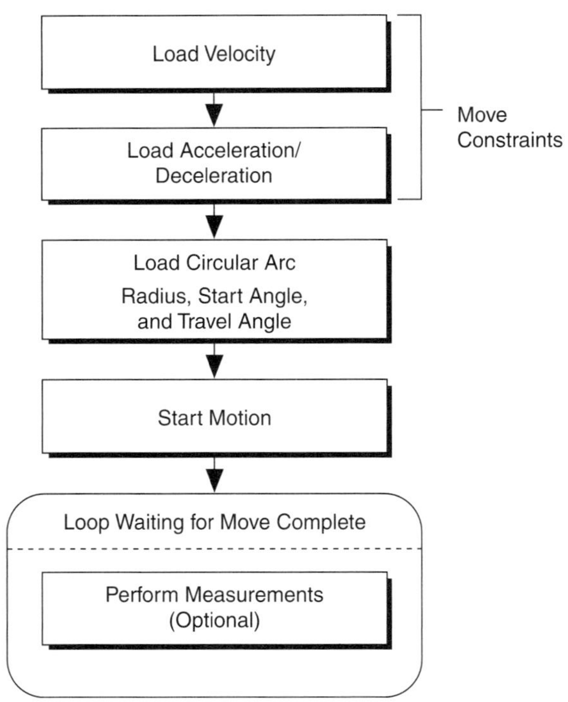

FIGURE 15.15 Circular arc move algorithm.

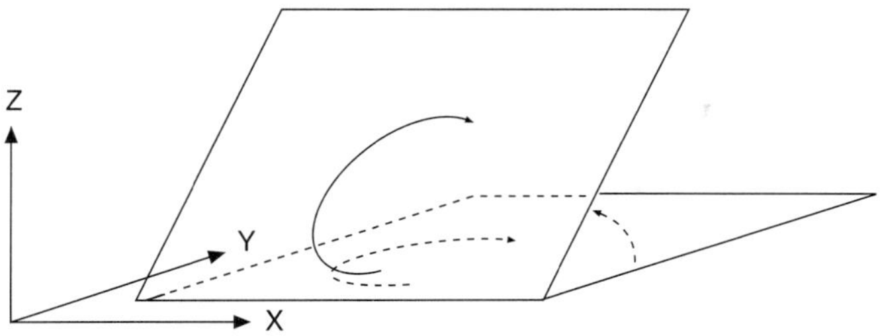

FIGURE 15.16 Changing pitch by rotating the X-axis.

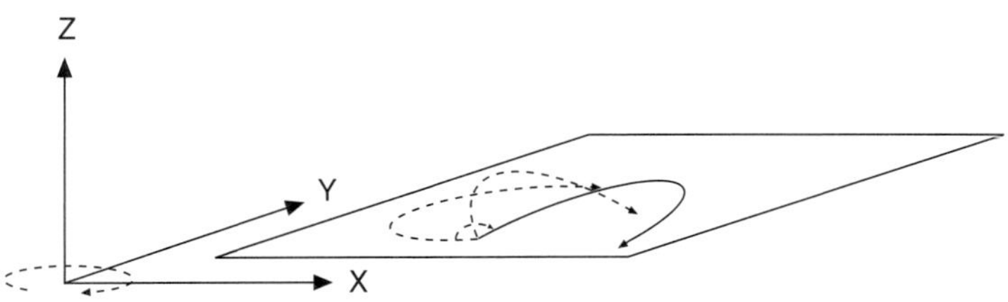

FIGURE 15.17 Changing yaw by rotating the Z-axis.

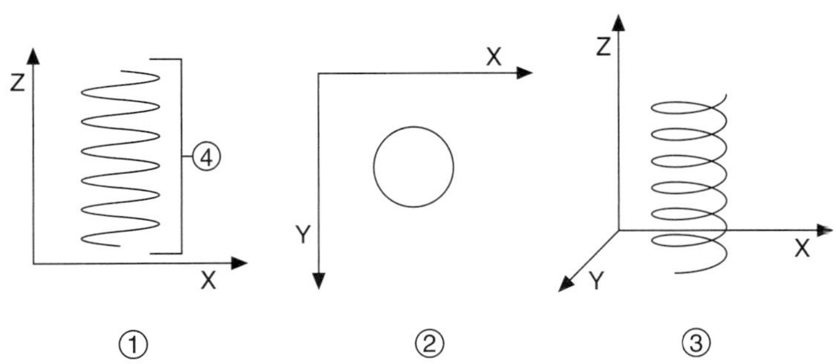

FIGURE 15.18 Helical Arc showing (1) side view, (2) top view, (3) isometric view, and (4) linear travel.

Spherical Arcs

A 3D spherical arc defines a 2D circular arc in the XY plane of a coordinate system that is transformed by rotation in pitch and yaw from the normal 3D coordinate space (XYZ), as shown in Figures 15.16 and 15.17.

In the transformed X′Y′Z′ space, the 3D arc is reduced to a simpler 2D arc. The 3D arc is defined as a 2D circular arc in the X′Y′ plane of a transformed vector space X′Y′Z′. This transformed vector space, X′Y′Z′, is defined in orientation only, with no absolute position offset. Its orientation is relative to the XYZ vector space, and is defined in terms of pitch and yaw angles.When rotating through the pitch angle, the Y and Y′ axes stay aligned with each other while the X′Z′ plane rotates around them. When rotating through the yaw angle, the Y′ axis never leaves the original XY plane, as the newly-defined X′Y′Z′ vector space rotates around the original Z-axis. The radius, start angle, and travel angle parameters also apply to a spherical arc that defines the arc in two dimensions.

Helical Arc

Ahelical arc defines an arc in a 3Dcoordinate space that consists of a circle in the XY plane and synchronized linear travel in the Z-axis. The arc is specified by a radius, start angle, travel angle, and Z-axis linear travel. Linear travel is the linear distance traversed by the helical arc on the Z-axis, as shown in Figure 15.18.

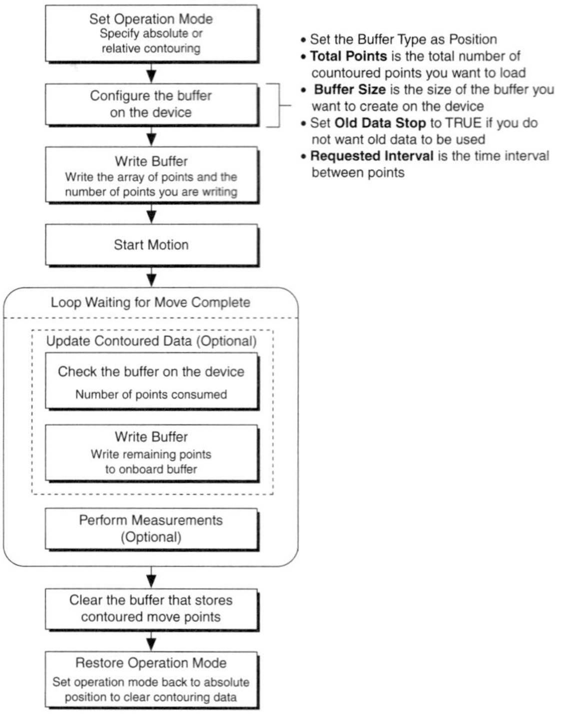

FIGURE 15.19 Contoured move algorithm.

15.3.2.3 Contoured Moves

A contoured move moves an axis or a coordinate space of axes in a pattern that you define. The trajectory generator on the motion controller is not used during a contoured move. The controller takes position data in the form of an array, and splines the data before outputting it to the control loop.

Contoured moves are useful when you want to generate a trajectory that cannot be constructed from straight lines and arcs. To ensure that the motion is smooth with minimum jerk, the motion controller creates intermediate points using a cubic spline algorithm. The move constraints commonly used to limit other types of moves, such as maximum velocity, maximum acceleration, maximum deceleration, and maximum jerk, have no effect on contoured moves. These need to be embedded in the series of contoured points provided (by spacing the points) (Figure 15.19).

15.3.2.4 Position Velocity Time Profiles

Position velocity time moves give you maximum control over the velocity of the move at certain times throughout the move. PVT moves specify a series of move segments by giving several move points for each axis. In addition, PVT moves specify both the time, in ms, each move segment takes and the velocity at which all axes are traveling at the end of each segment.

The following table demonstrates typical PVT inputs:

X	Y	Z	V	T
4317	4317	4317	1000	750
12245	11245	13245	500	1500
15173	15173	15173	500	2500

T: Time begins at 0 for each PVT move, and is cumulative through the end of the move. In this example, there are three segments of duration 750, 750, and 1000.

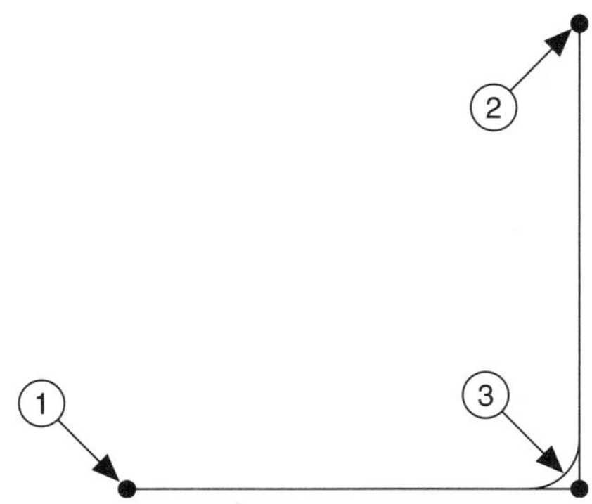

FIGURE 15.20 Two blended straight-linemoves showing (1) starting position, (2) end point, and (3) corner rounded by blending.

V: PVT moves assume a beginning velocity of zero. Therefore, you must bring each axis to a complete stop in the previous move before beginning a PVT move.

X, Y, Z: PVT moves start a position counter for each axis at 0, independent of where you are in relation to the original reference move. The numbers in the spreadsheet or text file are absolute in relation to the new counter. In this example, there are three move segments of lengths 4317, 7928, and 2928, respectively.

15.3.2.5 Blending

Blending, also called velocity blending, superimposes the velocity profiles of two moves to maintain continuous motion. Blending is useful when continuous motion between concatenated move segments is important. Examples of some applications that can use blending are scanning, welding, inspection, and fluid dispensing. Blending must occur on velocity profiles of two move segments, so the end positions of each move segment may or may not be reached. For example, if you are blending two straight-line moves that form a 90° angle, the blended move must round the corner to make the move continuous. In this case, the move never reaches the exact position where the two straight lines meet, but instead follows the rounded corner, as shown in Figure 15.20.

Motion controllers can perform blending between two straight-line moves, between two arc moves, or between straight-line and arc moves. Blending typically is not supported for reference and contoured moves.

Superimposing Two Moves

Superimposing two moves is the most common use of blending. In this case, the motion controller tries to maintain continuous motion by superimposing the two move segments such that the second move segment starts its profile while the first move is decelerating, as shown in Figure 15.21.

The velocity during the superimposition depends on the cruising velocity, deceleration, and jerk of the first move segment, and the jerk, acceleration, and cruising velocity of the second move segment.

Blending Moves after the First Move Is Complete

Blending moves after the first move is complete; this causes the first move segment to come to a complete stop before starting the profile of the second segment, as shown in Figure 15.22. This type of blending is useful if you want to start two move segments, one after the other, with no delay between them.

Blend after Delay

You can blend two moves after a delay at the end of the first move. Blending in this manner is useful if you want to start two move segments after a deterministic delay. The two move segments can be either straight-line moves or arc moves.

Because blending occurs on velocity profiles, the effect of reaching the end positions of the move segments and the maximum velocity depends on the velocity, acceleration, deceleration, and jerk loaded

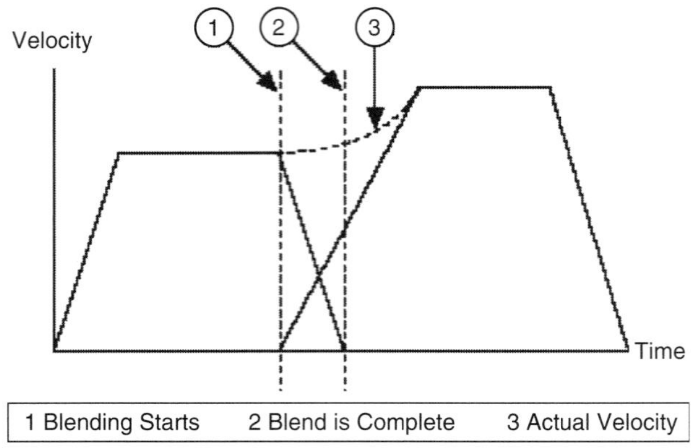

FIGURE 15.21 Superimposing two moves.

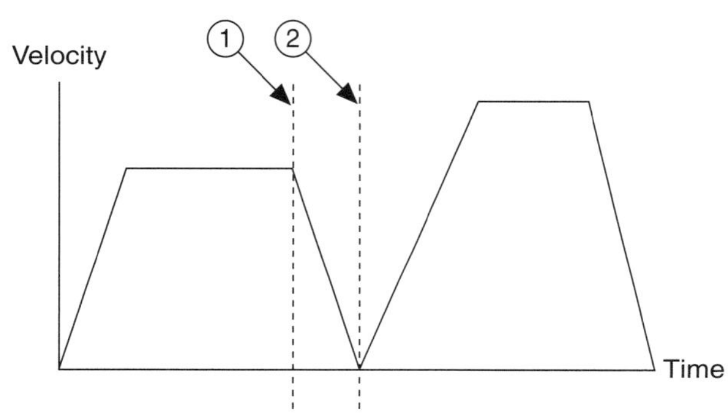

FIGURE 15.22 Blending after move complete showing (1) blending starting point and (2) blending complete.

for the two move segments. Because two move segments are always used while blending, it is very important that you wait for the blend to complete before loading the next move segment you want to blend.

15.3.2.6 Second- and Third-Order Profiles (Trapezoidal/s-Curve)

The trajectory generator uses the commanded parameters of position, acceleration, and velocity for a particular move to determine how much time it spends in the three primary move segments—acceleration, constant velocity, and deceleration—that define a velocity profile.

Trapezoidal Velocity Profile

When you use a trapezoidal profile, the axes accelerate at the acceleration value you specify, and then cruise at the maximum velocity you load. On the basis of the type of move and the distance being covered, it may be impossible to reach the maximum velocity you set. The velocity of the axis, or axes, in a coordinate space never exceeds the maximum velocity loaded. The axes decelerate to a stop at their final position, as shown in Figure 15.23.

S-curve Velocity Profile

The acceleration and deceleration portions of an s-curve motion profile are smooth, resulting in less abrupt transitions, as shown in Figure 15.24. This limits the jerk in the motion control system, but increases cycle time. The value by which the profile is smoothed is called the maximum jerk or s-curve value.

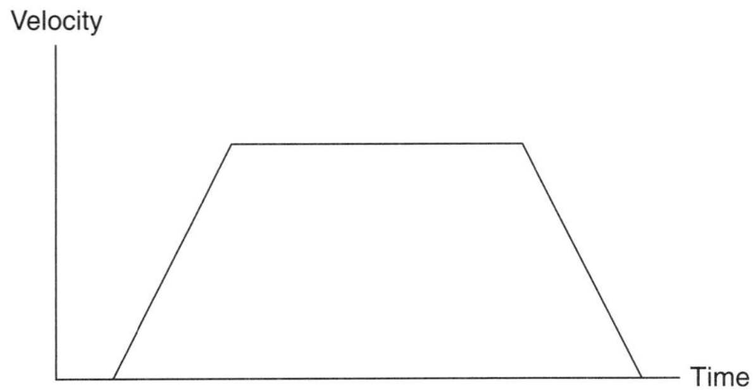

FIGURE 15.23 Trapezoidal move profile.

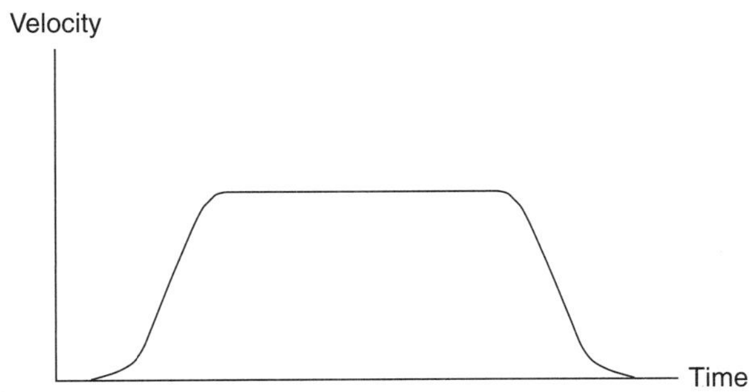

FIGURE 15.24 S-curve move profile.

15.3.2.7 Multiple Axis Synchronization (Coordinate/Vector Spaces)

With the exception of the arc move, you can execute all basic moves on either a single axis or on a coordinate space. A coordinate space is a logical grouping of axes, such as the XYZ axes Arc moves always execute on a coordinate space. If you are performing a move that uses more than one axis, you must specify a coordinate space made up of the axes the move will use.

Typically, there exists a function in the motion controller to configure a coordinate space. This function creates a logical mapping of axes and treats the axes as part of a coordinate space. The function then executes the move generated by the trajectory generator on the vector, and treats all the move constraints as vector values.

Multistarts versus Coordinate Spaces

Coordinate spaces always start and end the motion of all axes simultaneously. You can use multistarts to create a similar effect without grouping axes into coordinate spaces. Using a multistart automatically starts all axes virtually simultaneously. To simultaneously end the moves, you must calculate the move constraints to end travel at the same time. In coordinate spaces, this behavior is calculated automatically.

15.3.3 Control Loop

15.3.3.1 Spline Interpolation

The trajectory generator and the proportional, integral, derivative (PID) control loop need not execute at the same loop rates.While the PID control loop needs to run at microseconds, the trajectory generator may generate a new set point every millisecond. To maintain the steady flow of set points from the trajectory generator to the PID control loop, a motion controller needs a spline interpolator that interpolates between set points given by the trajectory generator to create finer set points to be acted on by the control loop (Figure 15.25).

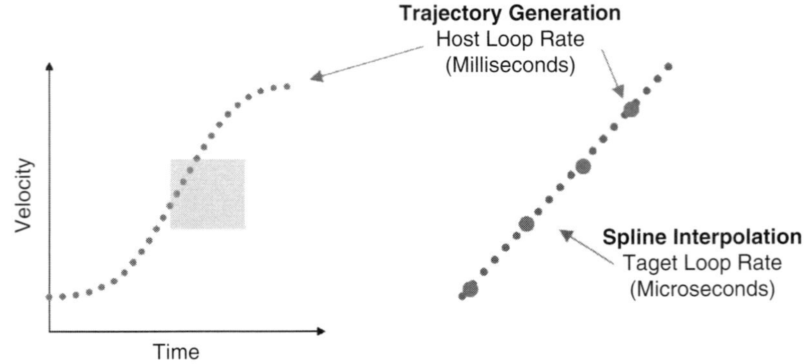

FIGURE 15.25 Spline engine function uses cubic spline algorithm to calculate interpolated positions between two positions from the trajectory generator.

An example of a spline interpolator is a cubic spline algorithm that uses four set points to calculate interpolated positions between two positions calculated by the trajectory generator. The polynomial used to calculate interpolated points can be represented as:

$$P(t) = a_3t^3 \times a_2t^2 \times a_1t^1 \times a_0t^0$$

Given four set points, P_0, P_1, P_2, and P_3, the spline coefficients are determined in real time as:

$$\begin{bmatrix} a_0 \\ a_1 \\ a_2 \\ a_3 \end{bmatrix} = \frac{1}{6}\begin{bmatrix} 1 & 4 & 1 & 0 \\ -3 & 0 & 3 & 0 \\ 3 & -6 & 3 & 0 \\ -1 & 3 & -3 & 1 \end{bmatrix}\begin{bmatrix} P_0 \\ P_1 \\ P_2 \\ P_3 \end{bmatrix}$$

The number of interpolated points created depends on the rate of the trajectory generator loop and control loop. For example, if the trajectory generator loop runs at 1 ms and the control loop runs at 100 μs, the spline engine calculates 1 ms/100 μs $=10$ interpolated points. Using the spline interpolator results in smoother motion and allows you to run the trajectory generator loop slower than the control loop.

15.3.3.2 Position and Velocity Loops

The control loop creates the command signal on the basis of the set point provided by the trajectory generator. In most cases, the control loop includes both a position and a velocity loop, but in some cases the control loop may include only a position loop. The position is typically read from encoders, but also may be read from analog inputs. The velocity is calculated from the position values, and may be read directly from a velocity sensor, such as a tachometer. Because feedback is not required for stepper motors, the control loop concerts the set point generated by the trajectory generator into stepper signals (step/direction).

The position and velocity control loops on the motion system correct for errors and maintain tight control of the trajectory. The position and velocity loops typically incorporate the PIDalgorithm. The PID algorithm consists of three basic coefficients: proportional, integral, and derivative, which are varied to get optimal response (Figure 15.26).

In addition to parameters such as proportional gain, integral gain, and derivative gain, the PID algorithm-implemented part of a motion control system includes additional parameters such as velocity feedback, velocity feed forward, and acceleration feed forward for optimal position control.

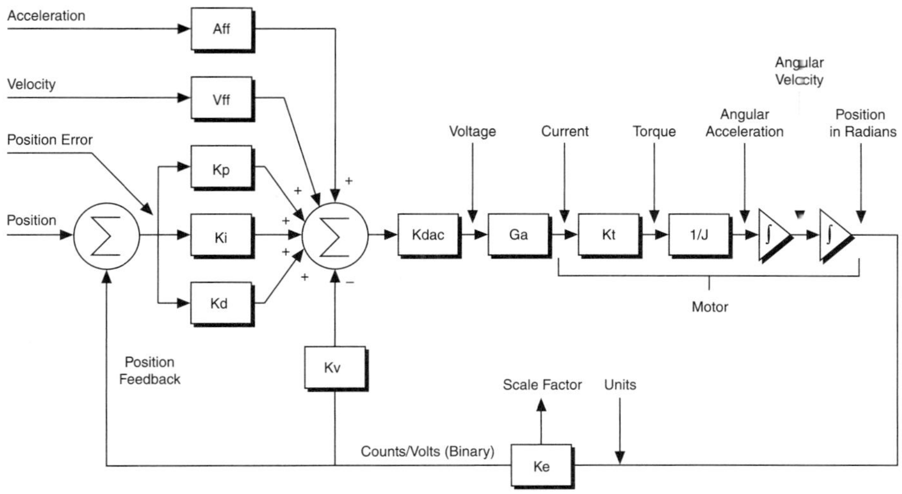

FIGURE 15.26 PID servo loop.

Kp (Proportional Gain)

The proportional gain (Kp) determines the contribution of restoring force that is directly proportional to the position error. This restoring force functions in much the same way as a spring in a mechanical system. An axis with too small a value of Kp is unable to hold the axis in position and is very soft. Increasing Kp stiffens the axis and improves its disturbance torque rejection. However, too large a value of Kp often results in instability.

Ki (Integral Gain)

The integral gain (Ki) determines the contribution of restoring force that increases with time, ensuring that the static position error in the servo loop is forced to zero. This restoring force works against constant torque loads to help achieve zero position error when the axis is stopped. In applications with small static torque loads, this value can be left at its default value of zero (0). For systems having high static torque loads, this value should be tuned to minimize position error when the axis is stopped. Although nonzero values of Ki cause reduced static position error, they tend to cause increased position error during acceleration and deceleration. This effect can be mitigated through the use of the integration limit parameter. Too high a value of Ki often results in servo loop instability.

Kd (Derivative Gain)

The derivative gain (Kd) determines the contribution of restoring force proportional to the rate of change (derivative) of position error. This force acts much like viscous damping in a damped spring and mass mechanical system. A shock absorber is an example of this effect. A nonzero value of Kd is required for all systems that use torque block amplifiers, where the command output is proportional to motor torque, for the servo loop operation to be stable. Too small a Kd value results in servo loop instability. With velocity block amplifiers, where the command output is proportional to motor velocity, it is typical to set Kd to zero or a very small positive value.

Kv (Velocity Feedback)

You can use a primary or secondary feedback encoder for velocity feedback. Setting the velocity feedback gain (Kv) to a value other than zero (0) enables velocity feedback using the secondary encoder, if configured, or the primary encoder if a secondary encoder is not configured. The derivative gain scales the derivative of the position error, which is the difference between the instantaneous trajectory position and the primary feedback position. Velocity feedback is estimated through a combination of speed-dependent algorithms.

Vff (Velocity Feedforward)

The velocity feedforward gain (Vff) determines the contribution in the command output that is directly proportional to the instantaneous trajectory velocity. This value is used to minimize following error during the constant velocity portion of a move and can be changed at any time to tune the PID loop. Velocity feedforward is an open-loop compensation technique and cannot affect the stability of the system. However, if you use too large a value for Vff, following error can reverse during the constant velocity portion, thus degrading performance, rather than improving it. Velocity feedforward is typically used when operating in PIVff mode with either a velocity block amplifier or substantial amount of velocity feedback. In these cases, the uncompensated following error is directly proportional to the desired velocity.

Aff (Acceleration Feedforward)

The acceleration feedforward gain (Aff) determines the contribution in the command output that is directly proportional to the instantaneous trajectory acceleration. Aff is used to minimize following error (position error) during acceleration and deceleration and can be changed at any time to tune the PID loop. Acceleration feedforward is an open-loop compensation technique and cannot affect the stability of the system. However, if you use too large a value of Aff, following error can reverse during acceleration and deceleration, thus degrading performance, rather than improving it.

Dual-Loop Feedback

Motion control systems often use gears to increase output torque, increase resolution, or convert rotary motion to linear motion. The main disadvantage of using gears is the backlash created between the motor and the load. This backlash can cause a loss of position accuracy and system instability. The control loop on the motion system corrects for errors and maintains tight control over the trajectory. The control loop consists of three main parts—proportional, integral, and derivative—known as PID parameters. The derivative part estimates motor velocity by differentiating the following error (position error) signal. This velocity signal adds, to the loop, damping and stability. If backlash is present between the motor and the position sensor, the positions of the motor and the sensor are no longer the same. This difference causes the derived velocity to become ineffective for loop damping purposes, which creates inaccuracy in position and system instability.

Using two position sensors for an axis can help solve the problems caused by backlash. As shown in Figure 15.27, one position sensor resides on the load and the other on the motor before the gears. The motor sensor is used to generate the required damping and the load sensor for position feedback. The mix of these two signals provides the correct position feedback with damping and stability.

15.3.3.3 Commutation

When a current-carrying conductor is placed in a magnetic field, it experiences a force. Motors are based on this principle of physics. Motors comprise of permanent magnets and current-carrying conductors in a mechanical assembly. Maintaining the correct direction of current and magnetic fields is critical to proper operation of the motor. This functionality is called commutation.

In case of brushed DC servo motors, a physical brush performs commutation by switching the current between different conductors. However, in case of brushless servo motors, three phases of current are

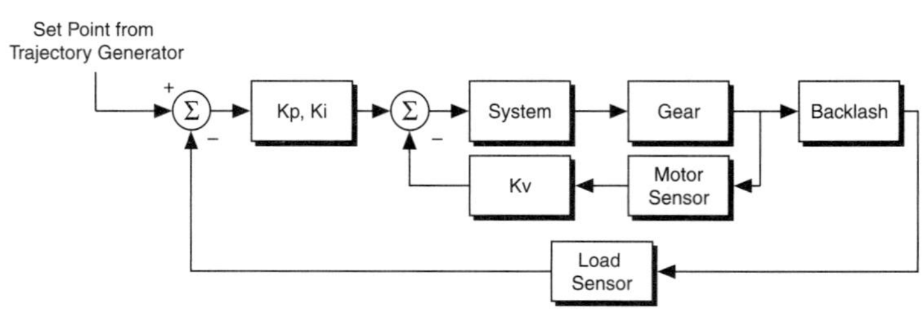

FIGURE 15.27 Dual-loop feedback.

electrically commutated. This requires commutation to be performed either by the motion controller or the drive.

15.3.3.4 Current Loop (Torque Loop)

A current loop takes a set point from the position and velocity loop and uses feedback from a current sensor to generate a voltage output signal. This voltage is used to control the power transistors in the drive to ensure that the right amount of torque is being generated. While in brushed DC servo motors the current command is directly proportional to the current in one winding in the motor, in brushless servo motors it is dependent on current in three windings. As a result, brushless servo motors often come paired with drives that include commutation and a pretuned current loop.

15.4 Motion Controller Hardware

A motion controller can be implemented on dedicated hardware connected to an analog drive, independent of the PC processor that hosts the user application. It can also be designed to share resources on the PC processor with the user application with some functions being executed within a digital drive, thus eliminating the need for dedicated motion controller hardware. These two architectures for implementing a motion controller have their advantages and disadvantages.

15.4.1 Architecture of an Analog Motion Controller (Dedicated Processor)

A motion controller is responsible for executing supervisory control, trajectory generation, and control-loop functions. In case of an analog motion controller, these functions are executed on dedicated hardware that communicates with the host PC via an interface such as the PCI bus, USB, Ethernet, Serial RS 232, and so forth. Commands from the user application running on the host computer are sent to the motion controller via this interface. The motion controller performs supervisory control, trajectory generation, and I/O synchronization, and executes position and velocity control loops. The motion controller then sends commands in the form of 10V analog voltage signals to an analog drive that closes the current loop. There are four main hardware components in a typical analog motion controller.

15.4.1.1 Microprocessor

A microprocessor runs amultitasking, real-time operating system, and handles host communication with the PC and supervisory control including command processing and fault handling.

15.4.1.2 Digital Signal Processor

A digital signal processor is used for math-intensive operations such as trajectory generation and position and velocity loops.

In some motion controllers, all the supervisory control, trajectory generation, and control-loop functions are performed on a single microcontroller that includes I/O interfaces, memory, and clock in addition to the CPU in an integrated circuit.

15.4.1.3 Field-Programmable Gate Array

Afield-programmable gate array (FPGA) is used to perform I/O synchronization functions such as encoder decoding, position capture and position compare, motion I/O processing, and stepper pulse generation.

15.4.1.4 Motion I/O

An analog motion controller connects to different types of motors as follows:

1. Brushed DC servo motor—An analog motion controller provides a ± 10 V analog output signal to a servo drive that closes the current loop to move a brushed DC servo motor.
2. Brushless servo motor—An analog motion controller provides a ± 10 V analog output signal to a servo drive that performs sinusoidal commutation and closes the current loop to move a brushless

servo motor. In case the drive does not perform sinusoidal commutation, the analog motion controller provides two analog output signals to the drive. The motion controller in this case performs sinusoidal commutation in addition to closing the position and velocity loops.

3. Stepper motor—An analog motion controller provides step and direction signals to a stepper drive to move a stepper motor. The motion controller performs step generation to output step and direction signals.

15.4.2 Architecture of a SoftMotion-Based Motion Controller (Shared Processor with User Application)

In case of a soft motion-based motion controller, functions such as supervisory control and trajectory generation execute on a real-time operating system shared with the operating system used for the user interface on the same processor. Control loops are executed within the drive, thus eliminating the need for dedicated motion controller hardware. Some configurations allow for the control loops to be also closed on the shared processor.

Common operating systems that are used for soft motion-based motion controllers include off-the shelf RTOSes such as QNX, Pharlap ETS, RTX, VxWorks, or Windows CE or, in many cases, proprietary implementations by motion control vendors.

15.4.2.1 PC/PLC-Based with Analog Interface (Motion I/O Connected to the PC/PLC via PCI or ISA Bus)

A PLC-based SoftMotion controller runs on the real-time operating system on the PLC. It connects to analog drives with a ± 10 V analog interface. In this case, the motion controller performs supervisory control, trajectory generation on the PLC processor, and closes position and velocity control loops on the PLC motion control modules. These modules connect to analog drives that handle the task of commutation, closing the current loops. In this case, the motion controller tasks run at the highest priority on the PLC processor and coexist with the user's sequential logic tasks.

A PC-based SoftMotion controller runs on a real-time operating system (VxWin, VxCE, InTime) or real-time extension (RTX, TwinCAT) that runs on a PC and shares the processor with Windows or any desktop OS. In this case, the motion controller performs supervisory control, trajectory generation on the PC processor, and closes position and velocity control loops on the PCI/ISA-based motion control card plugged into the PC. These cards connect to analog drives that handle the task of commutation, closing the current loops. In this case, since the motion controller runs on the real-time OS on the PC, it is guaranteed bandwidth and priority over Windows or any desktop OS sharing the processor with it.

15.4.2.2 PC/PLC-Based with Deterministic Digital Network (Distributed Motion, I/O Connected to the PC/PLC via a Deterministic Network)

As discussed in Section 15.4.2.1, a SoftMotion controller shares the processor with either the PLC tasks or tasks running on the Windows or desktop OS on a PC. The primary difference between this mode of operation and the former is that in this mode the communication of the motion controller with the drives is via a deterministic bus. This solves a number of issues faced in Section 15.4.2.1 such as connectivity and signal integrity. This architecture is also termed as "DistributedMotion." Similar to the previous case, the motion controller performs supervisory control, trajectory generation, and in some cases also closes the position and velocity control loops. The drives close handle the commutation and close the current loops. They also support closing the position and the velocity loops that allow for lower bandwidth utilization of the deterministic network.

A primary requirement for such a configuration is that the digital network connecting the motion controller to the digital drive is deterministic and that it is able to transmit data deterministically. Emerging networks include IEEE 1394, Ethernet PowerLink, EtherCAT, SERCOS III, PROFInet (IRT), EtherNET/IP, CANopen, and PROFIbus (MC).

IEEE 1394

IEEE 1394 (isochronous) (www.1394ta.org) is a high speed, real-time digital network based on the serial bus. It requires an IEEE 1394 interface on the host PC. It offers submillisecond response times and submicrosecond jitter for use with soft motion-based motion controllers.

Ethernet PowerLink

Ethernet PowerLink (www.ethernet-powerlink.org) is an Ethernet-based digital network based on TCP/IP and isochronous slot communication network management (SCNM). It offers submillisecond response times and submicrosecond jitter for use with soft motion-based motion controllers.

EtherCAT

EtherCAT (www.ethercat.org) is an Ethernet-based digital network based on the principle of reading and writing EtherCAT telegram packets as they pass through the different slave devices in cyclic order. It offers submillisecond response times and submicrosecond jitter for use with soft motion-based motion controllers.

SERCOS III

SERCOS-III (www.sercos.com) is an Ethernet-based digital network based on TCP/IP and the principle of cyclic data transfer with an exact time pattern. It requires the use of a master to control the timing behavior of the data transfer to the devices on the network. It provides submillisecond response rates for use with soft motion-based motion controllers.

PROFInet (IRT)

PROFInet (isochronous real time) (www.profinet.com) is an Ethernet-based digital network that leverages TCP/IP and DCOM standards, and works with internal switches in all devices to handle synchronization. It provides submillisecond response rates for use with soft motion-based motion controllers.

EtherNET/IP

EtherNET/IP (www.odva.org) is an Ethernet-based digital network that is based on TCP/IP and Control switches and uses time stamps for synchronization. It provides clock-cycle synchronous data transmission, and Information Protocol (CIP) standards. It supports a star topologywith devices connected via managed and millisecond response rates for use with soft motion-based motion controllers that performsupervisory control and trajectory generation.

CANopen

CANopen (www.canopen.de) is a layer on CAN that is based on the DS301 communication profile and the DS402 device profile for motion control. It supports deterministic data transmission, and millisecond response rates for use with soft motion-based motion controllers that perform supervisory control and trajectory generation. The control loops are closed on the digital drive.

PROFIbus (MC)

PROFIbus (Motion Control) (www.profibus.org) is based on the PROFIbus digital network with enhancements to response times and clock synchronization. PROFIbus (MC) has response rates of several milliseconds making it suitable for soft motion-based motion controllers that perform supervisory control and trajectory generation. The control loops are closed on the digital drive.

15.5 Summary

Motion control systems are used to control the position, velocity, and torque of a rotational or linear electromechanical device. A typical motion control system consists of a variety of components including user interface, motion controller, drive, motor, feedback device, and motion I/O The functions of a motion controller include trajectory generation, supervisory control, and a position and velocity feedback control loop.

16

Real-Time Monitoring and Control

Gerardo Garcia
National Instruments, Inc.

16.1 Introduction to Real-Time Systems

Real-time operating systems originated with the need to solve two main types of applications: event response and closed-loop control systems. Event response applications require a response to a stimulus in a determined amount of time; an example of such a system is an automotive airbag system. Closed-loop control systems continuously process feedback in order to adjust an output; an automotive cruise control system is an example of a closed-loop control system. Both of these types of systems require the completion of an operation within a specific deadline. This type of performance is referred to as determinism.

Real-time systems are sometimes classified as "soft" or "hard." Soft real-time typically means the utility of a system is inversely proportionate to how long it has been since a deadline is missed. For example, when pressing a cell phone button to answer an incoming call, the connection must be established soon after the button has been pressed. However, the deadline is not mission-critical and small delays can be tolerated. Hard real-time systems are those in which the utility of a system becomes zero in the event of a missed deadline. An automotive engine control unit (ECU) must process incoming signals and calculate spark plug timing within a deadline. If the deadline is missed, the engine will fail to operate correctly. The usefulness of a task after a deadline is missed depends on whether the system is a soft real-time or a hard real-time system, as shown in Figure 16.1.

Operating systems such as Microsoft Windows and Mac OS provide an excellent platform for developing and running your noncritical measurement and control applications. However, because these operating

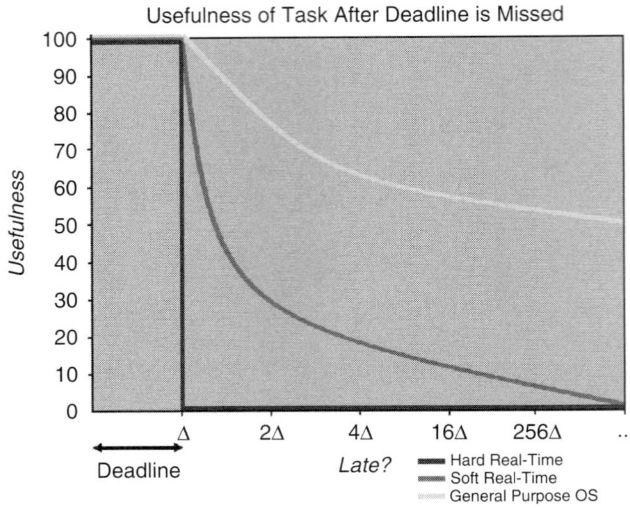

FIGURE 16.1 Differences between hard and soft real-time.

systems are designed for general purpose use, they are not the ideal platforms for running applications that require deterministic performance or extended uptime.

General-purpose operating systems are optimized to run a variety of applications simultaneously, ensuring that all applications receive some processing time. These operating systems must also respond to interrupts from peripherals such as the mouse and keyboard. The user has limited control regarding how these tasks are handled by the processor. As a result, high-priority tasks can be preempted by lower priority tasks, making it impossible to guarantee a response time for your critical applications.

In contrast, real-time operating systems give users the ability to prioritize tasks so that the most critical task can always take control of the processor when needed. This property enables you to program an application with predictable results.

Real-time operating systems are required when the processor is involved in operations such as closed loop control and time-critical decision making. These applications require timely decisions to be made on the basis of incoming data. For example, an I/O device samples an input signal and sends it directly to memory. Then, the processor must analyze the signal and send the appropriate response to the I/O device. In this application, the software must be involved in the loop; therefore, you need a real-time operating system to guarantee response within a fixed amount of time. In addition, applications requiring extended run-times or stand-alone operation are often implemented with real-time operating systems.

16.2 Real-Time Development Tools

Traditional real-time software development requires a code editor to develop code, a compiler to generate embedded code, and a compatible debugger to validate the application, as well as tools to profile the application performance, track memory usage, and trace system thread execution. Many software development packages integrate these tools into one environment. An example of such an environment is the LabVIEW Real-Time Development Module. Figure 16.2 depicts the real-time development environment, including the host computer for software development and the real-time target for execution.

System profiling tools are important for optimizing the execution of a real-time system. The LabVIEW Execution Trace Toolkit is an example of a tracing tool that captures low-level execution details of a realtime application. Figure 16.3 shows a typical trace capture that displays when application tasks executed in time.

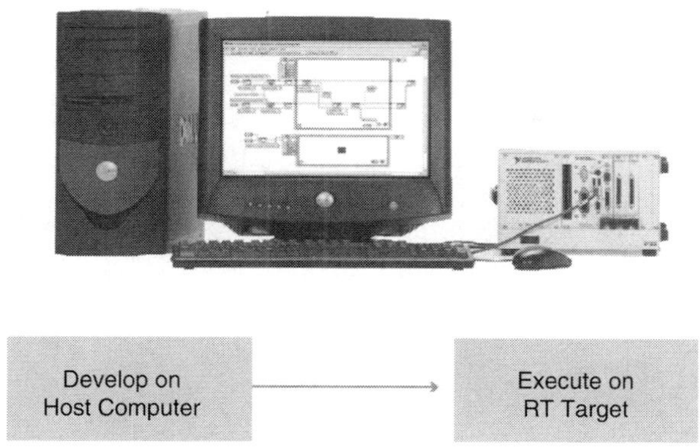

FIGURE 16.2 Real-time development process.

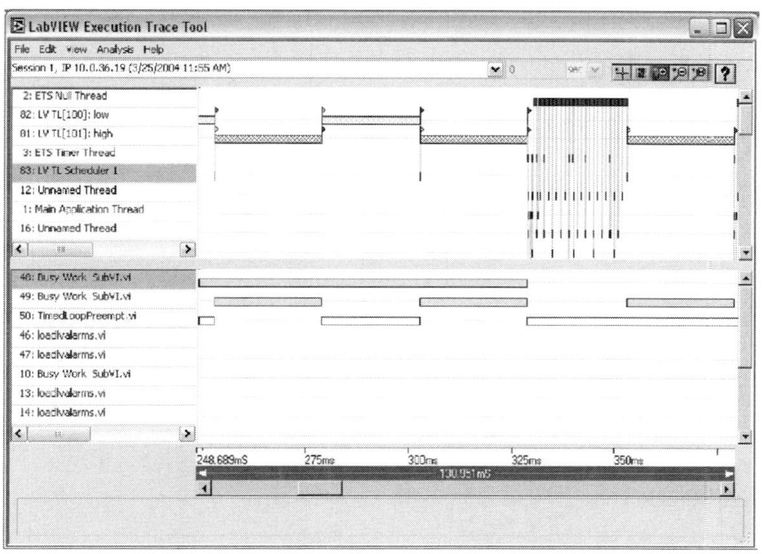

FIGURE 16.3 Application tracing tool.

Hardware deployment platforms vary depending on the application. Real-time applications use desktop PCs, modular hardware platforms such as PXI, off-the-shelf rugged controllers such as programmable logic controllers (PLCs) and programmable automation controllers (PACs), single-board computers, and embedded microprocessors on custom hardware. The popular NI CompactRIO PAC is shown in Figure 16.4. An appropriate real-time operating system is used in these hardware platforms to execute the application software.

16.2.1 Operating System Scheduler

All real-time operating systems have a scheduler that coordinates the execution of tasks (threads) by the CPU. There are various scheduling methods that are used by operating systems. Round-robin scheduling shares processor time between threads based on equal slices of time.

To illustrate round-robin scheduling, we consider the analogy of an automobile repair shop. The mechanic represents the CPU, the shop manager represents the scheduler, and multiple cars represent the multiple threads of our system. Using round-robin scheduling, the mechanic will cycle between each car for a set period of time. Round-robin scheduling guarantees that each thread will have some time with the processor.

FIGURE 16.4 NI CompactRIO programmable automation controller.

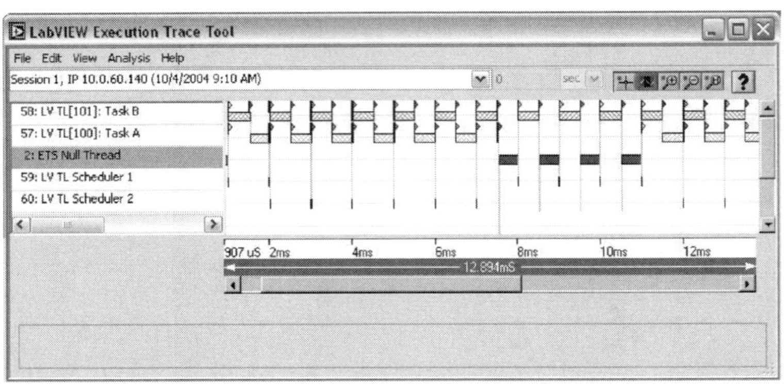

FIGURE 16.5 Example trace of preemptive scheduling.

Another method of scheduling is called preemptive scheduling, where priority can be given to the time-critical tasks.When a time-critical thread needs time from the processor, the other threads will wait until the time-critical thread is finished. In the repair shop example, an ambulance is assigned time-critical priority. As a result, it will be serviced as soon as it arrives. All other cars will be put on hold until the ambulance service is complete. Once the ambulance service is complete, the other cars will continue to share time with the mechanic. Figure 16.5 shows an example of two tasks running at different priority levels. Task B has a higher priority and runs every half millisecond, while Task A runs every 10 ms. The red flags indicate when Task A is preempted so that the higher priority Task B can be executed.

16.3 Real-Time Software Architecture

When developing a real-time application, you must decide which operations need to have real-time performance and which operations do not. The components that need real-time performance are those that need to occur within a specific period of time, meaning they are deterministic. These time-critical tasks should be assigned the highest priority in the application.

Operations that are of lower priority will run only when the time-critical operations are suspended. Examples of low priority operations are communication with a host computer, reading and writing to files, and nondeterministic communication with external instruments. These operations are sometimes called the house-keeping tasks.

16.4 Deterministic Timing

Determinism is how consistently a systemis able to perform operations within a fixed period of time. This is the most fundamental component of all real-time systems. Jitter is the variance between the expected execution time of a real-time task and the actual execution time. All real-time systems have some jitter; however, the jitter is bound and can be quantified. Systems that are not real time have very large or unbounded jitter. Figure 16.6 depicts a jitter diagram.

The most basic source of jitter is variation in the clock. However, you can induce additional jitter if you do not follow certain programming practices. To reduce jitter, tasks are programmed to run periodically on the basis of a hardware clock that can be derived from an I/O signal or a precise CPU timer. Figure 16.7 shows the configuration of a real-time task set to run every 500 ms.

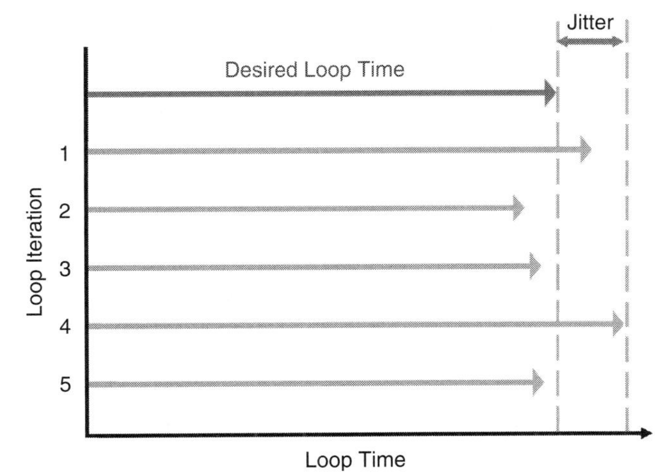

FIGURE 16.6 A jitter diagram.

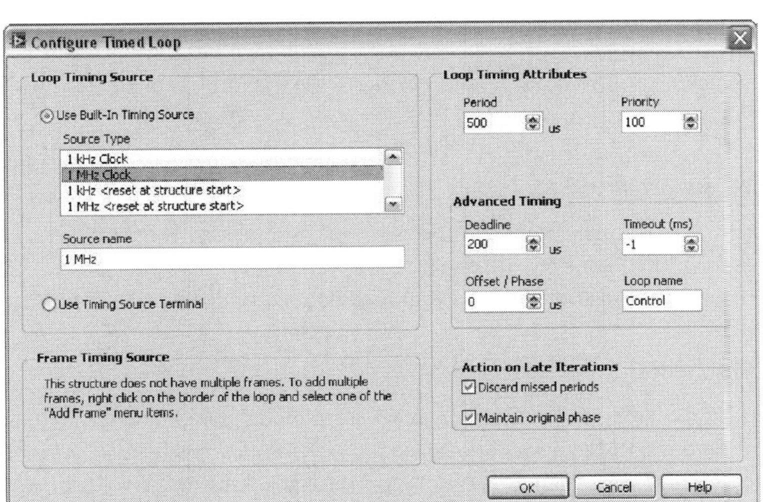

FIGURE 16.7 Real-time task configuration.

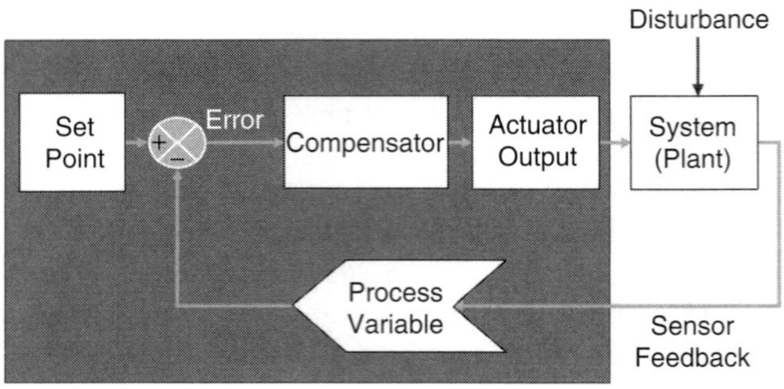

FIGURE 16.8 The block diagram of a closed-loop control system.

16.5 Implementing Real-Time Control

Control algorithm implementation is an important part of real-time system development. Control strategies can vary in complexity, from simple digital logic to advanced mathematical models that are carefully tuned to control dynamic systems. A typical block diagram of a closed-loop control system is shown in Figure 16.8.

16.5.1 Proportional-Integral-Derivative Control

The proportional-integral-derivative (PID) algorithm is the most common control algorithm used in industry. With PID, you specify a process variable and a set point. The process variable is the system parameter you want to control, such as temperature, pressure, or flow rate, and the set point is the desired value for the parameter you are controlling.

The proportional gain determines the ratio of output response to the error. The error is the difference between the set point and the process variable. In general, increasing the proportional gain will increase the speed of the control system response. However, if the proportional gain is too large, the process variable will begin to oscillate.

The integral component sums the error term over time. The integral response will continually increase over time unless the error is zero, so the effect is to drive the steady state error to zero.

The derivative component causes the output to decrease if the process variable is increasing rapidly. The derivative response is proportional to the rate of change of the process variable.Most practical control systems use very small derivative time, because the derivative response is highly sensitive to noise in the process variable signal.

16.5.2 Fuzzy Logic Control

Fuzzy logic control is most often used for expert systems and process control. It is a method of rule-based decision making that emulates the rule of thumb thought process of humans. If an expert can qualitatively describe a control strategy, fuzzy logic can be used to develop the controller.

16.5.3 Model-Based Control Design

Another method for designing control systems is by using model-based design methods. This involves using software tools to model the behavior of dynamic systems completely in software. Then, designers can use mathematical software tools to develop control algorithms and test them with the software simulation of the dynamic system. Figure 16.9 shows a simulation diagram created in LabVIEW.

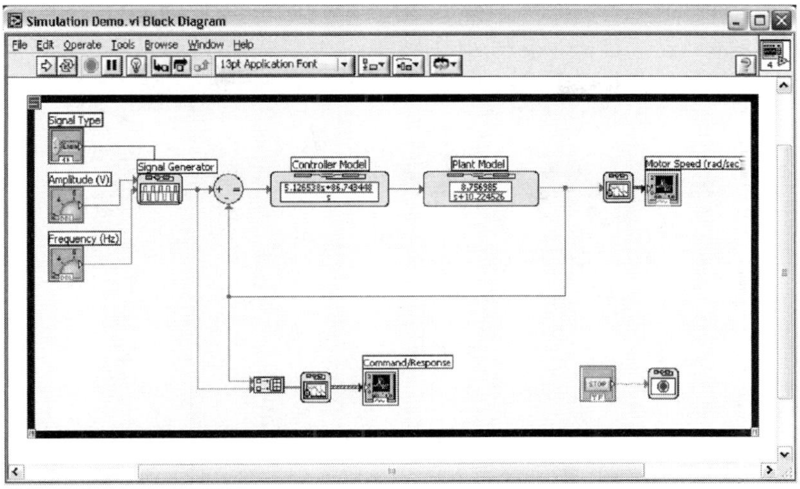

FIGURE 16.9 Simulation diagram.

16.6 Monitoring Systems

Real-time systems are usually monitored by one or more human–machine interfaces (HMI) that send and receive data from the real-time controllers. These HMIs can be in the form of powerful desktop servers, laptops, PDAs, web browser interfaces, or touch-panel displays. Communication is done through a variety of media and protocols. Ethernet is widely used as a media of data transfer with differing protocols depending on the application. These include TCP/IP for point-to-point streaming, OLE for Process Control (OPC) for distributed control systems, or web servers for remote viewing. Other methods for communicating are serial interfaces, fieldbuses such as PROFIBUS and DeviceNet, and wireless media such as WIFI and Bluetooth.

Large distributed real-time systems typically use a Supervisory Control and Data Acquisition (SCADA) system for monitoring. These systems have visualization features and can display data from real-time controllers, show alarm conditions, and log data to databases. An example of a SCADA software is the LabVIEW Datalogging and Supervisory Control (DSC) Module. A SCADA visualization of a heat exchanger is shown in Figure 16.10.

For distributed real-time control systems, data can be stored on multiple computers and monitored centrally, or it can be stored on one central server. The most difficult challenge is to communicate with live data. OPC is an industry standard interface through which software and hardware can communicate irrespective of the manufacturer. Controller hardwares from different manufacturers communicate with their own OPC server on a host system and publish I/O and calculated values as tags. SCADA systems act as OPC clients that subscribe to tags from all the individual OPC servers. Figure 16.11 shows a LabVIEW project with a list of tags that can be read from OPC servers to develop an operator interface.

For some applications, OPC data transfer may not be fast enough. In these cases, using lower-level protocols such as TCP/IP and UDP allows you to send data at faster rates.

16.7 Summary

In summary, real-time systems solve a variety of applications that require determinism and reliability. To achieve such requirements, real-time operating systems define strict scheduling behavior, which all

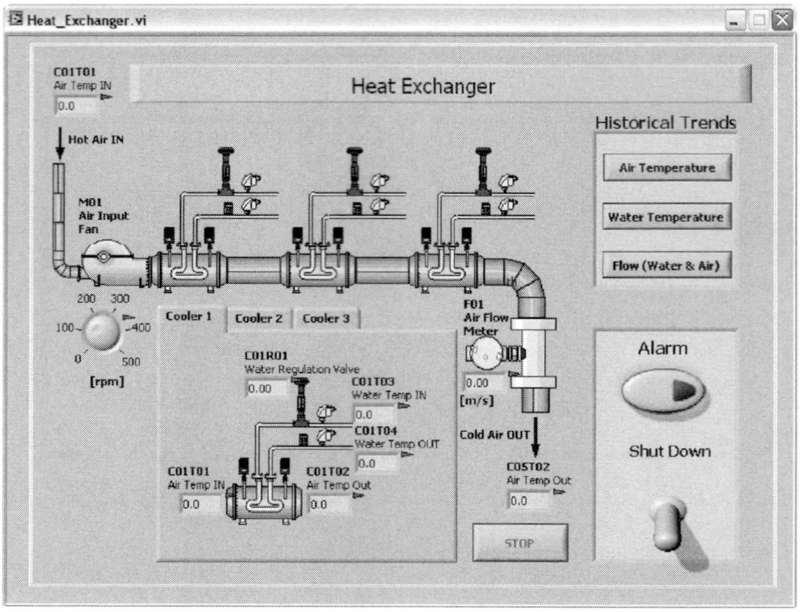

FIGURE 16.10 SCADA visualization.

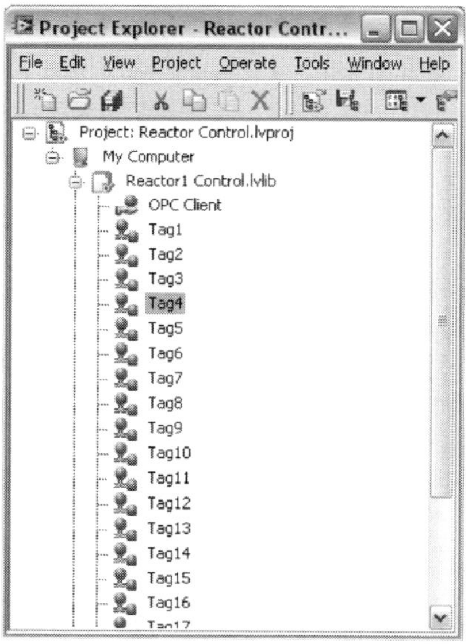

FIGURE 16.11 OPC tags.

software abides by. Due to the complex nature of real-time applications and the small margin for error, developers must employ good programming practices in order to be successful.

Further Information

Web sites

1. http://www.omimo.be/encyc/publications/faq/rtfaq.htm
2. http://www.ni.com/realtime

3. http://zone.ni.com/devzone/cda/tut/p/id/3938
4. http://zone.ni.com/devzone/cda/tut/p/id/4040

References

1. J.E. Cooling, *Software Design for Real-time Systems*, ISBN 0-412-34180-8, Chapman & Hall, London, 1991.
2. Stuart Bennett, *Real-Time Computer Control*, ISBN 0-13-764176-1, Prentice Hall, Upper Saddle River, NJ 2006.

17

Micromechatronics and Microelectromechanical Motion Devices*

Sergey Edward Lyshevski
Rochester Institute of Technology

17.1 Micromechatronic Systems Design

Micromechatronic systems integrate actuators, sensors, power electronics, and ICs. Within a focus on high-performance micromechatronic systems, books [1–3] cover the general issues in the systems design. Important topics and issues are outlined and covered in this section. One of the major components of micromechatronic systems is a high-performance microelectromechanical motion device that (1) converts physical stimuli to electrical or mechanical signals and vice versa, and, (2) performs actuation and sensing. These motion devices are controlled by ICs. Electromagnetic and electromechanical features of microelectromechanical motion devices are basics of their operation, design, analysis, and fabrication. Correspondingly, motion devices and ICs are designed taking into account possible system architectures [1–3]. The baseline step-by-step procedure in the design of micromechatronic systems is

1. Define application and environmental requirements
2. Specify performance specifications

*Some results reported in this chapter were published in [1–3].

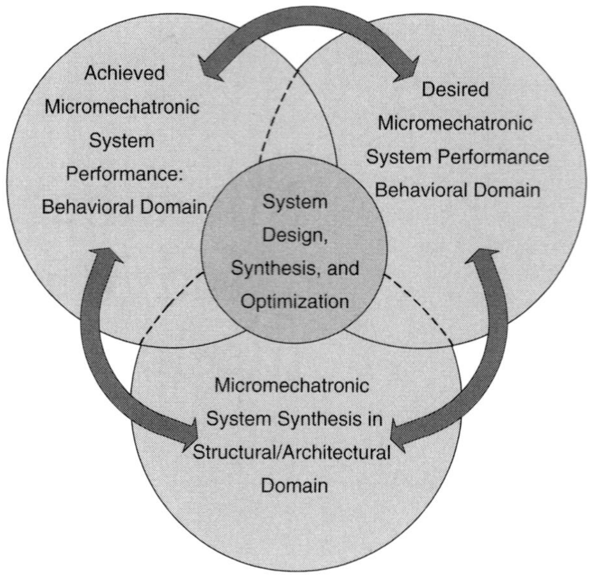

FIGURE 17.1 Design flow in synthesis of micromechatronic systems.

3. Devise (synthesize) microelectromechanical motion devices (actuators and sensors) researching operating principles, topologies, configurations, geometry, electromagnetic systems, and so forth
4. Perform electromagnetic, mechanical, and sizing-dimension estimates
5. Define technologies, techniques, processes, and materials to fabricate electromechanical motion devices
6. Design ICs to control electromechanical motion devices
7. Develop high-fidelity mathematical modelswith minimumlevel of simplifications and assumptions to examine integrated electromagnetic–mechanical phenomena and effects
8. Perform coherent electromagnetic and mechanical analysis
9. Modify and refine the design optimizing performance
10. Design control laws to control motion devices and implement these controllers using ICs
11. Integrate micromechatronic system

One of the most challenging design problems is the system architecture synthesis, system integration, optimization, and hardware (actuators, sensors, power electronics, ICs, microcontrollers, and DSPs) selection. The design starts with a given set of requirements and specifications. High-level functional design is performed first in order to produce detailed design at the subsystem and component level. Using the advanced subsystems and components, the initial design is performed, and the closed-loop electromechanical system performance is tested against the requirements. At each level of the design hierarchy, the system performance in the behavioral domain is used to evaluate and refine the design. The design flow is illustrated in Figure 17.1.

The micromechatronic system is illustrated in Figure 17.2. The systemperformance is measured against many criteria, for example, stability, robustness, transient behavior, accuracy, disturbance attenuation, and so forth. In the behavioral domain, the designer can examine and optimize the input–output transient dynamics using the tracking error $e(t) = y(t) - r(t)$, where $y(t)$ and $r(t)$ are the output and reference (command) variables. The tracking error $e(t)$ is minimized and dynamics is optimized using different performance criteria. For example,

$$J = \min_{t,e} \int_0^\infty t\,|\,e\,|\,dt, \quad J = \min_{e} \int_0^\infty |\,e\,|\,dt, \quad \text{or} \quad J = \min_{e} \int_0^\infty e^2 dt.$$

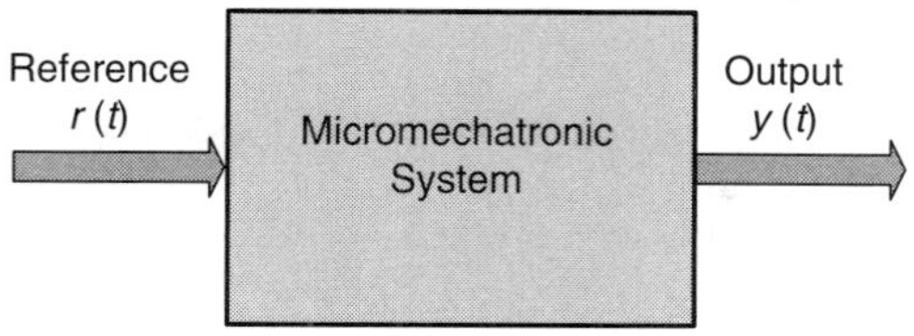

FIGURE 17.2 Dynamic micromechatronic system.

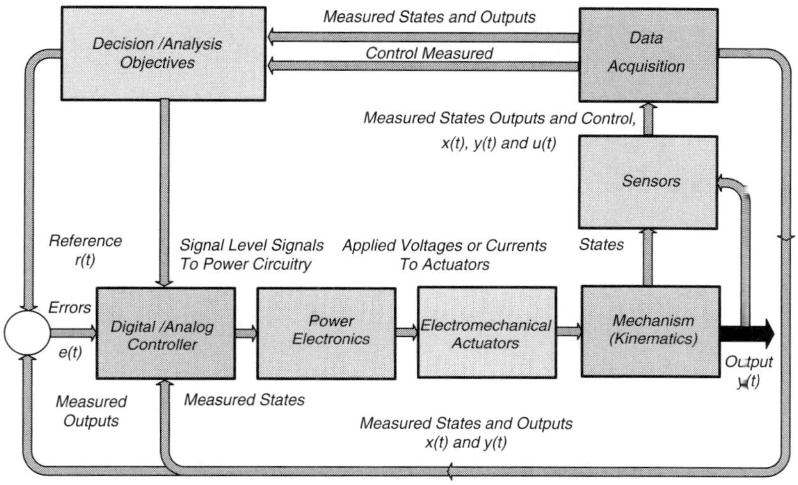

FIGURE 17.3 High-level functional block diagram of the closed-loop micromechatronic system.

Mechatronic systems (robots, electric drives, servomechanisms, pointing systems, assembles, etc.) are actuated by electromechanical motion devices that are highly nonlinear systems. Therefore, accurate actuator actuation and control are very important problems. Depending on operation and functionality of motion devices, onemust derive the expressions for the voltages applied to the phase windings in order to maximize the electromagnetic torque and ensure optimality. Actuators and sensors must be designed and integrated with the corresponding power electronics and ICs. The control laws are implemented using analog controllers, which can be integrated with power electronics (PWM amplifiers), as well as digital controllers (microcontrollers and DSPs). The principles of matching and compliance are general design principles that require that the system architectures should be synthesized integrating all subsystems and components. The matching conditions have to be determined and guaranteed, and actuators— sensors—power electronics—ICs—controller compliance must be ensured.

A functional block diagram of a controlled micromechatronic system is illustrated in Figure 17.3. Different techniques to design optimal controllers are reported in the next section. Those controllers should be derived using the baseline electromagnetic laws that define the operating principles of electromechanical motion devices. It was emphasized that depending on the motion devices, distinct power electronics must be utilized. High-performance micromechatronic systems can be designed using permanent-magnet electromechanical motion devices that ensure high torque and power densities. Radial and axial technology permanent-magnet motion devices are covered in the following sections.

17.2 Tracking Control of Micromechatronic Systems

In general, mechatronic systems are nonlinear [1–3]. Using the Hamilton–Jacobi theory and Lyapunov concepts, design of control algorithms for nonlinear systems is reported in References 1–3. To introduce the design, we consider linear and nonlinear models. A linear model is given as

$$x^{sys}(t) = Ax^{sys} + Bu \qquad y = Hx^{sys} \qquad x^{sys}(t_0) = x_0^{sys}, \tag{17.1}$$

where $x^{sys} \in X^{sys} \subset \mathbb{R}^n$ is the state vector with auxiliary conditions; $u \in U \subset \mathbb{R}^m$ is the control vector; $y \in Y \subset \mathbb{R}^b$ is the output vector; and $A^{sys} \in \mathbb{R}^{n \times n}$ and $B^{sys} \in \mathbb{R}^{n \times m}$ are the constant-coefficient matrices. Define the tracking error vector as

$$e(t) = Nr(t) - y(t) = Nr(t) - Hx^{sys}(t). \tag{17.2}$$

Therefore, from Equations 17.1 and 17.2, one finds

$$\dot{e}(t) = N\dot{r}(t) - \dot{y}(t) = N\dot{r}(t) - H\dot{x}^{sys}(t) = N\dot{r}(t) - HA^{sys}x^{sys} - HB^{sys}u. \tag{17.3}$$

Using the expanded state vector $x(t) = \begin{bmatrix} x^{sys}(t) \\ e(t) \end{bmatrix}$, we have

$$\begin{aligned} \dot{x}(t) &= \begin{bmatrix} \dot{x}^{sys}(t) \\ \dot{e}(t) \end{bmatrix} = \begin{bmatrix} A^{sys} & 0 \\ -HA^{sys} & 0 \end{bmatrix} \begin{bmatrix} x^{sys} \\ e \end{bmatrix} + \begin{bmatrix} B^{sys} \\ -HB^{sys} \end{bmatrix} u + \begin{bmatrix} 0 \\ N \end{bmatrix} \dot{r} \\ &= Ax + Bu + \begin{bmatrix} 0 \\ N \end{bmatrix} \dot{r} \quad y = Hx^{sys}. \end{aligned} \tag{17.4}$$

The *space transformation* method is based on the use of the following z and v vectors: $z = \begin{bmatrix} x \\ u \end{bmatrix}$ and $v = \dot{u}$. Thus, from Equation 17.4, we obtain the system

$$\dot{z}(t) = \begin{bmatrix} A & B \\ 0 & 0 \end{bmatrix} z + \begin{bmatrix} 0 \\ I \end{bmatrix} v = A_z z + B_z v, \;\; y = Hx^{sys} \qquad z(t_0) = z_0. \tag{17.5}$$

Minimizing the quadratic performance functional

$$J = \int_{t_0}^{t_f} \left(z^T Q_z z + v^T G_z v \right) dt, \; Q_z \in \mathbb{R}^{(n+m)\times(n+m)}, \; Q_z \geq 0, \; G \in \mathbb{R}^{m\times m} \quad G > 0, \tag{17.6}$$

with respect to the system dynamics (Equation 17.5), the application of the first-order necessary condition for optimality gives

$$v = -G_z^{-1} B_z^T K z. \tag{17.7}$$

The Riccati equation

$$-\dot{K} = KA_z + A_z^T K - KB_z G_z^{-1} B_z^T K + Q_z, \; K(t_f) = K_f \tag{17.8}$$

is solved to find the unknown matrix $K \in \mathbb{R}^{(n+m)\times(n+m)}$. Taking note of Equation 17.7, one has

$$\begin{aligned} \dot{u}(t) &= -G_z^{-1} B_z^T K z = -G_z^{-1} \begin{bmatrix} 0 \\ I \end{bmatrix}^T \begin{bmatrix} K_{11} & K_{21}^T \\ K_{21} & K_{22} \end{bmatrix} \begin{bmatrix} x \\ u \end{bmatrix} \\ &= -G_z^{-1} K_{21} x - G_z^{-1} K_{22} u = K_{f1} x + K_{f2} u. \end{aligned} \tag{17.9}$$

Using $\dot{x}(t) = Ax + Bu$, wehave $u = B^{-1}(\dot{x}(t) - Ax) = (BTB)^{-1} BT(\dot{x}(t) - Ax)$. From Equation 17.9, one obtains

$$\begin{aligned}\dot{u}(t) &= K_{f1}x + K_{f2}u = K_{f1}x + K_{f2}\left(B^T B\right)^{-1} B^T(\dot{x}(t) - Ax) \\ &= \left[K_{f1} - K_{f2}\left(B^T B\right)^{-1} B^T A\right]x(t) + K_{f2}\left(B^T B\right)^{-1} B^T \dot{x}(t) \\ &= \left(K_{f1} - K_{F1}A\right)x(t) + K_{F1}\dot{x}(t) = K_{F2}x(t) + K_{F1}\dot{x}(t).\end{aligned} \tag{17.10}$$

Using Equation 17.10, and taking note of $x(t) = \begin{bmatrix} x^{sys}(t) \\ e(t) \end{bmatrix}$, one finally derives the proportional-integral controller in the following form

$$\begin{aligned}u(t) &= K_{F1}x(t) - K_{F1}x_0 + \int K_{F2}x(\tau)d\tau + u_0 \\ &= K_{F1}\begin{bmatrix} x^{sys}(t) \\ e(t) \end{bmatrix} - K_{F1}\begin{bmatrix} x_0^{sys}(t) \\ e_0(t) \end{bmatrix} + \int K_{F2}\begin{bmatrix} x^{sys}(\tau) \\ e(\tau) \end{bmatrix} d\tau + u_0.\end{aligned} \tag{17.11}$$

For nonlinear micromechatronic systems, as given by

$$\dot{x}^{sys}(t) = F(x^{sys}) + B(x^{sys})u, \qquad y = Hx^{sys}, \qquad x^{sys}(t_0) = x_0^{sys}, \tag{17.12}$$

the proposed procedure can be straightforwardly used. In Equation 17.12), $F(\cdot)$: $\mathbb{R}^n \to \mathbb{R}^n$ and $B(\cdot)$: $\mathbb{R}^n \to \mathbb{R}^{n\times m}$ are the smooth Lipschitz maps; $H \in \mathbb{R}^{b\times c}$ is the output matrix with constant coefficients.

The proportional-integral controller for Equation 17.12 is given as

$$\dot{u}(t) = -G_z^{-1}B_z^T\frac{\partial V}{\partial z} = -G_z^{-1}\begin{bmatrix} 0 \\ I \end{bmatrix}^T \frac{\partial V(x,u)}{\partial [x\ u]^T}, \tag{17.13}$$

where $V(x, u)$ is the return function.

The control law is bounded, and $u_{min} \leq u \geq u_{max}$. To design the *admissible* control laws, we minimize

$$J = \int_{t_0}^{t_f}\left[z^T Q_z z + \int \left(\Phi^{-1}(v)\right)^T G_z dv\right]dt, \tag{17.14}$$

where $\Phi(\cdot)$: $\mathbb{R}^n \to \mathbb{R}^m$ is the bounded, integrable, one-to-one, real-analytic, globally Lipschitz continuous function, $\Phi \in U \subset \mathbb{R}^m$.

Minimizing Equation 17.14 with $u_{min} \leq u \geq u_{max}$, $u \{ U$, for linear and nonlinear systems, we have

$$-\frac{\partial V}{\partial t} = \min_{u\in U}\left\{z^T Q_z z + \int \left(\Phi^{-1}(v)\right)^T G_z dv + \frac{\partial V}{\partial z}^T (A_z z + B_z v)\right\}, \tag{17.15}$$

$$\text{and } -\frac{\partial V}{\partial t} = \min_{u\in U}\left\{z^T Q_z z + \int \left(\Phi^{-1}(v)\right)^T G_z dv + \frac{\partial V}{\partial z}^T [F_z(z) + B_z(z)v]\right\} \tag{17.16}$$

Using the first-order necessary condition for optimality, the minimization of Equations 17.15 and 17.16 results in the *admissible* controllers

$$\dot{u}(t) = -\Phi\left(G_z^{-1}B_z^T\frac{\partial V}{\partial z}\right) \quad u \in U. \tag{17.17}$$

Assuming that the solution of the Hamilton-Jacobi equation is approximated by the quadratic return function, from Equation 17.17, one obtains

$$u(t) = \Phi\left(K_{F1}x(t) + \int K_{F2}x(\tau)d\tau\right) = \Phi\left(K_{F1}\begin{bmatrix} x^{\text{sys}}(t) \\ e(t) \end{bmatrix} + \int K_{F2}\begin{bmatrix} x^{\text{sys}}(\tau) \\ e(\tau) \end{bmatrix} d\tau\right) \quad u \in U. \tag{17.18}$$

The *admissible* control law synthesized is bounded, and the second-order necessary condition for optimality is satisfied. However, the sufficient conditions must be examined. Using the *admissibility* concept, it is demonstrated in References 1–3 that a system with bounded control (Equation 17.18) is stable in $X(X_0,U)$, and robust tracking is guaranteed in the convex and compact set $E(E_0, Y, R)$ if for the reference input $r \in R$ there exists a positive-definite function $V(e,x)$ such that $[dV(e, x)]/dt \leq 0$. If the criteria imposed on the Lyapunov pair, $V(e, x) > 0$ and $[dV(e, x)]/dt \leq 0$, are guaranteed for all $x_0 \in X_0$, $e_0 \in E_0$, $u \in U$, and $r \in R$, then, the closed-loop system is robustly stable in $X(X_0,U)$, and robust tracking is guaranteed in the convex and compact set $E(E_0, Y, R)$. By computing the derivative of the $V(e, x)$, the unknown feedback gains can be found to satisfy the sufficient conditions for stability.

The designed analog controllers may be implemented using microcontrollers. Therefore, analog controllers are discretized. The feedback gains k_{dp} and k_{di} of the digital control law are found using the proportional, integral, and derivative coefficients of the analog PID controller (k_p and k_i) as well as the sampling period Ts. We have $k_{dp} = k_p \cdot \frac{1}{2} k_{di}$ and $k_{di} = T_s k_i$. In particular, the controller can be implemented as $u(kT_s) = k_p e(kT_s) + \frac{1}{2} k_i T_s \sum_{i=1}^{k} [e((i-1)T_s) + e(iT_s)]$. The proportional-integral controllers with the state feedbacks are implemented in the similar way, and the state variables are sampled at the specified sampling period T_s.

17.3 Synthesis of Microelectromechanical Motion Devices

The designer synthesizes microelectromechanical motion devices by devising and examining baseline physics and operational principles. Axial and radial topology, ac and dc, electromagnetic and electrostatic, and other actuators are covered in [1–3]. To introduce the synthesis taxonomy, we consider radial topology two-phase permanent-magnet synchronous motion devices as shown in Figure 17.4. The electromagnetic system is *endless*, and different geometries can be utilized. In contrast, the translational (linear) synchronous machines have the *open-ended* electromagnetic system. Thus, the device geometry and electromagnetic systems can be integrated within the synthesis, classification, analysis, design, and optimization taxonomy. In particular, motion devices can have different geometries (plate, spherical, torroidal, conical, cylindrical, and asymmetrical) and electromagnetic systems. Using these distinct features, one can classify motion devices.

The basic types of electromagnetic motion devices are induction, synchronous, while topologies are rotational and translational (linear). That is, devices can be classified using a type classifier as given by $Y = \{y : y \in Y\}$. As illustrated in Table 17.1, electromechanical motion devices are categorized using a geometric classifier (plate P, spherical S, torroidal T, conical N, cylindrical C, or asymmetrical A geometry), and an electromagnetic system classifier (*endless E*, *open-ended O*, or *integrated I*). The classifier, as documented in Table 17.1, is partitioned into 3 horizontal and 6 vertical strips, and contains 18 sections, each identified by ordered pairs of characters, such as (E, P) or (O, C). In each ordered pair, the first entry is a letter chosen from the bounded electromagnetic system set $M=\{E, O, I\}$. The second entry is a letter chosen from the geometric set $G=\{P, S, T, N, C, A\}$. That is, the electromagnetic system—geometric set is

$$M \times G = \{(E, F), (E, S), (E, T), \cdots, (I, N), (I, C), (I, A)\}.$$

In general, we have $M \times G = \{(m, g) : m \in M \text{ and } g \in G\}$.

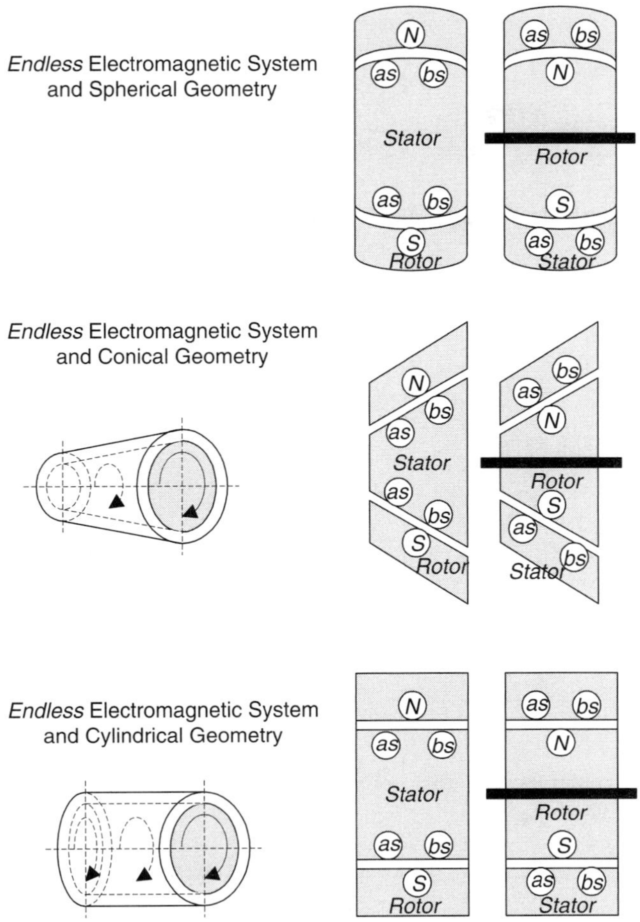

FIGURE 17.4 Radial topology permanent-magnet synchronous motion devices with *endless* electromagnetic system and different geometries.

Other categorization can be applied. For example, multiphase motion devices are classified using a phase classifier $H = \{h : h \in H\}$. Therefore, we have

$$Y \times M \times G \times H = \{(y,\ m,\ g,\ h) : y \in Y,\ m \in M,\ g, \in G \text{ and } h \in H\}$$

Topology (radial, axial, or integrated), permanent magnet shapes (strip, arc, disk, rectangular, rhomb, triangular, etc.), permanent magnet characteristics (*BH*demagnetization curve, energy product, hysteresis minor loop, etc.), *electromotive force* distribution, cooling, power, torque, size, torque-speed characteristics, bearing, packaging, as well as other distinct features can be classified. Hence, microelectromechanical motion devices can be devised and classified by an *N*-tuple as: motion device type, electromagnetic system, geometry, topology, phase, winding, sizing, bearing, cooling, fabrication, materials, packaging, and so forth.

By using distinct geometry, electromagnetic systems (*endless*, *open ended*, and *integrated*), topologies, and other features, novel high-performance motion devices can be synthesized by utilizing the Synthesis and Classification Solver. For example, the spherical, conical, and cylindrical geometry of two-phase permanent-magnet synchronous motion devices are illustrated in Figure 17.5. The cross-section of the slotless radial-topology synchronous microactuator, fabricated on the silicon substrate with polysilicon stator (with deposited windings), polysilicon rotor (with deposited permanent magnets), and contact

TABLE 17.1 Classification of Microelectromechanical Motion Devices Using the Electromagnetic System—Geometry Using Synthesis and Classification Solver

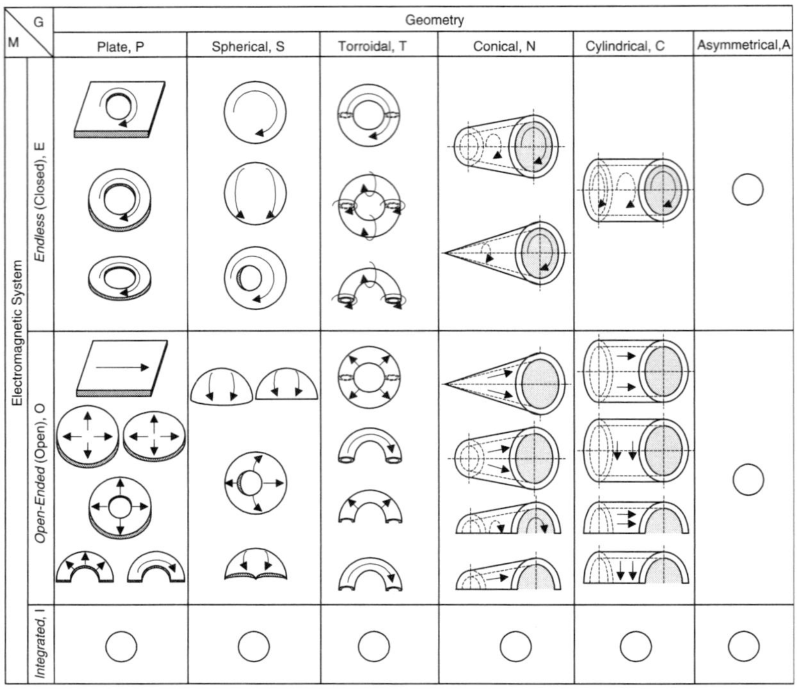

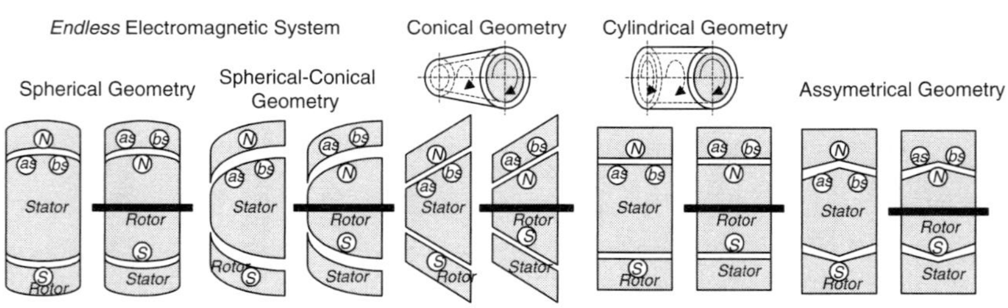

FIGURE 17.5 Two-phase permanent-magnet synchronous motion devices with *endless* electromagnetic system and distinct geometries.

bearing is illustrated in Figure 17.6. The fabrication of this microactuator and the processes are reported in References 1–3.

The major task is to devise high-performance motion devices relaxing fabrication difficulties guaranteeing affordability. The electrostatic and planar motion devices fabricated and tested to date are found to be inadequate for a wide range of applications owing to specifications imposed on performance [1–3]. Figure 17.7 illustrates the axial topology motion device. The stator is made on the substrate with deposited windings (surface micromachined or printed coils can be made using the fabrication processes as well as using double-sided substrate, applying conventional photolithography and etching processes [1–3]). The bearing post is fabricated on the stator substrate and the bearing hold is a part of the rotor microstructure. The rotor with permanent-magnet thin films rotates because of the electromagnetic torque developed. It is important to emphasize that stator and rotor are made using the conventional well-developed processes and materials.

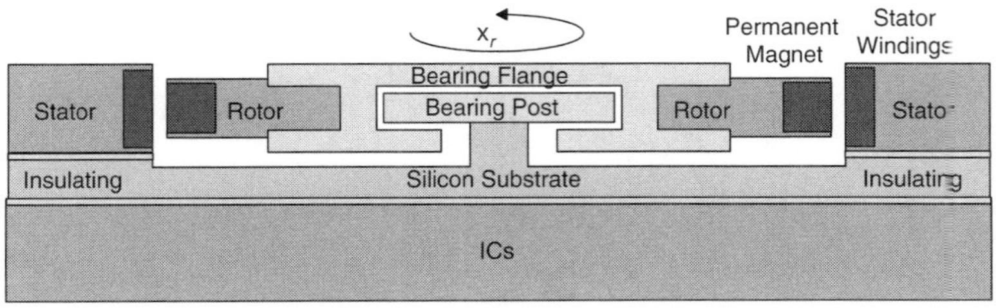

FIGURE 17.6 Cross-section schematics of a slotless radial-topology permanent-magnet brushless microactuator with controlling ICs.

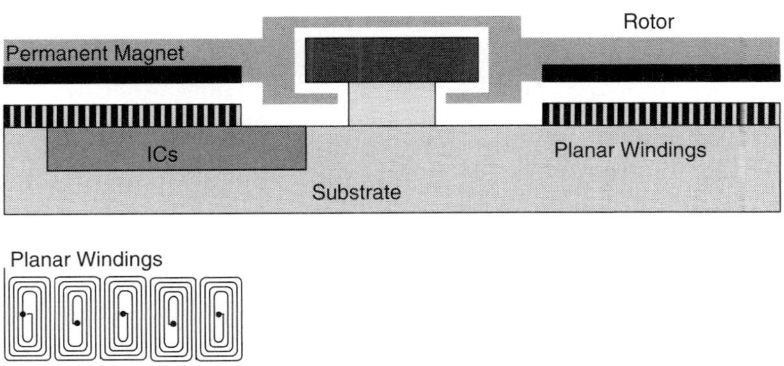

FIGURE 17.7 Cross-section of an axial topology motion device with controlling ICs.

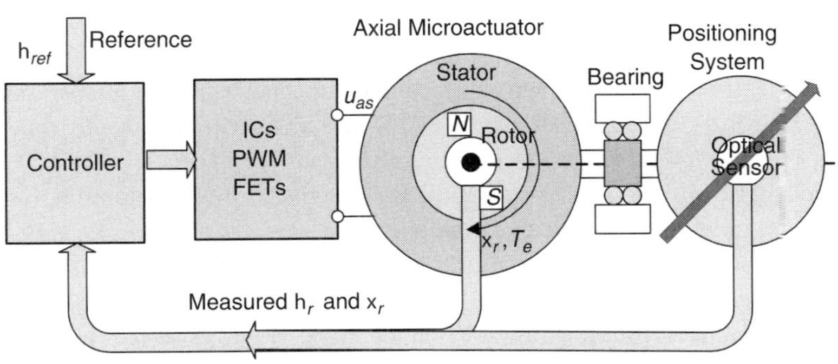

FIGURE 17.8 Schematic diagram of a servosystem: controller; ICs; actuator/sensors; and mechanism.

17.4 Micromechatronic System with an Axial Topology Motion Device

High-torque-density permanent-magnet axial-topology actuators with high switching-frequency power electronics and ICs, controlled by controllers (microcontroller or analog controllers), are found to be a very promising solution. High performance is achieved by utilizing: (1) high-torque-density brushless motion devices; (2) designing tracking controllers implemented utilizing controllers that integrate filters, IO devices, and advanced converters; (3) high-performance two- and four-quadrant power stages. The control laws are designed to satisfy the specifications imposed. Consider a servosystem as depicted in Figure 17.8. The reference signal is the angular displacement. Using the reference θ_{ref} and actual θ_r angular displacements (measured by the high-accuracy optical sensor), the controller develops PWM signals to

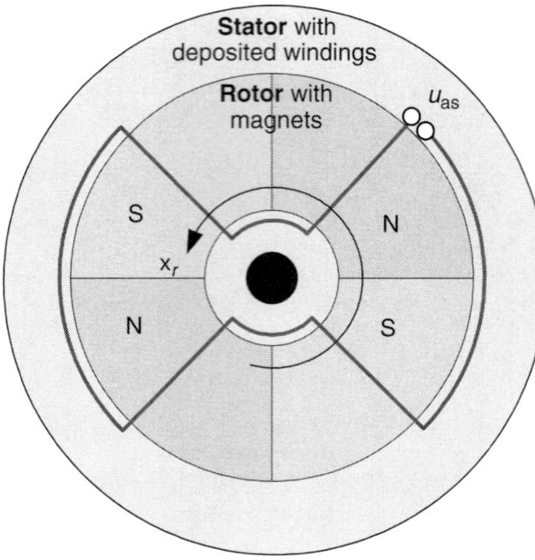

FIGURE 17.9 Axial topology actuator.

drive high-frequency field effect transistors in the power stage. The output voltage u_{as} is controlled by changing the duty cycle.

A limited-angle axial-topology permanent-magnet microactuator is documented in Figure 17.9. This device is used in the studied hard drive reported in Figure 17.8. To rotate the disk and displace the pointer, we utilize axial topology direct-drive synchronous actuators with the segmented array of the permanent-magnet strips. These high-torque-density permanent-magnet actuators exceed performance characteristics achieved by other servomotors [1–3]. The mathematical model must be found and used in the design of controllers as well as numerical analysis. The magnetic field developed by the magnets is approximated as continuous or discontinuous functions, for example, $B(\theta_r) = B_{max}\,\mathrm{sgn}(\sin(\frac{1}{2}N_m\theta_r))$, where B_{max} is the magnetic flux density from the permanent magnets as viewed at the coils; Nm is the number of magnets. For the limited-angle actuator, depending on the geometry, magnetization, and separations of magnets, one can approximate $B(\theta_r)$ as $B(\theta_r) = B_{max}\tanh^q(k\theta_r)$, where q is the technologydependent integer, and usually one finds $q = 1$ or $q = 3$; k is the technology-dependent coefficient.

The electromagnetic torque is derived using the magnetic dipole moment **m**, for example, $T_e = \mathbf{m}\times\mathbf{B}$. That is, for the symmetric two-filament limited angle actuator, for $q = 1$ we obtain $T_e = 2NlB_{max}|\tanh(k\theta_r)|\,i_{as}$, where N is the number turns in filaments; l is the equivalent coil active length; i_{as} is the current in the coil.

The mathematical model is found by using Kirchhoff's and Newton's second laws. In particular, the following circuitry and *torsional–mechanical* equations are used.

$$u_{as} = r_s i_{as} + \frac{d\psi_{as}}{dt},$$

$$T_e - B_m\omega_r - T_L = J\frac{d^2\theta_r}{dt^2},$$

where u_{as} is the applied voltage; r_s is coil resistance; ψ_{as} is the flux linkages as viewed from the winding; B_m is the friction coefficient; T_L is the load torque; J is the equivalent moment of inertia.

The flux linkage is expressed as

$$\psi_{as} = L_{as}i_{as} + A_eB(\theta_r),$$

where L_{as} is the self-inductance, and A_e is the equivalent cross-sectional area.

A set of nonlinear differential equations that models an axial topology microactuator is

$$\frac{di_{as}}{dt} = \frac{1}{L_{as}}\left(-r_s i_{as} - \underbrace{A_e \frac{dB(\theta_r)}{dt}}_{\text{back emf}} + u_{as}\right) = \frac{1}{L_{as}}\left(-r_s i_{as} - A_e B_{max} k \operatorname{sech}^2(k\theta_r)\omega_r + u_{as}\right),$$

$$\frac{d\omega_r}{dt} = \frac{1}{J}\left[2NlB_{max}\left|\tanh(k\theta_r)\right| i_{as} - B_m\omega_r - T_L\right],$$

$$\frac{d\theta_r}{dt} = \omega_r.$$

The bounded proportional-integral tracking controller with state feedbacks (Equation 17.18) is designed using the *state transformation* method. We use the expanded state vector as $z = \begin{bmatrix} x \\ u \end{bmatrix}$, $x(t) = \begin{bmatrix} x^{sys}(t) \\ e(t) \end{bmatrix}$, where $x^{sys}(t) = \begin{bmatrix} i_{as}(t) \\ \omega_r(t) \\ \theta_r(t) \end{bmatrix}$. Using different weighting matrices Q_z and G_z in Equation 17.14, distinct feedback gains are obtained. Letting $Q_z = I$ and $G_z = 1$, for the bounded controller (Equation 17.18) synthesized we have $k_p = 31$ and $k_i = 24$. The output dynamics for $\theta_{ref} = 0.349$ rad is shown in Figure 17.10 for $T_s = 0.0014$ s. For $k_p = 31$ and $k_i = 24$ with the state feedbacks, the settling time is 0.027 s, and the overshoot is 27%. It is possible to minimize the overshoot to 13%. The application of proportional-integral controllersmay not ensure the desired transient behavior as reported in Figure 17.10, when $k_p = 20$ and $k_i = 19$. We tested the servo without a return spring that is commonly used to ensure mechanical damping. Thus, the most challenging problem is considered. Pointer acceleration and deceleration are achieved by properly controlling the actuator that operates at very high efficiency and develops high electromagnetic torque.We conclude that optimal performance is achieved.

17.5 Synchronous Micromachines

In micromechatronics systems, rotational and translational motion devices (actuators and sensors), controlled by ICs, are widely used. Among various devices, synchronous permanent-magnet motion devices exhibit superior performance. The advantages of axial topology motion devices are feasibility, efficiency, and reliability. Fabrication simplicity results because: (1) magnets are flat; (2) there are no strict shape requirements imposed on magnets; (3) rotor back ferromagnetic material is not required; (4) it is easy to deposit planar windings on the flat stator. An axial topology permanent-magnet two-phase synchronous motion device is documented in Figure 17.11.

For two-phase actuator, one supplies two phase voltages u_{as} and u_{bs} in order to develop an electromagnetic torque. Assuming that the magnetic flux is constant through the magnetic plane (current loop), the torque on a planar current loop of any size and shape in the uniformmagnetic field is $\mathbf{T} = i\mathbf{s} \times \mathbf{B} = \mathbf{m} \times \mathbf{B}$, where i is the current in the loop (winding); $\mathbf{m}$ is the magnetic dipole moment. The electromagnetic force is given as $\mathbf{F} = \oint i d\mathbf{l} \times \mathbf{B}$. The interaction of the current (in windings) and magnets will produce the rotating electromagnetic torque. Microscale synchronous transducers can be used as motors (actuators) and generators (sensors). Microgenerators (velocity and position sensors) convert mechanical energy into electrical energy, while micromotors (actuators) convert electrical energy into mechanical energy. A broad spectrum of synchronous microtransducers can be used as actuators, for example, microscale drives, servos, and power systems.

Distinct electrostatic and electromagnetic motion devices can be designed and fabricated using the surface micromachining technology. However, all high-performance, high-power, and torque density devices are electromagnetic, and these devices use permanent magnets. Therefore, the issues of fabrication and analysis of magnets are of great interest. Different soft and hard magnets can be fabricated using surface micromachining. In particular, Ni, Fe, Co, NiFe, and other magnets can be made. In general, four types of

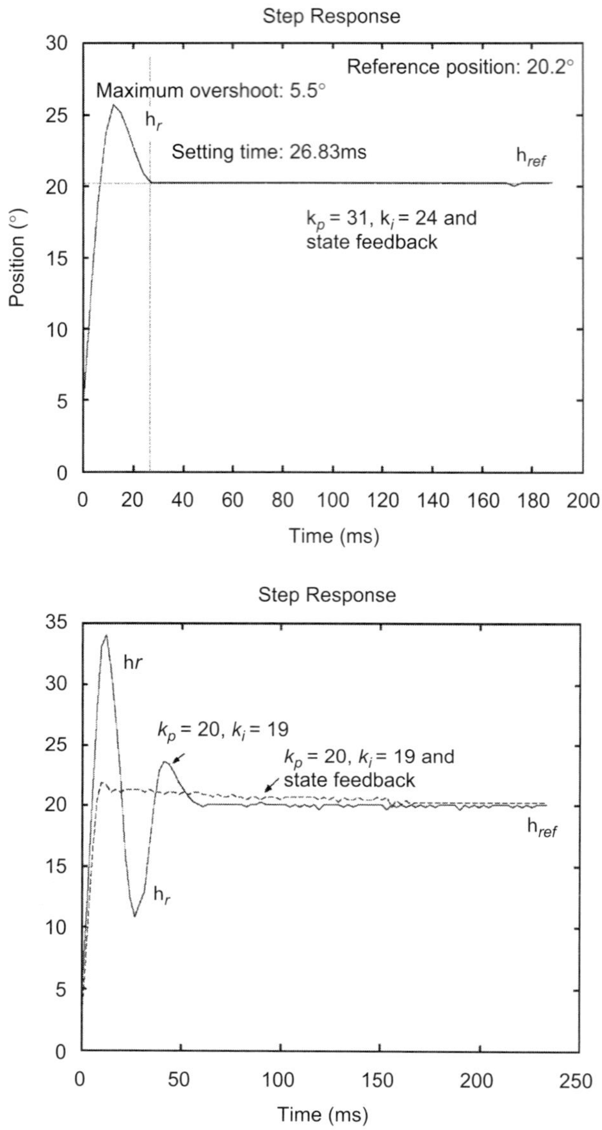

FIGURE 17.10 Transient dynamics of the closed-loop servo with the bounded proportional-integral controller with state feedback.

high-performance magnets have been commonly used in electromechanical motion devices, for example, neodymium iron boron ($Nd_2Fe_{14}B$), samarium cobalt (usually Sm_1Co_5 and Sm_2Co_{17}), ceramic (ferrite), and alnico (AlNiCo). The term soft is used to refer to magnets that have high saturation magnetization and a low coercivity (narrow *BH* curve). Another property of these magnets is their low magnetostriction. The soft magnets have been widely used in magnetic recording heads, and, in particular, NiFe thin films with the desired properties have been utilized. Hard magnets have wide *BH* curves (high coercivity) and, therefore, high-energy storage capacity. These magnets should be used in devices in order to attain high force, torque, and power densities. Unfortunately, limited progress has been made in the fabrication of hard permanent magnets.

The energy density is given as the area enclosed by the *BH* curve, for example, $w_m = \frac{1}{2}\mathbf{B} \cdot \mathbf{H}$. When magnets are used in motion devices, the demagnetization curve (second quadrant of the BH curve) is of

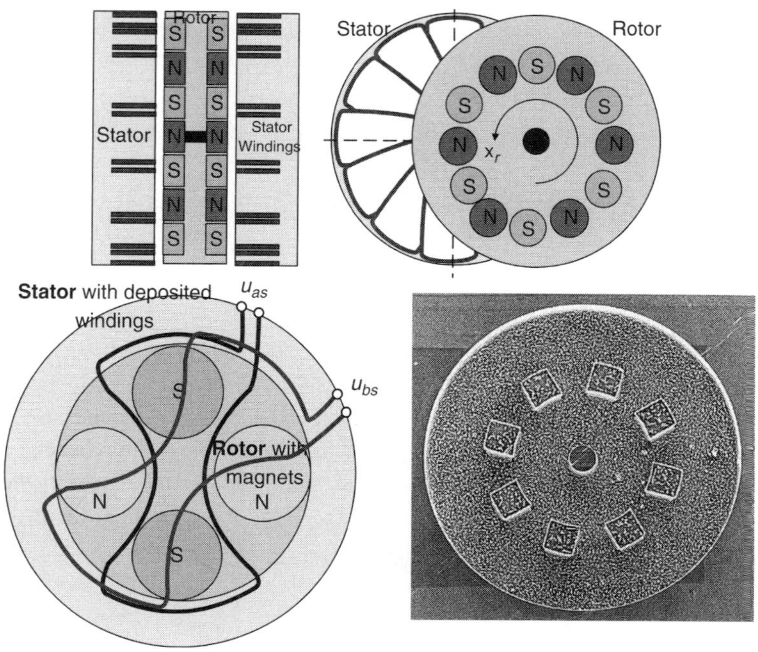

FIGURE 17.11 Axial permanent-magnet synchronous motion device.

interest. Permanent magnets store, exchange, and convert energy. In particular, permanent magnets produce stationary magnetic field without external energy sources or radiation devices.We apply Kirchhoff's and Newton's laws to derive the equations of motion for axial topology permanent-magnet synchronous motion devices. Using an axial permanent-magnet synchronous device documented in Figure 17.11, we assume that this variation of the flux density is sinusoidal, for example, $B(\theta_r) = B_{max}\sin^n(\frac{1}{2}N_m\theta_r)$, $n = 1, 3, 5, \cdots$, where B_{max} is the maximum flux density in the airgap produced by the magnet as viewed from the winding; n is the integer that is a function of the magnet geometry and the waveform of the airgap B. It should be emphasized that B_{max} depends on the magnets used, airgap length, temperature, and so forth.

The electromagnetic torque, developed by a single-phase axial topology permanent-magnet synchronous actuator, is found using the expression for the coenergy $W_c(i_{as}, \theta_r)$ that is given as $W_c(i_{as}, \theta_r) = NA_eB(\theta_r)i_{as}$. Assuming that the airgap flux density obeys $B(\theta_r) = B_{max}\sin(\frac{1}{2}N_m\theta_r)$, we have $W_c(i_{as}, \theta_r) = NA_eB_{max}\sin(\frac{1}{2}N_m\theta_r)i_{as}$, and the electromagnetic torque is

$$T_e = \frac{\partial W_c(i_{as}, \theta_r)}{\partial \theta_r} = \frac{\partial\left(NA_eB_{max}\sin(\frac{1}{2}N_m\theta_r)i_{as}\right)}{\partial\theta_r} = \frac{1}{2}N_mNA_eB_{max}\cos\left(\frac{1}{2}N_m\theta_r\right)i_{as}.$$

As we fed the following currentwith the magnitude i_M to the micromotorwinding $i_{as} = i_M\cos(\frac{1}{2}N_m\theta_r)$, the electromagnetic torque is $T_e = \frac{1}{2}N_mNA_eB_{max}i_M\cos^2(\frac{1}{2}N_m\theta_r) \neq 0$. The mathematical model of the single-phase permanent-magnet micromotor is found by using Kirchhoff's and Newton's second laws. In particular, one obtains

$$u_{as} = r_s i_{as} + \frac{d\psi_{as}}{dt} \text{ (circuitry equation)},$$

$$T_e - B_m\omega_r - T_L = J\frac{d^2\theta_r}{dt^2} \text{ (torsional–mechanical equation)}.$$

Taking note of the flux linkage equation $\psi_{as} = L_{as}i_{as} + NA_eB(\theta_r)$, we have a set of three first-order nonlinear differential equations that models a single-phase axial topology permanent-magnet synchronous motion device

$$\frac{di_{as}}{dt} = \frac{1}{L_{as}}\left(-r_s i_{as} - \frac{1}{2}N_m NA_e B_{max}\cos\left(\frac{1}{2}N_m\theta_r\right)\omega_r + u_{as}\right),$$

$$\frac{d\omega_r}{dt} = \frac{1}{J}\left(\frac{1}{2}N_m NA_e B_{max}\cos\left(\frac{1}{2}N_m\theta_r\right)i_{as} - B_m\omega_r - T_L\right),$$

$$\frac{d\theta_r}{dt} = \omega_r.$$

Single-phase axial and radial topologies permanent-magnet synchronous actuators have a torque ripple. The torque ripple can be minimized feeding $i_{as} = i_M\left[\cos(\frac{1}{2}N_m\theta_r) + \varepsilon\right]^{-1}$, $|i_{as}| \le i_M$, where $\varepsilon > 0$. To overcome this deficiency (torque ripple and velocity chattering) and increase torque density, two-phase motion devices are used. For two-phase actuators, the coenergy is given as $W_c(i_{as}, \theta_r) = NA_{ag}B_{max}\left(\sin(\frac{1}{2}N_m\theta_r)i_{as} - \cos(\frac{1}{2}N_m\theta_r)i_{bs}\right)$, and, hence, the electromagnetic torque is $T_e = \frac{\partial W_c(i_{as},\theta_r)}{\partial\theta_r} = \frac{1}{2}N_m NA_e B_{max}\left(\cos(\frac{1}{2}N_m\theta_r)i_{as} + \sin(\frac{1}{2}N_m\theta_r)i_{bs}\right)$. Thus, feeding the phase currents i_{as} and i_{bs} as $i_{as} = i_M\cos(\frac{1}{2}N_m\theta_r)$ and $i_{bs} = i_M\sin(\frac{1}{2}N_m\theta_r)$, we maximize the electromagnetic torque because $T_e = \frac{1}{2}N_m NA_e B_{max} i_M$.

17.6 Fabrication Aspects

Mini and microscale electromechanical motion devices can be fabricated by creating (etching or depositing) of conductors (coils and windings), ferromagnetic core, magnets, insulating layers, as well as other microstructures (radiating energy devices, movable and stationarymembers, and their components including bearings and posts). The subsequent order of the processes, sequential steps, and materials are different depending on the devised, designed, analyzed, and optimized devices. Several textbooks [1–6] provide the reader with the basic fabrication features and processes. In partcilular, the electrostatic devices are covered in References 1–6, while the electromagnetic motion devices are reported in References 1–3. Complementary metal oxide semiconductor (CMOS), high aspect ratio (LIGA and LIGA-like), and surface micromachining technologies are key features for fabrication of microdevices and structures. The LIGA (Lithography–Galvanoforming–Molding, or in German Lithografie–Galvanik–Abformung) technology allows one to make three-dimensional microstructures with the aspect ratio (depth versus lateral dimension is more than 100). The LIGA technology is based on the x-ray lithography that ensures short wavelength (from few to 10 Å), which leads to negligible diffraction effects and larger depth of focus compared with photolithography. The major processes in the machine's microfabrication are diffusion, deposition, patterning, lithography, etching, metallization, planarization, assembling, and packaging. Thin film fabrication processes were developed and used for polysilicon, silicon dioxide, silicon nitride, and other different materials (e.g., metals, alloys, composites, etc.). The basic steps are lithography, deposition of thin films and materials (electroplating, chemical vapor deposition, plasma-enhanced chemical vapor deposition, evaporation, sputtering, spraying, screen printing, etc.), removal of material (patterning) by wet or dry techniques, etching (plasma etching, reactive ion etching, laser etching, etc.), doping, bonding (fusion, anodic, and other), and planarization.

To fabricate motion and radiating energy microscale structures and devices, different fabrication technologies are used [1–6]. New processes were developed and novel materials were applied modifying the CMOS, surface micromachining, and LIGA technologies. Currently, the state-of-the-art technology in microfabrication has been progressed to the nanometer minimal features. Motion devices and structures (stator, rotor, bearing, coils, etc.) are defined photographically, and the high-resolution photolithography is applied to define two- (planar) and three-dimensional shapes (geometry). Deep ultraviolet lithography processes were developed to decrease the feature sizes of microstructures to 0.1 fÊm and less. Different

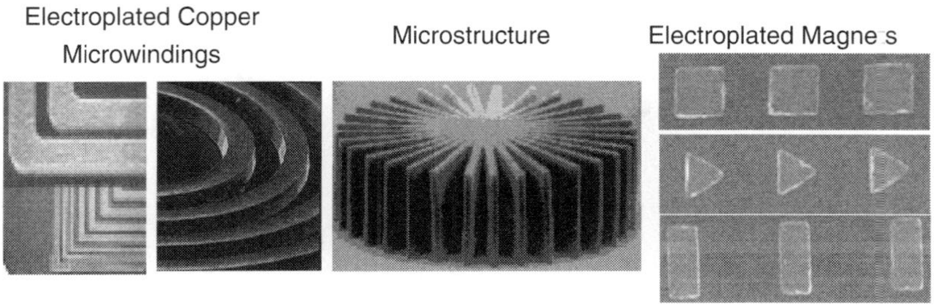

FIGURE 17.12 Deposited copper microwindings, microstructure, and permanent magnets.

exposure wavelengths λ (435, 365, 248, 193 nm) are used. Using the Rayleigh model for image resolution, the expressions for image resolution i_R and the depth of focus d_F are $i_R = k_i \frac{\lambda}{N_A}$ and $d_F = k_d \frac{\lambda}{N_A^2}$, where k_i and k_d are the lithographic process constants; λ is the exposure wavelength; N_A is the numerical aperture coefficient, and for high-numerical aperture we have $N_A = 0.5–0.6$.

The g- and i-line IBM lithography processes (with wavelengths of 435 and 365 nm, respectively) allow one to attain minimal features in the range of 0.35 μm. The deep ultraviolet light sources (mercury source or excimer lasers) with 248-nm wavelength enable the industry to achieve 0.25 μm resolution, leading to the minimal features in the range of 0.13 μm. The changes to short exposure wavelength possess challenges and present new highly desired possibilities. Using advanced lithography, 90-nm features were achieved. Different lithography processes commonly applied are photolithography, screen printing, electron-beam lithography, x-ray lithography (high-aspect ratio technology), and so forth. Althoughmachine topologies and configurations vary, magnetic and insulating materials, magnets, and windings are used in all motion devices. Figure 17.12 illustrates the electroplated 10 μm wide and thick with 10 μm spaced insulated copper microwindings (deposited on ferromagnetic cores), microstructures, and permanent magnets (electroplated NiFe alloy). Finally, it must be emphasized that usually the size of microelectromechanical motion devices is defined by the torque and power densities specified. Those requirements, as well as physical limits, define the device dimensionality.

Acknowledgments

The author sincerely acknowledges the support from the NSF under the NSF awards DUE 0311588 and EEC 0407281. *Disclaimer*—Any opinion, findings, and conclusions or recommendations expressed in this paper are those of the author and do not necessarily reflect the views of the National Science Foundation.

References

1. V. Giurgiutiu and S. E. Lyshevski, *Micromechatronics: Modeling, Analysis and Design with MATLAB*, CRC Press, Boca Raton, FL, 2003.
2. S. E. Lyshevski, *MEMS and NEMS: Systems, Devices, and Structures*, CRC Press, Boca Raton, FL, 2001.
3. S. E. Lyshevski, *Nano- and Micro-Electromechanical Systems: Fundamental of Micro- and Nano-Engineering*, CRC Press, Boca Raton, FL, 2005.
4. S. A. Campbell, *The Science and Engineering of Microelectronic Fabrication*, Oxford University Press, New York, 2001.
5. G. T. A. Kovacs, *Micromachined Transducers Sourcebook*,WCB McGraw-Hill, Boston, MA, 1998.
6. M. J. Madou, *Fundamentals of Microfabrication*, CRC Press, Boca Raton, FL, 2001.

II

Computers and Logic Systems

18

Introduction to Computers and Logic Systems

Kevin C. Craig
Fred Stolfi
Rensselaer Polytechnic Institute

18.1 Introduction: The Mechatronic Use of Computers

Mechatronics is the synergistic combination of mechanical engineering, electronics, control systems, and computers. The key element in mechatronics is the integration of these areas through the design process. Synergism and integration in design set a mechatronic system apart from a traditional, multidisciplinary system. In a mechatronic system, computer, electronic, and control technology allow changes in design philosophy, which lead to better performance at lower cost: accuracy and speed from controls, efficiency and reliability from electronics, and functionality and flexibility from computers. Automotive engine-control systems are a good example. Here a multitude of sensors measure various temperatures, pressures, flow rates, rotary speeds, and chemical composition and send this information to a microcomputer. The computer integrates all this data with preprogrammed engine models and control laws and sends commands to various valves, actuators, fuel injectors, and ignition systems so as to manage the engine's operation for an optimum combination of acceleration, fuel economy, and pollution emissions.

In mechatronics, balance is paramount. The essential characteristic of a mechatronics engineer and the key to success in mechatronics design is a balance between two sets of skills:

- Modeling (physical and mathematical), analysis (closed-form and numerical simulation), and control design (analog and digital) of dynamic physical systems
- Experimental validation of models and analysis and understanding the key issues in hardware implementation of designs

In mechatronic systems, computers play a variety of roles. First, computers are used to model, analyze, and simulate mechatronic systems and mechatronic system components and, as such, are useful for control design. Second, computers, as part of measurement systems, are used to measure the performance

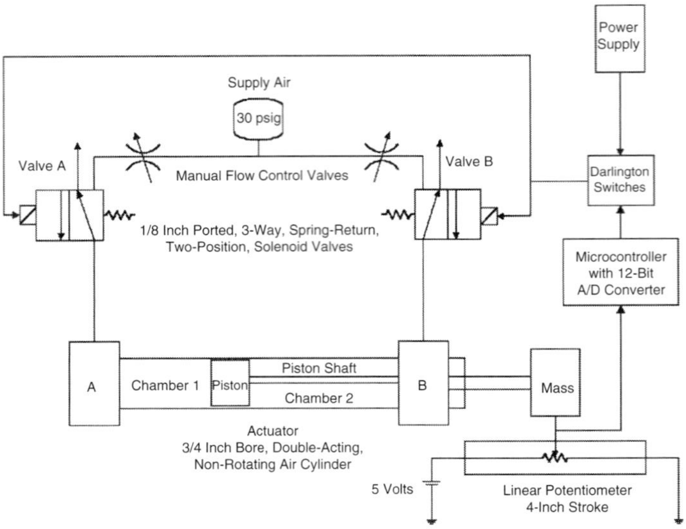

FIGURE 18.1 Pneumatic servomechanism.

of mechatronic systems, to determine the value of component parameters, and to experimentally validate models. Finally, computers or microcomputers form the central component in digital control systems for mechatronic designs. Thus, computers play an essential role in the two essential characteristics of the mechatronics balance and comprise a key component to mechatronic system designs. This is illustrated by the following example.

Consider the schematic of a pneumatic servomechanism, a computer-controlled, closed-loop positioning system, shown in Figure 18.1. Pneumatic servomechanisms have the advantages of low cost, high power-to-weight ratio, ease of maintenance, cleanliness, and a readily-available and cheap power source. However, the disadvantages are high, nonlinear friction forces, deadband due to stiction, and dead time due to the compressibility of air. The design goal is to implement a fast, accurate, and inexpensive pneumatic-actuator system using inexpensive on/off solenoid valves, rather than expensive continuously-variable servo valves. To accomplish this task, one must completely understand the physical system, develop a physical model on which to base analysis and design, and experimentally determine and/or validate model parameters. One must then develop a mathematical model of the system, analyze the system, and compare the results of the analysis to experimental measurements to validate the model. One must then design a closed-loop position control system utilizing on/off, modified on/off, or pulse-width modulated control. Finally, one must implement the control system and experimentally validate its predicted performance.

A MatLab/Simulink model of this system is shown in Figure 18.2. The mathematical model is highly nonlinear, as are the various control schemes. A computer numerical simulation is needed to understand the behavior of the system and the various control schemes. A data acquisition system is needed to take measurements of the various system inputs and outputs and validate the numerical simulation. And, a computer (a microcontroller in this case) is needed for the real-time implementation of the various control schemes. There are a variety of computer numerical simulation tools available, some requiring the detailed mathematical model while others enable virtual prototyping where the various system components are assembled on the computer screen with the component mathematical models given hidden in the background. There are also a variety of computer platforms on which to run the control algorithm, e.g., high-end PC using a DSP board and a real-time control-code generator; a microcontroller programmable in C or Basic with an analog-to-digital (A/D) converter and numerous digital input/output (I/O) ports; and a microchip implementation needed for product development.

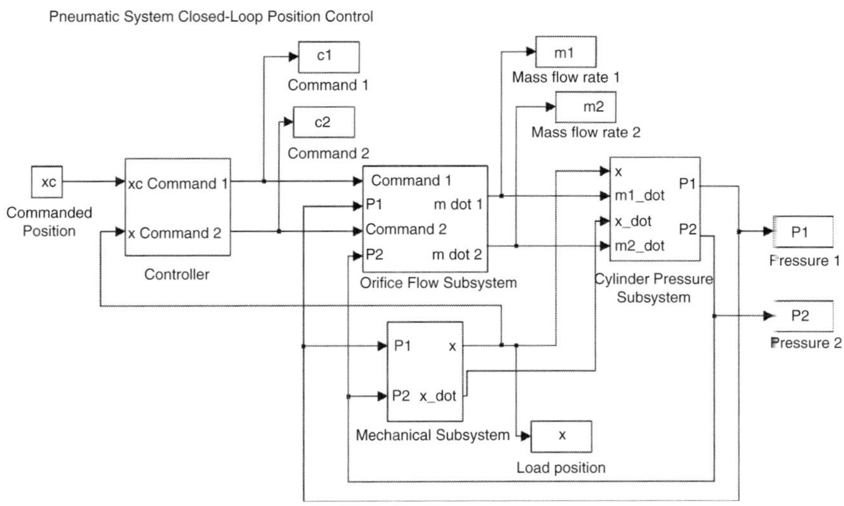

FIGURE 18.2 MatLab/Simulink model of the pneumatic servomechanism.

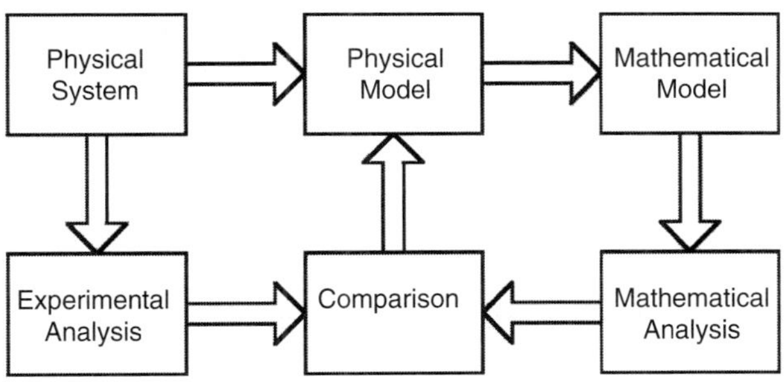

FIGURE 18.3 Dynamic system investigation process.

18.2 Mechatronics and Computer Modeling and Simulation

In design, balance is the key to success, i.e., balance between theory and practice and balance between modeling/analysis skills and hardware-implementation/measurement skills. Figure 18.3 illustrates the steps in a dynamic system investigation, which is the process that would be utilized to design a mechatronic system. The distinction between physical modeling and mathematical modeling is emphasized, as is the importance of both analytical and numerical solutions to the model equations. To generate a physical model, approximations must be made to the actual physical system. Small effects are neglected. The influence of the environment is ignored. Elements are assumed to be lumped instead of distributed. The dynamics are assumed to be linear. Parameters are assumed to be constant. Noise and uncertainty is ignored. These approximations have a direct influence on the mathematical model. Neglecting small effects limits the number of equations. Environmental independence reduces the complexity of the equations. Other approximations result in linear ordinary differential equations with constant coefficients. Neglecting uncertainty avoids the use of statistics in the model. In most cases, a design consideration is to develop the simplest model which adequately depicts the complexity of the system dynamics.

The predicted dynamic behavior of the model is only half the story, for these results, without experimental verification, are at best questionable, and at worst useless. Comparing the predicted dynamic behavior with the actual measured dynamic behavior is the key step in the dynamic system investigation process.

The steps in the dynamic system investigation process should be applied not only when an actual physical system exists (as in reverse engineering) and one desires to understand and predict its behavior, but also when the physical system is a concept in the design process that needs to be analyzed and evaluated. After recognizing a need for a new product or service, one uses past experience (personal and vicarious), awareness of existing hardware, understanding of physical laws, and creativity to generate design concepts. *The importance of modeling and analysis in the design process has never been more important than in this situation.* These design concepts can no longer be evaluated by the build-and-test approach because it is too costly and time consuming. Validating the predicted dynamic behavior in this case, when no actual physical system exists, becomes even more dependent on one's past hardware and experimental experience.

In physical modeling, one first specifies the physical system to be studied along with the system boundaries, input variables, and output variables. In modeling dynamic systems, we use engineering judgment and simplifying assumptions to develop a physical model. The complexity of the physical model depends on the particular need, e.g., system design iteration, control system design, control design verification, physical understanding. The intelligent use of simple physical models requires that we have some understanding of what we are missing when we choose the simpler model over the more complex model. The astuteness with which these approximations are made at the onset of an investigation is the very crux of engineering analysis. A variety of engineering models may be developed based on the particular need. Always ask the question: "Why am I modeling the physical system and what is the range of operation that I wish my model to be valid for?" If the need is system-design iteration or control-system design, then a "*design model*" is needed, i.e., a physical model whose mathematical model is a linear ordinary differential equation with constant coefficients and, therefore, useful with a broad, highly-developed assortment of linear design techniques. If the need is design verification before actual hardware implementation, then a "*truth model*" is needed, i.e., a physical model that is as close to reality as possible; with nonlinear simulation tools available, almost any mathematical model can now be simulated. Iterations can then be performed using, as a starting point, the results of the work performed with the design model. Models only need to be valid for the particular range of operation of interest; low-order models then can often represent very complex, higher-order models very effectively. In practice, you may need a hierarchy of models of varying complexity: a very detailed truth model for final performance evaluation before hardware implementation, several less complex truth models for use in evaluating particular effects, and one or more design models.

18.3 Mechatronics, Computers, and Measurement Systems

Measurement systems or data acquisition systems may be used for a variety of purposes, and a computer plays an integral role in each.

1. *Monitoring of Processes and Operations.* Certain applications of measuring instruments may be characterized as having essentially a monitoring function, e.g., thermometers, barometers, and water, gas, and electric meters.
2. *Control of Processes and Operations.* An instrument can serve as a component of a control system. To control any variable in a feedback control system, it is first necessary to measure it. A single control system may require information from many measuring instruments, e.g., industrial machine and process controllers, aircraft control systems.
3. *Experimental Engineering Analysis.* In solving engineering problems, two general methods are available: theoretical and experimental. Many problems require the application of both methods and theory and experiment should be thought of as complementing each other. Further, all models need validation, and measurement systems offer a means to collect the data required for model validation.

The distinction among monitoring, control, and analysis functions is not clear-cut; the category that a given application may fit may depend somewhat on the engineer's point of view and the apparent looseness of the classifications should not cause any difficulty. Rather it should be realized that computers,

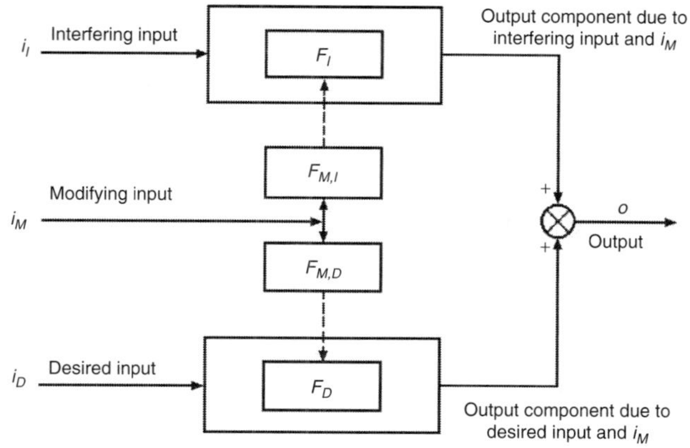

FIGURE 18.4 Input–output configuration of a measurement system.

as general purpose processing elements, can serve many functions in the processing of measured parameters from mechatronic systems and that these processing functions can be related to or unrelated to the modeling and control of such systems. Special purpose digital signal processing electronics are also used in measurement systems. High-speed digital signal processors (DSPs), for example, are used to collect input and output signals in the determination of transfer functions for mechatronic systems. The high speed allows the processing of simultaneous samples of the input and output for minimal phase error. The primary application for DSPs in mechatronic systems, however, is real-time control, discussed below.

Figure 18.4 is the input–output configuration of a measurement system. Input quantities are classified into three categories:

1. *Desired Inputs.* These are quantities that the instrument is specifically intended to measure.
2. *Interfering Inputs.* These are quantities to which the instrument is unintentionally sensitive. F_D and F_I are input–output relations, i.e., the mathematical operations necessary to obtain the output from the input. They represent different concepts depending on the particular input–output characteristic being described, e.g., a constant, a mathematical function, a differential equation, a statistical distribution function.
3. *Modifying Inputs.* These are quantities that cause a change in the input–output relations for the desired and interfering inputs, i.e., they cause a change in F_D and/or F_I. $F_{M,I}$ and $F_{M,D}$ represent the specific manner in which i_M affects F_I and F_D, respectively.

There are several methods for canceling or reducing the effects of spurious inputs. One method which relies upon computer processing of the signals is the method of calculated output corrections. This method requires one to measure or estimate the magnitudes of the interfering and/or modifying inputs and to know quantitatively how they affect the output. Then it is possible to calculate corrections, which may be added to or subtracted from the indicated output so as to leave (ideally) only that component associated with the desired input. Since many measurement systems today can afford to include a computer to carry out various functions, if sensors for the spurious inputs are provided, the computer can implement the method of calculated output corrections on an automatic basis.

18.4 Mechatronics and the Real-Time Use of Computers

We turn to the field of closed-loop control using a digital computer as the controller. Several comments are in order. First, a mechatronic system typically involves continuous variables. Elements rotate or translate in space. Fluids or gasses flow. Heat or energy is transferred. Computers are, by their nature, digital elements. Variables are represented in a computer by discrete values or simply by collections of zeroes and ones. For a computer to be used as the controller for a mechatronic system, therefore, the continuous variables must be converted to discrete variables for processing and then back again to continuous variables. This might seem obvious. What is not so apparent is that the computer algorithm forms an inherent separation between the processing of the signals and the signals themselves, which is not true of other mechatronic system components. Even if digital logic elements are used (as discussed in this chapter) the signals are converted to discrete form, but the flow of information is still continuous through the elements. When a computer is used for the control element, this information flow is broken and buried in the computer algorithm. As an example, computer algorithms sometimes mimic continuous proportional-integral-derivative (PID) control laws. When the execution of this algorithm is analyzed, even if the effects of sampling and quantization are included, it is assumed that the signals are processed just as if they were being determined by continuous processing elements. In reality, if the computer code is examined at the machine level (i.e., not in the high level language in which it may be written), it would bear very little resemblance to a differential equation representation of the PID algorithm. This has practical implications both for modeling the exact operation of the computer as a control element and for validating that the computer code actually produces the desired response to signals.

Other issues are involved when the mechatronic system controller is implemented in software. Software execution is often asynchronous to the other time constants in the system (i.e., the software execution and system response are often not synchronized). Software can be made synchronous by syncing it to the sampler period, but this typically limits performance and is difficult if the computer is to be used for other tasks than control. Once a computer is contained as an element in a mechatronic system, there is a tendency to use some of the processing power to provide additional functionality or ease of use for the product. This additional code can affect, sometimes adversely, the operation of the real time controller execution. Testing of the code and safety of the code are also issues. The engineer has to determine that his system operates deterministically and safely for all possible combinations of input signals and for all possible states in the execution of the algorithm. For real-time systems, execution order for the code is often not predictable since it can be dependent on the particular combination of input signals. Simplicity of the code, providing for testability of the code, using established software quality assurance practices, and developing extensive documentation are ways to achieve system determinism and safety. Often, a hardware interlock, that is, a safety system utilizing electronic or mechanical hardware, is often included in software controlled systems.

Code operation has to be further verified as the code is modified and as the code is reused for systems other than that for which it was developed. Unlike other controllers, computer code is portable, but this requires more thought for its possible reuse. Using standard software packages, standard processors, modular code, and commercial real-time environments increases the possibility for reuse.

Besides the issues inherent in using computer code as the controller, there are issues involved whenever a digital processing component is incorporated into a mechatronic system. Further, there are considerations that must be taken into account whenever digital signals are processed. Figure 18.5 shows a configuration useful for this discussion. The computer is important, but the computer "component" of many mechatronic machines and processes is often not the critical system element in terms of either technical or economic factors. Rather, components external to the computer, the actuators and sensors, the sampling system, and the anti-aliasing filter are more often the limiting factors in the system design.

Since both continuous (analog) and digital signals exist in computer-controlled systems, the signals in such a system can be classified as shown in the table below.

Signal Classification	Discrete in Time	Continuous in Time
Discrete in amplitude	D-D (digital)	D-C
Continuous in amplitude	C-D	C-C (analog)

For analog signals, the precise value of the quantity (voltage, rotation angle, etc.) carrying the information is significant, meaning that the specific waveform of input and output signals is of vital importance. Conversely, digital signals are binary (on/off) in nature, and variations in numerical value are associated

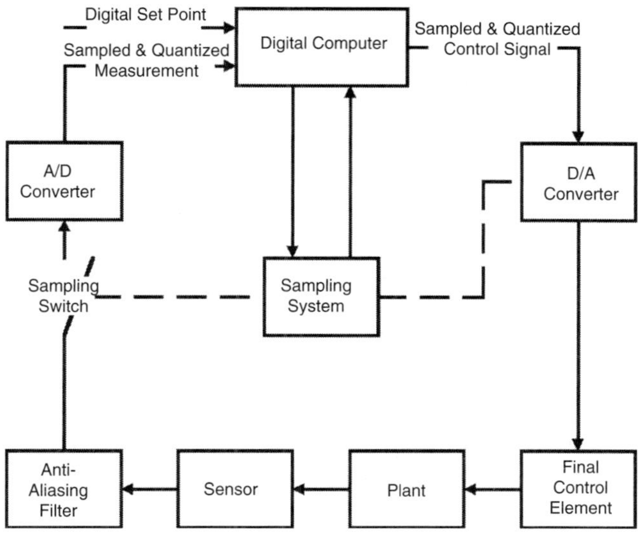

FIGURE 18.5 General computer-control configuration.

with changes in the logical state (true/false) of some combination of switches, e.g., +2 V to +5 V represents ON state, 0 V to +0.8 V represents OFF state.

In digital devices, it is simply the presence (logical 1) or absence (logical 0) of a voltage within some wide range that matters; the precise value of the signal is of no consequence. Digital devices are therefore very tolerant of noise voltages and need not be individually very accurate, even though the overall system can be extremely accurate. When combined analog/digital systems are used, the digital portions need not limit system accuracy; these limitations generally are associated with analog portions and/or the analog-to-digital (A/D) conversion devices. Since most mechatronic systems are analog in nature, it is necessary to have both A/D converters and digital-to-analog (D/A) converters, which serve as translators that enable the computer to communicate with the outside analog world.

In most cases, the sensor and the final control element are analog devices, requiring, respectively, A/D and D/A conversion at the computer input and output. There are, of course, exceptions, e.g., stepper motor and optical encoder. In most cases, however, the sensors can be thought of as providing analog voltage output and the final control element will accept an analog voltage input.

The current trend toward using dedicated, computer-based, and often decentralized (distributed) digital control systems in mechatronic applications can be rationalized in terms of the major advantages of digital control:

- Digital control is less susceptible to noise or parameter variation in instrumentation because data can be represented, generated, transmitted, and processed as binary words, with bits possessing two identifiable states.
- Very high accuracy and speed are possible through digital processing. However, hardware implementation is usually faster than software implementation. Determining the time required to develop a system in software is notoriously difficult to estimate.
- Digital control can handle repetitive tasks extremely well, through programming.
- Complex control laws and signal conditioning methods that might be impractical to implement using analog devices can be programmed. Very sophisticated algorithms can be implemented digitally.
- High product reliability can be achieved by minimizing analog hardware components and through decentralization using dedicated computers for various control tasks.

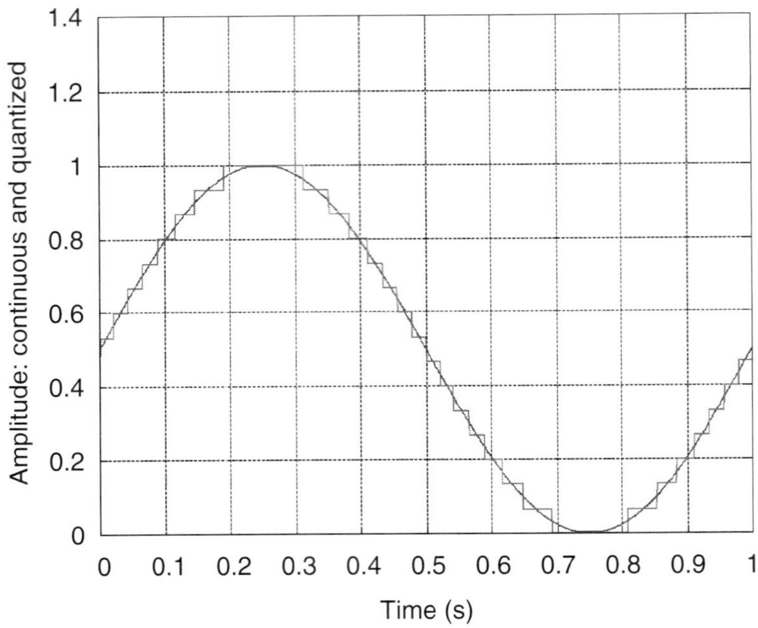

FIGURE 18.6 Simulation of a continuous and 4-bit quantized signal.

- Digital systems are more easily "programmed" and offer the ability to time-share a single processing unit among a number of different functions.
- Large amounts of data can be stored using compact high-density data storage methods.
- Data can be stored or maintained for very long periods of time without drift and without being affected by adverse environmental conditions. Digital control has easy and fast data retrieval capabilities.
- Fast data transmission is possible over long distances without introducing dynamic delays, as in analog systems.
- Digital processing uses low operational voltages (e.g., 0–12 V DC).
- Digital control has low overall component cost.

Further, from the standpoint of the mechatronic product, the inclusion of a computer means that additional system functions can be provided. The user can select from a range of operations. Additional features can be included. A user interface providing indications of operation can be added with minimal cost.

In a real sense, some of the problems of analysis and design of digital control systems (beyond the issues associated with software) are concerned with taking into account the effects of the sampling period, T, and the quantization size, q. If both T and q are extremely small (i.e., sampling frequency 50 or more times the system bandwidth with a 32-bit word size), digital signals are nearly continuous, and continuous methods of analysis and design can be used. It is most important to understand the *effects of all sample rates*, fast and slow, and the *effects of quantization* for large and small word sizes. Lower cost computers are typically slower and have a smaller word size. Figure 18.6 shows the effects of having too few quantization levels, i.e., too small a word size. The signal that will be processed by the controller has large errors over the original analog signal.

Figure 18.7 shows the effects of sampling. It is worthy to note that the *single most important impact* of implementing a control system digitally is often the delay associated with the D/A converter, i.e., $T/2$. This pure delay results in a substantial phase shift in the closed-loop feedback system and often limits the control operation.

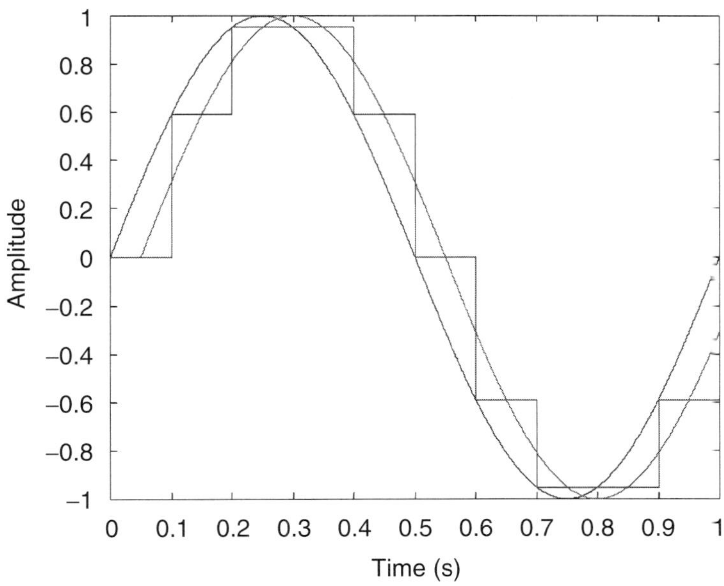

FIGURE 18.7 Continuous and D/A converter output.

In a feedback system, the analog signal coming from the sensor contains useful information related to controllable disturbances (relatively low frequency), but also may often include higher frequency "noise" due to uncontrollable disturbances (too fast for control system correction), measurement noise, and stray electrical pickup. Such noise signals cause difficulties in analog systems and low-pass filtering is often needed to allow good control performance. The phase shift from this filter also adversely affects control system stability.

Finally, in digital systems, a phenomenon called *aliasing* introduces some new aspects to the area of noise problems. If a signal containing high frequencies is sampled too infrequently, the output signal of the sampler contains low-frequency ("aliased") components not present in the signal before sampling. This is illustrated in Figure 18.8. If the higher frequency signal is sampled too infrequently, the result will be exactly the same values as the low frequency signal. From the standpoint of the controller, there is no way for the system to distinguish which signal is present. If we base our control actions on these false low-frequency components, they will, of course, result in poor control. The theoretical absolute minimum sampling rate to prevent aliasing is two samples per cycle; however, in practice, rates of about 10 are more commonly used. A high-frequency signal, inadequately sampled, can produce a reconstructed function of a much lower frequency, which cannot be distinguished from that produced by adequate sampling of a low-frequency function.

In all of the above, the word computer was used for the digital processing element. In electronics literature, a distinction is usually drawn between a microprocessor, microcomputer, DSP, and computer. There is no standard for what each of these terms can mean, but some insight can be gained by examining Figure 18.9, which is a general block diagram for a computer. All computers have a means of getting input, a means of generating output, a means of controlling the flow of signals and operations, memory for data storage, and an arithmetic logic unit (ALU) which executes the instructions. The ALU and control elements are often called the central processing unit (CPU). Small computers, which just contain a CPU, are often called microprocessors. Memory for these computers is often attached to the microprocessor but in distinct electronic packages. Input and output to the microprocessor is often handled by electronics called peripherals. If the memory is included in the same package, the computer is called either a micro-computer or computer depending on its physical size. CPU and memory on a single electronics chip is often

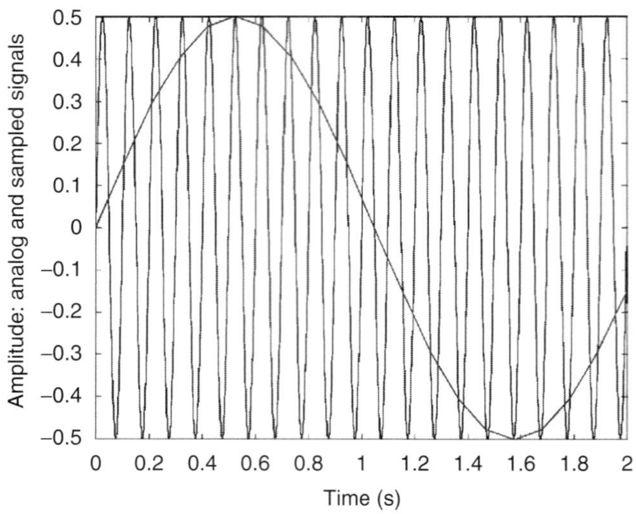

FIGURE 18.8 Simulation of continuous and sampled signal: aliasing.

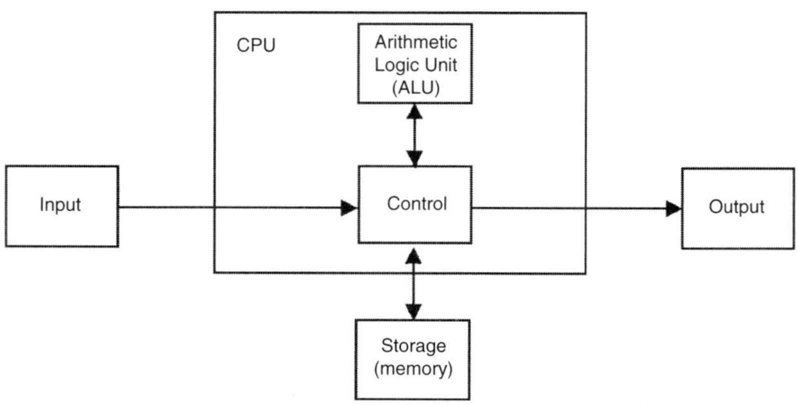

FIGURE 18.9 Elements of a computer.

called a microcomputer. The reader should be aware that a single electronics package can contain many "chips," which are connected by fine wires within the package. The overall package is still called a chip. Finally, if the A/D and D/A functions are provided in the same package, the computer is often called a DSP. However, these functions can also be contained in something which is called a microcomputer. DSPs are also computers which have a special instruction in the ALU called a multiply-accumulate (MAC) instruction even if the A/D and D/A are not present. Digital signal processing algorithms often involve MAC instructions and a computer, which can execute this instruction very effectively (in one instruction cycle of the computer), and are often called DSPs. To further complicate the situation, electronic devices called application specific integrated circuits (ASICs) exist. These devices can be custom made to perform a specific operation (such as a PID algorithm). ASICs can contain a CPU or memory or peripheral functions or even a MAC cell as part of its makeup. If the reader is thoroughly confused by this explanation, he probably has the proper grasp of the situation. However, he should be aware that diagrams like the one shown in Figure 18.9 often accompany the electronic component so the internal capabilities can be determined.

Before leaving computers, one final point will be made. Memory in a computer can often be divided between program space and data space, as shown in Figure 18.10. This representation is meant to be

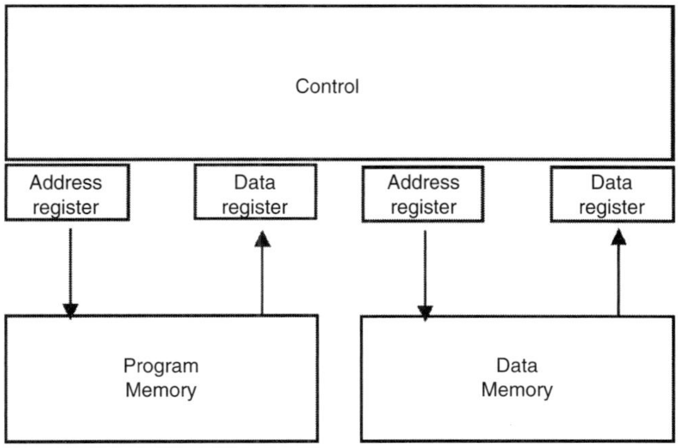

FIGURE 18.10 Computer memory organization.

pictorial rather than to define a specific computer architecture. In a von Neumann architecture, for example, the program memory and data memory share the same space and information busses. Whereas in a Harvard architecture, program memory and data memory are distinct (looking more like the figure). In either case, for a mechatronic system, one can think of the program (in program memory) as the set of instructions which tells the CPU how to manipulate data (in data memory) to produce an output. This view should emphasize the earlier point that the flow of signals in a mechatronic system becomes confused if a computer is to be used for real-time control.

Because of the low cost of modern microcomputers, the use of logic elements as discrete components in a mechatronic system has diminished. Microcomputers are often programmed to perform logic functions, which has the advantage that the operation can be altered in software rather than requiring electronic hardware changes. In analyzing this logic, of course, any of the traditional methods can be employed. The logic can be minimized via Karnaugh maps, for example. The only difference lies in the implementation of the algorithm. ASICs are also used to implement logic functions.

18.5 The Synergy of Mechatronics

As stated at the beginning of this section, mechatronics is the synergistic combination of mechanical engineering, electronics, control systems, and computers and the key element in mechatronics is the integration of these areas through the design process. The use of computers and logic elements as components in mechatronic systems will produce successful designs only if this synergy is achieved. The system must be designed as a system. Computers should never be an add-on component included when the design is complete. When computers are synergistically incorporated in the system, the power of the mechatronics approach to design is realized.

19

Digital Logic Concepts and Combinational Logic Design

George I. Cohn
California State University

19.1 Introduction

Digital logic deals with the representation, transmission, manipulation, and storage of digital information. A digital quantity has only certain discrete values in contrast with an analog quantity, which can have any value in an allowed continuum. The enormous advantage digital has over analog is its immunity to degradation by noise, if that noise does not exceed a tolerance threshold.

19.2 Digital Information Representation

Information can be characterized as qualitative or quantitative. Quantitative information requires a number system for its representation. Qualitative does not. In either case, however, digitalized information is represented by a finite set of different characters. Each character is a discrete quanta of information. The set of characters used constitutes the alphabet.

TABLE 19.1 Notation for Numbers

	Juxtaposition	Polynomial
Integer	$N = N_{n-1}N_{n-2} \ldots N_1N_0$	$N = \sum_{k=0}^{n-1} N_k R^k$
Fraction	$F = F_{-1}F_{-2} \ldots F_{-m+1}F_{-m}$	$F = \sum_{k=-m}^{-1} F_k R^k$
Real	$X = X_{n-1}X_{n-2} \ldots X_1X_0 \cdot X_{-1}X_{-1} \ldots X_{-m+1}X_{-m}$	$X = \sum_{k=-m}^{n-1} X_k R^k$

19.3 Number Systems

Quantitative information is represented by a number system. A character that represents quantitative information is called a **digit**. The number of different values which a digit may have is called the **radix**, designated by *R*. The symbols that designate the different values a digit can have are called numeric characters. The most conventionally used numeric characters are 0, 1, 2,..., etc., with 0 representing the smallest value. The largest value that a digit may have in a number system is the **reduced radix**, $r = R - 1$. Different radix values characterize different number systems: with *R* different numeric character values the number system is *R*nary, with 2 it is binary, with 3 it is ternary, with 8 it is octal, with 10 it is decimal, and with 16 it is hexadecimal.

Any value that can be expressed solely in terms of digits is an **integer**. A negative integer is any integer obtained by subtracting a positive integer from a smaller integer. Any number obtained by dividing a number by a larger number is a **fraction**. A number that has both an integer part and a fraction part is a **real number**.

All of the digits in a *number system* have the same radix. The radix is the **base** of the number system. Presumably, the possession of 10 fingers has made the decimal number system the most convenient for humans to use. The characters representing the 10 values a decimal digit can have are 0, 1, 2, 3, 4, 5, 6, 7, 8, and 9. The binary number system is the most natural for digital electronic systems because a desired reliability for a system can be most economically achieved using elements with two stable states. The characters normally used to represent the two values a binary digit may have are 0 and 1. The hexadecimal number system (0, 1, 2, 3, 4, 5, 6, 7, 8, 9, *A*, *B*, *C*, *D*, *E*, and *F*; $R = 16$) is of importance because it shortens by a factor of four the string of digits representing the binary information stored and manipulated in digital computers.

19.4 Number Representation

Numbers that require more than one digit can be represented in different formats, as shown in Table 19.1. Different formats facilitate execution of different procedures. Arithmetic is most conveniently done with the juxtaposition format. Theoretical developments are facilitated by the polynomial format.

19.5 Arithmetic

The most common arithmetic processes, addition, subtraction, multiplication, and division are conveniently implemented using multidiget notation. Development of formulation procedures is facilitated using the polynomial notation. Since the numbers are digital representations, the logic used to maniputate the numbers is *digital logic.* However, this is different than the logic of boolean algebra, which is what is usually meant by the term digital logic. The logic of the former is implemented in hardware by using the logic of the latter. The four basic arithmetic operations can be represented as functional procedures in equation form or in arithmetic manipulation form, as shown in Table 19.2.

The arithmetic processes in the binary system are based on the binary addition and multiplication tables given in Table 19.3.

Table 19.4 gives binary examples for each of the basic arithmetic operations.

TABLE 19.2 Arithmetic Operations

	Algebraic Form	Arithmetic Form
Addition	Sum = Augend + Addend	Augend + Addend Sum
Subtraction	Difference = Minuend–Subtrahend	Minuend – Subtrahend Difference
Multiplication	Product = Multiplicand × Multiplier	Multiplicand × Multiplier Product
Division	Dividend/Divisor = Quotient + Remainder/Divisor	Quotient Remainder Divisor |Dividend

TABLE 19.3 Single Digit Binary Arithmetic Table

(a) Addition

	0	1
0	0	1
1	1	10

(b) Multiplication

	0	1
0	0	0
1	0	1

TABLE 19.4 Binary Arithmetic Operation Examples

Addition:

1 1 0 0	carries
1 1 1 0 0	augend
+ 1 1 0 1	addend
1 0 1 0 0 1	sum

Multiplication:

1 1 0 1	multiplicand
×1 1 0	multiplier
0 0 0 0	partial product 1
1 1 0 1	partial product 2
1 1 0 1	partial product 3
1 0 0 1 1 1 0	product

Subtraction, borrow method:

10	borrows
⁄101	minuend
– 10	subtrahend
11	difference

Division, with fraction remainder:

quotient 1 0 0 1 0; remainder $\frac{111}{1011}$

1 0 1 1 | 1 1 0 0 1 1 0 1 dividend
divisor 1 0 1 1
0 1 1 1 0
1 0 1 1
0 1 1 1

Subtraction, payback method:

10	borrows
101	minuend
–1	payback
– 10	subtrahend
11	difference

Division, with remainder in quotient:

quotient
1 0 1 1 | 1 1 1.0 0 0 0 0 0 0 dividend
divisor 1 0 1 1
1 1 0 0
1 0 1 1
0 1 0 0 0 0
1 0 1 1

19.6 Number Conversion from One Base to Another

The method of using series polynomial expansions for converting numbers from one base to another is illustrated in Table 19.5.

Evaluation of polynomials is more efficiently done with the *nested form*. The nested form is obtained from the series form by successive factoring of the variable from all terms in which it appears as shown in Table 19.6. The number of multiplications to evaluate the nested form increases linearly with the order of the polynomial, whereas the number of multiplications to evaluate the series form increases with the square of the order.

Conversion of integers between bases is more easily done using the lower order polynomials, Table 19.6b, obtained by nesting. The least significant digit of the number in the new base is the remainder obtained after dividing the number in the old base by the new radix. The next least significant digit is the remainder obtained by dividing the first reduced polynomial by the new radix. The process is repeated until the most significant digit in the new base is obtained as the remainder, when the new radix no longer fits into the last reduced polynomial. This is more compactly represented with the arithmetic notation shown in Table 19.7 along with the same examples used to illustrate the polynomial series method.

TABLE 19.5 Series Polynomial Method for Converting Numbers between Bases

Method	Sample Conversion From: A Lower to a Higher Base	Sample Conversion From: A Higher to a Lower Base
1. Express number in polynomial form in the given base	$101.1_2 = 1\times2^2+0\times2^1+1\times2^0+1\times2^{-1}$	$19.5_{10} = 3\times10^1+6\times10^0+5\times10^{-1}$
2. Convert radix and coefficients to the new base	$= 1\times4+0\times2+1\times1+1\times0.5$	$= 11\times1010^1+110\times1010^0+101\times1010^{-1}$
3. Evaluate terms in the new base	$= 4+0+1+0.5$	$= 11\times1010+110+101/1010$ $= 11110+110+.1$
4. Add the terms	$101.1_2 = 5.5_{10}$	$19.5_{10} = 100100.1_2$

TABLE 19.6 Nested Polynomials

(a) Nested Polynomial via Iterated Factoring	(b) Lower Order Polynomials
$N = N_{n-1}R^{n-1} + N_{n-2}R^{n-2} + \cdots + N_2R^2 + N_1R + N0$	$N = N^{(1)}R + N_0$
$N = (N_{n-1}R^{n-2} + N_{n-2}R^{n-3} + \cdots + N_2R + N_1)R + N0$	$N^{(1)} = N^{(2)}R + N_1$
$N = ((N_{n-1}R^{n-3} + N_{n-2}R^{n-4} + \cdots + N_2)R + N_1)R + N0$	$N^{(2)} = N^{(3)}R + N_2$
$\vdots$	$\vdots$
	$N^{(n-2)} = N^{(n-3)}R + N_{n-2}$
$N = (\cdots(N_{n-1}R + N_{n-2})R + \cdots + N_2)R + N_1)R + N0$	$N^{(n-1)} = N_{n-1}$

TABLE 19.7 Radix Divide Method for Converting Numbers between Bases

Method		Sample Conversion From: A Higher to a Lower Base		Sample Conversion From: A Lower to a Higher Base	
$R\,/\,N$		$2\,/\,36_{10}$			
$R\,/\,N^{(1)}$	N_0	$2\,/\,18$	0		
$R\,/\,N^{(2)}$	N_1	$2\,/\,9$	0	$12\,/\,2012_3$	
$R\,/\,N^{(3)}$	N_2	$2\,/\,4$	1	$12\,/\,102$	11
$\vdots$		$2\,/\,2$	0	$12\,/\,2$	1
		$2\,/\,1$	0	$12\,/\,0$	2
$R\,/\,N^{(n-1)}$	N_{n-2}	0	1		
$R\,/\,0$	N_{n-1}	$36_{10} = 100100_2$		$2012_3 = 214_5$	

Conversion of a fraction from one base to another can be done by successive multiplications of the fraction by the radix of the number system to which the fraction is to be converted. Each multiplication by the radix gives a product that has the digits shifted to the left by one position. This moves the most significant digit of the fraction to the left of the radix point, placing that digit in the integer portion of the product, thereby isolating it from the fraction. This process is illustrated in algebraic form in the left column of Table 19.8 and in arithmetic form in the next column. Two sample numeric conversions are shown in the next two columns of Table 19.8.

Table 19.8 deals only with terminating fractions, that is, the remaining fractional part vanishes after a finite number of steps. For a nonterminating case the procedure is continued until a sufficient number of digits have been obtained to give the desired accuracy. A nonterminating case is illustrated in Table 19.9. A set of digits, which repeat ad infinitum, are designated by an underscore, as shown in Table 19.9.

TABLE 19.8 Radix Multiply Number Convertion Method (Terminating Case)

Formalism		Sample Conversion between Bases	
Algebraic	Arithmetic	Higher to Lower	Lower to Higher
$F = F_{-1}F_{-2}F_{-3}\cdots F_{-m}$	$F_{-1}\,F_{-2}\,F_{-3}\cdots F_{-m}$ $\times R$	0.125_{10} $\times 2$	0.100101_2 $\times 1010$
$R^*F = F_{-1}\cdot F_{-2}F_{-3}\cdots F_{-m}$ $= F_{-1}\cdot F^{(1)}$	$F_{-1}\rfloor\cdot F_{-2}\,F_{-3}\cdots F_{-m}$ $\times R$	$0\rfloor.25$ $\times 2$	$101\rfloor.11001$ $\times 1010$
$R^*F^{(1)} = F_{-2}\cdot F_{-3}F_{-4}$ $\cdots F_{-m} = F_{-2}\cdot F^{(2)}$	$F_{-2}\rfloor\cdot F_{-3}\cdots F_{-m}$ $\times R$	$0\rfloor.5$ $\times 2$ $1\rfloor.0$	$111\rfloor.1101$ $\times 1010$ $1000\rfloor.001$ $\times 1010$
$R^*F^{(2)} = F_{-3}\cdot F_{-4}F_{-5}$ $\cdots F_{-m} = F_{-2}\cdot F^{(3)}$ $\vdots$	$\vdots$		$1\rfloor.10$ $\times 1010$ $10\rfloor.1$ $\times 1010$
$R^*F^{(m-2)} = F_{-m+1}\cdot F_{-m}$ $= F_{-m+1}\cdot F^{(m-1)}$	$F_{-m+1}\rfloor\cdot F_{-m}$ $\times R$		$101\rfloor.$
$R^*F^{(m-1)} = F_{-m}$	$F_{-m}\rfloor\cdot$	$.125_{10} = .001_2$	$.100101_2 = 578125_{10}$

TABLE 19.9 Nonterminating Fraction Conversion Example

$0.1_{10} = 0.000110011\ldots_2$	
$\times 2$	
$0\rfloor.2$	or more compactly
$\times 2$	
$0\rfloor.4$	$0.1_{10} = 0.0\underline{0011}_2$
$\times 2$	
$0\rfloor.8$	
$\times 2$	
$1\rfloor.6$	
$\times 2$	
$1\rfloor.2$	
$\times 2$	
$0\rfloor.4$	
$\times 2$	
$0\rfloor.8$	
$\times 2$	
$1\rfloor.6$	

TABLE 19.10 Conversions between Systems Where One Base Is an Integer Power of the Other Base

(a) Conversion from high base to lower base
$B2.C5_{16} = 1011 \quad 0010 \quad .1100 \quad 0101_2$
$62.75_8 = 110 \quad 010 \quad .111 \quad 101_2$
(b) Conversion from lower base to high base
$11 \quad 0010 \quad 0100.0001 \quad 1100 \quad 01_2 = 324.1C4_{16}$
$10 \quad 110 \quad 001.011111 \quad 01_2 = 261.372_8$

Conversion to base 2 from a base, which is an integer power of 2, can be most simply accomplished by independent conversion of each successive digit, as illustrated in Table 19.10a. Inversely, conversion from base 2 to a base 2^k can be simply accomplished by grouping the bits into sets of k bits, each starting with the least significant bit for the integer portion and with the most significant bit for the fraction portion, as shown by the examples in Table 19.10b.

19.7 Complements

Each number system has two conventionally used **complements:**

$$\text{Radix complement of } N = N^{RC} = R^n - N$$
$$\text{Reduced radix complement of } N = N^{rC} = N^{RC} - 1$$

where R is the radix and n is the number of digits in the number N. These equations provide complements for numbers having the magnitude N.

A positive number can be represented by a code in the two character machine language alphabet, 0 and 1, which is simply the positive number expressed in the base 2, that is, the code for the number is the number itself. A negative number requires that the sign be coded in the binary alphabet. This can be done by separately coding the sign and the magnitude or by coding the negative number as a single entity. Table 19.11 illustrates four different code types for negative numbers. Negative numbers can be represented in the sign magnitude form by using the leftmost digit as the code for the sign (0 for + and 1 for −) and the rest of the digits as the code for the magnitude. Complements and biasing provide the means for coding the negative number as a single entity instead of having a discrete separate coding for the sign. The use of complements provides for essentially equal ranges of values for both positive and negative numbers. The biased representation can also provide essentially equal ranges for positive and negative values by choosing the biasing value to be essentially half of the largest binary number that could be represented with the available number of digits. The bias code is obtained by subtracting the biasing value from the code considered as a positive number, as shown in the rightmost column of Table 19.11.

Complements enable subtraction to be done by addition of the complement. If the result fits into the available field size the result is automatically correct. A diagnostic must be provided to show that the result is incorrect if **overflow** occurs, that is, the number does not fit in the available field. Table 19.12 illustrates arithmetic operations with and without complements. The two rightmost columns illustrate cases where the result overflows the 3-b field size for the magnitude. The overflow condition can be represented in terms of two carry parameters:

- C_0, the output carry from the leftmost digit position
- C_1, the output carry from the second leftmost digit position (the output carry from the magnitude field if sign magnitude representation is used)

If both of these carries are coincident (i.e., have the same value) the result fits in the available field and, hence, is correct. If these two carries are not coincident, the result is incorrect.

TABLE 19.11 Number Representations

		Signed Numbers Having the Specified Codes			
Available Codes	Positive Numbers	Sign Magnitude	One's Complement	Two's Complement	111 bias
1111	1111	−111	−000	−001	+1000
1110	1110	−110	−001	−010	+111
1101	1101	−101	−010	−011	+110
1100	1100	−100	−011	−100	+101
1011	1011	−011	−100	−101	+100
1010	1010	−010	−101	−110	+011
1001	1001	−001	−110	−111	+010
1000	1000	−000	−111	−1000	+001
0111	0111	+111	+111	+111	000
0110	0110	+110	+110	+110	−001
0101	0101	+101	+101	+101	−010
0100	0100	+100	+100	+100	−011
0011	0011	+011	+011	+011	−100
0010	0010	+010	+010	+010	−101
0001	0001	+001	+001	+001	−110
0000	0000	+000	+000	+000	−111

TABLE 19.12 Comparison of Arithmetic with and without Complements

Sample Illustrations	$N = 7 - 5$ $= 2$	$N = 5 - 7$ $= -2$	$N = 5 + 7$ $= 12$	$N = -5 - 7$ $= -12$
Pencil and paper arithmetic (without complements)	111 −101 10	(−111) −(−101) −10	101 +111 1100	(−101) +(−111) −1100
Computer arithmetic with 2's complement	0111 +1011 ⁄10010	0101 +1001 1110	0101 +0111 1100	1011 +1001 ⁄10100
4-binary digit working eld (accommodates 3-b magnitude)	↓ Designates positive number	↓ Designates negative number	↓ Designates negative number	↓ Designates positive number
Result Veracity	+010 True	−010 True	−100 False	−100 False
Significant carries	$C_0 = 1$ $C_1 = 1$	$C_0 = 0$ $C_1 = 0$	$C_0 = 0$ $C_1 = 1$	$C_0 = 1$ $C_1 = 0$
Veracity condition or equivalently	$C_0 \equiv C_1$ $C_0 \odot C_1 = 1$	$C_0 \equiv C_1$ $C_0 \odot C_1 = 1$	$C_0 \not\equiv C_1$ $C_0 \odot C_1 = 0$	$C_0 \not\equiv C_1$ $C_0 \odot C_1 = 0$

19.8 Codes

Various types of **codes** have been developed for serving different purposes. There are codes that enable characters in an alphabet to be individually expressed in terms of codes in a smaller alphabet. For example, the alphabet of decimal numeric symbols can be expressed in terms of the binary alphabet by the **binary-coded decimal** (BCD) or 8421 code shown in Table 19.13. The 8421 designation represents the weight given to each of the binary digits in the coding process.

There are codes that facilitate doing arithmetic. The 2421 code can also be used to represent the decimal numeric symbols. The 2421 code has the advantage that the code for the reduced radix complement is

TABLE 19.13 Sample Codes

Decimal	BCD		2-out-	Parity		Gray		
Digits	8421	2421	of-5	Even	Odd	1-bit	2-bit	3-bit
0	0000	0000	00011	0000 0	0000 1	0	00	000
1	0001	0001	00101	0001 1	0001 0	1	01	001
2	0010	0010	00110	0010 1	0010 0		11	011
3	0011	0011	01001	0011 0	0011 1		10	010
4	0100	0100	01010	0100 1	0100 0			110
5	0101	1011	01100	0101 0	0101 1			111
6	0110	1100	10001	0100 1	0110 0			101
7	0111	1101	10010	0111 1	0111 0			100
8	1000	1110	10100	1000 1	1000 0			
9	1001	1111	11000	1001 0	1001 1			

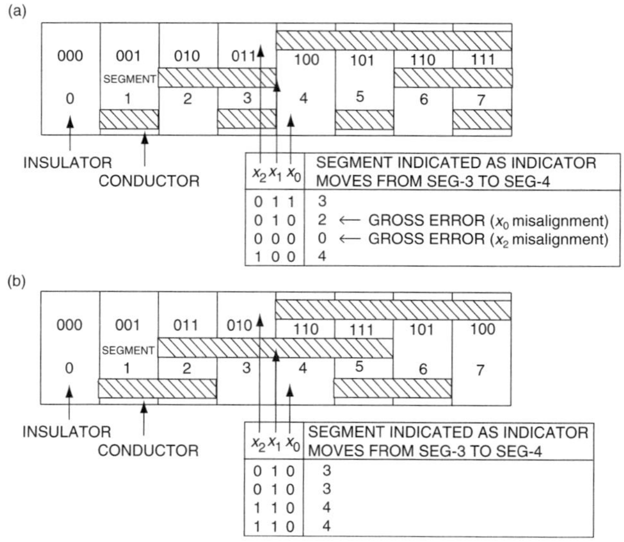

FIGURE 19.1 Eight-segment position sensor with slightly misaligned contacts: (a) binary code physical configuration, (b) gray code physical configuration.

the same as the reduced radix complement of the code, and this is not true of the BCD code. Thus, the 2421 code facilitates arithmetic with multiple individually coded digit numbers.

There are codes designed to detect errors that may occur in storage or transmission. Examples are the even and odd parity codes and the 2-out-of-5 code shown in Table 19.13. The 2-out-of-5 error detection code is such that each decimal value has exactly two high digit values. Parity code attaches an extra bit having a value such that the total number of high bits is odd if odd parity is used, and the total number of high bits is even if even parity is used. An even number of bit errors are not detectable by a single bit parity code. Hence, single bit parity codes are adequate only for sufficiently low bit error rates. Including a sufficient number of **parity bits** enables the detection and correction of all errors.

There are codes designed to prevent measurement misrepresentation due to small errors in sensor alignment. **Gray codes** are used for this purpose. A Gray code is one in which the codes for physically adjacent positions are also **logically adjacent**, that is, they differ in only one binary digit. Gray codes can be generated for any number of digits by reflecting the Gray code for the case with one less digit, as shown in Table 19.13, for the case of 1, 2, and 3-bit codes. The advantage of a Gray scale coded lineal position sensor is illustrated in Figure 19.1 for the eight-segment case.

19.9 Boolean Algebra

Boolean algebra provides a means to analyze and design binary systems and is based on the seven postulates given in Table 19.14. All other Boolean relationships are derived from these seven postulates. Expressed in graphical form, called *Venn diagrams*, the postulates appear more natural and logical. This benefit results from the two-dimensional pictorial representation freeing the expressions from the one-dimensional constraints imposed by lineal language format.

The OR and AND operations are normally designated by the arithmetic operator symbols + and · and referred to as sum and product operators in basic digital logic literature. However, in digital systems that perform arithmetic operations this notation is ambiguous and the symbols $\vee$ for OR and $\wedge$ for AND eliminates the ambiguity between arithmetic and boolean operators. Understanding the conceptual meaning of these boolean operations is probably best provided by set theory, which uses the union operator $\cup$ for OR and the intersection operator $\cap$ for AND. An element in a set that is the union of sets is a member of one set OR another of the sets in the union. An element in a set that is the intersection of sets is a member of one set AND a member of the other set in the intersection.

A set of theorems derived from the postulates facilitates further developments. The theorms are summarized in Table 19.15. Use of the postulates is illustrated by the proof of a theorem in Figure 19.2.

TABLE 19.14 Boolean Postulates

Postulate	Name	Meaning	Forms (a)	Forms (b)
1	De nition	$\exists$ a set $\{K\} = \{a,b,...\}$ of two or more elements and two binary operators $\ni \{K\} = \{a, b, a+b, a\cdot b, \ldots\}$	**OR** $+$ $\vee$ $\cup$	**AND** $\cdot$ $\wedge$ $\cap$
2	Substitution Law	$\text{expression}_1 = \text{expression}_2$ If one replaced by the other does not alter the value		
3	Identity Element	$\exists$ identity elements for each operator	$a + 0 = a$	$a \cdot 1 = a$
4	Commutativity	For every *a* and *b* in *K*	$a + b = b + a$	$a \cdot b = b \cdot a$
5	Associativity	For every *a*, *b*, and *c* in *K*	$a + (b + c) = (a + b) + c$	$a \cdot (b \cdot c) = (a \cdot b) \cdot c$
6	Distributivity	For every *a*, *b*, and *c* in *K*	$a + (b \cdot c) = (a + b) \cdot (a + c)$	$a \cdot (b + c) = (a \cdot b) + (a \cdot c)$
7	Complement	For every *a* in *K* $\exists$ a complement in $K \ni$	$a + \bar{a} = 1$	$a \cdot \bar{a} = 0$

TABLE 19.15 Boolean Theorems

Theorem		Forms (a)	Forms (b)
8	Idempotency	$a + a = a$	$a \cdot a = a$
9	Complement Theorem	$a + 1 = 1$	$a \cdot 0 = 0$
10	Absorption	$a + ab = a$	$a(a + b) = a$
11	Extra Element Elimination	$a + \bar{a}b = a + b$	$a(\bar{a} + b) = ab$
12	De Morgan's Theorem	$\overline{a + b} = \bar{a} \cdot \bar{b}$	$\overline{ab} = \bar{a} + \bar{b}$
13	Concensus	$ab + \bar{a}c + bc = ab + \bar{a}c$	$(a + b)(\bar{a} + c)(b + c) = (a + b)(\bar{a} + c)$
14	Complement Theorem 2	$ab + a\bar{b} = a$	$(a + b)(a + \bar{b}) = a$
15	Concensus 2	$ab + a\bar{b}c = ab + ac$	$(a + b)(a + \bar{b} + c) = (a + b)(a + c)$
16	Concensus 3	$ab + \bar{a}c = (a + c)(\bar{a} + b)$	$(a + b)(\bar{a} + c) = ac + \bar{a}b$

1)	$x + x = x + x$	IDENTITY
2)	$= (x + x) \cdot 1$	P3b IDENTITY ELEMENT EXISTENCE
3)	$= (x + x) \cdot (x + \bar{x})$	P7a COMPLEMENT EXISTENCE
4)	$= x + x \cdot \bar{x}$	P6a DISTRIBUTIVITY
5)	$= x + 0$	P7b COMPLEMENT EXISTENCE
6)	$= x$	P3a IDENTITY ELEMENT EXISTENCE

FIGURE 19.2 Proof of Theorem 8: Idempotency (a): $x + x = x$.

(a) $f(A, B, C) = AB + A\bar{C} + \bar{A}C$

(b) $f(0, 0, 1) = 0 \cdot 0 + 0 \cdot \bar{1} + \bar{0} \cdot 1$
$= 0 \cdot 0 + 0 \cdot 0 + 1 \cdot 1$
$= 0 + 0 + 1$
$= 1$

(c)

A	B	C	f(A, B, C)
0	0	0	0
0	0	1	1
0	1	0	0
0	1	1	1
1	0	0	1
1	0	1	0
1	1	0	1
1	1	1	1

FIGURE 19.3 Example of forms for defining boolean functions: (a) boolean expression definition, (b) boolean expression evaluation, (c) truth table definition.

TABLE 19.16 Number of Different Boolean Functions

Variables n	Arguments 2^n	Functions 2^{2^n}
0	1	2
1	2	4
2	4	16
3	8	256
4	16	65,536
5	32	4,194,304
⋮	⋮	⋮

19.10 Boolean Functions

Boolean functions can be defined and represented in terms of boolean expressions and in terms of **truth tables** as illustrated in Figure 19.3a,c. Each form can be converted into the other form. The function values needed for the construction of the truth table can be obtained by evaluating the function as illustrated in Figure 19.3b. The reverse conversion will be illustrated subsequently.

For a given number of variables there are a finite number of boolean functions. Since each boolean variable can have two values, 0 or 1, a set of n variables has 2^n different values. A boolean function has a specific value for each of the possible values that the independent variables can have. Since there are two possible values for each value of the independent variables there are 2^{2^n} different boolean functions of n variables. The number of functions increases very rapidly with the number of independent variables, as shown in Table 19.16.

The 16 different boolean functions of the two independent variables are defined in algebraic form in Table 19.17 and in truth table form in Table 19.18.

19.11 Switching Circuits

Boolean functions can be performed by digital circuits. Circuits that perform complicated boolean functions can be subdivided into simpler circuits that perform simpler boolean functions. The circuits that perform the simplest boolean functions are taken as basic elements, called *gates* and are represented

TABLE 19.17 Functions of Two Variables De ned as Boolean Expressions

	Name	Expression	Circuit Representation
0	ALWAYS	1	
1	NEVER	0	
2	1st Var	a	
3	2nd Var	b	
4	NOT 1st Var	$\bar{a}$	
5	NOT 2nd Var	$\bar{b}$	
6	MIN-0/NOR	$\bar{a}\wedge\bar{b} = a \downarrow b$	
7	MIN-1	$a\wedge\bar{b}$	
8	MIN-2	$\bar{a}\wedge b$	
9	MIN-3/AND	$a\wedge b$	
10	MAX-0/OR	$a\vee b$	
11	MAX-1	$a\vee\bar{b}$	
12	MAX-2	$\bar{a}\vee b$	
13	MAX-3/NAND	$\bar{a}\vee\bar{b} = a \uparrow b$	
14	EXOR	$A \oplus b = a\wedge\bar{b}\vee\bar{a}\wedge b$	
15	COIN	$a \ominus b = \overline{a \oplus b}$	

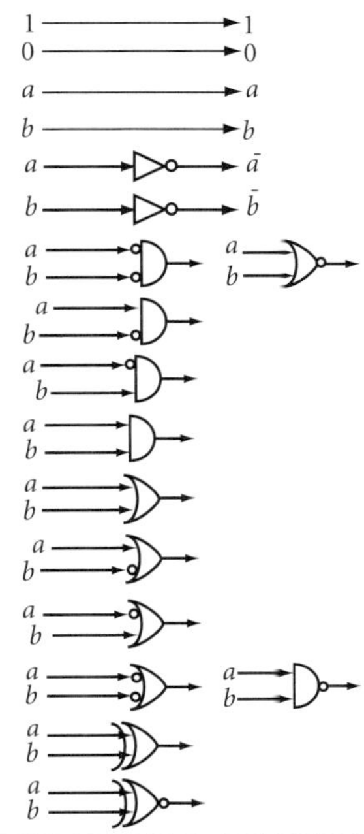

TABLE 19.18 Truth Tables for Two Variable Functions

a	b	NOR m_0	m_1	m_2	AND m_3	OR M_0	M_1	M_2	NAND M_3	XOR	COIN	a	b	$\bar{a}$	$\bar{b}$	LO	HI
0	0	1	0	0	0	0	1	1	1	0	1	0	0	1	1	0	1
0	1	0	1	0	0	1	0	1	1	1	0	0	1	1	0	0	1
1	0	0	0	1	0	1	1	0	1	1	0	1	0	0	1	0	1
1	1	0	0	0	1	1	1	1	0	0	1	1	1	0	0	0	1

by specialized symbols. The circuit symbols for the gates that perform functions of two independent variables are shown in Table 19.17. The gates are identified by the adjective representing the operation they perform. The most common gates are the AND, OR, NAND, NOR, XOR, and COIN gates. The only nontrivial single input gate is the invertor or NOT gate. Gates are the basic elements from which more complicated digital logic circuits are constructed.

A logic circuit whose steady-state outputs depend only on the present steady-state inputs (and not on any prior inputs) is called a *combinational logic circuit.* To depend on previous inputs would require memory, thus a combinational logic circuit has no memory elements.

Boolean algebra allows any combinational logic circuit to be constructed soley with AND, OR, and NOT gates. Any combinational logic circuit may also be constructed solely with NAND gates, as well as solely with NOR gates.

19.12 Expansion Forms

The **sum of products (SP)** is a basic form in which all boolean functions can be expressed. The **product of sums (PS)** is another basic form in which all boolean functions can be expressed. An illustrative example is given in Figures 19.4b,c for the example given in Figure 19.4a.

(a)

$$f(A, B, C) = A(B + \overline{C}) + (\overline{A} + B)C$$

TRUTH TABLE

A	B	C	f(A, B, C)
0	0	0	0
0	0	1	1
0	1	0	0
0	1	1	1
1	0	0	1
1	0	1	0
1	1	0	1
1	1	1	1

(b)

P6b → $f(A, B, C) = AB + A\overline{C} + \overline{A}C + BC$

T13a → $f(A, B, C) = AB + A\overline{C} + \overline{A}C$

(c)

P7b → $f(A, B, C) = A(\overline{A} + B + \overline{C}) + (\overline{A} + B + \overline{C})C$

P6b → $f(A, B, C) = (A + C)(\overline{A} + B + \overline{C})$

(d)

P7a →

$f(A, B, C) = AB(C + \overline{C}) + A(B + \overline{B})\overline{C} + \overline{A}(B + \overline{B})C$

P6b →

$f(A, B, C) = ABC + AB\overline{C} + AB\overline{C} + A\overline{B}\overline{C} + \overline{A}BC + \overline{A}\overline{B}C$

T8a →

$f(A, B, C) = ABC + AB\overline{C} + A\overline{B}\overline{C} + \overline{A}BC + \overline{A}\overline{B}C$

(e)

P7b → $f(A, B, C) = (A + C + B\overline{B})(\overline{A} + B + \overline{C})$

P6a → $f(A, B, C) = (A + B + C)(A + \overline{B} + C)(\overline{A} + B + \overline{C})$

(f)

$f(A, B, C) = m_{111} + m_{110} + m_{100} + m_{011} + m_{001}$

$f(A, B, C) = \Sigma m(1, 3, 4, 6, 7)$

(g)

$f(A, B, C) = M_{000} + M_{010} + M_{101}$

$f(A, B, C) = \Pi(0, 2, 5)$

FIGURE 19.4 Examples of converting boolean functions between forms: (a) given example, (b) conversion to SP form, (c) conversion to PS form, (d) conversion to canonical SP form, (e) conversion to canonical PS form, (f) minterm notation/canoncial SP form, (g) maxterm notation/canonical PS form.

Minterms are a special set of functions, none of which can be expressed in terms of the others. Each minterm has each of the variables in the complemented or the uncomplemented form ANDed together. An SP expansion in which only minterms appear is a canonical SP expansion. Figure 19.4d shows the development of the canonical SP expansion for the previous example. The canonical SP expansion may also be simply expressed by enumerating the minterms as shown in Figure 19.4f. Comparison of the truth table with the minterm expansion shows that each function value of 1 represents a minterm of the function and vice versa. All other function values are 0.

Maxterms are a special set of functions, none of which can be expressed in terms of the others. Each maxterm has each of the variables in the complemented or the uncomplemented form ORed together. A PS expansion in which only maxterms appear is a canonical PS expansion. Figure 19.4e shows the development of the canonical PS expansion for the previous example. The canonical PS expansion may also be simply expressed by enumerating the maxterms as shown in Figure 19.4g. Comparison of the truth table with the maxterm expansion shows that each function value of 0 represents a maxterm of the function and vice versa. All other function values are 1.

19.13 Realization

The different types of boolean expansions provide different circuits for implementing the generation of the function. A function expressed in the SP form is directly realized as an AND–OR **realization**, as illustrated in Figure 19.5a. A function expressed in the PS form is directly realized as an OR–AND realization as illustrated in Figure 19.5b. By using involution and deMorgan's theorm the SP expansion can be expressed in terms of NAND–NAND and the PS expansion can be expressed in terms of NOR–NOR, as shown in Figures 19.5c,d. The variable inversions specified in the inputs can be supplied by either NAND or NOR gates, as shown in Figures 19.5g,h, which then provide the NAND–NAND–NAND and the NOR–NOR–NOR circuits shown in Figures 19.5i,j.

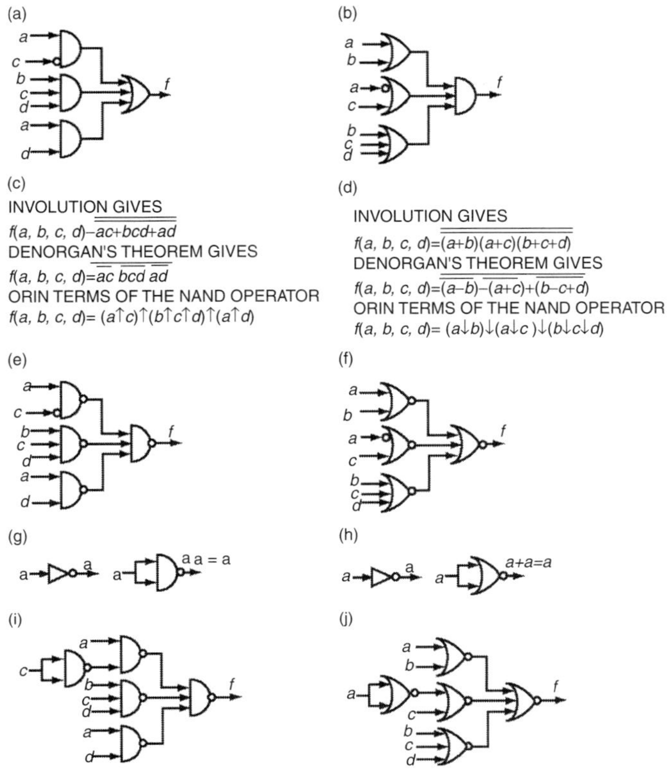

FIGURE 19.5 Examples of realizations based on various expansion forms: (a) AND-OR realization of $f(a, b, c, d) = a\bar{c} + bcd + ad$, (b) OR-AND realization of $f(a, b, c, d) = (a + b)(\bar{a} + c)(b + c + d)$, (c) AND-OR conversion to NAND-NAND, (d) OR-AND conversion to NOR-NOR, (e) NAND-NAND realization of $f(a, b, c, d) = a\bar{c} + bcd + ad$, (f) NOR-NOR realization of $f(a, b, c, d) = (a + b)(\bar{a} + c)(b + c + d)$, (g) NAND gate realization of NOT gate, (h) NOR gate realization of NOT gate, (i) NAND-NAND-NAND realization of $f(a, b, c, d) = a\bar{c} + bcd + ad$, (j) NOR-NOR-NOR realization of $f(a, b, c, d) = (a + b)(\bar{a} + c)(b + c + d)$.

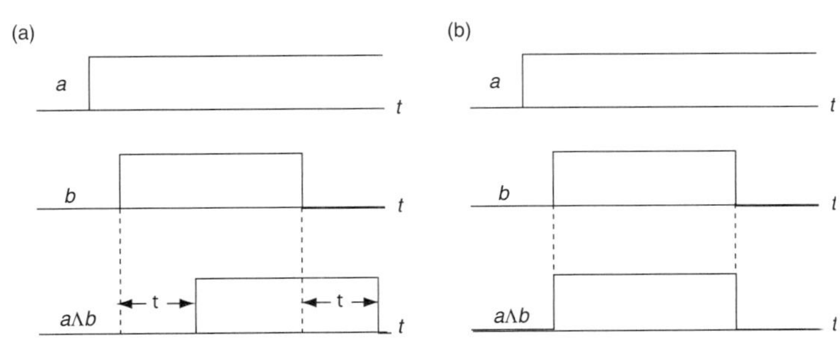

FIGURE 19.6 Timing diagrams for the AND gate circuit: (a) microtiming diagram, (b) macrotiming diagram.

19.14 Timing Diagrams

Timing diagrams are of two major types. A **microtiming diagram** has a time scale sufficiently expanded in space to display clearly the gate delay, such as shown in Figure 19.6a for an AND gate. A **macrotiming diagram** has a time scale sufficiently contracted in space so that the gate delay is not noticeable, as shown in Figure 19.6b for an AND gate.

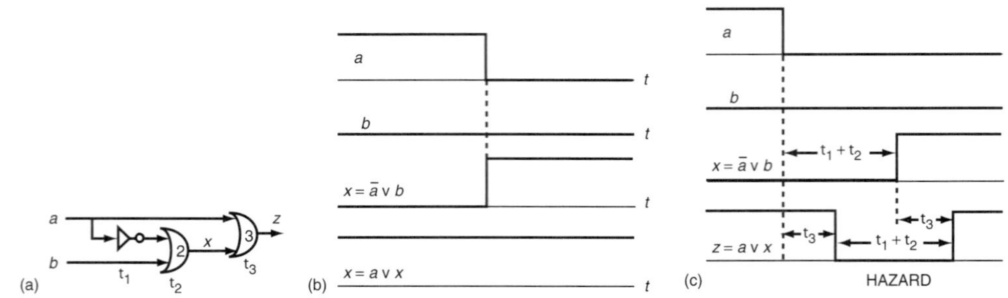

FIGURE 19.7 Example of a hazard (output variation) caused by unequal delay paths: (a) circuit for illustrating a hazard, (b) ideal case, no delays, $\tau_1 = \tau_2 = \tau_3 = 0$, no hazard introduced; (c) signal paths with different delays, $\tau_1 + \tau_2 + \tau_3 > \tau_3$, hazard introduced.

The advatange of the macrotiming diagram is that larger time intervals can be represented in a given spatial size and they can be more quickly developed. The disadvantage is that they do not display the information required for speed limitation considerations.

19.15 Hazards

The variation in signal delays through different circuit elements in different paths may cause the output signal to fluctuate from that predicted by non-time-dependent truth tables for the elements. This fluctuation can cause an undesired result and, hence, is a *hazard.* This is illustrated in Figure 19.7.

19.16 *K*-Map Formats

In a truth table the values of a boolean function are displayed in a one-dimensional array. A *K*-**map** contains the same information arranged in as many effective dimensions as there are independent variables in the function. The special form of the representation provides a simple procedure for minimizing the expression and, hence, the number of components required for realizing the function in a given form. The function is represented in a space that in a Venn diagram is called the *universal set.* The *K*-map is a special form of Venn diagram. The space is divided into two halves for each of the independent variables. The division of the space into halves is different for each independent variable. For one independent variable the space is divided into two different identical size regions, each of which represents a minterm of the function. For n independent variables the space is divided into 2^n different identical size regions, one for each of the 2^n minterms of the function. This and associated considerations are illustrated in a sequence of figures from Figures 19.8–19.15.

Figure 19.8 shows one-variable *K*-map formats. Figure 19.8a shows the space divided into two equal areas, one for each of the two minterms possible for a single variable. The squares could also be identified by the variable placed external to the space to designate the region that is the domain of the variable, with the unlabeled space being the domain of the complement of the variable, as shown in Figure 19.8b. Another way of identifying the regions is by means of the minterm number the area is for, as shown in Figure 19.8c. Still another way is to place the values the variable can have as a scale alongside the space, as shown in Figure 19.8d. The composite labeling, shown in Figure 19.8e, appears redundant but is often useful because of the different modes of thought used with the different label types. Putting the actual minterm expressions inside each square is too cluttering and is rarely used except as an aid in teaching the theory of *K*-maps. The use of minterm numbers, although widely used, also clutters up the diagram, and the methodology presented here makes their use superfluous once the concepts are understood.

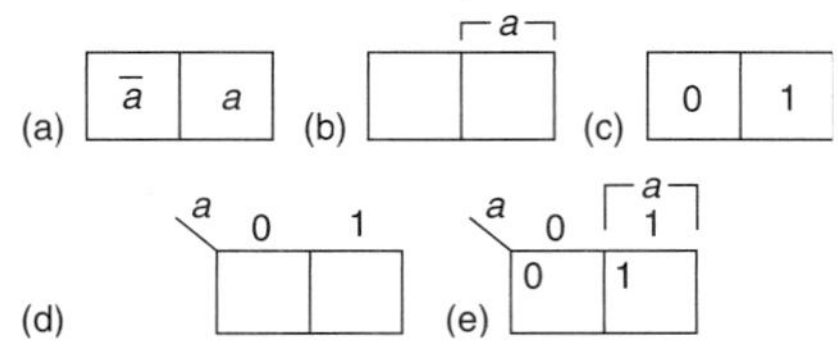

FIGURE 19.8 One vartiable *K*-map forms: (a) internal minterm labels, (b) external domain label, (c) internal minterm number labels, (d) external scale label, (e) composite labeling.

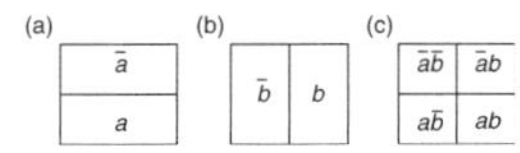

FIGURE 19.9 Two-variable *K*-map format construction: (a) domains for ***a***, (b) domains for ***b***, composite.

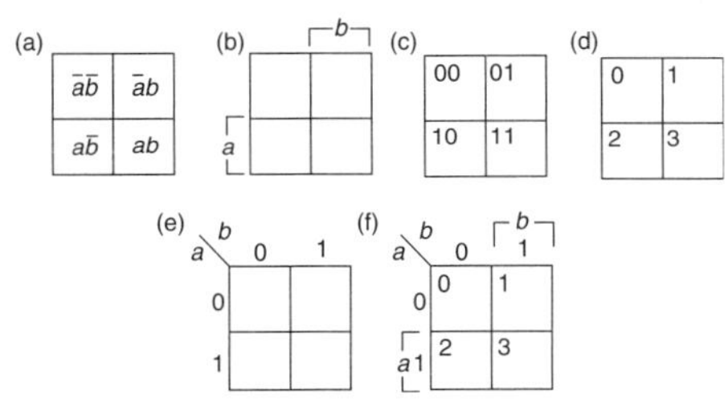

FIGURE 19.10 Two variable *K*-map formats: (a) minterm labels, (b) domain labels, (c) minterm binary labels, (d) minterm decimal labels, (e) scale labels, (f) composite labeling.

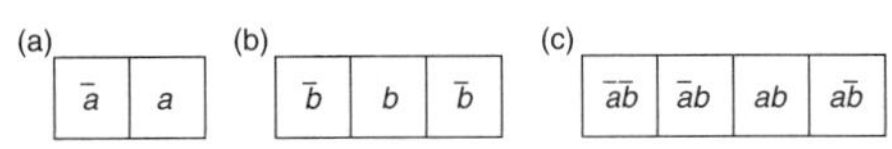

FIGURE 19.11 Two-variable *K*-map alternate format construction: (a) domains for ***a***, (b) domains for ***b***, (c) composite.

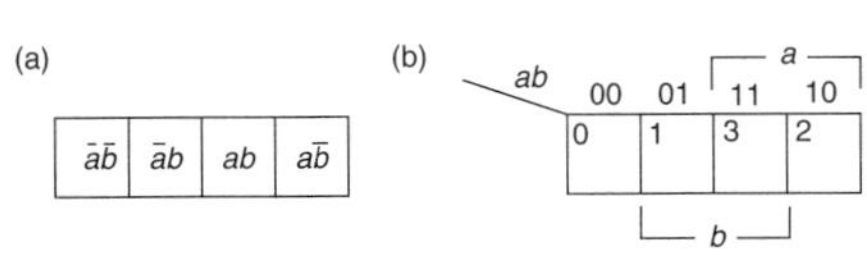

FIGURE 19.12 Two-variable *K*-map alternate format: (a) minterm labels, (b) composite labeling.

The organization of a two-variable *K*-map format is illustrated in Figure 19.9. The space is subdivided vertically into two domains for the variable ***a*** and its complement, and is subdivided horizontally for the variable ***b*** and its complement, as shown in Figures 19.9a,b. The composite subdivisions for both variables together with the expressions for the two-variable minterms are shown in Figure 19.9c.

The different formats for identifying the areas for the two-variable case are shown in Figure 19.10. Of particular interest is the comparison of the binary and decimal minterm number labels. The binary minterm number is simply the **catenation** of the vertical and horizontal scale number for the position. It is the use of this identity that makes internal labels in the square totally superfluous.

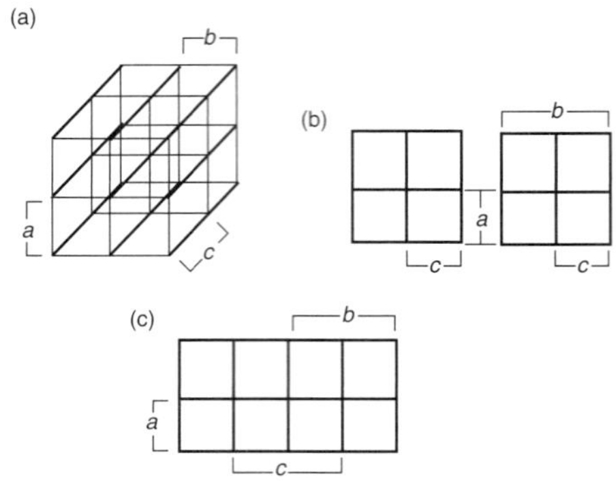

FIGURE 19.13 Three-variable *K*-map formats: (a) three-dimensional, (b) three-dimensional left and right halves, (c) two-dimensional.

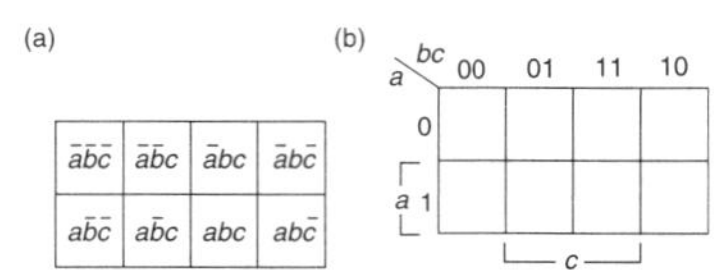

FIGURE 19.14 Three variable two-dimensional *K*-map formats: (a) minterm labels, (b) composite scale labels.

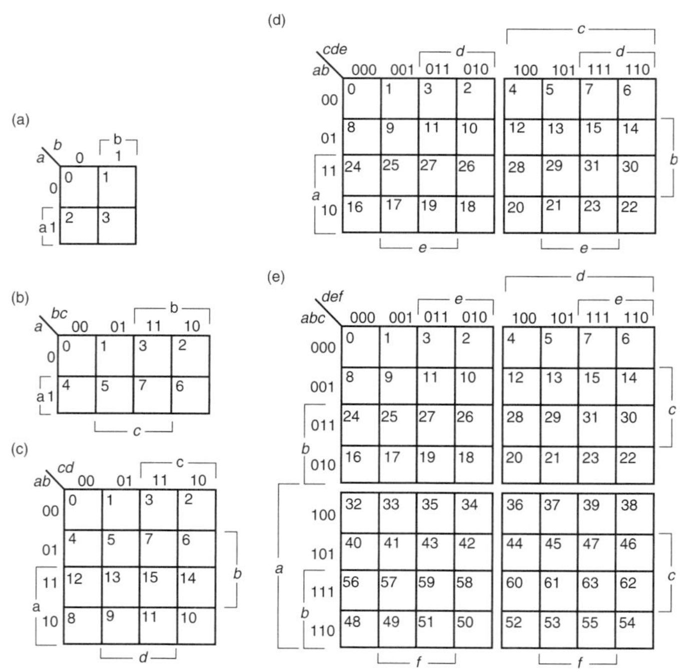

FIGURE 19.15 Formats for *K*-maps for functions of 2–6 independent variables with conformal coordinate scales: (a) two-variable case $f(a, b)$, three-variable case $f(a, b, c)$; (c) four-variable case $f(a, b, c, d)$; (d) five-variable case $f(a, b, c, d, e)$; (e) six-variable case $f(a, b, c, d, e, f)$.

An alternate way of subdividing the space for the two-variable case is illustrated in Figure 19.11 and labeling alternatives in Figure 19.12. The configuration employed in Figure 19.9 uses two dimensions for two variables, whereas the configuration employed in Figure 19.11 uses one-dimension for two variables. The two-dimensional configuration appears to be more logical and is more convenient to use than the one-dimensional configuration for the case of two variables. For the case of a larger number of variables the configuration in Figure 19.12 offers special advantages, as will be shown.

The organization of three-variable *K*-map formats is illustrated in Figure 19.13. It is logical to introduce an additional space dimension for each additional independent variable as depicted in Figure 19.13a; however, the excessive inconvenience of working with such formats makes them impractical. To make the mapping process practical it must be laid out in two dimensions. This can be done in two ways. One way is to take the individual slices of the three-dimensional configuration and place them adjacent to each other as illustrated in Figure 19.13b. Another way is to use the one-dimensional form for two variables, illustrated in Figure 19.12, as shown in Figure 19.13c. For the case of three and four independent variables the format given in Figure 19.13c is more convenient and for further independent variables that of Figure 19.13b is convenient. These are all illustrated in Figure 19.15. Labeling for three independent variables is given in Figure 19.14.

The independent boolean variables in **conformal coordinate scales** have exactly the same order as in the boolean function argument list, as depicted in Figure 19.15. Conformal assignment of the independent variables to the *K*-map coordinate scales makes the catenated position coordinates for a minterm (or maxterm) identical to the minterm (or max-term) number. Utilization of this identity eliminates the need for the placement of minterm identification numbers in each square or for a separate position identification table. This significantly decreases the time required to construct *K*-maps and makes their construction less error prone. The minterm number, given by the catenation of the vertical and horizontal coordinate numbers, is obvious if the binary or octal number system is used.

19.17 *K*-Maps and Minimization

A function is mapped into the *K*-map format by entering the value for each of the minterms in the space for that minterm. The function values can be obtained in various ways such as from the truth table for the function, the Boolean expression for the function, or from other means by which the function may be defined. An example is given for the truth table given in Figure 19.4a, which is repeated here in Figure 19.16a and whose *K*-map is shown in various forms in Figures 19.16b–d.

The function can also be mapped into a conformally scaled *K*-map directly from the canonical expansion, this being essentially the same process as entering the minterms (or maxterms) from the truth table. The function may also be directly mapped from any PS or SP expansion form. Another means of obtaining the *K*-map is to formulate it as a function table, as illustrated for the multiplication of 1- and 2-b numbers in Figure 19.17.

(a)

A	*B*	*C*	*F*(*A*, *B*, *C*)
0	0	0	0
0	0	1	1
0	1	0	0
0	1	1	1
1	0	0	1
1	0	1	0
1	1	0	1
1	1	1	1

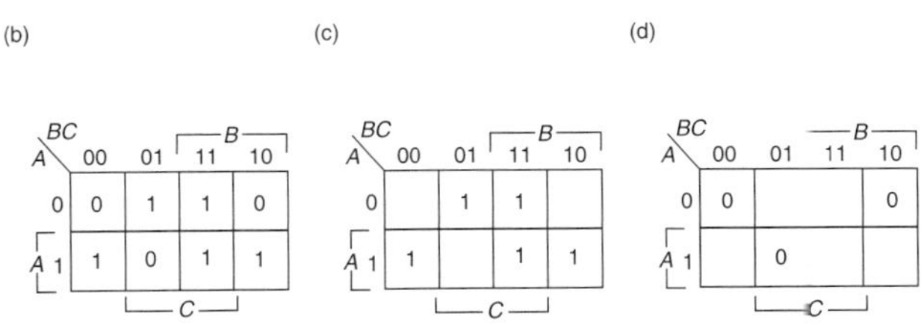

FIGURE 19.16 Three-variable *K*-map example: (a) truth table, (b) *K*-map with all values shown, (c) minterm *K*-map, (d) maxterm *K*-map.

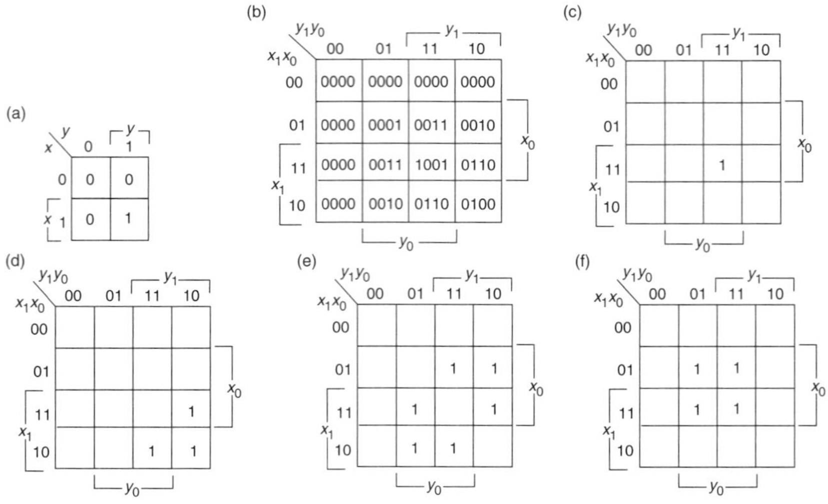

FIGURE 19.17 Examples of *K*-maps formulated as function tables: (a) *K*-map for the product of two 1-b numbers $P = x * y$; (b) composite *K*-map for the product of two 2-b numbers, $P_3\ P_2\ P_1\ P_0 = x_1\ x_0 * y_1\ y_0$; (c) *K*-map for the digit P_3 of the product of two 2-b numbers; (d) *K*-map for the digit P_2 of the product of two 2-b numbers; (e) *K*-map for the digit P_1 of the product of two 2-b numbers; (f) *K*-map for the digit P_0 of the product of two 2-b numbers.

19.18 Minimization with *K*-Maps

The key feature of *K*-maps that renders them convenient for minimization is that minterms, which are spatially adjacent in the horizontal or vertical directions, are logically adjacent. Logically adjacent minterms are identical in all variables except one. This allows the two minterms to be combined into a single terms with one less variable. This is illustrated in Figure 19.18. Two adjacent minterms combine into a first-order **implicant**. A first-order implicant is the combination of all of the independent variables but one. In this example, the first-order implicant expressed in terms of minterms contains eight literals but the minimized expression contains only three literals. The circuit realization for the OR combination of the two minterms has two AND gates and one OR gate, whereas the realization for the equivalent implicant requires only a single AND gate.

The combination of minterms into first-order implicants can be represented more compactly by using the single symbol minterm notation with the subscript that identifies the particular minterm expressed in binary, as illustrated in Figure 19.18d.

Two adjacent first-order implicants can be combined into a second-order implicant as illustrated in Figure 19.19. A second-order implicant contains all of the independent variables except two. In general, an *n*th-order implicant contains all of the variables except *n* and requires an appropriately grouped set of 2^n minterms.

Minterms that are at opposite edges of the same row or column are logically adjacent since they differ in only one variable. If the plane space is rolled into a cylinder with opposite edges touching, then the logically adjacent edge pairs become physically adjacent. For larger numbers of variables using *K*-maps with parallel sheets, the corresponding positions on different sheets are logically adjacent. If the sheets are considered as overlaid, the corresponding squares are physcally adjacent. The minimized expression is obtained by covering all of the minterms with the fewest number of largest possible implicants. A minterm is a zero-order implicant. Figure 19.20 illustrates a variety of examples. A **don't care** is a value that never occurs or if it does occur it is not used, and hence, it does not matter what its value is. Don't cares are also included to illustrate that they can be used to simplify expressions by taking their values to maximize the order of the implicants.

Maxterm *K*-maps can also be utilized to obtain minimized expressions by combining maxterms into higher order implicants, as illustrated for the example in Figure 19.21.

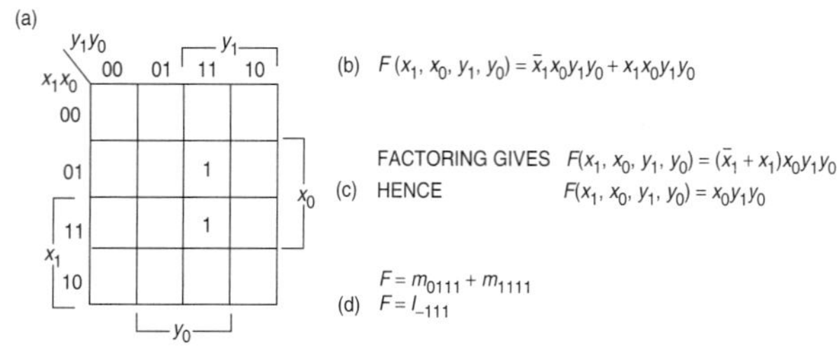

FIGURE 19.18 Example of minimization with a *K*-map: (a) sample *K*-map, (b) expression in minterms of function definition in (a), (c) simplification of the expression in (b), (d) simplification of the expression in (b) using single symbol minterm and implicant notation.

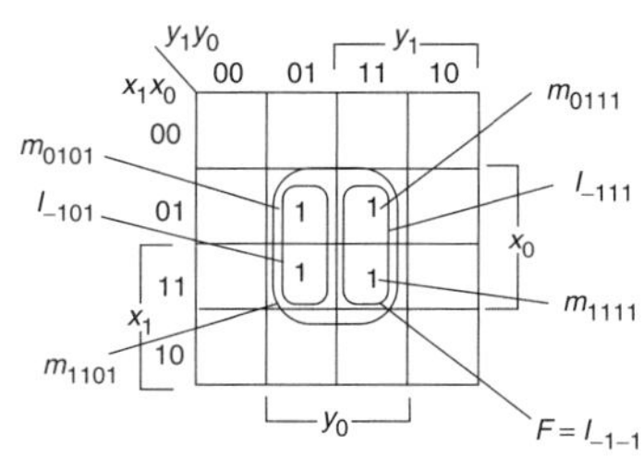

FIGURE 19.19 Example of minimization with *K*-map.

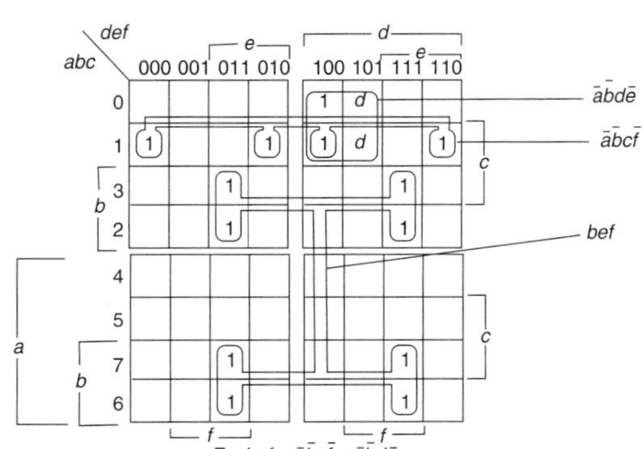

FIGURE 19.20 Example of minimization with six-variable *K*-map.

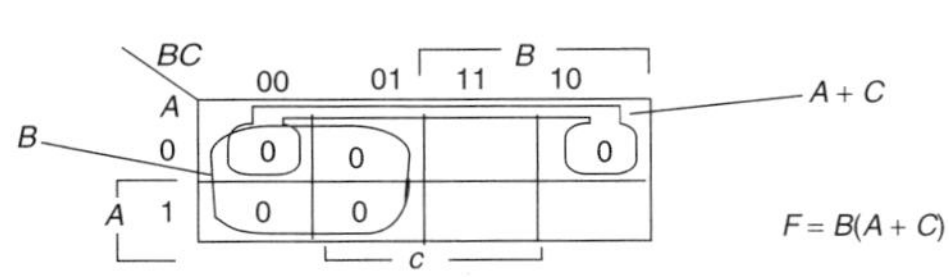

FIGURE 19.21 Three-variable maxterm *K*-map example.

GIVEN $F(A, B, C, D) = \Sigma m(2, 6, 7, 8) + d(0, 4, 5, 12, 13)$ and $G(A, B, C, D) = \Sigma m(2, 4, 5) + d(6, 7, 8, 10)$

ZERO-ORDER IMPLICANT LIST

NO. OF 1s	MIN-TERM	CODE *ABCD*	FLAGS	PI
0	0	0000	F-	√
1	2	0010	FG	√
	4	0100	FG	√
	8	1000	FG	√
2	5	0101	FG	√
	6	0110	FG	√
	10	1010	-G	√
	12	1100	F-	√
3	7	0111	FG	√
	13	1101	F-	√

FIRST-ORDER IMPLICANT LIST

IMPLICANTS	CODE *ABCD*	FLAGS	PI
0, 2	00-0	F-	√
0, 4	0-00	F-	√
0, 8	-000	F-	√
2, 6	0-10	FG	5
2, 10	-010	-G	6
4, 5	010-	FG	√
4, 6	01-0	FG	√
4, 12	-100	F-	√
8, 10	10-0	-G	7
8, 12	1-00	F-	√
5, 7	01-1	FG	√
5, 13	-101	F-	√
6, 7	011-	FG	√
12, 13	110-	F-	√

SECOND-ORDER IMPLICANT LIST

IMPLICANTS	CODE *ABCD*	FLAGS	PI
0, 2, 4, 6	0-0	F-	1
0, 4, 8, 12	-00	F-	2
4, 5, 6, 7	01-	FG	3
4, 5, 12, 13	-10-	F-	4

PRIME IMPLICANT CHART

		F				G		
	PI \ m	2	6	7	8	2	4	5
	1	X	X					
*	2				X			
*	3		X	X			X	X
	4							
**	5	X	X			X		
	6					X		
	7							

MINIMUM SP EXPANSIONS

$F = PI_2 + PI_3 + PI_5$
$= \overline{C}\overline{D} + \overline{A}B + \overline{A}C\overline{D}$

$G = PI_3 + PI_5$
$= \overline{A}B + \overline{A}C\overline{D}$

FIGURE 19.22 Illustration of the Quine–McClusky method of simultaneous minimization.

19.19 Quine–McCluskey Tabular Minimization

The *K*-map minimization method is too cumbersome for more than six variables and does not readily lend itself to computerization. A tabular method, which can be implemented for any number of variables and which lends itself to computer program implementation, consists of the following steps:

1. List all the minterms in the boolean function (with their binary code) organized into groups having the same number of 1s. The groups must be listed in consecutive order of the number of 1s.
2. Construct the list of first-order implicants. Use flags to indicate which minterms, don't cares, or implicants go with which functions. (Only minterms in adjacent groups have the possibility of being adjacent and, hence, this ordering method significantly reduces the labor of compiling the implicants.)
3. Construct the list of second-order implicants and the lists of all higher order implicants, until no higher order implicants can be constructed.
4. Construct the *prime implicant chart.* The prime implicant chart shows what prime implicants cover which minterms.
5. Select the minimum number of largest prime implicants that cover the minterms.

This procedure is illustrated in Figure 19.22 for the simultaneous minimization of two boolean functions.

Defining Terms

Base: The number of different values a single digit may have. The number a digit must be multiplied by to move it one digit to the left, also called the radix.

Binary-coded decimal (BCD): Each decimal digit is expressed individually in binary form.

Catenation: Symbols strung together to form a larger sequence, as the characters in a word and the digits in a number.

Code: The representation in one alphabet of something in another alphabet.

Complement: The quantity obtained by subtracting a number from the largest quantity that can be expressed in the specified number of digits in a given number system.

Conformal: The same arrangment of a set of quantities in two different contexts.

Digit: A character that represents quantitative information.

Don't care: A value that can be represented either as a minterm or a maxterm.

Fraction: Any number divided by a larger number.

Gray code: A set of codes having the property of logical adjacency.

Implicant: A first-order implicant is a pair of logically adjacent minterms. A second-order implicant is a set of logically adjacent first-order implicants and so on.

Integer: Any number that can be expressed solely in terms of digits.

***K*-map:** An arrangment of space into equal size units, each of which represents a minterm (or maxterm) such that each physically adjacent square is also logically adjacent.

Logically adjacent: Any two codes having the same number of digits for which they differ in the value of only one of the digits.

Macrotiming diagram: A graphical display showing how the waveforms vary with time, but with a time scale that does not have sufficient resolution to display the delays introduced by the individual basic elements of the digital circuit.

Maxterm: A function of a set of boolean variables that has a low value for only one combination of variable values and has a high value for all other combinations of the variable values.

Microtiming diagram: A graphical display showing how the waveforms vary with time, but with a time scale that has sufficient resolution to display clearly the delays introduced by the individual basic elements of the digital circuit.

Minterm: A function of a set of boolean variables that has a high value for only one combination of variable values and has a low value for all other combinations of the variable values.

Overflow: That part of a numerical operation result that does not fit into the allocated field.

Parity bit: An extra bit catenated to a code and given a value such that the total number of high bits is even for even parity and odd for odd parity.

Prime implicant: An implicant that is not part of a larger implicant.

Product of sums (PS): The AND combination of terms, which are OR combinations of boolean variables.

Radix: The number of different values that a digit can have in a number system.

Realization: A circuit that can produce the value of a function.

Real number: A number that has a fractional part and an integer part.

Reduced radix: The largest value a digit can have in a number system. It is one less than the radix.

Sum of products (SP): The OR combination of terms, which are AND combinations of Boolean variables.

Truth table: The table of values that a boolean function can have for which the independent variables considered as a multidigit number are arranged in consecutive order.

References

Hayes, J.P. 1993. *Introduction of Digital Logic Design.* Addison-Wesley, Reading, MA.

Humphrey, W.S., Jr. 1958. *Switching Circuits with Computer Applications.* McGraw-Hill, New York.

Hill and Peterson. 1974. *Introduction to Switching Theory and Logical Design,* 2nd ed. Wiley, New York.

Johnson and Karim. 1987. *Digital Design a Pragmatic Approach.* Prindle, Weber and Schmidt, Boston.

Karnaugh, M. 1953. The map method for synthesis of combinational logic circuits. *AIEE Trans. Comm. Elec.* 72 (Nov.): 593–599.

Mano, M.M. 1991. *Digital Design.* Prentice-Hall, Englewood Cliffs, NJ.

McClusky, E.J. 1986. *Logic Design Principles.* Prentice-Hall, Englewood Cliffs, NJ.

Mowle, F.J. 1976. *A Systematic Approach to Digital Logic Design.* Addison-Wesley, Reading, MA.

Nagle, Carrol, and Irwin. 1975. *An Introduction to Computer Logic,* 2nd ed. Prentice-Hall, Englewood Cliffs, NJ.

Pappas, N.L. 1994. *Digital Design* West, St. Paul, MN.

Roth, C.H., Jr. 1985. *Fundamentals of Logic Design,* 3rd ed. West, St. Paul, MN.

Sandige, R.S. 1990. *Modern Digital Design.* McGraw-Hill, New York.

Shaw, A.W. 1991. *Logic Circuit Design*. Saunders, Fort Worth, TX.
Wakerly, 1990. *Digital Design Principles and Practices*. Prentice-Hall, Englewood Cliffs, NJ.

Further Information

Further information on basic logic concepts and combinational logic design can be found in occasional articles in the following journals:

Lecture Notes in Computer Science (annual)
International Journal of Electronics (monthly)
IEEE Transactions on Education (quarterly)
IEEE Transactions on Computers (monthly)
IEEE Transactions on Software Engineering (monthly)
IEEE Transactions on Circuits and Systems 1. Fundamental Theory and Applications (monthly)

20

System Interfaces

Michael J. Tordon
Jayantha Katupitiya
The University of New South Wales

This chapter deals with asynchronous serial interfaces described by interface standards RS-232, RS-422, and RS-485 and with the general-purpose parallel interface bus described by IEEE-488 standard. The chapter also provides background information, terminology and parameters, which are important in the design of system interfaces for mechatronic systems.

20.1 Background

Modern mechatronic systems comprise a number of subsystems, which rely heavily on digital data communications. Different levels of complexity of these systems means that the requirement for data communications range from a simple communication between two devices to systems with a large number of subsystems, where each subsystem communicates directly or indirectly with other subsystems using a communication network. Depending on the proximity of subsystems, different requirements are placed on data communication channels, the physical implementation of channels, and interfaces between these devices. Figure 20.1 shows a schematic diagram of a simple data communication system connecting two devices.

A data source creates the data to be transmitted to the destination system and may convert the data into a specific form. The originating system usually does not create the data in a form suitable for transmission over transmission lines. This is left to the transmitter, which transforms the data into a signal suitable for transmission over a specific type of transmission line. The transmission line is generally implemented using electrical wiring but can involve a variety of physical medium including radio frequency, infrared, and sound

FIGURE 20.1 A schematic diagram of a simple data communication system.

signals. A transmission line provides a physical medium connecting the two systems. A receiver accepts the signal and converts it to the data form suitable to be passed onto the destination system. A data destination processes the data in order to recover the original information. From the previous information, it follows that even in the case of a simple data communication system a number of subsystems is involved in the communication task.

20.1.1 Terminology and Definitions

Interface: The common boundary between two subsystems is called an interface, and as can be seen in Figure 20.1, a number of interfaces can be involved even in a simple communication system.

Bit: The simplest form of data is one bit, which can take one of the two values 0 or 1, and hence is called binary data. All information in modern digital computers is stored in binary form.

Byte: A fixed number of bits (usually 8), which can be treated by a computer as a unit.

Character: Historically, the information is expressed in terms of characters. A character is a member of a character set. An example of a character set is the set of characters in the English language.

Character code: Individual characters from the selected character set are encoded in digital computers as binary numbers. One of the most widely used character set codes is the American Standard Code for Information Interchange (ASCII).

20.1.2 Serial versus Parallel

The basic unit of information to be transferred between subsystems is usually a character. For short distances, multiple parallel lines can be used to carry out simultaneous transmission of all the bits of a character. For the transmission of data over long distances, the cost of multiple data lines is often prohibitive and it is normal to serialize the data so that it can be passed over a single data path as a stream of bits.

20.1.3 Bit Rate versus Baud Rate

The speed of data transmission is usually expressed as a number of data bits transmitted per second and is called an effective bit rate with a unit bps. Larger units like kbps (1,000 bps) and Mbps (1,000,000 bps) are commonly used. The baud rate is a signaling rate and is expressed as a number of times per second that the signal transmitted over a data transmission line changes state. For systems using only two states, the signaling bit rate is equivalent to the baud rate. Distinction should be made between the effective data transmission bit rate and the signaling bit rate. In asynchronous serial communications, the effective data transmission bit rate can be significantly lower than the signaling bit rate because of the inclusion of start, stop, and parity bits. To maximize the transmission speed over a serial line, modern communication systems use signals with more than two states, thus achieving higher signaling bit rates. For example, if the transmission signal uses 16 states, then the signaling bit rate is four times higher than the baud rate. The terms baud rate, signaling bit rate, and effective data transmission rate are often used interchangeably which leads to confusion.

20.1.4 Synchronous versus Asynchronous

For both parallel and serial interface, the problem of synchronization must be solved. The communication over a transmission line can be done either in synchronous or asynchronous communication mode. In synchronous communication mode, the transmission of data is synchronized with a clock; thus, the transmission is occurring at regular time intervals. Since the data transmission takes place at fixed times, the completion of data transfer does not have to be acknowledged. In asynchronous transmission mode, the two systems are using clocks, that are not synchronized and may run at frequencies slightly out of step. Thus, for asynchronous systems, data validation requires a separate scheme called handshaking.

20.1.5 Data Flow-Control

Another problem in asynchronous communication systems is the speed of data processing. If one system is significantly slower in processing the data, a flow-control must be implemented to avoid data loss. Data flow-control may require additional handshaking. Similar problems may arise in multitasking systems in which, due to other tasks, the system is unable to handle incoming data during the period of high workload.

20.1.6 Handshaking

In order to ensure efficient transmission of data without errors, the sending system will use a separate signal to indicate that valid data has been presented to the interface. Because the instant at which the receiving device can process the data is not known, the sending device must wait for an acknowledgment signal before presenting new data to the interface. The handshaking can be implemented in either hardware or software.

20.1.7 Communication Protocol

Operation of a communication system is governed by a set of rules which must ensure reliable data transfer without errors and data loss. Such a set of rules is called a communication protocol.

20.1.8 Error Handling

Data transmitted over a communication line are subjected to noise and can thus be corrupted. Since it is essential to maintain the integrity of data, a number of different schemes for error detection have been developed. The simplest remedy after error detection is retransmission of the corrupted data. More sophisticated communication protocols can involve complex error correction schemes implemented at protocol level.

20.1.9 Simplex, Half-Duplex, Full-Duplex

In its simplest form, communication can be established with a single pair of wires. The data transmission mode, in which data can pass in one direction only, is called simplex or unidirectional channel. In most applications it is required that the communication takes place in both directions. If the cost of the data transmission line is high, it can be arranged that signals can pass in either direction over a single transmission line using additional circuitry on both ends of the transmission line but only in one direction at a time. This type of data communication mode is called half-duplex. Additional handshaking is required to implement the time sharing of the transmission line.

If signals can pass in either direction over a single transmission line simultaneously, the data communication mode is called full-duplex. An example of a full-duplex is a telephone line where the two channels are created as separate frequency bands. Cost permitting, two separate transmission lines can be established in which case the full-duplex communication is conducted over two simplex channels. This requires duplication of all the functions of a simple data communication system as shown in Figure 20.1.

20.1.10 Unbalanced versus Balanced Transmission

Implementation of the electrical transmission line can take two basic forms, unbalanced (single-ended) or balanced (differential). For unbalanced operation, a single conductor is used to carry the signal voltage, which is referenced to a signal ground. The signal ground is usually common return for all signals in the interface. Figure 20.2 shows an example of an unbalanced data transmission system with two channels and three wires. Symbol D represents driver and symbol R receiver. Unbalanced data transmission is

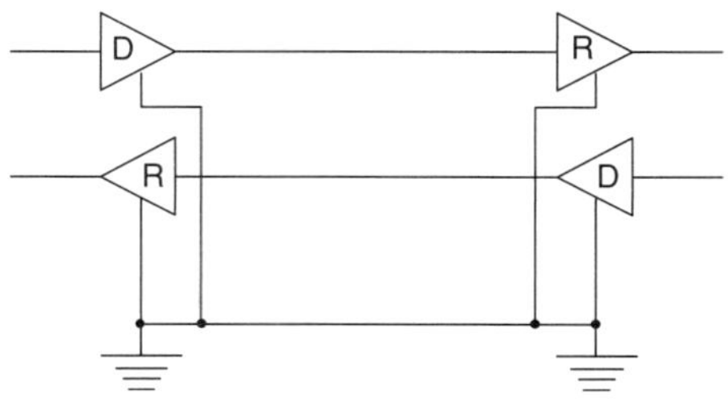

FIGURE 20.2 Example of an unbalanced data transmission.

FIGURE 20.3 Example of a balanced data transmission.

relatively inexpensive because, for multiple signal lines, only one common line is required; however, this type of interface is susceptible to induced and ground noise and is not suitable for high-speed communication over long distances. The ground noise is associated with voltage drop in a common return line, while the induced noise comes from interfering electromagnetic fields. Both types of noise can come from external sources or from neighboring transmission circuits. A remedy can be the use of coaxial cable, shielded cable, and/or the use of separate return lines for individual signals. These additional measures tend to increase the cost of the interface.

The balanced (differential) transmission mode has much better noise immunity than the unbalanced mode. Two complementary signal lines carry the data signal. The implementation often involves two single-ended drivers driving a twisted-pair transmission line. Figure 20.3 shows an example of balanced data transmission with two channels and five wires. As in Figure 20.2, symbol D represents driver and symbol R receiver. Symbol T represents termination resistor. Use of a termination resistor at the receiver end of the transmission line is critical for high-speed communications over long distances as unterminated transmission lines can cause severe distortion of signals. Both induced and ground noises appear on both conductors as common-mode signals that are rejected by the differential receiver. The differential signals carrying data are amplified while the common-mode noise signals are suppressed. As a result, the balanced data transmission lines can be used for longer distances with higher transmission rates. Both unbalanced and balanced interfaces shown in Figures 20.2 and 20.3 represent two simplex interfaces, which can form one full-duplex point-to-point (see below) communication channel.

A good source of information on individual drivers and receivers is provided in the data sheets and application notes of semiconductor manufacturers [1,2].

20.1.11 Point-to-Point versus Multi-Point

If communication takes place between two devices, we call such a communication link a point-to-point link. In mechatronic systems, it is often required for the master system to communicate with a number of subsystems. Cost permitting, a number of point-to-point data transmission lines can be implemented. In a point-to-point arrangement, the master system has a point-to-point connection to each individual subsystem, i.e., there is a separate port and communication line for each subsystem. This type of arrangement is shown in Figure 20.4. The connection can also be arranged as a multi-point connection in which

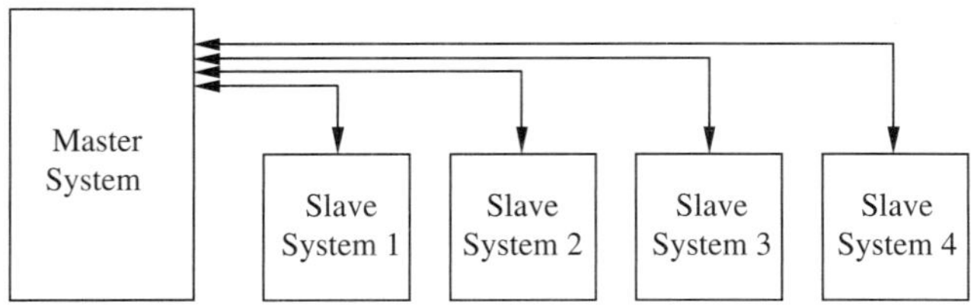

FIGURE 20.4 A point-to-point communication system with four subsystems.

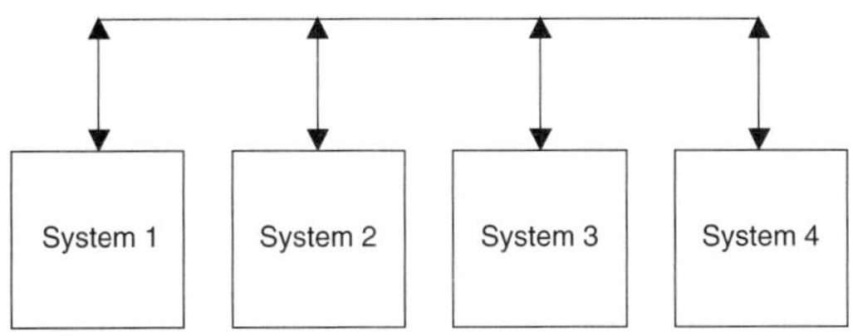

FIGURE 20.5 A multi-point communication system.

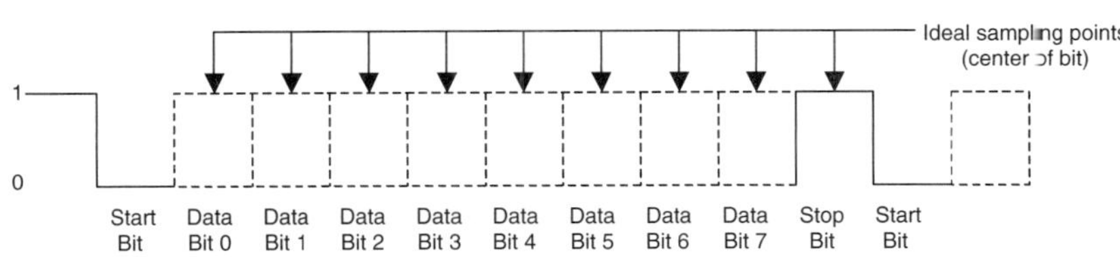

FIGURE 20.6 Asynchronous serial data format.

all devices are connected to a single transmission line, as shown in Figure 20.5. This arrangement is a data communication network arrangement where data can be transmitted from any device to any other device on the network. All devices on the network must be equipped with a receiver and a transmitter. Transmitters must have a tri-state (high output impedance) capability so that they do not provide additional load to the line. When transmitters are not transmitting, they are virtually disconnected from the transmission line. Complex communication protocol is required to manage individual transmitters on the network. The major advantage of a multi-point arrangement is usually the lower cost of the network compared to the individual communication links. The disadvantage is a more complex communication protocol (which must deal with the identity of the transmitting and receiving devices) and a more complex interface.

20.1.12 Serial Asynchronous Communications

In asynchronous serial communications, the data are transmitted at irregular intervals as a bit stream. Individual characters coded as binary numbers are converted to serial data streams, which are framed with start and stop bits. Optionally, a parity bit is added to the stream. In general, a computer represents information in parallel form such as bytes and words while the majority of communications with external devices takes place serially. The task of the parallel-to-serial and serial-to-parallel conversion is performed by a special integrated circuit called a universal asynchronous receiver transmitter (UART) as described later (see Figure 20.7).

Figure 20.6 shows an example of a typical data stream for asynchronous transmission. During idle time the line is in logical state 1 (for historical reasons also called "MARK"). The start of the data stream is always indicated by the start bit, which has logical value 0 (also called "SPACE"). The start bit is followed by 5–8 data bits representing a character. The data bits are followed by an optional parity bit. The stream is terminated by one or two stop bits with logical value 1, which can be followed by idle line or the start bit of the next character. The idle line corresponds to logical state 1. A parity bit is an extra bit inserted after the data bits and before the stop bit(s). It is set according to the parity information of the data in the stream. For example, if an even parity is used, the parity bit is set such that total number of ones in the data stream including the parity bit is even. The parity bit is used by the receiver for error checking. The task of the receiver is to detect the start of the data stream and to correctly sample individual bits in the stream. After the detection of the start bit the receiver should sample individual bits, ideally at the mid point of each bit, as shown in Figure 20.6. In the case of an ideal sampling, as shown in Figure 20.6, the receiver is said to have distortion tolerance of 50%. In practice, the receiver of a UART is sampling incoming signals using the Baud Rate Generator frequency, which is 16 times higher than the corresponding baud rate used for transmission. The uncertainty in the detection of the start bit will reduce the distortion tolerance by 6.25% (1/16) to 43.75% [3].

If, for example, the receiver clock is 1% slower than the clock of the corresponding transmitter, the sampling time of the first data bit will be delayed by 1.5% of the bit time and the sampling time of the stop bit will be delayed by 9.5%. In this case, the distortion tolerance would be further reduced to 34.25%. If the receiver clock is slower by 5%, then the receiver may detect the start bit of the next character instead of the stop bit of the current character. This results in a framing error. The above example shows the significance of the accuracy of the clock speed and the reason why the data stream must be kept short in asynchronous transmission.

Other factors affecting the error-free communications include length and type of transmission line, speed of communications, parameters of line drivers, termination of transmission line, and the level of noise in the communication system.

20.1.13 The Universal Asynchronous Receiver Transmitter

The basic function of the UART is to facilitate parallel-to-serial and serial-to-parallel data conversion. The UART usually contains one transmitter and one receiver. The receiver and transmitter can operate simultaneously and independently. The UART can operate in full-duplex or half-duplex mode.

Parallel data from the host computer are converted to an asynchronous serial bit stream. The UART automatically adds a start bit, an optional parity bit, and the programmed number of stop bits, and sends the stream out through the transmitter serial data output (TxD) output pin. The parallel data are converted to a serial stream with the least significant bit shifted out first. Figure 20.7 shows a typical arrangement for UART. As can be seen, the UART uses TTL (transistor transistor logic) compatible interface. The TIA/EIA (see later) transmission line drivers and receivers are specific to a particular interface; thus, changing system interface means changing the transmission medium and the relevant drivers and receivers. The use of UART is independent of the transmission medium.

Serial data received on the receiver serial data input (RxD) pin is converted to parallel data. In the process the UART checks the start bit, parity bit (if any), and stop bit and reports any error conditions. Note that the UART is capable of generating all signals required for successful bit-serial asynchronous communications.

The UART can also report a number of error conditions, including receiver overrun, parity error, framing error, and break error. Receiver overrun error occurs when the bytes are received faster than the computer

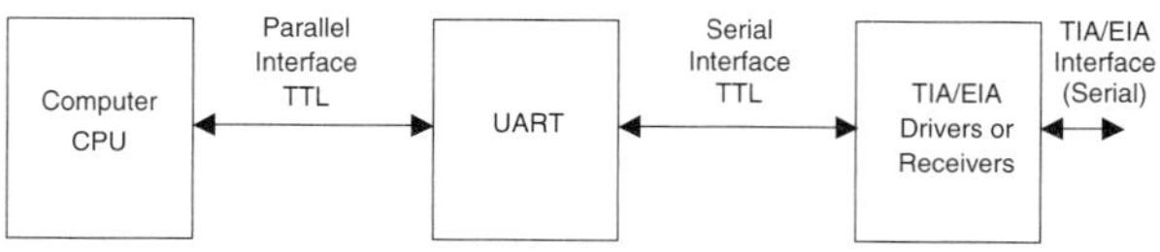

FIGURE 20.7 Typical arrangement for the UART.

processes them. The parity error is indicated if the parity of the bit stream changed during the communication process. Framing error is reported if the sampled stop bit is not at logic 1 level. The break error is reported if the communication line is idle for the time equivalent to the duration of at least one character.

Older types of UART devices such as 8250, 16450 had only one byte FIFO (first in first out) buffer and thus it was easy to overrun the receiver buffer. More recent devices are equipped with larger buffers providing more efficient communications. For example, device 16550D from National Semiconductor has a 16-byte receiver buffer and a 16-byte transmitter buffer and can operate at speeds up to 1.5 Mbps. Modern UARTs can also automatically handle tasks pertaining to multi-drop systems on a network.

20.2 TIA/EIA Serial Interface Standards

20.2.1 RS-232 Serial Interface

The RS-232 (Recommended Standard) was originally developed in 1962 by the Electronic Industries Association (EIA) as an interface between a computer and communication equipment. It is now jointly maintained by the Telecommunication Industries Association (TIA) and the EIA. The current version is designated as TIA/EIA 232-F (sixth revision) [4]. The Consulting Committee for International Telegraphs and Telephones (CCITT) issues recommendations that cover interfaces equivalent or similar to those issued by TIA/EIA.

Rapid development of computers created a demand for computer-to-computer communications over long distances. The switched public telephone network provided a readily available infrastructure for the communication task. Because computers generate digital data while the telephone network was designed for the transmission of voice signal, the digital signals from the computer had to be converted to a modulated signal which can be transmitted over the analog network. Modems (modulator/demodulator) are used to convert the digital signal into a modulated analog signal that is transmitted over the telephone line and converted back to digital signal by the modem at the other end of the telephone line. The RS-232 was designed as an interface between a computer and a modem. The formal name of the RS-232 standard is "Interface Between Data Terminal Equipment and Data Communication Equipment Employing Serial Binary Data Interchange," in which the Data Terminal Equipment (DTE) represents the computer and the Data Communication Equipment (DCE) represents the modem. Figure 20.8 shows an example of the RS-232 interface in the system providing computer-to-computer communication over the switched telephone network. The computers at each end represent DTE and the modems represent DCE.

The RS-232 interface standard specifies mechanical, electrical, and functional characteristics of the DTE/DCE interface. The CCITT V.24 interface describes equivalent functional characteristics and relies on other standards for mechanical and electrical characteristics of the interface. The RS-232 standard is widely used in applications where it provides a direct point-to-point connection between two computers or computers and field elements of mechatronic systems in which case we are dealing with DTE to DTE interface. As this is a situation where a modem is not required, the cable used to connect a DTE to another DTE is called a "null modem" cable, which has internal built-in connections to fake the presence of a modem.

The mechanical characteristic is concerned with the actual physical connection of the DTE and DCE and involves specification of pin assignments and genders of the connectors. The RS-232 standard does not specify a connector type, but it is customary to use either 25-pin D-type (DB-25) connector, which

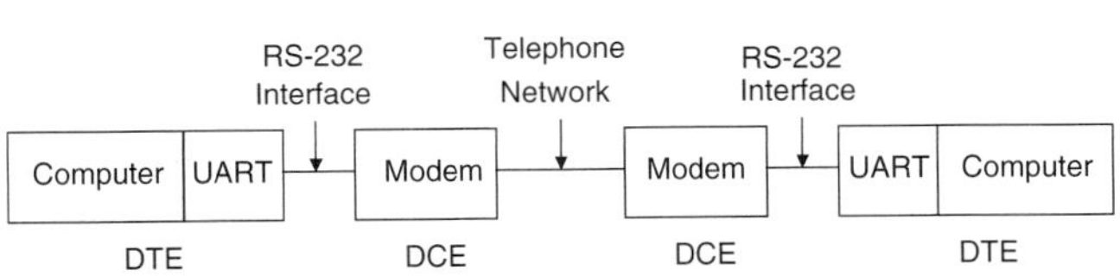

FIGURE 20.8 Data communication over a telephone network.

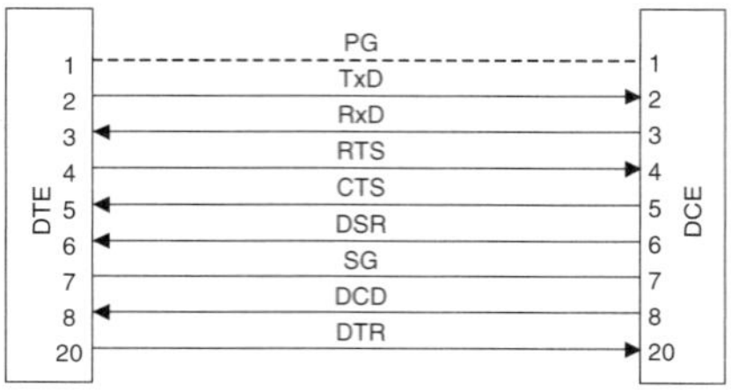

FIGURE 20.9 Pin assignment between a DTE and DCE.

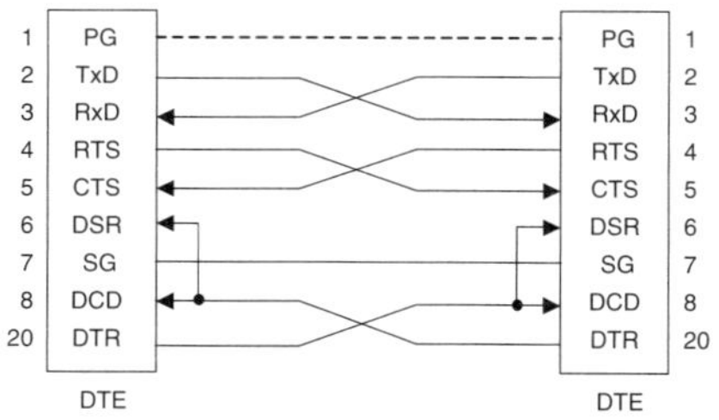

FIGURE 20.10 Example of a null modem cable pin assignment.

can accommodate all 25 pins, listed in the standard. In practice, a smaller number of pins are used; thus, as an alternative, a 9-pin D-type connector (DB-9) is often used. Please note that the pin assignment for DB-9 connector is not specified by RS-232 and is different from DB-25 pin assignment.

Figure 20.9 shows DB-25 connector pin assignments and the interconnection of selected circuits between a DTE and a DCE. Figure 20.10, on the other hand, shows an example of the DB-25 connector pin assignments and interconnection of selected circuits between two computers, i.e., two DTEs.

20.2.2 Functional Description of Selected Interchange Circuits

A full description of all signals as specified by the RS-232 standard is beyond the scope of this chapter. The reader is referred to the relevant standard [4]. We will describe the most common signals used in DTE/DCE and DTE/DTE interface. Please note that with the exception of the Protective Ground circuit and the Signal Ground circuit the circuits carry signals unidirectionally, as shown by arrows in Figure 20.9. Functional characteristics specify the functions that are performed by individual interchange circuits.

Protective Ground (PG). This line ensures that the chassis of the DTE and DCE are on the same potential.

Transmitted Data (TxD). Transmission line-signal originating from the DTE propagates to DCE.

Received Data (RxD). Receiver line-signal originating from the DCE propagates to DTE.

Request to Send (RTS). This signal is used to condition DCE for data transmission. On a half-duplex channel the signal controls the direction of data transmission of the DCE (transmit or receive). On a one-way-only channel (simplex) and on the full-duplex channels this signal controls the transmit state of the DCE (transmit or nontransmit state). A signal originating from the DTE propagates to DCE.

Clear to Send (CTS). This signal indicates that the DCE is ready to receive and is the response to the asserted RTS signal. The signal is originating from the DCE and propagates to DTE.

Data Set Ready (DSR). This signal indicates that the DCE is ready to operate. The signal is originating from the DCE and propagates to DTE.

Signal Ground (SG). This line is a common ground return line for all other signals.

Data Carrier Detect (DCD). This signal indicates that the DCE is receiving a valid modulated signal from the DCE at the other end. The signal is originating from the DCE and propagates to the DTE.

Data Terminal Ready (DTR). This signal indicates that the DTE is powered up and ready to operate. This signal is originating from the DTE and propagates to DCE.

The RS-232 specifies unbalanced, unidirectional, point-to-point interface. The interconnection is done over a set of wires referred to as interchange circuits. The electrical characteristics specify voltage levels of signals, rate of change of signals, and line impedance of interchange circuits. The standard specifies Nonreturn to Zero (NRZ) coding of digital signals.

The standard requires that the drivers be designed such that for the terminator load resistance between 3 and 7 kΩ the drivers should be capable of delivering high-level voltages between +5 and +15 V and low voltages between −5 and −15 V. The electrical signals are designed to provide a 2 V margin in signaling levels. The receiver signals are defined as +3 to +15 V for high voltage and as −3 to −15 V for low voltage.

It should be noted that for the data interchange circuit the high level voltage is defined as logic 0 (SPACE), while the low level voltage is defined as logic 1 (MARK). For control signals, on the other hand, the high level voltage defines the ON state while the low level voltage defines the OFF state. The maximum rate of change of signal allowed on both data and signal lines is 30 V/μs.

The original standard has also specified the maximum length of cable as 15 m. This specification was replaced by the specification of the maximum allowed capacitive load of 2500 pF in the EIA/TIA-232-D. The maximum cable length is determined by the capacitance of the cable per unit length; thus, this parameter now defines indirectly the length of the interface cable. The RS-232 interface is rated at signaling rates in the range from 0 to 20 kbps. It should be noted that in practice a good design would allow greater distances and greater data rates than the ones specified by the standard.

20.2.3 RS-422 and RS-485 Interfaces

The TIA/EIA-422-B standard "Electrical Characteristics of Balanced Voltage Digital Interface Circuits" [5] defines electrical characteristics of RS-422 interface. The RS-422 specifies a unidirectional, single driver, terminated balanced interface. The standard allows multiple receivers (up to 10) on one line. Figure 20.3 illustrates a typical point-to-point application of RS-422. As a result of improved noise immunity the RS-422 interface supports data rates up to 10 Mbps and cable length up to 1200 m, although not simultaneously. The maximum data rate of 10 Mbps is supported on a cable length up to 12 m, while a cable length of 1200 m supports data rates up to 100 kbps. Observe that the product of cable length and data rate is a limiting parameter of the interface. The transmission medium is a twisted-pair transmission line.

The TIA/EIA-485-A standard "Standard for Electrical Characteristics of Generators and Receivers for Use in Digital Multipoint Systems" [6] defines electrical characteristics of RS-485 interface. The RS-485 is a unique standard, which allows multiple nodes to communicate bidirectionally over a single twisted-pair transmission line. The RS-485 standard defines a low cost, multipoint balanced interface with electrical characteristic, supported cable types, cable length and data rates equivalent to those specified by the RS-422 standard. RS-485 parts are backward compatible and interchangeable with their equivalent RS-422 parts; however, RS-422 parts should not be used in RS-485 systems. RS-422 is usually used in point-to-point full-duplex communication systems, while RS-485 is used in multipoint half-duplex communication systems. The distinguishing feature of RS-485 drivers is their TRI-STATE capability, which allows the use of multiple drivers. The RS-485 has improved driver capability and input voltage range and supports up to 32 devices (drivers and/or receivers) on a single transmission line. Figure 20.11 shows a typical RS-485 multipoint application.

It should be noted that both RS-422 and RS-485 are electrical standards only. They do not specify mechanical or functional requirements.

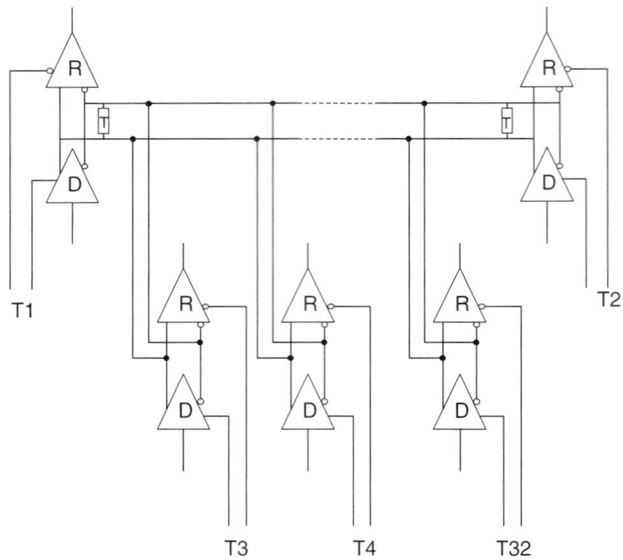

FIGURE 20.11 Example of an RS-485 multipoint application.

As mentioned earlier, data sheets and application notes of component suppliers provide an excellent source of information [1,2]. An excellent overview of practical data communications and interfacing for instrumentation and control is provided in reference 7. Detailed discussion of design aspects of serial communications and interfacing based on RS-232 and RS-485 is given in reference 8 It is recommended that for the full specification the designer should consult the relevant standards [4–6]. Additional information including design recommendations can be found in [9–11]. A good introduction and background theory to data communication and computer networks can be found in reference 12.

20.3 IEEE 488—The General Purpose Interface Bus

20.3.1 Introduction

The interface described by IEEE 488 standard, which will be referred to as general purpose interface bus (GPIB) in this chapter, is used to connect instruments to test and measurement systems. Examples of such instruments are digital voltmeters, storage oscilloscopes, printers, and plotters. In general, these instruments are called GPIB devices. These devices operate under the coordination of a controller. Most modern systems consist of a cluster of such devices connected to one or more computers. In such a system, one of the computers will become the controller.

Historically, the interface was developed by Hewlett–Packard in 1965. At that time, the interface was called HPIB, and a general standard did not exist. In 1975, it was formulated as IEEE 488 and was called IEEE Standard Digital Interface for Programmable Instrumentation. The standard specified the electrical, mechanical, and hardware aspects, i.e., the signals, their functioning, and purpose. Instrument manufacturers used the interface freely without adhering to a standard protocol in communicating with instruments. Instruments meant for the same purpose, yet manufactured by different manufacturers required widely varied commands. Some instruments made measurements in response to a command, while some other instruments of similar type made measurements without a command at all. Further, there were no agreed data formats between instruments sending data and instruments receiving data. This situation led to the development of an extension to the IEEE 488 standard. The new standard was published in 1987 and was called IEEE 488.2 Standard Codes, Formats, Protocols, and Common Commands for Use with IEEE 488.1 (1987) [13], where IEEE 488.1 is the new name for the original IEEE 488 standard.

The IEEE 488.2 compliant devices must present data through data formats and codes specified in the standard. The standard also specifies a minimum set of mandatory control sequences or commands and suggests a few other optional commands. It also provides a standard status-reporting model that must be implemented by the instrument manufacturers so that determining the status of instruments will be easier for the instrument programmers.

Although not yet a standard, The Standard Commands for Programmable Instrumentation (SCPI) put together in 1990 agrees upon a standard set of commands for various instrument categories. Accordingly, all digital voltmeters manufactured by different manufacturers will respond to the same GPIB command.

20.3.2 GPIB Hardware

This section describes the electrical and mechanical specifications of the GPIB interface as well as the signal description and their purpose.

All GPIB devices are connected using a special cable with each end having the male as well as the female ends of the connector. This permits piggyback connections of cables. The devices can be connected either in a chained manner (i.e., device B connected to device A, device C connected to device B, etc.) or in a star configuration (i.e., device A, B, C, etc. connected to a common node). The connection configurations are shown in Figure 20.12.

A maximum of 15 devices can be connected to the bus. The maximum separation between two devices is 4 m with an average separation of not more than 2 m. At least two-thirds of the devices connected must be powered on.

The GPIB cable consists of 24 wires. Eight of these lines are data lines, while three lines are used for handshaking. Another five lines are used for interface management and the remaining eight lines are ground lines. Among the ground lines are a cable shield line, a signal ground line, three ground return lines for the handshaking signals, and three other ground return lines for three of the interface management lines. All signals used are standard TTL signal levels with negative logic. The handshake lines and interface management lines are given in Table 20.1.

The operation of the individual lines is not important to the average user or programmer as their usage is taken care of by the controllers and the instruments that comply with IEEE 488.2 standard.

TABLE 20.1 Handshaking and Interface Management Lines

Handshaking Lines		Interface Management Lines	
NRFD	Not ready for data	ATN	Attention
NDAC	Not data accepted	IFC	Interface clear
DAV	Data valid	REN	Remote enable
		SRQ	Service request
		EOI	End of identify

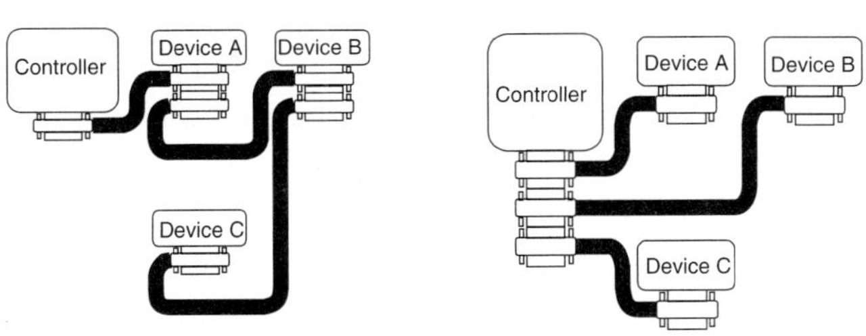

FIGURE 20.12 Linear and star configurations of connecting GPIB devices to a controller.

20.3.3 Controllers, Talkers, and Listeners

The controller carries out the general management of the bus. While there can be many controllers connected to the GPIB network, there can be only one controller-in-charge (CIC) which manages the bus at that given time. All information sent out by the controller on the data lines are called "commands" and all information sent out by other devices are termed "data." The GPIB devices that send data any time are called "talkers" and the devices that receive data are called "listeners." While there can be more than one listener operating at any given time, there can be only one talker operating at any given time. A system can have permanent talkers and permanent listeners; however, if the capability exists, a GPIB device can be a listener at one time and a talker at another time. A brief explanation of the signal lines is given below, as it would enhance our understanding of the operation of GPIB interface.

20.3.4 Interface Management Lines

20.3.4.1 Attention (ATN)

The ATN line is controlled by the CIC. When asserted, the signals on the data lines constitute a command signal and all devices must listen. When unasserted, the signals on the data lines represent data and are generally sent by a talker to one or more listeners.

20.3.4.2 Interface Clear (IFC)

The IFC line is asserted by the CIC to reset the GPIB bus. Upon receipt of this signal, all GPIB devices on the bus will initialize themselves.

20.3.4.3 Remote Enable (REN)

GPIB devices can be controlled either locally or remotely. The CIC asserts the REN line to bring all GPIB devices under remote programming mode. Thus, for example, the change of scale of a DVM can be carried out by a GPIB command instead of a front panel control.

20.3.4.4 Service Request (SRQ)

Any device other than a controller can asynchronously assert the SRQ line requesting service from the controller. The controller monitors the SRQ line and polls all devices to determine the device or devices requiring service.

20.3.4.5 End of Identity (EOI)

The EOI signal is used by a talker to indicate the end of the data message of the talker. It indicates to the listener(s) the end of the receiving data record.

20.3.5 Handshake Lines

In general, a data transfer with complete handshake gets through three stages: request or preparedness, data transfer, and acknowledgment. On some systems, where the stability of data on the data bus is questionable, a data valid signal may also be provided. On the GPIB bus, when a talker has to send data to a listener, the controller must address a device and instruct it to be the talker and then address one or more other devices and instruct them to be listeners. See Section 20.3.7 for addressing of GPIB devices.

20.3.5.1 Not Ready for Data (NRFD)

The NRFD line is controlled by the controller when sending commands or by the talker when sending data. A device that has been instructed to be a listener will unassert NRFD to indicate to the talker that it is ready to receive data. Of all the listeners, the slowest device will be the last to unassert NRFD and thus control the speed of data transfer.

20.3.5.2 Data Valid (DAV)

When all listeners have indicated their readiness to receive data by unasserting NRFD, the talker (or the controller when sending commands) will assert a DAV signal to indicte to all listeners that the data on the data lines DIO1-DIO8 are stable and may be read by the listeners. In response to a DAV signal, the listeners may assert NRFD to halt any further data transmission by talkers until the data already transmitted has been received.

20.3.5.3 Not Data Accepted (NDAC)

The NDAC line driven by all listeners is the acknowledgment signal. When data has been received by all listeners, the NDAC line will be unasserted. The talker can then remove the data and unassert the DAV signal.

20.3.6 Data Lines DIO1–DIO8 (8 lines)

The data lines are controlled by the controller when issuing commands or by the talker. As soon as the controller instructs a particular device to be a talker, that device will place data on the lines DIO1–DIO8 and will wait for at least T1 seconds.

20.3.7 Addressing of GPIB Devices

All GPIB devices connected to a GPIB bus must have a unique GPIB address. A device can have a primary address as well as a secondary address. Most devices use a primary address only. The addresses are in the range 0–30 decimal. In general, the addresses on the instruments as well as the controller are set using switches. The controller instructs a device with a particular address to be a talker or a listener by sending a bit pattern on the data bus. The bit pattern is formed according to Table 20.2. The data bits are numbered from D7 to D0. The value of each bit is listed straight underneath. The letter A signifies 0 or 1. The five bits D4–D0 will form a bit pattern representing the address of the device. The letter X signifies a "don't care" bit, which is not used. TA will be set to 1 if the controller is instructing the device to be a talker. LA will be set to 1 if the controller is instructing the device to be a listener. For example, if a particular device has the address 15 (decimal) and if the controller is instructing that device to be the talker, then the controller must send the following bit pattern over the data lines DIO8-DIO1.

Device with address 15 be the talker 0 1 0 0 1 1 1 1 = 4F (hex)

Similarly,

Device with address 0 be the listener 0 0 1 0 0 0 0 0 = 20 (hex)

"Untalk" the current talker 0 1 0 1 1 1 1 1 = 5F (hex)

"Unlisten" all listeners 0 0 1 1 1 1 1 1 = 3F (hex)

Note that Untalk and Unlisten commands look very similar to Talk and Listen commands; however, the address 31 (decimal) does not exist. Therefore, the address 31 is used to affect Untalk and Unlisten.

For the controllers, IEEE 488.2 Standard provides the Required and Optional Control Sequences. All controllers that comply with IEEE 488.2 must support all mandatory commands. The standard also provides the "Controller Protocols." Protocols are formed by combining a set of control sequences. For example, FINDLSTN command will issue a set of control sequences to determine the existing listeners.

For the instruments, IEEE 488.2 specifies a set of mandatory commands and queries. For example, when the command "*RST" is received by IEEE 488.2 compliant instruments, they all must carry out an instrument reset. Similarly, upon receipt of "*STB?" command the instrument will send the status byte to the controller.

TABLE 20.2 Talker/Listener Addressing Commands

D7	D6	D5	D4	D3	D2	D1	D0
X	TA	LA	A	A	A	A	A

All mandatory common commands and queries, all required and optional control sequences as well as the controller protocols can be found in references 14 and 15.

References

1. Goldie, J., Summary of well known interface standards, Application note AN-216, National Semiconductor, 1998, www.national.com.
2. Goldie, J., Comparing EIA-485 and EIA-422-A line drivers and receivers in multipoint applications, Application note AN-759, National Semiconductor, 1998.
3. McNamara, J.E., *Technical Aspects of Data Communication,* 3rd ed., Digital Press, 1988.
4. TIA/EIA-232-F, Interface between data terminal equipment and data communication equipment employing serial binary data interchange, TIA, EIA, 1997.
5. TIA/ EIA-422-B, Electrical characteristics of balanced voltage digital interface circuits, TIA, EIA, 1995.
6. TIA/EIA-485-A, Standard for electrical characteristics of generators and receivers for use in digital multipoint systems, TEI, EIA, 1998.
7. Mackay, S.G., et al., *Data Communications for Instrumentation and Control,* IDC Techbooks, 2000.
8. Axelson, J., *Serial Port Complete,* Lakeview Research, Madison, 1998.
9. Goldie, J., Ten ways to bulletproof RS-485 interfaces, Application note AN-1057, National Semiconductor, 1996.
10. RS-422 and RS-485 Application Note, B&B Electronics Manufacturing Co., 1997, www.bb-elec.com.
11. DALLAS SEMICONDUCTOR, Application Note 83, Fundamentals of RS-232 Serial Communications, 1998.
12. Stallings, W., *Data and Computer Communications,* 6th ed., Prentice-Hall, Upper Saddle River, NJ, 2000.
13. ANSI/IEEE 488.1-1987 IEEE standard digital interface for programmable instrumentation. institution of Electrical and Electronic Engineers, New York, 1987.
14. ANSI/IEEE 488.2-1987 IEEE standard codes, formats, protocols and common commands. institution of Electrical and Electronic Engineers, New York, 1987.
15. ANSI/IEEE 488.2-1992 IEEE standard codes, formats, protocols and common commands, and standard commands for programmable instruments. Institution of Electrical and Electronic Engineers, New York, 1992.

21

Communications and Computer Networks

Mohammad Ilyas
Florida Atlantic University

The field of communications and computer networks deals with efficient and reliable transfer of information from one point to another. The need to exchange information is not new but the techniques employed to achieve information exchange have been steadily improving. During the past few decades, these techniques have experienced an unprecedented and innovative growth. Several factors have been and continue to be responsible for this growth. The Internet is the most visible product of this growth and it has impacted the life of each and every one of us. This chapter describes salient features and operational details of communications and computer networks.

The contents of this chapter are organized in several sections. Section 21.1 describes a brief history of the field of communications. Section 21.2 deals with the introduction of communication and computer networks. Section 21.3 describes operational details of computer networks. Section 21.4 discusses resource allocation mechanisms. Section 21.5 briefly describes the challenges and issues in communication and computer networks that are still to be overcome. Finally, Section 21.6 summarizes the chapter.

21.1 A Brief History

Exchange of information (communications) between two or more entities has been a necessity since the existence of human life. It started with some form and shape of human voice that one entity can create and other(s) can listen to and interpret. Over a period of several centuries, these voices evolved into languages. As the population of the world grew, more and more languages were born. For a long time, languages were used for face-to-face communications. If there were ever a need to convey some information (a message) over a distance, someone would be briefed and sent to deliver the message to a distant site. Gradually, additional methods were developed to represent and exchange the information. These methods included symbols, shapes, and eventually alphabets. This development facilitated information recording and use of nonvocal means for exchanging information. Hence, preservation, dissemination, sharing, and communication of knowledge became easier.

Until about 150 years ago, all communication was via wireless means and included smoke signals, beating of drums, and use of reflective surfaces for reflecting light signals (optical wireless). Efficiency of these techniques was heavily influenced by environmental conditions. For instance, smoke signals were not very effective in windy conditions. In any case, as we will note later, some of the techniques that were in use centuries ago for conveying information over a distance were similar to the techniques that we currently use. The only difference is that the implementation of those techniques is exceedingly more sophisticated now than it was centuries ago.

As the technological progress continued and electronic devices started appearing on the surface, the field of communication also started making use of the innovative technologies. Alphabets were translated into their electronic representations so that information could be electronically transmitted. Morse code was developed for telegraphic exchange of information. Further developments led to the use of the telephone. It is important to note that in earlier days of technological masterpieces, users would go to a common site where one could send a telegraphic message over a distance or could have a telephonic conversation with a person at a remote location. This was a classic example of resource sharing. Of course, human help was needed to establish a connection with remote sites.

As the benefits of the advances in communication technologies were being harvested, electronic computers were also emerging and making the news. Earlier computers were not only expensive and less reliable, they were also huge in size. For instance, the computers that used vacuum tubes were of the size of a large room and used roughly about 10,000 vacuum tubes. These computers would stop working if a vacuum tube burned out, and the tube would need to be replaced by using a ladder. On average, those computers would function for a few minutes before another vacuum tube's replacement was necessary. A few minutes of computer time was not enough to execute a large computer program. With the advent of transistors, computers not only became smaller in size and less expensive, but also more reliable. These aspects of computers resulted in their widespread applications. With the development of personal computers, there is hardly any side of our lives that has not been impacted by the use of computers. The field of communications is no exception and the use of computers has escalated our communication capabilities to new heights.

21.2 Introduction

Communication of information from one point to another in an efficient and reliable manner has always been a necessity. A typical communication system consists of the following components as shown in Figure 21.1:

- Source that generates or has the information to be transported
- Transmitter that prepares the information for transportation
- Transmission medium that carries the information from one end to the other
- Receiver that receives the information and prepares it for delivering to the receiver
- Destination that takes the information from receiver and utilizes it as necessary

The information can be generated in analog or digital form. Analog information is represented as a continuous signal that varies smoothly in time. As one speaks in a microphone, an analog voice signal is generated. Digital information is represented by a signal that stays at some fixed level for some duration of time followed by a change to another fixed level. A computer works with digital information that has two levels (binary digital signals). Figure 21.2 shows an example of analog and digital signals. Transmission of

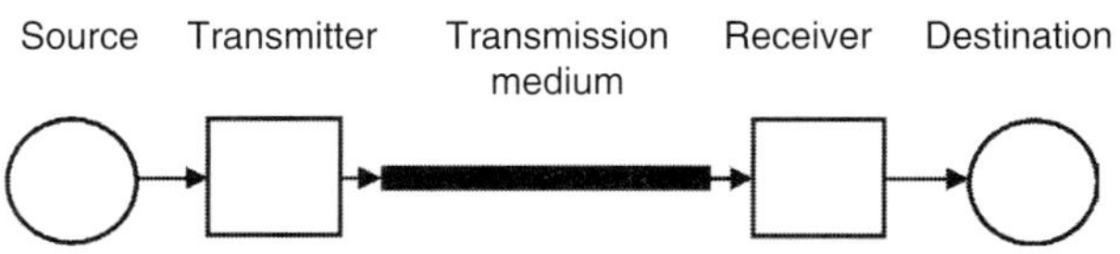

FIGURE 21.1 A typical communication system.

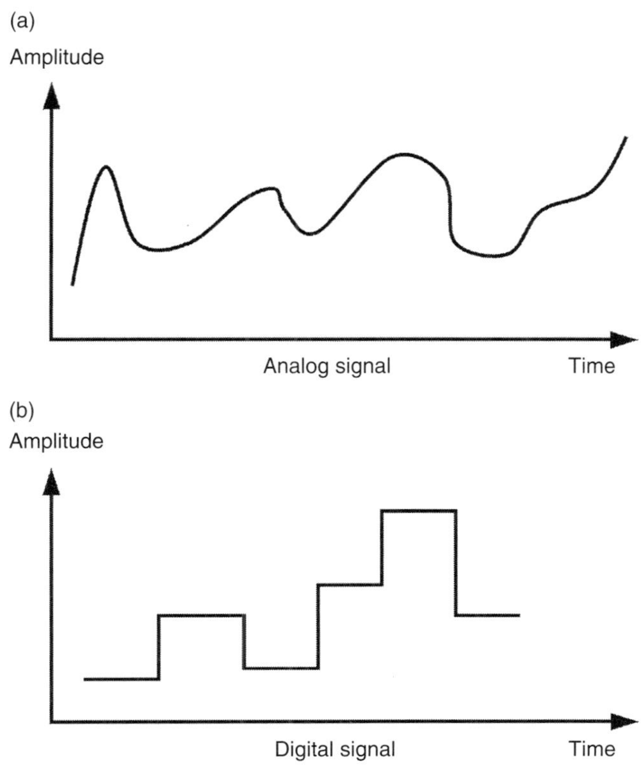

FIGURE 21.2 Typical analog (a) and digital (b) signals.

information can also be in analog or digital form. Therefore, we have the following four possibilities in a communication system [21]:

- Analog information transmitted as an analog signal
- Analog information transmitted as a digital signal
- Digital information transmitted as an analog signal
- Digital information transmitted as a digital signal

There may not be a choice regarding the form (analog or digital) of information being generated by a device. For instance, a voice signal as one speaks, a video signal as generated by a camera, a speed signal generated by a moving vehicle, and an altitude signal generated by the equipment in a plane will always be analog in nature; however, there is a choice regarding the form (analog or digital) of information being transmitted over a transmission medium. Transmitted information could be analog or digital in nature and information can be easily converted from one form to another.

Each of these possibilities has its pros and cons. When a signal carrying information is transmitted, it looses its energy and strength and gathers some interference (noise) as it propagates away from the transmitter. If the energy of the signal is not boosted at some intermediate point, it may attenuate beyond recognition before it reaches its intended destination. That will certainly be a wasted effort. In order to boost energy and strength of a signal, it must be amplified (in case of analog signals) and rebuilt (in case of digital signals). When an analog signal is amplified, the noise also becomes amplified and that certainly lowers expectations about receiving the signal at its destination in its original (or close to it) form. On the other hand, digital signals can be processed and reconstructed at any intermediate point and, therefore, the noise can essentially be filtered out. Moreover, transmission of information in digital form has many other advantages including processing of information for error detection and correction, applying encryption and decryption techniques to sensitive information, and many more. Thus, digital information transmission technology has become the dominant technology in the field of communications [9,18].

As indicated earlier, communication technology has experienced phenomenal growth over the past several decades. The following two factors have always played a critical role in shaping the future of communications [20]:

- Severity of user needs to exchange information
- State of the technology related to communications

Historically, inventions have always been triggered by the severity of needs. It has been very true for the field of communications as well. In addition, there is always an urge and curiosity to make things happen faster. When electricity was discovered and people (scattered around the globe) wanted to exchange information over longer distances and in less time, the telegraph was invented. Morse code was developed with shorter sequences (of dots and dashes) for more frequent alphabets. That resulted in transmission of messages in a shorter duration of time. The presence of electricity and the capability of wires to carry information over longer distances led to the development of devices that converted human voice into electrical signal, and thus led to the development of telephone systems. Behind this invention was also a need/desire to establish full-duplex (two-way simultaneous) communication in human voice. As use of the telephone became widespread, there was a need for a telephone user to be connected to any other user, and that led to the development of switching offices. In the early days, the switching offices were operated manually. As the state of the technology improved, the manual switching offices were replaced by automatic switching offices. Each telephone user was assigned a telephone number for identification purposes and a user able to dial the number for the purpose of establishing a connection with the called party. As the computer technology improved and the computers became easier to afford and smaller in size, they found countless uses including their use in communications. The computers not only replaced the automatic (electromechanical) switching offices, they were also employed in many other aspects of communication systems. Examples include conversion of information from analog to digital and vice versa, processing of information for error detection and/or correction, compression of information, and encryption/decryption of information.

As computers became more powerful, there were many other applications that surfaced. The most visible application was the amount of information users started sharing among themselves. The volume of information being exchanged among users has been growing exponentially over the last three decades. As users needed to exchange such a mammoth amount of information, new techniques were invented to facilitate the process. There was not only a need for users to exchange information with others in an asynchronous fashion, there was also a need for computers to exchange information among themselves. The information being exchanged in this fashion has different characteristics than the information being exchanged through the telephone systems. This need led to the interconnection of computers with each other and that is what is called computer networks.

21.3 Computer Networks

A computer network is an interconnection of computers. The interconnection forms a facility that provides reliable and efficient means of communication among users and other devices. User communication in computer networks is assisted by computers, and the facility also provides communication among computers. Computer networks are also referred to as computer communication networks. Interconnection among computers may be via wired or wireless transmission medium [5,6,10,13,18].

There are two broad categories of computer networks:

- Wide area networks
- Local/metropolitan area networks

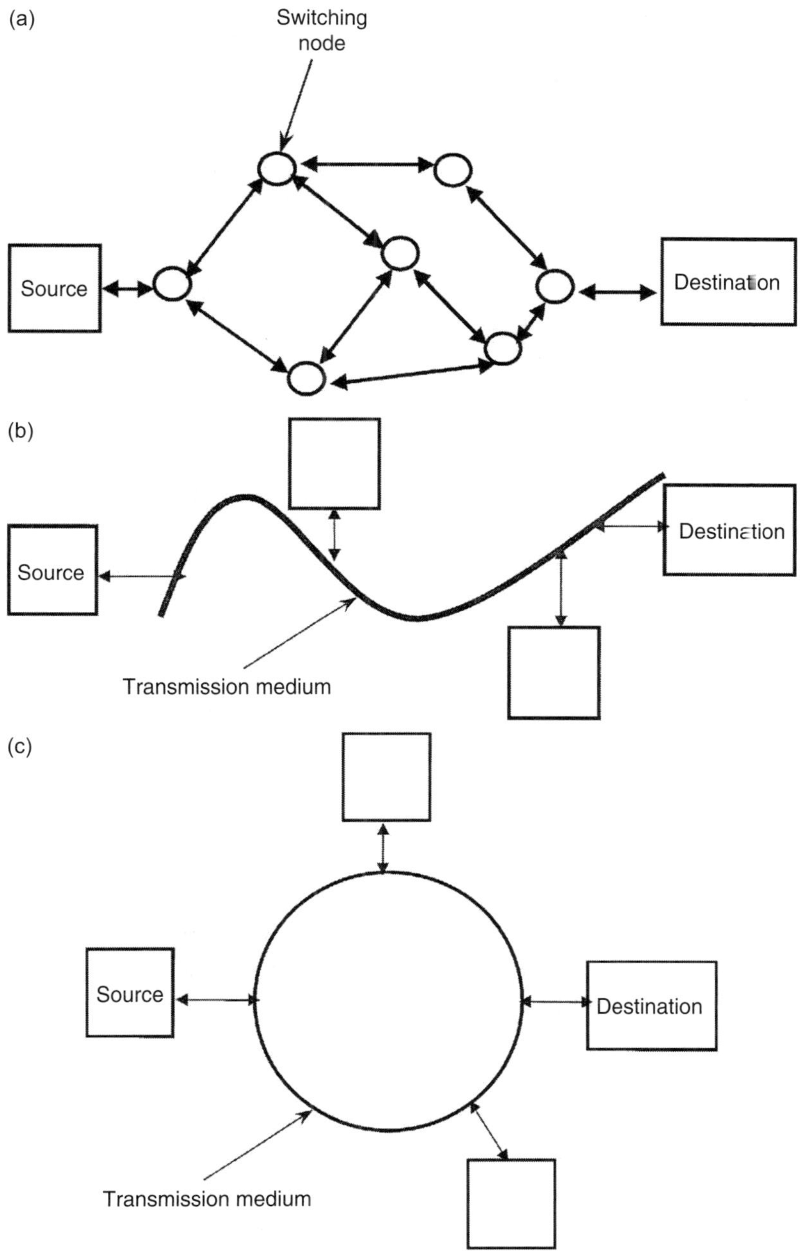

FIGURE 21.3 (a) A typical wide area computer communication network. (b) A typical local/metropolitan area communication bus network. (c) A typical local/metropolitan area communication ring network.

Wide area computer networks, as the name suggests, span a wider geographical area and essentially have a global scope. On the other hand, local/metropolitan area networks span a limited distance. Local area networks are generally confined to an industrial building or an academic institution. Metropolitan area networks also have limited geographical scope but it is relatively larger than that of the local area networks [19]. Typical wide and local/metropolitan area networks are shown in Figure 21.3.

Once a user is connected to a computer network, that user can communicate with any other user also connected to the network at some point. It is not required for a user to be connected directly to another user in order to communicate. In fact, in wide area networks, two communicating users will rarely be directly connected with each other. This implies that the users will be sharing the transmission links for exchanging their information. This is one of the most important aspects of computer networks. Sharing of resources improves utilization of the resources and is, of course, cost-effective as well. In addition to sharing the transmission links, the users will also share the processing power of the computers at the switching nodes, buffering capacity to store the information at the switching nodes, and any other resources that are connected to the computer network. A user who is connected to a computer network at any switching node will have immediate access to all the resources (databases, research articles, surveys, and much more) that are connected to the network as well. Of course, access to specific information may be restricted and a user may require appropriate authorization to access the information.

The information from one user to another may need to pass through several switching nodes and transmission links before reaching its destination. This implies that a user may have many options available to select one out of many sequences of transmission links and switching nodes to exchange information. That adds to the reliability of the information exchange process. If one path is not available, not feasible, or not functional, some other path may be used. In addition, for better and effective sharing of resources among several users, it is not appropriate to let any user exchange a large quantity of information at a time; however, it is not uncommon that some users may have a large quantity of information to exchange. In that case, the information is broken into smaller units known as packets of information. Each packet is sent toward its destination as a separate entity and then all packets are assembled together at the destination side to re-create the original piece of information [2].

Due to the resource sharing environment, users may not be able to exchange their information at any time they wish because the resources (switching nodes, transmission links) may be busy serving other users. In that case, some users may have to wait for some time before they begin their communication. Designers of computer networks should design the network so that the total delay (including wait time) is as brief as possible and that the total amount of information successfully exchanged (throughput) is as large as possible.

Many aspects must be addressed for enabling networks to transport users' information from one point to another. The major aspects are:

- Addressing mechanism to identify users
- Addressing mechanism for information packets to identify their source and destination
- Establishing a connection between sender and receiver and maintaining it
- Choosing a path or a route (sequence of switching nodes and transmission links) to carry the information from a sender to a receiver
- Implementing a selected route or path
- Checking information packets for errors and recovering from errors
- Encryption and decryption of information
- Controlling the flow of information so that shared resources are not over-taxed
- Informing the sender that the information has been successfully delivered to the intended destination (acknowledgment)
- Billing for the use of resources
- Ensuring that different computers running different applications and operating systems can exchange information
- Preparing information appropriately for transmission over a given transmission medium

This is not an exhaustive list of items that need to be addressed in computer networks. In any case, all such issues are addressed by very systematic and detailed procedures. The procedures are called communication protocols. The protocols are implemented at the switching nodes by a combination of hardware and software. It is not advisable to implement all these features in one module of hardware or software because that will become very difficult to manage. It is a standard practice that these features be divided into different smaller modules and then these modules can be interfaced together to collectively provide implementation of these features. International Standards Organization (ISO) has suggested dividing these features into seven distinct modules called layers. The proposed model is referred to as Open System Interconnection (OSI) reference model. The seven layers proposed in the OSI reference model are [2]:

- Application layer
- Presentation layer
- Session layer
- Transport layer
- Network layer
- Data link layer
- Physical layer

The physical layer deals with the transmission of information on the transmission medium. The data link layer handles the information on a single link. The network layer deals with the path or route of information from the switching node where the source is connected to the switching node where the receiver is connected. It also monitors end-to-end information flow. The remaining four layers reside with the user equipment. The transport layer deals with the information exchange from the source to the sender. The session layer handles the establishment of a session between the source and the receiver and maintains it. The presentation layer deals with the form in which information is presented to the lower layer. Encryption/decryption of information can also be performed at this layer. The application layer deals with the application that generates the information at the source side and what happens to it when it is delivered at the receiver side.

As the information begins from the application layer at the sender side, it is processed at every layer according to the specific protocols implemented at that layer. Each layer processes the information and appends a header and/or a trailer with the information before passing it on to the next layer. The headers and trailers appended by various layers contribute to the overhead and are necessary for transportation of the information. Finally, at the physical layer, the bits of information packets are converted to an appropriate signal and transmitted over the transmission medium. At the destination side, the physical layer receives the information packets from the transmission medium and prepares them for passing these to the next higher layer. As a packet is processed by the protocol layers at the destination side, its headers and trailers are stripped off before it is passed to the next layer. By the time information reaches the application layer, it should be in the same form as it was transmitted by the source.

Once a user is ready to send information to another user, he or she has two options. He or she can establish a communication with the destination prior to exchanging information or he can just give the information to the network node and let the network deliver the information to its destination. If communication is established prior to exchanging the information, the process is referred to as connection-oriented service and is implemented by using virtual circuit connections. On the other hand, if no communication is established prior to sending the information, the process is called connectionless service. This is implemented by using a datagram environment. In connection-oriented (virtual circuit) service, all packets between two users travel over the same path through a computer network and, hence, arrive at their destination in the same order as they were sent by the source. In connectionless service, however, each packet finds its own path through the network while traveling towards its destination. Each packet will therefore experience a different delay and the packets may arrive at their destination out of sequence. In that case, the destination will be required to put all the packets in proper sequence before assembling them [2,10,13].

As in all resource sharing systems, allocation of resources in computer networks requires careful attention. The main idea is that the resources should be shared among users of a computer network as fairly as possible. At the same, it is desired to maintain the network performance as close to its optimal level as possible. The fairness definition, however, varies from one individual to another and depends upon how one is associated with a computer network. Although fairness of resource sharing is being evaluated, two performance parameters—delay and throughput—for computer networks are considered. The delay is the duration of time from the moment information is submitted by a user for transmission to the moment it is successfully delivered to its destination. The throughput is the amount of information successfully delivered to its intended destination per unit time. Due to the resource sharing environment in computer networks, these two performance parameters are contradictory. It is desired to have the delay as small as possible and the throughput as large as possible. For increasing throughput, a computer network must handle increased information traffic, but the increased level of information traffic also causes higher buffer occupancy at the switching nodes and, hence, more waiting time for information packets. This results in an increase in delay. On the other hand, if information traffic is reduced to reduce the delay, that will adversely affect the throughput. A reasonable compromise between throughput and delay is necessary for the satisfactory operation of a computer network [10,11].

21.3.1 Wide Area Computer Networks

A wide area network consists of switching nodes and transmission links as shown in Figure 21.3a. Layout of switching nodes and transmission links is based on the traffic patterns and expected volume of traffic flow from one site to another site. Switching nodes provide the users access to a computer network and implement communication protocols. When a user is ready to transmit his or her information, the switching node, to which the user is connected, will establish a connection if a connection-oriented service has been opted. Otherwise, the information will be transmitted in a connectionless environment. In either case, switching nodes play a key role in determining the path of the information flow according to some well-established routing criteria. The criteria include performance (delay and throughput) objectives among other factors based on user needs. For keeping the network traffic within a reasonable range, some traffic flow control mechanisms are necessary. In late 1960s and early 1970s, when data rates of transmission media used in computer networks were low (a few thousand bits per second), these mechanisms were fairly simple. A common method used for controlling traffic over a transmission link or a path was an understanding that the sender would continue sending information until the receiver sent a request to stop. The information flow would resume as soon as the receiver sent another request to resume transmission. Basically the receiver side had the final say in controlling the flow of information over a link or a path. As the data rates of transmission media started increasing, this method was not deemed efficient. To control the flow of information in relatively faster transmission media, a sliding window scheme was used. According to this scheme, the sender will continuously send information packets but no more than a certain limit. Once the limit is reached, the sender will stop sending the information packets and will wait for the acknowledgment that the packets have been transmitted. As soon as an acknowledgment is received, the sender may send another packet. This method ensures that there are no more than a certain specific number of packets in transit from sender to receiver at any given time. Again, the receiver has control over the amount of information that the sender can transmit. These techniques for controlling the information traffic are referred to as reactive- or feedback-based techniques because the decision to transmit or not to transmit is based on the current traffic conditions.

Reactive techniques are acceptable in low to moderate data rates of transmission media. As the data rates increase from kilobits per second to megabits and gigabits per second, the situation changes. Over the past several years, there has been a manifold increase in data rates. Optical fibers provide enormously high data rates. Size of the computer networks has also experienced tremendous increase. The amount of traffic flowing through these networks has been increasing exponentially. Given that, the traffic control techniques used in earlier networks are not quite effective anymore [11,12,22]. One more factor that has added to the complexity of the situation is that users are now exchanging different types of information through the same network. Consider the example of the Internet. The geographical scope of the Internet is essentially global. Extensive use of optical fiber as transmission media provides very high data rates for exchanging information. In addition, users are using the Internet for exchanging any type of information they come across, including voice, video, and data. All these factors have essentially necessitated the use of a modified approach for traffic management in computer networks. The main factor leading to this change is that the information packets are moving so fast through the computer networks that any feedback-based (or reactive) control will be too slow to be of any use. Therefore, some preventive mechanisms have been developed to maintain the information traffic inside a computer network to a comfortable level. Such techniques are implemented at the sender side by ensuring that only as much information traffic is allowed to enter the network as can be comfortably handled by the networks [1,20,22]. Based on the users' needs and state of the technology, providing faster communications for different types of services (voice, video, data, and others) in the same computer network in an integrated and unified manner has become a necessity. These computer networks are referred to as broadband integrated services digital networks (BISDNs). BISDNs provide end-to-end digital connectivity and users can access any type of communication service from a single point of access. Asynchronous transfer mode (ATM) is expected to be used as a transfer mechanism in BISDNs. ATM is essentially a fast packet switching technique where information is transmitted in the form of small fixed-size packets called cells. Each cell is 53 bytes long and includes a header of 5 bytes. The information is primarily transported using a connection-oriented (virtual circuit) environment [3,4,8,12,17].

Another aspect of wide area networks is the processing speed of switching nodes. As the data rates of transmission media increase, it is essential to have faster processing capability at the switching nodes. Otherwise, switching nodes become bottlenecks and faster transmission media cannot be fully utilized. When transmission media consist of optical fibers, the incoming information at a switching node is converted from optical form to electronic form so that it may be processed and appropriately switched to an outgoing link. Before it is transmitted, the information is again converted from electronic form to optical form. This slows down the information transfer process and increases the delay. To remedy this situation, research is being conducted to develop large optical switches to be used as switching nodes. Optical switches will not require conversion of information from optical to electronic and vice versa at the switching nodes; however, these switches must also possess the capability of optical processing of information. When reasonably sized optical switches become available, use of optical fiber as transmission media together with optical switches will lead to all-optical computer and communication networks. Information packets will not need to be stored for processing at the switching nodes and that will certainly improve the delay performance. In addition, wavelength division multiplexing techniques are rendering use of optical transmission media to its fullest capacity [14].

21.3.2 Local and Metropolitan Area Networks

A local area network has a limited geographical scope (no more than a few kilometers) and is generally limited to a building or an organization. It uses a single transmission medium and all users are connected to the same medium at various points. The transmission medium may be open-ended (bus) as shown in Figure 21.3b or it may be in the form of a loop (ring) as shown in Figure 21.3c. Metropolitan area networks also have a single transmission medium that is shared by all the users connected to the network, but the medium spans a relatively larger geographical area, up to 150 km. They also use a transmission medium with relatively higher data rates. Local and metropolitan area networks also use a layered implementation of communication protocols as needed in wide area networks; however, these protocols are relatively simpler because of simple topology, no switching nodes, and limited distance between the senders and the receivers. All users share the same transmission medium to exchange their information. Obviously, if two or more users transmit their information at the same time, the information from different users will interfere with each other and will cause a collision. In such cases, the information of all users involved in a collision will be destroyed and will need to be retransmitted. Therefore, there must be some well-defined procedures so that all users may share the same transmission medium in a civilized manner and have successful exchange of information. These procedures are called medium access control (MAC) protocols.

There are two broad categories of MAC protocols:

- Controlled access protocols
- Contention-based access protocols

In controlled access MAC protocols, users take turns transmitting their information and only one user is allowed to transmit information at a time. When one user has finished his or her transmission, the next user begins transmission. The control could be centralized or distributed. No information collisions occur and, hence, no information is lost due to two or more users transmitting information at the same time. Examples of controlled access MAC protocols include token-passing bus and token-passing ring local area networks. In both of these examples, a token (a small control packet) circulates among the stations. A station that has the token is allowed to transmit information, and other stations wait until they receive the token [19].

In contention-based MAC protocols, users do not take turns transmitting their information. A user makes his or her own decision to transmit and also faces a risk of becoming involved in a collision with another station that also decides to transmit at about the same time. If no collision occurs, the information may be successfully delivered to its destination. On the other hand, if a collision occurs, the information from all users involved in a collision will need to be retransmitted. An example of contention- based MAC protocols is carrier sense multiple access with collision detection (CSMA/CD), which is used in Ethernet. In CSMA/CD, a user senses the shared transmission medium prior to transmitting its information. If the medium is sensed as busy (someone is already transmitting the information), the user will refrain from transmitting the information; however, if the medium is sensed as free, the user transmits the information. Intuitively, this MAC protocol should be able to avoid collisions, but collisions still do take place. The reason is that transmissions travel along the transmission medium at a finite speed. If one user senses the medium at one point and finds it free, it does not mean that another user located at another point of the medium has not already begun its transmission. This is referred to as the effect of the finite propagation delay of electromagnetic signal along the transmission medium. This is the single most important parameter that causes deterioration of performance in contention-based local area networks [11,19].

Design of local area networks has also been significantly impacted by the availability of transmission media with higher data rates. As the data rate of a transmission medium increases, the effects of propagation delay become even more visible. In higher speed local area networks such as Gigabit Ethernet, and 100-BASE-FX, the medium access protocols are designed to reduce the effects of propagation delay. If special attention is not given to the effects of propagation delay, the performance of high-speed local area networks becomes very poor [15,19].

Metropolitan area networks essentially deal with the same issues as local area networks. These networks are generally used as backbones for interconnecting different local area networks. These are high-speed networks and span a relatively larger geographical area. MAC protocols for sharing the same trans-mission media are based on controlled access. The two most common examples of metropolitan area networks are fiber distributed data interface (FDDI) and distributed queue dual bus (DQDB). In FDDI, the transmission medium is in the form of two rings, whereas DQDB uses two buses. FDDI rings carry information in one but opposite directions and this arrangement improves reliability of communication. In DQDB, two buses also carry information in one but opposite directions. The MAC protocol for FDDI is based on token passing and supports voice and data communication among its users. DQDB uses a reservation-based access mechanism and also supports voice and data communication among its users [19].

21.3.3 Wireless and Mobile Communication Networks

Communication without being physically tied-up to wires has always been of interest and mobile and wireless communication networks promise that. The last few years have witnessed unprecedented growth in wireless communication networks. Significant advancements have been made in the technologies that support wireless communication environment and there is much more to come in the future. The devices used for wireless communication require certain features that wired communication devices may not necessarily need. These features include low power consumption, light weight, and worldwide communication ability.

In wireless and mobile communication networks, the access to a communication network is wireless so that the end users remain free to move. The rest of the communication path could be wired, wireless, or combination of the two. In general, a mobile user, while communicating, has a wireless connection with a fixed communication facility and rest of the communication path remains wired. The range of wireless communication is always limited and therefore the range of user mobility is also limited. To overcome this limitation, the cellular communication environment has been devised. In a cellular communication environment, a geographical region is divided into smaller regions called cells, thus the name cellular. Each cell has a fixed communication device that serves all mobile devices within that cell. However, as a mobile device, while in active communication, moves out of one cell and into another cell, service of that connection is transferred from one cell to another. This is called the handoff process [7,16].

The cellular arrangement has many attractive features. As the cell size is small, the mobile devices do not need very high transmitting power to communicate. This leads to smaller devices that consume less power. In addition, it is well known that the frequency spectrum that can be used for wireless communication is limited and can therefore support only a small number of wireless communication connections at a time. Dividing communication regions into cells allows the use of the same frequency in different cells as long as they are sufficiently far apart to avoid interference. This increases the number of mobile devices that can be supported. Advances in digital signal processing algorithms and faster electronics have led to very powerful, smaller, elegant, and versatile mobile communication devices. These devices have tremendous mobile communication abilities including wireless Internet access, wireless e-mail and news items, and wireless video (though limited) communication on handheld devices. Wireless telephones are already available and operate in different communication environments across the continents. The day is not far when a single communication number will be assigned to every newborn and will stay with that person irrespective of his/her location.

Another field that is emerging rapidly is the field of ad hoc wireless communication networks. These networks are of a temporary nature and are established for a certain need and for a certain duration. There is no elaborate setup needed to establish these networks. As a few mobile communication devices come in one another's proximity, they can establish a communication network among themselves. Typical situations where ad hoc wireless networks can be used are in the classroom environment, corporate meetings, conferences, disaster recovery situations, etc. Once the need for networking is satisfied, the ad hoc networking setup disappears.

21.4 Resource Allocation Techniques

As discussed earlier, computer networks are resource sharing systems. Users share the common resources as transmission media, processing power and buffering capacity at the switching nodes, and other resources that are part of the networks. A key to successful operation of computer networks is a fair and efficient allocation of resources among its users. Historically, there have been two approaches to allocation of resources to users in computer networks:

- Static allocation of resources
- Dynamic allocation of resources

Static allocation of resources means that a desired quantity of resources is allocated to each user who may use it whenever he or she needs to. If the user does not use his/her allocated resources, no one else can. On the other hand, dynamic allocation of resources means that a desired quantity of resources is allocated to users on the basis of their demands and for the duration of their need. Once the need is satisfied, the allocation is retrieved. In that case, someone else can use these resources if needed. Static allocation results in wastage of resources, but does not incur the overhead associated with dynamic allocation. Which technique should be used in a given situation is subject to the concept of supply and demand. If resources are abundant and demand is not too high, it may be better to have static allocation of resources; however, when the resources are scarce and demand is high, dynamic allocation is almost a necessity to avoid the wastage of resources.

Historically, communication and computer networks have dealt with both situations. Earlier communication environments used dynamic allocation of resources when users walked to a public call office to make a telephone call or send a telegraphic message. After a few years, static allocation of res-ources was adopted, when users were allocated their own dedicated communication channels and these were not shared among others. In the late 1960s, the era of computer networks dawned with dynamic allocation of resources and all communication and computer networks have continued with this tradition to date. With the advent of optical fiber, it was felt that the transmission resources are abundant and can satisfy any demand at any time. Many researchers and manufacturers were in favor of going back to the static allocation of resources, but a decision to continue with dynamic resource allocation was made and that is here to stay for many years to come [10].

21.5 Challenges and Issues

Many challenges and issues are related to communications and computer networks that are still to be overcome. Only the most important ones will be described in this section.

High data rates provided by optical fibers and high-speed processing available at the switching nodes has resulted in lower delay for transferring information from one point to another. However, the propagation delay (the time for a signal to propagate from one end to another) has essentially remained unchanged. This delay depends only on the distance and not on the data rate or the type of transmission medium. This issue is referred to as latency vs. delay issue [11]. In this situation, traditional feedback-based reactive traffic management techniques become ineffective. New preventive techniques for effective traffic management and control are essential for achieving the full potential of these communication and computer networks [22].

Integration of different services in the same networks has also posed new challenges. Each type of service has its own requirements for achieving a desired level of quality of service (QoS). Within the networks any attempt to satisfy QoS for a particular service will jeopardize the QoS requirements for other services. Therefore, any attempt to achieve a desired level of quality of service must be uniformly applied to the traffic inside a communication and computer network and should not be intended for any specific service or user. That is another challenge that needs to be carefully addressed and its solutions achieved [13].

Maintaining security and integrity of information is another continuing challenge. The threat of sensitive information passively or actively falling into unauthorized hands is very real. In addition, proactive and unauthorized attempts to gain access to secure databases are also very real. These issues need to be resolved to gain the confidence of consumers so that they may use the innovations in communications and computer networking technologies to their fullest [13].

21.6 Summary and Conclusions

This chapter discussed the fundamentals of communications and computer networks and the latest developments related to these fields. Communications and computer networks have witnessed tremendous growth and sophisticated improvements over the last several decades.

Computer networks are essentially resource sharing systems in which users share the transmission media and the switching nodes. These are used for exchanging information among users that are not necessarily connected directly. There has been a manifold increase in transmission rates of transmission media and the processing power of the switching nodes (which are essentially computers) has also been multiplied. The emerging computer networks are supporting communication of different types of services in an integrated fashion. All types of information, irrespective of type and source, are being transported in the form of packets (e.g., ATM cells). Resources are being allocated to users on a dynamic basis for better utilization. Wireless communication networks are emerging to provide worldwide connectivity and exchange of information at any time.

These developments have also posed some challenges. Effective traffic management techniques, meeting QoS requirements, and information security are the major challenges that need to be surmounted in order to win the confidence of users.

References

1. Bae, J., and Suda, T., "Survey of traffic control schemes and protocols in ATM networks," *Proceedings of the IEEE*, Vol. 79, No. 2, February 1991, pp. 170–189.
2. Beyda, W., *Data Communications from Basics to Broadband*, 3rd., 2000.
3. Black, U., ATM: *Foundation for Broadband Networks*, Prentice-Hall, Englewood Cliffs, NJ, 1995.
4. Black, U., Emerging Communications Technologies, 2nd ed., Prentice-Hall, Englewood Cliffs, NJ, 1997.
5. Chou, C., "Computer networks in communication survey research," *IEEE Transactions on Professional Communication*, Vol. 40, No. 3, September 1997, pp. 197–208.
6. Comer, D., *Computer Networks and Internets*, Prentice-Hall, Englewood Cliffs, NJ, 1999.
7. Goodman, D., *Wireless Personal Communication Systems*, Addison-Wesley, Reading, MA, 1999.
8. Goralski, W., *Introduction to ATM Networking*, McGraw-Hill, New York, 1995.
9. Freeman, R., *Fundamentals of Telecommunications*, John Wiley & Sons, New York, 1999.
10. Ilyas, M., and Mouftah, H.T., "Performance evaluation of computer communication networks," *IEEE Communications Magazine*, Vol. 23, No. 4, April 1985, pp. 18–29.
11. Kleinrock, L., "The latency/bandwidth tradeoff in gigabit networks," *IEEE Communications Magazine*, Vol. 30, No. 4, April 1992, pp. 36–40.
12. Kleinrock, L., "ISDN-The path to broadband networks," *Proceedings of the IEEE*, Vol. 79, No. 2, February 1991, pp. 112–117.
13. Leon-Garcia, A., and Widjaja, I., *Communication Networks, Fundamental Concepts and Key Architectures*, McGraw Hill, New York, 2000.
14. Mukherjee, B., *Optical Communication Networks*, McGraw-Hill, New York, 1997.
15. Partridge, C., *Gigabit Networking*, Addison-Wesley, Reading, MA, 1994.
16. Rappaport, T., *Wireless communications*, Prentice-Hall, Englewood Cliffs, NJ, 1996.
17. Schwartz, M., *Broadband Integrated Networks*, Prentice-Hall, Englewood Cliffs, NJ, 1996.
18. Shay, W., Understanding Communications and Networks, 2nd ed., PWS, 1999.
19. Stallings, W., *Local and Metropolitan Area Networks*, 6th ed., Prentice-Hall, Englewood Cliffs, NJ, 2000.
20. Stallings, W., *ISDN and Broadband ISDN with Frame Relay and ATM*, 4th ed., Prentice-Hall, Englewood Cliffs, NJ, 1999.
21. Stallings, W., *High-Speed Networks, TCP/IP and ATM Design Principles*, Prentice-Hall, Englewood Cliffs, NJ, 1998.
22. Yuan, X., A study of ATM multiplexing and threshold-based connection admission control in connection-oriented packet networks, Doctoral Dissertation, Department of Computer Science and Engineering, Florida Atlantic University, Boca Raton, FL 33431, August 2000.

22

Fault Analysis in Mechatronic Systems

Leila Notash

Thomas N. Moore

Queen's University

22.1 Introduction

As the degree of automation increases, particularly intelligent automation, high reliability, fail-safe and fault tolerance become an essential part of the mechatronic system design. A mechatronic system is reliable if no failure and malfunction could result in an unsafe system; is safe if it causes no injury or damage to the operator, environment and system itself; is fail-safe if the system could be stopped safely after the failure; and is fault tolerant if the system could complete its task safely after any failure.

Fault/failure corresponds to any condition or component/subsystem degradation (sharp or graceful degradation) that affects the performance of a system such that the system cannot function as it is required. As the application of the mechatronic systems expands to areas such as highly dynamic/unstructured or space/remote environments, medical and high-speed applications, the necessity for the system to be fail-safe (could stop with no harm to the environment, operator, and itself) and fault tolerant (tolerate the failure and complete the assigned task) increases.

A mechatronic system is called fault tolerant if after any failures there will be no interruption in the task/operation of the system. Fault tolerance and high reliability could be achieved by using high quality components, through design and robust control, and by incorporating redundancy in the design of mechatronic systems. A mechatronic system consists of mechanical, electrical, computer, and control (hardware and software) subsystems. Therefore, their redundancy could be in the form of hardware redundancy (redundancy in sensing, actuation, transmission, communication, and computing), software redundancy, analytical redundancy, information redundancy, and time redundancy.

22.2 Tools Used for Failure/Reliability Analysis

The failure analysis techniques could be classified as inductive techniques and deductive techniques (Wolfe, 1978). Inductive techniques, such as decision or event trees and failure modes and effects analysis (FMEA), consider the possible states of components/subsystems and determine their effects on the system, that is, identify the undesired state. Deductive analyses, such as fault tree analysis (FTA), involve investigation of possible desired state of the overall system and identify the component states that contribute to the occurrence of the undesired state, that is, describe how the undesired state is achieved.

The event tree method is a pictorial representation of all the events (success or failures) that can occur in a system. Similar to other techniques, the event tree method can be used for systems in which all subsystems/components are continuously operating. This method is also widely used for systems in which some or all of the subsystems/components are in a standby mode with sequential operational logic and switching, such as safety oriented systems (Billinton and Allan, 1983).

FMEA is a bottom-up qualitative technique used to evaluate a design by identifying possible failure modes and their effects on the system, occurrence of the failure modes, and detection techniques. The history of FMEA goes back to the early 1950s when the technique was utilized in the design and development of flight control systems (Dhillon, 1983). Since then it has been widely used in the industry for specific designed systems with known knowledge of their components, subsystems, functions, required performance and characteristics, and so on. Criticality analysis (CA) is a quantitative method used to rank critical failure mode effects by taking into consideration the probability of their occurrence. FMECA is a design technique composed of FMEA and CA and provides a systematic approach to clarify hardware failures.

Fault tree analysis (FTA) is a top-down procedure which considers components in working or failed states, and it has been proven difficult to handle degraded component states. FTA can be used to obtain minimum cut sets, which define the modes of system failures and identify critical components. The reliability measures for the top event of FTA can be obtained provided that the failure data on primary events/failures is available.

22.3 Failure Analysis of Mechatronic Systems

The failure modes of a mechatronic system include failure modes of mechanical, electrical, computer, and control subsystems, which could be classified as hardware and software failures. The failure analysis of mechatronic systems consists of hardware and software fault detection, identification (diagnosis), isolation, and recovery (immediate or graceful recovery), which requires intelligent control.

The hardware fault detection could be facilitated by redundant information on the system and/or by monitoring the performance of the system for a given/prescribed task. Information redundancy requires sensory system fusion and could provide information on the status of the system and its components, on the assigned task of the system, and the successful completion of the task in case of operator error or any unexpected change in the environment or for dynamic environment.

The simplest monitoring method identifies two conditions (normal and abnormal) using sensor information/signal: if the sensor signal is less than a threshold value, the condition is normal, otherwise it is abnormal. In most practical applications, this signal is sensitive to changes in the system/process working conditions and noise disturbances, and more effective decision-making methods are required. Generally, monitoring methods can be divided into two categories: model-based methods and feature-based methods. In model-based methods, monitoring is conducted on the basis of system modeling and model evaluation. Linear, time-invariant systems are well understood and can be described by a number of models such as state space model, input–output transfer function model, autoregressive model, and autoregressive moving average (ARMA) model. When a model is found, monitoring can be performed by detecting the changes of the model parameters (e.g., damping and natural frequency) and/or the changes of expected system response (e.g., prediction error). Model-based monitoring methods are also referred to as failure detection methods.

Model-based systems suffer from two significant limitations. First, many systems/processes are nonlinear, time-variant systems. Second, sensor signals are very often dependent on working conditions. Thus, it is difficult to identify whether a change in sensor signal is due either to the change of working conditions or to the deterioration of the process.

Feature-based monitoring methods use suitable features of the sensor signals to identify the operation conditions. The features of the sensor signal (often called the monitoring indices) could be time and/or frequency domain features of the sensor signal such as mean, variance, skewness, kurtosis, crest factor, or power in a specified frequency band. Choosing appropriate monitoring indices is crucial. Ideally the monitoring indices should be: (i) sensitive to the system/process health conditions, (ii) insensitive to the working conditions, and (iii) cost effective. Once a monitoring index is obtained, the monitoring function is accomplished by comparing the value obtained during system operation to a previously determined threshold, or baseline, value. In practice, this comparison process can be quite involved. There are a number of feature-based monitoring methods including pattern recognition, fuzzy systems, decision trees, expert systems, and neural networks.

Fault detection and identification (FDI) process in dynamic systems could be achieved by analytical methods such as detection filters, generalized likelihood ratio (which uses Kalman filter to sense discrepancies in system response), and multiple mode method (which requires dynamic model of the system and could be an issue due to uncertainty in the dynamic model) (Chow and Willsky, 1984).

As mentioned above, the system failures could be detected and identified by investigating the difference between various functions of the observed sensor information and the expected values of these functions. In case of failure, there will be a difference between the observed and the expected behavior of the system, otherwise they will be in agreement within a defined threshold. The threshold test could be performed on the instantaneous readings of sensors, or on the moving average of the readings to reduce noise.

In a sensor voting system, the difference of the outputs of several sensors and each component (sensor or actuator) is included in at least one algebraic relation. When a component fails, the relations including that component will not hold and the relations that exclude that component will hold. For a voting system to be fail-safe and detect the presence of a failure, at least two components are required. For a voting system to be fail-operational and identify the failure, at least three components are required, for example, three sensors to measure the same quantity (directly or indirectly). As Chow and Willsky (1984) pointed out, for the detection and identification of a single failure among m components at least $(m - 1)$ relations are required (more relations are preferred for better performance in the presence of noise).

22.4 Intelligent Fault Detection Techniques

The fault tolerant control (robust control and decision-making process) should include allowable performance degradation in the failed state, criticality and likelihood of the failure, urgency of response to failure, tradeoffs between correctness and speed of response, normal range of system uncertainty, disturbance environment, component reliability vs. redundancy, maintenance goals (mean-time-to-failure, mean-time-to-repair, maintenance-hour/operation-hour, etc.), system architecture, limits of manual intervention, and life-cycle costs (Stengel, 1991).

Fault detection could be achieved by redundancy in sensing (measurement) and actuation, parallel redundancy (e.g., dual sensors or actuators), analytical redundancy, and artificial intelligence (expert systems, artificial neural network, or integration of both techniques) combined with redundancy.

Stengel (1991) classified the analytical redundancy into direct and temporal redundancy. Direct redundancy consists of algebraic relationship among instantaneous outputs of sensors and is useful for sensor failure detection, but not for actuator failure detection. Temporal redundancy includes the relationship among histories of sensor outputs and actuator inputs (also comparison of the outputs of dissimilar sensors at different times). Temporal redundancy could be used for both sensor and actuator FDI, for example, a sensor voting system with mixed displacement and velocity sensors could detect failures of both types of sensors. The computational complexity of temporal redundancy is higher compared to the direct redundancy case as it requires the dynamics of the system.

An expert system embodies in a computer the knowledge-based component of an expert skill in such a manner that the system can generate intelligent actions and advice and can, when required, justify to the user its line of reasoning. In general, an expert system is composed of three parts: an inference engine, a human-machine interface, and a knowledge base. The inference engine is the knowledge processor and is modeled after the expert's reasoning. The engine works with available information on a particular problem, coupled with the knowledge stored in the knowledge base to draw conclusions or recommendations.

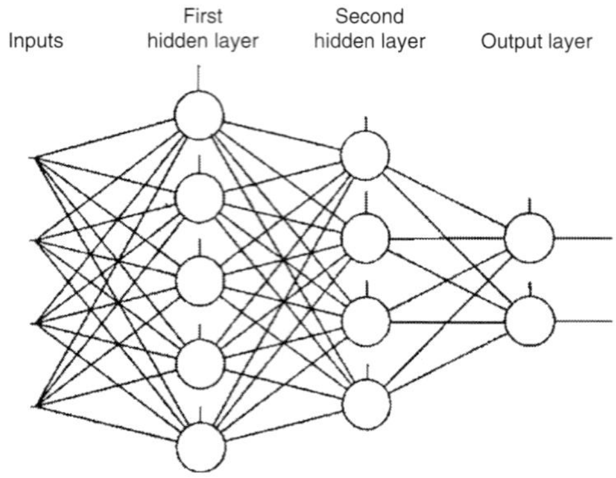

FIGURE 22.1 Architecture of a typical multilayer feedforward neural network.

The knowledge base contains highly specialized knowledge on the problem area as provided by the expert in the form of statistical analysis, empirical or semi-empirical rules, theoretic and computer simulation studies, and experimental testing. It includes problem facts, rules, concepts, and relationships.

Expert systems have obvious knowledge representation forms that make knowledge easy to manage, have the capability to explain their behavior, and can diagnose new faults using their knowledge bases. At the same time, self-learning is still a problem and computation time can be quite lengthy for difficult tasks.

A neural network is a highly nonlinear system with adaptation and generalization capabilities. There are many different architectures of neural networks; however, the multilayer feedforward neural network (refer to Figure 22.1) is one of the most popular ones. This is because of the simplicity, availability of efficient learning methods, generalization capabilities, and noise tolerance of these networks. This network is a collection of simple, interconnected nodes, also known as neurons, which operate in parallel and store knowledge on the strength of connections between the individual nodes. Such a parallel computing network, inspired by the computational architecture of the human brain, has been successfully applied to intelligent tasks such as learning, speech synthesis, and pattern recognition. The input vector feeds into each of the first layer neurons, the outputs of this layer feed into each of the second layer neurons, and so on. The last layer, which generates output to the external world, is called the output layer. The hidden layers are not connected to the external world. Often the neurons are fully connected between the layers, that is, every neuron in layer l is connected to every neuron in layer $l + 1$.

Training a neural network consists of the process of finding the set of interconnection weights (there is an interconnection weight associated with each neuron which modifies the input signal to that neuron in a specific manner), which results in a network output that satisfies a predefined criterion. Feedforward neural networks are trained using the backpropagation algorithm. This is a supervized training method. This means that the network will be presented with sample inputs and correct responses, called a training pattern. The network is then trained to reproduce the correct responses.

Neural networks have capabilities of association, memorization, error tolerance, self-adaptation, and multiple complex pattern processing. However, they cannot explain their own reasoning behavior and cannot diagnose new faults (those not already made available previously in training the network).

22.5 Problems in Intelligent Fault Detection

The fault detection scheme should be capable of detecting and identifying the failures correctly and promptly with minimum delays. This requires a reconfigurable robust controller. That is, the controller should distinguish between failures, uncertainties/inaccuracies in the model of the system, and disturbances such as sensor noise; and reduce the effect of measurement error and noise, uncertainties in the system model, and disturbances (even component failure) on the system output.

The sensor noise could be taken care of by statistical analysis on sensor readings. The uncertainties in the system model could be taken care of by estimating the effect of parameter uncertainties and compensating for it in the FDI system, or by minimizing the sensitivity of the FDI system to these uncertainties. The detection scheme should also be capable of monitoring the degradation of the system, as well as evolution and progress of failure over time (and predicting the failure), and responding to each accordingly.

22.6 Example Mechatronic System: Parallel Manipulators/Machine Tools

Parallel structured machine tools consist of multiple serial branches/legs acting in parallel on a common mobile platform with the spindle being connected to the mobile platform. Parallel manipulator-based devices have the advantages of not requiring actuation of base distal joints and of having their active joints acting in parallel on the mobile platform. These advantages can lead to parallel machine tools having desirable stiffness, accuracy, and dynamic characteristics, which, in turn, will provide high material removal rate (high product volume) with tight tolerances and in-process inspection capability (on-machine measurements of workpieces, fixtures, and tools during and after manufacturing process without breaking setups).

The failure analysis of parallel machine tools should include failures of parallel architecture, as well as failure of cutting tool, in addition to software failures.

22.6.1 Parallel Architecture Manipulators (Based on a Paper by Huang and Notash, 1999)

The following discussion will focus on the design orientated failure analysis of the mechanical system of parallel manipulators/machines.

Parallel manipulators consist of a base platform (stationary link), a mobile platform (end effector), and multiple branches/legs connecting the base and mobile platforms. Figure 22.2 depicts an example of a six-branch parallel manipulator.

The mechanical failure modes of manipulators could be classified as joint failure (component), link failure/breakage (component), branch failure (subsystem), end effector failure (subsystem), and device failure (system). Figure 22.3 represents the top level FTA of a three-branch parallel manipulator/machine.

22.6.1.1 Component Failures

Parallel (closed-loop) manipulators possess both active joints (joints that are sensed and actuated) and passive joints (unactuated joints which could be sensed or unsensed). Therefore, their failures could be due to the failures of active, passive sensed, or passive unsensed joints. The failure of any joint will cause

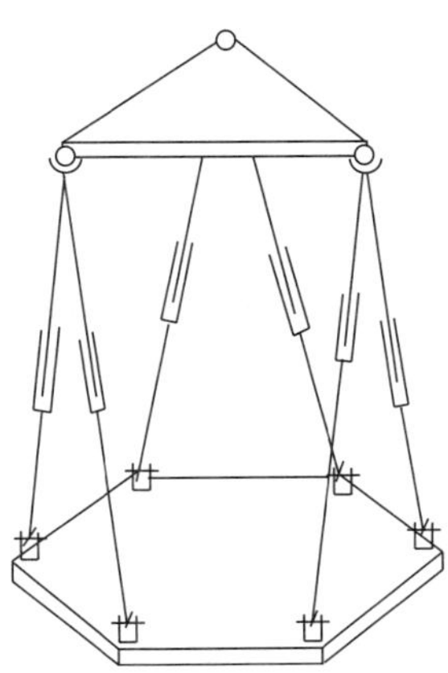

FIGURE 22.2 Example of a six-branch parallel manipulator/machine.

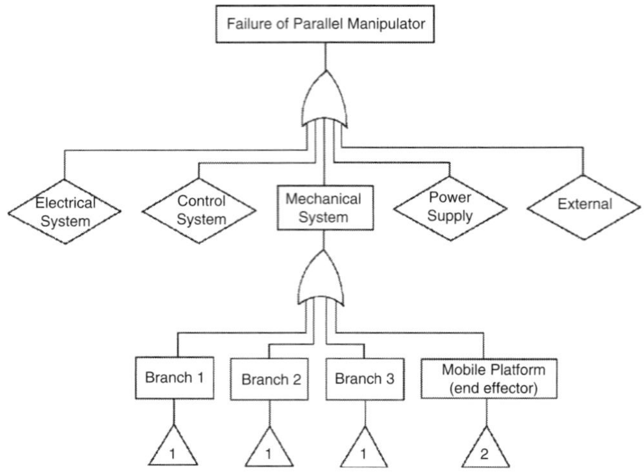

FIGURE 22.3 Top level fault tree for a three-branch parallel manipulator.

the failure of the parallel manipulator, unless the device is redundantly actuated/sensed or has a redundant unsensed joint(s) for the given task.

The common failure modes of active, passive sensed, and passive unsensed joints are the joint break and joint jam. The only failure modes of passive unsensed joints are the common joint failures. The failure modes of passive sensed joints include sensor failure, in addition to the common failure modes. In this case, the motion of the joint cannot be measured and the joint will be reduced to a passive unsensed joint. The major failure modes of active joints could be classified as actuator failure, transmission failure, and sensor failure. As a result of an actuator failure, the active joint degrades to a passive sensed joint, provided that the joint is back drivable; otherwise, the joint must be locked and the corresponding branch and the parallel manipulator will lose one DOF and an actuation. Because of a transmission failure, the actuator fails to drive the joint, and the active joint could only be used as a passive sensed joint. When the sensor of an active joint fails, although the actuator may operate properly, the motion of the joint cannot be controlled as there will be no reliable information available on the joint motion; hence, the active joint is degraded to a passive unsensed joint.

22.6.1.2 Subsystem Failures

The branches of a parallel device could be categorized as active or passive branches. An active branch possesses at least one active joint to provide a required force and to facilitate a suitable loci for the branch end location. A passive sensed branch has at least one sensed joint and its main function is to constrain the loci of the branch end position. Neither joint of a passive unsensed branch is sensed, and the branch is mainly used to constrain the motion of the mobile platform.

A branch of a parallel manipulator could fail because of component (link, joint) failures. As well, a branch will not follow its assigned path if it is in the workspace boundary, or at an internal singularity (where it loses one or more DOF). Therefore, the mechanical failure modes of a branch include branch break, loss of DOF, and loss of sensing/actuation.

22.6.1.3 Mechanical System Failures

A parallel manipulator could fail because of component and/or subsystem failures. Therefore, the mechanical failures of a parallel manipulator include loss of the DOF, loss of the actuation, loss of the motion constraint, and uncertainty configurations. A summary of the mechanical failure levels, modes, effects, and causes of parallel manipulators has been tabulated in Table 22.1.

TABLE 22.1 Failure Modes of Parallel Manipulators and Their Effects

Failure Levels			Failure Modes	Failure Causes	Effects
Components	Links		Break	Overload, fatigue, impact, material flaw	Reduction in number of branches
	Joints	Common	Joint break	Overload, fatigue, impact, material flaw	Reduction in number of branches
			Joint jam	Deterioration, external interference	Reduction in DOF of corresponding branch
		Active	Actuator failures	Depends on actuator type	Reduction in actuation, DOF if joint not back-drivable
			Transmission failures	Depends on transmission type	Reduction in actuation, DOF if joint not back-drivable
			Sensor failures	Depends on sensor type	Reduction in sensing, actuation, maybe DOF
		Passive sensed	Sensor failures	Depends on sensor type	Reduction in sensing
		Passive unsensed	Common failures (break, jam)	Overload, fatigue, impact, material flaw; deterioration, external interference	Reduction in number of branches; reduction in DOF
Branches	Common		Break	Joint/link break	Reduction in number of branches, maybe actuation and DOF, interference with other branches
			Loss of DOF	Joint jam, locked active joint, branch singularity	Reduction in DOF of manipulator
	Active		Loss of actuation	Active joint failure	Reduction in actuation, maybe DOF
			Loss of sensing	Sensor failure	Reduction in actuation, degradation to passive branch
	Passive sensed		Loss of sensing	Sensor failure	Reduction in sensing, degradation to passive unsensed branch
	Passive unsensed		Common failures (break, loss of DOF)	Joint/link break, joint jam, locked active joint, branch singularity	Reduction in constraint or DOF of manipulator
Manipulator			Loss of DOF	Joint jam, branch singularity, branch interference	Insufficient DOF
			Loss of actuation	Active joint/branch failure	Degradation in force and motion capabilities
			Loss of constraint	Reduction in number of active branches	Uncontrolled motion of manipulator
				Passive unsensed branch break	Uncontrolled motion of manipulator
				Uncertainty configuration	Instantaneous uncontrolled motion of manipulator

22.6.1.4 Failure Identification

A fault tolerant manipulator should be capable of identifying a failure, as well as tolerating the failure. The failed component (mechanical system) of a parallel manipulator, e.g., a failed joint sensor, could be identified via the manipulator controller using the information provided by the sensors of the device. A joint sensor fault detection scheme for a class of fault tolerant parallel manipulators, based on redundant sensing of joint displacements and the comparison of forward displacement solutions, was presented in (Notash, 2000).

While the failure of active joints could be identified based on the information provided by the sensor(s) on the corresponding joint, failure of passive joints could be identified by monitoring the overall performance of the manipulator in the software. For a given parallel manipulator, the criteria for failure should be incorporated in the simulation software. For example, the loss of DOF due to workspace boundary could be monitored (similar to the joint limits and branch interference) and the manipulator could be stopped before it reaches its envelope to prevent potential failure and damage to the device. As well, all of the potential special (uncertainty) configurations of the manipulator should be identified, and the closeness to these singularities should be monitored as the device moves around within its workspace.

22.6.1.5 Fault Tolerance through Redundancy

The fault tolerant capabilities of parallel manipulators could be improved by employing appropriate redundancies. Redundant sensing has been investigated for improving the fault tolerance capabilities of parallel manipulators, for simplifying the forward displacement analysis of these manipulators, and for facilitating fixtureless calibration of these devices. Redundancy in actuation has been considered for eliminating the uncertainty configurations of parallel manipulators. More work is required to develop methodologies for identifying the failed components of parallel manipulators with elements of redundancy, and compensating for their failures. For parallel manipulators, redundancy could be incorporated as redundant DOF (mobility), redundant sensing, and redundant actuation.

Redundant DOF could be achieved by incorporating additional joints into the parallel manipulator. A redundant DOF requires one more actuator on the parallel manipulator. This additional actuator is not considered as a redundant one because its failure will result in the failure of the parallel manipulator due to the loss of a required actuation. Redundancy in sensing could be obtained by sensing the existing passive unsensed joints of the manipulator, by adding a redundant passive sensed branch, or by using an external sensor such as a vision system. It should be noted that the information redundancy is achieved by redundant sensing, as well as by providing the task description of the manipulator, such as the Cartesian trajectory of the end effector (for robot path planning and machining operation). Redundancy in actuation could be accomplished by actuating the passive joints of the manipulator, or by adding an active branch (in addition to employing dual actuators).

22.6.2 Tool Condition Monitoring

An important element of the automated process control function is the real-time detection of cutting tool failure, including both wear and fracture mechanisms in machining operations. The ability to detect such failures online would allow remedial action to be undertaken in a timely fashion, thus ensuring consistently high product quality (quality of surface finish and dimensional precision) and preventing potential damage to the process machinery. The basic premise of any automated, real-time tool condition monitoring system is that there exists either a directly measurable, or a derived parameter, which can be related to advancing tool wear and/or breakage. Information about tool wear, if obtained online, can be used to establish tool change policy, adaptive control, economic optimization of machining processes, and full automation of machining processes.

In the ideal case, the system should be able to detect levels of wear well below those at which the tool would have to be replaced and should also be sensitive to relatively small changes in the level of wear. The latter characteristic would provide the system with the potential to "trend" the wear pattern and predict the amount of useful life left in the tool (allowable wear limit reached).

With respect to tool fracture, the system should be able to detect both small fractures, "chipping" phenomena, and catastrophic failure of a tool. Although prediction of such failures would be desirable, it is problematic whether this is a practical goal, at least in the near future. The number of variables, which determine the actual occurrence of tool fracture together with their complex interactions, and in many instances their underlying stochastic nature, make reliable prediction capabilities, at best, a long-term prospect in tool monitoring systems.

22.6.2.1 Cutting Tool Failure Monitoring Techniques

Tool condition monitoring systems are based upon either direct or indirect methods of quantifying the magnitude of tool failure.

The direct methods are those that utilize effects caused directly by tool failure. The direct methods, usually performed by means of optical, radiometric, pneumatic, or contact sensors can be effectively applied to the offline measurement of tool wear or breakage. However, such direct means of measuring tool failure have generally been found to be difficult to apply in practical shop floor applications. This is particularly true in those situations requiring online (real-time) monitoring capability.

Indirect methods of sensing tool failure depend upon the measurement of parameters, which are indirectly related to the condition of the cutting edge. For example, the cutting forces generated during a machining operation are dependent upon the condition of the tool's cutting edge. Generally, as the tool edge wears the generated cutting forces increase. Thus, measurement of the cutting forces present during a machining operation provides an indication of the tool condition, i.e., increasing cutting forces indicates increasing wear. In reality the relationship can be very complex. Other parameters that have been studied to determine their suitability as indicators of cutting tool failure include spindle motor current, acoustic emissions, cutting tool temperature, and noise and vibration signals. It is also possible to measure cutting forces directly and then relate these values to the condition of the cutting tool. In fact, this is one of the more common indirect tool wear monitoring methods. It has been reported that cutting force signals are more sensitive to tool wear than vibration or power measurements. The general reliability of force measurements is another reason for their popularity in tool condition monitoring applications. To use cutting force measurements for practical tool monitoring systems, there is a need to relate these forces to the state of tool condition online. However, since the measured cutting forces are affected by both cutting edge condition and changes in cutting conditions (feed rate, cutting speed, and depth of cut), the detection of tool failure using measurements of these forces becomes quite challenging in practice.

22.6.2.2 System Characteristics

Whether a tool condition monitoring system employs direct or indirect measures of tool failure (automated, computer-based system or not), it must include a number of common features if it is to be truly practical. Figure 22.4 shows the block diagram of a generalized tool condition monitoring system (Braun, 1986).

In the *measurement* section, the physical parameter (or possibly, parameters) of interest is converted to a form that is appropriate for further manipulation by the system (generally a digitized representation of an electric analog signal).

Within the *processing* section, various techniques are implemented in order to suppress noise, compress information, and emphasize important features of the acquired signal. Typical methods include analog

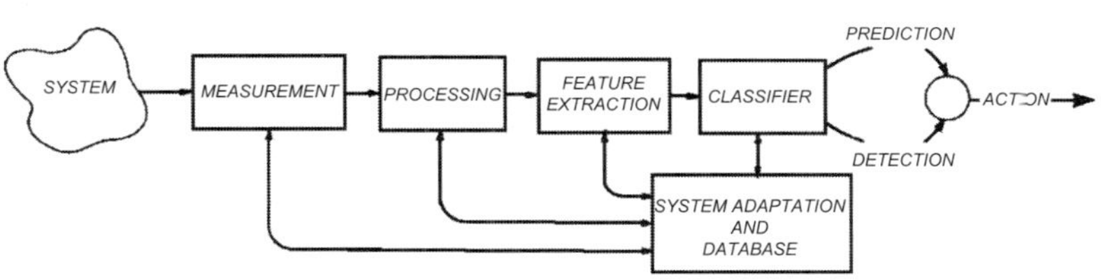

FIGURE 22.4 Block diagram of a generalized tool condition monitoring system.

or digital filtering, time domain averaging, Fourier transformation, parametric identification based on ARMA models, etc.

The purpose of the *feature extraction* section is to obtain a specific feature, or features, (often referred to as a feature vector), which can be used by the *classifier* to determine the specific type of failure and initiate appropriate corrective actions. Examples of features would include total power of the signal, crest factor value, power in a particular frequency range, frequency of the maximum peak, amplitude of the maximum peak, the autoregressive parameters of an ARMA model, etc. If multiple features are employed, they should be uncorrelated so that they provide independent indications of tool failure. When coupled with a hierarchical decision tree structure (or other appropriate structure) in the *classifier*, such multiple feature vectors can greatly improve the reliability of the tool monitoring system.

The *adaptation and database* section should not only efficiently manage all data storage and manipulation requirements but also provide the system, to as great a degree as possible, with the ability to learn from experience.

22.7 Concluding Remarks

It should be noted that the first and most practical step for increasing the reliability and improving the fault tolerance of mechatronic systems, for example, a parallel machine, is by enhancing the existing design, or by improving the robustness of the design, such as using coupled joints while designing the architecture of the manipulator. Redundancies through redesign are recommended for the applications where the fail-safe system could be very crucial, or the down time should be minimum and previously scheduled, such as the medical applications or space operation. It is also worth mentioning that not any redundancy could improve the fault tolerance of a system with no modification to the architecture of the device.

It is noteworthy to emphasize the importance of fail-safe simulation software and controller for a fault tolerant mechatronic system, which requires a robust software capable of monitoring the performance of system and responding to any system failures (including mechanical, electrical, and control systems).

References

Billinton, R., and Allan, R.N., *Reliability Evaluation of Engineering Systems: Concepts and Techniques,* Plenum Press, 1987.

Braun, S. ed., *Mechanical Signal Analysis—Theory and Applications,* Academic Press, 1986.

Chow, E.Y., and Willsky, A.S., "Analytical redundancy and the design of robust failure detection systems," *IEEE Trans. Automatic Control,* 29(7), 603–614, 1984.

Dhillon, B.S., *System Reliability, Maintainability and Management,* Petrocelli Books, 1983.

Huang, L., and Notash, L., "Failure analysis of parallel manipulators," *Proc. 10th IFToMM Congress on Theory of Machines and Mechanisms,* pp. 1027–1032, June, 1999.

Notash, L., "Joint sensor fault detection for fault tolerant parallel manipulators," *J. Robotic Systems,* 17(3), 149–157, 2000.

Stengel, R.F., "Intelligent failure-tolerant control," *IEEE Control Systems,* pp. 14–23, June 1991.

Wolfe, W.A., "Fault tree analysis," Atomic Energy of Canada, Report, 1978.

23

Logic System Design

M. K. Ramasubramanian
North Carolina State University

23.1 Introduction to Digital Logic

In analog electronics, voltages and current represent variables that vary continuously from the allowable minimum to the maximum. These variables are measured, amplified, added, and subtracted through analog circuits to achieve the desired results. For instance, measurement of temperature using thermocouples requires the amplification of voltages generated to a suitable range, calibration of the voltage with measured temperatures, and outputting the results on a voltmeter to indicate temperature. In this design, it may be necessary to subtract an offset voltage, multiply with a gain factor depending on the temperature range. The amplification of voltages and current are accomplished easily with operational amplifiers and transistors, respectively. The measured temperature can be used as the feedback signal in a control loop for a mechatronic temperature control system. In digital electronics, the variables assume a binary state, assuming a value of 0 or 1. In the above example, we might want to shut the solenoid valve down if the temperature was below desired value and open the valve if the temperature was above that value. In this case, we simply require a TRUE or FALSE input to the question "Is the temperature above or below the threshold?" The representation of these types of variables in circuits, which assume binary values, and their manipulation to achieve desired results is the topic of discussion in this chapter

23.1.1 Logic Switching Levels

In digital circuits, voltage levels indicate binary states where the HIGH or TRUE state is represented by the maximum voltage value, typically 5 V, and the LOW or FALSE state is represented by the minimum voltage value, typically 0 V. In Boolean logic,"1" represents TRUE and "0" represents FALSE. In practice, any voltage above a minimum input threshold, V_{IH}, is interpreted as logic HIGH and any voltage below

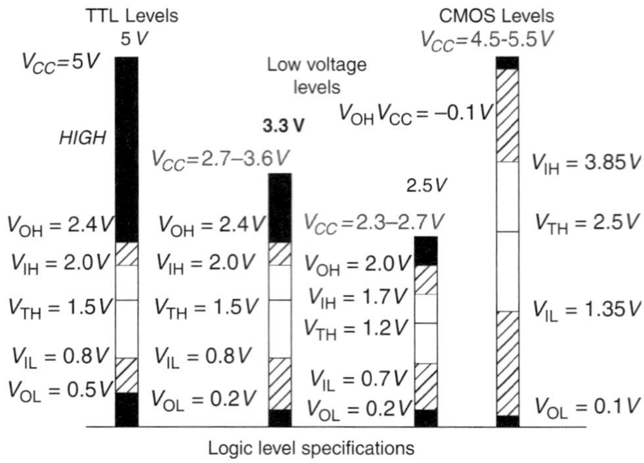

FIGURE 23.1 Switching levels for logic gates [1].

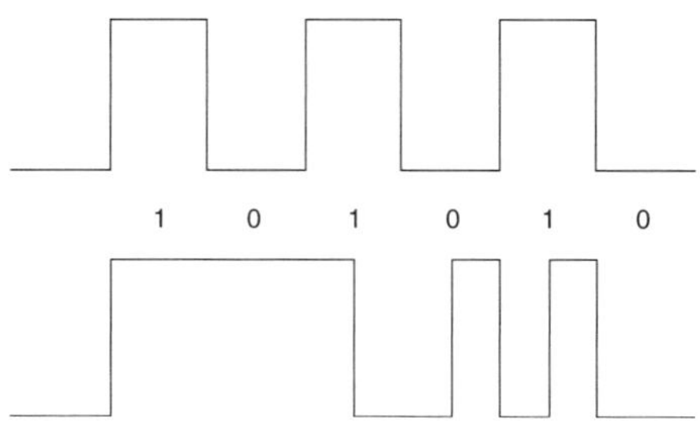

FIGURE 23.2 Periodic and nonperiodic logic level signals.

a maximum threshold, V_{IL}, is interpreted as logic LOW. The minimum output from a logic device for HIGH is represented by V_{OH} (different from V_{IH}), and maximum output level for a logic LOW is represented by V_{OL} (different from V_{IH}). These values depend on the type of logic device and a general chart of values for these parameters is shown in Figure 23.1 [1]. V_{CC} is the supply voltage. The difference between the V_{OH} and V_{IH}, or V_{OL} and V_{IL}, is called the noise margin. It is important to design the logic circuit with the constraint that voltages will never fall in the region between V_{IH} and V_{IL}, which is called the forbidden region where the logic device will fail to interpret signals. The differences between switching levels for different technologies such as 5-V logic, 3.3-V logic, CMOS (complementary metal oxide semiconductor), and TTL (transistor–transistor logic) should all be considered when interfacing these systems with each other.

A logic variable can rapidly change states as shown by an ideal pulse train in Figure 23.2. The variables can vary periodically or nonperiodically between 0 and 1. Logic gates read these signals as inputs, perform the appropriate Boolean operations among them, and generate the correct output at desired operating speeds. Robust design and use of logic functions and its implementation in circuits is an integral part of mechatronics design.

21.3.2 Logic Gate Application

Consider the example of an autonomous robot moving about on a table surface. The robot should move towards the destination denoted by a bright light source while avoiding obstacles and at the same time not falling off the edge. Assuming that we have three digital sensors, namely, obstacle detector,

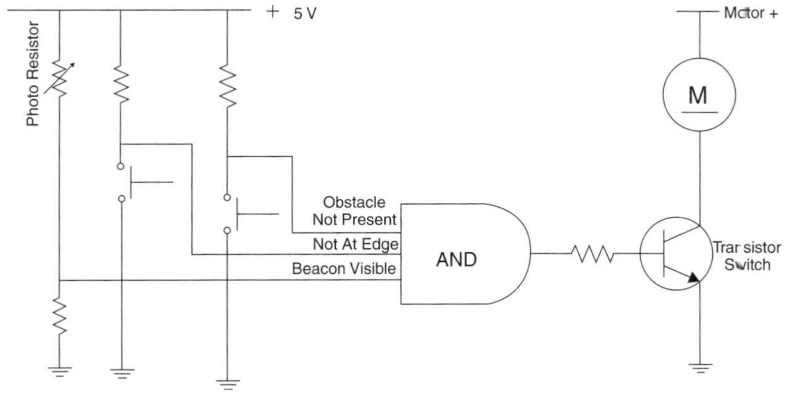

FIGURE 23.3 Forward motion logic implementation for a tabletop robot.

edge detector, and destination sensor, we can write a simple logic function for moving forward, as shown in Equation 23.1. Of course, this is not the complete logic required for the robot to function properly. However, we focus on one aspect of the problem to illustrate the use of logic functions.

$$\text{MOVE FORWARD} = (\text{OBSTACLE NOT DETECTED})\ \mathit{AND}\ (\text{EDGE NOT DETECTED})\ \mathit{AND}\ (\text{BEACON IS VISIBLE}) \quad (23.1)$$

The input from the three sensors is interfaced to a logic circuit consisting of logic gates, in this simple example, a three-input AND gate and the output drives the motors. Of course, other cases of behaviors for the robot where the edge is found or the beacon is not visible or an obstacle is detected have to be worked out to make this circuit robust and worthwhile. Figure 23.3 shows an implementation of the logical statement expressed in Equation 23.1.

23.2 Semiconductor Devices

23.2.1 Diode

In order to understand logic gates, it is important to develop a basic understanding of semiconductor devices, especially the diode and the transistor. A diode is a pn-junction, which means that the diode is made up of a p-type (electron deficient) material layer and an n-type (electron rich) material layer sandwiched together. When the positive terminal of a battery is connected to the p-side of the diode (anode) and the negative of the battery is connected to the n-side of the diode (cathode), then the diode is said to be forward biased as long as the voltage across the junction exceeds 0.7 V. When the terminals are reversed, the diode is said to be reverse biased and does not conduct until very high voltages are applied across the junction, known as the breakdown voltage. For all practical purposes, we can assume that a reverse-biased diode does not conduct. A schematic of a diode, its symbol, and a forward-biased circuit is shown in Figure 23.4. When forward biased, the diode can be treated as a simple closed switch with a 0.7 V drop across it and when the diode is reverse biased, the diode is an open switch.

23.2.2 Bipolar Transistor

A bipolar transistor has three semiconductor layers. In an npn-transistor, a very thin p-layer is sandwiched between two n-layers. Transistor types and their symbols are shown in Figure 23.5a,b.

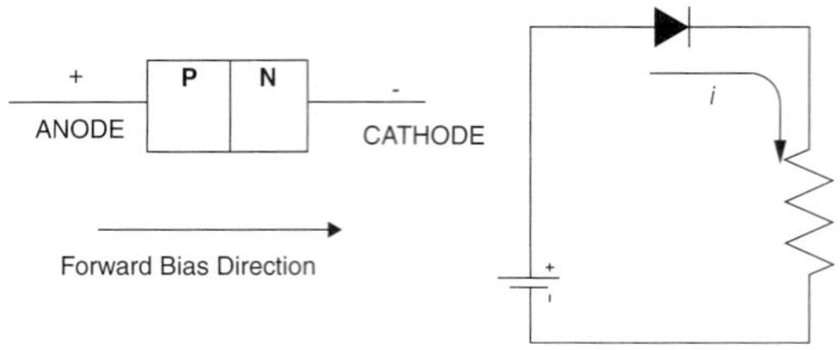

FIGURE 23.4 The diode and its behavior.

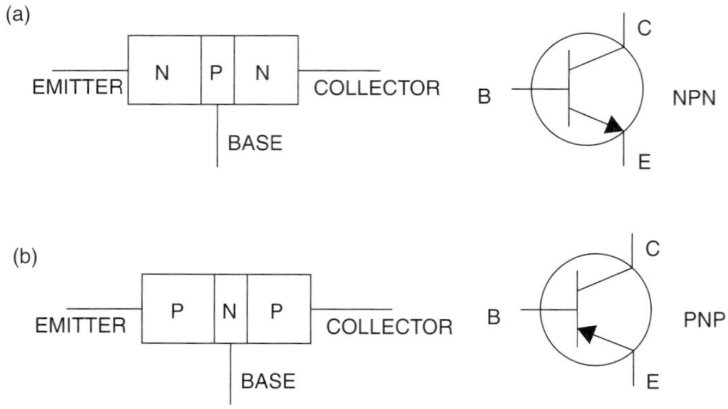

FIGURE 23.5 (a) npn-transistor symbol, and (b) pnp-transistor symbol.

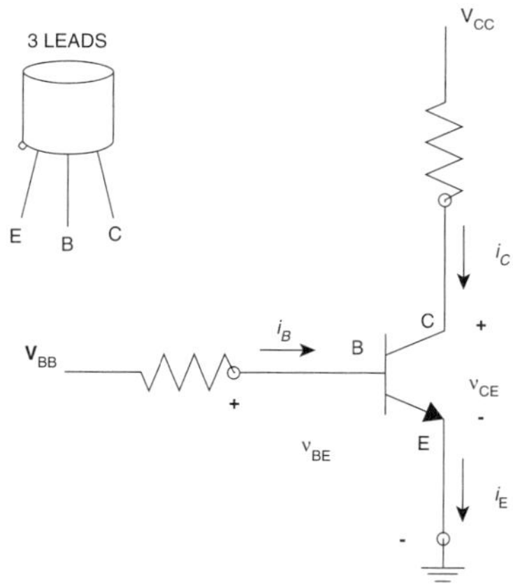

FIGURE 23.6 Schematic representation of the working of an npn-transistor.

There are three leads in a transistor, namely, the collector (C), emitter (E), and the base (B). For an npn-transistor in a circuit, as shown in Figure 23.6, the base–emitter junction is forward biased and will conduct if the voltage V_{BE} exceeds the forward bias voltage for the pn-junction, typically 0.7 V. V_{BE} is increased by increasing the voltage at B. However, the base–collector junction is reverse biased as the collector C is at a higher potential. As current flows in the base–emitter loop, the electrons from the emitter

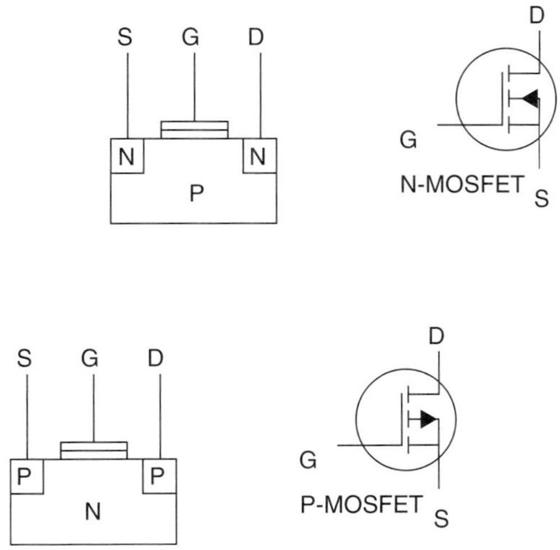

FIGURE 23.7 n- and p-channel MOSFETs and symbols.

flow into the base terminal by filling in the "holes" in the p-layer and subsequently releasing an electron from the p-layer out of the base terminal. However, because of a limited number of "holes" in the p-layer (which is very thin), the electrons from the emitter see a larger potential across the collector–emitter path and jump the junction. A large current, I_C, flows in the collector–emitter loop as a result. Thus, the transistor is a current amplifier. A small current flowing in the base–emitter loop, I_B, is amplified by typically a factor of about 100 in the collector–emitter path. As the current flow in the base–emitter is increased by increasing V_{BE}, the collector–emitter current increases by decreasing V_{CE}. Since the collector is connected to the power source, V_{CC}, and the emitter is connected to the ground, the device controls this current flow by controlling the drop in voltage across the collector–emitter junction, continuing to drop the voltage as the base–emitter current is increased. It is obvious that the voltage cannot drop below 0; in fact, it cannot drop below 0.2–0.35 V in a real device. Under these conditions, the transistor is said to be saturated and is acting as a closed switch. Circuits that are built with transistors in the saturating condition are called saturating circuits; for example, the TTL family of logic gates. Circuits that do not allow the transistor to saturate and find a stable operating point in the active region of the transistor are called nonsaturating circuits; for example, emitter-coupled logic (ECL) gates. The biggest advantage of a nonsaturating circuit is the speed with which states can be changed compared to a saturating circuit.

23.2.3 Field Effect Transistor

These devices are easier to make and uses less silicon. There are two major classes of field effect transistors (FETs), namely, the junction FET (JFET) and the metal oxide semiconductor FET (MOSFET). In both cases, a small input voltage controls the output current with practically no input current. The three terminals are called the source (S), drain (D), and gate (G). Figure 23.7 shows the symbols for the n- and p-channel enhancement type MOSFETs. MOSFET is the most popular of transistor technologies. A MOSFET gate has no electrical contact with the source and the drain. A silicon-dioxide layer insulates the gate. Electrical voltage applied at the gate attracts electrons to the region below the gate and provides an n-type channel in a p-type substrate for conduction between the drain and source. This is called the enhancement type of MOSFET. The other is the depletion-enhancement type where there is an n-channel present between the drain and source, but the channel resistance can be increased or decreased by applying either a negative or a positive voltage at the gate, respectively. Depletion-enhancement MOSFET symbols and function are described in Figure 23.8. MOSFET devices are slower than bipolar devices and are used in slower but high density circuits, due to ease of manufacture and use of less silicon.

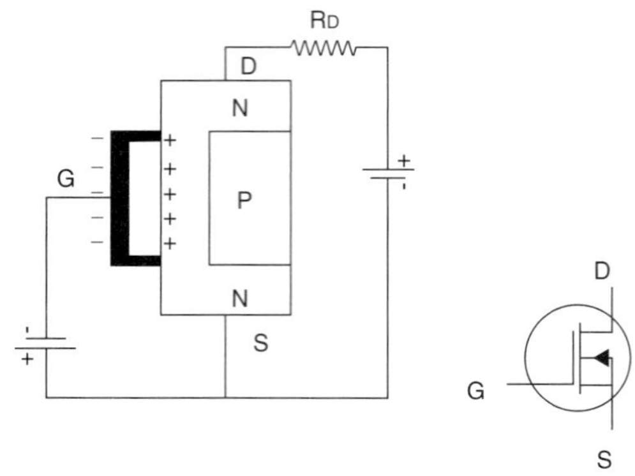

FIGURE 23.8 Depletion-enhancement type MOSFET.

23.3 Logic Gates

Logic gates are the basic building blocks of logic circuits and a computer. Mechatronic systems have a central computational element as well as specific logic functions implemented in hardware. A logic circuit consists of several logic gates working together. We will discuss the logic gates in general and as building blocks of mechatronic subsystems. Logic operations can be subdivided into two categories, namely, combinatorial and sequential. In the case of combinatorial logic circuits, the logic gates are used to produce an output based on instantaneous values of the inputs, whereas in the case of sequential logic circuits, the change in output depends on the present state as well as the state before the changes in input values, thus exhibiting memory behavior. Further, the sequential logic circuits can be synchronous or asynchronous. When the output changes synchronously with a clock input, it is said to be synchronous. When the inputs are read as soon as there is any change in it, it is called an asynchronous logic circuit.

There are three fundamental logic operations, namely the AND, OR, and NOT functions. Other logic operations are derived operations from these fundamental ones. The AND gate symbol and its truth table are shown in Figure 23.9. The AND gate can have more than two inputs.

Figure 23.10 shows an OR gate. Here the output is HIGH when either of the inputs or both the inputs are HIGH. The OR gate can also have more than two inputs. Figure 23.11 shows an inverter, also known as a NOT gate. This gate takes one input and simply inverts the logic, that is, a HIGH input is returned as LOW output and vice versa.

Other common logic gates that are derived from these fundamental ones are NAND, NOR, and Exclusive OR gates. NAND gate is a combination of AND and NOT gates; NOR is a combination of OR and NOT gates, and Exclusive OR can be generated with a combination of OR, NAND, and AND gates. Figures 23.12 through 23.14 show the derived gate types, namely the NAND, NOR, and XOR gates and their truth tables, respectively. The logic functions and their implementation into hardware using gates is the basic building block of a digital computer.

23.4 Logic Design

As in any design, it is important to keep it simple, robust, and cost effective. Mechatronics design or logic circuit design is no exception. When a logic function of a system is translated into relationships between inputs and outputs, it is not certain if the number of elements involved in realizing the design are the minimum or further simplification is possible. If the complexity is defined as the number of logic gates used, then the problem reduces to minimizing the logic function mathematically. However, if complexity is defined as the number of ICs used in the circuit (the amount of real estate occupied by the circuit), additional approaches have to be considered, namely using the same type of gate, as much

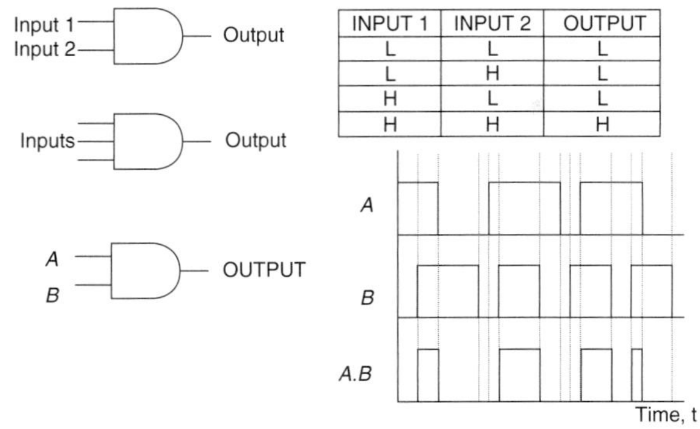

INPUT 1	INPUT 2	OUTPUT
L	L	L
L	H	L
H	L	L
H	H	H

FIGURE 23.9 AND gate, symbol, and behavior.

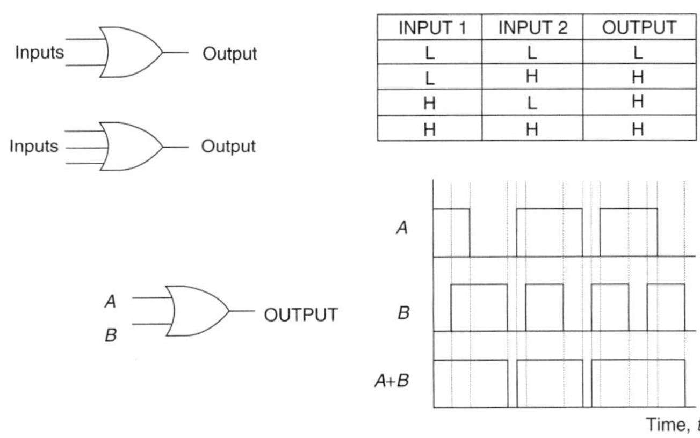

INPUT 1	INPUT 2	OUTPUT
L	L	L
L	H	H
H	L	H
H	H	H

FIGURE 23.10 OR gate, symbol, and behavior.

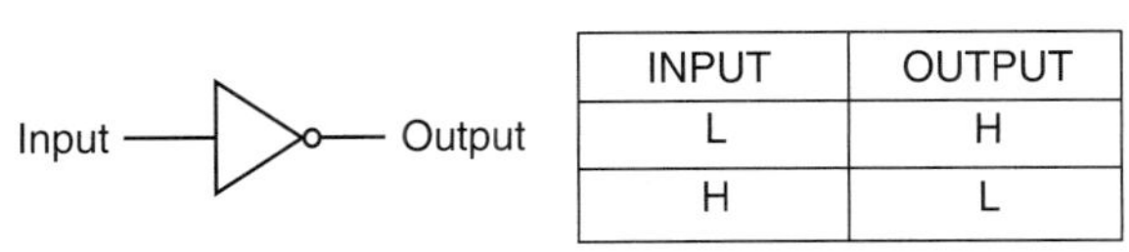

INPUT	OUTPUT
L	H
H	L

FIGURE 23.11 NOT gate or an inverter, symbol, and behavior.

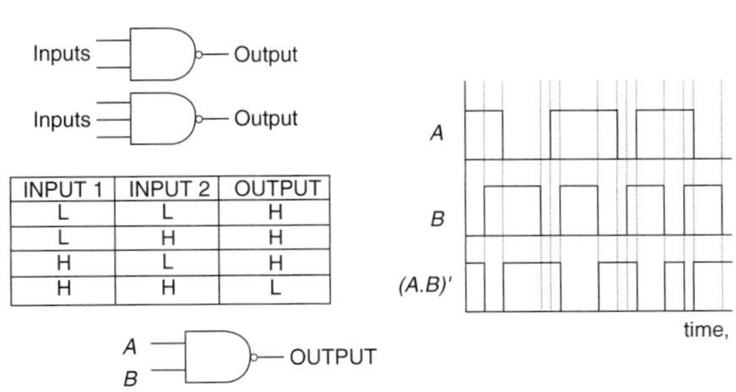

INPUT 1	INPUT 2	OUTPUT
L	L	H
L	H	H
H	L	H
H	H	L

FIGURE 23.12 NAND gate, symbol, and behavior.

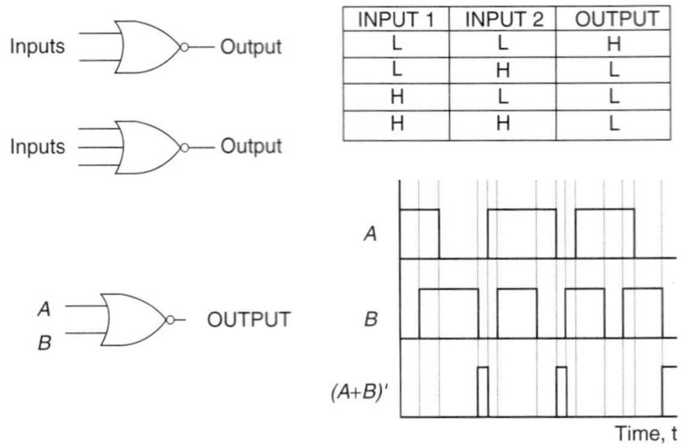

INPUT 1	INPUT 2	OUTPUT
L	L	H
L	H	L
H	L	L
H	H	L

FIGURE 23.13 NOR gate, symbol, and behavior.

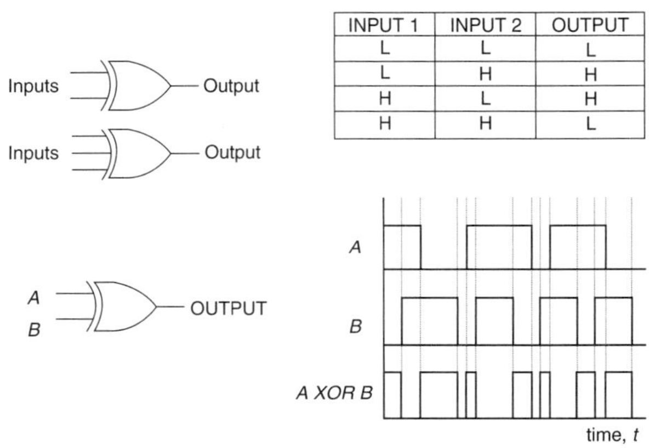

INPUT 1	INPUT 2	OUTPUT
L	L	L
L	H	H
H	L	H
H	H	L

FIGURE 23.14 XOR gate, symbol, and behavior.

as possible, although it may not be minimal in terms of number of gates. This will be preferred over using less different types of gates necessitating use of more ICs, in which some of the gates are unused.

23.4.1 Minimization

A method for minimizing Boolean functions is the Karnaugh map (K-map). From a physical description of the problem, logic statements are written as shown in Equation 23.1 for the tabletop robot problem. A truth table is generated showing the relationship between inputs and outputs. Let us take a truth table for a three-variable design, shown in Figure 23.15.

The logical function can be written as

$$X = A'B'C' + A'BC' + ABC' + A'BC \tag{23.2}$$

An implementation of the function without any further consideration will require four 3-input AND gates, one 4-input OR gate, and three inverters. If we assume that both complemented and uncomplemented forms of the signal for each variable are available, we still end up with a complex two-level circuit

INPUTS			OUTPUT
A	B	C	X
0	0	0	1
0	1	0	1
1	0	0	0
1	1	0	1
0	0	1	0
0	1	1	1
1	0	1	0
1	1	1	0

FIGURE 23.15 Truth table for a logic circuit design and minimization.

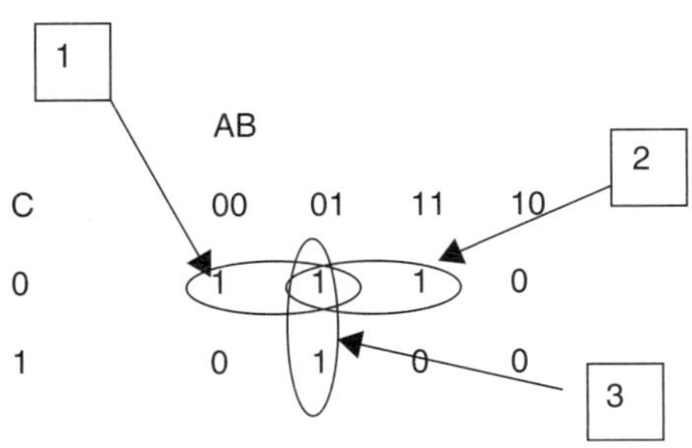

FIGURE 23.16 Karnaugh map for the logic design problem with three inputs and one output.

for what needs to be accomplished. Applying Karnaugh mapping, we can attempt to minimize the Boolean function and hence simplify the type and number of logic gates needed for circuit implementation.

The Karnaugh map is derived from the truth table shown in Figure 23.15. The two variables AB are grouped for column designations and the third variable provides the row designation. The values are arranged in such a way that adjacent columns or rows differ by only 1 bit.

Figure 23.16 essentially represents the logic described in the truth table in Figure 23.15. Because adjacent blocks in a K-map differ by 1 bit, the bit that changes is insignificant in a grouping of adjacent ones. In order to obtain the minimized function, adjacent ones on a K-map are identified by covering each one on the map at least once in a row or a column grouping, observing that in each case one variable is insignificant with respect to the value of *X*, the output. That variable is eliminated and the process is continued until all the groupings are evaluated. Finally, the reduced set of product terms is combined with an OR function to give the minimized function.

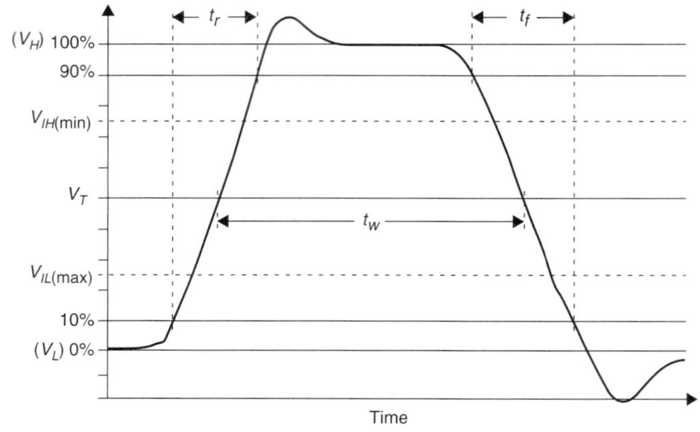

FIGURE 23.17 A real pulse and definition of characteristic parameters [2].

Figure 23.16 shows three sets of adjacent ones in rows and columns identified by circles around them and the following observations can be made:

Group 1. Only variable B changes states. Hence, it can be eliminated and the minimized form for the grouping is $A'C'$.

Group 2. Only variable A changes states. Hence, it can be eliminated and the minimized form of the grouping is BC'.

Group 3. Only variable C changes states. Hence, it can be eliminated and the minimized form of the grouping is $A'B$.

Hence, the minimized form for the logic function is

$$X = A'C' + BC' + A'B$$

This can be implemented with one 2-input AND gate IC and one 3-input OR gate. A K-map is helpful in minimizing up to six variables.

23.4.2 Dynamic Characteristics

Having studied the logic function and obtaining a minimum, we can build the logic circuit. However, in order to ensure that the circuit will work as intended over the entire operating range, dynamic characteristics of logic circuits must be considered. It was stated earlier that the input signal can change rapidly in a system and the logic circuit should perform as intended at frequencies at which the system is expected to operate.

The correct functioning of the logic circuit when the inputs are changing rapidly is an important consideration in design. In our discussion thus far, we have assumed that the logic signal is an ideal square wave and that the logic gates function without any delay. Let us examine the effects of relaxing these two assumptions to obtain some insight into the dynamic behavior of logic circuits.

A real pulse is shown in Figure 23.17 [2]. The rise time is denoted by t_r and fall time by t_f. The pulse further shows a settling time, overshoot, and undershoot when changing states. The signal amplitude is specified as the difference between the two stable signal levels for high (V_H) and low (V_L), that is, from 100% to 0%, and t_w is the pulse width of the signal measured at 50% of the amplitude. t_{THL}, t_{TLH} are the transition times for the output signal to go from high to low and low to high, respectively. t_{PHL} and t_{PLH} are propagation delay times for high to low and low to high transitions, respectively. For medium speed operation, t_{PHL}, and t_{PLH} are typically about 30 ns.

When an input to a logic gate changes states, the output lags behind by a characteristic time delay called the propagation delay, measured by the time difference between the input at 50% of the amplitude and the output at 50% of the amplitude. A simplified model of a real pulse for an inverter is shown in

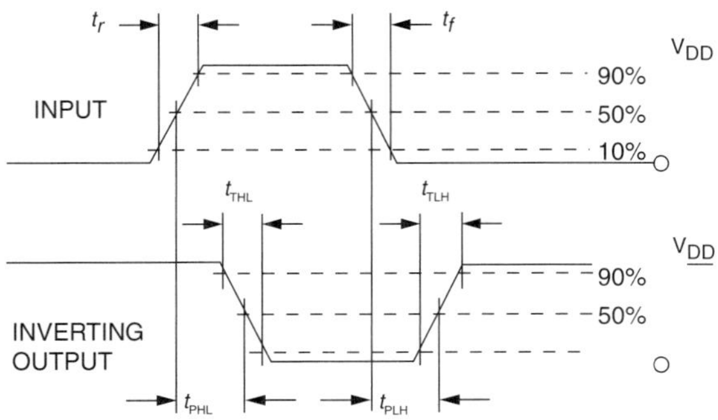

FIGURE 23.18 Propagation delay definition.

Figure 23.18 [3]. Values for the propagation, typically expressed in nanoseconds, are available in the datasheet for a device.

For a logic circuit, a propagation delay analysis is carried out by mapping out the total delay from input to output as the inputs change states, and identifying any static problems (frequency independent) and dynamic problems. Additional gates may be added to the circuit of the problem.

23.4.3 Other Design Considerations

The number of logic gates that a given gate can drive is called *fan-out* and the number of gates that can be connected to the input of a given gate is called *fan-in*. These data are given in the data sheet and should be adhered to. Further, minimizing the number of ICs needed in a logic circuit is an important consideration that might require modifying the design to use the same kind of gate although more numbers may be used than the minimum circuit identified with K-map analysis. Use of the same type of gates for compatibility between ICs is another design consideration in logic circuits.

23.5 Logic Gate Technologies

The first of the logic families that became commercially available was the resistor–transistor logic (RTL), where the transistor is used as a high-speed switch in circuits. Diode–transistor logic (DTL) and transistor–transistor logic (TTL) followed in the evolution. While the RTL and the DTL are obsolete, the TTL gates are still widely used. There are several variations of TTL logic, namely, the high-speed (H), low-power (L), Shottkey (S), and low-power Shottkey (LS). CMOS logic gates are an entirely different implementation of logic gates based on the complimentary metal-oxide semiconductor devices (CMOS) technology. These devices have low-power requirements and improved noise characteristics. These devices are extremely static sensitive and are easily damaged. A mixture of CMOS and bipolar processes resulted in the BiCMOS technology, using internal CMOS components and high-power bipolar outputs. Several different families evolved from the original BiCMOS processes [1].

23.5.1 Resistor–Transistor Logic

Figure 23.19a shows an resistor–transistor (RTL) inverter and Figure 23.19b shows a NAND gate. The transistor is assumed to operate in the saturation mode. Because of the nature of a transistor, there is a minimum voltage drop across the collector–emitter junction at saturation and there is a minimum current required in the base–emitter loop to saturate the transistor. In Figure 23.19a, the output voltage is between 0.3 and 5.0 V. The NAND gate shown has an output of 0.6 V for logic LOW as the collector–emitter drop for the two transistors in series has to be added up. If we add an additional transistor to add an input, we have additional power dissipation and the output voltage is at 0.9 V for a logic LOW causing problems

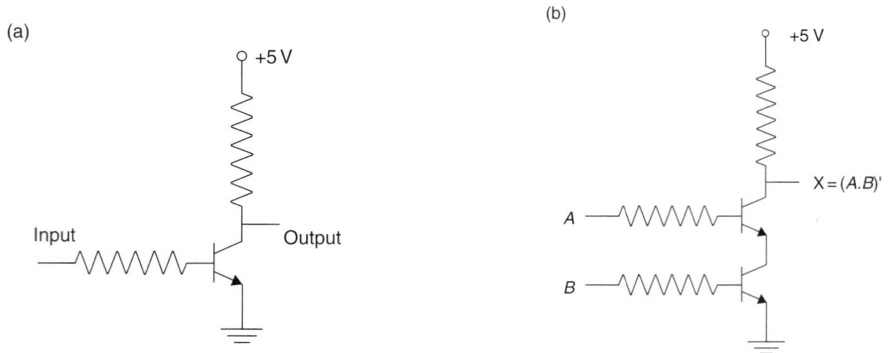

FIGURE 23.19 (a) Resistor–transistor NOT gate, (b) resistor–transistor NAND gate.

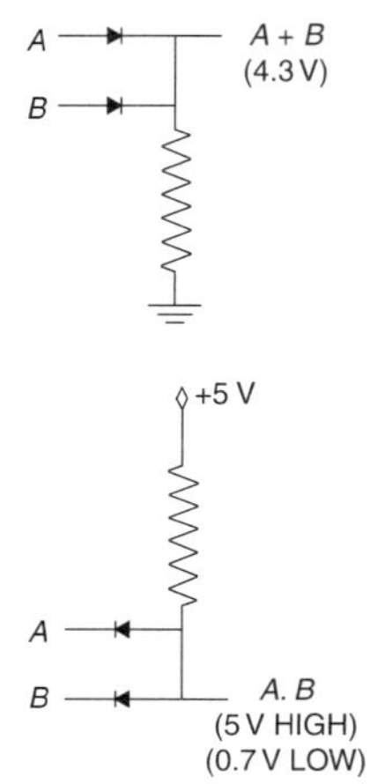

FIGURE 23.20 Diode–resistor logic.

with the logic devices that it is driving. The logic function operation becomes unreliable as output can get into the forbidden region. Further, the presence of resistors in the base–emitter loop tends to slow the device. Because of these limiting characteristics, RTL gates are obsolete.

23.5.2 Diode–Transistor Logic

Diodes themselves can be used to build logic gates for simple applications as shown in Figure 23.20. The diode drops the voltage by 0.7 V across the pn-junction when conducting, resulting in 0.7 V for a LOW and 4.3 V for a HIGH at the output. It is readily seen that the cascading of several of these circuits will push the circuit into the forbidden region, resulting in erroneous logic. Moreover, the diode resistor logic cannot implement an inverter (NOT) function, and it is not practical to produce high density ICs with diodes and resistors. Because of these shortcomings, the diode-resistor logic gate is obsolete.

A diode–transiistor logic (DTL) gate is shown in Figure 23.21. Here, the diodes are used for the OR function and the transistor is used for the NOT function to give a NOR gate. Still the presence of the resistor at the base of the transistor causes power dissipation and reduces the speed of operation. Figure 23.22 shows an improved diode–transistor design that eliminates the bias resistor, thereby improving the speed of operation. DTL devices are obsolete owing to the same limitations discussed earlier.

23.5.3 Transistor–Transistor Logic

In Figure 23.22, it can be observed that the diodes at the input are forward biased while the diode at the base of the transistor is reverse biased when any input is LOW. On the other hand, when all inputs are HIGH, the base diode is forward biased and the transistor conducts, giving the NAND function.

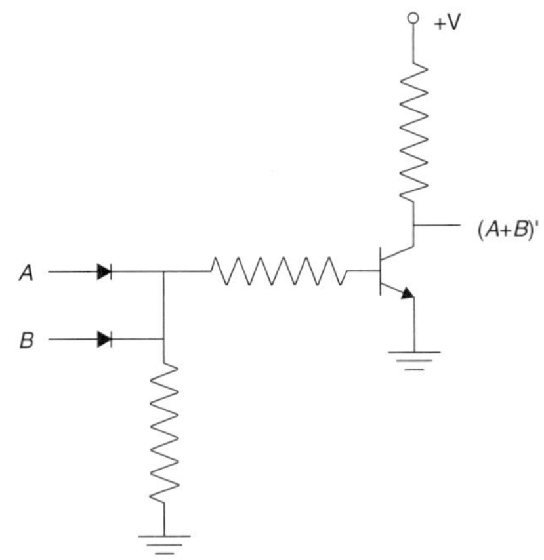

FIGURE 23.21 Diode–transistor logic NOR gate.

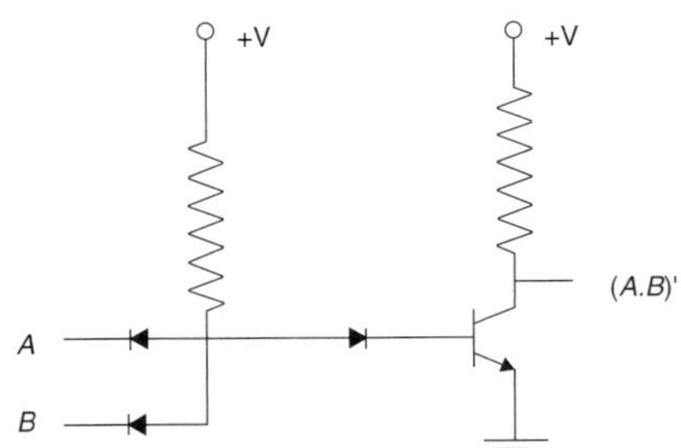

FIGURE 23.22 Improved diode–transistor NAND logic gate.

In a transistor npn, for example, the base–emitter is forward biased during conduction and collector–base is reverse biased. If we have a transistor with multiple emitter leads, then we can use the emitter–base junction for the input diodes. The base–collector junction is used for the base-diode in the DTL gate. The result is a transistor–transistor logic (TTL) gate implementation of a NAND function in Figure 23.23 [4].

Here, when any of the inputs is LOW, the base–emitter loop conducts and the emitter of the first stage transistor is at 0.2 V, giving HIGH for output at the inverter. When all the inputs are HIGH, the transistor multiple emitters (first stage) is cutoff. Therefore, all other transistors conduct with a logic LOW at output. The manufacturer's data sheet for each device provides circuit diagrams, and all technical data including maximum and minimum input values, propagation delay, rise and fall times, fan-out, fan-in limitations, power consumption, and application suggestions. These are excellent sources of information for the designer.

23.5.4 Emitter-Coupled Logic

Emitter-coupled logic (ECL) devices are bipolar devices in which the transistor is never saturated or completely shut off. The result is very high speed compared to TTL or CMOS implementations. The ECL gates are used in several applications where high speed is essential, for example, computer cache memory. Figure 23.24 shows a NOR/OR gate [5]. V_{CC} is connected to ground (0 V) while V_{EE} is connected to supply voltage, −5.2 V for better noise immunity. The transition time from one state to another is less than 1 ns, resulting in several gigahertz operating frequency when ECL gates are used.

Because the transistors are not fully saturated, the ECL gates output at HIGH and LOW are about −0.75 and −1.6 V, respectively. The bias voltage is set at the base of transistor Q_4, the value is the average

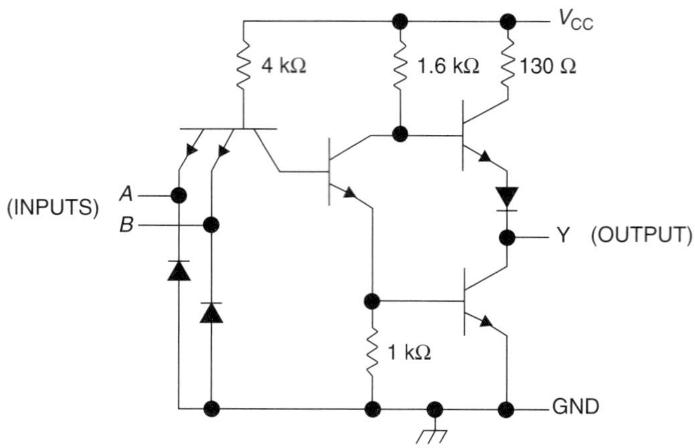

FIGURE 23.23 Transistor–transistor logic implementation of a NAND gate [4].

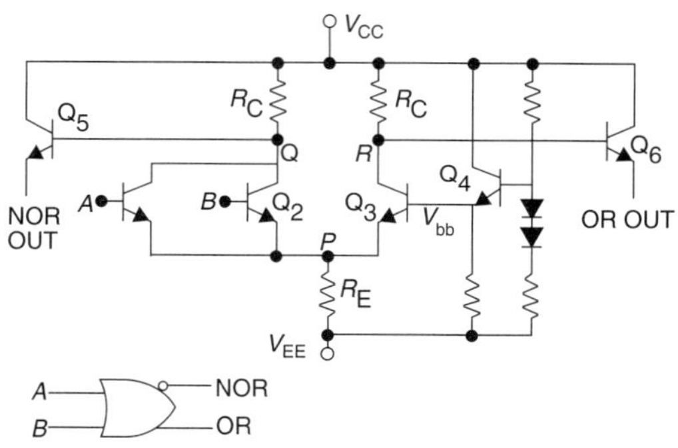

FIGURE 23.24 Emitter-coupled logic NOR/OR gate [5].

of HIGH and LOW values for the gate; in this case, the value would be –1.175 V. Next, the resistors are selected to control the current flow and prevent transistor saturation. When the signal at *A* or *B* is HIGH (–0.75 V), the transistors conduct. The voltage at point *P* becomes –1.5 V (V_{BE} of $Q_2 + V_A$). This reduces the difference between V_{bb} and the voltage at point *P* below the threshold for transistor Q_3 to conduct and hence it is off, raising the voltage at point *R* to 0 V. This turns transistor Q_6 on. With a V_{BE} threshold of 0.75 V, the measured OR output is –0.75 V, a logic HIGH. The value of resistance R_C and R_E are chosen so that the voltage at point Q is –0.85 V when transistor Q_1 or Q_2 is conducting. It is true when *B* is HIGH and *A* is LOW or when *A* and *B* are both HIGH. When both *A* and *B* are LOW, the transistors Q_1 and Q_2 are off and Q_3 conducts lowering the voltage at point *P* to –1.925 V. Voltage at point *R* is –0.85 V resulting in an OR output of –1.6 V at the OR output and –0.75 V at the NOR output.

Because of the constant operation of the transistors in the active region, there is continuous current draw and hence heat dissipation. The ECL devices draw four to five times the power of a comparable TTL device. Hence, this is used cautiously as front-end devices where speed is essential, while using HCMOS or TTL gates elsewhere. In order to mix ECL gates with TTL or CMOS devices, special level shifters are used, for example, National Semiconductor's 100325 Low Power ECL-to-TTL Translator. This device converts an ECL input (–0.75 (H) and –1.6 (L)) to a TTL output (2.4 V min for HIGH and 0.5 V max. for LOW), while maintaining a rise or fall time of less than 1 ns.

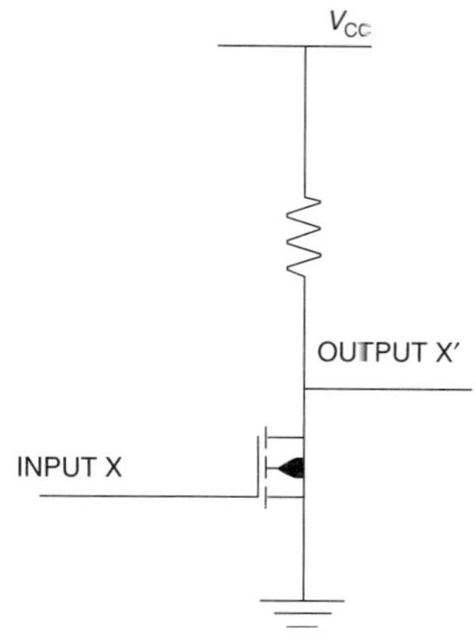

FIGURE 23.25 An NMOS inverter.

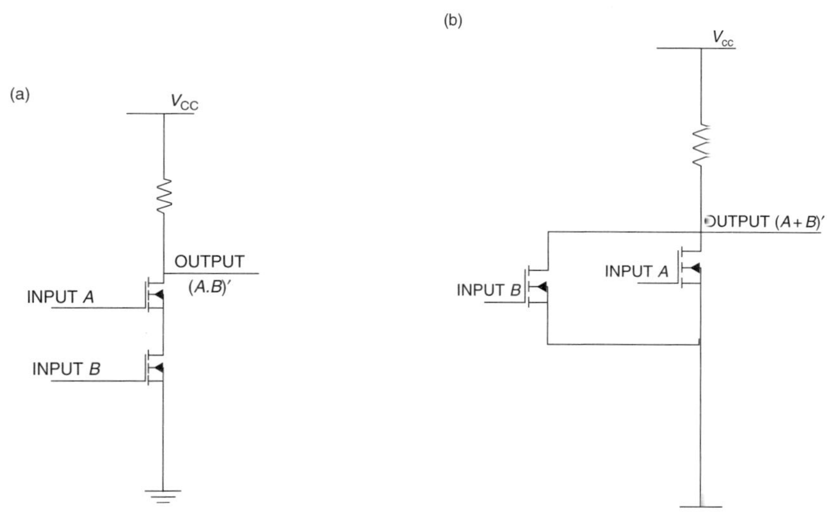

FIGURE 23.26 (a) An NMOS NAND gate, (b) an NMOS NOR gate.

23.5.5 CMOS Logic

As discussed earlier, MOSFET can be used as a transistor switch without significant power dissipation. NMOS logic gates are designed with n-MOSFETs and PMOS logic gates are designed with p-MOSFET transistors. As an example, an NMOS inverter is shown in Figure 23.25. Figure 23.26a shows a logic NAND function when two n-MOSFETs are connected in series and Figure 23.26b shows a parallel arrangement of two n-MOSFETs to give a NOR gate. The NMOS circuits shown have a pull-up resistor and a pull-down n-MOSFET. To eliminate the resistor, the pull-up side of the circuit is replaced with p-MOSFET. The modified NOT or inverter circuit is shown in a commercial implementation in Figure 23.27 [3]. Additional diodes are shown for static protection of the device. This is known as a CMOS circuit since the pull-down and pull-up parts of the circuit have complimentary MOSFET devices. When two n-MOSFETs are connected in the pull-down side of the circuit in series, the pull-up resistor is replaced by two p-MOSFETs in parallel, and vice versa. A CMOS implementation of the NOR gate is shown in Figure 23.28.

An important characteristic of CMOS gates is their low-power consumption as there is practically no current flow in both HIGH and LOW states. However, the device is slower than a bipolar transistor device.

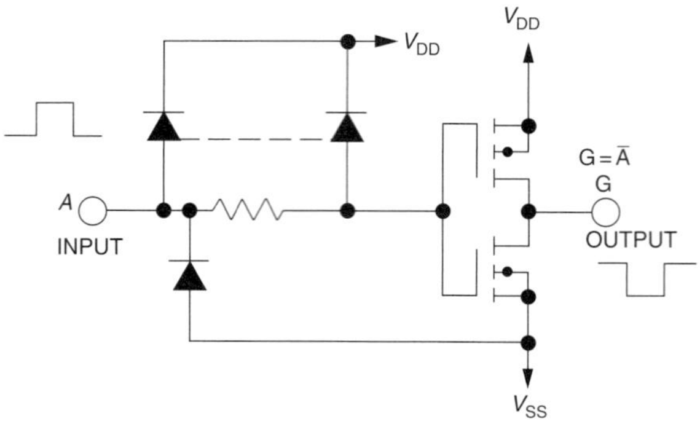

FIGURE 23.27 A CMOS inverter [3].

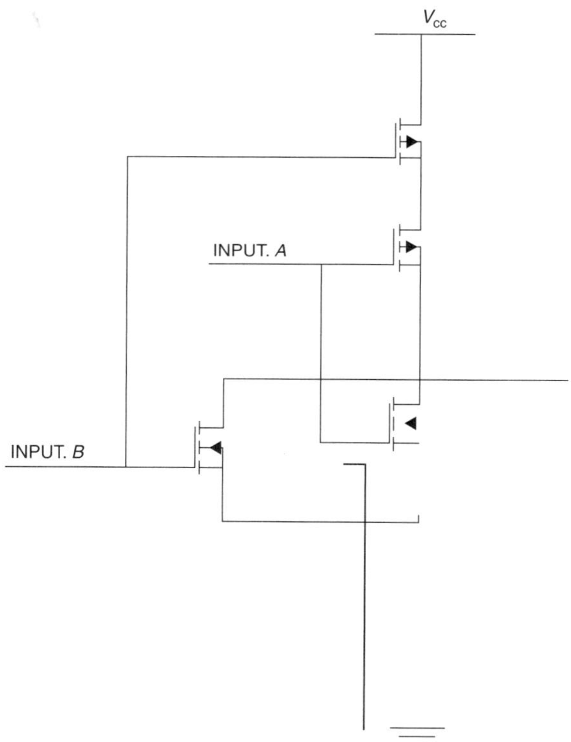

FIGURE 23.28 A CMOS NOR gate.

With decreasing transistor sizes due to advancements in fabrication technologies, the speed of CMOS devices continue to increase.

23.6 Logic Gate Integrated Circuits

A commercial logic gate ICs has several gates of the same type on it. For example, Figure 23.29 shows a commercial quad-AND gate IC. The chip itself is powered with V_{CC} and GND pins, *A* and *B* pins are inputs, and the *Y* pins are the corresponding outputs. You can use one or all of the gates on a chip as needed.

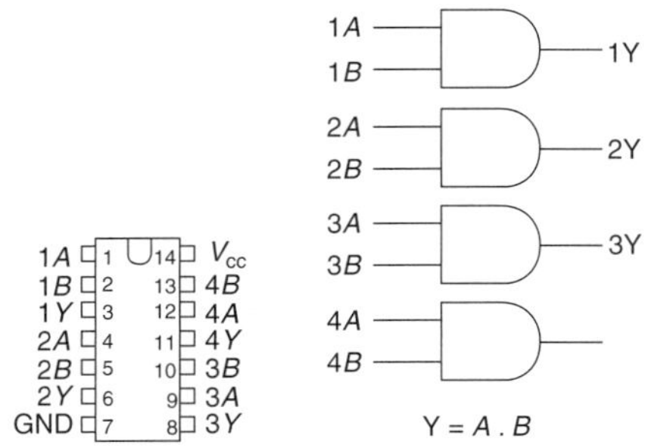

FIGURE 23.29 A DIP package of quad-AND gate IC.

The designation of the chips requires some attention. Let us take the Texas Instruments designations for an AND gate, namely, SN5408, SN54LS08, SN54S08, SN7408, SN74LS08, and SN74S08. While the designation shown on a device has much more information than what is shown here, the basic information that we should be aware of is the function designation (00 = NAND gate, 02 = NOR, 04 = Inverter, 08 = AND, etc.), and technology type (HC for high-speed CMOS, LS for low-power Schottky, etc.). For other notations used in chip designations, refer to the Texas Instruments Logic Selection Guide [1].

23.7 Programmable Logic Devices (PLD)

Programmable logic devices (PLDs) are ICs with several uncommitted logic gates in them, the connections among which are programmable based on the logic circuit design that needs to be implemented. This is especially helpful when very large circuits consisting of several thousands of logic gates have to be built and tested. For large circuit design and testing, it is not practical to use standard logic gate ICs since each IC has at most four or six logic gates on it, requiring large circuit boards and interconnects. The PLD consists of several hundred logic gates on it and the device design is programmable with a special programming hardware. When more than one PLD is used to implement a design, programmable interconnects are used between PLDs. One type of fully PLD, called the programmable logic array (PLA), consists of an AND level in the middle and an OR level at output, similar to a TTL single logic gate structure, with both layers being programmable. All input signals are connected to an inverter level, which provides both the normal and complemented values of input variables to the AND level. Appropriate connections are made at the AND level and at the OR level to produce the desired logic outputs. In this device all the levels are programmable.

A simpler version of PLD, called a programmable array logic (PAL) device, consists of a programmable AND layer and a fixed OR layer. This is easier and less expensive to manufacture, although it is not as flexible as a PLA. A variety of combinations is available to suit various needs A schematic of a PLA is shown in Figure 23.30 [6] where the connections to be made in the hardware are marked with an X. When programmed these connections will be made or "fused" and verified by the programming hardware.

23.8 Mechatronics Application Example

A driver circuit for a DC motor is a good example for the use of transistors and logic gates. The objectives of the design are the following:

1. The motor should drive forward and reverse at different speeds.
2. The motor should either coast to a stop or brake abruptly.
3. The motor should drive at different speeds, controllable by a microprocessor.

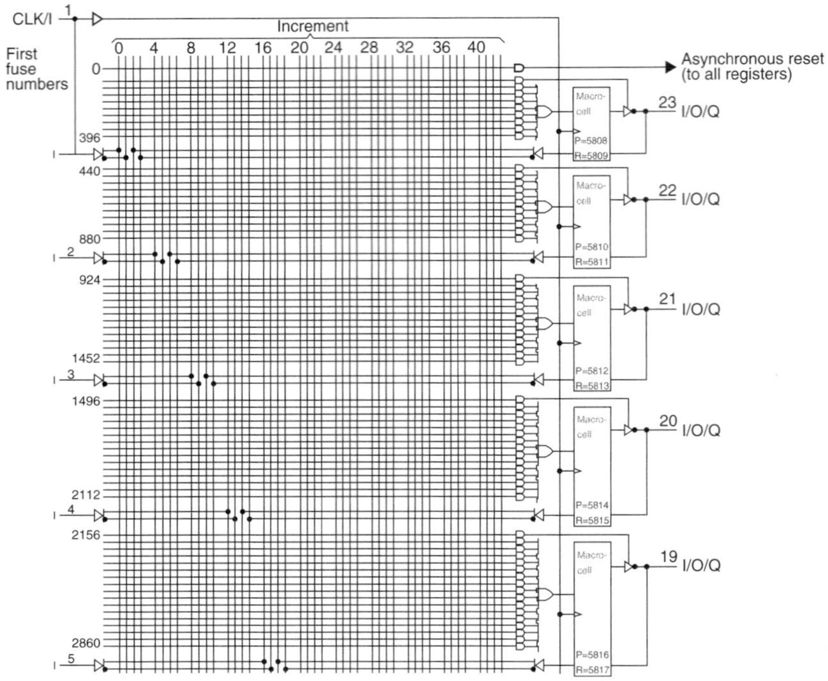

FIGURE 23.30 Programmable logic array [6].

The complete logic and power circuit implementation of the solution to this design problem is shown in Figure 23.31, which is known as the H-bridge. The motor is connected between the output pins (out1 and out2) [7]. The EN (enable) and IN1 (input 1) and IN2 (input 2) are the inputs. The behavior of the system is given by the adjacent table in Figure 23.31. When the enable signal is LOW, regardless of the input states, all the AND gates are LOW, and the power transistors are all off and the motor is off. If the motor is moving when the enable line switches to LOW, the motor coasts to a stop. When the enable input is HIGH, it can be seen that when IN1 is high and IN2 is LOW, transistors 1 and 4 are on, and 2 and 3 are off. This drives the motor one way as the current can flow through the motor to the ground through the two diagonal transistors. Since transistors 2 and 3 are off, short circuit from power to ground is prevented. This is designed by inputting the complements of IN1 and IN2 to the AND gates driving transistors 3 and 4, respectively. When IN2 is HIGH and IN1 is LOW, the motor runs in the opposite direction (while the enable is HIGH). Since transistors 2 and 4 are closed and 1 and 3 are open, current flows in the opposite direction through the motor.

When enable is HIGH, and the inputs IN1 and IN2 are either turned HIGH or LOW at the same time while the motor is moving, then the motor terminals are forced to V_{CC} or ground. However, the motor power is off since IN1 and IN2 are LOW. Now, the motor is a generator trying to maintain a potential difference across its terminal as the rotor moves in a magnetic field. The emf generated is forced to the source or sink potential. This brings the motor to a rapid stop, identified as the fast stop or the braking function. Further, the IN1 and IN2 lines can be used for direction and braking functions, while the enable can be pulsed at different duty cycle levels (pulse width modulation) to achieve different speeds. Since the motor is free running when enable is LOW regardless of input, as EN is switched rapidly, the inertia of the rotor helps smooth out the motion. The selection of pulse repetition time (PRT) and arrangement of pulses within the PRT in a uniform fashion to produce desired PWM signals should be done to fine-tune the performance of this system.

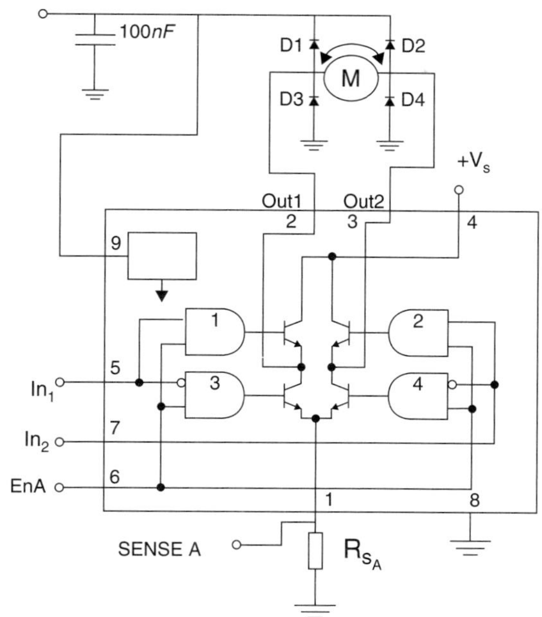

INPUTS		FUNCTION
ENA=H	In_1=H In_2=L	FORWARD
	In_1=L In_2=H	REVERSE
	In_1=In_2	FAST MOTOR STOP
ENA=L	In_1=X In_2=X	FREE RUNNING MOTOR STOP

FIGURE 23.31 H-bridge motor driver circuit [7].

References

1. "Logic Selection Guide, First Half 2001," Texas Instruments, Document sdyu001o.pdf. Source: www.ti.com.
2. "Designing with Logic," Texas Instruments, Document sdya009C.pdf. Source: www.ti.com.
3. "CD4069UB Types- Quad-Inverter," Texas Instruments, Datasheet, schs054.pdf, 1998. Source: www.ti.com.
4. "SN5400 Quadruple 2-Input Positive NAND-Gates," Texas Instruments, Datasheet, sdls025.pdf, March 1988, Source: www.ti.com.
5. Koga, R., Crain, W.R., Hansel, S.J., Crawford, K.B., Pinkerton, S.D., Peozin, S.H., Moses, S.C., and Maher, M., "Ion Induced Charge Collection and SEU Sensitivity of Emitter Coupled Logic (ECL) Devices," *IEEE Trans. on Nuclear Science,* 42(6), 1823–1828, 1995.
6. High-performance *Impact-X*™ Programmable Array Logic Circuits, TIBPAL22V10-7C, TI, 1995. Product datasheet.
7. "Dual Full-Bridge Driver L298," SGS Thomson Microelectronics Datasheet. Source: www.st.com.

24
Architecture

Daniel A. Connors
University of Colorado at Boulder

Wen-Mei W. Hwu
University of Illinois

The microprocessor industry is divided into the computer and embedded sectors. Both computer and embedded microprocessors share aspects of computer system design, instruction set architecture, organization, and hardware. The term **architecture** is used to describe these fundamental aspects, and more directly refers to the hardware components in a computer system and the flow of data and control information among them. In this section, various types of microprocessors will be described, fundamental architecture mechanisms relevant in the operation of all microprocessors will be presented, and microprocessor industry trends discussed.

24.1 Types of Microprocessors

Computer microprocessors are designed for use as the central processing units (CPU) of computer systems such as personal computers, workstations, servers, and supercomputers. Although microprocessors started as humble programmable controllers in the early 1970s, virtually all computer systems built after 1990 use microprocessors as their central processing units. The dominating architecture in the computer microprocessor domain today is the Intel 32-bit Architecture, also known as IA-32 or X86. Other high profile architectures in the computer microprocessor domain include Intel Itanium Processor Family (IPF), HP PA-RISC, SUN Microsystems SPARC, and IBM/Motorola PowerPC.

Embedded microprocessors are increasingly used in consumer and telecommunications products to satisfy the demands for quality and functionality. Major product areas that require embedded microprocessors include digital TV, DVD players, digital camera, network switches, high speed modems, digital cellular phones, video games, laser printers, and automobiles. Future improvements in energy consumption, fabrication cost, and performance will further enable new applications such as intelligent hearing aid. Many experts expect that embedded microprocessors will form the fastest growing sector of the semiconductor business in the next decade [1].

Embedded microprocessors have been categorized into DSP processors and Embedded CPUs owing to historic reasons. DSP processors have been designed and marketed as special purpose devices that are mostly programmed by hand to perform digital signal processing computations. A recent trend in the DSP market is to use compilers to alleviate the need for tedious hand coding in DSP development. Leading DSP microprocessor vendors include Texas Instruments, Lucent Technologies, Motorola, and ST.

Embedded CPUs are often used to execute operating system, networking, and user interface code in a consumer and telecommunications products. They have been traditionally derived from out-of-date computer microprocessors. Embedded CPUs often reuse the compiler and related software support developed for their computer cousins. Recycling the microprocessor design and compiler software minimizes engineering cost. Major vendors of embedded CPUs include IBM, Motorola, ARM, and MIPS.

An important recent trend in the embedded microprocessor market is toward integrating an embedded CPU, a DSP processor, and application-specific logic to form a single-chip solution. This approach is enabled by the ever-increasing transistor density achieved by the semiconductor fabrication technology. The major benefit is reduced system cost and energy consumption. In these designs, embedded CPU and DSP processor vendors no longer market microprocessor chips. Rather, they make their designs available for licensing by solution developers such as a cell phone vendor. The embedded CPU and the DSP processors thus incorporated into these single-chip solutions are called embedded CPU cores and DSP processor cores. For example, MIPS customized its embedded CPU core for use in Nitendo64. IBM, ARM, ST, NEC, and Hitachi offer similar products and services. Because of an increasing need to perform DSP computation in consumer and telecommunication products, an increasing number of embedded CPUs have extensions to enable more effective DSP computation.

There are several ways in which the needs of embedded computing differ from those of more traditional general-purpose computing systems. Constraints on the code size, weight, and power consumption place stringent requirements on embedded microprocessors and the software they execute. Also, constraints rooted in real-time requirements are often a significant consideration in many embedded systems. Furthermore, cost is a severe constraint in designing and manufacturing embedded processors.

In spite of the different constraints and product markets, both computer and embedded microprocessors share many main elements in their design. These main elements will be described. Additionally, over the past decade, substantial research has gone into the design of microprocessors embodying parallelism at the instruction level, as well as aggressive compiler optimization and analysis techniques for harnessing this opportunity. Exploitation of parallelism is a very effective approach to reducing the power consumption under given performance requirements. Much of this effort has since been validated through the proliferation of high-performance general purpose microprocessors, such as the Intel Itanium II processor, on the basis of these technologies. Nevertheless, growing demand for high performance in embedded computing systems is creating new opportunities to leverage these techniques in application-specific domains. The research of instruction-level parallelism (ILP) has developed distinct architecture methodologies referred to as very long instruction word (VLIW) and explicitly parallel instruction computing (EPIC) technology. Overall, these techniques represent fundamental, substantial changes in computer architecture.

24.2 Major Components of a Microprocessor

The main hardware of a microprocessor system can be divided into sections according to their functionalities. A popular approach is to divide a system into four subsystems: the central processor, the memory subsystem, the input/output (I/O) subsystem, and the system interconnect. Figure 24.1 shows

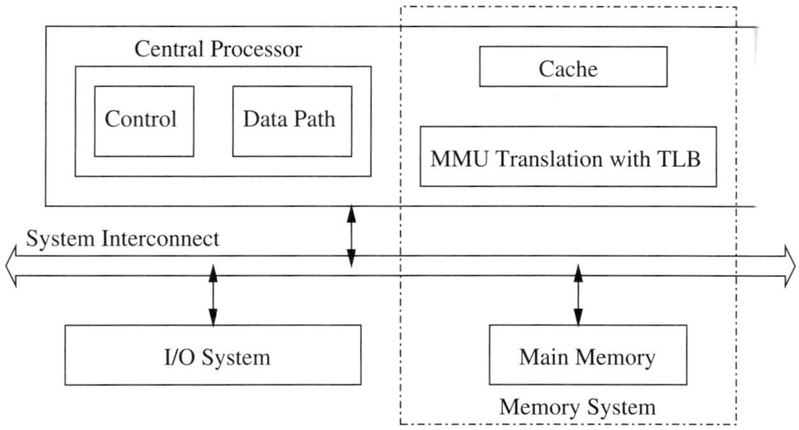

FIGURE 24.1 Subsystems of a computer system.

the connection between these subsystems. The main components and characteristics of these subsystems will be described.

24.2.1 Central Processor

A modern microprocessor system's central processor is typically further divided into control unit and data path.

24.2.1.1 Control Unit

The control unit of a microprocessor generates the control signals to orchestrate the activities in the data path. There are two major types of communication lines between the control unit and the data path: the control lines and the condition lines. The control lines deliver the control signals from the control unit to the data path. Different signal values on these lines trigger different actions in the data path. The condition lines carry the status of the execution from data path to the control unit. These lines are needed to test conditions involving the registers in the data path in order to make future control decisions. Note that the decision is made in the control unit but the registers are in the data path. Therefore, the conditions regarding the register contents are formed in the data path and then communicated to the control unit for decision making. A control unit can be implemented with hardwiring, microprogramming, or a combination of both.

A hardwired control unit is designed as a finite state machine that is realized with registers, combinational logic, and wires. Once constructed, the design can be changed only through physically rewiring the unit. Therefore, the resulting circuits are called hardwired control units. Design optimizations are typically performed on the combinational logic to minimize component count and maximize operation speed, which makes the resulting circuits exhibit little structure. The lack of structure makes it very difficult to design and debug complicated control units with this technique. Therefore, hardwiring is normally used when the control unit is relatively simple.

Most of the design difficulties in the hardwired control units are due to the effort of optimizing the combinational logic. One popular alternative to hardwired control unit design is to use either read only memory (ROM) or random access memory (RAM) to implement the combinational logic. A control unit whose combinational logic is realized by the use of ROM or RAM is called a *microprogrammed control unit*. The memory used is called *control memory* (CM). The practice of realizing the combinational circuit in a control unit with ROM/RAM is called *microprogramming*. The concept of microprogramming was first introduced by Maurice V. Wilkes.

The idea of using a memory to implement a combinational circuit can be illustrated with a simple example. Assume that we are to implement a logic function with three input variables, as described in the

(a)

A	B	C	F
0	0	0	0
0	0	1	0
0	1	0	1
0	1	1	0
1	0	0	1
1	0	1	1
1	1	0	0
1	1	1	0

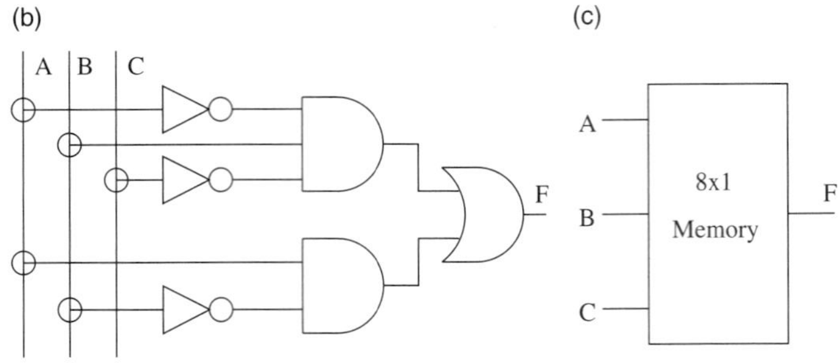

FIGURE 24.2 Using memory to simplify combinational logic design (a) truth table, (b) Karaugh map approach, and (c) memory approach.

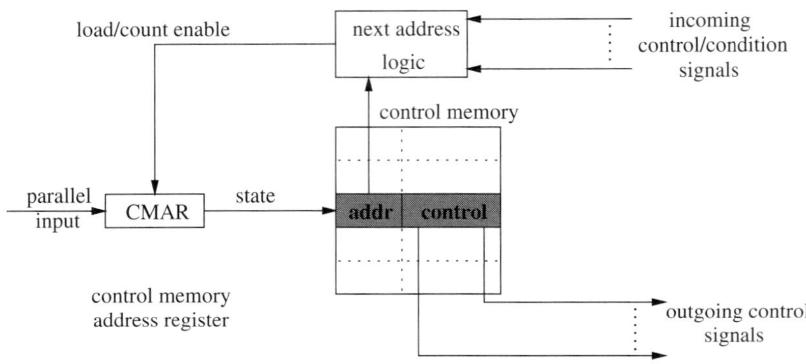

FIGURE 24.3 Basic model of microprogrammed control units.

truth table illustrated in Figure 24.2a. A common way to realize this function is to use Karnaugh maps to derive highly optimized logic and wiring. The result is shown in Figure 24.2b. The same function can also be realized using a memory with eight 1-bit locations to retain the eight possible combinations of the three input variables. Location i contains an F value corresponding to the *ith* input combination. For example, location 3 contains the F value (0) for the input combination 011. The three input variables are then connected to the address input of the memory to complete the design (Figure 24.2c). In essence, the memory contains the entire truth table. Considering the decoding logic and storage cells involved in a 8x1 memory, it is obvious that the memory approach uses a lot more hardware components than the Karnaugh map approach. However, the design is much simpler in the memory approach.

Figure 24.3 illustrates the general model of a microprogrammed control unit. Each control memory location consists of an address field and some control fields. The address field plus the next address logic implements the combinational logic for generating the next state value. The control fields implement the combinational logic for generating the control signals. The state register/counter is referred to as *control memory address register* (CMAR) for an obvious reason: the contents of the register are used as the address input to the control memory. An important insight is that the CMAR stores the state of the control unit.

In modern microprocessor designs, hardwiring and microprogramming are often used in conjunction with each other. In such a design, hardwiring is used to handle common, simpler cases of instruction execution whereas microprogramming is used to handle complex cases such as string move instructions in the IA-32 microprocessors.

24.2.1.2 Data Path

The data path of a microprocessor contains the main arithmetic and logic execution units required to execute instructions. Designing the data path involves analyzing the function(s) to be performed, then specifying a set of hardware registers to hold the computation state, and designing computation steps to transform the contents of these registers into the final result. In general, the functions to be performed are divided into steps, each of which can be done with a reasonable amount of logic in one clock cycle. Each step brings the contents of the registers closer to the final result. The data path must be equipped with sufficient amount of hardware to allow these computation steps in one clock cycle. The data path of a typical microprocessor contains integer and floating-point register files, ten ormore functional units for computation and memory access, and pipeline registers. The concept of pipelining and pipeline registers will be introduced later in this chapter.

24.2.2 Memory Subsystem

The memory system serves as a repository of information in a microprocessor system. The processing unit retrieves information stored in memory, operates on the information, and returns new information back to memory. The memory system is constructed from basic semiconductor DRAM units called modules or banks.

There are several properties of memory, including speed, capacity, and cost that play an important role in the overall system performance. The speed of a memory system is the key performance parameter in the design of the microprocessor system. The latency (L) of the memory is defined as the time delay from when the processor first requests data from memory until the processor receives the data. Bandwidth is defined as the rate at which information can be transferred to and from the memory system. Memory bandwidth and latency are related to the number of outstanding requests (R) that the memory system can service:

$$\mathrm{BW} = \frac{R}{L} \tag{24.1}$$

Bandwidth plays an important role in keeping the processor busy with work. However, technology tradeoffs to optimize latency and improve bandwidth often conflict with the need to increase the capacity and reduce the cost of the memory system.

24.2.2.1 Cache Memory

Cache memory, or simply cache, is a fast memory constructed using semiconductor SRAM. In modern computer systems, there is usually a hierarchy of cache memories. The top level (Level-1 or L1) cache is closest to the processor and the lowest level is closest to the main memory. Each lower level cache is about 3–5 times slower than its predecessor level. The purpose of a cache hierarchy is to satisfy most of the processor memory accesses in one or a small number of clock cycles. The L1 cache is often split into an instruction cache and a data cache to allow the processor to perform simultaneous accesses for instructions and data. Cache memories were first used in the IBM mainframe computers in the 1960s. Since 1985, cache memories have become a standard feature for virtually all microprocessors.

Cache memories exploit the principle of locality of reference. This principle dictates that some memory locations are referenced more frequently than others on the basis of two program properties. *Spatial locality* is the property that an access to a memory location increases the probability that the nearby memory location will also be accessed. Spatial locality is predominantly based on sequential access to program code and structured data. In order to exploit spatial locality, memory data are placed into cache inmultiple-word units called *cache lines*. *Temporal locality* is the property that access to amemory location greatly increases the probability that the same location will be accessed in the near future. Together, the two properties assure that most memory references will be satisfied by the cache memory.

Set associativity refers to the flexibility in placing a memory data into a cache memory. If a cache design allows each memory data to be placed into any of N cache locations, it is referred to as an N-way set-associative cache. A 1-way set-associative cache is also called a direct-mapped cache, as is shown in Figure 24.4a. A 2-way set-associative cache, as shown in Figure 24.4b, allows a memory data to reside in

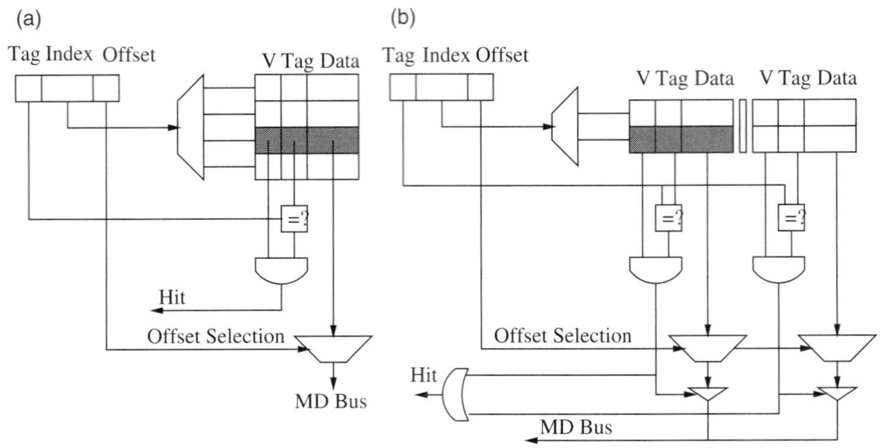

FIGURE 24.4 Cache memory (a) direct mapped design and (b) two-way set associative design.

one of two locations in the cache. Finally, an extreme design called fully associative cache allows amemory data to be placed anywhere in the cache.

Cache misses occur when the data requested does not reside in any of the possible cache locations. Misses in caches can be classified into three categories: conflict, compulsory, and capacity. Conflict misses are misses that would not occur for fully associative caches with least recently used (LRU) replacement. Compulsory misses are misses required in cache memories for initially referencing a memory location. Capacity misses occur when the cache size is not sufficient to keep data in the cache between references. Complete cache miss definitions are provided in Reference 2.

The latency in cache memories is not fixed and depends on the delay and frequency of cache misses. A performance metric that accounts for the penalty of cache misses is *effective latency*. Effective latency depends on the two possible latencies: hit latency (L_{HIT}), the latency experienced for accessing data residing in the cache; and miss latency (L_{MISS}), the latency experienced when accessing data not residing in the cache. Effective latency also depends on the *hit rate* (H), the percentage ofmemory accesses that are hits in the cache, and *miss rate* (M or 1 - H), the percentage of memory accesses that miss in the cache. Effective latency in a cache system is calculated as

$$L_{\mathrm{effective}} = L_{\mathrm{HIT}} \times H + L_{\mathrm{MISS}} \times (1 - H) \tag{24.2}$$

In addition to the base cache design and size issues, there are several other dimensions of cache design that affect the overall cache performance and miss rate in a system. The main memory update method dictates when the main memory will be updated by store operations. In a *write-through* cache, each write is immediately reflected to the main memory. In a *write-back* cache, the writes are reflected to main memory only when the respective cached data is purged from cache to make room for other memory data. Cache allocation designates whether cache locations are allocated on writes and/or reads. Finally, cache replacement algorithms for associative structures can be designed in various ways to extract additional cache performance. These include LRU, least frequently used (LFU), random, and first in, first out (FIFO). These cache management strategies attempt to exploit the properties of locality. Traditionally, when caches service misses they would *block* all new requests. However, *nonblocking* cache can be designed to service multiple miss requests simultaneously, thus alleviating delay in accessing memory data.

In addition to the multiple levels of cache hierarchy, additional memory buffers can be used to improve cache performance. Two such buffers are a streaming/prefetch buffer and a victim cache [3]. Figure 24.5 illustrates the relation of the streaming buffer and victim cache to the primary cache of a memory system. A streaming buffer is used as a prefetching mechanism for cache misses. When a cache miss occurs, the streaming buffer begins prefetching successive lines starting at the miss target. A victim cache is typically

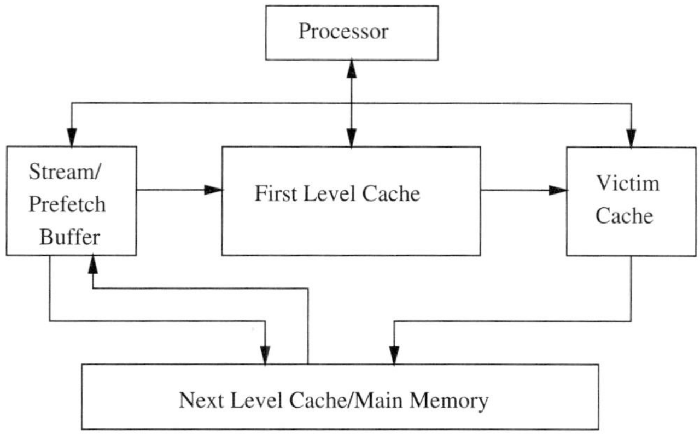

FIGURE 24.5 Advanced cache memory system.

a small fully associative cache loaded only with cache lines that are removed from the primary cache. In the case of a miss in the primary cache, the victim cache may hold the missed data. The use of a victim cache can improve performance by reducing the number of conflict misses. Figure 24.5 illustrates how cache accesses are processed through the streaming buffer into the primary cache on cache requests and from the primary cache through the victim cache to the secondary level of memory on cache misses.

Overall, cache memory is constructed to hold the most important portions of memory. Techniques using either hardware or software can be used to select which portions of main memory to store in cache. However, cache performance is strongly influenced by program behavior and numerous hardware design alternatives.

24.2.2.2 Virtual Memory

Cache memory illustrated the principle that the memory address of data can be separate froma particular storage location. Similar address abstractions exist in the two-level memory hierarchy of main memory and disk storage. An address generated by a program is called a *virtual address*, which needs to be translated into a *physical address* or location in main memory. Virtual memory management is a mechanism that provides the programmers with a simple, uniform method to access both main and secondary memories. With virtual memory management, the programmers are given a virtual space to hold all the instructions and data. The virtual space is organized as a linear array of locations. Each location has an address for convenient access. Instructions and data have to be stored somewhere in the real system; these virtual space locations must correspond to some physical locations in the main and secondary memory. Virtual memory management assigns (or maps) the virtual space locations into the main and secondary memory locations. The mapping of virtual space locations to the main and secondary memory is managed by the virtual memory management mechanism. The programmers are not concerned with the mapping.

The most popular memory management scheme today is demand paging virtual memory management, where each virtual space is divided into pages indexed by the page number (PN). Each page consists of several consecutive locations in the virtual space indexed by the page index (PI). The number of locations in each page is an important system design parameter called page size. Page size is usually defined as a power of two so that the virtual space can be divided into an integer number of pages Pages are the basic units of virtual memory management. If any location in a page is assigned to the main memory, the other locations in that page are also assigned to main memory. This reduces the size of the mapping information.

The part of the secondary memory to accommodate pages of the virtual space is called the swap space. Both the main memory and the swap space are divided into page frames. Each page frame can host a page of the virtual space. The mapping record in the virtual memory management keeps track of the association between pages and page frames.

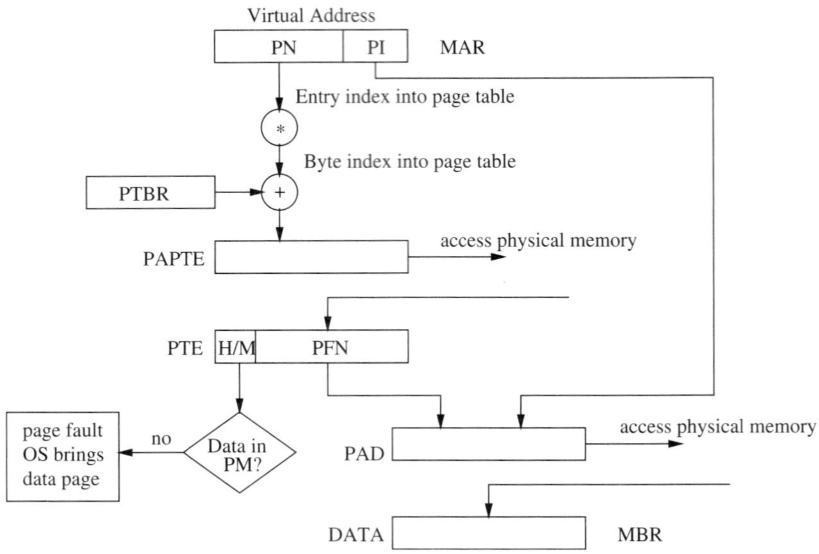

FIGURE 24.6 A simple virtual memory access algorithm.

When a virtual space location is requested, the virtual memory management looks up the mapping record. If the mapping record shows that the page containing requested virtual space location is in main memory, the management performs the access without any further complication. Otherwise a secondary memory access has to be performed. Accessing the secondary memory is a complicated task and is usually performed as an operating system service.When a page is mapped into the secondary memory, the virtual memory management has to request a service in the operating system to transfer the requested page into the main memory, update its mapping record, and then perform the access. The operating system service thus performed is called the page fault handler.

The core process of virtual memory management is a memory access algorithm. A one-level memory access algorithm is illustrated in Figure 24.6. At the start of the memory access, the algorithm receives a virtual address in a memory address register (MAR), looks up the mapping record, requests an operating system service to transfer the required page if necessary, and performs the main memory access. The mapping is recorded in a data structure called the page table located in main memory at a designated location marked by the page table base register (PTBR).

Each page is mapped by a page table entry (PTE) that occupies a fixed number of bytes in the page table. Thus, one can easily multiply the page number by the size of each PTE to form a byte index into the page table. The byte index of the PTE is then added to the PTBR to form the physical address (PAPTE) of the required PTE. Each PTE includes two fields: a hit/miss bit and a page frame number. If the hit/miss (H/M) bit is set (hit), the corresponding page is in main memory. In this case, the page frame hosting the requested page is pointed to by the page frame number (PFN). The final physical address (PAD) of the requested data is then formed by concatenating the PFN and PI. The data is returned and placed in the memory buffer register (MBR) and the processor is informed of the completed memory access. Otherwise (miss), a secondary memory access has to be performed. In this case, the page frame number should be ignored. The fault handler has to be invoked to access the secondary memory. The hardware component that performs the address translation part of the memory access algorithm is called the memory management unit (MMU).

The complexity of the algorithm depends on the mapping structure. A very simple mapping structure is used in this section to focus on the basic principles of the memory access algorithms. However, more complex two-level schemes are often used when the page table becomes a large portion of the main memory. There are also requirements for such designs in a multiprogramming system, where there are multiple processes active at the same time. Each processor has its own virtual space and therefore its own page table. As a result, these systems need to keep multiple page tables at the same time. It usually takes too much main memory to accommodate all the active page tables. Again, the natural solution to this problem is to provide additional levels of mapping where a second-level page table is used to map the main page table. In such designs, only the second-level page table is stored in a reserved region of main memory, while the first page table is mapped just like the code and data between the secondary storage and the main memory.

24.2.2.3 Translation Lookaside Buffer

Hardware support for virtual memory management generally includes a translation lookaside buffer (TLB) to accelerate the translation of virtual addresses into physical addresses. A TLB is a cache structure that contains the frequently used page table entries for address translation.With a TLB, address translation can be performed in a single clock cycle when TLB contains the required page table entries (TLB hit). The full address translation algorithm is performed only when the required page table entries are missing from the TLB (TLB miss).

Complexities arise when a system includes both virtual memory management and cache memory. The major issue is whether address translation is done before accessing the cache memory. In *virtual* cache systems, the virtual address directly accesses cache. In a *physical* cache system, the virtual address is translated into a physical address before cache access. Figure 24.7 illustrates the both the *virtual* and *physical* cache approaches.

A physical cache system typically overlaps the cache memory access and the access to the TLB. The overlap is possible when the virtual memory page size is larger than the cache capacity divided by the degree of cache associativity. Essentially, since the virtual page index is the same as the physical address index, no translation for the lower indexes of the virtual address is necessary. These lower index bits can be used to access the cache storage while the PN bits go through the TLB. The PFN bits that come out of the TLB are compared with the tag bits of the cache storage output, as shown in Figure 24.6, to determine the hit/miss status of cache memory access. Thus, the cache storage can be accessed in parallel with the TLB.

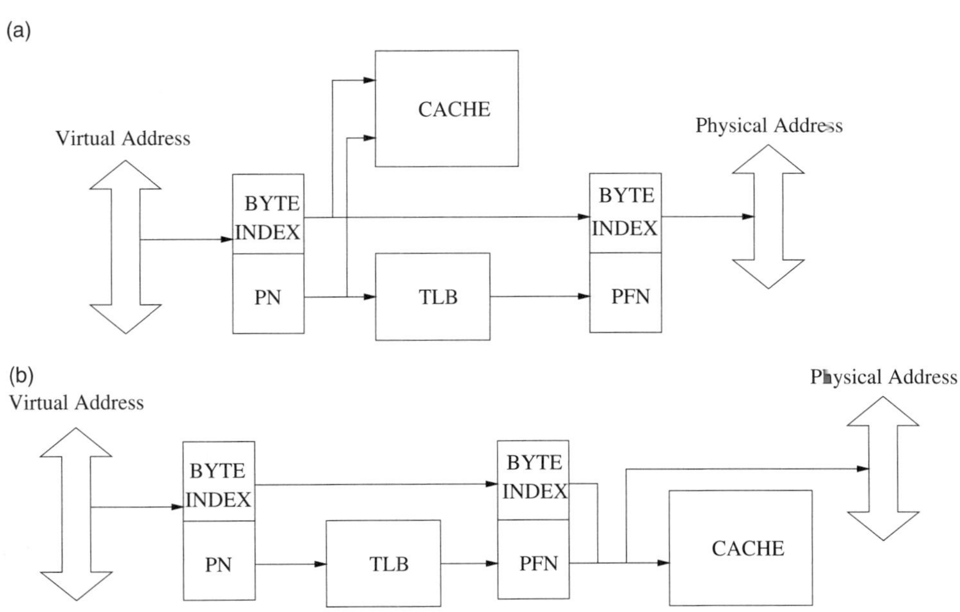

FIGURE 24.7 Translation lookaside buffer (TLB) architectures (a) virtual cache and (b) physical cache.

Virtual caches have their pros and cons. Typically, with no TLB logic between the processor and the cache, access to cache can be achieved at lower cost in virtual cache systems. This is particularly true in multiaccess-per-cycle cache systems where a multiported TLB is needed for a physical cache design. However, the virtual cache alternative introduces virtual memory consistency problems. The same virtual address from two different processes mean different physical memory locations. Solutions to this form of aliasing are to attach process identifier to the virtual address or to flush cache contents on context switches. Another potential alias problem is that different virtual addresses of the same process may be mapped into the same physical address. In general, there is no easy solution to this second form of aliasing; typical solutions involve reverse translation of physical addresses to virtual addresses.

Physical cache designs are not always limited by the delay of the TLB and cache access. In general, there are two solutions to allow large physical cache design. The first solution, employed by companies with past commitments to page size, is to increase the set associativity of cache. This allows the lower index portion of the address to be used immediately by the cache in parallel with virtual address translation. However, large set associativity is very difficult to implement in a cost-effective manner. The second solution, employed by companies without past commitment, is to use a larger page size. The cache can be accessed in parallel with the TLB access similar to the other solution. In this solution there are fewer address bits that are translated through the TLB, potentially reducing the overall delay. With larger page sizes, virtual caches do not have advantage over physical caches in terms of access time. With the advance of semiconductor fabrication processes, in particular the increasing levels of metals, it has become increasingly inexpensive to have highly set-associative caches and multiported TLB. As a result, physical caches have become much more favored solutions today.

24.2.3 Input/Output Subsystem

The input/output (I/O) subsystem transfers data between the internal components (CPU and main memory) and the external devices (disks, networks, printers, keyboards, pointing devices, and scanners).

24.2.3.1 Peripheral Controllers

The CPU usually controls the I/O subsystem by reading from and writing into the I/O (control) registers. There are two popular approaches for allowing the CPU to access these I/O registers: I/O instructions and memory-mapped I/O. In an I/O instruction approach, special instructions are added to the instruction set to access I/O status flags, control registers, and data buffer registers. In a memory-mapped I/O approach, the control registers, the status flags, and the data buffer registers are mapped as physical memory locations.

Due to the increasing availability of chip area and pins, microprocessors are increasingly including peripheral controllers on chip. This trend is especially clear for embedded microprocessors.

24.2.3.2 Direct Memory Access Controller

A direct memory access (DMA) controller is a peripheral controller that can directly drive the address lines of the system bus. The data is directly moved from the I/O data buffer registers to the main memory, rather than from the data buffer registers to a CPU register, then from CPU register to main memory. DMA controllers greatly increase the achieved transfer bandwidth of I/O operations and have become a standard feature of microprocessor systems today.

24.2.4 System Interconnect

System interconnect is the facility that allows the components within a computer system to communicate with each other. There are numerous logical organizations of these system interconnect facilities.

Dedicated links or point-to-point connections enable dedicated communication between components. There are different system interconnect configurations based on the connectivity of the system components. A complete connection configuration, requiring $N \cdot (N - 1)/2$ links, is created when there is one link between every possible pair of components. A *hypercube* configuration assigns a unique n-tuple {1,0} as the coordinate of each component and constructs a link between components whose coordinates differ only in one dimension, requiring $N \cdot \log N$ links. A *mesh* connection arranges the system components into a N-dimensional array and has connections between immediate neighbors, requiring $2 \cdot N$ links.

Switching networks are a group of switches that determine the existence of communication links among components. A cross-bar network is considered the most general form of switching network and uses a $N \times M$ two-dimensional array of switches to provide an arbitrary connection between N components switching network is the multistage network that employs multiple stages of shuffle networks to provide on one side and M components on another side using $N \cdot M$ switches and $N + M$ links. Another a permutation connection pattern between N components on each side by using $N \cdot \log N$ switches and $N \cdot \log N$ links.

Shared buses are single links that connect all components to all other components and are the most popular connection structures. The sharing of buses among the components of a system requires several aspects of bus control. First, there is a distinction between bus masters, the units controlling bus transfers (CPU, DMA), and bus slaves, the other units (memory, programmed I/O interface).

Bus interfacing and bus addressing are the means to connect and disconnect units on the bus. Bus arbitration is the process of granting the bus resource to one of the masters. Arbitration typically uses a selection scheme on the basis of some type of priority assignment. Fixed-priority arbitration gives every master a fixed priority, and dynamic-priority arbitration such as round-robin ensures every master become the most favorable at one point in time.

Bus timing refers to the method of communication among the system units and can be classified as either synchronous or asynchronous. Synchronous bus timing uses a shared clock that defines the time other bus signals change and stabilize. Clock sharing by all units allows the bus to be monitored at agreed time intervals and action taken accordingly. However, the synchronous system bus must operate at the speed of the slowest component. Asynchronous bus timing allows units to use different clocks, but the lack of a shared clock makes it necessary to use extra signals to determine the validity of bus signals. The use of these signals to determine the validity of bus signals is called hand-shaking protocols, which typically reduce the achievable transfer bandwidth via the bus.

24.3 Performance Enhancing Hardware Techniques

24.3.1 Pipelining

In the 1970s, only supercomputers and mainframe computers were pipelined. Today, virtually all microprocessors are pipelined. In fact, pipelining has been a major reason why microprocessors today outperform supercomputer CPU's built less than 10 years ago. Pipelining is a technique to coordinate parallel processing of operations [3]. This technique has been used in assembly lines of major industries for more than a century. The idea is to have a line of workers specializing in different pieces of work required to finish a product. A conveying belt carries products through the line of workers. Each worker performs a small piece of work on each product. Each product is finished after it is processed by all the workers in the assembly line.

The obvious advantage of pipelining is to allow one worker to immediately start working on a new product after finishing the work on a current product. The same methodology is applied to instruction processing in microprocessors. Figure 24.8a shows an example five-stage pipeline dividing instruction execution into fetch (F), decode (D), execute (E), memory (M), and write-back (W) operations, each requiring various stage-specific logic. Between each stage is a stage register (SR), or pipeline register, used to hold the instruction information necessary to control the instruction. A very basic principle of pipelining is that the work performed by each stage must take about the same amount of time. Otherwise, the efficiency will be significantly reduced because one stage becomes a bottleneck of the entire pipeline. That is, the time duration of the slowest pipeline stage determines the overall clock frequency of the processor. Because of this constraint and the often-time slower memory speeds, some of the principle five stages are often divided into smaller stages. For instance, the memory stage may be divided into three

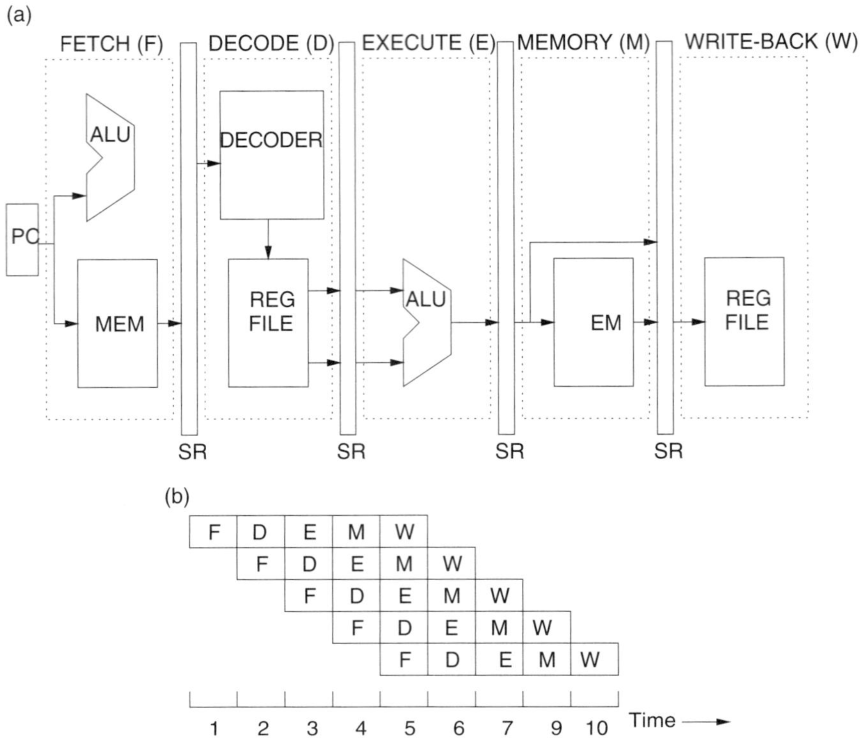

FIGURE 24.8 Pipeline architecture (a) machine and (b) overlapping instructions.

stages, allowing memory accesses to be pipelined and the overall processor clock speed to be a fraction of the memory access latency.

The time required to finish N instructions in a pipeline with K stages can be calculated. Assume a cycle time of T for the overall instruction completion, and an equal T/K processing delay at each stage. With a pipeline scheme, the first instruction completes the pipeline after T, and there will be a new instruction out of the pipeline per stage delay T/K. Therefore, the delays of executing N instructions without (1) and with (2) pipelining respectively are

$$T^{\times}(N) \tag{24.3}$$

$$T + (T/k)^{\times}(N-1) \tag{24.4}$$

There is an initial delay in the pipeline execution model before each stage has operations to execute. The initial delay is usually called *pipeline startup delay*, and is equal to total execution time of one instruction (T). The speedup of a pipelined machine relative to a nonpipelined machine is calculated as

$$\frac{T \times N}{T + (T/k) \times (N-1)} \tag{24.5}$$

When N is much larger than the number of pipe stages k, the ideal speedup approaches k. This is an intuitive result since there are k parts of the machine working in parallel, allowing the execution to go about k times faster in ideal conditions.

The overlap of sequential instructions in a processor pipeline is shown in Figure 24.8b. The processing of each instruction is shown as a five-stage process that goes from left to right; each stage is marked by the initial of its name: F(etch), D(ecode), E(xecute), M(emory), W(riteback). A new instruction is initiated in every clock cycle. The instruction pipeline becomes full after the pipeline delay of $k = 5$ cycles. Although the pipeline configuration executes operations in each stage of the processor, two important mechanisms are constructed to ensure correct functional operation between dependent instructions in the presence of data hazards. Data hazards occur when instructions in the pipeline generate results that are necessary for later instructions that are already started in the pipeline. In the pipeline configuration of Figure 24.8a, register operands are initially retrieved during the decode stage. However, the execute and memory stage can define register operands and contain the correct current value but are not able to update the register file until the later write-back execution stage. Forwarding (or register bypassing) is the action of retrieving the correct operand value for an executing instruction between the initial register file access and any pending instruction's register file updates. Interlocking is the action of stalling an operation in the pipeline when conditions cause necessary register operand results to be delayed. It is necessary to stall early stages of the machine so that the correct results are used, and the machine does not proceed with incorrect values for source operands. The primary causes of delay in pipeline execution are initiated due to instruction fetch delay and memory latency.

24.3.2 Branch Handling

Branch instructions pose serious problems for pipelined processors by causing hardware to fetch and execute incorrect instructions until the branch instructions are completed. Executing incorrect instructions can result in severe performance degradation through the introduction of wasted cycles into the execution process.

There are several methods for dealing with pipeline stalls caused by branch instructions. The simplest performance scheme handles branches by marking branch instructions as either *taken* or *not taken*. The compiler marks each branch by using a special branch opcode for each case. The designation allows the pipeline to fetch subsequent instructions according to the compiler's choice of opcode. However, the fetched instruction may need to be discarded and the instruction fetch restarted when the branch outcome is inconsistent with the compiler's designation.

Delayed branching is a scheme that treats the set of sequential instructions following a branch as delay slots. The delay-slot instructions are executed whether or not the branch instruction is taken. Limitations on delayed branches are caused by the compiler and program characteristics being unable to identify numerous instructions that execute independent of the branch direction. Improvements have been introduced to provide *nullifying* branches, which include a predicted direction for the branch.When the prediction is incorrect, the delay-slot instructions are nullified.

A more modern approach to reducing branch penalties uses hardware to dynamically predict the outcome of a branch. Branch prediction strategies reduce overall branch penalties by allowing the hardware to continue processing instructions along the predicted control path, thus eliminating wasted cycles. Efficient execution can be maintained while branch targets are correctly predicted. However, performance penalty is incurred when a branch is mispredicted. Branch target buffer is a cache structure that is accessed in parallel with the instruction fetch. It records the past history of branch instructions so that a prediction can be made while the branch is fetched again. The prediction method adapts the branch prediction to the run-time program behavior, generating a high prediction accuracy. The target address of each branch is also saved in the buffer so that the target instruction can be fetched immediately if a branch is predicted taken.

Several methodologies of branch target prediction have been constructed [4]. Figure 24.9 illustrates the relation of several general branch prediction schemes to the processor pipeline. The table in Figure 24.9a retains history information for each branch. The history includes the previous branch directions for making predictions on future branch directions. The simplest history is last taken, which uses 1 bit to recall whether the branch condition was taken or not taken during its most recent execution. A more effective branch predictor, as shown in Figure 24.9b, uses a 2-bit saturating state history counter to determine the future branch outcome. Two bits rather than one bit allow each branch to be tagged as strongly or weakly taken or not taken. Every correct prediction reinforces the prediction while an incorrect

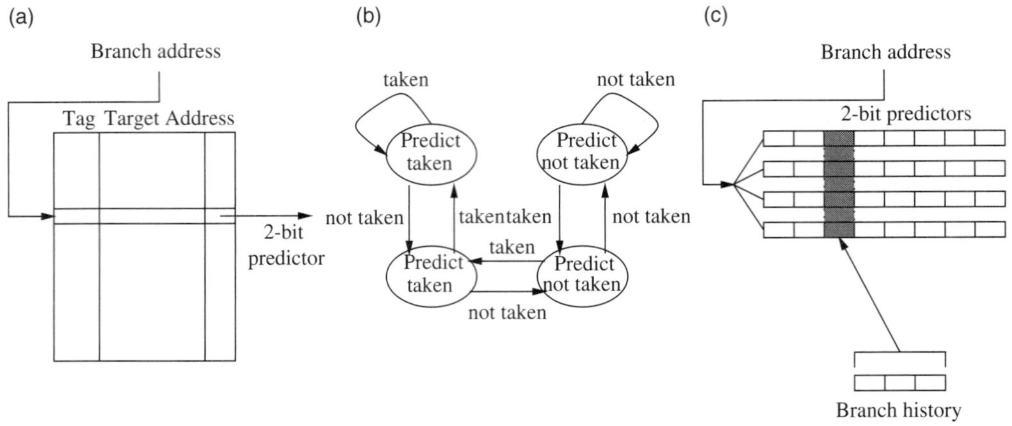

FIGURE 24.9 Branch prediction.

prediction weakens it. It takes two consecutive mispredictions to reverse the direction (whether taken or not taken) of the prediction.

Recently, more complex two-level adaptive branch prediction schemes have been built, which use two levels of branch history to make predictions [5]. An example of such predictor is shown in Figure 24.9c. The first level is the branch outcome history of the last branches encountered. The second level is the branch behavior for the last occurrences of a specific pattern of branch histories, commonly maintained as 2-bit saturating counters. There are alternative ways of constructing both levels of adaptive branch prediction schemes, the mechanisms can contain information that is either based on individual branches, groups (set-based), or all (global). Individual information contains the branch history for each branch instruction. Set-based information groups branch according to their instruction address, thereby forming sets of branch history. Global information uses a global history containing all branch outcomes. The second level containing branch behaviors can also be constructed using any of the three types. In general, the first-level branch history pattern is used as an index into the second-level branch history.

24.3.3 Dynamic Instruction Execution

A major limitation of pipelining techniques is the use of in-order instruction execution. When an instruction in the pipeline stalls, no further instructions are allowed to proceed to ensure proper execution of in-flight instruction. This problem is especially serious for multiple issue machines, where each stall cycle potentially costs work of multiple instructions. However, in many cases, an instruction could execute properly if no data dependence exists between the stalled instruction and the instruction waiting to execute. Dynamic instruction execution, also known as out-of-order execution, is an approach that uses hardware to rearrange the instruction execution to reduce the effect of stalls. The concept of dynamic execution uses hardware to detect dependences in the in-order instruction stream sequence and rearrange the instruction sequence in the presence of detected dependences and stalls.

Today, most modern superscalar microprocessors use dynamic execution to increase the number of instructions executed per cycle. Such microprocessors use basically the same concept: all instructions pass through an issue stage in order, are executed out of order, and are retired in order. There are several functional elements of this common sequence that have developed into computer architecture concepts. The first functional concept is *scoreboarding*. Scoreboarding is a technique for allowing instructions to execute out of order when there are available resources and no data dependences. Scoreboarding originates from the CDC 6600 machine's issue logic, named the scoreboard. The overall goal of scoreboarding is to execute every instruction as early as possible.

A more aggressive approach to dynamic execution is *Tomasulo's algorithm*. This scheme was employed in the IBM 360/91 processor. Although there are many variations on this scheme, the key concept of avoiding write-after-read (WAR) andwrite-after-write (WAW) dependences during dynamic execution is attributed to Tomasulo. In Tomasulo's scheme, the functionality of the scoreboarding is provided by *reservation stations*. Reservation stations buffer the operands of instructions waiting to issue as soon as they become available. The concept is to issue new instructions immediately when all source operands become available instead of accessing such operands through the register file. As such, waiting instructions designate the reservation station entry that will provide their input operands. This action removes WAW dependences caused by successive writes to the same register by forcing instructions to be related by dependences instead of by register specifiers. In general, renaming of register specifiers for pending operands to the reservation station entries is called *register renaming*. Overall, Tomasulo's scheme combines scoreboarding and register renaming. Tomasulo [6] provides complete details of his scheme in the context of a floating point unit design. Patt, et al. first described the extensions required to implement complete modern CPU's with the Tomasulo's Algorithm [7].

24.4 Instruction Set Architecture

There are several elements that characterize an instruction set architecture, word size, instruction encoding, and architecture style.

24.4.1 Word Size

Programs often differ in the size of data they prefer to manipulate. Word processing programs operate on 8-bit or 16-bit data that correspond to characters in text documents. Many applications require 32-bit integer data to avoid frequent overflow in arithmetic calculation. Scientific computations often require 64-bit floating point data to achieve desired accuracy. Operating systems and databases may require 64-bit integer data to represent a very large name space with integers. As a result, the processors are usually designed to access data of a variety of sizes from memory systems. This is a well-known source of complexity in microprocessor design.

The endian convention specifies the numbering of bytes within a memory word. In the little endian convention, the least significant byte in a word is numbered byte 0. The number increases as the positions increase in significance. The DEC VAX and X86 architectures follow the little endian convention. In the big endian convention, the most significant byte in a word is numbered 0. The number decreases as the positions decrease in significance. The IBM 360/370, HP PA-RISC, SUN SPARC, and Motorola 680X0 architectures follow the big endian convention. The endian convention determines how the word is stored in the address space or in a binary file. The difference usually manifests itself when users try to transfer binary files between machines using different endian conventions.

24.4.2 Instruction Encoding

Instruction encoding plays an important role in the code density and performance of microprocessors. Traditionally, the cost of memory capacity was the determining factor in designing either a fixed-length or variable-length instruction set. Fixed-length instruction encoding assigns the same encoding size to all instructions. Fixed-length encoding is generally a product of the increasing advancements in memory capacity.

Variable-length instruction set is the termused to describe the style of instruction encoding that uses different instruction lengths according to addressing modes of operands. Common addressing modes include either register or methods of indexing memory. Figure 24.10 illustrates two potential designs found in modern use of decoding variable-length instructions. The first alternative in Figure 24.10a involves an additional instruction decode stage in the original pipeline design. In this model, the first stage is used to determine instruction lengths and steer the instructions to the second stage where the actual instruction

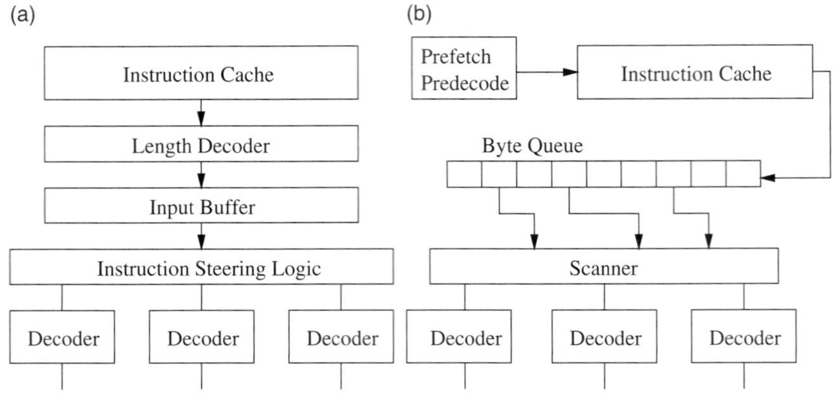

FIGURE 24.10 Variable-sized instruction decoding (a) staging and (b) predecoding.

decoding is performed. The second alternative in Figure 24.10b involves predecoding and marking instruction lengths in the instruction cache. The primary advantage of this scheme is the simplification of the number of decode stages in the pipeline design. However, the method requires a larger instruction cache structure for holding the resolved instruction information. Both design methodologies have been effectively used in decoding X86 variable instructions [8].

24.4.3 Architecture Style

Several instruction set architecture styles have existed over the past three decades of computing. First, complex instruction set computers (CISC) characterized designswith variable instruction formats, numerous memory addressing modes, and large numbers of instruction types. The original CISC philosophy was to create instruction sets that resembled high-level programming languages in an effort to simplify compiler technology. In addition, the design constraint of small memory capacity also led to the development of CISC. Two examples of the CISC model are the Digital Equipment Corporation VAX and Intel X86 architecture families.

Reduced instruction set computers (RISC) gained favor with the philosophy of uniform instruction lengths, load-store instruction sets, limited addressing modes, and reduced number of operation types. RISC concepts allow the microarchitecture design of machines to be more easily pipelined, reducing the processor clock cycle frequency and the overall speed of a machine. The RISC concept resulted from improvements in compiler technology and memory size. The HP PA-RISC, Sun SPARC, IBM Power PC, MIPS, and DEC Alpha machines are examples of RISC architectures.

Architecture models that specify multiple operations to issue in a clock cycle are very long instruction word (VLIW). VLIWs issue a fixed number of operations conveyed as a single long instruction and place the responsibility of creating the parallel instruction packet on the compiler. Early VLIW processor suffered from code expansion due to unfilled operation slots in the long instructions. Examples of VLIW technology are the Multiflow Trace, Cydrome Cydra machines, and TI-C6X. Explicitly parallel instruction computing (EPIC) is similar in concept to VLIW in that both use the compiler to explicitly group instructions for parallel execution. In fact, many of the ideas for EPIC architectures come from previous RISC and VLIW machines. In general, the EPIC concept solves the excessive code expansion and scalability problems associated with VLIW models by providing encoding mechanisms to reduce the need to represent unfilled operation slots in long instructions. Also, the trend of compiler-controlled architecture mechanisms such as predicated execution and speculative execution to be described later in this section are generally considered part of the EPIC-style architecture domain. The Intel IA-64, Philips Trimedia, and Lucent/Motorola StarCore are examples of EPIC machines.

24.5 Instruction Level Parallelism

Explicitly parallel instruction computing processors are equipped with instruction set architecture mechanisms designed to facilitate compiler's effort to arrange for the parallel execution of many operations in each clock cycle. These mechanisms, referred to as instruction level parallelism (ILP) features, are new instruction set architecture concepts for improving microprocessor performance.

24.5.1 Predicated Execution

Branch instructions are recognized as a major impediment to exploiting ILP. Branches force the compiler and hardware to make frequent predictions of branch directions in an attempt to find sufficient parallelism. Branch prediction strategies reduce this problem by allowing the compiler and hardware to continue processing instructions along the predicted control path, thus eliminating these wasted cycles. However, misprediction of these branches can result in severe performance degradation through the introduction of wasted cycles into the instruction stream.

Predicated execution provides an effective means to eliminate branches from an instruction stream. Predicated execution refers to the conditional execution of an instruction based on the value of a boolean source operand, referred to as the predicate of the instruction. This architectural support allows the compiler to use an *if-conversion* algorithm to convert conditional branches into predicate defining instructions, and instructions along alternative paths of each branch into predicated instructions [9]. Predicated instructions are fetched regardless of their predicate operand value. Instructions whose predicate operands are true are executed normally. Conversely, instructions whose predicate operands are false are nullified, and thus are prevented from modifying the processor state. Predicated execution allows the compiler to trade instruction fetch efficiency for the capability to expose ILP to the hardware along multiple execution paths.

Predicated execution offers the opportunity to improve branch handling in microprocessors. Eliminating frequently mispredicted branches may lead to a substantial reduction in branch prediction misses. As a result, the performance penalties associated with mispredicting these branches are removed. Eliminating branches also reduces the need to handle multiple branches per cycle for wide issue processors. Finally, predicated execution provides an efficient interface for the compiler to expose multiple execution paths to the hardware. Without compiler support, the cost of maintaining multiple execution paths in hardware can be prohibitive.

The essence of predicated execution is the ability to suppress the modification of the processor state on the basis of some execution condition. There must be a way to express this condition and a way to express when the condition should affect execution. Full predication cleanly supports this through a combination of instruction set and microarchitecture extensions. These extensions can be classified as support for suppression of execution and expression of condition. The result of the condition that determines if an instruction should modify state is stored in a set of 1-bit registers. These registers are collectively referred to as the predicate register file. The values in the predicate register file are associated with each instruction in the extended instruction set through the use of an additional source operand. This operand specifies which predicate register will determine whether the operation should modify processor state. If the value in the specified predicate register is 1, or true, the instruction is executed normally; if the value is 0, or false, the instruction is suppressed.

Predicate register values may be set using predicate define instructions, as described in the HPL Playdoh architecture [10]. There is a predicate define instruction for each comparison opcode in the original instruction set. The major difference with conventional comparison instructions is that these predicate defines have up to two destination registers and that their destination registers are predicate registers. The instruction format of a predicate define is shown below.

$$\text{pred\_}\langle cmp\rangle\ \text{Pout1}_{\langle type\rangle},\ \text{Pout2}_{\langle type\rangle},\ \text{src1},\ \text{src2}\ (\text{P}_{in})$$

This instruction assigns values to *Pout1* and *Pout2* according to a comparison of *src1* and *src2* specified by <*cmp*>. The comparison <*cmp*> can be: equal (eq), not equal (ne), greater than (gt), and so forth. A predicate <*type*> is specified for each destination predicate. Predicate defining instructions are also predicated, as specified by P_{in}.

The predicate <*type*> determines the value written to the destination predicate register given the result of the comparison and of the input predicate, P_{in}. For each combination of comparison result and P_{in}, one of three actions may be performed on the destination predicate. It can write 1, write 0, or leave it unchanged. There are six predicate types that are particularly useful, the unconditional (*U*), *OR*, and *AND* type predicates and their complements. Table 24.1 contains the truth table for these predicate definition types.

Unconditional destination predicate registers are always defined, regardless of the value of P_{in} and the result of the comparison. If the value of P_{in} is 1, the result of the comparison is placed in the predicate register (or its compliment for $\overline{U}$). Otherwise, a 0 is written to the predicate register. Unconditional predicates are utilized for blocks that are executed along only one path in the if-converted code region.

The *OR* type predicates are useful when execution of a block can be enabled along multiple paths, such as logical AND (&&) and OR (||) constructs in C. *OR* type destination predicate registers are set if P_{in} is 1 and the result of the comparison is 1 (0 for $\overline{OR}$), otherwise the destination predicate register is unchanged. Note that *OR* type predicates must be explicitly initialized to 0 before they are defined and used. However, after they are initialized, multiple *OR* type predicate defines may be issued simultaneously and in any order on the same predicate register. This is true since the *OR* type predicate either writes a 1 or leaves the register unchanged, which allows implementation as a wired logical *OR* condition. *AND* type predicates are analogous to the *OR* type predicate. *AND* type destination predicate registers are cleared if P_{in} is 1 and the result of the comparison is 0 (1 for $\overline{AND}$), otherwise the destination predicate register is unchanged.

Figure 24.11 contains a simple example illustrating the concept of predicated execution. Figure 24.11a shows a common programming if-then-else construction. The related control flow representation of that programming code is illustrated in Figure 24.11b. Using if-conversion, the code in Figure 24.11b is then transformed into the code shown in Figure 24.11c. The original conditional branch is translated into *pred_eq* instructions. Predicate register *p1* is set to indicate if the condition *(A = B)* is true, and *p2* is set if the condition is false. The "then" part of the if statement is predicated on *p1* and the "else" part is

TABLE 24.1 Predicate Definition Truth Table

		P_{out}					
P_{in}	Comparison	*U*	$\overline{U}$	*OR*	$\overline{OR}$	*AND*	$\overline{AND}$
0	0	0	0	—	—	—	—
0	1	0	0	—	—	—	—
1	0	0	1	—	1	0	—
1	1	1	0	1	—	—	0

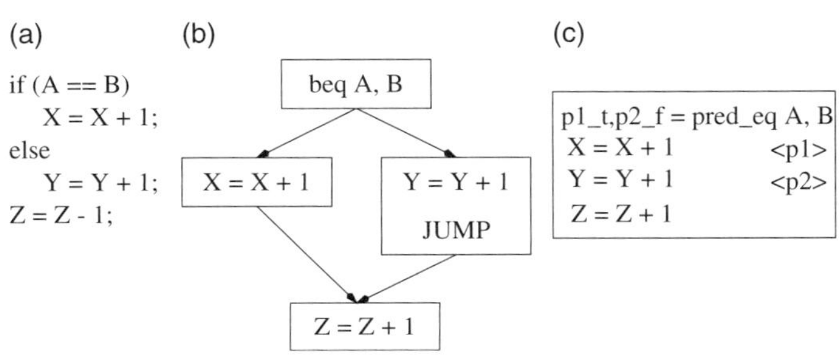

FIGURE 24.11 Instruction sequence (a) program code, (b) traditional execution, and (c) predicated execution.

predicated on *p2*. The *pred_eq* simply decides whether the addition or subtraction instruction is performed and ensures that one of the two parts is not executed.

24.5.2 Speculative Execution

The amount of ILP available within basic blocks, defined as consecutive instruction sequences without branching in or out, is extremely limited in most programs. As such, processors must optimize and schedule instructions across basic block boundaries to achieve higher performance. In addition, future processors must contend with both long latency load operations and long latency cache misses. When load data is needed by subsequent dependent instructions, the processor execution must wait until the cache access is complete.

In these situations, our-of-order processors dynamically perform branch prediction and reorder the instruction stream to execute nondependent instructions. As a result, they have ability of exploiting parallelism between instructions before and after a correctly predicted branch instruction. However, this approach requires complex circuitry at the cost of chip die space as well as additional power consumption. Similar performance gains can be achieved using static compile-time speculation methods without complex, power hungry out-of-order logic. Speculative execution, a technique for executing an instruction before knowing that its execution is required, is an important technique for exploiting ILP in programs. Speculative execution is best known for its use in hiding memory latency. A compiler can utilize speculative code motion to achieve higher performance in several ways. First, in regions of code where insufficient ILP exists to fully utilize the processor resources, useful instructions from other regions may be executed. Second, instructions at the beginning of long dependence chains may be executed early to reduce the computation's critical path. Finally, long latency instructions may be initiated early to overlap their execution with other useful operations. Figure 24.12 illustrates a simple example of code before and after a speculative compile-time transformation is performed to execute a load instruction above a conditional branch.

Figure 24.12a shows how the branch instruction and its implied control flow define a control dependence that restricts the load operation from being scheduled earlier in the code. Cache miss latencies would halt the processor unless out-of-order execution mechanisms were used. However, with speculation support, Figure 24.12b can be used to hide the latency of the load operation.

The solution requires the load to be speculative or nonexcepting. A speculative load will not signal exceptions such as address alignment errors or address space access violations. Essentially, the load remains silent for these exceptions. The additional check instruction in Figure 24.12b enables these signals to be detected when the execution does reach the original location of the load. When the other path of branch's execution is taken, such silent signals are meaningless and can be ignored. Using this mechanism, the load can be placed above all existing control dependences, providing the compiler with the ability to hide load latency. Details of compiler speculation can be found in Reference 11.

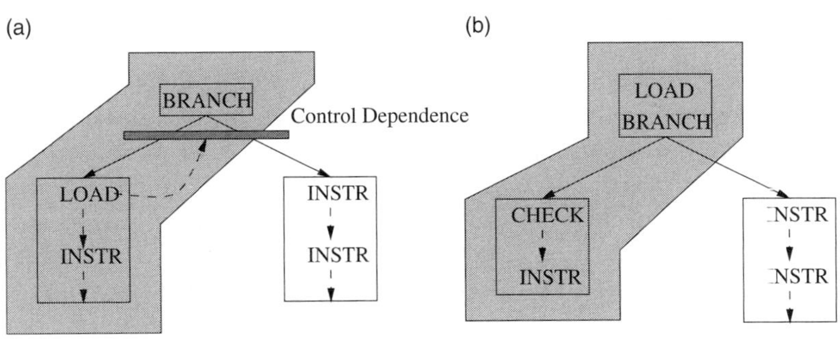

FIGURE 24.12 Instruction sequence (a) traditional execution and (b) speculative execution.

24.6 Industry Trends

The microprocessor industry is one of the fastest moving industry today. Healthy demands from the market place have stimulated strong competition, which in turn resulted in great technical innovations.

24.6.1 Computer Microprocessor Trends

Recent trends of computer microprocessors include deep pipelining, high clock frequency, wide instruction issue, speculative and out-of-order execution, predicated execution, multimedia data types, large on-chip caches, floating point capabilities, and multiprocessor support. In the area of pipelining, the Intel Pentium 4 processor is pipelined approximated twice as deeply as its predecessor Pentium 3. The deep pipeline has allowed the clock of Pentium 4 processor to run at a much higher clock frequency than Pentium 3. This trend has, however, been reversed in 2005 owing to power budget limitations. The pipeline depth of the Intel IA32 microprocessors that succeeded Pentium 4 have been reduced toward that of Pentium 3.

In the area of wide instruction issue, the Pentium 4 processor can decode and issue up to three X86 instructions per clock cycle, compared to the two-instruction issue bandwidth of Pentium. More recently, the Intel Itanium I and Itanium II processors can issue up to six instructions per clock cycle. Wide instruction issue, however, requires multiported register and multiple cache access ports that can significantly increase power consumption. As a result, future microprocessors will likely maintain or reduce the issue width compared to their recent predecessors.

Pentium 4 has dedicated a very significant amount of chip area to Branch History Table, Branch Target Buffer, Reservation Stations, Load-store Queue, and Reorder Buffer to support speculative and out-of-order execution. These structures together allow the Pentium 4 processor to maintain a large instruction window within which it performs aggressive speculative and out-of-order execution. All these structures, however, consume power in an intensive manner. As a result, the trend of larger instruction windows has also slowed owing to power budget limitations.

One important trend of the computer microprocessors in general is the slow down of the increase in complexity, size, and clock frequency of processor cores. Rather, the industry is moving into incorporating multiple processor cores on the same chip. If all the cores can be productively used, such model can achieve much higher performance than a single core given the same power and chip area budget. This, however, places much more burden on the programmer, compiler, and operating system than traditional single-core models.

In the area of predicated execution, Pentium 4 supports a conditional move instruction that was not available in Pentium. This trend is furthered by the next generation IA-64 architecture where all instructions can be conditionally executed under the control of predicate registers. This ability will allow future microprocessors to execute control-intensive programs much more efficiently than their predecessors.

In the area of data types, the multimedia instructions from Intel and AMD have become a standard feature of all X86 microprocessors. These instructions take advantage of the fact that multimedia data items are typically represented with a smaller number of bits (8–16 bits) than the width of an integer data path today (32–64 bits). On the basis of an observation, the same operation is often repeated on all data items in multimedia applications; the architects of multimedia instructions specify that each such instruction performs the same operation on several multimedia data items packed into one register word. Intel first proposed MMX instructions that process several integer data items simultaneously to achieve significant speedup in targeted applications. In 1998, AMD proposed the 3DNow! instructions to address the performance needs of 3D graphics applications. The 3DNow! instructions are designed on the basis of the concept that 3D graphics data items are often represented in single precision floating-point format, and they do not require the sophisticated rounding and exception handling capabilities specified in the IEEE Standard format. Thus, one can pack two graphics floating-point data into one double-precision floating-point register for more efficient floating-point processing of graphics applications. Note that

MMX and 3DNow! are similar in concepts applied to integer and floating-point domains. More recently, Intel proposed the SSE instructions to compete with AMD 3DNow!.

In the area of large on-chip caches, the popular strategies used in computer microprocessors are either to enlarge the first-level caches or to incorporate second-level and sometimes third-level caches on chip. For example, the AMD K7 microprocessor has a 64 KB first-level instruction cache and a 64 KB first-level data cache. These first-level caches are significantly larger than those found in the previous generations. For another example, the Intel Celeron microprocessor has a 128 KB second-level combined instruction and data cache. These large caches are enabled by the increased chip density that allows much more transistors on the chip. The Compaq Alpha 21364 microprocessor has a 64 KB first-level instruction cache, a 64 KB first-level data cache, and a 1.5 MB second-level combined cache. The recent Intel Itanium processors have up to 9 MB third-level combined cache on chip.

In the area of floating-point capabilities, the computer microprocessors in general have much stronger floating-point performance than their predecessors. For example, the Intel Pentium 4 processor achieves several times of floating-point performance improvements of the Pentium processor. For another example, most RISC and EPIC microprocessors now have floating-point performance that rival supercomputer CPUs built just a few years ago.

Owing to the increasing demand of multiprocessor enterprise computing servers, many computer microprocessors now seamlessly support cache coherence protocols. For example, the AMDK7 microprocessor provides direct support for seamless multiprocessor operation when multiple K7 microprocessors are connected to a system bus. This capability was not available in its predecessor AMD K6. The more recent AMD Opteron processors further support HyperTransport protocol to allow each processor in a multiprocessor system to have much higher communication bandwidth than what the traditional memory controllers can support.

24.6.2 Embedded Microprocessor Trends

There are three clear trends in embedded microprocessors. The first trend is to integrate a DSP core with an embedded CPU/controller core. Embedded applications increasingly require DSP functionalities such as data encoding in disk drives and signal equalization for wireless communications. These functionalities enhance the quality of services of their end consumer products. At the 1999 Embedded Microprocessor Forum, ARM, Hitachi, and Siemens all announced products with both DSP and embedded microprocessors [12].

Three approaches exist in the integration of DSP and embedded CPUs. One approach is to simply have two separate units placed on a single chip. The advantage of this approach is that it simplifies the development of the microprocessor. The two units are usually taken from existing designs. The software development tools can be directly taken from each unit's respective software support environments. The disadvantage is that the application developer needs to deal with two independent hardware units and two software development environments. This usually complicates software development and verification.

An alternative approach to integrating DSP and embedded CPUs is to add the DSP as a coprocessor of the CPU. This CPU fetches all instructions and forwards the DSP instructions to the coprocessor. The hardware design is more complicated than the first approach owing to the need to more closely interface the two units, especially in the area of memory accesses. The software development environment also needs to be modified to support the coprocessor interaction model. The advantage is that the software developers now deal with a much more coherent environment.

The third approach to integrating DSP and embedded CPUs is to add DSP instructions to a CPU instruction set architecture. This usually requires brand new designs to implement the fully integrated instruction set architecture. The benefit is that software developers need to deal with just one development environment.

The second trend in embedded microprocessors is to support the development of single-chip solutions for large volume markets. Many embedded microprocessor vendors offer designs that can be licensed and incorporated into a larger chip design that includes the desired input/output peripheral devices and application specific integrated circuit (ASIC), and field programmable gate array (FPGA) design. This paradigm is referred to as system-on-a-chip design. A microprocessor that is designed to function in such a system is often referred to as a licensable core.

The third major trend in embedded microprocessors is aggressive adoption of high-performance techniques. Traditionally, embedded microprocessors are slow to adopt high-performance architecture and implementation techniques. They also tend to reuse software development tools such as compilers from the computer microprocessor domain. However, owing to the rapid increase of required performance in embedded markets, the embedded microprocessor vendors are now making fast moves in adopting high-performance techniques. This trend is especially clear in the DSP microprocessors. Texas Instruments, Motorola/Lucent, and Analog Devices all ship aggressive EPIC/VLIW DSP microprocessors and associated compilers.

24.6.3 Microprocessor Market Trends

Readers who are interested in market trends for microprocessors are referred to Microprocessor Report, a periodic publication by MicroDesign Resources (www.MDRonline.com). In every issue, there is a summary of microarchitecture features, physical characteristics, availability, and pricing of microprocessors.

References

1. J. Turley, "RISC volume gains but 68K stll reigns,"*Microprocessor Report*, vol. 12, pp. 14–18, January 1998.
2. W. W. Hwu and T. M. Conte, "The susceptibility of programs to context switching," *IEEE Transactions on Computers*, vol. C-43, pp. 993–1003, September 1994.
3. J. L. Hennessy and D. A. Patterson, *Computer Architecture A Quantitative Approach*. San Francisco, CA: Morgan Kaufman, 1990.
4. J. E. Smith, "A study of branch prediction strategies," in *Proceedings of the 8th International Symposium on Computer Architecture*, pp. 135–148, May 1981.
5. Y. N. Patt,W.-M. Hwu, and M. Shebanow, "HPS, a new microarchitecture: rationale and introduction," in *Proceedings of the 18th AnnualWorkshop on onMicroprogramming*, pp. 103–106, December 1985.
6. L. Gwennap, "Klamath extends P6 family," *Microprocessor Report*, vol. 1, pp. 1–9, February 1997.
7. T. Y. Yeh and Y. N. Patt, "A comprehensive instruction fetch mechanism for a processor supporting speculative execution," in *Proceedings of the 25th International Symposium on Microarchitecture*, pp. 129–139, December 1992.
8. R. M. Tomasulo, "An efficient algorithm for exploiting multiple arithmetic units," *IBM Journal of Research and Development*, vol. 11, pp. 25–33, January 1967.
9. J. R. Allen, et al. "Conversion of control dependence to data dependence," in *Proceedings of the 10th ACM Symposium on Principles of Programming Languages*, pp. 177–189, January 1983.
10. V. Kathail, M. S. Schlansker, and B. R. Rau, "HPL PlayDoh architecture specification: Version 1.0," Tech. Rep. HPL-93-80, Hewlett-Packard Laboratories, Palo Alto, CA, February 1994.
11. S. A. Mahlke, et al. "Sentinel scheduling: A model for compiler-controlled speculative execution," *ACM Transactions on Computer Systems*, vol. 11, November 1993.
12. *Embedded Microprocessor Forum*, (San Jose, CA), October 1998.

25

Control with Embedded Computers and Programmable Logic Controllers

Hugh Jack
Andrew Sterian
Grand Valley State University

25.1 Introduction

Modern control systems include some form of computer, most often an embedded computer or programmable logic controller (PLC). An embedded computer is a microprocessor- or microcontroller-based system used for a specific task rather than general-purpose computing. It is normally hidden from the user, except for a control interface. A PLC is a form of embedded controller that has been designed for the control of industrial machinery (see Figure 25.1).

A block diagram of a typical control system is shown in Figure 25.2. The controller monitors a process with sensors and affects it with actuators. A user interface allows a user or operator to direct and monitor the control system. Interfaces to other computers are used for purposes such as programming, remote monitoring, or coordination with another controller.

When a computer is applied to a control application, there are a few required specifications. The system must always remain responsive and in control of the process. This requires that the control software be real-time so that it will respond to events within a given period of time, or at regular intervals. The systems are also required to fail safely. This is done with thermal monitoring for overheating, power level detection for imminent power loss, or with watchdog timers for unresponsive programs.

25.2 Embedded Computers

An embedded computer is a microprocessor- or microcontroller-based system designed for dedicated functionality in a specialized (i.e., nongeneral-purpose) electronic device. Common examples of embedded computers can be found in cell phones, microwave ovens, handheld computing devices, automotive systems, answering machines, and many other systems.

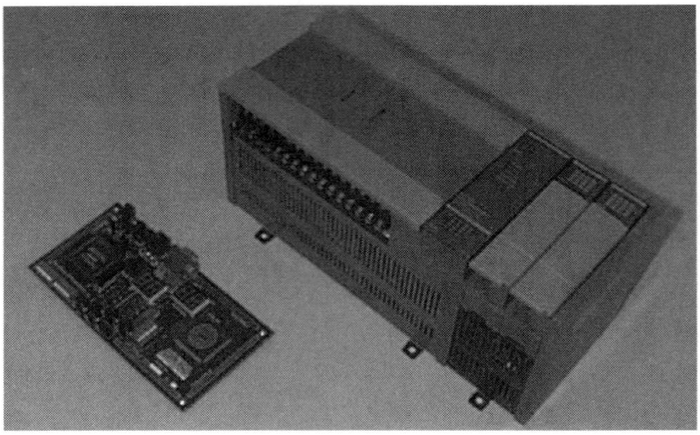

FIGURE 25.1 An embedded computer with an Altera FPGA (front-left) and an Allen Bradley SLC500 programmable logic controller (top-right).

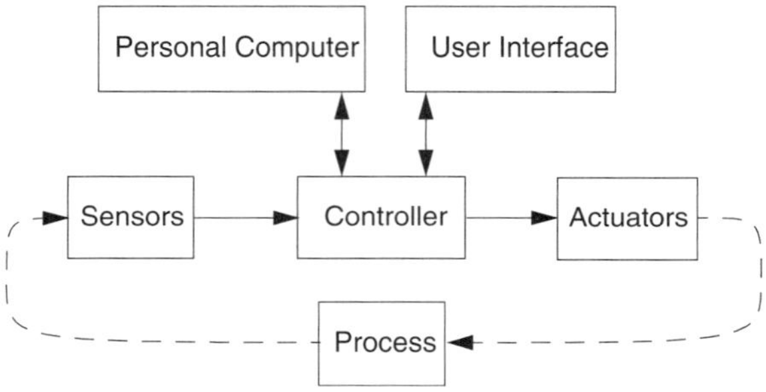

FIGURE 25.2 An example block diagram of a computer controlled application.

The design constraints and parameters for an embedded computer are usually different from those of a general-purpose computer. Although the latter is designed for maximum computing power and support for the latest interconnection and peripheral standards, an embedded computer is designed to be just powerful enough and to support only the interfaces and protocols that are specifically required. The constraints of an embedded computer design often include size, power consumption and heat dissipation, and cost.

25.2.1 Hardware Platforms

25.2.1.1 Microcontroller-Based Systems

Microcontrollers are closely related to the microprocessors that power today's general-purpose computers. They differ from microprocessors, in general, by being highly integrated, with built-in peripherals that minimize total system part count, having low power consumption, providing a small amount of on-chip RAM and ROM, and having several general-purpose input/output (I/O) lines available for instrument sensors and control. For this reason, a microcontroller-based embedded system may be designed with very few external components. In contrast, a microprocessor-based system requires external RAM, external peripherals, and I/O interfaces, and often dissipates so much heat that active cooling is required for proper operation.

The peripherals built into many microcontrollers include serial-line interfaces (such as RS232), timers, pulse generators, event counters, etc. These peripherals support many sensor and actuator control functions.

For example, pulse generators and timers can be used to construct stepper motor drive sequences. Microcontrollers are becoming increasingly specialized with respect to the data communication interfaces they support. While many support the ubiquitous RS232, SPI, and I^2C protocols, recent microcontrollers have built-in support for interfaces such as USB.

In order to minimize power consumption, most microcontrollers have a special sleep or standby mode in which no instructions are executed and very little power is consumed. Microcontrollers can be programmed to awaken in response to an external event so that the program code is executed, and power consumed, only when necessary.

Microcontrollers are a very large semiconductor market due to the wide range and high volume of devices that use them. There are many manufacturers and models of microcontrollers, ranging from tiny 8-pin devices with minimal functionality and costing mere pennies, to large devices with hundreds of pins, many features, and much higher cost. This broad spectrum reflects the highly specific nature of an embedded computer and its design.

25.2.1.2 FPLD-Based Systems

Field-programmable logic devices (FPLDs) such as complex programmable logic devices (CPLDs) and-field-programmable gate arrays (FPGAs) are a more recent alternative to microcontrollers for embedded computer design. An FPLD represents a programmable hardware device; the actual hardware functionality of the device is what is being designed. A microcontroller, in contrast, has fixed hardware functionality and is programmed with software. It is possible, however, to design an FPLD that behaves as a microcontroller, and is further programmed in software. The programmable hardware functionality, however, affords the designer a much greater degree of flexibility over a fixed hardware solution. The price for this flexibility, however, is complexity.

FPLDs may be designed from the ground up or may be composed of one or more predesigned core and peripheral blocks. It is possible, for example, to purchase microcontroller core functions, peripheral functions, etc. and assemble them to form a customized microcontroller on an FPLD with non-recurring engineering (NRE) costs that are much lower than a full custom chip design.

FPLDs can often be programmed "on-the-fly," allowing for reconfigurable computing. This is a computing paradigm that reprograms a system at the hardware level while it is in operation, according to system demands. This means, for example, that the same hardware device can implement multiple bus protocols, interfaces, or algorithms as needed, rather than requiring a larger and more expensive device that supports all of the necessary functions but only uses one at a time.

25.2.1.3 Digital Signal Processing Systems

Digital signal processing (DSP) devices are in many ways similar to microcontrollers with respect to peripheral integration, power consumption, etc. but also have specialized hardware support for common DSP operations, such as filtering. DSP devices are ideal for use in systems that process speech and music, or for robust control and communications applications. The specialized hardware support of these devices means that they are capable of sustaining much higher effective computation rates (on signal processing tasks), but at the same clock speed and power dissipation as the more general-purpose microcontrollers.

25.2.1.4 Real-Time Systems

Most embedded systems must operate in *real time*, that is, they must respond in a timely fashion to external events such as user commands and sensor readings. When an absolute upper limit on response time is required (and guaranteed), the system is a *hard real-time system*; otherwise, it is a soft real-time system. Systems that have safety constraints, such as automotive and industrial control systems, are often hard real-time systems so that absolute maximum time delays can be computed and verified for safety-critical events.

Real-time computation is effected using *interrupts*. These are mechanisms supported by all common microcontrollers that cause a change in the flow of execution of the program when the interrupt occurs. The program that is executed in response to the interrupt is expected to respond in some way to the interrupt.

For example, an interrupt may occur when a digital logic level changes at a device pin, indicating a sensor condition, or it may occur when the user presses a button on a keypad, indicating that an action is desired and is to be performed immediately.

The implementation of a real-time software system may either be custom designed or may make use of a commercial real-time operating system (RTOS). Since the design of an interrupt-driven real-time system has many potential pitfalls, the usage of a mature RTOS can greatly speed development time.

25.2.1.5 Embedded Modules

The functionality of commercially available embedded computing modules has been steadily increasing. It is common to find powerful microcontrollers, an Ethernet interface, and basic Internet protocol support all combined in a very small form factor for less than $50 in single quantities. This level of integration can greatly speed development time for network-enabled control or remote sensing applications.

25.2.2 Hardware Interfacing

25.2.2.1 Mechanical Switches

Switches are easily interfaced to digital logic with a resistor as shown in Figure 25.3. The mechanical nature of the switch may lead to *bounce* or oscillation of the digital signal for a brief period during the switch opening/closing action. This bounce may be eliminated in the software or with a small amount of additional hardware.

25.2.2.2 Analog Inputs

Analog inputs that indicate one of the two conditions can be interfaced to a digital logic input with a simple comparator (Figure 25.4). A threshold voltage is set with a resistor divider. The comparator generates a digital signal, which indicates whether the analog input voltage is above or below the threshold voltage. This approach can be used for sensors such as optical interrupters (for part counting, motor movement detection, etc.), temperature limit sensors, and many others.

When the analog voltage itself is of interest (as in, for example, temperature measurements), an analog-to-digital converter (ADC) can be used to provide either a serial or a parallel representation of the voltage with a precision ranging anywhere from 8 bits to 16 bits and above. A serial ADC may require as few as two digital I/O pins on a microcontroller for transferring data, while a parallel ADC requires at least as

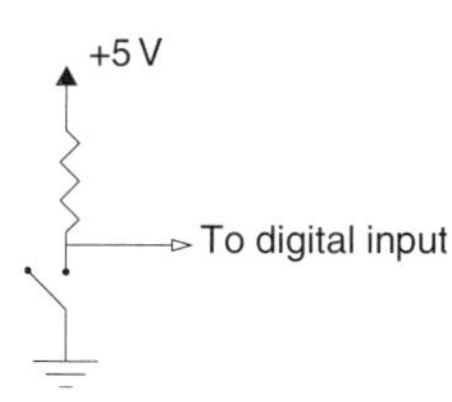

FIGURE 25.3 A mechanical switch is easily interfaced to a digital input on a microcontroller using a single resistor.

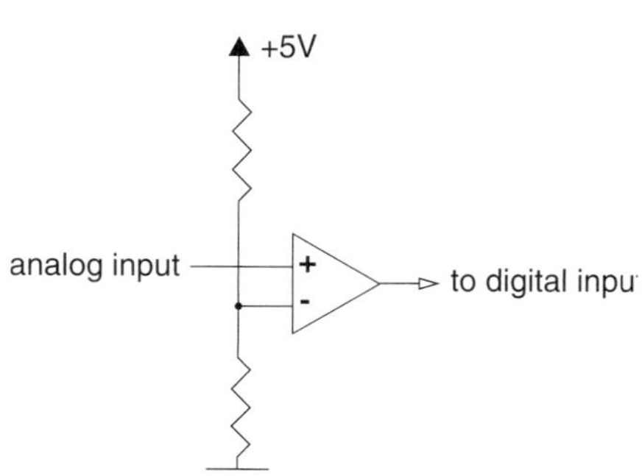

FIGURE 25.4 A digital input driven by a comparator detects whether an analog voltage signal is above or below a threshold voltage (set with a resistor divider network).

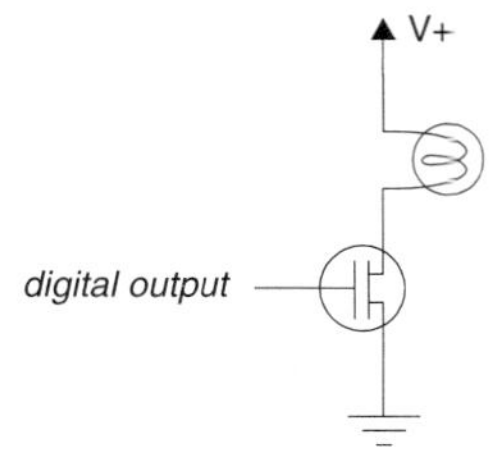

FIGURE 25.5 A digital output can control a high-current device (such as a lamp pictured to the right) using a transistor as a switch.

many pins as there are bits of resolution, but can transfer an entire analog reading in one transaction for faster throughput.

Some microcontrollers have built-in ADC peripherals with multiple input channels enabling highly integrated low-cost analog sensor systems.

Analog voltages that are very small may be amplified with an instrumentation amplifier prior to analog-to-digital conversion. Instrumentation amplifiers have very high gain and high impedance, and hence, are suitable for sensors with very weak driving voltages and currents.

25.2.2.3 Simple Actuators

Embedded computers are not usually capable of driving most practical actuators directly, since the latter often require voltages and currents not compatible with digital circuitry. As with sensors, however, some simple interface circuitry is all that is required. Simple on/off actuators such as lamps, LEDs, relay coils, etc. can be driven from a digital output, using a transistor as a switch, as shown in Figure 25.5. The digital output controls the on/off state of the transistor, which, in turn, either allows or does not allow current to flow through the actuator.

Motors can also be controlled using transistors as interfaces between digital outputs and the high-current motor coils. The lamp in Figure 25.5 can be replaced with a DC motor to allow simple on/off control of the motor. A set of four transistors arranged in an H-bridge configuration allows such a motor to rotate in either direction. Two H-bridge configurations can be used to control a stepper motor. In all cases, the speed and direction of rotation are under direct control of the embedded computer through its digital outputs.

25.2.2.4 Analog Outputs

For actuators that require a variable analog voltage or current, a digital-to-analog converter (DAC) can be used as an interface between the embedded computer and the actuator. As with ADCs, DACs are available in a variety of bit widths, conversion speeds, number of channels, etc. Often, the current driving capacity of these devices is not sufficient and an additional buffer amplifier is required to meet the current demands of the actuator.

25.2.3 Programming Languages

Embedded computers are most commonly programmed in low-level languages for maximum control over the hardware resources. The most time-critical sections of the code are generally programmed in assembly language, which is the lowest-level language understood by a microcontroller. The C language is generally used for higher-level structured programming. Even higher-level languages, such as C++ or Java, are not well suited for embedded programming as they require larger amounts of memory and are not designed for low-level access to hardware resources.

Figure 25.6 shows a fragment of an assembly language program written for the Microchip PIC 16F84A microcontroller. It enables power to a DC motor (through an external interface circuit) when two digital inputs are both at a logic 1 level. The code runs continuously, always checking for the status of the two digital inputs (which may be manual switches, current sensors, etc.).

The same code fragment written in C is shown in Figure 25.7. The code is more compact and easier to read since C is a higher-level language than assembly.

```
loop:
     btfss   PORTA,0      Check digital input bit 0 of Port A
     goto    turnoff          and disable motor if not 1
     btfss   PORTA,1      Check digital input bit 1 of Port A
     goto    turnoff          and disable motor if not 1

     bsf     PORTB,5      Enable motor by setting bit 5 of Port B
     goto    loop             and check inputs again

turnoff:
     bcf     PORTB,5      Disable motor by clearing bit 5 of Port B
     goto    loop             and check inputs again
```

FIGURE 25.6 Fragment of assembly code for Microchip PIC 16F84A microcontroller. This code fragment examines two digital inputs (bits 0 and 1 of input Port A) and sets bit 5 of output Port B if both inputs are at a logic 1 level. The output can be used to enable or disable a DC motor with appropriate interface circuitry.

```
while (1) {
        if (PA0 && PA1) {// Check status of bits 0 and 1 in Port A
                PB5 = 1;      // Set bit 5 of Port B
        } else {
                PB5 = 0;      // Clear bit 5 of Port B
        }
}
```

FIGURE 25.7 Fragment of C code to effect the same functionalilty as the code in Figure 25.6.

25.3 Programmable Logic Controllers

The modern programmable logic controller (PLC) is the successor of relay-based controls. The technological shift began in the 1960s, when the limitations of electromechanical relay-based controllers drove General Motors to search for electronic alternatives. The answer was provided in 1970 by Modicon, who provided a microprocessor-based control system. The programming language was modeled after relay ladder logic diagrams to ease the transition of designers, builders, and maintainers to these new controllers. Throughout the 1970s the technology was refined and proven, and since the early 1980s they have become ubiquitous on the factory floor.

Most PLC components are in card form that can be interchanged quickly in the event of a failure. A typical PLC application has about one hundred inputs and outputs, but the scale of the applications varies widely. A small PLC costing \$200 might have six inputs and four outputs. A large application might involve multiple PLCs working together over an entire plant and collectively have tens of thousands of inputs and outputs. In general, the aggregated cost of PLC hardware per input and output is approximately \$10–\$50. This does not include the cost of sensors (typically \$50–\$100), actuators (typically \$50–\$200), installation (typically \$10–\$100), design, or programming.

Manufacturing control systems always require logical control and sometimes continuous control. Logical control involves the examination of binary inputs (on or off) from sensors and setting binary outputs to drive actuators. A simple example is a photosensor that detects a box on a conveyor and actuates an air cylinder to divert the box. Continuous control systems are used less frequently because of their higher costs and increased complexity. A typical continuous controller might use an analog output card (\$1000) to output a voltage to a variable frequency motor driver (\$1000) to control the velocity of a conveyor.

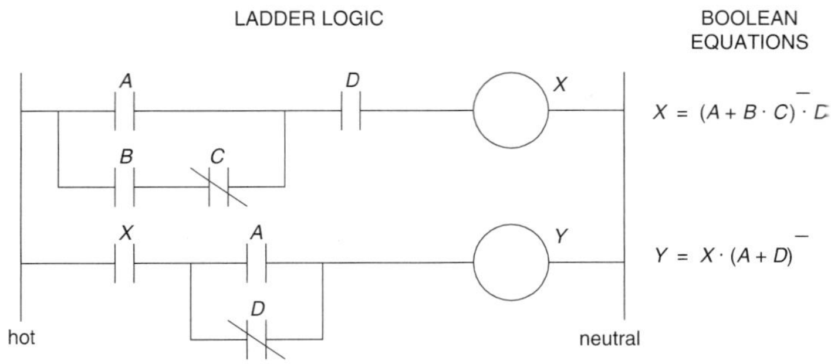

FIGURE 25.8 A simple ladder logic program with equivalent Boolean equations.

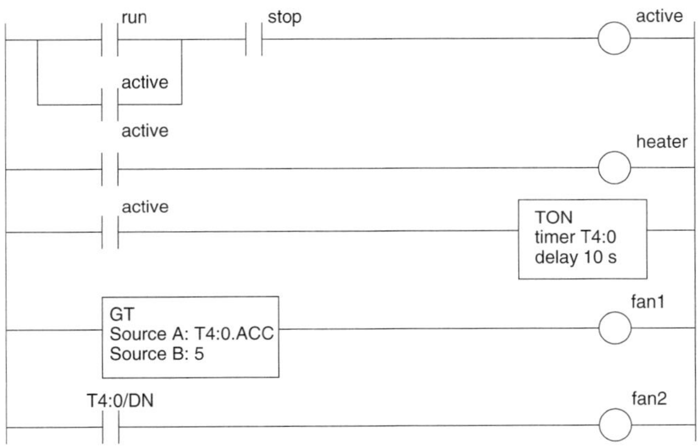

FIGURE 25.9 A complex ladder logic example.

25.3.1 Programming Languages

Every PLC can be programmed with ladder logic. Ladder logic uses input contacts (shown with two vertical lines) and output coils (shown with a circle). A contact with a slash through it represents a normally closed contact. In ladder logic, the left-hand rail is energized. When the contacts are closed in the right combinations, power can flow through the coil to the right-hand neutral rail.

Consider the ladder logic example in Figure 25.8. It is assumed that the *hot* rail at the left side has power, and the right side rail is *neutral.* When the contacts are opened and closed in the right combinations they allow power to flow through the output coils, thus actuating them. The program logic is interpreted by working from the left side of the ladder. In the first rung if *A* and *D* are on, the output *X* will be turned on. This can also be accomplished by turning *B* on, turning *C* off, and turning *D* on. In the second, the output *Y* will be on if *X* is on and *A* is on, or *D* is off. Notice that the branches behave as OR functions and the contacts in line act as an AND function. It is possible to write ladder logic rungs as Boolean equations, as shown on the right-hand side of the figure.

The example in Figure 25.8 contains only conditional logic, but Figure 25.9 shows a more complex example of a ladder logic program that uses timers and memory values. When the *run* input is active, output *heater* will turn on, 5 s later *fan1* will turn on, followed by *fan2* at 10 s. The first rung of the program will allow the system to be started with a normally open *run* push button input, or stopped with a normally closed push button *stop.* All stop inputs are normally closed switches, so the contact in this rung needs to be normally open to reverse the logic. The output *active* is also used to branch around the *run* to *seal-in* the run state. The next line of ladder logic turns on an output *heater* when the system is active. The third line will run a timer when *active* is on. When the input to the *TON* timer goes on, the timer *T4:0* will begin counting, and the timer element *T4:0.ACC* will begin to increment until the delay value of 10 s is reached, at this point the timer done bit *T4:0/DN* bit will turn on and stay on until the input to the timer is turned off. The fourth rung will compare the accumulated time of the timer and if it is greater than 5 the output *fan1* will be turned on. The final rung of the program will turn on *fan2* after the timer has delayed 10 s.

A PLC scans (executes) a ladder logic program many times per second. Typical execution times range from 5 to 100 ms. Faster execution times are required for processes operating at a higher speed.

The notations and function formats used in Figure 25.9 are based on those developed by a PLC manufacturer. In actuality, every vendor has developed a different version of ladder logic.

25.3.1.1 IEC 61131-3 Programming Languages

The IEC 61131 standards (formerly IEC 1131) have been created to unify PLCs [3,5]. The major portions of the standard are listed below.

IEC 61131-1 Overview
IEC 61131-2 Requirements and Test Procedures
IEC 61131-3 Data Types and Programming
IEC 61131-4 User Guidelines
IEC 61131-5 Communications
IEC 61131-7 Fuzzy Control

The most popular part of the standard is the programming specification, IEC 61131-3. It describes five basic programming models including ladder diagrams (LD), instruction list (IL), structured text (ST), sequential function charts (SFC), and function block diagrams (FBD). These languages have been designed to work together. It is possible to implement a system using a combination of the languages, or to implement the same function in different languages. A discussion of ST, SFC, and FDB programs follows.

Structured Text

A structured text program is shown in Figure 25.10. This program has the same function as the previous ladder logic example. The first line defines the program name. This is followed by variable definitions. The variables *run* and *stop* are inputs to the controller from sensors and switches. The variables *heater*,

```
PROGRAM example
VAR_INPUT
      run : BOOL ;
      stop : BOOL ;
END_VAR
VAR_OUTPUT
      heater : BOOL ;
      fan1 : BOOL ;
      fan2 : BOOL ;
END_VAR
VAR
      active :BOOL ;
      delay : TON ;
END_VAR
      active :=(run OR active) & stop ;
      heater := active ;
      delay(EN :=active,PRE := 10) ;
      IF ( delay.ACC >5 ) THEN
                        fan1 :=1 ;
      ELSE
                        fan1 := 0 ;
      END_IF ;
      fan2 :=delay.DN ;
END_PROGRAM
```

FIGURE 25.10 A structured text program equivalent to Figure 25.9.

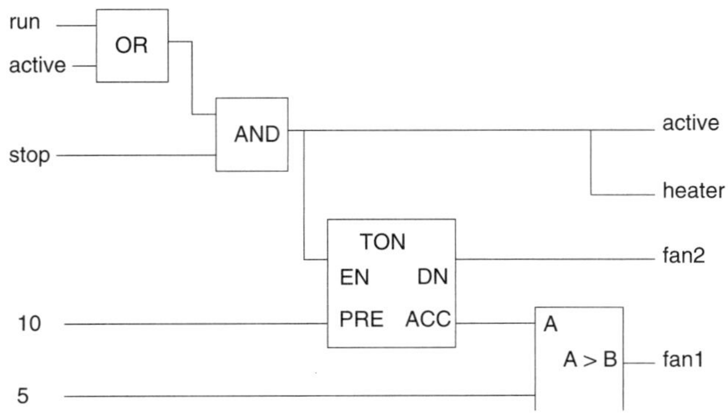

FIGURE 25.11 A FBD program equivalent to Figure 25.9.

fan1, and *fan2* are outputs to the actuators in the system. The variables *active* and *delay* are internal to the program only.

The program section immediately follows the variable declarations. In the program the first two lines set the values of *active* and *heater.* The instruction *delay* (....) calls the instantiated timer. The argument *EN* := *active* sets the timer to run, and *PRE* := 10 sets the timer delay to 10 s. The following lines use an "if" statement to set the value of *fan1*, using the accumulated timer value *delay.ACC.* The value of *fan2* is then set when the timer accumulator has reached the delay time and set the done bit *delay.DN.*

Structured text is popular and shows potential for eventually replacing ladder logic as the most popular programming language.

Function Block Diagrams

A data flow model is the basis of function block diagrams. In these programs, the data flows from the inputs on the left to the outputs on the right. The example in Figure 25.11 is equivalent to the previous ladder logic example. The *OR* and *AND* functions are used to set the values of *active* and *heater.* The *TON* timer uses the enable *EN* and delay *PRE* inputs to drive the accumulator *ACC* and *DN* outputs. The *DN* output drives *fan2* while the *ACC* value is compared to the value of 5 to set the output *fan1.*

Data flow diagrams can be very useful for doing a high-level design of a control system.

Sequential Function Charts

An SFC is used to describe a system in terms of *steps* and *transitions.* A step describes a mode of operation or state in which some *action* is performed, normally setting outputs. Transitions determine the change of states, normally by examining inputs. (Note: Some readers may notice that SFCs are based on Petri nets.)

Figure 25.12 shows an example of an SFC to control storage tanks. When the controller is started and the *power* input goes true, it will empty the tanks. After that the *run* input will start cycles where both the tanks are filled and then emptied repeatedly.

In this example, the flow of control begins at the initial step *start*, and then moves to step *S1.* The action associated with the step is *R*, which will reset, or turn off the outputs *in_valve1* and *in_valve2.* The system will remain in step *S1* until the transition is fired by input *power.* After this there are two possible paths. If *empty1* and *empty2* are both true, the left-hand branch will be followed, otherwise the right-hand transition will fire and that branch will be followed. The left-hand branch sets the *run_light* on (with *S*), and turns off the *outlet_valve.* The right-hand branch will turn on *outlet_valves* until the inputs *empty1* and *empty2* are both on. At that point *run_light* will be turned on, and the *out_valves* turned off. Regardless of which branch was followed, the flow of execution will pause at the following transition until the input *run* becomes true.

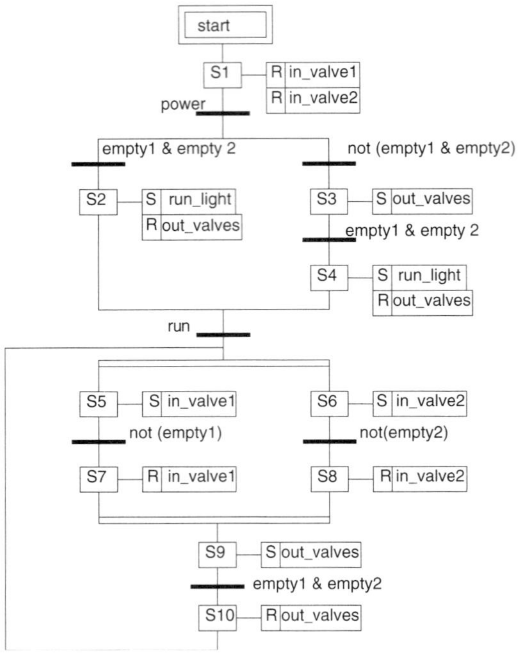

FIGURE 25.12 An SFC program for tank level control.

After the *run* transition is fired the flow of execution splits into both the left and right branches, as indicated by the two horizontal lines. The left branch fills one tank, while the right branch fills the other tank independently. When both branches are complete the flow of execution rejoins at the second set of horizontal lines, and then activates step *S9*. After step *S10* the flow of execution returns to the point after the *run* transition.

The SFC programming method differs from other programming methods in that the program is not expected to run completely in a single scan, while all others must run completely in each scan.

25.3.2 Interfacing

The installation and interfacing requirements for PLCs are driven by the need to protect people and equipment by failing safely. A typical wiring diagram for a PLC application is shown in Figure 25.13. At the top of the diagram a transformer is used to step down a higher supply voltage. This is immediately followed by a power disconnect and fuses. The power is then split into left and right rails, much like the ladder diagrams discussed earlier. Line 10 shows a master power control for the system. This includes a normally open start button and a normally closed stop button. These switches control a master control relay (MCR) *C1*. Notice that if power is supplied to the coil *C1*, it will close the contacts *C1* on the same run and hold *C1* on until the stop button is pushed. Another set of contacts is used on the left rail to disconnect power from the inputs to the PLC and the DC power supply. This control circuitry external to the PLC is required so that the stop buttons of a control system are able to directly disconnect the power. This is often required by law.

In this example the PLC is powered with 120 V AC, connected between the power rails. There are two 120 V AC inputs from normally open push buttons. The 24 V DC power supply is input to the *V+* on the output terminals of the PLC, which will then switch output power to solenoid *S1* and indicator light *L1*.

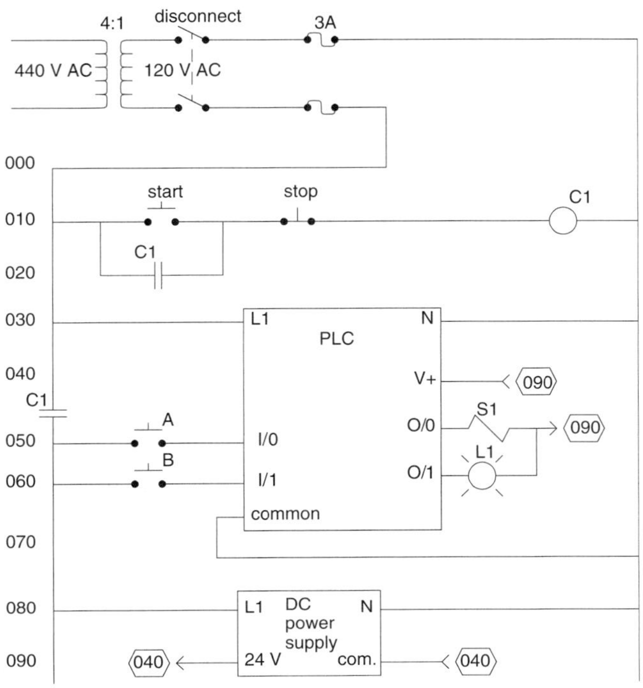

FIGURE 25.13 A PLC wiring example.

25.3.3 Advanced Capabilities

PLCs are often used in applications that go beyond basic logic solving. Some advanced programming and input/output (I/O) functions are listed below:

Calculations—The ability to do basic scientific calculations. Lower end PLCs only use integer math, while higher end PLCs also provide floating point math.
Analog I/O—Continuous voltage and current values can be input and output.
Feedback control—Proportional integral derivative (PID) controller calculations are provided as function blocks and can be used with analog I/O.
Communications—The ability to transmit data as strings over serial ports or to transfer parts of the PLC memory using proprietary protocols.
ASCII strings—Functions to manipulate ASCII strings.
System—Fault detection, status monitoring, interrupt routines, etc.
Fuzzy logic—Some PLCs include fuzzy set functions for nonlinear control problems.

At a minimum PLCs use communications for programming. But in many applications PLCs are used to communicate with other devices. In the past, most communications were based on proprietary, or closed, standards. More recently a few open communication standards have been developed and are supported by many vendors; these include Profibus, DeviceNet, CanBus, and ModBus. There has also been a trend to use more universal communication standards such as RS-232, RS-422, RS-485, and Ethernet. An example of an automation system is shown in Figure 25.14. An RS-232 connection is used between a laptop computer (e.g., COM1) and PLC1 for programming. DH+ is used to connect PLC1, PLC2, and the HMI; it is a proprietary communication standard developed by Allen-Bradley. An operator can use the human machine interface (HMI) to display data and accept operator input and communicate these values directly to both PLCs. Devicenet, an open automation standard, is used to connect PLC2 to a welding controller.

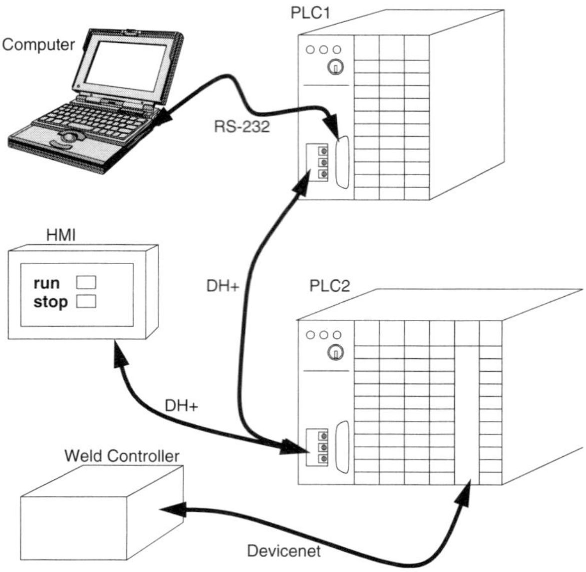

FIGURE 25.14 PLC communication example.

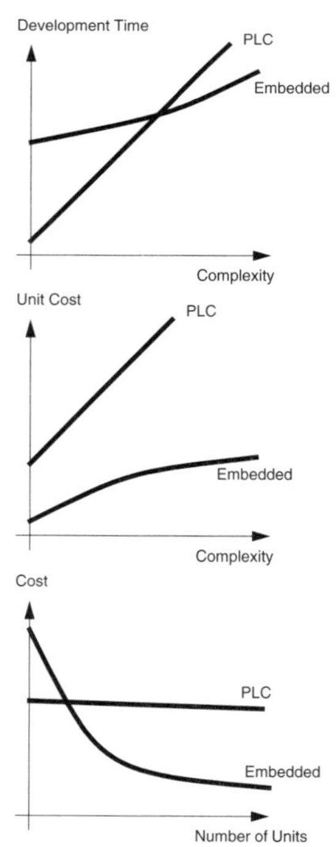

FIGURE 25.15 Relative trade-offs between control solutions.

25.4 Conclusion

PLCs and embedded controllers are complementary technologies and, when applied strategically, they will both provide low cost and reliable solutions to control problems. Figure 25.15 shows the relative trade-offs between the controllers. In general, an embedded controller requires more initial development time than a PLC for a simple system. As the system grows more complex, the embedded controller benefits from the existence of software libraries and design tools. When using a PLC the cost of the purchased hardware will always be higher per unit. The development costs for an embedded computer will usually be higher, but these become minimal when amortized over a large number of units. As a result, embedded controllers are typically selected for applications that will be mass-produced and allow a greater development time, such as a toy robot. PLCs are often selected for applications that only require a few controllers and are to be completed in a relatively short time, such as the production machines to make a toy.

References

1. Bryan, L.A., Bryan, E.A., *Programmable Controllers,* Industrial Text and VideoCompany, 1997.
2. Filer, R., Leinonen, G., *Programmable Controllers and Designing Sequential Logic,* Saunders College Publishing, 1992.
3. Lewis, R.W., *Programming Industrial Control Systems using IES1131-3,* The Institution of Electrical Engineers, 1998.
4. Petruzella, F., *Programmable Logic Controllers,* 2nd Ed., McGraw-Hill, 1998.
5. *Programmable Controllers—Part 3: Programming Languages,* IEC 61131-3 Ed. 1.0, 1993.
6. Stenerson, J., *Fundamentals of Programmable Logic Controllers, Sensors and Communications,* Prentice-Hall, 1998.
7. Webb, J.W., Reis, R.A., *Programmable Logic Controllers, Principles and Applications,* Prentice-Hall, 1995.

26

Graphical System Design for Embedded Systems

Shelley Gretlein
National Instruments, Inc.

26.1 Introduction

Nearly 50% of embedded system designs are late or never to market and nearly 30% fail after release [1]. Many of the challenges are a direct result of embedded systems becoming more complex as average code size has increased nearly an order of magnitude over the past 5 years [2]. In addition, with embedded systems becoming more common place, many domain experts, such as machine builders, test engineers, or control engineers, need embedded technologies to develop their systems, yet do not currently have the expertise required to develop embedded systems.With the increase in system complexity and with an ever-increasing number of nonembedded experts requiring this technology, there is a strong need for a new approach to embedded design. Graphical system design is a revolutionary approach to solving design challenges that blends intuitive graphical programming and flexible commercial off-the-shelf (COTS) hardware to help engineers and scientists more efficiently design, prototype, and deploy embedded systems. Figure 26.1 illustrates the embedded system development process. The graphical system design streamlines the embedded development process from design to prototype to deployment. Using the graphical system design approach will enable you to use a single environment across all stages of design to increase productivity, save money, and bring embedded technology to the domain expert.

26.2 Graphical Programming for Embedded Design

Many embedded systems run autonomously and need to execute many tasks in parallel with specific timing requirements. Consider a machine control system that needs to control a linear stage, rotate multiple shafts, control lighting, and read in video data; in a system like this, there are multiple processes that must happen deterministically, in real-time, and in parallel. Using traditional, text-based tools such

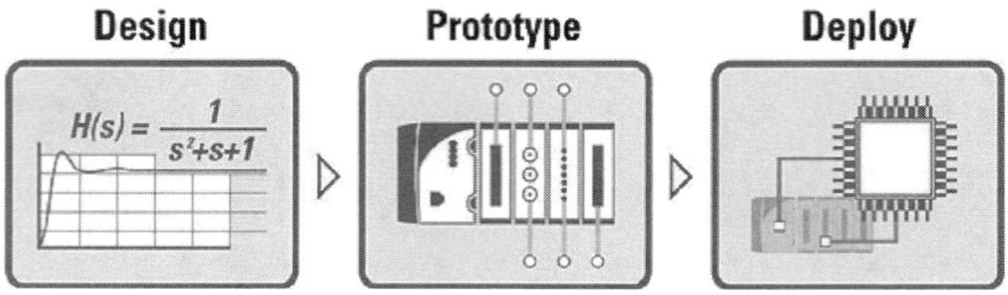

FIGURE 26.1 Graphical system design streamlines the embedded development process from design to prototype to deployment.

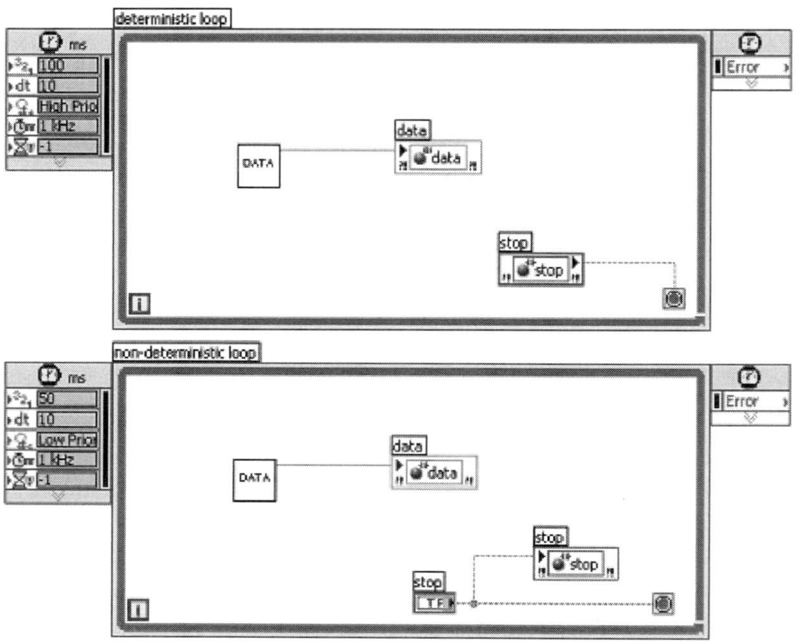

FIGURE 26.2 Parallel timed loops in the graphical programming environment of LabVIEW intuitively display parallel tasks.

as C for these applications quickly escalates in complexity. Graphical programming facilitates complex programming and timing models easily because of native functionality. More than 20 years ago, NI created components and technology in the formof the LabVIEW graphical development environment. LabVIEW natively incorporates timing into code using coding structures to implement timing, and representing parallelism is simply a matter of drawing another loop as shown in Figure 26.2.

To implement this level of timing and parallelism in text code is a barrier to a broader audience of domain experts, where the graphical representation is simply much more clear and accessible to scientists and engineers. If the graphical programming paradigm was extended to chips including FPGAs and microprocessors, it would be relatively painless to manage the parallel architecture of the silicon using the same consistency and scalability. Another key requirement for embedded system design is that the software platform addresses the various algorithm design views common in real-time embedded design. Various design views as models of computation for embedded software platforms are described in Reference 3. These models of computation match the way system designers view their system to help

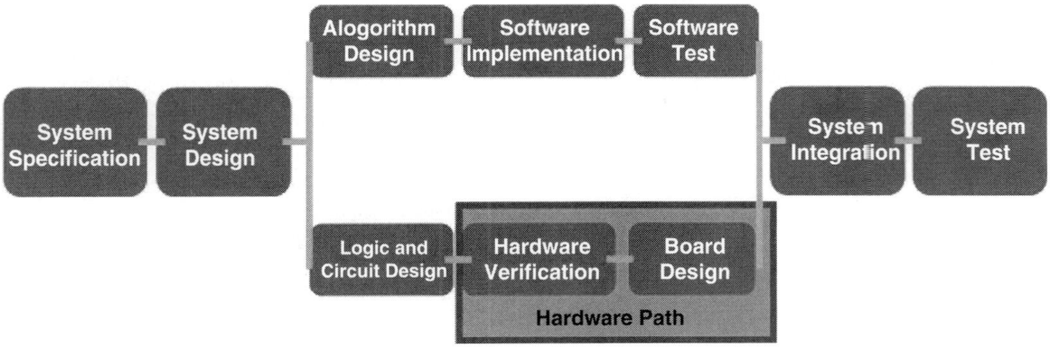

FIGURE 26.3 Typical embedded system software and hardware design flow showing separate software and hardware design flows.

minimize the complexity of translating system requirements into a software design. Over the past several years, LabVIEW has expanded to incorporate multiple models of computation to better meet the needs of embedded system designers and their varied skill sets. LabVIEW now includes text-based math, continuous time simulation, state diagrams, and graphical dataflow models to represent a variety of algorithms. LabVIEW also includes interactive tools to enhance the design experience for digital filters, control models, and digital signal processing algorithms to further ease designs in these vertical applications. You can implement algorithms on flexible COTS hardware platforms to significantly decrease your time to first prototype.

26.3 Customizable Off-the-Shelf Prototyping Platforms

As previously mentioned, with the many designs that are late or never released to market or worse, designs failing after release, something needs to be done to get more high-quality products out sooner. One way to address both of these issues is to prototype the systems better by integrating real-world signals and real hardware into the design process sooner so that high-quality designs are iterated on and problems are found (and fixed) earlier.

Consider the design flow in Figure 26.3. The hardware path is a viable location to search for efficiencies and ways to shorten the design process. Today, if you are creating custom hardware for final deployment, it is difficult to have the software and hardware developed in parallel, as the software is never tested on representative hardware until the process reaches the system integration step. In addition, you do not want the software development to be purely theoretical, because waiting until the system integration test to include I/O and test the design with real signals may mean that you discover problems too late to meet design deadlines.

Most designers currently use a solution such as an evaluation board to prototype their systems. However, these boards often only include a few analog and digital I/O channels and rarely include vision, motion, or ability to synchronize I/O. In addition, designers often have to waste time developing custom boards for sensors or specialized I/O, just to complete a proof of concept. Using flexible COTS prototyping platforms instead can truly streamline this process, as shown in Figure 26.4, and can eliminate much of the work required for hardware verification and board design. Much like PCs today, where anyone can go to an electronics store and plug components such as memory, motherboards, and peripheral together to create a PC, graphical system design strives to achieve the same standardization for prototyping platforms.

For most systems, a prototyping platform must incorporate the same components of the final deployed system. These components are often a real-time processor for executing algorithms deterministically, programmable digital logic for high-speed processing or interfacing the real-time processor to other

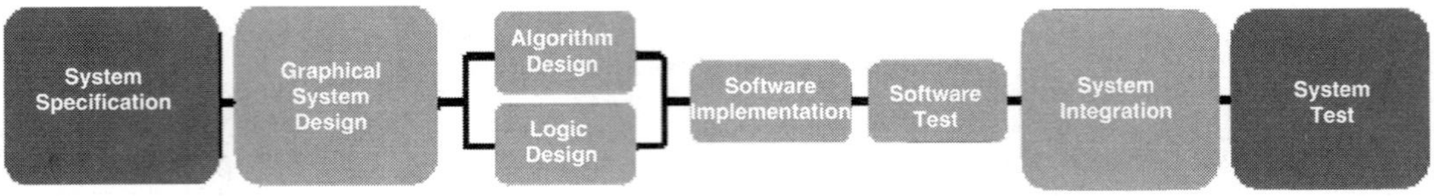

FIGURE 26.4 Stream-lined development flow with graphical system design.

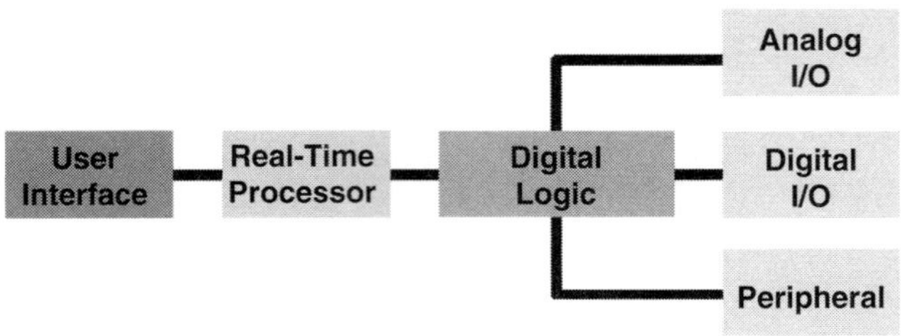

FIGURE 26.5 Typical components of an embedded system.

components, and varied types of I/O and peripherals, as shown in Figure 26.5. Finally, as with any system, if the off-the-shelf I/O does not serve all your needs, the platform should also be extensible and customizable when needed.

National Instruments offers several types of prototyping platforms including NI CompactRIO that contain all the basic building blocks of an embedded system. The controller contains a 32-bit processor running a real-time operating system. The CompactRIO backplane contains an FPGA that can implement high-speed processing and actually configure and provide interfaces to the I/O modules, which include options for analog input and output, digital input and output, and counter/timer functionally. Each of the modules includes direct connectivity to sensors and actuators as well as built-in signal conditioning and isolation. A module development kit is also available, which allows developers to expand the platform to include custom modules with all plugging into this COTS framework. In addition, CompactRIO is industrially packaged (−40°C to 70°C, 50G shock) with a small footprint (3.5 in. × 3.5 in. × 7.1 in.) and low power requirements (7–10 W typical), making it ideal for not only prototyping but also deployment of in-vehicle, machine control, and on-board predictive maintenance applications.

26.4 Custom Deployment Options

As previously mentioned, hardware devices, such as NI CompactRIO, can be employed for prototyping. It can be used for deployment because of its packaging, durability, and cost. However, there may be times when size or power may drive the need for smaller, custom board designs. To address that need, designers can preserve their software investment by using the LabVIEW Embedded Development Module to deploy code to any 32-bit processor. The LabVIEW Embedded Development Module combines all the graphical development benefits with the out-of-the-box analysis functions, integrated I/O, and interactive graphical debugging. This module is able to target any 32-bit microprocessor and provides a framework that can openly integrate a variety of existing C-based, third party tool chains and operating systems to target custom board designs. Once integrated, users can develop 100% graphically and interactively debug their application. By generating code to integrate with all existing targets in the market, users can have ultimate flexibility and breadth in target options. This new technology empowers a broader range of scientists, engineers, and domain experts to design their algorithms, develop the application, program the logic, prototype the system, and then deploy it to the target of choice.

26.5 Conclusion

The time has come for a new approach to electronic system design. Graphical system design brings a software platform combined with a hardware platform that offers the opportunity to significantly reduce development cost and time to market. A software platform that integrates multiple models of computation minimizes the time to implement specifications into a design. A flexible COTS hardware prototyping platform that supports the software platform and offers customizable components minimizes the time to first prototype, because the time and money required for designing custom hardware is eliminated. In addition, prototyping with real I/O results in a higher quality design—eliminating the design failures we see today. Finally, consistent graphical software from your design to your prototyping platform to your final deployment target maximizes code reuse and eases the transition to final deployment.

References

1. Embedded Software Development: Issues and Challenges. July 2003.
2. Rich Sevcik, "Taking the Frustration Out of Embedded Design," TechOnline, 2002, http://www.techonline.com/community/related_content/21543
3. Edward A. Lee, *Advances in Computers,* M. Zelkowitz, Ed., Vol. 56, Academic Press, London, 2002

27

Field-Programmable Gate Arrays

Daniel R. Fay
Daniel A. Connors
University of Colorado at Boulder

27.1 Introduction

Field-programmable Gate Arrays, or FPGAs, are a type of integrated circuit (IC) that contain logic components and interconnects that can be programmed after the device is manufactured in a semiconductor fabrication laboratory [1]. In short, an FPGA is a type of gate array that is programmable multiple times in the field rather than once at manufacturing time. FPGAs can host complex logic designs that include both combinational logic (AND, OR, NOT, XOR, etc.) as well as sequential logic (latches, flip–flops, etc). Most FPGAs also contain memory elements in the form of flip–flops or block memories. Modern FPGA-based chips contain up to hundreds of thousands of gates, and there are a variety of FPGA architectures on the market. In addition, there are emerging FPGA design trends to include not only programmable logic blocks but also programmable interconnects and switches between the blocks, and fixed hardware such as block memories, digital signal processing (DSP) cores, and processors. Overall, FPGA programmability allows them to be used in a variety of applications: prototyping ICs, in low-volume applications where the nonrecurring engineering (NRE) cost of designing and manufacturing a custom IC is too high, as glue logic for linking together other ICs, and for reconfigurable computing, which uses FPGAs as optimized computing engines. FPGAs provide this versatility through configurable logic and a hierarchy of flexible, programmable interconnects.

FPGAs [2] trace their roots to programmable logic devices, or PLDs, that emerged during the 1980s. These devices, known as programmable logic arrays (PLAs) and programmable array logics (PALs), allowed the designer to implement simple combinational and sequential logic. The PLA is composed of a set of programmable AND planes that connect to a set of programmable OR planes. The resulting connection of AND and OR planes creates the potential for any logic functions to be synthesized as its

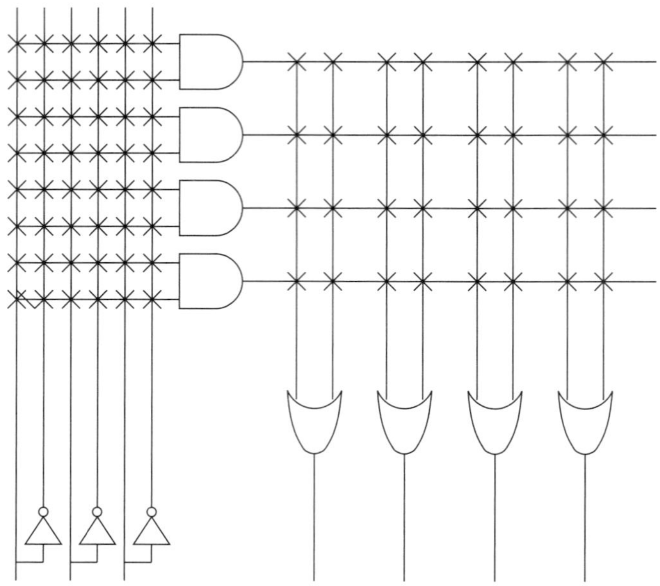

FIGURE 27.1 The respective AND and OR planes of a programmable logic array design.

sum of products (or alternatively the product of sums) form. Figure 27.1 displays the respective AND and OR planes of a PLA design. The original PLA-based designs were far more limited than FPGAs, with their functionality restricted to registering a few levels of simple sum-of-products Boolean expressions. Moreover, FPGA-generation of programmable logic, including Complex PLDs, or CPLDs, expanded these limited capabilities by adding more programmable logic and by allowing more flexible interconnect. As such, CPLDs can be used to implement functionality like feedback loops between logic blocks and even simple microcontrollers.

27.2 Field-Programmable Gate Array Architecture

At the heart of every FPGA architecture lies two basic resources: configurable logic blocks (CLBs) and routing resources. FPGAs organize the CLBs into a two-dimensional array with rows and columns of CLBs separated by routing channels. While the CLBs in modern FPGAs are often quite complicated, they all share similar basic functionality. To implement arbitrary combinational logic, they contain a lookup table (LUT), which is an $N \times N$ memory array that can implement any N-input Boolean function. The second component is a storage element such as a D flip–flop. Finally, there is also at least a multiplexer (MUX) that allows the storage element to be bypassed. These final two components allow the CLB to implement sequential logic like state machines. Figure 27.2 displays the basic view of a LUT-based CLB in modern FPGA systems.

Commercial FPGAs typically include more complex and specialized logic within their CLBs. Manufacturers often augment CLBs with multiple LUTs, fast carry chains for implementing low-latency arithmetic, and/or more sophisticated storage capabilities: some vendors' CLBs, for example, allow the LUT memory to be used as conventional read/write memory. Many FPGAs also contain various special–purpose blocks, such as memory, fast multipliers, DSP blocks, high-speed I/O, and even full-fledged microprocessors.

FPGAs can hold their configuration in different ways; however, the most common way for them to hold their configuration is using static RAM (SRAM). As compared to other configuration storage techniques, SRAM-programmed FPGAs enjoy high logic densities and can be rapidly reprogrammed

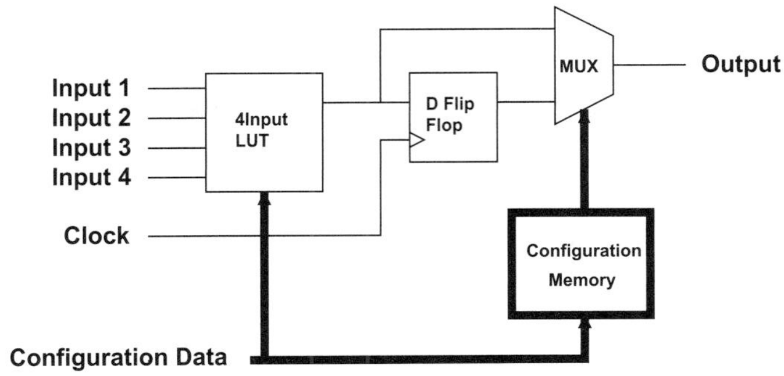

FIGURE 27.2 The basic view of a LUT-based CLB in modern FPGA systems.

multiple times; however, they lose their configuration if the FPGA loses power. Some FPGAs hold their configuration information using some kind of electrically erasable programmable read-only memory (EEPROM). FPGAs tend to have a lower logic density than SRAM-programmed FPGAs, and can only be reprogrammed a limited number (about 10,000–1,000,000) of times, but they can lose power and still retain their programming information. The final way FPGAs can store their configuration information is through either fuses or antifuses. Like EEPROM-programmed FPGAs, they retain their configuration information even after losing power; however, they can be programmed only once.

Most FPGAs have abundant clock resources consisting of dedicated low-skew logic and routing for distributing multiple clock signals around the FPGA. In addition, many FPGAs contain embedded clock managers that contain phase-locked loops (PLLs) for generating multiple arbitrary clock signals from a single external clock source.

27.3 Field-Programmable Gate Array Design Flow

To design for an FPGA, the designer first describes the behavior of the circuit. Once entered, the designer, using an electronic design automation (EDA) tool, generates a technology-mapped netlist. Place-androute tools then take this netlist and fit it to the target FPGA. After place-and-route is complete, the designer then uses the tools to generate a programming file to load on the FPGA. Timing analysis, simulation, and in-circuit logic analysis are used to debug the design throughout this process. Figure 27.3 illustrates a high-level overview of the design process using FPGAs including components of design entry, synthesis, verification, and simulation.

Design entry for FPGAs can be done in any number of ways [3]. One way is for the designer to enter the FPGA's design as a circuit schematic. Schematic entry is useful for designs that can be easily described using basic logic elements such as logic gates, flip–flops, multiplexers, encoders, and decoders. For larger projects, hardware description language (HDL) entry is usually a better choice. These languages allow the designer to describe an FPGA's functionality at a much higher level of abstraction. HDLs are like standard programming languages such as C/C++ and Java, with added extensions to support hardware-specific notations such as timesteps, interconnects, and concurrency. While there exist a variety of HDLs, the most widely used HDLs (and the ones the FPGA tools typically support) are VHDL or Verilog HDL. Many newer FPGA tools help increase designer productivity by allowing the designer to describe the design's higher levels of abstraction than those afforded by conventional HDLs. Some tools, such as Impulse-C, Handel-C, NAPA-C, and SystemC, allow the designer to describe the hardware in a modified version of the widely used C programming language. Other languages are used as well, including extensions to existing HDLs such as SystemVerilog and SystemVHDL. Tools also exist for converting DSP algorithms programmed in MATLAB into FPGA-synthesizable hardware.

Many toolchains provide libraries of predefined circuits, called IP cores. These cores are pretested and preoptimized for the target FPGA, which facilitates rapid development of complex FPGA designs. When

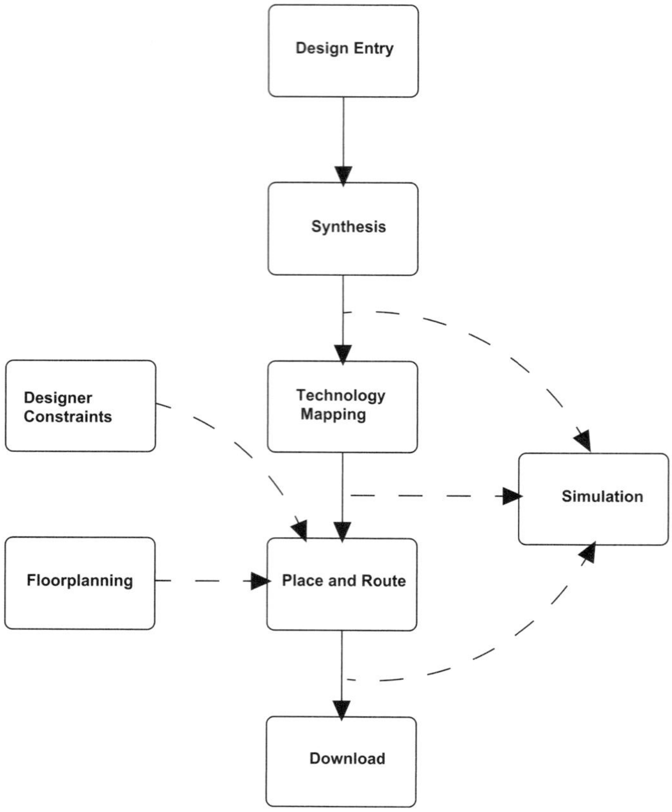

FIGURE 27.3 High-level overview of the design process using FPGAs.

the majority of the design consists of linking up predefined IP cores using standardized interconnect, the resulting design is called a system on a programmable chip (SoPC), which is analogous to the system on a chip (SoC) design flow for application-specific integrated circuits (ASICs).

Before testing the design while it is running on an actual FPGA, the FPGA designer debugs the design using a software simulator. Software simulation is first done after entering the design, to locate errors in the behavior of the circuit. After synthesis turns the circuit's behavioral description into a lower-level description, the simulation is rerun on the mapped netlist to look for bugs the synthesis tools might produce. After place and route, the design is tested again for tool-induced bugs. The final stage of testing involves testing the design while it actually runs on the FPGA. Like any other IC, FPGAs can be debugged with tools like logic analyzers and oscilloscopes. Many FPGA tools, however, also allow the designer to synthesize designs with full-speed logic analyzers. These tools synthesize logic analyzers onto the FPGA logic, allowing for full-speed, detailed analysis of the FPGA's internal circuitry.

The synthesis process outlined above consists of several key stages. First, the tools parse design entry information and convert it to an intermediate format understandable by the rest of the synthesis tools. Synthesis tools usually provide syntax checking at this stage. The tools then map the design to the functionality of the FPGA's logic blocks by packing Boolean logic functions to fit within the LUTs, by inferring arrays of storage as memory blocks, and so forth. Once the design is mapped to a technologyspecific netlist, the place and route phase begins. During this stage, the tools decide where to place the mapped blocks, with the route stage subsequently deciding how to interconnect the placed CLBs. During place and route, the tools can make tradeoffs between fast logic or logic that uses as few CLBs as possible. The designer can provide timing and area constraints to help guide the tools. It is also frequently possible for the designer to intervene at this stage to provide custom placement of the CLB mappings. Known as floorplanning, this technique allows the designer to manually place certain functionality to either improve synthesis time (the tools only need to place and route parts of the overall design), speed, or FPGA utilization. After completing the place and route stage, the tools proceed to generate a programming bitstream to load onto the FPGA. To protect the designer's intellectual property, some FPGA vendors allow this bitstream to be encrypted.

Once the tools finish generating the programming bitstream, it is then loaded onto the FPGA. FPGAs can be programmed in any of three ways. The first is to load the bitstream using a programming cable that connects the FPGA to computer that loads in the bitstream onto the array. This programming method is most frequently used during development/prototyping. The second method, which is used-more often for shipping products, is to have the FPGA read its configuration from an external PROM or Flash memory chip on power up. Finally, on FPGAs with internal nonvolatile storage, the FPGA can be programmed to hold a configuration indefinitely.

27.4 Different Uses for Field-Programmable Gate Arrays

Field-programmable gate arrays can be put to a variety of uses. One of these is for ASIC prototyping. For ASIC prototyping, a high-level description of the ASIC is synthesized to the FPGA to check for functionality bugs as well as for basic timing checks. For simple ASICs, the entire design can be synthesized onto a single FPGA. More complicated designs, however, may require multiple FPGAs Products such as QuickTurn and IKOS provide large-scale hardware emulation systems made up of many interconnected FPGAs for full circuit emulation of a large design.

FPGAs are also commonly used as ASIC replacements. They are desirable in low-volume applications where the high NRE costs cannot be amortized over enough units to justify the ASIC's lower per-unit cost over an FPGA. FPGAs are also useful in situations where the ASIC's design is subject to change; instead of having to qualify and manufacture an entire new set of ASICs to implement a change, the FPGA can be modified simply by reprogramming it.

FPGAs are also very useful in education of computer engineering.With the inherent attribute of being programmable, an FPGA can be used as a "programmable breadboard" that allows students to try out different designs. Their abundant logic resources allow students to implement elaborate projects with complexity ranging from hundreds of logic gates to embedded computer systems replete with multiple microprocessors, external I/O, and large external memories. A large selection of boards designed for education exist, which are made by the major FPGA vendors as well as by third-party companies. These boards usually contain an FPGA with access to external memory, switches and push buttons, light emitting diodes (LEDs), display screens, as well as expansion ports for attaching other hardware.

A growing use of FPGAs is to use them for reconfigurable computing. Reconfigurable computing leverages the vast parallel computational resources of FPGAs to accelerate certain classes of computational problems. These include DSP, bioinformatics, graphics/visualization, simulation, searching, video compression/decompression, and encryption/decryption. In general, any application that involves large amounts of computation but does not have very complex control flow (e.g., has few if-else statements and typically does not call a lot of procedures) is a candidate for FPGA acceleration. Most FPGAs do not implement floating-point arithmetic well, so applications that do a lot of floating-point calculations need to be reimplemented to only use integer arithmetic if they are to be efficiently implemented on an FPGA.

While it is possible for a reconfigurable computer to consist solely of FPGAs, virtually all reconfigurable computers contain one or more conventional microprocessors for running parts of the program that are ill-suited for running on the FPGA. A major way reconfigurable computers differentiate themselves is by how tightly coupled the FPGA logic is to the CPU. FPGA logic that is more closely coupled with the CPU logic reduces the communication latency between the CPU and the FPGA; however, these setups often have less FPGA logic available than more-decoupled designs. Generally, low-latency CPU–FPGA communication is essential when the FPGA part of the algorithm is relatively small or requires frequent communication with the CPU part of the algorithm. Production reconfigurable computers can take on a variety of configurations. In one configuration, the FPGA fabric is completely integrated with the processor itself. In this configuration, programs implement special complex instructions on the FPGA fabric. Another type of reconfigurable computer treats the FPGA as a tightly-coupled coprocessor. In this setup, the FPGA coprocessor shares memory with the main processor, allowing fast communication via direct memory access (DMA). Other systems take a traditional computer system and add FPGAs as expansion cards in the same way as someone might add a network card or video card.

27.5 Embedding Microprocessors and Microcontrollers onto Field-Programmable Gate Arrays

In recent years, it has become possible for designers to integrate microprocessor/microcontroller designs onto an FPGA. Many FPGAs now contain enough logic resources to implement a full-fledged microprocessor/microcontroller onto the FPGA. Known as "Soft CPU Cores," these parts are a standard part of many FPGA vendors' toolchains and contain microprocessor cores specially optimized for efficient implementation on the FPGA. Designers can typically tailor the soft microprocessor for its intended use by adding/removing resource-intensive hardware components such as caches, floating-point units, and barrel shifters. In addition to soft microprocessor cores, some FPGAs actually contain embedded microprocessor blocks. Since these microprocessor blocks are entirely custom logic, they can perform much faster, have more features, and use less FPGA area than a typical soft microprocessor core; however, the designer is limited to the hard microprocessor cores available within a given FPGA family.

Designers use integrated microprocessors/microcontrollers in a variety of ways. One way is to use them to implement complicated statemachines, as it is often easier towrite a statemachine as a software program than it is to express the statemachine as a state diagramand then synthesize a custom finite statemachine. Integrated microprocessor/microcontroller cores on FPGAs are also useful as an area-saving feature; instead of spending precious FPGA hardware resources implementing infrequently used, performance insensitive functionality, these features can instead be implemented as a software program running on the microprocessor/microcontroller. An example of thiswould be awireless router; while the signal processing and packet processing features are implemented on the FPGA fabric for high performance, the web server used for configuration purposes is run as a software program on the microprocessor/microcontroller.

FPGA-integrated microprocessors/microcontrollers usually have various interfaces that allow the core to interface with the rest of the FPGA. Many contain special high-speed interface blocks that allow for connecting custom accelerator blocks and coprocessors, such as fast Fourier transform (FFT) blocks or floating-point units. In addition, these cores usually have slower buses for interfacing with memory and/or I/O. Figure 27.4 illustrates the integration of programmable components with design blocks in a modern FPGA.

27.6 Reliability and Power Considerations

Serious and growing concerns with digital logic, particularly that used in extreme environments such as space, are soft error upsets, or SEUs. SEUs occur most often when a high-energy particle, such as a cosmic ray, hits a transistor and reverses its state. While the SEU does not permanently damage the transistor, it can lead to incorrect behavior in the circuit. While SEUs can occur in any kind of modern digital logic, they are of a special concern when using SRAM-based FPGAs. If an SEU flips one of the configuration bits on an SRAM-based FPGA, it can seriously alter the functionality of the FPGA's logic, with the error persisting until the FPGA gets reconfigured.

There are several ways to mitigate these problems. First, all critical state storage and interconnect buses should be protected with at least parity to detect the errors. If on-the-fly error detection *and correction* is desired, the circuits should also have some sort of error-correcting code (ECC) implemented on the state and interconnects. Another redundancy technique that can be used in lieu of or in conjunction with

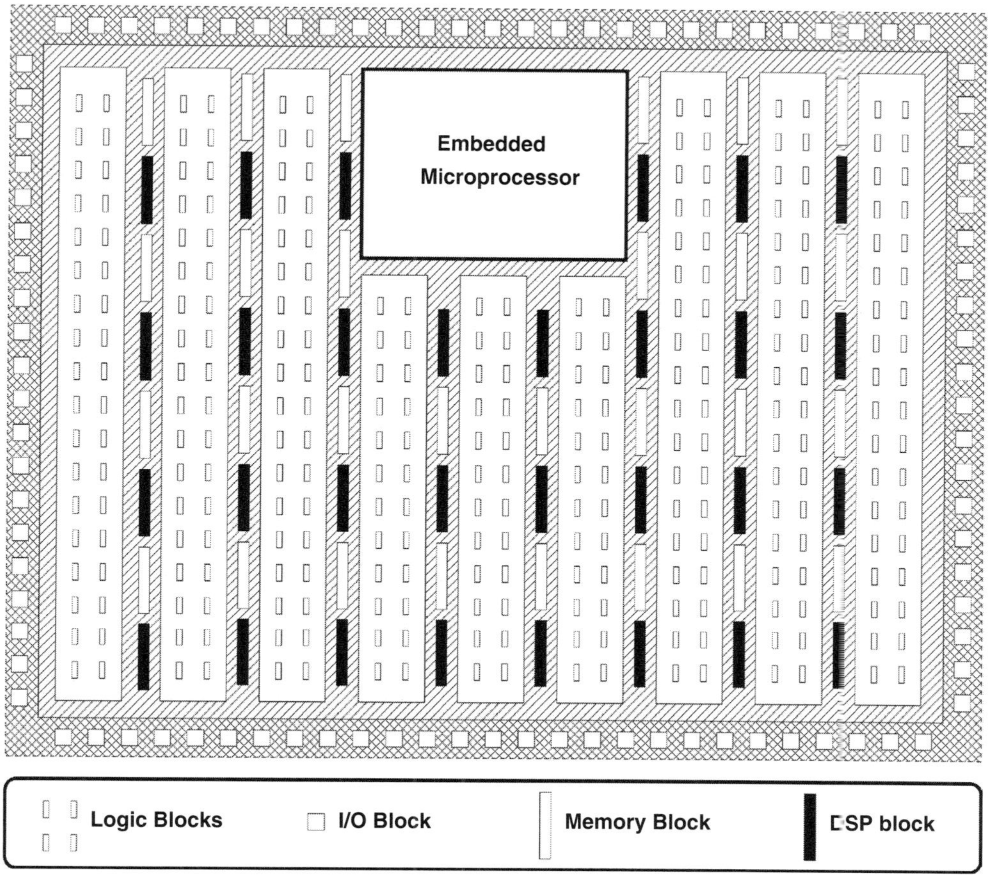

FIGURE 27.4 Integration of programmable components with design blocks in a modern FPGA.

the previously stated techniques is triple modular redundancy (TMR). TMR consists of triplicating all the on-chip logic. At regular stages throughout the design, the results of all three copies are compared using a minority voter scheme. If one of the results differs from the other two, the two matching results are then passed on downstream.

All the above techniques are effective ways to deal with SEU errors on any form of digital logic. One problem remains for SRAM-based FPGAs, however; as time goes on, the number of errors within the configuration storage grows until even ECC and TMR techniques cannot correct for the faulty logic. To mitigate this, the FPGA's configuration must be periodically refreshed to scrub out built-up configuration errors.

Power consumption is also an important consideration when designing with FPGAs. Overall, an FPGA's power needs tend to be significantly higher than for a custom ASIC; however, an FPGA typically uses less power than the equivalent algorithm implemented as a software program on a microprocessor or microcontroller. Relative to an ASIC, an FPGA design's power consumption can be easily estimated because of its premade, precharacterized logic blocks. Tools that will estimate power consumption using a simulation of the FPGA exist.

27.7 Similar Devices

Besides FPGAs and traditional PLDs, there exist a variety of other quasiprogrammable ICs; one of these is a mask programmable gate array, or MPGA. These parts start out as partially finished ICs; they come out of the factory with their transistors finished in a regular array as well as specifying the first few interconnect layers. To finalize the product, the customer connects the finished transistors in a desired pattern by specifying the remaining interconnect layers. Such a chip plays the middle ground between the low NRE/high unit cost of FPGAs and the high NRE/low unit cost of ASICs. They accomplish this by achieving densities close to ASICs; however, the customer only needs to design a few of the uppermost (and typically the lowest cost) of the masks in the mask set.

Structured ASICs are closely related to MPGAs. Structured ASICs bring the advantages of IP blockbased SoC design to MPGAs by combining mask-programmable logic with nonprogrammable blocks such as microprocessors, embedded memories, memory controllers, and high-speed I/O blocks. Some FPGA vendors provide special structured ASICs that exactly replicate the functionality of their FPGAs, to make it easier for customers to port their design from an FPGA over to a structured ASIC.

There is an analog equivalent to digital FPGAs: field-programmable analog arrays (FPAAs). These parts contain configurable analog blocks interfaced with programmable interconnect. Each configurable analog block typically contains at least one operational amplifier ("Op-Amp") along with various passive electronic components such as resistors and capacitors.

References

1. Brown, S.D., Francis, R.J., Rose, J., and Vranesic, S.G., *Field Programmable Gate Arrays*, Kluwer Academic Publishers, 1992.
2. Trimberger, S.M., *Field-Programmable gate Array Technology*, Kluwer Academic Publishers, 1994.
3. Salcic Z. and Smailagic, A., *Digital Systems Design and Prototyping Using Field Programmable Logic*, Kluwer Academic Publishers, 1997.

28

Graphical Programming for Field-Programmable Gate Arrays: Applications in Control and Mechatronics

Jeannie Sullivan Falcon

Michael Trimborn

National Instruments, Inc.

28.1 Introduction

Embedded control systems require a high degree of reliability and determinism. However, the costs of designing and testing these control systems can be very high because of application complexity and stringent performance requirements. In addition, reducing time-to-market for product development is essential for many companies. For researchers, the use of high-level software tools in a unified environment can abstract away many implementation details. This facilitates the rapid investigation of advanced algorithms and control approaches on real-world systems.

Field-programmable gate arrays (FPGAs) are being used for an increasing number of embedded control and monitoring systems. They are less expensive than ASICs in terms of development and can also be reprogrammed in the field. FPGAs also have advantages over microprocessors in that all of the control logic is implemented in hardware. This leads to a significant advantage in terms of speed. Also, multiple tasks can run in parallel because each task will utilize a different set of gates on the FPGA. FPGAs can be used for control, signal processing, digital communication, health monitoring, and custom timing and triggering.

Programming for FPGAs has previously required knowledge of VHDL or Verilog [1]. VHDL and Verilog are taught in electrical engineering programs but not in mechanical, aerospace, biomedical, or chemical engineering programs. This has limited the proliferation of FPGAs in control and mechatronic systems. Figure 28.1 shows a LabVIEW FPGA example for pulse width modulation (PWM) output generation. Figure 28.2 shows VHDL for the same task [2].

National Instruments (NI) has recently developed several new design tools and a hardware architecture that can be used to speed the design and implementation of embedded control systems utilizing FPGAs. The design tools are based on LabVIEWFPGA—agraphical programming language for FPGAs. The hardware architecture combines an FPGA with a microprocessor and is available for desktop computers, PXI/CompactPCI industrial computers, as well as an NI embedded control system called CompactRIO.

Built-in LabVIEW FPGA functions are available for control and signal processing. Control systems with digital inputs (such as encoders) and digital outputs (such as PWM signals) can be built on the FPGA with rates up to 20 MHz. Control systems with analog inputs and outputs can be built with rates up to 500 kHz. Parallelism is inherent because separate FPGA gates are used for each loop on a LabVIEW FPGA diagram. Thus, the addition of control or monitoring tasks will not slow the overall FPGA application. The total number of tasks is limited, however, by the number of gates of the FPGA.

This paper will present a background on FPGAs, discuss programming methods and system architectures, and describe the LabVIEW FPGA module for graphical programming of FPGAs. Several applications in control and mechatronics will also be discussed.

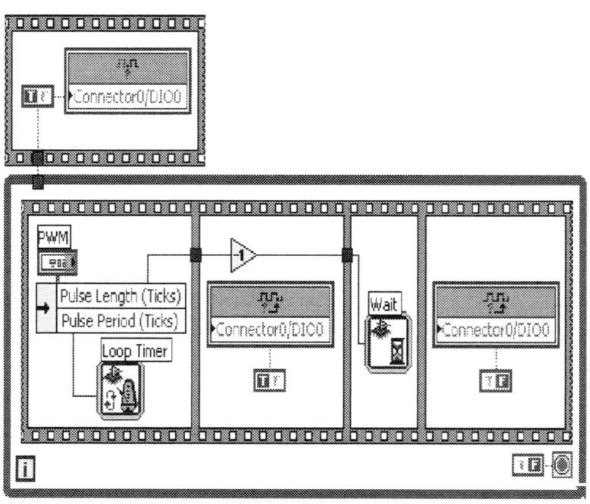

FIGURE 28.1 LabVIEW FPGA program for PWM output.

```vhdl
library IEEE;
use IEEE.std_logic_1164.all;

library SYNOPSYS;
use IEEE.std_logic_arith.all;
use IEEE.std_logic_unsigned.all;

-- Interface definition
entity pwm is
   port (
         data: in STD_LOGIC_VECTOR (7 downto 0);
      load: in STD_LOGIC;
         clk: in STD_LOGIC;
         enable: in STD_LOGIC
      outpwm: out STD_LOGIC
         );
end pwm;

-- Architecture definition with local variables and processes
architecture pwm_arch of pwm is
            signal value: STD_LOGIC_VECTOR (7 downto 0);
            signal count: STD_LOGIC_VECTOR (7 downto 0);
            signal tc: STD_LOGIC;
            signal cdir: STD_LOGIC;
begin

   -- Process A: Load new counter value
   process(clk)
         begin
            if (clk'event and clk = '1') then --CLK rising edge
                  if (load = '1') then
                        value <= data;
                  end if;
            end if;
         end process;

   -- Process B: Counter, Create PWM output signal
   process (clk)
         begin
            if (clk'event and clk = '1') then --CLK rising edge
               if (enable = '1') then
                  if (tc = '1') then
                        count <= value;
                  elsif (cdir = '1') then
                        count <= count + 1;
                  else
                        count <= count - 1;
                  end if;
                  if (cdir = '1') then
                        if (count = 255) then
                              tc <= '1';
                        else
                              tc <= '0';
                        end if;
                  else
                        if (count = 0) then
                              tc <= '1';
                        else
                              tc <= '0';
                        end if;
                  end if;
               end if;
            end if;
   end process;

   -- Process C: Change counter direction
   process (CLK)
   begin
      if CLK'event and CLK='1' then --CLK rising edge
         if ( enable = '1' ) then
                  if (TC = '1') then
                        cdir <= not cdir;
                  end if;
         end if;
      end if;
   end process;

   -- Pocess D: Output H-Bridge control signals
   process (CLK)
   begin
      if CLK'event and CLK='1' then --CLK rising edge
               outpwm <= cdir;
      end if;
   end process;

end pwm_arch;
```

FIGURE 28.2 VHDL example for PWM output.

28.2 Field-Programmable Gate Array (FPGA) Background

28.2.1 FPGA Definition and Market

An FPGA is a chip that consists of many unconfigured logic gates. Unlike the fixed, vendor-defined functionality of an application-specific integrated circuit (ASIC) chip, an FPGA can be configured and reconfigured for different applications. FPGAs are used in applications where the cost of developing and fabricating an ASIC is prohibitive, or the hardware must be reconfigured after being placed into service. Because FPGAs can be used for implementation of custom algorithms in hardware, they offer benefits such as precise timing and synchronization, rapid decision making, and simultaneous execution of parallel tasks. Today, FPGAs appear in such devices as electronic instruments, consumer electronics, automobiles, aircraft, copy machines, and application-specific computer hardware.

The major manufacturers of FPGAs are Xilinx [3] and Altera [4] and, together, they comprise more than 80% of the FPGA market [5].

28.2.2 Programming Methods

Historically, programming FPGAs has been limited to engineers who have in-depth knowledge of VHDL or other low-level design tools, and requires overcoming a very steep learning curve.

Other FPGA design options are available. Celoxica [6] offers comprehensive C/C++ design tools for FPGAs. Mentor Graphics [7] creates FPGA code using C++.

AccelChip [8] offers MATLAB® programming capability for FPGAs including tools for floating-point to fixed-point conversion. Xilinx [3] provides a graphical programming option using Simulink® as part of their Xilinx System Generator for DSP.

28.2.3 LabVIEW FPGA Graphical Programming

28.2.3.1 LabVIEW FPGA Introduction

The NI LabVIEW FPGA Module extends the LabVIEW graphical development environment to reconfigurable FPGAs on various hardware platforms [9]. With the LabVIEW FPGA Module, the operation of the FPGA device as well as the connected analog and digital I/O can be configured by programming in LabVIEW. The graphical dataflow and parallelism inherent in LabVIEW are useful for intuitively programming control tasks, algorithms, signal processing, and logic.

Once the FPGA is targeted in the LabVIEW development environment, LabVIEW displays only the functions that can be implemented in the FPGA (Figure 28.3). The LabVIEW FPGA Module functions palette includes typical LabVIEWstructures and functions, such as while loops, for loops, case structures, and sequence structures, as well as a dedicated set of LabVIEW FPGA-specific functions for math, signal generation and analysis, simple control (including PID), comparison logic, array and cluster manipulation, occurrences, analog and digital I/O, and timing. A combination of these functions can be used to embed logic and intelligence onto the FPGA-based device.Math and control in FPGA are limited to integer-math operations. However, simple control blocks such as PID have been developed with pseudo-floating point capability.

The LabVIEW block diagram is implemented in hardware, which provides direct control over the reconfigurable hardware I/O. The I/O signals can be analyzed and manipulated in ways that are not possible with fixed I/O hardware. Figure 28.4 shows the block diagram for a simple rising-edge counter. Following the data flow makes it very easy to understand the behavior of this counter. A while loop

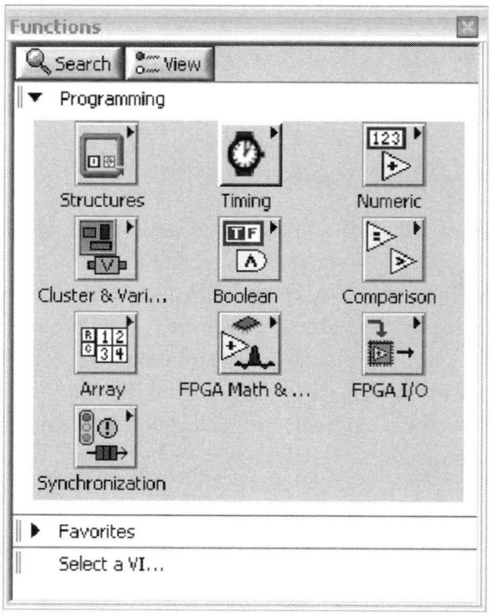

FIGURE 28.3 LabVIEW FPGA Functions Palette showing programming structures, timing, math, logic, control, synchronization, and I/O functions.

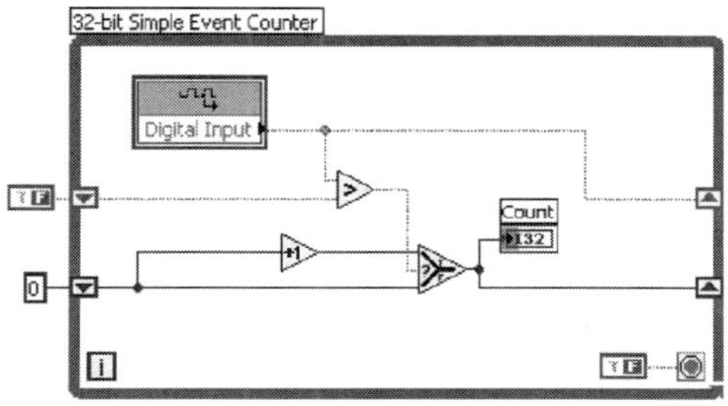

FIGURE 28.4 LabVIEW FPGA program for 32-bit counter.

structure is used to continuously read from a digital line and compare the line status from one iteration to the next. When the digital line changes state, the counter is incremented by one. The counter size is defined by the I32 data type of the counter indicator, making this a 32-bit counter.

The uniqueness of LabVIEW FPGA is thatmultiple, parallel counters can be created through additional while loops placed on the LabVIEW block diagram. These will consume completely different sections of the FPGA, resulting in true parallel process. In addition, the hardware can be reconfigured as needed. For example, the digital I/O lines connected to the FPGA can be apportioned as needed into any combination of encoder inputs, counter inputs, PWM outputs, or serial communication protocols. If different digital I/O is needed for a new application, the FPGA is simply reprogrammed with LabVIEW-FPGA to provide different I/O configuration.

The hardware configuration on the FPGA can be augmented with in-line signal processing, health monitoring, fault detection, logic, and feedback control. Each additional task consumes additional FPGA gates until the total number of gates on the FPGA device is consumed.

28.2.4 Development Flow

After creating the LabVIEW FPGA VI, the code must be compiled. Similar to other FPGA development tools, compile time for an FPGA VI can range fromminutes to several hours, depending on the complexity of the code and the specifications of the development system. To maximize development productivity, a bit-accurate emulation mode can be used, so the logic of the design can be verified before initiating the compile process. When targeting the FPGA Device Emulator, LabVIEW accesses I/O from the device and executes the VI logic on the Windows development computer. In this mode, it is possible to use the same debugging tools available in LabVIEW for Windows, such as execution highlighting, probes, and breakpoints.

Once the LabVIEW FPGA code is compiled, a LabVIEW host VI can be created to integrate the hardware into the rest of the test and control system. Figure 28.5 illustrates the development process for creating an FPGA application. The host VI uses controls and indicators on the FPGA VI front panel to transfer data between the FPGA on the RIO device and the host processing engine These front panel objects are represented as data registers within the FPGA. The host computer can be either a PC or PXI controller running Windows or a PC, PXI controller, Compact Vision System, or CompactRIO controller running a real-time operating system (RTOS).

28.2.5 Combined FPGA/MicroProcessor Architectures

National Instruments provides multiple hardware platforms with FPGA technology Figure 28.6 shows the PXI/cPCI Industrial Computer platform, with plug-in FPGA boards as an option, as well as the CompactRIO programmable automation controller with an FPGA on the backplane of the chassis. NI also provides programmable plug-in FPGA boards for desktop computers. An additional hardware platform with a programmable FPGA is the NI Compact Vision System with distributed machine vision and control capability.

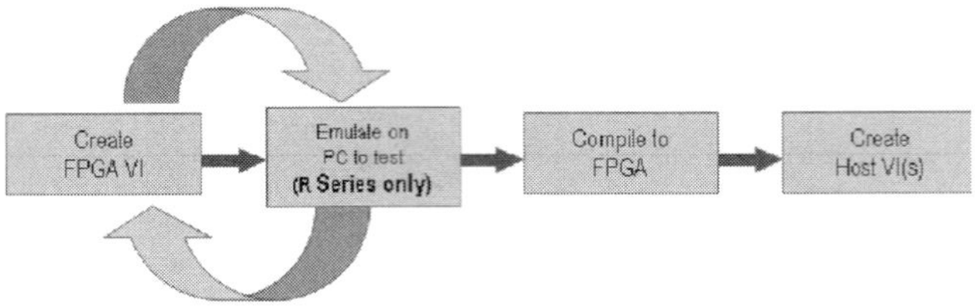

FIGURE 28.5 Application development flow showing FPGA compilation step.

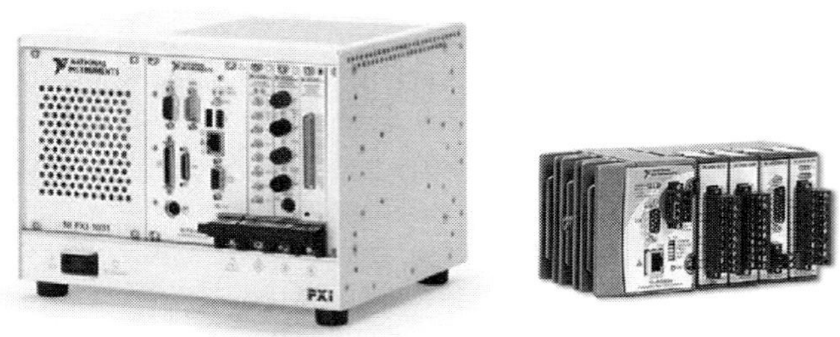

FIGURE 28.6 PXI/cPCI Industrial Computer and NI CompactRIO automation controller.

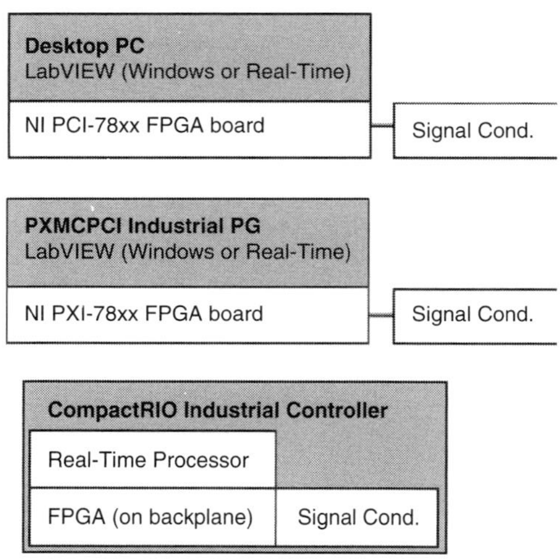

FIGURE 28.7 Hardware architecture diagram for FPGA/processor platforms.

The shared hardware architecture among the platforms is illustrated in Figure 28.7. The desktop PC and the PXI/cPCI Industrial PC can run Windows or LabVIEW Real-Time on the main floating-point processor. LabVIEW Real-Time allows LabVIEW diagrams to run on top of a commercial real-time operating system for multiple types of targets. In both the desktop and the PXI system, the plug-in FPGA board is connected to the processor via the PCI bus. Plug-in FPGA boards are available with both balanced analog I/O and multiple digital lines.

The CompactRIO architecture shown in Figure 28.7 is similar to that of the desktop and the PXI system with FPGA technology. The floating-point processor in the CompactRIO controller is tied to the FPGA on the backplane via an internal PCI bus. The FPGA is, in turn, connected to the analog and digital I/O modules in a star topology. Analog modules contain both conversion circuitry and signal conditioning while digital modules have signal conditioning alone.

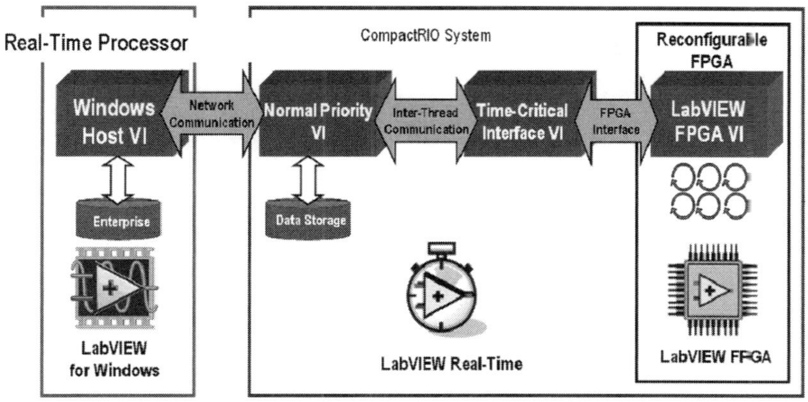

FIGURE 28.8 Software application architecture diagram for FPGA/processor platforms.

Because of the similar hardware architecture, a common software application architecture can be used for all three platforms. This software architecture is shown in Figure 28.8 and includes a host VI (program) running on a Windows PC, a real-time VI running on the processor, and an FPGA VI running completely in hardware on the FPGA.

With this software and hardware architecture, it is possible to quickly develop various hierarchical control strategies. Some lower level control loops can take place completely in the FPGA. Higher level control loops can be written for the floating-point processor in the system. In addition, various tasks such as custom startup and shutdown behavior, in-line signal processing, and custom timing, synchronization, and digital communication can be implemented in the FPGA. Time-critical parallel tasks such as health monitoring and fault detection can be developed on the FPGA as well.

While processors run computations in sequence, FPGAs can run computations in parallel. New processor architectures involving multiple cores are being released, but the FPGA is still the best choice for parallel computation. If multiple control loops or tasks are implemented in a processor, then the overall loop rate will decrease. Multiple control loops and tasks that are implemented in an FPGA can execute independently so that extremely high loop rates can be achieved. The limiting factor for FPGAs is the space available or the number of gates.

28.2.6 Integration of FPGA Graphical Programming with Simulation and Control

A simple analog input and analog output operation written with LabVIEWFPGA is shown in Figure 28.9. The analog I/O will run at the maximum rate allowed by the FPGA since a timing VI has not been added to the program. The timing is then limited only by the conversion rate for I/O on the FPGA-based device.

Figure 28.10 shows the front panel for this VI when the analog output is wrapped back to the analog input. The user can adjust the slider for the analog output on the front panel and see the result in the graph of the analog input.

A simple control program can be developed for the real-time processor using the LabVIEW Simulation Module. Figure 28.11 shows the simulation loop with a signal generator VI writing a sine wave to the analog output. The analog input is read back through the FPGA and communicated to the processor.

Many other dynamic system modeling and control blocks are available in LabVIEW Simulation Module including linear continuous and discrete-time models, as well as both graphical and textual blocks for nonlinear systems.

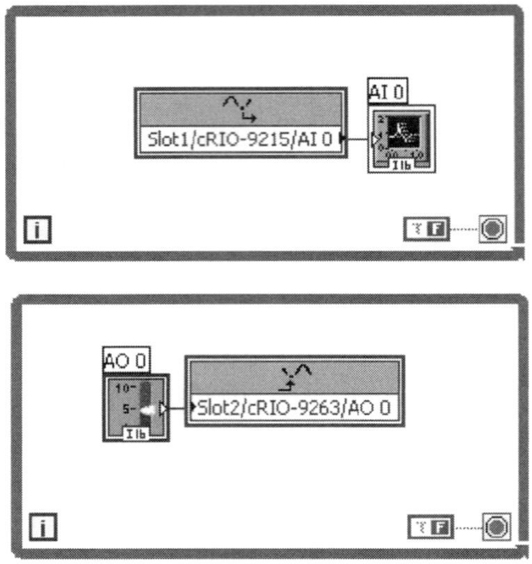

FIGURE 28.9 Simple analog input and output VI written in LabVIEW FPGA for single-point I/O.

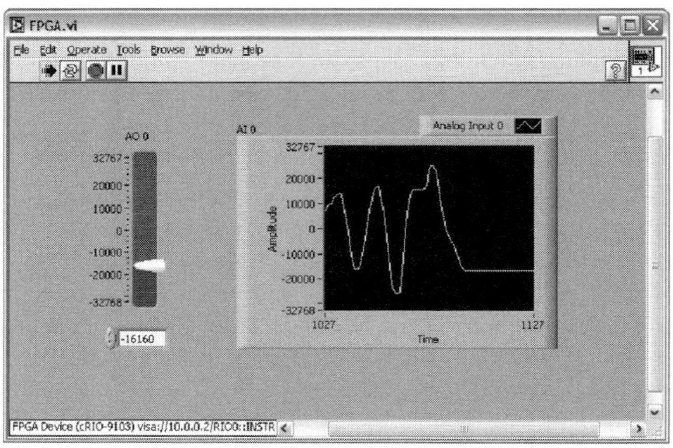

FIGURE 28.10 Front panel of analog input and output VI written in LabVIEW FPGA.

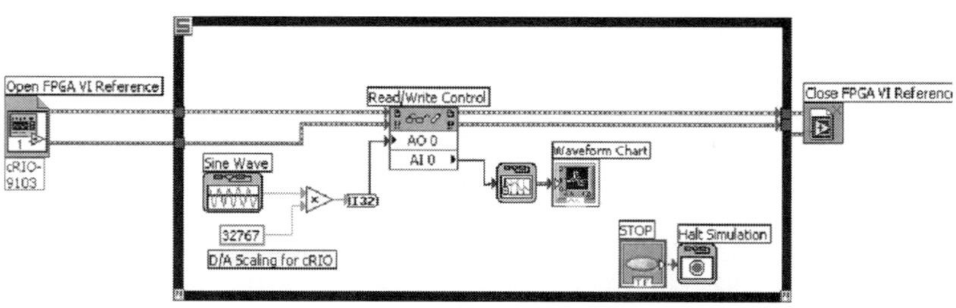

FIGURE 28.11 Simple LabVIEW Simulation Module VI that communicates with LabVIEW FPGA VI through an interface node.

28.3 Applications

28.3.1 High-Speed Control on a FPGA for Micro-Electromechanical Systems Device

Using LabVIEW FPGA and the NI PXI-7831R module, Kurfess and Degertekin implemented parallel, extremely high speed control (120 kHz) of an array of micro-electromechanical systems (MEMS) devices [10]. MEMS devices are produced in large quantities on a single silicon wafer, and, even when packaged or built into another device, they are often spatially arranged in arrays. Kurfess and Degertekin developed a novel measurement system that can test multiple MEMS devices in an array simultaneously rather than stepping and scanning a single measurement device over the MEMS array. The new measurement system is itself an array of MEMS microinterferometers. Each microinterferometer in the array is a position-sensing grating interferometer (μPSGI) capable of measuring out-of-plane distances with subnanometer accuracy at frequencies more than 20 MHz. The microinterferometer is 1000 times smaller than typical, bench-sized laser interferometers. Initial results demonstrate mapping of microscale membranes vibrating in the 2 MHz range (Figure 28.12).

Kurfess and Degertekin implemented the microinterferometer in both single and array units using the LabVIEW FPGA Module and NI reconfigurable I/O hardware, which makes the transition from monitoring and controlling a single interferometer to an array of interferometers a simple task. They used the NI PXI-7831R reconfigurable I/O module to provide analog I/O along with high-speed PID control. They applied the control to a deformable grating within the interferometer sensor to minimize the effects of sample misalignment and external disturbances and improve the sensitivity of the microinterferometer during operation. They used the pseudo–floating-point PID example in the LabVIEW FPGA Module as a basis for the control implementation. With the high-speed hardware timing capabilities of the LabVIEW FPGA Module, Kurfess and Degertekin easily achieved the required 120 kHz bandwidth for the control loop.

In this application, the maximum bandwidth was determined by the conversion speed of the A/D and D/A circuitry on the PXI-7831R module. Higher bandwidth control can be achieved if the digital lines on the plug-in FPGA board are used for both sensors and actuators. An example would be encoders or tachometers for sensors and PWM commands to amplifiers or drives for actuators. Also, the CompactRIO system includes options for high bandwidth analog-to-digital and digital-to-analog conversion circuitry.

28.3.2 Single-Point I/O on FPGA for Haptics

Kopp and O'Malley [11] used LabVIEW Real-Time and LabVIEW FPGA for control and high-speed single-point I/O in a haptics experiment focused on creating stiff "virtual walls." A limiting factor for

FIGURE 28.12 SEM image of MEMS microinterferometer.

stiffness in virtual walls is the sampling rate of the control system for the haptic device. If the desired stiffness is too high in relation to the sampling rate, then the virtual wall ceases to be passive and exhibits active behavior in the form of vibrations and, at too high of a stiffness, instability. The device used in the experiment was the Phantom Premium 1.0 haptic device from SensAble Technologies [12]. Using LabVIEW FPGA and a PXI-7831R FPGA board, the researchers were able to achieve a sampling rate of 20 kHz. The Windows/GHOST software and hardware platform that shipped with the Phantom device had a maximum sampling rate of 1 kHz. The increased sampling rate for the FPGA-based system resulted in a maximum wall stiffness of 500 N/m compared to a maximum wall stiffness of 25 N/m for the Windows/GHOST system.

28.3.3 Phase-Locked Loop for Scanning Probe Microscope

Rychen and Vancura [13] developed a compact, high-speed control system for a scanning probe microscope (SPM) using LabVIEW FPGA, LabVIEW Real-Time, and a PXI system. The 1.26 GHz real-time processor ran several algorithms such as raster scan generators, the tip-sample distance controller, and auxiliary feedback controllers. A digital phase-locked loop (PLL) was implemented with LabVIEWFPGA resulting in a demodulation bandwidth of more than 10 kHz, using the analog I/O lines to interface to the external components. With a redesign involving the use of digital I/O lines, 24 MS/s is possible with a frequency resolution of 0.1 μHz. The fully digital case will yield high-phase accuracy compared to the analog circuitry. In addition, the demodulation bandwidth may be adjusted into the millihertz range for the highest frequency resolution.

28.3.4 Model-Free Adaptive Control for FPGA

CyboSoft [14] has developed a model-free adaptive (MFA) control toolset for LabVIEW. Used in conjunction with NI SoftMotion on CompactRIO, one option in the toolset provides high-speed positioning control using LabVIEW FPGA with a minimum sampling period of 25 μs. The toolset has been successfully demonstrated with high precision piezoelectric stages with significant load changes.

28.3.5 Motorcycle Engine Control Prototyping

Dase [15] from Drivven Inc. [16] used LabVIEW FPGA, CompactRIO, and custom I/O modules to build an engine control prototyping system. Engine control is a challenging application involving loop times on the order of milliseconds and precision fuel and spark timing on the order of microseconds. Dase was able to completely replace the ECU on a 2004 Yamaha YZF-R6 motorcycle using the CompactRIO system such that experienced riders could not identify significant differences between the factory ECU control and the prototype control.

28.4 Additional Tools for FPGA Development

28.4.1 SoftMotion Position Control Loop on FPGA

The NI SoftMotion Development Module, shown in Figure 28.13, includes enhanced real-time and FPGA algorithms for motion control and built-in inputs for a variety of feedback devices including quadrature encoders and sine cosine encoders. Table 28.1 shows the servo update rates that can be obtained using NI SoftMotion on a variety of hardware targets including FPGA targets.

28.4.2 Digital Filter Design Toolkit

The NI Digital Filter Design Toolkit provides functions and interactive tools for design, analysis, and implementation of digital filters within LabVIEW FPGA. The toolkit includes interactive design methods for both classical filter design and filter design based on pole-zero placement (Figure 28.14). A control

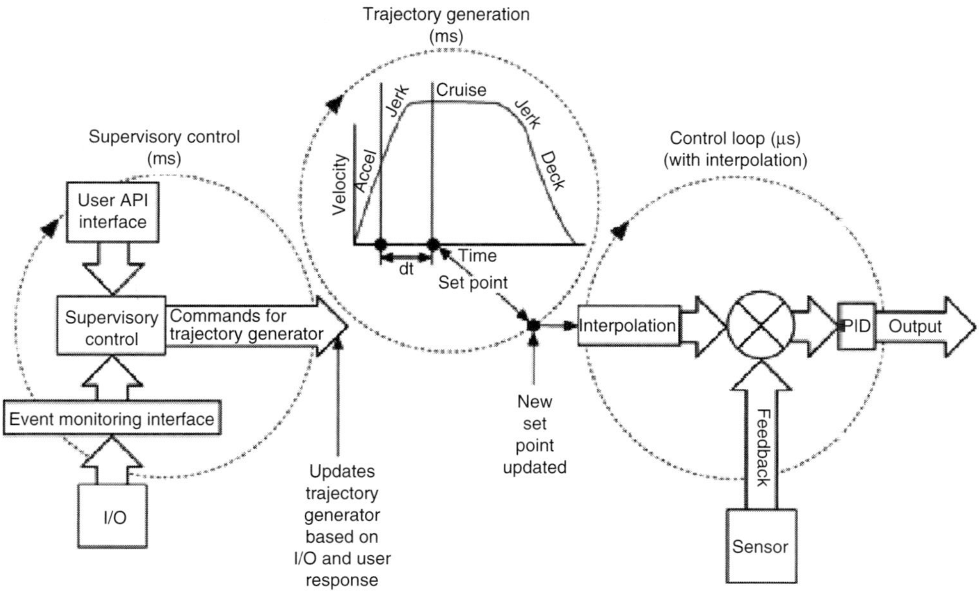

FIGURE 28.13 Architecture of NI SoftMotion Development Module.

TABLE 28.1 Servo Update Rates for NI Softmotion on Various Platforms

Hardware Platform	Servo Update Rate
CompactRIO	5 μS
PCI/PXI FPGA	5 μS
PCI/PXI Motion Control	62.5 μS
PCI/PXI M Series DAQ	1 ms
Compact FieldPoint	5 ms

law that can be represented as a discrete-time transfer function can be implemented on the FPGA using LabVIEW FPGA code generated from the Digital Filter Design Toolkit.

28.4.3 State Diagram Editor on FPGA

The LabVIEW State Diagram Toolkit can be used to help build state machines on an FPGA. The toolkit includes a state diagram editor for drawing the logic that defines an application. An example of this is shown in Figure 28.15. As states and transitions are added to the state machine diagram, LabVIEW generates code to match each change. The logic for the statemachines is represented in the LabVIEW code by a series of while loops and case statements. The toolkit can also insert comments in the LabVIEW block diagram to indicate where additional code can be added. The entire LabVIEW block diagram can then be implemented on an FPGA using the LabVIEW FPGA module.

28.5 Conclusion

Graphical programming for FPGAs can be used by control and mechatronic system designers and researchers to quickly develop robust and high-performance applications. The combination of FPGAs with floating-point processors allows engineers to intelligently distribute computation, control, and signal

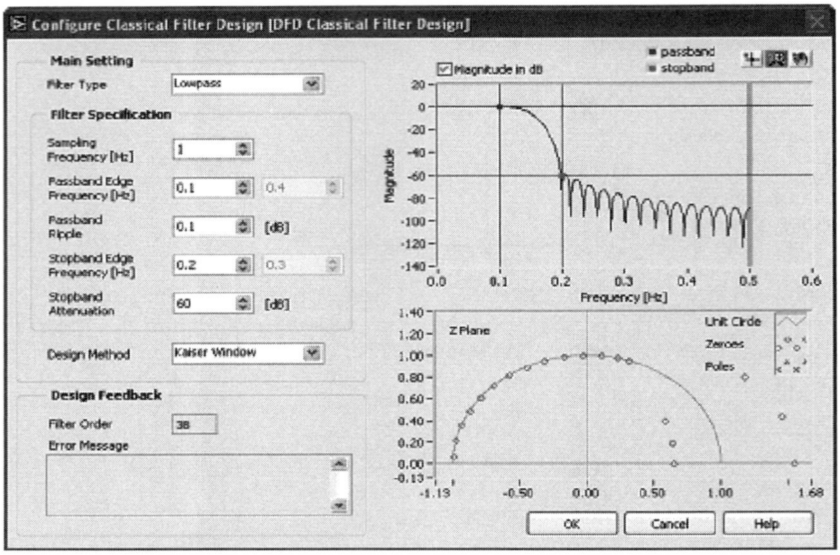

FIGURE 28.14 Interactive filter design using LabVIEW Digital Filter Design Toolkit.

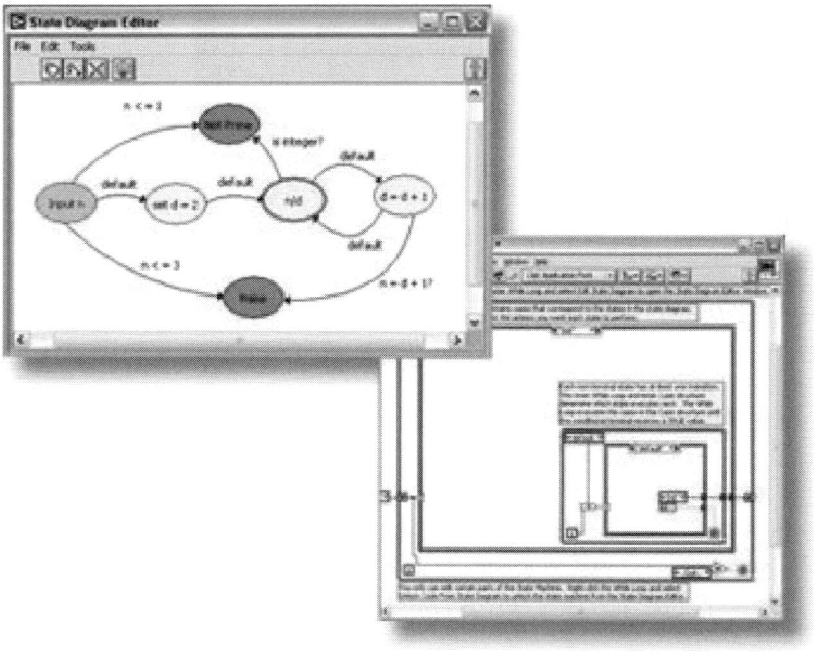

FIGURE 28.15 Architecture of NI SoftMotion Development Module.

processing functions. LabVIEW FPGA is a comprehensive tool for graphical programming FPGAs and includes functions for timing, logic, synchronization, high-speed control, analog, and digital I/O. The NI SoftMotion Module provides motion-specific functionality for combined real-time processor/FPGA systems. The NI Digital Filter Design Toolkit can be used in conjunction with LabVIEW FPGA to quickly prototype and implement digital filters on an FPGA.

References

1. C.Maxfield, *The DesignWarrior's Guide to FPGA's – Devices, Tools and FLows*, Oxford, UK: Newnes Press, 2004, pp. 153–172.
2. T. Salcudean, Project Integrated Program, Department of Electrical and Computer Engineering, University of British Columbia, http://www.ece.ubc.ca/~tims.
3. Xilinx, Inc., San Jose, CA, http://www.xilinx.com
4. Altera Corp., San Jose, CA, http://www.altera.com
5. M. Santarini (November 2005) "Xilinx keeps pace through innovation," *Movers and Shakers 2005*, 6th Edn., Reed Business Information, Available: http://www.reed-electronics.com/moversandshakers/article/CA6277496.html
6. Celoxica Ltd., Oxfordshire, UK, http://www.celoxica.com
7. Mentor Graphics Corp.,Wilsonville, OR, http://www.mentor.com
8. AccelChip, Milpitas, CA, http://www.accelchip.com
9. National Instruments Corp., FPGA portal, http://www.ni.com/fpga
10. B. Kim, M. Schmittdiel, F. L. Degertekin, and T. Kurfess, "Deformable diffraction grating for scanning micro interferometer arrays," *SPIE Photonic West Conference*, 2004.
11. E. Kopp and M. O'Malley, "Improving haptic simulations using LabVIEW FPGA," *Proceedings of the 23rd Annual Houston Conference on Biomedical Engineering Research*, February 9–10, 2006.
12. Sensable Technologies,Woburn, MA, http://www.sensable.com
13. J. Rychen and T. Vancura, Nanonis Gmbh, Switzerland, "Building high-precision control systems with LabVIEW7 Express," www.ni.com/pdf/csma/us/jorg_rychen_labviewsevenexpress.pdf
14. CyboSoft, General Cybernation Group, Inc., Rancho Cordova, CA, www.cybosoft.com
15. C. Dase, "Drivven Prototypes FPGA-Based Engine Control System Using NI CompactRIO," http://sine.ni.com/csol/cds/item/vw/p/id/538/nid/124100
16. Drivven Inc., San Antonio, TX, http://www.drivven.com

III

Software and Data Acquisition

29

Introduction to Data Acquisition

Craig Anderson
National Instruments, Inc.

29.1 Introduction

Data acquisition is the gathering of signals from real-world measurement sources and the digitizing of those signals for storage, analysis, and presentation on a personal computer (PC). Light, temperature, pressure, and torque are a few of the many different types of signals that can interface to a data acquisition system. In addition to acquiring data, data acquisition systems also are used to generate electrical signals.

These signals can either intelligently control mechanical systems or provide a stimulus so that the data acquisition system can measure the response. A data acquisition system provides a way to empirically test designs, theories, and real-world systems for validation or research. Figure 29.1 illustrates a typical computer-based data acquisition module.

The design and the production of amodern car, for instance, rely heavily on data acquisition. Engineers will first use data acquisition to test the design of the car's components. The frame can be monitored for mechanical stress, wind noise, and durability. The vibration and temperature of the engine can be acquired to evaluate the design quality. The researchers and engineers can then use this data to optimize the design of the first prototype of the car. The prototype can then bemonitored under many different conditions on a test track while information is collected through data acquisition. After required iterations of design change and data acquisition, the car is ready for production. Data acquisition devices can monitor the machines that assemble the car, and they can ensure that the assembled car meets the necessary specifications.

At first, data acquisition devices stood alone and were manually controlled by an operator. When the PC emerged, data acquisition devices and instruments could be connected to the computer through a serial port, parallel port, or some custom interface. A computer program could control the device automatically and retrieve data from the device for storage, analysis, or presentation. Now, instruments

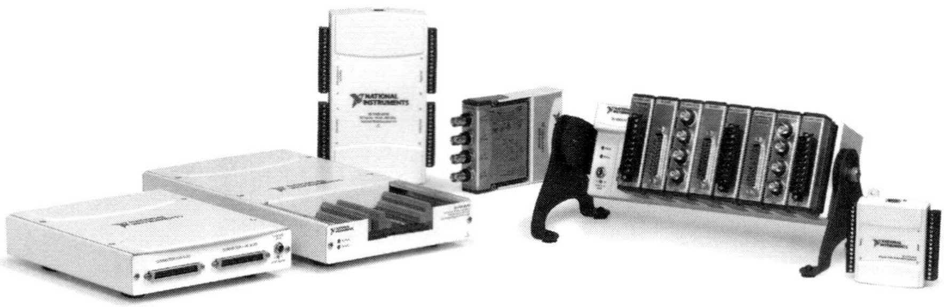

FIGURE 29.1 Examples of typical, computer-based data acquisition devices. Plug-in devices are also commonly available for USB, PCI, and PCI Express buses.

and data acquisition devices can be integrated into a computer through high-speed communication links such as PCI, PCI Express, Ethernet, IEEE 1394, or USB busses for tighter integration between the power and .exibility of the computer and the instrument or device.

Modern data acquisition systems are composed of transducers and sensors to measure a physical phenomenon, the resulting electrical signals, hardware to prepare the signals for measurement, measurement hardware, and data acquisition software.

29.2 Transducers and Sensors

Data acquisition begins with the physical phenomenon to be measured. This physical phenomenon could be the temperature of a room, the intensity of a light source, the pressure inside a chamber, the force applied to an object, or many other things. An effective data acquisition system can measure all these different phenomena.

A transducer is a device that converts a physical phenomenon into a measurable electrical signal, such as voltage or current. The ability of a data acquisition systemto measure different phenomena depends on the transducers that convert the physical phenomena into signals measurable by the data acquisition hardware. Transducers are synonymous with sensors in data acquisition systems. There are specific transducers for many different applications, such as measuring temperature, pressure, or fluid flow.

A common example of a transducer is a thermocouple. A thermocouple uses the material properties of dissimilar metals to convert a temperature into a voltage. As the temperature increases, the voltage produced by the thermocouple increases. A software program can then convert the voltage reading back into a temperature for analysis, presentation, and data logging. Many sensors produce currents instead of voltages. A current is often advantageous because the signal will not be corrupted by small amounts of resistance in the wires connecting the transducer to the data acquisition device. A disadvantage of current-producing transducers, though, is that most data acquisition devices measure voltage, not current.

Generally, the data acquisition devices that can measure current use a very small resistance of a known value to convert the known current into a readable voltage. Ultimately, the device isthen still acquiring a voltage.

29.3 Signals

The appropriate transducer converts the physical phenomena into measurable signals. However, different signals need to be measured in different ways. For this reason, it is important to understand the different types of signals and their corresponding attributes. Signals can be categorized into two groups; analog and digital.

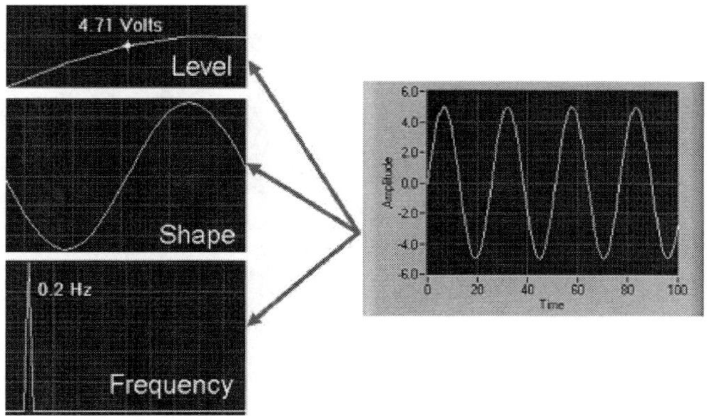

FIGURE 29.2 Primary characteristics of an analog signal.

29.3.1 Analog Signals

An analog signal can be at any value with respect to time. A few examples of analog signals include voltage, temperature, pressure, sound, and load. The three primary characteristics of an analog signal are level, shape, and frequency (Figure 29.2).

29.3.1.1 Level

Because analog signals can take on any value, the level gives vital information about the measured analog signal. The intensity of a light source, the temperature in a room, and the pressure inside a chamber are all examples that demonstrate the importance of a signal's level. When measuring a signal's level, the signal generally doesnot change quickly with respect to time. The accuracy of the measurement, however, is very important. A data acquisition system that yields maximum accuracy should be chosen to aid in analog level measurements.

29.3.1.2 Shape

Some signals are named after their speci.c shape—sine, square, sawtooth, and triangle. The shape of an analog signal can be as important as the level, because measuring the shape of an analog signal allows further analysis of the signal, including peak values, dc values, and slope. Signals where shape is of interest generally change rapidly with respect to time and therefore require faster sampling of the analog signal than is required for a level measurement. The analysis of heartbeats, video signals, sounds, vibrations, and circuit responses are some applications involving shape measurements.

29.3.1.3 Frequency

Analog signals can be categorized by their frequency. Unlike the level or shape of the signal, frequency cannot be directly measured. The signal must be analyzed using software to determine the frequency information. Thisanalys isisus ually done using an algorithm known asthe Fourier transform.

When frequency is the most important piece of information, it is important to consider both accuracy and acquisition speed. Although the acquisition speed for acquiring the frequency of a signal is less than the speed required for obtaining the shape of a signal, the signal must still be acquired fast enough so that the pertinent information is not lost while the analog signal is being acquired. The Nyquist sampling theorem states that an analog signalmust be sampled two times faster than the highest frequency component in the signal source in order to preserve frequency information. To preserve shape and frequency information, sampling should be performed 10 times faster than the highest frequency component in the signal source. Speech analysis, telecommunication, and earthquake analysis are some examples of common applications where the frequency of the signal must be known.

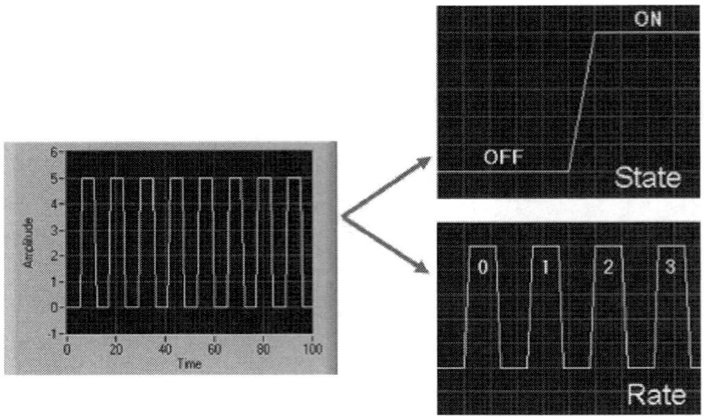

FIGURE 29.3 Primary characteristics of a digital signal.

29.3.2 Digital Signals

A digital signal cannot take on any value with respect to time. Instead, a digital signal has two possible levels—high and low. Digital signals generally conform to certain speci.cations that de.ne the characteristics of the signal. Digital signals are commonly referred to as transistor–transistor logic (TTL). TTL speci.cationsdictat e that a digital signal islo w when the level fallsw ithin 0–0.8 V, and high when it falls between 2 and 5 V. The useful information that can be measured from a digital signal includes the state and the rate, ass een in Figure 29.3.

29.3.2.1 State

Digital signals cannot take on any value with respect to time. The state of a digital signal is essentially the level of the signal—on or off, high or low.Monitoring the state of a switch—open or closed—is a common application showing the importance of knowing the state of a digital signal. The most commonly used digital levels are TTL and TTL-compatible CMOS. These are both very common 5-V standards for digital hardware.

Digital transfer rates to and from data acquisition hardware vary from unstrobed to high speed. Unstrobed digital input and output involves setting digital lines and monitoring states by software command. Thisformof digital input and output isals o known ass tatic or immediate digital I/O.The maximum speed of an unstrobed I/O is highly dependent on the computer hardware, the operating system, and the application program. Pattern digital I/O referst o inputsand outputsof digital patternsunder the control of a clock signal. The speed at which the data can be sent or received depends on the amount of data, the characteristics of the data acquisition hardware, and the computer speed.

29.3.2.2 Rate

The rate of a digital signal de.nes how it changes state with respect to time. One example of measuring a digital signal's rate is determining how fast a motor shaft spins. Unlike frequency, the rate of a digital signal measures how often a portion of the signal occurs. A software algorithm is not required to determine the rate of a signal. Instead, a digital signal's rate is typically calculated using counter/timers. Counter/timers are capable of measuring or producing time-critical digital pulses. These pulses, such asthe digital input and output, are generally TTL or TTL-compatible CMOS. These components are used for measuring or producing a number of time-critical signals, including event counting, pulse train generation, frequency-shift keying, and quadrature encoder monitoring. The two main characteristics of a counter/timer are counter size and maximum source frequency. The counter size is generally represented in bitsand determinesho w high a counter can count. For instance, a 32-bit counter can count $2^{32} - 1 = 4{,}294{,}967{,}295$ eventsbefor e it returnsthe count value back to zero. The maximum source frequency represents the speed of the fastest signal the counter can count. An 80-MHz counter can count events that are asfas t as12.5 nsapar t. An "event" isactually the rising or falling edge of a digital signal.

29.4 Signal Conditioning

No real situation will have perfect signals, be completely free of noise, or be guaranteed to be electrically safe. Most real signals require some sort of preparation (conditioning) before being gathered by a data acquisition system. Signal conditioning hardware provides a method to remove—as much as possible—unwanted componentsof a digital or analog signal.

Measurement hardware, particularly for high-frequency analog signals, is usually equipped with an antialiasing filter. Thisisa low-pass.lt er that blocks frequencies above the desired frequency range and increasesthe measurement accuracy. Digital and counter/timer lines are also commonly fitted with filters that remove signal spikes that could otherwise be mistakenly counted as a rising or falling edge. Isolation is another type of signal conditioning that separates the measurement hardware circuitry from the signal being measured. Thisisdone to remove large differencesin electric potential between the measurement hardware and the signal, and it protectsoper atorsand measurement hardware from damage, given a large surge in voltage or current. Other types of signal conditioning include ampli.cation, attenuation, multiplexing, excitation, and cold-junction compensation.

29.5 Data Acquisition Hardware

The heart of a data acquisition device is a digital-to-analog converter (DAC), an analog-to-digital converter (ADC), or some combinationsof the two. An ADC hasa .nite list of binary valuesthat represent analog voltage ranges. The ADC's purpose is to select a value from this list that is closest to an actual voltage at a specified time. The value is then transferred in binary format to a computer. Alternatively, a DAC can produce an analog voltage froma list of binary values. The voltage generated by a basicDAC stays the same until it receivesanother value from the computer. In order to acquire and produce analog waveforms, the DAC and ADC must activate at precise intervals. Consequently, measurement hardware has timing circuitry to produce a pulse train of a constant frequency to control the ADC and DAC.

The data that ist ransferred from the ADC and to the DAC travelst o the computer over a bus. A bus isa group of electrical conductorsthat transfer information inside a computer. Some common examples of a busar e PCI and USB. The buscan carry both control information and binary measurement data to and from measurement hardware. One of the most important considerations in selecting a bus is bus transfer rate, usually expressed in megabytes per second (Mb/s). A single analog value could require less than 1 byte or asm uch as4 bytes, depending on the type of measurement hardware. The busiss hared among multiple devices, so data acquisition devices often have onboard memory to serve as a holding place for data when the bus is not available. In very fast data acquisition routines, the memory can hold all the data, and at the end of the acquisition, all the data can be transferred to the computer for processing.

When data isac quired at high speedson multiple channels, it isoft en important to understand the phase relationship from one signal to the next. If the signals need to be generated or acquired on a single data acquisition device, simultaneously sampling data acquisition hardware provides a dedicated ADC or DAC and a shared clock for all channels. If the signals are generated or acquired on multiple data acquisition devices, there are a number of ways to synchronize the systems and preserve relative phase relationships. One way is to share the ADC and DAC clock between the data acquisition devices. The real-time system integration (RTSI) bus can connect multiple devices together to share timing circuitry among multiple devices. Phase-lock looping (PLL) is a more sophisticated synchronization method. A reference signal is supplied to all the data acquisition devices, and the internal clocks stay in phase with the reference signal. Consequently, the phase relationship can be reserved even if different measurement

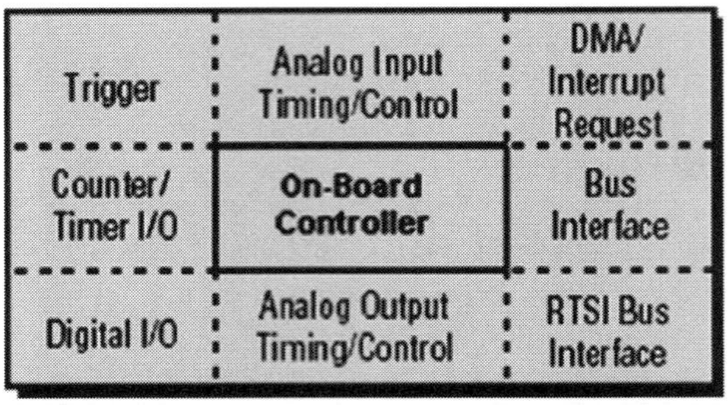

FIGURE 29.4 The components of a typical data acquisition system.

hardware is using different sampling or update speeds. Figure 29.4 is a diagram showing the components of a typical data acquisition system.

The difference between an actual analog voltage and the closest voltage from the list of binary values is called the quantization error. In a perfect digitizing measurement system that is free of noise, the quantization errorwould solely explain any difference between the actual voltage and the measured voltage. No measurement hardware and no environment, however, are perfect. The accuracy of an instrument describes the amount of uncertainty when considering quantization error, unavoidable system noise, and hardware imperfections. Accuracy is sometimes confused with precision. Precision refers to the amount of deviation inmultiple measurements connected to a constant and level signal source. Even if an instrument is precise, it could still be inaccurate if the readings were consistent but signi.cantly different than the actual value of the signal.

The accuracy of a data acquisition system can change with temperature, time, and usage. Data acquisition hardware can store onboard correction constants for offset and gain errors. An offset error is a constant difference between the measured and actual voltage, regardless of the voltage level. A gain error increases linearly as the measured voltage increases. Some data acquisition hardware also includes an accurate voltage source on board that can be periodically used as a reference to correct the gain and offset error parameters.

29.6 Data Acquisition Software

The final piece of a data acquisition systemto understand is the software. Software transforms the PC and the data acquisition hardware into a complete data acquisition, analysis, and presentation tool. Without software to control or drive the hardware, the data acquisition device will not function properly. The driver software is a set of commands that a programmer can incorporate into a program. The driver software is usually supplied by the manufacturer of the hardware and can be used in a variety of programming languages. Driver software is the layer of software that allows easy communication with the hardware. It forms the middle layer between the application software and the hardware. Driver software also prevents a programmer from having to do register-level programming or complicated commands in order to access the hardware functions.

Application software adds analysis and presentation capabilities to the driver software. To choose the right application software, evaluate the complexity of the application, the availability of con.gurationbased software that fits the application, and the amount of time available to develop the application. If the application isc omplex or there isno existing program, use a development environment. A programmer can use a programming language and an application development environment to build an application from the driver software such as the one shown in Figure 29.5, or use a con.guration-based programwith preset functionality. The application is then ready for an end user to easily control and acquire data from the hardware—a custom instrument built specifically for the user's needs.

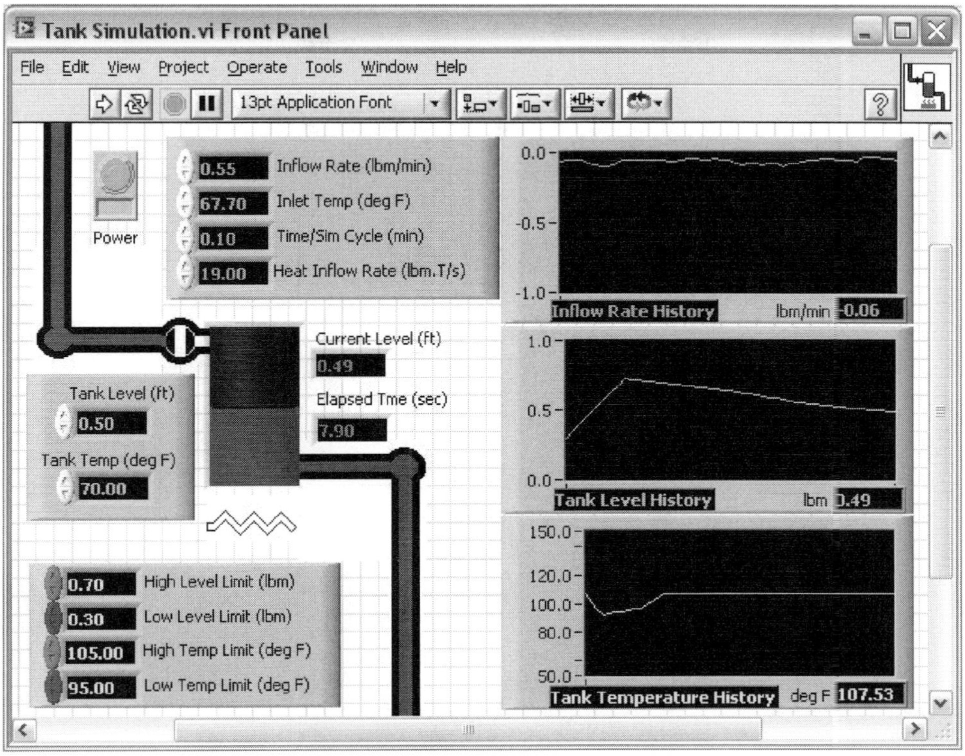

FIGURE 29.5 An example of an application built from LabVIEWsoftware.

29.7 Summary

Data acquisition is the process of collecting signals fromreal-world measurement sources and the digitizing of those signals for storage, analysis, and presentation. Data acquisition systems can also be used to generate signals that either intelligently control mechanical systems or provide a stimulus so that the data acquisition system can measure the response. Initially, data acquisition devices were stand-alone and controlled manually by an operator. Today, data acquisition devices and instruments are connected to computers and are composed of transducers and sensors for measuring physical phenomena, hardware to prepare the signals for measurement, measurement hardware, and data acquisition software.

29.8 Related Information

More information about PC-based data acquisition systems is available from National Instruments in the form of white papers, application notes, customer solutions, and product information. Visit http://www.ni.com and search for "data acquisition" to view available information.

30

Measurement Techniques: Sensors and Transducers

Cecil Harrison
University of Southern Mississippi

30.1 Introduction

An automatic control system is said to be *error actuated* because the **forward path** components (*comparator, controller, actuator,* and *plant* or *process*) respond to the error signal (Figure 30.1). The error signal is developed by comparing the measured value of the **controlled output** to some **reference input**, and so the accuracy and precision of the controlled output are largely dependent on the accuracy and precision with which the controlled output is measured. It follows then that measurement of the controlled output, accomplished by a system component called the **transducer**, is arguably the single most important function in an automatic control system.

A transducer senses the magnitude or intensity of the controlled output and produces a proportional signal in an energy form suitable for transmission along the feedback path to the comparator. [The term proportional is used loosely here because the output of the transducer may not always be directly proportional to the controlled output; that is, the transducer may not be a linear component. In linear systems, if the output of the transducer (the measurement) is not linear, it is linearized by the signal conditioner.] The element of the transducer which senses the controlled output is called the *sensor*; the remaining elements of a transducer serve to convert the sensor output to the energy form required by the **feedback path**. Possible configurations of the feedback path include:

- Mechanical linkage
- Fluid power (pneumatic or hydraulic)
- Electrical, including optical coupling, RF propagation, magnetic coupling, or acoustic propagation

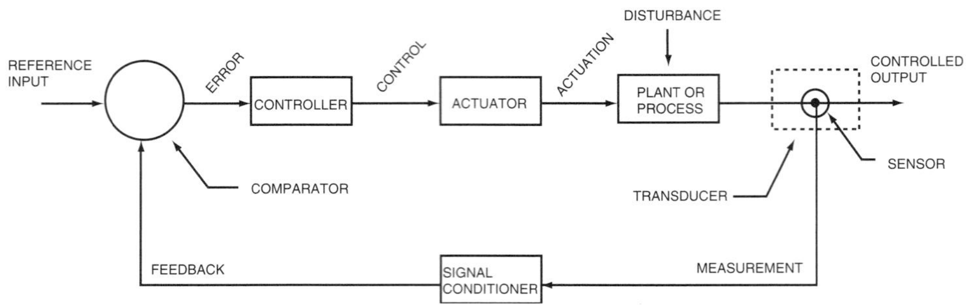

FIGURE 30.1 Functional block diagram of a canonical (standard) automatic control system.

Electrical signals suitable for representing measurement results include:

- DC voltage or current amplitude
- AC voltage or current amplitude, frequency, or phase (CW modulated)
- Voltage or current pulses (digital)

In some cases, representation may change (e.g., from a DC amplitude to digital pulses) along the feedback path.

The remainder of this discussion pertains to a large number of automatic control systems in which the feedback signal is electrical and the feedback path consists of wire or cable connections between the feedback path components. The transducers considered hereafter sense the controlled output and produce an electrical signal representative of the magnitude, intensity, or direction of the controlled output.

The **signal conditioner** accepts the electrical output of the transducer and transmits the signal to the comparator in a form compatible with the reference input. The functions of the signal conditioner include:

- Amplification/attenuation (scaling)
- Isolation
- Sampling
- Noise elimination
- Linearization
- Span and reference shifting
- Mathematical manipulation (e.g., differentiation, division, integration, multiplication, root finding, squaring, subtraction, or summation)
- Signal conversion (e.g., DC–AC, AC–DC, frequency–voltage, voltage–frequency, digital–analog, analog–digital, etc.)
- Buffering
- Digitizing
- Filtering
- Impedance matching
- Wave shaping
- Phase shifting

In cases in which part or all of the required signal conditioning is accomplished within the transducer, the transducer output may be connected directly to the comparator. [Connection of the transducer output directly to the comparator should not be confused with unity feedback. Unity feedback occurs when the cascaded components of the feedback path (transducer and signal conditioner) have a combined transfer function equal to 1 (unity).] In a digital control system, many of the signal conditioning functions listed here can also be accomplished by software.

Transducers are usually considered in two groups:

- **Motion** and **force transducers**, which are mainly associated with **servomechanisms**
- **Process transducers**, which are mainly associated with **process control** systems

As will be seen, most process transducers incorporate some sort of motion transducer.

30.2 Motion and Force Transducers

This section discusses those transducers used in systems that control motion (i.e., *displacement, velocity,* and *acceleration*). Force is closely associated with motion, because motion is the result of unbalanced forces, and so force transducers are discussed concurrently. The discussion is limited to those transducers that measure *rectilinear* motion (straight line motion within a stationary frame of reference) or *angular* motion (circular motion about a fixed axis). Rectilinear motion is sometimes called *linear* motion, but this leads to confusion in situations where the motion, though along a straight line, really represents a mathematically nonlinear response to input forces. Angular motion is also called *rotation* or *rotary motion* without ambiguity.

The primary theoretical basis for motion transducers is found in rigid-body mechanics. From the equations of motion for rigid-bodies (Table 30.1), it is clear that if any one of displacement, velocity, or

TABLE 30.1 Equations of Motion

Continuous	Discrete $\Delta t = t_i - t_{i-1}$
Rectilinear displacement:	
$x(t) = \int v(t)\,dt$	$x_i = x_{i-1} + \frac{v_i + v_{i-1}}{2} \cdot (\Delta t)$
$= \iint a(t)\,dt$	$= 2x_{i-1} - x_{i-2} + \frac{a_i + 2a_{i-1} + a_{i-2}}{2} \cdot \frac{(\Delta t)}{2}$
Angular displacement:	
$\theta(t) = \int \omega(t)\,dt$	$\theta_i = \theta_{i-1} + \frac{\omega_i + \omega_{i-1}}{2} \cdot (\Delta t)$
$= \iint \alpha(t)\,dt$	$= 2\theta_{i-1} - \theta_{i-2} + \frac{\alpha_i + 2\alpha_{i-1} + \alpha_{i-2}}{2} \cdot \frac{(\Delta t)}{2}$
Rectilinear velocity:	
$v(t) = \frac{d}{dt}x(t)$	$v_i = \frac{x_i - x_{i-1}}{\Delta t}$
$= \int a(t)\,dt$	$= v_{i-1} + \frac{a_i + a_{i-1}}{2} \cdot (\Delta t)$
Angular velocity:	
$\omega(t) = \frac{d}{dt}\theta(t)$	$\omega_i = \frac{\theta_i - \theta_{i-1}}{\Delta t}$
$= \int \alpha(t)\,dt$	$= \omega_{i-1} + \frac{\alpha_i + \alpha_{i-1}}{2} \cdot (\Delta t)$
Rectilinear acceleration:	
$a(t) = \frac{d}{dt}v(t)$	$a_i = \frac{v_i - v_{i-1}}{\Delta t}$
$= \frac{d^2}{dt^2}x(t)$	$= \frac{x_i - 2x_{i-1} + x_{i-2}}{(\Delta t)^2}$
Angular acceleration:	
$\alpha(t) = \frac{d}{dt}\omega(t)$	$\alpha_i = \frac{\omega_i - \omega_{i-1}}{\Delta t}$
$= \frac{d^2}{dt^2}\theta(t)$	$= \frac{\theta_i - 2\theta_{i-1} + \theta_{i-2}}{(\Delta t)^2}$

acceleration is measured, the other two can be derived by mathematical manipulation of the signal within an analog signal conditioner or within the controller software of a digital control system.

Position is simply a location within a frame of reference; thus, any measurement of displacement relative to the frame is a measurement of position, and any displacement transducer whose input is referenced to the frame can be used as a position transducer.

30.2.1 Displacement (Position) Transducers

Displacement transducers may be considered according to application as *gross* (large) displacement transducers or sensitive (small) displacement transducers. The demarcation between gross and sensitive displacement is somewhat arbitrary, but may be conveniently taken as approximately 1 mm for rectilinear displacement and approximately 10′ arc (1/6°) for angular displacement. The predominant types of gross displacement transducers (Figure 30.2) are:

- Potentiometers (Figure 30.2a)
- Variable differential transformers (VDT) (Figure 30.2b)
- Synchros (Figure 30.2c)
- Resolvers (Figure 30.2d)
- Position encoders (Figure 30.2e)

Potentiometer-based transducers are simple to implement and require the least signal conditioning, but potentiometers are subject to wear due to sliding contact between the wiper and the resistance element and may produce noise due to wiper bounce (Figure 30.2a). Potentiometers are available with strokes ranging from less than 1 cm to more than 50 cm (rectilinear) and from a few degrees to more than 50 turns (rotary).

VDTs are not as subject to wear as potentiometers, but the maximum length of the stroke is small, approximately 25 cm or less for a linear VDT (LVDT) and approximately 60° or less for a rotary VDT (RVDT). VDTs require extensive signal conditioning in the form of phase-sensitive demodulation of the AC signal; however, the availability of dedicated VDT demodulators in integrated circuit (IC) packages mitigates this disadvantage of the VDT.

Synchros are rather complex and expensive three-phase AC machines, which are constructed to be precise and rugged. Synchros are capable of measuring angular differences in the positions (up to ±180°) of two continuously rotating shafts. In addition, synchros may function simultaneously as reference input, output measurement device, feedback path, and comparator (Figure 30.2c).

Resolvers are simpler and less expensive than synchros, and they have an advantage over RVDTs in their ability to measure angular displacement throughout 360° of rotation. In Figure 30.2d, which represents one of several possibilities for utilizing a resolver, the signal amplitude is proportional to the cosine of the measured angle at one output coil and the sine of the measured angle at the other. Dedicated ICs are available for signal conditioning and for conversion of resolver output to digital format. The same IC, when used with a *Scott-T transformer*, can be used to convert synchro output to digital format.

Position encoders are highly adaptable to digital control schemes because they eliminate the requirement for digital-to-analog conversion (DAC) of the feedback signal. The code tracks are read by track sensors, usually wipers or electro-optical devices (typically infrared or laser). Position encoders are available for both rectilinear and rotary applications, but are probably more commonly found as shaft encoders in rotary applications. Signal conditioning is straightforward for *absolute* encoders (Figure 30.2e), requiring only a decoder, but position resolution depends on the number of tracks, and increasing the number of tracks increases the complexity of the decoder. *Incremental* encoders require more complex signal conditioning, in the form of counters and a processor for computing position. The number of tracks, however, is fixed at three (Figure 30.2f). Position resolution is limited only by the ability to render finer divisions of the code track on the moving surface.

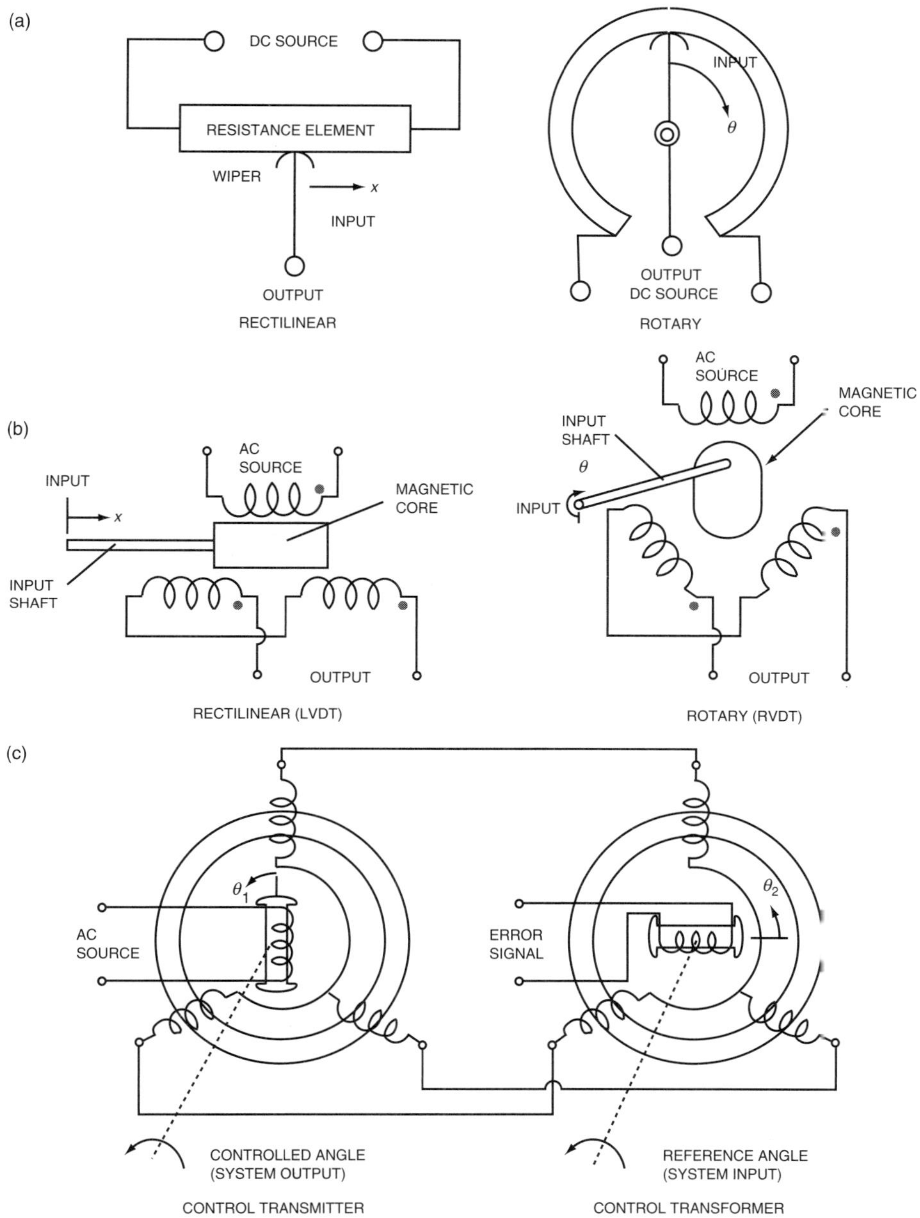

FIGURE 30.2 Gross displacement transducers: (a) potentiometers, (b) variable differential transformers (VDT), (c) synchros (typical connection), (d) resolvers (typical connection), (e) absolute position encoders, (f) code track for incremental position encoder.

Although gross displacement transducers are designed specifically for either rectilinear or rotary motion, a *rack* and *pinion*, or a similar motion converter, is often used to adapt transducers designed for rectilinear motion to the measurement of rotary motion, and vice versa.

The predominant types of *sensitive* (small) displacement transducers (Figure 30.3) are:

- Differential capacitors
- Strain gauge resistors
- Piezoelectric crystals

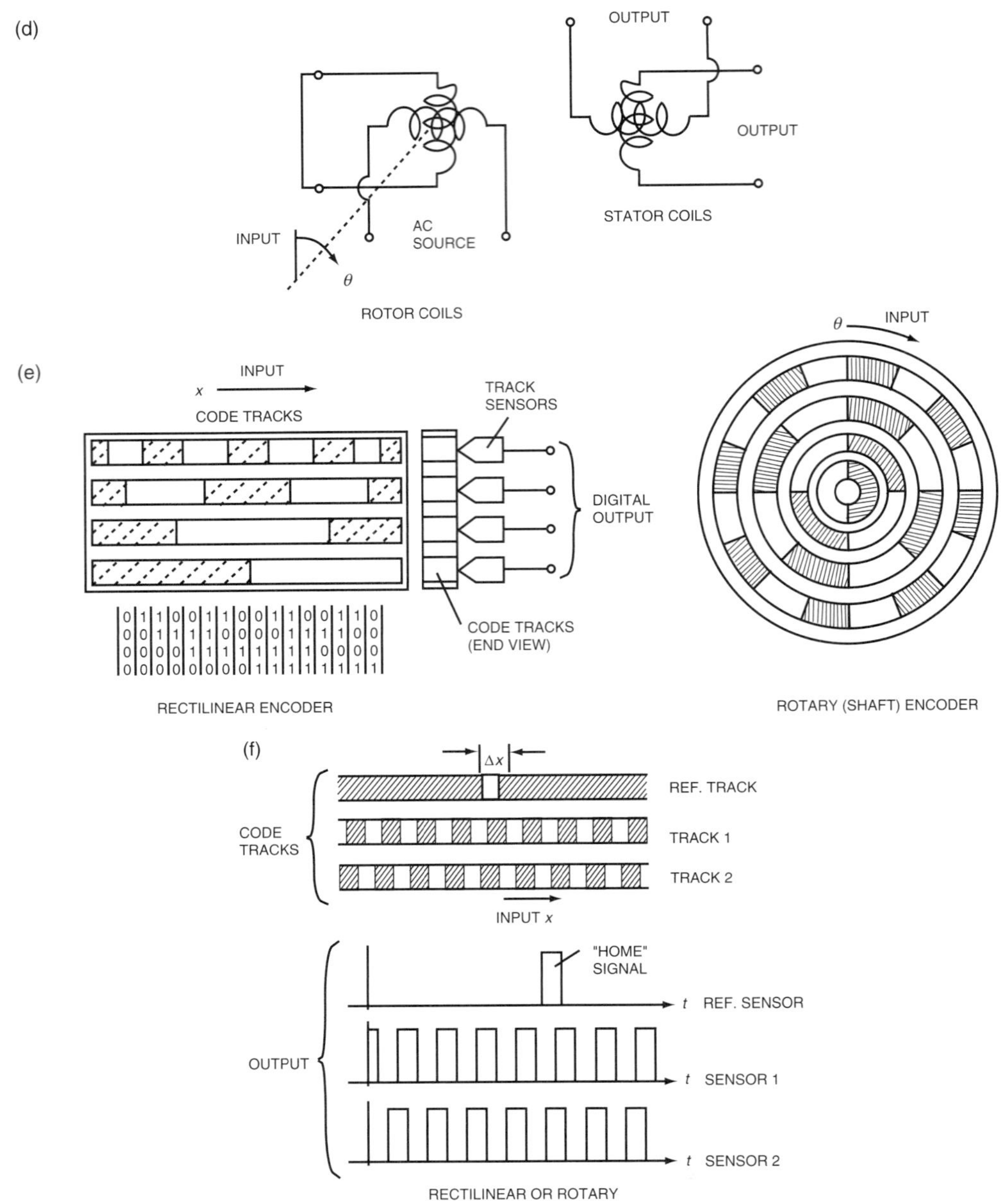

FIGURE 30.2 *(Continued)*

Figure 30.3a provides a simplified depiction of a differential capacitor used for sensitive displacement measurements. The motion of the input rod flexes the common plate, which increases the capacitance of one capacitor and decreases the capacitance of the other. In one measurement technique, the two capacitors are made part of an impedance bridge (such as a Schering bridge), and the change in the bridge output is an indication of displacement of the common plate. In another technique, each capacitor is connected to serve as tuning capacitor for an oscillator, and the difference in frequency between the two oscillators is an indication of displacement.

A strain gauge resistor is used to measure elastic deformation (strain) of materials by bonding the resistor to the material (Figure 30.3b) so that it undergoes the same strain as the material. The resistor is

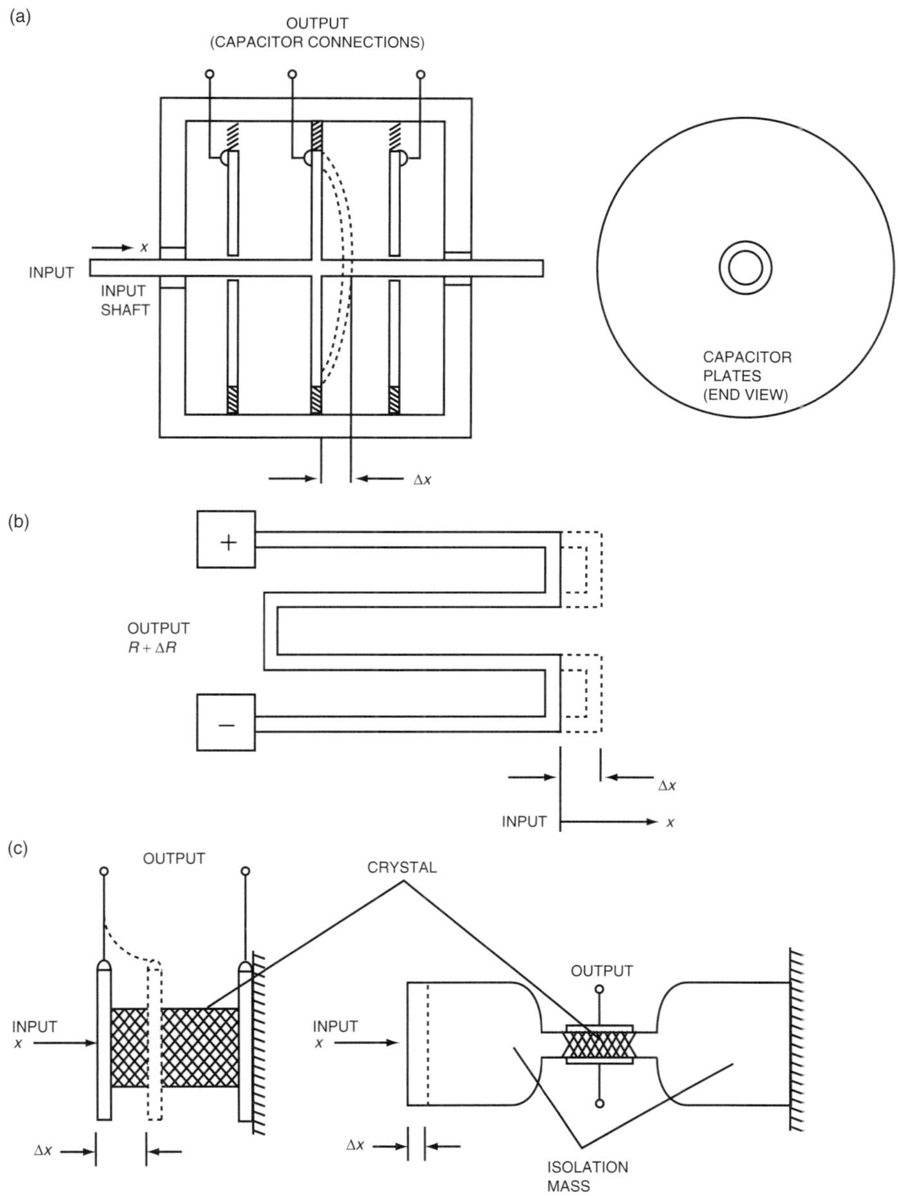

FIGURE 30.3 Sensitive displacement transducers: (a) differential capacitor, (b) strain gauge resistor, (c) piezoelectric crystals.

usually incorporated into one of the several bridge circuits, and the output of the bridge is taken as an indication of strain.

The piezoelectric effect is used in several techniques for sensitive displacement measurements (Figure 30.3c). In one technique, the input motion deforms the crystal by acting directly on one electrode. In another technique, the crystal is fabricated as part of a larger structure, which is oriented so that input motion bends the structure and deforms the crystal. Deformation of the crystal produces a small output voltage and also alters the resonant frequency of the crystal. In a few situations, the output voltage is taken directly as an indication of motion, but more frequently the crystal is used to control an oscillator, and the oscillator frequency is taken as the indication of strain.

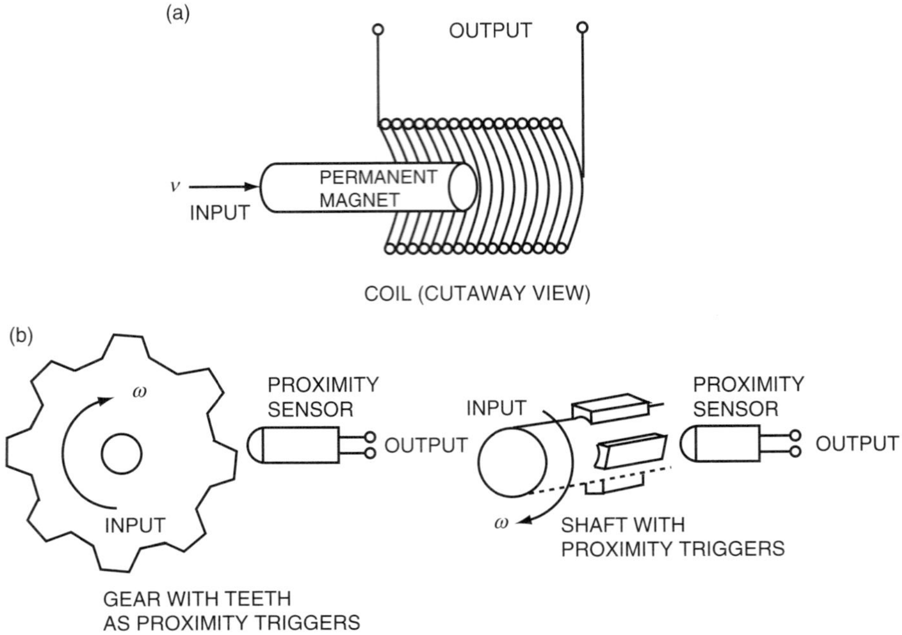

FIGURE 30.4 Velocity transducers: (a) magnet and coil, (b) proximity sensors.

30.2.2 Velocity Transducers

As stated previously, signal conditioning techniques make it possible to derive all motion measurements—displacement, velocity, or acceleration—from a measurement of any one of the three. Nevertheless, it is sometimes advantageous to measure velocity directly, particularly in the cases of short-stroke rectilinear motion or high-speed shaft rotation. The analog transducers frequently used to meet these two requirements are:

- Magnet-and-coil velocity transducers (Figure 30.4a)
- Tachometer generators

A third category of velocity transducers, *counter-type velocity transducers* (Figure 30.4b), is simple to implement and is directly compatible with digital controllers.

The operation of magnet-and-coil velocity transducers is based on Faraday's law of induction. For a solenoidal coil with a high length-to-diameter ratio made with closely spaced turns of fine wire, the voltage induced into the coil is proportional to the velocity of the magnet. Magnet-and-coil velocity transducers are available with strokes ranging from less than 10 mm to approximately 0.5 m.

A tachometer generator is, as the name implies, a small AC or DC generator whose output voltage is directly proportional to the angular velocity of its rotor, which is driven by the controlled output shaft. Tachometer generators are available for shaft speeds of 5000 rpm, or greater, but the output may be nonlinear and there may be an unacceptable output voltage ripple at low speeds.

AC tachometer generators are less expensive and easier to maintain than DC tachometer generators, but DC tachometer generators are directly compatible with analog controllers and the polarity of the output is a direct indication of the direction of rotation. The output of an AC tachometer generator must be demodulated (i.e., rectified and filtered), and the demodulator must be phase sensitive in order to indicate direction of rotation.

Counter-type velocity transducers operate on the principle of counting electrical pulses for a fixed amount of time, then converting the count per unit time to velocity. Counter-type velocity transducers rely on the use of a proximity sensor (*pickup*) or an incremental encoder (Figure 30.2f). Proximity sensors may be one of the following types:

- Electro-optic
- Variable reluctance
- Hall effect
- Inductance
- Capacitance

Two typical applications of counter-type velocity transducers are shown in Figure 30.4b.

Since a digital controller necessarily includes a very accurate electronic clock, both pulse counting and conversion to velocity can be implemented in software (i.e., made a part of the controller program). Hardware implementation of pulse counting may be necessary if time-intensive counting would divert the controller from other necessary control functions. A special-purpose IC, known as a *quadrature decoder/counter interface*, can perform the decoding and counting functions and transmit the count to the controller as a data word.

30.2.3 Acceleration Transducers

As with velocity measurements, it is sometimes preferable to measure acceleration directly rather than derive acceleration from a displacement or velocity measurement. The majority of acceleration transducers may be categorized as *seismic* accelerometers because the measurement of acceleration is based on measuring the displacement of a mass called the *seismic element* (Figure 30.5). The configurations shown in Figures 30.5a,b require a rather precise arrangement of springs for suspension and centering of the seismic mass. One of the disadvantages of a seismic accelerometer is that the seismic mass is displaced during acceleration, and this displacement introduces nonlinearity and bias into the measurement. The force-balance configuration shown in Figure 30.5c uses the core of an electromagnet as the seismic element. A sensitive displacement sensor detects displacement of the core and uses the displacement signal in a negative feedback arrangement to drive the coil, which returns the core to its center position. The output of the force-balance accelerometer is the feedback required to prevent displacement rather than displacement per se.

A simpler seismic accelerometer utilizes one electrode of a piezoelectric crystal as the seismic element (Figure 30.5d). Similarly, another simple accelerometer utilizes the common plate of a differential capacitor (Figure 30.3a) as the seismic element.

30.2.4 Force Transducers

Force measurements are usually based on a measurement of the motion, which results from the applied force. If the applied force results in gross motion of the controlled output, and the mass of the output element is known, then any appropriate accelerometer attached to the controlled output produces an output proportional to the applied force ($F = Ma$). A simple spring-balance scale (Figure 30.6a) relies on measurement of displacement, which results from the applied force (weight) extending the spring.

Highly precise force measurements in high-value servomechanisms, such as those used in pointing and tracking devices, frequently rely on gyroscope precession as an indication of the applied force. The scheme is shown in Figure 30.6b for a gyroscope with gimbals and a spin element. A motion transducer (either displacement or velocity) on the precession axis provides an output proportional to the applied force. Other types of gyroscopes and precession sensors are also used to implement this force measurement technique.

Static force measurements (in which there is no apparent motion) usually rely on measurement of strain due to the applied force. Figure 30.6c illustrates the typical construction of a common force transducer called a *load cell*. The applied force produces a proportional strain in the S-shaped structural member, which is measured with a sensitive displacement transducer, usually a strain gauge resistor or a piezoelectric crystal.

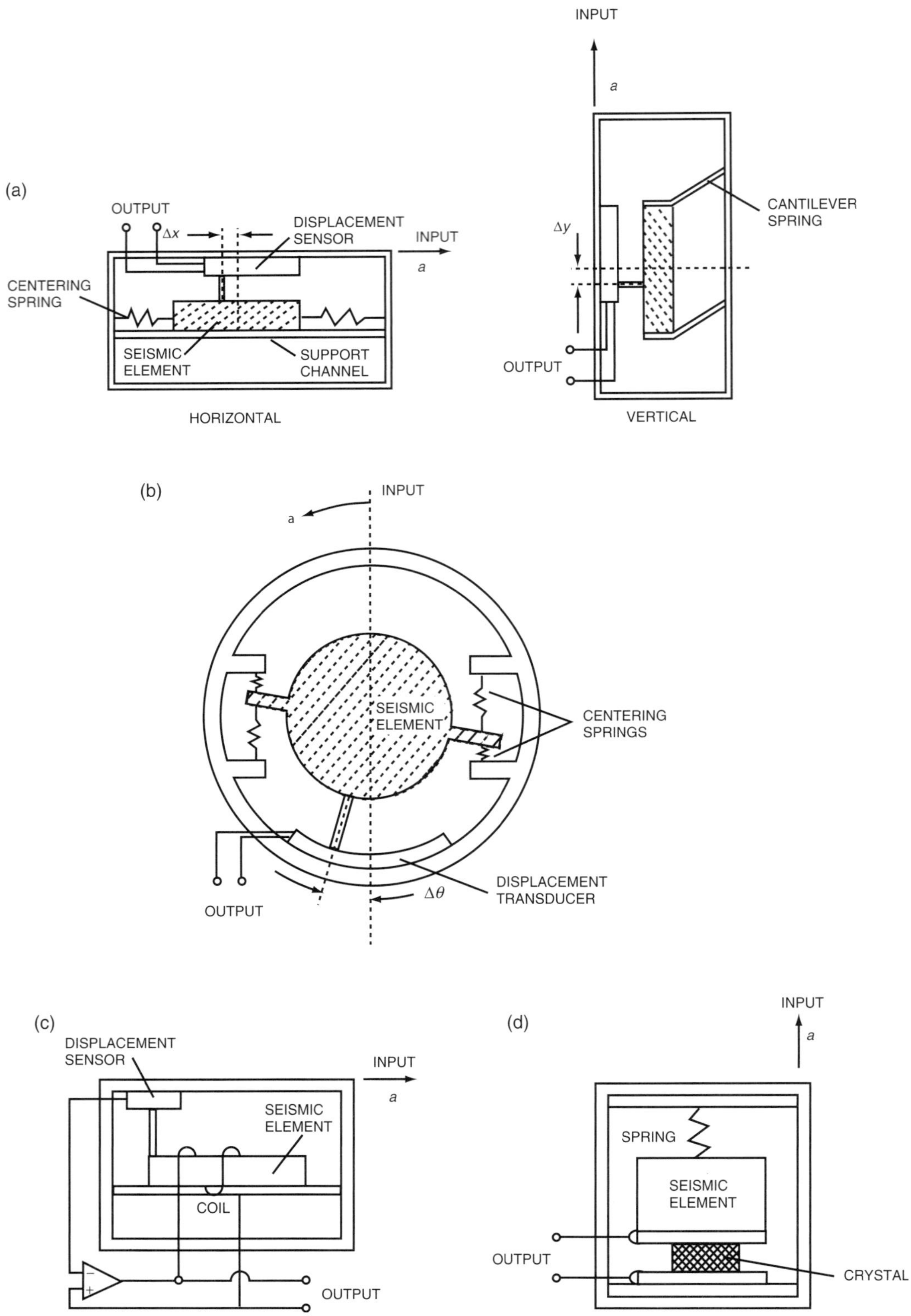

FIGURE 30.5 Seismic accelerometers: (a) rectilinear acceleration transducers, (b) rotary accelerometer, (c) force-balance accelerometer, (d) piezoelectric accelerometer.

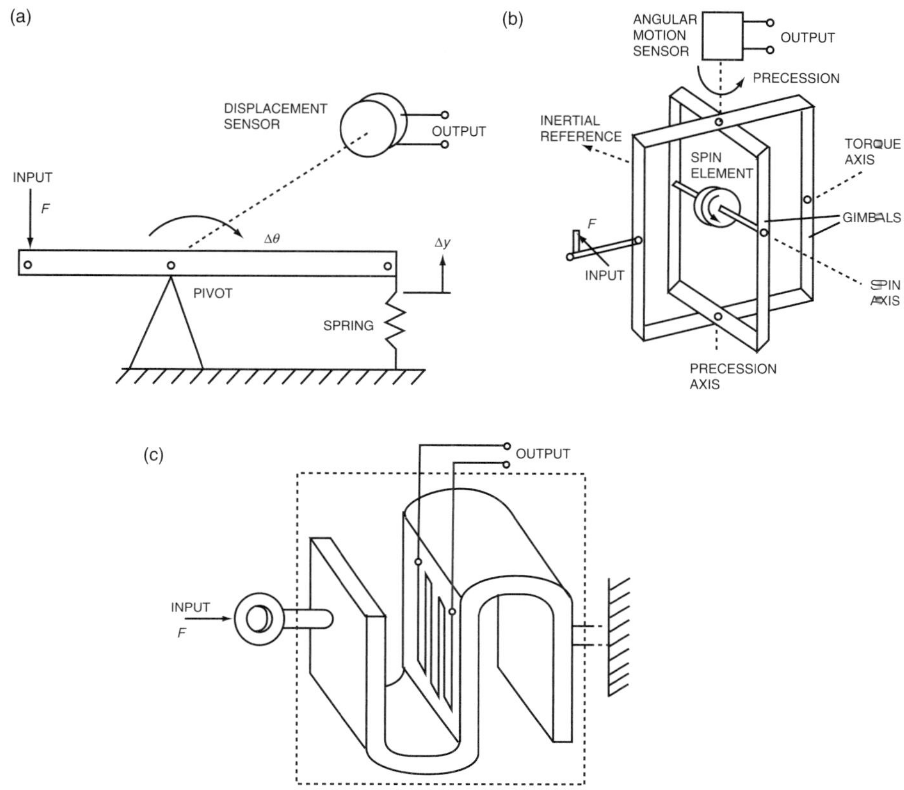

FIGURE 30.6 Force transducers: (a) spring scale, (b) gyroscope, (c) load cell.

30.3 Process Transducers

This section discusses transducers used in measuring and controlling the *process variables* most frequently encountered in industrial processes, namely,

- Fluid pressure
- Fluid flow
- Liquid level
- Temperature

30.3.1 Fluid Pressure Transducers

Most fluid pressure transducers are of the *elastic* type, in which the fluid is confined in a chamber with at least one elastic wall, and the deflection of the elastic wall is taken as an indication of the pressure. The *Bourdon tube* and the *bellows* are examples of elastic pressure transducers, which are used in laboratory-grade transducers and in some industrial process control applications. The fluid pressure transducer depicted in Figure 30.7, which uses an elastic *diaphragm* to separate two chambers, is the type most frequently encountered in industrial process control. Diaphragms are constructed from one of a variety of elastic materials ranging from thin metal to polymerized fabric.

For gross pressure measurements, the displacement of the diaphragm is sensed by a potentiometer or LVDT; for more sensitive pressure measurements, any one of the three sensitive displacement sensors described earlier is used. In the most common configuration for sensitive pressure transducers, a strain gauge resistor with a rosette pattern is bonded to the diaphragm. In another configuration, the outer

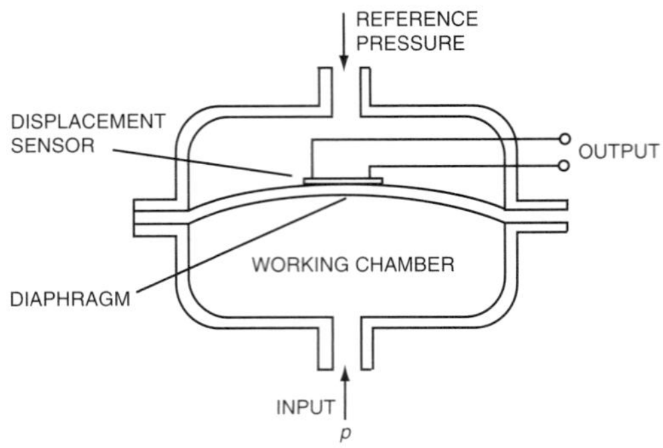

FIGURE 30.7 Diaphragm pressure transducer.

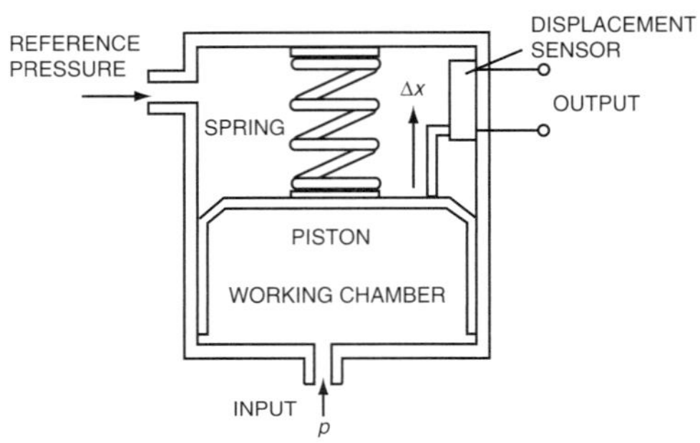

FIGURE 30.8 Piston-and-spring transducer.

walls of the pressure sensor serve as capacitor plates and the diaphragam serves as the common plate of a differential capacitor. In a very sensitive and highly integrated configuration, the diaphragm is a silicon wafer with a piezoresistive strain gauge and signal conditioning circuits integrated into the silicon.

High-vacuum (very low pressure) measurements, usually based on observations of viscosity, thermal conductivity, acoustic properties, or ionization potential of the fluid, will not be included in this discussion. Transducers used in high-pressure hydraulic systems [70 MPa (10,000 psi) or greater] are usually of the *piston and spring* type (Figure 30.8).

In either of the pressure transducers, the output is actually a measure of the difference in pressure between the working chamber and the reference chamber of the transducer (i.e., $p_{\mathrm{OUT}} = p - p_{\mathrm{REF}}$). The measurement is called:

- An *absolute pressure* if the reference chamber is sealed and evacuated (i.e., $p_{\mathrm{REF}} = 0$ and $p_{\mathrm{OUT}} = p$)
- A *gauge pressure* if the reference chamber is vented to the atmosphere (i.e., $p_{\mathrm{OUT}} = p - p_{\mathrm{ATM}}$)
- A *differential pressure* if any other pressure is applied to the reference chamber

30.3.2 Fluid Flow Transducers (Flowmeters)

Flowmetering, because of the number of variables involved, encompasses a wide range of measurement technology and applications. In industrial processes, the term *fluid* is applied not only to gases and liquids, but also to flowable mixtures (often called *slurries* or *sludges*) such as concrete, sewage, or wood pulp. Control of a fluid flow, and hence the type of measurement required, may involve *volumetric* flow rate, *mass* flow rate, or flow direction. Gas flows may be *compressible*, which also influences the measurement technique. In addition, the condition of the flow—whether or not it is homogenous and clean (free of suspended particles)—has a bearing on flowmeter technology. Another factor to be considered is flow velocity; slow moving laminar flows of viscous material require different measurement techniques than those used for high-velocity turbulent flows. Still another consideration is confinement of the flow. Whereas most fluid flow measurements are concerned with *full flow* through *closed channels* such as ducts and pipes, some applications require measurements of *partial flow* through *open channels* such as troughs and flumes. Only the most widely used flowmeters are considered here.

The major categories of flowmeters are:

- Differential pressure, constriction-type (venturi, orifice, flow nozzle, elbow (or pipe bend), and pitot static) (Figure 30.9)
- Fluid-power (gear motors, turbines, and paddle wheels) (Figure 30.10)

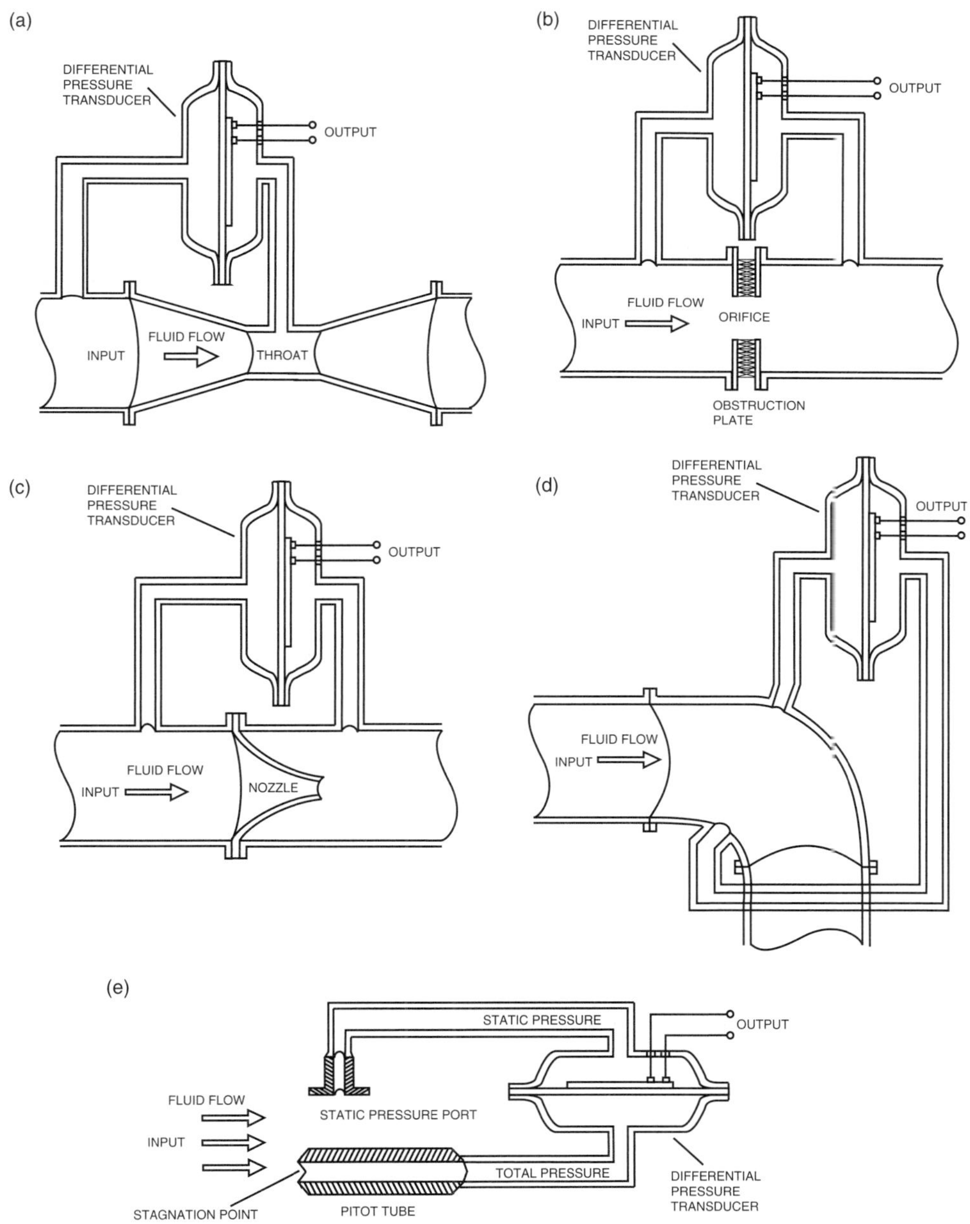

FIGURE 30.9 Differential pressure flowmeters: (a) Venturi flowmeter, (b) orifice flowmeter, (c) nozzle flowmeter, (d) pipebend (elbow) flowmeter, (e) pitot-static-flowmeter.

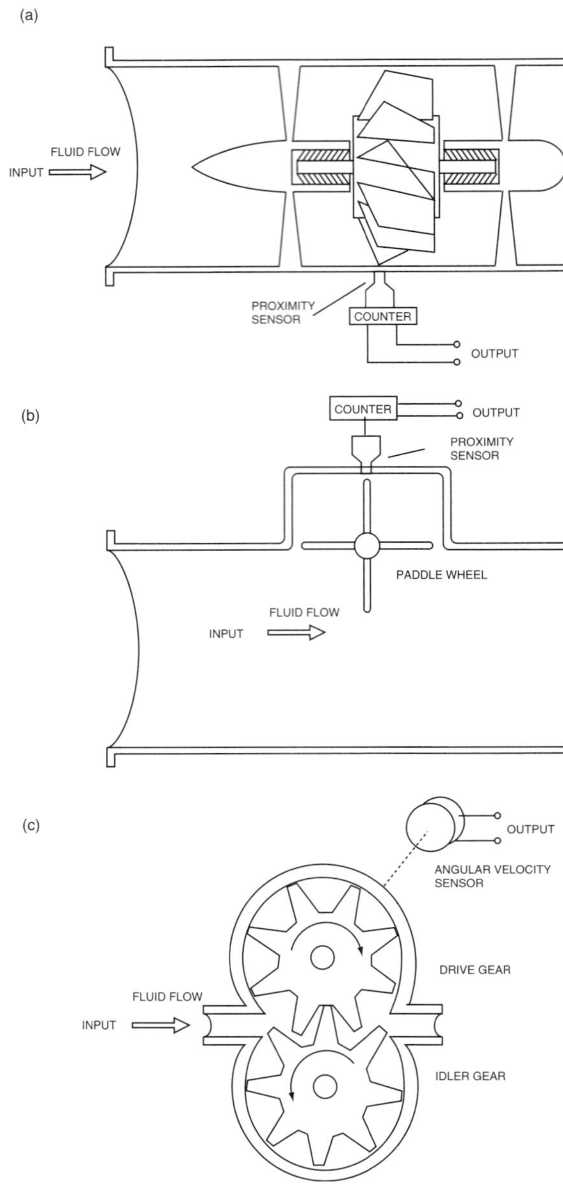

FIGURE 30.10 Fluid power flowmeters: (a) turbine flowmeter, (b) paddle wheel flowmeter, (c) gear motor flowmeter.

- Ultrasound (Figure 30.11)
- Vortex shedding (Figure 30.12)
- Thermal anemometer (Figure 30.13)
- Electromagnetic (Figure 30.14)
- Rotameter (variable-area in-line flowmeter) (Figure 30.15)

Differential pressure flowmeters are suited to high- and moderate-velocity flow of gas and clean, low-viscosity liquids. Venturi flowmeters (Figure 30.9a) are the most accurate, but they are large and expensive. Orifice flowmeters (Figure 30.9b) are smaller, less expensive, and much less accurate than venturi flowmeters.

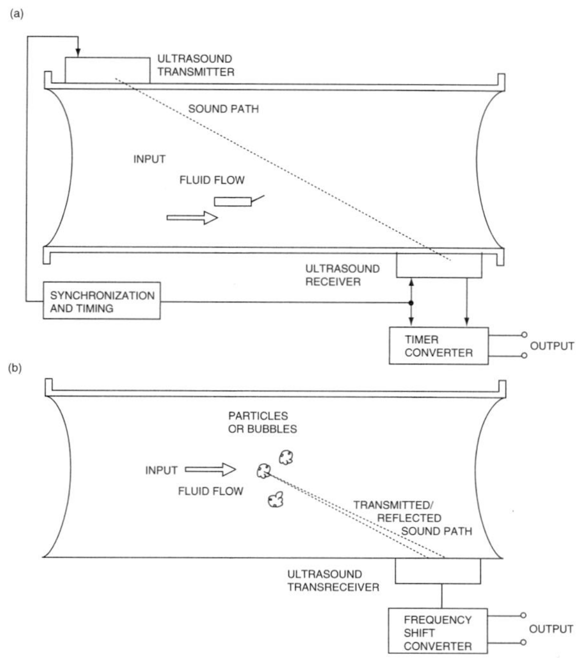

FIGURE 30.11 Ultrasound flowmeters: (a) transmission-type ultrasound flowmeter, (b) doppler ultrasound flowmeter.

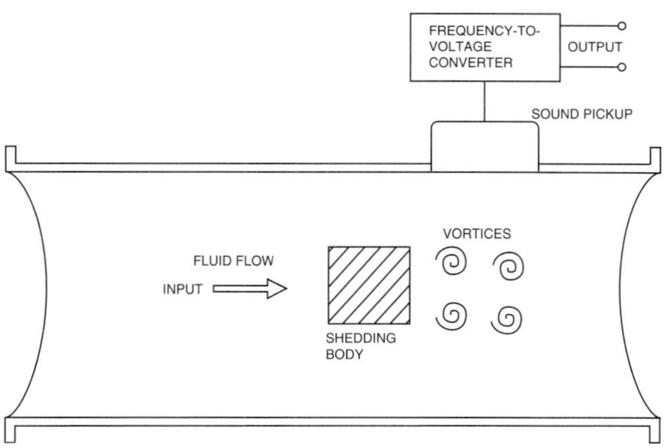

FIGURE 30.12 Vortex-shedding flowmeter.

Nozzle flowmeters (Figure 30.9c) are a compromise between venturi and orifice flowmeters. Pipe-bend flowmeters (Figure 30.9d), which can essentially be installed in any bend in an existing piping system, are used primarily for gross flow rate measurements. Pitot-static flowmeters (Figure 30.9e) are used in flows which have a large cross-sectional area, such as in wind tunnels. Pitot-static flowmeters are also used in freestream applications such as airspeed indicators for aircraft.

Fluid-power flowmeters are used in low-velocity, moderately viscous flows. In addition to industrial control applications, turbine flowmeters (Figure 30.10a) are sometime used as speed indicators for ships or boats. Paddle wheel flowmeters (Figure 30.10b) are used both in closed- and open-flow applications such as liquid flow in flumes. Since a fluid-power gear motor (Figure 30.10c) is a constant volume device, motor shaft speed is always a direct indication of fluid flow rate.

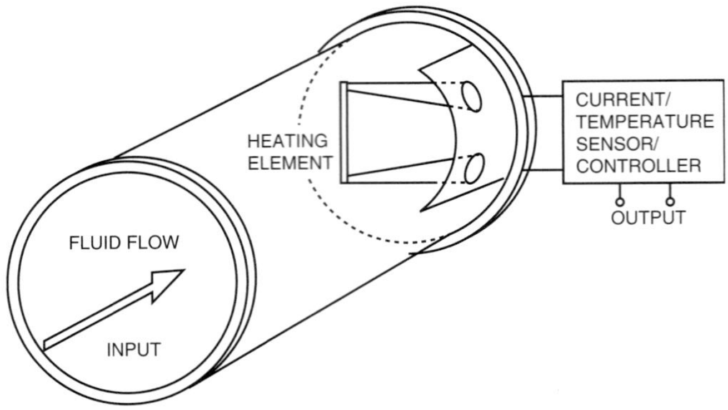

FIGURE 30.13 Thermal anemometer.

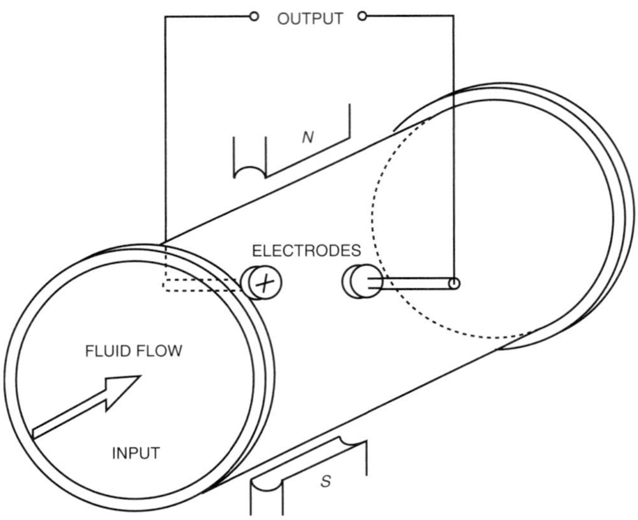

FIGURE 30.14 Electromagnetic flowmeter.

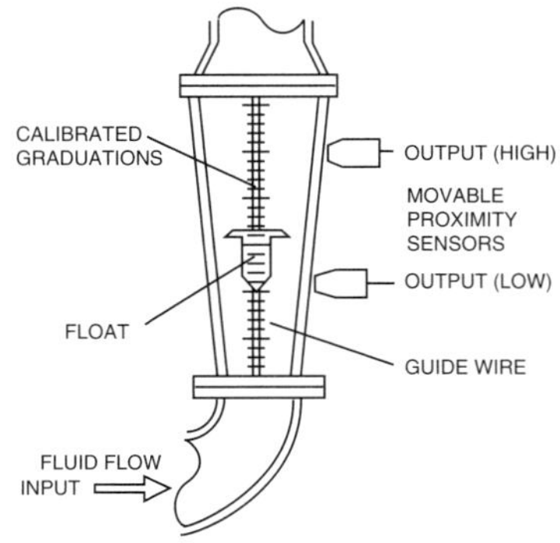

FIGURE 30.15 Variable-area in-line flowmeter (rotameter).

Ultrasound flowmeters of the transmission type (Figure 30.11a), which are based on the principle that the sound transmission speed will be increased by the flow rate of the fluid, are used in all types of clean, subsonic flows. Doppler flowmeters (Figure 30.11b) rely on echoes from within the fluid, and are thus only useful in dirty flows that carry suspended particles or turbulent flows that produce bubbles. Ultrasound flowmeters are nonintrusive devices, which can often be retrofitted to existing duct or pipe systems.

Vortex shedding flowmeters (Figure 30.12) introduce a *shedding body* into the flow to cause production (*shedding*) of *vortices*. The sound accompanying the production and collapse of the vortices is monitored and analyzed. The dominant frequency of the sound is indicative of the rate of vortex production and collapse, and hence an indication of flow rate. Vortex shedding flowmeters are useful in low-velocity, nonturbulent flows.

Thermal anemometers (Figure 30.13) are used in low-velocity gas flows with large cross-sectional area, such as in heating, ventilation, and air conditioning (HVAC) ducts. Convection cooling of the heating element is related to flow rate. The flow rate measurement is based either on the current required to maintain a constant temperature in the heating element, or alternatively on the change in temperature when the current is held constant.

Electromagnetic flowmeters (Figure 30.14) are useful for slow moving flows of liquids, sludges, or slurries. The flow material must support electrical conduction between the electrodes, and so in some cases it is necessary to ionize the flow upstream from the measurement point in order to use an electromagnetic flowmeter.

Variable-area in-line flowmeters (Figure 30.15), or *rotameters*, are sometimes referred to as sight gauges because they provide a visible indication of the flow rate. These devices, when fitted with proximity sensors (such as capacitive pickups) that sense the presence of the float, can be used in on–off control applications.

30.3.3 Liquid Level Transducers

Liquid-level measurements are relatively straightforward, and the transducers fall into the categories of *contact* or *noncontact*. Measurements may be *continuous*, in which the liquid level is monitored continuously throughout its operating range, or *point*, in which the liquid level is determined to be above or below some predetermined level.

The contact transducers encountered most frequently are:

- Float
- Hydrostatic pressure
- Electrical capacitance
- Ultrasound

The noncontact transducers encountered most frequently are:

- Capacitive proximity sensors
- Ultrasound
- Radio frequency
- Electro-optical

Float-type liquid level transducers are available in a wide variety of configurations for both continuous and point measurements. One possible configuration is depicted in Figure 30.16 for continuous measurement and for both single- and dual-point measurements.

Hydrostatic pressure liquid level transducers may be used in either vented or pressurized applications (Figure 30.17). In either case the differential pressure is directly proportional to the weight of the liquid column, since the differential pressure transducer accounts for surface pressure.

Capacitance probes (Figure 30.18a) are widely used in liquid level measurements. It is possible, when the tank walls are metal, to use a single bare or insulated metal rod as one capacitor plate and the tank walls as the other. More frequently, capacitance probes consist of a metal rod within a concentric cylinder

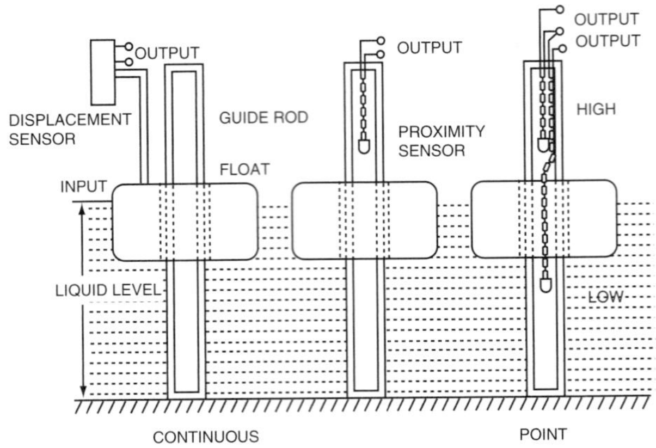

FIGURE 30.16 Float-type liquid level transducers.

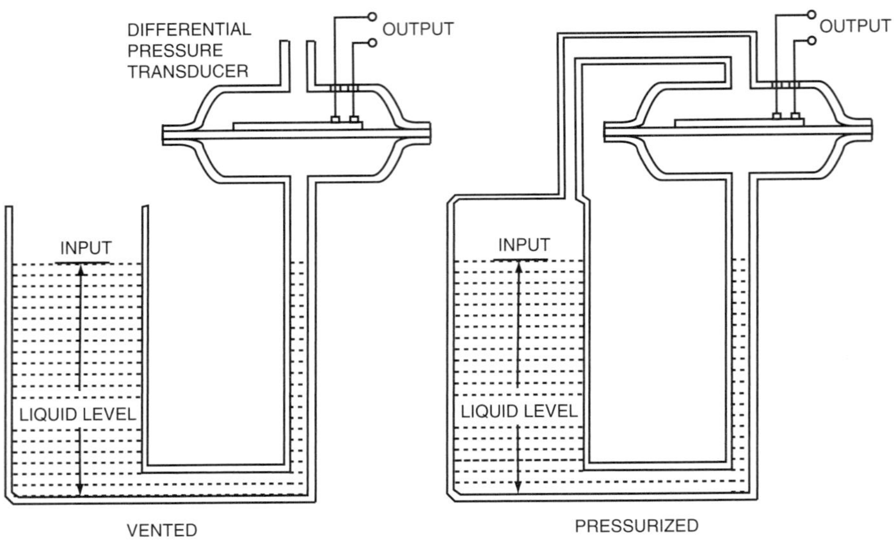

FIGURE 30.17 Hydrostatic pressure liquid level transducers.

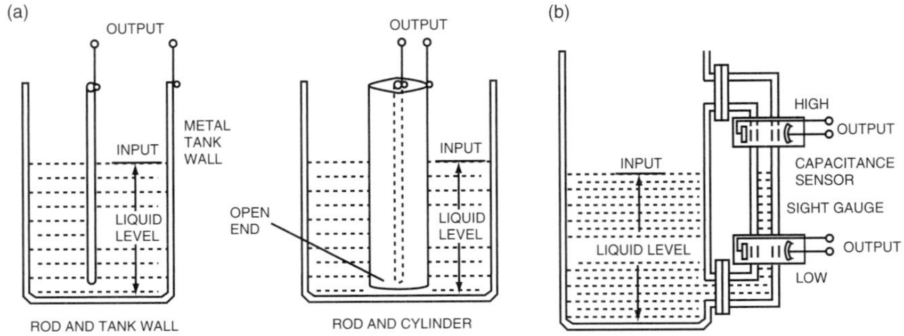

FIGURE 30.18 Capacitive-type liquid level transducers: (a) capacitive probes, (b) capacitive switches.

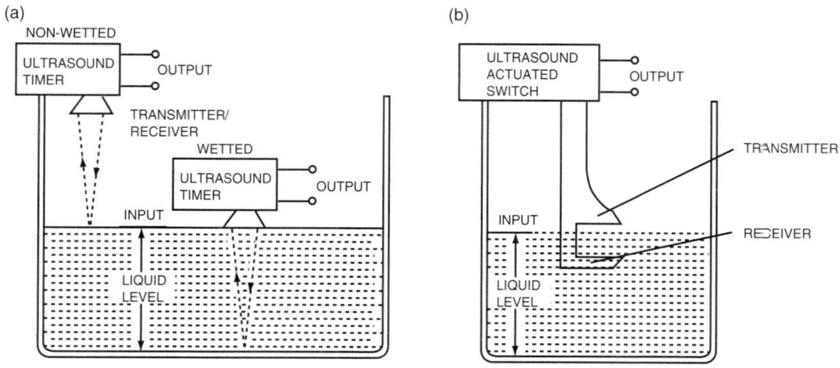

FIGURE 30.19 Ultrasound liquid level transducers: (a) echo-ranging liquid level transducer, (b) ultrasound switch.

open at the ends, which makes the transducer independent of the tank construction. An interesting application of this type of capacitance probe is as aircraft fuel quantity indicators. Capacitance switches can be utilizes as depicted in Figure 30.18b to provide noncontact point measurements of liquid level.

Ultrasound echo ranging transducers can be used in either *wetted* (contact) or *nonwetted* (noncontact) configurations for continuous measurement of liquid level (Figure 30.19a). An interesting application of wetted transducers is as depth finders and fish finders for ships and boats. Nonwetted transducers can also be used with bulk materials such as grains and powders. Radio-frequency and electro-optic liquid level transducers are usually noncontact, echo ranging devices that are similar in principle and application to the nonwetted ultrasound transducer.

Ultrasonic transducers can also be adapted to point measurements by locating the transmitter and the receiver opposite one another across a gap (Figure 30.19b). When liquid fills the gap, attenuation of the ultrasound energy is markedly less than when air fills the gap. The signal conditioning circuits utilize this sharp increase in the level of ultrasound energy detected by the receiver to activate a switch.

30.3.4 Temperature Transducers

Temperature measurement is generally based on one of the following physical principles:

- Thermal expansion
- Thermoelectric phenomena
- Thermal effect on electrical resistance
- Thermal effect on conductance of semiconductor junctions
- Thermal radiation

(Strictly speaking, any device used to measure temperature may be called a thermometer, but more descriptive terms are applied to devices used in temperature control.)

Bimetallic switches (Figure 30.20) are widely used in on–off temperature control systems. If two metal strips with different *coefficients of thermal expansion* are bonded together while both strips are at the same temperature, the bimetallic structure will bend when the temperature is changed. Although these devices are often called *thermal cutouts*, implying that they are used in normally closed switches, they can be fabricated in either normally closed or normally open configurations. The bimetallic elements can also be fabricated in coil or helical configurations to extend the range of motion due to thermal expansion.

Thermocouples are rugged and versatile temperature sensors frequently found in industrial control systems. A thermocouple consists of a pair of dissimilar metal wires twisted or otherwise bonded at one end. The *Seebeck effect* is the physical phenomena that accounts for thermocouple operation, so thermocouples are known alternatively as *Seebeck junctions*. The potential difference (*Seebeck voltage*) between the free ends of the wire is proportional to the difference between the temperature at the junction and

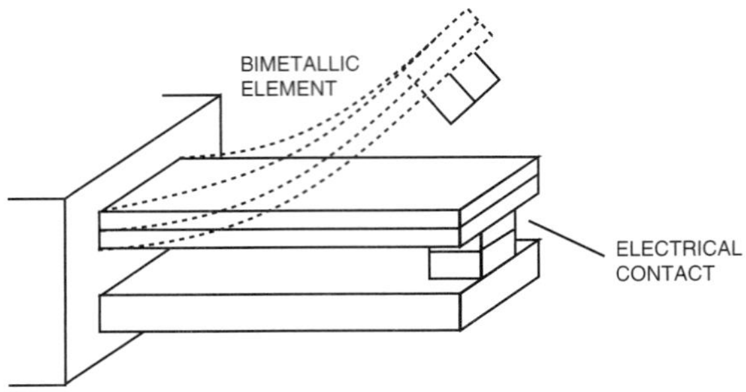

FIGURE 30.20 Bimetallic thermal switch.

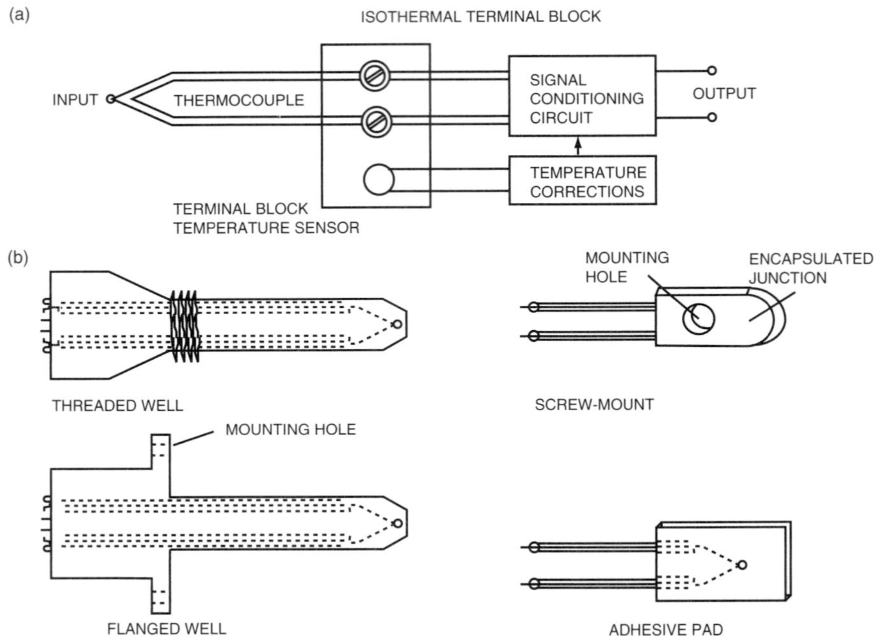

FIGURE 30.21 Thermocouples: (a) typical thermocouple connection, (b) some thermocouple accessories and configurations.

the temperature at the free ends. Thermocouples are available for measurement of temperature as low as −270°C and as high as 2300°C, although no single thermocouple covers this entire range. Thermocouples are identified as type B, C, D, E, G, J, K, N, R, S, or T, according to the metals used in the wire.

Signal conditioning and amplification of the relatively small Seebeck voltage dictates that the thermocouple wires must be connected to the terminals of a signal conditioning circuit. These connections create two additional Seebeck junctions, each of which generates its own Seebeck voltage, which must be canceled in the signal conditioning circuit. To implement cancellation to the corrections the following are necessary (Figure 30.21a):

- The input terminals of the signal conditioning circuit must be made of the same metal.
- The two terminals must be on an *isothermal terminal block* so that each Seebeck junction created by the connection is at the same temperature.
- The temperature of the terminal block must be known.

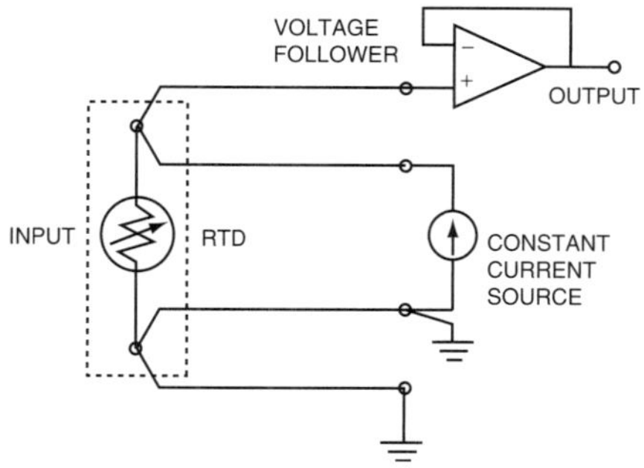

FIGURE 30.22 Typical resistance temperature detector (RTD) application.

The first two requirements are met by appropriate construction of the signal conditioning circuit. The third requirement is met by using a reference temperature sensor, probably an IC temperature transducer of the type described later.

Thermocouple and thermocouple accessories are fabricated for a variety of applications (Figure 30.21b). Protective shields (thermowells) are used to protect thermocouple junctions in corrosive environments or where conducting liquids can short circuit the thermocouple voltage; however, exposed (bare) junctions are used wherever possible, particularly when a fast response is essential.

Resistance temperature detectors (RTD) are based on the principle that the electrical *resistivity* of most metals increases predictably with temperature. Platinum is the preferred metal for RTDs, although other less expensive metals are used in some applications. The resistivity of platinum is one of the standards by which temperature is measured. The relatively good linearity of the resistivity of platinum over a wide temperature range (–200 to 800°C) makes platinum RTDs suitable for stable, accurate temperature transducers, which are easily adapted to control systems applications.

The disadvantage of the RTD is that the temperature-sensitive element is a rather fragile metal filament wound on a ceramic bobbin or a thin metal film deposited on a ceramic substrate. RTD elements are usually encapsulated and are rarely used as bare elements. The accessories and application packages used with RTDs are similar to those used with thermocouples (Figure 30.21b).

Most platinum RTDs are fabricated so as to have a nominal resistance of 100 Ω at 0°C. The *resistance temperature coefficient* of platinum is approximately 3–4 mΩ/Ω/°C, so resolution of the temperature to within 1°C for a nominal 100-Ω RTD element requires resolution of the absolute resistance within 0.3–0.4 Ω. These resistance resolution requirements dictate use of special signal conditioning techniques to cancel the lead and contact resistance of the RTD element (Figure 30.22). The circuit depicted in Figure 30.22 is a variation of a 4-*wire ohmmeter*. Most RTDs are manufactured with four leads to be compatible with such circuits.

Thermistors are specially prepared metal oxide semiconductors that exhibit a strong *negative* temperature coefficient, in sharp contrast to the weak positive temperature coefficient of RTDs. Nominal thermistor resistance, usually specified for 25°C, ranges from less than 1000 Ω to more than 1 MΩ, with sensitivities greater than 100 Ω/°C. Thus, the thermistor is the basis for temperature sensors that are much more sensitive and require less special signal conditioning than either thermocouples or RTDs. The tradeoff is the marked nonlinearity of the resistance-temperature characteristic. To minimize this problem, manufacturers provide packages in which the thermistor has been connected into a resistor network chosen to provide a relatively linear resistance-temperature characteristic over a nominal temperature range.

The development of thermistor technology has lead to the IC temperature sensor in which the temperature-sensitive junction(s) and the required signal conditioning circuits are provided in a monolithic package. The user is only required to provide a supply voltage (typically 5 V DC) to the IC in order to obtain an analog output voltage proportional to temperature. Thermistors and IC temperature sensors

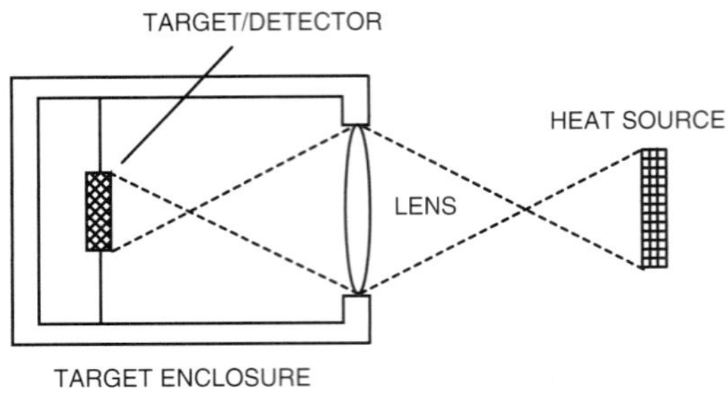

FIGURE 30.23 Schematic of the radiation thermometry scheme.

can be produced in very small packages, which permit highly localized temperature measurements. Some thermistors designed for biological research are mounted in the tip of a hypodermic needle. The shortcomings of both thermistor and IC temperature sensors are that they are not rugged, cannot be used in caustic environments, and are limited to temperatures below approximately 200°C.

Radiation thermometers are used for remote (noncontact) sensing of temperature in situations where contact sensors cannot be used. Operation is based on the principles of heat transfer through thermal radiation. Radiation thermometers focus the *infrared* energy from a heat source onto a *black body* (target) within the radiation thermometer enclosure (Figure 30.23). One of the contact temperature sensors described previously is incorporated into the target to measure the target temperature. The rise in temperature at the target is related to the source temperature. Typical radiation thermometers have standoff ranges (focal lengths) of 0.5–1.5 m, but instruments with focal length as short as 1 cm or as long as 10 m are available. Radiation thermometers are available for broadband, monochromatic, or two-color thermometry.

30.4 Transducer Performance

The operation of a transducer within a control system can be described in terms of its *static performance* and its *dynamic performance.* The static characteristics of greatest interest are:

- Scale factor (or sensitivity)
- Accuracy, uncertainty, precision, and system error (or bias)
- Threshold, resolution, dead band, and hysteresis
- Linearity
- Analog drift

The dynamic characteristics of greatest interest are:

- Time constant, response time, and rise time
- Overshoot, settling time, and damped frequency
- Frequency response

Static performance is documented through *calibration*, which consists of applying a known input (quantity or phenomenon to be measured) and observing and recording the transducer output. In a typical calibration procedure, the input is increased in increments from the lower range limit to the upper range limit of the transducer, then decreased to the lower range limit. The range of a component consists of all allowable input values. The difference between the upper and lower range limits is the *input span* of the component; the difference between the output at the upper range limit and the output at the lower range limit is the *output span.*

Dynamic performance is documented by applying a known change, usually a step, in the input and observing and recording the transducer output, usually with a strip recorder or a storage oscilloscope.

30.5 Loading and Transducer Compliance

A prime requirement for an appropriate transducer is that it be *compliant* at its input. Compliance in this sense means that the input energy required for proper operation of the transducer, and hence a correct measurement of the controlled output, does not significantly alter the controlled output. A transducer that does not have this compliance is said to *load* the controlled output. For example, a voltmeter must have a high-impedance input in order that the voltage measurement does not significantly alter circuit current and, hence, alter the voltage being measured.

Defining Terms

Controlled output: The principal product of an automatic control system; the quantity or physical activity to be measured for automatic control.

Feedback path: The cascaded connection of transducer and signal conditioning components in an automatic control system.

Forward path: The cascaded connection of controller, actuator, and plant or process in an automatic control system.

Motion transducer: A transducer used to measure the controlled output of a servomechanism; usually understood to include transducers for static force measurements.

Plant or process: The controlled device that produces the principal output in an automatic control system.

Process control: The term used to refer to the control of industrial processes; most frequently used in reference to control of temperature, fluid pressure, fluid flow, and liquid level.

Process transducer: A transducer used to measure the controlled output of an automatic control system used in process control.

Reference input: The signal provided to an automatic control system to establish the required controlled output; also called *setpoint.*

Servomechanism: A system in which some form of motion is the controlled output

Signal conditioning: In this context, the term used to refer to the modification of the signal in the feedback path of an automatic control system; signal conditioning converts the sensor output to an electrical signal suitable for comparison to the reference input (setpoint); the term can also be applied to modification of forward path signals.

Transducer: The device used to measure the controlled output in an automatic control system; usually consists of a sensor or pickup and signal conditioning components.

References

Bateson, R.N. 1993. *Introduction to Control System Technology,* 4th ed. Merrill, Columbus, OH.

Berlin, H.M. and Getz, F.C., Jr. 1988. *Principles of Electronic Instrumentation and Measurement.* Merrill, Columbus, OH.

Buchla, D. and McLachlan, W. 1992. *Applied Electronic Instrumentation and Measurement.* Macmillan, New York.

Chaplin, J.W. 1992. *Instrumentation and Automation for Manufacturing.* Delmar, Albany, NY.

Doeblin, E.O. 1990. *Measurement Systems Application and Design,* 4th ed. McGraw-Hill, New York.

Dorf, R.C. and Bishop, R.H. 1995. *Modern Control Systems,* 7th ed. Addison-Wesley, Reading, MA.

O'Dell, T.H. 1991. *Circuits for Electronic Instrumentation.* Cambridge Univ. Press, Cambridge, England, UK.

Seippel, R.G. 1983. *Transducers, Sensors, and Detectors.* Reston Pub., Reston, VA.

Webb, J. and Greshock, K. 1993. *Industrial Control Electronics,* 2nd ed. Macmillan, New York.

Further Information

Manufacturers and vendors catalogs, data documents, handbooks, and applications notes, particularly the handbook series (current year) by Omega Engineering, Inc.:

The Flow and Level Handbook
The Pressure, Strain, and Force Handbook
The Temperature Handbook

Trade journals, magazines, and newsletters, particularly:

Instrumentation Newsletter (National Instruments)
Personal Engineering and Instrumentation News
Test and Measurement News (Hewlett Packard)
Test and Measurement World

31

A/D and D/A Conversion

Brian Betts
National Instruments, Inc.

31.1 Introduction

As computers began to gain popularity, engineers and scientists realized that computers could become a powerful tool for data acquisition and control. However, almost all real-world phenomena (such as light, pressure, velocity, temperature, etc.) are measured with sensors that produce analog signals, and computers, on the other hand, rely on digital signals. Therefore, many companies began to invest in advancements in analog-to-digital and digital-to-analog converters (ADC and DAC). These devices have become the keystone in every measurement device. This chapter will examine the ADC and DAC on a functional level as well as discuss important specifications of each.

31.2 Sampling

In order to convert an analog signal into a digital signal, the analog signal must first be sampled. Sampling involves converting one value of a signal at a particular interval of time. Generally, conversions happen uniformly in time. For example, a digitizing system may convert a signal every 5 ms, or sample at 200 k S/s. Although it is not necessary to uniformly sample a signal, doing so provides certain benefits that will be discussed later.

A typical sampling circuit contains two major components: a track-and-hold (T/H) circuit and the ADC. Since the actual conversion in the ADC takes some amount of time, it is necessary to hold constant the value of the signal being converted. At the instance the sample is to be taken, the T/H holds the sample value even if the signal is still changing. Once the conversion has been completed, the T/H releases the value it is currently storing and is ready to track the next value.

One aspect of sampling that cannot be avoided is that some information is lost because of the fact that an analog waveform actually has an infinite number of samples and there is no way to capture every value. The major pitfall associated with this fact is called undersampling or sampling too slow. If a 10-kHz sine wave is to be acquired and sampling only occurs at 5 kS/s, the true waveform will not be preserved. In fact, a waveform of a different frequency will result. The result of undersampling is often referred to as aliasing. According to the Nyquist theory, which deals with sampling, sampling should occur at a rate twice as high as the highest frequency component of the signal. In general, this theory just preserves the frequency of the signal; so, if the shape of the waveform is desired, sampling should probably be at least ten times as fast as the signal.

31.3 Analog-to-Digital Converter Specifications

31.3.1 Range

The input range of an ADC is the span of voltages over which the ADC can make a conversion. For example, a common range for ADC is 0–5 V, meaning that the ADC can convert an input that is within 0–5 V. The end points of the low and high end of the range are called –full-scale and +full-scale (they are also referred to as rails). If the –full-scale is equal to 0 V, then the range is referred to as unipolar, and if the two full-scale values have the same magnitude, for example, −5 to +5 V, then the range of the ADC is referred to as bipolar. If an input voltage falls outside the range, the ADC is said to be overranged. In this case, most ADCs will return a value of the endpoint closest to the sampled signal.

31.3.2 Resolution

The resolution of a digitizer is the smallest detectable change in voltage; however, the resolution of an ADC usually refers to the number of binary bits it produces. For example, a 12-bit ADC represents a converted analog value, using 12 digital bits. This same 12-bit ADC can resolve a value to one of $4096 (= 2^{12})$ different levels. Another common way to specify resolution is by decimal digits. A 6-digit voltmeter measuring on a 1-V scale could measure in 0.000001 V steps from −0.999999 to 0.999999 V.

31.3.3 Coding Convention

The different formats an ADC can use to represent its output are known as coding conventions. An ADC using binary coding produces all 0s at –full-scale and all 1s at +full-scale (e.g., a 3-bit converter would produce 000 through 111).

31.3.4 Linear Errors

Linear errors are the largest and most common errors in an ADC, and are easily corrected by simple calibration or by additions and multiplications by correction constants. Although linear errors do not distort the ADC transfer function, they can change the range over which the ADC correctly operates.

31.3.5 Nonlinear Errors

Unlike linear errors, nonlinear errors are more difficult to compensate for in either the analog or digital signal. The best way to mitigate nonlinear error is to choose a well-designed, well-specified ADC. Nonlinear errors are characterized by two different specifications: differential nonlinearity (DNL) and integral nonlinearity (INL).

DNL measures any irregularity in the code width (smallest detectable change) by comparing the actual change in value to the ideal value of one code width (or 1 LSB). INL measures the deviation from an ideal transfer line of the code transitions.

Another important specification of an ADC in regards to DNL is if any codes are missing. A missing code can be thought of as a code with a width of 0 LSB (or a DNL of -1). If a code is missing, the step size at that point in the transfer function is doubled, effectively cutting the local resolution of the ADC in half. Therefore, ADC datasheets will specify if the ADC has no missing codes.

Another way to capture the same information included in INL is a measurement called relative accuracy. Relative accuracy indicates how far away fromthe ideal the code transitions are (which is INL), and also includes how far any part of the transfer function, including quantization "staircase" error, deviates fromideal. In an ideal noiseless ADC, the worse case relative accuracy is always greater than the INL.

31.3.6 Aperture Errors

Aperture errors deal with the timing of the conversions themselves. All ADCs require some signal, generally a pulse train clock, to tell the ADC when to start a conversion. Inherently, some small amount of time will elapse between the ADC receiving this convert signal and the sample being held. This amount of time is called the aperture delay. Most ADCs have an aperture delay of just a few nanoseconds. However, most measurement devices have additional circuitry in front of the ADC, such as amplifiers, which have the effect of negating the aperture delay caused by the ADC. For example, if the ADC has a delay of 10 ns and the amplifier has a delay of 160 ns, the effective aperture delay of the system is -150 ns.

Another important time specification is jitter. Jitter (or aperture jitter) measures the difference in the amount of time between each sample. If a signal is sampled at 1 million samples per second (1 MS/s), the expected period between each sample would be exactly 1μ s. Jitter caused by the clock source, digital clock circuitry, or S/H circuitry could cause this 1μ s sample time to vary by as much as a few picoseconds to a nanosecond fromcy cle to cycle.

31.3.7 Noise

Noise limits the ADC resolution because an interfering waveform is present in the input signal as it is being converted. The most common source of noise in a signal is thermal noise. Thermal noise is caused by the random nature of electrical components. With higher temperatures and resistances in components, the thermal noise will increase. Other common sources of noise are electromagnetically coupled signals from nearby components, such as logic circuits and clocks. Generally, noise is specified in volts peak-to-peak or rms, or LSBs rms or peak-to-peak.

Quantization error, resulting from the bit-resolution limitation of the ADC, is sometimes referred to as quantization noise. Although quantization error is perfectly predictable with respect to the input signal, when a signal is fairly "busy" (meaning that each consecutive conversions do not result in many common bits of data) the quantization error becomes chaotic. When this occurs, the quantization error can be thought of as another source of random noise, whose statistical distribution is uniform from -0.5 to 0.5 LSB. In spectral analysis, this is sometimes the dominant source of noise.

Once noise reaches the ADC, there are ways to process the noise out of the signal, provided that the noise is an independent signal. One of the most common ways to decrease noise in a dc measurement is to acquire a number of points and average the values. If the noise is white random noise, which has equal energy density at all frequencies, averaging will reduce the amount of noise by the square root of the number of samples averaged. If the noise is interfering with a repetitive waveform, the noise can be reduced by measuring a number of waveforms, using a level trigger and then averaging the waveforms. Most noise specifications for an ADC are for quiet, low-impedance signals. To preserve the noise performance of the ADC, the user must connect signals to the inputs with shielded cabling that keeps signals away from any electromagnetic interference.

31.3.8 Dynamic Range

Dynamic range is the ratio of largest to smallest signal the ADC can represent. The dynamic range is found by taking a full-scale signal value and comparing that to the smallest detectable noise level of the ADC. The dynamic range is usually expressed in decibels (dB) and can be found by the following formula:

$$\text{Dynamic Range} = 20\log(S/N)$$

where S is large signal level and N is noise level. The noise level includes quantization noise of the ADC, which for an ideal ADC is equal to LSB rms. A full-scale sine wave has an amplitude of $2n - 1$ LSB or (n = number of bits of the ADC) . Therefore, an ideal ADC has a dynamic range of

$$\begin{aligned}\text{Dynamic Range} &= 20\log(2^{n-1}/\sqrt{2} \times 1/\sqrt{12}) \\ &= 6.0206n + 1.7609\end{aligned}$$

Since no ADC is ideal, the effective number of bits (ENOB) of an actual ADC can be calculated using the above equation. The ENOB represents the real-world resolution of an ADC, and can be found with the following equation using dynamic range:

$$\text{ENOB} = (\text{Dynamic Range} - 1.7609)/6.0206$$

For example, a 12-bit ADC with a dynamic range of 69 dB has an ENOB of 11.17 bits.

31.3.9 Types of Analog-to-Digital Converters

All ADCs accomplish the same fundamental task of taking an analog signal and converting it into a digital representation. Two crucial characteristics of an ADC are the speed at which conversions can be made and the resolution of the conversions. Because it takes an ADC longer to resolve input signals to higher resolutions, designing or specifying an ADC often becomes a trade-off of speed versus resolution. This makes it important to understand how the ADC will be used in order to match the correct converter for the application.

Despite the many different types of ADCs available, they all share some common characteristics. The heart of any ADC converter is the comparator. A comparator is a simple 1-bit ADC that has two analog inputs and one digital output. One of the analog input signals is a reference voltage that has some known value. The other inputs will either be greater than or less than the known input value, and that will turn into a digital value of 1 or 0. Some ADCs are actually composed of multiple comparators, but the basic theory for each one is the same.

31.3.10 Flash

Flash ADCs are the fastest ADCs available, obtaining speeds in multiple gigasamples per second. However, true to the speed versus resolution trade-off discussed, the flash converters generally have resolutions of 10 bits and below. A flash converter with n bits of resolution is composed of $2n - 1$ high-speed comparators operating in parallel, see Figure 31.1. A string of $2n - 1$ resistors between two voltage references supplies a set of uniformly spaced voltages that span the input range, one for each comparator. The input voltage is then compared to each level simultaneously. The comparators then output a 1 for all voltages below the input voltage, and a 0 for all voltages above the input voltage. These resulting digital values are then fed into a logic convert to output an n-bit value.

Because of the simplicity of the design, flash converters are fast, but as the resolution of the converter is increased, the number of comparators and resistors needed increases exponentially. Both the size and power needed to operate the converter also increase exponentially as a result of increased resolution, and

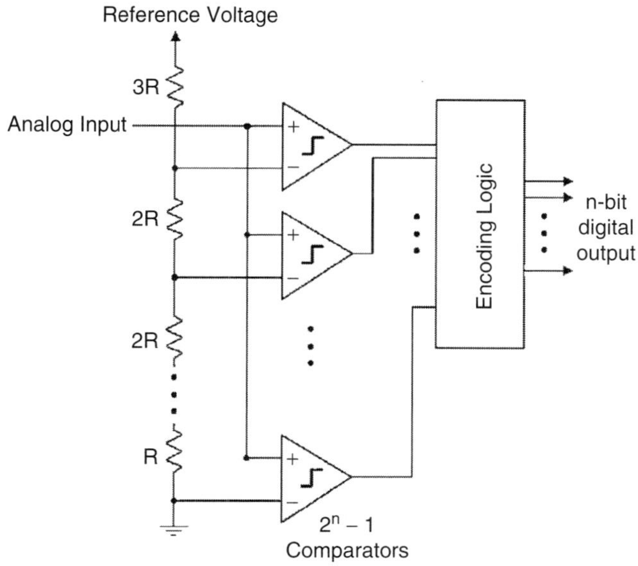

FIGURE 31.1 Flash ADC—a flash converter has $2n - 1$ comparators operating in parallel. It relies on the uniformity of the resistors for linearity.

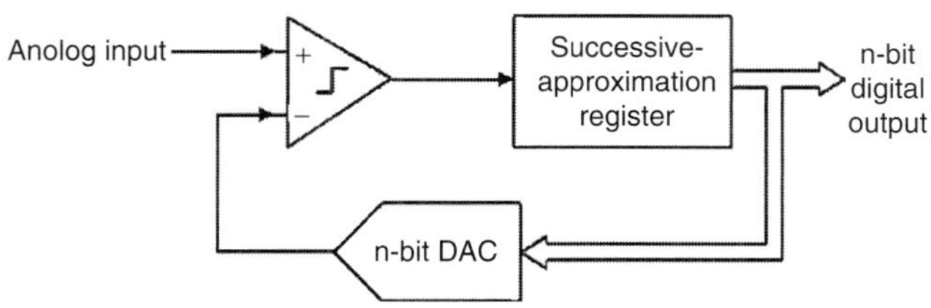

FIGURE 31.2 SAR ADC—a successive-approximation (SAR) converter has one comparator that iterates through a series of "guesses" to determine a digital representation of the signal.

this way the converters are limited in their resolution. However, because string resistors' values typically vary only a few percent fromone another, the differential linearity of the flash ADC is quite good.

31.3.11 Successive-Approximation Register

Successive-approximation register (SAR) ADCs are the most common ADCs, having resolutions of 8–18 bits and speeds up to 10 MS/s. These ADCs have a low cost, and generally have good integral linearity. The SAR ADC architecture contains a high-speed DAC in a feedback loop, see Figure 31.2. The SAR iterates the DAC through a series of levels, which are then compared to the input voltage. As the conversion progresses, the SAR builds the n-bit digital output as a result of these comparisons. When the SAR has finished, the output of the DAC is as close to the input signal as possible, and the digital input of the DAC becomes the output of the SAR ADC.

A good real-world analogy to an SAR is a balance scale. If an object of unknown mass is placed on one side and continues to test a combination of weights until the scale is balanced, the weight of the object can be obtained.

The speed of the SAR ADC is limited by the rate at which the DAC can settle inside the feedback loop. In fact, the DAC must settle n times for every n bits of resolution desired in the ADC. In order to achieve faster rates, the SAR architecture can be used as the basis for a different ADC, the multistage.

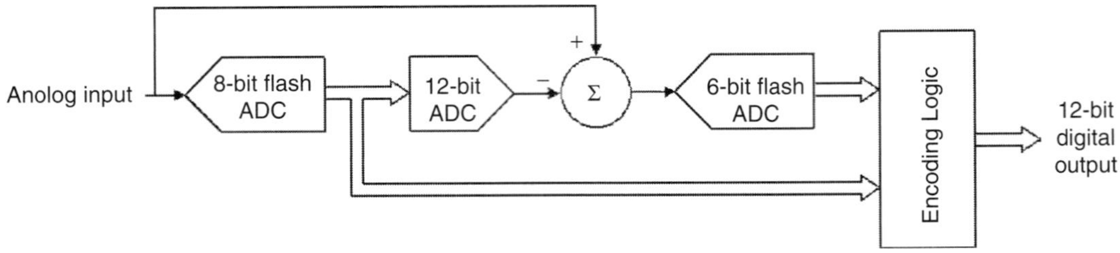

FIGURE 31.3 Multistage ADC—a multistage converter is a combination of the SAR and flash converters to provide faster sampling than the SARs and at a higher resolution than the flash converters could provide.

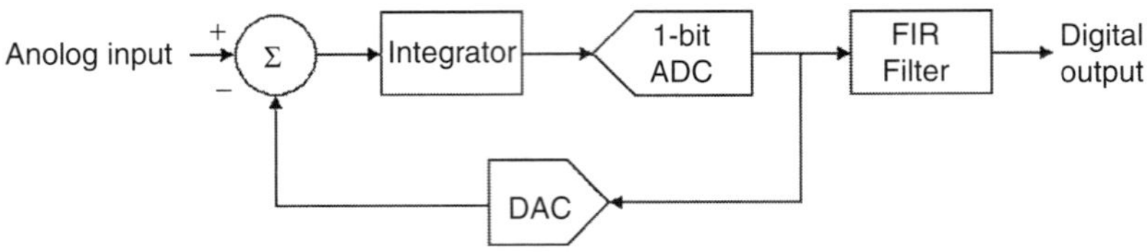

FIGURE 31.4 SDADC—asigma-delta converter uses a 1-bit comparator to determine the signal value. SD converters have great linearity by design, because the 1-bit ADC is perfectly linear, theoretically, since it can assume only one of two values.

31.3.12 Multistage

In order to achieve higher rates than the SAR, multistage ADCs use the iterative approach of the SAR but reduce the number of comparisons needed to complete the conversion. In addition to the comparator, the multistage ADC uses low-resolution flash converters, see Figure 31.3. In the figure, the 6-bit flash is used to convert the residual errors from the 8-bit flash. These two outputs fromthe ADCs are then combined using digital logic to produce a 12-bit output.

Most multistage ADCs are actually pipelined ADC. Pipelined ADCs have the same architecture as a multistage ADC, but each flash converter contains a T/H at the input. This allows each stage to convert the residual error while the previous stage has moved on to the next sample. This way the whole converter can operate at the speed of the slowest stage, as opposed to the multistage that operates at a speed equal to the sum of all the stages.

31.3.13 Integrating

Integrating ADCs are designed to return very high-resolution readings. As a trade-off, they operate at slower speeds. It is a very simple design; the integrating ADCs consist of an integrating amplifier, a comparator, a digital counter, and an extremely stable capacitor. The most common integrating ADC is the dual slope. In this architecture, the capacitor is initially discharged to have no potential across it. At a set time, the input is applied across the capacitor and it begins to charge for a set period of time T1. Because of the properties of a capacitor, the rate of charge is proportional to the input voltage. After T1, the capacitor is switched to a negative reference voltage and begins to discharge at a rate proportional to the reference. The digital counter simply measures how much time it took for the capacitor to completely discharge T2. Since T1 and the rate at which the capacitor discharges are both known values, the voltage of the input can be obtained by a simple ratio.

It is important to note that the converter is not actually measuring the input voltage itself. Instead, the ADC obtains the voltage by measuring time and using digital logic to calculate the input voltage. This method has the advantage of rejecting noise, such as periodic noise, to which other ADCs are susceptible. In addition, most integrating ADCs operate on a multiple of an AC line period (1/60 or 1/50 s) so that stray electromagnetic fields caused by power systems are cancelled.

31.3.14 Sigma Delta

The sigma-delta (SD) ADC is one of the most popular types of ADCs because of its fit on the speed versus resolution curve. SD ADCs can provide 16–24 bits of resolution at sample rates of up to hundreds of thousands of samples per second. This speed and resolution makes them ideal for certain applications such as vibration and audio analysis; however, the process of integration causes the SD ADC to have poor DC accuracy. Figure 31.4 shows the design of an SD ADC. The heart of an SD ADC is actually a 1-bit ADC that samples at incredibly high rates. Typically, these 1-bit ADCs sample at 64 or 128 times the eventual sample rate, which is a process known as oversampling. In addition to the high-speed ADC, an SD architecture consists of an analog low-pass filter and a DAC all together in a feedback loop. The result forces otherwise unavoidable quantization noise into higher frequency bands. This resulting spectrum of the noise is part of a process called noise shaping. The output of this feedback loop, which is actually just a streamof 1-bit conversions, is then fed to a digital filter. The digital filter then increases the resolution, reduces the data rate, and applies a low-pass digital filter to the data coming out of the feedback loop. After this process, the SD ADC has an output with high resolution and signals only in the frequency band of interest, eliminating most of the inherent electronic noise.

31.3.15 Digital-to-Analog Converters

The opposite of an ADC, which takes an analog value and produces a digital value would be a device that takes digital values and creates analog values. A DAC is a device that, given a digital representation of a signal, can create an analog signal at a specific voltage level. Although much of the theory behind ADCs discussed previously applies to DACs, a unique set of terms and phenomena do exist.

31.3.16 Updating

Updating can be thought of as the DAC equivalent to sampling. If a DAC is to generate a sine wave from a group of digital values, we need some way to specify how this waveform is to be generated. Simply put, the update rate is how many points per second that a DAC can output an analog value, generally given in samples per second, kilosamples per second, or million samples per second.

31.4 Digital-to-Analog Converter Specifications

31.4.1 Range

The range of a DAC is identical to the de.nition of the range of an ADC. This refers to the voltage range of values that the DAC can output.

31.4.2 Resolution

The resolution of a DAC is specified identical to the ADC; however, the perspective is reversed. In an ADC, resolution defines how many digital bits would represent an analog value, thus giving us a level of granularity we could acquire. With a DAC, the resolution indicates how many digital bits need to be supplied to the DAC to operate and what analog signal granularity we can produce.

31.4.3 Monotonicity

One of the most useful specifications of a DAC is the monotonicity. If a DAC is monotonic, this implies that as the digital value increases, the analog output value will also increase or at least stay the same. Conversely, a device is said to be nonmonotonic if one or more values of the analog output may actually be less than the values corresponding to codes having smaller weight. Many applications are sensitive to fine changes in output value; therefore, any DAC used needs to be monotonic on all bits.

31.4.4 Settling Time and Slew Rate

Settling time and slew rate together determine how rapidly a DAC can change the analog value it is outputting. Settling time refers to the amount of time it takes the output of the DAC to reach a specified accuracy level. Most DACs specify settling time as a full-scale change in voltage, from the smallest output value to the largest. Slew rate, specified in volt per second, is the maximum rate of change of the output of the DAC. Therefore, a DAC with a fast slew rate and a small settling time can generate high-frequency signals because an accurate voltage level can be obtained in a very small amount of time.

31.4.5 Offset Error and Gain Error

Offset error refers to the transfer characteristic of the DAC not outputting an analog value of 0 when the digital value of 0 is applied. The range fromz ero to full would be offset fromthe specified value because the offset would carry throughout the transfer function. The offset error can be thought of as a translation in the transfer line either up or down fromthe ideal. Gain error indicates a linear deviation fromthe ideal transfer line of a DAC. This can be caused by a variety of factors, which results in a change of slope from the ideal.

31.4.6 Architecture of Digital-to-Analog Converters

Unlike ADCs, DACs do not implement a wide range of approaches to convert a digital input code to an analog value. Instead, almost all DACs use some combination of a switch network, resistive network, and summing amplifier. This is not to say that all DACs have the same design, but they are all based on the principle of switching.

31.4.7 Switching Network

The switching network of a DAC can be thought of as the heart of the conversion. Since digital bits are either on or off, these bits can be used to control single poll switches. These switches are then used to direct some formof analog circuitry to develop an analog value. For example, a 3-bit DAC would comprise three switches, one for each bit of input data. Depending on the code given, these switches would close in such a way to develop an analog value froma reference source that is equal to the digital representation, see Figure 31.5. Depending on the design of the analog circuitry in the DAC, the switches may be connecting current or voltage references to a resistive network.

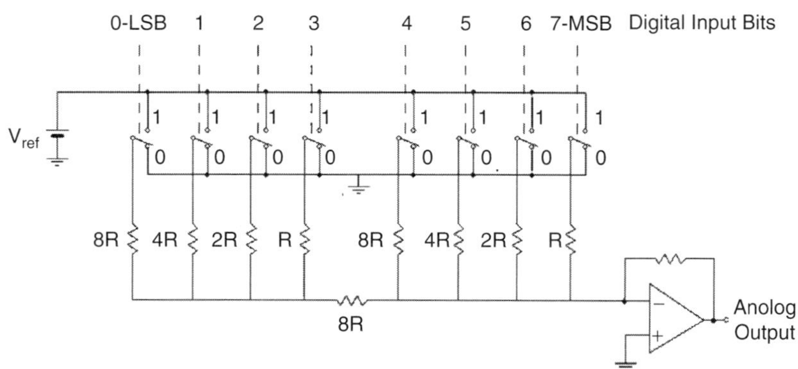

FIGURE 31.5 DAC architecture—most digital-to-analog converters (DACs) follow a standard architecture of a switch network, a resistive network, and an amplifier.

31.4.8 Resistive Networks

Resistive networks in a DAC provide the varying levels of analog output voltage, which will form the output of the DAC. Although many different resistive schemes are used in DAC design, the basic principle is common in all of them. The one shown in Figure 31.5 uses a dual resistor quad approach. In the figure, bits 0–3 and 4–7 are separated by a single resistor. These two independent groups are each a resistor quad with resistor values of 1R-2R-4R-8R where R is equal to 10 k. If any of the switches is in the 1 position, a current will develop across the resistor proportional to the resistors' value. Therefore, if bit 0 is on, a current proportional to 1/1R is generated; whereas, if switch 2 is on, a current proportional to 1/4R would be generated. The resistor between the two quads has the effect of a 16:1 current attenuator, so that even though bit 4 would generate a current proportional to 1/1R, once it gets to the amplifier it would appear to have a current proportional to 1/16R. In this case, bit 0 would be the most signi.cant bit (MSB), and bit 7 would be the least significant bit (LSB).

31.4.9 Summing Amplifier

The operational amplifier (op-amp) used in the DAC circuit of Figure 31.5 is acting as a summing amplifier. As the different bits generate a particular current, the op-amp is designed to collect the total current and would generate an output voltage. This output voltage of the op-amp is now an analog representation of the digital code that was fed to the DAC.

32

Signal Conditioning

Stephen A. Dyer
Kansas State University

Kelvin's first rule of instrumentation states, in essence, that the measuring instrument must not alter the event being measured. For the present purposes, we can consider the instrument to consist of an input transducer followed by a signal-conditioning section, which in turn drives the data-processing and display section (the remainder of the instrument). We are using the term *instrument* in the broad sense, with the understanding that it may actually be a measurement subsystem within virtually any type of system.

Certain requirements are imposed upon the transducer if it is to reproduce an event faithfully: It must exhibit amplitude linearity, phase linearity, and adequate frequency response. But it is the task of the signal conditioner to accept the output signal from the transducer and from it produce a signal in the form appropriate for introduction to the remainder of the instrument.

Analog signal conditioning can involve strictly *linear* operations, strictly *nonlinear* operations, or some combination of the two. In addition, the signal conditioner may be called upon to provide auxiliary services, such as introducing electrical isolation, providing a reference of some sort for the transducer, or producing an excitation signal for the transducer.

Important examples of linear operations include *amplitude scaling, impedance transformation, linear filtering,* and *modulation.*

A few examples of nonlinear operations include obtaining the *root-mean-square (rms) value, square root, absolute value,* or *logarithm* of the input signal.

There is a wide variety of building blocks available in either modular or integrated-circuit (IC) form for accomplishing analog signal conditioning. Such building blocks include operational amplifiers, instrumentation amplifiers, isolation amplifiers, and a plethora of nonlinear processing circuits such as comparators, analog multiplier/dividers, log/antilog amplifiers, rms-to-DC converters, and trigonometric function generators.

Also available are complete signal-conditioning subsystems consisting of various plug-in input and output modules that can be interconnected via universal backplanes that can be either chassis- or rack-mounted.

32.1 Linear Operations

Three categories of linear operations important to signal conditioning are amplitude scaling, impedance transformation, and linear filtering.

32.1.1 Amplitude Scaling

The amplitude of the signal output from a transducer must typically be scaled—either amplified or attenuated—before the signal can be processed.

32.1.1.1 Amplification

Amplification is generally accomplished by an *operational amplifier*, an *instrumentation amplifier*, or an *isolation amplifier*.

Operational Amplifiers

A conventional operational amplifier (op amp) has a differential input and a single-ended output. An *ideal* op amp, used often as a first approximation to model a real op amp, has infinite gain, infinite bandwidth, infinite differential input impedance, infinite slew rate, and infinite **common-mode rejection ratio (CMRR)**. It also has zero output impedance, zero noise, zero bias currents, and zero input offset voltage. Real op amps, of course, fall short of the ideal in all regards.

Important parameters to consider when selecting an op amp include:

1. DC voltage gain K_0.
2. Small-signal **gain-bandwidth product** (GBWP) f_T, which for most op amps is $f_T \approx K_0 f_1$, where f_1 is the lower break frequency in the op amp's transfer function. The GBWP characterizes the closed-loop, high-frequency response of an op-amp circuit.
3. **Slew rate**, which governs the large-signal behavior of an op amp. Slew rates range from less than 1 V/μs to several thousand volts per microsecond.

Other parameters, such as input and output impedances, DC offset voltage, DC bias current, drift voltages and currents, noise characteristics, and so forth, must be considered when selecting an op amp for a particular application.

There are several categories of operational amplifiers. In addition to "garden-variety" op amps there are many op amps whose characteristics are optimized for one or more classes of use. Some categories of op amps include:

1. *Low-noise* op amps, which are useful in the portions of signal conditioners required to amplify very-low-level signals.
2. *Chopper-stabilized* op amps, which are useful in applications requiring extreme DC stability.
3. *Fast* op amps, which are useful when large slew rates and large GBWPs are required.
4. *Power* op amps, which are useful when currents of greater than a few mA must be provided to the op amp's load.
5. *Electrometer* op amps, which are used when very high ($>10^{13}$ Ω) input resistances and very low (<1 pA) input bias currents are required.

An introduction to op amps and basic circuit configurations occurs in essentially any modern text on circuit theory or electronics, and the reader can find detailed theoretical developments and many useful configurations and applications in Roberge (1975), Graeme et al. (1971), Graeme (1973, 1977), Horowitz and Hill (1989), and Stout and Kaufman (1976).

Instrumentation Amplifiers

Instrumentation amplifiers (IAs) are gain blocks optimized to provide high input impedance, low output impedance, stable gain, relatively high **common-mode rejection** (CMR), and relatively low offset and drift. They are well suited for amplification of outputs from various types of transducers such as strain gages, for amplification of low-level signals occurring in the presence of high-level common-mode voltages, and for situations in which some degree of isolation is needed between the transducer and the remainder of the instrument.

Although instrumentation amplifiers can be constructed from conventional op amps [a three-op-amp configuration is typically discussed; see, for example, Stout and Kaufman (1976)], they are readily available and relatively inexpensive in IC form. Some IAs have digitally programmable gains, whereas others are programmable by interconnecting resistors internal to the IA via external pins. More-basic IAs have their gains set by connecting external resistors.

Isolation Amplifiers

Isolation amplifiers are useful in applications in which a voltage or current occurring in the presence of a high common-mode voltage must be measured safely, accurately, and with a high CMR. They are also useful when safety from DC and line-frequency leakage currents must be ensured, such as in biomedical instrumentation.

The isolation amplifier can be thought of as consisting of three sections: an input stage, an output stage, and a power circuit. All isolation amplifiers have their input stages galvanically isolated from their output stages. Communication between the input and output stages is accomplished by modulation/demodulation.

An isolation amplifier is said to provide two-port isolation if there is a DC connection between its power circuit and its output stage. If its power circuit is isolated from its output stage as well as its input stage, then the amplifier is said to provide three-port isolation. Isolation impedances on the order of 10^{10} Ω are not atypical.

Isolation amplifiers are available in modular form with either two-port or three-port isolation. Both single-channel and multichannel modules are offered.

32.1.1.2 Attenuation

Although the majority of transducers are low-level devices such as thermocouples, thermistors, resistance temperature detectors (RTDs), strain gages, and so forth, whose outputs require amplification, there are many measurement situations in which the input signal must be attenuated before introducing it to the remainder of the system.

Voltage Scaling

Most typically, the signals to be attenuated take the form of voltages. Broadly, the attenuation is accomplished by either a *voltage divider* or a *voltage transformer.*

Voltage Dividers

In many cases a simple chain divider proves adequate. The transfer function of a two-element chain of impedances $Z_1(s)$ and $Z_2(s)$ is

$$\frac{V_o(s)}{V_{\text{in}}(s)} = \frac{Z_1(s)}{Z_1(s) + Z_2(s)}$$

where the output voltage $V_o(s)$ is the voltage across $Z_1(s)$ and the input voltage V_{in} is the voltage across the two-element combination.

Of course, the impedances of the source (transducer) and the load (the remainder of the system) must be taken into account when designing the divider network.

Resistive Dividers

If the elements in the chain are resistors, then the divider is useful from DC up through the frequencies for which the impedances of the resistors have no significant reactive components. For $Z_1(s) = R_1$ and $Z_2(s) = R_2$,

$$\frac{V_o(s)}{V_{\text{in}}(s)} = \frac{R_1}{R_1 + R_2}$$

Other configurations are available for resistive dividers. One example is the Kelvin–Varley divider, which has several advantages that make it useful in situations requiring high accuracy. For a detailed description, see Gregory (1973).

Capacitive Dividers

If the elements in the chain divider are capacitors, then the divider has as its transfer function

$$\frac{V_o(s)}{V_{\text{in}}(s)} = \frac{C_2}{C_1 + C_2}$$

This form of divider is useful from low frequencies up through frequencies of several megahertz. A common application is in the scaling of large voltages.

Inductive Dividers

If the elements in the chain divider are inductors, then an autotransformer results. Inductive dividers are useful over frequencies from a few hertz to several hundred kilohertz. Errors in the parts-per-billion range are achievable.

Voltage Transformers

Voltage transformers constitute one of the most common means of accomplishing voltage scaling at line frequencies. Standard double-wound configurations are useful unless voltages above about 200 kV are to be monitored. For very high voltages, alternative configurations such as the *capacitor voltage transformer* and the *cascade voltage transformer* are employed (Gregory, 1973).

Current Scaling

Current scaling is typically accomplished via either a current shunt or a current transformer.

A *current shunt* is essentially an accurately known resistance through which the current to be measured is passed. The voltage developed across the shunt as a result of the current is the quantity measured. Shunts are useful at DC and frequencies through the audio range. Two disadvantages are (1) the shunt consumes power, and (2) the measurement circuitry must be operated at the same potential as the shunt.

The *current transformer* overcomes the mentioned disadvantages of the current shunt. Typically, the current transformer consists of a specially constructed toroidal core upon which the secondary (sense) winding is wrapped and through which the primary winding is passed. A single-turn primary is commonly used, although mutiturn primaries are available.

Other Attenuators

In addition to the aforementioned means of voltage and current scaling are attenuator pads, which provide, in addition to voltage or power reduction, the ability to be matched in impedance to the source and load circuits between which it is connected. The common pads include the T, L, and Π types, either balanced or unbalanced. Resistive attenuator pads are discussed in most textbooks on circuit design (e.g., Cuthbert, 1983). They are useful from DC through several hundred megahertz.

32.1.2 Impedance Transformation

Oftentimes the impedance of the transducer must be transformed to a value more acceptable to the remainder of the measurement system. In many cases maximum power must be transferred from the transducer's output signal to the remaining circuitry. In other cases it is sufficient to provide buffering that presents a very high impedance to the transducer, a very low impedance to the rest of the system, and a voltage gain of unity.

Matching transformers, passive matching networks such as attenuator pads, and unity-gain buffers are standard means of accomplishing impedance transformation. Unity-gain buffers are available in IC form.

32.1.3 Linear Filtering

Although, in general, digital signal processing offers many advantages over analog techniques for filtering signals, there are many relatively simple applications for which *frequency-selective analog filtering* is well suited.

Filters are used within signal conditioners (1) to reduce the effects of noise that corrupts the input signal, (2) as part of a demodulator, (3) to limit signal bandwidth, or (4) if the signal is to be sampled, to limit its bandwidth in order to prevent aliasing. These filters can be built either entirely of passive components or based on active devices such as op amps.

There are many good references that discuss methods of characterizing, specifying, and implementing frequency-selective analog filters. See Van Valkenburg (1960) for design of passive filters; for the design of active-RC filters, see Sedra and Brackett (1978) and Stephenson (1985).

32.2 Nonlinear Operations

There is a wide variety of nonlinear operations useful to signal-conditioning tasks. Listed below are some typical nonlinear blocks along with brief descriptions. Most of the blocks are available as ICs.

1. *Comparator.* A comparator is a two-input device whose output voltage, V_o, takes on one of two stable values, V_{o0} and V_{o1}, as follows:

$$V_o = \begin{cases} V_{o0}, & \text{if } V_2 < V_1 \\ V_{o1}, & \text{otherwise} \end{cases}$$

   where V_1 and V_2 are the voltages at the two inputs.
2. *Schmitt trigger.* A Schmitt trigger is a comparator with hysteresis. It can be constructed from a comparator by applying positive feedback.
3. *Multiplier.* A two-input multiplier supplies an output voltage that is proportional to the product of its input voltages.
4. *Divider.* A two-input divider has as its output a voltage proportional to the ratio of its input voltages. The functions of multiplication and division are usually combined within a single device.
5. *Squarer.* A squarer has as its output a voltage proportional to the square of its input. Squarers can be constructed by a number of means: from multipliers, based on diode-resistor networks, based on FETs, and so forth.
6. *Square-rooter.* A square-rooter has as its output a voltage proportional to the square root of its input. A square-rooter can be built most easily from either a divider or a log/antilog amplifier.
7. *Logarithmic/antilogarithmic amplifier.* A log/antilog amplifier produces an output voltage proportional to the logarithm or the antilogarithm of its input voltage.
8. *True RMS-to-DC converter.* A true RMS-to-DC converter computes the square root of the average, over some interval of time, of the instantaneous square of the input signal. The averaging operation is generally accomplished via a simple low-pass filter whose capacitor is selected to give the desired interval.
9. *Trigonometric function generator.* Generators are available in IC form that produce as their outputs any of the standard trigonometric functions or their inverses, taken as functions of the differential voltage at the generator's inputs.
10. *Sample-and-hold and track-and-hold amplifiers.* A sample-and-hold amplifier (SHA) is a device that samples the signal at its input and holds the instantaneous value whenever commanded by a logic control signal. A track-and-hold amplifier is identical to an SHA but is used in applications where it spends most of its time tracking the input signal (i.e., in "sample" or "track" mode), in contrast to the SHA, which spends most of its time in "hold" mode.
11. *Precision diode-based circuits.* Circuits such as precision half-wave rectifiers, absolute-value circuits, precision peak detectors, and precision limiters are relatively easy to design and implement based on diodes and op amps. See Horowitz and Hill (1989), Stout and Kaufman (1976), and Graeme (1977).

A detailed description of these and other nonlinear circuit blocks can be found in Sheingold (1976).

Example

We provide briefly an example of a device that has embedded within it several signal-conditioning circuits. Figure 32.1 shows the basic block diagram of a therapeutic ultrasound unit, which finds widespread use in physical medicine.

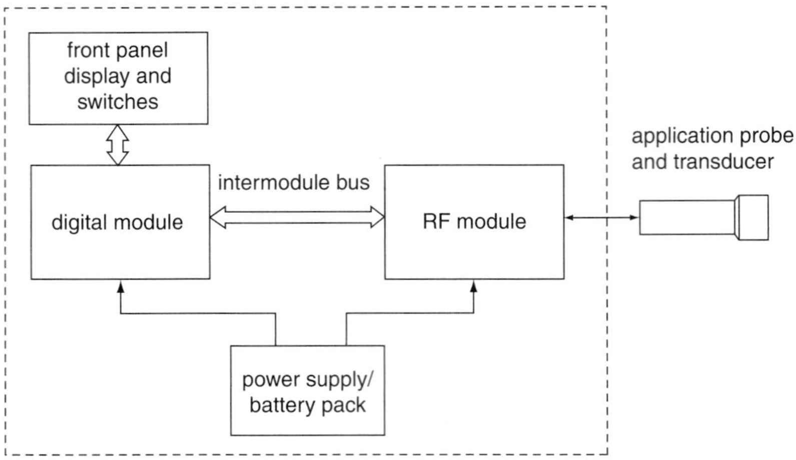

FIGURE 32.1 Basic block diagram of the therapeutic ultrasound unit discussed as an example.

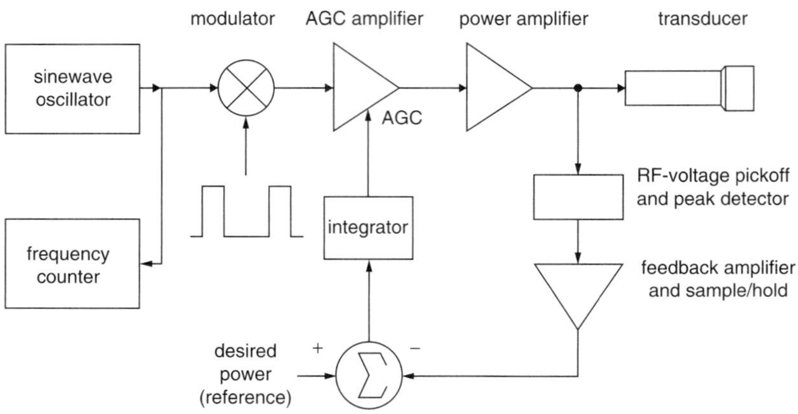

FIGURE 32.2 Simplified block diagram of the RF module used in the ultrasound unit of Figure 32.1.

The particular unit being discussed consists of five principal subsystems:

1. An application probe and ultrasound transducer, which imparts ultrasonic energy to the tissue being treated. *Note that this transducer is NOT an input transducer such as has been discussed in relation to signal conditioners.*
2. A radio-frequency (RF) module, which provides electrical excitation to the ultrasound transducer.
3. Front-panel display and switches, which allow communication between the unit and its operator.
4. A microprocessor-based digital module, which orchestrates the overall control of the ultrasound unit.
5. A power supply/battery pack, which provides operating power to the unit.

We focus now on the RF module, whose basic block diagram is shown in Figure 32.2. The module consists of a sine-wave oscillator that produces a signal at the resonant frequency of the transducer, a modulator that allows that signal to be pulse-modulated, and an amplifier with RF-voltage feedback. Incorporated in the amplifier are a power amplifier capable of driving the transducer and automatic-gain-control (AGC) circuitry required to adjust the output power to coincide with that selected by the operator. The AGC uses a standard feedback-control loop to maintain a constant-voltage envelope on the RF signal output from the power amplifier.

Some of the signal conditioners employed within the RF module include the following:

1. The RF-voltage pickoff at the output of the power amplifier. The pickoff employs a half-wave rectifier, followed by a simple capacitive chain divider for voltage scaling.
2. A precision peak detector, which obtains the peak value of the output from the voltage divider during a modulation cycle and presents that value to the feedback loop.
3. An amplifier, having digitally selectable gain, which amplifies the output of the peak detector.
4. A sample-and-hold amplifier, used to hold the amplified output from the peak detector during the "off-time" of the modulator. The SHA is needed since the time constant of the peak detector is not sufficient to prevent significant "droop" during the off-time of the modulator.
5. An integrator (an example of frequency-selective filtering), which develops the control voltage for the AGC loop from the output of the differencer.
6. A current shunt, not shown in Figure 32.2, which is used to monitor the DC current supplied to the power amplifier.

As can be seen from this simple example, several signal-conditioning functions may be employed within a single system, and the system itself might not even be an instrument!

Defining Terms

Common-mode rejection (CMR): CMRR given in decibels. CMR = 20 log|CMRR|. CMR is a nonlinear function of common-mode voltage and depends on other factors such as temperature.

Common-mode rejection ratio (CMRR): The ratio of the differential gain to the common-mode gain of an amplifier.

Gain-bandwidth product (GBWP): The product of an amplifier's highest gain and its corresponding bandwidth. Used as a rough figure of merit for bandwidth.

Slew rate: The maximum attainable time rate of change of an amplifier's output voltage in response to a large step change in input voltage.

References

Cuthbert, T. R. 1983. *Circuit Design Using Personal Computers.* John Wiley & Sons, New York.

Graeme, J. G. 1973. *Applications of Operational Amplifiers.* McGraw-Hill, New York.

Graeme, J. G. 1977. *Designing with Operational Amplifiers.* McGraw-Hill, New York.

Graeme, J. G., Tobey, G. E., and Huelsman, L. P. (Ed.) 1971. *Operational Amplifiers.* McGraw-Hill, New York.

Gregory, B. A. 1973. *An Introduction to Electrical Instrumentation.* Macmillan, London.

Horowitz, P. and Hill, W. 1989. *The Art of Electronics,* 2nd ed. Cambridge University Press, New York.

Roberge, J. K. 1975. *Operational Amplifiers.* John Wiley & Sons, New York.

Sedra, A. S. and Brackett, P. O. 1978. *Filter Theory and Design: Active and Passive.* Matrix, Beaverton, OR.

Sheingold, D. H. (Ed.) 1976. *Nonlinear Circuits Handbook.* Analog Devices, Norwood, MA.

Stephenson, F. W. 1985. *RC Active Filter Design Handbook.* John Wiley & Sons, New York.

Stout, D. F. and Kaufman, M. (Ed.) 1976. *Handbook of Operational Amplifier Circuit Design.* McGraw-Hill, New York.

Van Valkenburg, M. E. 1960. *Introduction to Modern Network Synthesis.* John Wiley & Sons, New York.

Further Information

IEEE Transactions on Instrumentation and Measurement. Published bimonthly by the Institute of Electrical and Electronics Engineers.

IEEE Transactions on Circuits and Systems—II: Analog and Digital Signal Processing. Published monthly by the Institute of Electrical and Electronics Engineers.

The Best of Analog Dialogue, *1967–1991*. 1991. Analog Devices, Norwood, MA. A collection of practical articles covering circuits, systems, and software for signal processing.

Analog Devices Special Linear Reference Manual and *Analog Devices Amplifier Reference Manual.* Presents an extensive selection of ICs, modules, and subsystems for signal conditioning.

Pallás-Areny, R. and Webster, J. G. 1991. *Sensors and Signal Conditioning.* John Wiley & Sons, New York. Provides an excellent introduction to sensors and signal-conditioning circuits required by them.

Sheingold, D. H. (Ed.) 1980. *Transducer Interfacing Handbook.* Analog Devices, Norwood, MA. Covers signal-conditioning techniques applicable to temperature, pressure, force, level, and flow transducers.

33 Virtual Instrumentation Systems

Darcy Dement
National Instruments, Inc.

33.1 Introduction to Instrumentation

Like instrumentation systems of the past, today's instrumentation systems build on widely accepted contemporary technologies. In the early days of instrumentation, the jeweled movement of clocks in the nineteenth century was first adapted to build analog meters. In the 1930s, when engineers began to accept the variable capacitor, variable resistor, and vacuum tubes as pieces of the radio, the first electronic instruments were introduced using the same components. As display technologies improved for use on the first televisions, oscilloscopes and analyzers began using the same technology to display the user's measurements. These first steps of integrating relevant technologies into instrumentation set the stage for today's virtual instrumentation.

33.2 Approaches to Instrumentation: Virtual and Traditional

Fundamentally, there are two types of instrumentation today: virtual and traditional. Figure 33.1 illustrates the architectures of these approaches.

The diagrams show the similarities in these two approaches. Both have measurement hardware, a chassis, a power supply, a bus, a processor, an OS, and a user interface. Because the approaches use the same basic components, the most obvious difference from a purely hardware standpoint is how the components are packaged. A traditional, or stand-alone, instrument puts all the components in the same

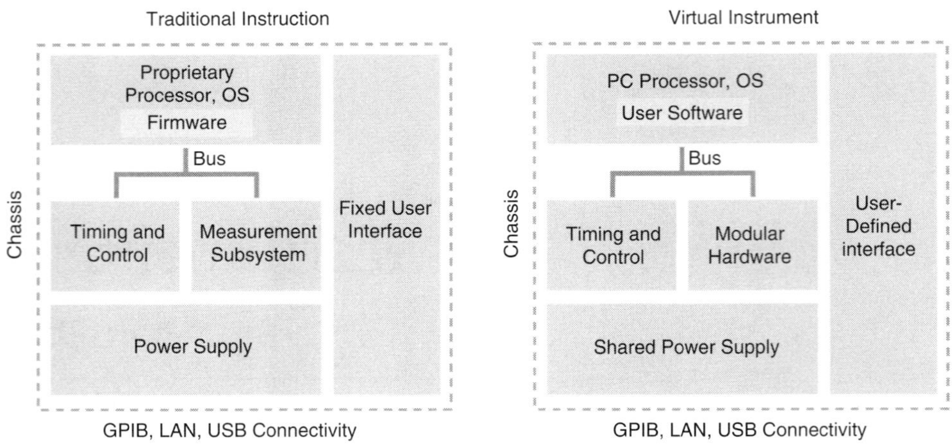

FIGURE 33.1 Traditional versus virtual instrumentation architectures. Both share similar hardware components; the primary difference between the architectures is where the software resides and whether it is user accessible.

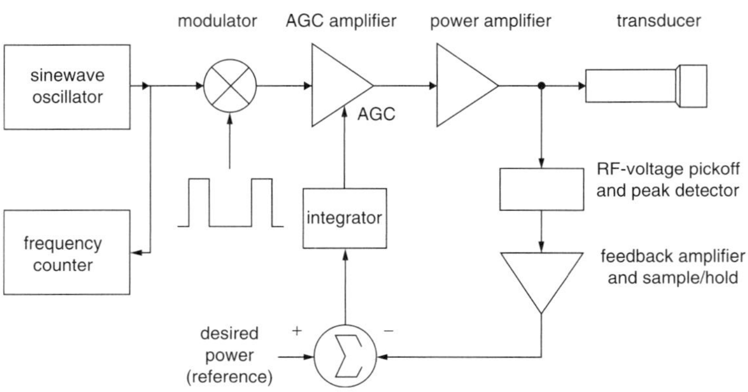

FIGURE 33.2 Example of a virtual instrumentation system using PXI hardware and National Instruments LabVIEW graphical development software.

box for every discrete instrument. When automated, a traditional instrument is often controlled with GPIB, USB, or LAN/Ethernet. While there are a large number of traditional instruments, the software and user interface are fixed and can only be updated when and how the vendor chooses (e.g., through a firmware update). This makes it impossible for the user to perform measurements not included in the traditional instrument's function list, and it makes it challenging to perform measurements for new standards or to modify the system if needs change.

By contrast, a virtual instrument makes the raw data from the hardware available to the user to define their own measurements and user interface. This software-defined approach lets users make custom measurements, perform measurements for emerging standards, or modify the system if requirements change (e.g., to add instruments, channels, or measurements).

A virtual instrument can take several forms. Referring again to the architecture of a virtual instrumentation system in Figure 33.1, many of the components are shared across instrument modules—such as the chassis and power supply—instead of duplicating these components for every instrument function. These instrument modules can also include different types of hardware, including oscilloscopes, function generators, digital, and RF. In some cases, such as PXI (PCI eXtensions for Instrumentation)—a rugged platform for test, measurement, and control supported by over 70 member companies—the measurement hardware is housed in an industrial chassis (see Figure 33.2).

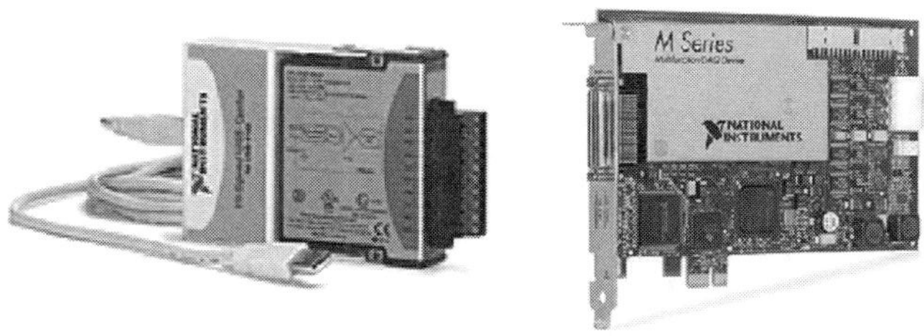

FIGURE 33.3 Examples of other measurement hardware choices for virtual instrumentation. On the left is a USB peripheral module and on the right, a PCI Express plug-in module.

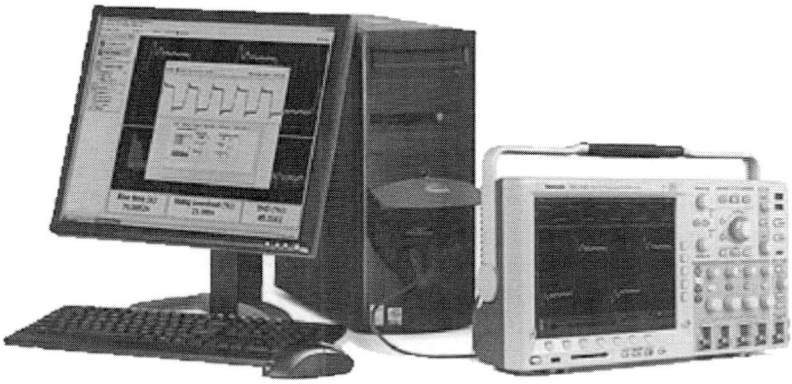

FIGURE 33.4 Virtual instrumentation implementation using a traditional instrument as the measurement hardware.

In those cases, the host computer can be embedded in the chassis (as shown in Figure 33.2) or it can be a separate desktop PC or server-class machine that controls the measurement hardware through a cable. In other cases, as shown in Figure 33.3, the measurement hardware is simply a peripheral that is installed in one of the host computer's peripheral ports or peripheral slots.

A virtual instrument can even use a traditional instrument as the measurement hardware (see Figure 33.4). While modular hardware provides other benefits described in the next section, fundamentally, a virtual instrument system can use any type of instrument as the measurement hardware, provided the user has access to the raw data to define custom measurements. A virtual instrumentation system can also use different measurement hardware—whether PXI, USB, plug-in peripheral, or traditional instrument—in the same system. (This is termed a "hybrid" system.) Strictly speaking, a virtual instrumentation system is defined as follows:

- Virtual instrument system (*n.*): Software-defined system, where software based on user requirements defines the functionality of generic measurement hardware
- Virtual instrument (*n.*): Diminutive form of a virtual instrument system

By any definition, the measurement software that resides on the host computer is fundamental to a virtual instrument. This user-defined software is what converts the raw data into the measurement of interest. It could be a common task such as a simple voltage measurement or a fast Fourier transform, or it could be a custom application such as a flyback measurement, octave analysis, or brain activity mapping.

Note that while the term "computer based" is often applied to virtual instruments, ultimately many modern traditional instruments are also computer based. Modern traditional instruments have all the same components of a PC, including a processor, memory, and a bus to move data internally, as shown in Figure 33.1. Thus, the distinction between virtual and traditional instruments is determined by whether the software is exposed to the user for custom measurements and tasks.

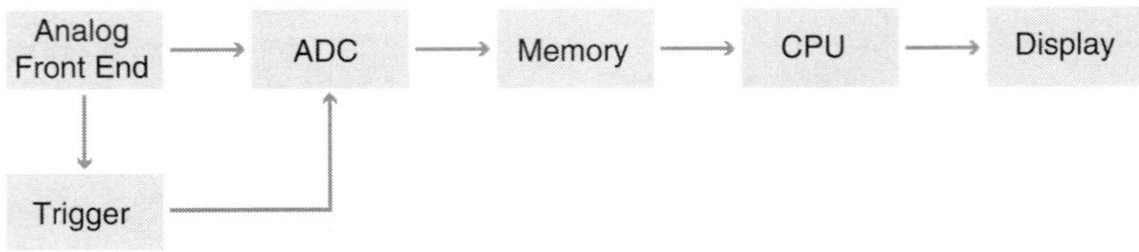

FIGURE 33.5 Typical digitizer architecture. In a virtual instrument implementation, most of the memory, the CPU, and the display are provided by the host PC.

33.3 Modular Hardware for System Scalability

Modular hardware uses commercial semiconductor, processor, and bus technologies to create virtual instruments with high performance at a low cost. The explosion of widely used commercial technologies such as analog-to-digital converters (ADCs), digital-to-analog converters (DACs), field-programmable gate arrays (FPGAs), digital signal processors (DSPs) has resulted in rapid growth of functionality and performance in modular I/O.

The most common instruments in use today digitize an analog signal into bits that can be interpreted by a processor, for example, digitizing a high-frequency signal (as in the case of an oscilloscope) or digitizing a slower signal with higher resolution (as with a digital multimeter). A greatly simplified digitizer architecture is shown in Figure 33.5.

The analog input path attenuates, amplifies, filters, and/or couples the signal to optimize the digitization by the ADC. (For some applications—e.g., with many physical measurements via transducers—additional signal conditioning is housed externally and connected to the digitizer via cable or direct mount.) The ADC samples the conditioned waveform and converts the analog input signal to digital values that represent the conditioned input signal. The data is then stored in onboard memory, where it is transferred across a bus (e.g., PCI Express) to the processor. (In some applications, there can be very little onboard memory if the bus throughput is high enough to stream the data continuously from the measurement hardware to the host PC.) Then, the software (in the CPU for a virtual instrument) makes the proper measurement and presents the data to the user interface, sends it to a database across the network, publishes it to the Internet, or whatever the instrument user chooses.

Analog input is presented here as an example, but many other measurement types can be inferred from this case. For example, engineers can build signal generation (analog output), digital I/O, counter/timers, RF, switching, and other instrument functions using the same virtual approach with different front-end architectures. Because the system is software defined, engineers can combine any of these elements into an integrated system with very little additional work—a unique property of virtual instrumentation. This property also means that the system developer can add new instruments, measurements, or channels over time as requirements change—a concept referred to as "scalability." Although not discussed in detail here, these modular I/O components also typically include timing and synchronization resources for sharing clocks, triggers, and events between hardware I/O modules. Engineers can use these shared resources to create a 1000-channel digitizer, a mixed-signal system with 10 ps skew between modules, or a fast, hardware-handshaked digital multimeter/switch system.

33.4 Software for Flexible, Custom Measurements

The role of software in virtual instrumentation cannot be overstated. Software converts the raw bitstream from hardware into a useful measurement. A well-architected virtual instrumentation system considers multiple layers of software, including I/O drivers, application development, and test management, as shown in Figure 33.6.

The bottom layer, measurement and control services, is one of the most crucial elements of a virtual instrumentation system, though often overlooked. This layer represents the driver I/O software and

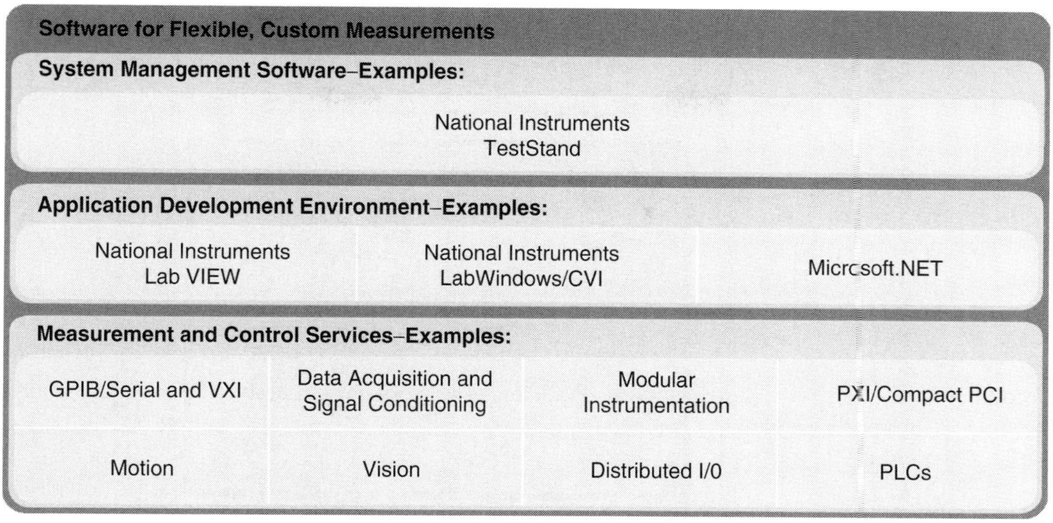

FIGURE 33.6 Software layers in a virtual instrumentation system.

hardware configuration tools. This driver software is critical, because it provides the connectivity between the test development software and the hardware for measurement and control.

Instrument drivers provide a set of high-level, human-readable functions for interfacing with instruments. Each instrument driver is specifically tailored to a particular model of instrument to provide an interface to its unique capabilities. Of particular importance in an instrument driver is its integration with the development environment so that the instrument commands are a seamless part of the application development. System developers need instrument driver interfaces optimized for their development environment of choice, for example, NI LabVIEW, C, C++, or Microsoft .NET.

Also represented in measurement and control services are configuration tools. These configuration tools include resources for configuring and testing I/O, as well as storing scaling, calibration, and channel aliasing information. These tools are important for quickly building, troubleshooting, and maintaining an instrumentation system.

The software in the application development environment layer provides the tools to develop the code or procedure for the application. Although graphical programming is not a requirement of a virtual instrument system, these systems often use graphical tools for their ease of use and rapid development time. Graphical programming uses "icons" or symbolic functions that pictorially represent the action to be performed, as shown in Figure 33.7. These symbols are connected together through "wires" that pass data and determine order of execution. NI LabVIEW provides the industry's most used and most complete graphical development environment.

Some applications also require an additional layer of software management, either for test execution or visibility into test data. These are represented in the system management software layer. For highly automated test systems, test management software provides a framework for sequencing, branching/looping, report generation, and database integration. The test management toolmust also provide tight integration into the development environments where the application-specific code is created. National Instruments TestStand, for example, provides this framework for sequencing, branching, report generation, and database integration and includes connectivity to all common development environments. In other applications that need visibility into large quantities of test data, other tools may be useful. These needs include quick access to large volumes of scattered data, consistent reporting, and data visualization. These software tools aid in managing, analyzing, and reporting data collected during data acquisition and/or generated during simulations.

Every layer of this software architecture should be considered for a virtual instrumentation system.

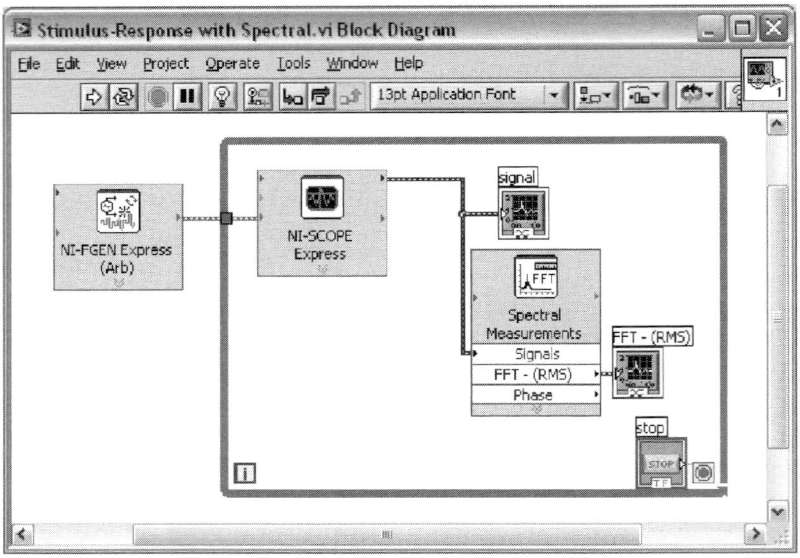

FIGURE 33.7 Code for a typical stimulus/response application using virtual instrumentation. This code is written in National Instruments LabVIEW for (1) generating a signal from an arbitrary waveform generator; (2) acquiring the signal with a digitizer/oscilloscope; (3) performing a fast Fourier transform; and (4) graphing the result of the FFT on the user interface (front panel).

33.5 Extensions beyond Test and Measurement

While virtual instrumentation has roots in test and measurement, these concepts can extend to industrial control and design prototyping. In fact, engineers and scientists who adopt a virtual instrumentation approach see large efficiency gains from using this approach across different types of applications.

In control and industrial applications, PCs and PLCs both play important roles. PCs bring great software flexibility and capability, while PLCs deliver outstanding ruggedness and reliability. A programmable automation controller (PAC) is an example of how virtual instrument techniques can be applied to industrial control; a PAC combines elements of both PCs and PLCs to accommodate more complex, dynamic, adaptive, and algorithm-based control while retaining its ruggedness and reliability. Other industrial-focused hardware might include logic, motion, machine vision, and other functions. In all these cases, the software environment can be the same one used for a virtual instrumentation system for test and measurement, bringing great efficiency gains to applications combining test, measurement, and industrial control.

For design prototyping, virtual instrumentation technology offers an architecture for designing custom measurement hardware circuitry using reconfigurable FPGA chips and NI LabVIEW graphical development tools. This architecture facilitates the customization of data acquisition, communication, and control hardware using the same software tools for test and measurement or industrial control. These users can quickly create custom hardware circuitry with high-performance I/O and unprecedented flexibility in system timing control without actually building custom circuitry.

33.6 Benefits of Virtual Instrumentation Systems

The software-defined nature of virtual instrumentation brings several benefits to engineers and scientists, which are discussed in the following sections.

33.6.1 Adaptability and Longevity

Only with virtual instrumentation can engineers and scientists create the user-defined instruments that nimbly adapt to a changing design or a multifunctional device under test. As the device under test gains more features and new measurements or tasks are added, only a software-defined architecture can adapt at the speed of the device under test itself. Because virtual instrumentation uses highly productive software, modular I/O, and commercial platforms, it is uniquely positioned to keep pace with the product development rate. Engineers can modify the software to add new functionality or support new devices or tests, and the modular I/O hardware can be rapidly combined in any order or quantity to performany I/O task required. These same properties can accommodate a wide variety of test measurement, or control tasks required for a single device or process, or disparate tasks across multiple devices.

33.6.2 Greater Speed

Virtual instrumentation applications that use modular I/O are capable of high-speed measurement and high throughput to PCmemory because of the use of bus and processor technologies. The PCI Express bus, for example, is commonly available today with a throughput of 1 GB/s—1000 times faster than the GPIB bus used to connect most traditional instruments to a PC, 16 times faster than USB 2.0, 80 times faster than 100-Mbps Ethernet, and even 8 times as fast as emerging Gbps Ethernet. For high-streaming applications, PCI Express slots are available today with 4 GB/s bandwidth, and plans exist for higher throughputs in the future. In virtual instrumentation systems, GHz PC processors analyze the data and make measurements using software. The result is measurements at 10–100 times the throughput of a test system built solely on traditional instruments that use built-in vendor-defined firmware and application-specific processors.

33.6.3 Lower Cost and Size

A virtual instrumentation system using modular I/O shares resources across the instruments, such as the processor, the bus, and the memory (see Figure 33.1). By sharing these resources across the instruments in the system, the integrated solution is much smaller and much less expensive than the traditional solution. The benefits of greater throughput, smaller size, and the ability to only change or upgrade specific components of the system when necessary also lead to lower total cost of ownership.

It is worth noting that these benefits do not sacrifice measurement performance. Today, the instruments designed with a virtual instrumentation approach include the industry's highest-resolution digitizer, the highest-bandwidth arbitrary waveform generator, and the most accurate 71/2-digit digital multimeter.

Further Reading

Allan Hambley, *Electrical Engineering: Principles and Applications*, 3rd Edn., Upper Saddle River, NJ: Prentice Hall (copyright 2005).

Gary W. Johnson and Richard Jennings, *LabVIEW Graphical Programming*, 4th Edn., New York:McGraw-Hill (copyright 2007).

David G. Alciatore and Michael B. Histand, *Introduction to Mechatronics and Measurement Systems*, 3rd Edn., New York: McGraw-Hill (copyright 2007).

34

Software Design and Development*

Margaret H. Hamilton
Hamilton Technologies, Inc.

A software-based system can be neatly compared with a biological entity called a superorganism. Comprising software, hardware, peopleware and their interconnectivity (such as the Internet), and requiring all to survive, the silicon superorganism is itself a part of a larger superorganism—for example, a medical system including patients, drugs, drug companies, doctors, hospitals, and health care centers; a space mission including the spacecraft, the laws of the universe, mission control, and the astronauts; a system for researching genes including funding organizations, funds, researchers, research subjects, and genes; a financial system including investors, money, politics, financial institutions, stock markets, and the health of the world economy; or it could be just the business itself.

Whether that business be government, academic, or commercial, the software-based system, like its biological counterpart, must grow and adapt to meet rapidly changing requirements. And, like other organisms, the business has both physical infrastructure and operational policies, which guide and occasionally constrain its direction and the rate of evolution, which it can tolerate without becoming dysfunctional.

Compared to a biological superorganism, which may take many generations to effect even a minor hereditary modification, software can be modified immediately. This makes it far superior in this respect to the biological entity in terms of its evolutionary adaptability. Continuity of business rules and/or the physical infrastructure provides a natural tension between "how fast the software can change" and "how rapidly the overall system can accept change." Software, the brain of the silicon superorganism, controls the action of the entire entity. Keep in mind, however, it was a human being that created the software.

In this chapter we will discuss the tenets of software, what it is and how it is developed, as well as the precepts of software engineering, which are the methodologies by which ideas are turned into software.[†]

* Parts of this chapter were taken from *Object Thinking: Development Before the Fact*, M. H. Hamilton and W. R. Hackler, in press.

[†] 001, 001 Tool Suite, DBTF, Development Before the Fact, SOO, and System Oriented Objects are all trademarks of Hamilton Technologies, Inc.

34.1 The Notion of Software

Software is the embodiment of logical processes, whether in support of business functions or in control of physical devices. The nature of software as an instantiation of process can apply very broadly, when modeling complex organizations, or very narrowly as when implementing a discrete numerical algorithm. In the former case, there can be significant linkages between reengineering businesses to accelerate the rate of evolution—even to the point of building operational models, which then transition into application suites and thence into narrowly focused implementations of algorithms, as above. Software thus has a potentially wide range of application, and that well designed has a potentially long period of utilization.

While some would define software as solely the code that a programming language generates from the compilation process, a broader and more precise definition includes requirements, specifications, designs, program listings, documentation, procedures, rules, measurements, and data as well as the tools used to create, test, optimize, and implement the software.

That there is more than one definition of software is a direct result of the confusion about the very process of software development itself. A 1991 study by the Software Engineering Institute (SEI) [1] amplifies this rather startling problem. SEI developed a methodology for classifying an organization's "software process maturity" into one of five levels which range from Level 1, the initial level (where there is no formalization of the software process), to Level 5, the optimizing level where methods, procedures, and metrics are in place with a focus toward continuous improvement in software reliability. The result of this study showed that fully 86% of organizations surveyed in the United States were at Level 1 where the terms "ad-hoc," "dependent on heroes," and "chaotic" are commonly applied. And, given the complexity of today's Internet-based applications, it would not be surprising to see the percentage of Level 1 organizations increase.

Creating order from this chaos requires an insightful understanding into the component parts of software as well as the development process. Borrowing again from the world of natural science, an entelechy is something complex that emerges when a large number of simple objects are put together. For example, one molecule of water is rather boring in its utter lack of activity. But pour a bunch of these molecules into a glass and there is a ring of ripples on the water's surface. If many of these molecules are combined, the result is an ocean. So too software. By itself, a line of code is a rather simple item. But combine many lines and the result is a complex program. Add additional programs and the result could be a system that can put a person on the moon.

Although the whole is indeed bigger than the sum of its parts, one must still understand those parts if the whole is to work in an orderly and controlled fashion. Like a physical entity, software can "wear" as a result of maintenance, changes in the underlying system, and updates made to accommodate the requirements of the ongoing user community. Entropy is a significant phenomenon in software, especially for Level 1 organizations.

Software at the lowest programming level is termed source code. This differs from executable code (i.e., which can be executed by the hardware to perform one or more specified functions) in that software is written in one or more programming languages and cannot, by itself, be executed by the hardware. A programming language is a set of words, letters, numerals, and abbreviated mnemonics, regulated by a specific syntax, used to describe a program to a computer. There are a wide variety of programming languages, many of them tailored for a specific type of application. C, one of today's more popular programming languages, is used in engineering as well as business environments while object-oriented languages such as C++ [2] and Smalltalk have been gaining acceptance in both of these environments. More recently, Java [3] has been gaining acceptance. In fact, it has become a language of choice for Internet-based applications. In the recent past, engineering applications have often used programming languages such as FORTRAN, HAL (or HAL/s) for NASA space applications, and Ada for government applications while commercial business applications have favored COBOL (COmmon Business Oriented Language). For the most part one finds that in any given organization there are no prescribed rules, other than those related to what is most popular at the moment, which dictate which languages are to be used. And, as one might expect, a wide diversity of languages is being deployed.

The programming language, whether it be C++, Java, Visual BASIC, C, FORTRAN, HAL/s, COBOL, or something else, provides the capability to code such logical constructs as that having to do with:

- *User Interface.* Provides a mechanism whereby the ultimate end-user can input, view, manipulate, and query information contained in an organization's computer systems. Studies have shown that productivity increases dramatically when visual user interfaces are provided. Known as GUIs (graphical user interfaces), each operating system provides its own variation. Some common graphical standards are Motif for UNIX systems and Microsoft Windows for PC-based systems.
- *Model Calculations.* Perform the calculations or algorithms (step-by-step procedures for solving a problem) intended by a program, e.g., process control, payroll calculations, or a Kalman filter.
- *Program Control.* Exerts control in the form of comparisons, branching, calling other programs, and iteration to carry out the logic of the program.
- *Message Processing.* There are several varieties of message processing. Help-message processing is the construct by which the program responds to requests for help from the end-user. Error-message processing is the automatic capability of the program to notify and then recover from an error during input, output, calculations, reporting, communications, etc. And, in object-oriented development environments, message processing implies the ability of program objects to pass information to other program objects.
- *Moving Data.* Programs store data in a data structure. Data can be moved between data structures within a program, moved from an external database or file to an internal data structure or from user input to a program's internal data structure. Alternatively, data can be moved from an internal data structure to a database or even to the user interface of an end-user. Sorting and formatting are data moving operations used to prepare the data for further operations.
- *Database.* A collection of data (objects[1]) or information about a subject or related subjects, or a system (for example, an engine in a truck or a personnel department in an organization). A database can include objects, such as forms and reports or a set of facts about the system (for example, the information in the personnel department needed about the employees in the company). A database is organized in such a way so as to be easily accessible to computer users. Its data is a representation of facts, concepts, or instructions in a manner suitable for processing by computers. It can be displayed, updated, queried, printed, and reports can be produced from it. A database can organize data in several ways including in a relational, hierarchical, network, or object-oriented format.
- *Data Declaration.* It describes data and data structures to a program. An example would be associating a particular data structure with its type (for example, data about a particular employee might be of type person).
- *Object.* A person, place, or thing, which could be physical or abstract. An object contains other more primitive objects (or data) and a set of operations to manipulate objects (or data). When brought to life, it knows things (called attributes) and can do things (to change itself or interact with other objects). For example, in a robotics system a robot object may contain the functions to move its own armature to the right, while it is coordinating with another robot to transfer yet another object. Objects can communicate with each other through a communications medium (e.g., message passing, radio waves, Internet).
- *Real-Time.* A software system that satisfies critical timing requirements. The correctness of the software depends on the results of computation, as well as on the time at which the results are produced. Real-time systems can have varied requirements such as performing a task within a specific deadline and processing data in connection with another process outside of the computer.

*Data and object are used interchangeably throughout this chapter to define information in a software program.

Applications such as transaction processing, avionics, interactive office management, automobile systems, and video games are examples of real-time systems.

- *Distributed.* Any system in which a number of independent, interconnected processes can cooperate. The client/server model is one of the most popular forms of distribution in use today. In this model, a client initiates a distributed activity and a server carries out that activity.
- *Simulation.* The representation of selected characteristics of the behavior of one physical or abstract system by another system. For example, a software program can simulate an airplane or an organization or another software program.
- *Documentation.* It includes the description of requirements, specification, and design as well as written or generated documentation, which describe how each program within the larger system operates and can be used; and comments which describe the operation of the program that are stored internally in the program.
- *Tools.* The software programs used to design, develop, test, analyze, or maintain system designs or another software program and its documentation. They include code generators, compilers, editors, database management systems (DBMS), GUI builders, debuggers, operating systems, and software development and systems engineering tools referred to in the 1990s as computer-aided software engineering (CASE) tools and now often referred to as life-cycle design or life-cycle development environments, which combine a set of tools, including some of those listed above.

Although the reader should by now understand the dynamics of a line of source code, where that line of source code fits into the superorganism of software is dependent upon many variables. This includes the industry the reader hails from as well as the software development paradigm used by the organization.

As a base unit, a line of code can be joined with other lines of code to form many things. In a traditional software environment many lines of code form a program, sometimes referred to as an application program or just plain application. But lines of source code by themselves cannot be executed. First, source code must be run through what is called a compiler to create an object code. Next, the object code is run through a linker which is used to construct an executable code. Compilers are programs themselves. Their function is twofold. The compiler first checks the source code for obvious syntax errors and then, if it finds none, creates object code for a specific operating system. UNIX, Linux (a spinoff of UNIX), and NT are all examples of operating systems. An operating system can be thought of as a supervising program that controls the application programs that run under its control. Since operating systems (as well as computer architectures) can be different from each other, the object code resulting from the source code compiled for one operating system cannot be executed under a different kind of operating system—without a recompilation.

Solving a complex business or engineering problem often requires more than one program. One or more programs that run in tandem to solve a common problem is known collectively as a system. The more modern technique of object-oriented development dispenses with the notion of the program altogether and replaces it with a classification-oriented concept of an object.

Where a program can be considered a critical mass of code, which performs many functions in the attempt to solve a problem with little consideration for object boundaries, an object is associated with the code to solve a particular set of functions having to do with just that type of object. By combining objects, like molecules, it is possible to create more organized systems than those created by traditional means. Software development becomes a speedier and less error-prone process as well. Since objects can be reused, once tested and implemented, they can be placed in a library for other developers to reuse. The more objects in the library, the easier and quicker it is to develop new systems. And since the objects being reused have, in theory, already been warranted (i.e., they've been tested and made error-free), there is less possibility that object-oriented systems will have major defects.

The process of writing programs and/or objects is known as software development, or software engineering. It is composed of a series of steps or phases, collectively referred to as a development life cycle. The phases include (at a bare minimum) the following: an analysis or requirements phase, where the business problem is dissected and understood; a specification phase, where decisions are made as to how the requirements will be fulfilled (e.g., deciding what functions are allocated to software and what functions are allocated to hardware); a design phase, where everything from the GUI to the database to the output is designed or selected as part of a design; an implementation or programming phase, where one or more tools are used to write and/or generate code; a testing (debugging) phase, where the code is tested against a business test case and errors in the program are found and corrected; an installation phase, where the systems are placed in production; and a maintenance phase, where modifications are made to the system. But different people develop systems in different ways. These different paradigms make up the opposing viewpoints of software engineering.

34.2 The Nature of Software Engineering

Engineers often use the term "systems engineering" to refer to the tasks of specifying, designing, and simulating a non-software system such as a bridge or electronic component. Although software may be used for simulation purposes, it is but one part of the systems engineering process. Software engineering, on the other hand, is concerned with the production of nothing but software.

In the 1970s industry pundits began to notice that the cost of producing large-scale systems was growing at a high rate and that many projects were failing or, at the very least, resulting in unreliable products. Dubbed the software crisis, its manifestations were legion and the most important include the following:

- *Programmer Productivity.* In government in the 1980s, an average developer using C was expected to produce 10 lines of code per day (an average developer within a commercial organization was expected to produce 30 lines a month); today the benchmark in the government is more like 2 to 5 lines a day while at the same time the need is dramatically higher than that, perhaps by several orders of magnitude, ending up with a huge backlog. Programmer productivity is dependent upon a plethora of vagaries—from expertise to complexity of the problem to be coded to the size of the program that is generated. The science of measuring the productivity of the software engineering process is called metrics. Just as there are many diverse paradigms in software engineering itself, there are many paradigms of software measurement. Today's metric formulas are complex and often take into consideration the following: cost, time to market, productivity on prior projects, data communications, distributed functions, performance, heavily used configuration, transaction rate, online data entry, end-user efficiency, online update, complex processing, reusability, installation ease, operational ease, and multiplicity of operational sites.
- *Defect Removal Costs.* The same variables that affect programmer productivity affect the cost of "debugging" the programs and/or objects generated by those programmers. It has been observed that the testing and correcting of programs consumes a large share of the overall effort.
- *Development Environment.* Development tools and development practices greatly affect the quantity and quality of software. Most of today's design and programming environments contain only a fragment of what is really needed to develop a complete system. Life-cycle development environments provide a good example of this phenomena. Most of these tools can be described either as addressing the upper part of the life cycle (i.e., they handle the analysis and design) or the lower part of the life cycle (i.e., they handle code generation). There are few integrated tools on the market (i.e., that seamlessly handle both upper and lower functionalities). There are even fewer tools that add simulation, testing, and cross-platform generation to the mix. Rare are the tools that seamlessly integrate system design to software development.
- *GUI Development.* Developing GUIs is a difficult and expensive process unless the proper tools are used. The movement of systems from a host-based environment to the workstation and/or PC saw the entry of countless GUI development programs onto the marketplace. But the vast majority of these GUI-based tools do not have the capability of developing the entire system (i.e., the processing component as opposed to merely the front-end). This leads to fragmented and error-prone systems. To be efficient, the GUI builder must be well integrated into the software development environment.

The result of these problems is that most of today's systems require more resources allocated to maintenance than to the original development of that system. Lientz and Swanson [4] demonstrate that the problem is, in fact, larger than the one originally discerned during the 1970s. Software development is indeed complex, and the limitations on what can be produced by teams of software engineers given finite amounts of time, budgeted dollars, and talent have been amply documented by Jones [5].

Essentially the many paradigms of software engineering attempt to rectify the causes of declining productivity and quality. Unfortunately, this fails because current paradigms treat symptoms rather than the root problem. In fact, software engineering is itself extremely dependent upon both the software and hardware as well as the business environments upon which they sit [6].

SEI's process maturity grid very accurately pinpoints the root of most of our software development problems. The fact that a full 86% of organizations studied remain at the ad hoc or chaotic level indicate that only a few organizations (the remaining 14%) have adopted any formal process for software engineering. Simply put, 86% of all organizations react to a business problem by just writing codes. If they do employ a software engineering discipline, in all likelihood it is one that no longer fits the requirements of the ever-evolving business environment.

In the 1970s, the "structured methodology" was popularized. Although there were variations on the theme (i.e., different versions of the structured technique included the popular Gane–Sarson method and Yourdon method), for the most part, it provided a methodology to develop usable systems in an era of batch computing. In those days, online systems with even the dumbest of terminals were a radical concept and GUIs were as unthinkable as the fall of the Berlin Wall.

Although times have changed and today's hardware is one thousand times more powerful than when structured techniques were introduced, this technique still survives. And it survives in spite of the fact that the authors of these techniques have moved on to more adaptable paradigms, and more modern software development and systems engineering environments have entered the market.

In 1981, Finkelstein and Martin popularized "information engineering" [7] for the more commercially oriented users (i.e., those whose problems to be solved tended to be more database centered) which, to this day, is quite popular among mainframe developers with an investment in CASE strategies of the 1990s. Information engineering is essentially a refinement of the structured approach. However, instead of focusing on the data so preeminent in the structured approach, information engineering focuses on the information needs of the entire organization. Here business experts define high-level information models, as well as detailed data models. Ultimately, the system is designed from these models.

Both structured and information engineering methodologies have their roots in mainframe-oriented commercial applications. Today's migration to client/server technologies (where the organization's data can be spread across one or more geographically distributed servers while the end-user uses his or her GUI of choice to perform local processing), disables most of the utility of these methodologies. In fact, many issues now surfacing in more commercial applications are not unlike those that needed to be addressed earlier in the more engineering-oriented environments such as telecommunications and avionics.

Client/server environments are characterized by their diversity. One organization may store its data on multiple databases, program in several programming languages, and use more than one operating system, and hence, different GUIs. Since software development complexity is increased 100-fold in this new environment, a better methodology is required. Today's object-oriented techniques solve some of the problems. Given the complexity of the client/server environment, code trapped in programs is not flexible enough to meet the needs of this type of environment. We have already discussed how coding via objects rather than large programs engenders flexibility as well as productivity and quality through reusability. But object-oriented development is a double-edged sword.

While it is true that to master this technique is to provide dramatic increases in productivity, the sad fact of the matter is that object-oriented development, if done inappropriately, can cause problems far greater than problems generated from structured techniques. The reason for this is simple. The stakes are higher. Object-oriented environments are more complex than any other, the business problems chosen to be solved by object-oriented techniques are far more complex than other types of problems, and there are few if any conventional object-oriented methodologies and corollary tools to help the development team develop good systems. There are many flavors of object orientation. But with this diversity comes some very real risks. As a result, the following developmental issues must be considered before the computer is even turned on.

- Integration is a challenge and needs to be considered at the onset. With traditional systems, developers rely on mismatched modeling methods to capture aspects of even a single definition. Whether it be integration of object to object, module to module, phase to phase, or type of application to type of application, the process can be an arduous one. The mismatch of products used in design and development compounds the issue. Integration is usually left to the devices of myriad developers well into development. The resulting system is sometimes hard to understand and objects are difficult to trace. The biggest danger is there is little correspondence to the real world. Interfaces are often incompatible and errors usually propagate throughout development. As a result, systems defined in this manner can be ambiguous and just plain incorrect.
- Errors need to be minimized. Traditional methods including those that are object oriented can actually encourage the propagation of errors, such as propagating errors through the reuse of objects with embedded and inherited errors throughout the development process. Errors must be eliminated from the very onset of the development process before they take on a life of their own.
- Languages need to be more formal.*[2] Although some languages are formal and others are friendly, it is hard to find languages both *formal* and *friendly.* Within environments where more informal approaches are used, lack of traceability and an overabundance of interface errors are a common occurrence. Recently, more modern software requirements languages have been introduced (for example, the Unified Modeling Language, UML [8]), most of which are informal (or semi-formal); some of these languages were created by "integrating" several languages into one. Unfortunately, the bad comes with the good—often, more of what is not needed and less of what is needed; and since the formal part is missing, common semantics need to exist to reconcile differences and eliminate redundancies.
- The syndrome of locked-in design needs to be eliminated. Often, developers are forced to develop in terms of an implementation technology that does not have an open architecture, such as a specific database schema or a GUI. Bad enough is to attempt an evolution of such a system; worse yet is to use parts of it as reusables for a system that does not rely on those technologies. Well thought-out and formal business practices and their implementation will help minimize this problem within an organization.
- Flexibility for change and handling the unpredictable must be dealt with up front. Too often it is forgotten that the building of an application must take into account its evolution. Users change their minds, software development environments change, and technologies change. Definitions of requirements in traditional development scenarios concentrate on the application needs of the user, but without consideration of the potential for the user's needs or environment to change. Porting to a new environment becomes a new development for each new architecture, operating system, database, graphics environment, or language. Because of this, critical functionality is often avoided for fear of the unknown, and maintenance, the most risky and expensive part of a system's life cycle, is left unaccounted for during development. To address these issues, tools and techniques must be used to allow cross technology and changing technology, as well as provide for changing and evolving architectures.
- Developers must prepare ahead of time for parallelism and distributed environments. Often, when it is known that a system is targeted for a distributed environment, it is first defined and developed for a single processor environment and then redeveloped for a distributed environment—an unproductive use of resources. Parallelism and distribution must be dealt with at the very start of the project.

---

*See Defining Terms for definition of formal.

- Resource allocation should be transparent to the user. Whether or not a system is allocated to distributed, asynchronous, or synchronous processors and whether or not two or ten processors are selected, with traditional methods, it is still up to the designer and developer to be concerned with incorporating such detail into the application. There is no separation between the specification of what the system is to do vs. how the system does it. This results in far too much implementation detail to be included at the level of design. Once such a resource architecture becomes obsolete, it is necessary to redesign and redevelop those applications which have old designs embedded within them.
- Automation that minimizes manual work needs to replace "make work" automated solutions. In fact, automation itself is an inherently reusable process. If a system does not exist for reuse, it certainly does not exist for automation. But most of today's development process is needlessly manual. Today's systems are defined with insufficient intelligence for automated tools to use them as input. In fact, automated tools concentrate on supporting the manual process instead of doing the real work. Typically, developers receive definitions, which they manually turn into code. A process that could have been mechanized once for reuse is performed manually again and again. Under this scenario, even when automation attempts to do the real work, it is often incomplete across application domains or even within a domain, resulting in incomplete code such as shell code. The generated code is often inefficient or hardwired to a particular kind of algorithm, an architecture, a language, or even a version of a language. Often partial automations need to be integrated with incompatible partial automations or manual processes. Manual processes are needed to complete unfinished automations.
- Run-time performance analysis (decisions between algorithms or architectures) should be based on formal definitions. Conventional system definitions contain insufficient information about a system's run-time performance, including that concerning the decisions between algorithms or architectures. System definitions must consider how to separate the system from its target environment. Design decisions, where this separation is not taken into account, thus depend on analysis of results from ad hoc "trial and error" implementations and associated testing scenarios.
- The creation of reliable reusable definitions must be promoted, especially those that are inherently provided. Conventional requirements definitions lack the facilities to help find, create, use, and ensure commonality in systems. Modelers are forced to use informal and manual methods to find ways to divide a system into components natural for reuse. These components do not lend themselves to integration and, as a result, they tend to be error-prone. Because these systems are not portable or adaptable, there is little incentive for reuse. In conventional methodologies, redundancy becomes a way of doing business. Even when methods are object oriented, developers are often left to their own devices to explicitly make their applications object oriented. This is because these methods do not support all that is inherent to the process of object orientation.
- Design integrity is the first step to usable systems. Using traditional methods, it is not known if a design is a good one until its implementation has failed or succeeded. Usually, a system design is based on short-term considerations because knowledge is not reused from previous lessons learned. Development, ultimately, is driven towards failure. The solution is to have an inherent means to build reliable, reusable definitions.

Once these issues are addressed, software will cost less and take less time to develop. But time is of the essence. These issues are becoming compounded and even more critical as developers prepare for the distributed environments that go hand in hand with the increasing predominance of Internet applications.

With respect to the challenges described above, an organization has several options, ranging from one extreme to the other. The options include: (1) keep things the same; (2) add tools and techniques that support business as usual, but provide relief in selected areas; (3) bring in more modern but traditional tools and techniques to replace existing ones; (4) use a new paradigm with the most advanced tools and techniques that formalizes the process of software development, while at the same time capitalizing on software already developed; or (5) completely start over with a new paradigm that formalizes the process of software development and uses the most-advanced tools and techniques.

34.3 Development before the Fact

Thus far, this chapter has explained the derivation of software and attempted to show how it has evolved over time to become the true "brains" of any automated system. But, like a human brain, this software brain must be carefully architected to promote productivity, foster quality, and enforce control and reusability.

Traditional software engineering paradigms fail to see the software development process from the larger perspective of the superorganism described at the beginning of this chapter. It is only when we see the software development process as made of discrete, but well-integrated, components can we begin to develop a methodology that can produce the very benefits that have been promised by the advent of software decades ago.

Software engineering, from this perspective, consists of a methodology as well as a series of tools with which to implement the solution to the business problem at hand. But even before the first tool can be applied, the software engineering methodology must be deployed to assist in specifying the requirements of the problem. How can this be accomplished successfully in the face of the issues needed to be addressed outlined in the last section? How can this be accomplished in situations where organizations must develop systems that run across diverse and distributed hardware platforms, databases, programming languages, and GUIs when traditional methodologies make no provision for such diversity? And how can software be developed without having to fix or "cure" those myriad of problems, which result "after the fact" of that software's development?

What is required is a radical revision of the way we build software, an approach that understands how to build systems using the right techniques at the right time. First and foremost, it is a preventative approach. This means it provides a framework for doing things right the first time. Problems associated with traditional methods of design and development are prevented "before the fact" just by the way a system is defined. Such an approach would concentrate on preventing problems of development from even happening rather than letting them happen "after the fact," and fixing them after they have surfaced at the most inopportune and expensive point in time.

Consider such an approach in its application to a human system. To fill a tooth before it reaches the stage of a root canal is curative with respect to the cavity, but preventive with respect to the root canal. Preventing the cavity by proper diet prevents not only the root canal, but the cavity as well. To follow a cavity with a root canal is the most expensive alternative, to fill a cavity on time is the next most expensive, and to prevent these cavities in the first place is the least expensive option.

Preventiveness is a relative concept. For any given system, be it human or software, one goal is to prevent, to the greatest extent and as early as possible, anything that could go wrong in the life cycle process.

With a preventative philosophy, systems would be carefully constructed to minimize development problems from the very outset. A system could be developed with properties that controlled its very own design and development. One result would be reusable systems that promote automation. Each system definition would model both its application and its life cycle with built-in constraints—constraints that protect the developer, but yet do not take away his flexibility.

The philosophy behind preventative systems is that reliable systems are defined in terms of reliable systems. Only reliable systems are used as building blocks, and only reliable systems are used as mechanisms to integrate these building blocks to form a new system. The new system becomes reusable for building other systems.

Effective reuse is a preventative concept. That is, reusing something (e.g., requirements or code) that contains no errors to obtain a desired functionality avoids both the errors and the cost of developing a new system. It allows one to solve a given problem as early as possible, not at the last moment. But to make a system truly reusable, one must start not from the customary end of a life cycle, during the implementation or maintenance phase, but from the very beginning.

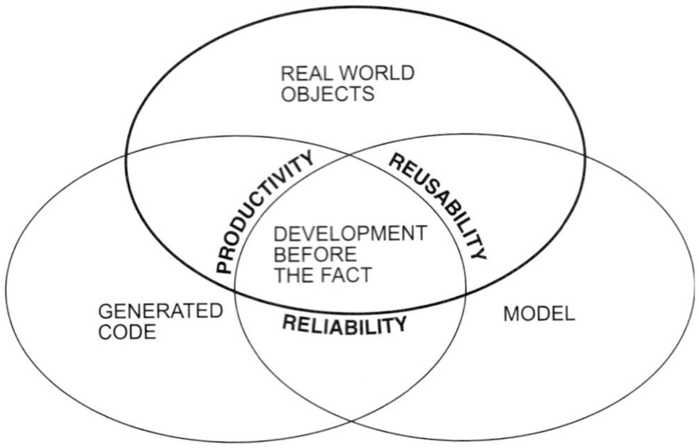

FIGURE 34.1 The development before the fact paradigm.

Preventative systems are the true realization of the entelechy construct where molecules of software naturally combine to form a whole much greater than the sum of its parts. Or one can think of constructing systems from the tinker toys of our youth. One recalls that the child never errs in building magnificent structures from these tinker toys. Indeed, tinker toys are built from blocks that are architected to be perpetually reusable, perfectly integratable, and infinitely user-friendly.

One approach that follows this preventative philosophy is development before the fact (DBTF), as shown in Figure 34.1. Not yet in the mainstream, it has been used successfully by research and "trail blazer" organizations and is now being adopted for more commercial use. This technology is described in order to illustrate, by example, the potential that preventative approaches have.

Where traditional approaches begin the process of developing software after the fact, the DBTF paradigm is very much about beginnings. It was derived from the combination of steps taken to solve the problems of traditional systems engineering and software development. DBTF includes a technology, a language, and a process (or methodology) based on a formal theory.

34.3.1 Language

Once understood, the characteristics of good design can be reused by incorporating them into a language for defining any system (i.e., not just a software system). One language based on DBTF is a formalism for representing the mathematics of systems. A system defined with this language has properties that come along "for the ride" that in essence control its own destiny. Based on a theory (DBTF) that extends traditional mathematics of systems with a unique concept of control, this formal, but friendly language has embodied within it a natural representation of the physics of time and space. With this language, every object is a system-oriented object (SOO), an integration that includes aspects of being function oriented (including dynamics) and object oriented. Instead of systems being object oriented, objects are systems oriented. All systems are objects and all objects are systems.

Because of this, many things heretofore not believed possible with traditional methods are possible. A DBTF system inherently integrates all of its own objects (and all aspects, relationships, and viewpoints of these objects) and the combinations of functionality, including timing, using these objects; maximizes its own reliability and flexibility to change (including the change of target requirements, static and dynamic architectures, and processes and as well reconfiguration in real time); capitalizes on its own parallelism and traceability; supports its own run-time performance analysis; and maximizes the potential for its own reuse (providing inherent resource allocation and reuse without need for the designer's intervention); and it provides the ability to automate design and development wherever and whenever possible. Each DBTF system is defined with built-in quality, built-in productivity, and built-in control.

The language—meta-language, really—is the key to DBTF. Its main attribute is to help the designer reduce the complexity and bring clarity into his thinking process, turning it into the ultimate reusable, which is wisdom itself. It can be used to define any aspect of any system and integrate it with any other aspect. The crucial point is that these aspects are directly related to the real world and, therefore the same language can be used to define system requirements, specifications, design, and detailed design for functional, resource, and resource allocation architectures throughout all levels and layers of seamless definition, including hardware, software, and peopleware.

This language based on DBTF can be used to define organizations of people, missile or banking systems, cognitive systems, as well as real-time or database environments and is, therefore, appropriate across industries, academia, or government.

34.3.2 Technology

Real-world experience sets the stage for the DBTF technology. Having evolved over three decades, the theory has roots in the worlds of systems theory, formal methods, and object technology. The DBTF technology embodies the theory, the language supports its representation, and its automation supports its application and use. Each is evolutionary (in fact, recursively so), with experience feeding the theory and the theory feeding the language, which in turn feeds the automation. All are used, in concert, to design systems and build software.

The DBTF approach had its beginnings in 1968 with an empirical analysis of the Apollo space missions. A better way was needed to define and develop systems than the ones being used and available because the existing ones (just like the traditional ones today) did not solve the pressing problems. Research for developing software for man-rated missions led to the finding that interface errors accounted for approximately 75% of all errors found in the flight software during final testing (in traditional development, the figure is as high as 90%). Such errors include data flow, priority, and timing errors from the highest levels of a system to the lowest level of detail. Each error was categorized according to how it could be prevented just by the way a system is defined. This work led to a theory and methodology for defining a system that would eliminate all interface errors.

The first technology derived from this theory concentrated on defining and building reliable systems. Having realized the benefits of addressing one major issue, such as reliability, research continued to evolve by addressing other major issues the same way, that is, just by the way a system is defined [9–11].

DBTF is a function- and object-oriented approach based on a unique concept of control, which is lacking in any other software engineering paradigm. The foundations are based on a set of axioms and on the assumption of a universal set of objects. Each axiom defines a relation of immediate domination. The union of the relations defined by the axioms is control. Among other things, the axioms establish the relationships of an object for invocation, input and output, input and output access rights, error detection and recovery, and ordering during its developmental and operational states. Table 34.1 summarizes some of the properties of objects within DBTF systems.

34.3.3 Process

Where software engineering fails is in its inability to grasp that not only the right paradigm (out of many paradigms) must be selected, but that the paradigm must be part of an environment that provides an integrated automated means to solve the problem at hand. What this means is that the paradigm must be coupled with an integrated system of tools with which to implement the results of utilizing that paradigm to develop the model of the system.

Essentially, the paradigm generates the model and a toolset must be provided to generate the system. DBTF provides this next-generation capability.

This DBTF approach is used throughout a life cycle, starting with requirements and continuing with functional analysis, simulation, specification, analysis, design, system architecture design, algorithm development, implementation, configuration management, testing, maintenance, and reverse engineering. Its users include end users, managers, system engineers, software engineers, and test engineers.

TABLE 34.1 System Oriented Object Properties of Development before the Fact

Quality (better, faster, cheaper)
- Reliable
- Affordable

Reliable (better)
- In control and under control
- Based on a set of axioms
  - –domain identification (intended, unintended)
  - –ordering (priority and timing)
  - –access rights: Incoming object (or relation), outgoing object (or relation)
  - –replacement
- Formal
  - –consistent, logically complete
  - –necessary and sufficient
  - –common semantic base
  - –unique state identification
- Error free (based on formal definition of "error")
  - –always gets the right answer at the right time and in the right place
  - –satisfies users and developers intent
- Handles the unpredictable
- Predictable

Affordable (faster, cheaper)
- Reusable
- Optimizes resources in operation and development
  - –in minimum time and space
  - –with best fit of objects to resources

Reusable
- *Understandable*, integratable and maintainable
- Flexible
- Follows standards
- Automation
- Common definitions
  - –natural modularity
    - -natural separation (e.g., functional architecture from its resource architectures);
    - -dumb modules
    - -an object is integrated with respect to structure, behavior and properties of control
  - –integration in terms of structure and behavior
  - –type of mechanisms
    - -function maps (relate an object's function to other functions)
    - -object type maps (relate objects to objects)
    - -structures of functions and types
  - –category
    - -relativity
      - instantiation
      - polymorphism
      - parent/child
      - being/doing
      - having/not having
    - -abstraction
      - encapsulation
      - replacement

Handles the unpredictable
- throughout development and operation
- Without affecting unintended areas
- Error detect and recover from the unexpected
- Interface with, change and reconfigure in asynchronous, distributed, real-time environment

Flexible
- Changeable without side effects
- Evolvable
- Durable
- Reliable
- Extensible
- Ability to break up and put together
  - –one object to many: modularity, decomposition, instantiation
  - –many objects to one: composition, applicative operators, integration, abstraction
- Portable
  - –secure
  - –diverse and changing layered developments
  - –open architecture (implementation, resource allocation, and execution independence)
  - –plug-in (or be plugged into) or reconfiguration of different modules
  - –adaptable for different organizations, applications, functionality, people, products

Automation
- the ultimate form of reusable
- formalize, mechanize, then automate
  - –it
  - –its development
  - –that which automates its development

Understandable, integratable and maintainable
- Reliable
- A measurable history
- Natural correspondence to real world
  - –persistence, create and delete
  - –appear and disappear
  - –accessibility
  - –reference
  - –assumes existence of objects
  - –real time and space constraints
  - –representation
  - –relativity, abstraction, derivation
- Provides user friendly definitions
  - –recognizes that one user's friendliness is another user's nightmare
  - –hides unnecessary detail (abstraction)
  - –variable, user selected syntax
  - –self teaching
  - –derived from a common semantic base
  - –common definition mechanisms
- Communicates with common semantics to all entities
- Defined to be simple as possible but not simpler

(continued)

TABLE 34.1 (Continued)

relation including function typing including classification form including both structure and behavior (for object types and functions) -derivation deduction inference inheritance	• Defined with integration of all of its objects (and all aspects of these objects) • Traceability of behavior and structure and their changes (maintenance) throughout its birth, life and death • Knows and able to reach the state of completion –definition –development of itself and that which develops it -analysis -design -implementation -instantiation -testing -maintenance

Note: all underlined words point to a reusable.

Source: Hamilton, M., "Software Design and Development," *The Electronics Handbook,* CRC Press, Boca Raton, FL, 1996. With permission.

The DBTF process combines mathematical perfection with engineering precision. Its purpose is to facilitate the "doing things right in the first place" development style, avoiding the "fixing wrong things up" traditional approach. Its automation is developed with the following considerations: error prevention from the early stage of system definition, life cycle control of the system under development, and inherent reuse of highly reliable systems. The development life cycle is divided into a sequence of stages, including requirements and design modeling by formal specification and analysis, automatic code generation based on consistent and logically complete models, test and execution, and simulation.

The first step in building a DBTF system is to define a model with the language. This process could be in any phase of the developmental life cycle, including problem analysis, operational scenarios, and design. The model is automatically analyzed to ensure it was defined properly. This includes static analysis for preventive properties and dynamic analysis for user-intent properties.

In the next stage, the generic source code generator automatically generates a fully production-ready and fully integrated software implementation for any kind of application, consistent with the model, for a selected target environment in the language and architecture of choice. If the selected environment has already been configured, the generator selects that environment directly; otherwise, the generator is first configured for a new language and architecture.

Because of its open architecture, the generator can be configured to reside on any new architecture (or interface to any outside environment), e.g., to a language, communications package, an Internet interface, a database package, or an operating system of choice; or it can be configured to interface to the users own legacy code. Once configured for a new environment, an existing system can be automatically regenerated to reside on that new environment. This open architecture approach, which lends itself to true component-based development, provides more flexibility to the user when changing requirements or architectures, or when moving from an older technology to a newer one.

It then becomes possible to execute the resulting system. If it is software, the system can undergo testing for further user-intent errors. It becomes operational after testing. Application changes are always made to the requirements/specification definition—not to the code (the developer does not even need to change the code). Target architecture changes are made to the configuration of the generator environment (which generates one of a possible set of implementations from the model)—not to the code. If the real system is hardware or peopleware, the software system serves as a simulation upon which the real system can be based. Once a system has been developed, the system and the process used to develop it are analyzed to understand how to improve the next round of system development.

Seamless integration is provided throughout from systems to software, requirements to design to code to tests to other requirements and back again; level to level and layer to layer. The developer is able to trace from requirements to code and back again.

TABLE 34.2 A Comparison

Traditional (After the Fact)	DBTF (Before the Fact)
Interface errors (over 75% of all errors)	*No interface errors*
Most found after implementation	All found before implementation
Some found manually	All found by automatic and static analysis
Some found by dynamic runs analysis	
Some never found	Always found
Ambiguous requirements	*Unambiguous requirements*
Informal or semiformal language	formal, but friendly language
Different phases, languages, and tools	All phases, same language and tools
Different language for other systems than for software	Same language for software, hardware and any other system
Automation supports manual process	*Automation does real work*
Mostly manual documentation, programming, test generation, traceability, etc.	Automatic documentation, programming, test generation, traceability, etc.
	100% code automatically generated for any kind of software
No guarantee of function integrity after implementation	*Guarantee of function integrity after implementation*
Systems not traceable or evolvable	*Systems traceable and evolvable*
Locked in products, architectures, etc.	Open architecture
Painful transition from legacy	Smooth transition from legacy
Maintenance performed at code level	Maintenance performed at spec level
Reuse not inherent	*Inherent reuse*
Reuse is adhoc	Every object a candidate for reuse
Customization and reuse are mutually exclusive	Customization increases reuse pool
Mismatched objects, phases, products, architectures and environment	*Integrated & seamless objects, phases, products, architectures, and environment*
System not integrated with software	System integrated with software
Function oriented <u>or</u> object oriented	System oriented objects: integration of function, timing, <u>and</u> object oriented
GUI not integrated with application	GUI integrated with application
Simulation not integrated with software code	Simulation integrated with software code
Automation not defined and developed with itself	*Automation defined with and generated by itself*
	#1 in all evaluations
Dollars wasted, error prone systems	*Better, faster, cheaper systems*
Not cost-effective	10 to 1, 20 to 1, 50 to 1...dollars saved
Difficult to meet schedules	Minimum time to complete
Less of what you need and more of what you don't need	No more, no less of what you need

Given an automation that has these capabilities, it should be of no surprise that an automation of DBTF has been defined with itself and that it continues to automatically generate itself as it evolves with changing architectures and changing technologies. Table 34.2 contains a summary of some of the differences between the more modern preventative paradigm and the traditional approach.

A relatively small set of things is needed to master the concepts behind DBTF. Everything else can be derived, leading to powerful reuse capabilities for building systems. It quickly becomes clear why it is no longer necessary to add features to the language or changes to a developed application in an ad hoc fashion, since each new aspect is ultimately and inherently derived from its mathematical foundations.

34.4 Experience with DBTF

That preventative development is a superior alternative has been proven rather dramatically in several experiments. DBTF has been through many evaluations and competitions conducted and sponsored by leading academic institutions, government agencies, and commercial organizations. In every evaluation and competition this alternative came out on top. What set this alternative apart from the others was that it provided a totally integrated system design and development environment, whereas the traditional methods resulted in an informal, difficult to integrate (including application modules as well as the products used to implement them), fragmented, more manual, and "after the fact" life-cycle process.

The National Test Bed of the U.S. Department of Defense sponsored an experiment in which it provided a development problem to each of three contractor/vendor teams chosen from a large pool of vendors and development environments, based upon a well-defined set of requirements. The application was a real-time, distributed, multiuser, client server system, which needed to be defined and developed under the government 2167A guidelines.

All teams were able to complete the first part, the definition of preliminary requirements. Two teams completed the detailed design. But only one team was able to generate complete, integrated, and fully production-ready code automatically; a major portion of this code was running in both C and Ada at the end of the experiment [12]. The team that was able to generate the production-ready code was using the 001 Tool Suite, a development environment based on the DBTF methodology.

34.5 Conclusion

Businesses that expected a big productivity payoff from investing in technology are, in many cases, still waiting to collect. A substantial part of the problem stems from the manner in which organizations are building their automated systems. While hardware capabilities have increased dramatically, organizations are still mired in the same old methodologies that saw the rise of the behemoth mainframes. Old methodologies simply cannot build the new systems.

There are other changes as well. Users demand much more functionality and flexibility in their systems. And given the nature of many of the problems to be solved by this new technology, these systems must also be error-free as well.

Where the biological superorganism has built-in control mechanisms fostering quality and productivity, until now the silicon superorganism has had none. Hence, the productivity paradox.

Often, the only way to solve major issues or to survive tough times is through nontraditional paths or innovation. One must create new methods or new environments for using new methods.

Innovation for success often starts with a look at mistakes from traditional systems. The first step is to recognize the true root problems, then categorize them according to how they might be prevented. Derivation of practical solutions is a logical next step. Iterations of the process entail looking for new problem areas in terms of the new solution environment and repeating the scenario. That is how DBTF came into being.

With DBTF all aspects of system design and development are integrated with one systems language and its associated automation. Reuse naturally takes place throughout the life cycle. Objects, no matter how complex, can be reused and integrated. Environment configurations for different kinds of architectures can be reused. A newly developed system can be safely reused to increase even further the productivity of the systems developed with it.

The paradigm shift occurs once a designer realizes that many of the old tools are no longer needed to design and develop a system. For example, with one formal semantic language to define and integrate all aspects of a system, diverse modeling languages (and methodologies for using them), each of which defines only part of a system, are no longer necessary. There is no longer a need to reconcile multiple techniques with semantics that interfere with each other.

DBTF can support a user in addressing many of the challenges presented in today's software development environments. There will, however, always be more to do to capitalize on this technology. That is part of what makes a technology like this so interesting to work with. Because it is based on a different premise or set of assumptions (set of axioms), a significant number of things can and will change because of it. There is the continuing opportunity for new research projects and new products. Some problems can be solved, because of the language, that could not be solved before. Software development as we know it will never be the same. Many things will no longer need to exist—they, in fact, will be rendered extinct, just as that phenomenon occurs with the process of natural selection in the biological system. Techniques for bridging the gap from one phase of the life cycle to another become obsolete. Testing procedures and tools for finding most errors are no longer needed because those errors no longer exist. Tools to support programming as a manual process are no longer needed.

Compared to the development using traditional techniques, the productivity of DBTF developed systems has been shown to be significantly greater. Upon further analysis, it was discovered that the larger and more complex the system, the greater the productivity—the opposite of what one finds with traditional systems development. This is, in part, because of the high degree of DBTF's support of reuse. The larger a system, the more it has the opportunity to capitalize on reuse. As more reuse is employed, productivity continues to increase. Measuring productivity becomes a process of relativity—that is, relative to the last system developed.

Capitalizing on reusables within a DBTF environment is an ongoing area of research interest. An example is understanding the relationship between types of reusables and metrics. This takes into consideration that a reusable can be categorized in many ways. One is according to the manner in which its use saves time (which translates to how it impacts cost and schedules). More intelligent tradeoffs can then be made. The more we know about how some kinds of reusables are used, the more information we have to estimate costs for an overall system. Keep in mind also that the traditional methods for estimating time and costs for developing software are no longer valid for estimating systems developed with preventative techniques.

There are other reasons for this higher productivity as well, such as the savings realized and time saved due to tasks and processes that are no longer necessary with the use of this preventative approach. There is less to learn and less to do—less analysis, little or no implementation, less testing, less to manage, less to document, less to maintain, and less to integrate. This is because a major part of these areas has been automated or because of what inherently take place because of the nature of DBTF's formal systems language.

In the end, it is the combination of the technology and that which executes it that forms the foundation of successful software. Software is so ingrained in our society that its success or failure will dramatically influence both the operation and the success of an organization. For that reason, today's decisions about systems engineering and software development have far-reaching effects.

Software is a relatively young technological field that is still in a constant state of change. Changing from a traditional software environment to a preventative one is like going from the typewriter to the word processor. Whenever there is any major change, there is always the initial overhead needed for learning the new way of doing things. But, as with the word processor, progress begets progress.

Collective experience strongly confirms that quality and productivity increase with the increased use of properties of preventative systems. In contrast to the "better late than never" after the fact philosophy, the preventive philosophy behind DBTF is to solve—or if possible, prevent—a given problem as early as possible. Finding a problem statically is better than finding it dynamically. Preventing it by the way a system is defined is even better. Better yet is not having to define (and build) it at all.

Reusing a reliable system is better than reusing one that is not reliable. Automated reuse is better than manual reuse. Inherent reuse is better than automated reuse. Reuse that can evolve is better than one that cannot evolve. Best of all is reuse that ultimately approaches wisdom itself. Then, have the wisdom to use it.

The answer continues to be in the results just as in the biological system; and the goal is that the systems of tomorrow will inherit the best of the systems of today.

Defining Terms

Data Base Management System (DBMS): The computer program that is used to control and provide rapid access to a database. A language is used with the DBMS to control the functions that a DBMS provides. For example, SQL is the language that is used to control all of the functions that a relational architecture-based DBMS provides for its users, including data definition, data retrieval, data manipulation, access control, data sharing, and data integrity.

Graphical User Interface (GUI): The ultimate user interface, by which the deployed system interfaces with the computer most productively, using visual means. Graphical user interfaces provide a series of intuitive, colorful, and graphical mechanisms that enable the end-user to view, update, and manipulate information.

Interface: A point of access in a boundary between objects or programs or systems. It is at this juncture that many errors surface. Software can interface with hardware, humans, and other software.
Methodology: A set of procedures, precepts, and constructs for the construction of software.
Metrics: A series of formulas that measure such things as quality and productivity.
Software Architecture: The structure and relationships among the components of software.
Formal: A system defined in terms of a known set of axioms (or assumptions); it is therefore, mathematically based (e.g., a DBTF system is based on a set of axioms of control). Some of its properties are that it is consistent and logically complete. A system is consistent if it can be shown that no assumption of the system contradicts any other assumption of that system. A system is logically complete if the assumptions of the method completely define a given set of properties. This assures that a model of the method has that set of properties. Other properties of the models defined with the method may not be provable from the method's assumptions. A logically complete system has a semantic basis (i.e., a way of expressing the meaning of that system's objects). In terms of the semantics of a DBTF system, this means it has no interface errors and is unambiguous, contains what is necessary and sufficient, and has a unique state identification.

Further Information

Hamilton, M. and Hackler, W. R., Object Thinking: Development Before the Fact, In Press.

Krut, Jr., B. "Integrating 001 Tool Support in the Feature-Oriented Domain Analysis Methodology" (CMU/SEI-93-TR-11, ESC-TR-93-188). Pittsburgh, PA: Software Engineering Institute, Carnegie-Mellon University, 1993.

Ouyang, M. and Golay, M. W., "An Integrated Formal Approach for Developing High Quality Software of Safety-Critical Systems," Massachusetts Institute of Technology, Cambridge, MA, Report No. MIT-ANP-TR-035., September, 1995.

McCauley, B. "Software Development Tools in the 1990s," AIS Security Technology for Space Operations Conference, July 1993, Houston, TX.

Hamilton, M. and Hackler, W. R., Towards Cost Effective and Timely End-to-End Testing, HTI, prepared for Army Research Laboratory, Contract No. DAKF11-99-P-1236, July 17, 2000.

Keyes, J., *Internet Management,* Chapters 30–33, on 001-developed systems for the Internet, Auerbach, Boca Raton, FL, 2000.

References

1. Software Engineering Institute. *Capability Maturity Model,* Pittsburgh, PA: Carnegie, Mellon University, 1991.
2. Stroustrup, B., *The C++ Programming Language,* Reading, MA: Addison-Wesley, 1997.
3. Gosling, J., Joy, B., and Steele, G., *The Java Language Specification,* Reading, MA: Addison-Wesley, 1996.
4. Lientz, B.P., and Swanson, E.B., *Software Maintenance Management,* Reading, MA: Addison-Wesley, 1980.
5. Jones, T.C., *Program Quality and Programmer Productivity,* IBM Tech. Report TR02.764 January: 80, San Jose, CA: Santa Teresa Labs, 1977.
6. Keyes, J., *Handbook of E-Business,* Chapter F5, Hamilton, M., Defining e...com for e-Profits, RIA, 2000.
7. Martin, J., and Finkelstein, C.B., *Information Engineering,* Carnforth, Lancs, U.K.: Savant Institute, 1981.
8. Booch, G., Rumbaugh, J., and Jacobson, I., *The Unified Modeling Language User Guide,* Addison-Wesley, 1999.
9. Hamilton, M., "*Inside Development Before the Fact,*" Electronic Design, April 4, 1994, ES.
10. Hamilton, M., *"Development Before the Fact in Action,"* Electronic Design, June 13, 1994, ES.
11. Keyes, J., *The Ultimate Internet Developers Sourcebook,* AMACOM, to be published Fall 2001.
12. Software Engineering Tools Experiment-Final Report, Vol. 1, Experiment Summary, Table 1, Page 9, Department of Defense, Strategic Defense Initiative, Washington, D.C., 20301-7100, October 1992.

35

Data Recording and Logging

Craig Anderson
National Instruments, Inc.

35.1 Overview

Data logging and recording is a very common measurement application. In its most basic form, data logging is the measurement and recording of physical or electrical parameters over a period of time. These parameters can be temperature, strain, displacement, .ow, pressure, voltage, current, resistance, power, or any of a wide range of other measurement types. Real-world data-logging applications are typically more involved than just acquiring and recording signals, as illustrated in Figure 35.1. They usually involve some combination of online analysis, offline analysis, display, report generation, and data sharing. In addition, many data-logging applications are beginning to require the acquisition and storage of other types of data. One example would be recording sound and video in conjunction with the other parameters measured during an automobile crash test.

Data logging is used in a broad spectrum of applications. Chemists record data such as temperature, pH, and pressure when performing experiments in a lab. Design engineers log performance parameters such as vibration, temperature, and battery level to evaluate product designs. Civil engineers record strain and load on bridges over time to evaluate safety. Geologists use data logging to determine mineral formations when drilling for oil. Breweries log the conditions of their storage and brewing facilities to maintain quality.

The list of applications for data logging goes on and on, but all these applications have similar requirements. The purpose of this chapter is to provide some general background on data logging, discuss the various functional requirements that are common to most data-logging applications, and examine some

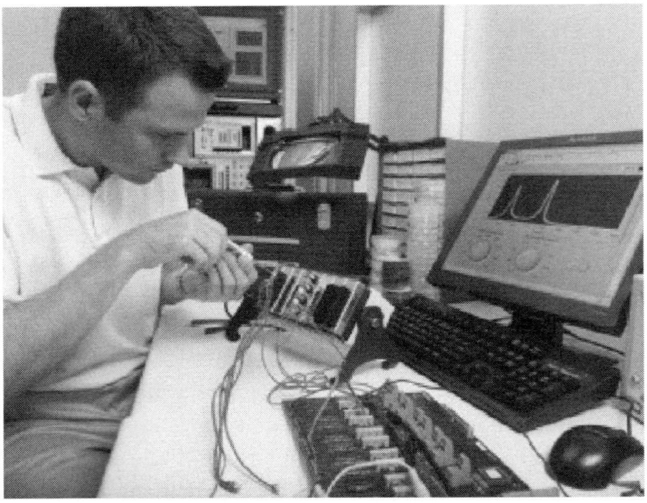

FIGURE 35.1 PC-based data-logging application.

of the modern hardware and software options available that allow scientists and engineers to implement powerful PC-based data-logging systems.

35.2 Historical Background

The earliest form of data logging involved taking manual measurements from analog instruments, such as thermometers and manometers. These measurements were recorded in a written log, along with the time of observation. To view trends over time, people manually plotted their measurements on paper. In the late-nineteenth century, it became possible to begin automating this process with machines, and strip chart recorders evolved. Strip chart recorders are analog instruments that translate electrical impulses from sensors into mechanical movement of an arm. A pen is attached to the arm, and long rolls of paper aremoved at a constant rate under the pen. The result is a paper chart displaying the parameters measured over the course of time. Strip chart recorders were a great improvement over manual data logging, but they still had drawbacks. For example, translating the traces on the paper into meaningful engineering measurements was tedious at best, and the data that was recorded took up reams and reams of paper.

With the development of the personal computer in the 1970s and 1980s, people began to leverage computers for analysis of data, data storage, and report generation. The need to bring data into the PC brought about a new type of equipment—the data logger. Data loggers are stand-alone box instruments that measure signals, convert them to digital data, and store the data internally. This data must be transferred to the PC for analysis, permanent storage, and report generation. Data is typically transferred either by manually moving a storage device, such as a .oppy disk, from the data logger to the computer, or by connecting the data logger to the PC through some communications link, such as serial or Ethernet.

In the 1990s, a further evolution in data logging took place as people began to create PC-based data-logging systems. These systems combined the acquisition and storage capabilities of stand-alone data loggers with the archiving, analyzing, reporting, and displaying capabilities of modern PCs. PC-based logging systems finally enabled full automation of the data-logging process. The move to PC-based data-logging systems was made possible by the following three technological enhancements:

1. Increasing PC reliability
2. Steadily decreasing cost of PC hard drive space
3. PC-based measurement hardware that could meet or exceed stand-alone data logger measurement capabilities

Today, PC-based logging systems provide the widest range of measurement types, analysis capabilities, and reporting tools. The remainder of this chapter focuses on the functionality necessary to implement a PC-based data-logging system.

35.3 Data Logging Functional Requirements

Every data-logging application—from .fteenth-century monks manually recording weather patterns to twenty-first-century physicists logging the experimental parameters of a fusion reactor test—can be broken down into a set of .ve commonfunctional requirements, illustrated in Figure 35.2. Acquiring is the process of actually measuring the physical parameters and bringing them into your logging system. Online analysis consists of any processing done to the data while it is being acquired including alarms, data scaling, and sometimes control. Logging or storing the data is an obvious requirement of every data-logging system.

Offline analysis is everything that is done with the data after it has been acquired in order to extract useful information from it. The final functional block is made up of display, reporting and data sharing. These are all the "miscellaneous" requirements that fill out the functionality of a data-logging system. Let us examine how each of these functional blocks is addressed with modern PC-based data-logging systems.

35.3.1 Acquisition

The acquisition function is one of the most critical components of every data-logging system. In a PCbased system, the acquisition is accomplished by the measurement hardware, which can be broken down further into sensors, signal connectivity, signal conditioning, and analog-to-digital (A/D) conversion, as shown in Figure 35.3. Each of these topics is covered in more detail in other sections of this chapter, so only a high-level overview is given here.

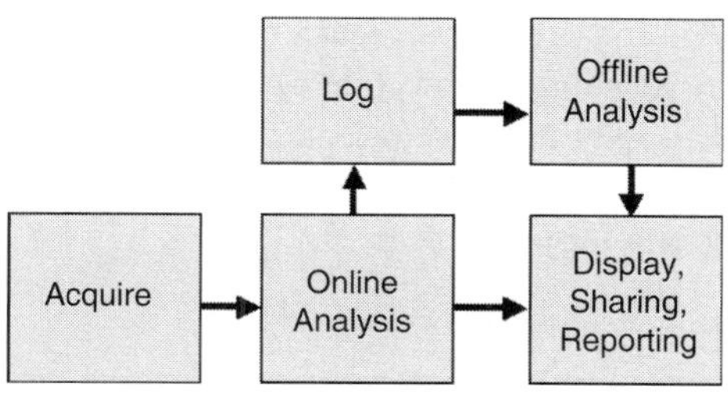

FIGURE 35.2 Basic elements of a data-logging system.

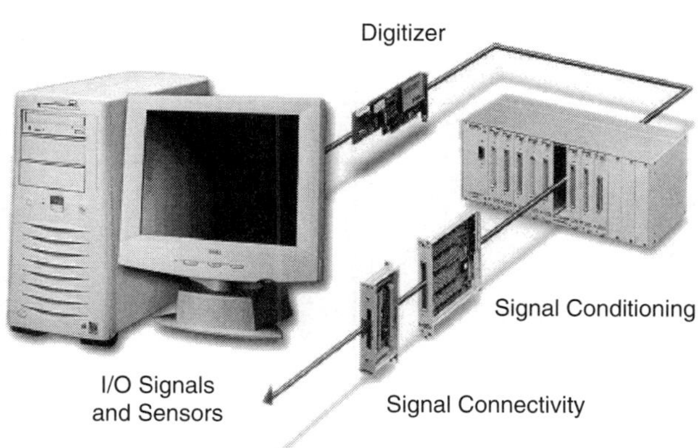

FIGURE 35.3 Measurement hardware components of a PC-based data-logging system.

35.3.2 Sensors

A wide variety of sensors are used to convert physical parameters into electrical signals. Temperature sensors such as thermocouples, RTDs, or thermistors are some of the most common sensors used in data-logging applications. Other widespread sensors are .owmeters, pressure transducers, strain gauges, accelerometers, and microphones, to name a few. Proper sensor selection and installation is beyond the scope of this chapter.

35.3.3 Signal Connectivity

After sensors are installed, they must be connected to the data-logging system. Signal connectivity is the component of the measurement hardware that allows you to connect your sensors to your logging system. Screw terminals, which allow you to connect bare wires from sensors directly to your logging system, are the most basic form of connectivity. Screw terminals are a good choice for general-purpose use, particularly when you need to connect a large number of signals into a small amount of space. The disadvantage of screw terminals is that they are time-consuming to connect and dif.cult to recon.gure. Figure 35.4 shows some other standard connectivity options that are designed to make connecting and disconnecting sensors less labor intensive. Minithermocouple connectors are a widely used connectivity option for thermocouples. BNC and SMB connectors are commonly used when electrical shielding is required for noise immunity. Banana jacks are often used when measuring current, resistance, or higher voltages. The sensor provider typically defines the connectivity options available for a sensor, and it is up toy ou toc hoose measurement hardware that can accept that connectivity.

35.3.4 Signal Conditioning

Signal conditioning is one of the most important and most overlooked components of a PC-based datalogging system. Many sensors require special signal conditioning technology. For example, signals from a thermocouple have very low voltage and require amplification, filtering, and linearization. Other sensors, such as strain gauges and accelerometers, require power in addition to amplification and filtering, while other signals may require isolation to protect the system from high voltages. No single stand-alone data logger can provide the flexibility required to make all these measurements. However, with front-end signal conditioning, you can combine the necessary technologies to bring these various types of signals into a single PC-based data-logging system.

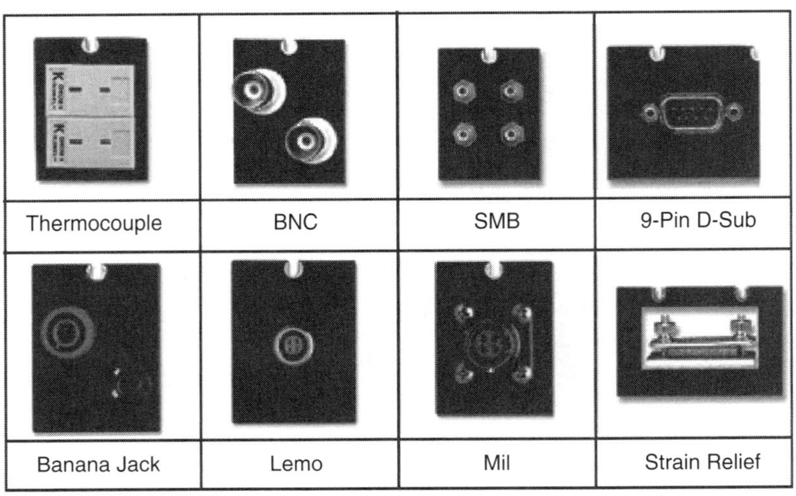

FIGURE 35.4 Examples of signal connectivity options.

Because of the vast array of signal-conditioning technologies, the role and need for each technology can quickly become confusing. A list of common types of signal conditioning, their functionality, and examples of when they are needed is given below.

35.3.4.1 Amplification

When the voltage levels being measured are very small, amplification is used to maximize the effectiveness of the digitizer. By amplifying the input signal, the conditioned signal uses more of the effective range of the A/D converter. This allows better accuracy and resolution of the measurement. Typical sensors that require amplification are thermocouples and strain gauges.

35.3.4.2 Attenuation

Attenuation is the opposite of amplification. It is necessary when the voltages to be digitized are outside the input range of the digitizer. This form of signal conditioning divides the input signal so that the conditioned signal is within the range of the A/D converter. Attenuation is necessary for measuring high voltages.

35.3.4.3 Isolation

Voltage signals outside the range of the digitizer can damage the measurement systemand harm the operator. For that reason, isolation is usually required in conjunction with attenuation to protect the system and the user from dangerous voltages or voltage spikes. Isolation may also be required when the sensor is on a different electrical ground plane from the measurement sensor (such as a thermocouple mounted on an engine).

35.3.4.4 Multiplexing

Typically, the digitizer is the most expensive part of a data acquisition system. Multiplexing allows you to automatically routemultiple signals into a single digitizer, providing a cost-effective way to greatly expand the signal count of your system. Multiplexing is necessary for any high-channel-count application.

35.3.4.5 Filtering

Filtering is required to remove unwanted frequency components from a signal. This prevents aliasing and reduces signal noise. Thermocouple measurements typically require a low-pass filter to remove power-line noise from the signals. Vibration measurements normally require a higher-frequency low-pass filter to remove high-frequency signal components that are above the range of the acquisition system.

35.3.4.6 Excitation

Many sensor types, including RTDs, strain gauges, and accelerometers, require some form of power to make a measurement. Excitation is the signal conditioning technology required to provide this power. This excitation can be a voltage or current source, depending on the sensor type.

35.3.4.7 Linearization

Some types of sensors produce voltage signals that are not linearly related to the physical quantity they are measuring. Linearization is the process of interpreting the signal from the sensor as a physical measurement. This can be done either with signal conditioning or through software. Thermocouples are the classic examples of sensors that require linearization.

35.3.4.8 Cold-Junction Compensation

Another technology that is required for thermocouple measurements is cold-junction compensation (CJC). Any time a thermocouple is connected to a data acquisition system, the temperature of the connection must be known in order to calculate the true temperature the thermocouple is measuring. A built-in CJC sensor must be present at the location of the connections.

35.3.4.9 Simultaneous Sampling

When it is critical tomeasur emultiple signals at exactly the samemoment in time, simultaneous sampling is required. Front-end signal conditioning can provide amuchmore cost-effective simultaneous sampling solution than purchasing a digitizer with those capabilities. Typical applications that might require simultaneous sampling include vibration measurements and phase-difference measurements.

Most sensors require a combination of the above signal conditioning technologies. Again, the thermocouple is the classic example because it requires amplification, linearization, CJC, filtering, and sometimes isolation. Ideally, a good PC-based data-logging platform should give you the ability to select the type of signal conditioning that you need for your application. In some systems, front-end signal conditioning is an option, but in others, front-end signal conditioning is a necessity to make the required measurements. As a rule of thumb, your measurement system should include front-end signal conditioning if you are planning to use any of the following: thermocouples, RTDs, thermistors, strain gauges, LVDTs, accelerometers, switching, multiplexing, mixed low-voltage/high-voltage signals, current inputs, or resistance inputs.

35.3.5 Conversion

After physical parameters have been converted into electrical signals and properly conditioned, it is time toc onvert the analog electrical signals intodig ital values and pass those values back tothe computer. The A/D conversion can be accomplished with either a plug-in data acquisition (DAQ) board, or it can be integrated into a single package with the conditioning and connectivity.

The combination of sensors, signal connectivity, signal conditioning, and A/D conversion makes up the measurement hardware portion of a data-logging system. In a PC-based system, the measurement hardware is con.gured and controlled through software, and it is critical to use software that is designed to integrate smoothly with all components of your data-logging system.

35.3.6 Online Analysis

The next functional component in a typical data-logging system is online analysis. In PC-based systems, online analysis is accomplished through software. Different data-logging applications require different forms of online analysis.We will discuss some of the most common ones here.

Channel scaling is the conversion of the raw binary values returned by the acquisition system into properly scaled measurementswith appropriate engineering units. One example is computing temperature (in ¢ªC) from a thermocouple reading. The digitizer returns binary measurements of the thermocouple voltage and the cold-junction sensor voltage. The software converts the binary measurements into voltages, and then uses a thermocouple conversion formula to compute temperature. Similar channel scaling routines are used for strain gauges, RTDs, accelerometers, and others. Fortunately, modern PC-based measurement software handles most scaling functions automatically.

Another important online analysis function is alarming and event management. This includes monitoring a channel and providing some noti.cation if preset limits are exceeded. This noti.cation can be as basic as turning on a warning light, or as complex as paging someone with information about the problem. Alarming can also include an automated response to certain events. For example, a data-logging system could shutdown a machine being monitored if the oil temperature exceeded a certain limit.

A wide range of online analysis functionality can be required in different data-logging applications. This functionality could include feedback control systems or advanced signal analysis. Only PC-based data-logging systems have the flexibility to implement these differing requirements.

35.3.7 Logging and Storage

The logging (or storage) functional block is, by de.nition, required in every data-logging system.Methods of storing data varywidely across different systems. Strip chart recorders use paper; traditional data loggers can use internal nonvolatile memory, .oppy disks, or a variety of other mediums. PC-based data-logging systems typically use the hard drive of the PC, although they can also use tape drives, network drives, RAID drives, and other more exotic options.

Software is of critical importance in PC-based data-logging systems because well-written logging software determines how data is stored, how quickly data can be written to disk, and how efficiently disk space is used. Logging software also gives you data management capabilities, such as changing data formats, archiving data, and connecting to databases.

The data storage format has a strong link to the performance and ease of use of your data-logging system. Three general formats are commonly used for storage in data-logging systems—ASCII text files, binary files, and databases.

ASCII text files are the most common and .exible form of data storage. Text files for data-logging applications are typically made up of a header section and columns of data. The header section provides information that includes channel names, units, test equipment, and user comments. The first data column is usually the time stamp of each sample, and it is followed by another column for each channel being logged. Text files are useful because they can be opened or imported into almost any software package, and they are easily transferred between operating systems. Some disadvantages of text files are that they use disk space inefficiently, and that they require additional processing overhead to write and read from files. ASCII text files are commonly used when the acquisition speed is slow (<1000 samples per second), the total amount of data to log is not large, or the user needs to share data between different software applications.

Binary files are the most efficient method of data storage. With binary files, the raw bytes that the computer uses to store data in memory are written directly to the file. This data takes up considerably less space than the same information written in ASCII text format, and it requires much less processor overhead than formatting into text. Binary files cannot be viewed in common software applications like Microsoft Excel. Instead, they must be translated by a software routine into meaningful data. With PCbased data-logging systems, you can log scaled data that is already processed into correct engineering units, or you can log the raw binary values returned by the digitizer. The raw binary values representing the A/D conversions of each sample returned from a 16-bit data acquisition device take up 16 bits, or 2 bytes, of memory. The channel scaling routines in your logging software automatically convert this raw data intoa real number that represents the physical value you measured. Scaled data is typically handled inside your data-logging software as a double precision .oating-point value that refers to a data type taking up 8 bytes of memory on most computer systems.

For performance reasons, some high-speed data-logging systems might log the raw binary values to disk, along with the necessary scaling constants to convert them to scaled data at a later time. Figure 35.5 shows the relationship between logging raw binary data, scaled binary data, and ASCII text. Binary files take up less space and allow greatly improved stream-to-disk speed. Raw binary files can be less than one-tenth the size of a text file containing the same information. The disadvantage of binary files is that they typicallymust be translated to another format before they can be shared between different application types.

Many data-logging software packages log data into databases. Databases are typically binary files that provide a structured format for inserting and retrieving data. They are optimized for efficiently handling large amounts of data and for searching through information in the database without loading everything intomemo ry. Often, databases are alsodesig ned toallo w easy backup and archiving of data, and multiple user access. They usually have software methods tomak e it easy toimpo rt data into differ ent software packages for analysis and report generation. In many ways, databases are the ideal storage format for PC-based data-logging systems. Two disadvantages of using databases for storage are that they add increased complexity, and that they can be dif.cult to implement if starting from scratch.

Many different storage media types are used for data logging. Stand-alone data loggers can use onboard nonvolatile memory, .oppy disks, PCMCIA memory cards, tapes, or a variety of other options. PC-based data-logging systems usually rely on the computer's internal hard disk. This is possible because of the trends toward more reliable and higher-capacity hard drives. The 200 GB (and larger) hard drives that are

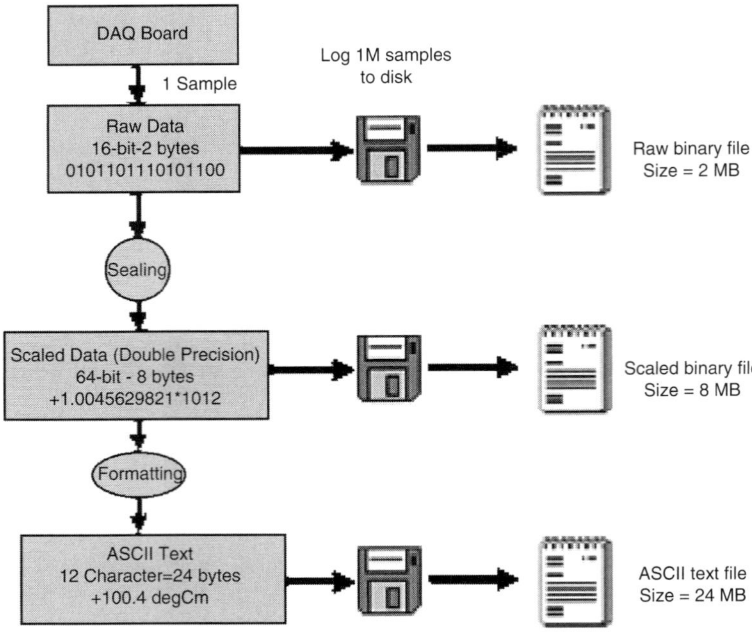

FIGURE 35.5 Data storage file size examples.

readily available today make hard drives one of the most economical storage devices. It is still advisable to periodically back up or archive data stored on a local hard drive.

High-speed data-logging applications (more than 1 MS/s) can start to exceed the write-to-disk speeds of normal PC hard drives. One of the advantages of PC-based logging systems is that you can move to more high-performance storage devices and higher-performance computers, often with little or no modification to your logging software or measurement hardware. One type of high-performance storage device is the redundant array of independent disks (RAID) controller. RAID controllers usemultiple hard drives in concert to greatly enhance the combined stream-to-disk speed and to provide improved data integrity. Audiovisual (AV) drives are another type of storage device used for high-speed data logging.

AV drives are optimized for streaming large amounts of audio and video information to disk, and this optimization also makes them well suited for high-performance data-logging applications. Finally, some companies make custom hardware that allow data acquisition devices to stream data across the computer's PCI or PCI Express bus directly into device storage. The stream-to-disk rates of these devices are limited by the available bandwidth of the data transfer bus.

35.3.8 Offline Analysis

Offline analysis is performing mathematical functions on data after it has been acquired in order to extract important information. Types of Offline analysis can include computing basic statistics of measured parameters as well as more advanced functions such as the frequency content of signals and order analysis. Offline analysis can be integrated with the rest of the data-logging application, or it can occur separately through stand-alone analysis software packages. Often, offline analysis is combined with the report generation, historical display, and data-sharing functions.

35.3.9 Display

Most data-logging applications require some form of display to view the measurements being recorded. The display function can be broken down further to view live data and historical data. Live data display

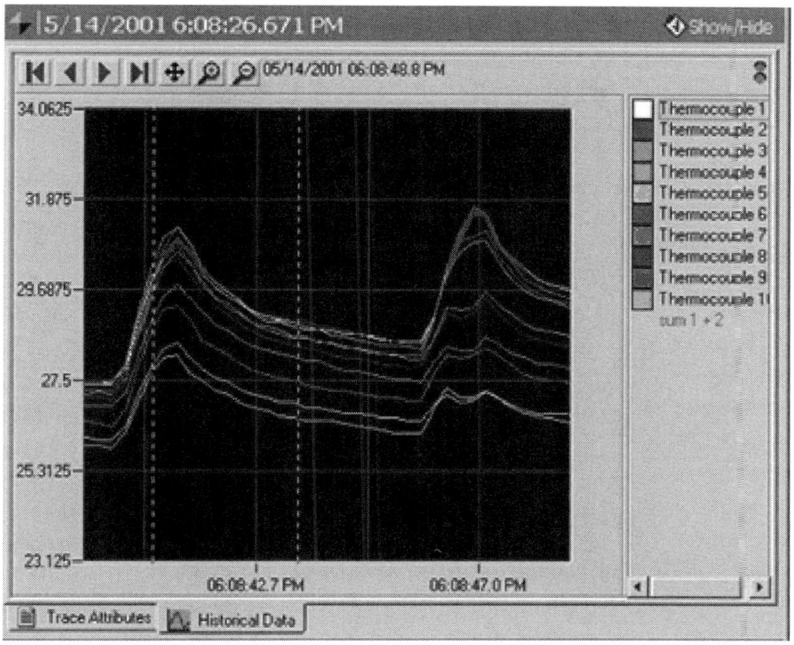

FIGURE 35.6 Example of historical data display from National Instruments VI logger software.

is necessary if you need to view data as it is being acquired. Many stand-alone data loggers have a live data display integrated into the box with them. Historical display lets you view data that was previously acquired. Most stand-alone data loggers require you to move the data to a PC for historical viewing. PC-based data-logging applications let you combine both live display and historical display into the same user interface. Data-viewing utilities should provide an intuitive user interface, scrolling and zooming capabilities, cursors, and general customization features. Figure 35.6 is an example of a typical historical data display found with commercially available software.

35.3.10 Report Generation

Report generation is a function that is often not considered part of the data-logging application. In reality, almost every data-logging application requires some form of reporting capability, for the simple reason that if data is being recorded, someone needs to see it in a presentable format. Report generation can be integrated into PC-based data-logging applications for increased ef.ciency. The logging application can be set up toper iodically generate prede.ned reports and distribute them tothe appropriate people. Powerful commercial software designed to give you advanced capabilities for analyzing data and generating reports from your measurements is available. Figure 35.7 shows an example of some of the report generation capabilities possible with commercially available packages.When choosing software for report generation, it is critical that it integrates smoothly with the rest of your data-logging software. Ideally, the logging software should be able to pass data directly to the report generation application and trigger automatic reports.

35.3.11 Data Sharing and Publishing

In order for logged data to be useful, it must be available to the right people. With the networking capabilities found in modern data-logging software, sharing data and publishing it to the network no longer requires a degree in computer science. Logging applications can be set up to publish live data to

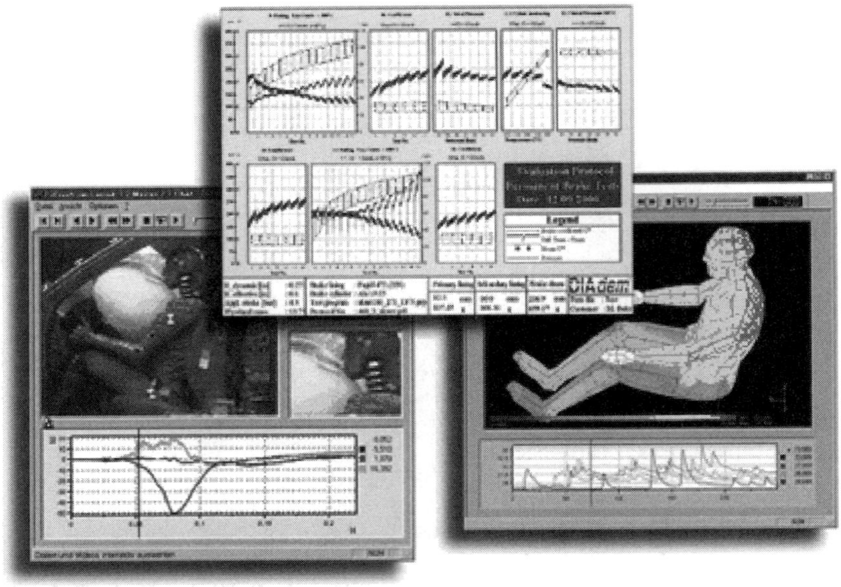

FIGURE 35.7 Advanced report generation capabilities with National Instruments DIAdem software.

the network as it is acquired, periodically e-mail both raw data and analyzed results to key personnel, or automatically post reports to aWeb page.

In widely distributed data-logging applications, each logging node can publish its measurements to the network, and a main computer can serve as the central collection facility. The central computer retrieves the measurements from each node, combines them for further analysis, logs the results for permanent archiving, and periodically generates reports analyzing the data.

35.4 Data-Logging Systems

Nowthatwe have seen the functional components of a data-logging system, let us examine howthese components can be implemented in real systems. All PC-based data-logging systems are made up of hardware and software. The measurement hardware handles the acquisition portion of the logging application, and the hardware choice defines channel count, sensor type, acquisition speed, and measurement accuracy. The measurement software, in addition to controlling the hardware, also handles the online analysis, logging, offline analysis, display, reporting, and data sharing.

35.4.1 Software Options

Choosing software is one of the most critical steps when defining a PC-based data-logging system. Your logging system depends on software to give you a productive, .exible solution. The measurement software must be designed toint egrate seamlessly with your hardware. In addition tothe basic task of acquiring data and logging it to disk, your software should provide tools to handle configuration of measurement hardware, scaling of data from channels, and calibration of your system. The software should allow you to complete your entire application—including report generation, analysis, archiving, and sharing. There are two general categories of software that can be used for PC-based data-logging applications—turnkey software, also known as configuration-based software, and application development environments.

Turnkey packages are ready-to-run data-logging software applications that interface with your measurement hardware to acquire and log data. These applications provide a user-friendly environment for

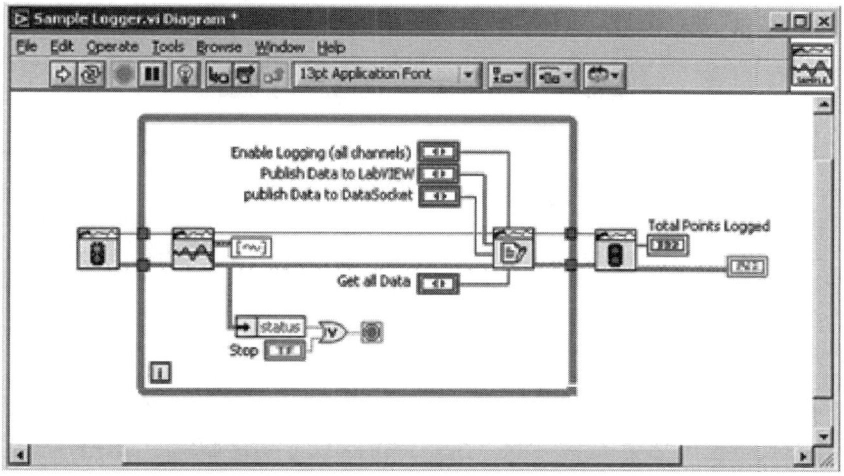

FIGURE 35.8 Example of graphical programming of data-logging applicationswithNational Instruments LabVIEW.

configuring your logging task and getting up and running quickly. An example of graphical programming of data-logging is shown in Figure 35.8.

A good configuration-based data-logging software package provides the features discussed below.

35.4.1.1 An Intuitive User Interface

The software configuration should be through a window-based, menu-driven interface with easily accessible help functions and tutorials.

35.4.1.2 Automatic Data Storage and Archiving

One of the primary functions of any data-logging software package is to handle the storage of the data. It should automatically store the data in an efficient manner, and the software should provide a method for backing up and archiving data.

35.4.1.3 Capability to Export Data

At a minimum, the software package should allow you to export data to ASCII text files, so you can import it into other packages. More advanced data-logging software packages let you automatically transfer data into common databases and analysis programs.

35.4.1.4 Alarming and Event Management

The data-logging softwaremust provide the capability to handle alarms and events. This includes detecting if a signal is over or under a limit, outside of a range, or inside of a range. If an alarm occurs, the software should allow a range of actions, such as sending pages or performing some type of digital or analog output.

35.4.1.5 Display and Trending Tools

All turnkey logging software packages need to have a good interface for viewing both live and historical data. This interface must let you scroll through data, zoom in on regions of interest, and see long-term data trends.

A disadvantage of configuration-based applications is that unless there is a method for customization, you are locked into the functionality provided by the manufacturer. If your measurement needs to change, and you need to add a different type of signal, you might be out of luck if your turnkey software does not support that measurement type. In addition, if you want to integrate offline analysis, report generation, and network connectivity into your data-logging application, a closed turnkey software application can make that dif.cult. On the other hand, there are turnkey software applications that provide methods for customizationwith popular application development tools. These customizable logging software packages provide the best of both worlds, allowing you to get up and run quickly with your logging application, and also giving you a method for integrating more advanced functionality at a later date.

Application development tools are the other option available for developing PC-based data-logging systems. Development tools can range fromtext-based programming languages to graphical programming environments. Figure 35.8 is an example of the software code for a data-logging application developed in a graphical programming environment. Development tools let you build your own customized data-logging application to do exactly what you need. Application development tools let you modify your application as your needs change, integrate customized analysis and report capabilities with your logging application, and fully automate your data-logging system.

When developing data-logging applications, it is advisable to choose a development environment with productivity features that let you create powerful PC-based logging systems. Some features to look for when evaluating application development tools are as follows:

A Wide Range of Graphical User Interface Components
Developing user interface components such as graphs, displays, and controls from scratch is extremely time-consuming. You should choose a development environment that contains high-quality user interface components.

Tight Integration with Measurement Hardware
It is critical that you use software designed to work with your measurement hardware. Not only does proper software integration result in signi.cantly shorter development times, but it also helps ensure that you get measurements you can trust.

Analysis Functions
One of the primary reasons for custom developing a data-logging software application is to integrate advanced analysis functions. A good application development environment provides a wide range of analysis functions to handle almost any need.

Network Connectivity
In today's networked environment, the ability to connect your data-logging application to a network or publish data to the Web can be very important. Your application development software should provide tools tomak e publishing results tothe network a trouble-free process.

Report Generation
Your application development environment should either allow you to generate reports automatically or allow programmatic control of external report generation packages. The choice between turnkey software and development tools depends on the complexity of your datalogging application and the amount of customization required.With either choice, it is important to use a software vendor that specializes in connecting measurements to computers and that provides high-quality service and support.

35.4.2 Hardware Options

Many different hardware platforms are available for data-logging systems. Platform choice depends on your size, operating environment, and installation requirements. Although the combinations are nearly endless, platforms for PC-based data logging can be broken down broadly into four categories—portable, desktop, rack mount/industrial, and distributed. One of the key benefits of PC-based data-logging systems is that the same data-logging software scales across all these platforms.

Portable data-logging solutions are necessary in a variety of applications, such as in-vehicle data logging and field-testing of equipment. Portable PC-based solutions use laptop computers and measurement hardware designed to be easily portable. Figure 35.9 shows a portable, PC-based data-logging system from National Instruments. The digitizer is a USB-based data acquisition system with integrated signal

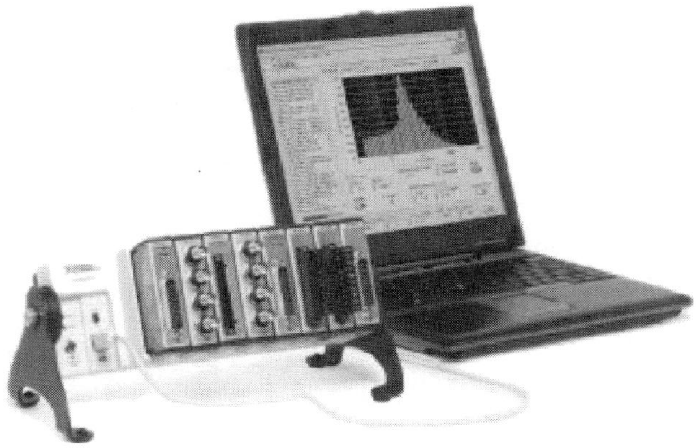

FIGURE 35.9 Portable PC-based data-logging system.

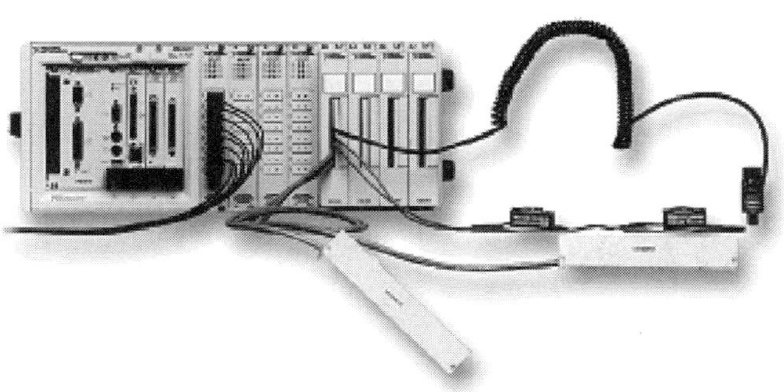

FIGURE 35.10 Rack-mount industrial PC-based data-logging system.

conditioning and connectivity. However, portable data-logging systems are frequently limited to less than 40 channels, purely owing to size constraints.

Desktop systems, such as the one shown in Figure 35.3, use measurement hardware designed to work with standard desktop PCs. Desktop systems are ideal for a wide range of laboratory-based data-logging applications, such as validation testing of new product designs. Because fixed desktop systems are not as constrained by size, the signal connectivity and conditioning functions are typically accomplished by a modular front-end signal conditioning system that provides the capability to measure a wide range of sensor and signal types and toeasily expand tolo g hundreds of channels.

Often, desktop systems take up toomuch space or donot fit well in environments such as large laboratories or manufacturing facilities. In these cases, the more compact and clean solution of amodular industrial PC, based on the PXI or CompactPCI standard, might be a more appropriate data-logging solution. Figure 35.10 is an example of a PXI-based data-logging system. The PC, data acquisition board, signal conditioning, and connectivity functions are all comprised in one modular system. These systems are designed tobe rack mountable, sothe y can be cleanly installed intoan industrial or laboratory environment.

Finally, some data-logging systems need to be distributed away from the PC. This is the case when you need to log data from multiple locations around a facility, such as when logging the performance parameters of a chemical plant. Distributed logging systems should be compact so that they can be mounted unobtrusively, and they typicallymust operate in extended temperature ranges. With distributed logging systems, you typically have multiple measurement nodes that communicate back to a central computer through a communications link such as RS485 or Ethernet. Figure 35.11 presents two examples of distributed logging systems.

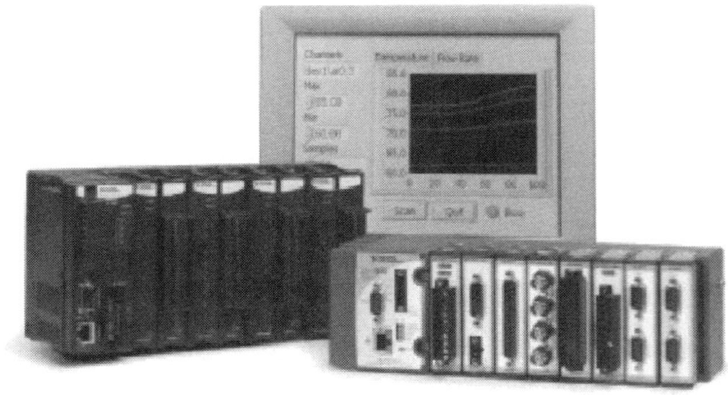

FIGURE 35.11 Examples of distributed data-logging systems.

The choice of logging platform depends on the requirements of your data-logging system, and some systems might require using multiple platforms together.With properly designed hardware and software, data-logging systems can scale from simple, low-channel-count laboratory systems up to very highchannel-count, distributed industrial logging systems.

35.5 Conclusion

Data logging lets scientists and engineers evaluate a variety of phenomena, fromweather patterns to factory performance. PC-based data-logging systems provide the most flexibility, customization, and integration. To define a data-logging system, you must evaluate your requirements for acquisition, online analysis, logging, offline analysis, display, report generation, and data sharing. On the basis of these requirements, you can choose data-logging software and hardware to meet your needs.

35.6 Related Information

More information about PC-based data-logging systems is available fromNational Instruments in the form of white papers, application notes, customer solutions, and product information. Visit http://www.ni.com and search for "data logging" to view available information.

Index

A

B

D

E

F

K

L

M

N

Q

R

S

T

U

V

W

Z

L'autonomie des collectivités territoriales dans l'espace ouest africain francophone

Logiques Juridiques

Fondée par Gérard Marcou
Dirigée par Jean-Claude Némery et Thomas Perroud

Le droit n'est pas seulement un savoir, il est d'abord un ensemble de rapports et pratiques que l'on rencontre dans presque toutes les formes de sociétés. C'est pourquoi il a toujours donné lieu à la fois à une littérature de juristes professionnels, produisant le savoir juridique, et à une littérature sur le droit, produite par des philosophes, des sociologues ou des économistes notamment.

Parce que le domaine du droit s'étend sans cesse et rend de plus en plus souvent nécessaire le recours au savoir juridique spécialisé, même dans des matières où il n'avait jadis qu'une importance secondaire, les ouvrages juridiques à caractère professionnel ou pédagogique dominent l'édition, et ils tendent à réduire la recherche en droit à sa seule dimension positive. À l'inverse de cette tendance, la collection « Logiques juridiques » des éditions L'Harmattan est ouverte à toutes les approches du droit. Tout en publiant aussi des ouvrages à vocation professionnelle ou pédagogique, elle se fixe avant tout pour but de contribuer à la publication et à la diffusion des recherches en droit, ainsi qu'au dialogue scientifique sur le droit. Comme son nom l'indique, elle se veut plurielle.

Dernières parutions

Alex ATANASOV, *La machine judiciaire. Analyse économique et comparative du fonctionnement du système judiciaire*, 2023.
Mohamed KETATA, *Droit du marché financier et droit des contrats. Contribution à l'étude de l'opération de marché en droits tunisien et français*, 2023.
Van Anh LY, *La codification du droit international privé vietnamien à la lumière de la codification québécoise*, 2023.
Ruben AKONO MINLO, *Introduction au droit minier applicable au Cameroun. Une vue synoptique des aspects juridiques de l'exploitation minière*, 2023.
Yao AGBETSE, *La supervision de la mise en œuvre des droits de l'homme à l'épreuve de la volonté politique des États africains. La Commission africaine des droits de l'homme et des peuples dans tous ses états*, 2023.
Golnaz KARIMZADEH MEIBODI, *La subrogation réelle en droit patrimonial de la famille*, 2023.

Balakiyém Gnazouyoufei

L'autonomie des collectivités territoriales dans l'espace ouest-africain francophone

Préface de Jean-François Brisson

L'Harmattan

5-7, rue de l'École polytechnique - 75005 Paris

www.editions-harmattan.fr
ISBN : 978-2-336-58013-5
EAN : 9782336580135
Dépôt légal : janvier 2026

À mes parents

À mes frères et sœurs

SOMMAIRE

PREFACE

Diplômé de l'Université de Kara au Togo, inscrit à l'Université de Bordeaux dès la première année de Master, Monsieur Balakiyém GNAZOUYOUFEI nous livre avec ses *« Recherches sur l'autonomie des collectivités territoriales dans l'espace ouest-africain francophone »*, une étude tout à fait originale et d'une grande modernité.

Caractérisée par le souci de se démarquer des travaux déjà existants sur la décentralisation en Afrique, l'approche retenue par Balakiyém GNAZOUYOUFEI fait honneur à la jeune doctrine africaine. Elle repose sur une approche critique des dispositifs institutionnels retenus par les États de l'espace ouest-africain francophone. A cet égard, elle témoigne de la hauteur de vue de l'auteur, lequel a su inscrire sa recherche dans le temps long, adoptant ainsi une approche de type historique l'amenant à exploiter des travaux, habituellement peu sollicités par les juristes, lui permettant au final d'identifier les causes profondes de la situation actuelle. Approche d'une grande modernité aussi dans la mesure aussi où Monsieur Balakiyém GNAZOUYOUFEI invite les législateurs africains à se départir de leur tendance à voir dans les systèmes juridiques occidentaux une sorte de modèle de référence ; et ce, afin de proposer des solutions africaines pour l'Afrique.

Interroger au regard des standards juridiques internationaux l'autonomie des collectivités territoriales dans l'organisation des États de l'espace ouest africain francophone, mesurer le décalage entre les énoncés normatifs des textes constitutionnels et la réalité vécue par les acteurs publics locaux, en rechercher les causes profondes relevaient d'un parti pris scientifique audacieux, c'est à quoi s'est attelé Monsieur Balakiyém GNAZOUYOUFEI à l'occasion d'un travail de cinq années de recherche distingué par le jury appelé à l'évaluer.

Il fallait du courage pour s'engager dans cette recherche tant les difficultés étaient nombreuse. L'absence de sources documentaires notamment de jurisprudences, susceptibles d'intéresser les relations entre l'État et les collectivités territoriales, l'obligea ainsi à s'en tenir pour l'essentiel à l'étude des textes constitutionnels, législatifs ou réglementaires, le privant d'éléments d'analyse ailleurs utiles à une juste compréhension des systèmes juridiques étudiés. La rareté des recherches doctorales relative à la décentralisation en Afrique constitua l'autre écueil, et avec elle, le renvoi nécessaire aux travaux de la doctrine française, eux-mêmes construits à partir des solutions propres à l'organisation territoriale française, exposant alors l'analyse à un risque de superposition artificielle entre le discours doctrinal de référence et la situation spécifique des organisations territoriales africaines. Les concepts analytiques n'offraient eux-mêmes que peu de garanties.

L'auteur finira, après de longues hésitations, à privilégier la notion d'« autonomie » par rapport à d'autres concepts approchants, mais jugées moins signifiants, tels que la décentralisation ou la libre administration. Un choix judicieux dès lors que l'autonomie, débarrassée de la charge politique qu'elle a dans la doctrine publiciste française, correspond désormais aux standards internationaux prônés par la Charte européenne de l'autonomie locale qui en donne une définition générale appuyée sur une liste de critères et conditions permettant d'en apprécier la portée juridique dans chacun des Etats étudiés. Les conclusions de la recherche, enfin, allaient elles-mêmes s'avéraient peu concluantes. M. Balakiyém GNAZOUYOUFEI dressant le constat, sans appel, de l'ineffectivité du concept d'autonomie locale dans les systèmes juridiques de l'espace ouest-africain francophone.

Mais l'intérêt de ce travail n'est pas dans ses conclusions, il est dans les explications renouvelées qu'en donnent Balakiyém GNAZOUYOUFEI dont l'esprit critique et la profondeur des analyses n'échapperont pas au lecteur. C'est à une véritable entreprise de déconstruction que se livre l'auteur quand il cherche à en découvrir les causes profondes. Des causes qui tiennent à l'importation d'un modèle territorial inadapté aux réalités africaines, à la manière dont a été conçue la décentralisation en Afrique après la période coloniale, aux comportements de dirigeants africains en quelque sorte piégés par l'héritage institutionnel issu de la colonisation et par les exigences liées à la conditionnalité de l'aide au développement.

Ce faisant, la recherche s'inscrit dans le débat actuel de déconstruction des modèles juridiques établis à partir de matériaux tirés du droit français et contribue grandement à la compréhension des stagnations des processus de décentralisation en Afrique de l'Ouest francophone. La greffe initiale conjuguée au prêt à penser juridique amènent encore aujourd'hui les juristes africains à s'en remettre aux constructions juridiques élaborées en France, ce qui contribue à l'ineffectivité de la régulation par le droit. Les institutions et les conceptions importées ne sauraient rendre compte de la réalité politique et sociale des États d'Afrique noire francophone. Pour autant, l'idée de décentralisation s'est imposée dans les esprits. Nul ne peut nier que les énoncés textuels ont joué un rôle décisif et le retour en arrière ne parait plus possible. Comme le martèle l'auteur, la décentralisation a vocation à accompagner le développement d'États qui sont encore en construction démocratique. Fidèle aux enseignements d'Alexis de Tocqueville, il souligne combien le rôle joué par les collectivités territoriales est un moyen de limiter l'hégémonie du pouvoir central en mettant en exergue à l'égard des acteurs publics des exigences nouvelles de transparence et de redevabilité. La difficulté rencontrée par les États africains tient toutefois à ce que le modèle d'administration décentralisée retenu, parce qu'imposé, d'abord par la puissance coloniale, puis par les bailleurs de fonds dans le cadre de l'aide au développement, ne permet pas, en raison de son inadaptation à la réalité africaine, de répondre aux objectifs affichés, tant d'amélioration de la démocratie que de lutte contre la pauvreté. Balakiyém GNAZOUYOUFEI insiste pour admettre que la responsabilité n'en revient que partiellement aux dirigeants africains : d'abord, parce que les choix d'organisation territoriale sont encore aujourd'hui dictés de l'extérieur (rôle des organisations internationales et experts) ; ensuite, parce qu'il s'avère difficile de se défaire de l'influence de la culture juridique française qui, pour beaucoup encore en Afrique, constitue le modèle de référence auquel on saurait renoncer sans régresser. Or le modèle institutionnel français connaît lui aussi ses propres difficultés en témoignent la multiplication des textes de lois intervenus ces dernières années, la complexité des solutions retenues et les hésitations du législateur français contemporains. Dans ces conditions, l'auteur parvient sans difficulté à nous convaincre qu'il appartient aux États africains de rechercher des solutions propres et adaptées aux réalités locales africaines.

Balakiyém GNAZOUYOUFEI se propose alors d'interroger le « tabou » que constitue l'autonomie des collectivités territoriales en Afrique. Car si l'idée de décentralisation semble désormais admise dans son principe, la question de l'autonomie réelle à accorder des collectivités territoriales reste en suspens. Fort de ce constat, il restait à l'auteur à mesurer l'ampleur du déficit d'autonomie locale en Afrique de l'ouest. C'est à partir des standards internationaux que nous offre la Charte européenne de l'autonomie locale que Balakiyém GNAZOUYOUFEI a bâti sa grille d'analyse. Elle lui permet de mettre en évidence les trop larges possibilités laissées au pouvoir central d'intervenir dans les moindres aspects de la gestion des collectivités territoriales, faisant de la libre administration, proclamée dans les textes, une sorte de leurre au service d'une organisation territoriale purement artificielle.

De ce tableau critique, deux pistes de réflexion émergent. L'une consiste à en finir avec le principe, hérité du droit français, d'unité catégorielle des communes. M. Balakiyem GNAZOUYOUFEI propose de favoriser la différenciation statutaire de manière à tenir compte des spécificités de la ruralité et de la diversité ethnique. L'autre proposition consiste dans ce cadre rénové à reconnaître institutionnellement la place qu'occupe dans l'organisation sociale la chefferie traditionnelle. Les chefferies traditionnelles ont perduré malgré la période coloniale. Elles correspondent à une réalité sociale forte. L'auteur se défend d'y voir un retour au tribalisme et la division ethnique. D'après lui, l'institutionnalisation de la chefferie, reposant sur une légitimité coutumière distincte, serait d'autant plus possible qu'elle repose sur un mode d'organisation proche du modèle communal.

Sans doute, la lecture de ce travail, documenté et argumenté-, peut laisser la place à l'expression d'un regret : l'auteur pouvait-il, devait-il, aller plus loin dans l'entreprise de reconstruction, après s'être méthodiquement livré à sa déconstruction, et proposer un autre modèle ? Un modèle africain d'organisation territoriale débarrassé de l'héritage colonial et post-colonial. A sa décharge, la prospective n'est jamais sans risque dans une recherche juridique. Celui, qui s'y risque, encourt le reproche de faire de la « science-fiction » et l'ingénierie juridique n'est pas nécessairement ce qui est attendu d'une recherche académique. En outre, il s'agissait d'un autre projet de thèse que les travaux de M. GNAZOUYOUFEI rendent, sans doute, aujourd'hui, possible.

C'est tout le mérite de cet important travail de recherche que d'ouvrir la voie à d'autres chercheurs pour jeter les bases conceptuelles d'un système d'administration territoriale adapté aux spécificités des États d'Afrique.

Jean-François Brisson,
Fait à Bordeaux le 18 octobre 2025

REMERCIEMENTS

Toute ma reconnaissance va, en tout premier lieu, au professeur Jean-François Brisson. Si l'exercice d'écriture d'une thèse peut parfois tourmenter celui qui s'y essaie, mon directeur de thèse a participé à ce que ce sentiment m'atteigne le moins possible. Par sa disponibilité à me lire et à m'écouter, le professeur Brisson m'a permis de nourrir ma réflexion, en l'affinant ou en l'amendant.

Je tiens également à remercier Mme Malyis Douence ainsi que Mme et M. les professeurs Laetitia Janicot, Nadjombé Gbeou-Kpayile et Pascal Combeau. Je suis très sensible à l'honneur qu'ils m'ont fait de s'être intéressés à ce travail et d'en avoir apprécié la teneur, notamment en acceptant de constituer mon jury de thèse.

Je ne saurais exprimer avec suffisamment de mots l'étendue de ma reconnaissance, pour leurs regards et conseils précieux, envers Clothilde, Manon, Mattéo, Paul, Deyana, Élodie, Yaodia, Frédéric. Qu'ils sachent combien la force de leur amitié, qui n'a jamais compté le temps qu'ils ont bien voulu me consacrer m'est cher et m'honore.

Cette thèse n'est certainement pas l'œuvre d'une vie. Mais elle est au moins l'œuvre d'une jeunesse. Que tous ceux qui ont croisé ma route au cours de ces vingt-sept années s'en trouvent remerciés. Ils ont tous contribué, à leur façon, à forger l'homme qui a écrit ces lignes. Je pense, notamment à mes vieux amis d'ici et d'ailleurs : Edwige, Rébecca, Rufus, Fernanda, Hannah, Melchior, Rodrigue, Hervé, Samuel, Jonas, Marius, Tchasse. Votre présence, parfois malgré la distance, a été déterminante.

Enfin, qu'il me soit permis de réserver mes derniers remerciements, les plus affectueux, à mes frères et sœurs ainsi qu'à mes parents, pour leur compréhension, et parfois pour leur incompréhension, pour leur entière confiance, leur indéfectible soutien et leur extrême patience.

PRINCIPALES ABREVIATIONS

AFD – Agence française de développement
AJCT – Actualité juridique, collectivités territoriales
AJDA – Actualité juridique, droit administratif
AJFP – Actualité juridique, fonction publique
al. – alinéa
APD – Aide publique au développement
Art. – article

CAA – Cour administrative d'appel
CE – Conseil d'État
Cf. – *Confer,* (se reporter à)
CGCT – Code général des collectivités territoriales
chron. – chronique
CODESRIA – Conseil pour le développement de la recherche en sciences sociales en Afrique
Coll. – collection
comm. – commentaire
cons. – considérant

D. – Recueil Dalloz
dactyl. – dactylographiée (pour les thèses non publiées)
DC – Type de décision du Conseil constitutionnel portant sur la conformité a priori de la loi à la Constitution
Dir. – sous la direction de
Dr. Adm. – Droit administratif

éd. – édition
EDCE – Études et documents du Conseil d'État
s. – et [pages] suivante
Fasc. – Fascicule
Gaz. Pal. – Gazette du Palais
Idem – exactement la même référence que la note précédente
Ibid. – Ibidem - la même référence que la note précédente, excepté la page
Infra – ci-dessous

JCP A – La semaine juridique, administrations et collectivités territoriales
JCP G – La semaine juridique, édition générale
LGDJ – Librairie générale de droit et de jurisprudence
LPA – Les petites affiches

n° – numéro
not. – notamment
OCDE – Organisation pour la coopération et le développement économique
op. cit. – *opus citatum*, ouvrage/article cité
p. – pages
par ex. – par exemple
PDM – Partenariat pour le développement municipal
PUAM – Presses universitaires d'Aix-Marseille
PUF – Presses universitaires de France
PUG – Presses universitaires de Grenoble
QPC – Question prioritaire de constitutionnalité
RDP – Revue de droit public
RDSS – Revue de droit sanitaire et social
RFAP – Revue française d'administration publique
RFDA – Revue française de droit administratif
RFDC – Revue française de droit constitutionnel
RFFP – Revue française de finances publiques
RFSP – Revue française de science politique
RGCT – Revue générale des collectivités territoriales
RIDC – Revue internationale de droit comparé

spé. – spécialement, plus particulièrement (en général, n° de page exact d'une citation ou d'une référence précise)
Supra – ci-dessus

TA – Tribunal administratif
trad. – traduit par

UEMOA – Union économique et monétaire ouest africaine

v. – voir, se référer à
vol. – volume

INTRODUCTION

> « Le règne du droit tient […] du rêve, si l'on admet que l'effectivité de la règle de droit est la condition nécessaire de l'existence de l'État de droit »[1].

1. - Entre 1960, année d'accession à l'indépendance de nombreux États d'Afrique subsaharienne et le début des années 90, la plupart des pays de la sous-région ouest-africaine ont connu une longue période de monopartisme ou de domination militaire du pouvoir politique. Les conférences nationales souveraines qui se sont tenues à partir de 1990 marquent le commencement d'une nouvelle ère. Décentraliser, tel est le nouveau mot d'ordre sur le continent[2]. Les dirigeants des États africains affirment leur engagement à promouvoir la démocratie et la décentralisation conformément aux recommandations issues des conférences nationales souveraines. Ces recommandations sont reprises et traduites dans les Constitutions. La Constitution de la IIIe République du Niger du 26 décembre 1992 dispose dans son article 115 que « *l'administration territoriale repose sur les principes de la décentralisation et de la déconcentration* ». Celle de la IVe République togolaise adoptée en octobre de la même année dispose à l'alinéa 1 de l'article 141 que « *la République togolaise est organisée en collectivités territoriales sur la base du principe de décentralisation, dans le respect de l'unité nationale* ». Le même processus est enclenché au Burkina Faso avec l'adoption par référendum de la Constitution du 2 juin 1991 qui consacre l'organisation du pays « *en collectivités territoriales* » en son article 143, et énonce « *la participation démocratique des populations à la libre administration des collectivités territoriales* » à

---

[1] M. KAMTO, *Pouvoir et droit en Afrique noire. Essai sur le fondement du constitutionnalisme dans les États d'Afrique noire francophone*, thèse, LGDJ, 1986, p. 441.

[2] C'est d'ailleurs ce qui justifie le choix opéré par Charles Nach Mback d'étudier dans sa thèse le lien entre l'adoption des politiques de développement et le processus généralisé de démocratisation sur le continent avec comme repère temporel l'année 1990. C. NACH MBACK, *Démocratisation et décentralisation. Genèse et dynamiques comparés des processus de décentralisation en Afrique subsaharienne*, thèse, Karthala-PDM, 2003,

l'article 145. Dans le même sens, la Constitution malienne du 25 septembre 1992 consacre à nouveau en son article 97 l'existence des collectivités territoriales.

2. - Le choix de décentralisation à ce moment de l'histoire de la construction des jeunes États africains devait répondre à un triple objectif : approfondir le processus démocratique, responsabiliser les populations dans la gestion de leur développement et promouvoir des services de proximité sur la base d'une nouvelle gouvernance locale. Les collectivités territoriales étaient considérées comme des institutions capables de contribuer à la consolidation de l'unité nationale et à la promotion d'un développement local durable axé notamment sur la réduction de la pauvreté et la délivrance des services sociaux de base. La Communauté économique des États de l'Afrique de l'Ouest (CEDEAO) réaffirme dans son protocole sur la démocratie et la bonne gouvernance adopté en 2001 à l'article premier que constitue un principe constitutionnel commun à tous les États de la communauté « *la participation populaire aux prises de décision, le strict respect des principes démocratiques, et la décentralisation du pouvoir à tous les niveaux de gouvernement* »[3]. Les espoirs placés dans le processus de décentralisation se matérialisent par l'adoption en 2014 d'une Charte africaine des valeurs et des principes de la décentralisation, de la gouvernance locale et du développement local par les États membres de l'Union africaine (UA)[4]. Les États réaffirment leur volonté collective d'approfondir la démocratie participative et s'engagent à respecter un noyau dur de principes fondamentaux, notamment l'autonomisation des citoyens et des communautés, de promouvoir l'obligation de rendre compte et la transparence dans les institutions publiques, de promouvoir et de protéger la diversité culturelle. Ils reconnaissent le rôle des collectivités territoriales comme premier niveau d'exercice de la démocratie[5].

3. - Cependant, après des décennies de mise en œuvre, les modalités d'organisation administrative et territoriale de l'État suscitent encore des interrogations. C'est une antienne classique dans l'étude du droit en

---

[3] Protocole A/SP1/12/01 sur la démocratie et la bonne gouvernance additionnel au protocole relatif au mécanisme de prévention, de gestion, de règlement des conflits, de maintien de la paix et de la sécurité adopté à Dakar le 21 décembre 2001.

[4] La Charte a été adoptée le 27 juin 2014 à la vingt troisième session ordinaire de la conférence de l'UA à Malabo en Guinée Équatoriale.

[5] Voir le préambule de la Charte.

Afrique que d'affirmer qu'entre le texte et la pratique du texte peuvent exister d'importantes divergences[6]. L'analyse des politiques institutionnelles de décentralisation[7] sur le continent africain semble confirmer ce propos et offre ainsi l'occasion de vérifier l'effectivité du droit de la décentralisation en Afrique. Rien n'est plus au cœur de la relation entre droit et société que la question de l'effectivité[8]. Elle permet de s'attacher à une description du fonctionnement d'un système juridique et de ses liens avec le réel[9]. Ainsi, il est possible d'observer les écarts entre la consécration de règles en faveur de la décentralisation territoriale et le comportement des acteurs qui traduisent une continuité de la centralisation. La force du droit est alors démystifiée par sa faible capacité à assurer son observance[10]. L'ineffectivité met en doute la capacité du droit de la décentralisation à remplir objectivement sa fonction au sein de la société. Les objectifs[11] de la décentralisation paraissent à cet effet compromis non seulement par les conditions de sa production, mais aussi par les modalités de sa mise en œuvre en Afrique.

4. - Si à la question « décentraliser, qu'est-ce à dire ? », Émile Marcère répondait qu'il s'agit d'« un terme un peu vague et qu'il faut conserver pourtant puisqu'il exprime un vœu intense, un sentiment général : c'est comme un cri sorti de milliers de poitrines oppressées, mais c'est un mot dont le sens n'est pas défini »[12] ; ce « sentiment général », ce « cri sorti de

---

6 L. MATALA-TALA, « L'ineffectivité du droit positif en Afrique subsaharienne », *Civitas Europa*, vol. 31, n° 2, 2013, pp. 239-260.

7 Sur la nature institutionnelle de la décentralisation, v., J.-L. QUERMONNE, « Les politiques institutionnelles », in M. GRAWITZ, J. LECA (dir.), *Traité de science politique*, PUF, vol. 4, 1985, p. 61, spé., pp. 77-78.

8 J. COMMAILLE, « Effectivité », in D. ALLAN et S. RIALS (dir.), *Dictionnaire de la culture juridique*, PUF, 2003, p. 583. Pour une argumentation détaillée sur le sens de cette acception, voir les développements *infra*.

9 J. BETAILIE, « Le concept d'effectivité-action », in L. MIGUEZ MACHO, M. ALMEIDA CERREDA (coord.), Los retos actuales del derecho administrativo en el Estado autonomico : estudios en humenaje al profesor José Luiz Carro Fernández-Valmayor, Andavira, Fundación Democracia y Gobierno Local, vol. 2, 2017, p. 367-383.

10 R. M. INGHLITERRA, La réalisation du droit de la non-discrimination, thèse, LGDJ, 2022, p. 2.

11 Efficacité administrative, développement, démocratie. Pour une argumentation détaillée, voir les développements *infra* n° 274 et s.

12 E. MARCERE, « Lettre sur la décentralisation », *Revue politique et parlementaire*, 1895, p. 1, cité par S. REGOURD, « La génétique de la décentralisation. Tabous, métaphores et autres occultations », in N. KADA (dir.), *Les tabous de la décentralisation*, Berger-Levrault, coll. « Au fil du débat », 2015, p. 15.

milliers » de voix est d'abord en Afrique l'expression d'un « vœu intense » des organisations internationales pourvoyeurs de l'Aide publique au développement (APD). La Banque mondiale dans son rapport sur le développement de 1989 recommandait aux États africains de renouveler leurs institutions pour garantir une utilisation efficace des ressources[13]. La décentralisation devient à cet effet, un axe majeur des réformes institutionnelles sur le continent africain. Elle prend ainsi, toute sa place dans les exigences de restructuration et de modernisation des administrations publiques africaines comme le souligne Charles Nach Mback[14]. « L'appui aux politiques de décentralisation et à la gestion des municipalités africaines est devenu une des priorités importantes de la communauté internationale, et les bailleurs de fonds multilatéraux tels que la Banque mondiale, la CEE ou la BAD, ou bilatéraux et notamment la France, y consacrent chaque année une part croissante de leur aide »[15].

5. - Cet attrait pour la décentralisation trouve une première explication dans l'échec des politiques d'ajustement structurel[16]. Il tient à deux préoccupations majeures, l'instauration d'un « bon » gouvernement et la promotion du développement[17]. La décentralisation est perçue par les organisations internationales comme un rempart efficace contre les excès du pouvoir central[18] et son incapacité à produire le développement local[19]. Le « bon » gouvernement serait alors celui qui contribue à la libéralisation des énergies locales à travers la décentralisation du pouvoir de décision, la promotion de la participation des populations au processus décisionnel et

---

[13] Cf. WORLD BANK, *Sub-saharan Africa: from crisis to substainable growth*, 1989.

[14] C. NACH MBACK, *Démocratisation et décentralisation. Genèse et dynamiques comparés des processus de décentralisation en Afrique subsaharienne*, thèse, Karthala-PDM, 2003, p. 27.

[15] J.-L. VENARD, « Bailleurs de fonds et développement local », in S. JAGLIN, A. DUBRESSON (dir.), *Pouvoirs et cités d'Afrique noire : décentralisations en question*, Karthala, 1993, pp. 20-33.

[16] *Cf.* P. HUGON, « Les effets des politiques d'ajustement sur les structures politiques africaines », in G. CONAC (dir.), *L'Afrique en transition vers la démocratie*, Economica, 1993, p. 89 et s.

[17] Plusieurs épithètes étaient accolées à cette notion de développement : durable, participatif, local, etc.

[18] D. G. LAVROFF, « Conclusion », in A. PARIENTE (dir.), *La séparation des pouvoirs. Théorie contestée, pratique renouvelée*, Dalloz, 2007, p. 150.

[19] C. NANAKO, *La libre administration des collectivités territoriales au Bénin et au Niger*, thèse dactyl., Université d'Abomey-Calavi (Bénin), 2016, p. 2.

un rétrécissement du rôle de l'État sur les missions essentielles[20]. Face aux exigences externes des partenaires au développement et à la résurgence des replis identitaires avec des revendications tendant à la prise en compte des particularités locales, l'État en Afrique se décentralise après des années de centralisation. En proclamant sa « *profession de foi décentralisatrice* »[21], l'État cherche d'abord à assurer sa pérennité et à garantir son idéal d'unité nationale.

6. - La reconnaissance par les constituants africains de l'existence aux côtés de l'État d'autres centres de décision — les collectivités territoriales — dans le système d'administration du territoire conduit à la restauration du pluralisme territorial[22]. Cela est vrai au sens où la décentralisation suppose l'existence d'une pluralité de centres autonomes de décision. Les « organes locaux » ont dans cet ordre d'idée, la maitrise juridique de leur activité c'est-à-dire qu'ils sont libres de prendre, dans le respect des lois et règlements, la décision qu'ils veulent[23]. Partant de ce postulat, on peut se demander si la multiplication des textes et des discours en faveur de la décentralisation a suffi pour changer les comportements et les mentalités afin de provoquer le développement tant souhaité. La recomposition de l'armature institutionnelle a-t-elle vraiment permis aux États africains de changer de religion en adoptant avec une ferveur nouvelle la foi décentralisatrice, après s'être, pendant longtemps, adonnée avec passion à la vénération centralisatrice[24] ?

---

20 Voir sur le changement de la nature de l'État, J. CHEVALLIER, *L'État post-moderne*, LGDJ, coll. « Droit et société », 2017.
21 C. NACH MBACK, *Démocratisation et décentralisation. Genèse et dynamiques comparés des processus de décentralisation en Afrique subsaharienne, op.cit.*, p. 28.
22Sur le pluralisme territorial, v. not, G. SAEZ, « Recompositions de l'action publique et pluralisme territorial », *Annuaire des collectivités locales*, t. 19, 1999. pp. 17-26 ; J.-B. AUBY, *La décentralisation et la droit*, LGDJ, coll. « Systèmes », 2006 ; Conseil d'État, *Décentralisation et ordre juridique*, La documentation française, 1993 ;
23 C. EISENMANN, *Centralisation et décentralisation. Esquisse d'une théorie générale*, LGDJ, 1948. Voir pour une systématisation, E. THALINEAU, Essai sur la centralisation et la décentralisation. Réflexions à partir de la théorie de Charles Eisenmann, thèse dactyl. Université de Tours, 1994 ; J. CAILLOSSE, Les « mises en scènes » juridiques de la décentralisation. Sur la question du territoire en droit public français, LGDJ, coll. « Droit et société », 2009, pp. 53-77.
24 Nous reprenons pour notre compte cette interrogation formulée par Jacques Baguenard à propos des réformes décentralisatrices en France. V. en ce sens, J. BAGUENARD, *La décentralisation*, PUF, coll. « Que sais-je ? », 2004, pp. 3-22.

7. - La réponse ne peut être absolue tant il est difficile, encore aujourd'hui, de « théoriser » la décentralisation[25]. Si l'on s'accorde sur le fait que la décentralisation dans un État unitaire se mesure à la liberté dont disposent les collectivités territoriales, en la matière, tout est une question de nuances et de degrés[26]. Jacques Baguenard relève à ce propos qu'« *elle constitue nécessairement un point d'équilibre, instable et fragile, résultant d'un compromis dynamique entre des forces centripètes qui poussent au renforcement de l'unité étatique et des forces centrifuges qui stimulent l'épanouissement de la diversité locale* »[27]. Toutefois, décentraliser suppose que des conditions minimales soient réunies. En effet, la décentralisation ne saurait exister sans une redistribution des compétences entre l'État et les collectivités territoriales, une indépendance des organes locaux par rapport au pouvoir central et une gestion autonome par les autorités locales des affaires relevant de leurs compétences. La réformation de l'organisation administrative et territoriale de l'État conduit ainsi à une répartition des compétences entre un organe central et des organes périphériques relativement autonomes[28]. Par conséquent, la réalisation de la décentralisation s'apprécie dans un État en fonction du degré de liberté reconnue aux collectivités territoriales. L'étendue des compétences ainsi que la liberté d'initiative et de gestion dont bénéficient les collectivités territoriales permettent de déterminer l'effectivité ou non de la décentralisation. Après trois décennies de mise en œuvre des politiques décentralisatrices en Afrique, l'idée de décentralisation semble désormais admise dans son principe, mais la question de l'autonomie des collectivités territoriales constitue toujours un tabou comme c'est d'ailleurs le cas en droit français[29].

8. - Notre recherche se propose de s'interroger sur ce tabou en Afrique. La proclamation de la décentralisation apparait insuffisante pour la réalisation des fonctions sociales qu'on lui assigne. L'attention doit désormais se porter sur les modalités susceptibles de favoriser son effectivité. Tel est le sens de la consécration de l'autonomie des collectivités territoriales dans les États d'Afrique noire francophone. Or,

---

[25] J. BENOIT, *Théorie juridique de la décentralisation administrative*, thèse, dactyl., Université de Paris, 1990.

[26] F. DELPERE, « Organisation territoriale et exception française », *Cahiers français*, n° 318, janvier-février 2004, p. 15.

[27] J. BAGUENARD, *La décentralisation*, PUF, coll. « Que sais-je ? », 2004, pp. 3-22.

[28] C. ROIG, « Théorie et réalité de la décentralisation », *RFSP*, n° 3, 1966, p. 452.

[29] N. KADA (dir.), *Les tabous de la décentralisation*, Berger-Levrault, coll. « Au fil du débat », 2015.

l'autonomie locale qui ne se laisse pas facilement appréhender est susceptible d'être éprouvée par le constat de son ineffectivité. Entreprendre une recherche sur ce tabou du droit de la décentralisation nécessite de préciser l'objet de l'étude marqué par une imprécision du vocabulaire (**§1**). Il sera alors possible d'expliciter davantage la problématique de l'étude (**§2**) et les méthodes utilisées (**§3**) pour y répondre. Enfin, la thèse soutenue sera présentée (**§4**).

§1.- L'autonomie des collectivités territoriales comme objet de la recherche

9. - L'histoire des institutions politiques et administratives des États d'Afrique noire francophone est liée à celle de la France du moins à partir du XIXe siècle[30]. Pour appréhender le concept d'autonomie locale en Afrique, il importe de revenir sur les conditions historiques de son émergence. Ce détour par l'histoire nous permettra dans un premier temps de montrer qu'en raison de la perpétuation du legs colonial, l'autonomie locale en Afrique porte en elle, les germes de son ineffectivité (**A**). Dans un second temps, nous préciserons ce que l'on entend par autonomie locale dans la présente recherche (**B**).

A- L'émergence historique du concept d'autonomie locale

10. - Il n'est pas évident de parler d'autonomie locale dans un système juridique marqué par une centralisation administrative. Cette dernière constitue sans aucun doute la particularité du modèle français d'administration publique[31] hérité par les anciennes colonies françaises d'Afrique[32]. Alexis de Tocqueville dans *L'Ancien régime et la Révolution*,

---

30 Voir en ce sens, G. CONAC (dir.), *Les institutions administratives des États francophones d'Afrique noire*, Economica, 1979.

31 G. BIGOT, *L'administration française. Politique, droit et société*, 1789-1870, LexisNexis, 2014, p. 158.

32 Les rapports entre le centre et la périphérie tels que décrit par Jean Chaptal sont marquées par leur étroitesse. Il s'agit pour le pouvoir central de s'assurer de la transmission de « *la loi et* [des] *ordres du gouvernement jusqu'aux dernières ramifications de l'ordre social avec la rapidité du fluide électrique* ». Cité par L. JAUME, S. SOLEIL, « Centralisation/Décentralisation. Retour sur quelques certitudes historiques », *AJDA*, 2005, p. 760.

ouvrage publié en 1856, montrait d'ailleurs que la centralisation n'est pas une création de la Révolution, mais un héritage de l'Ancien Régime[33]. Cette volonté de centralisation qui caractérise l'histoire de l'administration française[34] et mise en œuvre dans les colonies africaines permet d'expliquer la rupture du lien originel entre commune et autonomie dans les États d'Afrique noire francophone. En reconfigurant l'État en Afrique sur le modèle occidental, la colonisation a détruit l'autonomie initiale des communautés de base (**1°**). L'absorption de ces communautés par le pouvoir étatique permet de dresser le constat de l'ineffectivité de l'autonomie locale (**2°**).

1. La rupture du lien originel entre commune et autonomie dans le contexte colonial

11. - La période féodale est marquée sur le continent africain par l'existence de communautés politiques constitutives d'entités préexistantes à l'État[35]. Ces « *structures naturelles de peuplement* »[36] ont une existence autonome, « *c'est-à-dire une légitimité antérieure à leur consécration institutionnelle* »[37]. Cette autonomie des communautés traditionnelles africaines disparait progressivement sous l'influence de la pensée coloniale française. Il est important pour comprendre le cheminement de l'autonomie locale en Afrique de se référer aux différentes évolutions institutionnelles en France. À la différence des communes françaises qui sont nées de la reconnaissance par le pouvoir central des groupements préexistants[38], la commune dans les pays d'Afrique subsaharienne nait de l'initiative du pouvoir colonial et se confine sur l'aire géographique des

---

33 Pour une approche critique de cette assertion, v. L. JAUME, S. SOLEIL, « Centralisation/Décentralisation. Retour sur quelques certitudes historiques », *op.cit.*, p. 760 et s.

34 Cf. P. LEGENDRE, Trésor historique de l'État en France. L'administration classique, Fayard, 1992.

35 En France, les communautés d'habitants ont aussi préexisté à l'État, v., S. REGOURD, « La génétique de la décentralisation. Tabous, métaphores et autres occultations », *op.cit.*, p. 19.

36 P. LEGENDRE, Trésor historique de l'État en France, op.cit., p. 61.

37 S. REGOURD, « L'autonomie communale : des vicissitudes de la théorie juridique à l'enseignement de l'histoire », *Pouvoirs locaux*, n° 1, 1989, p. 73.

38 F. ZERARI, « L'évaluation de l'autonomie des collectivités territoriales dans les systèmes juridiques d'inspiration française », *Les Annales de droit*, n° 12, 2018, pp. 211-227.

circonscriptions administratives qu'il découpe[39]. En effet, s'il est possible d'établir une filiation directe entre les 44 000 paroisses de l'Ancien régime français et les 36 000 communes actuelles[40], une telle opération est impossible dans le cas des communes africaines.

12. - L'existence d'une autonomie communale originelle des communautés est fondée en France sur le serment juratoire[41]. Les communautés d'habitants ainsi formées obtenaient de leurs féodaux une concession dans le cadre d'une charte reconnaissant à la commune certains droits et privilèges en contrepartie du versement d'une somme et de la reconnaissance des obligations de vassalité[42]. Les franchises dont disposaient les communautés d'habitants variaient d'une ville ou d'un bourg à l'autre. La période d'affirmation de ces institutions locales connaitra son apogée entre le XIIIe et XIVe siècle. L'autonomie initiale des communes va se déliter progressivement pour des raisons économiques, mais surtout politique et juridique. Le développement des oligarchies bourgeoises conduit à la confiscation par ces dernières des privilèges octroyés par la charte « *réduisant à néant la relation de solidarité constitutive de la commune dans le serment initial d'aide mutuelle* »[43].

---

39 L. NGOMO TSIMI, *L'autonomie administrative et financière des collectivités territoriales décentralisées : l'exemple du Cameroun*, thèse, dactyl., Université Paris-Est, 2010, p. 26.

40 S. REGOURD, « La génétique de la décentralisation. Tabous, métaphores et autres occultations », *op.cit.*, p. 15.

41 Ce serment permettait de vivre en paix dans une même Cité et de se porter mutuellement assistance. Il constitue l'essence de la commune française. Charles Petit-Dutaillis, affirme à ce propos que : « *sans association par serment, il n'y avait pas de commune, et cette association suffisait pour qu'il y ait commune* », la « *commune a exactement le même sens que serment communal* ». La commune d'alors se manifestait sous la forme de groupements ou d'associations entre les habitants, unis par une solidarité de voisinage, inscrite sur le périmètre d'un territoire. Ainsi, la période féodale fut marquée par l'existence d'institutions, ancêtres des institutions locales d'aujourd'hui. De cette période, on retiendra volontiers que les traits dominants de la société féodale ne sont pas seulement ceux connus d'une société stratifiée où le roi cherche à affirmer son hégémonie, mais aussi ceux d'une société imprégnée de fortes valeurs humaines. En effet, de nombreux indices portent à croire qu'il s'agit d'une société fortement dominée par l'expression des groupements humains autonomes rattachés à leur territoire (la commune) et qui n'aspirent qu'à une seule chose, vivre en paix et en solidarité, conformément à un serment juré.

42 J.-B. AUBY, J.-F. AUBY, R. NOGUELLOU, *Droit des collectivités locales*, PUF, coll. « Thémis », 2015, p. 3 et s.

43 S. REGOURD, « La génétique de la décentralisation. Tabous, métaphores et autres occultations », *op.cit.*, p. 21.

13. - La dénaturation du concept d'autonomie intervient dès lors que se multiplient les opinions en faveur des vertus de la centralisation[44]. L'idée centralisatrice connait son heure de gloire sous l'administration napoléonienne avec le triomphe du modèle autoritaire du Consulat et de l'Empire[45]. La centralisation de l'an VIII s'est construite autour de deux éléments : le renforcement de l'autorité centrale et l'inféodation des autorités locales. La loi du 28 pluviôse an VIII consacre la mise sous tutelle des autorités locales. Ce texte du 17 février 1800 réorganise l'administration de la nation sur la base d'une hiérarchisation entre trois degrés constitués par les départements, les arrondissements et les municipalités. Grégoire Bigot évoquant cette loi estime qu'on pourrait parler à propos de cette organisation, de « constitution administrative » de la France pour au moins trois raisons. D'abord, elle est marquée selon l'auteur par la philosophie constitutionnelle d'un exécutif ramassé sur un seul homme qui inféode les assemblées qui ne garantissent *in fine* qu'un faux semblant de représentativité. Ensuite, les trois niveaux de l'administration locale sont placés sous l'autorité directe du chef de l'État, responsable du gouvernement de la Nation. La « chaine de dépendance » est maintenue par la nomination par le Consul ou l'Empereur des préfets, sous-préfets et maires. Enfin, la loi du 28 pluviôse an VIII est le reflet de la secondarisation des intérêts locaux par rapport à ceux de l'État. Le Consulat ne nie pas la spécificité des collectivités locales, mais simplement l'administration agissant en son sein est « *d'abord entendue comme une courroie de transmission de l'administration générale* »[46].

14. - Cette conception de l'administration locale qui prend forme sous le régime napoléonien va être éprouvée dans les colonies françaises d'Afrique à partir du XIXe siècle. Ainsi, la IIIe République marque une étape décisive en faveur des collectivités territoriales en métropole et dans les colonies africaines. En France, elle se caractérise par des réformes qui sont à l'origine de deux grandes lois demeurées célèbres : la loi

---

[44] G. BIGOT, *L'administration française. Politique, droit et société*, *op.cit.*, n° 126.

[45] Ce modèle autoritaire et centralisateur est lui-même inspiré des *précé*dents révolutionnaires de de la Convention et du Directoire (1792-1795). V. not., P. LEGENDRE, *Trésor historique de l'État en France. L'administration classique*, Fayard, 1992 ; J. MOREAU, M. VERPEAUX, *Révolution et décentralisation. Le système administratif français et les principes révolutionnaires de 1789*, Economica, coll. « Droit public positif », 1992.

[46] G. BIGOT, *L'administration française. Politique, droit et société, 1789-1870*, *op.cit.*, n° 127 et s.

départementale du 10 août 1871 et la loi municipale du 5 avril 1884. Cette dernière consacre une véritable clause de compétence générale[47] au profit du conseil municipal, qui désormais « *règle par ses délibérations les affaires de la commune* ». En Afrique, cette période symbolise la poursuite de la politique d'assimilation avec la création des premières communes dans la colonie du Sénégal[48].

15. - La commune africaine issue de cette époque coloniale se caractérise par la rupture de l'équation commune égale communauté traditionnelle préexistante. Cette absence de conformité entre la commune en Afrique et les communautés traditionnelles de base traduit la volonté du pouvoir colonial de rompre le lien entre commune et autonomie. L'objectif étant de briser toute tentative de révolte des populations locales. Ce faisant, le pouvoir colonial divise les légitimités traditionnelles et affermit sa position au sein des colonies. Les chefs des communautés traditionnelles devenus des agents auxiliaires de l'administration coloniale, l'idée d'une existence de communautés autonomes disparait définitivement[49].

16. - À partir de la IVe République, malgré des avancées remarquées, le droit de la décentralisation demeure marqué par une méfiance de l'État à l'égard des collectivités territoriales et de l'autonomie locale. Même dans les colonies françaises, aussi éloignées qu'elles soient, la France coloniale continue par pratiquer une politique d'assimilation et non d'autonomisation progressive considérée comme une atteinte à l'unité de la République formée de toutes ces colonies[50]. Les groupements humains reconnus en tant que collectivités territoriales sont alors mis au service de l'appareil étatique colonial. Les communes ayant perdu leur légitimité

---

47 Sur la clause générale de compétence, voir nos développements *infra ;* lire utilement : M. VERPEAUX, « Pavane pour une notion défunte. La clause générale de compétence », *RFDA*, 2014, p. 457 ; G. Le CHATELIER, « Le débat sur la clause générale de compétence est-il vraiment utile ? », *AJDA*, 2009, p. 186 ; J.-M. PONTIER, « Semper Manet. Sur une clause générale de compétence », *RDP*, 1984, p. 1443.

48 M. DIOUF, « L'idée municipale : une idée neuve en Afrique », *Politique africaine* n° 74, 1999, p. 13 et s., Voir les développements *infra* n° 226 et s.

49 Voir les développements *infra* n° 71 et s.

50 Sur une exception de l'apparition de l'autonomie dans la Ve République à propos des colonies : Titre XIII, art. 77, abrogé par la révision constitutionnelle d'août 1995 : « *Dans la Communauté instituée par la présente Constitution, les États jouissent de l'autonomie ; ils s'administrent eux-mêmes et gèrent démocratiquement leurs propres affaires* ».

territoriale n'ont plus vocation à revendiquer l'exercice d'un pouvoir originaire par lequel leurs habitants géraient en commun des intérêts. On assiste à une personnification[51] progressive des collectivités territoriales[52]. L'autonomie locale, désormais redoutée, est réduite dans son entendement philosophique. Le rapport État-collectivités territoriales ne désigne plus qu'une soumission des seconds envers les premiers[53].

17. - En somme, deux constats peuvent être faits sur l'autonomie locale en Afrique. D'une part, les trajectoires de naissance du mouvement communal en Afrique et en France sont liées par la préexistence de communautés autonomes. D'autre part, si dans le premier cas, les communautés d'habitants entretiennent un lien de filiation direct avec les communes actuelles, il en va autrement de la commune africaine. Cette dernière a été créée sur des espaces territoriaux qui ne correspondent pas à celles des communautés traditionnelles afin de briser les légitimités locales et conforter le pouvoir colonial. À ce stade, l'ineffectivité de l'autonomie locale s'expliquerait donc par la rupture de l'équation commune égale communauté traditionnelle préexistante. L'autonomie locale est, par ailleurs, soumise à une rationalité centralisatrice. Elle est toujours considérée comme un obstacle à l'unité nationale et à l'indivisibilité de la République. C'est la raison pour laquelle les autorités normatives africaines en font un usage parcimonieux depuis les indépendances.

2. Le constat de l'ineffectivité de l'autonomie locale dans l'État postcolonial africain

18. - La mise en œuvre de l'autonomie locale ne suffit pas pour assurer dans les faits l'exercice effectif des compétences des collectivités territoriales afin de répondre aux besoins des populations. Si l'institutionnalisation de l'autonomie locale dans les systèmes juridiques

---

[51] Sur la personnalité morale en droit public, voir not. : J.-B. AUBY, *La notion de personne publique en droit administratif français*, thèse, dactyl., Université de Bordeaux, 1979 ; F. LINDITCH, *Recherche sur la personnalité morale en droit administratif*, thèse, LGDJ, 1997. Et plus spécifiquement sur l'émergence de la personnalité morale en droit des collectivités territoriales, F. MUGNIER, *La personnalité juridique des collectivités territoriales. Genèse et développement d'une personne morale dans l'État*, thèse, Dalloz, 2022.

[52] B. FAURE, *Droit des collectivités territoriales*, Dalloz, Coll. « Précis », 2021, n° 4.

[53] L. NGOMO TSIMI, L'autonomie administrative et financière des collectivités territoriales décentralisées : l'exemple du Cameroun, op.cit., p. 24.

africains semble arriver à maturité, le décalage entre les textes et les intentions réelles des autorités normatives révèle une contradiction qui nuit à son effectivité.

19. - Les années 1960 sont marquées par l'accession à la souveraineté internationale des États en Afrique subsaharienne. Cette période correspond à l'adoption de la Constitution française de la Ve République qui introduit de profonds changements en matière d'autonomie des collectivités territoriales. D'abord, l'adoption de la loi Deffere du 2 mars 1982 constitue un tournant dans l'histoire moderne de l'organisation administrative française. L'autonomie locale fut modifiée par la réforme des modalités du contrôle administratif exercé par le représentant de l'État sur les décisions prises par les collectivités locales. Ce changement est d'ailleurs considéré par la doctrine comme un bouleversement majeur[54] dans le droit de la décentralisation. Ensuite, la révision constitutionnelle du 28 mars 2003[55] marque une étape supplémentaire et décisive dans la reconnaissance de l'autonomie des collectivités territoriales sur le plan financier et matériel. Au titre des innovations introduites, l'organisation de la République française est désormais dite « décentralisée »[56] et les collectivités territoriales peuvent déroger, à titre expérimental et sous réserve du respect de nombreuses conditions, aux dispositions législatives ou règlementaires qui régissent l'exercice de leurs compétences. Aussi, l'entrée en vigueur en 2010 de la question prioritaire de constitutionnalité a dans un premier temps été perçue comme un moyen de renforcement de la garantie juridictionnelle du droit à l'autonomie locale[57]. Le bilan de l'application de la QPC en matière de décentralisation après quelques années de mise en œuvre apparait pour le moins décevant. Elle n'a pas contribué à renforcer l'autonomie des collectivités territoriales[58].

---

[54] Cf. M. VERPEAUX, L. JANICOT, *Droit des collectivités territoriales, op.cit.*, p. 67-68.

[55] Loi n° 2003-276 du 28 mars 2003.

[56] J.-F. BRISSON, « La France est une République indivisible… son organisation est décentralisée », *RDP*, 2003, p. 111.

[57] Sur ce thème, voir, M. VERPEAUX, « Libre administration des collectivités territoriales et QPC », in X. BIOY, W. MASTOR, S. MOUTON, X. MAGNON (dir.), *Le réflexe constitutionnel : question sur la QPC*, Bruylant, 2013, p. 73 ; P. DE MONTALIVET, « QPC et droit des collectivités territoriales », *AJDA*, 2016, p. 586.

[58] Voir not., G. DRAGO, « QPC et droit des collectivités territoriales : premier bilan », *JCP A*, 2011, p. 221 ; M. DOUENCE, « Où en est la jurisprudence du conseil constitutionnel sur le principe de libre administration des collectivités territoriales ? », in *Mélanges Gérard Marcou*, IRJS Éditions, 2017, p. 195.

20. - Dans un élan de « modernité », tous les États nouvellement indépendants se dotent d'une Constitution écrite qui accorde une place pour le moins mitigée à l'autonomie locale. À première vue, les premières Constitutions adoptées par les États africains sont toutes empreintes de libéralisme. La Constitution de la première République du Togo consacre, par exemple, son titre VII composé d'un seul article aux collectivités territoriales (art 58) : « *la République une et indivisible, reconnait l'existence des collectivités territoriales. Ces collectivités sont les circonscriptions et communes. Elles s'administrent librement et démocratiquement dans les conditions prévues par la loi* ». On pourrait donc croire à la lecture d'une telle disposition constitutionnelle que le Togo indépendant s'est engagé en faveur de la décentralisation territoriale et reconnait aux collectivités une autonomie dans la conduite de leurs affaires tout en restant dans le cadre défini par la loi. Il ne faut pas se méprendre sur les intentions du constituant togolais pour au moins deux raisons. La première est que les constituants africains se sont tous inspirés des Constitutions françaises de 1946 et 1958. Ils ont en ce sens repris à leur compte les articles 87 de la première, 34 et 72 de la seconde. La deuxième raison, la plus importante, c'est qu'au moment des indépendances, les dirigeants africains héritent de pays profondément divisés et des populations épuisées par près d'un siècle de colonisation. Dans ces conditions, l'heure n'est pas à la décentralisation et à la reconnaissance de collectivités territoriales autonomes, mais plutôt à la centralisation administrative. Reproduisant le schéma colonial, les États africains vont chercher à asseoir leur domination et renforcer le pouvoir du nouvel État naissant afin de réunir toutes les forces vives de la Nation. Ces dispositions constitutionnelles ne feront l'objet d'aucune mesure d'application, car elles ne reflétaient pas la réalité des préoccupations internes de l'époque[59].

---

[59] Cf. L. NGOMO TSIMI, *L'autonomie administrative et financière des collectivités territoriales décentralisées : l'exemple du Cameroun, op.cit.*, p. 24 et s. ; I. DIALLO, *Le droit des collectivités locales au Sénégal*, thèse, L'Harmattan, 2007 ; P. AGNAMA-EBOUMI, *La décentralisation territoriale et le développement local au Gabon*, thèse, Université de Toulouse, 2004 ; A. CABANIS, « Le dialogue entre civilisation et culture : le nouveau cadre constitutionnel en Afrique francophone », article consulté sur le site http://www.cmiesi.ma, rubrique notes et analyses, p. 236 : « Les premières constitutions dont se sont dotés les États de l'Afrique francophone aux lendemains des indépendances recouvrées, sont très naturellement placées sous le signe de la construction de l'État-Nation. Dans cette perspective, tout ce qui pouvait affaiblir la domination du pouvoir central devait être banni [...] ».

21. - L'État en Afrique s'opposait alors fortement à toute idée d'autonomie locale[60]. Les termes « décentralisation » et « collectivités territoriales » étaient insignifiants et ne pouvaient tout au plus qu'exprimer une continuité du pouvoir central. Le pouvoir est alors incarné « *par une même personne, le parti unique et l'État avaient fait collusion en une manière d'alliance hégémonique qui contrôle tous les pôles du pouvoir : État-parti, parti-État, deux puissances coalisées qui font main-basse sur la société par fractionnement et prolifération des monolithismes, parti unique bien sûr, mais aussi syndicat unique, organisation unique des femmes, organisation unique des jeunes, etc. autant de chapes coulées sur les pôles actifs de la société qui avaient transformé nos villes et nos campagnes en univers sans aération* »[61]. La perpétuation par les États d'Afrique noire francophone de l'héritage jacobin a fini par exacerber les tensions au sein des populations. Les premières élections municipales qui se feront au cours de la décennie 80 toujours sous la stricte direction du parti unique avec une liste unique présentée par le parti unique cherchent à contenir les revendications sociales[62]. La reconnaissance d'une autonomie aux collectivités territoriales ne constituait pas une préoccupation des gouvernements tant elles sont perçues comme dérangeantes et susceptibles à la fois de nuire à l'autorité du chef de l'État, fondateur et chef du parti unique et d'entraver la marche vers l'unité nationale.

22. - L'avènement du multipartisme[63] et des élections pluralistes en 1990 constitue le point de rupture avec les anciennes pratiques institutionnelles. La conjonction de deux éléments, l'un endogène et l'autre exogène, a participé à ce changement de paradigme dans la conception de l'État en Afrique. De l'intérieur, les États africains font face à une crise due à la baisse de l'exportation des matières premières agricoles et minières, provoquant un désastre sur le plan social, économique et politique. Selon Maurice Kamto, cette crise combinée à « l'individualisme » introduit par la colonisation a entrainé l'effondrement de l'organisation traditionnelle et l'affirmation du pouvoir de l'individu sur le groupe social.

---

60 *Cf.* T. MICHALON, « Autoritarisme et hypercentralisation : l'État africain contre les libertés », *Revue africaine de droit international comparé*, vol. 3, 1991, pp. 76-82.

61 M. KAMTO, L'urgence de la pensée. Réflexion sur une précondition du développement en Afrique, Mandara, 1993, p. 26.

62 J.-P. DUPRAT, « Togo », *Annuaire de législation française et étrangère*, n° 32, 1985, p. 513.

63 V. not., G. CONAC (dir.), L'Afrique en transition vers le pluralisme politique, Economica, 1993 ; M. C. DIOP, M. DIOUF (dir.), Les figures du politique en Afrique. Des pouvoirs hérités aux pouvoirs élus, Codesria-Karthala, 1999,

Cela a facilité l'expression du mécontentement généralisé qui s'est par la suite répandu d'une capitale africaine à l'autre[64]. Sur le plan extérieur, la fin de la guerre froide marque la promotion du modèle libéral par les puissances occidentales[65] en Afrique subsaharienne à travers la pratique des conditionnalités politiques[66]. L'aide au développement sera « tiède » dans les pays qui refusent de s'ouvrir au pluralisme et de reconnaitre les vertus de la décentralisation en tant que catalyseur du développement et de la démocratie. De nouvelles lois concrétisent, enfin les dispositions constitutionnelles relatives à la décentralisation territoriale. La libre administration des collectivités territoriales par des conseils élus est désormais formellement protégée par le législateur[67]. Pour certains auteurs, ce changement de cap dans les États africains s'analyse davantage comme un moyen de préservation de la paix sociale qu'une modalité d'organisation administrative de l'État visant à construire des collectivités territoriales viables[68]. L'autonomie accordée aux collectivités territoriales sert la stratégie de l'État. En répondant aux velléités sécessionnistes, il évite l'implosion de l'État unitaire et apparait comme réformateur aux yeux des acteurs de la coopération internationale pourvoyeurs de l'APD.

23. - L'autonomie locale intègre à nouveau le droit positif sous l'influence des puissances occidentales combinées aux tensions sociales suscitées par la crise généralisée que traverse le continent. Toutefois, il faut constater que depuis la période coloniale, l'autonomie locale et la démocratisation de vie publique sont principalement le résultat des forces extérieures. Cette situation est symptomatique des difficultés qu'il y a dans

---

64 M. KAMTO, *L'urgence de la pensée. Réflexion sur une précondition du développement en Afrique, op.cit.*, pp. 121-136.

65 *Cf.* F. FUKUYAMA, *La fin de l'histoire et le dernier homme*, Flammarion, coll. « champs essais », 2009.

66 *Cf.* M. FAU-NOUGARET, *La conditionnalité démocratique. Étude de l'action des organisations internationales*, thèse, dactyl., Université de Bordeaux, 2004 ; J.-L. ATANGANA AMOUGOU, « La conditionnalité juridique des aides et respect des droits fondamentaux », *Revue numérique Afrilex*, n° 2, 2001, p. 1.

67 Ainsi au Togo, intervient en 1998, la loi n° 98-006 du 11 février 1998 fixant le cadre juridique de la décentralisation qui fera l'objet d'une modification en 2007 puis en 2019.

68 C. NACH MBACK, *Démocratisation et décentralisation. Genèse et dynamiques comparés des processus de décentralisation en Afrique subsaharienne, op.cit.*, p. 36 et s. ; R. OTAYEK, « La décentralisation comme mode de redéfinition de la domination autoritaire. Quelques exemples à partir des situations africaines », http://www.afsp.msh-paris.fr/archives/congreslyon2005/communications/tr4/otayek.pdf consulté le 22 octobre 2019.

les États d'Afrique noire francophone à donner un sens à l'autonomie des collectivités territoriales. Le passage de la norme aux comportements correspondants des destinataires permettant la réalisation de l'autonomie locale mobilise peu les acteurs. En dépit, de sa proclamation par les textes, il existe un contraste entre l'importance accordée en théorie à l'autonomie locale dans les discours et le résultat de la pratique marquant ainsi son ineffectivité. Quoi qu'il en soit, le nouvel État unitaire africain ne peut qu'être décentralisé. Telle est sa nouvelle « manière d'être ». La décentralisation en Afrique intègre le concept de l'autonomie des collectivités territoriales au sein des principes fondamentaux qui régissent l'action publique locale. C'est d'ailleurs ce qui justifie la liaison entre l'effectivité du droit et l'objet de la présente recherche qui mérite d'être davantage précisé.

B- Le concept d'autonomie dans le droit de la décentralisation

24. - L'autonomie est un concept qui sert plusieurs discours[69]. Bien connue des juristes, elle est un attribut de la personnalité juridique[70]. Lorsqu'on lui adjoint l'épithète « locale », l'autonomie devient une formule syncrétique qui rassemble une variété de termes susceptibles de l'exprimer, renvoyant concomitamment à un nombre d'objets connus du juriste[71] notamment, la décentralisation et la libre administration. Quelle consistance juridique faut-il alors reconnaitre à l'autonomie locale ? D'abord, il convient de préciser que l'autonomie locale dans cette étude s'inscrit dans le cadre de la décentralisation territoriale. Il y a donc ici un lien entre autonomie locale et décentralisation territoriale (**1°**). Ensuite, nous procéderons à la définition de l'autonomie locale comme une liberté d'action limitée des collectivités dans un État unitaire (**2°**).

---

69 Sur la notion de concept, v. X. BIOY, « Notions et concepts en droit : interrogations sur l'intérêt d'une distinction… », in G. TUSSEAU (dir.), *Les notions juridiques*, Economica, 2010, p. 21 ; « L'usage du concept de "personne" en droit », *Revue de la recherche juridique-Droit prospectif*, vol. 5, n° 26, 2012, p. 2271.

70 Qui implique une autonomie administrative et financière.

71 Le même constat est fait à propos de l'intérêt public local. V. en ce sens, T. ROMBAUTS-CHABROL, *L'intérêt public local*, thèse, Dalloz, 2016, p. 5 et s.

1. Autonomie locale et décentralisation : deux concepts liés

25. - La décentralisation et l'autonomie sont deux concepts différents, mais liés. Dans un État unitaire de tradition centralisatrice comme c'est le cas des États d'Afrique noire francophone, la décentralisation territoriale est autant un programme politique qu'une technique administrative[72]. Elle constitue l'expression de la volonté des gouvernants d'un État unitaire, de consacrer des « *libertés et responsabilités locales* »[73], de donner plus de respiration et d'espace aux administrations territoriales[74]. Dans ce sens, la décentralisation est le résultat d'un long processus. C'est un mouvement qui vise à donner un second souffle à la démocratie en responsabilisant davantage les collectivités territoriales. Cela suppose que les collectivités territoriales aient la capacité d'agir pour répondre aux besoins des populations, donc qu'elles soient autonomes. John Keudjeu de Keudjeu relève que le concept d'autonomie locale ne peut être défini par rapport à lui-même « *mais plutôt en fonction de la modalité d'organisation administrative et politique étudiée.* [...] *Suivant que l'on se trouve dans le cadre de la décentralisation, de la régionalisation ou de la fédération, l'appréhension qui en sera donnée va varier* »[75]. C'est ce qui explique que, pour rendre compte de l'autonomie dans un État unitaire, l'on recourt à la notion de décentralisation.

26. - Cette situation traduit aussi la difficulté à saisir le terme « décentralisation » qui en définitive obéit à deux fonctions différentes. D'abord, ainsi que le relève Bertrand Faure, « *la décentralisation sert d'élément de construction du droit, car elle suppose pour se réaliser, l'autonomie organique des autorités locales par l'élection et leur autonomie fonctionnelle par le transfert de compétences et de moyens par l'État, celui-ci n'assurant plus qu'un contrôle de tutelle apte à contenir, sans annihiler, l'exercice de cette autonomie* ». Ensuite, *« le concept*

---

[72] Maurice Hauriou écrivait à ce propos que : « s'il ne s'agissait que du point de vue administratif, la centralisation assurerait au pays une administration plus habile, plus impartiale, plus intègre et plus économe que la décentralisation. Mais les pays modernes n'ont pas seulement besoin d'une bonne administration, ils ont besoin aussi de liberté politique ». M. HAURIOU, Précis de droit administratif et de droit public, réimpression de la 12e éd., de 1933, Dalloz, 2002, p. 86.

[73] Voir en ce sens l'analyse de : P.-Y. CHICOT, « Autonomie locale : liberté et responsabilité », in S. LARCHER, *Un kaléidoscope de l'autonomie locale : théorie, pratique institutionnelle et déclinaisons ultramarines,* Rapport d'information du Sénat, n° 452, 2014, p. 17.

[74] B. PLESSIX, *Droit administratif général,* LexisNexis, 2020, p. 299 et s.

[75] J. R. KEUDJEU DE KEUDJEU, *Recherche sur l'autonomie des collectivités territoriales décentralisées au Cameroun,* thèse, dactyl., Université de Douala, 2011, p. 25.

sert également d'élément de <u>*connaissance du droit*</u> *en jouant le rôle de principe de classification des régimes d'autonomie locale. La décentralisation caractérise alors un régime intermédiaire entre celui de la centralisation et celui du fédéralisme* »[76]. Si la décentralisation est un élément de classification des États qui se situe à l'intermédiaire de la centralisation et de la fédération, c'est qu'elle n'épuise pas toutes les formes d'autonomies. Mais, de quelle décentralisation parle-t-on exactement ? Certaines Constitutions africaines[77] utilisent expressément le mot « décentralisation » sans pourtant le définir. Toutefois, elle ne manque pas de définitions[78]. En tant que phénomène juridique, elle a fait l'objet de nombreuses études doctrinales[79].

27. - Maurice Hauriou définit la décentralisation comme « une manière d'être de l'État caractérisée par le fait que l'État se résout en un certain nombre de personnes administratives qui ont la jouissance de droits de puissance publique et qui assurent le fonctionnement des services publics en exerçant ces droits c'est-à-dire en faisant des actes d'administration »[80]. Selon René Chapus, ce qui traduit l'existence de la décentralisation c'est « le transfert d'attributions de l'État à des institutions [territoriales ou non] juridiquement distinctes de lui et bénéficiant, sous la surveillance de l'État, d'une certaine autonomie de gestion »[81]. À la lecture de ces deux définitions classiques de la décentralisation, on se rend compte qu'au sens

---

76 B. FAURE, *Droit des collectivités territoriales, op.cit.*, n° 3.

77 *Cf.* art 141 Constitution du Togo du 14 octobre 1992 ; art. 164, Constitution du Niger du 25 novembre 2010 ; art 1 al. 2 de la loi constitutionnelle du 18 janvier 1996 au Cameroun. De façon générale l'expression fait partie du bloc de constitutionnalité des États ouest africains puisque le protocole de la CEDEAO de 2001 dispose en son article 1er d) que la « *décentralisation du pouvoir à tous les niveaux de gouvernement* » constitue un principe constitutionnel. *Cf.* Protocole A/SP1/12/01 sur la démocratie et la bonne gouvernance, additionnel au protocole relatif au mécanisme de prévention, de gestion et de règlement des conflits, de maintien de la paix et de la sécurité. Sur la constitutionnalisation de la notion en droit français, Voir, F. PRIET, « Sur une notion désormais constitutionnalisée : la décentralisation », in, *Terres du droit, Mélanges en l'honneur d'Y Jégouzo,* Dalloz, 2009, p. 249.

78 J.-B. AUBY, *La décentralisation et le droit, op.cit.*, p. 13.

79 La littérature juridique est particulièrement abondante sur le sujet. V. not. C. EISENMANN, Centralisation et décentralisation. Esquisse d'une théorie générale, LGDJ, 1948 ; S. FLOGAÏTIS, *La notion de décentralisation en France, en Allemagne et en Italie,* thèse, LGDJ, 1979 ; J. BENOIT, Théorie juridique de la décentralisation administrative en France, thèse, dactyl., Université Paris 2, 1990 ; F. FOURNIE, Recherche sur la décentralisation dans l'œuvre de Maurice HAURIOU, thèse, LGDJ, 2005.

80 M. HAURIOU, « Décentralisation », in *Répertoire du droit administratif,* Paul Dumont, 1891, p. 482.

81 R. CHAPUS, *Droit administratif général,* Montchrestien, 2004, p. 403.

juridique, elle peut à la fois concerner le territoire, on parle alors de décentralisation territoriale et les services qu'on désigne par décentralisation technique ou fonctionnelle. Retenons dans le cadre de cette étude que la décentralisation pose le problème de l'organisation administrative des structures territoriales[82].

28. - La décentralisation engage la question du territoire[83]. Charles Eisenmann estime qu'il ne peut exister qu'une seule décentralisation : celle qui est territoriale[84]. Aux antipodes des conceptions dominantes, il refuse de compromettre la notion de décentralisation en parlant de décentralisation fonctionnelle. Pour cet auteur, la structuration territoriale de l'administration et la distribution fonctionnelle des tâches administratives sont deux modes séparées de l'organisation qui ne procèdent pas de la même logique. Les confondre conduit inévitablement à brouiller les figures où prend forme le système administratif[85]. Il peut, certes, exister des similitudes entre les résultats institutionnels de la division fonctionnelle et ceux de la division territoriale, mais il faut, soutient Charles Eisenmann, éviter d'assimiler ces deux phénomènes, car « *ils n'affectent pas la même sorte de structure de l'administration* »[86].

29. - Il faut également noter que la décentralisation territoriale dans le cadre d'un État unitaire s'inscrit dans une « dialectique de l'Un et du Multiple »[87] qui conduit à la recherche d'un équilibre entre la préservation de l'unité nationale et la reconnaissance de libertés locales. En ce sens, l'approche théorique de la centralisation et de la décentralisation esquissée par Charles Eisenmann constitue une clé d'analyse de la décentralisation territoriale dans un système juridique unitaire à travers des phénomènes de conciliation et de coopération administrative qui vont au-delà de la

---

82 C. EISENMANN, *Cours de droit administration*, LGDJ, coll. « Anthologie du droit », 2014, p. 155 et s.

83 Cf. J. CAILLOSSE, *Les « mises en scènes » juridiques de la décentralisation. Sur la question du territoire en droit public français*, LGDJ, coll. « Droit et société », 2009.

84 Même si nous partageons cette position, il faut relever qu'elle mérite d'être nuancée aujourd'hui tant la frontière entre la collectivité territoriale et l'établissement public s'amenuise et tend à disparaitre.

85 J. CAILLOSSE, Les « mises en scènes » juridiques de la décentralisation. Sur la question du territoire en droit public français, op.cit., p. 61.

86 C. EISENMANN, « Les structures de l'administration », in *Traité de science administrative*, Mouton, 1966, pp. 317-318.

87 J.-A. MAZERES, « La dialectique de l'un et du multiple », *La Décentralisation, Les cahiers français*, n° 204, 1982, p. 4.

simple contradiction entre la décentralisation et la centralisation[88]. L'analyse faite par Charles Eisenmann de la décentralisation et de la centralisation conduit de façon globale à nier l'existence d'une forme pure de centralisation ou de décentralisation. L'auteur propose alors un niveau intermédiaire, la « semi-décentralisation » qui permet aujourd'hui à une partie de la doctrine de repenser autrement la décentralisation autour de ces manifestations effectives, c'est-à-dire une co-administration entre l'État et les collectivités territoriales[89]. En ce sens, le constat des difficultés de réalisation de l'autonomie locale en Afrique invite à repenser la décentralisation plutôt que de *« tenter des justifications fondées sur des constructions hypothétiques »*[90].

30. - Dans sa dimension territoriale, la décentralisation se définit « comme un processus de dévolution de compétences administratives au profit de collectivités publiques élues qui les exercent en leur nom et pour leur compte dans le respect des finalités et des modalités d'un contrôle, d'ordinaire qualifié de "tutelle" »[91]. Dès lors, « la collectivité cesse d'être

---

[88] « Saisie au niveau élémentaire, microscopique, la semi-décentralisation consiste à attribuer le pouvoir de poser des décisions (des normes) intéressant une circonscription territoriale, c'est-à-dire une collectivité locale, conjointement à une autorité d'État et une autorité propre à cette circonscription ou collectivité locale, – donc : à une autorité mixte, composée d'un organe d'État et d'un "organe local". Ce qui signifie que l'adoption de ces décisions ; l'édiction de ces normes, requièrent l'accord des deux organes, disons : leur double volonté, leur double consentement. C'est donc une formule d'association et participation de deux autorités au pouvoir de décision ; elle fait des deux autorités (…) des co-auteurs juridiques, des co-créateurs de normes édictées. Sur un plan global ou macroscopique, on pourra dire que la semi-décentralisation consiste à remettre l'administration de chacune des collectivités territoriales semblables que l'on a constituées au sein de la collectivité étatique (nationale) totale à des autorités formées d'une autorité propre à chacune et d'une autorité d'État », C. EISENMANN, Cours de droit administratif, op.cit., p. 287.

[89] J.-F. BRISSON, « Décentraliser autrement le territoire », in *Révolution, constitution, décentralisation, Mélanges en l'honneur de Michel Verpeaux,* Dalloz, 2020, p. 539 ; « les compétences des collectivités territoriales », *RFDA*, 2022, p. 291 ; Pour une étude plus ancienne, J.-M. PONTIER, *L'État et les collectivités locales. La répartition des compétences*, thèse, LGDJ, 1978.

[90] C. NANAKO, *La libre administration des collectivités territoriales au Bénin et au Niger, op.cit.*, p. 359.

[91] B. FAURE, *Droit des collectivités territoriales, op.cit.*, n° 6. Voir aussi pour une définition proche, M. E. SOHOUENOU, « Contribution à l'élaboration d'une échelle de la décentralisation territoriale en Afrique noire francophone », *Revue béninoise des sciences juridiques et administratives*, n° 22, 2009, p. 148 ; C. NANAKO, *La libre administration des collectivités territoriales au Bénin et au Niger, op.cit.*, p. 8.

une circonscription administrative de l'État pour devenir une entité relativement autonome. Elle possède ainsi des organes distincts de ceux de l'État, dispose de compétences propres, est propriétaire de biens, recrute du personnel, est titulaire d'un patrimoine et établit son propre budget ; bref, elle peut s'administrer librement »[92]. La décentralisation territoriale porte des valeurs de démocratie et de proximité. Elle correspond alors d'après Benoit Plessix « à la nécessité de créer une indispensable soupape de sureté pour répondre au besoin des individus de se rattacher à quelque chose de moins vaste et intimidant que le seul sentiment national, de s'accrocher au cocon d'une partie plus intime, de renouer le lien avec leur région originelle et leur village natal. Elle peut aussi résulter de la volonté des gouvernants d'instituer une vie locale démocratique, de diffuser l'idée d'un self-government à l'anglaise »[93]. Cette prise de position contemporaine n'est pas sans rappeler les professions de foi des partisans de la diversité et de la liberté, à l'image des libéraux pour qui la liberté locale est le fondement de toute liberté politique, selon Dareste, la sève de la démocratie, selon Laboulaye, l'école primaire de la liberté, selon l'image restée fameuse d'Alexis de Tocqueville[94].

31. - En définitive, ramenée à son contenu essentiel, la décentralisation s'applique à une collectivité territoriale qui malgré son appartenance à une collectivité plus grande qu'est l'État s'administre elle-même et gère ces propres affaires[95]. L'idée d'une certaine autonomie locale constitue alors la base de la décentralisation[96]. Sans elle, la décentralisation se viderait de son sens puisqu'elle constitue un élément de sa définition. Lorsqu'on évoque l'autonomie locale, on cherche à saisir les libertés dont disposent les collectivités territoriales sur le territoire de l'État, et dans le cadre de ce

---

92 P. BODINEAU, M. VERPEAUX, *Histoire de la décentralisation,* PUF, 1997.

93 B. PLESSIX, *Droit administratif général, op.cit.*, p. 300.

94 Pour un aperçu de la pensée de ces auteurs libéraux, G. BIGOT, *L'administration française. Politique, droit et société,1789-1870, op.cit.,* p. 158 et s ; G. BACOT, « L'apport de Tocqueville aux idées décentralisatrices », in L. GUELLEC (dir.), *Tocqueville et l'esprit de la démocratie,* Presses de Sciences Po, 2005, pp. 203-229.

95 A. DE LAUBADERE, J.-C. VENEZIA, Y. GAUDEMET, *Traité de droit administratif,* LGDJ, 1990, p. 98. La Charte africaine des valeurs et des principes de la décentralisation, de la gouvernance locale et du développement local utilise l'expression « gouvernement local » pour désigner les autorités locales. L'utilisation d'une telle expression lui permet de couvrir la diversité des situations sur le continent notamment les différences entre les systèmes d'inspiration français et anglais.

96 *mm.*

qu'il est convenu d'appeler la décentralisation[97]. Parce que la décentralisation est un processus et que l'autonomie locale est un état qui traduit un rapport de forces entre l'État et les collectivités territoriales qu'on peut chercher à mesurer, le concept d'autonomie offre une perspective d'analyse intéressante du droit de la décentralisation en Afrique.

2. L'autonomie locale : une liberté d'agir des collectivités territoriales ?

32. - La liberté au sens kantien[98] ne renvoie pas à l'indépendance, mais à l'autonomie, c'est-à-dire la capacité à se donner à soi-même ses propres lois. Dans le cadre de la décentralisation territoriale, l'autonomie renvoie à l'idée d'une certaine liberté des collectivités territoriales. Toutefois, l'autonomie locale demeure un concept flou et de faible densité[99] dans les États unitaires de tradition centralisatrice. « [...] *Par son étymologie* [l'autonomie] *a une portée d'apparence absolue et désigne le pouvoir de se diriger librement en droit et par extension, en politique ou en gestion, tandis que dans ses expressions réelles elle est toujours relative et limitée* »[100]. Dans ces conditions, entreprendre une recherche sur l'autonomie des collectivités territoriales peut paraitre périlleux en raison de l'imprécision du terme autonomie[101]. Jean-Yves Vincent dans sa préface à la thèse précitée constatait que, « *en ce domaine aux confins du politique et du juridique, le mot autonomie conserve un relent de soufre ; l'autonomie est soupçonnée d'être le fourrier de l'indépendance, de frayer le chemin aux micro-souverainetés, de préparer insidieusement l'implosion de l'État unitaire et de la République indivisible* »[102]. Voilà, qui justifie, dans les États

---

97 J. CAILLOSSE, « Les libertés locales : bréviaire de quelle décentralisation ? », in N. KADA (dir.), *Les tabous de la décentralisation*, Berger-Levrault, coll. « Au fil du débat », 2015, p. 61.
98 E. KANT, *Critique de la raison pratique*, Flammarion, 2003, p. 473.
99 J.-F. BRISSON, « L'autonomie locale au sens de la Charte européenne : quel apport concret à la spécificité des collectivités territoriales ?», in *Peut-on encore définir la collectivité territoriale ?*, Colloque AFDCL, 2021. https://www.youtube.com/watch?v=sCZzLmU0118&t=2s. Consulté le 31 octobre 2021.
100 R. HERTZOG, « L'autonomie en droit : trop de sens, trop peu de signification », in *Mélanges offerts à Paul Amselek*, Bruylant, 2005, p. 446.
101 M. JOYAU, *De l'autonomie des collectivités territoriales françaises. Essai sur la liberté du pouvoir normatif*, thèse, LGDJ, 1998, p. 1.
102 J.-Y. VINCENT, Préface à M. JOYAU, *De l'autonomie des collectivités territoriales françaises. Essai sur la liberté du pouvoir normatif, op.cit.*, p. XIX.

unitaires de l'Afrique noire francophone très attachés à la préservation de l'intégrité du territoire, que le terme autonomie soit peu employé dans les Constitutions. L'aversion pour les formes d'État dit « autonomique » ou « régional »[103] ou encore pour la fédération[104] conduit à une utilisation parcimonieuse de l'autonomie par le constituant et le législateur. L'autonomie signifie que « la volonté est législatrice », qu'elle « va poser par elle-même ses lois »[105]. Or, la centralisation des États africains ne s'accommode guère d'une telle conception dès lors qu'il est question des collectivités territoriales. La collectivité territoriale ne peut s'auto-organiser. Elle ne dispose pas de prérogatives d'ordre politique et ne peut exprimer des intérêts particuliers que s'ils sont conformes aux orientations nationales. L'autonomie des collectivités territoriales ne peut, par conséquent, être qu'une autonomie administrative.

33. - La difficulté à définir le concept d'autonomie locale tient à cette ambiguïté inhérente à l'étymologie même du mot autonomie. Selon Charles Cadoux, le statut d'autonomie se situe quelque part entre « *le fédéralisme et la décentralisation puisqu'il évoque l'idée de liberté d'administration voire de gouvernement, mais toujours sous un certain contrôle du pouvoir central* ». Il ajoute que, « *on ne peut aller au-delà de cette approximation* » parce qu'il « *n'y a pas "un" degré précis de liberté politique caractérisant une fois pour toutes le concept d'autonomie, mais des "seuils" possibles de plus ou moins grande autonomie* »[106]. Le droit supranational de la décentralisation en Afrique a hissé l'autonomie locale au rang des valeurs démocratiques universelles[107] sans pour autant en préciser le contenu. C'est l'article 3, de la Charte européenne de

---

103 K. BLAIRON, *Le régionalisme et le fédéralisme dans les réformes de l'État italien. Contribution à la définition de la forme d'État italienne*, thèse dactyl., Université de Nancy 2, 2005.

104 J. RIVERO, « Fédéralisme et décentralisation : harmonie ou contradiction ? », in *Pages de doctrine*, t. 1, LGDJ, 1980, pp. 213-221.

105 Définition étymologique que l'on retrouve dans l'ensemble des dictionnaires. Pour une approche détaillée en philosophie, E. GAZIAUX, *L'autonomie en morale : au croisement de la philosophie et de la théologie*, Presse universitaire de Louvain, 1998, p. 360.

106 C. CADOUX, « La conception française de l'État unitaire (comparaisons) », *Revue de science administrative de la Méditerranée occidentale*, 1990, p. 26.

107 M. E. SOHOUENOU, « L'administration territoriale béninoise et le droit supranational de la décentralisation », in *Leçon inaugurale, Colloque international sur l'effectivité de la décentralisation au Bénin*, janvier 2015, p. 2 ; C. NANAKO, *La libre administration des collectivités territoriales au Bénin et au Niger*, *op.cit.*, p. 32.

l'autonomie locale,[108] qui dans un effort intéressant de conceptualisation juridique, précise dans son premier alinéa, ce qu'il faut entendre par autonomie locale. L'autonomie locale est « *le droit et la capacité effective pour les collectivités de régler et de gérer, dans le cadre de la loi, sous leur propre responsabilité et au profit de leurs populations, une part importante des affaires publiques* ». Cette définition qui constitue, à ce jour, la plus aboutie en la matière, semble-t-il[109], guidera notre réflexion sur l'autonomie locale dans les États d'Afrique noire francophone.

34. - Dans ce contexte, l'autonomie locale suppose une absence de contraintes excessives. Son étude se résume donc à la mise en évidence des degrés variables qui caractérise la nature des relations entre l'État et les collectivités territoriales. L'autonomie locale ne se laisse pas enfermer dans une vue unique. La relativité du concept est parfaitement résumée par Jean-Marie Pontier à travers la métaphore suivante, « *l'autonomie locale est peut-être comme une gemme, pour laquelle l'art de l'orfèvre est de tailler de telle sorte que la pierre puisse renvoyer le plus d'éclat possible, afin que l'on puisse admirer et s'extasier. S'il n'est guère possible d'affirmer qu'elle comporte, comme un diamant taillé de manière classique, cinquante-deux facettes, l'autonomie locale en présente de nombreuses, et la manière dont on la regarde produit des effets variables* »[110]

Depuis 1990, les révisions des textes constitutionnels des pays francophones d'Afrique noire ont singulièrement bouleversé la conception de l'autonomie en matière de décentralisation territoriale dans un État unitaire de type classique. On est passé d'une « *décentralisation-soumission* » c'est-à-dire d'une décentralisation dans laquelle les collectivités territoriales ne disposaient d'aucun pouvoir propre à l'idée d'une

---

108 P. WILLIAMS-RIQUIER, « La charte européenne de l'autonomie locale : un instrument juridique international pour la décentralisation », *Revue française d'administration publique*, n° 121-122, 2007, p. 191.

109 M. VERPEAUX, Préface à V. KALIMERI, *Le Conseil de l'Europe et les collectivités territoriales. Contribution à l'étude de l'autonomie locale*, thèse, L'Harmattan, 2017, p. 13 : « La Charte européenne de l'autonomie locale constitue, encore aujourd'hui, le seul traité international qui consacre les droits des collectivités territoriales. Malgré quelques lacunes, elle constitue un guide indispensable pour tous les pays qui veulent établir un régime démocratique au niveau local »

110 J.-M. PONTIER, « Autonomie locale, libre administration, décentralisation », in *Itinéraire d'un constitutionnaliste, Mélanges en l'honneur de Francis DELPÉRÉE*, Bruylant, 2007, p. 1221.

« *décentralisation-liberté* »[111] qui accorde une marge de manœuvre aux collectivités. Ainsi, on peut lire, par exemple, à l'article 141 de la Constitution togolaise que « *les collectivités territoriales s'administrent librement par des conseils élus au suffrage universel, dans les conditions prévues par la loi* »[112]. En réalité, il s'agit simplement d'une restauration à rebours après une longue période marquée par le monopartisme puisque, comme nous l'avons expliqué ci-dessus, les premières Constitutions africaines étaient empreintes d'un libéralisme rêveur. L'autonomie locale dans les nouvelles Constitutions des États d'Afrique francophone trouve ainsi à s'exprimer à travers la libre administration. Mais alors, pourquoi avoir choisi de formuler cette recherche autour du concept d'autonomie et non de libre administration ?

35. - Le choix de l'autonomie au détriment de la libre administration tient en ce que la nature juridique de la libre administration des collectivités territoriales prête à discussion. En effet, il existe un débat doctrinal sur la question de savoir si la libre administration est une véritable « liberté » ou s'il s'agit simplement d'un « principe » d'organisation de l'État, duquel découleraient certains droits et libertés[113]. Michel Verpeaux, dans son commentaire de l'arrêt du Conseil d'État du 18 janvier 2001, *Commune de*

---

[111] Cf. L. NGOMO TSIMI, *L'autonomie administrative et financière des collectivités territoriales décentralisées : l'exemple du Cameroun, op.cit.*, p. 52 et s. ; C. ROIG, « Théorie et réalité de la décentralisation », *op.cit.*, p. 447.

[112] Voir aussi, art 164 Const. Niger ; Burkina Faso, art. 143 et 145 ; Côte d'Ivoire, art. 119 ; Mali, art. 98 ; Sénégal, art. 102 ; Tchad, art. 205.

[113] La littérature juridique est abondante sur le sujet. V. not., C. NANAKO, *La libre administration des collectivités territoriales au Bénin et au Niger*, thèse dactyl. Université d'Abomey-Calavi (Bénin), 2016 ; C. BACOYANNIS, *Le principe constitutionnel de libre administration des collectivités territoriales*, thèse, Economica-PUAM, 1993 ; A. ROUX, « Le principe constitutionnel de la libre administration des collectivités territoriales », *RFDA*, 1992, p. 435 ; M. BOURJOL, « Libre administration et décentralisation », in CRERAL, *Vertus et limites de la décentralisation*, Colloque des 2 et 3 mai 1985, *Les cahiers du droit public*, éd. Bergeret, p. 61 ; L. FAVOREU, A. ROUX, « La libre administration des collectivités territoriales est-elle une liberté fondamentale ? », *Les cahiers du Conseil constitutionnel*, 2002, n° 12, p. 88 ; C.-A. DUBREUIL, « Libre administration des collectivités territoriales et liberté fondamentale », *JCP G*, 2012, n° 46, p. 2048 ; M. VERPEAUX, « Question prioritaire de constitutionnalité et libre administration des collectivités territoriales », *AJDA*, 2010, n° 28, p. 1595 ; M. TROPER, « Libre administration et théorie générale du droit », in J. MOREAU, G. DARCY (dir.), *La libre administration des collectivités territoriales. Réflexion sur la décentralisation*, Economica-PUAM, 1993, p. 59.

Venelles c/M. Morbelli[114], considère que « *le principe de libre administration des collectivités territoriales constitue… une garantie, au même titre que le principe de la séparation des pouvoirs.* […] *La libre administration peut d'ailleurs être conçue comme une forme de séparation verticale des pouvoirs tandis que la forme habituelle de la séparation serait horizontale. L'une comme l'autre ne sont pas des droits, mais des moyens d'asseoir des droits ou des libertés, ils sont des moyens, ils ne constituent pas des buts* »[115]. Constantinos Bacoyannis dans sa thèse[116] défend en revanche l'idée selon laquelle la libre administration ne se réduit pas à un simple principe d'organisation, mais constitue bien une liberté. Il précise que le droit de s'administrer librement n'est pas conféré à la personne morale « collectivité territoriale », mais au groupement naturel qui est délimité grâce à son rattachement à un territoire et qui préexistait à sa reconnaissance par l'État[117]. Or, comme nous l'avons montré précédemment, l'équation commune égale communauté traditionnelle préexistante a été rompue par le pouvoir colonial en Afrique. Les collectivités territoriales africaines telles qu'elles sont organisées aujourd'hui n'ont jamais existé « *en tant que groupements naturels territoriaux vivant indépendamment de l'existence et de la volonté de l'État* »[118]. Elles sont dépourvues de réalité sociologique et historique. La qualification de la liberté appliquée à la libre administration ne peut donc se justifier en réalité dans les États d'Afrique noire francophone d'où la préférence à l'autonomie des collectivités territoriales. Cette autonomie n'est pas naturelle, mais construite, elle est une concession de l'État, qu'il accorde plus ou moins selon les exigences de rationalité administrative qui prévalent à un moment donné[119].

36. - Pour l'essentiel, l'autonomie locale peut être définie comme « une liberté reconnue à une collectivité instituée, ayant pour objet la satisfaction de besoins collectifs, et qui s'exerce dans le cadre d'un système d'action

---

114 M. VERPEAUX, « Libre administration, liberté fondamentale, référé-liberté » *RFDA*, 2001, p. 681.

115 *Ibid.*, p. 684.

116 C. BACOYANNIS, *Le principe constitutionnel de libre administration des collectivités territoriales*, thèse, Economica-PUAM, 1993, p. 99 et s.

117 *Ibid.*, p. 100.

118 *Idem* ; voir aussi, L. FAVOREU, A. ROUX, « La libre administration des collectivités territoriales est-elle une liberté fondamentale ? », *Les cahiers du Conseil constitutionnel*, n° 12, 2002.

119 *Cf.* J.-B. AUBY, J.-F. AUBY, R. NOGUELLOU, *Droit des collectivités locales*, *op.cit.*, p. 69.

publique. En termes juridiques, cela se manifeste par un droit de libre administration, des institutions démocratiques et des compétences effectives au sein de l'État »[120]. Elle est à la fois une autonomie institutionnelle et fonctionnelle. L'autonomie des collectivités territoriales se traduit institutionnellement par l'élection des organes des collectivités territoriales au suffrage universel. Cette situation contraste avec la période post-indépendance, où prétexte pris de la construction de l'unité nationale, les autorités locales étaient nommées par le pouvoir central[121]. Avec l'élection, les autorités administratives locales ne sont plus soumises au pouvoir hiérarchique des autorités centrales, lesquelles ne peuvent, en principe, leur imposer ou leur opposer leur pouvoir propre de décision. La décentralisation se différencie ainsi de la déconcentration, en coupant précisément le lien hiérarchique qui relie les autorités locales au gouvernement central[122]. Sur le plan fonctionnel, la libre administration suppose l'attribution des compétences effectives et l'existence des moyens nécessaires pour une libre gestion par les collectivités territoriales. En exigeant que les organes locaux ne soient pas nommés par le pouvoir central et soumis à son pouvoir hiérarchique, la mise en œuvre de la décentralisation « implique le principe de l'autonomie des personnes morales. Cela signifie que, à moins de règles expresses en sens contraire, les personnes morales décentralisées sont libres d'agir à leur guise »[123].

37. - Cependant, la décentralisation territoriale n'aboutit pas à faire des autorités locales des entités politiques disposant en vertu de la Constitution des pouvoirs législatif[124], exécutif et judiciaire, et habilitées à

---

[120] G. MARCOU, « L'autonomie communale : étude comparative », *Pouvoirs*, n° 95, 2000, p. 69.

[121] C. NACH MBACK, *Démocratisation et décentralisation. Genèse et dynamiques comparés des processus de décentralisation en Afrique subsaharienne, op.cit.*, p. 33.

[122] B. PLESSIX, *Droit administratif général, op.cit.*, p. 300.

[123] J. BAGUENARD, *La décentralisation*, PUF, coll. « Que sais-je ? », 2004, pp. 3-22.

[124] Les autorités locales disposent d'un pouvoir règlementaire résiduel et subordonné au pouvoir règlementaire national. V. not., B. Faure, « Règlements locaux et règlements nationaux », *Les nouveaux cahiers du Conseil constitutionnel*, n° 42, 2014, p. 43 ; *Le pouvoir règlementaire des collectivités locales*, thèse, LGDJ, 1998 ; G. CHAVRIER, *Le pouvoir normatif local : enjeux et débats*, LGDJ, 2011 ; L. JANICOT, « Le pouvoir normatif des régions », *RFDA*, 2016, p. 664 ; C. MONEMBOU, « Le pouvoir règlementaire des collectivités locales dans les États d'Afrique noire francophone », *Revue CAMES/SJP*, n° 002, 2015, p. 79.

représenter seules des intérêts propres, distincts de ceux de l'État[125]. Les collectivités territoriales ne disposent pas de la « compétence de la compétence »[126]. En ce sens, l'État assure sur les actes des collectivités territoriales un contrôle nécessaire à la préservation de la légalité et de la forme unitaire de l'État. Ce contrôle prend, dans la plupart des États d'Afrique noire francophone, la forme d'une tutelle administrative[127] et d'un contrôle juridictionnel. Dans le premier cas, la tutelle consiste en la soumission des actes pris par les collectivités territoriales à un contrôle *a priori* exercé par le représentant de l'État qui dispose du pouvoir de les réformer et de les annuler. Le contrôle juridictionnel permet quant à lui de faire sanctionner l'illégalité des actes par le juge administratif. La décentralisation territoriale en tant que mode d'organisation territoriale de l'État unitaire en Afrique suppose donc la constitution de collectivités territoriales sous la forme de personnes morales de droit public[128] à qui l'État décide de confier l'exercice d'un certain nombre de compétences qu'elles exerceront sous son contrôle. Le cadre unitaire dans lequel se déploie la décentralisation implique que cette dernière repose sur un équilibre entre l'autonomie administrative et le contrôle de tutelle exercé par l'État. Le niveau symétrique entre les deux permet de déterminer l'étendue de l'autonomie des collectivités territoriales et par extension le niveau de centralisation de l'État.

38. - La fonction de l'autonomie locale consacrée peut être atteinte dès lors que les conditions d'exercice de la liberté d'action recherchée s'accompagnent d'un ensemble de règles et de pratiques harmonisées qui tendent à conférer plus d'épanouissement aux collectivités territoriales décentralisées dans leurs rapports avec l'État[129]. Au-delà des considérations administrative et financière, l'autonomie locale entretient dans le discours politico-administratif un lien étroit avec la

---

125 J.-A. MAZERES, « Les collectivités locales et la représentation. Essai de problématique élémentaire », *RDP*, 1990, p. 607.

126 R. CARRE DE MALBERG, *Contribution à la théorie générale de l'État*, Sirey, 1920, p. 170.

127 Sur la tutelle, l'œuvre pionnière de : R. MASPETIOL, P. LAROQUE, *La tutelle administrative, le contrôle des administrations départementales, coloniales et des établissements publics*, Sirey, 1930. Voir aussi, S. REGOURD, *L'acte de tutelle en droit administratif français*, thèse, LGDJ, 1982.

128 Pour une approche historique et critique de la personnalité morale des collectivités, voir, F. MUGNIER, *La personnalité juridique des collectivités territoriales. Genèse et développement d'une personne morale dans l'État*, thèse, Dalloz, 2022.

129 L. NGOMO TSIMI, *L'autonomie administrative et financière des collectivités territoriales décentralisées : l'exemple du Cameroun, op.cit.*, p. 7.

démocratisation de la vie publique[130]. Les conditions essentielles à l'effectivité de cette « décentralisation-liberté » se résument en l'habilitation à l'exercice de compétences effectives et en la disponibilité des moyens[131]. Si l'autonomie n'est pas un « concept magique »[132], en raison de sa valeur relative, elle « *se mesure plus qu'elle ne se définit* »[133]. Sous cet angle, l'autonomie locale se prête à une analyse sur le plan de l'effectivité. Un concept polysémique qu'il convient aussi de préciser.

§2.- L'effectivité de l'autonomie des collectivités territoriales comme problématique de la recherche

39. - La décentralisation est susceptible de degrés, c'est-à-dire que le point où elle peut être poussée dépend de diverses circonstances de temps et de lieu[134]. Dans les États d'Afrique noire francophone, la décentralisation c'est-à-dire « *le transfert des pouvoirs, des responsabilités, des capacités et des ressources* » du pouvoir central vers les autorités infranationales vise le renforcement des capacités de ces dernières afin de « *promouvoir la participation des populations et la fourniture de services de qualité* »[135]. Pour ce faire, elle consacre le droit des collectivités territoriales à s'administrer librement et à gérer des affaires propres[136]. L'analyse du droit des collectivités territoriales en Afrique et du droit en général ne saurait se limiter au premier degré de l'énoncé du droit. Il n'est pas suffisant que les textes constitutionnels ou législatifs reconnaissent l'existence des collectivités

---

130 M. HAURIOU, « Décentralisation », in *Répertoire de droit administratif*, *op.cit.*, p. 479 : « une organisation décentralisatrice est particulièrement nécessaire dans un pays, lorsque le pouvoir central lui-même est démocratiquement organisé » ; v. aussi S. REGOURD, « De la décentralisation dans ses rapports avec la démocratie. Genèse d'une problématique », RDP, 1990, p. 964. « On n'imagine guère de progrès de la décentralisation sans démocratie […] ».

131 C. NANAKO, *La libre administration des collectivités territoriales au Bénin et au Niger*, *op.cit.*, p. 27.

132 A. MOYRAND, « De l'autonomie administrative à l'autonomie politique », in *Le statut du territoire de Polynésie française. Bilan de dix ans d'application : 1984-1994*, Economica-PUAM, 1996, p. 148.

133 V. DUSSART, *L'autonomie des pouvoirs publics constitutionnels*, CNRS, 2000, p. 13.

134 *Cf.* F. FOURNIE, *Recherche sur la décentralisation dans l'œuvre de Maurice HAURIOU*, thèse, LGDJ, 2005, p. 267.

135 Art 1er de la Charte africaine des valeurs et des principes de la décentralisation… *préc.*

136 *Cf.* art. 2 du CGCT au Burkina Faso.

territoriales et affirment leur libre administration[137]. De ce point de vue, étudier le droit de la décentralisation « *par le truchement des analyses tendant à relever l'ineffectivité ou l'inefficacité des lois et du droit* » permet « *de dresser le procès d'un système et d'une certaine conception du droit* »[138] en Afrique. Il convient toutefois, dans un premier temps, en raison de la polysémie du concept d'effectivité d'en préciser les contours[139] (**A**) puis dans un second temps, de présenter les facteurs qui mettent à mal l'autonomie des collectivités territoriales en Afrique à travers une analyse critique des modes d'élaboration, de diffusion et d'application du droit de la décentralisation (**B**).

A- Une définition stipulative du concept d'effectivité

40. - L'effectivité est un concept particulièrement fuyant. Paul Amselek affirmait d'ailleurs que l'effectivité est trop mal définie pour être retenue par la science du droit[140]. Sa définition varie selon les champs disciplinaires et les différents courants de pensée. Sans qu'il soit nécessaire de revenir ici en détail sur l'ensemble des acceptions doctrinales de l'effectivité, il faut préciser que ces premières occurrences s'inscrivent dans un courant sociologique. En effet, le concept d'effectivité fut mobilisé afin de mesurer les effets des normes juridiques. Il s'agissait alors d'analyser l'ensemble des effets de la règle de droit, quel qu'il soit, c'est-à-dire intentionnels ou non intentionnels, directs ou indirects, immédiats ou différés, conformes ou non conformes au « *devoir être* » ou au « *pouvoir être* »[141]. Dans une approche restrictive, le rapprochement du concept d'effectivité aux effets du droit

---

137 C. NANAKO, *La libre administration des collectivités territoriales au Bénin et au Niger, op.cit.*, p. 32.

138 V. RICHARD, *Le droit et l'effectivité – contribution à l'étude d'une notion*, thèse, dactly., Université Paris II, 2003, p. 430.

139 Les travaux de Julien Betaille sur « les conditions juridiques de l'effectivité en droit public interne » constitue en ce sens une œuvre utile. J. BETAILLE, *Les conditions juridiques de l'effectivité de la norme en droit public interne : illustrations en droit de l'urbanisme et en droit de l'environnement*, thèse, dactly., Université de Limoges, 2012.

140 P. AMSELEK, *Perspectives critiques d'une réflexion épistémologique sur la théorie du droit (Essai de phénoménologie juridique)*, thèse, 1962, p. 334.

141 R. M. INGHLITERRA, *La réalisation du droit de la non-discrimination*, thèse, LGDJ, 2022, p. 12. Pour un usage de cette approche, V. not., G. ROCHER, « L'effectivité du droit », in A. LAJOIE, R.-A. MACDONALD, R. JANDA, G. ROCHER (dir.), *Théorie et émergence du droit : pluralisme, surdétermination et effectivité*, Montréal, Thémis, 2004, p. 135 ; V. DEMERS, *Le contrôle des fumeurs. Une étude d'effectivité du droit*, Montréal, coll. « Thémis », 1996.

conduit à ne prendre en compte que la finalité des règles de droit. Telle est la conception défendue par Yann Leroy dans sa thèse[142]. L'auteur définit l'effectivité comme « *la qualité d'une norme qui produit des effets* ». Ces derniers sont « *tout à la fois les effets concrets et symboliques, les effets juridiques, économiques, sociaux ou de quelque autre nature, les effets désirés ou non voulus, prévus ou non intentionnels, immédiats ou différés, à la seule condition qu'ils n'entrent pas en contradiction avec les finalités des règles de droit évaluées* »[143]. Ainsi, pour Yann Leroy l'effectivité comprend « *les effets qui sont en adéquation avec la finalité de la règle de droit qui les produit, qu'il s'agisse d'effets voulus ou d'effets non désirés, mais désirables, ou même d'effets non prévus tant qu'ils ne sont pas contradictoires avec ladite finalité* »[144]. Cette acception de l'effectivité présente l'inconvénient majeur d'exclure de l'effectivité les effets pervers de la règle de droit. Par conséquent, une norme juridique qui produit des effets pervers ne peut être effective. Aussi, cette acception, dans la logique binaire qu'elle introduit, entretient une confusion non souhaitable entre effectivité et efficacité.

41. - Il est important de distinguer entre effectivité et efficacité[145] or, la frontière conceptuelle entre les deux se révèle poreuse[146]. La notion d'efficacité est en règle générale définie comme le caractère de ce qui est efficace, c'est-à-dire « *qui produit, dans de bonnes conditions et sans autre aide, l'effet attendu* »[147]. L'efficacité fait donc référence à l'atteinte d'un effet, un effet que l'on attend. Il s'agit d'une adéquation des objectifs assignés à la règle de droit par son auteur et des résultats obtenus[148]. L'effet « attendu » est donc celui, qui non seulement concourt à la finalité de la norme, mais surtout atteint cette finalité, l'objectif posé par l'auteur de la norme[149]. On parlerait alors d'inefficacité « *lorsqu'on considère que les résultats attendus de telle*

---

142 Y. LEROY, L'*effectivité du droit au travers d'un questionnement en droit du travail*, thèse, LGDJ, 2011.

143 *Ibid.*, p. 339.

144 Y. LEROY, « La notion d'effectivité en droit », *Droit et société*, n° 70, 3/2011, p. 730.

145 Le constat est souvent dressé sans pourtant conduire à une clarification. R. M. INGHLITERRA, *La réalisation du droit de la non-discrimination*, *op.cit.*, p. 14.

146 *Ibid.*, p. 15. Cette porosité conduit certains auteurs à affirmer que « *l'effectivité et l'efficacité sont identiques* » : M.-A. FRISON-ROCHE, « L'efficacité des décisions en matière de concurrence : notions, critères, typologie », *LPA*, n° 259, 2000, p. 4 et s.

147 Entrée « Efficace », Cntrl, Sens 1.

148 R. M. INGHLITERRA, *La réalisation du droit de la non-discrimination, op.cit.*, p. 15.

149 J. BETAILLE, *Les conditions juridiques de l'effectivité de la norme en droit public interne, op.cit.*, p. 18.

ou telle règlementation ne sont pas obtenus »[150]. Pour Alexandre Flückiger, « *une politique ou une norme sont efficaces si les résultats correspondent à leurs objectifs* »[151]. Alors que les contours de l'efficacité semblent relativement simples, ceux de l'effectivité fluctuent[152] au point de constituer un embarras certain pour la doctrine la plus illustre[153]. Dans la doctrine juridique contemporaine, l'effectivité est en général définie comme le « *caractère d'une règle de droit qui produit l'effet voulu, qui est appliquée réellement* »[154]. Ainsi, dans le *Vocabulaire juridique* dirigé par Gérard Cornu, l'adjectif effectif désigne à la fois ce « *qui produit l'effet recherché* » que ce « *qui correspond à la réalité* »[155].

42. - Une partie de la doctrine présente l'effectivité comme ayant une « vocation pratique dans la mesure où elle vise à évaluer les degrés d'application du droit, à préciser les mécanismes de pénétration du droit dans la société »[156]. Dans ce cadre, l'effectivité est « l'instrument conceptuel d'évaluation [du] degré de réception [de la norme], le moyen de mesurer des "écarts" entre pratique et droit »[157]. La mesure concrète de cet écart observe Julien Betaille relève en grande partie des techniques propres à la sociologie, mais l'idée d'évaluer sur le plan de « degrés » l'effectivité d'une règle de droit constitue une richesse de la notion, qui doit rejaillir sur toutes les autres définitions du concept[158]. Ainsi Pierre Lascoumes définit l'effectivité comme le « degré de réalisation [du droit] dans les pratiques

---

150 P. LASCOUMES, E. SERVERIN, « Théories et pratiques de l'effectivité du droit », *Droit et Société*, n° 2, 1986, p. 146.
151 A. FLÜCKIGER, « L'évaluation législative ou comment mesurer l'efficacité des lois », *Revue européenne des sciences sociales*, 2007, n° 138, p. 86.
152 R. M. INGHLITERRA, *La réalisation du droit de la non-discrimination*, thèse, LGDJ, 2022, p. 15.
153 Voir, en ce sens, E. MILLARD, E. MATZNER, « Note des traducteurs », in A. Ross, *Introduction à l'empirisme juridique*, trad. E. Millard et E. Matzner, présenté et annoté par É. Millard, LGDJ, 2004, p. 11-12 ; E. MILLARD, « Deux lectures critiques d'Alf Ross », in JOUANTAN, O. (édit.), *Théories réalistes du droit*, Presses universitaires de Strasbourg, 2000, pp. 13-14.
154 G. CORNU (dir.), *Vocabulaire juridique*, PUF, coll. « Quadrige », 2018, p. 388.
155 *Idem.*
156 F. RANGEON, « Réflexions sur l'effectivité du droit », in D. LOCHACK (dir.), *Les usages sociaux du droit*, PUF, 1989, p. 128.
157 P. LASCOUMES, E. SERVERIN, « Théories et pratiques de l'effectivité du droit », *Droit et Société*, n° 2, 1986, p. 127.
158 J. BETAILLE, *Les conditions juridiques de l'effectivité de la norme en droit public interne, op.cit.*, p. 15.

sociales »[159]. L'apport principal de la sociologie juridique se trouve dans cette idée de degré de l'effectivité. Elle va constituer la base de notre définition du concept dans le cadre de cette étude sur l'autonomie des collectivités territoriales en Afrique.

43. - À ce stade de notre réflexion, il convient de préciser qu'aucune acception de l'effectivité ne peut s'imposer avec autorité. Ce qui permet d'envisager une définition purement « stipulative » du concept d'effectivité c'est-à-dire une définition qui « *ne sera ni vraie ni fausse, mais seulement opératoire pour un problème spécifique* »[160] et répondant aux finalités de la recherche. Cela implique également de procéder à des exclusions[161] en distinguant l'effectivité des notions voisines.

François Rangeon fait remarquer que « la confusion entre l'effectivité [et des notions voisines comme l'efficacité ou l'efficience] présente [...] plus d'inconvénients que d'avantages [...] et surtout elle ne permet pas de comprendre les raisons profondes de l'effectivité »[162]. Cette remarque est d'autant plus vraie lorsqu'il est question de l'autonomie locale. Sa mesure ne peut s'établir sur le plan de l'efficacité puisque, l'autonomie locale ne peut se définir de façon universelle. Elle s'apprécie en fonction des circonstances de temps et de lieu qui font qu'elle se prête davantage à une analyse juridique de son effectivité. Jacques Commaille considère « que poser la question de l'effectivité du droit, c'est donc bien se préoccuper de son adéquation avec les comportements sociaux et les écarts éventuels (comme autant de manifestations d'ineffectivité) entre les normes juridiques et la réalité sociale qu'elles sont censées régir »[163]. L'effectivité se réfère selon Robin Médard Inghliterra à « une appréciation pragmatique de

---

[159] P. LASCOUMES, « effectivité », in A.-J. ARNAUD (dir.), *Dictionnaire encyclopédique de théorie et de sociologie du droit*, LGDJ, coll. « Anthologie du droit », 2e éd., corrigée et augmentée, 2018, p. 217. Cette conception est partagée par : E. MILLARD, « Effectivité des droits de l'Homme », in J. ANDRIANTSIMBAZOVINA, H. GAUDIN, J.-P. MARGUENAUD, S. RIALS, F. SUDRE (dir.), *Dictionnaire des Droits de l'Homme*, PUF, 2008, p. 349 ; L. HEUSCHLING, « Effectivité, efficacité, efficience et qualité », in M. FATIN-ROUGE STEFANINI, L. GAY, A. VIDAL-NAQUET (dir.), *L'efficacité de la norme juridique*, Bruylant, 2012, p. 59. Ces deux auteurs pensent l'effectivité en termes de niveau de « réalisation sociale ».

[160] M. TROPER, « Pour une définition stipulative du droit », *Droits*, n° 10, 1989, p. 102.

[161] *Cf.* C. EISENMANN, « Quelques problèmes de méthodologie des définitions et classifications en science juridique », *Archives de philosophie du droit*, 1966, p. 26.

[162] F. RANGEON, « Réflexions sur l'effectivité du droit », *op.cit.*, p. 127.

[163] J. COMMAILLE, « Effectivité », in S. RIALS, *Dictionnaire de culture juridique*, PUF, 2003.

la correspondance entre la signification de la norme et les comportements de ces destinataires » alors que « l'efficacité se réfère à une approche téléologique de la correspondance entre la volonté de l'auteur de la norme et les faits sociaux supposément produits par elle »[164]. À l'appréciation téléologique du second concept, « qui repose sur l'idée de finalité, qui constitue un rapport de finalité », répond le caractère pragmatique du premier, « qui concerne les faits réels, l'action et le comportement que leur observation et leur étude enseignent »[165].

44. - Sur la base de ces différentes considérations, l'effectivité, dans le cadre de notre étude sera entendue comme le degré d'influence qu'exerce la consécration de l'autonomie des collectivités territoriales sur le comportement des acteurs, notamment l'État. Autrement dit, l'analyse en termes d'effectivité permet de saisir les comportements des acteurs en tant qu'ils permettent ou non aux collectivités de prendre en charge les affaires relevant de leur compétence en vue de promouvoir le développement à la base et de renforcer la gouvernance locale. Cette définition retenue, il convient d'analyser les facteurs de réalisation de l'autonomie des collectivités territoriales.

B- L'identification doctrinale des conditions juridiques de l'effectivité du droit

45. - Plusieurs auteurs ont entrepris un travail doctrinal d'identification des conditions ou facteurs pouvant affecter l'effectivité d'une règle de droit[166]. La multiplicité des facteurs impose de recentrer le propos sur les éléments propres à l'ordre juridique interne. Le travail réalisé par Julien Betaille dans sa thèse de doctorat[167] constitue en ce sens une ressource utile. L'auteur présente une typologie des « conditions juridiques d'effectivité de la norme en droit public interne » c'est-à-dire de *« l'ensemble des éléments du système juridique qui contribuent à rendre possible l'effectivité de la norme juridique »*[168]. Julien Betaille distingue dans le système juridique, un

164 R. M. INGHLITERRA, *La réalisation du droit de la non-discrimination, op.cit.*, p. 17.
165 *Idem.*
166 Pour une présentation synthétique des contributions doctrinales, V. not. R. M. INGHLITERRA, *La réalisation du droit de la non-discrimination, op.cit.*, § 43 et s.
167 J. BETAILLE, *Les conditions juridiques de l'effectivité de la norme en droit public interne : illustrations en droit de l'urbanisme et en droit de l'environnement*, thèse, dactly., Université de Limoges, 2012.
168 *Ibid.*, p. 63.

certain nombre d'éléments qui ont une influence sur l'effectivité de la norme juridique qu'il range d'une part, dans les conditions « classiques » et d'autre part, dans les conditions « ampliatives ». Dans le premier cas, il renvoie notamment à la cohérence de l'ordre juridique (réception des normes juridiques externes) et à la sanction. Dans le second, il s'agit plutôt des conditions relatives à la conception (la production normative) et à la réception de la norme (par les sujets et les juges). Ces deux facteurs sont à la fois susceptibles d'influencer, voire d'amplifier l'effectivité de la norme juridique[169].

46. - Partant de ces conditions, il est possible de retenir que l'effectivité d'une règle de droit se situe essentiellement à deux niveaux. Primo, en amont du processus même de l'élaboration c'est-à-dire dès le moment de sa conception. Secundo, en aval, c'est-à-dire dans le comportement des acteurs à l'étape de la mise en œuvre. Il est alors possible de retenir dans le cadre de la présente étude deux idées qui influencent la réalisation de l'autonomie locale. D'un côté, les conditions liées à la conception du droit de la décentralisation en Afrique, car la phase d'élaboration de la norme influence par la force des choses son effectivité dans la mesure où elle lui confère une certaine légitimité ainsi qu'un certain nombre de caractères susceptibles de favoriser ou d'entraver son appropriation. Il s'agit donc de la cohérence interne de l'ordre juridique. De l'autre, les conditions liées à la mise en œuvre de la décentralisation censée contribuer à une autonomie suffisamment large des collectivités territoriales permettant de rénover la démocratie locale et de satisfaire qualitativement aux besoins des populations.

47. - En partant de ces éléments, notre recherche peut être ainsi problématisée : *l'autonomie des collectivités territoriales est-elle effective ou n'est-elle qu'un symbole qui cache une (re) centralisation* ? Le libéralisme dont a fait preuve le constituant africain en affirmant l'autonomie locale n'est-il pas, en pratique, remis en cause par les contraintes qui pèsent sur les collectivités territoriales ? Comment la décentralisation territoriale peut-elle s'inscrire dans un système hérité de la colonisation et qui trouve son identité dans la centralisation ? En somme, *quelle est l'étendue réelle de l'autonomie des collectivités territoriales dans les États ouest-africain francophone* ?

---

169 *Ibid.*, p. 65.

48. - L'hypothèse qui sous-tend ce questionnement est la suivante : la décentralisation territoriale peut malgré son institutionnalisation dans les textes n'avoir guère d'effectivité[170]. En effet, malgré la consécration textuelle de l'autonomie des collectivités territoriales et les discours en faveur de la décentralisation territoriale, la plupart des collectivités territoriales africaines ne disposent pas de la « capacité effective » de « gérer et de régler » les affaires locales. Les États africains peuvent à peine justifier aujourd'hui un retour en arrière en matière de décentralisation, mais elle demeure dans une large mesure encore perfectible. Les obstacles auxquels elle se heurte justifient notre recherche.

La décentralisation en Afrique[171] a retenu l'attention de la doctrine africaine, mais très rarement dans la période récente elle fut étudiée sous l'angle de l'autonomie[172]. La dernière étude, à notre connaissance, qui traite de la libre administration des collectivités territoriales dans l'espace ouest-africain, est celle de Cossaba Nanako en 2016[173]. L'auteur soulignait déjà le fait que « *les libertés locales ne bénéficient que d'une place congrue dans les recherches et colloques universitaires aussi bien en droit qu'en science politique, comme si les chantiers y afférant étaient achevés ou que la décentralisation n'était pas en prise directe avec les questions majeures de développement socio-économique des États* »[174]. En Afrique centrale, notamment au Cameroun, la question de l'autonomie n'a également été que très peu étudiée[175]. Lorsqu'on analyse ces travaux, on se rend compte que leurs auteurs cherchent, soit à définir l'autonomie des

---

170 R. CHAPUS, *Droit administratif général, op.cit.*, p. 403.

171 V. not., K. APPIA., *Recherche sur le contenu de l'autonomie des organes décentralisés. Étude appliquée à l'expérience ivoirienne*, thèse, dacyl., Université Aix-Marseille, 1985. ; T. HOLO, « La décentralisation au Bénin : mythe ou réalité ? », *Revue béninoise des sciences juridiques et administratives*, n° 7, 1986 ; NLEP R. G., *L'administration publique Camerounaise : Contribution à l'étude des systèmes africains d'administration publique*, thèse, LGDJ, 1986 ; C. NACH MBACK, *Démocratisation et décentralisation. Genèse et dynamiques comparés des processus de décentralisation en Afrique subsaharienne*, thèse, Karthala-PDM, 2003.

172 Voir en ce sens, N. MEDE ; « L'autonomie "retenue" : étude sur le principe de libre administration des collectivités territoriales en Afrique de l'Ouest francophone », *Revue juridique et politique des États francophones*, vol. 62, n° 2, 2008, pp. 188-208, S. OUATTARA, *Gouvernance et libertés locales pour une renaissance de l'Afrique*, Karthala, 2007 ; M. TOTTE (dir.), *La décentralisation en Afrique de l'Ouest : entre politique et développement*, Cota- Karthala-ENDA GRAF, 2003.

173 C. NANAKO, *La libre administration des collectivités territoriales au Bénin et au Niger*, thèse dactyl. Université d'Abomey-Calavi (Bénin), 2016.

174 *Ibid.*, pp.19-20.

175 L. NGOMO TSIMI, *L'autonomie administrative et financière des collectivités territoriales décentralisées…*, *op.cit.* ; J. R. KEUDJEU DE KEUDJEU, *Recherche sur l'autonomie des collectivités territoriales décentralisées au Cameroun*, thèse, dactly., Université de Douala, 2012.

collectivités territoriales[176], soit à mesurer le degré d'emprise de l'État sur les collectivités territoriales[177]. Notre recherche se situe dans cette deuxième catégorie. Toutefois, dans l'ensemble de ces travaux, la dimension historique ainsi que les contraintes externes dans la construction de l'État décentralisé en Afrique n'ont pas suffisamment été étudiées. Ainsi, dépassant les conclusions auxquelles ces travaux aboutissent, notre recherche se propose d'enrichir l'étude de l'autonomie des collectivités territoriales en Afrique, notamment à travers l'élargissement du cadre spatio-temporel[178].

En effet, il nous semble que le rejet par les auteurs des thèses précitées du poids de l'histoire des institutions et des conséquences du mimétisme sur l'effectivité de l'autonomie locale tient en l'absence d'une analyse du mode de fonctionnement des sociétés précoloniales. Or, l'effectivité d'une norme est conditionnée par sa cohérence à l'ordre juridique. L'introduction de normes institutionnelles étrangères à celles pratiquées sur le continent africain avant la période coloniale perturbe cette cohérence. Ainsi, nous nous éloignons des positions de ces auteurs qui visent à rejeter les conditions de conception et d'élaboration de la norme comme un facteur juridique clé de la réalisation de l'autonomie locale. L'introduction de l'État colonial a permis d'absorber les structures territoriales préexistantes. L'organisation territoriale n'est plus fondée sur le divin, mais sur l'attribution de la personnalité juridique qui permet au nouvel État d'encadrer les collectivités qui perdent ainsi leur « liberté d'être ». Aussi, dans les travaux précédents, les enjeux de la décentralisation en matière de démocratisation de la vie publique locale n'ont bénéficié que d'une place congrue.

49. - Notre étude s'efforce donc d'apprécier le degré de liberté dont jouissent les collectivités territoriales dans les nouveaux schémas d'aménagement du pouvoir au sein des pays de l'Afrique de l'Ouest francophone. Il s'agit, d'un point de vue heuristique, d'examiner le contenu des textes relatifs à la décentralisation, d'en évaluer l'origine et la consistance comme un révélateur de la consistance par symétrie de l'autonomie des collectivités territoriales. L'approfondissement historique

---

176 L. NGOMO TSIMI, *L'autonomie administrative et financière des collectivités territoriales décentralisées : l'exemple du Cameroun, op.cit.*

177 C. NANAKO, *La libre administration des collectivités territoriales au Bénin et au Niger, op.cit.* ; J. R. KEUDJEU DE KEUDJEU, *Recherche sur l'autonomie des collectivités territoriales décentralisées au Cameroun, op.cit.*

178 Voir les développements *infra* n°63 et s.

s'avère en ce sens opportun afin de démontrer le décalage existant. Dépassant le cadre des énoncés, que Luc Sindjoun désigne comme une « *révolution passive* »[179], c'est sur l'observation des faits sociaux que les rapports entre l'État et les collectivités locales seront étudiés. Avant de revenir plus en détail sur ce questionnement, il convient de clarifier le cadre méthodologique qui nous permettra de vérifier cette hypothèse de recherche.

§3.- Méthodologie de la recherche

50. - Afin de conduire notre réflexion, il s'avère indispensable de fixer le cadre épistémologique de notre travail de recherche. Après avoir précisé la méthode ainsi que les éléments de théorie qui nous permettront d'atteindre l'objectif de la présente recherche (**A**), nous procéderons à une délimitation du cadre spatio-temporel de l'étude (**B**).

A- La démarche méthodologique retenue

51. - Le droit de la décentralisation a pendant longtemps été posé en des termes techniques. Jean-Arnaud Mazères précise par ailleurs qu'il convient désormais de poser des questions d'ordre épistémologique lorsque l'on aborde la décentralisation[180]. La présente recherche s'inscrira dans la pratique habituelle des juristes à savoir la dogmatique juridique[181].

---

179 C'est-à-dire un « changement bricolé par lequel le régime et ses acteurs dominants, confrontés à la "revanche de la société", des dominés, parviennent à la relative émasculation révolutionnaire de celle-ci ou de ceux-ci à travers les réformes politiques ». L. SINDJOUN, « Présentation générale : éléments pour une problématique de la révolution passive », in L. SINDJOUN, (dir.), *La révolution passive*, Dakar, CODESRIA, 1999, p. 3.

180 J.-A. MAZERES, « Débat : le processus de décentralisation », *AJDA*, 1992, p. 6 : « Il semble qu'en droit administratif il y ait en quelque sorte un "noyau dur", le domaine des obligations, des actes et du contentieux, autour duquel se développerait une zone périphérique, plus incertaine, soumise à une "scientificité" moindre : les institutions, les aspects statutaires, les organes, notamment. Il est d'ailleurs très difficile de donner des définitions proprement juridiques des concepts utilisés dans ces domaines, par exemple pour ce qui est de la décentralisation, du principe hiérarchique, ou de la tutelle (le mot exprime d'ailleurs le malaise conceptuel que les juristes éprouvent face à ce phénomène) ».

181 Étude du droit consacrée « à l'interprétation et à la systématisation des normes juridiques ». A.-J. ARNAUD (dir.), *Dictionnaire encyclopédique de théorie et de sociologie du droit*, *op.cit.*, p. 188.

Il s'agit d'une combinaison de l'évaluation et de la synthèse afin de favoriser la connaissance des systèmes juridiques africains. Ainsi, nous serons en mesure de statuer sur la qualité des protections qu'ils énoncent en matière d'autonomie des collectivités territoriales.

52. - En travaillant sur des aspects institutionnels en droit, il est impossible de respecter la frontière épistémologique entre l'*être* et le *devoir être*[182]. La démarche épistémologique adéquate impose le dépassement du positivisme, spécifiquement celui normativiste, car si l'on s'en tient à la définition du droit telle que posée par Hans Kelsen dans la « Théorie pure du droit »[183], le droit appartient à la sphère du *Sollen* c'est-à-dire du *devoir être.* Ainsi, la décentralisation ne saurait faire l'objet d'un travail de recherche juridique ou bien elle serait soumise à une « scientificité » moindre dans la mesure où elle porte pour une grande part sur des aspects institutionnels[184]. Aussi, l'appréciation de l'effectivité se situe au cœur de la tension entre l'*être* et le *devoir-être*, entre la réalité et la norme. Ce qui permet d'« *analyser l'écart qui existe entre droit et application du droit afin d'essayer de comprendre quels sont les facteurs qui le déterminent, et quels rapports dynamiques peuvent exister entre eux* »[185]. Pour Georges Vedel, *« le juriste doit se mouvoir à la fois dans le "Sein" et dans le "Sollen" et cet exercice, tout d'illogique équilibre, lui est imposé sous peine pour lui de se transformer soit en algébriste des normes, soit en chroniqueur des phénomènes juridiques* »[186]. En ce sens, la méthode classique proposée par Kelsen ne peut être utilisée dans la présente étude d'où le recours à une approche analytique[187].

---

182 M. DOAT, *Recherches sur la notion de collectivité locales en droit administratif français,* thèse, LGDJ, 2003, p. 5.
183 H. KELSEN, *Théorie pure du droit*, LGDJ, coll. « La pensée juridique », 1999.
184 J.-A. MAZERES, « Débat : le processus de décentralisation », *op.cit.,* p. 6.
185 M.-A. COHENDET, *L'épreuve de la cohabitation (mars 1986 – mai 1988),* thèse, dactyl., Université Lyon 3, 1991, p. 9.
186 G. VEDEL, Préface à J-J. ISRAEL, *La régularisation en droit administratif français,* thèse, LGDJ, 1981, p. II, cité par J. BETAILLE, *Les conditions juridiques de l'effectivité de la norme en droit public interne, op.cit.,* p. 34 ; v. aussi G. Vedel, « Topologie et recherche politique », in Pages de doctrine, t. 1, LGDJ, 1980, p. 415. Georges Vedel considérait que « l'effort du professeur est de rechercher perpétuellement la conciliation entre l'intelligible et le réel. Sacrifier le second à la première, c'est tricher, mais faire l'inverse, c'est manquer le but de tout enseignement ».
187 V. CHAMPEIL-DESPLATS, *Méthodologies du droit et des sciences du droit,* Dalloz, coll. « Méthodes du droit », 2016, p. 255 et s.

53. - La méthode analytique procède d'un discours ayant pour fonction pratique l'évaluation par l'analyse des textes, l'approfondissement de leurs contenus, leurs interprétations, leurs explications et leur mise en relation en vue de coordonner leurs significations[188]. La dogmatique juridique dans sa fonction évaluative permet la réalisation d'un travail de systématisation, de rationalisation et de mise en cohérence du droit[189]. Ce travail de mise en cohérence est d'autant plus utile que le droit de la décentralisation s'est construit par strates[190] progressives depuis la période coloniale. Dans une telle démarche, on « *s'attache, plus qu'à autre chose, à la simple description et juxtaposition des divers éléments* »[191] d'un objet tendant « *davantage la mise en ordre que la mise en perspective* [...] »[192]. Ainsi, ceux qui utilisent cette approche pour saisir un objet défini, « les descripteurs »[193] comme les nomme Théodore Fortsakis, « *visent exclusivement à servir la pratique* »[194]. L'analyse, l'explication, les commentaires, les réflexions, les suggestions liées à cette démarche laissent transparaitre un propos doctrinal possiblement prescriptif[195]. Les fonctions évaluative et prescriptive de la dogmatique juridique permettent d'apprécier le contenu du droit. Le travail mené ici confirme « *la perméabilité des statuts descriptif et prescriptif des discours méthodologiques* »[196]. Cette posture méthodologique nous permettra de vérifier si les pratiques sont conformes aux références établies en naviguant entre l'intelligible et le réel.

B- La délimitation du champ de la recherche

54. - Le continent africain qui constitue le champ géographique de la présente étude est riche de sa diversité. Contrairement aux usages courants qui semblent la présenter comme un espace unifié, le continent africain recoupe des réalités complexes et diverses pour être appréhendé dans sa

---

188 *Ibid.*, pp. 91-93, 257 et s.

189 *Idem.*

190 J.-A. MAZERES, « Essai d'analyse archéologique de la décentralisation », *Les Cahiers du L.E.R.A.S.S.* n° 21, 1990, pp. 91-115.

191 T. FORTSAKIS, *Conceptualisme et empirisme en droit administratif français*, thèse, LGDJ, 1987, p. 36.

192 T. ROMBAUTS-CHABROL, *L'intérêt public local*, thèse, Dalloz, 2016, p. 18.

193 T. FORTSAKIS, *Conceptualisme et empirisme en droit administratif français, op.cit.*, p. 36 et s.

194 *Idem.*

195 R. M. INGHLITERRA, *La réalisation du droit de la non-discrimination, op.cit.*, p. 27.

196 V. CHAMPEIL-DESPLATS, *Méthodologies du droit et des sciences du droit, op.cit.*, p. 14.

globalité[197]. Par conséquent, il est apparu évident que les ambitions de notre recherche ne peuvent se déployer sur l'ensemble du continent africain. Ainsi, nous avons décidé de centrer notre étude sur les États d'Afrique subsaharienne de succession française. Ces États, à quelques variations près, ont en commun une tradition politique et administrative marquée par une forte concentration des pouvoirs, une centralisation administrative et une volonté nouvelle de décentraliser l'État. Aussi, avons-nous choisi par souci d'efficacité, d'orienter notre réflexion sur un petit nombre de pays témoins à savoir : le Burkina Faso, le Mali, le Niger et le Togo.

55. - Ce choix obéit au fait que ces pays nous ont semblé pouvoir servir d'exemple puisque leurs expériences permettent d'envisager la nécessité de rendre effective l'autonomie des collectivités territoriales. Tous ces pays ont connu ces dernières années une instabilité politique marquée par des coups d'État à répétition en plus de la menace terroriste à laquelle ils sont confrontés. Cette dernière trouvant un terreau fertile sur les relents identitaires et la grande vulnérabilité économique et sociale qui sévit dans certaines régions. La généralisation de la crise a permis de replacer la question de la décentralisation et de la gouvernance locale au cœur des préoccupations nationales et sous régionales, car les populations sont non seulement en demande de prestation en matière d'accès aux services sociaux de base, mais aussi de transparence, de redevabilité et de participation aux choix des politiques publiques qui touchent directement à leur vie quotidienne. Quant au Togo, il est vrai que la décentralisation est encore neuve. Le pays n'ayant connu que deux élections municipales depuis son accession à l'indépendance. Les premières furent organisées en 1987 et les secondes en 2019. Il nous a paru déterminant de vérifier si les changements intervenus en matière de décentralisation intègrent les expériences tirées de l'échec des politiques précédentes en mettant à la disposition des collectivités les moyens nécessaires à leur opérationnalisation. En outre, ces pays témoins font partie des organisations politique et économique ouest-africain tels que l'UEMOA et la CEDEAO. Ces dernières ont fait de la démocratie et de la décentralisation un axe important de pacification et de développement de la sous-région ouest-africaine. Par ailleurs, nous ne manquerons pas de

---

[197] Aujourd'hui, la doctrine africaniste notamment dans le domaine des sciences politiques utilise le pluriel « Les Afriques ». La création du laboratoire LAM (Les Afriques dans le Monde) à Bordeaux en est une illustration. https://www.lam.sciencespobordeaux.fr/projet-scientifique/

recourir à des exemples issus d'autres États d'Afrique subsaharienne chaque fois que cela sera nécessaire.

56. - Au regard de la diversité des pays entrant dans le champ de notre étude, la comparaison sera utilisée. Il est vrai que la comparaison dans la science juridique ne fait pas l'objet d'une adhésion unanime en raison de sa fonction subversive[198]. Toutefois, dans la présente recherche, elle ne sera pas utilisée dans une perspective différentielle et culturaliste[199] ou même de modélisation[200]. Le recours à la comparaison n'est pas davantage justifié par le but de contribuer à la concurrence des systèmes juridiques en déterminant celui qui produirait les meilleurs résultats sociaux et économiques. L'approche comparatiste[201] vise ici à éclairer, enrichir et soutenir l'argumentation. On ne peut que regretter, encore une fois, l'absence d'ouvrages, d'écrits scientifiques substantiels, et d'une jurisprudence abondante sur la question de la décentralisation et de l'autonomie locale en Afrique.

57. - En analysant les institutions et les « infrastructures conceptuelles »[202] dans les différents pays, il s'agit de mettre en relief de façon globale l'état actuel du droit de la décentralisation et de l'autonomie locale en Afrique. Ainsi, l'objectif de la comparaison dans notre recherche consiste en une description à des fins de connaissances[203]. La comparaison sera également faite par rapport à la France, tant il est incontestable que les réformes institutionnelles en France continuent d'avoir une emprise sur l'organisation administrative et politique des États étudiés Il s'agit donc de mesurer les dynamiques juridiques à l'œuvre dans l'espace ouest-africain à la suite des réformes adoptées ces trente dernières années en matière de décentralisation. En outre, l'influence exercée par les organisations internationales avec la transmission « clef en main » de modèle institutionnel incite à une approche comparative afin de déterminer son ampleur sur les droits nationaux.

---

198 H. MUIR-WATT, « La fonction subversive du droit comparé », *RIDC*, n° 3, 2000, p. 503.

199 V. CHAMPEIL-DESPLATS, *Méthodologies du droit et des sciences du droit, op.cit.*, p. 224.

200 *Ibid.*, p. 225.

201 M.-C. PONTHOREAU, « Le droit comparé en question(s). Entre pragmatisme et outil épistémologique », *RIDC*, n° 1, 2005, p. 8.

202 C. GREWE, H. RUIZ-FABRI, *Droits constitutionnels européens*, PUF, 1995, p. 11.

203 V. CHAMPEIL-DESPLATS, *Méthodologies du droit et des sciences du droit, op.cit.*, p. 226.

58. - Par ailleurs, la connaissance de l'administration et de la vie administrative africaine exige de recourir à d'autres disciplines. Bien que relevant résolument du champ du droit public, du droit administratif de façon générale, et en particulier du droit de la décentralisation, notre recherche sur l'autonomie des collectivités territoriales s'inscrivant dans le triple héritage précolonial, colonial et postcolonial de l'État ne pourrait pas se passer de l'apport d'autres disciplines scientifiques. « *Une articulation de savoirs entre les disciplines qui développent "des problématiques [se] recoupant partiellement"* »[204] s'impose. Même si le droit des collectivités territoriales reste *lato sensu* une branche du droit administratif[205], on ne peut ignorer l'importance du droit constitutionnel sur les pratiques administratives. Il existe un rapport d'articulation entre les matières constitutionnelles et administratives puisque le premier permet, restreint, ou garantit les secondes[206]. En témoigne d'ailleurs l'apparition d'un « *droit constitutionnel local* »[207]. Ainsi, il sera convoqué, au premier plan, dans la présente recherche, le droit constitutionnel. En outre, d'autres disciplines telles la science administrative, l'anthropologie juridique, la sociologie, l'histoire du droit, et même l'histoire seront mobilisés *« ceci sans nier les difficultés et même*

---

204 F. OST, M. VAN DE KERVOCHE, *Jalons pour une théorie critique du droit*, Bruxelles, Publications des Facultés universitaire Saint-Louis, 1987, p. 70.

205 Voir, B. FAURE, « Peut-on ne pas affirmer un particularisme du Droit des collectivités locales ? » in *Les collectivités locales et le droit. Les mutations actuelles*, PUF, 2001, p. 13.

206 G. VEDEL, « Les bases constitutionnelles du droit administratif », *EDCE*, 1954, n° 8, in *Pages de doctrine*, t. II, LGDJ, 1980 p. 129 ; *Cf.* égal. L. FAVOREU, « Les bases constitutionnelles de la décentralisation », in *Révolution et décentralisation. Le système administratif français et les principes révolutionnaires de 1989*, *op.cit.*, p. 83 ; B. STIRN, *Les sources constitutionnelles du droit administratif*, LCDJ, coll. « Systèmes », 2011 ; B. FAURE, « L'apparition d'un droit constitutionnel des collectivités territoriales », in S. REGOURD, J. CARLES, D. GUIGNARD (dir.), *La décentralisation 30 ans après*, *op.cit.*, p. 73.

207 *Cf.* A. ROUX, *Droit constitutionnel local*, Economica, coll. « Poche droit public », 1995 ; A.-M. Le POURHIET, *Droit constitutionnel local*, colloque, Martinique, déc. 1997, coll. « Travaux de l'AFDC », Economica-PUAM, 1999 ; J.-C. DOUENCE « Statut constitutionnel des collectivités locales », in *Encyclopédie des collectivités locales*, Dalloz, folio n° 63 ; B. MATHIEU, M. VERPEAUX, *Les mutations constitutionnelles des collectivités territoriales*, Dalloz, coll. « Thèmes & commentaires », 2014 ; P. TURK (dir.), *Pour un droit constitutionnel des collectivités territoriales : évolutions et débats*, L'Harmattan, 2020.

les limites inhérentes à ce type d'exercice, notamment la nécessité de respecter le "génie propre" »[208] de chaque discipline.

59. - L'objectif principal de cette thèse consiste à procéder à « *un test de vérification* »[209] d'un énoncé juridique, celui de la consécration de l'autonomie des collectivités territoriales en Afrique. Si la finalité principale de cette thèse n'est « *pas de prôner ou de condamner* » un système donné, « *mais de chercher à en rendre compte* », on ne peut s'« *abriter derrière une axiologie de façade* »[210] ni en ce qui concerne l'objet de recherche ni sur la neutralité de la description[211]. L'attitude évaluative liée au discours doctrinal qui consiste à soulever parfois des incohérences, à saluer des avancées ou encore à avancer des suggestions comporte une part de prescription[212]. Aussi, concevoir le droit de la décentralisation en Afrique comme la consécration de collectivités territoriales qui s'administrent librement pour la concrétisation des ambitions de développement local et de démocratie locale implique *ipso facto* de lui concéder une part d'axiologie. Relayer le contenu de ce droit revient non seulement à faire écho des valeurs qui le sous-tendent[213] (développement, démocratie, participation citoyenne…), mais aussi à les légitimer. Même si on partage ces valeurs, on peut néanmoins regretter l'acharnement sur la décentralisation et les constructions juridiques hypothétiques qui ne conduisent pas à leurs réalisations.

---

208 F. OST, « Science du droit », in A.-J. ARNAUD (dir.), *Dictionnaire encyclopédique de théorie et de sociologie du droit*, LGDJ, coll. « Anthologie du droit », 2e éd., corrigée et augmentée, 2018, p. 543 : Sur l'interdisciplinarité « […] *la recherche s'opère à partir du champ théorique d'une des disciplines en présence qui développe des problématiques et des hypo*thèses *qui recoupent partiellement celles qu'élabore, de son côté, l'autre discipline. Il s'agit cette fois d'une articulation de savoirs qui entraine, par approches successives, comme dans un dialogue, des réorganisations partielles des champs théoriques en présence. On pourrait dire que, dans ce cas, on procède à la traduction d'un jeu de langage dans un autre, ceci sans nier les difficultés et même les limites inhérentes à ce type d'exercice, notamment la nécessité de respecter le "génie propre" de chaque langue* ».

209 V. CHAMPEIL-DESPLATS, *Méthodologies du droit et des sciences du droit, op.cit.*, p. 115.

210 D. LOCHAK, « La neutralité de la dogmatique juridique : mythe ou réalité », in P. AMSELEK (dir.), *Théorie du droit et science*, PUF, 1994, p. 308.

211 Cf. V. FORRAY, S. PIMONT, *Décrire le droit… et le transformer. Essai sur la décriture du droit*, Dalloz, 2017, pp. 159-161.

212 Voir sur les énoncés descriptifs et prescriptifs, V. CHAMPEIL-DESPLATS, *Méthodologies du droit et des sciences du droit, op.cit.*, pp. 257-298.

213 D. LOCHAK, « La neutralité de la dogmatique juridique : mythe ou réalité », *op.cit.*, p. 294.

§4.-Annonce de la thèse et du plan

60. - Cette thèse prend pour point de départ l'hypothèse selon laquelle la décentralisation territoriale peut être consacrée par les textes, mais n'avoir aucune effectivité c'est-à-dire aucune influence sur les comportements des acteurs. Au bout de ce cheminement introductif, le lecteur ne sera plus surpris, peut-être moins, de la thèse soutenue dans la présente étude : *la décentralisation territoriale dans les États d'Afrique considérés dans cette étude n'est qu'une forme de centralisation.* Sur la base, de l'identification doctrinale des facteurs de l'effectivité d'un énoncé juridique, l'évaluation, à travers une analyse méthodique, permet d'affirmer *l'ineffectivité de l'autonomie des collectivités territoriales dans les États ouest-africains francophone.* Le « local » est une question qui concerne au premier plan l'État. La réalité institutionnelle renvoie aujourd'hui au constat d'une instrumentalisation des libertés locales dont bénéficient les collectivités territoriales. Par la force des choses, elles se retrouvent impliquées dans la mise en œuvre de politique publique entièrement définie par l'État[214]. L'ineffectivité de l'autonomie locale est un révélateur de ce que la décentralisation comme la centralisation est une manière d'être de l'État[215]. Cet état de fait invite à repenser la décentralisation dans les États d'Afrique noire francophone dans une logique plus fonctionnelle associant l'État et les collectivités territoriales dans la conduite des politiques publiques.

61. - Les causes profondes de cette situation d'ineffectivité se trouvent, d'une part dans le bouleversement des formes traditionnelles d'organisation africaine par le fait colonial et d'autre part, par le refus de l'État « moderne » africain de remettre en cause l'héritage institutionnel de la colonisation. Les États occidentaux, en particulier la France ainsi que les organisations internationales, y ont joué un rôle important. Le premier à travers notamment l'instrumentalisation de la coopération décentralisée au service d'une « diplomatie démultipliée » portée par le Quai d'Orsay. Les seconds, par leurs programmes d'aide au développement conduisant à une

---

[214] Le constat est d'ailleurs partagé en Europe, voir en ce sens, Y. MÉNY, Y. SUREL, *Politique comparée. Les démocraties : Allemagne, Etats-Unis, France, Grande-Bretagne, Italie*, LGDJ, 2009, p. 466 et s. Voir aussi, Y. MÉNY, *Centralisation et décentralisation dans le débat politique français (1945-1969)*, thèse, LGDJ, 1974.

[215] M. HAURIOU, « Décentralisation », *op.cit.* pp. 471-491.

élaboration et à la transmission de modèles institutionnels[216]. La réception des normes ou des institutions d'origine externe dans l'ordre juridique interne constitue le premier foyer d'incohérences susceptibles de limiter l'effectivité d'un énoncé juridique. C'est dire, que dès l'étape même de la conception, les autorités normatives africaines n'ont pas suffisamment apprécié les facteurs de réalisation de l'autonomie des collectivités territoriales, soit à dessein, soit par omission (**Partie 1**).

62. - Quoi qu'il en soit, les collectivités territoriales décentralisées sont perçues comme des autorités publiques assujetties au pouvoir de l'État et gérant pour son compte les compétences que ce dernier veut bien leur accorder selon une rationalité dont il est le seul à fixer les déterminants. L'autonomie locale censée permettre à ces autorités d'agir concrètement demeure très limitée dans son effectivité, la puissance d'État pouvant toujours intervenir sur elles par des moyens statutaires ou financiers par exemple[217]. Dans ces conditions, la notion de compétence ou d'« attributions effectives » demeure assez floue et entièrement définie par l'État central[218]. Par ailleurs, la tutelle conserve tout son sens avec un accroissement des pouvoirs de l'autorité de tutelle qui pèse sur la vie locale. L'État est installé au cœur du local[219]. En même temps qu'il le fonde, il le prive des moyens d'une affirmation au sein de l'État unitaire. En Afrique francophone, malgré une décentralisation apparente, le pouvoir de décision appartient principalement à des autorités administratives centrales[220]. La production des textes juridiques par les autorités administratives décentralisées reste très pauvre ou quasiment inexistante[221]. La décentralisation territoriale apparait alors comme une stratégie de réforme de l'État pour l'État[222]. L'autonomie locale bénéficie,

---

216 Sur les stratégies d'import-export institutionnel, voir, Y. MÉNY (dir.), *Les politiques de mimétisme institutionnel. La greffe et le rejet*, L'Harmattan, coll. « Logiques politiques », 1993.

217 Pour un exposé de la situation française qui n'est pas si différente, voir, C. CHABROT, *La centralisation territoriale. Fondements et continuité en droit public français*, thèse, dactyl., Université de Montpellier I, 1997.

218 *Cf.* en droit français, J.-F. BRISSON, « Les compétences des collectivités territoriales », *RFDA*, 2022, p. 291.

219 J. CAILLOSSE, Les « mises en scènes » juridiques de la décentralisation. Sur la question du territoire en droit public français, LGDJ, coll. « Droit et société », 2009, pp. 29-30.

220 U. NGAMPIO-OBELE-BELE, « Brèves réflexions sur la gestion administrative de la crise du Coronavirus en Afrique francophone », *RDSS*, 2022, p. 73.

221 *Idem.*

222 Cf. C. DEBBASCH, *La décentralisation pour la rénovation de l'État*, PUF, 1976.

dans ces conditions, d'un faible degré de contrainte sur les comportements des acteurs dans les États d'Afrique francophone (**Partie 2**).

Partie 1 – L'État décentralisé : l'importation d'un modèle territorial inadapté
Partie 2 – L'État décentralisé : une construction juridique artificielle

PARTIE 1 – L'ÉTAT DÉCENTRALISÉ : L'IMPORTATION D'UN MODELE TERRITORIAL INADAPTÉ

63. - « *Rien de plus dangereux que les emprunts au droit étranger pour le droit public* ». Cette mise en garde de Ferdinand Larnaude[223] ne semble guère faire recette sur le continent africain un siècle après avoir été prononcée. N'en déplaise à Montesquieu pour qui « [les lois] *doivent être tellement propres au peuple pour lequel elles sont faites que c'est un grand hasard si celles d'une nation peuvent convenir à une autre* »[224], de nombreux États exportent et importent leur droit ainsi que leurs institutions. L'organisation de l'administration publique africaine dans son ensemble n'est que le résultat de multiples emprunts au modèle[225] administratif français. Or, la réception des normes d'origine externe dans l'ordre juridique interne constitue le premier foyer d'incohérences susceptibles de limiter l'effectivité de la norme[226].

64. - Il y a plus de trois siècles, Jean Rivero proposait dans un article resté célèbre[227] une grille d'analyse de ce phénomène d'emprunt de modèles. L'auteur s'interroge sur les causes du phénomène d'imitation[228]. S'agit-il de choix imposés, conditionnés ou libres ? Si en ce qui concerne les emprunts réalisés par le droit administratif français, le problème parait simple selon Fabrice Melleray[229], il en va autrement pour les États d'Afrique noire francophone, car l'idée de décentralisation fut imposée. Les autorités africaines ont été contraintes de reprendre à leur compte une

---

223 « Droit comparé et droit public », *RDP*, 1902, tome XVII, p. 5 cité par F. MELLERAY « L'imitation des modèles étrangers en droit administratif », *AJDA*, 2004, p.1224.

224 *L'Esprit des lois*, I. 3.

225 Le modèle est ce que l'on reproduit par imitation, ce qui sert de référence, que ce soit « *dans les arts* », « *les choses d'esprit ou pour les choses morales* ». Le modèle est également une construction théorique, une « *représentation simplifiée d'un objet ou d'un processus qui existe ou doit exister* ».

226 J. BETAILLE, *Les conditions juridiques de l'effectivité de la norme en droit public interne : illustrations en droit de l'urbanisme et en droit de l'environnement*, thèse, dactly., Université de Limoges, 2012, p. 73 et s.

227 J. RIVERO, « Les phénomènes d'imitation des modèles étrangers en droit administratif », *in Miscellanea W. J. Ganshof Van Der Meersch*, Bruylant, 1972, t. 3, p. 619

228 Il analyse ensuite les modalités de l'emprunt (suivant leur étendue, leur objet et leur technique), les résultats de l'emprunt (rejet, réussite ou cas douteux) et enfin les limites du phénomène.

229 F. MELLERAY « L'imitation des modèles étrangers en droit administratif », *op.cit.*, p.1224.

construction administrative développée dans un autre pays, la France, ex-puissance coloniale. Le système administratif français, en raison de son attachement à une histoire nationale particulière, parait inexportable en tant que système[230]. Toutefois, dans nombre de ces aspects, les institutions administratives françaises essaiment et irriguent les systèmes juridiques des pays de l'Afrique noire francophone[231].

65. - L'histoire des idées politiques et des institutions administratives en Afrique noire francophone a été fortement marquée par la colonisation française. L'administration a joué un rôle important, dans la soumission et la domination des peuples colonisés. La maitrise des territoires conquis ne pouvait se passer d'une administration servant de relais entre la métropole et les peuples colonisés. La puissance coloniale fait la loi, elle gouverne et administre à la fois. Les réflexes centralisateurs de l'administration française conduisent les autorités coloniales à partir du milieu du XIXe siècle à se substituer de plus en plus aux autorités locales, détruisant ainsi les fondements de l'organisation sociale des peuples colonisés. L'administration coloniale se structure autour d'une direction centralisée dont la verticalité permet de réduire les relais et les détours inutiles pour réduire le coût de la colonisation. L'enjeu d'une telle administration est alors de bien administrer, c'est-à-dire en situation coloniale gouverner, mais aussi et surtout de légitimer la domination sur les sociétés colonisées. La colonisation française aura non seulement détruit les formes d'organisation traditionnelle, mais également légué aux pays d'Afrique noire francophone un appareil administratif fortement centralisé, autoritaire, voire despotique. Le processus de décolonisation ayant conduit à l'accession à la souveraineté internationale des États d'Afrique n'a pas entrainé une remise en cause des principes de centralisation et d'unité sur lesquels repose le système administratif français.

66. - Bien que les États africains aient opéré des choix différents en matière d'organisation administrative et territoriale, tous se rejoignent en un point qui est la « parenté » de leurs systèmes administratifs avec le modèle français érigé en référence. Il est admis communément que le mimétisme qui passe par des emprunts à des systèmes étrangers n'est que

---

230 Y. GAUDEMET, « L'exportation du droit administratif français. Brèves remarques en forme de paradoxe », *in Mélanges Philippe Ardant*, LGDJ, 1999, p. 432.
231 *Ibid.*, p. 437.

la couverture formelle de rapports de domination[232]. Le « *transfert de technologies institutionnelles* »[233] résulte d'une combinaison à la fois de l'aptitude des pays « exportateurs » à imposer leurs vues et de l'incapacité des pays « importateurs » à faire prévaloir une conception de l'organisation politico-administrative qui leur est propre. C'est ce qui explique l'adoption par de nombreux pays d'Afrique subsaharienne francophone d'une organisation politique administrative autour des principes d'uniformité et de hiérarchie dans le but de réaliser l'unité nationale. Et ce, en dépit, de leur inadaptation aux réalités sociales composites du continent. Que ce soit la construction de circonscriptions administratives ou l'institution du préfet à la tête de l'administration déconcentrée, le système administratif français a été largement réceptionné par l'« Afrique napoléonienne »[234].

67. - L'étude de l'organisation administrative des pays de l'Afrique noire francophone constitue un terrain propice à l'analyse des conséquences de la colonisation sur les modes d'organisation traditionnelle des sociétés africaines (**Titre 1**). Aussi, l'héritage institutionnel colonial continue d'imprégner l'État postcolonial africain à travers les emprunts voulus ou forcés aux institutions administratives étrangères (**Titre 2**) malgré leur inadéquation.

Titre 1 — La réception forcée d'un modèle territorial étranger

Titre 2 — La reconstruction sous influence des structures territoriales de l'État

---

232 J. CHEVALLIER, *Institutions politiques*, LGDJ, coll. « Systèmes », 1996, p. 124.

233 Y. MENY (dir.), *Les politiques de mimétisme institutionnel. La greffe et le rejet*, L'Harmattan, coll. « Logiques politiques », 1993.

234 Y. GAUDEMET, « L'exportation du droit administratif français. Brèves remarques en forme de paradoxe », *op.cit.*, p. 438.

TITRE 1 — LA RÉCEPTION FORCÉE D'UN MODELE TERRITORIAL ÉTRANGER

> « C'était là l'œuvre des Blancs : abattre ces chefs, puis les restaurer pour maintenir les morceaux d'une communauté ou ce qui en restait une fois que les Blancs avaient enlevé le meilleur »[235].

68. - La conquête européenne de l'Afrique a laissé au continent de profonds traumatismes. La colonisation française a modifié les modes d'organisation des sociétés africaines. L'administration coloniale a introduit de nouvelles normes institutionnelles qui continuent d'imprégner ces sociétés. Elle marque un point de rupture dans l'évolution des structures sociopolitiques précoloniales. Ces dernières ont fait l'objet de nombreux travaux anthropologiques et historiques dont toute réflexion sur l'État ou plus largement sur les modalités d'organisation du territoire en Afrique noire ne peut se passer. Certains auteurs affirment, notamment en raison de la prédominance du mode de production paysan, que l'Afrique subsaharienne précoloniale n'a connu avant l'invasion occidentale ni l'idée de politique[236] ni de véritables structures étatiques[237]. Une telle posture nous semble excessive et sert dans une certaine mesure le discours légitimant de la « mission civilisatrice » que les pays européens se sont donnés pour justifier l'expansion coloniale.

69. - S'il est vrai que l'État moderne, c'est-à-dire dans sa forme actuelle, ne pouvait exister à cette période, il y'a eu en Afrique noire précoloniale une diversité de formes d'organisation sociopolitique dont l'Histoire

---

235 A. PATON, *Pleure, ô pays bien aimé*, Alban Michel, 1999.

236 P. GESCHIERE, « Le poids de l'histoire », *in* C. COULON, D.-C. MARTIN (dir.), *Les Afriques politiques*, La découverte, 1991, p. 29-43.

237 D. BOURMAUD, *La politique en Afrique*, Montchrestien, coll. « Clefs », 1997, p. 13.

témoigne de la richesse[238]. À la veille de la colonisation, le phénomène politique dans les sociétés africaines ne pouvait être dissocié de l'organisation sociale de la société qui l'environne. En introduisant des structures administratives étrangères à la conception de l'autorité en Afrique noire précoloniale, la colonisation française a bouleversé leur système de valeurs. Au-delà de la littérature scientifique, un détour par la littérature romanesque, ainsi que le recommande Mamoudou Gazibo, permet de mieux saisir les conséquences de ce bouleversement[239].

70. - L'impact le plus évident des siècles de contact des sociétés africaines avec les sociétés occidentales a été « de reconfigurer la conception de l'organisation territoriale et de l'exercice de l'autorité. De ce contact entre les sociétés africaines et occidentales sont nés des États et des formes institutionnelles dont l'extériorité a pu amener certains auteurs à les qualifier de produits de pure importation »[240]. Actuellement, les rapports qui sont établis entre les structures anciennes et nouvelles issues de la colonisation s'orientent davantage vers un phénomène d'hybridation. Pour saisir l'impact du phénomène colonial dans sa globalité, nous nous intéresserons à la particularité des modes d'organisation sociopolitique des sociétés précoloniales africaines (**Chapitre 1**) puis nous constaterons l'ampleur des conséquences de la colonisation dans l'effondrement des modes d'organisation traditionnelle au profit des formes occidentales de domination (**Chapitre 2**).

Chapitre 1 – Les prémices d'une conception décentralisée du pouvoir dans les sociétés africaines précoloniales

Chapitre 2 – Les reconfigurations institutionnelles introduites par la colonisation.

---

238 Voir en ce sens, *Histoire générale de l'Afrique*, UNESCO, 8 vol., 1998.

239 V. not., C. HAMIDOU KANE, *L'aventure ambiguë*, Poche, 2003, F. OYONO, *Une vie de Boy*, Ed., Julliard, Paris, 1956. *Le vieux nègre et la médaille*, Poche, 2005.

240 M. GAZIBO, *Introduction à la politique africaine,* Presses de l'Université de Montréal, 2010, p. 63 et s.

Chapitre 1 – Les prémices d'une conception décentralisée du pouvoir dans les sociétés africaines précoloniales

71. - L'objet de ce chapitre est de démontrer qu'il a existé en Afrique précoloniale une variété de formes d'organisation sociopolitiques qui régissait la vie en société. Les systèmes politiques africains ont déjà fait l'objet de nombreux travaux restituant l'histoire générale du monde noir[241]. Dès lors, notre démarche consistera à présenter succinctement en l'adaptant à notre objet d'étude, l'évolution des formes d'organisation sociale en Afrique précoloniale indispensable à la compréhension des défis actuels. L'enjeu est d'autant plus important, car la réussite des réformes territoriales sur le continent africain ne peut se passer d'une approche historique de l'organisation sociale des sociétés précoloniales.

72. - Aborder la question de l'organisation sociopolitique des sociétés précoloniales africaines reste encore aujourd'hui une entreprise délicate. La difficulté inhérente à cette entreprise a conduit à des solutions nihilistes et simplistes niant l'existence de toute forme d'organisation sociale dans les sociétés précoloniales africaines. Il n'aurait pas pu en être autrement, puisqu'on observe généralement les sociétés précoloniales avec une focale occidentale. Étienne Le Roy note à ce propos que « *le droit africain comme droit des institutions africaines est rarement observé en tant que tel. On s'intéresse d'abord à ce que l'on pourrait appeler folklore : ce sont les modes de conduite étrange, les pratiques sociales ou religieuses extraordinaires qui retiennent l'attention* »[242]. Comment admettre qu'il ait pu exister sur une aussi longue période de l'histoire des sociétés humaines sans la moindre manifestation du phénomène politique et donc sans aucune forme d'organisation administrative et territoriale ?

73. - L'Afrique précoloniale parait sous le regard de l'observateur avisé comme une mosaïque[243]. Les sociétés précoloniales de l'Afrique ont en

---

[241] Voir not., P.-F. GONIDEC, *L'État africain : évolution, fédéralisme, centralisation et décentralisation*, 2e éd., remanié, LGDJ, 1985 ; *Les systèmes politiques africains*, LGDJ, 1971 ; G. BALANDIER, « Réflexions sur le fait politique : le cas des sociétés africaines », *Cahiers internationaux de sociologie*, vol. 37, 1964, pp. 23-50 ; P. DIAGNE, « Le pouvoir en Afrique », in *Le concept de pouvoir en Afrique*, Les presses de l'Unesco, 1981, p. 28.

[242] E. LE ROY, *Cours d'histoire des institutions d'Afrique noire*, Séminaire d'Anthropologie juridique, 1976-1977, p.14, cité par R.-G. NLEP, *L'administration publique camerounaise : Contribution à l'étude des systèmes africains d'administration publique*, LGDJ,1986, p. 19.

[243] D. BOURMAUD, La politique en Afrique, op.cit., p. 9.

effet développé plusieurs systèmes politiques. Dans tous les pays d'Afrique, les peuples se sont toujours organisés en communautés[244]. Ces dernières pouvaient se bâtir autour d'une structure lignagère, villageoise, ou même étatique. Le fonctionnement de ces communautés était régi par des règles claires. Les rapports sociaux se construisaient aux jeux des alliances voulues ou contraintes, entre autonomie et dépendance. L'organisation du territoire dans les sociétés précoloniales africaines était loin d'être aussi autocratique que l'on est enclin à le penser. En réalité, il existait déjà dans ces sociétés, une certaine « décentralisation » avant l'heure, car on assistait à une multiplicité des lieux de responsabilités chacun avec son pouvoir propre. Sur ce dernier point, les avis divergent, car pour certains auteurs[245], la décentralisation administrative est un produit d'importation. Il ne pouvait donc exister dans les sociétés précoloniales une quelconque décentralisation.

74. - À l'inverse de cette tendance, nous démontrerons que non seulement il a existé des États en Afrique précoloniale, mais aussi que le pouvoir dans ces sociétés était décentralisé. Nous le ferons, en donnant aux notions d'« État » et de « décentralisation » une signification adaptée[246] au vécu social des sociétés précoloniales. Vu sous cet angle, l'Afrique précoloniale offre à l'observateur une diversité de systèmes d'organisation politico-administrative (**Section 1**). Néanmoins, indépendamment de la multiplicité des formes d'organisation dans les sociétés précoloniales, il existe des éléments constituants communs à toutes ces sociétés (**Section 2**).

---

244 V. P. DIAGNE, « Le pouvoir en Afrique », *op.cit.*, p. 28, M. NDIAYE, « La démocratie en Afrique subsaharienne précoloniale », in *Constitution, justice, démocratie, Mélanges en l'honneur de Dominique Rousseau*, LGDJ, 2020, p. 946.

245 C. NACH MBACK, *Démocratisation et décentralisation. Genèse et dynamiques comparés des processus de décentralisation en Afrique subsaharienne*, thèse, Karthala-PDM, 2003, p. 258.

246 Il ne s'agit pas comme le font certains auteurs d'appliquer les critères modernes de reconnaissance de la décentralisation aux sociétés précoloniales. B. BERIDOGO, « Processus de décentralisation et pluralité de logiques des acteurs au Mali », *in* C. FAT, Y. F. KONE, C. QUIMINAL (dir.), *Décentralisation et pouvoirs en Afrique : En contrepoint, modèles territoriaux français* [en ligne]. Marseille : IRD Éditions, 2006 (généré le 28 avril 2021) : « *Mais quels que soient les types d'organisation de la société à ces époques, et contrairement à un discours étatique actuel qui veut y repérer les prémisses de la décentralisation, on n'y trouvait nullement une administration décentralisée dans le sens moderne, qui implique la possibilité de l'accès de tous au pouvoir par le jeu de la compétition électorale, de l'égalité des voix et partant, des hommes* ».

Section 1 : Une approche dominante malgré l'hétérogénéité des structures politiques en Afrique précoloniale

75. - Les structures politiques de l'Afrique subsaharienne précoloniale étaient diverses et variées. Rien que dans sa partie occidentale, l'Afrique a vu naitre une variété de structures politiques allant du lignage à des ensembles plus structurés comme l'État doté d'une véritable organisation politique et administrative. Face à cette multiplicité des formes d'organisation, construire une typologie des entités politiques négro-africaines précoloniales constitue une entreprise difficile. Le chercheur doit s'abstenir d'établir des typologies arbitraires, trop générales ou trop complexes[247]. L'hétérogénéité des sociétés politiques précoloniales pose donc un véritable problème de classification. Maurice Duverger faisait déjà remarquer, en 1959, l'inexistence d'« *aucune typologie des groupes sociaux* [...] *acceptée de façon générale par les spécialistes.* [Chaque chercheur] *doit encore établir sa propre typologie pour servir de cadre à ses recherches* »[248]. L'actualité de cette observation demeure certaine. Nous montrerons d'abord, l'inopérance des typologies généralement admises (**§1**) puis nous proposerons ensuite une classification pouvant permettre de mieux saisir notre objet d'étude (**§2**).

§1.- La difficulté à saisir l'organisation politico-administrative de l'Afrique précoloniale : l'inopérance partielle des classifications traditionnelles

76. - L'usage de la technique classificatoire sert les desseins de la science juridique depuis bien longtemps. Elle offre la possibilité sur la base de critères préalablement définis de rapprocher des situations ou des objets présentant une homogénéité afin de les dissocier d'autres qui sont

---

247 G. CONAC, « Le développement administratif des États d'Afrique noire », in G. CONAC (dir.), *Les institutions administratives des États francophones d'Afrique noire*, Economica, 1979, p. VIII.
248 M. DUVERGER, *Méthodes de la science politique*, PUF, 1959, p. 432.

jugés hétérogènes[249]. La valeur de la classification ainsi obtenue s'apprécie à l'aune d'une double exigence. Elle doit à la fois avoir des qualités logiques et présenter un intérêt scientifique[250]. En somme, la valeur d'une opposition dépendra de sa capacité à établir, non seulement des classes qui s'opposent trait pour trait selon le principe de non-contradiction, mais aussi de son caractère opératoire. Le critère de la classification doit permettre par sa présence d'identifier clairement l'une des classes et son absence l'autre. L'analyse des différentes formes de pouvoir qui se manifestent dans les sociétés précoloniales africaines constitue une pierre d'achoppement pour le juriste. Dans son œuvre de systématisation, il doit s'abstenir de dénaturer la réalité des phénomènes observés à travers un choix consciencieux des matériaux de classification.

77. - La recherche des critères de différenciation permettant de regrouper et de classer les sociétés africaines avant la période coloniale a donné lieu à de nombreux travaux d'historiens, d'anthropologues et de sociologues qui ont permis de dégager une typologie devenue classique dans les études africanistes. Cette typologie permet de classer les sociétés précoloniales africaines à l'aune du concept d'État. L'État étant entendu ici comme la première forme d'organisation administrative territoriale. Bien que cette typologie ne soit pas suffisamment opératoire pour saisir précisément notre champ d'analyse, elle permet de rendre compte des obstacles inhérents à l'élaboration d'une classification fonctionnelle. Les premières tentatives de classifications en ce sens se sont révélées peu opératoires (**A**). Aussi, la perspective holistique des études anthropologiques conduites sur les formes d'organisation des sociétés africaines précoloniales ne conduit qu'à saisir approximativement l'idée de décentralisation dans ces sociétés (**B**).

A- L'échec des tentatives de classification des entités politico-administratives négro-africaines

78. - Les structures des sociétés précoloniales étant très variées, les ethnologues, les anthropologues, les sociologues, les historiens, les politistes et dans une moindre mesure, les juristes, se sont livrés à la

249 Faire une classification, c'est réaliser « une opération de partitionnement et d'ordonnancement d'une collection donnée d'objets aboutissant à leur regroupement exhaustif en classes d'équivalence mutuellement exclusives définie par les prédicats d'appartenance », in Encyclopédie philosophique universelle, 2e éd., PUF, 1998.

250 F. HAMON, M. TROPER, *Droit constitutionnel, op.cit.*, p. 88.

recherche de critères permettant de classifier les différentes formes d'organisation sociopolitiques observées en Afrique précoloniale. Si, ainsi que le rappelle Charles Eisenmann, le classificateur dispose d'une liberté absolue pour faire parler son imagination, la répartition des objets observés doit se réaliser sur la base d'un critère de jugement pertinent[251]. Dans son ouvrage sur les civilisations noires[252], Jacques Maquet répartit les sociétés africaines précoloniales en cinq (5) catégories avec pour critères de répartition l'économie, la culture et la nature de l'organisation sociale. Ainsi, sur la base de ces critères, l'auteur distingue la civilisation de l'arc propre aux peuples chasseurs et récolteurs tels les pygmées ou Bochimans, la civilisation des clairières regroupant les agriculteurs itinérants de la forêt humide, la civilisation des guerriers englobant les agriculteurs de la savane méridionale, tandis que les pasteurs des hauts plateaux de l'Est africain relèveraient de la civilisation de la lance, les artisans et les marchands de l'Afrique de l'Ouest composant la civilisation des cités. Une telle répartition semble *a priori*, inopérante pour expliquer les formes d'organisation du pouvoir en Afrique précoloniale en raison du caractère statique des descriptions.

79. - Reprenant alors les travaux de Jacques Maquet, Daniel Bourmaud va dans son ouvrage dédié à la politique africaine, élaborer une typologie des organisations sociopolitiques précoloniales. En partant des civilisations guerrières et de cités notamment, l'auteur explique comment le risque de disette, la nécessité de constituer et de gérer un stock de surplus de production a conduit à l'émergence d'une chefferie en tant que structure politique. Ces chefferies fonctionnent soit sur un territoire isolé, limité à un ou quelques villages, soit au contraire elles se solidarisent pour donner naissance à une structure de type fédéral. Cette dernière hypothèse étant le plus souvent consécutive à un processus de conquête. Quant aux civilisations des cités dont les plus grandes manifestations ont été les empires du Ghana, du Mali et du Songhaï, elles disposent d'une économie fondée sur l'exploitation des ressources naturelles et leur commercialisation vers l'extérieur, elles montrent une gradation des systèmes politiques reposant sur une intégration poussée ou, au contraire, sur une indépendance affirmée des composantes urbaines et paysannes les

[251] C. EISENMANN, « Essai d'une classification théorique des formes politiques », *Politique*, n° 41-44, 1968, p.5.

[252] J. MAQUET, *Les civilisations africaines*, Marabout Université, 1962, repris et systématisé par D. Bourmaud, *La politique en Afrique*, Montchrestien, coll. « Clefs », p. 9-10.

unes par rapport aux autres[253]. On constate que la classification ainsi opérée permet de déduire l'existence de formes complexes d'organisation des sociétés précoloniales. Nous pouvons également en déduire que la notion d'État est *in fine* le critère principal de la classification.

80. - Hubert Deschamps[254] va quant à lui énumérer trois types de systèmes politiques. Les anarchies qui regroupent les sociétés dans lesquelles il n'existe pas de différenciation entre les gouvernants et les gouvernés, mais où il peut exister des institutions collectives. Les chefferies soumises à l'autorité d'un chef qui appartient à la famille traditionnellement détentrice du pouvoir et les États au sein desquelles il existe une véritable structure administrative. Cette classification issue de la littérature ethnographique s'établit sur la base du degré de structuration en prenant en compte diverses institutions sociales africaines tels, la famille, le lignage, le clan, la tribu. Bien qu'elle tende à s'imposer, elle a fait l'objet de vives critiques tant du point de vue des expressions qu'elles utilisent[255] que, et c'est la critique principale, de son caractère statique[256]. En effet, cette classification comporte en elle-même des éléments de confusion conduisant à considérer certaines sociétés tels les Dogons au Mali, qui sont pourtant parvenues à créer des ensembles pourvus à la fois d'autorité centrale et de projet politique de sociétés anarchiques c'est-à-dire sans chefs[257].

---

253 D. BOURMAUD, *La politique en Afrique, op.cit.*, p. 9 et s.

254 H. DESCHAMPS, *L'Afrique noire précoloniale*, PUF, coll. « Que sais-je ? », 1962.

255 G. CONAC estime que les expressions d'anarchie et de chefferie utilisées dans cette classification peuvent prêter à confusion. G. CONAC (dir.), *Les institutions administratives des États francophones d'Afrique noire*, Economica, 1979.

256 Le professeur Théophile Obenga rejette cette classification issue des travaux ethnographiques, car elle soulève beaucoup de problèmes. « *Étant descriptive, la méthode ethnographique aboutit à des résultats qui relèvent des études de simple distribution géographique. Le caractère linéaire et statique des morphologies décrites surprend : anarchie, chefferie et État. Un type spécifique de structure sociale correspond-il à une période historique précise ? Si oui, à quelle coupure historique, à quel temps précis correspondrait l'anarchie en tant que système politique africain traditionnel ? Les Dogons, qui sont décrits comme vivant dans un système politique anarchique, seraient-ils plus anciens dans l'histoire que les peuples qui créèrent les Empires de Ghana, de Mali et de Songhaï ? Les chefferies mendés en Sierra Leone sont-elles plus anciennes que l'État de Kouch ou l'État pharaonique ?* ». T. OBENGA, « De l'État dans l'Afrique *pré*coloniale : Le cas du royaume de Kouch dans la Nubie ancienne », *Présence africaine*, n° 127-128, 1983, p. 128. Voir aussi du même auteur, *L'Afrique dans l'Antiquité. Égypte pharaonique, Afrique noire*, Présence africaine, 1973.

257 *Idem.*

81. - Des constitutionnalistes, à l'instar de Gérard Conac, se sont aussi livrés à l'exercice de la classification des sociétés politiques précoloniales. En effet, cet auteur distingue au titre des structures politiques différentes des formes constitutionnelles : les sociétés ne disposant pas d'un personnel politique nettement différencié, les sociétés dotées de gouvernants et enfin, les sociétés complexes avec une spécialisation poussée des fonctions dotées d'un personnel administratif, civil et militaire soumis à l'autorité du chef[258].

82. - La classification des structures politiques en Afrique est, on le voit, on ne peut plus compliquée. En témoigne, si besoin en est, le recours quasi systématique à la tripartition. Or, on sait que le recours aux classifications tripartite, en Droit, conduit souvent à de fausses notes. De telles classifications existent dans le droit positif, mais elles constituent une entorse aux qualités logiques de la classification[259]. Charles Eisenmann considérait d'ailleurs à ce propos qu'il était inadmissible « *de poser, en considération annoncée d'un même trait [...] une division tripartite* »[260]. Pour étudier donc les sociétés politiques africaines précoloniales, nous devons nous résoudre à la classification binaire, devenue classique dans l'étude de l'Afrique précoloniale, entre d'un côté les sociétés sans État considérées comme des sociétés « anarchiques » et de l'autre les sociétés avec État. La parenté va jouer un rôle déterminant dans cette classification. Elle permettra d'établir l'existence ou non d'un pouvoir détaché de la société servant d'instrument de domination à une catégorie sociale précise[261].

---

258 G. CONAC, « Le développement administratif des États d'Afrique noire », in G. CONAC (dir.), *Les institutions administratives des États francophones d'Afrique noire*, Economica, 1979, p. IX.
259 M. BARTOLUCCI, *L'acte plurilatéral en droit public*, thèse, Dalloz, 2022, pp. 300-301.
260 C. EISENMANN, « Quelques problèmes de méthodologie des définitions et des classifications en sciences juridiques », *Archives de philosophie du droit*, vol. XI, 1966, p. 37. Prenant en exemple les fonctions juridiques de l'État réparties selon le triptyque fonction législative, fonction administrative et fonction juridictionnelle, l'auteur propose pour plus de rigueur, de les regrouper en fonction de création (législative) et fonction d'application du droit (administrative et juridictionnelle). Si cette classification binaire est pertinente, car répondant aux exigences logiques, elle n'est pas exempte de critiques sur le fond.
261 P.-F. GONIDEC, *L'État africain : évolution, fédéralisme, centralisation et décentralisation*, *op.cit.*, p. 30.

B- L'insuffisance explicative de la distinction classique entre sociétés sans État et sociétés étatiques

83. - L'Afrique n'est pas réfractaire à l'État. Établir une distinction binaire en prenant pour critère la notion d'État suppose le parti pris des différents travaux de reconnaitre qu'il a existé sur le continent noir avant la colonisation, un État africain précolonial distinct de l'État occidental et postcolonial. Il est donc possible d'opposer trait pour trait[262] les sociétés politiques africaines dotées de structures étatiques et celles qui fonctionnent sans État. Nous devons cette classification à Meyer Fortes et Edward Evans-Pritchard, anthropologues et politistes, qui, dans leur ouvrage *Systèmes politiques africains*[263] devenu aujourd'hui un classique des études africanistes, distinguent ces deux types de systèmes politiques dans les sociétés précoloniales. Les sociétés à État territorialement unifiées sont placées sous la domination d'une autorité centrale disposant de forces armées. Elles connaissent une organisation basée sur les inégalités de classes, de richesses et de privilèges. À l'inverse, trait pour trait, les sociétés sans État sont dépourvues de toute autorité centrale en raison de l'inexistence des divisions sociales de rang ou de statut[264]. Malgré le caractère simpliste de cette distinction, elle conduit à une exagération de l'opposition dans la mesure où le simple recours au concept d'État emporte un parti pris ethnocentrique qui consiste à penser l'État au sens d'État occidental. Pour s'en convaincre, il suffit de se référer à l'expression de sociétés « anarchiques » retenue pour désigner les sociétés dépourvues d'appareils étatiques. Or, l'absence d'une structure formelle d'État n'implique ni l'anarchie ni le désordre, mais l'existence de structures politiques régulatrices parfois très précisément règlementées[265]. Nous évoquerons successivement les sociétés sans État dans lesquelles l'autonomie locale semble être la règle (**1°**) puis les sociétés avec État davantage centralisées (**2°**).

---

262 C. EISENMANN, « Quelques problèmes de méthodologie des définitions et des classifications en sciences juridiques », *op.cit.*, p. 37.

263 M. FORTES, E. EVANS-PRICHARD, *Systèmes politiques africains*, PUF, 1964.

264 J.W. LAPIERRE, Vivre sans État ? Essai sur le pouvoir politique et l'innovation sociale, Esprit-Le Seuil, 1977, p. 70.

265 C. COQUERY-VIDROVITCH, « Les structures du pouvoir et la communauté rurale précoloniale », *Revue française d'histoire d'outre-mer*, 1981, p. 54.

1. L'exercice d'un pouvoir décentralisé dans les sociétés sans État

84. - On sait désormais grâce aux travaux d'historiens et d'ethnologues qu'il a existé dans l'Afrique précoloniale des sociétés sans chefs institutionnalisées. Cette réalité a pu surprendre auparavant les juristes occidentaux qui ne pouvaient concevoir l'existence de telles sociétés que sur le plan purement théorique[266]. D'ailleurs encore aujourd'hui, la science du droit constitutionnel n'étudie le phénomène du pouvoir que dans le cadre de l'État[267]. Loin du mirage, l'Afrique précoloniale offre donc l'occasion de revenir sur les ressorts et les mécanismes qui président à l'organisation des sociétés dites « anarchiques ». Cette catégorie regroupe par ailleurs de nombreuses variantes construites sur la base d'entités sociales africaines que sont : le lignage, le clan et la tribu. Il n'existe pas dans ces entités de pouvoir organisé, le pouvoir y est diffus (**a**). La régulation sociale est assurée à travers des mécanismes participatifs et communautaires (**b**).

a) L'absence d'institutionnalisation du pouvoir

85. - Les sociétés précoloniales africaines non étatiques ne résistent pas à la grille d'analyse du pouvoir exprimé dans sa forme juridique qu'est l'État au sens moderne. Elles correspondent dans la typologie établie par Hubert Deschamps aux anarchies ; c'est-à-dire des organisations politiques caractérisées par un pouvoir diffus exercé par l'ensemble de la communauté sans que l'on puisse désigner expressément son titulaire. Les sociétés non étatiques sont ainsi dépourvues des différents organes qui donnent naissance à l'appareil d'État. Elles ne disposent pas d'organes dédiés à la création de la loi ni d'exécution ni de justice. On note par exemple chez les Konkomba au nord du Togo l'absence d'institution juridictionnelle et de cadre formel de législation ainsi que d'agents chargés de l'application de la loi[268]. En revanche, en raison de l'existence d'un corps de règles coutumier[269], on peut affirmer avec certitude qu'il existe une

---

266 G. CONAC, « Le développement administratif des États d'Afrique noire », *op.cit.*, p. IX.
267 F. HAMON, M. TROPER, *Droit constitutionnel*, *op.cit.*, p. 88 et s.
268 P.-F. GONIDEC, *L'État africain : évolution, fédéralisme, centralisation et décentralisation*, *op.cit.*, p. 31.
269 Ces règles coutumières ne sont pas le fruit d'un travail législatif, ne sont pas interprétées par des tribunaux ni mises en application par un exécutif.

forme d'organisation politique. L'ordre social est en l'espèce assuré par des rapports de parenté puisqu'il n'existe aucun pouvoir d'État.

86. - Ainsi, réunie par des liens de parenté, la tribu est fixée sur un territoire bien défini correspondant à un groupement humain déterminé. Le territoire joue un rôle fondamental[270] dans ces sociétés, car les communautés humaines reliées par des liens de parenté forment, *in fine*, une unité politique. Les travaux de sociologues et d'ethnologues ont permis de prouver que ces sociétés sans État étaient des sociétés globales définies comme des entités spatiales et humaines au sein desquelles, le pouvoir politique est assuré[271]. Lorsque la famille prend des proportions encore plus étendues, elle devient un clan c'est-à-dire une unité sociale dont l'ensemble des membres se reconnaissent un ancêtre commun et respectent les mêmes interdits religieux comme le totem. Plusieurs clans ayant en partage la même langue peuvent constituer une tribu. « *En principe, le chef du clan ou de la famille est le membre de chaque génération le plus proche de l'ancêtre. Mais il faut tenir compte du conseil de famille. Cette assemblée, qui rassemble tous les adultes des deux sexes ou les hommes seuls, a son mot à dire dans chaque affaire importante. Elle peut décider par exemple de refuser l'héritage de l'autorité clanique au candidat "légal" parce qu'il est reconnu malade, inactif ou mauvais administrateur. Par ailleurs, à côté et au-dessus de l'autorité familiale et économique du patriarche, il y a l'autorité politico-administrative de la communauté territoriale et du village* »[272].

87. - L'absence d'une institution formelle disposant seule de l'exercice du pouvoir témoigne de la démarche participative qui préside à l'organisation de ces sociétés. Chaque membre de la société étant appelé à jouer un rôle dans la conduite des affaires qui touchent à la survie du groupe social. Le caractère diffus du pouvoir renforce l'autonomie décisionnelle et empêche la concentration du pouvoir aux mains d'une aristocratie dominante. Par conséquent, dans les sociétés considérées comme anarchiques, l'exercice du pouvoir est diffus et l'appartenance au groupe se fait sentir partout. Ainsi, chaque individu en son domaine permet de pousser plus loin les vertus de la « décentralisation ». Les

---

270 Sur cet aspect Luc Sindjoun montre à la suite de J. Herbst que « le royaume Bamoun [au Cameroun], par exemple, au moment de la rencontre coloniale au début du XXe siècle a un roi cartographe, le sultan Njoya, qui dessine les limites de son autorité et représente le centre du pouvoir ».

271 P.-F. GONIDEC, *L'État africain : évolution, fédéralisme, centralisation et décentralisation*, *op.cit.*, p. 33.

272 J. KI-ZERBO, *Le Monde noir*, Abidjan, CEDA, 1963, p. 47.

aspirations à la décentralisation ne s'apprécient que dans la mesure où, il existe un accroissement des individus capables d'assumer en toute autonomie leurs responsabilités dans la construction d'un groupe social interdépendant.

b) L'originalité des modes d'exercice du pouvoir

88. - La souveraineté dans les sociétés sans État est complètement diffuse de sorte qu'il n'existe pas d'institutions de domination telles qu'on peut le concevoir pour les sociétés politiques actuelles. En raison de l'absence d'institutionnalisation de l'autorité dans les sociétés sans État, c'est la communauté tout entière qui décide. Cela conduit à admettre que les sociétés sans État en Afrique précoloniale pratiquaient pour une large part la démocratie directe ou populaire. L'ordre social est entièrement régulé à l'intérieur des structures parentales. Cependant, la fixation du lignage, du clan, de la tribu sur un territoire entraine souvent des conflits d'ordre interne préalable à une différenciation et éventuellement à la naissance d'une autorité. En réalité, dans ces sociétés, le maintien ou le rétablissement de l'ordre social implique le recours à des pratiques comme le droit d'aînesse. Ce qui suppose que dans le lignage, l'autorité du rétablissement de l'ordre appartient à l'aîné c'est-à-dire le doyen d'âge.

89. - Le lignage se reconnait à travers un ancêtre commun avec sa tête un chef qui, comme le note Maurice Delafosse, est un patriarche qui « *tient le plus près à l'ancêtre de la famille globale, c'est-à-dire au fondateur de la première famille réduite primitive* »[273]. Dans un tel schéma, renchérit Daniel Bourmaud, il n'y a pas, « *à proprement parler d'institutionnalisation permanente du politique, mais l'exercice ponctuel de la contrainte constitue un attribut réservé à ceux qui remplissent les conditions socialement définies* »[274]. Le chef ici n'est pas un gouvernant, mais simplement un leader. Le caractère ponctuel de la tâche qu'il assume préserve de toute tentation à l'autoritarisme. Aussi, comme nous l'avons montré précédemment, le conseil de famille joue un rôle important dans la prise de décision. Ce conseil constitue le garant de l'exercice partagé du pouvoir et un contre-pouvoir à l'expansion excessive de l'autorité du chef de famille.

---

[273] Cf. J. VILLANDRE, *Les chefferies traditionnelles en Afrique occidentale française*, thèse, Université de Paris, 1950, p. 5.
[274] D. BOURMAUD, *La politique en Afrique*, *op.cit.*, p. 17.

90. - Dans les sociétés précoloniales africaines dépourvues d'un pouvoir institutionnalisé, la collectivité prime sur l'individu. Ce dernier n'existe que parce qu'il appartient à un groupe social auquel il s'identifie c'est-à-dire l'ensemble des descendants d'un ancêtre commun et soudé entre eux par les liens du sang et du sol. L'individu n'est pas un être abstrait, mais un être concret qui prend corps avec la collectivité au nom de laquelle il exerce ses fonctions[275]. L'autonomie dont il dispose à cet effet n'est pas une autonomie personnelle, mais elle le prédispose à agir constamment dans l'intérêt et pour la prospérité du groupe. Le pouvoir n'a de sens que dans la mesure où il agit dans l'intérêt exclusif de toute la collectivité. Ces caractères propres aux sociétés politiques africaines non étatiques faiblement institutionnalisées consacrent le primat de la parenté et de l'âge dans la régulation de l'ordre social[276]. Le pouvoir politique institutionnalisé demeure résiduel[277].

91. - Tout à l'opposé de ces sociétés fondées sur la limitation du politique et le primat de la collectivité, apparait un phénomène étatique. Il est fondé sur la différenciation sociale entre gouvernés et gouvernants ainsi que le monopole de la contrainte armée par un groupe d'individus. L'État se présente alors comme un instrument de domination de la majorité par une minorité gouvernante. Le passage d'une société non étatique à une société étatique conduit à une centralisation du pouvoir.

2. L'exercice d'un pouvoir centralisé dans les sociétés avec État

92. - Les sociétés sans État se transforment progressivement en des sociétés politiques étatiques à la faveur des conquêtes et de l'accumulation des richesses par une catégorie de personnes au sein de la société. La politisation croissante des sociétés sans État en raison de leur rapport au territoire a conduit progressivement à une différenciation sociale et à des évolutions politiques aboutissant à une institutionnalisation du pouvoir. L'Afrique occidentale précoloniale a constitué un champ de cohabitation de ces deux systèmes politiques. À la faveur des évolutions sociales, des

---

275 M. NDIAYE, « La démocratie en Afrique subsaharienne précoloniale », in *Constitution, justice, démocratie,* Mélanges en l'honneur de Dominique Rousseau, LGDJ, 2020, p. 946 : « il y a une antériorité et primauté du groupe sur l'individu. L'unité de base de la société n'est pas l'individu mais le lignage ».

276 P. DIAGNE, *Pouvoir politique traditionnel en Afrique occidentale. Essai sur les institutions politiques précoloniales*, Présence africaine, 1967.

277 D. BOURMAUD, *La politique en Afrique, op.cit.*, p. 17.

liens territoriaux se sont formés pour donner corps à de grandes unités politiques. Des royaumes sont nés du fait de conquêtes entreprises étendant ainsi leurs frontières tout en englobant les sociétés sans État[278]. Le pouvoir politique dans ces nouvelles entités est détenu par un individu ou un groupe d'individus disposant d'un véritable pouvoir de coercition. L'État, instrument de domination de la masse, prend ainsi forme avec une hiérarchisation des individus en fonction de leurs possessions.

93. - La naissance de l'État précolonial s'explique en premier lieu par des facteurs endogènes que nous avons commencés par décrire dans les sociétés non étatiques. Des regroupements volontaires ont pu être observés entre deux ou plusieurs tribus pour donner naissance à un village. Ces alliances volontaires traduisent la volonté des uns et des autres de disposer du surplus de production des tribus voisines. Ainsi, le niveau de développement des sociétés a contribué à la formation de certaines confédérations. Jacques Maquet dans son ouvrage sur les civilisations noires précité nous renseigne sur l'état d'esprit des chefs. En effet, souligne-t-il, « *l'existence d'un surplus à s'approprier sous forme de tributs saisonniers était un puissant stimulant pour les chefs* »[279]. La volonté des chefs de constituer de grands ensembles étatiques était essentiellement mue par les revenus que leurs sujets pouvaient leur fournir. Sinon, quel intérêt y verraient-ils à constituer un État unissant des villages qui, pour leur besoin de subsistance, doivent consommer tout ce qu'ils produisent[280] ?

94. - Le développement économique va conduire à une stratification sociale avec pour conséquence l'établissement d'une distinction entre gouvernés et gouvernants. Ainsi, la civilisation des guerriers décrite par Jacques Maquet, composée d'agriculteurs de la savane, a bâti leur organisation sociale sur la base de leur mode de production. La gestion du surplus de production stocké dans les greniers pour faire face aux aléas relatifs aux changements climatiques a permis l'émergence de la chefferie. L'autorité du chef est fortement liée à l'existence de ce surplus de production. La conclusion à laquelle parvient Jacques Maquet est la suivante : « *parce qu'un homme peut disposer de la surproduction de la communauté,*

---

278 M. SOME, « État, pouvoir et démocratie en Afrique de l'Ouest contemporaine : les héritages du passé », in A. LOADA, J. WHEATLEY (dir.), *Transitions démocratiques en Afrique de l'Ouest. Processus constitutionnels, société civile et institutions démocratiques*, L'Harmattan, coll. « Série politique », 2014, p. 53.

279 J. MAQUET, *Les civilisations noires, op.cit.*, p. 139.

280 *Idem.*

il a les moyens de pression et de coercition, c'est-à-dire le pouvoir politique, et ce pouvoir lui permet de revendiquer le surplus d'autres agriculteurs. Et ainsi, continue la spirale des accroissements alternatifs et cumulatifs du revenu et du pouvoir». Cette accumulation des richesses permet au chef de s'attacher les services de nombreux courtisans et agents d'exécution pour asseoir son autorité. Le chef se constitue alors progressivement un appareil d'État.

95. - Certains auteurs nient aux chefferies précoloniales leur qualité d'État en raison de leur taille. Hubert Deschamps dont on a évoqué la classification précédemment s'efforce en ce sens d'établir une distinction entre les « anarchies », les « chefferies » et les « États »[281]. La taille des entités observées constitue sans aucun doute un enjeu important pour l'analyse du point de vue de la population, du territoire des sociétés considérées ainsi que de l'organisation politique dont le degré de complexité en dépend. Toutefois, on retrouve aussi bien dans la plus petite unité politique qui puisse exister tout comme dans la plus grande des phénomènes de même nature. Une idée de domination basée sur l'accumulation du pouvoir économique et politique qui est irréductible au déploiement de l'État[282]. Les chefferies précoloniales constituent quoiqu'en miniature des États. Refuserait-on, ainsi que s'interroge Pierre François Gonidec, à la Principauté de Liechtenstein le caractère d'État sous prétexte de son exiguïté ? La réponse ne peut être que négative.

L'Afrique précoloniale a connu de grands empires et royaumes. Ces grands ensembles se sont principalement construits à travers des conquêtes. Les chefferies précoloniales ont évolué grâce à ces conquêtes tout au long de l'histoire. C'est ainsi que se sont constitués les grands royaumes sur fond de lutte et de conquête[283]. De cette manière, se sont succédé en Afrique subsaharienne des États de type primaire, les chefferies et des États de type secondaire, les royaumes et les empires. Si les premiers sont le fruit de facteurs endogènes, les seconds relèvent davantage de facteurs exogènes[284].

96. - Dans le Soudan occidental, des structures étatiques territorialement très étendues ont pu exister. C'est le cas des royaumes

---

281 H. DESCHAMPS, *L'Afrique noire précoloniale*, PUF, coll. « Que sais-je ? », 1962.

282 P.-F. GONIDEC, *L'État africain : évolution, fédéralisme, centralisation et décentralisation, op.cit.*, p. 39.

283 D. BOURMAUD, *La politique en Afrique, op.cit.*, p.11.

284 P.-F. GONIDEC, *L'État africain : évolution, fédéralisme, centralisation et décentralisation, op.cit.*, p. 39.

Mossi de Ouagadougou, des royaumes d'Abomey et de l'empire Mandingue du Mali qui englobait le royaume du Ghana vassalisé. Ces structures étatiques disposaient du pouvoir de contrainte physique pour assurer le maintien de l'ordre social. Dans l'empire Mandingue du Mali, le pouvoir central était assuré par le roi des rois, le *Mansa*, chef de gouvernement entouré pour l'exercice de ces fonctions de hauts dignitaires. Avec les descendants des généraux, ils formaient une aristocratie militaire chargée de conseiller le *Mansa*. Le griot occupait les fonctions de porte-parole du *Mansa*. Il était chargé du protocole. On avait à côté le précepteur des princes et le majordome de la cour impériale[285]. Ainsi s'institutionnalise progressivement le pouvoir avec la construction d'une élite politico-administrative. Le souverain dispose désormais d'un véritable appareil administratif et militaire pour s'imposer à ces sujets. Les grands royaumes de l'Afrique subsaharienne s'identifient à travers leur mode de fonctionnement. Selon que leur formation s'est réalisée sur la base de guerre de conquête ou par association volontaire, l'État dans l'exercice du pouvoir pouvait prendre la forme d'un État de type unitaire ou d'un État de type fédéral[286]. Toutefois, une diversité de systèmes politiques verra le jour grâce à des combinaisons sociopolitiques variées, preuve d'ailleurs, de la complexité des sociétés africaines précoloniales.

97. - Les différentes tentatives de classification des entités politiques africaines se sont fondées essentiellement sur la notion d'État et sur le degré de différenciation qui pouvait s'établir entre les gouvernants et les gouvernés. Si ces typologies ont permis de démontrer au-delà de la critique qu'il existait un État précolonial africain, ils ne nous éclairent en revanche que très peu sur la réalité même de l'exercice du pouvoir politique. Au-delà des développements qui précèdent, il parait peu probable, peu importe la taille des entités politiques africaines précoloniales, que les activités essentielles à la vie politique et administrative de la communauté soient exercées par un organe unique. La reconnaissance d'un tel postulat nous permet, par ailleurs, d'esquisser une nouvelle typologie des sociétés africaines précoloniales qui fait fi de l'origine du pouvoir, des formes d'autorité pour s'intéresser uniquement à la manière dont le pouvoir se traduit *in concreto*.

---

285 M. SOME, « État, pouvoir et démocratie en Afrique de l'Ouest contemporaine : les héritages du passé », *op.cit.*, p. 53 et s.

286 P.-F. GONIDEC, *L'État africain : évolution, fédéralisme, centralisation et décentralisation*, *op.cit.*, p. 39. Voir aussi du même auteur, *Les systèmes politiques africains*, LGDJ, 1971.

§2.- Esquisse d'une classification renouvelée des sociétés précoloniales

98. - La typologie que nous proposons se base sur le couple unité/pluralité des centres de décisions dans les sociétés africaines précoloniales. Existait-il un centre unique de décision ou plusieurs centres qui participaient à l'exercice des fonctions politiques et administratives au sein des sociétés africaines précoloniales ? La réponse à cette interrogation déterminera la tendance à la centralisation ou non du pouvoir politique. Il s'agit concrètement de dépasser le caractère général des classifications classiques en choisissant comme matériau de base de la nouvelle classification le mode d'exercice des différentes fonctions administratives dans les sociétés africaines précoloniales. Cela permettra donc de construire des catégories de sociétés en s'appuyant sur le degré d'unification de l'exercice des fonctions administratives (**B**). Il convient, avant, de vérifier l'existence même du phénomène administratif et ces manifestations dans les entités précoloniales africaines (**A**).

A- Une classification fondée sur l'exercice des fonctions administratives dans la société précoloniale

99. - L'analyse de la répartition des taches dans les sociétés africaines peut être un révélateur du modèle d'organisation de ces sociétés. Il existe aujourd'hui un consensus entre les différents observateurs des réalités précoloniales africaines autour de l'existence d'un « chef » dans les sociétés de l'époque. Nous avons démontré que même dans les sociétés sans pouvoir institutionnalisé il existe sur la base de l'ancienneté ou de l'apparition de certaines situations d'inégalités des chefs leaders chargés d'organiser et de diriger les activités essentielles à la survie de la communauté. Si l'existence d'un chef dans l'organisation administrative et politique de ces sociétés parait primordiale au sens où il constitue un rouage politique et administratif important dans la structure des sociétés précoloniales, il ne dispose pas, en revanche, du monopole de l'exercice de toutes les activités. En effet, malgré son importance, il n'est pas l'organe unique et ne peut exercer seul l'ensemble des fonctions nécessaire à la survie du groupe[287]. Le processus de structuration des entités précoloniales

[287] *Idem.*

partant des chefferies primaires à des ensembles plus grands tels que les royaumes ou les empires nous permet d'observer les tendances à l'unité ou à la multiplicité des organes qui se répartissent les fonctions essentielles à la survie du groupe dans ces sociétés.

100. - Il est admis depuis 1748 une conception de l'État que l'on doit à l'œuvre de Montesquieu. Dans *L'Esprit des lois*, Montesquieu affirme qu'il a dans chaque État trois types de puissances à savoir la puissance législative, la puissance exécutrice et la puissance de juger[288]. La possession d'un pouvoir suppose l'exercice d'une fonction spécifique dévolue à un organe particulier. « Pouvoir » et « fonction » sont appréhendés par les interprètes de la pensée de Montesquieu comme les réalités d'une même pièce permettant de distinguer les fonctions étatiques[289]. La séparation des pouvoirs telle qu'elle a été théorisée pour les systèmes politiques modernes perd toute sa substance dans son application aux sociétés africaines précoloniales. Elle demeure en ce sens qu'une vue de l'esprit, car il est impossible d'isoler à part dans ces sociétés une fonction législative. Il n'est pas question ici de dépouiller le chef africain de toute activité normative, mais c'est presque un non-sens de parler de l'activité législative du chef dans une société dont les règles régissant l'organisation sociale et la conduite des individus sont le fruit de « *normes fondamentales transcendantales* »[290]. Pour accomplir sa fonction gouvernementale, le chef sera amené à prendre des mesures à caractère générale ou individuelle pour renforcer, adapter ou appliquer la « *norme fondamentale transcendantale* » mettant ainsi en œuvre ce que le droit public moderne désigne comme son « pouvoir règlementaire »[291].

101. - Par suite, en sa qualité d'organe principal du gouvernement, le chef organise et dirige les activités de la société. Sa mission consiste donc à défendre le territoire contre toute agression extérieure. Il s'occupe des affaires étrangères et de la résolution des conflits intéressants des catégories de la population. L'organisation des activités comme la chasse,

---

[288] C. MONTESQUIEU, *De l'Esprit des lois*, Librairie Ch. Delagrave, 2e éd., 1892.

[289] Pouvoir et fonction entretiennent des rapports de réciprocité mais ne se coïncident pas. Voir, G. TIMSIT, *Le rôle de la notion de fonction administrative en droit administratif français*, thèse, LGDJ, 1963, p. 5 et s.

[290] M. KAMTO, *Pouvoir et droit en Afrique noire. Essai sur les fondements du constitutionnalisme dans les États d'Afrique noire francophone*, thèse, LGDJ, 1987.

[291] E. KATE KWAYEB, *Les institutions de droit public du pays Bamiléké*, LGDJ, 1960, p. 26 et s.

le déplacement des populations d'un endroit à un autre, le commerce relève en principe de l'autorité du chef. Ces activités communes existent dans toutes les sociétés africaines précoloniales. L'administration s'assure donc de la paix et de la sécurité ainsi que de la réalisation du bien-être économique et social des populations. Accomplies dans l'intérêt de la collectivité, ces activités permettent de remplir les fonctions d'une administration classique à savoir : police et service public[292]. C'est donc à partir de l'analyse des modes d'exercice des fonctions administratives que le juriste devra établir une classification *in concreto* des entités précoloniales africaines[293].

102. - La reconnaissance de l'existence d'une fonction administrative comme donnée constante de toutes les sociétés africaines précoloniales permet d'entrevoir autrement les rapports entre centre et périphérie. La répartition de l'exercice de l'action administrative entre plusieurs organes pour répondre aux besoins des populations renseigne sur la nature des entités politiques africaines précoloniales. En effet, elle permet d'aborder plus aisément la question de l'existence ou non d'une conception décentralisée du pouvoir en Afrique noire précoloniale. Les modalités d'exercice de ces fonctions varient d'une société à l'autre en fonction de la place faite aux institutions administratives locales dans la conduite des affaires du groupe social. Elles se résument, *in fine*, pour reprendre les propos de Charles Eisenmann « *au problème de l'unité ou de la division corrélative de l'appareil étatique et de la collectivité étatique. C'est le problème des formes d'États et autres collectivités politiques, de leur structure unitaire (simple) ou composée (divisée)* »[294].

103. - La question de la centralisation et de la décentralisation se résume ainsi à celle des formes de l'État — l'unitaire et la composée[295].

---

292 C. EISENMANN, Préface à G. TIMSIT, *Le rôle de la notion de fonction administrative en droit administratif français, op.cit.*, p. VI.

293 Il ne s'agit pas dans le cadre de notre étude de refaire le débat sur l'existence ou non d'une fonction administrative. Voir en ce sens l'ouvrage de F.-P. BENOIT, *Le droit administratif français*, Dalloz, 1968, spéc., Livre II, p. 27 et s. ; G. TIMSIT, *Le rôle de la notion de fonction administrative en droit administratif français, op.cit.* ; R.-G. NLEP, *L'administration publique camerounaise. Contribution à l'étude des systèmes africains d'administration publique*, LGDJ, 1986, J. BENOIT, *Théorie juridique de la décentralisation administrative*, thèse, dactyl., Université de Paris, 1990.

294 C. EISENMANN, *Centralisation et décentralisation. Esquisse d'une théorie générale*, LGDJ, 1948.

295 *Idem.*

Pierre-François Gonidec adopte la même démarche lorsqu'il distingue dans les sociétés précoloniales africaines deux formes d'État à savoir les États unitaires et les États de type de fédéral. La reconnaissance de l'existence d'un fédéralisme dans les sociétés précoloniales peut de l'aveu même de l'auteur surprendre. Cependant, poursuit-il, « *réduit à sa nudité essentielle, le fédéralisme répond partout et toujours à la même préoccupation. Il permet de concilier la nécessité de reconnaitre aux collectivités composantes une certaine autonomie et la volonté de créer une communauté globale regroupant ces collectivités* »[296]. Ce que décrit ici le professeur Gonidec n'est rien d'autre que la décentralisation. La reconnaissance de l'existence de la décentralisation ne peut se faire qu'à travers l'étude dans le temps des sociétés précoloniales marquées par leur dynamisme constant. N'étant pas figée, l'approche historique des formes d'organisation et des manifestations du pouvoir est indispensable. Jean Rivero souligne d'ailleurs que « *telle forme de concentration du pouvoir n'est pas significative en elle-même, elle ne l'est que par rapport à l'état qui la* précède *ou par rapport à l'état qui la suit. Il faut l'interpréter en fonction d'un ensemble historique ; c'est dans évolution, dans un dynamisme qu'il faut demander à chacune des formes… leur sens* »[297].

De ce point de vue, la décentralisation du pouvoir au profit de structures administratives locales exige non seulement que les autorités locales représentent effectivement la population, mais aussi qu'elles disposent d'une marge de choix réelle. Bien entendu, la représentation au sens moderne, basée sur le principe de l'élection, ne peut s'appliquer aux sociétés précoloniales africaines dans lesquelles elle prend d'autres formes reconnues et acceptées telle la transmission héréditaire ou clanique du pouvoir marqué par son auréole de sacralité. C'est donc à l'aune de ces différentes considérations que sera établie une nouvelle typologie des sociétés africaines précoloniales.

---

296 P.-F. GONIDEC, *L'État africain : évolution, fédéralisme, centralisation et décentralisation*, *op.cit.*, p. 41.
297 J. RIVERO, *Fédéralisme et décentralisation dans la structure de l'État moderne*, Cours polycopié, 1949, p. 319, cité par C. ROIG, « Théorie et réalité de la décentralisation », *Revue française de science politique*, vol. 16 n° 3, 1966, p. 452.

B- Pour une nouvelle classification des sociétés précoloniales africaines autour de l'opposition entre unité et pluralité institutionnelle

104. - Comment étaient gouvernées les sociétés africaines précoloniales ? Telle est la question sous-jacente à laquelle nous essayerons d'apporter une réponse par un examen des modes d'exercice du pouvoir dans les entités politiques précoloniales en Afrique. À la suite des travaux généraux entrepris par Charles Eisenmann sur la théorie générale de la centralisation et de la décentralisation, il nous parait opportun de résoudre la question de la répartition du pouvoir dans les sociétés précoloniales à partir des rapports entre centre et périphérie, unité et pluralisme. Pour ce dernier en effet, « *centralisation signifie centre unique, unité de centre ; décentralisation, centres multiples, pluralité de centres…* »[298].

105. - Dans les sociétés précoloniales en Afrique, la fragmentation du pouvoir a permis à la société d'assurer sa prééminence. En effet, l'existence des autorités locales était un moyen de réduire et d'encadrer le pouvoir du chef. La décentralisation permettait par ailleurs d'instituer un véritable contre-pouvoir à l'autorité du chef. Selon Paul Mercier, dans la configuration des sociétés africaines précoloniales l'autorité est à la fois pyramidale et horizontale. Il relève en ce sens que l'organisation d'un système politique répétitif constitue la caractéristique des sociétés précoloniales en Afrique. « *C'est-à-dire que plusieurs niveaux de foyers subordonnés du pouvoir peuvent être distingués, organisés en pyramide par rapport à l'autorité centrale. Les autorités centrales et périphériques reflètent le même modèle, les dernières étant les images en réduction de la première. Des pouvoirs semblables sont répétés à chaque niveau, avec une étendue décroissante. Chaque autorité a certains pouvoirs reconnus sur les autorités subordonnées qui lui sont articulés* »[299]. Cependant, la verticalité des relations entre le centre et la périphérie doit être fortement nuancée. Les autorités périphériques ont tendance à affirmer leur caractère naturel c'est-à-dire d'entités qui sociologiquement s'imposent par leur attitude profondément autonomique. Ce mode d'organisation bien que

298 C. EISENMANN, *Centralisation et décentralisation. Esquisse d'une théorie générale*, LGDJ, 1948, p. 17.

299 P. MERCIER, « L'organisation sociale et les formes de groupement. Problèmes de classification », in J. POIRIER (dir.), *Ethnologie régionale*, Gallimard, 1972, p. 958. Dans le Cameroun oriental, chez les Bamilékés, la fragmentation du pouvoir va consister en une superposition institutionnelle mettant en concurrence plusieurs autorités. V. en ce sens P. BOURMAUD, *La politique en Afrique*, *op.cit.*, p. 21.

répandu n'est pas exclusif. Les modalités d'exercice du pouvoir varient selon les sociétés. Ainsi, en fonction du degré d'unité ou non réalisé dans les sociétés précoloniales, on peut les classer en deux catégories à savoir d'un côté, les sociétés à pouvoir diversifié de tendance démocratiques (**1°**) et de l'autre, les sociétés à pouvoir unifié de tendance despotique (**2°**).

1. Le modèle prédominant : les sociétés à pouvoir diversifié de tendance démocratique

106. - Le processus de formation de l'État africain précolonial permet d'observer les mécanismes d'agrégation des unités politiques primaires. L'accès aux ressources économiques dans l'espace ouest-africain considéré comme plaque tournante du commerce de longue distance a permis d'asseoir de grands royaumes tout au long de l'histoire[300]. Les travaux réalisés par les historiens ont permis de comprendre comment par la conquête, le pays des Sarakolés a soumis sur la côte occidentale les Mandingues et les Wolofs, les intégrant ainsi à son ressort territorial. Cependant, contrairement au royaume d'Abomey sur lequel nous reviendrons plus loin, le monarque ne réussit pas à créer un État unitaire centralisé. Dans les sociétés politiques plurales, l'exercice du pouvoir est un reflet des communautés constitutives de la société[301].

107. - L'empire du Ghana s'est ainsi organisé sous la forme d'un État composé de type confédéral. Les États conquis gardaient leur organisation propre tout en reconnaissant l'autorité de *Koumbi Saleh* — capitale de l'empire du Ghana — sur leurs ressorts territoriaux. Si dans le fonctionnement interne du pays des Sarakolés, le pouvoir politique était centralisé, il n'est pas de même dans ces relations avec les minorités soumises qui conservaient une personnalité politique propre. Il existait donc dans cet ensemble que formait l'empire du Ghana, une pluralité de centres de décisions politiques. Cet empire se présentait comme une forme de confédération très « lâche », pour reprendre les propos de Pierre François Gonidec, dont la politique de gouvernance pourrait se résumer en un seul mot : « la décentralisation »[302]. Chacun des États conquis

---

300 J. F. ADE AJAYI (dir.), *Histoire générale de l'Afrique, L'Afrique au XIXe siècle jusque vers les années 1880*, vol. VI, Ed., Unesco, 1996.

301 M. NDIAYE, « La démocratie en Afrique subsaharienne précoloniale », *op.cit.*, p. 948.

302 P.-F. GONIDEC, *L'État africain : évolution, fédéralisme, centralisation et décentralisation, op.cit.*, p. 41.

devenant désormais province de l'empire disposait d'une large autonomie. Le chef représentant sa population prenait en charge l'administration générale et disposait d'une armée propre placée sous son commandement. L'autorité centrale symbolisait l'unité de l'empire en arbitrant les conflits entre ses vassaux lui ayant prêté serment d'allégeance. Toutefois, le pouvoir réside toujours dans l'assemblée représentative des communautés[303].

108. - La confédération mise en place sous l'empire du Ghana connut son déclin avec l'invasion des Almoravides vers la fin du XIe siècle. Sur les cendres du pouvoir de *Koumbi Saleh*, naquit au milieu du XIIIe siècle, l'empire du Mali qui pour l'essentiel à reproduit le système de fonctionnement de l'empire du Ghana. Magloire Somé affirme que si le nouvel État formé s'est doté d'une constitution, dite charte de « *Kurukan Fuga* », fondement de la cohésion entre les différents peuples confédérés sous l'autorité de la capitale *Niani* ; elle distingue deux types de provinces. Chaque province garde sa personnalité propre. Cependant, elles ne bénéficient pas du même traitement selon qu'elles se sont ralliées d'elles-mêmes ou conquises par le nouvel empire. Dans les premières, les rois conservent leurs titres. En revanche, dans les secondes, à côté du pouvoir traditionnel, le *Mansa* Soundjata Keita procédait à la nomination d'un *farin* — gouverneur — qui représentait son pouvoir[304]. Il s'agit là d'un compromis permettant au Mansa de garantir l'unité de l'empire en introduisant une souplesse dans l'administration des provinces. En respectant les autorités traditionnelles préétablies, le nouvel empire permettait aux différentes entités qui la composent de conserver leur autonomie, multipliant ainsi les centres de décisions. L'organisation d'une administration locale permettait de contrebalancer le pouvoir suprême du roi.

109. - La décentralisation du pouvoir, du centre vers une périphérie à laquelle le pouvoir central reconnait une personnalité politique et culturelle, est la caractéristique principale des deux plus grands ensembles étatiques qui se sont succédé en Afrique occidentale pendant la période précoloniale. Chaque État/chefferie associé à l'empire ou conquis dispose d'une totale autonomie dans son organisation interne. Les institutions du

---

303 M. NDIAYE, « La démocratie en Afrique subsaharienne précoloniale », *op.cit.*, p. 949.
304 M. SOME, « État, pouvoir et démocratie en Afrique de l'Ouest contemporaine : les héritages du passé », *op.cit.*, p. 57.

pouvoir central se juxtaposent à celles des pouvoirs locaux. Les empires, en réalité, n'ont été rien d'autre que des espaces d'hégémonie de conquérants plus forts qui ont réussi à s'imposer en obtenant des allégeances à leur autorité, que des États centralisés où le contrôle du pouvoir était vertical[305].

2. L'exception : les sociétés à pouvoir unifié de tendance despotique

110. - La volonté d'unification de l'exercice du pouvoir dans certaines sociétés africaines précoloniales renseigne sur la nature du régime politique existant. Le régime aura tendance à être despotique lorsque la volonté d'unifier et de concentrer l'exercice du pouvoir sera d'autant plus grande. Les sociétés de cette catégorie recoupent celles dites à fonctions administratives différenciées décrites par Roger-Gabriel Nlep. Il s'agit d'entités dans lesquelles « *les fonctions de maintien de l'ordre, d'expression symbolique de la société et la règlementation de la production (centralisation, redistribution d'un certain produit) peuvent donner naissance à des rôles spécifiques qui peuvent être l'apanage de groupes familiaux, ou d'individus prééminents au sein de ces mêmes groupes* »[306]. La division de la société consacre une hiérarchisation des classes sur la base des fonctions assumées ou du statut économique. À travers cette organisation de la société en classe, « *les mécanismes permettant l'exercice de la fonction administrative se trouvent enrayés, et l'idée même de démocratie s'estompe, au risque de glisser vers le despotisme* »[307].

111. - En Afrique subsaharienne, le royaume d'Abomey offre une excellente illustration de ce type de société. Le roi d'Abomey avait réussi à se débarrasser de toute forme de limitation du pouvoir pour s'assurer un contrôle absolu du pouvoir étatique. L'administration du royaume était fortement centralisée. Maurice Glélé évoque à ce propos « *un système de décentralisation administrative, d'encadrement au profit du pouvoir central fortement structuré, hiérarchisé* »[308]. Malgré la division du royaume en province, ces

---

[305] *Ibid.*, p. 58.

[306] J. COPANS, *Anthropologie, science des sociétés primitives*, Ed. Denoël, 1972, p. 133. Cité par R-G. NLEP, *L'administration publique camerounaise. Contribution à l'étude des systèmes africains d'administration publique, op.cit.*, p. 25.

[307] *Ibid.*, p. 26.

[308] M. A. GLELE, *Naissance d'un État noir*, LGDJ, 1969, p. 345. En réalité, il ne s'agit pas d'une décentralisation administrative mais d'un procédé de déconcentration.

dernières ne jouissaient d'aucune autonomie. Les chefs placés à la tête des provinces dépendaient strictement de l'autorité du roi et du *Migan*, ministre de l'Intérieur chargé de l'administration territoriale[309]. La décentralisation était quasi inexistante dans le royaume d'Abomey qui constitue sans aucun doute la forme la plus achevée d'État unitaire centralisé dans les sociétés ouest-africaines précoloniales.

112. - Ce royaume fait en ce sens figure d'exception puisque rares sont les monarques qui ont réussi à imposer une centralisation aussi aboutie que celle du royaume d'Abomey. Cependant, les chefs placés à la tête des subdivisions administratives n'étant pas toujours nommés, ces derniers se considèrent davantage comme des représentants de la population que de simples substituts du pouvoir central. On constate également que si l'on s'éloigne du centre vers la périphérie, l'intensité de la centralisation diminue considérablement de sorte qu'à *« la centralisation de droit se substitue une autonomie de fait »*[310]. La tendance des régimes despotiques à l'unité du pouvoir est fortement amenuisée par le principe d'autonomie et de diversification du pouvoir cher aux sociétés précoloniales. L'idée de participation des différentes strates de la société à l'exercice du pouvoir y est prépondérante[311]. Les sociétés précoloniales africaines sont ainsi diversement organisées, mais elles possèdent aussi des traits communs.

Section 2 : Les éléments constituants du modèle d'organisation territoriale des sociétés africaines précoloniales

113. - Les sociétés africaines précoloniales sont hétérogènes. Elles offrent à cet effet à l'observateur une variété d'organisations sociopolitiques. De la petite chefferie de lignage aux grands ensembles étatiques, on retrouve plusieurs éléments de différenciation entre ces sociétés. Néanmoins, quelles que soient leurs formes d'organisation, elles

---

309 P.-F. GONIDEC, *L'État africain : évolution, fédéralisme, centralisation et décentralisation, op.cit.*, p.42.
310 *Ibid.*, p. 43.
311 J. F. ADE AJAYI (dir.), *Histoire générale de l'Afrique, l'Afrique au XIXe siècle jusque vers les années 1880*, vol. VI, Ed., Unesco, 1996.

étaient toutes avant l'invasion occidentale[312] des sociétés sacrales[313]. La diversité des formes d'organisation du pouvoir politique dans cette période sur le continent africain cache à peine le caractère sacré que ces sociétés ont en commun (**§1**). La religion est intimement liée au politique. La vie sociale est alors fortement imprégnée de valeur religieuse qui crée un lien fondamental unissant ces sociétés.

114. - Ce lien formé autour de la combinaison d'éléments mystiques et religieux ne peut être étudié qu'à travers une analyse des rapports que les hommes entretiennent avec le pouvoir. Cela conduit à s'intéresser à la cellule de base au sein de laquelle se manifeste ce type relation à savoir le village[314]. Celle-ci est l'unité rurale de base de production dans les sociétés africaines précoloniales. Chaque village a développé, qu'il fasse partie d'un ensemble étatique ou non, une conscience de lui-même et se reconnait comme une entité socialement indépendante et distincte des autres. Ce caractère « autonomique » des sociétés africaines précoloniales (**§2**) a permis de donner naissance à un sentiment ethnique, que l'on pourrait qualifier de sentiment national avec la précaution que l'existence d'une nation ne doit pas être confondue à celle de l'État.

§1.- L'essence sacrée du pouvoir dans les sociétés d'Afrique noire précoloniale

115. - Dans les sociétés d'Afrique noire précoloniale, que ce soit dans le domaine social, dans le domaine politique ou même en matière économique, aucune institution ne peut véritablement exister si elle ne repose sur un concept religieux ou si elle n'a pas la religion comme pierre angulaire[315]. Le couple religion-politique est consubstantiel à toutes les sociétés humaines. Dans son *Traité de science politique*, Georges Burdeau

---

312 J. F. ADE AJAYI, « L'Afrique à la veille de la conquête européenne », in *Histoire générale de l'Afrique*, vol. 6, éd., Unesco, 1996, p. 825.
313 M. KAMTO, *Pouvoir et droit en Afrique noire. Essai sur les fondements du constitutionnalisme dans les États d'Afrique noire francophone*, thèse, LGDJ, 1987.
314 Sur l'exercice communautaire du pouvoir dans les sociétés africaines précoloniale, v., M. NDIAYE, « La démocratie en Afrique subsaharienne précoloniale », *op.cit.*, pp. 946-952.
315 Cf. P.-F. GONIDEC, *L'État africain : évolution, fédéralisme, centralisation et décentralisation*, *op.cit.*, p. 36.

affirme à cet effet que « *la sacralisation du pouvoir est un phénomène trop constant* [de l'histoire des sociétés humaines] *pour être considéré comme un accident* »[316]. Aujourd'hui, malgré l'affirmation du principe de laïcité dans les constitutions des États postcoloniaux, on note encore la prégnance du fait religieux[317] dans les stratégies de conquête et de conservation du pouvoir en Afrique. C'est en ce sens que Luc de Heusch affirme que « *La science politique* [africaine] *relève de l'histoire comparée des religions* »[318].

116. - Dans l'Afrique précoloniale, le pouvoir est sacré non seulement du fait de ses origines, mais aussi parce que les modalités de contrôle de son exercice sont organisées autour du sacré. Toute activité politique et sociale trouve son essence et sa finitude dans le sacré. L'« union congénitale » entre le sacré et le pouvoir n'est pas nouvelle. Il s'agira donc de restituer les travaux sur l'origine sacrée du pouvoir dans les sociétés africaines précoloniales qui a la particularité de ne pas avoir des origines divines, mais mythiques et ancestrales basées sur la croyance dans l'existence d'un au-delà, car les morts ne sont pas morts[319]. Le sacré apparait comme fondement et source de légitimité du pouvoir dans les sociétés africaines précoloniales (**A**) en même temps qu'il limite le pouvoir et donne les arguments de sa contestation (**B**).

A- La légitimation du pouvoir par le sacré dans les sociétés africaines précoloniales

117. - Le caractère religieux qui s'attache au pouvoir politique dans les sociétés d'Afrique noire précoloniale fonde la légitimité de ces titulaires. Le religieux étant intimement lié au politique, « *essayer de comprendre l'Afrique et l'Africain sans l'apport des religions traditionnelles serait ouvrir une gigantesque armoire vidée de son contenu le plus* précieux », écrivait Amadou Hampâté Ba[320]. Ainsi, toute pensée politique en Afrique noire précoloniale ne peut échapper aux considérations religieuses. Cela dit, en Afrique noire précoloniale, relève

---

316 G. BURDEAU, *Traité de science politique*, t. I, LGDJ, 3e éd., revue et augmentée, 1980.
317 L'évocation de Dieu, également dans les préambules des constitutions africaines (Togo, Niger…) En France notamment, la séparation de l'État et de l'Église est récente (placée dans le temps long de l'histoire) mais continue de rythmer les débats politiques.
318 Pour une dialectique de la sacralité du pouvoir, L. DE HEUSCH, *Le Pouvoir et le sacré*, Bruxelles, Annales du Centre d'étude des religions, 1962.
319 M. KAMTO, *Pouvoir et droit en Afrique. Essai sur les fondements du constitutionnalisme dans les États d'Afrique noire francophone*, thèse, LGDJ, 1987.
320 Textes sacrés d'Afrique noire, Gallimard, 2005.

Maurice Kamto, le roi n'est pas un Dieu, pas plus que le pouvoir n'a une origine divine. La croyance en la survie des morts dans l'au-delà a donné naissance à la notion fondamentale d'ancêtre sur laquelle repose la sacralité du pouvoir dans les sociétés africaines précoloniales (**1°**). Le culte des ancêtres permet d'assurer la sacralisation du pouvoir. Le chef est le seul capable de créer ce lien entre le monde des vivants et celui des ancêtres, sacralisé et idéalisé. Par ce procédé d'intermédiation, le chef se trouve dans une position de supériorité par rapport à l'ensemble du groupe. Cette participation à l'ordre du sacré et la singularisation de la position du chef va se manifester à travers des rituels sacrés qui vont donner sens à son pouvoir (**2°**).

1. Le pouvoir comme expression mythique imprégné de sacralité

118. - Le sacré dans les sociétés africaines précoloniales relève d'une conception particulière de l'univers. Dans cet univers mouvant et dynamique se mêlent différents éléments issus du monde visible et invisible qu'il faut appréhender afin de comprendre les réalités sociopolitiques de ces sociétés. Le pouvoir s'inscrit dans une tentative d'explication du monde dans lequel tous les éléments qui le composent sont mus par la force vitale des dieux et des morts[321]. Le titulaire du pouvoir politique dans ces sociétés est une personnification de l'ensemble des phénomènes transcendants qui trouvent une explication dans le recours au mythe. Discours logique et explicatif, nous dit Maurice Kamto, le mythe permet une explication des origines du groupe et de la création du monde. Le mythe explique l'origine du pouvoir et du droit. Cet ordre mythique est par essence sacré. La légalité et la légitimité du pouvoir dépendent ainsi de la conformité à cet ordre. La reconnaissance du pouvoir par le corps social suppose une ferme fidélité aux valeurs sacrées qui fondent le groupe.

119. - Le mythe apparait comme l'élément fondamental de la littérature sacrée, ésotérique et profonde, jouant dans ces civilisations marquées par l'oralité un rôle identique au dogme des religions liées à l'écriture[322]. Le sacré est par essence lié au mythe. Il constitue sa caractéristique principale.

---

[321] M. KAMTO, *Pouvoir et droit en Afrique. Essai sur les fondements du constitutionnalisme dans les États d'Afrique noire francophone, op.cit.*, p.71 ; « Les morts ne sont pas morts » disait le poète. B. DIOP, *Leurres et lueurs*, Présence Africaine, 2000.

[322] L. V. THOMAS, « Réflexions à propos des mythes d'Afrique noire », in *L'autre et l'ailleurs. Hommages à Roger Bastide*, coll. « IDERIC », 1976, p. 313.

Par son intermédiaire, la société est saisie en tant qu'unité, en tant qu'ordre et permanence. Elle est éprouvée sous l'aspect d'une valeur suprême et contraignante qui constitue ainsi que le souligne l'anthropologue Georges Balandier, la matérialisation d'une transcendance s'imposant aux individus et aux groupes. Le rapport du pouvoir à la société n'est pas essentiellement différent du rapport établi entre le « totem » et le clan. Et cette relation est évidemment chargée de sacralité[323]. Le pouvoir politique issu du mythe est, par conséquent, un pouvoir sacré qui marque par son attachement à travers les rituels, le renouvellement sans cesse des valeurs du passé. Ainsi, chez les Mossis de la Haute-Volta, en matière politique, le concept de « *naan* » se rapporte au pouvoir du temps des origines, celui que les fondateurs ont mis en œuvre pour construire l'État et à la force reçue de Dieu qui permet de dominer autrui. Le souverain — *Mogho-Naba* — procède ainsi d'une double origine à la fois divine et historique. Cette double origine fait du souverain une puissance sacrée et lui accorde la capacité d'agir sur le groupe. Le pouvoir politique confondu au sacré en Afrique noire précoloniale est le symbole de la perpétuation continue du passé dans le présent. Le chef ou le monarque qui incarne le sacré est au carrefour de tous les temps, du temps psychologique qui lui est individuel, du temps mythique en tant qu'incarnation du héros fondateur du groupe et du temps social puisqu'il amine par sa présence le groupe[324].

120. - Le mythe dans les sociétés africaines précoloniales ne peut donc être dissocié du sacré. Il épouse en ce sens les caractères de ce dernier tels que la transcendance, la valence affective qui sous-tend le jeu des images-analogies et l'exemplarité ; le mythe est le modèle de tout ce qui est et doit être, c'est le principe structurant de l'être et du phénomène[325]. C'est un récit du monde qui porte en lui le religieux et véhicule les valeurs sacrées qui donne au pouvoir politique toute sa légitimité[326]. Un pouvoir qui ne procède pas du sacré ne peut donc exister dans les sociétés d'Afrique noire précoloniale. Or, comme nous l'avons évoqué *supra,* les grands royaumes qui ont pu exister dans le Soudan occidental ont été dans leur majorité le

---

323 G. BALANDIER, « Réflexions sur le fait politique : le cas des sociétés africaines », *Cahiers internationaux de sociologie*, vol. 37, 1964, pp. 23-50.

324 M. KAMTO, *Pouvoir et droit en Afrique. Essai sur les fondements du constitutionnalisme dans les États d'Afrique noire francophone, op.cit.*, p. 76.

325 L. V. THOMAS, « Réflexions à propos des mythes d'Afrique noire », *op.cit.*, pp. 320-321.

326 *Idem.*

fait de conquête militaire. De quelle légitimité pourraient alors se réclamer de tels pouvoirs ?

121. - Conscient que pour asseoir son autorité, il doit justifier du respect de l'ordre social issu du mythe et bénéficier de l'onction du sacré, le pouvoir conquérant consent à transmettre le pouvoir à la famille à laquelle il revient normalement suivant l'histoire mythique du groupe dans le respect des procédures normales de dévolution du pouvoir. C'est pour cette raison que dans l'empire Mandingue du Mali, le *Mansa* ne touchait pas au pouvoir traditionnel dans les provinces conquises[327]. Parfois, le recours au mariage avec l'une des filles du souverain accorde au conquérant militaire le bénéfice de la sacralité. Ainsi donc, le recours au sacré met fin à l'illégalité et donne au pouvoir toute sa légitimité[328]. Le pouvoir du souverain désormais reconnu sera empreint de sacralité à tous les moments de son exercice.

2. L'empreinte permanente du sacré dans la vie politique

122. - Le chef en Afrique noire précoloniale est sacré. La double origine historique et divine de son pouvoir en fait un personnage à part placé au-dessus du groupe social. La sacralité de son pouvoir lui accorde la légitimité nécessaire pour asseoir son autorité sur le groupe qu'il anime. Les rites d'intronisation révèlent, une fois de plus, le caractère sacré du pouvoir comme fondement de l'organisation sociopolitique des sociétés africaines précoloniales. Le souverain séparé du groupe après les rituels et le cérémonial d'investiture devient l'homologue, le médiateur des dieux. L'intronisation officialise le règne et constitue la marque de l'institutionnalisation du pouvoir. Cette intronisation qui marque la prise de pouvoir effective est précédée d'une période d'interrègne qui correspond à la période de renouvellement du pouvoir après le décès du précédent roi au cours de laquelle l'héritier du trône subit une initiation rituelle.

---

327 Voir *supra* n° 110 et s.

328 Dans les sociétés modernes, ce procédé se retrouve dans la légitimation par le recours à la constitution à la suite d'une révolution ayant suspendue l'application de la loi fondamentale.

123. - Chez les Bassar[329], le nouveau chef est présenté dans le secret absolu aux grandes divinités et aux ancêtres par un Conseil afin de recevoir leur souffle et leurs bénédictions. À la fin de cette période initiatique qui dure deux semaines, le nouveau chef est présenté à la population lors d'une cérémonie publique[330]. Le chef ayant reçu les bénédictions des dieux et des ancêtres, communie à l'occasion du cérémonial d'investiture avec son peuple sans le soutien duquel il ne peut gouverner. L'intronisation prend ainsi la forme symbolique d'une double investiture d'abord par les forces surnaturelles ancestrales puis par les hommes[331]. La permanence du caractère sacré du pouvoir du chef est ainsi affirmée par le biais de la mythologie l'inscrivant dans la continuité historique du pouvoir des pères fondateurs du groupe.

124. - Tout au long du règne du souverain, la sacralité du pouvoir s'exprime à travers son activité religieuse. Prêtre du groupe, le chef assure le culte aux ancêtres. Il est le point de liaison entre le groupe actuel, constitué par les vivants, et le groupe idéalisé, sacralisé, symbolisé par la totalité des ancêtres. Il transmet la parole des ancêtres aux vivants, celle des vivants aux ancêtres[332]. La sacralité du pouvoir s'affirme également pendant le règne du souverain dans ses rapports avec ses sujets. Le peuple doit vénération et totale soumission au pouvoir sacré du souverain[333]. Toute transgression étant perçue comme un sacrilège. Ainsi, au Mali, rapporte l'historien Joseph Ki-Zerbo, le roi était salué comme un « dieu », dans la position de l'homme qui se prosterne pour la prière, les sujets frappaient la terre avec leurs coudes et jetaient la poussière sur leurs dos[334]. Dans la Haute-Volta, le *Mogho Naba* n'étant plus après son intronisation un homme comme les autres, il perd définitivement son ancien nom pour

---

329 Les Bassars sont une ethnie du nord du Togo.

330 D. F. GBIKPI-BENISSAN, *Pouvoirs politiques anciens et pouvoir moderne au Togo-La chefferie dans la nation contemporaine. Essai de sociologie politique sur la chefferie en pays Bassar, Akposso et Mina*, thèse de sociologie, université de Lomé, 1976, p. 151 ; S. DUGAST, « Déterminations économiques versus fondements symboliques : la chefferie de Bassar », Cahiers d'études africaines, vol. 23, n° 2, 1988, p. 265.

331 M. KAMTO, *Pouvoir et droit en Afrique, op.cit.*, p. 92.

332 G. BALANDIER, « Réflexions sur le fait politique : le cas des sociétés africaines », *Cahiers internationaux de sociologie*, vol. 37, juillet-décembre 1964, pp. 23-50.

333 G. BALANDIER, *Anthropologie politique*, PUF, coll. « Quadrige », 2013, p.119.

334 J. KI-ZERBO, Histoire de l'Afrique d'hier à demain, Hatier, 1972, p. 138. Cité par M. KAMTO, *Pouvoir et droit en Afrique, op.cit.*, p. 84.

prendre un nouveau nom royal. Le simple rappel de son ancien nom est considéré comme un crime de lèse-majesté[335].

125. - Le pouvoir dans les sociétés africaines précoloniales trouve ainsi son sens dans le mythe fondateur qui constitue d'une certaine manière la norme fondamentale, la « *Grundnorm* » kelsénienne qui donne au pouvoir politique sa légitimité et ses attributs de sacralité. Si nous avons écarté de l'analyse les sociétés africaines précoloniales pénétrées par l'islam[336], il faut néanmoins reconnaitre que, malgré la pénétration des religions monothéistes, les sociétés africaines restent fortement attachées au sacré qui en même temps qu'il fonde le pouvoir le limite. L'introduction de l'islam ne dénature pas ce caractère sacré et renforce par ailleurs, en raison de sa nature décentralisée par opposition à la hiérarchie chrétienne centralisée, l'autonomie du pouvoir.

B- La limitation du pouvoir par le sacré dans les sociétés africaines précoloniales

126. - Si pour Montesquieu, « *il faut que par la disposition des choses, le pouvoir arrête le pouvoir* »[337], en Afrique noire précoloniale, il faut que par la disposition des choses, le sacré arrête le sacré. La religion en Afrique noire précoloniale constitue la principale source du droit. Le pouvoir et le droit trouvent leur essence dans une charte mythique transcendante ayant valeur constitutionnelle. L'autorité issue de cette « constitution » est imprégnée de sacralité et est toujours respectée, conservée et entretenue comme un impératif juridique. Le pouvoir a toujours été investi de cette auréole de sacralité. Le sacré a fondé le pouvoir, mais que l'on ne se trompe point, rappelle Bernard Asso, si le sacré fonde le pouvoir, il détermine également les modalités de contrôle de ce pouvoir[338]. Le plus puissant des chefs dans les sociétés africaines précoloniales connait des limites à son pouvoir. Le caractère sacré du pouvoir qui établit un rapport de vénération et de soumission entre le chef et ses sujets n'est ni absolu ni inconditionnel. Malgré la diversité des formes d'organisation sociopolitiques qu'on rencontre dans ces sociétés, le rejet de l'absolutisme est une constante qu'il

---

335 M. KAMTO, *Pouvoir et droit en Afrique, op.cit.*, p. 85.
336 *Idem.*
337 *De l'esprit des lois.* Livre XI, Chap. IV, p. 117.
338 B. ASSO, « De la sacralisation du pouvoir : essai sur l'Afrique noire animiste », *Annales méditerranéennes d'Histoire et d'Ethnologie*, 1980, p. 90, cité par M. KAMTO, *Pouvoir et droit en Afrique, op.cit.*, p. 110.

s'agisse des sociétés à pouvoir unifié ou des sociétés à pouvoir diversifié. Le non-respect des obligations issues de la charte fondamentale par le chef entraine une contestation de son pouvoir par le corps social.

127. - La sacralité du pouvoir s'accorde avec l'idée de participation de l'ensemble des individus composant la société à l'exercice de la fonction politique. Les systèmes politiques en Afrique noire précoloniale comme l'avons montré sont variés bien que dans la majorité des sociétés observées par les historiens les gouvernements étaient quasiment tous de type monarchique[339]. Ces monarchies africaines ne pouvaient être des monocraties autocratiques, car le pouvoir tel qu'il est conçu en Afrique noire précoloniale annule les possibilités d'émergence de tels régimes[340]. Cette observation s'applique également aux sociétés dans lesquelles le pouvoir est diffus. En effet, l'absence d'une autorité établie distincte de l'ensemble du corps social dans ces sociétés ne doit pas conduire l'observateur à la conclusion hâtive d'un pouvoir anonyme. L'absence de distinction entre gouvernés et gouvernants ne signifie pas que les sociétés africaines précoloniales sont dépourvues de chefs. À l'intérieur des familles, clans et tribus, le lien avec les ancêtres est assuré par la présence de chefs de famille, de clans et de tribus. Les réseaux de parenté remplissent ainsi les fonctions conservatrices qui sont dévolues ailleurs aux réseaux politiques[341]. Lorsqu'on s'intéresse de près au mode de gouvernance des sociétés à pouvoir diffus, on se rend compte que la décision était l'œuvre de l'ensemble du corps social qui, dans le même temps, exerçait un contrôle direct sur le pouvoir.

128. - Le contrôle direct du pouvoir par le corps social dans les sociétés africaines précoloniales prend la forme de la *palabre* considérée comme universelle en Afrique noire. L'institution de la *palabre* africaine pouvait prendre des formes variées, notamment celle d'un système de gouvernement par le verbe, la controverse et le dialogue[342]. En tant que telle, la *palabre* constitue un outil de contrôle du pouvoir, l'expression d'un

---

339 C. ANTA DIOP, *L'Afrique noire précoloniale*, Présence Africaine, 1960, p. 58. Avec une généralisation excessive, l'auteur affirme que : « À l'époque précoloniale, en effet, tout le continent était couvert de monarchies et d'empires. Aucun coin où vivent les hommes, fût-ce en forêt vierge, n'échappait à une autorité monarchique ».

340 M. KAMTO, *Pouvoir et droit en Afrique, op.cit.*, p. 112.

341 *Ibid.*, p. 113.

342 L. SYLLA, « Démocratie de l'arbre à palabre et bois sacré ». Contribution au colloque du LAJ, 1980, p. 3 repris par M. KAMTO, *Pouvoir et droit en Afrique, op.cit.*, p. 114.

gouvernement direct où la parole est érigée en technique de gouvernement. Elle permettait au chef de prendre à la fois connaissance des problèmes spécifiques, de mesurer les tendances de l'opinion du groupe social et d'être à l'écoute de la population. C'est ainsi que se manifeste l'exercice concret du pouvoir dont le ressort fondamental est l'échange entre la base et le sommet[343]. Le chef africain imprégné de toutes les valeurs sacrées ne pouvait agir sans prendre en compte l'opinion publique. La pratique de la *palabre* ainsi qu'elle conduit à la recherche d'un consensus dans le processus décisionnel a conduit à l'affirmation par certains politologues que les sociétés africaines précoloniales étaient des « sociétés unanimistes »[344]. La discussion « palabrique » est l'essence de la démocratie dans les sociétés africaines précoloniales[345], mais il serait faux de conclure que ces sociétés étaient monolithiques et unanimistes. La tradition africaine de la décision unanime ne peut aucunement servir de prétexte pour la condamnation du pluralisme[346].

129. - Dans les structures de type étatiques, le roi était en plus de l'utilisation de la palabre entourée de conseillers sans lesquels il ne pouvait gouverner. Les conseillers sont pour la plupart issus de la noblesse et appartiennent à la sphère du pouvoir en raison de leur position sociale. Dans l'empire Mossi de la Haute-Volta, le *Mogho Naba* était entouré d'un Conseil du roi et d'un Conseil des ministres. Le premier se réunissait tous les jours pour informer le roi de l'état du groupe. Le second composé de l'ensemble des dignitaires de la cour royale se réunissait en séance publique. Les décisions prises par ce dernier après une discussion générale constituent la manifestation de la volonté générale[347]. Il en est de même pour l'empire du Ghana et du Mali qui, comme nous l'avons rappelé, était très décentralisé dans leur organisation. Du fait de cette décentralisation,

---

343 *Ibid.*, p. 115.

344 J. BUSCHMANN, *L'Afrique noire indépendante*, LGDJ, 1962. Cette thèse est largement critiquée par Maurice Kamto. Voir, M. KAMTO, *Pouvoir et Droit en Afrique, op.cit.*, pp. 131-147.

345 M. NDIAYE, « La démocratie en Afrique subsaharienne *préc*oloniale », in *Mélanges en l'honneur de Dominique Rousseau, Constitution, justice, démocratie,* LGDJ, 2020, p. 945.

346 P.-F. GONIDEC, *Les systèmes politiques africains,* LGDJ, 1971, p. 33. On note à ce propos une évolution dans la pensée de cet auteur, car en 1966 dans un article, « Traditionalisme et modernisme dans les institutions publiques de l'Afrique noire » reprenait la thèse de l'unanimisme des sociétés traditionnelles africaines développée par le professeur Jean Buschmann.

347 G.-V. KABORE, *Organisation politique traditionnelle et évolution politique des Mossi de Ouagadougou,* 1966, cité par P.-F. GONIDEC, *Les systèmes politiques africains, op. cit.*, p. 39.

les conseils de clans disposaient d'un véritable pouvoir de contrôle sur l'autorité du roi.

130. - Le chef dont le pouvoir est établi sur la base de la charte mythique du groupe n'exerce pas seul ce pouvoir. Soumis aux pressions du corps social et de ses conseillers, il est obligé de tenir compte des différents avis pour éviter toute contestation de son pouvoir. Parfois, en raison de la structure politique de la société, le pouvoir du souverain est fortement limité par l'autonomie des entités politiques qui forment l'État. Par la disposition des choses, le sacré limite le sacré.

§2.-Le village, communauté rurale de base dans les sociétés précoloniales africaines

131. - Le village est dans l'Afrique noire précoloniale la matrice de l'organisation de l'espace sociopolitique. Cette affirmation peut surprendre quand on sait que le besoin de vivre ensemble des différentes populations africaines a donné naissance à une variété de structures politiques. Cependant, quelles que soient les modalités de regroupement de ces sociétés, on ne peut nier le fait que c'est le village qui constitue la structure sociopolitique la plus répandue en Afrique précoloniale[348] avant l'invasion européenne[349]. Les sociétés étatiques précoloniales ne sont d'ailleurs, dans leur majorité, rien d'autre qu'une juxtaposition de communautés villageoises. C'est à l'intérieur de cette cellule de base que s'exprime toute la vitalité des sociétés africaines. Dans l'expérience du village africain précolonial, on observe le développement d'un sentiment d'appartenance à une communauté propre, distincte des autres qui par la participation de l'ensemble de ces membres, assure de manière autonome sa survie (**A**). Noyau principal et autonome de production agricole et de subsistance, le village n'est pas pour autant fermé à des échanges extérieurs que ce soit en matière économique ou politique (**B**).

---

[348] Cette affirmation mérite d'être nuancée. Le village est particulièrement présent en Afrique occidentale mais la partie centrale et australe du continent ignore plus ou moins cette réalité.

[349] J. F. ADE AJAYI, « L'Afrique à la veille de la conquête européenne », in *Histoire générale de l'Afrique*, vol. 6, éd., Unesco, 1996, p. 825.

A- Le village, une communauté à caractère autonomique

132. - Le village est la structure par excellence au sein de laquelle et autour de laquelle est organisée la totalité de la vie sociale et politique. Toutefois, la notion de village, regroupement des membres d'une famille étendue ou d'un clan élargi à des familles alliées, se laisse difficilement saisir dans une seule définition. En tant qu'unité rurale de production, le village se présente sous la forme d'un *« groupe plus ou moins restreint de familles qui tirent la quasi-totalité de leur subsistance du travail de la terre, dans le cadre d'une division du travail peu différencié, où même les individus ou groupes spécialisés ou privilégiés — artisans, chefs, prêtres... — ont besoin, pour vivre, de participer ou de faire participer leurs proches ou leurs dépendants à la production agricole »*[350]. Pour Henri Labouret, le village se caractérise par son aspect en ce qu'il *« possède identité et personnalité, fonde les relations entre ses occupants sur la parenté, plus souvent sur le voisinage et la solidarité... »*[351].

133. - Ainsi, nous pouvons dire du village qu'il constitue la plus petite unité organique autonome dans les sociétés négro-africaines précoloniales. Il ne s'agit pas d'une simple agrégation d'individus, mais d'une organisation communautaire[352] qui possède une vie intérieure institutionnelle et économique qui lui est propre. Les communautés villageoises d'Afrique noire précoloniale sont autonomiques c'est-à-dire que chaque communauté à une conscience propre d'elle-même en tant qu'entité socialement distincte des autres et indépendante. C'est d ailleurs au nom de cette organisation de la société dans le cadre sociopolitique particulier qu'est le village, qu'en Afrique de l'Ouest les populations manifestent une certaine hostilité envers toute forme de centralisation excessive[353].

---

[350] C. COQUERY-VIDROVITCH, « Les structures du pouvoir et la communauté rurale précoloniale », *Revue française d'histoire d'outre-mer*, n° 250-253, 1981, p. 54.

[351] H. LABOURET, *Paysans d'Afrique occidentale*, Gallimard, 1941, p. 106 et s.

[352] Sur l'organisation « communaucratique » du pouvoir, v., P. DIAGNE, « Le pouvoir en Afrique », in *Le concept de pouvoir en Afrique*, Les presses de l'Unesco, 1981, p. 28 ; M. NDIAYE, « La démocratie en Afrique subsaharienne précoloniale *op.cit.*, p. 946.

[353] C. SAVONNET-GUYOT, « La communauté villageoise comme système politique : un modèle ouest-africain », *RFSP*, n° 6, 1975, p. 1114. Elles sont « *réputées pour leurs techniques agricoles, leur amour de la terre, leur stabilité et, de manière presque unanime, leur caractère réfractaire à toute organisation centralisée* ».

134. - Chez les *Bwas* et les *Samos* de la Haute-Volta, le village apparait comme une unité politique tout à fait autonome au-dessus de laquelle plus rien ne mériterait en réalité leur loyalisme[354]. Bien entendu, cette organisation autour du village ne supprime pas totalement les liens de famille qui donnent à la société son caractère lignager qui s'avère important dans la répartition des tâches dans le cadre du village. Le premier lignage installé conserve des privilèges politiques importants. Il possède les terres les plus vastes du village sur lesquelles il procède à l'accueil des étrangers ou en confie la gestion aux familles qui n'en possèdent pas. On peut dans ce cadre s'interroger sur les mécanismes qui conduisent à la création d'un sentiment d'appartenance à la communauté villageoise dans la mesure où les liens familiaux demeurent prégnants. Le caractère authentique du village comme modèle autonomique d'organisation de l'espace politique tient en ce qu'il n'est pas une *« simple juxtaposition de lignages exploitant en commun un terroir, généralement indivis, c'est l'expression spatiale d'une volonté de vivre ensemble »*[355].

135. - Cette volonté de vivre ensemble se manifeste à travers des gestes rituels et des sacrifices réalisés d'une part, par le chef de village sur l'autel fondateur du village conférant ainsi aux nouveaux arrivants la « citoyenneté villageoise ». D'autre part, pour montrer leur attachement à leur nouvel espace d'accueil, le chef du lignage immigré dépose dans la maison des ancêtres du village d'accueil une pierre en *banco* volé dans la maison des ancêtres du village d'origine. La communauté villageoise repose de ce fait sur un socle socioreligieux qui implique une attitude de loyauté. Ce loyalisme villageois suppose l'effacement progressif au niveau des relations extérieures des anciens liens d'appartenance lignagère. De même, il suppose la réalisation en chaque individu « citoyen de son village » d'un équilibre entre son appartenance à une cellule lignagère et son intégration au village[356].

136. - Le village apparait dans les sociétés ouest-africaines précoloniales comme un véritable cadre de solidarité et d'échange. S'il constitue une unité de production autonome, au sens où il assure l'essentiel de la production destinée à la consommation, l'espace villageois n'est pas replié

---

354 *Idem.*

355 C. SAVONNET-GUYOT, « La communauté villageoise comme système politique : un modèle ouest-africain », *op.cit.*, p. 1115.

356 *Ibid.*, p. 1117.

sur lui-même. En effet, il entretient avec le monde extérieur une multiplicité de relations conduisant au rejet des théories faisant du village une entité autarcique. Le caractère autonomique du village n'exclut pas la possibilité d'alliances villageoises pouvant naitre d'un service rendu, d'un accord politique, d'un pacte de défense mutuelle ou d'un courant d'échanges matrimoniaux[357].

B- Une communauté villageoise ouverte au monde extérieur

137. - Le caractère autonomique des communautés villageoises en Afrique noire précoloniale a pu, à tort, conduire à les définir comme des sociétés closes, fermées sur elles-mêmes et vivant en parfaite autarcie sans aucun rapport avec l'extérieur. Le village, s'il constitue le principal noyau de production et de subsistance, ne peut être considéré comme une entité autarcique repliée sur elle-même qu'il soit ou non intégré dans un État.

138. - La « mythisation » de la communauté villageoise perçue comme un ensemble « intouché et intouchable », vivant en vase clos et en parfait équilibre avec elle-même, apparait dans la littérature ethnologique de la seconde moitié du XVIIIe siècle faisant ainsi du village dans les sociétés négro-africaines un conservatoire des vertus du passé à l'abri de la civilisation urbaine et industrielle[358]. Dans cette optique, « *les communautés villageoises disposeraient par elles-mêmes de tout le nécessaire, et se passeraient presque entièrement de rapports avec l'extérieur. Cette union indestructible de la communauté villageoise, qui constitue à elle seule un petit État, donne lieu de croire que, grâce à elle, les habitants vivent heureux et jouissent d'une liberté réelle. La ruine ou le partage des royaumes n'inquièteraient guère les villageois ; tant que le village est entier, peu leur importe sous quel pouvoir il passe… ; son économie interne ne s'en trouve pas changée* »[359]. Cette vision autarcique de l'unité rurale de production bien qu'erronée s'est épanouie depuis lors au point d'intégrer même dans les années 1980 les considérations économiques des bailleurs de fonds. Ces derniers, ignorant le fait que le continent africain n'a jamais été

---

357 *Ibid.*, p. 1118.

358 C. COQUERY-VIDROVITCH, « Les structures du pouvoir et la communauté rurale précoloniale », *op.cit.*, p. 55.

359 House of Commons: 5th Report of the Select Committee on the Affairs of the East India Company (1812). Et. Ch. Metcalfe, Minute dated 7th november 1830, Report from the Select Committee. Traduit par J. Pouchepadass et cité par C. Coquery-Vidrovitch, « Les structures du pouvoir et la communauté rurale *préc*oloniale », *op.cit.*, p. 56.

protectionniste ni vécu en autarcie, ont incité les pays nouvellement indépendants à démanteler leurs frontières douanières[360].

139. - L'image idéalisée du village comme lieu communautaire ignore ainsi que le souligne Catherine Coquery-Vidrovitch, la subordination aussi bien sur le plan politique qu'économique de la collectivité vis-à-vis de l'extérieur. Il existe une réelle extraversion de la communauté villageoise qui par le jeu d'un réseau de connexions entretient avec l'extérieur des échanges économiques, politiques et sociaux prenant diverses formes[361]. Le continent africain dans son ensemble a toujours été ouvert aux échanges commerciaux. L'Afrique occidentale précoloniale fait à cet effet figure d'une véritable plaque tournante du commerce de longue distance[362]. L'or étant échangé depuis la période médiévale contre le sel du désert dans l'ancien Soudan occidental. Par ailleurs, l'existence dans cette partie de l'Afrique de sociétés dotées de véritables appareils étatiques conduits à replacer le village dans son contexte pour saisir la nature des relations que ce dernier entretient avec l'extérieur. La reconnaissance d'un pouvoir supérieur se traduisant *in fine* par des rapports de domination et de dépendance. Dans les sociétés à pouvoir diversifié de type fédéral notamment, ces relations de dépendance et d'autonomie se manifestent avec une certaine acuité. En effet, la réalité du pouvoir politique dans ces sociétés réside à l'intérieur des communautés villageoises qui résistent à toute tentative de concentration excessive du pouvoir. Ainsi que le souligne Maurice Kamto, le plus petit segment lignager ou villageois était tout aussi que le plus grand, jaloux de ses traditions. Le pouvoir fédéral ne prend ici tout son sens que lors d'une menace extérieure. L'ensemble des communautés se réunissant alors à cette occasion pour faire face au danger[363]. Le caractère autonomique de la communauté rurale précoloniale en Afrique n'exclut pas non plus la possibilité d'alliances villageoises

---

360 B. HIBOU, *L'Afrique est-elle protectionniste ? Les chemins buissonniers de la libéralisation extérieure,* Karthala, 1996, 354 p. J.-F. BAYART, B. HIBOU, B. SAMUEL, « L'Afrique "cent ans après les indépendances" : vers quel gouvernement politique ? », *Politique africaine,* n° 119, 2010, pp. 129-157.

361 C. COQUERY-VIDROVITCH, « Les structures du pouvoir et la communauté rurale précoloniale », *op.cit.,* p. 56.

362 Voir en ce sens, J. F. ADE AJAYI (dir.), Histoire générale de l'Afrique, L'Afrique au XIXe siècle jusque vers les années 1880, vol. VI, Ed., Unesco, 1996.

363 M. KAMTO, *Pouvoir et droit en Afrique, op.cit.,* p. 121. Ce constat est fait également par J. Capron dans le cadre de ces travaux sur les Bwas du Mali et de la Haute-Volta : Associations d'âge, économie et pouvoir chez les populations Bwa Pweysa, *in* D. PAULME (dir.), *Classes et Associations d'âge en Afrique de l'Ouest,* Plon, 1971, p. 33.

pouvant naitre d'un service rendu ou d'un courant d'échanges matrimoniaux[364].

Cette dernière hypothèse est assez courante dans toutes les sociétés africaines précoloniales, car, qu'il ait ou non État, les échanges matrimoniaux permettent de renforcer la communauté villageoise[365]. Cellule de base, la communauté rurale est au cœur des structures politiques dans les sociétés africaines précoloniales. Elle configure l'espace politique et assure la survie et la subsistance du groupe familial.

---

364 *Idem.*

365 Le mariage dans les sociétés africaines précoloniales est un acte collectif qui engage l'ensemble du groupe social, C. COQUERY-VIDROVITCH, « Les structures du pouvoir et la communauté rurale précoloniale », *op.cit.*, p. 58.

Conclusion du chapitre 1

140. - L'Afrique noire précoloniale a connu une variété de systèmes politiques faisant osciller leur architecture entre deux extrêmes. D'un côté, selon la typologie de Fortes et Evans-Pritchard[366], il existe des sociétés à faible institutionnalisation du pouvoir dites « anarchiques » et de l'autre, des sociétés ayant connu une forte institutionnalisation du pouvoir impliquant l'existence d'une structure étatique. Une approche organique de ces sociétés précoloniales a permis d'esquisser une nouvelle typologie qui traduit la réalité de l'exercice du pouvoir. En effet, dans la période précoloniale malgré la permanence de l'autorité du chef comme rouage principal de la structure de ces sociétés bâties sur les valeurs sacrées, ce dernier n'est pas l'organe unique de l'exercice du pouvoir. L'organisation du pouvoir autour du sacré induit une multiplicité des centres de décisions qui, dans le même temps, servent de limitation au pouvoir du chef. Le chef en Afrique précoloniale n'est donc pas un autocrate. On est en présence dans les sociétés précoloniales africaines d'une administration véritablement décentralisée. Certes, il ne s'agit pas d'une décentralisation telle que le droit positif donne à la concevoir à ce jour. Cependant, l'hostilité à toute forme de centralisation excessive du pouvoir dans ces sociétés conforte l'idée d'une organisation décentralisée du pouvoir qui s'appuie sur le village comme facteur de cohésion du groupe social. Cette conception de l'organisation des sociétés africaines va être remise en cause pendant la période coloniale. À l'autonomie qui caractérisait ces sociétés succédera une organisation centralisée.

---

366 M. FORTES, E. EVANS-PRICHARD, *Systèmes politiques africains*, PUF, 1964.

Chapitre 2 — Les reconfigurations institutionnelles imposées par la colonisation

141. - L'expansion coloniale a porté une atteinte considérable aux structures traditionnelles africaines. La période coloniale marque le début de la rupture du lien originel entre autonomie et structure de peuplement avec la consécration du monopole de l'exercice du pouvoir politique par une autorité centrale. La colonisation implique dans son essence une situation de dépendance totale[367]. Elle est un phénomène global de domination multiforme dont l'ensemble des aspects sont solidaires et complémentaires[368]. L'étude d'un aspect particulier de la colonisation ne peut être envisagée sans le recours au tout. L'interprétation de l'impact de la colonisation reste encore aujourd'hui une question délicate tant elle divise et soulève controverses et passions[369]. Ainsi, pour l'étudier, il faut comme le rappelle Marcel Prélot à propos de l'histoire constitutionnelle du Second Empire français « *en parler non plus en vocables politique et militant, mais en termes de science* […] *avec compréhension et objectivité* »[370]. Bien qu'elle ait été de courte durée[371], la colonisation n'a pas été qu'une simple parenthèse dans la trajectoire des pays africains. Elle a bouleversé les mentalités, mis à mal la culture indigène et provoqué la destruction des systèmes politiques précoloniaux par l'introduction des idées et des valeurs occidentales. La colonisation a constitué une période de rupture. Il faut néanmoins s'abstenir de la penser en des termes manichéens. La rupture introduite par la période coloniale n'a été ni systématique ni totale[372]. En revanche, son influence sur l'organisation des structures sociales en Afrique noire est certaine et continue à produire des effets dans l'Afrique actuelle.

La colonisation s'est traduite sur le plan politique par une confiscation du pouvoir des sociétés précoloniales. Par la contrainte, les colonisateurs vont imposer aux pays colonisés des pratiques sociales spécifiques qui vont conduire à la mutation des sociétés anciennes. La colonisation va

---

367 La dépendance pouvant être à la fois économique, culturelle, religieuse, politique, juridique etc.

368 P.-F. GONIDEC, *L'État africain : évolution, fédéralisme, centralisation et décentralisation, op.cit.*, p. 59.

369 M. KAMTO, *Pouvoir et droit en Afrique, op.cit.*, p. 207.

370 M. PRELOT, « La signification constitutionnelle du second empire », *Revue française de science politique*, vol. 3, n° 1, 1953, p. 33.

371 La colonisation française de l'Afrique de l'Ouest a commencé en 1895 et s'est achevé approximativement à partir de 1958.

372 M. GAZIBO, *Introduction à la politique africaine, op.cit.*, §25 ; M. KAMTO, *Pouvoir et droit en Afrique, op.cit.*, p. 207.

conduire à la mise en place d'une superstructure administrative empreinte de bureaucratie affaiblissant ainsi le pouvoir africain traditionnel. L'instauration de l'« État administratif colonial » va ainsi transformer l'organisation territoriale des sociétés colonisées pour répondre à la volonté de domination des colonisateurs français. La période coloniale correspond ainsi à l'introduction dans tous les territoires occupés d'un modèle d'administration identique à celui de la métropole (**Section 1**). La pénétration européenne et la structuration du pouvoir colonial ont conduit, par ailleurs, à la désacralisation du pouvoir africain. Le chef dépossédé de son auréole de sacralité perd son influence en devenant un simple rouage de l'administration coloniale (**Section 2**).

Section 1 : L'implémentation d'un modèle administratif de type métropolitain

142. - L'administration en Afrique noire est un fait colonial total. L'essence profonde de l'entreprise coloniale est le profit. La colonisation est intrinsèquement liée à l'évolution sociale et économique de l'Europe puisque le motif déterminant de l'expansion coloniale est d'ordre économique. Cette affirmation n'entraine pas la négation de l'existence d'autres motifs qui ont pu inspirer la colonisation. Toutefois, ces motifs demeurent secondaires dans la mesure où les artisans les plus purs de la colonisation « *ne furent eux-mêmes pas toujours des saints, mais plutôt des conquérants porteurs de foi, chercheurs de gain, donnant le salut, prenant le profit* »[373]. Pour atteindre cet objectif, le colonisateur français a utilisé dans les territoires colonisés des recettes issues de son expérience par la diffusion des principes de fonctionnement identiques.

Ainsi, la doctrine coloniale de la France fut celle de l'administration directe assimilant totalement les populations et les territoires conquis outre-mer au modèle métropolitain centralisé. Le pouvoir colonial entend ainsi asseoir sa domination sur les sociétés africaines. La mise en place d'une superstructure administrative calquée sur le modèle français sert les desseins de cette volonté de domination et permet de préserver les intérêts de la métropole. De ce fait, la colonisation correspond à la reconfiguration des structures institutionnelles africaines avec l'introduction de l'appareil étatique moderne, amorçant dans les territoires occupés « *le processus de*

---

[373] P.-F. GONIDEC, *L'État africain : évolution, fédéralisme, centralisation et décentralisation*, *op.cit.*, p. 59.

développement administratif»[374], c'est-à-dire la mise en place d'un véritable réseau administratif au sens où on l'entend aujourd'hui. La doctrine coloniale française est donc marquée par l'omniprésence de la métropole (**§1**) et la réorganisation de la conception territoriale des sociétés africaines autour d'une idée nouvelle d'État (**§2**).

§1.- L'omniprésence de la puissance coloniale dans l'organisation administrative

143. - La prépondérance métropolitaine est assurée par la centralisation. Les colonies sont placées dans la même situation que les collectivités métropolitaines. À travers la duplication de son modèle de centralisation, la France consacre le monopole de l'exercice du pouvoir politique dans les colonies. Le pouvoir métropolitain en tant que centre unique d'impulsion politique, légifère, gouverne et administre à la fois. La centralisation du pouvoir par la transposition dans les colonies des règles issues des réalités métropolitaines permettait aux autorités françaises de renforcer le lien colonial[375]. Cette orientation de la politique coloniale française vise à affaiblir les formes d'autorités traditionnelles afin d'y substituer le pouvoir colonial. Malgré son caractère empirique, l'administration coloniale repose sur une théorie de la colonisation qui fonde son action (**A**) et qui permet de la distinguer d'autres modes de gouvernance coloniale (**B**).

A- Les fondements théoriques de la politique administrative coloniale

144. - La colonisation est au service d'un dessein commun. Cependant, sa nature uniforme interroge. A-t-elle été la même partout ? Ou bien, est-il possible qu'elle produise des effets différents selon les choix opérés par la puissance colonisatrice ? À ce propos, il est convenu d'opposer de façon

---

374 G. CONAC, « Le développement administratif des États d'Afrique noire », in G. CONAC (dir.), *Les institutions administratives des États francophones d'Afrique noire*, Economica, coll. « La vie du Droit en Afrique », 1979, p. X. L'auteur souligne qu'a l'exception de l'Éthiopie et de Madagascar le processus de développement d'un appareil administratif moderne relève entièrement du phénomène colonial.

375 P. LEGENDRE, *Histoire de l'administration de 1750 à nos jours*, PUF, coll. « Thémis », 1968, p. 162.

péremptoire le système français d'administration coloniale, le « *direct rule* », à l'administration indirecte, l'« *indirect rule* » attribué à la Grande-Bretagne. Cette opposition se base essentiellement sur la place réservée aux autorités traditionnelles dans chacun des deux systèmes. En effet, la distinction entre la politique coloniale française et celle des Britanniques repose sur la volonté supposée de la première de détruire les systèmes politiques précoloniaux alors que la seconde s'évertuait à les intégrer dans sa gouvernance. En réalité, le monopole de cette politique d'intégration des pouvoirs traditionnels africains ne peut être attribué à la Grande-Bretagne, car elle fut dans une moindre mesure, certes, pratiquée aussi par la France. À l'inverse, la Grande-Bretagne n'a pas manqué de neutraliser les chefferies traditionnelles chaque fois que celles-ci constituaient un danger pouvant entraver la prospérité de l'Empire colonial britannique. Jacques Lombard constate à ce propos que l'opposition entre le système français d'administration coloniale et la conception britannique de l'administration indirecte a été trop exacerbée[376]. Ces deux politiques bien qu'étant apparemment divergentes produisent *in fine* le même résultat : l'affaiblissement du pouvoir traditionnel africain[377]. Nous évoquerons dans cette logique les données de l'opposition entre les politiques coloniales française et anglaise (**1°**) puis la relativité de l'opposition induite par la finalité commune de deux politiques coloniales (**2°**).

1. L'opposition classique des politiques coloniales française et anglaise

145. - La colonisation a pris des formes différentes à la fois selon les pays colonisateurs et les contextes locaux. Néanmoins, malgré cette pluralité de situations coloniales, l'administration a été partout appelée à jouer un rôle primordial dans le processus de domination des peuples colonisés. L'organisation administrative devait, dans cette optique, répondre aux besoins et aux objectifs de la colonisation. La France et la Grande-Bretagne ont pris possession d'immenses territoires en y imposant une puissante administration permettant ainsi d'établir une opposition binaire entre les modes de gouvernance de ces deux puissances coloniales.

---

[376] J. LOMBARD, *Autorités traditionnelles et pouvoirs européens en Afrique noire*, Armand Colin, 1967, p. 95.
[377] *Idem.*

146. - L'entreprise coloniale française pourrait se résumer en ces mots : conquérir des peuples étrangers qui viendront grossir l'effectif de sa population et la maintenir au rang des grandes puissances[378]. La politique coloniale de la France s'inscrit dans une dialectique de destruction-création. La réalisation du « dessein colonial » français suppose ainsi un aménagement du territoire colonial de sorte que les objectifs économique et politique puissent être atteints. Le système colonial se donnait alors pour but d'effacer de la mémoire collective tout ce qui pouvait rattacher les peuples colonisés à leur terroir, à leur histoire afin de leur faire perdre tout repère de manière à pouvoir disposer des consciences et des personnes sans aucune résistance[379]. Même si les procédés français vont connaitre des évolutions selon la nature du régime en place au sein de la métropole, il se dégage au gré des expériences un style colonial français marqué par une action directe de l'administration. Seule au sommet, l'administration dirige toute l'entreprise coloniale visant, par ailleurs, à assimiler les peuples colonisés à la métropole. Ainsi, comme le relève Pierre Legendre, ce style particulier typique de la France tend à réaliser une « contagion culturelle » ou plus exactement une « *infusion des mœurs* »[380].

147. - Les peuples noirs colonisés, étant issus de race inférieure, doivent pour accéder à la civilisation s'élever au rang des Blancs. La métropole, convaincue de sa supériorité civilisationnelle et du caractère universel de la culture française, s'assure le monopole du savoir et de la connaissance. Elle se donne à travers son administration la mission de dispenser les Lumières en apportant aux peuples colonisés les « bienfaits » de la civilisation occidentale[381]. Au cours de ce processus d'élévation vers le « monde des Idées », la politique d'assimilation française va provoquer le démantèlement de l'organisation précoloniale et l'inculcation aux sociétés colonisées de techniques considérées comme modernes[382]. L'Empire colonial français se devait de transformer les quelque cent millions de sujets de ces colonies en autant de citoyens rompus aux

---

378 P. LEGENDRE, *Histoire de l'administration de 1750 à nos jours, op.cit.*, p. 159.

379 P. LOUM, *La politique de décentralisation et les nouvelles stratégies de développement*, thèse, dactyl., Université de Montpellier, 2013, p. 49.

380 P. LEGENDRE, *Histoire de l'administration de 1750 à nos jours, op.cit.*, p. 174.

381 S. EL MECHAT (dir.), *Les administrations coloniales, XIXe-XXe siècles : Esquisse d'une histoire comparée,* Presses universitaires de Rennes, 2009, p. 14.

382 Voir en ce sens, V. FOUCHER, E. SMITH, « Les aventures ambiguës du pouvoir traditionnel dans l'Afrique contemporaine », *Revue internationale et stratégique*, vol. 81, n° 1, 2011, pp. 30-43.

valeurs, aux méthodes et aux droits métropolitains. La conséquence principale de cette politique d'assimilation qui permettait de transcender la contradiction entre démocratie et colonisation était la reconnaissance d'un droit à la participation des peuples colonisés aux instances représentatives du gouvernement central de la métropole[383]. Paradoxalement, le pouvoir colonial n'était pas disposé à assumer pleinement les conséquences d'une telle politique d'où la substitution à cette dernière d'une politique dite d'association n'impliquant pas l'octroi systématique de la citoyenneté française aux peuples colonisés. Cette politique présentait d'ailleurs, sans modifier son essence centralisatrice et assimilatrice quelques traits communs avec l'« *indirect rule* » anglais[384].

148. - À l'opposé de la conception française se dresse une politique d'administration indirecte pratiquée par les Britanniques dans leurs colonies. L'*indirect rule* anglais se présente comme la doctrine coloniale la plus conciliante avec les formes traditionnelles du pouvoir. Le système colonial britannique intègre les peuples colonisés dans la gestion administrative du territoire colonial. Ce mode de gouvernance permet d'asseoir la gestion des populations autochtones sur leurs propres institutions, leurs propres habitudes de pensée, leurs us et coutumes, le tout adapté malgré tout à certaines exigences du pouvoir colonial. La différence avec le système colonial français se dessine déjà à ce niveau. En cherchant à s'adapter aux différentes formes de pouvoir qui ont précédé son occupation, la politique coloniale britannique se caractérise par l'absence d'uniformité dans son système de gouvernance. Débarrassés des élans centralisateurs de la politique coloniale française, les territoires placés sous la domination anglaise disposaient de statuts juridiques hétérogènes. À la différence de la politique coloniale française, « *le système d'administration britannique respecta les pouvoirs des anciens souverains, s'efforçant de ne détruire aucune monarchie et de conserver l'unité des États, si vastes soient-ils* »[385].

149. - Cette politique a été rendue possible car le colonisateur britannique fut moins soucieux de l'unification de son système juridique

---

383 Dans cette optique, les quatre communes du Sénégal pouvaient élire un député à l'Assemblée française depuis 1848.
384 C. COQUERY-VIDROVITCH, « Les systèmes politiques précoloniaux jusqu'en 1945 », in C. COQUERY-VIDROVITCH, H. MONIOT (dir.), *L'Afrique noire de 1800 à nos jours*, PUF, 2005, p. 67.
385 J. LOMBARD, *Autorités traditionnelles et pouvoirs européens en Afrique noire*, *op.cit.*, p. 165.

tout comme de l'unification territoriale de ces colonies[386]. Partout où il y'avait un chef, l'administration coloniale britannique s'assurait de son maintien au pouvoir. Cette politique a permis d'accompagner la reconnaissance effective du rôle des « *native authorities* » en tant que piliers de l'administration indirecte et du *self-government*. Les pouvoirs traditionnels ont continué ici malgré le dessein colonial à être des acteurs de premier plan dans la gestion quotidienne du territoire[387]. L'adaptation de la politique coloniale aux contextes locaux a provoqué une différenciation dans l'approche des problèmes liés à la chefferie traditionnelle même pour des pays situés dans un même espace géographique. La question de la chefferie ne fut ainsi pas résolue de la même façon au Ghana qu'au Nigéria du Nord par exemple.

150. - Si tout semble, au regard des développements qui précèdent, faire du colonisateur anglais le protecteur des pouvoirs anciens en Afrique, il ne faut pas perdre de vue que le but de toute entreprise coloniale est la domination. En réalité, la période coloniale montre une ambivalence de l'attitude des pays colonisateurs à l'égard des pouvoirs anciens précoloniaux. En même temps que leurs interventions visaient à les transformer, à leur priver de leur essence, les pays colonisateurs souhaitaient les maintenir pour en faire des relais[388]. Ce constat est valable aussi bien pour le colon français qu'anglais d'où la nécessité de relativiser l'opposition classique des deux doctrines coloniales.

2. Une opposition à relativiser

151. - Le caractère relatif de l'opposition des politiques coloniales française et anglaise tient en l'argument selon lequel la domination coloniale qu'elle soit directe ou indirecte a profondément porté atteinte aux structures politiques préexistantes. Malgré la permanence de cette opposition reprise systématiquement par les historiens européens, il est difficile de s'en tenir à pareille opposition sans tomber dans le simplisme. En réalité, les deux puissances coloniales ont indistinctement pratiqué le « *direct rule* » et l'« *indirect rule* ». La France a mis en œuvre une politique d'administration indirecte, notamment au Maroc et en Tunisie. Le facteur

---

[386] *Idem.*
[387] *Ibid.*, p. 166.
[388] V. FOUCHER, E. SMITH, « Les aventures ambiguës du pouvoir traditionnel dans l'Afrique contemporaine », *Revue internationale et stratégique, op.cit.*, p. 33.

le plus important qui conforte l'idée de la relativité de l'opposition c'est que les systèmes traditionnels sont mis à mal chaque fois qu'ils constituent une menace pour l'autorité coloniale dans les deux systèmes[389]. Cette affirmation doit être nuancée en ce que les Anglais ont une préférence pour la neutralisation des structures politiques précoloniales plutôt que leur destruction pure et simple[390].

152. - Il faut néanmoins relever qu'il s'agisse des Anglais ou des Français, leurs actions conduisent aux mêmes résultats. La stratégie consistait alors à conserver les institutions que le pouvoir estimait utiles à son action tout en les privant de toute autonomie de décision et de possibilité d'action[391]. Les chefferies n'étaient maintenues que tant qu'elles ne portaient pas atteinte aux intérêts et à la prospérité de l'entreprise coloniale. Les exigences fondamentales du système colonial eurent partout pour effet de modifier les structures sociopolitiques, de les affaiblir et donc de gauchir les fonctions des institutions africaines de base[392]. Les chefferies les plus revendicatives avaient été déchues et remplacées par des personnages plus conciliants. Le chef africain, que ce soit dans le système britannique ou français, n'était rien d'autre qu'un subordonné du pouvoir colonial. Lord Lugard, principal théoricien de l'*indirect rule*, déclarait à ce propos qu'« *il n'existe pas deux ordres de gouvernements — britanniques et indigènes — travaillant séparément ou en coordination, mais un gouvernement unique dans lequel les chefs indigènes ont des devoirs bien définis, et un statut reconnu tout comme les fonctionnaires britanniques. Leurs attributions ne doivent pas entrer en conflit et se recouvrir le moins possible. Elles doivent être complémentaires et le chef lui-même doit comprendre qu'il n'a droit à sa place et à ses pouvoirs que pour autant qu'il rende à l'État les services qu'on attend de lui* »[393].

153. - Somme toute, cette idée correspond à celle exprimée par les autorités coloniales françaises. Les deux puissances coloniales ont un objectif commun qui consiste à faire à tout prix de la chefferie traditionnelle un instrument de l'affermissement du pouvoir colonial. Il ne

---

389 J. LOMBARD, *Autorités traditionnelles et pouvoirs européens en Afrique noire, op.cit.*, p. 95.
390 P.-F. GONIDEC, *Les systèmes politiques africains,* LGDJ, 1971, p. 49.
391 R. F. BETTS, « La domination européenne : méthodes et institutions », in *Histoire générale de l'Afrique. L'Afrique sous domination coloniale 1800-1935, op.cit.*, p. 339.
392 *Ibid.*, p. 334.
393 F. D. LUGARD, *The dual mandate in British tropical Africa*, 1922, cité par P.-F. GONIDEC, *Les systèmes politiques africains, op.cit.*, p. 50.

faut pas s'y tromper[394]. Pour les masses paysannes africaines que l'on se trouve dans l'indirect rule anglais ou la politique d'assimilation française, la pénétration européenne se traduit partout par les mêmes exigences en termes d'accroissement des taxes et des impôts. Qu'importe les modalités d'exercice du pouvoir colonial, les « vrais chefs de l'Empire »[395] étaient les Blancs[396]. Cela étant, quels furent les principaux éléments structurants de la politique coloniale française en Afrique noire francophone ?

B- Les éléments caractéristiques de l'administration coloniale française

154. - La centralisation du pouvoir et l'émiettement des structures administratives caractérisent l'administration coloniale française. L'État colonial reproduit partout les règles d'administration inspirées de son propre système. Dans toutes les colonies françaises, l'ossature institutionnelle est la même. La politique coloniale française a consisté à étendre au-delà des limites de la métropole un système administratif centralisé. Ce système s'accommode, par ailleurs, très mal avec l'idée d'existence de collectivités locales qui pourraient du fait de leur autonomie influencer la politique du gouvernement central colonial[397]. Le refus de reconnaitre une quelconque autonomie d'action à des pouvoirs non centraux oblige l'administration coloniale française à intervenir directement dans les affaires locales. Cette démarche, loin de paraitre surprenante, est en phase avec le rejet de toute idée de différenciation dans l'administration des colonies françaises.

En effet, les autorités coloniales ont voulu bâtir un système d'administration uniformisé et applicable indistinctement à l'ensemble de l'Empire colonial. La législation applicable aux colonies devait en ce sens se rapprocher davantage de celle applicable dans la métropole. Sans surprise, l'administration coloniale fut ainsi marquée par son caractère hiérarchique et fortement centralisé (**1°**). Pour atteindre ces objectifs et asseoir son autorité, la maitrise des territoires conquis s'avère également primordiale pour le pouvoir colonial. La France va alors procéder au découpage systématique des territoires contrôlés en plusieurs entités. Ainsi, l'espace politique africain se voit remodeler de sorte à permettre à

---

394 Voir les développements *infra* n° 188 et s.

395 R. DELAVIGNETTE, *Les vrais chefs de l'Empire*, Gallimard, 1939.

396 C. COQUERY-VIDROVITCH, « Les systèmes politiques précoloniaux jusqu'en 1945 », *op.cit.*, p. 73.

397 J. LOMBARD, *Autorités traditionnelles et pouvoirs européens en Afrique noire*, *op.cit.*, p. 101.

l'administration coloniale d'assurer le contrôle des pouvoirs anciens. Cette réorganisation territoriale provoqua de nombreuses tensions, car certains chefs ont vu leurs pouvoirs s'amenuiser et se confiner dans de petites unités administratives. C'est par ce nouveau découpage arbitraire que l'administration coloniale va imposer de manière autoritaire son pouvoir (2°).

1. Un système administratif colonial centralisé

155. - L'idée de centralisation est indissociable de la politique coloniale française. Dans les colonies, la métropole gouverne et administre à la fois[398]. La verticalité de l'exercice du pouvoir colonial conduit à la suppression et à la substitution des autorités locales traditionnelles par l'administration coloniale elle-même. Cette volonté centralisatrice se traduit dans la concentration du pouvoir au sein d'une instance unique de direction. L'organisation administrative coloniale fut assurée par le ministère de la Marine jusqu'en 1881[399]. Pierre Legendre note à ce propos que la question coloniale depuis l'Ancien Régime ne pouvait être dissociée de celle du commerce maritime[400]. L'évolution des idées et la transformation de la perception de la colonisation en une « industrie capitaliste » vont susciter la création par Gambetta d'un sous-secrétariat d'État aux colonies rattaché au Commerce qui ne durera que deux ans en raison de sa contestation. La croissance de l'expansion coloniale combinée à l'autonomisation du sous-secrétariat qui se libère de la tutelle du ministère de la Marine conduit à la création législative en 1894 d'un ministère des Colonies. L'influence de l'administration centrale dans la gestion des colonies se voit renforcer à travers la création de ce ministère. La IIIe République française comme ses devancières s'est engagée dans la voie d'une politique coloniale centralisée et politiquement dirigée par l'administration[401].

156. - Une unité de direction se construit au sommet de la machine coloniale française avec à sa tête les gouverneurs. On distingue parmi ces

---

[398] *Cf.* P. LEGENDRE, *Trésor historique de l'État en France*, Fayard, 1992.

[399] *Cf.* P. LEGENDRE, *Histoire de l'administration de 1750 à nos jours,* PUF, coll. « Thémis », 1968.

[400] L'auteur précise que seule une brève interruption de 1858 à 1860 a permis d'instituer un ministère de l'Algérie et des colonies. P. LEGENDRE, *Histoire de l'administration de 1750 à nos jours, op.cit.*, p. 166 et s.

[401] *Ibid.*, p. 167.

derniers le gouverneur général et les gouverneurs de colonies. Le premier est placé à la tête du gouvernement central de l'Afrique-Occidentale Française (AOF) créé en 1895 et qui regroupe huit colonies[402]. Cette centralisation fédérale fut imposée par la métropole. Alors qu'on aurait pu penser qu'elle permettrait de réaliser une intégration interafricaine, elle servait plutôt les intérêts de la France à travers la création d'une unité franco-africaine destinée à garantir la sécurité des relations politiques, commerciales et financières de la zone franc CFA[403]. Les seconds, placés sous la responsabilité du gouverneur général s'occupaient de l'administration directe des colonies. Le développement de l'institution de gouverneur tel que le remarque Pierre Legendre est une constante de tous les régimes politiques qui se sont succédé dans la métropole[404].

Clé de voute du système colonial, cette institution (de gouverneur) constitue une « *véritable charnière du système centralisé* »[405]. Si l'orientation de la politique coloniale est faite depuis la métropole, le contexte local impose en revanche des adaptations. Le législateur métropolitain agissant très souvent en ce qui concerne l'administration des colonies en deçà de ces compétences accorde une liberté d'action totale aux gouverneurs. Cette pratique justifie le foisonnement des décrets-lois pris par les gouverneurs dans les territoires coloniaux. Les gouverneurs disposent ainsi de pouvoirs considérables qui font d'eux à la fois des chefs d'État locaux, chefs de l'administration civile, chefs militaires et magistrats. L'administration coloniale fait ainsi preuve d'une concentration excessive du pouvoir aux mains des gouverneurs. Ces derniers, personnages de confiance, représentants de l'État français sont à la fois des autorités administratives et politiques[406]. Ainsi, le gouverneur « *quoique responsable en dernier ressort devant le gouvernement de la métropole jouissait des pouvoirs d'un souverain* »[407]. De ce fait, la colonisation française est dirigée par l'administration dont l'unité de commandement est garantie par les gouverneurs. Le schéma général de l'entreprise coloniale française pourrait se résumer par les nombreux

---

402 L'AEF fut créée plus tard en 1908 pour regrouper les colonies françaises de la partie équatoriale de l'Afrique.

403 C. COQUERY-VIDROVITCH, « Les systèmes politiques précoloniaux jusqu'en 1945 », *op.cit.*, pp. 67-68.

404 P. LEGENDRE, *Histoire de l'administration de 1750 à nos jours, op.cit.*, p. 167.

405 *Idem.*

406 B. FISCHER, *Les relations entre l'administration et les administrés au Mali : contribution à l'étude du droit administratif des États d'Afrique subsaharienne de tradition juridique française*, thèse, dactyl., Université de Grenoble, 2011, p. 75.

407 R. F. BETTS, « La domination européenne : méthodes et institutions », *op.cit.*, p. 344.

efforts consentis pour réaliser au sommet une unité de direction et à la base, la mise en place d'une monarchie déléguée[408].

157. - Ces éléments contribuent à expliquer le caractère autoritaire de l'administration coloniale. Les objectifs de la colonisation française ne sauraient être atteints sans une administration capable d'endiguer toute aspiration des peuples colonisés à la liberté et à l'égalité. En tant que rouage principal de la stratégie de domination pensée par le colonisateur, elle doit anéantir tout esprit d'initiative chez le colonisé. C'est d'ailleurs à cet objectif que répond la politique administrative qui s'exprime dans la création arbitraire de nouveaux découpages territoriaux avec en prime l'émiettement des structures administratives.

2. L'émiettement des structures administratives

158. - L'administration directe pratiquée par la France aboutit à une stratification des organes administratifs qui doivent garantir l'ordre public colonial afin de s'assurer de l'efficacité des décisions économiques et politiques. L'application de la politique métropolitaine et celle des gouverneurs nécessite la mise en place d'une administration territoriale dotée d'un pouvoir conséquent de commandement. La réorganisation territoriale a permis d'introduire arbitrairement sur les anciens espaces africains constitués de royaumes et d'empires de nouvelles circonscriptions administratives. En ce sens, le système colonial français introduit dans les colonies africaines le cercle en tant qu'échelon principal de commandement[409]. La naissance du cercle en tant que structure coloniale remonte à son intégration dans la constitution du royaume Walo du 10 octobre 1856[410]. La colonie du Sénégal a servi de terrain d'expérimentation de cette technique de maillage territorial avant d'être étendue à l'ensemble des colonies de l'AOF[411].

159. - L'extension de la structure du cercle à toutes les colonies fait écho aux motifs sous-jacents qui ont conduit à la création des départements pendant la Révolution française de 1790. Il s'agissait de doter la colonie d'une structure nationale unique servant de base à la vie

---

408 P. LEGENDRE, *Histoire de l'administration de 1750 à nos jours, op.cit.*, pp. 166-169.

409 Le cercle est équivalent du département.

410 P. LOUM, *La politique de décentralisation et les nouvelles stratégies de développement, op.cit.*, pp. 46-59.

411 *Ibid.*, p. 47 et s.

politique et administrative. Le cercle comprend au moins deux subdivisions avec à leurs têtes des administrateurs blancs placés sous la responsabilité d'un commandant de cercle. Le cercle devient l'unité territoriale principale de l'administration coloniale. Ce découpage inquiète tant par l'arbitraire des arbitrages territoriaux effectués que de la scission « *des entités écologiques ou des groupes humains homogènes* »[412] qu'il provoque.

160. - Le commandant de cercle dispose d'un pouvoir absolu au sein de sa circonscription. Il détenait *de jure* ou *de facto* tous les pouvoirs sur les milliers d'individus dont il constituait l'unique lien avec l'autorité supérieure[413]. Le commandant du cercle inspirait crainte et obéissance chez le paysan africain[414]. Le commandant de cercle était selon le propos de l'historien Joseph Ki-Zerbo « *la cheville ouvrière de tout le système, l'homme-orchestre, le Maitre-Jacques chargé de préparer les décisions et de les exécuter. C'est le Dieu de la brousse* »[415]. L'administration coloniale revêt les apparences d'une administration militaire. En témoigne, la connotation militaire des expressions de « cercle » et de « commandant » qui rappellent la nature de l'acquisition coloniale[416]. Des mobiles militaires ont sans aucun doute présidé au découpage territorial des colonies. Le système colonial est ainsi profondément marqué par une administration de commandement. Elle englobe ici les structures de pouvoir et de coercition, les instruments et les agents de leur mise en œuvre, un style de rapport entre ceux qui donnent des ordres et ceux qui sont supposés obéir sans naturellement les discuter. Cette notion renvoie *in fine* au caractère autoritaire par excellence du

---

412 J. BUGNICOURT, « Le mimétisme institutionnel en Afrique : obstacle majeur au développement », *RFSP*, vol. 6, 1973, p. 1245.

413 C. COQUERY-VIDROVITCH, « Les systèmes politiques précoloniaux jusqu'en 1945 », *op.cit.*, p. 74 ; R. F. BETTS, « La domination européenne : méthodes et institutions », *op.cit.*, p. 344.

414 À cet égard, la littérature romanesque offre un excellent aperçu du sentiment de crainte qu'inspire le commandant de cercle. Voir notamment, F. OYONO, *Une vie de Boy*, Ed., Julliard, 1956 ; *Le vieux nègre et la médaille*, Poche, 2005.

415 J. KI-ZERBO, Histoire de l'Afrique noire, Hatier, 1978, pp. 429-430 cité par M. KAMTO, Pouvoir et Droit en Afrique noire. Essai sur les fondements du constitutionnalisme dans les États d'Afrique noire francophone, op.cit., p 231 ; R. DELAVIGNETTE, « Des commandants français aux préfets africains », RFFPA, n° 2, 1966, p. 30.

416 R. F. BETTS, « La domination européenne : méthodes et institutions », *op.cit.* pp. 344-345.

pouvoir colonial[417]. Cet héritage colonial marque encore aujourd'hui de son empreinte les relations entre l'administration et les administrés[418]. Alors que certaines études sur l'impact historique de la colonisation tendent à minimiser l'influence coloniale sur le caractère autoritaire du pouvoir postcolonial[419], Catherine Coquery-Vidrovitch soulignait déjà, dans un article paru en 1987, son opposition à cette idée en relevant que « *le dictateur militaire autocrate d'aujourd'hui doit davantage au "commandant blanc" qu'au chef coutumier d'antan, d'ailleurs dénaturé et transformé depuis longtemps en "chef administratif", c'est-à-dire en rouage subalterne de la bureaucratie coloniale* »[420].

161. - D'un point de vue juridique, les colonies ne peuvent être considérées comme des entités politiques. Il s'agit simplement de circonscriptions administratives rattachées à l'État colonisateur. La duplication du modèle administratif métropolitain dans les colonies témoigne aussi de la nécessité d'y construire un véritable pouvoir colonial autonome vis-à-vis du pouvoir métropolitain. Si, relève Maurice Kamto, l'appareil politique n'avait été qu'une simple structure administrative, cela signifierait que la moindre décision appliquée dans les colonies eut été prise à Paris[421]. Les gouverneurs disposaient d'une considérable liberté d'action qui leur permettait, tout en suivant les orientations générales de Paris, de développer leurs propres politiques pour les adapter à chaque colonie[422]. L'érection des cercles comme unité de base de la réorganisation du territoire colonial avec à leurs têtes les commandants de cercle assure l'implantation effective du pouvoir colonial. L'introduction de nouvelles exigences sociales par la colonisation s'est faite sans aucun effort de

---

417 A. MBEMBE, *De la postcolonie. Essai sur l'imagination politique dans l'Afrique contemporaine*, La découverte, 2000, pp. 41-94 et 141.

418 B. FISCHER, *Les relations entre l'administration et les administrés au Mali. Contribution à l'étude du droit administratif des États d'Afrique subsaharienne de tradition juridique française, op.cit.*, p. 81.

419 M. GAZIBO, C. THIRIOT, « Le politique en Afrique dans la longue durée : historicité et héritages », in M. GAZIBO, C. THIRIOT (dir.), *Le politique en Afrique : état des débats et pistes de recherches*, Karthala, 2009, p. 27.

420 C. COQUERY-VIDROVITCH, « Les débats actuels en histoire de la colonisation », *Revue du Tiers monde*, vol. 28, n° 112, 1987, p. 784.

421 M. KAMTO, *Pouvoir et droit en Afrique noire. Essai sur les fondements du constitutionnalisme dans les États d'Afrique noire francophone, op.cit.*, p. 224.

422 R. DELAVIGNETTE, *Les vrais chefs de l'Empire, op.cit.*, p. 122 rapporte à ce propos : « Nous n'étions pas les auxiliaires situés à la périphérie d'une sorte de commandement qui nous dictait les ordres ».

conciliation avec les règles coutumières africaines. Leur intégration n'a été faite qu'au prix d'une distorsion des institutions endogènes.

§2.- L'introduction de l'État « moderne » et de ses techniques en Afrique

162. - Le continent africain a connu avant la colonisation des structures étatiques[423]. Le passé du continent est en effet marqué, dans sa partie occidentale, notamment d'empires et de royaumes. Toutefois, la question à laquelle nous tenterons de répondre ici est celle de savoir si la domination coloniale a eu un impact sur l'idée d'État. Un nouvel État africain — moderne — est-il né de la colonisation ? Une réponse positive s'impose, car l'on ne peut nier les évidences.

Les territoires des États contemporains de l'Afrique noire francophone n'ont pas les mêmes configurations que ceux de l'État précolonial. Le droit moderne procède à une définition purement formelle et descriptive de l'État moderne comme étant l'exercice d'un pouvoir politique organisé sur une population bien définie, dans le cadre d'un territoire délimité. C'est cette conception de l'État qui a prévalu lors du découpage du continent africain à la conférence de Berlin de 1885. Le droit s'est érigé en « répartiteur d'espaces » pendant la période coloniale en permettant de considérer que « *toute terre sur le globe est soit territoire d'États européens ou d'États mis sur le même pied, soit terre encore librement occupable, c'est-à-dire étatique potentiel ou colonie potentielle* »[424]. Les territoires sans maitres (terra nullius) sont ainsi, selon les règles de droit international, ouverts à la conquête coloniale. Des procédés juridiques ont ainsi légitimé la conquête coloniale qui va indirectement contribuer à la naissance d'un État moderne (**A**) à l'image de l'État-nation occidental (**B**).

A- Le fait colonial et la genèse de l'État moderne

163. - La pénétration européenne dès le début du XVIe siècle a contribué à l'insertion de l'Afrique dans le système européen. Cette insertion du continent africain dans cet ordre nouveau s'est faite de façon progressive et constante grâce notamment à l'institutionnalisation de la

---

[423] Voir *supra* n° 71 et s.

[424] C. SCHMITT, *Le nomos de la terre,* PUF, coll. « Quadrige », 2012, p. 171.

traite négrière et aux nombreuses missions d'explorateurs ayant précédé la colonisation[425]. Phénomène historique de domination, la colonisation européenne a fortement impacté les structures politiques et juridiques des sociétés africaines en établissant des liens de dépendance. Si, la domination coloniale s'est établie dans un premier temps par la force des armes, elle a fini par trouver une justification de sa raison d'être dans la force du droit. Les différents procédés de légitimation de la domination coloniale laissent, en revanche, apparaitre quelques contradictions (**1°**). La colonisation a participé à la destruction des États africains précoloniaux tout en cristallisant les éléments susceptibles de favoriser la naissance d'un nouvel État en tant que support juridique à la nation (**2°**).

1. Une domination coloniale légitimée par le droit

164. - Les modes d'établissement de la domination coloniale reposent sur des critères juridiques élaborés par le droit international. Le droit des gens prévoyait la possibilité d'acquisition de territoires nouveaux à travers deux procédés : l'annexion ou le protectorat[426]. L'annexion est un procédé radical qui bien que présentée comme première formule d'occupation des territoires n'intervient en réalité que dans un second temps. En effet, pour faciliter l'acceptation de la domination coloniale dans les sociétés indigènes, les nations colonisatrices procèdent d'abord à la signature des traités de protectorat[427]. Toutefois, la différence entre ces deux modes de domination coloniale n'est que juridique, car l'un ou l'autre aboutissent de fait à une occupation effective[428]. Sur la base de considérations d'ordre politique, le droit international a contribué à poser le principe selon lequel les territoires africains sont des territoires sans maitres non soumis à une quelque conque autorité politique souveraine. L'occupation de tels territoires donnait ainsi aux nouveaux occupants un titre juridique valable[429]. Les puissances colonisatrices avaient alors tout intérêt à

---

425 M. GAZIBO, *Introduction à la politique africaine*, Presses de l'Université de Montréal, 2010.

426 P.-F. GONIDEC, *L'État africain : évolution, fédéralisme, centralisation et décentralisation, op.cit.*, p. 64 et s.

427 L'occupation allemande du Togo fut en ce sens précédé de la signature d'un traité de protectorat avec le roi Mlapa III de Togoville en 1884.

428 B. TRAORE, « De la genèse de la nation et de l'État en Afrique noire », *Présence Africaine*, vol. 3, n° 127-128, 1983, p. 157.

429 P.-F. GONIDEC, *L'État africain : évolution, fédéralisme, centralisation et décentralisation, op.cit.*, p. 64.

considérer les sociétés politiques africaines comme des anarchies. À ce titre, elles sont dépourvues de toute forme d'organisation politique, d'États au sein desquels ne s'exerce encore aucune souveraineté organisée même lorsqu'on y rencontre une population indigène sans organisation politique suffisante[430]. Georges Scelles dans son ouvrage dédié au droit des gens évoque, pour parler des territoires africains, des « *territoires exotiques peuplés de collectivités primitives arriérés ou décadentes* »[431]. Les États coloniaux se considérant comme plus avancés se sont alors donné pour mission d'apporter civilisation et protection internationale aux peuplades africaines enfermées dans l'immobilisme[432].

165. - Le système de protectorat international permettait ainsi à deux États de nouer une relation particulière dans laquelle l'État protecteur voit ses compétences et son autorité s'accroitre au détriment de l'État protégé qui perd sa liberté d'action dans un certain nombre de domaines. Il en ressort que le régime de protectorat ne peut s'appliquer qu'en présence de deux États. La signature des traités de protectorat valait ainsi reconnaissance de la qualité d'États aux sociétés signataires. Les États africains conservent dans le cadre du système de protectorat leur indépendance. L'État protecteur ne bénéficie que d'une délégation de compétence en matière internationale. Il faut néanmoins relever que de façon de tout à fait curieuse, les juristes issues des pays colonisateurs ont développé une théorie du protectorat colonial qu'il faut distinguer du protectorat de droit international. Le protectorat colonial s'applique aux pays non civilisés dans lesquels il n'existe aucune organisation durable des pouvoirs publics, où l'on pourrait vainement rechercher les éléments de stabilité et d'unité qui fondent les sociétés politiques indépendantes. Ces pays sont constitués de peuplades avec à leur tête des chefs, mais ne disposent pas d'un peuple avec un gouvernement. Par conséquent, ils ne peuvent se voir appliquer le régime de protectorat de droit international. Ainsi, il revient à toute puissance colonisatrice désirant étendre son pouvoir sur ces territoires sans personnalité internationale d'apporter les éléments nécessaires pour jeter les bases d'une souveraineté, c'est-à-dire un appareil administratif capable de préserver les intérêts publics et privés,

---

430 M. SIBERT, *Traité de droit international public*, Dalloz, 1951, p. 864.
431 G. SCELLES, *Principes de droit des gens. Principe et systématique*, Dalloz, 2008, Réimpression de l'édition de 1932.
432 R. F. BETTS, « La domination européenne : méthodes et institutions », *op. cit.*, p. 339.

sous une direction commune[433]. À l'égard des territoires africains, la puissance colonisatrice ne dispose que d'une possibilité d'acquisition de la souveraineté puisqu'il ne peut se constituer protecteur d'un État qui n'existe pas. Les traités de protectorat signés avec les chefs de peuplades africains ne constituent pas au sens du droit international des actes de cession de souveraineté. Ils témoignent plutôt du devoir « d'humanité et de justice »[434] des pays colonisateurs et d'une stratégie de pacification afin d'éviter les conflits au moment de la possession.

166. - Le protectorat colonial administratif a servi de prétexte à la violation des règles de droit international. Ce régime a facilité la destruction des pouvoirs publics locaux en les neutralisant dans un premier temps. La fin des structures politiques traditionnelles se matérialise dans la décision d'annexion pure et simple du territoire nouvellement conquis. C'est ainsi que, rappelle Pierre-François Gonidec, la monarchie du Dahomey fut supprimée par arrêté du gouverneur français en date du 12 février 1900 après une période de six ans au cours de laquelle le royaume placé sous protectorat français avait conservé ses institutions politiques[435]. La stratégie fut la même à Madagascar. Sur le

---

433 F. GERAL DE SEREZIN, *Le protectorat international : le protectorat sauvegarde, le protectorat de droit des gens, le protectorat colonial*, Pédone, 1896, p. 267.

434 *Idem.*

435 L'arrêté d'annexion se présente comme suit : « Considérant que depuis plusieurs années, Agoli-Agbo n'a pas cessé d'annihiler par sa mauvaise volonté et ses intrigues les efforts accomplis par les résidents d'Abomey pour la mise en valeur de cette province ;
Considérant que Agoli-Agbo s'est rendu coupable de nombreux attentats contre la vie, la liberté et les biens de ses sujets ;
Considérant qu'il importe de prendre d'urgence des mesures de rigueur pour mettre fin à une situation préjudiciable aux intérêts généraux de la colonie de Dahomey et Dépendances ;
Considérant enfin que la population du Dahomey a exprimé à l'unanimité le désir de relever directement de l'autorité du résident français à Abomey ;
Vu le câblogramme du ministre des Colonies en date du 3 février 1900, prescrivant d'une part la suppression de la souveraineté d'Abomey, la destitution d'Agoli-Agbo et l'internement de ce chef à Porto-Novo et d'autre part la division du royaume d'Abomey en cantons indépendants placés sous l'autorité du résident :
ARRETE
Article 1 : La souveraineté d'Abomey telle qu'elle a été constitué par le traité susvisé du 29 janvier 1894 est supprimée ; les territoires formant le royaume d'Abomey seront divisés, par un arrêté spécial, en cantons indépendants placés sous l'autorité du résident.
Article 2 : Le roi Agoli-Agbo est déchu du trône d'Abomey. Ce prince sera jusqu'à nouvel ordre interné à Porto-Novo. »

point de réaliser l'unification politique de l'ile au XIXe siècle, le royaume de Mérina fut placé sous le protectorat français grâce au traité de 1885 en laissant néanmoins subsister le système politique en vigueur à cette époque. Une décennie plus tard, par la loi du 6 août 1896, l'ile était annexée et la monarchie abolie forçant la reine Ranavalona à l'exil[436]. Le protectorat se présente ainsi comme une conquête suivie d'annexion. Il ne s'agit que d'un voile pudiquement jeté sur l'annexion, un régime circonstanciel que le colonisateur n'avait aucune volonté de respecter. Les territoires annexés deviennent ainsi des colonies.

167. - L'État africain dont l'existence a été niée par les colons européens devient, du point de vue du droit international comme du droit public interne, une collectivité administrative, une simple subdivision territoriale de la métropole colonisatrice. La colonisation va alors permettre au pays colonisateur disposant de la pleine souveraineté l'institutionnalisation d'un État moderne fait à son image en tant que support juridique des nations nouvellement créées.

2. La constitution d'un État moderne de type occidental

168. - La colonisation a contribué à la création d'un nouvel État bâti autour du concept de nation. Les puissances colonisatrices ont fait l'effort de transposer en Afrique noire le modèle d'État moderne de type occidental construit essentiellement autour de ce concept très discuté. Ils s'efforcent de faire coïncider deux entités *a priori* différentes que sont l'État, entité politique par excellence et la nation qui est une réalité sociale et historique construite autour de l'unité culturelle d'un peuple. La théorie de l'État-nation apparait dès le début XVe siècle comme un moyen de légitimation du pouvoir politique avant d'être consacrée dans son acception moderne en Europe au XVIIIe siècle. Pour Carré de Malberg, le concept de nation « *désigne non pas une masse amorphe d'individus, mais bien la collectivité organisée des nationaux, en tant que cette collectivité se trouve constituée par le fait même de son organisation en une unité indivisible* »[437]. L'État n'étant selon

---

[436] F. GERAL DE SEREZIN, *Le protectorat international : le protectorat sauvegarde, le protectorat de droit des gens, le protectorat colonial, op.cit.*, p. 267 et s.
[437] Cf. R. CARRE DE MALBERG, *Contribution à la théorie générale de l'État*, Sirey, 1920.

cet auteur que « *la personnification juridique de la nation* »[438]. Ainsi, l'État se caractérise selon cette conception par le caractère indivisible de la collectivité nationale, l'égalité abstraite de ces membres et la nationalité. Il se cristallise dans le « *vouloir-vivre collectif* »[439].

169. - L'expansion coloniale qui s'est réalisée sur la négation de l'État africain précolonial a été à l'origine des nations africaines. Il faut noter que cette idée ne fait pas l'unanimité, car pour certains auteurs[440], la nation africaine ne peut tenir son identité de l'accident historique que constitue la colonisation. Le régime colonial a simplement remodelé les nations africaines en agissant comme un agent de consolidation et non de fondation[441]. L'Afrique noire portait en elle les aptitudes nécessaires à l'édification des nations et des États nationaux à travers les différentes agrégations d'ethnies qui s'opèrent par les relations interpersonnelles qu'elles entretiennent. Cependant, la coupure profonde provoquée par la colonisation a permis d'introduire une nouvelle conception de l'État-nation née avec les Lumières. Les nations modernes ont été créées de façon tout à fait artificielle et par la force grâce au pouvoir politique. Ce constat est vrai aussi bien pour l'Afrique noire que pour l'Europe. Les grandes nations européennes ne sont rien d'autre que le résultat de l'extension forcée par l'État a tous les citoyens d'une langue unique et de l'idée qu'il existerait une unicité des mœurs[442]. Pour s'en convaincre, il suffit de se référer aux velléités constantes de reconnaissance des particularismes, corses, basques, bretons, etc.

170. - En Afrique noire, la colonisation a créé le contexte favorable à l'éclosion d'une nation au sens où on l'entend aujourd'hui. La construction de nations africaines n'était pas l'objectif de la colonisation, mais par ces procédés de domination et de soumission, le pouvoir colonial a facilité en dépit parfois du bon sens et souvent de façon arbitraire la volonté de vivre ensemble des communautés africaines. C'est à l'intérieur des frontières

---

438 Pour une critique de cette conception de l'État-nation : L. DUGUIT, *Traité de droit constitutionnel*, Paris, 3è éd., tome 2, 1930 ; M. HAURIOU, *Précis de droit constitutionnel*, Sirey, 1929.

439 E. RENAN, « Qu'est-ce que la nation », Conférence donnée à la Sorbonne le 11 Mars 1882.

440 B. TRAORE, « De la genèse de la nation et de l'État en Afrique noire », *op.cit.*, p. 159.

441 *Idem.*

442 J. W. LAPIERRE, *Le Pouvoir politique et la langue. Babel et Léviathan*, PUF, coll. « La politique éclatée », 1988, pp. 13-130.

artificielles dont les contours ont été fixés par le colonisateur que les populations africaines parfois éloignées les unes des autres ont été obligées de vivre ensemble. C'est ainsi que se produisit un processus d'acculturation et d'assimilation des cultures des groupes apparentés qui étaient mises en contact. La multiplication des échanges de biens favorisés par la mise en place d'un réseau de communications à l'intérieur des nouvelles frontières pour les besoins de la domination coloniale a indirectement renforcé la coopération ethnique.

171. - L'État dans la pensée occidentale répond à une logique de l'unité par l'unicité de l'État et par l'uniformité qu'il impose aux individus. Le système colonial a ainsi réussi par son pouvoir centralisateur, son ordre administratif et ses enseignements, à faire naitre une culture européenne aux côtés des cultures traditionnelles. Malgré le fait que, seule une minorité d'individus se réclament de cette culture européenne, qui se manifeste à travers l'usage du français et l'accès à une instruction de type occidental, elle renforce les liens de solidarité. La langue française aujourd'hui utilisée partout dans la sphère publique participe au rayonnement de la France et la création d'un sentiment d'appartenance entre les utilisateurs de cette langue commune. L'administration coloniale a ainsi cristallisé par ses méthodes les éléments constitutifs de la nation dans son acception moderne, telle qu'issue de l'Europe du XVIIIe siècle. Les nations africaines nouvellement créées vont constituer les bases de l'État moderne qui s'accompagne du transfert d'un système juridique bouleversant ainsi les méthodes de production du droit en Afrique noire francophone[443].

B- La déformation de l'ordre juridique traditionnel

172. - La colonisation s'est établie sur un postulat simple, mais aux conséquences lourdes pour les pays de l'Afrique noire francophones. Les nations économiquement et industriellement avancées se sentant investies d'une mission civilisatrice se devaient de disposer des pays restés en marge de l'histoire et de l'évolution sociale et politique en leur apportant les traits culturels nécessaires à leurs passages de « sociétés froides » sans histoire à

[443] P. O. SECK, « La singularité du pouvoir politique traditionnel en Afrique : exemples comparés des sociétés Wolof et Bantu », *Revue numérique Afrilex*, 2014, p. 1.

des « sociétés chaudes » historicisées[444]. L'introduction de l'État moderne en Afrique résulte de cette conception dans laquelle le colonisateur fera du droit le pilier de son action. On assiste à une « francisation » de l'ordre juridique traditionnel selon le nouveau schéma imposé par le colonisateur. La colonisation française va ainsi profondément modifier les techniques de production du droit dans les territoires africains (**1°**) par la diffusion du système normatif français (**2°**).

1. Le bouleversement des techniques de production du droit

173. - Le Droit est une traduction de la représentation du monde propre à chaque société. Les sociétés humaines ne partageant pas les mêmes conceptions du monde ni les mêmes valeurs. Le contenu du droit qu'elles produisent est par conséquent indéniablement marqué par la différence des logiques qui le sous-tendent. Le Droit en tant qu'objet scientifique est un discours sur le monde. Ainsi, les sociétés humaines construisent leurs institutions selon des représentations du droit qui sont difficilement transposables d'une société à une autre. L'hégémonie culturelle de l'occident semble ne pas tenir compte de cette réalité. En effet, l'introduction par le colonisateur français des principes et techniques de l'État moderne a profondément impacté les méthodes de production et le sens du droit dans les territoires colonisés. Pour en saisir toute la dimension, il faudra restituer les fondements de l'ordonnancement juridique dans les sociétés africaines précoloniales afin de les différencier des représentations occidentales.

174. - Jusqu'à une période récente de l'histoire, les auteurs ne reconnaissaient ni l'existence d'une quelque conque forme d'organisation politique dans les sociétés précoloniales africaines ni l'existence du Droit. Pour René David, il n'existait aucun véritable système de droit dans les pays africains. Selon cet auteur, la structure tribale des sociétés africaines très fragmentée a amenuisé toutes les chances d'éclosion d'un système juridique. La notion même de droit méconnue de ces sociétés y est

---

[444] M. KAMTO, *Pouvoir et droit en Afrique. Essai sur les fondements du constitutionnalisme dans les États d'Afrique noire francophone, op.cit.* ; R. F. BETTS, « La domination européenne : méthodes et institutions », op.cit., p. 335 ; A. SAURRAUT, La mise en valeur des colonies françaises, Payot, 1923, p. 29 : « Le seul droit qu'elle veuille connaitre, c'est le droit du plus fort à protéger le plus faible », ajoutant qu'elle garantissait « l'accroissement économique et le développement humain » de ses colonies.

importée grâce à la colonisation entreprise par les nations occidentales[445]. Contrairement à ces affirmations, Henri Lévy-Bruhl soulignait qu'« *aucune société ne peut exister sans une discipline qu'elle impose à ces membres, sans un minimum d'organisation, c'est-à-dire sans droit. Toute société est donc par définition une société juridique* »[446]. À l'évidence, les sociétés africaines disposaient bel et bien de systèmes juridiques. La sacralité qui rythme le quotidien des sociétés précoloniales africaines imprègne également le Droit dont il constitue le fondement[447].

175. - Tout système juridique répond à une logique qui le dépasse et le détermine, mais cette logique constitue dans le même temps permission et limitation de l'intervention. Les systèmes juridiques des sociétés traditionnelles africaines reposent sur une logique cosmogonique qui influence la manière dont elles se pensent. Le mythe joue un rôle prépondérant dans la conception et la représentation du Droit des sociétés africaines. Il permet d'expliquer l'origine et le fondement de la sacralité qui imprègne le pouvoir politique. En ce sens, le mythe est la source fondamentale du Droit en Afrique. Il prescrit les comportements, fixe les interdits, crée des obligations, bref, le mythe régule l'ordre du groupe[448]. Si le mythe constitue la clé de voute du phénomène juridique en Afrique noire précoloniale, il ne crée pas le Droit des sociétés traditionnelles. Il permet simplement de saisir le Droit à travers la description des structures originelles de la société. Le mythe apparait alors, selon Maurice Kamto, comme le socle sur lequel repose l'ordre juridique des sociétés africaines précoloniales construit autour de l'héritage normatif originel des ancêtres fondateurs permettant l'esquisse d'une théorie du droit africain[449].

176. - Maurice Kamto montre dans sa thèse le lien qui existe entre le mythe et la construction de l'ordre juridique en Afrique noire traditionnelle. Le mythe est marqué par sa transcendance en tant qu'il

---

445 R. DAVID, *Les grands systèmes de droit contemporains*, Dalloz, 1964, p. 73.

446 Cité par M. KAMTO, *Pouvoir et droit en Afrique. Essai sur les fondements du constitutionnalisme dans les États d'Afrique noire francophone, op.cit.*, p. 152. Traduction de l'adage latin *Ubi societas, ibi jus.*

447 *Ibid.*, pp. 149-199.

448 Ce caractère prescripteur du mythe a fait dire, à tort à certains auteurs que le mythe dans les sociétés africaines constitue un législateur rationnel absolu. Les hommes ne peuvent que dire le droit selon cette conception sans le créer. Voir en ce sens, N. ROULAND, *Anthropologie juridique*, PUF, 1988, p. 184 et s.

449 M. KAMTO, *Pouvoir et droit en Afrique. Essai sur les fondements du constitutionnalisme dans les États d'Afrique noire francophone, op.cit.*, p.153 et s.

constitue le fondement axiologique des sociétés bâties sur la symbiose entre le monde des vivants et celui des morts. Il ne faut point se méprendre sur le rôle du mythe dans les sociétés traditionnelles africaines. Cadre cosmogonique de l'ordre juridique, le mythe n'est pas un législateur. Il crédite du « coefficient de sacralité » les normes originaires et affecte par son caractère transcendantal tout ce qui lui est attaché. L'ordre juridique traditionnel n'est donc pas une émanation de l'ordre mythique au sens où il serait créé par le mythe, mais il se structure autour de deux types de règles juridiques. L'auteur distingue sur cette base d'un côté les « normes transcendantes » qui révèlent le mythe des ancêtres fondateurs et de l'autre, les « règles temporelles » qui sont l'œuvre du temps et des hommes[450].

Toute la théorie du droit africain repose sur ces deux catégories de normes juridiques. Le droit dans les sociétés traditionnelles africaines serait ainsi que le définit Maurice Kamto, un « *ensemble de normes créées ou constatées qui cherchent à provoquer des conduites humaines en attachant aux conduites contraires des actes de contrainte socialement organisés* »[451]. La distinction entre ces deux normes se fait au regard de la force sociale de la règle envisagée sous l'angle cosmogonique, c'est-à-dire en fonction de sa transcendance ou de sa non-transcendance, suivant qu'elle soit temporelle ou au contraire a-temporelle. Ainsi, le mythe fonde le droit originel africain auquel les pouvoirs qui se succèdent doivent se conformer. La validité des règles édictées par ces derniers dépend de leur conformité à la « norme fondamentale transcendante ». En somme, l'ordre juridique des sociétés traditionnelles africaines repose sur la soumission du pouvoir à la norme transcendante scrupuleusement gardée par l'ensemble de la société sans qu'il soit besoin d'en confier la mission à un organe particulier ou à un seul être[452].

177. - Ce rapport constant entre le droit et le mythe va être remis en cause par les puissances colonisatrices. Le mythe est alors perçu par les colonisateurs occidentaux comme une « maladie d'enfance » du droit. Ce dernier ne pourra attendra l'âge adulte que lorsqu'il parviendra à accéder à

---

450 *Idem.*

451 Souligné par l'auteur. Les normes créées renvoient aux règles de droit issues de l'activité normative du pourvoir politique traditionnel, tandis que les normes constatées sont celles posées par le pouvoir originel. M. KAMTO, *Pouvoir et droit en Afrique. Essai sur les fondements du constitutionnalisme dans les États d'Afrique noire francophone*, *op.cit.*, p.170.

452 B. FISCHER, *Les relations entre l'administration et les administrés au Mali. Contribution à l'étude du droit administratif des États d'Afrique subsaharienne de tradition juridique française*, *op.cit.*, p. 92.

la rationalité c'est-à-dire débarrassée de toutes les scories métaphysiques[453]. Par un processus d'acculturation juridique[454], les sociétés traditionnelles africaines considérées comme primitives vont passer de l'univers du mythe structurant leurs sociétés à celui de la loi suivant une évolution progressive, linéaire et univoque. Michel Alliot démontre que ce processus d'acculturation ayant pour point de départ le mythe de la « société archaïque » et pour point d'aboutissement optimum et inéluctable la loi dans la société « évoluée » répond à une logique préétablie par les nations colonisatrices de « progrès juridique »[455]. Cette vision du droit européocentriste hérité d'un Occident « *colonialiste scientiste et positiviste* » convaincu de sa « *mission civilisatrice* »[456], construite sur des représentations du droit différentes de celles des sociétés négro-africaines va s'ériger en règle.

178. - Le discours juridique occidental trouve sa cohérence dans sa vision judéo-chrétienne. Dieu est Tout ! On pourrait tenter de rapprocher cet occident bâti autour de valeurs chrétiennes des sociétés africaines islamisées qui sont-elles aussi organisées autour l'existence d'un Dieu préexistant créateur du monde. Toutefois, le rapprochement s'arrête là, car dans les sociétés islamisées, Dieu est sacré. Or, il est profane dans la tradition occidentale[457]. La conséquence de cette différenciation se trouve dans le fait dans les premières, le droit est sacré et tous les pouvoirs publics y sont soumis. À l'inverse, dans les sociétés occidentales, la source du droit étant profane, les pouvoirs publics ne sont pas soumis aux lois religieuses, au droit divin. À l'image de Dieu, l'Occident chrétien fait des pouvoirs publics les seules forces de création et d'abrogation du Droit[458]. Pour Michel Alliot, le droit français ne peut se défaire d'une telle conception.

---

453 D. MANAÏ, J. PERRIN, « Le droit entre le mythe, la raison et la réalité sociale », *L'Année Sociologique (1940/1948-)*, Troisième série, vol. 31, 1981, p. 403.
454 On aurait même pu parler de déculturation juridique au regard de l'ampleur du phénomène.
455 M. ALLIOT, « L'acculturation juridique », in *Ethnologie générale*, *Encyclopédie de la Pléiade*, Gallimard,1968, pp. 1180-1230.
456 D. MANAÏ, J. PERRIN, « Le droit entre le mythe, la raison et la réalité sociale », *op.cit.*, p. 403 ; R. F. BETTS, « La domination européenne : méthodes et institutions », in *Histoire générale de l'Afrique. L'Afrique sous domination coloniale 1800-1935*, *op.cit.*, p. 339.
457 M. ALLIOT, « La coutume dans les droits originellement africains », *Bulletin de liaison du LAJP*, n° 7-8, 1985, pp. 79-100.
458 B. FISCHER, *Les relations entre l'administration et les administrés au Mali. Contribution à l'étude du droit administratif des États d'Afrique subsaharienne de tradition juridique française*, *op.cit.*, p. 92.

L'État, « *avatar laïcisé du Dieu chrétien* » agit comme Dieu « *unique, centralisé, extérieur aux citoyens, leur accordant à chaque instant la personnalité qui leur permet d'être et les droits qui leur permettent d'agir et les gouvernants souverainement par la contrainte uniforme des lois et décrets : hors de l'État et de ses lois uniformes point de Droit* »[459].

179. - Ainsi, la période de colonisation française a vu naitre un pouvoir souverain d'édiction des lois renforcé par ailleurs, par la désacralisation du pouvoir traditionnel[460]. Par l'intermédiaire du droit dont il est l'unique créateur, l'État colonial français instille sur les territoires africains un système normatif propre au monde occidental.

2. La diffusion du système juridique français d'origine romano-germanique

180. - « *Le Droit, c'est l'esprit d'un peuple* », disait Savigny[461]. Il faut donc pour assurer son effectivité que le peuple s'y reconnaisse, qu'il n'en soit pas froissé, qu'il n'en soit pas brûlé, qu'il ne se sente pas étranger dans le droit qu'on lui tisse au quotidien[462]. Cette affirmation aussi pertinente soit elle fut mise à rude épreuve dans les territoires africains. En effet, l'introduction de l'État moderne dans les colonies africaines marque aussi celle d'un droit nouveau propre à la métropole. Ce nouveau droit sera d'un atout considérable pour consolider la pénétration occidentale en Afrique. Tout dispositif institutionnel opère au service de certaines conjonctions de forces et d'intérêts. Le Droit continental dans lequel s'insère le droit français va servir en situation coloniale les intérêts de la politique

---

459 M. ALLIOT, « La coutume dans les droits originellement africains », *op.cit.*, p. 79 et s.
460 Voir à ce propos les développements *infra* n° 187 et s.
461 Savigny, juriste, fondateur de l'École historique selon laquelle une organisation politique exprime « l'esprit d'un peuple », comme le fait son Droit. L'on retrouve cela encore dans la position de l'Allemagne en matière européenne. Il ne faut donc pas sous-estimer la portée du Droit et de la pensée de tel ou tel juriste dans les constructions en cours. V. par ex. Savigny, Puchta, et Rückert, *L'esprit de l'école historique du droit*, trad. franç. Olivier Jouanjan, 2004.
462 M.-A. FRISON-ROCHE, « Le système juridique français constitue-t-il un atout ou un handicap pour nos entreprises et nos territoires ? », in M. PEBEREAU (dir.), *Réformes et transformations*, PUF, 2018. Disponible en ligne https://mafr.fr/fr/article/le-droit-est-il-un-atout-ou-un-handicap-pour-nos-e/ consulté le 5 juillet 2021.

métropolitaine[463] en s'assurant de la diffusion d'un « *système de normes à la française* »[464]. Les conquêtes militaires et la colonisation vont assurer au droit français une large diffusion sur les terres africaines en raison principalement de son caractère écrit[465].

181. - Le droit africain — ou devrons nous dire les droits africains — est marqué par sa diversité. L'unité dans les sociétés d'Afrique noire se présente dans une logique complexe de différenciations. Les différences produites par les mythes fondateurs constituent des facteurs de solidarité dans la mesure où la société trouve son essence à la fois dans l'autonomie de chaque communauté et dans l'interdépendance que crée la spécialisation fonctionnelle[466]. Le système juridique français est aux antipodes de cette logique d'autonomie et d'interdépendance. La Révolution française de 1789, en mettant fin à l'Ancien Droit, a construit le droit français sur un système légaliste basé sur des textes uniformisants et juridiquement égalitaires[467]. De la territorialité des coutumes africaines marquées par l'oralité, on passe à une codification systématique. Puisque le droit africain précolonial est perçu comme un « droit de l'enfance » appelé à grandir, la codification lui permettra d'atteindre la rationalité nécessaire à l'âge adulte[468]. Le système juridique français depuis le XVIIIe siècle à ériger l'écriture en une forme de progrès juridique. La loi, expression de la volonté générale et principale source du droit, doit impérativement figurer dans un texte. Le droit français va s'imprègne de ce dogme de l'écriture et en faire une parfaite application sur les territoires africains colonisés puisqu'« *aucun droit ne peut naitre sans un texte qui le consacre* » ni s'éteindre sans une abrogation expresse[469].

---

463 E. LE ROY, « Les usages politiques du droit », in C. COULON, D.-C. MARTIN (dir.), *Les Afriques politiques*, La Découverte, 1991, p. 112 : « en dehors de la force des armes, plus ou moins brutalement utilisée, la colonisation ne dispose que de la force des mots et singulièrement du droit ».

464 P. LEGENDRE, *Histoire de l'administration de 1750 à nos jours*, PUF, coll. « Thémis », 1968, p. 182.

465 « Ce qui convient aux Français convient à tous (car) il y'a bien peu de différence entre un peuple et un autre » disait Napoléon, cité par G. LEFBEVRE, *Napoléon*, Poche, 2012, p. 443.

466 M. ALLIOT, « L'État et la société en Afrique noire, greffes et rejets », *Revue française d'histoire d'outre-mer*, n° 250- 253, 1981, pp. 95-99.

467 N. ROULAND, *Introduction historique au droit*, PUF, coll. « Droit fondamental », 1998, p 36 et s.

468 M. ALLIOT, « L'État et la société en Afrique noire, greffes et rejets », *op.cit.*, p. 95.

469 P. LEGENDRE, *Histoire de l'administration de 1750 à nos jours, op.cit.*, p. 459.

182. - En réalité, c'est la question de la nature du droit et donc de ses sources dans les sociétés dites « sans écriture » qui se pose. Le fait que l'oralité soit la principale caractéristique des droits africains suffit-il à justifier leur qualification en « droit coutumier » ? Ou au contraire ainsi que s'interroge le professeur Maurice Kamto, le droit africain serait à l'instar du droit de toutes les sociétés une combinaison de règles d'origine coutumière et de règles édictées par un organe spécialisé ayant le pouvoir de créer le droit ? Ces interrogations bien que légitimes n'ont pas été l'objet de préoccupations de la part des nations colonisatrices. L'absence d'écriture a servi de prétexte à la qualification du droit des sociétés précoloniales de « droit coutumier ». L'introduction de cette terminologie pendant la colonisation avait pour but d'établir une dichotomie des droits de sorte à distinguer d'un côté le droit non écrit des non évolués et de l'autre, le droit écrit de la métropole, des nations civilisées. Or, malgré l'absence de codification, le droit des sociétés précoloniales reconnait l'existence d'organes habilités à créer des normes juridiques. Les chefs africains légiféraient en assemblée publique et imposaient des règles de conduite générales. Ainsi, l'argument de l'oralité parait insuffisant et rend inexacte la qualification de « droit coutumier », car il existait à côté des coutumes, un droit législatif et jurisprudentiel de sorte que les Africains avaient sous des formes différentes une connaissance des procédés modernes de création de Droit[470].

183. - La recension et la codification des droits africains entreprise par le colonisateur au début du XXe siècle en Afrique-Occidentale Française n'ont été qu'« *un artefact colonial et postcolonial qui* [a permis] *d'aménager la transition dans un processus de soumission aux modèles du pouvoir et de l'économie ; de l'État moderne et du marché capitaliste* »[471]. La retranscription des droits africains a servi les intérêts de l'Empire colonial français dans la mesure où le processus de recension fut tronqué[472], les traditions et coutumes sont inventées pour permettre à l'administration d'en disposer comme il

---

470 P.-F. GONIDEC, *Les droits africains. Évolutions et sources*, LGDJ, 1968, p. 224.

471 E. LE ROY, *Les Africains et l'institution de la justice : entre mimétismes et métissages*, Dalloz, coll. « Regards sur la justice », 2004, p. 112.

472 E. SAADA, « Citoyens et sujets de l'Empire français. Les usages du droit en situation coloniale », *Genèses*, n° 53, 2003, pp. 13-14.

l'entend afin de contrôler le comportement des populations[473]. Ainsi est née, la distinction droits traditionnels — droit moderne. Les premiers n'étant conservés que dans la mesure où ils étaient conformes au second réinterprétés et codifiés selon les *desiderata* du colonisateur.

184. - Le droit écrit a pris le contrôle de toutes les prescriptions pour ce qui est de l'organisation administrative des territoires. Le système juridique français se définit et exprime par lui-même une culture qui se traduit dans l'usage des mots qui donnent une idée de la façon dont les réalités sont saisies et construites. Le primat du droit écrit va bouleverser l'organisation sociale construite autour des chefs et des aînés[474] qui devaient conserver et transmettre par l'oralité la mémoire collective du groupe. L'État et le droit moderne vont permettre la naissance d'une élite alphabétisée capable de comprendre et d'interpréter le langage du droit formulé dans une langue étrangère : le français. C'est tout le système social des sociétés africaines précoloniales qui se trouve ainsi modifié. L'oralité du fait de la proximité que suppose l'échange de paroles entre individus et groupes renforçait les relations interpersonnelles. L'anonymat du message écrit met fin au modèle d'organisation sociale communautaire[475].

185. - Le droit écrit dans son expression législative est instrumentalisé par le colonisateur pour servir ses intérêts. Alors qu'il était dans les sociétés traditionnelles déclaratoire et conforme aux normes transcendantes, le droit devient prospectif et un vecteur essentiel de l'action civilisatrice des nations colonisatrices[476]. Le colonisateur va faire du droit un objet de légitimation de son action puisqu'il le présente comme seule condition de l'accès à la modernité par les populations. Le droit devient donc l'outil du politique et est utilisé au fur et à mesure de l'occupation coloniale comme un instrument de domination puis peu avant les indépendances comme un vecteur de développement[477].

---

473 B. FISCHER, *Les relations entre l'administration et les administrés au Mali : contribution à l'étude du droit administratif des États d'Afrique subsaharienne de tradition juridique française, op.cit.*, p. 97.
474 M. NDIAYE, « La démocratie en Afrique subsaharienne précoloniale », in *Mélanges en l'honneur de Dominique Rousseau, Constitution, justice, démocratie,* LGDJ, 2020, p. 945.
475 B. FISCHER, *Les relations entre l'administration et les administrés au Mali : contribution à l'étude du droit administratif des États d'Afrique subsaharienne de tradition juridique française, op.cit.*, p. 97.
476 *Ibid.*, p. 101.
477 *Ibid.*, p. 102.

186. - La politique coloniale française visait donc à supplanter des droits africains par le droit moderne à travers l'importation pure et simple des règles applicables dans la métropole. C'est en cela que se situe le particularisme de la tradition juridique française dont ont hérité les pays de l'Afrique noire francophones. La colonisation française marque l'ébranlement de l'ordre juridique traditionnel africain sur les cendres duquel va émerger un nouvel ordre juridique et politique dans lequel le pouvoir traditionnel africain perd son essence sacrée. Il s'agissait essentiellement d'adapter les réalités africaines aux objectifs européens en dépit des conséquences sur les modes d'organisations traditionnelles.

Section 2 : La chefferie traditionnelle, un rouage de l'administration coloniale

187. - La conquête coloniale a affaibli les pouvoirs traditionnels africains. Malgré la volonté de mettre en place une administration centralisée avec un pouvoir concentré selon les principes métropolitains, le pouvoir colonial a dû pour son implantation effective s'assurer le soutien des chefs locaux[478]. Contrainte par des nécessités pratiques dues à l'étendue des territoires ainsi qu'aux moyens de communication limitée, l'intégration de la chefferie traditionnelle dans l'administration coloniale s'est imposée. L'effectivité du pouvoir colonial exige l'établissement d'un lien de proximité entre l'administration et les populations. La politique assimilationniste de la France se mue alors en une collaboration avec un déséquilibre des rapports de forces entre l'administration coloniale et les autorités traditionnelles. Si, «*par tempérament, les coloniaux français penchaient pour l'administration directe ; par nécessité, ils devaient s'appuyer sur les chefferies plus ou moins réorganisées par eux*»[479]. Ainsi, par l'intermédiaire de la chefferie traditionnelle dont l'existence est maintenue au gré des circonstances, l'administration coloniale assure le maillage de l'ensemble de son territoire. La sauvegarde de la chefferie traditionnelle relève d'une stratégie administrative du pouvoir colonial (**§1**). Au séisme institutionnel provoqué par la période coloniale suivront, après la Seconde Guerre mondiale, des

---

478 R. F. BETTS, « La domination européenne : méthodes et institutions », in *Histoire générale de l'Afrique. L'Afrique sous domination coloniale 1800-1935, op.cit.*, pp. 339-344.

479 G. CONAC, « Le développement administratif des États d'Afrique noire », in G. CONAC (dir.), *Les institutions administratives des États francophones d'Afrique noire*, Economica, coll. « La vie du Droit en Afrique », 1979, p. XII.

tentatives de démocratisation en trompe-l'œil des institutions coloniales axées sur le développement des territoires (§2).

§1.- L'instrumentalisation de la chefferie traditionnelle par le pouvoir colonial

188. - L'attitude du colonisateur français à l'égard des pouvoirs traditionnels en Afrique noire francophone est pour le moins ambivalente et contradictoire[480]. Son approche doctrinaire qui a consisté aux premières heures de la colonisation à mettre fin aux traités de protectorats pour s'assurer la pleine souveraineté sur les territoires colonisés conduit nécessairement à l'anéantissement de la chefferie traditionnelle. Toutefois, l'effet cumulé des deux grandes guerres ainsi que l'éveil de conscience qu'elles provoquent en Afrique, amène la France colonisatrice, soucieuse de l'application à l'identique des principes d'organisation administrative de la métropole, à admettre l'importance de la chefferie dans le cadre de son administration[481]. La politique coloniale française balbutie entre idéologie et empirisme. Toutefois, les exigences pratiques penchent davantage vers une attitude empirique qui se concrétise par une évolution du rôle de la chefferie traditionnelle dans l'appareil administratif colonial[482]. Cette évolution entamera de façon définitive le pouvoir traditionnel déjà affaibli par la pénétration des valeurs idéologiques et religieuses occidentales (**A**). Le chef traditionnel perd de ce fait, son autorité et sa proximité avec le groupe social. Il tire désormais sa légitimité de la loi métropolitaine. Le chef devient un « agent auxiliaire » inféodé à l'administration coloniale (**B**).

A- L'affaiblissement progressif de la chefferie traditionnelle dans le système colonial

189. - L'objet de toute politique coloniale est d'établir une situation de dépendance sur les plans culturel, économique et politique. Dans cette optique, peu importe la nature du système colonial, il porte atteinte à

---

480 V. FOUCHER, E. SMITH, « Les aventures ambiguës du pouvoir traditionnel dans l'Afrique contemporaine », *Revue internationale et stratégique*, *op.cit.*, p. 34.

481 M. KAMTO, *Pouvoir et droit en Afrique. Essai sur les fondements du constitutionnalisme dans les États d'Afrique noire francophone*, *op.cit.*, p. 226.

482 J. LOMBARD, *Autorités traditionnelles et pouvoirs européens en Afrique noire*, *op.cit.*, p .128 et s.

l'organisation sociale et à l'autorité des pouvoirs investis traditionnellement[483]. La seule présence européenne sur le « continent noir » a suffi à influencer irrémédiablement les fondements traditionnels du pouvoir. La tutelle coloniale a contraint les populations africaines à adopter des traits de civilisation européenne et française. Cette contrainte s'est progressivement installée sous l'action de l'Église. Les missionnaires religieux ont combattu les coutumes africaines au nom de l'universalisme chrétien, bouleversant l'ordre religieux des sociétés négro-africaines (**1°**). La pénétration des valeurs de l'Occident chrétien participe du mouvement de désacralisation du pouvoir traditionnel africain par la superposition de deux systèmes de références culturelles contradictoires (**2°**).

1. La désacralisation du pouvoir traditionnel africain par la pénétration des valeurs religieuses et idéologiques occidentales

190. - La désacralisation prive les chefs africains de leur autorité sur les populations. Aucun pouvoir ne peut exister dans les sociétés africaines précoloniales s'il ne trouve son fondement dans les valeurs sacrées du groupe social. La sacralité accorde au pouvoir légitimité et proximité avec le groupe dans la gestion des affaires publiques. Ce caractère sacré du pouvoir traditionnel africain va souffrir de l'introduction pendant la période coloniale non seulement de la conception laïque du pouvoir, mais aussi des croyances religieuses chrétiennes. Dans les sociétés où « *l'animisme seul formait le ciment de la chefferie* » écrit Jacques Lombard, « *la pénétration des valeurs chrétiennes ou laïques a désacralisé le pouvoir* »[484] en éloignant par la même occasion du groupe social. La coïncidence entre le début de la colonisation européenne au XIXe siècle et l'arrivée significative des missionnaires chrétiens en Afrique noire n'est pas le fruit du hasard. Les missionnaires religieux avaient pour mission de faciliter l'occupation coloniale par l'évangélisation des indigènes. Ils concouraient ainsi au succès de l'entreprise coloniale[485]. L'Europe, seule détentrice des savoirs techniques modernes, des croyances universelles, avait l'obligation morale et civique d'instruire les peuplades noires d'Afrique. Cette obligation s'est

---

[483] R. F. BETTS, « La domination européenne : méthodes et institutions », *op.cit.*, pp. 358-359.

[484] J. LOMBARD, *Autorités traditionnelles et pouvoirs européens en Afrique noire, op.cit.*, p. 78.

[485] M. KAMTO, *Pouvoir et droit en Afrique. Essai sur les fondements du constitutionnalisme dans les États d'Afrique noire francophone, op.cit.*, p. 209.

construite sur la base d'un raisonnement qui se résume dans les propos d'Alexis Gochet qui affirmait que : « *c'est par le christianisme que l'Europe a été civilisée… c'est aussi par la Religion que l'Afrique sortira de la Barbarie païenne* »[486].

191. - Cette idéologie chrétienne malgré l'« offense métaphysique »[487] qu'elle constitue pour les sociétés précoloniales africaines fait des émules dans le rang des indigènes qui se christianisent en recevant le sacrement du baptême. L'Africain ainsi christianisé perd son attachement à sa communauté et à son univers culturel. La culture traditionnelle perdue, le rapport d'inégalités introduit par les colons conduit à une acculturation voire une déculturation des Africains. À la recherche de cet élément qui lui manque pour accéder à l'Humanité aux yeux du colonisateur, le colonisé se précipite vers la culture et la religion étrangère comme un moyen d'émancipation[488]. L'institutionnalisation des écoles catholique et protestante dans les colonies[489] sert en effet, ce dessein. Ouverte seulement à une minorité, l'école chrétienne occidentale a permis de former une élite contestataire. Le développement du christianisme a exacerbé les mouvements d'opposition à l'encontre du pouvoir traditionnel sacré. L'effort missionnaire se révèle alors être un soutien de l'action de l'État colonial.

Il est vrai, certains auteurs, notamment Jacques Lombard, insistent sur le fait que l'action de l'église missionnaire en Afrique, tout comme la politique coloniale française, ne procède pas d'une stratégie globale de colonisation mise en place par l'État métropolitain. Qu'il soit permis d'en douter, car le missionnaire apparait comme un auxiliaire au service de l'entreprise coloniale française. Partout, il accompagne le militaire et l'administrateur. En réalité, il est assez difficile de mesurer l'ampleur réelle de ce phénomène de désacralisation du pouvoir dans un monde où le pouvoir spirituel s'exprimait tout entier dans la personne du chef traditionnel. Son pouvoir et sa légitimité se fondent sur les origines mythiques du groupe, ce qui en fait un personnage à part. Avec la nouvelle explication du monde telle qu'elle se raconte dans la Bible, le chef perd

---

486 Cité par M. KAMTO, *Pouvoir et droit en Afrique. Essai sur les fondements du constitutionnalisme dans les États d'Afrique noire francophone*, *op.cit.*, p. 212.

487 *Idem.*

488 Frantz Fanon écrivait dans le compte rendu du premier congrès des écrivains et artistes noirs en 1956 que : « *L'opprimé avec l'énergie et la ténacité du naufragé se rue vers la culture imposée* ».

489 L'école chrétienne constituait une institution clé de l'entreprise coloniale.

toute crédibilité. Il est dépouillé de son aura réduit à un rôle instrumental[490].

192. - Le pouvoir africain est déstructuré et désacralisé à travers l'action de l'église missionnaire qui déstabilise l'ordre politique traditionnel en altérant toutes les valeurs traditionnelles. Il perd l'avantage de la proximité avec les ancêtres fondateurs et les membres du groupe social. Cependant, il faut préciser que la propagation du modèle européen d'éducation culturelle et religieuse ne se déploie pas sur une *tabula rasa* sur laquelle on pouvait inscrire tout nouveau système de valeurs. En effet, le système européen ne s'est pas complètement substitué aux systèmes traditionnels. Ainsi, la coexistence entre ces deux systèmes de valeurs contradictoires dans leur essence va s'avérer complexe rendant plus précaire la position du chef traditionnel.

2. La difficile coexistence d'un double système de valeurs

193. - La situation coloniale met systématiquement en présence deux sociétés hétérogènes que tout oppose que ce soit dans leurs représentations sociales ou dans leurs systèmes de pensées. De cette différence nait un rapport de subordination et de domination contraint qui engendre de nombreux conflits[491]. Les pouvoirs coutumiers africains vont subir les conséquences de la double cohabitation difficile de systèmes construits sur la base de références culturelles qui s'opposent en permanence. Le chef africain se retrouve avec la pénétration des valeurs idéologiques et religieuses occidentales dans une position complexe. Comment juger les comportements des individus lorsque ceux-ci répondent à des systèmes culturels par principe contradictoires ? Un même comportement peut selon le cadre de référence culturelle qu'on applique avoir un fondement légal ou non. Cette situation va être une source de malentendus entre d'une part le chef africain et les populations et d'autre part entre celui-ci et l'administration coloniale. Le chef agissant conformément aux normes fondamentales traditionnelles de son groupe encourt les sanctions de la part du pouvoir colonial tandis que, lorsqu'il se

---

490 M. KAMTO, *Pouvoir et droit en Afrique. Essai sur les fondements du constitutionnalisme dans les États d'Afrique noire francophone*, *op.cit.*, p. 212 et s.

491 C. COQUERY-VIDROVITCH, H. MONIOT, *L'Afrique noire, de 1800 à nos jours*, PUF, coll. « Nouvelle Clio », 2005, p. 237.

réfère aux normes objectives établies par la métropole, c'est le soutien de son groupe social qu'il perd[492].

194. - Le pouvoir colonial a introduit sur les territoires africains de nouvelles normes de comportement sans pour autant suspendre définitivement l'application des coutumes ancestrales[493]. Les règles issues des principes traditionnels ne peuvent être appliquées que dans l'hypothèse de leur non-contrariété avec l'ordre public colonial mouvant dont la définition relève de la seule appréciation de l'autorité coloniale. Le recours à la notion d'ordre public entraine ainsi l'application systématique des règles issues de la métropole dès lors que les pratiques traditionnelles sont considérées comme contraires à cet ordre « *ou prévaut l'hypothèse de leur silence réel ou supposé* »[494]. Aussi, malgré les volontés d'assimilation des colonies à la métropole, le pouvoir colonial met en place une politique « indigène »[495]. Les populations colonisées sont soumises non seulement au respect des règles du droit pénal de la métropole, mais aussi à un ensemble de règles dérogatoires qui sont par ailleurs doublement contradictoire par rapport au droit métropolitain et aux normes traditionnelles en vigueur dans les colonies[496].

195. - Le régime de l'indigénat marque de façon décisive la domination coloniale française ainsi que la perte du prestige de la chefferie traditionnelle dans les sociétés africaines. La contradiction introduite par ce pluralisme normatif laisse la possibilité à chaque individu de choisir selon les circonstances de se référer aux systèmes de valeurs et de normes qui lui est favorable. Le vol du trône du roi d'Ashanti illustre parfaitement les conséquences de cette contradiction ainsi que la désacralisation de l'autorité du chef en Afrique[497]. Le siège d'or, symbole de l'unité du royaume Ashanti, fut à la suite de l'invasion britannique enterré dans un

---

492 J. LOMBARD, *Autorités traditionnelles et pouvoirs européens en Afrique noire*, *op.cit.*, pp. 87-88.

493 Notons qu'il s'agit ici des coutumes telles qu'elles ont été réinterprétées par le pouvoir colonial à l'occasion du processus de codification des coutumes.

494 S. BELLINA, *Droit public et institutionnalisation en situation de pluralisme normatif : le cas de l'État malien*, thèse, dactyl., Université Pierre Mendès France, Grenoble II, 2001, p. 218.

495 R. F. BETTS, « La domination européenne : méthodes et institutions », *op.cit.*, pp. 339-344.

496 I. MERLE, « De la "légalisation" de la violence en contexte colonial. Le régime de l'indigénat en question », *Revue des sciences sociales du politique*, vol. 17, n° 66, 2004, pp. 137-162.

497 J. LOMBARD, *Autorités traditionnelles et pouvoirs européens en Afrique noire*, *op.cit.*, p. 79.

village. À l'occasion des travaux de construction d'une route, la caisse qui servait d'emballage au trône fut découverte et pillée de ses ornements d'or. Les chefs traditionnels ont instruit le procès en condamnant à mort les coupables. Il s'agissait selon les lois coutumières d'un crime. L'administration coloniale en application de la législation européenne a procédé à une requalification de ce qui constituait un crime pour la population en un délit de vol. La dépendance à la fois politique et juridique issue de la colonisation a privé les chefs traditionnels de pouvoir. Ils n'avaient plus aucune emprise sur le groupe social qu'ils dirigent.

196. - Le système coutumier qui permettait d'assurer la régularisation de l'ordre social et la gestion participative des affaires est soumis au contrôle de la puissance coloniale. Le chef africain perd toute initiative règlementaire ou législative en raison de la subordination au droit européen. Les pouvoirs judiciaires reconnus autrefois aux chefs traditionnels dans toutes les sociétés précoloniales leur furent retirés. La politique coloniale française a durablement altéré l'autorité des élites traditionnelles. Toutefois, la France va renouer avec la chefferie traditionnelle dans l'entre-deux-guerres. Le chef traditionnel devient alors un instrument de l'administration coloniale.

B- L'intégration de la chefferie traditionnelle dans les structures du pouvoir colonial

197. - Le pouvoir colonial adopte à l'égard de la chefferie traditionnelle une démarche faite de tâtonnements. Cette attitude marquée par des avancées et des reculs témoigne de l'absence d'une doctrine absolue[498]. Le colonisateur français a préféré procéder à l'expérimentation progressive de différentes politiques selon des conceptions propres à la métropole. Ainsi, après des années passées à vouloir anéantir le pouvoir traditionnel sur les territoires africains, le pouvoir colonial va en faire un allié pour l'efficacité de son administration. La modification de l'attitude du colonisateur vis-à-vis de la chefferie n'implique pas une restauration effective de cette institution traditionnelle. Tout en conservant sa défiance contre les pouvoirs traditionnels, le pouvoir colonial réalise la nécessité de sauvegarder la légitimité traditionnelle du commandement indigène (**1°**). Il va ainsi chercher à renforcer l'autorité des chefs traditionnels dans une

---

[498] R. F. BETTS, « La domination européenne : méthodes et institutions », *op.cit.*, pp. 339-344.

logique purement utilitariste pour l'amélioration de sa gestion administrative[499]. Le pouvoir colonial ne voit dans la chefferie traditionnelle qu'un instrument auxiliaire inféodé à son administration (**2°**).

1. La nécessité d'une sauvegarde de la légitimité traditionnelle du chef

198. - Le pouvoir colonial va tenter de restaurer la chefferie traditionnelle en tant qu'institution représentative du groupe social. La politique coloniale française semble alors se mouvoir entre la volonté de mettre en place une administration directe et celle d'une administration indirecte induite par la nécessité de collaborer avec les chefs traditionnels. Malgré son affaiblissement, la chefferie traditionnelle bénéficie encore d'une certaine influence au sein des populations. C'est ce qui justifie le changement d'attitude du pouvoir colonial à son égard. Pour répondre aux exigences liées à une bonne administration du territoire, la France va opter pour une véritable politique en faveur des chefs. De l'aveu même de Robert Delavignette, directeur de l'École nationale de la France d'outre-mer (ENFOM), il ne peut y avoir « *de colonisation sans politique indigène ; pas de politique indigène sans commandement territorial ; et pas de commandement territorial sans chefs indigènes* »[500].

199. - La revalorisation de l'autorité du chef va se traduire par une série de circulaires adoptées par les gouverneurs généraux au début du XXe siècle. C'est d'abord, sous l'observation avisée du gouverneur général Van Vollenhoven qu'en 1917 est posé le constat de l'échec d'une stratégie de contact direct entre les administrateurs coloniaux et les populations. Le commandant du cercle doit disposer pour son action des intermédiaires légitimes qui bénéficient de la confiance de la population, car « *on peut dire sans aucun paradoxe, que le commandant de cercle perd toujours le contact, quand il cherche à l'établir directement* »[501]. L'administration directe semble dans ce contexte quasiment impossible d'où la nécessité de recourir à des intermédiaires indigènes : les chefs traditionnels. Ces derniers doivent être

---

499 J. LOMBARD, *Autorités traditionnelles et pouvoirs européens en Afrique noire, op.cit*, p. 125.
500 R. DELAVIGNETTE, *Les vrais chefs de l'Empire*, Gallimard, 1939, p. 121.
501 Circulaire du gouverneur général Van Vollenhoven, 18 aout 1917, J.O. AOF, 1917, p. 467.

issus de lignées familiales disposant selon la coutume du pouvoir de régner.

200. - Dans sa circulaire fixant la politique générale au sujet des chefs indigènes, Van Vollenhoven précise que « si on veut bien se souvenir que le lien qui unit en pays noir le chef à ses administrés est un lien infiniment complexe, moral plus qu'administratif ; si on veut se convaincre qu'auprès de ces populations, on ne fait rien, absolument rien, si elles ne sont en confiance, on conclura nécessairement qu'il faut choisir un chef accepté volontiers par les populations et désiré par elles »[502]. Aussi, la restauration de la chefferie traditionnelle dans son rôle d'organe représentatif de la population, passe par le renforcement de son autorité par le pouvoir colonial au regard des populations. Cela passe par des gestes simples de courtoisie, notamment l'interdiction de faire patienter un chef durant de longues heures avant de le recevoir[503].

201. - Derrière cette réintégration de la chefferie traditionnelle conduisant à une amélioration de la condition des pouvoirs coutumiers se cache le paternalisme de l'administration coloniale. L'interruption du mouvement de neutralisation du pouvoir des chefs traditionnels ne met pas fin à la politique coloniale française qui consiste à faire de l'administration indigène en Afrique un simple prolongement de l'administration française. Les chefs ne disposaient d'aucun pouvoir propre. La reconstitution d'un commandement indigène n'avait de sens que dans la mesure où elle facilitait l'action coloniale sur les territoires africains. L'objectif n'était pas de rétablir les chefs en leur donnant des pouvoirs bien qu'ils soient un rouage important de l'administration. La chefferie traditionnelle est inféodée au pouvoir colonial. Ne disposant plus de son autonomie d'antan, la chefferie sert d'instrument de consolidation de la souveraineté de la puissance coloniale.

2. L'inféodation de la chefferie traditionnelle à l'administration coloniale

202. - La logique coloniale française se résume en une captation des structures traditionnelles africaines. Pour consolider l'emprise du pouvoir

---

502 *Idem.*

503 M. KAMTO, *Pouvoir et droit en Afrique. Essai sur les fondements du constitutionnalisme dans les États d'Afrique noire francophone, op.cit.*, p. 229.

colonial, il était indispensable de mettre en place un appareil administratif qui favorise l'intégration des chefs dans le système. À travers cette institutionnalisation du rôle de la chefferie dans l'administration coloniale, le colonisateur exerce une influence croissante sur une institution « cheffale » qui tend à devenir plus bureaucratique que coutumière. Non seulement la chefferie va servir les besoins de l'administration coloniale, mais elle va se révéler également utile au maintien de la situation coloniale. En effet, elle demeure une institution provisoire appelée à disparaitre dans l'approche moderne de l'organisation administrative voulue par la France colonisatrice. La superposition de deux autorités de commandement distinct était inenvisageable dans la politique coloniale française. Elle n'admet qu'un seul et unique commandement : l'autorité française. Les chefferies étaient alors condamnées à disparaitre ou à devenir de simples axillaires de l'administration coloniale.

203. - L'assouplissement du système d'administration directe ne permet pas de reconnaitre au chef, quel que soit son rang, un autre rôle que celui d'exécutant[504]. Les chefs, disait le gouverneur général Van Vollenhoven dans sa circulaire précitée, « *n'ont aucun pouvoir d'espèce, car il n'y a pas deux autorités dans le cercle : l'autorité française et l'autorité indigène ; il n'y en a qu'une ! Seul, le commandant de cercle commande, seul il est responsable. Le chef indigène n'est qu'un instrument, un auxiliaire de transmission* ». S'il était nécessaire de reconstituer les hiérarchies indigènes affaiblies par la pénétration coloniale, il ne pouvait l'être que pour servir le dessein de l'efficacité administrative. L'intégration de la chefferie répond de façon pragmatique à une logique de maitrise administrative et territoriale des colonies par le pouvoir colonial. Il s'agit d'une stratégie administrative du pouvoir colonial[505] qui conduit à institutionnaliser un rapport de subordination entre les chefs et l'administration coloniale.

204. - Le chef relève dans l'exercice de ses fonctions de l'autorité coloniale. Sa docilité est récompensée par son maintien en poste et son insubordination punie d'une destitution[506]. Le bon chef serait donc le chef corvéable qui se cantonne à l'exercice avec zèle des fonctions

---

504 J. LOMBARD, *Autorités traditionnelles et pouvoirs européens en Afrique noire*, *op.cit.*, p. 129.
505 M. KAMTO, *Pouvoir et droit en Afrique. Essai sur les fondements du constitutionnalisme dans les États d'Afrique noire francophone*, *op.cit.*, p. 226.
506 C. NACH MBACK, « La chefferie traditionnelle au Cameroun : Ambiguïtés juridiques et dérives politiques ». *Africa Development / Afrique Et Développement*, vol. 25, n° 3-4, 2000, pp. 77-118.

limitativement énumérées par le colonisateur. Cette position de dépendance vis-à-vis de l'administrateur colonial renforce la tendance à la fonctionnarisation des chefs. Le colonisateur ne semble pas s'attarder sur l'aspect coutumier de l'institution, pas plus que sur son apport réel à l'édification de l'État moderne qu'il prône. Le chef ne peut prétendre jouer un rôle autonome. Il n'est pas une autorité administrative, mais un agent auxiliaire de l'administration coloniale. Les fonctions d'agent d'information, d'exécution et de propagande qui lui sont dévolues s'exercent sous le contrôle d'un administrateur territorial. À ce titre, il ne dispose d'aucune marge de manœuvre[507].

205. - La résistance des chefs face à l'attitude condescendante du pouvoir colonial à leur égard amène ce dernier à créer des chefferies administratives. La chefferie administrative permet à l'administration coloniale de disposer de chefs totalement acquis à sa cause. Le chef administratif nommé par le gouverneur sur proposition du commandant de cercle ne dispose d'aucune légitimité coutumière. Il tire son autorité de la loi européenne. Ainsi, les fondements mêmes du pouvoir traditionnel en Afrique noire francophone sont en permanence remis en cause par la création des « chefs de paille »[508] c'est-à-dire des « chefs sans lignée ». On pouvait donc distinguer dans les colonies un chef traditionnel issu des règles coutumières et un chef administratif émanant du pouvoir colonial ne disposant d'aucune assise traditionnelle[509].

206. - *In fine*, l'enrôlement politique de la chefferie traditionnelle dans l'administration coloniale a ruiné son autonomie et sa proximité avec les populations qu'il gouvernait autrefois. Le pouvoir indigène bien qu'utile à la maitrise du territoire colonial n'était rien d'autre qu'un simple rouage de l'administration. Les véritables maitres demeurent les commandants de cercle. Inféodée au pouvoir colonial, la chefferie ne jouit plus d'aucune autorité. L'autonomie dont bénéficiaient les communautés est bafouée sur

---

507 L'inspecteur des colonies Manet écrivait à ce propos : « Le chef (…) fut il descendant d'un roi avec lequel nous avons traité, ne détient aucun pouvoir propre » cité par M. KAMTO, *Pouvoir et droit en Afrique. Essai sur les fondements du constitutionnalisme dans les États d'Afrique noire francophone, op.cit.*, p. 230.

508 R. DELAVIGNETTE, *Les vrais chefs de l'Empire*, Gallimard, 1939, p. 128 : « Les chefs comme celui que je vis la première fois dans le village où je fis école, ce sont en quelque sorte des hommes de paille, et je les appellerai des chefs de paille ».

509 M. KAMTO, *Pouvoir et droit en Afrique. Essai sur les fondements du constitutionnalisme dans les États d'Afrique noire francophone, op.cit.*, p. 227.

l'autel de la centralisation administrative en dépit des tentatives de démocratisation des institutions coloniales.

§2.- Les tentatives de démocratisation des institutions coloniales

207. - La Seconde Guerre mondiale a constitué un déclic dans la conception des rapports entre la puissance coloniale et les populations en Afrique noire. Dévasté par six années de guerre atroce, l'empire colonial français se retrouve ébranlé. Dans les possessions coloniales françaises, des voix s'élèvent contre la politique coloniale. La France métropolitaine se voit obligée de satisfaire à la demande d'évolution juridique et politique des colonies réclamée par les élites contestataires en leur octroyant des droits nouveaux. L'adoption de la Constitution de la IVe République en France achève le processus de structuration d'un pouvoir nouveau. L'administration coloniale est amenée à faire des concessions de nature juridique et politique qui dénaturent sans y mettre fin la centralisation du pouvoir métropolitain. Cela passe d'abord par la mise en place des corps municipaux, organes chargés de canaliser les aspirations locales qui revendiquent plus d'autonomie (**A**). Puis, à la faveur des principes issus de la constitution de la IVe République, le coup de grâce est porté à la chefferie traditionnelle, remplacée par une nouvelle élite locale dérivée de la colonisation (**B**). Cette dernière fera de la centralisation administrative son cheval de bataille.

A- La municipalisation comme stratégie de contrôle des aspirations locales

208. - Les dernières années de la présence coloniale française sur les territoires d'Afrique noire occidentale sont marquées par la mise en œuvre d'une politique de décentralisation. En réalité, il s'agit d'une politique en trompe-l'œil pour canaliser les revendications à plus d'autonomie des populations colonisées. La décentralisation mise en œuvre à ce moment visait davantage à réaffirmer le contrôle de la puissance coloniale qu'à assurer le partage du pouvoir décisionnel qu'on pouvait légitimement attendre d'une telle politique. Cette volonté politique de changement de la France se manifeste ainsi que le souligne Gérard Conac, par la construction hâtive d'un réseau de collectivités préfabriquées permettant essentiellement à la métropole de faire preuve de son esprit d'ouverture en exhibant comme un parangon l'association des indigènes au processus décisionnel[510].

209. - Un processus d'émancipation politique s'enclenche dès 1946 avec pour conséquence logique l'accession massive à la souveraineté internationale des États de l'Afrique noire francophone. Une série de textes législatifs et règlementaires adoptée cette année va améliorer la condition des indigènes à travers la suppression, notamment des peines de l'indigénat[511], de la justice indigène en matière pénale[512] et du travail forcé. Elle reconnait aux ressortissants africains la citoyenneté avec pour corollaire un droit de vote restreint qui permettra à l'élite politique indigène d'acquérir influence et prestige par sa participation aux assemblées fédérales de l'Union française créée en application de la Constitution de la IVe République. Le nouveau pouvoir qui nait va progressivement se substituer aux chefferies dans le cadre de l'État nouveau. L'instauration de l'élection des conseillers généraux et des parlementaires dans les rangs indigènes apparait comme un nouvel affront pour les autorités traditionnelles. La stratégie coloniale est simple. Il s'agit de réaliser une mutation contrôlée du pouvoir colonial en un pouvoir africain moderne. C'est dans cet ordre d'idée que l'institution municipale apparait comme un bon compromis entre la volonté coloniale de garder un contrôle sur ces possessions et le désir de la nouvelle élite de pouvoir opérer en toute liberté.

210. - L'instauration du système municipal dans l'empire colonial français est très variable. En Afrique occidentale, le Sénégal a été le premier à expérimenter la municipalisation au milieu du XIXe siècle. Pour son extension à l'Afrique noire française, l'administration coloniale a dû faire face à une série de difficultés liées à la soumission des populations aux formes d'organisations traditionnelles du pouvoir. La puissance coloniale fut obligée dans un premier temps de combiner les institutions modernes issues de la métropole avec les chefferies traditionnelles. Des communes mixtes rurales et des communes de moyen exercice sont créées et dotées de conseils municipaux au sein desquels siègent les représentants des chefs traditionnels. Toutefois, il ne faut pas se méprendre sur la portée

---

[510] G. CONAC, « Le développement administratif des États d'Afrique noire », *op.cit.*, p. XVII.

[511] Décret du 20 février 1946 portant suppression des peines de l'indigénat en AEF, AOF, au Togo et au Cameroun, JORF, février 1946, p. 1848.

[512] Décret du 30 avril 1946.

de cette démarche. En effet, dans ce processus de municipalisation, l'administration coloniale continue de régner en maître. Elle procède dans la plupart des communes à la nomination des maires et des représentants des notabilités traditionnelles[513]. Malgré l'érection de l'élection, l'administration coloniale dispose toujours d'un pouvoir de nommer les membres du conseil municipal.

211. - L'absence totale d'autonomie des instances mise en place par la puissance coloniale conforte l'idée selon laquelle il n'a jamais été question pour la France d'octroyer un quelconque pouvoir de décision aux indigènes. Il était surtout question pour la puissance coloniale de transplanter « *des institutions organisées d'en haut pour servir de canaux au développement politique* ». La France colonisatrice cherche à travers la décentralisation du système colonial à se constituer un pouvoir moderne acquis à sa cause[514]. La participation à la vie politique timidement introduite par la création des organes municipaux brise les hiérarchies traditionnelles au nom de la liberté et de l'égalité républicaine. Les conséquences de la substitution des anciennes aristocraties par une nouvelle élite issue des rangs de l'administration coloniale consacrent le triomphe du modèle d'administration centralisé de la France. La municipalisation conduit progressivement les communautés traditionnelles autonomes à l'agonie.

B- Les facteurs favorables au maintien de la centralisation politico-administrative

212. - Nous avons jusqu'ici présenté la colonisation comme un phénomène de domination et de soumission. Aussi, il importe de revenir sur les différents facteurs qui ont facilité la pénétration et la prospérité européenne sur les territoires africains. Ce retour aux éléments psychologiques permet de comprendre pourquoi malgré les revendications d'autonomie, la situation coloniale demeure présente dans ses principales caractéristiques.

---

513 C. NACH MBACK, « La chefferie traditionnelle au Cameroun : Ambiguïtés juridiques et dérives politiques ». *op.cit.,* p. 77 et s.

514 G. CONAC, « Le développement administratif des États d'Afrique noire », *op.cit.,* p. XVIII.

213. - La rencontre entre la civilisation européenne et africaine s'est établie dans un rapport de supériorité et d'infériorité qui hante encore aujourd'hui les populations africaines. La prospérité qu'a connue l'œuvre coloniale nait du complexe de supériorité de l'Européen sur le colonisé. À travers son discours civilisationnel, le colonisateur a réussi à persuader les peuples colonisés de leur infériorité congénitale de sorte à créer autour de lui un mythe qui suscite une crainte révérencielle à l'égard de son pouvoir[515]. Au-delà des mouvements de résistance contre la pénétration européenne, les populations noires vont rechercher dans le mythe l'origine de la domination occidentale. Cette attitude fataliste conduit progressivement à l'intériorisation pour le paysan africain d'une situation de domination sur laquelle il estime n'avoir aucune prise. En somme, ainsi que le montre Véronique Gorog, l'Africain va essayer de donner un sens à ce phénomène de domination en intégrant l'homme blanc dans l'univers mythique traditionnel. La force du mythe dans l'explication du monde des sociétés négro-africaines précoloniales étant connue, la réalité de la domination s'explique par le détour du mythe. Le complexe d'infériorité est inscrit dans la structure mentale des Africains dès le début de la colonisation[516].

Ce processus d'intériorisation de la domination à la suite de l'échec des résistances a conduit certains auteurs à affirmer que la soumission des colonisées à la domination occidentale était congénitale[517]. Octave Mannoni explique alors qu'il y'a chez les Africains un besoin de protection et donc de dépendance, car « *tous les peuples ne sont pas aptes à être colonisés, seuls le sont ceux qui possèdent ce besoin* »[518] de dépendance. Nous ne reviendrons pas sur le mépris qu'exprime une telle affirmation qui a fait d'ailleurs l'objet de très vives critiques dans la littérature sur la colonisation[519]. L'Européen a l'image du Dieu chrétien, se doit de protéger et de conduire ses brebis africaines vers de beaux pâturages où elles pourront enfin s'élever vers la Civilisation. Par une sorte de pédagogie institutionnelle s'installe un paternalisme « bienveillant » du colonisateur à l'égard des peuples africains. Trop fier au départ, ce paternalisme va se

---

515 M. KAMTO, *Pouvoir et droit en Afrique. Essai sur les fondements du constitutionnalisme dans les États d'Afrique noire francophone*, *op.cit.*, p. 253.

516 V. GOROG, *L'origine de l'inégalité des races. Études des trente-sept contes africains*, cité par M. KAMTO, *Pouvoir et droit en Afrique*, *op.cit.*, p. 256.

517 O. MANNONI, *Psychologie de la colonisation*, Seuil, 1950, p. 15.

518 *Ibid.*, p. 87.

519 Voir en ce sens, F. FANON, *Peaux noires masques blancs*, Points, 2015 ; *Les damnées de la terre*, La découverte, 2019.

structurer en une méthode politique de gouvernement pour retirer toute assurance à l'indigène pour lui faire accepter son infériorité et par conséquent la supériorité du colonisateur ainsi que le modèle d'organisation du territoire qu'il prône[520].

214. - Pour mener à bien cette mission d'élévation des populations primitives vers les Lumières, il a fallu mettre en place une politique ambitieuse d'instruction de la jeunesse. L'enseignement obligatoire du français et son érection en langue officielle de l'administration coloniale participent de ce mouvement de transformation. Le programme scolaire vante non seulement les mérites de la colonisation, mais aussi pointait le caractère absurde des cultures traditionnelles africaines. Le gouverneur Roumel instruisait ainsi ses collaborateurs : *« il faut amener l'indigène, par un enseignement bien conduit, à situer convenablement sa race et sa civilisation au regard des autres races et civilisations passées et présentes. C'est un excellent moyen d'atténuer cette vanité native qu'on lui reproche, de le rendre modeste, tout en lui inculquant un loyalisme solide et raisonné »*[521]. L'heureuse diffusion du savoir se trouve ainsi prise dans le piège originel de la supériorité de la culture occidentale française. On était loin de l'idéal républicain proclamé par la Déclaration des droits de l'homme et du citoyen à la suite de la Révolution française. « *Les hommes naissent et demeurent libres et égaux en droits…* »[522]. Tout porte à croire que les idéaux de Liberté et d'Égalité s'estompent au-delà des frontières métropolitaines.

215. - Pour toutes ces raisons, la colonisation constitue pour les peuples colonisés une véritable offense aussi bien sur le plan métaphysique que sur les plans physique et psychologique. La colonisation a détruit tout ce qui constituait l'essence de l'indigène africain. Cette perte de soi se manifeste différemment chez les masses paysannes accrochées au peu qu'il reste des formes d'organisations sociopolitiques traditionnelles et chez la nouvelle aristocratie intellectuelle profondément affectée par la situation coloniale. Alors que les premiers se sont résignés, les seconds expriment davantage le souhait de déposséder les détenteurs du pouvoir colonial. Ayant renoncé à sa personnalité, les rapports se définissent désormais en

---

520 M. KAMTO, *Pouvoir et Droit en Afrique. Essai sur les fondements du constitutionnalisme dans les États d'Afrique noire francophone*, *op.cit.*, p. 259.
521 Cité par M. KAMTO, *Pouvoir et droit en Afrique. Essai sur les fondements du constitutionnalisme dans les États d'Afrique noire francophone*, *op.cit.*, p. 260.
522 Extraits de l'article premier de la DDHC de 1789.

des termes nouveaux de « civilisés » et « non civilisés » en dehors de toute classification selon la couleur de la peau. L'élite politique et intellectuelle noire civilisée estime désormais recueillir les qualités nécessaires pour se substituer au pouvoir colonial. Frantz Fanon décrit avec exactitude ce procédé de substitution. Il écrit dans son ouvrage posthume *Les damnées de la terre* que « *cette catégorie de colonisés* [l'élite lettrée] *a déjà décidé de remplacer le colon, de prendre sa place* »,[523] car l'instruction coloniale a aiguisé son appétit pour le pouvoir et les « manières des Blancs ». Elle se dit civilisée[524]. La contestation par les colonisés lettrés du pouvoir colonial ne signifie donc pas la destruction du système colonial. Convaincus que leur salut ne viendra que de la politique d'assimilation menée par la France colonisatrice, ils acceptent et font siennes l'idéologie du colonisateur[525].

216. - En somme, la rencontre entre le pouvoir colonial et les populations indigènes s'est faite en Afrique noire sur fond de violence et de résignation. La combinaison de ces deux facteurs, soumission par la force et résignation a produit à terme chez le colonisé une attitude permanente de révérence à l'égard de tout pouvoir. L'intériorisation du système colonial suppose préalablement un rejet de soi et l'adulation de l'autre[526]. Ainsi, l'intégration active des catégories mentales métropolitaines produit chez la nouvelle élite politique titulaire du nouveau pouvoir une tentation à la reproduction de l'autoritarisme et de la centralisation qui a marqué la période de domination coloniale[527]. La masse africaine insuffisamment alphabétisée se soumet désormais au

---

523 F. FANON, *Les damnées de la terre*, La découverte, 2019, p. 13.

524 M. KAMTO, *Pouvoir et droit en Afrique. Essai sur les fondements du constitutionnalisme dans les États d'Afrique noire francophone, op.cit.*, p. 261.

525 C'est à Albert Memmi qu'il revient le mérite d'avoir dressé le portrait impeccable du colonisé civilisé. Par un mécanisme connu de mystification : *« l'idéologie d'une classe dirigeante se fait adopter dans une large mesure par les classes dirigées… En consentant à cette idéologie, les classes dominées confirment d'une certaine manière le rôle qu'on leur assigné. Ce qui explique, entre autres, la relative stabilité des sociétés ; l'oppression y est, bon gré mal gré, tolérée par les opprimés eux-mêmes… Il existe assurément — à un point de son évolution- une certaine adhésion à la colonisation. Mais cette adhésion est le résultat de la colonisation et non sa cause ; elle nait après et non avant l'occupation coloniale. Pour que le colonisateur soit complètement maitre, il ne suffit pas qu'il le soit objectivement, il faut qu'il croie en sa légitimité ; et pour que cette légitimité soit entière, il ne suffit pas que le colonisé soit objectivement esclave, il est nécessaire qu'il s'accepte tel* ». A. MEMMI, *Portait du colonisé*, Pauvert, 1966, pp. 125-126.

526 P.-F. GONIDEC, *Les systèmes politiques africains, op.cit.*, p. 65.

527 C. COQUERY-VIDROVITCH, « Les débats actuels en histoire de la colonisation », *Revue du Tiers monde*, vol. 28, n° 112, 1987, p. 784.

pouvoir de l'État nouveau de la même façon qu'il se soumettait hier encore au pouvoir colonial. Jaloux de ces nouvelles compétences, le noir civilisé en tant que nouveau chef n'entend pas les partager, reproduisant ainsi le système administratif hérité du pouvoir colonial.

217. - On comprend donc que l'administration locale n'est pas plus une priorité aussi bien pour les nouveaux dirigeants indigènes que pour les anciens colons pendant la période coloniale. Dans les pays où, la culture politico-administrative a été façonnée dans un contexte autoritaire, les nouveaux souverains n'entendent point accorder une quelconque autonomie à des organes municipaux. Au contraire, l'idée d'accorder des libertés locales suscite une profonde méfiance. La municipalisation entamée sous la colonisation est maintenue dans une constance qui pourrait surprendre l'observateur non avisé. Celle d'une émasculation progressive de l'autonomie des collectivités territoriales au profit de la centralisation étatique. À ce titre, Thierry Michalon souligne que les pouvoirs « *issus du démantèlement des empires coloniaux* [ont] *disloqué les solidarités locales, régionales, ethniques au profit de la formation d'une solidarité nationale* (…) *Autoritaires et hypercentralisées* [ces pouvoirs] *ont bien été jusqu'à présent ennemies des libertés tant politiques que locales* »[528]. Cela se traduit d'un côté, sur le plan juridique, par la consécration de l'exécutif comme lieu unique de pouvoir avec une prééminence du chef de l'État. De l'autre, sur le plan institutionnel, le monopartisme se révèle être l'outil indispensable en faveur de la tendance à la centralisation.

218. - Le passage de l'univers administratif contrôlé par la puissance coloniale à un univers soumis au contrôle d'une minorité indigène suffisamment « blanchie » n'a pas suffi à restaurer le pluralisme dans l'exercice des fonctions administratives.

---

528 T. MICHALON, « Autoritarisme et hypercentralisation : l'État africain contre les libertés », *Revue africaine de droit international comparé*, vol. 3, 1991, pp. 76-82.

Conclusion du chapitre 2

219. - Phénomène de domination, la colonisation a été à l'origine de nombreux bouleversements dans les sociétés négro-africaines précoloniales. La domination coloniale s'est traduite par des changements institutionnels dans les sociétés précoloniales africaines. Elle emporte également en tant que phénomène global des conséquences sur les plans culturel, économique, politique et juridique. La colonisation produit un discours sur les structures institutionnelles des sociétés précoloniales entièrement reconfigurées par l'implémentation d'un système d'administration ayant pour support l'État moderne. La France métropolitaine, modelée par le régime napoléonien centralisateur, a laissé aux pays de l'Afrique noire francophone les caractères les plus détestables de son système administratif ruinant ainsi l'éclosion entamée pendant la période précoloniale d'une idée de décentralisation et d'autonomie locale. Le bilan colonial est loin d'être élogieux de ce point de vue. Unité, centralisation, uniformisation, telles furent les grandes valeurs transmises par la France a ses possessions africaines et reçues par les élites locales comme les composants nécessaires d'un État moderne[529]. Ces valeurs justifient l'échec de la décentralisation aussi bien dans les anciennes colonies qu'en France[530].

---

[529] T. MICHALON, « Quel État pour l'Afrique ? », *Présence africaine*, vol. 3, n° 107, 1978, p. 18.

[530] B. FAURE, « L'échec de la décentralisation française : L'État, les élus et les règles », in F. CROUZATIER-DURAND, M. TOUZEIL-DIVINA (dir.), *40 regards sur 40 ans de décentralisation(s)*, Ed. L'Épitoge, 2022, p. 281.

Conclusion du titre 1

220. - La colonisation ne peut être mise hors de cause dans l'effondrement des formes d'organisation traditionnelles des sociétés africaines précoloniales. « *L'éthique coloniale* »[531] qui ignorait dans les territoires dépendants toute logique politique au profit d'un processus purement administratif va introduire un système administratif fortement centralisé. En mettant brutalement en contact au XIXe siècle un peuple européen conquérant avec des populations africaines péjorativement qualifiées de primitives, la colonisation a durablement affecté les institutions précoloniales. Le pouvoir né de cette période se distingue du pouvoir africain par son cadre explicatif : sacré et transcendantal, il devient légal et rationnel, car en désacralisant le pouvoir du chef africain l'administration coloniale l'a vidé de ses fondements. L'État moderne ainsi que le pouvoir qui s'y exerce sont des inventions de la colonisation.

221. - C'est la fin de la conception pragmatique de l'organisation sociale africaine dans laquelle les formes sont déterminées par le résultat à atteindre. Désormais, seule la beauté de l'épure institutionnelle[532] a un sens aux yeux des nouveaux dirigeants. L'espace politique va être recomposé sur la base de nouveaux découpages territoriaux dépourvus de base sociologique. L'observation de ce passé institutionnel nous prédispose à comprendre la nature des relations du couple colonisateurs-colonisés. La colonisation a produit des effets de masse et à très long terme qui se traduisent par la religiosité avec laquelle les concepts métropolitains sont intériorisés dans les consciences colonisées[533]. Elles sont encore historiquement « identifiées » aux métropoles en matière d'organisation politique et administrative. Le système colonial a fixé le cadre administratif général dans lequel s'insèrent les gouvernements nationaux depuis les indépendances. « *Cette amorce de normalisation de la vie politique dans le cadre d'une structure organisée à l'Européenne constitue l'aspect principal de la modernisation que les Européens ont alors introduite en Afrique, mais pour servir leurs propres desseins* »[534]. La « fin » de la colonisation n'a pas permis à l'Afrique de se

531 T. MICHALON, « Quel État pour l'Afrique ? », *op.cit.*, p. 18.
532 E. LE ROY, « Gouvernance et décentralisation ou le dilemme de la légitimité dans la réforme de l'État africain de la fin du xx siècle », in *L'État en Afrique, indigénisations et modernités*, Cahier du GEMDEV, n° 34, 1996, pp. 97-99.
533 P. LEGENDRE, Trésor historique de l'État en France. L'Administration classique, Fayard, coll. « Les savoirs », 1992, p. 185.
534 R. F. BETTS, « La domination européenne : méthodes et institutions », *op.cit.*, p. 359.

réinventer. Elle a de nouveau été soumise à d'autres contraintes dans la construction des structures territoriales.

TITRE 2 — LA RECONSTRUCTION SOUS INFLUENCE DES STRUCTURES TERRITORIALES DE L'ÉTAT

> « Il n'y a pas de chose plus difficile à entreprendre et plus incertaine à réussir, ni plus périlleuse à conduire, que de prendre l'initiative pour introduire de nouvelles institutions »[535].

222. - Les réformes des administrations territoriales en Afrique ont en commun leur caractère exogène. En effet, les réformes entreprises par les États d'Afrique subsaharienne ont été dictées par des impératifs extérieurs qui sont soit d'ordre financier soit lié à la mise en œuvre de plans de développement imposé par les organisations internationales. La recherche d'une politique capable de fournir à l'État un appareil administratif efficace rendant la gestion des affaires publiques possible devient ainsi une priorité[536]. Devant l'urgence de se doter d'institutions à la veille des indépendances, les leaders africains ont maintenu les structures administratives issues de la colonisation. Fortement centralisées, elles devaient conduire à la formation de l'unité nationale et favoriser le développement des jeunes États indépendants. Des décennies après les indépendances, il ne reste plus « *grand-chose des espoirs de développement qui avaient été mis dans les structures étatiques des pays africains.* [...] *Les économies des pays du continent sont à l'agonie et, la légitimité de leurs appareils de gestion sociale demeure toujours aussi faible. Le concept d'État gestionnaire ou promoteur du*

---

535 N. Machiavel, Discours (III, 30), in *Œuvres* (traduites par Ch. Bec), Laffont, 1969, p. 123.

536 Voir à ce propos, G. LANGROG, « Genèse et conséquences du mimétisme administratif en Afrique », *Revue internationale des sciences administratives*, vol. 59, n° 6, 1973, p. 119.

développement semble avoir complètement échoué en Afrique»[537]. C'est dans ce contexte général de remise en cause du modèle d'administration centralisé que les institutions financières internationales vont imposer aux États africains une série de réformes institutionnelles. Parmi ces réformes, la décentralisation, par son attractivité[538], occupe une place de choix dans les programmes de l'ensemble des bailleurs de fonds, qui vont en faire une conditionnalité de l'accès à l'Aide publique au développement (APD).

223. - Cette forte implication internationale en faveur de la décentralisation dans les États d'Afrique subsaharienne renforce par ailleurs les tentatives mimétiques. La quasi-totalité des pays francophones de l'Afrique de l'Ouest a hérité des institutions d'origine occidentale. Bien qu'une partie de la doctrine s'oppose à l'analyse des structures administratives africaines en termes de mimétisme[539], force est de constater que les anciennes colonies françaises d'Afrique noire francophone n'ont pas fait preuve d'innovation institutionnelle[540]. Si l'on admet, avec Marie-Claire Ponthoreau, que « *l'emprunt en droit est sans doute beaucoup plus fréquent que l'innovation*»[541], on constate par ailleurs que cet emprunt dans le cas des pays d'Afrique a été davantage contraint que voulu. Les autorités africaines se sont vues obligés d'intégrer dans leur droit public des constructions institutionnelles développées sous d'autres cieux. Sous l'influence conjuguée de l'attrait à la décentralisation et de l'assistance technique majoritairement française, des solutions importées, toutes faites, vont être imposées aux États africains. Ces derniers vont reproduire *in extenso* le modèle d'organisation territoriale de leur l'ex-puissance coloniale. C'est donc sans surprise qu'on retrouve dans les États d'Afrique noire francophone, à quelques exceptions près, les mêmes structures territoriales dont la construction s'inspire des évolutions institutionnelles en droit français (**Chapitre 1**). Le succès du modèle français d'administration territoriale en Afrique s'explique non seulement

---

537 D. DARBON, « L'État prédateur », *Politique africaine*, n° 39, 1990, p. 45.

538 J. DU BOIS DE GAUDUSSON, « La décentralisation menacée par la (bonne) gouvernance », *in Mélanges Franck Moderne*, Dalloz, 2004, pp. 995-1009.

539 J. DU BOIS DE GAUDUSSON, « Le mimétisme postcolonial, et après ?», *Pouvoirs*, vol. 2, n° 129, 2009, pp. 45-55.

540 E. MBONSI, Avant-Propos à C. NACH MBACK, *Démocratisation et décentralisation. Genèse et dynamiques comparés des processus de décentralisation en Afrique subsaharienne,* thèse, Karthala-PDM, 2003, p. 5 et s.

541 M.-C. PONTHOREAU, « Trois interprétations de la globalisation juridique. Approche critique des mutations du droit public », *AJDA*, 2006, p. 20.

en raison de l'histoire commune que la France partage avec les États d'Afrique noire francophone, mais aussi à cause des différents changements intervenus dans la dogmatique internationale (**Chapitre 2**).

Chapitre 1 — La perpétuation des cadres territoriaux hérités de la colonisation

Chapitre 2 — La décentralisation : une conditionnalité de l'aide au développement

Chapitre 1 – La perpétuation des cadres territoriaux hérités de la colonisation

224. - « Les configurations institutionnelles reflètent partout, quelle que soit la nature du régime, un entrecroisement complexe de biens importés et de biens conservés ou redécouverts »[542]. L'analyse de la structure territoriale d'un État est révélatrice du fonctionnement concret d'un pouvoir politique. Les observateurs des institutions politico-administratives africaines abordent quasi systématiquement l'analyse des structures territoriales en Afrique sous l'angle du mimétisme postcolonial. Peut-on encore au XXIe siècle, ainsi que s'interrogeait Jean du Bois de Gaudusson, analyser les démocraties en Afrique en termes de mimétisme[543] ? La réponse à cette question est loin d'être évidente tant la doctrine est divisée sur le contenu même du concept de mimétisme. Le mimétisme se traduit-il par la parfaite transposition du modèle exporté ou par une simple conformité dans la terminologie utilisée ? Ou encore dans la pratique institutionnelle parce que le texte n'est qu'un support et que le mimétisme dans le texte ne signifie pas mimétisme dans les comportements des acteurs institutionnels[544] ? Une partie de la doctrine africaniste estime qu'il est dépassé aujourd'hui d'analyser les institutions africaines en termes de mimétisme, car ce concept ne permet pas — plus — de rendre compte de la complexité des institutions et du contexte international dans lequel elles sont créées[545]. Dans le même ordre d'idée, André Cabanis et Babacar Guèye estiment que les élites africaines n'ont pas copié leurs homologues français, mais « tous ont puisé dans un fonds commun aux origines très variées et dont on décèle des traces dans à peu près tous les pays du monde »[546]. Bien que cette idée soit pertinente, car elle consiste in fine à rejeter la perspective condescendante dans les études portant sur l'Afrique, elle ne nous convainc pas.

---

542 Y. MENY (dir.), *Les politiques institutionnelles du mimétisme. La greffe et le rejet*, L'Harmattan, coll. « Logiques politiques », 1993, p. 8.

543 J. DU BOIS DE GAUDUSSON, « Le mimétisme postcolonial, et après ? », *op.cit.*, p. 45.

544 F. DARGENT, « Les échecs du mimétisme constitutionnel en Afrique noire francophone », *RDP*, n° 5, 2017, p. 1347.

545 J. DU BOIS DE GAUDUSSON, « Le mimétisme postcolonial, et après ? », *op.cit.*, p. 45.

546 A. CABANIS, B. GUEYE, « Dire le droit en Afrique francophone », *Droit sénégalais*, n° 11, 2013, p. 105.

225. - L'analyse des structures territoriales en Afrique noire francophone montre que leur existence relève pour l'essentiel de l'héritage colonial. La levée du joug colonial aurait pu susciter la tentation chez les élites africaines de transformer radicalement les institutions en place. Au contraire, la peur du chaos se mêlant à l'absence d'autres options réalistes ou raisonnables dans le temps court dont il disposait, celle de la continuité administrative est retenue par les dirigeants africains[547]. L'administration territoriale coloniale a établi des relais de commandement sur l'ensemble du territoire qui a constitué pour les nouveaux dirigeants nationaux « *un instrument d'action* [...] *précieux dans une période où ils devaient faire accepter leur propre autorité* »[548]. À cela s'ajoute l'influence de la formation des cadres et la coopération administrative entretenue dans les pays où l'indépendance a été obtenue pacifiquement avec l'accord de la puissance colonisatrice. Sans prétendre qu'il existe une homogénéité totale et parfaite dans tous les États d'Afrique noire francophone, il faut noter que cet héritage administratif est important et se retrouve dans l'ensemble des pays ayant en commun un passé colonial français. Pour répondre aux urgences post-indépendance l'administration africaine à commencer par « *se* différencier » et « *se rendre étrangère* » des structures traditionnelles « *en s'associant à l'administration génitrice métropolitaine coloniale, en se constituant héritière* »[549]. En perpétuant cet héritage colonial, les élites africaines perpétuent le déphasage entre le territoire des collectivités et les réalités locales. Ainsi, l'organisation territoriale en Afrique francophone s'inspirant des institutions françaises est organisée autour des communes (**Section 1**), des préfectures (**Section 2**) et des régions (**Section 3**).

---

547 B. FISCHER, Les relations entre l'administration et les administrés au Mali : contribution à l'étude du droit administratif des États d'Afrique subsaharienne de tradition juridique française, op.cit., p. 68.

548 G. CONAC, « Le développement administratif des États d'Afrique noire », in G. CONAC (dir.), *Les institutions administratives des États francophones d'Afrique noire*, Economica, Coll. « La vie du Droit en Afrique », 1979, p. XLVI.

549 D. DARBON, « L'"aventure ambiguë" des administrations africaines de succession française dites francophones », *Revue internationale de politique comparée,* vol. 1, n° 2, 1994, p. 211.

Section 1 : L'ancrage de l'institution communale

226. - L'introduction de « l'idée municipale »[550] et le développement de l'institution communale en Afrique subsaharienne sont liés à la présence européenne sur le continent. C'est dans la colonie du Sénégal que les premières communes ont été créées par l'administration coloniale. L'institution sur le continent africain d'un maire remonte au XVIIIe siècle dans les villes de Saint-Louis et de Gorée[551]. Plus tard, le développement économique de la ville de Rufisque et de Dakar à la fin du XIXe siècle va conduire l'administration coloniale à les ériger également en communes. Ces quatre communes appelées les « quatre vieilles » vont bénéficier d'un régime municipal identique à l'organisation métropolitaine notamment avec le décret du 1er août 1872[552]. Au Burkina Faso, les villes de Bobo-Dioulasso et de Ouagadougou ont été érigées en communes mixtes dans la typologie de l'administration coloniale respectivement par des arrêtés du 4 décembre 1926 et du 1er août 1952. Les maires à cette époque étaient nommés par le commandant du cercle. En 1956, les deux villes sont érigées en communes de plein exercice avec des organes élus au suffrage universel direct. Au Niger, l'administration coloniale a institué les premiers conseils de notables par les décrets du 21 mars 1919 et du 1er avril 1936. La réforme municipale en Afrique occidentale française de 1955 a vu la création de la commune de plein exercice de Niamey et les communes mixtes de Zinder et Maradi érigées en 1958 en communes de moyen exercice. Il en va de même au Mali et au Togo où les premières communes ont été créées respectivement en 1918 et en 1929 sous le statut de communes mixtes.

227. - Les premières occurrences de la commune en Afrique remontent donc à la période coloniale. Cette institution demeure depuis lors présente

---

550 M. DIOUF, « L'idée municipale en Afrique. Une idée neuve », *Politique africaine*, 1999, n° 74, p.13.

551 L. JORE, « Les établissements français sur la côte occidentale d'Afrique de 1758 à 1809 », *Revue française d'histoire d'Outre-mer*, 1964, p. 257, mentionne un maire à Saint-Louis en 1778. Le « Maire de ville » apparaît encore dans l'État général des dépenses à faire au Sénégal pour l'année 1783 (AN Colonies C8 22) et dans une lettre du Ministre de la Marine au Commandant du Sénégal, en date du 23 avril 1789 (AS 1 B 1, pi 212). Lire aussi, H. J. LEGIER, « Institutions municipales et politique coloniale : les Communes du Sénégal », *Revue française d'histoire d'outre-mer*, n° 201, 1968, pp. 414-464.

552 H. J. LÉGIER, « Institutions municipales et politique coloniale : les Communes du Sénégal », *Revue française d'histoire d'outre-mer*, n° 201, 1968, p. 415.

dans les institutions africaines. Elle peut prendre diverses formes. Malgré le recours permanent à l'institution communale dans l'organisation territoriale de l'État postcolonial, elle ne bénéficie pas d'une définition précise dans les textes qui la consacrent. Les lois de décentralisation dans la majorité des États d'Afrique noire francophone se contentent d'affirmer que la commune est une « collectivité territoriale de base »[553]. L'idée de la commune collectivité de base, école de la liberté au sens où l'entend Alexis de Tocqueville[554] est pourtant difficile à défendre dans les ex-colonies françaises. En effet, la caractéristique principale des communes héritées de la colonisation est leur inadéquation aux solidarités communautaires préexistantes. Dans le but d'assurer l'ordre public et d'encadrer les populations, l'administration coloniale, tout comme celle postcoloniale, n'a pas tenu compte dans le découpage communal ni du caractère communautaire de l'organisation sociale africaine, ni vraiment des potentialités locales (§**1**). Cette situation emporte des conséquences quant à la viabilité économique des communes (§**2**).

§1.- La commune, une collectivité territoriale de base

228. - La commune est la collectivité territoriale la mieux implantée dans le paysage administratif des États d'Afrique noire francophone. Elle est donc l'échelon administratif de base. Les législations nationales définissent d'ailleurs la commune comme « *la collectivité territoriale de base dotée de la personnalité morale et de l'autonomie financière. Elle regroupe les habitants d'un espace territorial continu* »[555]. Cette définition, qui fait de la place de la commune dans la hiérarchie des organisations décentralisées, son critère absolu de définition, est pour le moins réductrice et insuffisante. Réductrice et insuffisante d'abord parce que, ainsi que le souligne Henry

---

553 Art. 17 du CGCT au Burkina Faso ; art. 56 Loi n° 2019-006 relative à la décentralisation et aux libertés locales au Togo ; art. 20 du CGCT au Niger.

554 « C'est dans la commune que réside la force des peuples libres. Les institutions communales sont à la liberté ce que les écoles primaires sont à la science ; elles la mettent à la portée du peuple ; elles lui en font goûter l'usage paisible et l'habitue à s'en servir. Sans institutions communales, une nation peut se donner un gouvernement libre, mais elle n'a pas l'esprit de liberté » A. DE TOCQUEVILLE, *De la démocratie en Amérique*, Gallimard, coll. « Idées », 1968, p. 72.

555 Art. 56 de la Loi n° 2019-006 du 26 juin 2019 relative à la décentralisation et aux libertés locales ; art. 17 du CGCT du Burkina Faso ; Art. 20 du CGCT du Niger ; art. 1 du code des collectivités territoriales au Mali

Roussillon, il n'est pas intellectuellement satisfaisant de définir une collectivité locale par sa seule taille, alors même que celle-ci constituerait un caractère important[556]. Ensuite, il y a également un risque de confusion, en partant de ce critère, entre la commune « cellule de base » et les unités administratives qui se situent en dessous de la commune.

En effet, cette définition ne prend pas suffisamment en compte les réalités sociologiques de l'organisation communautaire et villageoise dans les sociétés africaines. Dans un appel à la nation, le président Modibo Keïta[557] déclarait que « *le village est chez nous la cellule de base et c'est la vitalité de cette cellule qui engendrera la vitalité de la nation tout entière* ». À ce titre, la Constitution malienne du 22 septembre 1960 disposait en son article 41 que « *les collectivités territoriales de la République du Mali sont : les régions, les cercles, les arrondissements, les tribus nomades, les communes, les villages et les fractions nomades* ». Cette disposition constitutionnelle transformait toutes les subdivisions administratives en collectivités qui s'administrent librement par des conseils élus. À l'exception des communes qui existaient déjà, aucune de ces collectivités territoriales ne verra le jour.

229. - La commune, telle qu'elle est définie dans le droit africain de la décentralisation, ne semble avoir aucun lien avec la communauté locale[558]. L'histoire du droit administratif français enseigne que la commune française trouve son origine dans la conjuration bourgeoise aux XIe et XIIe siècles à travers le serment de commune. Il s'agit selon le propos d'Esmein d'« *une association sous la foi du serment, entre habitants d'une ville, pour se défendre mutuellement contre les agressions et les oppressions et pour empêcher entre eux les désordres et les violences. Elle porte souvent le nom caractéristique de pax, amicitia, foedus pacis* »[559]. Il est donc clair que la commune à l'origine est un phénomène purement communautaire qui a pris tout son sens avec l'œuvre révolutionnaire. La loi du 14 et 22 décembre 1789 en son article 7, en instituant « *une municipalité dans chaque ville, bourg, paroisse ou communauté de compagne* », a conçu un modèle simple pour l'organisation de la société reposant sur le vécu social. Ce caractère communautaire et spontané est introuvable dans le cadre de la création de la « commune africaine » depuis

---

556 H. ROUSSILLON, *Les structures territoriales des communes*, thèse, LGDJ, p. 224.
557 Modibo Keïta est le premier président du Mali indépendant. Il a dirigé le pays entre 1960 et 1968.
558 A. T. KOKO, *Les chefferies traditionnelles africaines face à la dynamique des réformes territoriales. Contribution à l'étude des processus de décentralisation, op.cit.*, p. 85.
559 Cité par A. T. KOKO, *op.cit.*, p. 85.

l'époque coloniale. Il aurait fallu, si l'on avait souhaité faire de la commune la collectivité infraétatique par excellence, l'organiser autour des unités villageoises et des chefferies traditionnelles africaines. Mais tel n'a pas été le cas aussi bien dans la période coloniale que celle après les indépendances. Les communautés traditionnelles de base ont été écartées au profit de la création de nouvelles circonscriptions artificielle et technocratique.

230. - La loi française relative à la réorganisation municipale en Afrique noire de 1955[560], qui a généralisé la pratique communale par la création des communes de plein et moyen exercice[561] dans les colonies françaises, continue d'être une source d'inspiration pour les textes de loi organisant la commune au sein des États d'Afrique noire francophone. Les législations nationales organisent les institutions communales autour de deux organes principaux : une assemblée délibérante, le conseil municipal et un exécutif composé du maire et des adjoints. Les conseillers municipaux sont élus au suffrage universel direct, libre, égal et secret[562] pour une durée allant de cinq ans au Mali, au Niger et au Burkina Faso, à six ans au Togo. L'article 7 du code des collectivités territoriales au Mali prévoit que « *le mandat du conseil communal est de cinq (5) ans. Toutefois, il peut être prorogé de six (6) mois par décret motivé pris en conseil des ministres. En cas de nécessité, il peut faire l'objet d'une seconde prorogation de six (6) mois par décret motivé pris en conseil des ministres* ». Au Niger, la prorogation ne peut intervenir qu'une seule fois, pour une durée de six mois par décret pris en conseil des ministres[563]. Ces prorogations interviennent généralement en cas de difficultés à organiser des élections municipales dans les délais prévus par la loi. En l'absence de pareilles dispositions, toute prorogation du mandat des conseillers doit être approuvée par le parlement, comme ce fut le cas au Burkina Faso avec l'adoption le 29 avril 2021 du projet de loi portant prorogation du mandat des exécutifs locaux.

Le nombre des conseillers municipaux varie en fonction de la taille de la population. Le chiffre de la population pris en considération est celui

---

560 Loi n° 55-1489 du 19 novembre 1955 relative à la réorganisation municipale en Afrique occidentale française, en Afrique équatoriale française, au Togo, au Cameroun et à Madagascar.

561 Dans la commune de plein exercice, le maire est désigné au sein du conseil municipal, tandis que dans la commune dite de moyen exercice, le maire est nommé par le chef de la circonscription administrative parmi les fonctionnaires de sa localité.

562 Art. 25 du CGCT au Niger.

563 Art. 23 du CGCT au Niger.

du dernier recensement général publié avant la tenue des élections[564]. Au Togo, il varie entre onze, pour les communes dont la population est inférieure ou égale à 50 000 habitants, et vingt-trois pour les communes dont la population est supérieure à 150 000 habitants. Au Mali, le nombre de conseillers varie entre onze, pour les communes d'au plus 10 000 habitants, et 45 pour les communes de plus de 200 000 habitants. Le nombre exact de conseillers à élire est fixé par décret pris en conseil des ministres au Togo ainsi qu'au Niger, et par arrêté du ministre en charge de décentralisation au Mali.

Le conseil municipal se réunit en session ordinaire, sous la présidence du maire, au moins une fois par trimestre. La durée d'une session ne saurait excéder cinq jours. En cas de convocation d'une session extraordinaire à l'initiative du maire, soit à la demande d'un président de commission permanente, soit d'au moins un tiers des membres du conseil, la durée de la session est de trois jours[565]. Le conseil municipal règle par ses délibérations les affaires de la commune. Il est en outre chargé de programmer et de mettre en œuvre les actions de développement de la commune en harmonie avec les orientations nationales[566]. En cas de dissensions graves entrainant un mauvais fonctionnement du conseil municipal, la dissolution est prononcée par décret motivé pris en conseil des ministres. Elle entraine la mise en place d'une délégation spéciale remplaçant provisoirement le conseil municipal avant l'organisation des prochaines élections municipales.

231. - L'exécutif du conseil municipal est assuré par le maire assisté par des adjoints. Le maire et les adjoints sont élus pour un mandat équivalent à celui des conseillers municipaux parmi les membres du conseil municipal à la première séance. Au Mali, le maire de la commune est investi sur la base des résultats définitifs des élections communales. C'est donc le conseiller figurant à la tête de la liste ayant obtenu le plus grand nombre de sièges qui est désigné comme maire[567]. Le nombre d'adjoints est déterminé en fonction de la taille de la population et du nombre de sièges au sein du conseil municipal. Ainsi, au Togo, le maire est assisté « *d'un adjoint pour les communes dont la population est inférieure ou égale à*

---

564 *Cf.* not., art. 5 du code des collectivités territoriales au Mali ; art. 28 du CGCT au Niger.

565 *Cf.* not., art. 234 du CGCT au Burkina Faso ; art. 28 du code des collectivités territoriales au Mali.

566 Art. 88 Loi n° 2019-006 relative à la décentralisation et aux libertés locales au Togo.

567 Art. 50 du code des collectivités territoriales au Mali.

50 000 habitants ; deux adjoints pour les communes dont la population est comprise entre 50.001 et 100 000 habitants ; trois adjoints pour les communes dont la population est comprise entre 100.001 et 150 000 habitants ; quatre adjoints pour les communes dont la population est supérieure à 150 000 habitants»[568]. Au Niger, pour les communes dont la taille du conseil varie de onze (11) à quinze (15) sièges inclus, le maire est assisté d'un adjoint et de deux (2) adjoints pour les communes de seize (16) sièges et plus[569].

Le maire est chargé de la mise en œuvre de la politique municipale dans l'intérêt supérieur de la commune et dans le respect des règles de l'État. Il dispose d'importants pouvoirs en tant que premier magistrat de la commune et représente également l'État au niveau de sa collectivité. Il bénéficie donc d'un dédoublement fonctionnel. L'importance et la complexité de ces fonctions justifient que certains États subordonnent l'accès aux fonctions exécutives à la satisfaction de certaines conditions. Au Niger, le législateur impose une condition de diplôme. Ainsi, l'article 52 du CGCT prévoit que « *le maire ainsi que le ou les adjoint (s) sont élus parmi les conseillers élus titulaires d'au moins le Brevet d'études du premier cycle (BEPC) ou d'un diplôme équivalent* ». Au Burkina Faso, selon les dispositions de l'article 255, alinéa 2 du CGCT, « *tout candidat aux fonctions de maire ou d'adjoint au maire doit remplir les conditions d'aptitude et de capacité intellectuelle, physique et morale nécessaires au bon accomplissement des missions et tâches d'administration des affaires communales* ». Ces conditions peuvent se révéler difficiles à remplir dans les milieux ruraux où le niveau d'alphabétisation est moins élevé.

En tant qu'agent de l'État, le maire exécute les lois et règlements dans la commune. Il est aussi à ce titre, officier de police judiciaire et officier d'état civil[570]. Pour l'ensemble de ces fonctions, le maire est soumis à l'autorité du représentant de l'État qui exerce sur lui un contrôle hiérarchique. Lorsqu'il agit au nom de la commune, le maire est placé sous le contrôle du conseil municipal ainsi que sous la tutelle administrative assurée par le représentant de l'État dans le département. Ordonnateur principal du budget, le maire est chargé notamment de l'exécution des

---

568 Art. 124 Loi n° 2019-006 *préc.*

569 Art. 50 du CGCT au Niger.

570 Voir not., art. 263 & 264 du CGCT au Burkina Faso ; art. 69 & 70 du code des collectivités territoriales au Mali.

délibérations du conseil municipal[571] et du maintien de l'ordre public dans la commune[572]. Il peut être suspendu ou révoqué[573].

232. - La commune prend en charge la satisfaction des besoins de proximité en matière d'hygiène et de salubrité publique. Elle assure le maintien des services publics indispensables. L'État lui transfère également des compétences en matière d'éducation, de culture, de sport, etc. L'exercice de l'ensemble de ces compétences attribuées implique que les communes disposent des moyens nécessaires. C'est là que se pose à nouveau la question du choix des priorités pendant les découpages communaux. L'organisation actuelle de la commune en Afrique, consistant à poser des structures institutionnelles sur des espaces territoriaux sans une réelle prise en compte des réalités sociohistoriques et économiques, emporte des conséquences quant à la survie même des communes.

§2.- Une carte communale en manque de rationalisation : la nécessité d'une coopération intercommunale

233. - L'émergence des territoires communaux dans les États d'Afrique noire francophone pose la question de la détermination du territoire pertinent pour l'organisation de la commune. Cette question a donné lieu à des réponses diverses à travers le continent. Le Togo, avec une superficie globale de 56 600 km², compte 117 communes[574]. La région maritime compte à elle seule 32 communes réparties sur un territoire de 6100 km². L'exiguïté du territoire n'a pas manqué de soulever des problèmes de délimitation des frontières communales. Si cet émiettement communal peut être considéré comme satisfaisant sur le plan de la démocratie locale, il l'est moins sur celui de la rationalité de la gestion. En effet, la multiplication des entités économiquement peu viables hypothèque les espoirs de développement. Le rapport sur la mise en œuvre de la décentralisation au Niger a montré, en se basant sur deux critères

571 Voir, art. 257 du CGCT au Burkina-Faso.

572 Art. 152 Loi n° 2019-006 *préc.* ; art. 81 du CGCT au Niger ; art. 261 du CGCT au Burkina Faso.

573 Voir les développements *infra* n°371 et s.

574 Loi n° 2017-008 du 29 juin 2017 portant création de communes au Togo.

élémentaires, que sur deux cent cinquante-deux (252) communes quatre-vingt-seize (96) sont économiquement non viables[575]. Les deux critères utilisés sont la capacité des collectivités à mobiliser les ressources pour la prise en charge du personnel minimal de la commune ainsi que les charges de l'exécutif, soit les indemnités de fonction, de l'eau, de l'électricité et du téléphone et les charges afférents au fonctionnement des commissions. En ne prenant en compte que les dépenses relatives aux traitements et indemnités du personnel communal, on constate un nombre élevé de communes en incapacité de prendre en charge leurs dépenses de fonctionnement.

La situation est la même au Burkina Faso et au Mali. Dans ce dernier cas, le choix d'une communalisation intégrale du territoire a conduit à l'adoption d'une technique de découpage communal considérée au départ par beaucoup d'observateurs comme novatrice. La Mission de décentralisation a laissé le découpage au libre jeu du regroupement des différentes communautés avec toutes les difficultés que cette stratégie pouvait engendrer pour la viabilité des collectivités naissantes. Ces regroupements devraient se faire sur la base de quatre critères : le maintien et la dynamisation des solidarités sociales, la viabilité économique de chaque territoire érigé en entité administrative, la logique géographique et spatiale et enfin, dans une certaine mesure, la prise en compte du découpage existant. Les travaux de la Mission de décentralisation se sont achevés en 1996 avec la création de 703 communes, dont 19 communes urbaines. Plus de la moitié des communes ont une population inférieure à 18 000 habitants[576]. Malgré le libre choix laissé aux populations par la Mission de décentralisation, la création des communes au Mali a donné lieu à des situations de conflits débouchant sur le périclitemment de l'État en 2012. L'extrême pauvreté de certaines collectivités a servi de prétexte à un repli identitaire entrainant une crise politique et sécuritaire que l'État malien peine à résorber depuis lors.

234. - Ainsi, la création des communes économiquement non viables se révèle être un obstacle rédhibitoire au développement économique et social qui est censé être promu par la décentralisation. Les collectivités territoriales doivent, à l'instar des personnes physiques, développer entre elles un système relationnel pouvant faciliter l'exercice en commun de

---

575 DGDCT, *Rapport final sur l'évaluation de la mise en œuvre de la décentralisation au Niger*, 2020, p. 121 et s.
576 PNUD, *Rapport sur le profil de pauvreté des 703 communes du Mali*, 2018, p.18.

leurs compétences[577]. Dès lors, la coopération intercommunale s'est imposée comme une première étape au renforcement de l'action des collectivités territoriales[578]. L'incitation à la coopération intercommunale repose sur un postulat simple. Ainsi que l'écrivait un auteur français, l'intercommunalité permet « *de remédier à l'inconvénient venant de ce que beaucoup de communes sont trop petites, dépourvues de ressources en hommes et en argent* »[579]. Il s'agit de constater que la commune n'offre pas l'optimum géographique nécessaire par la mise en place et la gestion d'un certain nombre de services publics locaux. Par conséquent, toute modalité de coopération entre communes est bienvenue afin d'élargir leur périmètre d'intervention et d'engranger des économies d'échelle. L'intercommunalité en tant que de modalité de coopération entre au moins deux collectivités territoriales au sein de l'espace national permet une mutualisation des ressources des collectivités pour la réalisation d'action d'intérêt commun. Mais, pour être efficace, elle doit dépasser le cadre de la simple mutualisation des moyens pour la gestion de services communs pour devenir celle d'un projet de développement de territoires partagé par les collectivités territoriales membres. Les communes trouvent ainsi le moyen de faire « communauté »[580] en liant leur destin, tout en fixant elles-mêmes le rythme de leur construction communautaire[581].

D'une manière générale, la coopération entre communes trouve sa justification à travers deux approches différentes. Dans une approche purement fonctionnelle, l'intercommunalité est appréhendée comme un moyen d'améliorer la qualité du service public rendu. Elle peut aussi être considérée comme un levier de réforme de l'organisation territoriale de l'État et des collectivités décentralisées. Dans cette optique, l'intercommunalité constitue une étape sur la voie de la disparition

---

577 I. DIALLO, *Le droit des collectivités locales au Sénégal*, thèse, L'Harmattan, 2007, p. 121 et s.

578 Sur la critique de l'émiettement et la solution de l'intercommunalité en droit français, N. KADA, « L'intercommunalité sous les feux de la rampe », *RDP*, n° 5, 2015, p. 1187.

579 L. ROLLAND, *Précis de droit administratif*, Dalloz, 1934, p. 175.

580 Au sens philosophique du terme, voir notamment les travaux de l'universitaire et écrivain sénégalais F. SARR, *Afrotopia*, Ed. Philippe Rey, 2016.

581 N. GBEOU-KPAYILE, « Essai sur le financement des collectivités locales dans les États d'Afrique noire francophone : Les cas du Sénégal et du Togo », *Revue béninoise des sciences juridiques et administratives*, n° 33, 2014, p. 96.

programmée des communes actuelles[582]. À terme, l'intercommunalité peut conduire à repenser l'organisation territoriale de l'État à travers un découpage communal autour de véritables zones économiques plus larges qui favoriseraient une amélioration de l'assiette fiscale des collectivités. La coopération entre communes peut non seulement permettre une gestion commune des services publics locaux ; être un outil du développement local et de l'organisation rationnelle des territoires, mais aussi un instrument au service d'une plus grande solidarité territoriale[583]. Dans ce contexte, l'intérêt de la collaboration semble évident au regard de la mission de développement local confiée aux collectivités territoriales. Il ne s'agit pas de faire de la coopération intercommunale une panacée, mais elle peut contribuer à réduire les inégalités entre les communes et accroitre les capacités financières des collectivités. La coopération intercommunale apparait alors comme une source de marge de manœuvre financière[584] pour les collectivités africaines et potentiellement une source d'augmentation du niveau de satisfaction des besoins des populations.

235. - Les législations nationales semblent avoir pris en compte toutes ces considérations en fixant le cadre de coopération entre les communes. L'article 325 du CGCT au Niger prévoit que « *les collectivités territoriales peuvent s'associer pour l'exercice de leurs compétences en créant des organismes publics de coopération. Cette coopération peut se traduire par la création de structures appropriées de promotion et de coordination d'actions intercommunales (…) entrant dans leurs domaines de compétence* ». La coopération intercommunale se déroule sous le contrôle de l'État[585]. La loi togolaise précitée du 26 juin 2019 précise que « *l'État veille à cette coopération entre les collectivités territoriales afin de garantir la solidarité nationale* ». La recherche d'un équilibre entre les

---

582 *Cf* P. AGNAMA-EBOUMI, *La décentralisation territoriale et le développement local au Gabon*, thèse, Université des sciences sociales de Toulouse, 2004, p. 299 ; N. KADA, « L'intercommunalité sous les feux de la rampe », *op.cit.*, p. 1190. Voir aussi pour une analyse critique des dernières réformes sur la question, M. DOUENCE, « La confortation du *statu quo* de la coopération intercommunale ou "beaucoup de bruit pour rien" » ?, *RFDA*, 2020, p. 223.

583 P. AGNAMA-EBOUMI, *La décentralisation territoriale et le développement local au Gabon, op.cit.*, p. 300.

584 N. GBEOU-KPAYILE, « Essai sur le financement des collectivités locales dans les États d'Afrique noire francophone : Les cas du Sénégal et du Togo », *op.cit.*, p. 96 ; L. TARTOUR, « L'intercommunalité, source de marges de manœuvre financière », *RFFP*, n° 119, 2012, p. 35.

585 Art. 25, Loi n° 2017-052 du 02 octobre 2017 déterminant les conditions de la libre administration des collectivités territoriales au Mali.

potentialités communales au niveau de chaque préfecture constitue un objectif de l'intercommunalité. Les structures intercommunales sont essentiellement constituées sous la forme d'établissements publics de coopération intercommunale dotés de la personnalité morale et d'une autonomie de gestion[586].

La coopération intercommunale prend diverses formes. Elle peut consister en une simple gestion[587] en commun des services à la population ou prendre la forme d'un véritable projet de développement territorial[588]. Dans le premier cas, il est créé un syndicat de communes qui matérialise la naissance d'une intercommunalité de gestion. Le syndicat des communes a donc pour vocation la gestion des services publics communs à plusieurs communes. Cette forme d'intercommunalité consacrée par le législateur togolais n'est pas partagée par les autres législations qui elles préfèrent la communauté de communes. En effet, cette dernière correspond au second cas présenté ci-dessus c'est-à-dire l'intercommunalité de projet. La communauté de communes est un établissement public de coopération intercommunale à caractère fédératif. Elle est constituée par l'association volontaire de plusieurs communes afin de mutualiser des moyens et des services au profit d'un projet commun à l'ensemble du territoire de l'établissement. Selon le décret n° 2016-301 du 29 juin 2016 fixant le régime juridique de la coopération entre les collectivités territoriales en République du Niger, l'objet de la communauté de communes est la création d'un espace de solidarité en vue de la réalisation d'un projet commun de développement économique et d'aménagement de l'espace. La création d'une communauté de communes est organisée par le pouvoir législatif et règlementaire.

Si la loi prévoit que les communautés de communes sont créées sur l'initiative des communes concernées[589], elle doit faire l'objet d'une approbation par l'autorité de tutelle. Au Burkina Faso, le gouverneur de région est l'autorité compétente pour adopter l'acte constitutif de la communauté de communes. L'arrêté d'approbation est publié au Journal

---

586 Art. 132 du CGCT au Burkina Faso.

587 Art. 39, Loi n° 2019-006 préc. : « La coopération de gestion est la forme de coopération par laquelle des communes décident de la création d'un établissement public territorial ayant mission de gérer pour le compte des communes créatrices, un ou des services publics d'intérêt commun, sous la forme de syndicat de communes ».

588 Ibid., art 40 : « La coopération de projet consiste en l'élaboration d'un projet commun de développement et d'aménagement de l'espace intercommunal, pour un ensemble de collectivités intéressées ».

589 Art. 42 de la loi n° 2019-006 *préc.*

officiel du Faso avec une mention obligatoire de l'objet, de l'identité des membres de la communauté ainsi que son siège social. Au Niger, le représentant de l'État dans la circonscription assure l'instruction préliminaire du dossier et le transmet au ministre en charge de la décentralisation. L'approbation de la création de la communauté est adoptée en conseil des ministres après présentation du projet de décret par le ministre de tutelle. La communauté de communes est donc créée au Niger par décret pris en conseil des ministres sur proposition du ministre en charge de la tutelle des collectivités territoriales.

L'EPCI ainsi créé ne peut exercer que les compétences définies dans ses statuts. Cette règle de spécialité se conjugue avec celle de l'exclusivité. « *Les communautés sont encadrées par le principe de spécialité fonctionnelle des communautés et le principe d'exclusivité de la compétence transférée* » dispose l'article 131, alinéa 3 du CGCT au Burkina Faso. Une fois la compétence transférée à la communauté de communes, la commune ne peut plus intervenir. La communauté de communes est chargée uniquement de la gestion des affaires d'intérêt communautaire. L'intérêt communautaire s'analyse comme « *la ligne de partage au sein d'une compétence entre les domaines d'action transférés à la communauté et ceux qui demeurent au niveau communal* »[590]. Il s'agit là d'un élément de distinction entre l'EPCI et les collectivités territoriales. Les pouvoirs de l'EPCI sont limités, ses organes dirigeants ne bénéficient pas non plus de la même légitimité démocratique que ceux des collectivités territoriales. Il n'existe pas de lien direct et propre entre les administrés et les organes de direction de l'EPCI.

236. - Il faut ici signaler que si la création des communautés de communes relève principalement de la volonté des collectivités concernées, le législateur togolais prévoit une hypothèse d'intercommunalités contraintes. En effet, la loi n° 2019-006 précitée impose aux communes issues d'un même ressort préfectoral de créer un cadre de coopération dénommé l'intercommunalité de préfecture. C'est le sens de l'article 30 de la loi qui dispose que « *les communes d'une même préfecture mettent ensemble obligatoirement un cadre de coopération* ». L'intercommunalité de préfecture apparait comme une instance de substitution à la suppression de l'échelon préfectoral en tant que collectivité décentralisée. Elle est dotée d'un organe délibérant et d'un bureau. La désignation des conseillers de l'intercommunalité de préfecture se fait parmi les membres des conseils municipaux. La détermination du nombre de sièges des conseillers se fait

---

[590] *Ibid.*, art. 131 al. 2.

sur la base du nombre total de conseillers de la commune la plus grande majoré de 25 %. Le nombre ainsi obtenu est divisé par le nombre de communes pour obtenir le nombre de sièges par commune devant siéger au sein de l'intercommunalité. Le nombre de sièges restant est réparti entre les communes sur la base d'un critère démographique[591]. Le bureau est composé de trois membres[592]. L'intercommunalité de préfecture est dirigée à tour de rôle par un représentant de chaque commune pour un mandant d'un an. Les deux autres membres du bureau sont soumis à la même règle de rotation.

237. - Le droit institutionnel de la coopération intercommunale, c'est-à-dire l'ensemble des règles qui établissent la manière dont les structures intercommunales sont constituées, leurs organes, leurs modes de fonctionnement, reste encore à perfectionner dans le contexte africain pour prendre en compte, notamment, les exigences de démocratisation locale. L'expérience française montre que l'EPCI rencontre beaucoup de succès et « *bouleverse durablement le paysage administratif français en limitant très fortement les effets négatifs de l'émiettement communal excessif* »[593]. La France compte à ce jour 1 253 EPCI et seules 27 communes ne sont pas membres d'un EPCI. La doctrine africaine place beaucoup d'espoirs dans l'intercommunalité[594]. Toutefois, ce procédé est encore peu connu des autorités décentralisées et surtout, les « guerres de domination » entre les communes jalouses de leur territoire exigu empêche l'aboutissement du processus de coopération des collectivités africaines[595]. Malgré les critiques qu'on peut adresser à l'intercommunalité, l'expérience française montre tout de même qu'on peut en faire un outil de gestion de l'action publique locale. Encore faudrait-il, dans le cadre des États d'Afrique francophone, réussir non seulement le pari d'en faire un outil efficace de conception et de gestion des projets, mais aussi d'asseoir la légitimité démocratique des instances intercommunales en révisant le mode de désignation. Si la

---

[591] Art. 31, Loi n° 2019-006 *préc.*

[592] *Idem.*

[593] N. KADA, « L'intercommunalité sous les feux de la rampe », *op.cit.*, p. 1190 ; B. FAURE, *Droit des collectivités territoriales*, *op.cit.*, n° 486, p. 402.

[594] P. AGNAMA-EBOUMI, *La décentralisation territoriale et le développement local au Gabon*, *op.cit.*, p.304 et s. ; N. GBEOU-KPAYILE, « Essai sur le financement des collectivités locales dans les États d'Afrique noire francophone : Les cas du Sénégal et du Togo », *op.cit.*, p. 96 et s.

[595] Voir le rapport final du colloque des maires de l'espace UEMOA sur « la problématique de l'intercommunalité dans l'espace UEMOA », octobre 2011.

commune fait l'objet d'une attention particulière des acteurs de la réforme décentralisatrice en Afrique, la place de l'échelon préfectoral en tant que collectivité territoriale est de plus en plus contestée dans l'organisation administrative des États africains.

Section 2 : L'ambivalence assumée de l'institution préfectorale

238. - La préfecture équivalent français du département a gardé pendant longtemps la double nature de circonscription administrative et de collectivité décentralisée dans les États d'Afrique noire francophone. En même temps qu'elle relève de l'organisation déconcentrée de l'État à travers la figure du préfet, la préfecture dispose également d'un conseil préfectoral avec des conseillers élus. L'institution préfectorale a fait l'objet de nombreuses critiques doctrinales[596]. Elle est perçue comme un échelon de trop qui affecte l'efficacité de l'action publique locale en Afrique. Aussi, en raison de ces origines historiques et des justifications pratiques de son existence, la collectivité préfectorale n'a pas cessé d'essuyer les critiques entrainant sa disparition progressive dans les pays de la sous-région ouest-africaine (§**1**). Cependant, malgré l'effacement progressif de la préfecture dans les pays objet de cette recherche, elle semble avoir encore un avenir certain au Mali où elle porte la dénomination de Cercle (§**2**).

§1.- La suppression de la collectivité préfectorale

239. - Les interrogations sur la pertinence d'une architecture territoriale à trois niveaux ont très vite fait apparaitre l'échelon préfectoral comme le niveau de trop de l'organisation décentralisée. Plusieurs raisons ont pu être avancées pour justifier la suppression de ce niveau de collectivité territoriale : l'origine historique, la simplification des circuits administratifs

596 S. PAGNOU, « La place du contrôle de légalité dans les processus de bonne gouvernance liés à la décentralisation. Le cas du Togo », *Revue internationale des sciences administratives*, vol. 79, n° 3, 2013, p. 605 ; M. TOUZEIL-DIVINA, « Voyage départemental dans la doctrine publiciste et française », *Pouvoirs locaux*, n° 116, 2019, p. 40 ; N. FERREIRA, L. JANICOT (dir.), *Quel(s) avenir(s) pour les départements ?* LGDJ, coll. « LEJEP », 2017. Voir aussi, Dossier : Département : état des lieux, *AJCT*, 2019, p. 317 ; L. JANICOT, « Les départements : quel(s) avenir (s) », *AJDA*, 2019, p. 2417.

et surtout la réalisation d'économies ainsi que l'amélioration de l'efficacité de l'action publique.

240. - L'échelon préfectoral est un héritage de la colonisation française. En effet, le cercle était l'institution centrale de toute l'organisation coloniale en Afrique occidentale française où l'utilisation de ce terme servait à rappeler la nature militaire de l'acquisition coloniale[597]. La construction de l'empire colonial s'est donc réalisée à travers l'institution des cercles de 1895 à 1946. Devenu une subdivision politique équivalente au département au moment des indépendances, le nom de Cercle a été modifié par les États nouvellement indépendants pour en finir avec cette connotation impériale. Ainsi, les cercles ont été transformés en préfecture ou département selon les pays. À ce jour, le Mali est la seule nation de l'Afrique noire francophone à avoir maintenu la dénomination de cercle. Cet héritage colonial fait apparaitre l'institution préfectorale comme un outil de centralisation du pouvoir d'autant plus qu'après le changement de nom, cet échelon intermédiaire entre la commune et la région a continué par recouvrir deux réalités administratives distinctes. La préfecture était une circonscription administrative de l'État avec le préfet comme autorité responsable et une collectivité territoriale avec l'institution d'une assemblée délibérante et d'un président de conseil préfectoral.

Au Togo, les préfectures sont nées de la réforme administrative du 23 juin 1981. Cette réforme a conduit à la transformation des circonscriptions administratives coloniales en préfectures avec l'adoption de la loi n° 81-8 du 23 juin 1981 portant organisation territoriale. L'adoption de cette loi marque le début de la coexistence de la préfecture en tant que circonscription administrative et collectivité territoriale. L'article 10 de la loi du 23 juin 1981 prévoit que « *la préfecture est une collectivité territoriale dotée de la personnalité morale et de l'autonomie financière* ». Elle est administrée par un conseil de préfecture qui élit en son sein un président et un comité permanent[598]. Il faut néanmoins relever que, assez curieusement, malgré l'institution d'une assemblée délibérante et d'un président du conseil élu, c'est le préfet qui assurait en réalité les fonctions

---

597 R. F. BETTS, « La domination européenne : méthodes et institutions » in, A. ADU BOAHEN (dir.), *Histoire générale de l'Afrique, VII : L'Afrique sous domination coloniale, 1880-1935*, Éditions UNESCO, 1987, p. 344.

598 Art. 11 & 12 Loi n° 81-8 du 23 juin 1981 *préc.*

exécutives et administratives au sein de la collectivité[599]. Le préfet représentait à la fois l'État et les populations sur le territoire concerné. C'est le sens des dispositions de l'article 14 de la loi de 1981 : « *le préfet exerce dans la préfecture les pouvoirs de police administrative. Il peut prendre toute décision pour assurer l'ordre et la tranquillité publique. Il représente la population et fait connaitre ses besoins et aspirations à l'autorité hiérarchique* ».

Cette organisation superposée, considérée par certains auteurs comme innovante[600], entraine davantage de confusion dans la distinction de ces deux entités. Il ne s'agit pas non plus d'une particularité introduite par le législateur togolais pour prendre en compte les particularités locales, ainsi que semble l'expliquer Adiki Tovenim Koko dans sa thèse[601]. Cette confusion, voulue et souhaitée par le législateur, visait à promouvoir l'unité de l'État afin d'asseoir un système fortement hiérarchisé et centralisé comme ce fut le cas au Niger[602]. La Commission spéciale de « redécoupage administratif » mise en place par un arrêté ministériel de janvier 1995 au Niger a abouti à la construction d'un dispositif territorial à trois étages : commune, département et régions[603]. La loi n° 98-30 du 14 septembre 1998 qui crée 36 départements fixe leur limite et le nom de leurs chefs-lieux reconnait aux départements un double statut de circonscription administrative déconcentrée et de collectivité territoriale autonome. L'autorité du préfet déteint sur celle du conseil préfectoral et de son président. C'est l'une des raisons pour lesquelles des voix se sont élevées pour demander la suppression pure et simple de l'échelon départemental, qui n'est une collectivité territoriale que de nom.

241. - Une réorganisation de l'administration territoriale autour de deux niveaux de collectivités : commune et région se sont imposées dans la majorité des États de l'Afrique noire francophone. L'évolution des structures administratives est apparue comme la condition *sine qua non* de

---

599 M. S. AMLALO, « La décentralisation en marche au Togo », Article disponible sur www.afriquegouvernement.net, Mars 2007, p. 4. La loi de 1981 fait coexister « *la circonscription administrative dirigée par le préfet et la collectivité décentralisée disposant d'une assemblée délibérante élue (le Conseil de Préfecture) et d'un Président de Conseil élu, mais c'est le préfet qui est chargé des fonctions exécutives et administratives* ».

600 A. T. KOKO, *Les chefferies traditionnelles africaines face à la dynamique des réformes territoriales : contribution à l'étude des processus de décentralisation, op.cit.*, p. 99.

601 *Idem.*

602 M. DANDA, *Niger, une décentralisation importée. Une comparaison avec la France*, L'Harmattan, 2013, p. 76.

603 *Cf.* S. HALIDOU, « Une décentralisation inachevée au Niger », *Télescope*, vol. 11, n° 3, 2004, p. 9.

la réussite du processus de décentralisation[604]. Il est impératif de faire évoluer les structures administratives pour atteindre les objectifs d'efficacité de l'action publique portés par les réformes décentralisatrices. La suppression de la collectivité préfectorale a donc pour but de supprimer certains frais liés à la gestion de cette collectivité, notamment les indemnités des élus, les frais de personnel, etc. L'empilement des trois niveaux de collectivités est considéré comme une source d'accroissement inutile des dépenses de fonctionnement, en plus de l'absence d'opérationnalité des trois niveaux de collectivités du fait notamment de leur déconnexion du contexte socio-économique et de leur inadéquation par rapport à la réalité des pays[605]. Ainsi, la suppression de la préfecture devait permettre la réalisation des économies substantielles en réaffectant les compétences et les ressources préfectorales entre les communes et les régions. En effet, le personnel préfectoral devra en cas de suppression être redéployé au sein des communes et des régions renforçant ainsi les capacités en ressources humaines de ces dernières. La commune et la région pourront ainsi remplir leur vocation de contributeur au développement local et régional.

242. - C'est le choix opéré par le Burkina Faso dès 2004, au moment de l'adoption de la loi n° 055-2004 du 21 décembre 2004 portant code général des collectivités territoriales. Les réajustements introduits par le CGCT burkinabè ont conduit à la consécration de deux niveaux de collectivités : commune et région, et à la suppression de la province en tant que collectivité territoriale. Au Niger, l'ordonnance n° 2010-54 du 17 septembre 2010 portant code général des collectivités territoriales a également procédé à la suppression des départements en consacrant en son article 2 que « *les collectivités territoriales sont : la commune et la région* ». Les modifications de l'architecture de l'administration locale ainsi intervenues n'ont pas connu de résistance particulière en raison notamment de l'absence de constitutionnalisation de l'échelon intermédiaire dans ces pays.

604 S. PAGNOU, « La place du contrôle de légalité dans les processus de bonne gouvernance liés à la décentralisation. Le cas du Togo », *op.cit.*, p. 620.

605 Le Doyen Vignon a soulevé, il y a déjà plusieurs années, le problème de la viabilité des collectivités locales dans les mêmes termes. Y. B. VIGNON, « Les aspects structurels de la mal administration au Togo », *Annales de l'Université du Bénin*, Tome XVII, 1997-1998, p. 75. Cité par S. PAGNOU, « La place du contrôle de légalité dans les processus de bonne gouvernance liée à la décentralisation. Le cas du Togo », *op.cit.*, p. 620.

Il faut rappeler qu'en France, le débat sur la suppression du département s'est enlisé en raison notamment de l'inscription de l'institution départementale en tant que collectivité territoriale à l'article 72 de la Constitution. Sa suppression nécessiterait donc une révision constitutionnelle dont la mise en œuvre n'est jamais aisée, surtout au regard du contexte politique français. La fin annoncée du département depuis 2008 par le rapport Attali[606] n'aura pas eu lieu, car le gouvernement issu de l'alternance politique de 2012 n'a pas souhaité s'engager dans la voie de la révision. Le Premier ministre d'alors affirmait à ce propos que l'on avait encore « *besoin de cet échelon intermédiaire entre la région et la commune* »[607]. Le département en France est originellement conçu pour être au service de l'État et non directement des populations. Il est historiquement, écrit Jean-Marie Pontier, marqué par cette ambiguïté fondamentale, car il n'avait été pensé que pour être une simple circonscription administrative dont l'exercice des fonctions est placé sous l'autorité et l'inspection du Roi[608]. L'action départementale est alors conçue comme une collaboration à l'administration générale du Royaume d'où les soubresauts que connait l'institution[609]. Si les dernières réformes législatives notamment, la loi RCT de 2010, la loi MAPTAM de 2014 et la loi NOTRe de 2015 ont « dévitaliser » les départements ; tous les projets visant leur suppression n'ont pas abouti et semblent même abandonnés garantissant ainsi l'avenir des départements[610].

À la différence de la France, le Togo, qui se retrouvait dans la même situation, puisque la Constitution de 1992 avait inscrit en son article 141 la préfecture en tant que collectivité territoriale, la suppression de cette dernière s'est faite sans heurts. Le débat sur le projet de révision constitutionnelle s'étant porté davantage sur la question de la limitation du mandat présidentiel, la modification des structures administratives n'a retenu que très peu l'attention de la classe politique. Ainsi, la loi n° 2019-003 du 15 mai 2019, adopté à la majorité des quatre cinquièmes des députés composant l'Assemblée nationale, a procédé à la modification des dispositions de l'article 141 de la Constitution togolaise. Il dispose

---

606 Rapport de la commission pour la libération de la croissance française, présidée par Jacques Attali, 300 décisions pour changer la France, janv. 2008, décis. 19.

607 N. FERREIRA, L. JANICOT (dir.), *Quel(s) avenir(s) pour les départements ?* LGDJ, Coll. « LEJEP », 2017, p. 8.

608 J.-M. PONTIER, « Une approche historique », in N. FERREIRA, L. JANICOT (dir.), *Quel(s) avenir(s) pour les départements* ? LGDJ, Coll. « LEJEP », 2017, p. 14.

609 *Ibid.*, p. 16.

610 L. JANICOT, « Les départements : quel(s) avenir (s) », *op.cit.,* p. 2417.

désormais que « *la République togolaise est organisée en collectivités territoriales sur la base du principe de décentralisation, dans le respect de l'unité nationale. Ces collectivités territoriales sont : les communes et les régions* ». À la suite de l'adoption de la loi de révision constitutionnelle, le législateur a procédé à l'abrogation des dispositions relatives à l'organisation de la collectivité préfectorale, en adoptant la loi n° 2019-006 portant modification de la loi n° 2007-011 du 13 mars 2007 relative à la décentralisation et aux libertés locales modifiée par la loi n° 2018- 003 du 31 janvier 2018.

243. - Les origines historiques de l'échelon préfectoral, ainsi que la volonté d'amélioration de l'efficacité de l'action publique, ont renforcé les critiques à l'égard de cet échelon entrainant sa suppression dans de nombreux États de l'Afrique noire francophone. Toutefois, cet échelon intermédiaire entre la commune et la région continue d'exister dans certains États, notamment au Mali.

§2.- L'exception malienne du maintien d'un échelon intermédiaire : le cas des cercles

244. - Pendant la période coloniale, le cercle constituait l'échelon de base de l'administration coloniale dirigé par un officier de la métropole, le commandant de cercle. La vie politique et administrative de l'empire colonial français s'est développée très largement dans le cadre du cercle. En 1887 fut créé dans le Soudan français, l'actuel territoire du Mali, le premier cercle dans la localité de Bafoulabé. Dans les années qui ont suivi l'accession à la souveraineté internationale des pays de l'ancienne Afrique-Occidentale Française (AOF), le terme de « Cercle » a été remplacé par celui de « Préfecture » ou « Département ». Le Mali a quant à lui fait le choix non seulement du maintien de la dénomination coloniale, mais aussi de la préservation de cette institution en tant que collectivité territoriale. La loi n° 2017-051 du 2 octobre 2017 portant code des collectivités territoriales au Mali dispose à l'alinéa 1er de l'article 78 que « *le Cercle est une collectivité territoriale dotée de la personnalité morale et de l'autonomie financière. Il est composé de communes* ». Instrument de centralisation, les cercles issus de la période coloniale constitués d'un regroupement de cantons étaient conçus comme un outil administratif de gestion du territoire de manière uniforme. Le cercle au Mali est un cadre de mise en cohérence des stratégies et des actions de développement des communes qui le composent[611]. C'est ce caractère de pôle susceptible de se voir confier différentes missions en fonction des choix politiques qui en fait une institution pivot dans la stratégie gouvernementale.

245. - En tant que collectivité territoriale, le cercle bénéficie d'une organisation comparable à celle des communes avec un conseil de cercle et un bureau du conseil de cercle composé du président et des vice-présidents. Les conseillers de cercle sont élus au suffrage universel direct pour une durée de cinq (5) ans au scrutin de liste à la représentation proportionnelle sans panachage ni vote préférentiel[612]. Toutefois, le code des collectivités territoriales prévoit que le mandat des conseillers de cercle «*peut être prorogé de six (6) mois par décret motivé pris en conseil des ministres. En cas de nécessité, il peut faire l'objet d'une seconde prorogation de six (6) mois par décret motivé pris en conseil des ministres*»[613]. Le nombre de conseillers est déterminé en fonction d'un critère démographique en prenant en compte le chiffre de la population issue du dernier recensement administratif. Le nombre des conseillers de cercle varie entre 27 conseillers pour les cercles ayant au plus 100 000 habitants et 41 conseillers par les cercles de plus de 300 000 habitants[614]. Conformément à l'article 100 de la loi du 2 octobre 2017 portant code des collectivités territoriales, « *le conseil de cercle se réunit en session ordinaire une fois par trimestre sur convocation du président. Celui-ci peut, toutefois, le convoquer en session extraordinaire lorsque les circonstances l'exigent. Il est tenu, en outre, de le convoquer à la demande d'un tiers (1/3) des conseillers ou par le représentant de l'État dans le Cercle* ». Les délibérations du conseil du cercle sont exécutoires de plein droit dès leur publication et leur transmission au représentant de l'État dans le cercle. Toutefois, les délibérations portant sur : les budgets et le compte administratif, l'aliénation des biens du patrimoine et les emprunts de plus d'un an ne sont exécutoires qu'après approbation par le représentant de l'État dans le cercle[615]. Le conseil de cercle peut être suspendu par arrêté du ministre en charge de la décentralisation ou dissous par décret motivé pris en conseil des ministres[616].

---

611 Art. 78, al. 2 du code des collectivités territoriales au Mali.
612 Art. 198 Loi n° 2022-019 du 24 juin 2022 portant loi électorale au Mali.
613 Art. 80 du code des collectivités territoriales au Mali.
614 *Ibid.*, art. 79.
615 *Ibid.*, art. 96.
616 Sur la question du contrôle de l'État sur l'action des collectivités territoriales, voir les développements *infra* n°368 et s.

246. - Lors de la première séance suivant l'élection du conseil, il est procédé à l'installation du président du conseil de cercle sur la base des résultats définitifs des élections de cercle. Est investi président du conseil de cercle le conseiller figurant à la tête de la liste ayant obtenu le plus grand nombre de sièges. Une fois investi, le président du conseil de cercle assure la présidence du conseil de cercle qui procède à l'élection des vice-présidents au scrutin uninominal majoritaire. La différence entre le conseil de cercle et le conseil départemental en France va au-delà de la simple dénomination pour concerner les modalités mêmes de désignation du bureau du conseil. Le droit positif français consacre l'élection du président du conseil départemental à la majorité absolue des membres du conseil départemental pour les deux premiers tours et à la majorité relative pour le suivant. Aussitôt après l'élection du président et sous sa présidence, le conseil départemental fixe le nombre des vice-présidents et des autres membres de la commission permanente. Il est procédé à l'élection des membres de la commission permanente qui est composée du président du conseil départemental, de quatre à quinze vice-présidents et éventuellement d'un ou plusieurs autres membres au scrutin de liste[617].

Le président du conseil de cercle est le chef de l'exécutif et de l'administration de la collectivité territoriale de cercle. Il assume de multiples fonctions. Il prépare et exécute les délibérations du conseil de cercle. Il prépare et exécute le budget du cercle ; assure la gestion du personnel. Le président du conseil de cercle gère le domaine du cercle, à ce titre, il exerce les pouvoirs de police afférents à cette gestion[618]. Il peut également recevoir l'autorisation du conseil de cercle pour la durée de son mandat afin d'arrêter ou de modifier l'affectation des propriétés du cercle utilisées par les services du cercle ; de fixer, dans les limites déterminées par le conseil de cercle, les tarifs des redevances, etc.[619]

En ce qui concerne les compétences du cercle, la loi précitée du 2 octobre 2017 prévoit en son article 95 que « le conseil de cercle règle par ses délibérations les affaires du cercle, notamment celles relatives au développement économique, social et culturel ». Le cercle dispose de compétences dans le domaine de l'enseignement par la prise en charge des collèges relevant l'enseignement secondaire général, de l'apprentissage, de

---

617 *Cf.* art. L. 3122-1, L. 3122-4, L. 3122-5 du CGCT en France.
618 Art. 134 du code des collectivités territoriales au Mali.
619 *Ibid.*, art. 135 & 136.

la formation professionnelle… ; de la santé ; de la jeunesse, des sports, des arts et de la culture[620].

247. - L'échelon intermédiaire entre la commune et la région quelle qu'est la dénomination qu'on lui donne : « Préfecture », « département » ou « Cercle » connait une destinée singulière dans les États de l'espace ouest-africain francophone. Créée au cours de la période coloniale, elle a fait l'objet de critiques et de remises en cause permanentes en écho au débat sur le maintien ou non de la collectivité départementale en droit français. Beaucoup de pays africains, notamment le Burkina Faso, le Niger et le Togo, ont franchi le pas de la simple contestation en supprimant purement et simplement cet échelon de l'architecture de l'administration locale. « Vilipendé et méprisé »[621] ailleurs, cet échelon semble trouver une nouvelle jeunesse dans le droit des collectivités territoriales au Mali. L'absence de consensus sur le maintien ou la suppression de l'échelon départemental est contrebalancée par la volonté unanime des décideurs politiques des pays objet de la présente recherche de promouvoir l'institution régionale.

Section 3 : <u>La consécration progressive de l'échelon régional</u>

248. - La région occupe une place incontestable aujourd'hui dans l'administration territoriale des États d'Afrique subsaharienne francophone. Dénoncée au départ par certains États d'Afrique francophone[622], la régionalisation est de plus en plus perçue comme une « thérapie » pour calmer les courants centrifuges à portée autonomiste[623].

---

620 *Ibid.*, art. 95.

621 Ces expressions ont été utilisées par les auteurs Auby et Pontier pour désigner la situation du département en droit français. *Cf.* F. AUBY, J.-M. PONTIER, *Le département*, Economica, coll. « collectivités territoriales », 1988, p. 5.

622« Toute propagande particulariste de caractère régionaliste, raciale ou ethnique, toute manifestation de discrimination raciale, sociale, sexiste, ethnique, politique ou religieuse sont punies par la loi ». Cf. article 8 alinéa 4 de la Constitution nigérienne de 2010. Le Tchad dénonce après « le régionalisme », « le tribalisme, le népotisme, les inégalités sociales » condamnant encore « la propagande à caractère tribal et régional » (article 5).

623 E. MOUBITANG, « la régionalisation constitutionnel au Cameroun : étude d'une dynamique d'intégration nationale à l'épreuve du terrorisme international », *Revue numérique Afrilex*, 2021, p. 5.

Ainsi, la tendance à la régionalisation correspond en Afrique subsaharienne francophone à une «*philosophie ou une idéologie juridique par laquelle le bonheur devrait arriver c'est-à-dire la liberté et la paix*»[624]. Elle apparait désormais comme le pôle pertinent de développement et de restauration de la paix. L'objectif est aussi de faire de la région un nouvel espace de définition et de mise en œuvre des politiques publiques locales. L'échelon régional semble progressivement s'affirmer comme une autorité supérieure au sein des collectivités territoriales. Elle se voit confier la planification, l'adoption des plans de développement, du plan régional d'aménagement en plus du transfert des compétences étatiques en matière de sport, de tourisme et d'action culturelle. Cette orientation novatrice en faveur d'un leadership régional pour une administration décentralisée cohérente axée sur la recherche de la performance et de l'efficacité n'est pas sans rappeler les efforts entrepris par le législateur français en ce sens depuis la loi NOTRe de 2015[625]. La région, qui jusque-là n'avait occupé aucune place véritable dans l'ordre politique, devient un enjeu dans la relance du processus de décentralisation et de restauration de l'État (**§1**) avec le renforcement de ces organes (**§2**).

§1.- Le développement de la régionalisation en Afrique

249. - La région est aujourd'hui considérée comme l'échelon pertinent de mise en œuvre de projets d'envergure. D'une idée prosaïque associée à la crainte de l'État unitaire de voir son pouvoir et son autorité contestée, par des entités aux assises territoriales aussi grandes que les régions, s'est développée une idée plus conciliante. La régionalisation en Afrique est le fruit d'une longue évolution à la fois des esprits et des institutions. La notion de régionalisation, rappelle Jean-Marie Pontier, se situe à un double niveau, administratif et économique, non pas qu'elle ne soit pas politique, mais elle ne l'est que modestement[626]. Elle est, poursuit l'auteur, «*fréquemment comprise comme l'action du pouvoir central tendant à adapter certaines*

---

624 A. KPODAR, « Prolégomènes à une virée constitutionnelle en Afrique noire francophone : une approche de théorie juridique », in *Les voyages du droit, Mélanges en l'honneur de Dominique Breillat,* LGDJ, p. 331. Voir aussi, N. GBEOU-KPAYILE, « L'idée de Constitution en Afrique noire francophone », *Revue numérique Afrilex*, 2014, p. 4.
625 B. FAURE, « Le leadership régional : nouvelle orientation du droit des collectivités territoriales ? », *AJDA*, 2015, p. 1898.
626 J.-M. PONTIER, « Région : collectivité territoriale », in *Encyclopédie des collectivités locales*, folio n° 1710, actualisé en mai 2022.

de ses actions dans un cadre régional. Appliqué à l'ordre administratif et juridique, le régionalisme semble tout naturellement conduire à la décentralisation. La décentralisation par la création d'une collectivité régionale est logiquement la conséquence de la doctrine régionaliste. C'est ainsi d'ailleurs que cela sera entendu à notre époque »[627]. Cette doctrine régionaliste conciliante vis-à-vis de l'État unitaire et de sa refondation a fait progresser l'idée régionale en Afrique au fil de l'évolution des dispositions constitutionnelles et législatives (**A**) pour faire d'elle un échelon de valorisation des potentialités locales (**B**).

A- De l'idée régionale à la consécration de la collectivité régionale

250. - La régionalisation nait des revendications par certaines parties du territoire d'une reconnaissance de particularismes qui marquent l'identité d'une population. Parce qu'il traduit une volonté de remise en cause de l'unité républicaine, elle cristallise l'opposition entre ceux qui, d'un côté, craignent que la consécration des régions ne nuise à l'unité de la République et ceux qui, de l'autre côté, brandissent le fait régional pour appeler à une nouvelle organisation territoriale de la République. Certains auteurs voient d'ailleurs dans la régionalisation une avancée de l'État unitaire vers un État fédéral[628]. L'opposition entre la régionalisation et le fédéralisme n'étant plus en raison de la naissance de la collectivité régionale qu'une question de gradation[629]. Il suffit pour s'en convaincre de se référer aux revendications des indépendantistes corses qui ont obligé l'État français à y apporter des réponses institutionnelles[630] poussant le plus loin possible le régionalisme. Toutefois, il faut noter que l'essor de l'idée régionale est d'abord fortement marqué en France, comme dans les États de l'espace ouest-africain, par des préoccupations d'ordre économique (**1°**) avant son érection en tant que collectivité territoriale (**2°**).

1. Le développement historique de l'idée régionale

251. - La naissance de l'idée régionale en Afrique de l'Ouest est liée à son développement historique en France. La région apparait en France

627 *Idem.*

628 R. PASQUIER, « Les régions dans la réforme territoriale : vers un fédéralisme à la française », *AJCT*, 2016, p. 71.

629 J.-M. PONTIER, « Région : collectivité territoriale », in *Encyclopédie des collectivités locales, op.cit.*, § 22.

630 M. DOUENCE, « La Corse », *RFDA*, 2016, p. 645.

dans l'entre-deux-guerres comme un échelon pertinent pour la réforme économique de l'action de l'État. C'est par une circulaire du 25 août 1917 que sont mises en place les premières institutions régionales françaises. Du nom du ministre qui les a créées, les « régions Clémentel » sont tournées vers la coordination et l'interventionnisme économique de l'État[631]. Sur la base de cette circulaire, les compagnies consulaires furent organisées en regroupements volontaires autour des chambres de commerce[632]. Pour Joseph Berthélémy, ces groupements ont eu une grande utilité, car « *leur concours a permis de pratiquer une décentralisation, accueillie avec faveur dans tous les domaines. C'est ainsi qu'est organisée avec persévérance la mobilisation économique, c'est-à-dire la coordination des efforts industriels et commerciaux vers le même but : la meilleure répartition des produits de chaque région et l'utilisation plus rationnelle de sa puissance* »[633]. Toutefois, leurs faibles moyens limitaient sérieusement leur capacité d'action et d'intervention à grande échelle. Ces groupements se sont vus reconnaitre la qualité d'établissements publics avec le décret du 14 juin 1938. Mais, elles disparaitront en septembre 1938 avec la création de nouvelles régions dites « régions Gentin »[634] qui confirme la vocation économique des régions. Le gouvernement de Vichy mettra en place en 1941 des préfets de régions en charge des circonscriptions administratives d'action qui ne constituaient en rien des collectivités territoriales. C'est pendant la décennie 1950-1960 que la question régionale va ressurgir à travers le problème de spatialisation des activités économiques c'est-à-dire de l'aménagement du territoire.

252. - Les élites politiques et économiques vont alors se saisir de cette occasion pour faire de la planification un outil indispensable à l'aménagement du territoire[635]. Le processus de régionalisation va connaitre une ampleur considérable avec le passage de la IVe à la Ve République sous la présidence du Général de Gaulle. La planification régionale et la réduction des inégalités sont au cœur de son projet de modernisation. Il était question de faire de la région le niveau où l'administration pourra nouer et entretenir le dialogue avec les nouvelles couches sociales qui émergent c'est-à-dire les « forces vives » dont le

---

631 P. JAN, *Institutions administratives*, LexisNexis, coll. « Objectif cours », 2021 p. 103.
632 J.-M. PONTIER, « Région : collectivité territoriale », *op.cit.*, § 31.
633 J. BERTHELEMY, *Traité de droit administratif*, 10e éd., cité par J.-M. PONTIER, *op.cit.*, § 31.
634 J.-M. PONTIER, « Région : collectivité territoriale », *op.cit.*, § 34.
635 R. PASQUIER, « Les régions dans la réforme territoriale : vers un fédéralisme à la française », *op.cit.*, p. 71.

régime est soucieux de s'assurer le soutien. Il s'agissait de construire un espace fonctionnel d'étude et de délibération afin de construire des pôles de développement équilibrés au sein de l'espace national[636]. L'échec du projet de référendum de 1969 qui entendait faire de la région une collectivité territoriale à part entière dotée de moyens propres, avec une assemblée d'élus, marque la fin de la régionalisation fonctionnelle[637].

253. - Ce foisonnement d'idées autour du fait régional a reçu un écho favorable dans les anciennes colonies françaises d'Afrique occidentale. Ainsi, la régionalisation en Afrique suit dans les grands traits les mêmes évolutions qu'en France. L'idée régionale a progressé de sa conception en tant que circonscription administrative à vocation économique pour devenir une collectivité territoriale. Les États d'Afrique noire francophone vont très vite constitutionnaliser l'échelon régional même si son existence effective tarde à venir en l'absence d'élection des conseils régionaux. C'est le cas notamment du Mali qui dès son accession à la souveraineté internationale a entendu accorder à la région le statut de collectivité territoriale. Les pères de l'indépendance malienne s'étaient fermement engagés dans un projet ambitieux de décentralisation. Ainsi, la Constitution malienne du 22 septembre 1960 prévoit en son article 41 que *« les collectivités territoriales de la République du Mali sont : les régions, les cercles, les arrondissements, les tribus nomades, les communes, les villages et les fractions nomades. Toute autre collectivité territoriale est créée par la loi. Ces collectivités s'administrent librement par des conseils élus et dans les conditions prévues par la loi. Dans les collectivités territoriales, le délégué du gouvernement a la charge des intérêts de la République, du contrôle administratif et du respect des lois »*.

254. - La création de régions par la Constitution de la première République malienne intervient dans la continuité du projet de création de provinces entamé par les autorités coloniales en 1947 dans le but d'améliorer le fonctionnement des structures coloniales[638]. La mention de la région en tant que collectivité territoriale dans la Constitution malienne de la première République n'est ni anodine ni hasardeuse. Cette érection constitutionnelle de la région s'explique par deux raisons. D'abord, la prégnance continue des idées de l'ex-métropole et son impact sur les élites

---

636 *Idem.*

637 R. PASQUIER, « La régionalisation française revisitée. Fédéralistes, mouvement régional et élites modernisatrices, *RFSP,* vol. 53, n° 1, 2003, p. 101.

638 M. DJIRE, A. KEITA, *Régionalisation au Mali. Regards croisés*, Éditions Tombouctou, 2015, p. 38.

administratives et politiques africaines. Ensuite, et c'est la principale raison, qui n'est pas dépourvue de tout lien avec la première, c'est l'organisation en un État fédéral du Mali entre 1959 et 1960. En effet, la république soudanaise (actuel Mali) formait alors avec le Sénégal la fédération du Mali. Les deux textes législatifs relatifs à l'organisation administrative et territoriale adoptée par cette fédération consacrent les régions au titre de collectivité territoriale. La loi n° 60-3/AL-RS du 7 juin 1960 portant organisation territoriale de la République soudanaise, promulguée par le décret n° 30/PC du 21 juin 1960, a structuré le territoire en région, cercle, arrondissement, commune et village pour les populations sédentaires ; en tribu et fraction pour les populations nomades. La loi n° 60-5-AL-RS du 7 juin 1960 portant organisation des régions et des assemblées régionales consacre, quant à elle, le statut juridique des régions comme collectivités publiques dotées de la personnalité civile et de l'autonomie financière. Au titre des institutions de la région, cette loi prévoyait l'existence d'un gouverneur de région et d'une assemblée régionale. La région était alors organisée en cercles, arrondissements, communes de plein ou de moyen exercice, villages et fractions.

Cependant, malgré cette profondeur historique de la région, il n'y a jamais eu d'élection pour mettre en place les assemblées régionales. Les régions étaient dirigées par les gouverneurs et les assemblées régionales remplacés par des commissions spéciales composées majoritairement des représentants de l'administration, sans pouvoir réel[639]. Les régions ont davantage fonctionné comme des circonscriptions administratives déconcentrées que des collectivités territoriales. Les premiers responsables du Mali n'ont pas voulu ériger les régions en entités politiques par peur des velléités irrédentistes ; ils en ont fait des régions économiques[640].

255. - Au Togo, aucune des trois premières constitutions ne mentionne les régions. La Constitution de la première République de 1961 dispose que les collectivités territoriales sont les circonscriptions administratives[641] et les communes. Celle de la IIe République de 1963 reconnait l'existence des collectivités territoriales sans les énumérer expressément. La Constitution de la IIIe République sera quant à elle marquée par une régression en supprimant purement et simplement le titre réservé aux

---

639 *Ibid.*, p. 39.

640 *Idem.*

641 Ces circonscriptions administratives héritées de la période coloniale seront appelées plus tard transformées en préfectures.

collectivités territoriales. La première mention des régions figure à l'article 4 de la loi n° 60-4 du 10 février 1960 portant réorganisation administrative de la République du Togo qui dispose que les circonscriptions administratives sont regroupées en régions économiques[642] contrôlées par un inspecteur de région nommé[643]. La loi n° 81-8 du 23 juin 1981 portant organisation territoriale au Togo va diviser le territoire en région, préfecture, commune, sous-préfecture, canton et village, en accordant la personnalité morale et l'autonomie financière aux préfectures et aux communes uniquement. Cette loi de 1981 ainsi que son décret d'application[644] ont fait de la région une circonscription administrative à vocation économique dirigée par un chef de région dépositaire de l'autorité de l'État.

256. - Ainsi, dans ces deux pays, la région a été mise en place d'abord pour des raisons économiques. Il était exclu de faire d'elle une force politique concurrente à l'État unitaire. La décennie 90 marque la consécration de la région en tant que collectivité territoriale.

2. La consécration de la collectivité régionale

257. - La décentralisation régionale se traduit par la consécration de la région en tant que collectivité territoriale. Elle induit la mise en place d'une démocratie régionale. En droit français, l'échec de la réforme de 1969 a conduit à la démission du Général de Gaule sans pourtant que le débat sur la région s'estompe. Georges Pompidou qui le succède fait adopter la loi n° 72-619 du 5 juillet 1972 qui organise modestement la décentralisation régionale en créant de nouvelles régions sous la forme d'établissements publics régionaux. Ces établissements sont dotés de trois organes : l'assemblée délibérante, le comité économique et social et le préfet de région. L'année 1982 marque la consécration de la collectivité régionale qui se voit désormais appliquer le principe de libre administration des collectivités territoriales. Il faut toutefois relever que la collectivité régionale entre dans l'histoire avec l'organisation des premières élections régionales au suffrage universel direct en 1986, soit quatre ans après la création des régions par le législateur de 1982. Cette avancée du droit de

---

642 On dénombrait 4 régions économiques jusqu'en 1965 avec l'érection de Kara en région.

643 Art. 5, Loi n° 60-4 *préc.*

644 Décret n° 81-129 du 6 juillet 1981 portant modalités d'application de la loi n° 81-8 *préc.*

la décentralisation français va avoir un impact sur l'organisation administrative des États de l'espace ouest-africain francophone. En effet, dans les années qui ont suivi la consécration de l'échelon régional, les États africains ont procédé à une modification de leurs structures territoriales en y intégrant une décentralisation régionale.

258. - Le Togo qui, ainsi que nous l'avons montré précédemment, ne reconnaissait la région qu'en tant que circonscription administrative, va procéder à sa constitutionnalisation. Ainsi, la transformation de la région en collectivité territoriale s'est faite avec l'adoption de la Constitution de la IVe République de 1992. Malgré les multiples révisions constitutionnelles, la région conserve sa place au sein de la Loi fondamentale qui en son article 141 dispose que « *les collectivités territoriales sont : les communes et les régions.* […] *Les collectivités territoriales s'administrent librement par des conseils élus au suffrage universel, dans les conditions prévues par la loi* ». Le statut de la région n'est que partiellement constitutionnel, car la loi du 26 juin 2019 relatif à la décentralisation et aux libertés locales précise que « *les collectivités territoriales sont créées ou supprimées par la loi. La loi fixe la dénomination des collectivités territoriales* »[645]. Cependant, aucune élection des conseillers régionaux n'a été organisée au Togo depuis la consécration de la collectivité régionale. La région n'existe donc que de fait. Le ministère de l'Administration territoriale a entrepris des discussions avec la classe politique togolaise pour l'organisation prochaine d'une élection. Ces discussions ont abouti, le 3 mars 2022, à l'adoption de la loi portant création des nouvelles régions dite « loi de Kara ».

259. - Au Mali, après l'échec de la régionalisation proclamée dans la Constitution de la première République, les constitutions qui ont suivi ne mentionnent pas les catégories de collectivités. La région malienne a donc un statut originellement législatif qui découle de la loi n° 95-034 du 12 avril 1995 portant code des collectivités territoriales en République du Mali. Cette loi dispose en son article 122 que la région est une collectivité territoriale dotée de la personnalité morale et de l'autonomie financière. Les premiers conseils régionaux ont été installés à la suite de l'élection organisée en 1999. Toutefois, l'organisation régionale va être remise en cause à partir de 2011 avec la crise politico-sécuritaire qui a débouché sur la déliquescence de l'État malien en 2012. La crise malienne a permis,

---

645 Art. 2, al. 3 de la loi n° 2019-006 *préc.*

toutefois, de renforcer la place de la région dans l'organisation administrative en tant qu'acteur de pacification et de restauration de l'État.

260. - Au Burkina Faso, la loi n° 041/98-AN portant organisation de l'administration du territoire prévoyait deux collectivités territoriales : la commune et la province. En 2001, les textes d'orientation de décentralisation adoptés en 1998 ont été modifiés par la loi n° 013-2001/AN du 2 juillet 2001 afin d'introduire la région dans l'organisation administrative de l'État. Elle a fait l'objet d'une consécration à l'article 8 du CGCT qui dispose que « *la collectivité territoriale est une subdivision du territoire dotée de la personnalité juridique et de l'autonomie financière. Elle constitue une entité d'organisation et de coordination du développement. Les collectivités territoriales sont : la région et la commune* ». Au Niger, c'est la loi n° 98 -31 du 14 septembre 1998 qui crée les régions en transformant le statut des anciens départements. L'article 2 du CGCT au Niger reconnait à ce titre la région comme une collectivité territoriale.

261. - Dans l'ensemble de ces pays, l'organisation administrative et territoriale fait coexister la région circonscription administrative et la région collectivité territoriale. Les deux types de régions se confondent dans leur territorialité et dans leur dénomination, mais se distinguent par leur statut et leur régime juridique. En tant que collectivité, la région dispose de deux institutions : un organe délibérant, le conseil régional, et un organe exécutif, le président assisté de vice-présidents. La région fonctionne dans les mêmes conditions que la commune. Elle dispose de compétences d'attribution dans les domaines du développement économique, social, éducatif, sanitaire, scientifique, culturel et sportif. L'échelon régional bénéficie aujourd'hui d'une attention particulière dans les États africains. L'intérêt renouvelé pour la décentralisation régionale dans ces pays se justifie par le consensus formé autour de la nécessité de territorialiser les politiques publiques pour non seulement approfondir la décentralisation, mais aussi refonder l'État[646] afin de mieux satisfaire les besoins des populations et préserver l'unité de la République.

---

646 M. DJIRE, A. KEITA, *Régionalisation au Mali. Regards croisés*, Éditions Tombouctou, 2015 ; C. DEBBASCH, *La décentralisation pour la rénovation de l'État*, PUF, 1976 ; E. MOUBITANG, « la régionalisation constitutionnelle au Cameroun : étude d'une dynamique d'intégration nationale à l'épreuve du terrorisme international », *Revue numérique Afrilex*, 2021, p. 1.

B- La territorialisation des politiques publiques : une réponse à la crise de l'État

262. - Le territoire est au centre des enjeux de développement. La prise en compte du territoire, de son histoire, de sa culture est essentielle dans la conception du développement. La territorialisation est ainsi au centre des politiques de décentralisation qui vise entre autres à moderniser et à renforcer l'action publique. La territorialisation est une approche des politiques publiques qui s'inscrit au sein des territoires selon leurs spécificités avec une prise en charge plus localisée des besoins des populations. Territorialiser c'est donc promouvoir le transfert des politiques publiques sectorielles vers des politiques publiques locales. En intégrant davantage la prise en compte des niveaux infranationaux dans l'élaboration des politiques publiques, l'action publique locale en ressort plus efficace. La territorialisation passe par une réorganisation profonde de l'appareil d'État à travers une politique de décentralisation régionale qui confère aux régions des compétences plus étendues[647].

263. - La volonté de territorialiser prend son sens au regard des inégalités de développement régional. L'État doit pouvoir offrir à toutes les régions des chances égales de développement au nom du principe de solidarité[648]. L'égalité des chances sur l'intégralité du territoire national et donc entre les régions est comme le rappelle Odile Togolo, une affaire d'équilibre à entretenir en permanence. Tout déséquilibre accidentel ou voulu, c'est-à-dire toute tendance à privilégier ou à sous-estimer telle ou telle région par rapport à d'autres, peut déboucher sur des conflits affectant les efforts de développement[649]. En effet, les inégalités qui naissent du déséquilibre de développement constituent le fondement même des tensions sécuritaires que connaissent les États d'Afrique subsaharienne. La pauvreté et les déséquilibres régionaux servant de terreau aux groupes armés terroristes que ce soit au Burkina Faso, au Niger, au Mali ou encore récemment au Togo.

---

647 B. DIAWARA, *Territorialisation des politiques publiques et reconstruction de l'État au Mali en termes de gouvernance,* thèse, dactyl., Université Paris-Est, 2019, p.175.

648 Art. 142 de la Constitution togolaise : « L'État veille au développement harmonieux de toutes les collectivités territoriales sur la base de la solidarité nationale, des potentialités régionales et de l'équilibré inter-régional ».

649 O. TOGOLO, « La régionalisation : le défi de la souveraineté économique du pays », in M. KAMTO, S. DOUMBE-BILLE, B. M. METOU (dir.), *Regard sur le droit public en Afrique, Mélanges en l'honneur du Doyen Joseph-Marie Bipoun Woum,* L'Harmattan, p. 79 et s.

264. - Dans le cas du Mali par exemple, le déséquilibre entre le Nord et le Sud est alarmant. Les plus grandes villes du Sud concentrent plus de 80 % de la population et de la richesse totales du pays alors que le Nord représente plus de 80 % de la superficie non habitée. Les régions de Kidal, Gao et Tombouctou ont constitué le bastion des groupes rebelles lors de la crise de 2012. Il s'agit, désormais à la faveur de l'Accord d'Alger de 2015, de faire de la région un nouvel espace politique et un cadre de développement économique pour réduire les disparités régionales. La commune ne pouvant en raison de son exiguïté offrir un cadre propice pour la mise en œuvre des politiques de développement de grande ampleur, la région est plébiscitée. L'échelon régional est celui de la stratégie et de la mise en cohérence des politiques de l'État et des interventions des collectivités décentralisées. Les politiques de planification et d'aménagement du territoire doivent s'efforcer à prendre en compte les réalités et les potentialités régionales pour un développement harmonieux. Il ressort des entretiens réalisés dans le cadre d'une étude sur la régionalisation au Mali[650], que les acteurs, qu'ils soient issus de rang politique ou de la société civile, partagent la nécessité d'une territorialisation des politiques publiques pour sortir le Mali de l'impasse.

265. - Aucun développement ne peut réussir systématiquement par une approche sectorielle dans le contexte des États de l'Afrique subsaharienne. Pour éviter une juxtaposition des politiques sectorielles sans prise sur le réel, l'approche sectorielle et l'approche territoriale doivent être complémentaires. Faire le sectoriel sans le territorial ou le territorial sans le sectoriel n'a aucun sens[651]. L'absence de prise en compte des potentialités endogènes a abouti à ce que Jean Pierre Elong M'Bassi nomme « Le syndrome de Manantali »[652]. Cette ville malienne abrite l'un des plus grands barrages hydroélectriques d'Afrique de l'Ouest, mais se retrouve elle-même dépourvue de desserte et d'infrastructures adéquates.

266. - La régionalisation mue par cette volonté de prise en compte des réalités régionales dans la mise en œuvre des politiques publiques suscite des interrogations. Elles sont relatives non seulement à la préservation de

---

[650] M. DJIRE, A. KEITA, *Régionalisation au Mali. Regards croisés*, *op.cit.*, p. 10 et s.

[651] *Ibid.*, p. 72.

[652] Manantali est une ville située sur le fleuve Sénégal au Mali. Elle abrite l'un des plus grands barrages hydro- électriques appelé barrage de Manantali et fournit de l'électricité à 4 pays différents (Mali, Sénégal, Mauritanie et la Guinée).

l'intégrité territoriale et de la cohésion nationale[653], mais aussi à la définition de la nouvelle carte régionale. Le débat sur le découpage et le nombre de régions est sans fin et n'est pas sans rappeler celui en France avec le contentieux suscité par la politique de fusion des régions de 2015[654]. La réforme de l'organisation administrative et territoriale a été relancée en 2012 avec l'adoption de la loi n° 2012-017 du 2 mars 2012 qui prévoyait le passage de 8 à 19 régions de manière progressive sur une durée de 5 ans. Le même jour est adopté la loi n° 2012-018 qui crée deux nouvelles régions au nord du Mali : Taoudénit et Ménaka. À ce jour, seules dix régions sont opérationnelles sur le plan administratif avec à leur tête des autorités intérimaires en l'absence d'élection des conseils régionaux. Les lois de 2012 ont marqué une étape décisive dans la régionalisation au Mali. Toutefois, le processus de régionalisation a été interrompu en raison du coup d'État intervenu quelques jours après leur adoption le 22 mars 2012. Le retour à l'ordre constitutionnel en septembre 2013 a permis la tenue des « États généraux de la décentralisation » en octobre 2013. Cette rencontre a recommandé l'approfondissement de la régionalisation entamé par les lois de 2012 avec la reconnaissance de la capacité de l'échelon régional à mettre en cohérence les politiques et programmes nationaux et les priorités et programmes des collectivités territoriales et de coordonner la mise en œuvre du développement du territoire régional. La régionalisation ambitionne dans une logique de territorialisation de faire revivre les territoires régionaux en contribuant à un développement adapté et équilibré du territoire national qui tient compte de la diversité de ces territoires et des communautés qui y vivent tout en restant ancrées dans un ensemble national. Cependant, la situation se détériore à nouveau. Le dialogue entre l'État et les groupes rebelles s'enlise jusqu'à l'adoption des Accords d'Alger en 2015 qui place la région au cœur des enjeux de pacification.

267. - Si la régionalisation peut conduire au développement économique des territoires par la création de riches et d'emploi, l'on s'interroge également sur les réelles intentions des acteurs de l'accord pour la paix et la réconciliation issu des pourparlers d'Alger. En 1990, la décentralisation communale a été pensée pour résoudre la crise touarègue

---

653 E. MOUBITANG, « la régionalisation constitutionnelle au Cameroun : étude d'une dynamique d'intégration nationale à l'épreuve du terrorisme international », *op. cit.*, p. 1

654 V. not., F. HOURQUEBIE, « La nouvelle carte des régions : question de bon sens ou de baronnie », *AJDA*, 2015, p. 626.

du Nord Mali. Deux décennies plus tard, pour répondre encore une fois aux irrédentismes des rebelles, la régionalisation du pays leur est proposée comme la solution pour une sortie de crise. Le processus de régionalisation, tel que prévu dans les accords d'Alger, suscite des craintes dans son volet politique et institutionnel. Le texte de l'Accord d'Alger porte en lui-même les germes de sa contestation. En effet, le chapitre 3 du titre II de cet accord relatif au cadre institutionnel et à la réorganisation territoriale entend doter les régions d'assemblées régionales élues au suffrage universel direct. Le président de l'Assemblée régionale, chef de l'exécutif et de l'administration de la région, étant par ailleurs, lui aussi élu au suffrage universel direct. En prônant la mise en place des assemblées régionales et non des conseils régionaux et l'élection du président de région au suffrage universel direct comme pour le président de la République, les mouvements signataires de l'accord d'Alger cachent à peine le régionalisme ou le fédéralisme que l'on veut dissimuler sous le vocable de régionalisation[655].

Une telle conception de la régionalisation peut conduire à un éclatement de la nation au regard de l'état de déliquescence dans lequel se trouvent les États d'Afrique en proie aux contestations multiples. Le législateur malien ainsi que celui des pays de l'Afrique subsaharienne francophone en général, semble avoir pris conscience de ces craintes dans l'attribution des compétences à la région.

§2.- Le refus d'une décentralisation du pouvoir normatif au profit des régions

268. - L'institutionnalisation des régions est particulièrement lente parce que les régions portent en elles un handicap originel. Les pouvoirs publics sont en effet partagés entre deux visions contradictoires de l'espace régional. En même temps que la région constitue un espace fonctionnel important pour l'animation et la coordination des politiques publiques en faveur du développement ; ils craignent que l'autonomie régionale ne

---

[655] E. DIARRA, « Le processus de régionalisation au Mali », in *Fédéralisme, Décentralisation et Régionalisation de l'Europe : Perspectives comparatives*, Colloque de Rouen 2015 Disponible sur https://webtv.univ-rouen.fr/videos/federalisme-decentralisation-et-regionalisation-de-leurope-le-processus-de-regionalisation-au-mali-par-eloi-diarra/ consulté le 22 octobre 2021 ; M. DJIRÉ, A. KEITA, *Régionalisation au Mali. Regards croisés, op.cit.* p. 68.

conduise à l'éclatement de l'État unitaire[656]. Des exemples de droit étranger[657] confortent cette crainte dans la mesure où la régionalisation peut être perçue comme une phase de transition vers la construction d'un fédéralisme. Cette argumentation est suffisante à elle seule pour créer un climat de suspicions à l'égard des régions surtout dans les États d'Afrique noire francophone marqués par une instabilité croissante qui menace l'unité nationale. Par conséquent, les pouvoirs de l'échelon régional restent fortement limités et encadrés.

269. - Qu'il s'agisse d'une menace réelle ou supposée, les risques de la régionalisation pour l'unité nationale ne peuvent être ignorés dans l'étude de l'organisation administrative et territoriale de l'État en Afrique. Les mutations institutionnelles peuvent réveiller les vieux démons du nationalisme ambiant dans beaucoup de pays où les conditions de viabilité d'un État ne sont pas réunies[658]. Le maintien de l'unité nationale et de l'intégrité de l'État demeure depuis l'accession à la souveraineté internationale des pays d'Afrique noire francophones, une préoccupation majeure. La libre administration assortie de l'autonomie dont disposent les régions, couplée à la « *diversité géographique, culturelle et ethnique, crée une complexité susceptible de porter préjudice à l'unité nationale* »[659]. À ces facteurs viennent s'ajouter les conséquences négatives issues de la pénétration européenne en Afrique. La superposition d'un système d'administration moderne aux modes traditionnels de gouvernance ainsi que le dévoiement culturel constituent chacun un point de départ potentiel d'une série de tensions qui en se combinant et en se nourrissant réciproquement, hissent des obstacles sérieux sur le chemin de l'unité nationale[660] et par conséquent de la régionalisation.

---

656 E. MOUBITANG, « la régionalisation constitutionnelle au Cameroun : étude d'une dynamique d'intégration nationale à l'épreuve du terrorisme international », *op.cit.*, p. 1.
657 Voir les mutations de la forme d'État et les revendications vers plus d'autonomie en Italie, Espagne, Belgique, Grande-Bretagne.
658 O. TOGOLO, « La régionalisation : le défi de la souveraineté économique du pays », in M. KAMTO, S. DOUMBE-BILLE, M.-B. METOU (dir.), *Regards sur le droit public en Afrique, Mélanges en l'honneur du Doyen Joseph-Marie Bipoun WOUM*, L'Harmattan, p. 87.
659 J.-T. HOND, « État des lieux de la décentralisation territoriale au Cameroun », in M. ONDOA (dir.), *L'administration publique camerounaise à l'heure des reformes*, L'Harmattan, 2010, p. 101.
660 J.-P. FOGUI, *L'intégration politique au Cameroun : Analyse centre-périphérique*, LGDJ, 1990, p. 42 ; O. TOGOLO, « La régionalisation : le défi de la souveraineté économique du pays », *op.cit.*, p. 89.

270. - Ainsi, malgré la ferme conviction que les régions peuvent conduire au « bonheur » c'est-à-dire à la liberté et à la paix, l'État en Afrique n'entend pas leur accorder un pouvoir normatif. Or, l'autonomie d'une politique publique se mesure à l'aune des différentes ressources, qu'elles soient juridiques, financières ou organisationnelles, susceptibles d'être mobilisées pour sa mise en œuvre[661]. La région dans les États d'Afrique noire francophone ne dispose pas d'un pouvoir de premier rang, celui qu'elle choisirait librement et qu'elle exercerait d'initiative. Elle n'exerce que les attributions que l'État lui a consenties[662]. Héritier de la centralisation napoléonienne, l'État africain est à l'image de l'État français. Il dispose seul du pouvoir souverain, c'est-à-dire de la capacité à déterminer non seulement sa propre compétence, mais aussi celle des autres institutions publiques. La régionalisation dans les États de l'espace ouest-africain francophone est une régionalisation administrative identique à celle qui existe en France et qui ne reconnait aucune compétence normative aux régions[663] sous réserve des dérogations à titre expérimental[664].

271. - Les textes de loi relative à la décentralisation, s'inspirant du droit français, ne reconnaissent qu'un pouvoir règlementaire local résiduel[665] aux collectivités territoriales. Dans la conception française de l'administration territoriale dont a hérité l'État en Afrique, les collectivités territoriales ne sont que des relais de l'expression législative[666]. Elles sont partie intégrante

---

661 R. PASQUIER, « Les régions dans la réforme territoriale : vers un fédéralisme à la française », *op.cit.*, p. 79.

662 *Cf.* en droit français, F. DELPÉRÉE, *Le fédéralisme en Europe*, PUF, coll. « Que-Sais-je ? », 2000, p. 77.

663 *Cf.* G. CHAVRIER, *Le pouvoir normatif local : enjeux et débats*, LGDJ, 2011 ; L. JANICOT, « Le pouvoir normatif des régions », *RFDA*, 2016, p. 664.

664 L.O n° 2021-467 du 19 avril 2021 relative à la simplification des expérimentations mises en œuvre sur le fondement du quatrième alinéa de l'article 72 de la Constitution ; art. 72 al. 4 issu de la révision constitutionnelle de 2003. Voir sur la question l'étude de, L. JANICOT, M. VERPEAUX, « Réformer sans réviser », *AJDA*, 2021, p. 1375.

665 B. FAURE, *Le pouvoir règlementaire des collectivités locales*, thèse, LGDJ, 1998. Voir aussi du même auteur, *Droit des collectivités territoriales*, Dalloz, coll. « Précis », 2021.

666 Le schéma pensé par Sieyès avec la loi au centre rayonnant sur les citoyens placés en circonférence l'illustre parfaitement : « *Je me figure la loi au centre d'un globe immense ; tous les citoyens sans exception sont à même distance sur la circonférence et n'y occupent que des places égales ;*

de la chaine d'exécution qui descend sans interruption du ministre à l'administré par l'intermédiaire des régions et des communes. Cela se traduit notamment dans le rappel constant fait par le législateur africain aux collectivités territoriales qui doivent exercer leurs compétences en cohérence avec les orientations nationales. Les règles juridiques posées par l'État doivent donc iriser simultanément tous les territoires de la République afin de garantir l'égalité et l'indivisibilité de la République. L'hésitation des pouvoirs publics en Afrique à donner à la décentralisation régionale une portée normative trouve ainsi sa justification.

272. - Nonobstant ce refus de mettre en place des gouvernements régionaux, la région dispose en réalité de suffisamment de possibilités pour qu'elle puisse réaliser les progrès socio-économiques espérés à condition qu'elle sache saisir les opportunités qui lui sont offertes. En effet, la loi reconnait à la région de larges compétences en matière économique, sociale, culturelle et sportive de sorte que, le fait qu'elle n'ait pas reçu les attributs de souveraineté, notamment le pouvoir de légiférer, ne constitue pas un obstacle qui l'empêcherait de s'affirmer comme entité territoriale autonome[667]. Il est évident que cette autonomie sera moindre par rapport à celle dont bénéficient les régions italiennes, belges ou anglaises dont le fonctionnement se rapproche de celui d'un État fédéré. Il faudra néanmoins s'en contenter, car les risques de sécession et de fédéralisme sont à prendre en compte dans l'organisation administrative de l'État en Afrique. Il convient d'empêcher que le passage à la régionalisation ne constitue davantage une menace pour les jeunes États d'Afrique noire francophone. Les surenchères régionalistes affaiblissent davantage la capacité gouvernante de l'État. Le risque de basculement de l'édifice est important pour des États qui demeurent des géants au pied d'argile.

---

tous dépendent également de la loi, tous lui offrent leur liberté et leur propriété à protéger ; et c'est ce que j'appelle les droits communs des citoyens, par où ils se ressemblent tous. Tous ces individus correspondent entre eux, ils négocient, ils s'engagent les uns envers les autres, toujours sous la garantie commune de la loi » E. SIEYES, *Qu'est-ce que le tiers État ?* Paris, 1788, chap. VI « Ce qui reste à faire. Développement de quelques principes ».

[667] O. TOGOLO, « La régionalisation : le défi de la souveraineté économique du pays », *op.cit.*, p. 89.

Conclusion du chapitre 1

273. - La carte administrative des États de l'espace ouest-africain francophone est pour l'essentiel issue de la période coloniale. La colonisation européenne a d'abord été l'expression d'un impérialisme territorial. Pour assurer les desseins de l'entreprise coloniale, l'administration coloniale devait s'appuyer sur une organisation administrative et territoriale qui comprenait les communes, les cercles et les subdivisions administratives. Les premières expériences de décentralisation dans les États d'Afrique noire francophone nouvellement indépendants ont généralement conduit à la mise en place de structures territoriales identiques à celle de la période coloniale. Ainsi, les communes ont été maintenues dans l'ensemble des pays sans une remise en cause de leur ancrage territorial. Les cercles et les subdivisions administratives ont été transformés d'abord en circonscriptions administratives[668], avant d'être érigés plus tard en collectivités territoriales sous la dénomination de « Préfecture », « Département » ou « Cercle ». L'institution préfectorale a fait l'objet de nombreuses contestations qui ont conduit à sa suppression en tant que structure territoriale décentralisée. Néanmoins, elle continue d'exister au Mali qui d'ailleurs a choisi de préserver la dénomination coloniale de « Cercle ». Quant à la région, elle n'existait pas formellement pendant la colonisation. Toutefois, l'organisation décentralisée de la République française a continué d'être une source d'inspiration pour ces anciennes colonies qui placent désormais dans la région les espoirs de paix et de développement économique. Au-delà des structures territoriales, les pays africains ont aussi hérité des principes d'organisation administrative construite autour de l'unité républicaine.

---

[668] Loi n° 60-4 du 10 février 1960 portant réorganisation administrative de la République du Togo.

Chapitre 2 – La décentralisation : une conditionnalité de l'aide au développement

274. - Le phénomène d'internationalisation ou de globalisation du droit[669] est vécu différemment par les États selon leurs situations. Autrement dit, la question ne se pose pas dans les mêmes termes selon que l'on soit face à des pays en transition démocratique[670] comme ceux de l'Afrique ou des États ayant une démocratie consolidée[671]. À l'aune de la décennie 90, les pays d'Afrique se sont vus obligés de se conformer à de nouveaux standards de démocratie en échange de l'Aide publique au développement. La trilogie « droits de l'homme — démocratie pluraliste — État de droit »[672] constitue le nouveau sésame pour participer à de nombreuses organisations internationales ou pour bénéficier des aides de la Banque mondiale et d'autres bailleurs de fonds internationaux. Seuls les « bons élèves », qui promeuvent les valeurs de la démocratie occidentale, peuvent bénéficier de l'aide économique[673]. Dans le même sens, les institutions financières internationales vont axer leur discours sur la « bonne gouvernance » et imposer le « Consensus de Washington »[674] pour obliger les États africains à épurer leur dette. La décentralisation va dans ce contexte devenir le cheval de bataille des organisations internationales qui placent en elle les espoirs du renforcement de l'efficacité de l'action publique en Afrique.

275. - Imposés de l'extérieur sur l'action combinée des institutions financières internationales et des bailleurs de fonds, les États africains disposent de peu de marge de manœuvre dans la mise en œuvre de la décentralisation. Les « entrepreneurs institutionnels » occidentaux vont exporter la décentralisation comme un outil à part entière de

---

669 J.-B. AUBY, *La globalisation, le droit et l'État*, Montchrestien, 2010.

670 C. EUZET, Éléments pour une théorie générale des transitions démocratiques de la fin du XXème siècle, thèse, dactyl., Université de Toulouse I, 1997.

671 M.-C. PONTHOREAU, « Trois interprétations de la globalisation juridique. Approche critique des mutations du droit public », *op.cit.,* p. 20.

672 J. CHEVALLIER, « Mondialisation du droit ou droit de la mondialisation ? », in C.-A. MORAND (dir.), *Le droit saisi par la mondialisation*, Bruylant, 2001, p. 39.

673 S. BOLLE, « La conditionnalité démocratique dans la politique africaine de la France », *Revue numérique Afrilex*, 2001, p. 1.

674 Il s'agit d'un ensemble de règles édictés par John Williamson relatives à la bonne gestion macroéconomique, la bonne gouvernance institutionnelle et l'orientation vers le marché. Le respect de ces règles conditionne les aides financières aux pays en développement.

développement selon les dogmes occidentaux sans une réelle réflexion quant aux spécificités nationales des pays africains. La greffe institutionnelle qui en résulte empêche alors toute reconnaissance d'un droit à la différence. Or, un processus d'harmonisation fondée sur un consensus autour des valeurs communes de bonne gouvernance aurait pu permettre aux États africains de mettre en place des solutions endogènes s'inspirant des réalités sociologiques propres au continent. Mireille Delmas-Marty souligne à ce propos que c'est l'existence ou l'absence de marge d'appréciation nationale qui permet de distinguer l'unification de l'harmonisation[675]. La décentralisation intrinsèquement liée au processus de démocratisation[676] se répand en Afrique comme par une sorte de mimétisme[677] avec le concours des bailleurs de fonds internationaux.

276. - Un tel changement dans le fonctionnement des administrations appelle nécessairement une adaptation des normes juridiques qui régissent les rapports entre les différents acteurs de l'administration publique. La matérialisation de la décentralisation doit donc passer par l'élaboration d'un cadre légal et règlementaire pour sa mise en œuvre effective. L'ingénierie institutionnelle étant en panne[678], faute de formation adéquate des acteurs nationaux annihilant ainsi toute capacité à l'innovation, les gouvernements africains vont recourir à l'expertise des pays partenaires au développement, en particulier la France. Les experts sollicités dans le cadre de la coopération décentralisée ou de l'assistance technique vont concevoir les textes en se référant à leur modèle avec des critères de référence hors d'atteintes, conçus en d'autres lieux et tributaire de présupposés et de

---

[675] M. DELMAS-MARTY, « Les processus de mondialisation du droit », in C.-A. MORAND (dir.), *Le droit saisi par la mondialisation*, Bruylant, 2001, p. 72 : « alors que l'unification suppose des règles précises auxquelles les États sont tenus de se conformer à l'identique, en application d'une stricte hiérarchie des normes, l'harmonisation implique un affaiblissement de la hiérarchie, imposant seulement un rapprochement autour de principes communs, qui ménage une sorte de droit à la différence, chaque État gardant une marge pour mettre en œuvre ces principes, à condition que la proximité soit suffisante pour être jugée compatible. La détermination d'un seuil de compatibilité est donc le corollaire de l'apparition d'une marge nationale d'appréciation ».

[676] Cf. C. NACH-MBACK, *Démocratisation et décentralisation ; genèse et dynamiques comparés des processus de décentralisation en Afrique subsaharienne*, thèse, Karthala-PDM, 2003.

[677] Cf. F. PETITEVILLE, *La coopération décentralisée, les collectivités locales dans la coopération nord-sud*, coll. « Logiques politiques », L'Harmattan, 1995.

[678] E. MBONSI, Avant-propos à C. NACH-MBACK, *Démocratisation et décentralisation ; genèse et dynamiques comparés des processus de décentralisation en Afrique subsaharienne, op.cit.*, p. 6.

conceptions *a priori* idéalistes ou utopistes[679]. L'internationalisation du droit a transformé l'enjeu politique et culturel de son exportation[680]. L'ambition clairement affichée est d'influencer par la coopération internationale la formation du droit à travers la diffusion de modèle et des valeurs françaises. L'intégration de la décentralisation dans le discours international (**Section 1**) comme unique modèle de développement et de refondation de l'État en Afrique a contribué à asseoir l'influence occidentale et française en particulier en Afrique francophone subsaharienne (**Section 2**).

Section 1 : L'action internationale en faveur de la décentralisation

277. - La décentralisation a progressivement fait son chemin dans le discours international. Surfant sur l'érosion de l'État centralisé ainsi que dans la croyance du rôle des autorités locales dans la promotion du développement local, la décentralisation est érigée en modèle de développement. On attend de la décentralisation administrative qu'elle assure une meilleure efficacité de la gestion des affaires publiques locales. En effet, puisqu'elle permet un transfert de compétences des autorités centrales vers les collectivités territoriales, la décentralisation est présentée comme le moyen pertinent pour promouvoir la démocratie et la bonne gouvernance en Afrique. Par suite, les gouvernements africains vont opter, pas toujours de bon gré, pour un ajustement à l'ordre international en adoptant clairement la trilogie « droits de l'homme — démocratie pluraliste — État de droit » censée être favorisée par la décentralisation. Ainsi que le souligne René Otayek, « *la décentralisation s'impose comme une norme universelle, étroitement associée à l'universalisation d'une autre norme, la*

---

679 J. DU BOIS DE GAUDUSSON, Préface à C. NACH-MBACK, *Démocratisation et décentralisation ; genèse et dynamiques comparés des processus de décentralisation en Afrique subsaharienne, op.cit.*, p. 7.

680 O. DUTHEILLET DE LAMOTTE, M.-A. LOUTERNERIE, *L'influence internationale du droit français*, La Documentation française, 2001, p. 13 : « Il s'agit toujours, comme hier, au travers de la négociation de conventions internationales bilatérales et multilatérales, de l'envoi de missions d'expertise, […] de marquer la présence d'un État au travers de son droit et de ses juristes ».

démocratie de marché, unique produit désormais disponible sur le marché du design institutionnel et idéologique depuis la chute du modèle soviétique »[681].

278. - L'on ne peut s'empêcher de faire le lien entre le renouvellement de l'intérêt pour la décentralisation et les efforts entrepris par les libéraux en France au début du XIXe siècle pour lutter contre la centralisation administrative héritée de l'Ancien régime. Dès 1861, dans son ouvrage « *De la centralisation et de ses effets* »[682], Odilon Barrot dresse un inventaire des conséquences néfastes de la centralisation et appelle à un partage du pouvoir avec des institutions décentralisées pour lutter contre la tyrannie du pouvoir central. Aussi, Alexis de Tocqueville[683], à la suite des écrits de Benjamin Constant[684], montre que, sans la décentralisation, la liberté n'est qu'un leurre. Les communes sont alors considérées comme des structures capables d'entraver la toute-puissance de l'État et d'impulser le développement.

279. - Ce regain d'intérêt pour la décentralisation comme modèle absolu de démocratisation et de développement s'inscrit aussi à l'intérieur de nouveaux paradigmes de développement durable. L'Organisation des Nations unies, en élaborant les Objectifs de développement durable (anciennement Objectifs du millénaire pour le développement), et les institutions financières internationales avec à leur tête la Banque mondiale et le FMI, en faisant de la bonne gouvernance un axe majeur de leurs interventions, vont asséner davantage l'idée que, sans la décentralisation, l'on ne peut lutter efficacement contre la pauvreté dans le monde. Les premiers projets de décentralisation administrative élaborés pour les États africains étaient essentiellement conçus du point de vue de l'intérêt économique de l'État. Ils combinaient trois intentions principales, à savoir, décharger le gouvernement de responsabilités jugées secondaires, mais trop contraignantes et inutilement compromettantes pour son autorité ;

---

681 R. OTAYEK, « La décentralisation comme mode de redéfinition de la domination autoritaire ? Quelques réflexions à partir de situations africaines », Disponible en ligne http://www.afsp.mshparis.fr/archives/congreslyon2005/communications/tr4/otayek.pdf , consulté le 22 octobre 2019.

682 O. BARROT, *De la centralisation et de ses effets*, éd., H. Dumineray, 1861.

683 A. DE TOCQUEVILLE, *De la démocratie en Amérique, op.cit.*, pp. 58-67.

684 B. CONSTANT, « Principes de politique applicables à tous les gouvernements représentatifs et particulièrement à la Constitution actuelle de la France (1815) », in B. CONSTANT, *De la liberté chez les modernes. Écrits politiques*, Hachette, coll. « Pluriel », 1980 (Présenté par Marcel GAUCHET), pp. 263-427.

améliorer l'efficacité de la gestion locale en la rendant plus proche des administrés ; stimuler le dynamisme national en impliquant les citoyens dans la marche des affaires publiques. Ces différents arguments utilisés en France[685] à la fin du XVIIIe siècle seront repris en Afrique comme facteur indispensable de la bonne gouvernance de l'administration publique.

280. - Désormais, dans la sphère des acteurs de la coopération internationale, on parle d'une approche territoriale du développement local. Toutefois, derrière ces concepts se cache toujours la même réalité : celle de la croyance quelque peu « naïve » selon laquelle la décentralisation c'est le développement. Après des décennies d'application, on est toujours à la recherche de la démocratie et du développement témoignant du rejet de la greffe[686]. L'adoption de la décentralisation est en réalité le signe d'une croisade contre le « tout État » (**§1**). Elle appelle à un changement de paradigme dans la conception même de celui-ci. Il s'agit de substituer à l'État fort et centralisé, prônant l'unicité au nom de la construction de l'unité nationale, un État fort et décentralisé assurant la participation de tous à la gestion des affaires publiques (**§2**).

§1.- La remise en cause du modèle d'administration centralisé

281. - Les États d'Afrique à l'aube des indépendances ont largement fait le choix d'instituer des États de type unitaire, exception faite du Nigéria, de l'Éthiopie et de l'Afrique du Sud. Ce choix s'explique par des raisons historiques. Les États africains procèdent d'une agrégation de territoire dont la plupart ont été annexés à l'ensemble national au prix de l'usage de la force et de la contrainte[687]. Le tracé des frontières actuelles des États en Afrique s'est opéré de façon arbitraire. Cette situation a provoqué une dispersion des ethnies de part et d'autre des frontières les obligeant à cohabiter. Par conséquent, l'objectif des pays africains

---

685 G. BACOT, « L'apport de Tocqueville aux idées décentralisatrices », *in* L. GUELLEC (dir.), *Tocqueville et l'esprit de la démocratie, The Tocqueville Review/La Revue Tocqueville,* Presses de Sciences Po, 2005, pp. 203-239.

686 Y. MENY, « la greffe et le rejet. Les politiques du mimétisme institutionnel », in Y. MENY (dir.), *Les politiques de mimétisme institutionnel. La greffe et le rejet*, *op.cit.*, p. 7.

687 M. FAU NOUGARET, « Originalité et convergence des phénomènes de décentralisation en Afrique Sub-Saharienne », *Revue numérique Afrilex*, 2009, p 1.

nouvellement indépendants était de créer un sentiment national dans ces ensembles étatiques hérités de la colonisation. La volonté d'extinction de la multiplicité des différences a conduit à l'érection de l'unité en principe absolu de l'organisation politique et administrative du territoire (**A**). L'impératif du moment pour reprendre les mots de Pierre Rosanvallon, était de fondre « *l'esprit local et particulier dans l'esprit national et public* »[688], pour favoriser le développement. Cette conception ne résistera pas à l'épreuve du temps. L'incapacité de l'État à atteindre le développement va démontrer les limites de la centralisation administrative comme système d'organisation du pouvoir justifiant l'intervention des acteurs internationaux (**B**).

A- L'échec de la logique unitaire

282. - La recherche de la stabilité de l'État appelle à l'unité. Du moins, tel est l'argument des pères des indépendances africaines qui tirent parti de l'étymologie même du mot « État » « *status* ». En effet, dérivé du verbe « *stare* » qui signifie au sens premier « se tenir debout », il paraissait difficile de concevoir l'existence de l'État en Afrique sans cette idée de stabilité. La volonté de construire une unité nationale dans les ensembles hérités de la colonisation a conduit à la proclamation de l'unité et de l'indivisibilité des Républiques naissantes. Ainsi, pouvait-on lire dans article 2 de la première constitution togolaise de 1961, que « *La République togolaise est indivisible* […] »[689]. Pareille disposition se retrouve dans la quasi-totalité des constitutions africaines[690] inspirées de la constitution française du 4 octobre 1958. La diversité existante dans l'organisation des sociétés africaines précoloniales à travers la multiplicité des ethnies a été muselée dans le mythe de l'unité (**1°**) et le refus du pluralisme (**2°**) qui se sont révélés à l'usage inefficaces.

---

688 Cité par R. PASQUIER, « Une révolution territoriale silencieuse ? Les communes nouvelles entre européanisation et gouvernance territoriale », *Revue française d'administration publique*, vol. 162, n° 2, 2017, pp. 239-252.

689 La nouvelle constitution en vigueur depuis 1992 y ajoute l'unité. Art 1 : « La République togolaise est un État de droit, laïc, démocratique et social. Elle est une et indivisible ».

690 V. not., art 25 de la Constitution du Mali ; art. 3 de la Constitution du Niger.

1. Le principe de l'unité dans la construction de l'État-nation

283. - « L'indivisibilité est l'essence même de l'unité, le signe de son effectivité et de sa durée. Le caractère indivisible de la République est la garantie du maintien de son unité, la projection de cette dernière dans l'avenir »[691] écrivent Rolland Debbasch et André Roux. Les pays d'Afrique noire francophones sont restés fidèles à cette conception française de l'État et de la République. Les principes de l'unité et d'indivisibilité, repris dans la majorité des constitutions des pays de l'Afrique francophone, ont permis de conférer à ces États le caractère d'État unitaire. Ce choix institutionnel de la forme de l'État témoigne de l'organisation des rapports entre le centre et la périphérie et du rejet du fédéralisme. Le défi pour les pays nouvellement indépendants d'Afrique était donc de réussir à construire une nation. Certains auteurs relèvent d'ailleurs que la nécessité de construire une unité nationale constitue « la hantise » des gouvernements africains, car si « l'État existe, la nation est à créer. Or sans nation, l'État est fragile »[692]. Ainsi, le souci de l'unité a été et demeure aujourd'hui une préoccupation majeure sur le continent. La centralisation va permettre la mise en place d'un centre d'administration unique pour traiter des affaires nationales. Elle permet l'uniformité de l'action sur toute l'étendue du territoire et l'atteinte d'un niveau d'efficacité propice au développement. Pour rendre viable la centralisation, les États font recours au procédé de la déconcentration pour décongestionner les services centraux de l'État sans remettre en cause le modèle unitaire de l'État.

284. - Dans sa quête exacerbée de l'unité, l'État en Afrique se dresse contre toute tentative qui vise à reconnaitre des droits subjectifs à quelque groupe que ce soit en raison d'une quelconque spécificité régionale. Sékou Touré affirmait d'ailleurs en ce sens qu'« *il ne saurait être admis des groupes religieux ou ethniques qui disposent d'une personnalité propre* »[693] au sein de l'État. Tout comme en France, l'unité est construite autour de l'unicité du peuple. C'est le Conseil constitutionnel qui, à l'occasion de l'examen de la loi portant statut de la collectivité de Corse, a formulé expressément cette

---

691 R. DEBBASCH et A. ROUX, « L'indivisibilité de la République », in B. MATHIEU et M. VERPEAUX (dir.), *La République en droit français*, Economica, 1996, p. 75.

692 F. GALLETTI, *Les transformations du droit public africain francophone*, thèse, Bruylant, 2004, p. 225.

693 J.-P. FOGUI, *Autorités traditionnelles et intégration politique au Cameroun*, thèse, dactyl., Université de Bordeaux, 1980.

exigence d'unité. Le Conseil affirme, en se référant au principe de l'indivisibilité, que « *la constitution* […] *ne connait que le peuple français, composé de tous les citoyens français sans distinction d'origine, de race ou de religion* »[694]. Les principes d'indivisibilité et d'unicité constituent pour le Conseil constitutionnel des « principes fondamentaux » qui s'opposent à la reconnaissance des droits collectifs à quelque groupe que ce soit, défini par une communauté d'origine, de culture, de langue ou de croyance[695].

285. - Aussi, l'objectif d'unité nationale ne peut être dissocié de l'unité linguistique. Le nouvel État-nation a imposé aux peuples conquis et soumis l'assimilation linguistique. L'unité de la langue apparait comme une condition fondamentale de l'unité nationale[696]. Alors que pendant longtemps, les conquêtes militaires en Afrique précoloniale ont permis de préserver les identités culturelles des territoires conquis, l'État-nation a été construit en assemblant des nations (ethnies) annexées, dominées et mutilées de leur langue et de leur culture propre[697]. La langue devient une affaire d'État dès lors que le pouvoir politique prend la forme de l'État moderne[698]. Cette réalité est introduite sur le continent africain à travers la promotion de la langue française depuis la période coloniale. Il s'agissait alors de faire de la langue l'instrument ultime pour la réalisation de la mission coloniale. Dans ce contexte, les langues locales ont été reléguées au second plan. La langue française va alors être repris lors des indépendances par les gouvernements africains qui proclament, dès 1960,

---

694 Décision n° 91-290 DC du 9 mai 1991, relative à la loi portant sur le statut de la collectivité territoriale de corse, le Conseil constitutionnel refuse de reconnaître l'existence d'un « peuple corse » distinct du « peuple français » d'origine, de culture, de langue ou de croyance »

695 Sur l'application de ces principes, voir la décision n° 99-412 DC du 15 juin 1999, *Charte européenne des langues régionales ou minoritaire*s. Sur la promotion des langues régionales en droit français, voir : Loi n° 2021-641 du 21 mai 2021 *relative à la protection patrimoniale des langues régionales et à leur promotion* dite « Loi Molac » et la déc. n° 2021-818 DC du 21 mai 2021, *Loi relative à la protection patrimoniale des langues régionales et à leur promotion*. Lire aussi, M. VERPEAUX, « Défiance jacobine et réalités constitutionnelles », *AJDA*, 2021, p. 1750.

696 J. W. LAPIERRE, *Le Pouvoir politique et la langue. Babel et Léviathan*, PUF, coll. « La politique éclatée », 1988, p. 26.

697 *Ibid.*, p. 26.

698 *Ibid.*, p. 37.

dans leurs constitutions, la langue française comme langue officielle[699]. Cette solution de facilité permet aux gouvernants d'éviter les conflits ethniques qui pourraient naitre s'ils faisaient le choix d'ériger une langue nationale en langue officielle.

286. - Pour les jeunes États africains, l'adoption de la langue française comme langue officielle au détriment des langues nationales devait permettre la formation et la consolidation de l'unité nationale ainsi que le développement de l'État en l'ouvrant au reste du monde. Adopter la langue française était un signe de progrès et d'accès à la modernité. Le modèle de l'État-nation unitaire et centralisé s'accommode mal au multilinguisme. Néanmoins, aujourd'hui, les Constitutions de certains États reconnaissent l'existence de langues nationales qui peuvent être utilisées dans la sphère privée et qui constituent un facteur de cohésion sociale. Au Togo, depuis 1975, le Kabyè et l'Éwé sont considérés comme les langues nationales. Elles ont même intégré les programmes scolaires au titre des enseignements facultatifs. Au Niger, la Constitution reconnait l'existence de langues nationales. Elle dispose dans son article 5 que « *toutes les communautés composant la Nation nigérienne jouissent de la liberté d'utiliser leurs langues en respectant celles des autres. Ces langues ont, en toute égalité, le statut de langues nationales. L'État veille à la promotion et au développement des langues nationales. La loi fixe les modalités de leur promotion et de leur développement* ». L'appel à l'unité nationale, seul moyen de faire prospérer l'État unitaire et jacobin va se traduire sur le plan politique par une restriction des droits civils et politiques à travers la consécration du monopartisme.

2. L'unicité du parti comme condition nécessaire à la construction de l'État-nation

287. - Les États africains ont autorisé, lors de leur indépendance, le multipartisme affirmant ainsi de façon formelle leur caractère libéral. L'article 4 de la Constitution togolaise du 14 avril 1961 prévoit l'instauration d'un régime libéral et multipartisan en disposant que « *les partis et formations politiques concourent à l'expression du suffrage. Ils se constituent et exercent leur activité librement dans le cadre des lois et règlements. Ils doivent respecter*

---

699 Il a fallu attendre la révision constitutionnelle du 25 juin 1992 pour que le français soit érigé en langue de la République en France. La révision constitutionnelle du 23 juillet 2008 introduit un nouvel article 75-1 : « *Les langues régionales appartiennent au patrimoine de la France* ».

les principes de la souveraineté nationale et de la démocratie». Cependant, ce pluralisme formel n'empêchera pas la domination du parti gouvernemental et plus tard l'institutionnalisation de parti unifié à partir de 1980.

288. - Cet écart entre le texte et sa pratique trouve une explication à travers de deux arguments. D'une part, en raison de la diversité ethnique et linguistique des États d'Afrique noire francophone, il était nécessaire de créer un instrument politique susceptible de réunir tous les citoyens et capable de faire naitre une conscience nationale. D'autre part, face à la réalité du sous-développement, le multipartisme est considéré par les dirigeants africains comme une source d'exacerbation des tensions politiques entre les populations qui compromettrait les efforts entrepris en faveur du développement. La construction de l'unité nationale a donc servi de fondement idéologique au rejet du multipartisme. Selon les propos de Tom Mboya, « *un État a besoin de maintenir l'unité comme base de tout développement. [...]. Le pays doit être placé sous la direction d'une seule formation politique, un parti fort et efficace pour pouvoir mobiliser pleinement le potentiel de la population afin que l'indépendance se traduise en termes de développement significatif qui puisse satisfaire les espoirs du peuple. Ce but ne peut pas être atteint sous un régime de pluralisme politique* »[700].

289. - Le parti unique devient le seul instrument à la disposition des gouvernants capable de résorber les dérives régionalistes et tribalistes qui minent l'État en Afrique et d'amorcer le décollage économique[701]. Maurice Glélé soutient à ce propos que, pour réaliser l'unité nationale et se consacrer aux tâches de développement économique, face aux querelles stérilisantes des hommes politiques, la voie moderne la plus adéquate demeure le parti unique. Un parti unique bien structuré, ayant une doctrine, du moins un programme, c'est-à-dire « sachant où il va »[702]. La recherche d'un pouvoir efficace et stable conduit donc les constituants africains à faire valoir l'idée que dans les États nouveaux, « *qui ne correspondent pas toujours à des nations déjà intégrées, à une volonté préexistante dans les esprits* » ; l'État ne pourrait « *résister aux forces centrifuges des particularismes*

---

700 Cité par B. KOUMANTIGA, *Le parti unique et la question de l'unité nationale au Togo de 1961 à 1990,* Mémoire Maîtrise ès Lettre Sciences Humaines, Université de Kara, 2013. Voir aussi, B. YACINE-TOURÉ, *Afrique : l'épreuve de l'indépendance,* Genève, Graduate Institute Publications, 1983.

701 B. YACINE-TOURÉ, *Afrique : l'épreuve de l'indépendance, op.cit.*, pp. 46-68.

702 M.GLELE, *Naissance d'un État noir*, LGDJ, 1969, p. 345.

locaux ou ethniques, au "régionalisme" ou au "tribalisme", que par la force du pouvoir central, c'est-à-dire essentiellement du pouvoir gouvernemental»[703]. Ainsi, la transformation des conditions économiques nécessaires à des pays encore insuffisamment développés avait pour condition une discipline des activités de production, qui supposait elle-même une action ferme et continue, menée par un pouvoir assuré de perdurer[704].

290. - À partir de ce constat, l'existence d'un État-nation ne peut se penser en dehors de l'œuvre unitaire du parti unique. Tirant sa légitimité de la promesse d'un développement socio-économique, l'État jacobin postcolonial en Afrique a fini par s'imposer comme outil de la croissance économique et du développement. Cette position est largement partagée par les États occidentaux ainsi que par les institutions financières. En effet, ces derniers voient dans le discours des dirigeants africains sur la nécessaire construction d'États forts et stables un préalable au développement[705]. Pierre Lampué écrit en ce sens qu'il faut reconnaitre à « *la concentration du pouvoir un avantage majeur, qui est d'assurer la force et l'efficacité de l'action politique.* [...] *Un tel avantage est assurément précieux chez des peuples dont l'intégration nationale est encore imparfaite, qui sont* [...] *des "nations en voie de se faire", et qui aspirent, au surplus, à une modernisation et à un développement rapide de leur économie* »[706]. Ainsi, le règne de la dictature qui précède l'autoritarisme colonial est justifié par la recherche du développement économique, reléguant au second plan l'affirmation des libertés politiques et des libertés locales.

291. - Au début des années 1990, des voix s'élèvent contre l'unicité du parti et le caractère autocratique des régimes incapables de satisfaire aux besoins des populations. La contestation du parti unique, à la suite de l'aggravation des conditions sociales due à l'application des plans d'ajustement structurel des institutions de Bretton Woods à travers l'organisation des conférences nationales qui s'en est suivi, témoigne de l'échec du couple État-nation/Parti unique. Le mécontentement des populations, face à l'incapacité de l'État à réduire les disparités économiques, est relayé par une partie de la classe d'opposition en exil. Il

---

703 P. LAMPUÉ, *Les Constitutions des États africains d'expression française*, LGDJ, 1962, p. 2.
704 *Idem.*
705 J.-L. ATANGANA AMOUGOU, *L'État et les libertés publiques au Cameroun*, thèse, dactyl., Université Jean Moulin, Lyon 3, 1999.
706 P. LAMPUÉ, *Les Constitutions des États africains d'expression française, op.cit.*, p. 28.

trouvera un écho favorable auprès des pays partenaires au développement, qui ont étamé une révision de leur doctrine du développement en Afrique.

B- Une intervention internationale légitimée par l'échec du mythe de l'unité

292. - L'échec de la logique unitaire dans la construction de l'État-nation va constituer une source de légitimation pour l'intervention active des institutions internationales financières et des bailleurs de fonds occidentaux en Afrique. L'accès à l'aide publique au développement est subordonné à la mise en œuvre de conditionnalités économiques et politiques fixées par les pays pourvoyeurs de l'aide. Trois facteurs expliquent ce changement de paradigme dans l'attribution de l'aide publique aux pays en développement : la chute du mur de Berlin, l'affaiblissement des régimes autoritaires en Afrique et les revendications internes en faveur du multipartisme[707]. Cette crise de régime, qui, à cette époque, n'est pas propre au continent africain, va ouvrir la voie à une nouvelle politique internationale en faveur des pays du tiers-monde. Sensibles à la stabilité du continent, les acteurs de la coopération internationale font de la démocratie une condition d'accès à l'aide au développement. L'action publique internationale, qui pendant longtemps s'était refusée à une ingérence dans les affaires internes des États, vient d'institutionnaliser ce que l'on pourrait qualifier d'« ingérence démocratique »[708] (**1°**). Cependant, des inquiétudes persistent quant au lien factuel établi entre la promotion des valeurs démocratiques et le progrès économique (**2°**).

1. Un accès conditionné à l'aide publique au développement

293. - La fin de la bipolarisation mondiale après la chute du mur de Berlin va constituer une occasion unique pour la promotion des valeurs occidentales de démocratie. Les partenaires occidentaux des pays du tiers-monde vont intégrer les objectifs de démocratisation dans les accords de coopération bilatérale et multilatérale. La Commission européenne, en tant que principal pourvoyeur de l'aide au développement en Afrique

---

707 D. ÉTHIER, « La conditionnalité démocratique des Agences d'aide et de l'Union européenne ». *Études internationales*, vol. 32, n° 3, 2001, p. 495.

708 J. KEUTCHEU, « L'"ingérence démocratique" en Afrique comme institution : dispositif et scène », *Études internationales*, vol. 45, n° 3, 2014, pp. 425-451.

subsaharienne, va progressivement soumettre l'accès à l'aide à une conditionnalité démocratique. Il faut noter que l'exigence démocratique n'était pas à l'ordre du jour dans les premiers accords de Yaoundé[709] et de Lomé[710] entre la Communauté économique européenne et les pays de l'Afrique, des Caraïbes et du Pacifique (ACP). Dès 1985, la Communauté économique européenne tente une première approche de conditionnalité sans succès dans le cadre de la conclusion des accords ACP. Dans un contexte encore marqué par les séquelles de la colonisation, les pays de l'ACP dénoncent une ingérence qu'ils considèrent comme excessive dans les affaires internes des États. Ces accords font, aujourd'hui encore, à l'heure de leur renégociation, l'objet de vives critiques[711], car l'effet de levier effectif des dispositions juridiques et les conditionnalités de l'aide sont très limités comme l'illustre la persistance des cas de changements anticonstitutionnels dans les pays d'Afrique noire francophones[712].

294. - Finalement, seule la Charte des Nations unies a pu être invoquée dans le préambule de l'Accord de Lomé III[713]. La résistance ne fut pas malgré cela, que de courte durée, car la signature de la convention de Lomé IV en 1989[714] marque l'introduction expresse d'une « clause de Droits de l'Homme ». On pouvait lire à l'alinéa 1 de l'article 5 de cette convention que « *la coopération vise un développement centré sur l'homme, son acteur et bénéficiaire principal, et qui postule donc le respect et la promotion de l'ensemble des droits de celui-ci. Les actions de coopération s'inscrivent dans cette perspective, où le*

---

709 Convention d'association entre la Communauté économique européenne et les États africains et malgache de Yaoundé, 20 juillet 1963.

710 Convention I A.C.P.-C.E.E. (Lomé I), Lomé, 28 février 1975 et Convention II A.C.P.-C.E.E. (Lomé II), Lomé, 31 octobre 1979.

711 K. NUBUKPO, Une solution pour l'Afrique. Du néoprotectionnisme aux biens communs, Odile Jacob, 2022, p. 31 et s.

712 Les exemples sont légion pour ne retenir que les plus récents, citons, les cas du Burkina Faso, du Mali qui connait une situation d'instabilité avec une succession de coup d'état, le Niger avec la tentative de coup d'état manqué dans la nuit du 30 au 31 Mars 2021. Voir aussi sur la question de la renégociation des conventions ACP-UE, G. LAPORTE, « L'après Cotonou : de la coopération Nord-Sud traditionnelle à un partenariat d'intérêts partagés ? », *Revue de l'UE*, 2019, p. 401.

713 Convention 3 A.C.P.-C.E.E. (Lomé III), Lomé, 8 décembre 1984.

714 Convention 4 A.C.P.-C.E.E. (Lomé IV), Lomé, 15 décembre 1989. L'accord de Cotonou en 2000 va clore cette évolution de la conditionnalité en affirmant à l'article 33§ 3 que : « *la coopération appuie les efforts des États A.C.P. pour développer leurs institutions publiques comme facteur dynamique de croissance et de développement, et pour améliorer de manière significative l'efficacité et l'impact des services publics sur la vie quotidienne des citoyens. Dans ce contexte, la coopération soutient la réforme, la rationalisation et la modernisation du secteur public* ».

respect des droits de l'homme est reconnu comme un facteur fondamental d'un véritable développement et où la coopération elle-même est conçue comme une contribution à la promotion de ces droits ». Les stipulations de cet article montrent clairement le lien qui venait d'être établi entre les objectifs de coopération pour le développement et la promotion des droits humains. Ainsi, la communauté internationale découvrait subitement qu'il ne pouvait y avoir de développement sans un certain degré de démocratie, pas de démocratie sans respect des Droits de l'Homme et enfin, pas de démocratie sans développement[715].

295. - C'est dans ce contexte, que le président François Mitterrand prononcera un discours qui bouleversa à jamais les modalités de l'action publique internationale. La France, dit-il, *« liera tout son effort de contribution aux efforts qui seront accomplis pour aller vers plus de liberté* ». « *L'aide française sera plus tiède envers les régimes autoritaires* » et « *plus enthousiaste envers ceux qui franchiront le pas vers la démocratie* »[716]. Mitterrand souscrivait ainsi à la nouvelle trouvaille internationale qu'est la conditionnalité démocratique marquant un tournant pas très apprécié de tous[717] dans les relations entre la France et l'Afrique. Ce tournant sera perpétué par les présidents successifs de la France. François Hollande, en 2012, dans son discours à Dakar, affirmait que « *la démocratie vaut pour elle-même [...]. Mais elle vaut aussi pour ce qu'elle permet. Il n'y a pas de vrai développement économique ni de vrai progrès social sans pluralisme* ». Le parti unique défendu au début des années 1970 tombe en disgrâce aux yeux de la communauté internationale. Et pour cause, un lien quasi mécanique semble s'établir désormais entre la promotion des valeurs démocratiques et le progrès économique et social. L'Afrique, champ d'expérimentation privilégié des puissances occidentales, va subir conjointement l'ajustement économique et l'ajustement démocratique.

---

715 « Observations relatives aux droits de l'homme en ce qui concerne les activités de coopération au développement de la communauté européenne », in P. ALSTON (dir.), *L'Union européenne et les droits de l'Homme*, Bruylant, 2001, pp.596-597.

716 « Discours de François Mitterrand à La Baule du 20 juin 1990 », Politique étrangère de la France, mai-juin 1990, p. 130. Voir sur la politique étrangère française sous la présidence Mitterrand, L. DENIS, *La politique étrangère de la France : contribution à l'étude et analyse des politiques étrangères de la France sous la présidence de François Mitterrand*, thèse, dactyl., Université de Perpignan, 2010.

717 Le discours de la Baule a suscité de très vives réactions de la part de certains chefs d'États africains notamment Déby itno et Habiarinéa.

296. - Alors même que certains auteurs semblaient dire que « le but des conditionnalités économiques était [...] de restructurer les économies des pays africains que d'avoir des effets particuliers sur les systèmes politiques africains »[718], l'association développement et forme de gouvernement a pris de l'ampleur dans la littérature officielle des institutions financières internationales. Les questions de gouvernance sont présentées comme un remède aux différents obstacles à la qualité de la vie au niveau local[719]. Le concept de gouvernance va être formulé explicitement pour la première fois en 1989 dans un rapport de la Banque mondiale intitulé « Afrique subsaharienne : de la crise à une croissance durable ». Ce rapport met l'accent sur l'arbitrage des relations entre l'État et le marché pour atteindre l'objectif d'efficacité économique. Les programmes d'ajustement structurel paraissent à eux seuls insuffisants pour le développement. Ce rapport va donc introduire dans l'analyse de cet échec les dimensions politiques et culturelles, en dénonçant le mauvais fonctionnement des administrations, la corruption, le laxisme dans la gestion des affaires publiques. Dans ce rapport sont liées la libéralisation économique et la légitimité du pouvoir invitant ainsi à un renouveau politique dans les États d'Afrique[720]. Dès lors, la Banque mondiale va porter un intérêt considérable à la production d'une idéologie de l'État et accroitre dans le même temps l'intervention des services publics[721].

297. - Cependant, après des décennies de mise en œuvre de la conditionnalité politique, ni la démocratie ni le développement ne sont atteints par les pays du Sud. Le déficit démocratique est permanent. On assiste toujours à une centralisation du pouvoir au bénéfice du président, une captation de l'administration publique avec des inégalités croissantes qui sont source de conflits et de déstabilisation de l'État. Les conférences

---

718 A. WISEMAN, « Démocratisation, réforme économique et conditionnalités en Afrique subsaharienne : contradictions et convergences », in S. MAPPA (dir), *Développer par la démocratie ? Injonctions occidentales et exigences planétaires*, Karthala, 1995, p. 467.
719 OCDE, Les nouvelles formes de gouvernance et le développement économique, Éditions OCDE, 2005.
720 BANQUE MONDIALE, L'Afrique subsaharienne. De la crise à une croissance durable, Washington D.C., 1989. Voir également, A. SADIO, Conditionnalité politique de l'aide publique au développement des partenaires occidentaux à l'Afrique : analyse des actions françaises en Afrique subsaharienne, thèse, dactyl., Université de Rouen Normandie, 2019.
721 BANQUE MONDIALE, L'État dans un monde en mutation : rapport sur le développement dans le monde, Washington 1997.

nationales souveraines[722], organisées dans les pays d'Afrique noire francophones, ont toutes intégré dans les débats, la question de l'hyper centralisation du pouvoir, sans véritable succès. L'État africain entretient toujours une profonde méfiance envers la démocratie libérale occidentale et ses vertus en faveur du développement économique et social.

2. L'inefficacité de la conditionnalité politique à assurer le développement

298. - Trois décennies après le discours de la Baule et l'ouverture des régimes politiques africains au pluralisme politique, où en sont les pays d'Afrique noire francophones en matière de développement ? Le bilan des transitions démocratiques en Afrique est mitigé. Si l'introduction du pluralisme a permis de mettre fin, à court terme, à la crise que traversaient les gouvernements africains, elle n'a conduit ni à un enracinement de l'idéal démocratique, ni au développement économique et social tant espéré.

299. - La réticence des élites africaines à l'égard des politiques de conditionnalité, considérées comme portant atteinte à la souveraineté de leurs jeunes États indépendants, est motivée par les risques politiques qu'une telle démarche pourrait engendrer. En effet, les dirigeants africains craignent que l'instauration d'un véritable jeu démocratique conduise à la perte de leur pouvoir et des différents privilèges qui y sont attachés. Par conséquent, pour disposer de l'aide publique au développement, une ouverture démocratique de façade est organisée, avec une domination d'un seul parti dans les faits, malgré la multiplication des formations politiques qui ont accès aux élections. La démocratie dans les pays d'Afrique subsaharienne est constamment en recul. L'indice de démocratie développé par *l'Economist Intelligence Unit* montre que malgré l'organisation des élections à intervalles réguliers, l'Afrique subsaharienne fait figure de mauvaise élève[723]. En témoigne, le caractère prolifique du contentieux

[722] N. MEDE, « La conférence nationale souveraine, 30 après : quel bilan économique et social ? », *Revue numérique Afrilex*, 2022, p. 1.

[723] Les cinq critères d'évaluation : le processus électoral et le pluralisme, les libertés civiles, le fonctionnement du gouvernement, la participation politique, et la culture politique.

électoral en Afrique[724]. Même dans des pays comme le Bénin qui fait figure de bon élève en matière de démocratisation, la démocratie politique n'a pas permis de faire advenir le développement. Nicaise Médé écrit à ce propos que « *trente ans après* [les conférences nationales] *nous nous réveillons de notre sommeil d'espérance avec une gueule de bois et le sentiment désormais partagé que la démocratie politique ne suffit pas à accoucher le développement* »[725].

300. - Les agences d'aides ont échoué dans la mise en œuvre des politiques de conditionnalité dans les pays en développement sur le continent africain. L'efficacité potentielle des conditions politiques des programmes d'aide a fait l'objet de plusieurs études dont les conclusions sont convergentes[726]. Tous les auteurs estiment que les réformes demandées au titre du respect des droits de la personne et de la démocratisation ont fort peu de chance d'être réalisées, de manière satisfaisante, si les conditions internes, qui sont décisives en matière de réformes démocratiques, notamment la détermination des élites dirigeantes à procéder aux réformes n'existent pas. Ce qui est le cas des pays en développement d'Afrique noire francophones. Ces derniers sont réfractaires à la conditionnalité politique pour les raisons sus mentionnées. Par ailleurs, les sanctions prévues (réduction, suspension ou report de l'aide) ont peu d'influence sur le comportement des États receveurs et font l'objet d'une application parcimonieuse en fonction des intérêts des pays pourvoyeurs de l'Aide. La conditionnalité perd de sa substance lorsqu'il s'agit de sanctionner des pays « amis » violant *de facto* les principes démocratiques[727].

301. - Le parti unique n'a pas réussi le pari du développement économique et social. Il en est de même du multipartisme qui s'est traduit par une inflation de partis politiques. Cependant, la démocratie ne peut pas se résumer à l'organisation périodique d'élections entachées de fraudes

---

724 C'est l'un des rares domaines du droit public africain où les décisions de justice ne manquent pas. Les matériaux sont suffisants pour toute œuvre de recherche. Voir not., Y. NIANG, *Le contrôle juridictionnel du processus électoral en Afrique noire francophone : les exemples du Sénégal et du Bénin*, thèse, dactyl., Université de Bordeaux, 2018.
725 N. MEDE, « La conférence nationale souveraine, 30 après : quel bilan économique et social ? », *op.cit.*, p. 3.
726 D. ÉTHIER, « La conditionnalité démocratique des Agences d'aide et de l'Union européenne », *op.cit.*, p. 495 et s.
727 S. BOLLE, « La conditionnalité démocratique dans la politique africaine de la France », *Revue numérique Afrilex*, 2001, p.10.

trahissant la sincérité des scrutins[728]. Dans ces conditions, le développement économique tant attendu n'a pu advenir. Néanmoins, l'intérêt des acteurs de la coopération internationale ne faiblit pas. Dans la quête des moyens à mettre en œuvre pour accélérer et consolider les transitions démocratiques qui balbutient, la décentralisation va apparaitre comme une nouvelle composante de la mobilisation des forces endogènes face à la faillite de l'État postcolonial pour résorber le sous-développement. Parmi les mesures préconisées pour améliorer les conditions des plus pauvres, le niveau décentralisé est privilégié pour mettre en œuvre les programmes sociaux. La décentralisation ainsi que la participation de l'ensemble des bénéficiaires sont de nouveau à l'ordre du jour[729]. La lutte contre la pauvreté et la réduction des inégalités est alors inscrite au cœur du processus de développement. La dimension sociale devient une composante centrale du processus de développement.

§2.- Le recours au modèle d'administration décentralisé

302. - Les conditionnalités politiques et économiques mises en œuvre par les institutions de Bretton Woods et les agences d'aide internationale n'ont pas résolu les problèmes de sous-développement de l'Afrique. Au contraire, les plans d'ajustement structurel ont contribué à exacerber les déséquilibres et la qualité des services sociaux de base comme la santé et l'éducation. L'instauration du parti unique a également participé à la démobilisation et à la réduction de la place du citoyen dans la gestion des affaires publiques. Face aux inégalités sans cesse croissantes, s'est développé un sentiment d'hostilité vis-à-vis de l'État unitaire et centralisé. La Banque mondiale va alors déployer un plaidoyer en faveur de la décentralisation dans son rapport de 2000[730]. Il préconise à la suite des séries de conditionnalités, l'intégration de la décentralisation

728 Sur la question de la sincérité des élections, R. GHEVONTIAN, « La notion de sincérité du scrutin », *Les cahiers du Conseil constitutionnel,* n° 13, 2003 ; J. DU BOIS DE GAUDUSSON, « les élections à l'épreuve de l'Afrique », *Les cahiers du Conseil constitutionnel,* n° 13, 2003.

729 B. YACINE-TOURÉ, *Afrique : l'épreuve de l'indépendance,* Genève, Graduate Institute Publications, 1983, p. 48 et s.

730 BANQUE MONDIALE, *Le développement dans le monde au seuil du XXIe siècle*, éd. Eska, 2000.

administrative dans le champ global de refondation de l'État en Afrique subsaharienne[731].

303. - La pauvreté apparait aussi comme un problème d'ordre politique. La nécessité de procéder à un renouvellement des stratégies va conduire les institutions financières internationales à adopter de nouvelles politiques d'aide au développement avec l'entrée en lice d'un nouvel objectif de lutte contre la pauvreté. Les pays en développement désireux de bénéficier d'un allègement de la dette afin de relancer leurs économies doivent, dans le cadre de l'initiative Pays pauvres très endettés (PPTE), élaborer un programme de lutte contre la pauvreté prenant la forme d'un Document stratégique de réduction de pauvreté (DRSP). Cette stratégie a reçu l'assentiment des Nations Unies à travers notamment l'action du Programme des Nations unies pour le développement (PNUD). Convaincue du caractère universel des normes libérales et de la nécessité de réduire l'action de l'État pour relancer le développement, la critique de l'État autoritaire a puissamment alimenté le discours dominant sur les vertus supposées de la décentralisation et l'émergence de pouvoirs locaux censés favoriser la participation politique[732]. Pour agir efficacement contre la pauvreté, il faut repenser l'État en agissant non seulement sur son fonctionnement, mais aussi et surtout, sur sa structure. Le modèle décentralisé constitue pour les pourvoyeurs de l'aide le moyen privilégié de la lutte contre la pauvreté (**A**). Les collectivités décentralisées deviennent ainsi les nouveaux catalyseurs du développement (**B**).

A- L'apparition de la lutte contre la pauvreté comme nouveau paradigme de l'aide internationale au développement

304. - Les acteurs de l'aide au développement ont au fil des années repensé leurs stratégies d'intervention dans les pays en développement de l'Afrique subsaharienne. L'inefficacité des programmes d'ajustement structurel a conduit à replacer l'humain au centre de toute action de développement. Ils reconnaissent, ainsi que le souligne les Nations unies, que mettre fin à la pauvreté doit aller de pair avec des stratégies de développement de la croissance économique et qui répondent à une série

---

731 *Ibid.*, pp. 113-134.

732 R. OTAYEK, « La décentralisation comme mode de redéfinition de la domination autoritaire ? Quelques réflexions à partir de situations africaines », *op.cit.*, p. 3.

de besoins sociaux, notamment l'éducation, la santé, la protection sociale et les possibilités d'emploi tout en luttant contre le changement climatique et la protection de l'environnement[733]. Les partenaires au développement vont faire de la participation politique une nouvelle condition de réussite des programmes de lutte contre la pauvreté.

305. - La Charte africaine des valeurs et des principes de la décentralisation, de la gouvernance locale et du développement local, adoptée en 2014, entérine aussi le lien entre décentralisation et développement. Dans son préambule, elle se résout à promouvoir les valeurs et les principes de la décentralisation, de la gouvernance locale et du développement local en Afrique comme préalable à l'amélioration des conditions de vie de l'ensemble des peuples du continent. Ainsi, pour s'assurer de l'implication des populations dans l'élaboration et le suivi des DSRP, les politiques de décentralisation constituent des cadres de référence pour la mise en œuvre d'une démarche participative capable d'impulser le développement local (**1°**). Dans le même temps, les acteurs de la coopération internationale vont considérer que le développement local ne peut advenir sans la « bonne gouvernance » (**2°**).

1. L'émergence de la décentralisation dans l'élaboration des stratégies de réduction de la pauvreté

306. - L'organisation en communauté des populations pauvres constitue le fondement de la réussite des stratégies de lutte contre la pauvreté. Déjà, en 1999, dans son rapport sur le développement dans le monde, la Banque mondiale attirait l'attention des États sur la nécessité d'organiser et d'assurer la représentation des populations pauvres. Quelques années plus tard, on peut lire dans le rapport de 2002 de la Banque mondiale préparé pour la conférence des Nations unies sur le financement du développement à Monterrey, cette formule qui sonne comme un aveu d'échec : « *une leçon tirée de l'expérience passée est que la réforme ne réussit généralement pas sans une forte appropriation locale et une approche large, qui inclut la prise en compte des institutions, la gouvernance et la participation des acteurs — une leçon qui constitue le moteur du processus des documents stratégiques de réduction de la pauvreté* »[734].

---

733 www.un.org/sustainabledevelopment

734 *Cf.*, Rapport Banque mondiale 2002, p.46.

307. - La nouveauté introduite dans ces documents stratégiques se retrouve dans cette démarche participative qui impose que l'ensemble des acteurs soit associé à l'élaboration et au suivi des politiques. Une telle démarche vise à renforcer le débat démocratique et à s'assurer de l'efficacité de l'aide. La Charte africaine des valeurs et des principes de la décentralisation, de la gouvernance locale et du développement local stipule, en son article 12.3, que « *les gouvernements locaux ou les autorités locales encouragent la participation de tous les segments de la société aux activités de planification, de mise en œuvre, de suivi et d'évaluation des politiques, programmes et projets, à travers des plates-formes communautaires structurées et d'autres formes de participation, en vue d'assurer la fourniture de services de qualité* ».

Pour Payadowa Boukpessi, ministre de l'Administration territoriale du Togo, la décentralisation est un processus important dans la lutte contre la pauvreté. Elle permet aux élus locaux, qui sont plus proches des populations, de contribuer de façon plus efficace au règlement des difficultés d'accès aux services de base. Cette idée est une constante dans les orientations nationales de réduction de la pauvreté depuis l'adoption du DSRP en 2001 à l'élaboration de la politique nationale de développement pour la période 2018-2022. L'adoption par le gouvernement togolais de la stratégie intérimaire de réduction de la pauvreté en 2008 marque le renouement des relations entre le Togo et la coopération financière internationale. L'élaboration de cette stratégie au niveau national s'est déroulée dans un cadre participatif axé sur la consultation des acteurs relevant de toutes les composantes de la vie économique et sociale, de toutes les associations de développement, organisations communautaires, etc.[735]

308. - La réussite de la stratégie est subordonnée à son appropriation par les populations cibles. La prise en compte des besoins réels des populations, afin de favoriser le développement économique et social au niveau local, passe par une réforme de l'État centralisé. Dans sa résolution du 28 février 1995, l'Assemblée générale de l'ONU reconnait « *le rôle important que les gouvernements et les administrations publiques peuvent jouer en assumant les responsabilités nouvelles résultant de la poursuite d'une croissance économique soutenue et du développement durable dans tous les pays, dont notamment la mise en place d'infrastructures de base, la promotion du développement social, la lutte*

---

[735] *Cf.*, Gouvernement togolais DRSP-I 2008 p. 3 et s.

contre les disparités socio-économiques et la pauvreté, la création de conditions appropriées pour le secteur privé et la protection de l'environnement»[736].

Ainsi, c'est à travers l'adoption des DSRP que les institutions internationales entendent produire la démocratie locale, en donnant la parole aux populations les plus défavorisées. La décentralisation apparait alors comme une « *réappropriation du pouvoir par le peuple* »[737] qui permet de garantir la stabilité politique indispensable à toute politique de développement. Dans les pays minés par des conflits ethniques, la décentralisation peut offrir un cadre institutionnel de choix qui fédère l'ensemble des belligérants. C'est d'ailleurs en ce sens qu'elle a été le plus souvent utilisée dans les pays d'Afrique noire subsaharienne[738]. Les programmes de lutte contre la pauvreté et les politiques en faveur de la décentralisation deviennent, dans les pays d'Afrique subsaharienne, un prétexte pour un partage de pouvoirs, de ressources et de compétences entre l'échelon national et local pour la résolution de crise sociopolitique. Toutefois, cette démarche ne saurait aboutir sans une bonne gouvernance.

2. La bonne gouvernance, une clé de réduction de la pauvreté

309. - La gouvernance est « un mot-talisman paré de tous les fantasmes associés à l'action publique, tout en revêtant le vocabulaire rassurant de l'objectivité technique »[739]. L'imprécision s'accroit lorsqu'on lui adjoint l'adjectif « bonne »[740]. La bonne gouvernance n'est pas un concept juridique[741], elle renvoie à l'idée utopique du « meilleur gouvernement »[742] ainsi qu'aux idées de légitimité et de participation citoyenne. Elle apparait dans le discours international à la faveur des stratégies de lutte contre la pauvreté initiée par les institutions de Bretton Woods et les agences d'aide

---

736A/RES/49/136. Résolution adoptée par l'Assemblée générale [sur le rapport de la Deuxième Commission (A/49/726)] du 28 février 1995.

737 R. OTAYEK, « La décentralisation comme mode de redéfinition de la domination autoritaire ? Quelques réflexions à partir de situations africaines », *op.cit.*, p. 12.

738 Voir les développements *supra* n° 248 et s.

739 J. PITSEYS, « Le concept de gouvernance », *RIEJ*, vol. 65, n° 2, 2010, p. 207.

740 R. BOUSTA, « Le discours sur la "bonne gouvernance" : arrêt sur image », in G. KOUBI, W. TAMZINI (dir.), *Discours administratifs, droit (s) et transformations sociales*, éd. IRJS, coll. « Les voies du droit », 2020, p. 238.

741 J. CAILLOSSE, « Question sur l'identité juridique de la "gouvernance" », in R. PASQUIER, V. SIMOULIN, J. WEISBEIN, *La gouvernance territoriale. Pratique, discours et théories*, LGDJ, coll. « Droit et société », 2007, p. 52.

742 R. BOUSTA, *Essai sur la notion de bonne administration*, L'Harmattan, coll. « Logiques juridiques », 2010.

au développement. Tour à tour qualifié d'« attrape-tout », de « fluctuant », de « transversal », le concept de bonne gouvernance a fini par s'imposer comme un redoutable carrefour d'ambiguïtés qui mérite chaque fois des précautions dans son utilisation[743]. En effet, les facettes de la bonne gouvernance sont multiples en raison du caractère polysémique du concept. Cette polysémie n'est pas en soi une difficulté dans le cadre de notre étude, puisque l'essence même de cette notion est connue. La bonne gouvernance est la manifestation expresse du désaveu du modèle de politique traditionnel qui laisse aux autorités politiques la responsabilité de la gestion des affaires publiques pour mettre l'accent sur la multiplicité des acteurs pouvant intervenir dans cette gestion, moyennant l'instauration de processus de négociation entre cette diversité d'intervenants. En clair, la bonne gouvernance fait ici référence à l'exemplarité de l'action publique et à la participation des citoyens à la gestion des affaires publiques[744].

310. - Dans son analyse de la gouvernance, la Banque mondiale semble *a priori* faire clairement la distinction entre les dimensions politiques et économiques du concept. Elle la considère comme étant « *la manière dont le pouvoir est exercé pour gérer les ressources nationales, économiques et sociales consacrées au développement* »[745]. Ayant un mandat limité aux aspects économiques, elle devrait s'intéresser uniquement à ce que la bonne gouvernance contribue au développement et à la réduction de manière durable de la pauvreté dans le monde. L'échec des politiques d'ajustement structurel mis en place dans les pays d'Afrique subsaharienne amène la Banque mondiale et le FMI à remettre en question l'efficacité des modes des gestions des affaires publiques dans ces États et plus largement leur manière de gouverner, ce qui implique une redéfinition des rapports entre les pouvoirs publics et les administrés. Les agences des Nations unies, notamment le PNUD, développent une conception plus large de la bonne gouvernance englobant à la fois une dimension économique et politique. Pour cette dernière, la bonne gouvernance c'est la bonne gestion des affaires publiques et la démocratie, aussi bien sur le plan politique, économique et administratif, avec comme principaux indicateurs : la transparence, l'éthique et le respect de la déontologie dans la conduite des affaires ;

---

743 J. DU BOIS DE GAUDUSSON, « La décentralisation menacée par la (bonne) gouvernance », in *Mélanges Franck Moderne*, Dalloz, 2004, pp. 995-1009.
744 R. BOUSTA, « Le discours sur la "bonne gouvernance" : arrêt sur image », *op.cit.*, p. 241.
745 WORLD BANK, *Managing development – the governance dimension*, Washington, 1996.

l'indépendance et la crédibilité des juges et des tribunaux ; des élections libres et transparentes ; la décentralisation ; le renforcement des contre-pouvoirs[746].

311. - Ainsi, « la confiance et la transparence de l'action publique concrétisent un aspect normatif » du discours international sur la bonne gouvernance, « qui implique l'ouverture et l'accessibilité des politiques publiques aux citoyens »[747]. Les institutions de Bretton Woods et les agences internationales d'aides au développement sont convaincues que la cause de l'échec des politiques de développement entamées depuis 1980 réside dans la mauvaise gestion, la corruption, le clientélisme. Ces maux sont érigés en traits distinctifs de l'État dans les pays du Sud[748]. Ainsi, l'inclusion de la bonne gouvernance traduit selon un rapport du Fonds international de développement agricole (FIDA) de 1999, l'idée généralement acceptée que la qualité de la gouvernance influe sur les perspectives de croissance et l'évolution de la pauvreté dans un pays. Dans le même sens, l'association internationale de développement a estimé que la bonne gouvernance est cruciale pour le bon déroulement du processus de développement et pour l'efficacité de l'aide publique au développement[749]. C'est ainsi que la bonne gouvernance va, face à l'instabilité politique du début de la décennie 90, devenir une condition d'accès aux financements des bailleurs de fonds, qui la considèrent comme la clé de la réduction de la pauvreté en Afrique.

312. - Cependant, des études démontrent que la corrélation entre bonne gouvernance et développement n'est pas aussi évidente qu'on peut être amené à le penser. La bonne gouvernance tout comme la mauvaise gouvernance peuvent être un facteur de croissance économique[750]. Ainsi, la mauvaise gouvernance ne suffit pas à elle seule à expliquer le faible niveau de développement qui prévaut dans les pays du Sud[751]. La lutte

---

[746] PNUD, *Public Sector Management, Governance and sustainable Human Development,* New York, 1995.

[747] R. BOUSTA, « Le discours sur la "bonne gouvernance" : arrêt sur image », *op.cit.*, p. 247.

[748] J.-F. MEDARD, « La spécificité des pouvoirs africains », *Pouvoirs*, n °25, 1983, p. 5.

[749] FIDA, Rapport sur la bonne gouvernance : une mise au point, 1999.

[750] Pour une analyse complète de cette relation entre bonne gouvernance et développement : voir N. MEISEL, J. OULD AOUDIA, « L'insaisissable relation entre "bonne gouvernance" et développement », *Revue économique*, vol. 59, n° 6, 2008, p. 1159.

[751] *Idem.*

contre la pauvreté et la bonne gouvernance ont été les grandes dimensions du paradigme néolibéral qui s'est imposé au reste du monde grâce aux institutions financières internationales[752]. Mais plus de trente années après le début de leur intervention dans les pays en développement, les résultats manquent toujours à l'appel.

313. - Moins qu'un processus durable, la bonne gouvernance est davantage un phénomène de mode porté par la Banque mondiale et les institutions internationales pour assurer l'efficacité de l'action publique. Il faut néanmoins rappeler que cette démarche comporte un aspect politique et social non négligeable. Elle assure la diffusion de valeurs idéologiques de démocratisation et de limitation du rôle de l'État par l'intégration des institutions décentralisées capables d'assurer une mobilisation effective des populations locales. Aux plans politique et social, les collectivités territoriales ont vocation alors à restaurer la crédibilité et la légitimité de l'État et conduire à un développement économique harmonieux basé sur les initiatives communautaires et individuelles avec la mise en place d'un cadre incitatif qui promeut les dynamiques locales.

B- Le rôle renouvelé des collectivités territoriales dans les stratégies de lutte contre la pauvreté

314. - La Banque mondiale, dans son rapport de 1997 sur le développement dans le monde[753], place le processus de décentralisation au cœur de son concept de « bonne gouvernance ». La rencontre de la décentralisation, de la bonne gouvernance et de la lutte contre la pauvreté, ainsi que le relève Jean du Bois de Gaudusson, est la conséquence de l'incapacité de l'État à assurer les services essentiels à la population[754]. Alors, on a vu dans les collectivités locales un substitut pour résoudre la crise de l'État africain. Une crise qui affecte ce dernier dans ses différents aspects d'État-nation, d'État interventionniste et providence, d'État central[755]. Cela entraine un recentrage des missions de l'État et une redéfinition de son profil à travers le nouveau slogan « moins d'État, mieux

752 A. M. DIOP, « Les politiques de développement élaborées par la Banque Mondiale à l'intention des pays de l'hémisphère Sud de 1980 à nos jours : origines, variantes et implications », *Afrique et Développement*, vol. 41, n° 1, 2016, p. 180.
753 BANQUE MONDIALE, Rapport sur le développement dans le monde, 1997.
754 J. DU BOIS DE GAUDUSSON « La décentralisation menacée par la bonne gouvernance… », *op.cit.*, p. 997.
755 *Idem.*

d'État». Expliquant cette formule, Jacques Nzouankeu estime que le « moins d'État » se tourne vers le passé pour décrire ce qu'il faut changer. Le « moins » suggère que ce changement doit être réducteur, qu'il s'agit de décharger l'État de quelque chose. Le « mieux d'État » est quant à lui un regard sur l'avenir. Il décrit ce qui sera ou devra être la nouvelle image, le nouveau profil de l'État. Le « mieux » exprime en abrégé un jugement de valeur qui suggère qu'il s'agira d'un changement qualitatif, que le profil de l'État lorsqu'on l'aura déchargé sera préférable à ce qu'il est actuellement[756].

315. - Dans le but de rendre les collectivités territoriales capables de se prendre en charge afin de réduire la pauvreté et d'améliorer le bien-être de la population, notamment par la fourniture des services publics de base, les pays en développement de l'Afrique subsaharienne se sont engagés à aller vers une décentralisation réelle et effective qui suppose des collectivités territoriales fortes et opérationnelles. La vocation générale des collectivités territoriales est de contribuer à un développement économique et social plus harmonieux. Elles doivent mettre en place des dispositifs qui permettent de garantir la participation des habitants, en particulier des personnes pauvres, aux institutions afin de favoriser leur accès aux services sociaux de base[757]. Pour la réalisation de cet objectif, les États ont consenti, sous l'effet des pressions internes (mouvements sociaux) et externes (exigence des partenaires au développement), au transfert de certaines de leurs compétences au profit des collectivités territoriales, notamment en matière sociale.

Les collectivités territoriales se sont ainsi vues confier la charge de la promotion du développement local. Cela signifie qu'elles ont désormais la charge de la conception et de la mise en œuvre des projets de développement en matière d'éducation, de santé, d'accès aux infrastructures, etc. Les autorités locales doivent à cet effet s'engager avec les différents partenaires au développement sur des modes d'intervention qui privilégient la concertation, la participation, le consensus autour des

---

[756] J. M. NZOUANKEU, « La problématique de la gouvernance vue du Sud : aspects socioculturels », Communication à la table préparatoire n° 3 : *La bonne gouvernance : objet et condition du financement*, disponible en ligne https://base.afrique-gouvernance.net/docs/bonne_gouv.aspects_socio-culturels.pdf. Consulté le 22 aout 2021.

[757] Cf. Rapport sur le développement dans le monde 1997, préc.

axes majeurs du développement local[758]. Des comités de gestion ont été institués dans les différentes collectivités. C'est le cas des Comités villageois de Développement (CVD), des Associations des Parents d'élèves (APE), etc. qui collectent les informations, participent à la concertation et à l'évaluation des projets. En s'appuyant sur une décentralisation effective, l'espace local offre des possibilités pour réinventer et mettre en application de nouveaux modes de gestion participative des affaires publiques et du développement local. Les institutions décentralisées apparaissent comme des acteurs clés dans l'amélioration des conditions de vie des populations[759]. La proximité est considérée comme un facteur important de la bonne gouvernance en ce qu'elle met en relation directe le besoin social vécu par le citoyen et la décision publique locale qui est appelée à y répondre[760]. La décentralisation devient indispensable pour quiconque souhaite s'engager dans la voie de la bonne gouvernance.

316. - Toutefois, malgré ces efforts pour établir un cadre institutionnel de participation citoyenne pour lutter contre la pauvreté et réduire les inégalités entre les populations, l'évaluation de la décentralisation des services sociaux de base révèle de nombreuses insuffisances. La décentralisation, souligne Karine Tade, n'est pas suffisamment prise en compte dans les planifications nationales[761]. À cela s'ajoute la faiblesse des services déconcentrés censés assurer l'assistance et le conseil aux collectivités territoriales. Pour garantir l'effectivité de la décentralisation et en faire un véritable outil de développement, les autorités décentralisées doivent trouver aisément au même niveau territorial qu'elles, des autorités déconcentrées agissant au nom de l'État[762]. Les études de terrain ont montré que les services déconcentrés « *ne sont pas dotés conséquemment en ressources humaines et en matériels pour jouer pleinement leur partition dans*

---

758 M. NTOLO BEKOA, *Banque mondiale et développement des pays d'Afrique subsaharienne : l'impact des programmes mis en œuvre au Bénin, au Cameroun et au Togo,* thèse, dactyl., Université de Lyon, 2014.

759 G. ENEE, « Les enjeux des actions de coopération décentralisée en Afrique de l'ouest dans un contexte de décentralisation territoriale. L'exemple burkinabè », *Espaces et sociétés* n° 31, 2011, p. 95.

760 *Idem.*

761 K. TADE, *Impact des projets de coopération décentralisée sur la réduction des inégalités d'accès aux services sociaux de base : Cas des coopérations décentralisées franco-burkinabés et franco-béninoises,* thèse dactyl., Université Paris-Saclay, 2016.

762 N. KADA, « La réforme de l'état territorial », *RFAP,* vol. 1, n° 141, 2012, p. 109.

l'accompagnement des collectivités territoriales émergentes»[763]. Les collectivités territoriales et les services déconcentrés doivent former ensemble un dispositif cohérent et complémentaire pour la fourniture de services publics de qualité aux populations.

317. - Beaucoup restent donc à faire pour permettre l'ancrage du modèle d'administration décentralisé. Les élus locaux, tout comme les membres des services techniques déconcentrés, doivent s'approprier le modèle pour pouvoir exercer effectivement leurs nouvelles compétences. Les partenaires techniques et financiers vont, par ailleurs, se mobiliser pour assurer le « service après-vente » de la diffusion du modèle décentralisé qu'ils prônent avec vigueur, en renforçant les capacités des différents acteurs impliqués dans le processus de décentralisation. Dans les pays d'Afrique subsaharienne, en particulier en Afrique noire francophone, on note une présence importante de la coopération française qui n'hésite pas à mettre en scène ces propres collectivités territoriales comme nouvel outil d'aide au développement dans sa stratégie d'influence diplomatique et économique.

Section 2 : La coopération Nord-Sud, un vecteur de promotion de l'administration décentralisée en Afrique

318. - La coopération décentralisée, ou le jumelage, désigne les relations de toute forme, de collaboration ou d'actions que les collectivités territoriales africaines entretiennent avec des autorités locales étrangères de même niveau ou non[764]. Si les deux expressions sont aujourd'hui utilisées indistinctement par les acteurs de la coopération, la première

---

[763] K. TADE, *Impact des projets de coopération décentralisée sur la réduction des inégalités d'accès aux services sociaux de base : Cas des coopérations décentralisées franco-burkinabés et franco-béninoises*, *op.cit.*, p. 187.

[764] Pour une définition en droit français, B. FAURE, *Droit des collectivités territoriales*, Dalloz, coll. « Précis », 2021, n° 648.

semble être la terminologie la plus usitée dans les textes de loi[765] Face au désengagement de l'État et à la reconnaissance des collectivités territoriales comme acteur du développement économique local, les villes françaises sont appelées à faire valoir leur expertise en matière de décentralisation[766] auprès de leurs homologues africains. La coopération décentralisée rencontre ainsi les enjeux de développement dans les pays du Sud. Bien qu'elles soient restées longtemps en dehors de l'action extérieure au nom de la conception absolue du principe d'indivisibilité et de souveraineté de l'État, les collectivités territoriales sont désormais appelées à jouer un rôle dans l'exercice des compétences en matière de relations internationales. La pratique des jumelages entamée après la Seconde Guerre mondiale va ouvrir une brèche dans la reconnaissance par l'État de l'action extérieure des collectivités territoriales. À partir de 1980, l'action extérieure des collectivités territoriales va connaître une intégration significative dans le droit, marquant ainsi leur capacité à entreprendre des actions en dehors des frontières nationales. Cette irruption des collectivités territoriales sur la scène internationale (**§1**), dans un domaine autrefois monopolisé par l'État, ne semble pas heurter ce dernier qui par différentes stratégies l'instrumentalise (**§2**) pour accroitre sa présence sur le continent africain.

§1.- La coopération décentralisée au service d'une stratégie globale d'aide au développement

319. - L'affirmation des collectivités territoriales sur la scène internationale porte atteinte à la compétence régalienne de l'État dans la

---

765Voir la loi n° 2019-006 du 26 juin 2019 relative à la décentralisation et aux libertés locales au Togo, Chapitre V- Section II : Coopération entre les collectivités territoriales nationales et les collectivités territoriales étrangères ou coopération decentralisée (Articles 43 à 46) ; Ord. n° 2010-54 portant code général des collectivités territoriales au Niger, Livre III- Des relations entre les collectivités territoriales avec notamment l'article 329 qui institue une commission nationale de la coopération décentralisée. En France CGCT, Livre 1er, Titre unique, Chapitre V Action extérieure des collectivités territoriales (Articles L 1115-1 à L 1115-7). Au Burkina Faso, la loi consacre plutôt l'expression de jumelage : art. 127 à 129.

766 Cf. F. PETITEVILLE, *La coopération décentralisée. Les collectivités locales dans la coopération Nord-Sud*, L'Harmattan, 1995.

conduite des Affaires étrangères[767]. En effet, une telle action des collectivités est d'abord apparue comme contraire à la conception française de l'État-nation, uni et souverain. Les doutes relatifs au risque d'empiètement des collectivités territoriales sur le monopole de la conduite des affaires diplomatiques dont dispose l'État se sont dissipés au fil du temps. Les deux acteurs, bien qu'intervenant en dehors des frontières nationales, agissent avec des acteurs distincts. Démembrements internes de l'État, les collectivités territoriales se présentent comme des acteurs de la globalisation du droit[768]. La reconnaissance en France d'une capacité d'action à l'international des collectivités territoriales a eu des conséquences dans les anciennes colonies françaises d'Afrique. En effet, l'essor des entités décentralisées sur la scène internationale en a fait des acteurs importants du financement du développement dans les pays du Sud. Or, pour agir à l'international, les collectivités territoriales ne peuvent contracter qu'avec des acteurs de même rang qu'elles. La volonté de captation de cette nouvelle source de financement a conduit progressivement les États de l'espace francophone à s'engager vers une décentralisation effective. Le développement de l'action des collectivités territoriales s'insère dans un nouveau cadre juridique (**A**) et fait l'objet d'un encadrement qui se révèle à l'analyse limitée et pour cause, elle s'inscrit dans une stratégie globale de l'aide au développement (**B**).

A- Une reconnaissance progressive de l'action extérieure des collectivités territoriales

320. - Le droit de l'action extérieure des collectivités territoriales est d'appréhension difficile en raison de son caractère transnational. La transnationalité[769] pose la capacité des collectivités territoriales à établir des relations avec des collectivités étrangères. Les collectivités territoriales ne disposent pas de la compétence pour organiser elles-mêmes le cadre de leurs interventions extérieures. Les relations qu'entretiennent les collectivités françaises et leurs homologues étrangères sont soumises à un

---

767 *Cf.* A. SIMON, *Les compétences régaliennes et les collectivités territoriales*, thèse, dactly., Université Grenoble Alpes, 2018.

768 M. CHAMBON, « Les collectivités territoriales, acteurs de la globalisation du droit administratif ? », *RFDA*, 2019, p. 994.

769 La transnationalité désigne « des relations internationales qui mettent en présence deux êtres dont l'un au moins n'est pas un État ». J. COMBACAU et S. SUR, *Droit international public*, Montchrestien, 2014, p. 4.

régime juridique dont « *l'État français est l'élément principal et directeur* »[770]. Il définit le cadre juridique de la coopération décentralisée. Il faut néanmoins préciser qu'en réalité, ce droit se trouve au carrefour de plusieurs disciplines juridiques allant au-delà du droit administratif *stricto sensu*[771]. L'action extérieure des collectivités territoriales tire ses fondements dans un droit positif interne au départ « *flou et lacunaire* »[772] jalonné par une influence des normes européennes (**1°**) et faisant l'objet d'un encadrement par le législateur (**2°**). L'action des collectivités territoriales françaises hors des frontières nationales s'établit donc au cœur d'un dispositif complexe de normes qui n'a pas cessé de progresser et qu'il conviendrait d'expliciter.

1. L'influence des normes européennes dans la reconnaissance de l'action extérieure des collectivités territoriales

321. - Les faits ont précédé le droit dans le cadre de l'action extérieure des collectivités. Le développement de l'action des collectivités françaises en dehors du territoire national s'est fait en l'absence de texte. Malgré ce vide juridique, les collectivités ont développé de nombreuses relations avec des homologues étrangères insérant alors leur action à mi-chemin entre le droit international public et le droit interne[773] avant que le législateur ne se saisisse de la question. Ces premières relations transnationales correspondent à la pratique des jumelages qui avait pour vocation de rapprocher culturellement les peuples[774]. Au début des années 1980, l'action extérieure des collectivités territoriales va prendre un tournant décisif. À la faveur des proximités géographiques, de l'existence d'intérêts économiques communs, les collectivités vont multiplier les contacts entre elles. Sous l'impulsion du Conseil de l'Europe, les relations entretenues par les collectivités territoriales par-delà les frontières terrestres et maritimes ont donné naissance à une coopération dite transfrontière.

---

770 B. FAURE, *Droit des collectivités territoriales, op.cit.*, n° 650.

771 Il y'a un recours constant aux règles de droit international public.

772 F. PETITEVILLE, *La coopération décentralisée. Les collectivités locales dans la coopération Nord-Sud, op.cit.*, p. 25.

773 X. BARELLA, « La coopération décentralisée à la recherche d'une sécurité juridique renforcée », *AJDA*, 2008, p. 1580.

774 Sur les jumelages de coopération, d'amitié, F. PETITEVILLE, *La coopération décentralisée. Les collectivités locales dans la coopération Nord-Sud, op.cit.*, p. 12.

322. - La convention de Madrid[775] constitue le premier instrument juridique de promotion de la coopération transfrontière. Cette convention-cadre européenne, ratifiée par la France le 14 février 1984, va contribuer au développement de la coopération transfrontalière des collectivités territoriales. Toutefois, l'apport de la convention de Madrid va se révéler très modeste en pratique. En effet, la lecture combinée des articles 1 et 4 permet de constater que la convention n'entend donner aucun droit subjectif aux collectivités territoriales pour leurs actions extérieures[776]. Elle se borne simplement à « favoriser et promouvoir les initiatives » (art 1) et à « résoudre les difficultés d'ordre juridique, administratif ou technique » qui seraient « de nature à entraver les développements et le bon fonctionnement de la coopération transfrontalière » (art. 4). Ce faisant, la convention de Madrid n'œuvre pas de façon suffisamment précise et impérative à sa cause[777].

Néanmoins, sur la base de cette convention des accords bilatéraux et multilatéraux vont s'établir pour renforcer le droit de la coopération transfrontalière. La France va notamment conclure avec l'Italie[778] et l'Espagne[779] des accords de relations bilatérales et un accord plurilatéral[780] avec l'Allemagne, le Conseil Fédéral suisse et le Grand-duché du Luxembourg. Ces différents accords vont donner lieu à une diversité de situations rendant complexe l'appréhension de ce droit et appelant ainsi à un effort d'unification du régime. Bernard Dolez va encourager le Conseil des ministres du Conseil de l'Europe à élaborer un protocole additionnel qui, selon son rapport explicatif, « *expose des instruments juridiques éprouvés par l'expérience, unifie les principes fondamentaux de la coopération transfrontalière* [...] *et propose des solutions adaptées aux parties contractantes* »[781]. La convention de Madrid est alors complétée par trois protocoles additionnels dont le deuxième innove en élargissant le champ d'application du système

---

[775] La Convention-cadre européenne sur la coopération transfrontalière des collectivités ou autorités territoriales du 21 mai 1980.

[776] Les collectivités ne disposent de pouvoir en la matière que dans le cadre des dispositions du droit interne prévu à cet effet, P. COMBEAU, « Action extérieure des collectivités territoriales », *JCP- A*, Fasc. 129-40.

[777] B. FAURE, *Droit des collectivités territoriales, op.cit.*, n° 659.

[778] Accord franco-italien du 26 novembre 1993, entré en vigueur le 6 octobre 1995.

[779] Accord franco-espagnol conclu le 10 mars 1995, entré en vigueur le 24 février 1997.

[780] Accord de Karlsruhe conclu le 26 janvier 1996, entré en vigueur le 1 er septembre 1997 entre la France, partiellement complété entre les parties par un accord conclu à Bâle le 21 septembre 2000, entré en vigueur le 1er juin 2001.

[781] *Cf.* A. CABANES, A. ROBBES, « La coopération décentralisée : comment s'extraire des contingences de l'intérêt public local ? », *AJDA*, 2003, p. 593.

européen de la coopération au-delà des relations de voisinage[782]. Il introduit la notion de coopération interterritoriale qu'il définit en son article premier comme « *toute concertation visant à établir des rapports entre collectivités ou autorités territoriales de deux ou plusieurs Parties contractantes, autres que les rapports de coopération transfrontalière des collectivités voisines, y inclus la conclusion d'accords avec les collectivités ou autorités territoriales d'autres États* ». La coopération transfrontalière s'ouvre ainsi progressivement à la reconnaissance d'une forme de coopération plus large qu'est la coopération décentralisée.

323. - La Commission européenne est longtemps restée silencieuse sur la question de la coopération décentralisée transfrontalière. Le statut ambivalent des collectivités territoriales en droit de l'Union européenne l'empêche de se prononcer directement sur le cadre juridique de l'action extérieure des collectivités. Face au foisonnement des accords conclus par les États membres de l'Union, la commission va cependant mettre en place un cadre européen pour soutenir les initiatives de coopération décentralisée. Les projets de coopération décentralisée transfrontalière sont soutenus par le biais du Fonds européen de développement régional (FEDER) et le programme INTERREG des fonds structurels[783]. La mise en œuvre des jumelages entre les territoires des pays du Nord et ceux des pays du Sud a été également encouragée par l'UE en 1989 dans le cadre de la signature de la convention de Lomé IV avec les pays ACP. Dans ce texte apparaissait déjà le terme de « coopération décentralisée » pour désigner les relations de partenariats entre collectivités territoriales[784].

---

782 P. COMBEAU, « Action extérieure des collectivités territoriales », *op.cit.* ; M. VERPEAUX, L. JANICOT, *Droit des collectivités territoriales, op.cit.*, p. 496 et s.

783 B. FAURE, Droit des collectivités territoriales, op.cit., n° 659.

784 Convention Lomé IV 1989, art. 20 : « En conformité avec les articles 2, 3 et 13 et afin d'encourager l'épanouissement et la mobilisation des initiatives de tous les acteurs des États ACP et de la Communauté, susceptibles d'apporter leur contribution au développement autonome des États ACP, la coopération appuie également, dans les limites fixées par les États ACP intéressés, les actions de développement d'acteurs économiques, sociaux et culturels, **dans le cadre d'une coopération décentralisée**, notamment sous forme de conjonctions d'efforts et de moyens entre homologues des États ACP et de la Communauté. Cette forme de coopération vise en particulier à mettre au service du développement des États ACP les compétences, les modes d'action originaux et les ressources de ces acteurs. Les acteurs visés par le présent article sont les

324. - Les normes européennes ont jeté les bases de la coopération décentralisée entre les collectivités des pays du Nord et du Sud. Le Conseil de l'Europe à travers la convention de Madrid et ces protocoles additionnels n'a pas manqué l'occasion d'inviter les États à aménager la compétence extraterritoriale des collectivités territoriales. Le législateur français, longtemps resté coi face à cet appel, a fini par y répondre en donnant une base légale à l'action extérieure des collectivités territoriales.

2. La reconnaissance de l'action extérieure des collectivités territoriales par le législateur français

325. - Les collectivités territoriales sont reconnues comme jouant un rôle important dans la conduite de l'action publique au sein de l'État. En revanche, pour toutes les actions impliquant une intervention au-delà du cadre étatique, « *la quasi-exclusivité d'action du pouvoir central était soigneusement protégée* »[785]. La question de la compétence des collectivités territoriales françaises à nouer des relations transnationales a donné lieu à d'importants débats juridiques[786] relatifs à l'articulation possible entre le principe de libre administration des collectivités et la reconnaissance d'une compétence internationale aux entités infraétatiques. Selon une partie de la doctrine, le fait que l'article 72 de la Constitution dispose en son alinéa 3 que les « collectivités s'administrent librement », traduit une reconnaissance au profit de ces collectivités d'une compétence internationale qui peut servir de fondement juridique à la coopération décentralisée. Cette position n'a pas manqué d'être contestée[787], car cet article prévoit que les collectivités territoriales exercent librement leurs compétences « dans les conditions

---

pouvoirs publics décentralisés, les groupements ruraux et villageois, les coopératives, les entreprises, les syndicats, les centres d'enseignement et de recherche, les organisations non gouvernementales de développement, des associations diverses et tous groupes et acteurs capables et désireux d'apporter leur contribution spontanée et originale au développement des États ACP ».

785 P.-Y. CHICOT, « Les politiques internationales des collectivités territoriales françaises : une compétence locale atypique ? », in J.-F. BRISSON (dir.), *Les transferts de compétences de l'État aux collectivités locales*, L'Harmattan, 2009, p. 435 et s.

786 Cf., B. DOLEZ, Coopération décentralisée et souveraineté de l'État. Contribution à l'étude du régime juridique de l'action extérieure des collectivités territoriales, thèse, Université de Lille 2, 1993.

787 J. FERSTENBERT, « Droit et pratique de l'action extérieure des collectivités territoriales », *Revue du centre ouest* n° 1, 1988, pp.16-17.

prévues par la loi ». Or, ainsi que le relève Xavier Barella, avant 1982, aucun texte n'aborde expressément le sujet de la coopération décentralisée et, de ce fait, l'article 72 peut difficilement être considéré comme un fondement juridique suffisant de l'action internationale locale[788].

326. - Le régime juridique des relations entre les collectivités territoriales françaises et leurs homologues étrangères était particulièrement « flou et lacunaire » en raison du caractère fragmentaire et paradoxal des règles du droit positif[789]. La loi du 2 mars 1982 est la première à s'être saisie de la question de l'action extérieure des collectivités territoriales. Les dispositions de l'article 65 de cette loi autorisaient les collectivités à établir des relations uniquement avec les régions frontalières. Cet article constitue l'exemple type de disposition législative difficilement compréhensible ayant donné lieu à une série de circulaires[790] qui sont venues préciser son contenu et étendre son application à toutes les collectivités. Le Conseil d'État a dénoncé ce foisonnement de texte et d'instruments en matière de coopération décentralisée, en ce qu'il ne contribue pas à la lisibilité du système[791].

327. - Finalement, l'action extérieure des collectivités va bénéficier d'un cadre normatif avec l'adoption de la loi du 6 février 1992[792]. Le titre IV de cette loi relatif à la coopération décentralisée est repris par le code général des collectivités territoriales[793]. Il consacre expressément l'existence d'une compétence extérieure des collectivités territoriales[794]. L'article 131-1, alinéa 1 de la loi de 1992, dispose que « *les collectivités territoriales et leurs groupements peuvent conclure des conventions avec des collectivités étrangères et leurs groupements, dans les limites de leurs compétences et dans le respect*

---

788 X. BARELLA, « La coopération décentralisée à la recherche d'une sécurité juridique renforcée », *op.cit.,* p. 1580.

789 C. AUTEXIER, « De la coopération décentralisée », *RFDA*, 1993, p.411.

790 Solution insatisfaisante d'après C. AUTEXIER pour qui, le juriste attaché à la hiérarchie des normes, à moins de considérer que ces circulaires, loin de créer du droit, ne faisaient qu'expliciter une question déjà réglée par la Constitution, et notamment son article 72.

791 *Le cadre juridique de l'action extérieure des collectivités locales*, La Documentation française, 2006.

792 . La loi n° 92-125 dite « ATR » administration territoriale de la République

793 CGCT, art. L. 1115-1 et s.

794 J. DU BOIS DE GAUDUSSON, « L'État français et le contrôle des actions extérieures des collectivités territoriales », in *Le droit appliqué à la coopération interrégionale en Europe*. Colloque de Thessalonique 11-12 sept. 1993, LGDJ, 1995, p. 227.

des engagements internationaux de la France»[795]. Sur le fondement de ce texte, les collectivités territoriales disposent désormais d'une compétence générale en matière internationale qu'ils s'exercent au moyen des conventions de coopération[796]. Il faut reconnaitre avec Bertrand Faure que le statut légal fixé par la loi de 1992 est recognitif, car ainsi que l'avons montré précédemment, les collectivités territoriales n'ont pas attendu l'édiction de cette loi pour jeter les ponts vers les collectivités étrangères[797]. Placées désormais dans la sphère de la politique étrangère autrefois monopolisée par l'État, les collectivités territoriales doivent agir dans le strict respect des contraintes posées par la loi. C'est à cette seule condition qu'elles pourront participer à l'exercice des fonctions régaliennes[798].

B- Le partage de la compétence de l'aide au développement entre les collectivités territoriales et l'État en France

328. - Le cadre juridique de la coopération décentralisée connait depuis sa consécration par le législateur une avancée significative qui ne manque pas de susciter de nouvelles interrogations. Le monopole de l'État dans la conduite des affaires diplomatiques au nom du principe de souveraineté a toujours été évoqué comme un obstacle à l'action extérieure des collectivités territoriales. La coopération décentralisée invite à repenser autrement les ressorts de la souveraineté. Si, on reconnait aux collectivités territoriales la possibilité d'agir au-delà des frontières dans le cadre de la coopération décentralisée, elles doivent s'abstenir de faire concurrence à l'État. La primauté de l'État, détenteur absolu de la souveraineté, constitue

---

[795] La nouvelle version de l'article L. 1115-1 dispose « Dans le respect des engagements internationaux de la France, les collectivités territoriales et leurs groupements peuvent mettre en œuvre ou soutenir toute action internationale annuelle ou pluriannuelle de coopération, d'aide au développement ou à caractère humanitaire. [...] A cette fin, les collectivités territoriales et leurs groupements peuvent, le cas échéant, conclure des conventions avec des autorités locales étrangères. Ces conventions précisent l'objet des actions envisagées et le montant prévisionnel des engagements financiers. Elles entrent en vigueur dès leur transmission au représentant de l'État dans les conditions fixées aux articles L. 2131-1, L. 2131-2, L. 3131-1, L. 3131-2, L. 4141-1 et L. 4141-2. Les articles L. 2131-6, L. 3132-1 et L. 4142-1 leur sont applicables ». On note la suppression notamment de la limite relative aux compétences.

[796] X. BARELLA, « La coopération décentralisée à la recherche d'une sécurité juridique renforcée », *AJDA*, 2008, p. 1580.

[797] B. FAURE, *Droit des collectivités territoriales, op.cit.*, n° 653.

[798] J-M. PONTIER, « La notion de compétences régaliennes dans la problématique de la répartition des compétences entre les collectivités publiques », *RDP*, 2003, p. 193.

encore une limite empêchant les collectivités territoriales d'entretenir des relations avec des États étrangers (**1°**). Les collectivités territoriales deviennent des acteurs clés de la solidarité internationale qui ne peuvent contracter qu'avec des collectivités infranationales. Les États de l'Afrique subsaharienne, afin de pouvoir bénéficier des avantages de l'action extérieure des collectivités françaises, sont obligés de mettre en place des institutions décentralisées, seules capables contracter avec leurs homologues françaises. Aussi, l'on a pu constater que l'action des collectivités territoriales ne pouvait se limiter à un processus essentiellement technique, elle est aussi politique (**2°**).

1. L'interdiction faite aux collectivités de contracter avec les États

329. - L'action extérieure des collectivités territoriales porte atteinte à la compétence internationale de l'État. La souveraineté, fondement même de l'existence de l'État, se matérialise par le monopole que possède l'État dans la conduite des relations internationales. Carré de Malberg définit la notion de souveraineté comme « *le caractère suprême d'un pouvoir suprême en ce que ce pouvoir n'en admet aucun autre ni au-dessus de lui ni en concurrence avec lui* »[799]. Cette « *qualité essentielle* »[800] de l'État lui réserve l'absolu « *droit de représenter la Nation* »[801] et donc de conduire sa politique étrangère. Aussi, le droit international public est indifférent au sujet des collectivités territoriales. L'État, en tant que détenteur de la souveraineté, est l'unique titulaire de la personnalité internationale. À ce titre, il bénéficie seul de la qualité de sujet de droit international qui ne peut être reconnue à des collectivités infraétatiques[802]. Les États sont ainsi les seuls capables d'imputation. Ses sujets, les collectivités territoriales, ne sont ici que de simples organes de l'État[803]. Cette idée d'imputation fait de l'État, l'auteur et le responsable de la conclusion de tout acte qu'il estime nécessaire pour maintenir ses relations avec des États étrangers ou des organisations internationales.

---

799 R. CARRE DE MALBERG, *Contribution à la théorie générale de l'État*, Sirey, 1920, p. 70.
800 A. ESMEIN, *Éléments de droit constitutionnel français et comparé*, éd. Panthéon-Assas, 2001, p. 1.
801 *Idem.*
802 P. COMBEAU, « Action extérieure des collectivités territoriales », *op.cit.*, p 7.
803 J. DAVID, *Les sujets infra-étatiques et le droit supranational. Le cas des collectivités territoriales*, Mémoire M2 recherche Bordeaux IV, 2008, p. 21.

330. - Le Conseil d'État a rappelé, dans un avis du 16 mai 1980 rendu à l'occasion de la conclusion de la convention de Madrid, que les collectivités territoriales françaises n'ont pas de compétence en matière de droit international. La compétence diplomatique appartient uniquement à l'État[804]. Dans un second avis du 25 octobre 1994, le Conseil d'État a également précisé la nature des conventions passées entre les collectivités françaises et étrangères. Pour ce dernier, ces conventions ne sont que de simples contrats « *qui ne constituent pas des engagements internationaux au sens du titre VI de la Constitution* [et] *ne sauraient déroger aux règles de légalité interne* »[805]. Malgré la reconnaissance par le législateur de la coopération décentralisée, celle-ci ne peut être mise au même degré que la conduite des affaires étrangères. Le Conseil constitutionnel adopte un raisonnement identique à celui du juge administratif, en rappelant la subordination des actions de coopération décentralisée au respect des engagements internationaux de la France[806]. S'il reconnait aux collectivités territoriales une compétence internationale « *ce n'est qu'au travers d'un grand luxe de* préc*autions, en examinant que celles-ci ne portent pas atteinte aux compétences régaliennes de l'État* »[807].

331. - Les collectivités territoriales sont donc limitées dans leur action au-delà des frontières nationales. Loin d'être un inconvénient, cela constitue un avantage considérable dans la promotion du modèle occidental de développement. En effet, tout État africain, qui voudrait tirer profit de l'apport considérable des collectivités territoriales françaises dans le domaine de l'aide au développement, se voit contraint de façon incidente de réformer son administration pour consacrer des entités infranationales. En fait, la reconnaissance d'une compétence extérieure aux collectivités françaises et le rôle que l'État entend leur faire jouer[808] vont agir comme un catalyseur en Afrique. La décentralisation, présente dans les discours des dirigeants africains après les indépendances, mais jamais réalisée faute de volonté politique, va connaitre un essor considérable. Pour les bailleurs de fonds, la réussite de la coopération décentralisée passe impérativement par le respect de trois conditions préalables : une décentralisation effective, un engagement en faveur de la démocratie et des garanties de transparence de la gestion publique.

---

804 CE, avis 16 mai 1980, *EDCE*, 1986, p. 109.

805 CE, avis 25 octobre 1994, *EDCE*, 1994, n°46.

806 DC, 26 janvier 1995, *Aménagement et développement du territoire.*

807 X. BARELLA, « La coopération décentralisée à la recherche d'une sécurité juridique renforcée », *AJDA*, 2008, p.1580.

808 Voir les développements *supra* § 2.

332. - Apparue tardivement dans le contexte africain[809], la coopération décentralisée est aujourd'hui consacrée dans la quasi-totalité des États africains. Les collectivités territoriales africaines bénéficient d'un cadre juridique à l'action extérieure dans les textes de loi relatifs à la décentralisation. Les politiques de décentralisation sont utilisées comme des supports à l'essor de la coopération décentralisée. Ainsi, Franck Petiteville relève que, « *partout où elle s'est développée, l'action extérieure des collectivités locales a bénéficié d'un contexte déterminant de décentralisation territoriale, qu'il s'agisse de dispositions juridiques anciennes, de réformes récentes, ou d'un desserrement progressif du contrôle étatique sur l'action des collectivités locales* »[810]. Face au désenchantement de l'État africain à favoriser le développement, la consécration de l'action extérieure des collectivités territoriales est un moyen de captation de nouvelles ressources financières ainsi qu'une réponse aux logiques externes qui commandent le changement de la nature de l'État en Afrique. La coopération décentralisée a par ailleurs permis de porter au-devant de la scène les villes africaines. Le développement de leur action internationale a légitimé et renforcé dans une certaine mesure la place des collectivités territoriales africaines face à l'État.

333. - L'appui institutionnel des collectivités territoriales françaises aux institutions décentralisées dans les pays du Sud, par le biais de la coopération décentralisée, parait être le meilleur moyen de renforcement des processus de construction et de fonctionnement participatif de l'État[811]. On assiste à un remplacement de la coopération bilatérale, qui a contribué au développement de la centralisation administrative, par une coopération décentralisée s'appuyant sur les administrations locales. Ainsi, la coopération décentralisée favorise l'essor de la décentralisation dans les pays de l'Afrique subsaharienne. L'action extérieure des collectivités françaises en Afrique est en ce sens empreinte de mobile politique, car les collectivités ne font preuve d'aucune neutralité. Pour Pierre Yves Chicot, une telle attitude est moins le produit d'un militantisme idéologique que la

---

809 L'expression « coopération décentralisée » apparait pour la première fois en 1989 dans le cadre des conventions ACP-UE

810 Association des professionnels de l'action européenne et internationale des collectivités territoriales, *L'action internationale des collectivités territoriales*, éd. Cavalier Bleu, 2012, p. 27.

811 L. CARRINO, « Les institutions locales et l'État. Nouvelles approches pour la coopération au développement », in M. GALY, E. SANNELLA (dir.), *Les défis de l'État en Afrique*, L'Harmattan, 2007, p. 189.

culture d'un pragmatisme qui lui-même est le résultat de la prise de conscience de l'interconnexion entre le global et le local[812]. L'impératif de développement économique, qui naissait au début des années 1990, a contribué à faire passer la dimension technique de l'action extérieure des collectivités à celle politique dans le cadre de la coopération décentralisée. Les collectivités territoriales françaises deviennent des acteurs, aux côtés de l'État, de la mise en œuvre de la politique d'aide au développement.

2. L'extension du domaine d'intervention des collectivités territoriales

334. - L'action extérieure des collectivités territoriales peut être présentée dans ces modalités comme plus technique que politique, mais indiscutablement, elle est porteuse d'un message politique[813]. L'assemblée délibérante est à l'origine de la décision qui engage la collectivité dans une coopération décentralisée. Une telle décision de la majorité politique au niveau local ne peut être exempte de mobiles politiques. L'État semble s'accommoder de ce constat en prenant des mesures favorables à l'action extérieure des collectivités. Le législateur lève les limites juridiques à l'exercice de la compétence internationale des collectivités territoriales et facilite leur présence sur la scène internationale à travers une série de lois.

335. - La loi du 6 février 1992 précitée prévoyait que les collectivités ne pouvaient conclure des conventions que dans les limites de leurs compétences. Le juge administratif était alors amené à vérifier l'existence d'un intérêt public local pour valider l'action internationale des collectivités[814]. D'abord, deux lois intervenues respectivement en 2005[815] et en 2006[816] vont permettre aux communes et aux EPCI chargés de la

---

[812] P.-Y. CHICOT, « Les politiques internationales des collectivités territoriales françaises : une compétence locale atypique ? », in J.-F. BRISSON, *Les transferts de compétences de l'État aux collectivités locales*, L'Harmattan 2009, p. 437.
[813] *Ibid.*, pp. 437-438.
[814] CE, 28 juill. 1995, *Cne de Villeneuve d'Ascq*, n° 129838, *AJDA*, 1995, p. 834, concl. R. Schwartz ; TA Poitiers, 18 nov. 2004, *Charbonneau c/ Dpt Deux-Sèvres*, *AJDA*, 2005, p. 486, note Y. Gounin ; CAA Bordeaux, 30 oct. 2007, *Dpt Deux-Sèvres c/ Charbonneau*, n° 05BX00167, *AJDA*, 2008, p. 198, note Y. Gounin ; *JCP A* 2008, note P. Combeau.
[815] Loi n° 2005-95 du 9 février 2005 relative à la coopération internationale des collectivités territoriales et des agences de l'eau dans les domaines de l'alimentation en eau et de l'assainissement, codifié à l'article L. 1115-1-1 du CGCT.
[816] Loi n° 2006-1537 du 7 décembre 2006 *relative au secteur de l'énergie*, codifié à l'article L. 1115-2 du CGCT.

distribution d'eau potable et de l'assainissement ainsi que du traitement des déchets ménagers d'affecter 1 % de leur budget à des actions de coopération. Les dispositions résultant de ces deux démarches législatives sont respectivement codifiées aux articles L. 1115-1-1 et L. 1115-2 du CGCT. Ce dispositif a suscité un grand intérêt dans les rangs des observateurs étrangers et des organisations internationales qui y ont vu un modèle de financement innovant et une contribution significative à la réalisation des Objectifs de développement durable (ODD)[817]. Ensuite, la loi Thiollière de 2007[818] a contribué à la reconnaissance définitive de l'action extérieure des collectivités territoriales en faisant de la coopération décentralisée une compétence à part entière reconnue à toutes les collectivités territoriales et à leurs groupements. Désormais, les collectivités territoriales et leurs groupements n'ont plus besoin de démontrer la présence d'un intérêt public pour la réalisation d'une action de coopération décentralisée[819]. Avec l'adoption en 2014 de la loi d'orientation et de programmation relative à la politique de développement et de solidarité internationale[820], la coopération décentralisée va être intégrée au cœur de la politique d'aide aux pays du Sud. Cette aide publique correspond à l'aide financière fournie par les États dans le but de promouvoir le développement économique et d'améliorer les conditions de vie dans les pays en développement[821]. L'APD s'inscrit dans le cadre global de la lutte contre la pauvreté depuis l'adoption et la mise en place par les Nations unies des objectifs du millénaire pour le développement[822].

336. - L'APD qui relève en principe de la compétence exclusive de l'État devient ainsi une compétence partagée entre l'État et les collectivités territoriales. L'adoption de la loi d'orientation et de programmation relative à la politique de développement et de solidarité internationale souligne l'importance du rôle que l'État entend faire jouer aux collectivités

817 Ministère des Affaires étrangères, *Guide juridique de l'action extérieure des collectivités territoriales*, La Documentation française, 1er éd., 2019, p. 17.

818 Loi n° 2007-147 du 2 février 2007 relative à l'action extérieure des collectivités territoriales et de leurs groupements.

819 Ministère des Affaires étrangères, *Guide juridique de l'action extérieure des collectivités territoriales, op.cit.*, p. 17.

820 Loi n° 2014-773 du 7 juillet 2014 d'orientation et de programmation relative à la politique de développement et de solidarité internationale.

821 Définition donnée par le CAD de l'OCDE

822 En 2019, 3 461 220 euros de l'APD des collectivités territoriales ont pour objectif principal la réduction de la pauvreté soit l'objectif de développement durable

territoriales en intégrant la coopération décentralisée dans la politique publique d'aide au développement. Annick Girardin, alors secrétaire d'État, affirmait à propos de cette loi qu'elle tend à « *consacrer les collectivités territoriales comme des acteurs à part entière de la politique française de développement et de solidarité internationale* »[823]. Les relations de jumelages entretenues à la veille des indépendances avec les collectivités africaines, marquées par leur caractère humanitaire, se muent en des projets caractérisés par une approche d'aide au développement et d'appui institutionnel[824]. En effet, les premières formes de coopération des villes francophones du Nord avec les villes Sud sont marquées par l'assistance humanitaire. Il s'agissait surtout de soulager la misère des populations du Sud par l'envoi de médicaments et de produits de première nécessité. Cette forme de coopération n'a pas complètement disparu aujourd'hui et prend la forme de financement de petites infrastructures telles, les forages d'eau, la construction de blocs scolaires, etc.[825] Le deuxième temps de la coopération est marqué par une volonté d'agir pour le développement afin de valoriser les ressources locales, d'accompagner l'émergence d'organisations et de favoriser les échanges entre les peuples. Dans ce cadre, sont mis en place des projets de développement de plus grande envergure dans tous les domaines de la vie sociale[826].

337. - Aujourd'hui, la coopération décentralisée prend davantage la forme d'un appui au processus de la décentralisation avec une volonté plus ou moins affirmée de mettre en place un partenariat mutuel centré sur un développement économique et durable, prenant en compte les intérêts des collectivités territoriales françaises et celles de leurs partenaires. Cependant, les rapports des premières avec les anciens territoires colonisés demeurent des rapports de dominants à dominés. L'observation des dynamiques spatiales de l'action internationale des collectivités françaises permet de se rendre compte qu'il y'a une superposition quasi identique de la carte de la coopération décentralisée avec les pays composant l'ex-empire colonial. Le fait que ces pays aient en commun le partage de la même langue et soient membres de l'Organisation internationale de la Francophonie renforce également la présence française. Les collectivités

---

823 Déb. parl., Sénat, séance du 26 mai 2014, JORF Sénat, 27 mai 2016 p. 4220.

824 G. ENEE, « Les enjeux des actions de coopération décentralisée en Afrique de l'ouest dans un contexte de décentralisation territoriale. L'exemple burkinabè », *op.cit.*, p. 91.

825 Ce type de coopération est souvent appelée « coopération container ».

826 M. BONDO, M. BALIKWISHA NYONYO, *La coopération décentralisée dans l'espace francophone*, Rapport 2013, Assemblée parlementaire francophone, p. 5 et s.

françaises continuent de s'insérer davantage dans une logique d'aide plus que dans une logique de partenariat avec les collectivités du Sud. L'intégration de l'action internationale des collectivités territoriales dans le cadre de la loi de programmation sur le développement et la solidarité internationale témoigne de l'attachement de l'action extérieure des collectivités territoriales à l'aide publique au développement[827].

338. - L'aide au développement consenti par les collectivités territoriales françaises au profit des collectivités africaines se concentre sur les enjeux internationaux que sont la décentralisation territoriale et le développement local. L'appui institutionnel aux collectivités africaines passe alors essentiellement par le renforcement des capacités du personnel dans la gouvernance locale. Alors que les collectivités françaises bénéficient dans ce cadre d'un dispositif institutionnel d'appui à la coopération décentralisée avec des orientations clairement identifiées, on ne peut en dire autant des collectivités territoriales africaines qui sont dans une logique de réception passive. Les orientations et le contenu des projets de coopération correspondent davantage à une volonté d'habillage afin de les rendre conformes aux critères des collectivités financeurs qu'à la conviction d'une nécessité partagée[828].

339. - Le monopole traditionnel de l'État en matière internationale demeure-t-il incontestable ? Cette interrogation est d'autant plus légitime lorsque l'on observe l'accroissement, tant quantitatif que qualitatif, de l'action extérieure des collectivités territoriales. Ces dernières interviennent désormais en complément de l'État dans le domaine de la coopération internationale. Ce processus n'est pas désintéressé. Loin de là ! L'État, en acceptant le partage de sa souveraineté extérieure avec des entités infranationales, entend instrumentaliser à son profit l'accroissement de l'action extérieure des collectivités territoriales au moyen notamment des institutions dont il la dote.

---

[827] E. GARCIA, *L'action internationale des collectivités territoriales : un outil de développement des territoires français ?* thèse, dactyl., Université de Cergy Pontoise, 2013, p. 305.
[828] G. ENEE, « Les enjeux des actions de coopération décentralisée en Afrique de l'ouest dans un contexte de décentralisation territoriale. L'exemple burkinabè », *op.cit.*, p. 92. Les acteurs satisfont à la logique du « *tout financement est bon à prendre* ».

§2.- L'instrumentalisation de l'action extérieure des collectivités au service de la politique africaine de l'État français

340. - L'action extérieure des collectivités territoriales françaises n'a pas manqué de s'étendre depuis sa reconnaissance comme complément indispensable à l'aboutissement de la politique étrangère de l'État par le législateur. La coopération décentralisée ne se présente plus en termes de conflit de compétence entre l'État et les collectivités[829]. En 2001, le Conseil économique et social écrit dans un rapport que « *s'il y eut, peut-être une certaine méfiance au début de la part de l'État face à une coopération qu'il ne maitrisait pas, cet état d'esprit est aujourd'hui révolu. La conception à l'honneur est plutôt que la mobilisation des collectivités territoriales, des entreprises, des ONG, et autres acteurs de la société civile disposant d'un ancrage local, mais aussi de compétences et savoir-faire spécifiques, constitue une valeur ajoutée irremplaçable pour l'action internationale de la France, une action internationale dont la diversité constituerait ainsi un des principaux atouts devant les exigences d'adaptation à des situations complexes* »[830]. Cette longue citation traduit d'une certaine manière, le fait que la coopération décentralisée tend à devenir un rouage important de la diplomatie française. L'État s'efforce d'encourager cette coopération pour assurer sa présence et son influence dans le monde[831]. L'action extérieure des collectivités territoriales, déployée au moyen de la coopération décentralisée pour assister techniquement les pays en développement, est placée sous le contrôle de l'État (**A**) pour servir sa stratégie de domination et d'influence (**B**).

---

829 F. PETITEVILLE, La coopération décentralisée. Les collectivités locales dans la coopération Nord-Sud, op.cit., p. 50.

830 *La contribution de la France au progrès des pays en développement*, 2001, disponible en ligne https://www.lecese.fr/sites/default/files/pdf/Avis/2001/01022803.pdf, p.113. consulté le 22 aout 2021.

831 B. FAURE, Droit des collectivités territoriales, op.cit., n° 649.

A- Une coopération décentralisée contrôlée

341. - La capacité juridique des collectivités territoriales à agir à l'extérieur des frontières nationales en fait « l'humble supplétif »[832] de l'État en matière internationale. Elle ne fait pas obstacle à la politique étrangère de l'État puisque ce dernier, saisissant son fort potentiel, en a fait un allié. Il existe sur la scène internationale une convergence des actions des collectivités territoriales et de celles de l'État. Les premières sont subordonnées aux priorités définies par la seconde. L'État a mis en place des structures institutionnelles de concertation et de pilotage pour le soutien (**1°**) et le financement de l'action extérieure des collectivités territoriales (**2°**).

1. La mise en place d'un dispositif institutionnel de dialogue et de pilotage

342. - Les pouvoirs publics ayant cessé de voir dans la coopération décentralisée une menace à la souveraineté de l'État vont davantage chercher à l'accompagner par la mise en place d'institutions de dialogue et de pilotage. L'institutionnalisation de la coopération décentralisée s'inscrit dans un mouvement de globalisation des échanges et de décloisonnement des frontières. L'environnement juridique interne des États s'ouvre ainsi aux différentes mutations contemporaines. La loi du 6 février 1992, en même temps qu'elle reconnaissait la possibilité pour les collectivités territoriales de conclure des conventions de coopération, créait une Commission nationale de la coopération de décentralisée (CNCD)[833]. Il s'agit d'un organe interministériel compétent dans l'ensemble des domaines relevant de l'action extérieure des collectivités territoriales. Régie par l'article L. 1115-6 du CGCT et par les dispositions règlementaires codifiées aux articles R. 1115-8 et suivants du même code, modifié en dernier lieu par le décret n° 2017-939 du 10 mai 2017, la CNCD est chargée

---

[832] P.-Y. CHICOT, « Les politiques internationales des collectivités territoriales françaises : une compétence locale atypique ? », *op.cit.*, p. 448.

[833] Art. L. 1115-6 CGCT issu de la loi d'orientation de 2014 : « il est créé une Commission nationale de la coopération décentralisée qui établit et tient à jour un état de l'action extérieure des collectivités territoriales. Elle favorise la coordination entre l'État et les collectivités territoriales et entre les collectivités territoriales et peut formuler toute proposition relative à l'action extérieure des collectivités territoriales. Les collectivités territoriales et leurs groupements transmettent à la commission les informations nécessaires à l'accomplissement de ses missions ».

de faire l'état de l'action extérieure des collectivités territoriales et de leurs groupements. Elle tient également à jour l'Atlas de l'action extérieure qui recense plus de 12 000 projets de coopération décentralisée dans près de 150 pays. La DAECT, qui assure le secrétariat de la Commission nationale de la coopération décentralisée, procède à cet effet au nom de la CNCD à la collecte des données relatives à l'aide publique au développement des collectivités territoriales.

343. - En application de la loi d'orientation et de programmation du 7 juillet 2014 et de la primauté de principe de l'État en droit international, les principes dégagés par la communauté internationale en matière de développement s'appliquent aux actions extérieures entreprises par les collectivités territoriales françaises. La période actuelle voit donc l'accroissement du rôle des collectivités territoriales dans l'approche territoriale du développement économique local et l'effort d'organisation au niveau mondial des mouvements représentatifs des « pouvoirs locaux » depuis 2004 avec l'émergence de Cités et gouvernements locaux unis (CGLU). Pour prendre en compte ces acteurs, la CNCD a fait l'objet d'une réforme en 2017 à la suite du rapport Laignel[834] pour élargir sa composition à « *des représentants d'établissements publics, d'associations ou d'organismes ayant une activité en relation avec l'action extérieure des collectivités territoriales ou la Francophonie* »[835].

344. - Non seulement, les collectivités sont instrumentalisées dans la mise en œuvre de la politique étrangère, notamment en matière d'aide publique au développement, l'État les associe également à sa stratégie de reconstruction des États des pays du Sud en période post-conflit. En effet, il est reconnu aujourd'hui que les collectivités territoriales françaises jouent un rôle important dans la réponse aux crises qui secouent l'Afrique subsaharienne. Cela a été le cas au Mali où les collectivités sont devenues des acteurs de premier plan de la mise en œuvre de l'accord de paix d'Alger[836]. Les collectivités territoriales jouent en ce sens un rôle central dans la fourniture des services de base et l'assistance aux populations, en particulier lorsque les institutions de l'État sont affaiblies, voire

---

834 A. LAIGNEL, Rapport public, « l'action extérieure des collectivités territoriales. Nouvelles approches… Nouvelles ambitions », La documentation française, 2013.
835 Décret n° 2017-939 du 10 mai 2017 relatif à la Commission nationale de la coopération décentralisée.
836 A. SIMON, *Les compétences régaliennes et les collectivités territoriales, op.cit.*, pp. 480-482.

momentanément inexistantes. Cet état de faiblesse structurelle appelle à un renforcement de la coopération décentralisée sur l'ensemble du continuum de gestion de crise : prévention (risques naturels), réponse d'urgence (aide humanitaire) et post-crise (reconstruction/réconciliation)[837].

345. - Cette situation démontre la volonté de l'État à faire des collectivités territoriales des partenaires à l'appui de sa politique de domination et de rayonnement dans le monde. L'intervention des collectivités territoriales dans le domaine de l'action humanitaire énoncée par l'article L. 1115-1[838] va bien au-delà des catastrophes naturelles. L'État va chercher à tout prix à profiter de l'ampleur prise par cette action extérieure des collectivités territoriales en exerçant un contrôle très subtil sur la coopération décentralisée au travers de sa politique de cofinancement. Pour l'État, l'action extérieure des collectivités territoriales n'est qu'un instrument additionnel en termes d'échelle et d'expertise à la politique française de coopération internationale. Il formule à cet effet des orientations thématiques et géographiques qui sont traduites en instruments et dispositifs incitatifs. Les incitations de l'État en direction de l'action extérieure des collectivités territoriales s'exercent principalement dans le cas de l'accès aux financements, mais leurs effets doivent toutefois être relativisés. Il est en effet rare que les collectivités territoriales se voient limitées ou critiquées par l'État dans leurs initiatives internationales. En revanche, ces orientations qui bornent l'action internationale à une fonction d'aide au développement n'encouragent pas les autorités locales à tenir compte de leurs propres enjeux territoriaux et relèguent l'intérêt local de l'action extérieure à un plan secondaire[839]. Les intérêts étatiques prennent le pas sur les intérêts locaux.

2. La mise en place d'un soutien technique et financier

346. - L'État s'est érigé en mécène de la coopération décentralisée par la mise en place d'une diversité d'instruments financiers. À travers ces

---

837 *Guide juridique de l'action extérieure des collectivités territoriales, op.cit.*, p. 46.

838 « Dans le respect des engagements internationaux de la France, les collectivités territoriales et leurs groupements peuvent mettre en œuvre ou soutenir toute action internationale annuelle ou pluriannuelle de coopération, d'aide au développement ou à caractère humanitaire ».

839 E. GARCIA, *L'action internationale des collectivités territoriales : un outil de développement des territoires français ? op.cit.*, p. 306.

interventions financières, il guide et oriente la coopération décentralisée vers des domaines et des zones géographiques qu'il considère comme prioritaires dans sa stratégie d'intervention en faveur du développement[840]. La DAECT s'est dotée à partir de 2007-2008 d'une palette d'instruments rationalisée : les « appels à projets » dont l'effet levier et l'influence tiennent selon le ministère des Affaires étrangères « *essentiellement à leur fonction prescriptrice, contribuant à orienter les coopérations décentralisées des collectivités territoriales vers les thèmes et pays prioritaires* »[841].

347. - Si les collectivités territoriales assurent une partie du financement de leurs actions de coopération décentralisée, l'État intervient à travers les appels à projets dans le cofinancement de l'action extérieure. En effet, le ministère des Affaires étrangères[842] apporte, à travers ce mécanisme, un appui financier aux projets de coopération décentralisée présentés par les collectivités territoriales françaises. Ces cofinancements, qui s'élèvent à 8,5 millions d'euros par an[843], sont attribués selon des modalités précises. Le nouveau dispositif de cofinancement en soutien à la coopération décentralisée, qui a été réformé en 2006, s'appuie désormais sur des appels à projets annuels, triennaux ou des appels à projets thématiques[844] qui fixent les priorités de l'État au regard des concertations au sein de la Commission nationale de la coopération décentralisée. Le ministère des Affaires étrangères peut ainsi en fonction d'un certain nombre de critères apporter un soutien financier aux collectivités territoriales ou à leurs réseaux nationaux. On peut noter à ce titre, l'obligation faite à l'article L. 1115-6 du CGCT aux les collectivités

---

840 Les pays relevant de l'aide publique au développement sont également définis selon les critères de l'OCDE

841 MAEDI, *Diplomatie et territoire, Pour une action extérieure démultipliée*, La documentation française, 2017, p. 87 ; A. SIMON, *Les compétences régaliennes et les collectivités territoriales*, *op.cit.*, p. 484.

842 Il ne s'agit pas du seul acteur de cofinancement. On relève aussi l'intervention notamment de l'Agence française de développement, et d'Expertise France.

843 MAEDI, *Diplomatie et territoire, Pour une action extérieure démultipliée, La documentation française*, 2017, p. 87.

844 Les priorités thématiques sont : – la lutte contre le changement climatique ; – la jeunesse et sa mobilité internationale (notamment dans sa dimension « inclusion sociale» et « formation professionnelle») ; – le soutien à l'export des PME-ETI et à l'attractivité des territoires, que ce soit en matière de tourisme des visiteurs étrangers ou d'accueil des investissements directs étrangers ; – urbanisme et patrimoine ; – l'agriculture et la sécurité alimentaire ; – la gouvernance territoriale et la formation des cadres territoriaux, là où la décentralisation est en marche.

territoriales et leurs groupements de transmettre à la Commission nationale de la coopération décentralisée les informations nécessaires à l'accomplissement de ses missions. L'octroi du financement est également subordonné à une déclaration préalable à la Commission nationale de la coopération décentralisée.

348. - En dehors des appels à projets, les pouvoirs publics ont aussi mis en place des outils de financement spécifique pour renforcer l'action internationale des collectivités territoriales. C'est le cas notamment du fonds de stabilisation qui permet de financer des projets recourant à l'expertise des collectivités françaises en vue de renforcer le rôle des autorités locales dans la reconstruction et l'approfondissement de la décentralisation[845]. Notons également que la Commission européenne intervient dans le financement de l'action extérieure des collectivités territoriales. Dans le cadre de sa nouvelle approche territoriale du développement qui cherche à promouvoir un développement endogène « de la base vers le haut », les autorités locales sont au centre du dispositif. En témoigne, la création d'une ligne thématique « Organisations de la société civile et autorités locales »[846] dotée de 1,9 milliard d'euros pour 2014-2020 avec 450 millions consacrés à l'appui aux collectivités des pays en développement. La Commission montre ainsi qu'elle reconnait les autorités locales comme acteurs de la coopération décentralisée pour le développement. C'est principalement à travers l'appui-pays (financements administrés par les délégations de l'UE) que sont soutenues les actions de coopérations décentralisées.

349. - On pourrait croire que l'appel à projets et le recours au contrat seraient une garantie d'équilibre des parties. Cependant, en mettant en place une politique de labellisation à travers le programme de valorisation de l'expertise des collectivités territoriales[847], les pouvoirs publics exercent

---

845 MAEDI, *Diplomatie et territoire, Pour une action extérieure démultipliée, op.cit.*, p. 89.

846 OSC-AL, instrument de coopération au développement. Celles-ci sont financées directement « priorité 1 : appui pays, renforcement des AL aux processus de développement et de gouvernance » ou soutenues par le biais de leurs associations « priorité 2 : renforcement des réseaux au niveau régional, européen et mondial ».

847 La Délégation pour l'action extérieure des collectivités territoriales a lancé en juillet 2020 le programme de valorisation de l'expertise territoriale des collectivités territoriales, dénommé EXPE-CT. Ce programme permet de promouvoir à l'international les compétences et les savoir-faire des collectivités françaises et de leurs associations. Sur l'effet label, voir MAEDI, *Diplomatie et territoire, Pour une action extérieure démultipliée*, *op.cit.*

indéniablement une influence sur l'orientation de l'action extérieure des collectivités. Il est évident que la procédure de labellisation, tout comme celle d'attribution des financements de soutien à la coopération décentralisée, sont soumises à des impératifs de conformité aux priorités de l'État en matière de politique internationale[848]. Les discours tiers-mondistes qui consistent à venir en aide aux pays en développement ne sont en réalité que la face cachée du dispositif de coopération décentralisée. L'État et les collectivités territoriales agissent *in fine* dans une logique partenariale en vue de la réalisation d'un même objectif qui est celui de la diffusion du « design institutionnel » français.

B- Une coopération décentralisée détournée

350. - La contribution des collectivités territoriales à l'action internationale de la France prend de l'ampleur avec l'évolution des logiques qui président à leurs actions. Aujourd'hui, la majorité des collectivités territoriales françaises sont engagées dans l'action internationale dans un nombre croissant de pays et de secteurs. Elles sont devenues des actrices incontournables de la scène internationale. Il s'agit, selon le propos de Jean Marc Ayrault, « *d'un atout de tout premier plan pour l'attractivité de la France et pour la solidarité internationale* »[849]. L'action extérieure des collectivités territoriales va être mise au service du développement des territoires à l'extérieur des frontières nationales. Cette propension des collectivités territoriales françaises à transformer la coopération décentralisée pour la promotion et l'attractivité de leurs territoires en fait des acteurs de la diplomatie économique (**1°**) auxquelles le gouvernement n'hésite plus à faire appel pour la promotion de l'expertise française (**2°**) dans le cadre des politiques d'appui à la décentralisation dans les pays en développement.

1. Une coopération décentralisée au service d'une diplomatie des territoires

351. - La généralisation du modèle de l'économie libérale à l'échelle du globe induit par le phénomène de globalisation touche les collectivités

---

848 F. PETITEVILLE, *La coopération décentralisée. Les collectivités locales dans la coopération Nord-Sud, op.cit.*, p. 47.

849 MAEDI, *Diplomatie et territoire, Pour une action extérieure démultipliée*, La documentation française, 2017, p. 5.

territoriales[850]. L'impératif de développement économique guide leurs actions dans la sphère internationale. Les collectivités territoriales françaises intègrent désormais dans leurs partenariats de coopération décentralisée une stratégie de promotion et d'attractivité de leurs territoires. La capacité des collectivités territoriales à agir à l'international dans le domaine économique a été maintenue par le législateur, malgré la suppression en 2015 de la clause de compétence générale pour les départements et les régions[851].

352. - Le détournement de la coopération décentralisée pour le développement au profit de l'attractivité remonte aux années 1990. En effet, dans un rapport de 1996, Hubert Perrot, Délégué pour l'action extérieure des collectivités, affirmait que les liens établis entre des collectivités territoriales françaises et des collectivités territoriales étrangères par la conclusion de conventions de coopération décentralisée devraient pouvoir faciliter l'établissement de stratégies d'action, voire d'implantation, vers des pays où ces liens ont été établis. Une collectivité territoriale qui souhaiterait aider le développement des activités de PME/PMI de sa région vers un pays étranger pourrait favoriser le financement ou le cofinancement d'opérations de coopération décentralisée avec des collectivités territoriales de ce pays[852]. Le financement du développement par la coopération décentralisée ne s'inscrit donc pas dans une logique purement altruiste. Pour le Conseil d'analyse économique[853], l'influence et le rayonnement des territoires

---

850 M. CHAMBON, « Les collectivités territoriales, acteurs de la globalisation du droit administratif ? », *RFDA*, 2019, p. 994.

851 Loi n° 2015-991 du 7 août 2015 portant nouvelle organisation territoriale de la République a supprimé la clause de compétence générale pour les départements et les régions. Sur la clause générale de compétence, lire utilement : M. VERPEAUX, « Pavane pour une notion défunte. La clause générale de compétence », RFDA, 2014, p 457 ; G. Le CHATELIER, « Le débat sur la clause générale de compétence est-il vraiment utile ? », AJDA, 2009, p. 186 ; J.-M. PONTIER, « Semper Manet. Sur une clause générale de compétence », RDP, 1984, p. 1443.

852 H. PERROT, *De nouvelles perspectives pour la coopération décentralisée et l'action extérieure des collectivités locales,* La documentation française, 1996, p. 24.

853 Le Conseil d'analyse économique réalise, en toute indépendance, des analyses économiques pour le gouvernement et les rend publiques. Il examine les questions qui lui sont soumises par le Premier ministre et par le ministre chargé de l'économie et peut

n'apparaissent jamais comme des objectifs officiels alors qu'ils le sont dans les faits. L'APD accordée aux pays en voie de développement relève de la poursuite d'intérêts égoïstes par les donateurs[854]. L'action extérieure à des fins de promotion et d'attractivité parvient à coexister avec les partenariats de coopération décentralisée voire, parfois, se mêlent aux partenariats à des fins de développement de la collectivité étrangère, se transformant ainsi progressivement en partenariat à dimension économique[855] au profit des acteurs territoriaux des collectivités françaises. En effet, comme nous l'avons montré précédemment, il s'avère difficile de mutualiser les intérêts dans le cadre de rapports de domination.

353. - Ainsi, les collectivités territoriales françaises sont désormais les nouveaux acteurs de ce qu'il est convenu d'appeler une « diplomatie démultipliée ». L'État français encourage cette diplomatie qu'il considère comme complémentaire à son action. Sous des aspects de la coopération décentralisée, cette diplomatie cache un champ de lutte dans le cadre de la concurrence territoriale qui a désormais dépassé les frontières et voit les villes s'affronter entre elles au niveau mondial[856]. L'influence des logiques de mondialisation économique et d'internationalisation des politiques publiques locales contribue à faire changer la nature de l'action extérieure des collectivités territoriales. Longtemps cantonnée à la célébration de liens d'amitié et de solidarité entre les peuples, elle s'intègre désormais à des stratégies globales de projection et d'attractivité territoriale[857]. L'action internationale est ainsi devenue pour les élus et leurs équipes politico-administratives une ressource de leadership qu'ils utilisent à des fins de valorisation et de promotion des territoires dont ils ont la charge.

---

procéder de sa propre initiative à l'analyse prospective de questions économiques qu'il estime pertinentes pour la conduite de la politique économique du pays. Placé auprès du Premier ministre, le Conseil d'analyse économique a pour mission « d'éclairer, par la confrontation des points de vue et des analyses, les choix du gouvernement en matière économique ».

854 Conseil d'analyse économique, *La France et l'aide publique au développement*, La Documentation française, 2006.

855 A. SIMON, *Les compétences régaliennes et les collectivités territoriales, op.cit.*, p. 437.

856 *Idem.*

857 M. KARZAZI, *La coopération décentralisée franco-marocaine : entre coopération au développement et diplomatie économique d'influence*, thèse, dactyl., Université de Cergy Pontoise, 2013.

354. - La coopération décentralisée Nord-Sud répond donc à une logique de marché[858] qui a fait des collectivités territoriales françaises des acteurs incontournables du développement à travers le transfert du modèle d'administration décentralisé. L'action extérieure des collectivités territoriales constitue donc tout à la fois un acte de solidarité internationale et plus encore un enjeu économique pour les territoires. Cela va évidemment à l'encontre de l'esprit même de la coopération décentralisée qui doit être guidée par l'altruisme, la générosité et le désintérêt dans un cadre partenarial égalitaire[859].

2. Une coopération décentralisée au service de la diffusion de l'expertise française

355. - Les collectivités territoriales sont des acteurs essentiels de la promotion de l'expertise française à l'étranger. La coopération décentralisée joue un rôle de premier plan dans le développement municipal des pays en développement à travers l'appui institutionnel. L'action extérieure des collectivités territoriales « *se mesure moins en termes financiers qu'en termes d'influence et de rayonnement* ». Elle offre une vitrine des richesses et des potentiels des collectivités en « *valorisant l'expertise française à l'étranger* »[860]. Les partenariats de coopération décentralisée vont s'inscrire dans le nouveau paradigme international qui fait de la décentralisation une priorité politique pour assurer le développement local et la démocratie locale. D'une manière générale, de l'avis même du ministère des Affaires étrangères, dans les pays en développement, la coopération française appuie principalement les processus de décentralisation et considère l'échelon local comme un niveau très pertinent pour rendre compte de la diversité des enjeux et pour construire la coopération entre les différents acteurs dans une perspective à la fois de développement et de renforcement de la démocratie[861].

---

858 F. PETITEVILLE, *La coopération décentralisée. Les collectivités locales dans la coopération Nord-Sud*, L'Harmattan, 1995. Il démontre notamment l'ambivalence de la coopération décentralisée entre les logiques de coopération et celles du marché.

859 C. JEBEILI, « La coopération décentralisée », in S. REGOURD, J. CARLES, D. GUIGNARD (dir.), *La décentralisation 30 ans après*, *op.cit.*, pp. 307-322.

860 Réponse ministérielle du ministre des Affaires étrangères du 3 mai 2011 à la question nº 92170 posée par Philippe Vitel, *JO 3 mai 2011*, p. 4452.

861 Diplomatie et territoire, *Pour une action extérieure démultipliée, op.cit.*, p. 58.

356. - L'expertise des collectivités territoriales est au cœur de la problématique du développement. Il y a un véritable enjeu de rayonnement et d'influence qui se joue dans l'élaboration des normes juridiques, techniques et des « bonnes pratiques ». L'assistance technique française organisée et articulée autour de la coopération décentralisée et des accords bilatéraux participe à l'élaboration des textes fondamentaux dans le cadre de la mise en œuvre des politiques de décentralisation dans les pays d'Afrique subsaharienne. La France, en tant qu'acteur principal de la coopération décentralisée en Afrique noire francophone, dispose ainsi de moyens pour influencer la production des règles applicables au processus de la décentralisation. Cette capacité d'influence « para-normative » passe par son active participation à l'élaboration des « *standards de références, nomenclatures, lignes directrices, législations types, codes divers, mis à la disposition des États à titre purement indicatif, soit par voie de résolutions, soit mêmes, simplement par publications directes émanant du secrétariat des organisations* »[862]. On pourrait utilement s'interroger sur le caractère normatif, c'est-à-dire sur la force obligatoire, de ces différents actes. Bien qu'ils ne puissent pas bénéficier du caractère impératif des actes règlementaires ou législatifs, ils influent indirectement sur les systèmes juridiques des pays destinataires en orientant considérablement leur contenu[863].

357. - Dotés en pratique « d'une grande importance pour l'harmonisation progressive des comportements, et surtout, des législations nationales des États membres dans les domaines techniques les plus variés », les actes para-normatifs se retrouvent à mi-chemin entre l'acte politique et l'acte juridique. Ils relèvent de la soft Law et n'ont donc pas une valeur contraignante. Toutefois, avec le mouvement de globalisation du droit, leur effet sur les systèmes juridiques des États aidés est considérable. Ils constituent des actes prénormatifs dans la mesure où ils finiront par être adoptés les rendant ainsi obligatoires par l'effet de transposition au niveau national[864]. La portée de ces standards internationaux ainsi élaborés est encore renforcée par la pratique de la conditionnalité imposée par les bailleurs de fonds pour l'octroi de leur

---

862 P.-M. DUPUY, Y. KERBRAT, *Droit international public*, Dalloz, coll. « Précis », 2010, p. 217.

863 Voir en droit interne l'application des jurisprudences *Fairvesta* et *Numericable*. CE, 21 mars, 2016, *RFDA*, 2016, p. 497, concl. S. von Coester *; AJDA*, 2016, p. 717, chron. L. Dutheillet de Lamothe et G. Odinet *; AJDA, 2016*, p. 931, *Dr. adm. 2016,* comm. 34, note A. Sée.

864 P.-M. DUPUY, Y. KERBRAT, *Droit international public, op.cit.*, p. 217.

aide[865]. En invitant les collectivités territoriales à mettre leur expertise en matière de services aux citoyens et de gouvernance locale au profit des collectivités des pays en développement, la France entend influencer le modèle d'organisation territoriale des pays bénéficiaires de l'aide publique au développement.

358. - L'appui institutionnel implique toutes les actions de soutien au processus de décentralisation. Dans le cadre d'une convention de coopération décentralisée, cet appui peut prendre plusieurs formes : aide aux collectivités du Sud à renforcer leurs capacités de gestion des dépenses, des services publics, à connaitre leurs ressources humaines, leur potentiel économique, à échanger de bonnes pratiques, à mettre en œuvre des politiques de protection de l'environnement, du patrimoine, de planification rurale ou urbaine, etc. Selon le ministère des Affaires étrangères, 1 500 à 2 000 agents territoriaux sont directement investis dans la gestion de la coopération décentralisée[866] et environ 140 experts techniques français travaillent sur les questions de gouvernance locale, particulièrement en Afrique[867]. Cette présence d'experts emporte d'énormes conséquences dans la mesure où, participant à la mise en œuvre de la décentralisation, ils seront tentés de proposer des solutions dont ils ont une meilleure connaissance parce qu'expérimentés dans leur pays d'origine[868].

359. - Les collectivités territoriales sont ainsi devenues incontournables dans les stratégies de développement pensées pour les pays du Sud. Elles revendiquent d'ailleurs cette place auprès des Nations unies. Reconnaissant leur expertise, les partenaires au développement font de plus en plus appel aux entités infraétatiques pour mettre en œuvre les ODD, que ce soit pour le renforcement du processus de décentralisation et de la gouvernance locale, ou pour l'appui aux politiques locales sectorielles[869]. Près de 2000 partenariats de coopération décentralisée sont entretenus par les collectivités territoriales françaises en Afrique

---

865 B. FISCHER, *Les relations entre l'administration et les administrés au Mali : contribution à l'étude du droit administratif des États d'Afrique subsaharienne de tradition juridique française, op.cit.*, p. 130 et s.
866 Diplomatie et territoire, *Pour une action extérieure démultipliée, op.cit.*, p. 58.
867 Ministère des Affaires étrangères, *L'appui de la France aux processus de décentralisation et de gouvernance locale démocratique*, Rapport 2012.
868 C. NACH MBACK, *Décentralisation et démocratisation, op.cit.*, p. 228.
869 MAEDI, Diplomatie et territoire, *Pour une action extérieure démultipliée, op.cit.*, p. 58.

occidentale[870]. L'agence Cités unies France intervient pour animer et mutualiser l'action extérieure des collectivités. En 2018, avec l'appui de ses partenaires, l'Agence française de développement et la Délégation pour l'action extérieure des collectivités territoriales, elle a lancé un dispositif d'appui à l'action internationale des collectivités territoriales. Il s'agit d'un appui en ingénierie qui permet d'accompagner les collectivités dans le montage d'un projet ou la définition d'une stratégie à l'international ou encore dans la mise en place des recommandations d'une évaluation[871]. D'autres organisations proches des collectivités interviennent également dans l'appui aux politiques de décentralisation. C'est le cas notamment du Centre national de la fonction publique territoriale (CNFPT) qui organise des sessions de formations de courtes durées au profit des fonctionnaires territoriaux des pays en développement d'Afrique subsaharienne[872]. Il entretient par ailleurs, tout comme les collectivités territoriales, un partenariat avec France Expertise[873] pour mieux mobiliser leur expertise.

360. - L'analyse de ces politiques de coopération visant à renforcer les capacités des collectivités des pays en développement donne lieu à un bilan critique. Marie-Christine Kessler relève que la coopération décentralisée n'a pas permis d'éviter « l*es cercles vicieux universels de la coopération institutionnelle* »[874]. Elle s'est traduite par une transposition du modèle français d'administration décentralisée avec pour seul objectif de

---

870 Voir Atlas français de la coopération décentralisée et des autres actions extérieures.

871 www.cites-unies-france.org

872 L'AFD organise également depuis 2009, un parcours de formation sur le pilotage des collectivités locales (PCL), en appui aux opérations, destiné aux collectivités locales des zones d'intervention de l'AFD. Ce parcours a pour finalité de renforcer les compétences managériales des collectivités locales et s'adresse aux cadres des collectivités exerçant des responsabilités dans les domaines des finances locales, de l'aménagement du territoire, de la gestion des services publics urbains et du management et conduite de projets. Il est structuré en modules thématiques.

873 Expertise France est l'agence française d'expertise technique internationale, issue de la réforme du dispositif français de coopération technique. L'agence inscrit son action dans le nouvel agenda international 2015-2030 et les ODD. Elle intervient en particulier dans les domaines de la gouvernance démocratique, économique et financière, de la lutte contre le dérèglement climatique et du développement urbain, d'une agriculture respectueuse de l'environnement, de la santé, de la protection sociale et de l'emploi, ainsi que de la stabilisation des pays fragiles ou en crise et de la sécurité.

874 M.-C. KESSLER, « La coopération administrative et ses problèmes : réflexions à partir de l'expérience française », *RISA*, vol. 59, n°2, 1993, cité par F. PETITEVILLE, *La coopération décentralisée. Les collectivités locales dans la coopération Nord-Sud*, *op.cit.*, p. 120.

contribuer à consolider l'influence de la France dans le monde[875] et particulièrement dans ces anciennes colonies. Pour Mahamadou Danda, la coopération française qu'elle soit décentralisée ou bilatérale, a largement influencé le système de décentralisation et ses modes de fonctionnement en Afrique[876]. Cependant, l'impact des différentes pressions exogènes dont nous venons de parler doit être nuancé, car on constate également que les États et les collectivités du Sud les utilisent pour asseoir leur pouvoir de domination sur les populations.

---

875 *Idem.*

876 M. DANDA, Niger, une décentralisation importée : une comparaison avec la France, L'Harmattan 2013.

Conclusion du chapitre 2

361. - L'organisation de l'État en Afrique est en constante mutation. Depuis les indépendances, les pays d'Afrique subsaharienne peinent à trouver le modèle d'organisation qui sied à leurs réalités. Cette absence de modèle d'organisation propre au continent fait d'elle un champ d'expérimentation propice pour les pays du Nord. Ainsi, la décentralisation, modèle de gouvernance politique dans les sociétés occidentales, va se muer en outil de développement et de promotion de la démocratie en Afrique. À partir de 1990, les organisations internationales sont convaincues que la décentralisation est le remède absolu pour sortir l'Afrique du sous-développement et lutter efficacement contre la pauvreté. L'aide publique au développement est désormais conditionnée à la mise en œuvre des politiques de décentralisation. Ce nouveau dogme influe sur toute politique d'aide en faveur des pays en développement.

362. - La France va jouer un rôle important dans la diffusion en Afrique du modèle d'organisation décentralisé, n'hésitant pas à faire une place à ces collectivités territoriales sur la scène internationale. L'appui de la coopération française décentralisée pour la mise en œuvre des politiques de décentralisation va être instrumentalisé pour servir les stratégies d'influence et de rayonnement de l'État français à travers la diffusion de son modèle d'administration décentralisé. Portés constamment par des valeurs d'universalisme, les pays du Sud qui disposent de peu de marge de manœuvre sont coincés dans le discours international dont ils sont dépendants. Après des décennies d'essais, les résultats tardent à venir. L'appui international a, par ailleurs, renforcé les tentatives de mimétisme et de transposition d'un modèle d'administration considérée comme une panacée, parce qu'ayant fait ses preuves ailleurs.

Conclusion du titre 2

363. - La permanence de l'héritage institutionnel colonial sur les territoires africains est indéniable. Les institutions administratives territoriales en Afrique noire francophone ne reflètent pas les réalités sociologiques qu'elles sont censées encadrer. Construites sur la base des structures administratives coloniales, ces institutions portaient déjà en elles-mêmes les germes de leur propre destruction[877]. Or, l'effectivité de la décentralisation suppose une adhésion active de la population. En ignorant cette donnée fondamentale dans l'élaboration des politiques de décentralisation, l'action administrative se trouve paralysée au niveau local. Ce déphasage entre les institutions, les réalités sociales et les pratiques ne contribuent pas à la réalisation de l'autonomie des collectivités territoriales. À ce stade de notre étude, deux facteurs permettent de l'expliquer. Du point de vue interne, la construction des structures administratives par une élite africaine résignée ou réalisant l'impossibilité d'anéantir le système colonial a conduit à une contestation en demi-teinte du pouvoir colonial sans réelles intentions de le détruire. Cette attitude passive est facilitée par la transformation des enjeux mondiaux en faveur du développement et de la démocratisation sur le plan externe. Imposés de l'extérieur sur l'action combinée des institutions financières internationales et des bailleurs de fonds, les États africains disposent de peu de marge de manœuvre dans la mise en œuvre des réformes territoriales. Ainsi, la combinaison de la contrainte et l'opportunisme des choix opérés par l'élite africaine ont permis de maintenir l'architecture institutionnelle de l'administration territoriale coloniale faisant preuve de ce qu'il est convenu de désigner selon l'analyse faite par Georges Langrod de « *macro-mimétisme* »[878].

---

877 M. NEMBOT, Le glas de la fonction publique. Essai sur la signification d'une institution en quête de légitimité, thèse, L'Harmattan, 2000, p. 11.

878 G. LANGROD, « Genèse et conséquences du mimétisme administratif en Afrique », *op.cit.*, p. 121.

Conclusion de la partie 1

364. - L'histoire générale de colonisation éclaire l'histoire administrative des pays africains. Transposé d'autorité par le pouvoir colonial, le modèle d'administration territoriale français est reçu et maintenu en Afrique noire francophone. La réception du modèle administratif français dans les États africains postcoloniaux est le fait aussi bien de la colonisation, par la parenté institutionnelle qu'elle induit, que de l'influence juridico-économique préservée par l'ex-puissance coloniale. La France est restée longtemps après les indépendances, le principal pourvoyeur de capitaux en Afrique noire francophone. Sur la base des structures administratives coloniales, est échafaudée la nouvelle organisation administrative et territoriale des nouveaux États de l'Afrique noire francophone. Malgré les influences diverses et les besoins multiples qui ont amené les différents États ouest-africains à se doter d'un système administratif particulier, ces pays anciennement colonies françaises, souffrent tous d'une crise d'identité. L'idée de décentralisation et de recherche de l'autonomie locale ayant présidé à l'élaboration d'un corps de règles régissant les rapports entre le centre et la périphérie sur le continent, contraste avec les réalités socioculturelles qui animent les peuples en Afrique. Or, la cohérence de l'ordre juridique est l'une des conditions de l'effectivité. À ce titre, le continent africain apparait comme une terre de prédilection pour l'étude des conséquences de la circulation et de la réception des modèles d'administration étrangers.

365. - Ainsi, il est possible d'affirmer que la décolonisation est une gageure que l'on ne peut soutenir longtemps. Le constat est même devenu banal, selon Pierre Legendre. Cet auteur relève qu'après de nombreuses oppositions, entre des systèmes d'administration dans les organismes internationaux chargés de l'assistance technique ; les fonctionnaires internationaux transposent, sans aucune protestation apparente, les schémas nationaux comme à la belle époque coloniale[879]. Le développement institutionnel, priorité des programmes internationaux de développement, sert d'instrument de diffusion d'une culture administrative globale. Les systèmes d'administration dominants, comme celui de la France, deviennent après des années de domination culturelle non seulement subies, mais aussi désirées. De cette manière, la colonisation est un exemple de réception juridique et institutionnelle. Si

---

[879] P. LEGENDRE, Trésor historique de l'État en France, op.cit., p. 185.

l'État moderne s'est constitué par un processus d'unification de sa souveraineté territoriale, celle-ci doit aujourd'hui être repensée. Le territoire sur lequel l'État entend exercer sa souveraineté est revendiqué par d'autres acteurs en quête d'autonomie. Les réformes de l'organisation administrative et territoriale ne portent pas les marques de l'histoire profonde des peuples auxquels ils sont destinés. La décentralisation, professée à grand renfort de slogans et d'une rhétorique grandiloquente, n'a pas conduit à jeter les bases d'institutions décentralisées autonomes.

PARTIE 2 – L'ÉTAT DÉCENTRALISÉ : UNE CONSTRUCTION JURIDIQUE ARTIFICIELLE

366. - La décentralisation doit être repensée en Afrique. Il s'agit de la repenser, car la majeure partie des pays d'Afrique subsaharienne sont désormais résolus à faire de la décentralisation une réalité. Le foisonnement des textes fixant les conditions, les modalités et l'entendue de la décentralisation témoigne de l'intérêt pour la décentralisation en Afrique en dépit de son extranéité. Loin de l'hostilité qu'on pouvait constater dans les premières Constitutions au lendemain des indépendances à l'égard d'une quelconque affirmation des collectivités décentralisées ou *a fortiori* d'entités fédératives[880], la décentralisation est désormais sur le plan formel une réalité. Les Constitutions reconnaissent l'existence des collectivités territoriales et consacrent le principe selon lequel ces collectivités s'administrent librement. Cependant, malgré cette consécration constitutionnelle, plusieurs points continuent de susciter des débats. Les collectivités territoriales africaines continuent ainsi que l'écrit Bertrand Faure à propos de l'échec de la décentralisation française « *d'offrir bien de commodités pour le pouvoir central lui-même. Les résultats obtenus au titre de la décentralisation n'ont pas permis de freiner le développement de tous les vices auxquels cette décentralisation est spécialement sujette et les a poussés du côté même, ou en suivant une inclinaison naturelle, elle penchait déjà* »[881]. L'étude des relations entre l'État et les collectivités territoriales sans les systèmes juridiques d'inspiration française démontre la présence permanente du pouvoir central dans le fonctionnement des collectivités. La décentralisation est viciée par cette omniprésence de l'État : c'est l'État qui aide, autorise et interdit[882].

367. - Tout se passe comme si la décentralisation n'était qu'une autre manifestation de la centralisation. La décentralisation africaine est toujours marquée par une unité de commandement. L'administration d'État est omniprésente à travers ses règles et ces multiples contrôles de l'action publique locale. La faible autonomie offerte par le transfert des compétences aux collectivités territoriales disparait progressivement pour

---

880 A. CABANIS, M.-L. MARTIN, « Décentralisation et pouvoirs locaux dans les constitutions d'Afrique francophone », *Politeia,* n° 6, 2004, p. 75.
881 B. FAURE, « L'échec de la décentralisation française : L'État, les élus et les règles », in F. CROUZATIER-DURAND, M. TOUZEIL-DIVINA (dir.), *40 regards sur 40 ans de décentralisation(s),* Ed. L'Épitoge, 2022, p. 281.
882 *Idem.*

laisser place à une (re) centralisation de l'exercice des compétences décentralisées. Le contexte actuel marqué par de multiples crises sécuritaires et politiques en Afrique noire subsaharienne qui mettent à rude épreuve la stabilité des États déjà fragiles appelle à la prudence dans les évolutions institutionnelles. L'argument sécuritaire sert de prétexte à l'administration d'État pour renforcer son emprise sur les collectivités territoriales[883]. La constitutionnalisation de la libre administration au profit des collectivités territoriales africaines n'a donc pas permis de faire émerger des collectivités territoriales capables de construire des projets structurants locaux pour satisfaire les besoins des populations. Les moyens de mise en œuvre de la libre administration appellent à une réflexion sur le cadre juridique de l'action des collectivités territoriales c'est-à-dire sur l'existence effective d'une libre administration. En premier lieu, il s'agit de s'interroger sur les compétences des collectivités territoriales et les conditions de leur exercice. En second lieu, il s'agira de montrer le décalage entre la vision politique de la décentralisation attachée à promouvoir une certaine démocratie locale et les pratiques institutionnelles dans les États d'Afrique noire francophone. La décentralisation telle qu'elle est conduite aujourd'hui est artificielle, car elle ne garantit ni la libre administration des collectivités territoriales (**Titre 1**) ni l'émergence d'une véritable démocratie locale (**Titre 2**) présentée comme étant ces principaux enjeux. Cette affirmation sera vérifiée par l'analyse et la confrontation des textes et des pratiques africaines avec les standards juridiques internationaux en matière de décentralisation.

Titre 1 – Une décentralisation sans libre administration

Titre 2 – Une décentralisation sans démocratie locale

---

[883] Au Burkina Faso, à la suite du dernier coup d'État du 23 janvier 2022, la junte au pouvoir a prononcé la dissolution des assemblées délibérantes des collectivités territoriales et leur remplacement par des délégations spéciales.

TITRE 1 — UNE DÉCENTRALISATION SANS LIBRE ADMINISTRATION

> « Les collectivités territoriales sont des miroirs qui reflètent les exigences du groupement humain plus rapidement et plus efficacement que l'État »[884].

368. - La construction d'un État décentralisé implique la reconnaissance de l'autonomie des collectivités territoriales. L'autonomie est un concept juridique général, flou et de faible densité. Parce qu'il est un concept flou, il appelle par la force des choses des qualificatifs. On parlera alors d'autonomie administrative s'agissant des collectivités territoriales dans un État unitaire. L'autonomie fait donc référence dans les pays d'Afrique noire francophones à la libre administration des collectivités territoriales et ignore la dimension politique de l'autonomie locale. Le droit supranational de la décentralisation en Afrique a hissé l'autonomie locale au rang des valeurs démocratiques universelles[885] sans pour autant en préciser le contenu. Ainsi, la Charte africaine de la décentralisation se contente d'affirmer que les autorités locales doivent «*pouvoir gérer, de manière responsable et transparente, leur administration et leurs finances à travers des assemblées délibératives et des organes exécutifs démocratiquement élus*»[886]. Cette disposition est reprise dans les Constitutions africaines qui consacrent la libre administration des collectivités territoriales par des conseils élus dans les conditions prévues par la loi.

---

884 J.-M. PONTIER, *L'État et les collectivités locales, op.cit.*, p. 114.

885 M. E., SOHOUENOU « L'administration territoriale béninoise et le droit supranational de la décentralisation », Leçon inaugurale, *Colloque international sur l'effectivité de la décentralisation au Bénin*, 12 janvier 2015, p. 2.

886 Art. 5. 2 de la Charte africaine de la décentralisation… *préc.*

369. - Pour comprendre ce que recouvre le concept d'autonomie locale, il faut recourir à des standards juridiques internationaux. À cet effet, c'est l'article 3 de la Charte européenne de l'autonomie locale qui, dans un effort intéressant de conceptualisation juridique, précise dans son premier alinéa ce qu'il faut entendre par autonomie locale. Elle est définie comme « *le droit et la capacité effective pour les collectivités de régler et de gérer, dans le cadre de la loi, sous leur propre responsabilité et au profit de leurs populations, une part importante des affaires publiques* ». L'autonomie locale suppose donc le droit et la capacité effective des collectivités à gérer et régler leurs compétences. L'effectivité renvoie dans ce contexte à la situation dans laquelle les collectivités territoriales peuvent décider librement des orientations qu'elles souhaitent donner à leurs actions. En clair, la capacité effective des collectivités territoriales peut se résumer en une question de moyens. Parler de décentralisation dans ce contexte, c'est poser la question des compétences, mais aussi des ressources, c'est-à-dire des capacités de financement des collectivités. Les collectivités territoriales africaines disposent-elles des compétences et des moyens nécessaires pour décider pleinement de l'orientation de l'action publique locale ? La réponse à cette question est déterminante, car le principe de libre administration serait vidé de son sens si les collectivités étaient privées de compétences et de moyens pour les exercer, qu'il s'agisse de moyens financiers, humains ou juridiques.

370. - L'environnement juridique dans lequel les collectivités territoriales agissent et évoluent est particulièrement contraignant. Même si ce cadre est constitutionnel, c'est au législateur qu'il revient de le définir et de l'encadrer. La mise en œuvre de la libre administration appelle donc l'intervention de lois attributives de compétences dont l'exercice est par ailleurs placé sous le contrôle de l'État. En Afrique, si le législateur est formellement moins réticent lorsqu'il s'agit de transférer des compétences, il l'est davantage en ce qui concerne l'attribution des moyens nécessaires à leur exercice. Ainsi, non seulement les compétences des collectivités territoriales ne sont pas consolidées en raison de l'instabilité législative, mais elles ne disposent pas non plus de moyens suffisants à leur exercice. C'est tout l'enjeu des réformes décentralisatrices en cours sur le continent africain[887]. La question de l'attribution de compétences effectives et des

---

[887] C. NACH MBACK, *Démocratisation et décentralisation. Genèse et dynamiques comparés des processus de décentralisation en Afrique subsaharienne*, *op.cit.*, p. 397.

moyens de leur exercice n'a pas fini de tourmenter les acteurs et observateurs de la chose locale africaine[888]. Il semble qu'en la matière, il faudra donner du temps à la décentralisation en Afrique. Ainsi, les collectivités réputées être autonomes souffrent du brouillamini des compétences qui les empêchent d'assurer pleinement l'exercice des compétences locales. L'administration d'État continue d'investir le champ de compétences dévolu aux collectivités territoriales (**Chapitre 1**). Face à la complexité sans cesse croissante des compétences décentralisées et l'absence de ressources techniques et financières, l'autonomie des collectivités territoriales africaines s'avère fortement limitée (**Chapitre 2**). La décentralisation se réalise sans assurer aux collectivités territoriales une liberté d'action.

Chapitre 1 – L'omniprésence de l'État dans le fonctionnement des collectivités territoriales

Chapitre 2 – L'omnipotence de l'État dans l'attribution des moyens

888 Le constat est aussi vrai pour les pays d'Afrique francophone que pour la France. Voir en ce sens, J.-F. BRISSON, « Clarification des compétences et coordination des acteurs », *AJDA*, 2014, p. 605.

Chapitre 1 — L'omniprésence de l'État dans le fonctionnement des collectivités territoriales

371. - L'État en Afrique est systématiquement considéré comme incapable de satisfaire l'intérêt général[889]. C'est la raison pour laquelle les constituants africains de la décennie 1990 ont entendu reconnaitre l'existence aux côtés de l'État d'autres entités territoriales dotées de la personnalité morale de droit public. L'institutionnalisation de cette dualité de personnes publiques entraine nécessairement une dualité de compétences[890]. La différenciation organique entre l'État et les collectivités territoriales suppose que chaque entité dispose d'un domaine de compétences réservé et protégé contre d'éventuels empiètements. Une telle protection ne peut être effective que si des règles claires opèrent cette délimitation.

372. - Dès lors se pose la question de la répartition des compétences entre l'État et les collectivités territoriales et des modalités du partage. Cette question est au cœur du processus de décentralisation, qui ne peut se concevoir sans un transfert de compétences de l'État vers les collectivités territoriales. Les transferts de compétences sont un des éléments indispensables de la décentralisation. Il s'agit d'une question, écrit Jean-Marie Pontier, qui se pose « *dans tous les régimes, même les moins décentralisateurs, parce que la centralisation pure est impossible, elle aboutit à l'apoplexie de l'État et, donc, à la mort de celui-ci* »[891]. On pourrait objecter, dans ce cas, qu'il suffit d'une déconcentration pour donner un second souffle à l'État. Cependant, les peuples n'ont pas seulement besoin que d'une bonne

---

[889] R. A. SAWADOGO., *L'État africain face à la décentralisation*, Karthala-Club du Sahel et de l'Afrique de l'Ouest, p. 55 ; v. aussi C. NANAKO, *La libre administration des collectivités territoriales au Bénin et au Niger,* thèse, dactyl., Université d'Abomey-Calavi, p. 42 ; O. SY, *Reconstruire l'Afrique. Vers une nouvelle gouvernance fondée sur les dynamiques locales*, Ed. Charles Léopold Mayer, 2009, p. 146 pour qui « *la crise en Afrique est avant tout celle de l'État-nation centralisé et arrogant qui a été mis en œuvre après les indépendances formelles* » et C. SAVONNET-GUYOT, *État et sociétés au Burkina-Faso. Essai sur le politique africain,* Karthala, 1986, p. 141 : « *la crise de l'État* (...) *met en relief la difficulté d'intégrer à l'État national des formations sociales et des systèmes politiques qui lui préexistaient* ».

[890] H. K. NOUBISSI, *Décentralisation et centralisation au Cameroun. La répartition des compétences entre l'État et les collectivités locales,* thèse, L'Harmattan, 2013 ; J.-M. PONTIER, *L'État et les collectivités locales. La répartition des compétences*, thèse, PUAM, 1978.

[891] J.-M. PONTIER, « Actualité, continuité et difficultés des transferts de compétences entre l'État et les collectivités territoriales », *RFDA*, 2003, p. 35.

administration d'État, ils ont besoin de liberté et de démocratie[892]. Le sujet de la distribution des compétences devient un axe décisif du débat sur l'organisation politique et administrative de l'État en Afrique. Pourtant, cette question ne trouve pas de réponse simple rendant ainsi périlleuse toute tentative de construction d'une théorie générale des compétences décentralisées dans les États d'Afrique francophone[893].

373. - Néanmoins, dans une optique de clarification, il apparait nécessaire de relever que la construction juridique de la répartition des compétences en Afrique francophone est essentiellement le fait des textes constitutionnels et surtout législatifs. Il serait en effet vain d'espérer en raison de l'inopérance du contentieux des actes administratifs toute précision de la part du juge à ce sujet. La détermination des compétences locales relève donc de la seule volonté du pouvoir central qui accorde par la voie législative une liberté d'action aux collectivités territoriales dans un champ précisément délimité (**Section 1**). La décentralisation en Afrique est très étroite et n'accorde que des compétences très limitées aux collectivités territoriales. Par conséquent, il s'avère indispensable pour l'État de renforcer sa présence dans la gestion de ces compétences afin d'éviter tout débordement de la part des collectivités territoriales qui pourrait nuire à son autorité. À cet effet, le législateur a pris le soin de mettre en place des contrôles stricts et nombreux avec des procédures particulièrement lourdes pour les autorités décentralisées. C'est ce qui explique la multiplicité des procédés de tutelle dans les États de l'espace ouest-africain francophone (**Section 2**).

Section 1 : La détermination des compétences locales par le pouvoir central

374. - Dans l'étude du droit de la décentralisation, aborder la question de la distribution des compétences locales est l'une des tâches les plus délicates. La recherche des techniques de répartition des compétences entre l'État et les collectivités territoriales ne cesse de tourmenter les

---

892 Voir *supra* n° 573 et s.

893 Le constat est identique en France. Voir en ce sens, J.-C. DOUENCE, « Réflexions sur la vocation générale des collectivités locales à agir dans l'intérêt public local », *in Quel avenir pour l'autonomie des collectivités locales ?* Ed. de l'Aube, 1999, p. 335 ; J.-M. PONTIER, « La redéfinition des compétences : à la recherche de la cohérence », *RDP*, 2015, p. 1241.

juristes. Pourtant, la recherche de tels critères part d'une constatation somme toute banale : « *il y a indissociabilité entre l'organisation locale et l'existence de textes constitutionnels ou légaux* »[894]. En effet, la collectivité territoriale n'accède à l'existence juridique qu'à travers sa reconnaissance par l'État ; sans préjuger de l'opinion que l'on peut avoir quant à l'antériorité de la collectivité sur le texte qui l'organise[895]. On peut déduire de cette affirmation qu'il existe toujours un texte qui règle la question du mode de définition des compétences locales. Quel est donc l'acte adéquat pour la définition et la distribution des compétences entre l'État et les collectivités territoriales ? Faut-il recourir à la Constitution ou à la loi ?

375. - D'emblée, on pourrait souhaiter que ce soit la Constitution qui se prononce directement sur l'organisation des compétences locales. Cela offrirait, en raison de la place de la Constitution dans la hiérarchie normative, plus de garanties à l'action des autorités décentralisées. Pour autant, aucune répartition des compétences n'est organisée dans les Constitutions des États d'Afrique francophone. Il serait alors vain de rechercher dans les textes constitutionnels de ces États une théorie générale de la répartition des compétences[896]. Toutefois, les textes constitutionnels ne sont pas totalement muets au sujet des collectivités territoriales. Les Constitutions d'Afrique francophone font preuve de beaucoup de détermination lorsqu'il s'agit d'affirmer le principe de décentralisation, mais elles sont circonspectes quant aux précisions des conditions d'exercice des compétences décentralisées[897]. L'augmentation des références faites aux collectivités territoriales dans les Constitutions prouve l'intérêt du recours à la décentralisation. Cependant, la question de la distribution des compétences locales n'y trouve aucune réponse (**§1**). Dans une logique de complémentarité, les constituants africains s'accordent tous à renvoyer au législateur le soin de déterminer la

---

894 J.-M. PONTIER, *L'État et les collectivités locales. La répartition des compétences*, *op.cit.*, p. 64.
895 *Idem.*
896 Il s'agit d'une tradition commune partagée par les États unitaires par opposition aux États composés. Voir en ce sens, J.-F. BRISSON, « Constitution et compétences locales. Entre théorie introuvable et vaine praxis », in J.-M. PONTIER, L. RAPP (dir.), *Constitution et collectivités territoriales. Réflexions sur le 60e anniversaire de la Constitution*, LexisNexis, 2019, p. 51 ; G. MARCOU, « Décentralisation : quelle théorie de l'État ? », in *Réforme de la décentralisation, réforme de l'État. Régions et villes en Europe, Annuaire des collectivités locales*, t. 24, 2004, pp. 235-252.
897 A. CABANIS, M.-L. MARTIN, « Décentralisation et pouvoirs locaux dans les constitutions d'Afrique francophone », *op.cit.*, p. 79.

répartition des compétences. Ainsi, laissée à la discrétion du législateur, la question des compétences devient « *un sujet éminemment politique, fonction d'arbitrages successifs, de compromis où les constructions juridiques chimiquement pures cèdent devant toutes sortes de considérations* »[898] (**§2**).

§1.- L'indétermination constitutionnelle de la répartition des compétences locales

376. - Autant le dire tout de suite, l'affirmation par les constituants africains des principes de décentralisation ne traduit aucunement une vision d'ensemble cohérente pour l'avenir des collectivités territoriales. L'influence des organisations internationales et des nations occidentales qui pourvoient à l'Aide publique au développement en est pour beaucoup dans cette exaltation constitutionnelle de la décentralisation[899]. Malgré la permanence d'énoncés réservés aux collectivités territoriales, les bases constitutionnelles du droit des collectivités territoriales ont fait l'objet de peu d'études de la part de la doctrine[900]. La raison de ce désintérêt se trouve, notamment dans l'absence de constitutionnalisation de ce droit[901]. Dans la mesure où toutes les Constitutions des États d'Afrique francophone contiennent *peu ou prou* des articles relatifs aux collectivités

---

[898] *Cf.* J.-F. BRISSON, « Les compétences des collectivités territoriales », *RFDA*, 2022, p. 291.

[899] Voir les développements *supra* n° 274 et s. Notons aussi que les nouveaux États africains issues de la décolonisation ont pris pour modèle de référence la Constitution française de 1958.

[900] Pour quelques études qui abordent les fondements constitutionnels de la décentralisation en Afrique, H. K. NOUBISSI, *Décentralisation et centralisation au Cameroun. La répartition des compétences entre l'État et les collectivités locales*, thèse, L'Harmattan, 2013. Notons aussi que même en France, la place des collectivités territoriales n'a connu un regain d'intérêt qu'a la faveur de l'adoption de la loi constitutionnelle du 23 mars 2003. Pour un aperçu des études avant cette date : L FAVOREU, « Les bases constitutionnelles du droit des collectivités locales », in F. MODERNE (dir.), *La nouvelle décentralisation*, Sirey, 1983 ; C. AUTEXIER, « L'ancrage constitutionnel des collectivités de la République », *RDP*, 1981, p. 595 ; A. ROUX, *Droit constitutionnel local*, Economica, coll. « Droit Poche », 1995.

[901] « Pour qu'il existe un droit constitutionnel des collectivités territoriales (ou locales), il faut qu'il y ait une Constitution, si possible formelle. Ceci n'est cependant pas suffisant, car encore faut-il que ce droit soit constitutionnalisé ». M. VERPEAUX, « Droit constitutionnel des collectivités territoriales ou droit constitutionnel local ? », in P. TÜRK (dir.), *Pour un droit constitutionnel des collectivités territoriales*, L'Harmattan, 2020, p. 20.

territoriales, nous nous intéresserons à l'examen du contenu de ces dispositions constitutionnelles (**A**) et de leur impact sur la détermination des compétences des collectivités territoriales (**B**).

A- La Constitution et les collectivités territoriales

377. - Toutes les constitutions africaines accordent une place aux collectivités territoriales. Les institutions décentralisées sont prises en compte avec une certaine régularité malgré les nombreuses révisions constitutionnelles que connaissent les États dans l'espace africain francophone. La longueur et la taille des dispositions constitutionnelles réservées à ces collectivités peuvent différer d'une constitution à l'autre sans pour autant avoir des effets sur l'étendue de la décentralisation.

378. - La Constitution de la première République du Togo adoptée dès l'accession du pays à la souveraineté internationale consacre un titre composé d'un seul article aux collectivités territoriales (art 58) : « *la République une et indivisible, reconnait l'existence des collectivités territoriales. Ces collectivités sont les circonscriptions et communes. Elles s'administrent librement et démocratiquement dans les conditions prévues par la loi* »[902]. On retrouve à quelques exceptions près les mêmes dispositions dans la Constitution de la deuxième République adoptée deux ans plus tard. Dans son préambule, le constituant togolais proclame sa volonté de veiller au respect et à la garantie de la personne humaine, de la famille et des collectivités locales. À la différence de la première, la Constitution de 1963 ne mentionne pas les catégories de collectivités de territoriales[903]. Elle se contente de rappeler dans les deux premiers alinéas de l'article 84 que « *la République une et indivisible, reconnait l'existence des collectivités territoriales. Les collectivités territoriales sont créées par la loi* ».

Les innovations notionnelles introduites par la Constitution française du 27 octobre 1946 qui a inséré pour la première fois, expressément, dans le texte constitutionnel les concepts de « libre administration » et de « collectivités territoriales » aux articles 85 à 89 du Titre X[904] ont d'une certaine façon influencé les constituants africains. Toutefois, dans certains

---

902 Constitution togolaise de 1961, Titre VII « Des collectivités territoriales ».
903 Elles vont passer deux à trois avec la Constitution de 1992 : communes, préfectures et régions puis à nouveau à deux avec la loi constitutionnelle de 2019 qui supprime la collectivité territoriale préfectorale. Voir *supra* n° 224 et s.
904 Voir le Titre VIII intitulé « De l'Union française » dont la Section III était réservée aux départements et territoires d'Outre-mer.

pays, cette constitutionnalisation a été tardive. C'est le cas du Burkina Faso. La décentralisation ne sera consacrée constitutionnellement qu'avec l'adoption en 1991 d'un nouveau texte constitutionnel qui dispose que le territoire du Burkina Faso est subdivisé en collectivités territoriales décentralisées. L'article 143 de la Constitution burkinabè dispose à cet effet que « *le Burkina Faso est organisé en collectivités territoriales* ». On retrouve aujourd'hui dans toutes les Constitutions africaines une disposition qui organise l'administration territoriale. Ainsi, l'article 164, alinéas 2 et 3 de la Constitution du Niger dispose que « *l'administration territoriale repose sur les principes de la décentralisation et de la déconcentration. Les collectivités territoriales sont créées par une loi organique. Elles s'administrent librement par des conseils élus* ». Au Mali, l'article 97 dispose aussi que « *les collectivités territoriales sont créées et administrées dans les conditions définies par la loi* ».

379. - Dans l'histoire des institutions locales au Togo, la Constitution de 1980 fait figure d'exception. Il s'agit en effet de la seule Constitution qui ne réserve pas un titre spécifique aux collectivités territoriales. L'article 6 de cette Constitution affirme que « *la République togolaise assure à chaque citoyen le respect, conformément à la loi, des droits et libertés de la personne humaine, de la famille et des collectivités locales* ». Elle reprend les dispositions précitées du préambule de 1963. Paradoxalement, c'est sous cette République marquée par le règne du monopartisme que le Togo connaitra l'une des deux élections municipales de son histoire[905].

Sous la VIe République, le droit constitutionnel local va connaitre quelques modifications avec l'adoption de nouvelles lois constitutionnelles. Désormais, le Titre XII de la Constitution réservé aux collectivités territoriales est intitulé « Des collectivités territoriales et de la chefferie traditionnelle ». Le constituant de 1992 consacre à côté des collectivités territoriales[906] les autorités coutumières, qui accèdent ainsi à une reconnaissance constitutionnelle. La décennie 1990 correspond en Afrique francophone à un mouvement général de réhabilitation des valeurs traditionnelles : on affirme la volonté de protéger les spécificités

---

905 La deuxième élection des conseillers municipaux n'étant intervenue que le 30 juin 2019.

906 Art 141 : « La République Togolaise est organisée en collectivités territoriales sur la base du principe de décentralisation dans le respect de l'unité nationale. Ces collectivités territoriales sont : les communes, les préfectures et les régions. Toute autre collectivité territoriale est créée par la loi. Les collectivités territoriales s'administrent librement par des conseils élus au suffrage universel, dans les conditions prévues par la loi ».

culturelles ainsi que les langues nationales[907]. On est loin des craintes suscitées à la veille des indépendances à l'égard des autorités coutumières qui justifiaient la recherche effrénée de l'homogénéité la plus complète des populations par les élites africaines[908].

Ainsi, la recherche d'une légitimité rénovée et le sentiment que les autorités traditionnelles sont capables de jouer un rôle important dans le contrôle des populations face à l'éclatement de la société justifient en partie le retour en grâce des notabilités coutumières[909]. L'article 143 de la Constitution togolaise dispose à cet effet que « *l'État togolais reconnait la chefferie traditionnelle, gardienne des us et coutumes. La désignation et l'intronisation du chef traditionnel obéissent aux us et coutumes de la localité* ». S'il est rappelé que les chefferies traditionnelles sont gardiennes des *us et coutumes*, l'article ne donne aucune autre précision quant à son rôle exact dans l'organisation territoriale de la République. La Constitution nigérienne de 2010 est aussi peu loquace sur le rôle de la chefferie et se contente de renvoyer au législateur le soin d'en préciser le rôle. Elle dispose dans son article 167 que « *l'État reconnait la chefferie traditionnelle comme dépositaire de l'autorité coutumière. À ce titre, elle participe à l'administration du territoire de la République dans les conditions déterminées par la loi. La chefferie traditionnelle est tenue à une stricte obligation de neutralité et de réserve. Elle est protégée contre tout abus de pouvoir tendant à la détourner du rôle que lui confère la loi* ».

380. - En outre, la Constitution togolaise de 1992 et celle burkinabè issue de la révision du 11 juin 2012 ont introduit le bicaméralisme. Le parlement est depuis lors composé de deux chambres : l'Assemblée nationale et le Sénat[910]. Bien que consacrée constitutionnellement, l'institution sénatoriale tarde à voir le jour. Les articles 52, alinéa 6 de la Constitution togolaise et 80 al. 2 de la Constitution burkinabè intègrent les collectivités territoriales à l'exercice de la souveraineté de l'État. Véritables acteurs du jeu politique national, les autorités locales participent à l'élection des membres du Sénat qui assure leur représentation au sein de la République.

Aussi, on ne saurait passer sous silence l'ensemble des dispositions présentes dans les titres placés en tête de toutes les constitutions relatives

---

907 A. CABANIS, M.-L. MARTIN, « Décentralisation et pouvoirs locaux dans les constitutions d'Afrique francophone », *op.cit.*, p. 78.

908 *Ibid.*, p. 78 et s.

909 *Idem.*

910 Cette institution bien que consacrée constitutionnellement tarde à voir le jour.

à la souveraineté de l'État[911]. Ces dispositions figurant dans les tous premiers articles de la Constitution consacrent, conformément à la longue tradition constitutionnelle héritée de la France, l'indivisibilité de la République et le caractère unitaire de l'État. Bien que ces diverses dispositions ne concernent pas directement les collectivités territoriales, elles peuvent s'appliquer en matière locale. La préservation du caractère unitaire de l'État justifie que les Constitutions fixent les bases du contrôle de l'action des collectivités territoriales. Si les modalités du contrôle sont souvent renvoyées généralement à la loi, certaines constitutions mentionnent expressément l'institution d'un délégué de l'État. À ce titre, l'article 165, alinéa 2 de la Constitution nigérienne de 2010 dispose que *« le représentant de l'État veille au respect des intérêts nationaux »*.

381. - De façon générale, il est possible d'identifier dans les Constitutions africaines des références expresses à la notion de collectivité territoriale. Les Constitutions comportent sans grande discontinuité un titre relatif aux collectivités territoriales. Même s'il n'est pas expressément défini, on peut déduire de la rédaction de certaines dispositions constitutionnelles que les constituants ont entendu reconnaitre aux collectivités territoriales la gestion des affaires locales. La tendance générale en Afrique francophone consiste à reconnaitre deux niveaux de collectivités décentralisées : les communes et les régions[912]. En revanche, les Constitutions ne règlent pas la question de l'attribution des compétences entre ces deux échelons décentralisés et l'État.

B- La Constitution et les compétences locales

382. - Les compétences locales ne sont pas de nature constitutionnelle. La Constitution en Afrique subsaharienne a été le réceptacle des revendications politiques et sociales, notamment en matière de décentralisation[913]. Toutefois, il faudrait se méfier des slogans et des affichages constitutionnels dans la promotion des processus de

---

911 Au Burkina Faso, dans la Constitution de 1991, les questions relatives à la souveraineté nationale se retrouvent dans un Titre II intitulé « De L'État et de la souveraineté nationale ». La loi constitutionnelle du 27 janvier 1997 a modifié l'intitulé du Titre II en remplaçant « souveraineté du peuple » par « souveraineté nationale ».

912 Voir les développements *supra* Partie 1.

913 Voir en ce sens l'étude de, C. NACH MBACK, *Démocratisation et décentralisation. Genèse et dynamiques comparés des processus de décentralisation en Afrique subsaharienne, op.cit.*

décentralisation[914]. À l'issue de l'adoption de la loi constitutionnelle française du 28 mars 2003, Bertrand Faure, dans un article au titre évocateur : « Réforme constitutionnelle et décentralisation : des slogans font loi », faisait remarquer que les Constitutions introduisent souvent des références juridiques par leur support « *mais platoniques dans leur effet, de telle manière que les avancées concrètes devront encore tout au législateur* »[915]. Sont dans cette situation les Constitutions africaines. Elles introduisent des principes en matière de décentralisation sans en tirer directement les conséquences. Cette situation trouve une explication dans la tradition politique des systèmes juridiques d'inspiration française où la décentralisation dans sa vision politique est présentée en termes de « *libertés locales* »[916]. Les collectivités territoriales sont dressées comme un moyen de défense contre les dérives qui peuvent naitre de la centralisation de l'État[917]. C'est la raison pour laquelle les Constitutions africaines héritière de la tradition française de l'État unitaire font preuve d'une certaine timidité dans l'amorce constitutionnelle de la répartition des compétences décentralisées[918]. À travers l'énoncé du principe de la libre administration des collectivités territoriales, la Constitution donne une habilitation générale au législateur pour régler la question des compétences des collectivités territoriales[919].

---

914 H. K. NOUBISSI, *Décentralisation et centralisation au Cameroun. La répartition des compétences entre l'État et les collectivités locales*, thèse, L'Harmattan, 2013, p. 51.

915 B. FAURE, « Réforme constitutionnelle et décentralisation : des slogans font loi », *RDP*, n°1, 2003, p. 122.

916 Voir à ce sujet, J.-F. BRISSON, « Faut-il inscrire l'expression "*libertés locales*" dans la Constitution ? », in P. TÜRK (dir.), *Pour un droit constitutionnel des collectivités territoriales*, L'Harmattan 2020 p. 223 : « *La seule évocation des libertés locales suggère, en effet, que la réforme défendue plonge ses racines dans le mouvement politique engagé dès le XIXe siècle par les auteurs libéraux pour contrer les excès du gouvernement autoritaire. Elle tente de nous persuader que, sous d'autres formes, c'est le même combat qui, enjambant les siècles, continue. L'expression peut en effet se réclamer d'illustres promoteurs. Tocqueville, d'abord qui, dans "De la démocratie en Amérique", fait l'éloge et la promotion du self-government par opposition à la situation de la France ; sous cette réserve que Tocqueville député n'a jamais appuyé en France le moindre projet de décentralisation, comme si, dans son esprit, la centralisation était finalement plus conforme aux mœurs des français* ».

917 J.-M. PONTIER, *L'État et les collectivités locales. La répartition des compétences, op.cit.*, p. 78.

918 H. K. NOUBISSI, Décentralisation et centralisation au Cameroun. La répartition des compétences entre l'État et les collectivités locales, op.cit., p. 88.

919 Pour le professeur Jean-François Brisson, le fait que la question de la décentralisation relève en France (cela vaut pour les États d'Afrique francophone) de la loi et non du règlement alors qu'elle est fondamentalement une question d'organisation administrative est une survivance de l'approche philosophique de la décentralisation appréhendée sous l'angle de la liberté. J.-F. BRISSON, « Faut-il inscrire l'expression "*libertés locales*" dans la constitution ? », *op.cit.*, p. 228.

383. - Toutes les Constitutions d'Afrique francophone se rejoignent sur un point : elles renvoient au législateur le soin de décider de l'organisation et de l'étendue des compétences des collectivités territoriales. L'article 141, alinéa 3 de la Constitution togolaise du 14 octobre 1992 dispose que « *les collectivités territoriales s'administrent librement par des conseils élus au suffrage universel, dans les conditions prévues par la loi* ». Le constituant précise à l'article 84, alinéa 35, que la détermination des principes fondamentaux de la libre administration des collectivités territoriales, de leurs compétences et de leurs ressources relève du domaine du législateur. L'article 164, alinéa 3 de la constitution du Niger de 2010 renvoie également à la loi le soin de « *détermine* [r] *les principes fondamentaux de la libre administration des collectivités territoriales, leurs compétences et leurs ressources* ». On retrouve une disposition identique dans les constitutions burkinabè et malienne[920].

La Constitution exclut donc de son champ toute question relative à l'organisation et au fonctionnement des collectivités territoriales. Elle se contente de régler la distribution des compétences entre l'exécutif, le législatif et l'autorité judiciaire. En confiant la mission au législateur de fixer les bornes de la libre administration des collectivités territoriales, le constituant semble conforté dans l'idée que le parlement serait plus à même de garantir l'unité nationale. Ce faisant, le principe de libre administration, qui n'est pas défini par la Constitution, va fluctuer au gré des choix politiques opérés par le législateur. La conséquence immédiate étant l'imprécision du partage et le risque que le transfert des compétences ne se transforme en une opération de recentralisation des compétences au profit de l'État. La détermination des compétences des collectivités par la loi ordinaire pose ainsi un véritable problème. La loi dans les pays d'Afrique peut se révéler antilibérale[921] sans que les principes posés par la Constitution soient d'un grand secours en l'absence de tout contrôle de la part des juridictions constitutionnelles.

---

[920] Respectivement art. 101-2 : « la loi détermine les principes fondamentaux de la libre administration des collectivités territoriales, leurs compétences et leurs ressources » et art. 98 : « les collectivités s'administrent librement par des Conseils élus et dans les conditions fixées par la loi ».

[921] Si le constat est aussi vrai ailleurs, il l'est encore plus dans la situation africaine marqué par la prééminence de l'exécutif sur les autres pouvoirs. *Cf.*, B. FAURE, *Droit des collectivités territoriales, op.cit.*, n° 699, p. 594 et s.

384. - En somme, les textes constitutionnels dans les États « simples » ne renseignent pas sur la détermination matérielle des compétences entre l'État et les collectivités territoriales. Le droit constitutionnel habilite le législateur à fixer la ligne de partage entre l'administration étatique et l'administration décentralisée. Les communes et les régions n'exerceront que les compétences que le législateur voudra bien leur accorder. « *On aura beau souligner le perfectionnement de la discipline* [le droit de la décentralisation], *ce qui est décentralisé est un cadeau de la loi* »[922]. C'est donc dans les textes législatifs qu'il faut rechercher les clés et méthodes de répartition des compétences locales.

§2.- La prévalence de la loi dans l'attribution des compétences locales

385. - Les collectivités territoriales ne bénéficient d'aucune garantie en matière de répartition des compétences. La nature unitaire de l'État fait du législateur, autorité politique donc, le *« grand ordonnateur du système de compétences »*[923]. Les textes de loi relatifs à la décentralisation en Afrique posent des principes fondamentaux qui doivent orienter l'action du parlement dans la distribution des compétences. Bien que placés en tête du processus de décentralisation, ces principes ont la particularité de ne pas s'imposer au législateur de façon absolue. Il s'agit de principes ayant davantage une valeur politique que juridique. Puisque la Constitution se limite à une habilitation formelle du législateur pour la détermination des compétences décentralisées, l'idée même de décentralisation demeure attachée à la liberté du législateur. Ce dernier a pris soin d'énoncer un certain nombre de principes fixant le cadre de la répartition (**A**). Cependant, ces principes n'exercent qu'une influence relative sur la détermination matérielle des compétences des collectivités territoriales (**B**).

---

922 *Idem.*

923 J.-F. BRISSON, « Les compétences des collectivités territoriales », *op.cit.*, p 291.

A- La définition législative du cadre de répartition des compétences

386. - Le cadre de répartition des compétences renvoie à l'ensemble des principes énoncés par la loi dont l'objectif est de guider le législateur sur la manière de transférer les compétences. Ces principes restent toutefois très théoriques. Ils ne s'imposent pas au législateur de façon impérative. Ils constituent simplement un point de départ. On dénombre, à la lecture des différents textes de loi sur la décentralisation, quatre principes dont l'un marque la particularité de la décentralisation sur le continent africain : le principe de progressivité souvent lié à celui de la subsidiarité. Ainsi, seront étudiés dans les lignes qui suivent les principes : de subsidiarité (**1°**), de progressivité (**2°**), d'interdiction d'une tutelle entre collectivités (**3°**) et de compensation (**4°**).

1. Le principe de subsidiarité

387. - La subsidiarité suppose dans le cadre de la répartition des compétences de rechercher le niveau adéquat pour l'exercice d'une compétence, c'est-à-dire le niveau adapté pour répondre aux besoins des populations sans toutefois perdre de vue les impératifs d'efficacité financière. Autrement dit, les décisions doivent être prises à l'échelon le plus proche des citoyens sauf si elles sont susceptibles d'une meilleure mise en œuvre et à moindre coût à l'échelon supérieur. Concrètement, l'application de ce principe voudrait que le législateur distribue les compétences en partant du bas vers le haut. Les compétences doivent ainsi être confiées en priorité aux communes, puis aux régions et enfin à l'État. La mise en œuvre du principe de subsidiarité place l'administration décentralisée au premier plan et relègue l'État en seconde position dans la satisfaction des besoins des populations[924].

388. - Le principe de subsidiarité ne figure pas expressément dans les droits malien et togolais. L'article 3, alinéa 3 de la loi n° 2019-006 du 26 juin 2019 dispose que dans le cadre de la définition des compétences locales, « *les collectivités territoriales ont vocation à exercer les compétences qui peuvent, le mieux, être mises en œuvre à leur échelon* ». Le législateur togolais de 2019 s'est

---

924 Pour une étude d'ensemble, A. DURANTHON, *Subsidiarité et collectivités territoriales. Étude sur la subsidiarisation des rapports entre État et collectivités en droit public français*, thèse, Dalloz, 2017.

sans douter inspirer de l'article 72 al. 2 de la Constitution française issue de la révision constitutionnelle du 28 mars 2003 qui sans désigner formellement le principe de subsidiarité prévoit que, « *les collectivités territoriales ont vocation à prendre les décisions pour l'ensemble des compétences qui peuvent le mieux être mises en œuvre à leur échelon* ». À l'opposé de cette démarche, l'article 34 de loi n° 55-2004 du 21 décembre 2004 portant code général des collectivités territoriales au Burkina Faso dispose sans équivoque que « *la répartition des compétences entre l'État et les collectivités territoriales est régie par le principe de subsidiarité* ». Il en va de même pour le législateur nigérien qui à l'article 154 de l'ordonnance n° 2010-54 du 17 septembre 2010 portant code général des collectivités territoriales prévoit que « *la répartition des compétences entre l'État et les collectivités territoriales s'opère par blocs de compétences et selon le principe de subsidiarité* ». Si ces deux pays s'illustrent par la consécration formelle de la subsidiarité dans l'attribution des compétences, il faut reconnaitre avec Charles Nach Mback que, même non expressément formulé, le principe de subsidiarité semble s'applique à tout mécanisme de transfert de compétences dans les États africains[925].

389. - Dès 1997, le principe de subsidiarité était fortement recommandé dans le cadre des réformes décentralisatrices sur le continent africain par l'assistance technique française[926]. Cependant, malgré l'importance croissante qu'on accorde à ce principe dans les études sur la décentralisation, la notion de subsidiarité demeure insaisissable selon un constat partagé par l'ensemble de la doctrine. Guillaume Drago écrit à propos de la subsidiarité qu'elle n'est pas un principe cartésien et le juriste ne s'y retrouve pas aisément[927]. Pour Jean-François Brisson, l'énoncé du principe est à l'image du principe de libre administration : assez peu parlant[928].

390. - L'apparition de ce principe dans les sciences juridiques est relativement récente, bien que ces origines soient lointaines. Chantal Millon-Delsol démontre dans ses travaux consacrés au principe de

---

925 C. NACH MBACK, *Démocratisation et décentralisation. Genèse et dynamiques comparés des processus de décentralisation en Afrique subsaharienne, op.cit.*, p. 400.

926 Secrétariat d'État à la coopération/Direction du développement, *La décentralisation en Afrique subsaharienne*, Paris, 1997, p. 23 cité par C. NACH MBACK, *op.cit.*, p. 400.

927 G. DRAGO, « Le principe de subsidiarité comme principe de droit constitutionnel », *RIDC*, 1994, p. 583.

928 J.-F. BRISSON, « Les nouvelles clefs constitutionnelles de la répartition matérielle des compétences entre l'État et les collectivités territoriales », *AJDA*, 2003, p. 529.

subsidiarité que ce principe « *trouve ses fondements à la fois dans l'esprit grec, la philosophie chrétienne médiévale et dans la vision germanique de la société* »[929]. Certains auteurs font remonter la première consécration de la subsidiarité au droit canonique[930]. En effet, on retrouve dans l'Encyclique *Quadragesimo Anno*, publiée en 1931 sous l'égide du Pape Pie IX une exhortation à l'abandon aux citoyens des tâches qu'ils peuvent accomplir avec aisance à leur niveau. L'autorité publique doit ainsi abandonner « *aux groupements de rang inférieur le soin des affaires de moindre importance où se disperserait à l'excès son effort ; elle pourra dès lors assurer plus librement, plus puissamment, plus efficacement les fonctions qui n'appartiennent qu'à elle, parce qu'elle seule peut les remplir* »[931]. Autrement, « *ce serait commettre une injustice, en même temps que troubler d'une manière très dommageable l'ordre social, que de retirer aux groupements d'ordre inférieur, pour les confier à une collectivité plus vaste et d'un rang plus élevé, les fonctions qu'ils sont en mesure de remplir eux-mêmes* »[932]. En droit français, les auteurs font remonter la première définition du principe de subsidiarité au Rapport Guichard de 1976[933]. La subsidiarité, selon ce rapport, « *conduit à rechercher toujours le niveau adéquat d'exercice des compétences, un niveau supérieur n'étant appelé que dans les cas où les niveaux inférieurs ne peuvent pas exercer eux-mêmes les compétences correspondantes. L'État doit ainsi déléguer aux collectivités tous les pouvoirs qu'elles sont en mesure d'exercer* »[934].

391. - Dans les États d'Afrique francophone, la loi fait une place au principe de subsidiarité dans le droit positif, mais aucun texte n'en donne une définition. Il est possible de voir dans la consécration de ce principe que ce soit formellement ou non, la volonté du législateur africain de faire écho au texte de l'article 6 al.1 de la Charte africaine des valeurs et des

---

929 C. MILLON-DELSOL, *L'État subsidiaire*, PUF, 1992, p. 13.
930 J.-M. PONTIER, « Les principes de répartition des compétences », *op.cit.*, p. 7.
931 PIE XI, Encyclique « Ouadragesimo anno », sur la restauration de l'ordre social en pleine conformité avec les *préce*ptes de l'Évangile, à l'occasion du quarantième anniversaire de l'Encyclique « Rerum novarum », 15 mai 1931 (traduction officielle du latin, publiée par la Typographie vaticane), Paris, Bonne Presse, 1931, p. 31 cité par J.-P. DEROSIER, « La dialectique centralisation/décentralisation - Recherches sur la caractère dynamique du principe de subsidiarité » *Revue internationale de droit comparé*, vol. 59, n°1, 2007, p. 126.
932 *Ouadragesimo anno, Actae Apostolicae Sedis 23* (1931) p. 203, § 126, cité par J.-M. PONTIER, « Les principes de répartition des compétences », *op.cit.*, p. 7.
933 Voir en ce sens, A. DELCAMP, « Principe de subsidiarité et décentralisation », *RFDC*, n° 23, 1995, p. 616.
934 Rapport *Vivre ensemble* de la commission de développement des responsabilités locales, La documentation française, 1976.

principes de la décentralisation, de la gouvernance locale et du développement local adoptée par la conférence de l'Union Africaine (UA) en 2014. Cet article stipule que « *le gouvernement central crée les conditions propices à la prise de décisions, à l'élaboration, à l'adoption et à la mise en œuvre des programmes et des politiques aux niveaux inférieurs du gouvernement où les gouvernements locaux ou les autorités locales offrent une meilleure garantie de pertinence et d'efficacité* ». Contrairement à la Charte européenne de l'autonomie locale qui insiste à titre principal sur la proximité démocratique du pouvoir de décision, la Charte africaine opte pour une vision technocratique de la subsidiarité axée sur la recherche de l'efficacité de l'action publique[935].

392. - Toutefois, une telle définition ne renseigne pas non plus sur les critères d'application du principe. Qu'est-ce qui pourrait justifier que le législateur considère un échelon plutôt qu'un autre comme plus à même d'exercer une compétence ? Pour Elong Mbassi par exemple, la répartition des compétences devrait privilégier « *la prise en charge des problèmes locaux et quotidiens par les collectivités locales, sauf si cette prise en charge nécessite des moyens financiers ou techniques hors de portée de leurs capacités* »[936]. Le critère de subsidiarité en Afrique serait donc principalement déterminé par l'existence de moyens matériels, financiers et techniques nécessaires à l'exercice d'une compétence[937]. Il s'agit là de pures considérations factuelles qui font ressortir le principe de subsidiarité du seul domaine de l'opportunité administrative. Ce faisant, on ne peut exclure l'arbitraire de ceux qui s'en recommandent pour justifier une recentralisation, comme le note Bertrand Faure à propos de l'application du principe en droit français[938].

---

935 Ce qui contredit l'analyse faite par C. Nach Mback qui estime que les réformes décentralisatrices en Afrique tendent en application du principe de subsidiarité à remettre le pouvoir à la base, au plus des populations. C. NACH MBACK, *Démocratisation et décentralisation. Genèse et dynamiques comparés des processus de décentralisation en Afrique subsaharienne*, *op.cit.*, p. 402. Pour une analyse des formulations européenne et française de la subsidiarité, voir notamment, J.-F. BRISSON, « Les nouvelles clefs constitutionnelles de la répartition matérielle des compétences entre l'État et les collectivités territoriales », *op.cit.*, p. 529. L'auteur estime que la formulation issue de la révision de 2003 est en retrait par rapport à celle de la Charte européenne de l'autonomie locale.

936 J.-P. ELONG MBASSI, « Introduction », in *La décentralisation en Afrique de l'Ouest*, P.D.M., Cotonou, 1994, p. 16.

937 C. NACH MBACK, *Démocratisation et décentralisation. Genèse et dynamiques comparés des processus de décentralisation en Afrique subsaharienne*, *op.cit.*, p. 403.

938 B. FAURE, *Droit des collectivités territoriales*, *op.cit.*, n°688.

393. - En réalité, la subsidiarité n'est pas un véritable principe juridique. Il ne lie pas le législateur dans l'exercice de son rôle de répartiteur de compétences locales. Pour Frédéric Baudin-Cuillère, la subsidiarité n'est pas une norme juridique, mais un état d'esprit, une orientation[939]. Principe éthique[940], les termes de sa formulation ne permettent pas de l'opposé au législateur. Dans les termes de son énoncé — « le mieux », « vocation » —, il ne figure aucun impératif qui puisse être sanctionné. Le critère du « mieux » renvoie aux notions d'efficacité et de performance[941] tandis que celui de « vocation » en fait un simple objectif. Pour garantir efficacement le principe de subsidiarité, il est important non seulement de l'inscrire dans la Constitution, mais aussi d'en faire un véritable principe normatif. L'expérience de l'article 72, alinéa 3 de la Constitution française, montre qu'il ne faut pas simplement en faire un objectif à valeur constitutionnelle,[942] mais un principe normatif sur lequel pourront s'appuyer les collectivités territoriales. La liaison de la subsidiarité à la décentralisation et sa consécration constitutionnelle pourrait en effet conduire à l'invalidation des lois ne respectant pas le principe de subsidiarité[943].

394. - D'une certaine manière, le législateur africain avait bien conscience de la difficulté à mettre en œuvre le principe de subsidiarité au moins pour une raison. L'application de ce principe dans la répartition des compétences entre les collectivités publiques en Afrique est difficilement opératoire, car les collectivités territoriales n'ont pas, de fait, entre elles, le même niveau. Certaines seraient, en effet, inaptes à exercer les

---

939 F. BAUDIN-CUILLERE, *Principe de subsidiarité et administration locale*, L.G.D.J., 1995, p. 35.

940 C. MILLON-DELSOL, *Le principe de subsidiarité*, P.U.F. 1993. p. 55. Voir aussi, V. CONSTANTINESCO, « Le principe de subsidiarité : un passage obligé vers l'Union européenne », in *Mélanges Boulouis*, Dalloz, 1991, p. 38.

941 Ces préoccupations relèvent davantage de la gestion publique que du droit. J.-F. BRISSON, « Les nouvelles clefs constitutionnelles de la répartition matérielle des compétences entre l'État et les collectivités territoriales », *op.cit.*, p. 529 et s.

942 Voir en ce sens, B. FAURE, « Les objectifs à valeur constitutionnelle, une nouvelle catégorie juridique ? », *RFDC*, 1995, p. 47 ; P. DE MONTALIVET, *Les objectifs de valeur constitutionnelle*, Dalloz, 2006. Et aussi l'application qu'en fait le Conseil constitutionnel en matière de décentralisation : Cons. Cons. DC n° 2005-516 du 7 juill. 2005, Loi de programme fixant les orientations de la politique énergétique, JORF du 14 juill. 2005, p. 11589.

943 J.-P DEROSIER, « La dialectique centralisation/décentralisation. Recherches sur le caractère dynamique du principe de subsidiarité », *op.cit.*, p. 137 et s.

compétences qu'elles doivent en principe exercer. Il semble que c'est cette situation qui justifie dans les législations africaines la combinaison du principe de subsidiarité avec celui de la progressivité[944].

2. Le principe de progressivité

395. - La stratégie de transfert de compétences adoptée dans les textes africains relatifs à la décentralisation consiste en un transfert progressif des compétences aux communes et aux régions. Le législateur subordonne ainsi l'entièreté du transfert de compétences à la capacité des collectivités territoriales à les assumer. Selon l'article 5 du CGCT au Burkina Faso : « *la mise en œuvre de la décentralisation se fait selon la règle de la progressivité et le principe de subsidiarité* ». Dans l'exposé des motifs de l'ordonnance portant code général des collectivités territoriales au Niger, on peut lire qu'il est retenu au nombre de principes de base du nouveau cadre juridique la progressivité dans la mise en œuvre des différents paliers de décentralisation[945]. L'article 63 de la loi n° 2019-006 précitée au Togo prévoit que « *le transfert de compétences se fait suivant le principe de progressivité, en tenant compte de chaque niveau de décentralisation et de la capacité des collectivités territoriales à les assumer* ». Le principe de progressivité traduit l'idée selon laquelle l'État peut continuer d'intervenir de façon légale dans les domaines de compétences transférées par la loi aux collectivités territoriales tant qu'il estime qu'elles n'ont pas la capacité suffisante pour les exercer.

396. - Ainsi, le principe de progressivité dans la mise en œuvre de décentralisation est lié à la « capacité » des collectivités territoriales. Ce terme ne fait pas l'objet d'une définition de la part du législateur et pourtant il conditionne le transfert et l'élargissement des compétences des collectivités.

Le législateur togolais distingue trois catégories de compétences attribuées aux collectivités territoriales[946] : les compétences propres, les compétences partagées et les compétences transférées. Les compétences propres font référence à celles exercées par une collectivité territoriale

---

[944] H. K. NOUBISSI, *Décentralisation et centralisation au Cameroun. La répartition des compétences entre l'État et les collectivités locales*, *op.cit.*, p. 124.
[945] Exposé des motifs de l'ordonnance portant code général des collectivités territoriales au Niger repris dans le CGCT, p. 18.
[946] Loi n° 2019-006 *préc.*, art. 64.

décentralisée dans un domaine déterminé de manière à lui permettre d'accomplir les actes relatifs à ce domaine dans la limite de ses ressources propres[947]. Les compétences partagées renvoient quant à elles à des actions menées par les collectivités territoriales en complémentarité de celles menées par l'État[948]. Enfin, les compétences transférées regroupent la catégorie des compétences qui sont transférées par l'État aux collectivités territoriales de manière à permettre l'élargissement progressif des compétences propres[949]. La première catégorie d'attribution semble laisser aux collectivités territoriales une maitrise des éléments de décision sous réserve des conditions de ressources. Néanmoins, la frontière entre les compétences exclusives et les compétences partagées n'est pas toujours évidente en raison de l'enchevêtrement des compétences. Cela justifie la présence permanente de l'État aussi bien dans les compétences propres que celles partagées pour assurer la cohérence de l'action publique. C'est sur la dernière catégorie de compétence que le droit positif togolais fait peser le principe de progressivité. En réalité, l'utilisation de l'expression « compétences transférées » révèle une erreur dans le choix des mots dans la mesure où toutes les compétences décentralisées sont des compétences d'attribution et donc des compétences transférées. Les collectivités ne disposent pas de compétences propres par nature.

La progressivité dans le transfert des compétences traduit la réticence de l'État à faire aboutir totalement le processus de décentralisation. Conformément à ce principe, les autorités étatiques peuvent procéder au transfert effectif des compétences légalement attribuées aux collectivités locales selon un rythme totalement aléatoire en toute légalité[950]. Les titres légaux attribués aux collectivités territoriales peuvent être mis en échec par des décrets gouvernementaux. Le choix des modalités et les délais des transferts sont laissés à l'entière responsabilité du pouvoir règlementaire. C'est le sens de l'article de 77 du code général des collectivités territoriales burkinabè qui dispose que « *les modalités et les délais de transfert des compétences à la région et aux communes, nécessaires à l'effectivité* [...] *sont fixés par décret pris en conseil des ministres* ». Chaque domaine de compétence transférée par la loi aux collectivités territoriales doit faire « *l'objet de décret de transfert pris en conseil des ministres sur proposition du ministre en charge de la tutelle des collectivités*

---

947 *Ibid.*, art. 65.
948 *Ibid.*, art. 66.
949 *Ibid.*, art. 67.
950 H. K. NOUBISSI, *Décentralisation et centralisation au Cameroun. La répartition des compétences entre l'État et les collectivités locales, op.cit.*, p. 126.

territoriales en rapport avec les ministères sectoriels»[951]. À travers cette possibilité offerte par le législateur lui-même, le décret devient l'outil nécessaire pour la mise en œuvre effective de la décentralisation. Les décrets de transfert seront ainsi pris progressivement selon une chronologie définie par le gouvernement et son appréciation de la « capacité » des collectivités à exercer les compétences que la loi leur attribue.

397. - La décentralisation territoriale demeure ainsi dans une logique de centralisation administrative. Le décret donne le moyen à l'État de recentraliser les compétences et de demeurer l'acteur principal de la décentralisation. « *Il est tout même paradoxal de constater qu'à l'issue d'un processus de décentralisation qu'il a voulu, l'État ne se désengage pas au sens où, s'il n'est plus acteur, il reste régulateur*»[952]. Par ailleurs, au-delà de cette approche critique du principe de progressivité, il est difficile de distribuer concrètement des compétences identiques à des collectivités territoriales de même niveau sans tenir compte de moyens effectivement en leur possession pour en espérer un exercice au minimum effectif. En ce sens, l'attribution de compétences expérimentales aux collectivités territoriales en vue d'une application ultérieure aurait été préférable dans une conception libérale de la réforme décentralisatrice.

Quoi qu'il en soit, le législateur reste le principal organisateur de la décentralisation qui ne trouve son expression juridique et politique que dans sa volonté. La décentralisation est conçue, décidée et mise en œuvre par le pouvoir central. En somme, même s'il est vrai que la nouveauté du processus de décentralisation nécessite un effort d'apprentissage de la part des collectivités territoriales, on ne peut s'empêcher de constater que le principe de progressivité peut toujours servir de prétexte pour justifier les précautions et les lenteurs des autorités gouvernementales dans la mise en œuvre des politiques de décentralisation[953]. En dépit des disparités qui peuvent exister entre les différents échelons décentralisés, le législateur veille à assurer l'égalité entre toutes les collectivités territoriales. Cette égalité se traduit sur le plan normatif par la consécration du principe de l'interdiction de tutelle entre collectivités.

---

951 Art. 164 du CGCT au Niger.

952 B. FAURE, *Droit des collectivités territoriales, op.cit.*, n° 676.

953 H. K. NOUBISSI, *Décentralisation et centralisation au Cameroun. La répartition des compétences entre l'État et les collectivités locales, op.cit.*, p. 126.

3. Le principe de l'interdiction de tutelle entre collectivités

398. - La tutelle dans les législations des États d'Afrique noire francophone est une attribution exclusive de l'État. Une collectivité territoriale ne saurait exercer sur une autre un quelconque pouvoir de tutelle. Les relations horizontales entre les collectivités publiques secondaires doivent se faire dans le respect de l'autonomie de chacune d'elles. Le législateur malien[954], burkinabè,[955] nigérien[956] et togolais[957] affirme que, « *aucune collectivité territoriale ne peut établir ou exercer de tutelle sur une autre collectivité territoriale* ». Ce principe autrefois implicite a été également consacré explicitement en droit français à l'article 72 al. 5 de la Constitution issue de la révision du 28 mars 2003. Ce qui signifie qu'il s'impose désormais au législateur qui ne peut instituer une tutelle entre collectivités. L'égalité de droit institué entre les différentes collectivités territoriales n'est que la traduction du principe d'égalité républicaine garantie par les Constitutions africaines. Même si ces dernières n'évoquent que les personnes physiques, les collectivités territoriales, personnes morales de droit public, sont également des sujets de droit[958].

399. - L'étude de ce principe nécessite quelques précisions sur sa portée. La tutelle dans son acception large telle qu'elle est présentée ici par le législateur s'analyse comme une capacité d'orientation d'une collectivité par une autre par la pression que cette dernière peut exercer sur la précédente. Une collectivité territoriale suffisamment puissante peut constituer un centre de régulation des activités d'autres collectivités. Dans cette situation, l'interdiction de la tutelle est une garantie de bonne relation entre les collectivités territoriales. Le principe permet de préserver à la fois l'égalité des collectivités et leur libre administration. Néanmoins, les collectivités territoriales sont appelées à exercer en commun un certain nombre de compétences. On ne saurait interdire au nom du principe de l'interdiction d'une tutelle entre les collectivités cet exercice en commun des compétences. Le législateur précise à cet effet que la possibilité pour une collectivité territoriale d'accorder une aide financière à une autre ne

---

954 Art. 24 de la loi n°52-2017 du 2 octobre 2017 déterminant les conditions de la libre administration des collectivités territoriales au Mali.

955 Art. 42 du CGCT du Burkina Faso.

956 Art. 8 du CGCT du Niger.

957 Art. 5 de la loi n°2019-006 *préc.*,

958 Voir par exemple, F. MELIN-SOUCRAMANIEN, « Le principe d'égalité entre collectivités locales », *Les cahiers du Conseil constitutionnel*, 2002, n° 12.

peut donner lieu à l'établissement ou à l'exercice d'une quelconque tutelle[959]. En la matière, la jurisprudence française retient une interprétation souple du principe qui ne saurait interdire toute contrainte de fait relevant notamment des mécanismes d'incitation ou de dissuasion. Le Conseil d'État a jugé que l'octroi d'une subvention par le département aux communes sous réserve de la satisfaction d'un certain nombre de conditions n'est pas illicite[960].

400. - Il faut toutefois envisager le fait que la loi ne suffit pas, à elle seule, à régler la question des rapports inégalitaires entre les collectivités territoriales. Pour Charles Nach Mback, le pire est même à craindre pour les collectivités territoriales africaines. Les déséquilibres démographique, financier et technique entre collectivités sont si graves dans les pays africains que la faiblesse des moyens correctifs de l'État fait craindre à terme une « décentralisation malthusienne » dans laquelle les collectivités les plus démunies s'effaceront de fait progressivement devant les autres[961]. Il faudra s'assurer que l'évolution en pratique évite les pièges d'une hiérarchisation entre collectivités. Le législateur doit également s'assurer que les collectivités territoriales disposent des moyens nécessaires pour l'exercice de leurs compétences.

4. Le principe de compensation

401. - Le principe de compensation est une contrepartie à l'attribution des compétences. La consécration d'un tel principe part du constat que l'État a tendance sous le couvert du transfert des compétences aux collectivités territoriales à se débarrasser des charges qu'il n'arrive plus à exécuter ou qui deviennent très coûteuses. Cette pratique malheureuse conduit à annihiler l'amélioration de la qualité des services publics qu'on aurait pu espérer des réformes de décentralisation. Il est certain que les collectivités territoriales ne seraient, dans ces conditions, pas plus que l'État, en mesure de réaliser ces nouvelles missions. Le transfert des compétences doit se faire dans le respect des exigences d'équilibre entre les nouvelles compétences locales et les moyens dont disposent les

---

959 Art. 9 al. 2 de la loi n° 2019-006 *préc.*

960 CE, Ass., 12 décembre 2003, *Département des Landes, Lebon*, p. 502, *RFDA*, 2004, p. 518, concl. F. Seners et 225, note J.-C. Douence. Voir aussi, Cons. Const. 8 juillet 2011, n° 2011-146 QPC, *Département des Landes, AJDA*, 2011, p. 2067, note M. Verpeaux.

961 C. NACH MBACK, *Démocratisation et décentralisation. Genèse et dynamiques comparés des processus de décentralisation en Afrique subsaharienne, op.cit.*, p. 412.

collectivités territoriales pour les assurer. La loi impose à cet effet que tout transfert de compétence soit accompagné d'un transfert concomitant des ressources afin de donner les moyens aux collectivités territoriales de les gérer. En pratique, si les transferts de compétences sont généralement réalisés, le transfert des ressources financières qui devait être concomitant devient finalement progressif.

402. - Le législateur togolais dispose à l'article 61, alinéa 2 que, « tout transfert de compétences à une collectivité territoriale doit être accompagné de transfert concomitant, par l'État, à celle-ci, des ressources et des charges correspondantes, ainsi que du transfert des services, des biens meubles et immeubles et du personnel nécessaires à l'exercice normal de ces compétences ». L'alinéa 2 de l'article 4 de loi n° 52-2017 du 2 octobre 2017 déterminant les conditions de la libre administration des collectivités territoriales au Mali impose également que « tout transfert de compétences à une collectivité territoriale doit être accompagné du transfert concomitant par l'État à celle-ci, des ressources et moyens nécessaires à l'exercice normal de ces compétences ». Le transfert des compétences implique non seulement une compensation financière pour l'exercice des responsabilités, mais aussi une compensation matérielle c'est-à-dire la mise à disposition des biens, des services et des agents utilisés précédemment par l'État.

403. - La compensation financière est un droit des collectivités territoriales. Pour assurer son respect, le législateur a prévu un certain nombre de garanties qui se révèlent en pratique difficile à atteindre en raison, notamment des moyens limités dont dispose l'État en Afrique. La loi impose que l'État transfert concomitamment les ressources. Cela suppose qu'il soit procédé, avant toute opération, à une évaluation préalable des dépenses réalisées par l'État pour l'exercice des compétences transférées. La garantie d'une concomitance dans le transfert n'est pas véritablement assurée dans les États d'Afrique noire francophone. Le législateur burkinabè n'y fait d'ailleurs pas mention dans le code général des collectivités territoriales. L'article 36 du CGCT prévoit que « *les transferts de compétences par l'État doivent être accompagnés du transfert aux collectivités territoriales des moyens et des ressources nécessaires à l'exercice normal de ces compétences, dans les conditions prévues par la loi* », sans aucune précision, ni sur la concomitance, ni sur l'évaluation préalable des dépenses. Au Togo, le législateur se contente de prévoir que « *les charges correspondant à la mise en œuvre des compétences partagées attribuées aux collectivités territoriales font l'objet d'une* *évaluation préalable au transfert desdites compétences*». L'évaluation préalable n'est donc expressément prévue que pour les compétences partagées sans toutefois que la méthode de calcul de ces charges soit précisée. De fait, l'évaluation des dépenses est confiée aux ministères sectoriels qui doivent déterminer le montant des ressources financières correspondant aux compétences transférées dans leurs domaines afin d'alimenter l'adoption du projet de loi de finances. En réalité, les énormes difficultés auxquelles font face les collectivités territoriales trouvent un début d'explication dans l'inaccomplissement de cette formalité. Les ministères gardent toujours la main basse sur l'essentiel des compétences sectorielles dans la mesure où les transferts financiers sont réalisés de façon aléatoire en méconnaissance de l'exigence de concomitance.

404. - En dépit de cela, il faut noter que les législations nationales précisent la nature de la compensation financière. Elle peut se réaliser « *soit par transfert de fiscalité, soit par dotations ou par les deux à la fois* »[962]. Des deux modes de compensation, seul le transfert de fiscalité offre aux collectivités territoriales une véritable garantie d'indépendance en raison de l'augmentation des ressources propres qu'elle permet. Par ailleurs, « *toute charge nouvelle incombant aux collectivités territoriales, du fait de la modification par l'État des règles relatives à l'exercice des compétences transférées, doit être compensée par un transfert approprié de moyens* »[963]. Toutefois, les charges étant en permanente évolution, les ressources transférées ne correspondent quasiment jamais aux ressources effectivement engagées par la collectivité qui souhaite fournir un service de qualité. La question de la compensation financière reste donc entière dans les États d'Afrique noire francophone.

405. - En outre, la compensation n'est pas que financière. Elle concerne aussi les biens et services. Les compétences exercées par les services de l'État relevant désormais de l'administration décentralisée, il convient de placer ces services sous la responsabilité des autorités décentralisées. En la matière, rien n'est gagné d'avance. D'abord, lorsqu'on se réfère à la loi togolaise, le législateur semble poser la compensation matérielle comme une faculté et non une obligation. L'article 306 de la loi n° 2019-006 du 26 juin 2019 dispose à cet effet que « *dans le cadre de son accompagnement, l'État peut mettre à la disposition des collectivités territoriales les*

---

962 Art. 37 du CGCT au Burkina Faso.

963 *Ibid.*, al. 2 ; art. 310 al. 2 Loi n° 2019-006 *préc.* ; art. 159, al. 2 du CGCT au Niger.

ressources humaines rétribuées sur le budget de l'État»[964]. Le caractère immédiat du transfert des services est ici fortement remis en cause. En droit français, le juge constitutionnel sanctionne tout report du transfert des services en l'absence de « *caractéristiques et de circonstances particulières* »[965]. Ensuite, dans les pays où la loi impose clairement d'accompagner les transferts de compétences de tout ou partie des services correspondants, le législateur aménage deux modes de transfert des services. Il en est ainsi au Burkina Faso et au Niger où la loi prévoit que la mise à disposition peut être définitive ou temporaire. Dans le premier cas, les services sont durablement placés sous la responsabilité des autorités décentralisées. Dans le second cas, l'utilisation des services de l'État repose sur une convention signée entre le représentant de l'État et le président du conseil de la collectivité concernée. Cette convention est établie suivant des modèles types de convention fixés par le pouvoir règlementaire[966].

406. - En ce qui concerne les transferts de biens, le pouvoir central est aussi tenu de transférer les biens meubles et immeubles affectés aux services transférés. La mise à disposition des biens à la collectivité est de « plein droit » à la date du transfert. C'est le sens des dispositions de l'article 315 de la loi n° 2019-006 et des articles 41 et 161 des codes burkinabè et nigérien qui prévoit que « *le transfert de compétences entraine de plein droit la mise à la disposition de la collectivité territoriale bénéficiaire des biens meubles et immeubles utilisés à la date de ce transfert pour l'exercice de ces compétences* ». Le transfert est constaté par un décret de dévolution pris en conseil des ministres[967]. Il est précédé d'un inventaire exhaustif des biens meubles et immeubles[968] sanctionné par procès-verbal établi contradictoirement entre les représentants de l'État et ceux des organes exécutifs des collectivités territoriales[969].

407. - La loi fait par ailleurs référence à un « exercice normal » des compétences comme moyen d'évaluation de la qualité du service que les ressources et moyens transférés doivent permettre d'atteindre. Les

---

964 Sur la question du personnel, voir les développements *infra* § 506 et s.

965 Cons. Const. 12 aout 2004, n° 04-503 DC, *Loi relative aux libertés et responsabilités locales*, *AJDA*, 2004, p. 1960, note M. Verpeaux.

966 Voir respectivement pour le Niger et le Burkina Faso, les articles 40 et 160 du CGCT.

967 Voir les articles : 41 al. 2 CGCT du Burkina Faso ; 161 al. 2 du CGCT du Niger ; art. 314 al. 2 Loi n° 2019-006 *préc.*

968 Loi n° 2019-006 *préc.*, art. 314 al. 2.

969 art. 161 al. 2 du CGCT du Niger.

moyens transférés seraient donc jugés insuffisants s'ils ne permettent pas un exercice normal des compétences attribuées. Toutefois, cette notion d'« exercice normal » est difficile à apprécier en l'absence de l'intervention d'un juge. Quelle serait la jauge pour considérer qu'une collectivité territoriale exerce les compétences que la loi lui attribue dans des conditions normales ?

On peut considérer que l'exercice normal renvoie à l'équilibre parfait entre les charges et les ressources transférées. Cependant, la compensation parait parfois exiger de l'État plus qu'un simple équilibre. En effet, l'article 310 de la loi n° 2019-006 précitée dispose que, « *les ressources attribuées aux collectivités territoriales sont au moins équivalentes aux charges supportées par l'État à la date du transfert des compétences* ». Le fait pour le législateur de préciser que les ressources doivent être « au moins » équivalentes aux charges financières résultant du transfert sans imposer un seuil maximal laisse subsister la possibilité que l'exercice dit « normal » des compétences transférées par les collectivités ne coïncide pas avec ce qu'en faisait l'État avant le transfert. Le législateur semble ainsi valider l'idée que l'État puisse allouer des ressources permettant un exercice « optimal » des compétences transférées aux collectivités[970]. Il semble *a priori* impossible de déterminer avec précision le montant des ressources à transférer pour un exercice normal des compétences locales.

En tout état de cause, l'équilibre entre les charges et les ressources transférées voulu par le législateur parait difficile à atteindre. La mise en œuvre effective du principe de compensation suppose en amont que l'État puisse disposer des ressources budgétaires lui permettant d'assumer efficacement les compétences transférées. Or, l'État en Afrique est lui-même incapable de répondre aux besoins des populations faute de moyens suffisants. Dans ce contexte, l'équilibre entre les charges transférées et les ressources nécessaires pour les assumer n'est que pure illusion. À la limite, il est possible de constater que l'État se décharge sur les collectivités territoriales pour l'avenir de telle ou telle compétences[971]. Outre l'absence de véritables règles au contenu précisé par l'intervention du juge, l'impératif de compensation est marqué dans le droit des collectivités territoriales en Afrique par une souplesse favorisant un transfert de charges de plus en plus coûteux pour les collectivités. La verticalité du régime de distribution des compétences laisse l'État maitre du calendrier

970 H. K. NOUBISSI, *Décentralisation et centralisation au Cameroun. La répartition des compétences entre l'État et les collectivités locales, op.cit.*, p. 127.

971 *Ibid.*, p. 128.

des transferts de compétence ainsi que de la détermination de leur contenu.

B- La détermination matérielle des compétences des collectivités territoriales

408. - La détermination du champ d'intervention des collectivités territoriales est l'une des questions les plus aléatoires du droit de la décentralisation en Afrique. Loin d'être une spécificité africaine, ce caractère est le propre des États unitaires[972]. Comme indiqué précédemment, la Constitution ne détermine expressément aucune matière relevant de la compétence des collectivités. Elle est muette et ne crée par elle-même aucune dynamique de décentralisation. Les matières que l'État entend confier à l'administration décentralisée reposent principalement sur des choix d'opportunité opérés par le parlement. Les modalités de cette répartition des compétences sont pour le moins ambiguës. En effet, le législateur en Afrique semble partagé entre deux procédés de répartition. Il s'agit d'une question délicate qui se pose également avec une certaine acuité en droit français. Bertrand Faure écrit à ce propos que le droit français n'a jamais choisi entre les deux techniques de répartition de compétences : la compétence d'attribution et la compétence générale[973]. L'amalgame entre les deux techniques s'est construit, explique-t-il, sur la double fidélité au pragmatisme et à la tradition française. Le résultat de ce mélange révèle au regard de l'expérience juridique et pratique française que, « *ces deux techniques s'appellent et se conjuguent* [en] *offrant aux collectivités un pouvoir général d'administration au-delà de leurs missions spéciales confiées par l'État* »[974].

409. - Ce résultat ne peut toutefois pas être partagé dans le cadre des collectivités africaines. Il est exclu, en effet, que les collectivités territoriales puissent disposer d'un pouvoir d'administration au-delà des missions que la loi leur confie expressément. L'inspiration du droit français a conduit dans certains cas à l'affirmation de l'existence d'une clause de compétence générale au profit des collectivités africaines. Cependant, l'interprétation des textes relatifs à la distribution des compétences permet de lever

---

972 Voir en ce sens à propos de la France, B. FAURE, *Droit des collectivités territoriales, op.cit.*, n° 704.

973 *Ibid.*, n° 674.

974 *Idem.*

l'équivoque autour de la compétence générale (**1°**) et d'établir un consensus sur le procédé d'énumération des compétences (**2°**).

1. L'exclusion de la clause générale de compétence comme procédé de répartition

410. - Les différentes études sur le droit des collectivités territoriales en Afrique font apparaitre la clause générale de compétence comme une modalité classique de transferts de compétences aux collectivités[975]. Cependant, la clause de compétence générale reste fortement orientée par la conception que s'en fait le pouvoir central puisque les affaires locales n'existent pas par elles-mêmes. La médiation de la loi est donc nécessaire pour leur donner une forme. Aussi, le manque de ressource ainsi que le contexte politique dans lequel intervient la décentralisation en Afrique semblent défavorables à la reconnaissance d'une compétence générale qui accroitrait l'autonomie des collectivités territoriales. Ainsi, la clause générale de compétence n'a jamais été et ne peut être, du moins en l'état actuel, le système retenu pour déterminer les compétences des collectivités territoriales ouest-africaines.

411. - Le spectre de la clause générale de compétence issue de l'article 61 de loi communale française du 5 avril 1884 qui prévoit que, « *le conseil municipal règle par ses délibérations les affaires de la commune* »[976] continue de planer sur les législations africaines dans la définition des compétences locales. Cette disposition figure aujourd'hui à droit constant à l'alinéa premier de l'article L. 2121-29 du CGCT français. Elle est reprise dans certaines législations africaines et s'applique à l'ensemble des catégories de collectivités. C'est le cas au Burkina Faso, au Mali et au Niger. L'alinéa 3 de l'article 224 du CGCT burkinabè dispose que le conseil municipal « *règle par ses délibérations, les affaires de la commune et donne son avis sur toutes les questions qui lui sont soumises par l'État ou par d'autres collectivités territoriales* »[977]. Les codes malien et nigérien disposent respectivement à l'article 22 que, « *le*

975C. NACH MBACK, *Démocratisation et décentralisation. Genèse et dynamiques comparés des processus de décentralisation en Afrique subsaharienne*, *op.cit.*, pp. 412-420 ; H. K. NOUBISSI, *Décentralisation et centralisation au Cameroun. La répartition des compétences entre l'État et les collectivités locales*, *op.cit.*, p. 128.

976 Cette disposition est elle-même issue de la loi départementale du 10 aout 1871, art. 48 al. 5.

977 L'article 145 dispose dans les mêmes termes pour le conseil régional

conseil communal règle par ses délibérations les affaires de la commune, notamment celles relatives au développement économique, social et culturel » et à l'article 29 que, « *le conseil municipal règle par ses délibérations les affaires de la commune. À cet effet, il décide des mesures à prendre pour assurer le développement économique, social et culturel de la commune* ». Ces dispositions s'appliquent à tous les niveaux décentralisés. Il faut signaler, toutefois, une différence rédactionnelle dans la loi au Niger pour le cas des régions. L'article 104 prévoit que « *le conseil régional règle par ses délibérations les affaires de la région dans le respect des compétences des autres collectivités territoriales. À cet effet, il prend des mesures nécessaires pour promouvoir le développement économique, social, éducatif, sanitaire, scientifique, culturel et sportif de la région* ».

412. - La reprise de cette formule législative semble traduire la volonté de consacrer une clause générale de compétence qui reconnait aux collectivités territoriales notamment à leurs assemblées délibérantes, le pouvoir de gérer elles-mêmes en toute liberté leurs affaires propres. Pour le dire autrement, les collectivités territoriales ne seraient plus en application de la compétence générale subordonnées à l'État pour la détermination des compétences qu'elles souhaitent exercer. Elles sont libres de leurs initiatives, toute restriction n'étant qu'exceptionnelle. Lorsqu'on se réfère à la doctrine française sur le sujet, certains auteurs ont vu, dans la décision du Conseil constitutionnel du 9 décembre 2010 sur la loi de réforme des collectivités territoriales[978], un lien entre compétence générale et libre administration. La première étant une garantie essentielle de la seconde[979]. La libre administration serait donc la possibilité offerte aux collectivités de définir librement leur champ d'intervention. Le Sénat français voit d'ailleurs dans la clause générale de compétence « *ce qui*

---

[978] Cons. Const. DC n°2010-618, 9 décembre 2010, Loi de réforme des collectivités territoriales, consid. 54 & 55 : « Considérant que l'article 48 de la loi du 10 août 1871 susvisée précisait que le conseil général délibère "sur tous les objets d'intérêt départemental dont il est saisi, soit par une proposition du préfet, soit sur l'initiative d'un de ses membres" ; que ces dispositions n'ont eu ni pour objet ni pour effet de créer une "clause générale" rendant le département compétent pour traiter de toute affaire ayant un lien avec son territoire » ; « les dispositions critiquées permettent au conseil général ou au conseil régional, par délibération spécialement motivée, de se saisir respectivement de tout objet d'intérêt départemental ou régional pour lequel la loi n'a donné compétence à aucune autre personne publique ; (…) par suite, doit être écarté le grief tiré de ce que les dispositions critiquées seraient contraires au principe de libre administration des collectivités territoriales »

[979] B. FAURE, *Droit des collectivités territoriales, op.cit.*, n° 739.

représente une extension significative des attributions des communes » et qui « *exprime le principe, aujourd'hui constitutionnel, de la libre administration des collectivités locales* »[980].

La clause générale de compétence serait alors « *la traduction juridique de l'aptitude générale d'une collectivité à intervenir* »[981] dans un certain nombre de domaines qui ne se limiterait pas à ceux établis par la loi. Ainsi, les affaires propres de la collectivité regroupent l'ensemble des affaires qui présente un intérêt public local et qui répond aux besoins de la population. Les collectivités territoriales peuvent justifier une part de leurs interventions et de leurs décisions en s'appuyant sur l'existence d'un intérêt public local. Sur cette base, aucun domaine précis ne peut être réservé aux collectivités puisqu'elles ont en charge toute question qui pourrait revêtir un intérêt local.

413. - Or, dans le cadre d'un État unitaire, le domaine des affaires locales ne signifie pas que les collectivités territoriales possèdent une liberté d'initiative complète. En dehors des contrôles qui pèsent sur l'action des collectivités, l'administration décentralisée ne peut définir elle-même la liste des affaires locales ni définir les besoins spécifiquement locaux. C'est à la loi qu'il revient dans un État unitaire, comme c'est le cas dans les pays étudiés, de procéder à cette définition. Le législateur en Afrique ne se contente pas d'affirmer que l'assemblée délibérante « règle par ses délibérations les affaires » de la collectivité. Il délimite en plus le champ d'intervention possible des collectivités territoriales. L'assemblée locale règle par ses délibérations les affaires de la collectivité dans les domaines de compétences que la loi a entendu lui attribuer. Ainsi, le conseil ne peut régler par ses délibérations que les affaires relevant du domaine économique, social, culturel… dont la consistance est précisée par la suite. La position du législateur togolais offre sur cet aspect une perspective d'analyse intéressante.

414. - L'intérêt local est défini dans la loi n° 2019-006 précitée comme étant constitué de l'ensemble des « besoins et des projets spécifiques des habitants d'une collectivité territoriale liés par un destin commun et une

---

980 https://www.senat.fr/evenement/archives/D18/principes.html, consulté le 17 juillet 2022.

981 J.-M. PONTIER « Semper Manet. Sur une clause générale de compétence », *RDP*, 1984, p. 1443.

solidarité d'intérêts »[982]. L'alinéa premier de l'article 3 reconnait aux collectivités territoriales des « compétences spécifiques » qui constituent le domaine des affaires d'intérêt local sans toutefois préciser immédiatement le contenu de ces compétences dites spécifiques. Plus loin, dans le chapitre réservé à la libre administration, le législateur rejette toute idée d'existence d'une clause de compétence générale au profit des collectivités territoriales. Au sens de l'article 8 de la loi n° 2019-006 du 26 juin 2019, « les collectivités territoriales règlent, par les délibérations de leurs conseils, les affaires relevant de leurs domaines de compétence. La détermination des compétences des régions et des communes relève de la loi ». Pour chacune des catégories constitutionnelles de collectivités, le droit positif reste constant sur le rejet de la compétence générale. Pour les communes, on peut lire à l'article 88 que « le conseil municipal règle, par ses délibérations, les affaires de la commune, telles que définies par les articles 82, 83 et 84 de la présente loi ». Les articles 82, 83 et 84 renvoient respectivement aux compétences propres, partagées et transférées que la loi attribue aux communes. Pour les régions, l'article 240 dispose que « le conseil régional règle, par ses délibérations, les affaires de la région. Il programme et met en œuvre les opérations et les actions de développement de la région, conformément aux orientations nationales en la matière ». La lecture du premier alinéa de cette disposition aurait pu faire penser que le législateur consacre au profit des régions une clause de compétence générale. Cependant, l'alinéa 2 limite la compétence régionale à la seule promotion du développement dans le cadre des orientations définies par le pouvoir central.

415. - La technique de la clause de compétence générale n'existe pas dans le droit de la décentralisation en Afrique pour les communes ou les régions. Les collectivités territoriales ne disposent pas d'une compétence générale leur permettant de décider d'intervenir dans telle ou telle compétence qui ne leur aurait pas été attribuée par la loi. Même si, les dispositions législatives des pays de la présente étude ne sont pas aussi claires que les dispositions togolaises sur la question de la compétence générale, l'évolution jurisprudentielle[983] et doctrinale française permet de

---

982 Loi n° 2019-006 *préc.*, art. 3, al. 2.

983 Cons. Const., déc., n° 2016-565 QPC, 16 sept. 2016, *Assemblée des départements de France*, cons. 5 ; *AJDA,* 2016, n° 43, p. 2438-2444, note Faure ; *Dr. adm.* 2016, n° 12, p. 1-2, note Plessix ; *JCP* 2016, n° 43-44, p. 1971-1975, note Verpeaux ; *LPA*, 21 mars 2017, n° 57, p. 13-14, note Bourdon.

consolider la thèse de l'inexistence de la clause générale de compétence[984]. On ne saurait voir dans la formule législative, le conseil « *règle par ses délibérations les affaires* » locales, la consécration d'une clause de compétence générale. La vocation générale des collectivités territoriales n'a de sens que dans le cadre de l'exercice des compétences que la loi leur a explicitement attribuées. À travers l'alinéa 7 de l'article 70, le législateur togolais a entendu exclure du transfert de compétences de l'État aux collectivités territoriales toutes *« les matières qui ne sont pas dévolues expressément à la région et à la commune* [...] ». Les affaires locales ne sauraient donc avoir un quelconque fondement naturel, car c'est la norme d'État c'est-à-dire la loi qui détermine les affaires locales[985] et attribue telle ou telle compétence à telle ou telle catégorie de collectivités territoriales.

416. - On pourrait alors se demander si l'exclusion de la clause générale de compétence dans la répartition des compétences ne porte pas atteinte à la libre administration des collectivités territoriales. Le juge constitutionnel français donne une réponse qui s'inscrit dans la tradition française et de sa conception de l'organisation décentralisée de la République. Le rejet de la compétence générale ne remet pas en cause le principe de la libre administration. Dans sa décision « *Assemblée des départements de France* » précitée, le Conseil constitutionnel a jugé que le principe de libre administration *« n'implique pas, par lui-même, que les collectivités territoriales doivent pouvoir intervenir dans les domaines pour lesquels aucune autre personne publique ne dispose d'une compétence attribuée par la loi »* ; surtout *« compte tenu de l'étendue des attributions dévolues aux départements par les dispositions législatives en vigueur, qu'il s'agisse de compétences exclusives, de compétences partagées avec d'autres catégories de collectivités territoriales ou de compétences susceptibles d'être déléguées par d'autres collectivités territoriales, les dispositions contestées ne privent pas les départements d'attributions effectives »*[986]. Ainsi, dès lors que les collectivités disposent de « compétences effectives », dont la détermination par voie d'énumération relève du législateur, il n'y a

984 O. GOHIN, « Semper abset. La prétendue clause de compétence générale des collectivités territoriales », in P. TURK (dir.) *Pour un droit constitutionnel des collectivités territoriales. Évolutions et débats*, *op.cit.*, p. 214.
985 Sur cette question, voir notamment, J. CHAPUISAT, *La notion d'affaires locales en droit administratif français*, thèse dactyl., Université de Paris II, 1972 ; « Libertés locales et libertés publiques », *AJDA*, 1982, p. 354.
986 Cons. Const., déc. n° 2016-565 QPC, 16 septembre 2016, *Assemblée des départements de France*, *préc.*

pas d'atteinte à l'autonomie locale. Bien que controversée, cette solution pourrait à l'occasion inspirer les juges constitutionnelles africains qui n'hésitent pas d'ailleurs à prendre le parti de l'État.

417. - Il faut aussi rappeler que le rejet de la clause générale de compétence dans la définition des compétences décentralisées prend un sens particulier dans le contexte africain marqué par la pénurie de moyens des collectivités territoriales[987]. Cette situation de pénurie fonde la limitation des compétences locales puisque les ressources pour les exercer font défaut. Cet argument rapproché du principe de progressivité de la décentralisation permet d'en tirer toutes les conséquences, car l'élargissement des compétences propres/exclusives des collectivités territoriales dépendra de leur capacité à les exercer. La loi demeure le principal vecteur pour déterminer le contenu des compétences locales. L'étude des différentes compétences énumérées par le législateur permet donc de délimiter les matières dans lesquelles les autorités locales peuvent intervenir.

2. La détermination législative des compétences par voie d'énumération

418. - Les collectivités territoriales ne sont pas en mesure de développer des initiatives dans les domaines que ne leur sont pas attribués par la loi. La détermination des compétences décentralisées est une opération de l'État. Dans un État unitaire, les collectivités territoriales ne sont pas « maitresses de leur destin ». L'État unitaire décentralisée est « monoconstitutionnel » et « monolégislatif ». Il dispose seul de la compétence pour déterminer ce qui doit relever ou non de la responsabilité des collectivités territoriales. Les Constitutions réservent au parlement la charge de la répartition des compétences entre l'État et les collectivités territoriales. C'est le sens des dispositions des articles 84, 98, 101, 164 des constitutions togolaise, malienne, burkinabè et nigérienne qui proclament que la loi détermine « *les principes fondamentaux de la libre administration des collectivités territoriales, de leurs compétences et de leurs ressources* ». Pour définir une compétence décentralisée, il faut donc la médiation du législateur comme c'est le cas en droit français[988]. L'opération par laquelle

---

[987] C. NACH MBACK, *Démocratisation et décentralisation. Genèse et dynamiques comparés des processus de décentralisation en Afrique subsaharienne, op.cit.*, p. 414.
[988] B. FAURE, *Droit des collectivités territoriales, op.cit.*, n°673.

le législateur confère une compétence précédemment exercée par l'État à une collectivité territoriale caractérise l'expression d'attribution de compétences[989]. L'attribution des compétences est donc un processus vertical de transferts de compétences de l'État aux collectivités territoriales.

419. - Il n'y a pas une logique propre qui préside aux transferts de compétence dans États d'Afrique. C'est en se référant aux différents textes de loi que l'on peut dresser un état des lieux sommaire des domaines de compétence d'attribution. Pour le législateur togolais, la mission des collectivités territoriales consiste en : « *la conception, la programmation et l'exécution des actions de développement d'intérêt local de leur ressort territorial, en particulier dans les domaines économique, social et culturel* »[990]. Cette formulation somme toute assez générale est suivie d'une disposition qui spécifie les matières explicitement transférées aux collectivités territoriales sans toutefois préciser les compétences exercées par chaque catégorie de collectivité territoriale. Au sens de l'article 62, « *dans le cadre défini par la présente loi, l'État transfère aux collectivités territoriales, dans leur ressort territorial respectif, des compétences dans les matières suivantes : développement local et aménagement du territoire ; urbanisme et habitat ; infrastructures, équipements, transports et voies de communication - énergie et hydraulique ; assainissement, gestion des ressources naturelles et protection de l'environnement ; commerce et artisanat ; éducation et formation professionnelle ; santé, population, action sociale et protection civile ; sports, loisirs, tourisme et action culturelle* ».

Le droit positif togolais ne distingue pas explicitement les domaines de compétences dévolues à chacune des catégories constitutionnelles de collectivités. À la différence des autres législations africaines, il n'existe pas dans le droit togolais de disposition identique à celle qu'on retrouve à l'article 35 du CGCT du Burkina Faso qui prévoit que, « *la répartition des compétences entre les collectivités territoriales s'effectue, en distinguant celles qui sont dévolues aux régions et aux communes* » ou l'article 158 du CGCT du Niger qui dispose que, « *la répartition des compétences entre l'État et les collectivités territoriales s'opère par blocs de compétences…* ». Ces articles posent le principe d'une répartition par blocs de compétences également consacré dans le droit positif français à l'article L.1111-4 du code général des collectivités

---

989 *Ibid.*, n° 675.

990 Art. 7, Loi n° 2019-006 *préc.*

territoriales selon lequel les compétences doivent être transférées, dans la mesure du possible, en totalité à une catégorie précise de collectivités[991].

Ces dispositions portent ainsi le principe de spécialisation à son degré maximal. Le législateur souhaite dans cette optique instaurer une cohérence absolue dans la répartition des compétences entre les différents niveaux de collectivités. Ainsi, chaque niveau de collectivité se verrait attribuer à titre exclusif un domaine de compétence. Autrement dit, pour chaque matière, la loi habilite une seule et unique autorité à intervenir. Le pragmatisme de cette technique bute sur l'inextricable imbrication des compétences[992]. Cette logique de répartition s'est révélée impraticable. Les compétences (sociale, éducation, environnement, développement économique, santé, sport…) sont enchevêtrées, c'est-à-dire qu'il est impossible pour une matière transférée par la loi d'avoir un seul et unique niveau de collectivité qui dispose d'un titre à agir. Même dans les pays qui ont fait le choix d'une méthode de répartition par blocs de compétence, notamment le Burkina Faso et le Niger, plusieurs domaines de compétences demeurent partagés. Que l'on songe notamment à l'enseignement préscolaire et primaire ; à la protection civile, d'assistance et le secours ; l'aménagement du territoire ou encore le développement économique, on s'aperçoit qu'il est impossible d'opérer une répartition par blocs de compétences dans ces domaines. Aucune collectivité ne peut se réclamer d'une compétence exclusive dans la mesure ou pour l'exercice de cette compétence, la participation d'un autre niveau de collectivité est requise également. L'examen des compétences attribuées manifeste la porosité des frontières dans la détermination des compétences. Il n'est pas aisé de distinguer les compétences attribuées exclusivement aux communes de celles confiées aux régions.

420. - Or, la notion de compétence renvoie à la double idée de l'existence d'un domaine d'action propre et celle qui est conséquente d'une

---

991 Art. L. 1111-4 CGCT : « La répartition des compétences entre les collectivités territoriales et l'État s'effectue, dans la mesure du possible, en distinguant celles qui sont mises à la charge de l'État et celles qui sont dévolues aux communes, aux départements ou aux régions de telle sorte que chaque domaine de compétences ainsi que les ressources correspondantes soient affectés en totalité soit à l'État, soit aux communes, soit aux départements, soit aux régions ».

992 En France, face à l'illusion d'une distribution par blocs des compétences, le constituant de 2003 a promu le procédé de la « collectivité chef de file », voir art. 72 al. 5 de la Constitution française ; B. FAURE, *Droit des collectivités territoriales, op.cit.*, p. 578 et s.

limitation de ce domaine[993]. Dès lors, la différenciation des sphères d'action des collectivités communale et régionale n'est pas évidente. En effet, il n'existe aucun critère objectif de séparation. Toutefois, l'analyse des compétences propres attribuées par le législateur à l'échelon communal permet d'affirmer qu'il a entendu en faire un niveau de solidarité communautaire. La collectivité communale en tant que « collectivité territoriale de base » est le premier niveau d'encadrement et de gestion des populations locales naturellement voué à recevoir les compétences dites de « proximité ». À ce titre, la commune prend en charge les services publics traditionnels de collecte, de transport et de traitement des déchets, veille au bon état de la signalisation routière et assure le maintien de l'ordre public local, la construction et la gestion des infrastructures et équipements de proximités tels que les gares routières et les aires de stationnement, les marchés et abattoirs locaux. Elle dispose également de compétences renforcées en matière d'urbanisme à travers notamment l'établissement des plans de lotissement, la délivrance des permis de construire, etc. À y voir de près, quantitativement, dans les pays étudiés, c'est dans les domaines de l'assainissement, de la protection de l'environnement, de l'habitat et de l'urbanisme que le législateur a entendu accentuer le particularisme communal par rapport à la région.

421. - La région est davantage portée sur la planification et la coordination des actions de développement. Elle reçoit au sens de l'article 35 al. 2 du CGCT du Burkina Faso des compétences transversales. La collectivité régionale regroupe sur son territoire les communes. Elle apparait ainsi comme le niveau naturel de mise en cohérence de l'ensemble des politiques locales. La planification du développement économique ainsi que l'aménagement du territoire restent le cœur de compétences de la région. Dans ces matières, les autorités communales ne possèdent que des attributions consultatives. Tout porte à croire que le législateur a voulu faire de la région une puissance d'unification de l'action locale en lui confiant des missions d'encadrement, d'organisation et de coordination du développement local. La véritable difficulté comme le souligne Hilaire Noubissi réside dans le fait que l'attribution de telles compétences de coordination et de planification aux régions comporte des risques quant à l'avenir des régions et des communes situées sur leur ressort territorial. La « délicatesse » des compétences que la loi leur attribue, additionnée à la nouveauté de l'institution régionale, laisse entrevoir un risque de

[993] J.-M. PONTIER, *L'État et les collectivités locales. La répartition des compétences*, *op.cit.*, p. 35.

recentralisation des compétences par l'État en attendant que les régions se prévalent d'une certaine expérience dans la mise en œuvre d'autres compétences moins lourdes de conséquences[994].

422. - Il est difficile d'appréhender clairement la question des compétences locales. L'analyse des différents droits positifs montre que le législateur privilégie un cadre d'administration « multiniveaux » dans lequel chacun des niveaux est amené à coopérer avec les autres. En ce sens, il est possible de douter de la pertinence même de la notion de compétence en tant qu'elle renvoie aux idées d'autonomie et d'exclusivité. Elle n'est plus apte à décrire ce qui se joue dans les relations entre l'État et les différentes catégories de collectivités territoriales. Par ailleurs, en faisant reposer les systèmes de détermination des compétences sur la loi, le constituant crée une situation de précarité des compétences décentralisées ainsi qu'il a été dénoncé par la doctrine au sujet des collectivités territoriales en droit français[995]. L'absence d'intangibilité des compétences attribuées aux collectivités territoriales est une limite du processus de décentralisation. L'instabilité de la loi implique que les compétences dont disposent les collectivités territoriales peuvent être retirées par le législateur lui-même. Au Togo, un projet de loi portant modification de la loi n° 2019-006 relative à la décentralisation et aux libertés locales a été adopté par le gouvernement le 30 mai 2022 et prévoit de retirer notamment aux communes, au titre de leurs compétences propres, les compétences en matière de gestion du domaine foncier cédé par l'État à la commune et celle relative à l'adressage des voies communales. Après seulement trois années d'exercice, le pouvoir central tente de modifier les compétences décentralisées. Cette situation de mise à jour des compétences crée une instabilité qui n'est pas favorable aux collectivités territoriales et à une bonne administration.

Le transfert des compétences est un révélateur de la capacité de l'État en Afrique à se recomposer et à maintenir sa présence dans l'action des collectivités territoriales. L'exercice des compétences d'attribution est soumis à une forte tutelle de l'État qui caractérise la permanence de la centralisation administrative.

---

[994] H. K. NOUBISSI, *Décentralisation et centralisation au Cameroun. La répartition des compétences entre l'État et les collectivités locales, op.cit.*, p. 134.

[995] B. FAURE, *Droit des collectivités territoriales, op.cit.*, p. 593 et s.

Section 2 : La tutelle permanente de l'État sur les collectivités territoriales

423. - « Donner c'est donner, reprendre c'est voler ! ». L'adage est bien connu de tous. Pourtant, en ce qui concerne les relations entre l'État et les collectivités territoriales, il parait difficile de donner sans retenir. L'État dispose d'un droit de regard sur l'action des collectivités territoriales. Il exerce une tutelle sur les collectivités territoriales qui trouve son origine dans la volonté du pouvoir central de surveiller l'action des collectivités territoriales. À propos de cette tutelle sur les collectivités territoriales, on pourrait s'interroger avec Jean Waline sur le sens d'une telle pratique. Comment concevoir que l'État qui a procédé à la distribution des compétences au profit des entités décentralisées, indépendantes de lui, puisse reprendre d'une main ce qu'il a donné de l'autre, remettant ainsi en cause le principe même de la décentralisation[996] ?

424. - D'un point de vue juridique, la réponse est simple. Il ne saurait avoir de décentralisation sans un contrôle de l'État sur les collectivités territoriales. La tutelle a été d'ailleurs présentée comme *« la traduction juridique de la décentralisation »*[997]. Décentralisation et tutelle sont deux notions indissolublement liées malgré leurs caractères antagonistes[998]. Ainsi, les articles 98, 164, 141 des constitutions malienne, nigérienne et togolaise disposent en des termes quasi identiques que les collectivités territoriales s'administrent librement dans les conditions fixées par la loi. En outre, l'article 165, alinéa 2 de la Constitution du Niger dispose que *« le représentant de l'État veille au respect des intérêts nationaux »*. Le législateur malien dispose quant à lui à l'article 20 de la loi n° 2017-052 du 2 octobre 2017[999] que « *les collectivités territoriales exercent leurs activités sous le contrôle de l'État et dans les conditions définies par la loi* ». Le droit de la décentralisation dans les États d'Afrique noire francophone consacre ainsi le principe de libre administration et l'existence d'un pouvoir de tutelle de l'État sur les collectivités territoriales en vue de la sauvegarde de l'intérêt général et de

---

[996] J. WALINE, « A propos de la tutelle sur les collectivités territoriales », in *Mélanges en l'honneur de Jean-François Lachaume,* Dalloz, p. 1096.
[997] R. MAPESTIOL, P. LAROQUE, *La tutelle administrative,* Sirey, 1930, p. 24.
[998] G. MELLERAY, *La tutelle de l'État sur les communes,* thèse, Sirey, 1981, p. 19.
[999] Loi n°2017-052 du 02 octobre 2017 déterminant les conditions de la libre administration des collectivités territoriales au Mali.

la légalité (§ **1**). Sous la notion de tutelle se trouvent rangés des pouvoirs de portées différentes à savoir des pouvoirs de contrôle : l'autorisation, l'approbation et la substitution et des pouvoirs de direction : annulation, nomination, révocation. Cette conception large de la tutelle dans les législations africaines contrarie le jeu de la décentralisation. La diversité des moyens mis à disposition de l'autorité de tutelle constitue un indicateur « précieux »[1000] de mesure du degré d'autonomie dont bénéficient les collectivités territoriales (§ **2**).

§1.- La justification de l'existence d'un pouvoir de tutelle sur les collectivités territoriales

425. - Le contrôle de l'État sur les collectivités territoriales africaines est désigné par l'expression de « tutelle ». En France, la loi du 2 mars 1982 a entendu supprimer la tutelle administrative et la tutelle financière sur les collectivités. L'expression de « tutelle » disparait alors du vocabulaire officiel au profit des expressions de « contrôle administratif » ou de « contrôle budgétaire »[1001]. Cependant, pour une partie de la doctrine, le maintien du concept de tutelle pour décrire les rapports entre l'État et les collectivités n'a rien d'inapproprié[1002]. Au contraire, le concept de contrôle est trop vague et n'a aucun contenu juridique déterminé. Il ne permet donc pas de saisir véritablement la situation juridique des collectivités territoriales dans leurs rapports avec l'État[1003].

426. - Cette position doctrinale a eu un écho favorable auprès du législateur africain qui conserve dans les textes relatifs à la décentralisation l'expression de « tutelle ». Le pouvoir tutélaire qu'exerce l'État sur les collectivités apparait comme un outil de protection de la collectivité elle-même contre les erreurs de ces dirigeants, mais surtout des administrés. La tutelle permet de garantir le respect de la légalité et de protéger les

---

[1000] Y. VILLAN, *L'État et les collectivités locales en France et en Allemagne : étude comparée des limites constitutionnelles à la décentralisation*, thèse, dactyl., Université Paris I, 2018. p. 551.

[1001] *Cf.* J.-M. PONTIER, « Les contrôles et la tutelle », *AJDA*, 2022, p. 1311.

[1002] B. FAURE, *Droit des collectivités territoriales, op.cit.*, n° 838 ; J. WALINE, « A propos de la tutelle sur les collectivités territoriales », *op.cit.*, p. 1096.

[1003] *Idem.* ; Voir aussi, L. JANICOT, « La tutelle de l'État sur les collectivités territoriales trente après la loi du 2 mars 1982 », *AJDA,* 2012, p. 753 ; B. FAURE, « La glorieuse trentenaire. A propos du 30ᵉ anniversaire de la loi du 2 mars 1982 », *AJDA,* 2012, p. 738.

administrés de la collectivité[1004]. Si ces éléments permettent de justifier l'existence du pouvoir de tutelle, il faut relever que son existence est aussi rendue nécessaire du fait de l'imprécision de la notion de compétences dont nous avons parlé précédemment (**A**). Aussi, la préférence du législateur africain pour l'expression de tutelle permet de la distinguer d'autres formes de contrôle que possède l'État, notamment le pouvoir hiérarchique (**B**).

A- La nécessité du pouvoir de tutelle

427. - L'existence du pouvoir de tutelle est rendue nécessaire pour au moins deux raisons : la doctrine s'accorde d'une part sur le fait que la tutelle constitue un moyen de préservation de l'unité de l'État. Le contrôle de l'État sur l'action des collectivités territoriales et la préservation de l'unité de l'État permettent de garantir l'indivisibilité de l'État. Ainsi, la tutelle apparait comme un élément important de la définition de l'État unitaire. En outre, les États de l'espace francophone ouest-africain ont importé un modèle d'organisation administrative d'inspiration française avec tous ces défauts, notamment en ce qui concerne l'imprécision de la répartition des compétences. Cette imprécision permet de démontrer également la nécessité d'instituer un pouvoir de tutelle.

428. - Le maintien d'une tutelle étatique sur l'action des collectivités territoriales est une caractéristique propre de l'État unitaire qui permet de le différencier d'autres formes d'États[1005]. Léon Duguit fait partie des auteurs de la doctrine française qui ont démontré la corrélation entre la forme unitaire de l'État et l'existence d'un pouvoir de tutelle[1006]. En partant de la distinction entre gouvernants et agents décentralisés, l'auteur montre que la différence entre l'État fédéral et l'État unitaire réside dans le fait qu'il existe dans les États fédéraux deux groupes de gouvernants sur un même territoire. Les gouvernants des collectivités fédérées échappent, à la

---

1004 J. BENOIT, *Théorie juridique de la décentralisation, op.cit.* : « un ensemble de procédés prévus par la loi, distincts des pouvoirs hiérarchiques, organisant un contrôle par les autorités administratives de l'État des collectivités locales ou des établissements publics considérés comme décentralisés, constituant la limite de l'autonomie de ces personnes administratives dans l'intérêt national ou aussi dans leur intérêt propre » ; J. WALINE, « A propos de la tutelle sur les collectivités territoriales », op.cit., p. 1096.

1005 Y. VILLAN, *L'État et les collectivités locales en France et en Allemagne : étude comparée des limites constitutionnelles à la décentralisation, op.cit.*, p. 446.

1006 L. DUGUIT, *Traité de droit constitutionnel*, E. Boccard, t. 3, 1923, p. 74.

différence des agents décentralisés de l'État unitaire, dans leur champ de compétence, à tout contrôle des gouvernants fédéraux[1007]. Dès lors qu'un « *un tel contrôle* [existe], *on n'est plus en en présence d'un État fédéral, mais d'un État pseudo-fédéral* »[1008], ou pour le dire on ne peut plus clairement d'un État unitaire décentralisé. Maurice Hauriou ne disait pas autre chose lorsqu'il affirmait que la décentralisation est une « manière d'être de l'État ». Dans sa thèse sur l'œuvre du doyen de Toulouse, François Fournié montre que par cette formule, Maurice Hauriou distingue la décentralisation du fédéralisme[1009]. La première respecte l'unité de l'État en laissant subsister une tutelle[1010] pour « *protéger les intérêts de la personne en tutelle et* [...] *maintenir le bon ordre dans l'administration* »[1011].

429. - Cette position défendue par des auteurs classiques de la doctrine française permet d'expliquer les positions actuelles du législateur africain. En effet, les pays d'Afrique subsaharienne francophones ont rejeté l'idée de fédéralisme dans la construction de leurs États aussi bien au moment des indépendances que pendant la période des conférences nationales souveraines[1012]. Si la décentralisation est devenue nécessaire dans ces pays tant du point de vue politique qu'administratif afin de faciliter la participation des citoyens et accroitre l'efficacité de la machine administrative, elle ne peut conduire à changer la forme unitaire de l'État. En ce sens, tel que cela a été relevé par la doctrine française, la tutelle étatique « *reste le dernier rempart de protection et d'identification de l'État unitaire* »[1013]. Par conséquent, la notion de tutelle « *est inhérente à celle de collectivité territoriale : il ne peut y avoir collectivité territoriale sans contrôle. Si le*

1007 *Idem.* V. aussi, S. FLOGAÏTIS, *La notion de décentralisation en France, en Allemagne et en Italie, op.cit.*, pp. 45-49, Y. VILLAN, L'État et les collectivités locales en France et en Allemagne : étude comparée des limites constitutionnelles à la décentralisation, op.cit., p. 446.

1008 S. FLOGAÏTIS, *La notion de décentralisation en France, en Allemagne et en Italie, op.cit.*, p. 47.

1009 F. FOURNIÉ, *Recherches sur la décentralisation dans l'œuvre de Maurice Hauriou*, thèse, LGDJ, 2005, pp. 234-248.

1010 Sur la tutelle dans l'œuvre de Maurice Hauriou, *Ibid.*, pp. 252-262.

1011 *Ibid.*, p. 253.

1012 C. NACH MBACK, *Démocratisation et décentralisation. Genèse et dynamiques comparés des processus de décentralisation en Afrique subsaharienne, op.cit.*, pp. 65-178.

1013 P. COMBEAU, « Introduction », in P. COMBEAU (dir.), *Les contrôles de l'État sur les collectivités territoriales aujourd'hui, op.cit.*, p. 1.

contrôle disparait, la collectivité territoriale change de nature et s'identifie à la qualité de la collectivité composante d'un État quasi-fédéral »[1014].

430. - Il parait donc impossible pour l'État de laisser aux collectivités territoriales la liberté de gérer et de régler sans aucune intervention du pouvoir central les affaires qui les concernent. L'absence d'un pouvoir de tutelle est un danger pour l'État dans la mesure où, la revendication d'une plus grande autonomie des collectivités territoriales peut entrainer la dislocation de l'État. La décentralisation, en tant que phénomène purement administratif, permet de préserver l'unité et l'indivisibilité de l'État. La notion de tutelle qu'elle induit systématiquement a pour fonction de maintenir l'unité nationale et d'éviter l'éclatement en fédéralisme[1015]. La consécration de la libre administration des collectivités territoriales ne s'oppose donc pas, dans un État de type unitaire, à l'institution d'un pouvoir de tutelle[1016]. La libre administration ne signifie pas « libre gouvernement ». Les collectivités territoriales africaines à l'instar des collectivités territoriales françaises sont ainsi en permanence soumises à une tutelle, à une volonté d'une autorité étatique. Par conséquent, l'autonomie dont elles bénéficient ne peut être que relative et subordonnée. Les actes des collectivités territoriales peuvent à tout moment être contestés par le pouvoir central. Pour préserver l'unité nationale dans le cadre de la décentralisation, le pouvoir central dispose d'une « omnicompétence »[1017] sur toute l'étendue du territoire afin de protéger l'unité de l'État et l'intérêt général.

431. - Au-delà d'une garantie pour la préservation de l'unité de l'État, le pouvoir de tutelle est nécessaire pour faire face aux défauts de la décentralisation à la française importée dans les États d'Afrique francophone. Les défauts dont il s'agit sont ceux que nous avions mentionnés dans la section précédente et qui rendent impossible toute tentative de théorisation de la répartition des compétences. Les collectivités territoriales ne disposent pas de compétences propres. Il

---

1014 *Idem.*

1015 A. GRUBER, *La décentralisation et les institutions administratives*, Armand Colin, 1986, p. 23.

1016 G. MARCOU, « L'autonomie des collectivités locales en Europe : fondements, formes et limites » in, *Quel avenir pour l'autonomie des collectivités locales* ? Les Deuxièmes Entretiens de la Caisse des dépôts et consignations, GRALE-Éditions de l'Aube, p. 41 ; « L'autonomie communale : étude comparative », *Pouvoirs*, n° 95, 2000, pp. 69-85.

1017 O. BEAUD, *La puissance de l'État*, PUF, 1994, p. 145.

n'existe pas d'affaires propres qui relèveraient par essence de la compétence des collectivités territoriales. Les affaires propres sont celles que le législateur a décidé de considérer comme telles. S'il avait existé des affaires propres par essence que le pouvoir central ne faisait que restituer aux collectivités alors, souligne Guy Melleray, le pouvoir de tutelle perdrait par là même toute vocation à en surveiller l'exercice[1018]. Parce que, « *les affaires locales n'ont d'autre source institutive qu'un auto-dessaisissement de l'État* [qui] *juridiquement demeure relatif, temporaire et toujours révocable* »[1019], le pouvoir central conserve un pouvoir de surveillance sur les prérogatives qu'il reconnait aux collectivités territoriales. Les collectivités en même temps qu'elles constituent des personnes autonomes sont des organes de l'État. La tutelle administrative permet à l'État de vérifier l'exercice par les collectivités territoriales des compétences qu'il a entendu leur déléguer, de s'assurer du respect de la légalité, de l'intérêt général et de se substituer à elles en cas de manquement à leur obligation pour assurer une continuité dans la vie administrative.

432. - Le pouvoir de tutelle marque ainsi les limites de la liberté dont disposent les collectivités territoriales. Il témoigne de ce que la décentralisation ne peut être « pure de tout alliage »[1020] dans un État unitaire. La tutelle étatique est le symbole même de la permanence de la centralisation administrative dans la décentralisation. Dans son essence, l'existence de la tutelle ne peut être sérieusement contestée. Toutefois, une fois admise, tout est dans l'art de l'exécution et de l'équilibre. L'autonomie accordée aux collectivités territoriales bien que relative implique de distinguer le pouvoir de tutelle du pouvoir hiérarchique.

B- La nécessité de distinguer entre pouvoir de tutelle et pouvoir hiérarchique : le flou conceptuel

433. - La distinction entre hiérarchie et tutelle occupe une fonction explicative importante dans l'organisation administrative du territoire des États unitaires. Le pouvoir hiérarchique s'inscrit dans le cadre de la centralisation et de l'administration d'État tandis que, le pouvoir de tutelle serait l'emblème de la décentralisation et le cadre d'action des collectivités

---

1018 G. MELLERAY, *La tutelle de l'État sur les communes, op.cit.*, p. 29.
1019 J. CHAPUISAT, *La notion d'affaires locales en droit administratif français, op.cit.*, p. 338.
1020 G. MELLERAY, *La tutelle de l'État sur les communes, op.cit.*, p. 27.

territoriales[1021]. Cette présentation d'apparence simple se révèle en pratique complexe en raison du caractère équivoque de la notion de tutelle. En effet, si les fondements des deux pouvoirs sont distincts, ils utilisent les mêmes procédés conduisant à relativiser l'opposition entre ces pouvoirs.

434. - Sur le plan des fondements, l'idée qui préside à la distinction de la hiérarchie et de la tutelle est que la première se manifeste à l'intérieur d'une même personne publique alors que la seconde s'intéresse au rapport entre l'État — personne publique — et d'autres personnes morales — les collectivités territoriales — auquel le droit positif reconnait une autonomie administrative. Ainsi, le pouvoir hiérarchique « *doit être envisagé à l'aune de sa fonction pour les structures administratives : il se définit alors comme un moyen d'unification de la volonté et de l'action au sein de chaque personne publique. Il est un pouvoir inconditionné affectant l'ensemble de l'activité des subordonnés* »[1022]. Par opposition, le pouvoir de tutelle permet de concilier la nécessité d'un contrôle et la libre administration en supprimant tout lien de subordination entre l'État et les collectivités territoriales. C'est en ce sens que Charles Eisenmann écrit que l'opposition entre les organes centraux et décentralisés repose sur la notion de hiérarchie et corrélativement sur celle de subordination[1023].

435. - Dans une conception idéale, la tutelle traduit l'absence d'un rapport de supériorité entre l'autorité centrale et les autorités décentralisées. Ces dernières disposent d'un pouvoir de décision autonome qu'elle s'exerce sous la simple surveillance du représentant de l'État. Le pouvoir de surveillance serait alors exercé au moyen d'un contrôle de la légalité. Ce pouvoir de contrôle qui permet à l'autorité de tutelle de s'assurer de la régularité juridique des actes des collectivités territoriales ne limite pas leur indépendance. Une telle indépendance n'étant par ailleurs concevable que dans le respect de la légalité. Cependant, la tutelle étatique sur les collectivités territoriales africaines implique la mise à disposition de l'autorité de tutelle des pouvoirs conséquents telles l'annulation, la substitution d'action, la révocation, la nomination, etc. On peut lire par exemple à l'article 56 du code général des collectivités du Burkina Faso que « *le contrôle de tutelle comporte les*

1021 C. EISENMANN, *Cours de droit administratif*, *op.cit.*, p. 221 et s.
1022 C. CHAUVET, *Le pouvoir hiérarchique*, thèse, LGDJ, 2013.
1023 C. EISENMANN, *Cours de droit administratif*, *op.cit.*, p. 221.

fonctions : d'approbation ; d'autorisation préalable ; d'annulation ; de suspension ou de révocation ; de substitution ; d'inspection». Ce faisant, elle utilise les mêmes procédés que le pouvoir hiérarchique entrainant une confusion entre centralisation, déconcentration et décentralisation. En effet, la doctrine range au sein du pouvoir hiérarchique des instruments qui recoupent ceux de la tutelle, notamment la révocation et la suspension[1024].

436. - L'analyse doctrinale permet de montrer l'hétérogénéité des positions sur la distinction entre pouvoir de tutelle et pouvoir hiérarchique. La suspension d'un élu local est considérée par exemple comme l'exercice d'un pouvoir disciplinaire de la part du pouvoir central. Il faut toutefois relever que la doctrine semble partagée lorsqu'il s'agit de classer le pouvoir disciplinaire dans l'une des deux catégories fondamentales que sont la tutelle et la hiérarchie. En raison du dédoublement fonctionnel, le maire possède à la fois le statut d'élu local et de représentant de l'État. À ce titre, le maire en sa qualité d'agent de l'administration déconcentrée est placé sous l'autorité du préfet qui peut donc décider de prendre à son encontre des sanctions en cas de refus d'assurer des missions qui lui sont confiées en sa qualité d'agent de l'État. L'article 163 de la loi n° 2019-006 du 26 juin 2019 relative à la décentralisation et aux libertés locales dispose à cet effet que le représentant de l'État exerce « *son pouvoir hiérarchique sur les actes du maire lorsque celui-ci agit comme agent de l'État dans la commune* ». Cette situation traduit un rapport de hiérarchisation entre les deux autorités.

437. - Cependant, le recours à la notion de pouvoir hiérarchique semble insuffisant pour traduire l'entièreté des situations que recouvre le pouvoir de sanction disciplinaire de l'autorité étatique. L'autorité de tutelle peut en dehors de l'exercice des fonctions de l'État révoquer un maire pour des faits étrangers dès lors qu'ils conduisent à compromettre l'exercice satisfaisant des fonctions[1025]. Bertrand Seiller montre dans une étude relative à la suspension du maire d'une commune française[1026] que le pouvoir disciplinaire dépasse le domaine du pouvoir hiérarchique. Selon cet auteur, le pouvoir disciplinaire du « *gouvernement s'exerce indifféremment à l'égard de l'exercice des compétences déconcentrées et décentralisées dévolues aux organes*

1024 B. FAURE, *Droit des collectivités territoriales, op.cit.*, n° 883.

1025 B. SEILLER, « Le pouvoir disciplinaire sur les maires », *AJDA*, 2004, p. 1637.

1026 Il s'agit d'une suspension du maire de la commune de Bègles à la suite de la décision de ce dernier de marier deux personnes de même sexe.

exécutifs des communes. Il transcende la distinction classique entre le pouvoir hiérarchique — qui accompagne la déconcentration — et le pouvoir de tutelle — qui caractérise la décentralisation »[1027]. Dans le même sens, pour Léon Duguit, le pouvoir disciplinaire est « *la sanction du pouvoir de surveillance qui appartient à l'administration supérieure sur les fonctionnaires subordonnés pour garantir l'accomplissement par eux de leurs obligations fonctionnelles* ». Une telle surveillance précise, le doyen de Bordeaux peut *« s'exercer sur tous les agents sans distinction, quel que soit leur mode de nomination, quelles que soient la nature et l'étendue de leurs attributions, sur les agents nommés et les agents élus* »[1028]. Le pouvoir disciplinaire apparait ainsi dans toute institution primaire ou secondaire pour assurer le maintien de la discipline en son sein[1029]. Il constitue un pouvoir à part, que l'on ne peut faire entrer dans l'opposition binaire entre le pouvoir de tutelle et le pouvoir hiérarchique[1030].

438. - L'ambiguïté de ce pouvoir de sanction introduit une certaine confusion dans les catégories juridiques du droit administratif[1031]. Elle conduit à relativiser la classification des structures administratives territoriales sur la base de la division hiérarchie/tutelle. Ainsi, certains auteurs invitent à ne pas exagérer l'opposition entre les deux pouvoirs[1032]. La relativité de l'opposition entre tutelle et hiérarchie dans l'organisation administrative démontre le caractère perméable entre la déconcentration et la décentralisation. Le flou des contours se révèle être un avantage pour le pouvoir central, car l'État en Afrique s'est toujours efforcé de combiner la décentralisation avec des procédés de déconcentration au sein d'une même personne morale dans le but de limiter l'autonomie qui accompagne l'attribution de la personnalité juridique aux collectivités territoriales. La multiplicité des procédés dont dispose le pouvoir de tutelle pour interférer

---

1027 B. SEILLER, « Le pouvoir disciplinaire sur les maires », *op.cit.*, p. 1637.

1028 L. DUGUIT, *Traité de droit constitutionnel, op.cit.*, p. 272. Voir aussi, C. EISENMANN, *Cours de droit administratif, op.cit.*, p. 224-225 ; L. RICHER, « La notion de tutelle sur les personnes en droit administratif », *RDP*, 1979, p. 971-1008 : ce dernier se réfère en particulier aux travaux d'Hauriou et à sa vision englobante et se déclare favorable à un régime juridique répressif. Serge Regourd étudie également en détail cette question et plaide à cet égard en faveur d'une distinction entre les mesures de tutelle et celles relevant du pouvoir disciplinaire en raison des fonctions administratives distinctes de ces pouvoirs, récusant ainsi la distinction entre la tutelle sur les personnes et tutelle sur les actes : S. REGOUD, *L'acte de tutelle en droit administratif français, op.cit.*, p. 110-119.

1029 Cf. J. MOURGEON, *La répression administrative*, LGDJ, 1967.

1030 B. SEILLER, « Le pouvoir disciplinaire sur les maires », *op.cit.*, p. 1637.

1031 B. FAURE, *Droit des collectivités territoriales, op.cit.*, n° 883.

1032 *Ibid.*, n° 4.

dans la vie juridique locale conforte notre hypothèse d'une décentralisation sans libre administration.

§2.- La multiplicité des procédés d'exercice du pouvoir de tutelle

439. - Le pouvoir de tutelle renvoie aux différents procédés par lesquels l'État soumet à sa volonté l'action des collectivités territoriales. Approbation, autorisation, annulation, substitution, suspension, révocation, dissolution, démission d'office sont autant de procédés mis à la disposition de l'autorité de tutelle pour exercer un contrôle direct sur l'action des collectivités. Cette panoplie de procédés démontre la méfiance de l'État vis-à-vis des collectivités territoriales. L'État en Afrique ne fait pas suffisamment confiance aux collectivités territoriales pour exercer de manière satisfaisante les compétences qu'il a entendu leur accorder. Les procédés de la tutelle traduisent ainsi la subordination directe des collectivités territoriales au pouvoir central. Ils correspondent à la finalité première du pouvoir de tutelle qui est essentiellement négative[1033], les collectivités territoriales étant considérées comme des mineures incapables. Contrairement à la France où persiste un « malaise » autour de l'usage du mot de tutelle[1034], sa pertinence ne fait aucun doute pour le législateur dans les pays ouest-africains étudiés. Le législateur nigérien justifie en ce sens l'institution d'une tutelle *a priori* sur les actes des collectivités territoriales au regard des « *difficultés de l'expérience du premier mandat des communes* » qui « *ont montré de nombreuses illégalités de gestion, génératrices de* préc*édents fâcheux et difficilement rattrapables* »[1035] et donc de l'incapacité des collectivités à se gérer elles-mêmes. Le développement de multiples mécanismes de sanctions renforce l'emprise tutélaire de l'État sur les collectivités territoriales et limite considérablement leur autonomie. L'étude des procédés de la tutelle permet de mesurer l'étendue réelle des pouvoirs de l'autorité et, par conséquent, l'étendue de l'autonomie des collectivités territoriales. Pour aborder cette question, les développements qui vont suivre vont s'articuler autour des domaines fondamentaux que

---

1033 G. MELLERAY, *La tutelle de l'État sur les communes, op.cit.*, p. 160.

1034 J.-A. MAZÈRES, « Débat – le processus de décentralisation », *AJDA*, 1992, p. 6 ; B. FAURE, « Faut-il garder le mot "tutelle" en droit administratif ? », *AJDA*, 2008, p. 113.

1035 Exposé des motifs du projet d'adoption du code général des collectivités territoriales.

les législations africaines assignent au contrôle de tutelle : le contrôle des actes (**A**) et le contrôle des organes (**B**).

A- Le contrôle exercé par l'État sur les actes des collectivités territoriales

440. - L'État organise un contrôle sur les actes des collectivités territoriales. La tutelle sur les actes des collectivités territoriales présente un caractère uniforme c'est-à-dire que dans les législations ici étudiées, la loi prévoit un mécanisme de contrôle des actes identique pour toutes les collectivités territoriales. Suivant l'article 292 de loi n° 2017-051 du 2 octobre 2017 portant code des collectivités territoriales au Mali, ce « *contrôle des actes des collectivités territoriales consiste, à l'exclusion de toute appréciation d'opportunité, en la vérification de leur légalité. Il s'exerce sur les délibérations, arrêtées et autres actes des collectivités territoriales* ». Le législateur burkinabè dispose quant à lui à l'article 55 du code général des collectivités territoriales que, « *les actes et les délibérations des collectivités territoriales sont soumis au contrôle de légalité exercé par l'autorité de tutelle* ». Le contrôle de légalité des actes locaux est présenté par la doctrine comme l'instrument le plus compatible avec le principe de libre administration des collectivités territoriales. Pour Sasso Pagnou, dans la mesure où ce contrôle permet une conciliation du principe de légalité et du principe de la libre administration, il est un élément de la bonne gouvernance[1036]. Il constitue l'encadrement nécessaire à la réussite du processus de décentralisation dans les États africains. La mise en place de ce contrôle contribue à rétrécir le champ du contrôle *a priori* qui est l'une des « *formes énergiques du contrôle de tutelle* »[1037].

441. - S'il est vrai que le contrôle de légalité des actes semble s'imposer comme principale modalité de contrôle, le contrôle *a priori* ne disparait pas pour autant. Les législations ouest-africaines laissent subsister dans les textes qui organisent la décentralisation des mécanismes, notamment d'approbation préalable et d'annulation par le représentant de l'État. Il coexiste donc dans le droit positif des pays étudiés, une dualité de régimes de contrôle des actes locaux. Certains actes sont exécutoires de plein droit

---

1036 S. PAGNOU, « La place du contrôle de légalité dans les processus de bonne gouvernance liés à la décentralisation. Le cas du Togo », *Revue internationale des sciences administratives*, vol. 79, n° 3, 2013, p. 605.

1037 I. M. FALL, « Le contrôle de légalité des actes des collectivités locales au Sénégal », *Revue numérique Afrilex*, n° 5, p. 82.

tandis que d'autres doivent être approuvés par le représentant de l'État. Cela montre la persistance d'une tension entre la volonté d'un recul de l'administration étatique et la permanente volonté de surveillance étatique. Il convient d'exposer les modalités du *contrôle a posteriori* (**1°**) ainsi que l'étendue des pouvoirs du représentant de l'État qui symbolise la persistance d'une tutelle *a priori* (**2°**).

1. Les modalités de mise en œuvre du contrôle sur les actes

442. - L'exercice du contrôle de légalité des actes des collectivités territoriales est organisé autour de deux autorités : le représentant de l'État et le juge administratif. Le premier est chargé de constater les illégalités et le second de les sanctionner. Le contrôle des actes des collectivités territoriales comporte ainsi deux phases. La première, administrative au cours de laquelle un dialogue s'installe entre les services de la préfecture et la collectivité afin de corriger les illégalités constatées et la seconde, juridictionnelle qui consiste en la saisine par la voie du déféré préfectoral du juge administratif pour faire sanctionner l'illégalité. Ces deux phases ne sont pas toujours cumulées. Lorsque le représentant de l'État parvient dans la phase de dialogue à faire retirer ou modifier l'acte illégal, la phrase administrative clôt le contrôle. Le juge n'étant saisi qu'en cas d'échec du dialogue instauré dans la phase administrative. Seront donc examinées successivement la phase administrative du contrôle (**a**) puis la phase juridictionnelle dont on ne peut que déplorer l'ineffectivité dans les pays étudiés (**b**).

a) Une phase administrative déterminante

443. - La phase administrative du contrôle ne peut se réaliser qu'à partir du moment où le représentant de l'État prend connaissance des actes de la collectivité (**i**). Cette phase est aussi considérée comme celle permettant d'instituer un dialogue entre l'autorité de tutelle et la collectivité (**ii**).

i) La transmission des actes des collectivités territoriales au représentant de l'État

444. - Le contrôle *a posteriori* de la légalité repose sur la double condition de publicité et de transmission des actes des collectivités territoriales au représentant de l'État. Il est précisé à l'article 77 de la loi togolaise relative à la décentralisation et aux libertés locales que « *les actes pris par les autorités* *locales sont exécutoires* [...] *à compter de leur transmission à l'autorité de tutelle*»[1038]. La loi nigérienne consacre également ce principe à l'article 312 du CGCT : « *les actes pris par les autorités des collectivités territoriales sont exécutoires de plein droit dès qu'il a été procédé à leur publication ou à leur notification aux intéressés ainsi qu'à leur transmission à l'autorité de tutelle* ». Il en est de même au Burkina Faso[1039] et au Mali[1040]. Si la règle de publicité ou de notification est une condition classique de l'entrée en vigueur des actes administratifs, celle de la transmission apparait comme une exception conçue de façon restrictive[1041]. Toutefois, le législateur tente dans certains cas[1042] d'en limiter la portée en dressant la liste des actes soumis à l'obligation de transmission. Il s'agit pour le législateur nigérien des actes ci-après[1043] :

- Les budgets primitifs, les actes modificatifs et les comptes ;
- Les délibérations du conseil ou les décisions prises par délégation du conseil ;
- Les actes à caractère règlementaire pris dans tous les domaines qui relèvent de leurs compétences en application de la loi ;
- Les conventions relatives aux marchés publics ainsi que les conventions de concession ou d'affermage des services publics locaux et les contrats de partenariat ;
- Les décisions relatives au personnel (nomination, avancement et sanction) ;
- Les outils de planification et d'urbanisme ;
- Le règlement intérieur

445. - Les actes concernés par l'obligation de transmission sont ainsi principalement les actes unilatéraux et dans une moindre mesure les actes contractuels. En cas de non-transmission d'un acte figurant dans cette liste, celui-ci ne peut obtenir force exécutoire. L'alinéa 3 de l'article 77

1038 Art. 77 Loi n° 2019-006 *préc.*

1039 Art. 60 CGCT : « Les délibérations qui ne sont pas soumises à approbation ou à autorisation préalable deviennent exécutoires après leur transmission à l'autorité de tutelle, sous réserve du respect des conditions d'entrée en vigueur des actes des autorités locales ».

1040 Loi n° 2017-051 du 02 octobre 2017 *préc.*, L'obligation de transmission des actes pèse sur le chef de l'exécutif, art. 61 (pour les communes), art. 134 (pour les cercles) et art. 202 (pour la région)

1041 A. GRUBER, La décentralisation et les institutions administratives, op.cit., p. 275.

1042 Une telle liste n'existe pas dans toutes les législations. Au Togo par exemple, seuls les actes non soumis à l'obligation sont énumérés (art. 79 Loi n° 2019-006 *préc.*).

1043 Ces actes sont énumérés à l'article 313 du CGCT.

précité précise que, « *est nul et de nul effet, tout acte soumis à l'obligation de transmission et qui n'a pas été transmis effectivement à l'autorité de tutelle* ». En outre, il faut relever que la compétence du représentant de l'État ne se limite pas aux seuls actes soumis à l'obligation de transmission[1044], tous les actes administratifs même les moins importants peuvent faire l'objet d'un contrôle de légalité. Ainsi, pour le législateur togolais, bien qu'ils ne soient pas soumis à l'obligation de transmission « *les actes de gestion courante des services et du domaine public de la collectivité territoriale, les actes d'administration interne, notamment en matière de gestion du personnel de la collectivité territoriale ; les actes d'administration courante, notamment les correspondances échangées entre les élus locaux et leurs électeurs* »[1045] demeurent assujetties au contrôle de légalité. Seuls sont exclus les actes de droit de privé des collectivités territoriales et les actes relevant de l'exercice des missions déconcentrées par les collectivités territoriales.

446. - Une fois la transmission faite, la loi impose à l'autorité de tutelle d'en accuser réception dans un délai de 15 jours à compter de la date de réception de l'acte (art. 77 préc. Togo). À défaut d'un accusé de réception délivré immédiatement par l'autorité de tutelle, la preuve de la réception de l'acte peut être apportée pour tous moyens (art. 312 CGCT préc. Niger). Se pose alors la question du mode de transmission des actes et de son impact sur l'efficacité de l'action publique locale. L'écrit reste le mode d'expression normal de l'administration publique africaine. Les administrations locales se retrouvent ainsi dans l'obligation, afin d'obtenir un accusé de réception au plus vite, à titre de preuve, de se déplacer pour déposer une version papier des actes des autorités territoriales. Ceci coûte la mise à disposition d'une ressource humaine et d'un moyen de déplacement[1046]. Ces coûts pourraient être considérablement réduits et permettre d'économiser du personnel à des tâches répétitives d'enregistrement, de délivrance d'accusés de réception, si l'administration

---

1044 Voir en droit français la jurisprudence *Commune du Port*, CE, 28 février 1997, *Lebon*, p. 61, *RFDA*, 1997, p. 1190, concl. J. H. Stahl, note J.-C. Douence, *AJDA*, 1997, p. 46.
1045 Art. 79 de la loi n° 2019-006 *préc.*
1046 C. NANAKO, La libre administration des collectivités territoriales au Bénin et au Niger, op.cit., p. 170.

optait pour une dématérialisation et une télétransmission des actes[1047]. Mais il faudrait en amont résoudre la question de la numérisation et de la couverture par internet de l'administration publique[1048].

447. - Il convient aussi de relever une différence dans le régime d'entrée en vigueur des actes soumis à l'obligation de transmission. En effet, le législateur togolais introduit deux régimes différents d'entrée en vigueur des actes des collectivités territoriales. Seuls les actes qui ne sont pas soumis à l'obligation de transmission sont exécutoires de plein droit sous réserve du respect des conditions d'entrée en vigueur des actes locaux. En revanche, pour les actes soumis à l'obligation de transmission, donc l'ensemble des actes des collectivités à l'exception de ceux mentionnés à l'article 79 précité, l'exécution est subordonnée, après la transmission, à l'écoulement d'un délai de trente jours. « *Les actes pris par les autorités locales sont exécutoires dans un délai de trente jours à compter de leur transmission à l'autorité de tutelle* »[1049]. Ainsi, contrairement aux autres législations étudiées, la postériorité du contrôle n'implique pas dans la législation togolaise une exécution de plein droit après transmission, publication ou notification aux intéressés. Elle est soumise à une condition supplémentaire de délai.

448. - La transmission des actes constitue un préalable important pour l'exercice du contrôle de légalité par l'autorité de tutelle. C'est la réalisation de cette formalité qui ouvre véritablement la phase administrative du contrôle de légalité. Le représentant de l'État dispose dès lors de deux possibilités : la confirmation de régularité des actes transmis ou alors la saisine par la voie du déféré du juge administratif pour l'annulation des actes qu'il estime entachés d'illégalités. Le contrôle de légalité est effectué sans suite contentieuse en l'absence d'irrégularités constatées par l'autorité de tutelle. En revanche, lorsqu'il relève l'existence d'irrégularités, il est tenu de déférer l'acte à la juridiction administrative afin d'en obtenir

---

1047 En France, la question s'est également posée. Pour améliorer la performance du contrôle de légalité, le législateur a fait le choix de la dématérialisation : art. 139 de la loi n° 2004-809 relative aux libertés et responsabilités locales codifié à l'art. L. 2131-2 CGCT. Voir aussi, C. BRANQUART, « Contrôle de légalité : un réel renouveau ? », *AJDA*, 2011, p. 198 ; P. CADIEU, « La modernisation du contrôle de légalité », *RLCT*, mai 2006, p. 4 ; « le "résistible déclin" du contrôle de légalité », *RLCT*, mai 2008, p. 5.

1048 À cet effet nous ne pouvons que saluer le lancement le 15 juin 2022 au Togo de la plateforme national des services digitalisés de l'administration.

1049 Art. 77 de la loi n° 2019-006 *préc.*

l'annulation. Toutefois, le déféré tant à être substitué par l'instauration d'un dialogue entre l'autorité de tutelle et la collectivité.

ii) La place prépondérante du dialogue avant l'exercice du déféré préfectoral

449. - Le déféré préfectoral n'est pas systématique. Une fois l'irrégularité constatée dans cette phase du contrôle s'installe un dialogue entre l'autorité de tutelle et les autorités décentralisées. Cette étape préalable qui tend à se substituer au déféré constitue ce que l'on appelle la phase précontentieuse du contrôle de légalité. Cette phase de dialogue est qualifiée par certains auteurs d'étape « *inquisitoriale du contrôle administratif de légalité* »[1050]. La saisine du juge administratif manifeste alors pour les autorités préfectorales l'échec de l'usage de la négociation. Cela signifie qu'elles n'ont pas réussi à établir un dialogue constructif avec la collectivité. L'état du droit est établi en sorte que ce moment de dialogue soit ménager. Il est précisé en ce sens à l'article 295 du code des collectivités territoriales au Mali que « *le représentant de l'État communique à la collectivité territoriale ses observations sur les délibérations non soumises à approbation, les arrêtés et autres actes qu'il estime contraires à la légalité, dans les quinze (15) jours qui suivent leur transmission. La collectivité territoriale répond aux observations du représentant de l'État dans le délai de quinze (15) jours, à compter de la date de réception desdites observations. À défaut de réponse dans le délai imparti ou lorsque la collectivité n'apporte pas un traitement approprié aux observations formulées, le représentant de l'État défère l'acte incriminé au tribunal administratif* ». La loi nigérienne prévoit quant à elle un délai de deux mois au cours duquel l'autorité de tutelle peut adresser des observations aux autorités décentralisées. C'est le sens de l'article 314 du code général des collectivités territoriales qui prévoit que « *l'autorité de tutelle a compétence pour demander dans les deux (2) mois à dater de l'accusé de réception prévu à l'article 312 du présent code, à la collectivité territoriale concernée, le retrait ou la modification de l'acte transmis pour contrôle et qu'elle estime entaché d'illégalité* ».

450. - L'aménagement de ce délai a pour conséquence de proroger le délai de mise en œuvre du déféré préfectoral qui est habituellement de

---

[1050] E. AUBIN, « Contrôle administratif de légalité–Objet et effets », *JCl. Coll. terr.*, Fasc. 912, § 105.

deux mois[1051]. L'action du préfet est analysée comme un recours gracieux porté devant l'autorité décentralisée, auteur de l'acte contesté avec pour objet de demander à l'auteur la correction de l'illégalité relevée en réformant sa décision ou en l'abrogeant. En cas de refus, l'autorité de tutelle dispose d'un délai deux mois pour attaquer l'acte initial à compter de la décision de refus. Ainsi, l'article 296 du code des collectivités territoriales au Mali dispose que : « *le représentant de l'État défère au tribunal administratif les délibérations non soumises à approbation, les arrêtés et autres actes de la collectivité territoriale qu'il estime contraires à la légalité, dans les deux (2) mois qui suivent leur transmission. Ce délai est suspendu lorsque le représentant de l'État met en œuvre les dispositions prévues à l'article 295* ». C'est aussi le sens de l'alinéa 2 de l'article 314 du CGCT au Niger qui prévoit qu'en cas de refus, l'autorité de tutelle « *défère l'acte devant la juridiction administrative dans les deux (2) mois qui suivent le refus écrit de la collectivité ou le silence gardé par elle durant deux (2) mois* ». La conservation du délai du recours prouve que le préfet n'est pas tenu, en présence d'un acte suspect, de le déférer immédiatement devant la juridiction administrative[1052].

451. - La phase administrative du contrôle est marquée par deux étapes importantes. La première permet à travers l'obligation de transmission des actes d'informer l'autorité de tutelle. La seconde est caractérisée par le dialogue qui s'établit entre l'autorité de tutelle et les autorités décentralisées dès le moment où il constate une irrégularité et celui où il décide d'exercer le déféré préfectoral. Le représentant de l'État dispose dans cette dernière phase d'une marge de manœuvre importante. Toutefois, il ne peut en principe sanctionner de son propre gré l'acte considéré comme illégal[1053]. Mis dans une situation presque identique à celle du requérant ordinaire, le représentant de l'État doit se référer au juge administratif, en cas d'échec du dialogue avec la collectivité, pour obtenir l'annulation de l'acte. C'est alors que démarre la seconde phase dite juridictionnelle du contrôle de légalité qui s'avère *in fine* accessoire.

[1051] Ce délai est raccourci à 30 jours au Togo. Loi n° 2019-006 art. 164 al. 2 pour les communes et art. 298 pour les régions.
[1052] Voir les développements *infra* n° 454 et s.
[1053] Sur l'étendue des pouvoirs du représentant de l'État, voir les développements *infra* n° 467 et s.

b) Une phase juridictionnelle accessoire

452. - Cette phase du contrôle de la légalité des actes repose en principe sur un changement de paradigme. La sanction de l'illégalité des actes des autorités territoriales est désormais une compétence du juge administratif. La vérification de la légalité interne et externe des actes locaux par l'autorité de tutelle doit se faire en dehors de toute ap*préc*iation d'opportunité. La loi lui impose de déférer au juge tout acte qu'il estime irrégulier. Cette dernière situation est présentée par la doctrine comme une innovation en matière de contrôle des actes locaux[1054]. Il apparait ainsi à la lecture des lois nigérienne, malienne et togolaise notamment que le recours à la juridiction administrative est un facteur d'atténuation des pouvoirs de tutelle de l'administration. L'impression générale qui consiste à attribuer à la juridiction administrative un rôle charnière dans l'ap*préc*iation des rapports entre l'État et les collectivités territoriales doit cependant être nuancée pour au moins deux raisons. D'abord, non seulement le contrôle de légalité est marqué par des interférences politiques avec une saisine du juge soumise à la volonté du préfet, mais aussi, par l'ineffectivité des juridictions administratives. La situation togolaise nous permettra d'illustrer cet élément (**i**). Ensuite, l'institutionnalisation de l'assistance-conseil (**ii**) contribue à limiter fortement la fonction confiée au juge administratif.

i) L'inexistence d'un contrôle juridictionnel des actes locaux

453. - L'idée que nous souhaitons défendre ici est celle de la difficulté à voir émerger dans les États d'Afrique noire francophone un contentieux des actes des collectivités territoriales. L'enjeu étant que l'absence d'un contrôle juridictionnel prive les collectivités d'une garantie de leur autonomie. Nous montrerons que, d'une part, l'inexistence du contrôle juridictionnel est due au caractère hypothétique du contrôle du juge, car elle dépend de la seule volonté du représentant de l'État. Le fait que le représentant de l'État ap*préc*ie seul l'opportunité d'une saisine du juge, le contrôle juridictionnel des actes locaux devient une question de compromis. D'autre part, la judiciarisation du contentieux des actes tarde à venir en raison de la quasi-ineffectivité des juridictions administratives.

1054 I. M. FALL, « Le contrôle de légalité des actes des collectivités locales au Sénégal », *op.cit.*, p. 82 ; S. PAGNOU, « La place du contrôle de légalité dans les processus de bonne gouvernance liés à la décentralisation. Le cas du Togo », *op.cit.*, p. 605.

454. - Le législateur africain fait reposer la phase juridictionnelle du contrôle sur l'exercice du déféré. Théoriquement, l'édifice parait solide. Le représentant de l'État qui n'a pas réussi à faire retirer ou modifier un acte administratif local peut saisir le juge administratif. Lorsqu'il défère un acte illégal, l'autorité de tutelle est tenue d'en informer le chef de l'exécutif local. « *Le préfet défère devant la juridiction administrative compétente les délibérations, arrêtés, actes et conventions qu'il estime contraires à la légalité, dans les trente jours qui suivent leur transmission prévue à l'article 162. Il en informe le maire* »[1055] et « *lui communique toutes précisions sur les illégalités invoquées à l'endroit de l'acte concerné* »[1056]. À l'occasion d'un déféré à l'encontre d'un acte entaché d'illégalité, l'autorité de tutelle peut accessoirement demander la suspension temporaire de l'exécution de cet acte. Ce mécanisme dit du « référé-déféré » est prévu en des termes quasi identiques par les législations africaines. Le déféré assorti d'une demande de suspension s'assimile au référé-suspension du droit commun, à ceci près, que l'autorité de tutelle n'a pas à démontrer l'urgence de sa demande. L'acte contesté sera suspendu « *si l'un des moyens invoqués dans la requête est de nature à justifier l'annulation de la délibération, de l'arrêté, de la convention ou de l'acte attaqué* »[1057]. Cette procédure permet d'éviter que l'acte ne soit appliqué s'il apparait comme illégal. Le juge dispose d'un délai de huit jours pour se prononcer sur la demande de sursis sauf si « *l'acte attaqué est de nature à compromettre l'exercice d'une liberté publique ou individuelle ou lorsque les conséquences qui découleraient de son exécution peuvent s'avérer irréparables* »[1058]. Dans ce cas, le juge statue dans un délai de quarante-huit heures[1059].

455. - Le problème que pose le déféré préfectoral quant au bon fonctionnement du contrôle de légalité est relatif à sa nature juridique. S'agit-il d'une faculté ou d'une obligation ? Qu'adviendrait-il si le représentant de l'État s'abstient de saisir le juge administratif pour des actes qui seraient manifestement illégaux ? Les textes législatifs étudiés ici sont muets sur les conséquences d'un refus d'exercice du déféré par l'autorité de tutelle. Toutefois, un regard sur la jurisprudence française

---

1055 Loi n° 2019-006 *préc.*, art. 164 al. 2 pour les communes et art. 298 pour les régions.

1056 Art. 314 al. 3 CGCT au Niger.

1057 Loi n° 2019-006 *préc.*, art. 164 al. 3. Voir aussi, art. 314 al. 3 CGCT au Niger ; Loi n° 2017-051 du 2 octobre 2017 *préc.*, art. 297.

1058 Art. 315 CGCT au Niger

1059 Art. 297 al. 2 de la loi n° 2017-051 *préc.*

permet d'affirmer qu'en la matière le représentant de l'État dispose d'une marge de manœuvre qui lui permet de déclencher ou non le déféré. Le Conseil d'État a ainsi jugé que la décision de refus du préfet de déférer un acte, sur demande d'une personne lésée en l'espèce, ne constitue pas un acte susceptible de faire l'objet d'un recours pour excès de pouvoir[1060]. Cependant, en cas de refus répété du préfet d'exercer le déféré pour des actes dont l'illégalité est manifeste, la responsabilité pour faute lourde de l'État peut être engagée[1061]. Il est prévu en ce sens à l'article 311 du CGCT Niger que les décisions de l'autorité de tutelle « *sont aussi de nature à engager la responsabilité de l'État, y compris en cas de carence à agir dommageable pour la collectivité concernée ou pour les tiers* ».

456. - L'exercice du déféré préfectoral semble être une faculté. Il ne pèse sur l'autorité de tutelle aucune obligation à déclencher la procédure du déféré, y compris lorsque la demande de déféré est formulée par un administré. Cette possibilité dite du « déféré provoqué » est prévue à l'article 166 de la loi n° 2019-006 relative à la décentralisation et aux libertés locales au Togo qui prévoit que « *sans préjudice du recours direct dont elle dispose, si une personne physique ou morale est lésée par un acte d'une commune, elle peut demander au représentant de l'État dans la commune de mettre en œuvre la procédure prévue aux articles 164 et 165 de la présente loi* ». Cette demande doit porter sur les seuls actes attaquables par la voie du recours pour excès de pouvoir. Elle présente un double intérêt : elle permet d'une part, aux administrés d'informer l'autorité de tutelle de l'existence d'une éventuelle illégalité, et d'autre part, de faire supporter au représentant de l'État le coût du recours juridictionnel s'il met effectivement en œuvre le déféré préfectoral. Ici encore, ni les textes ni la jurisprudence ne permettent de déterminer les effets de l'exercice du déféré sur demande. En droit français, dans l'hypothèse où l'autorité de tutelle décide d'exercer le déféré préfectoral, son recours doit intervenir dans le temps restant à courir entre la demande et l'expiration du délai contentieux de deux mois. Si le préfet décide de ne pas déférer, le délai de recours est prorogé au profit de l'administré qui pourra agir lui-même dans les deux mois suivant le refus

---

1060 CE, 25 janvier 1991, *Brasseur, Lebon,* p. 23 ; concl. B. Stirn , *AJDA,* 1991, p. 351 ; voir aussi, B. SEILLER, « La contestation du refus d'exercer le déféré préfectoral », *RDP,* 1992, p. 1149.

1061 CE, 6 octobre 2000, Ministre de l'Intérieur c/Commune de Saint-Florent, Lebon., p. 395.

préfectoral au moyen d'un recours pour excès de pouvoir[1062]. Cette solution bien que pertinente semble difficilement applicable dans les pays d'Afrique noire subsaharienne en raison notamment de la réticence des administrés à attaquer devant les juridictions les actes de l'administration. Le respect de la légalité dépend donc de la seule tutelle de l'État.

Le préfet est la pièce maitresse de l'exercice du déféré préfectoral. Il semble donc qu'à tous les niveaux, le droit de la décentralisation offre au représentant de l'État la possibilité de transiger avec la loi en s'abstenant de déférer un acte dont il reconnait l'illégalité. Cette situation n'est pas une caractéristique du contrôle de légalité des actes en Afrique[1063]. Pour autant, une telle situation se révèle en pratique comme le souligne Sasso Pagnou, « *anti-pédagogique dans* [des] *pays où à quelque niveau que ce soit la légalité n'est pas encore entrée dans les mœurs et où le*[s] *gouvernement*[s] *déclare* [nt] *mettre tout en œuvre pour la réhabilitation de la règle de droit* »[1064]. On ne peut donc pas s'étonner de la banalisation et du désintérêt pour le recours au déféré préfectoral pour la contestation des actes locaux[1065].

457. - Les textes sont clairs, mais la pratique est tout autre. Dans les textes, le juge occupe une place importante dans le contentieux de l'annulation des actes administratifs locaux. Cependant, l'intervention du juge dans les relations entre l'État et les collectivités territoriales s'avère en pratique marginale. La possibilité offerte aux organes des collectivités de saisir le juge administratif ou de voir leurs actes annulés par l'unique intervention du juge aurait pu constituer un rempart contre la propension

---

1062 C.E. Sect., 25 janvier 1991, Brasseur, préc. ; B. FAURE, Droit des collectivités territoriales, op.cit., n° 870.

1063 J.-F. BRISSON, « Le contrôle de légalité dans la loi du 13 aout 2004. A la recherche des illusions perdues », *AJDA*, 2005, p. 126.

1064 S. PAGNOU, « La place du contrôle de légalité dans les processus de bonne gouvernance liés à la décentralisation. Le cas du Togo », *op.cit.*, p. 605.

1065 Pour un exemple en France mais qui vaut aussi pour les pays ici étudiés, I. DIALLO, « L'avenir du déféré préfectoral en droit français », *AJDA*, 2005, p. 2438 *« Le déféré préfectoral est non seulement banalisé mais aussi concurrencé par la procédure du référé liberté qui offre beaucoup plus de garanties au requérant. De plus, le préfet préfère la voie de la négociation à celle contentieuse. Dans ce contexte, le déféré préfectoral ne saurait prospérer, rendant son avenir incertain* » ; P. GERARD, « Décentralisation : le déféré délaissé », *Dr. Adm.*, n° 1, 1992, p. 1.

de l'État à tout contrôler[1066]. Cependant, la judiciarisation du contentieux des actes locaux tarde à venir. L'effectivité du contrôle de légalité des actes locaux dépend de la réussite du processus de réforme de la justice. Le plaidoyer de Palouki Massina en faveur de l'opérationnalisation de la juridiction administrative au Togo semble peu à peu aboutir[1067].

458. - Le Togo s'est doté d'un nouveau code de l'organisation judiciaire le 30 octobre 2019[1068]. Ce code vient en remplacement de l'ordonnance n° 78-35 du 7 septembre 1978. Parmi les innovations apportées par le nouveau code, on note l'institution des chambres administratives près les tribunaux de grande instance. La création des chambres administratives près les tribunaux de grande instance est censée révolutionner le contentieux administratif au Togo jusqu'alors ineffectif. Cette création semble s'être imposée par l'évolution du processus de décentralisation au Togo. La nuance est particulièrement importante ici, car le nouveau texte ne mentionne pas expressément la compétence des chambres administratives du tribunal de grande instance (TGI) en matière d'annulation des actes locaux. L'article 79 du code de l'organisation judiciaire dispose que, « *en matière administrative, les chambres administratives sont juges de droit commun, en premier ressort, du contentieux administratif, sous réserve des compétences attribuées à la chambre administrative de la Cour suprême* ». Puis le législateur procède à une énumération des matières relevant de la compétence des chambres administratives en premier ressort. Il mentionne principalement les litiges à caractère pécuniaire mettant en cause les agents des administrations à titre individuel, les litiges à caractère fiscal et le contentieux des marchés publics[1069]. Mais, l'interprétation de la réserve constitutionnelle de compétence au profit de la chambre administrative de la Cour suprême nous conduit à affirmer qu'il est possible d'ajouter à cette liste de matières la compétence en matière d'annulation des actes des collectivités territoriales dans le cadre du déféré

1066 B. COULIBALY, « Le juge administratif, rempart de protection des citoyens contre l'administration en Afrique noire francophone ? », *Revue numérique Afrilex,* 2013, p. 1 ; D. LOCHAK, «Le droit administratif, rempart contre l'arbitraire ?», *Pouvoirs*, n°46, 1988, pp. 44-55 : « *Le juge administratif, de son coté, garantit l'effectivité de ce dispositif d'encadrement de l'activité administrative par le droit, en annulant les actes illégaux qui lui sont déférés par le biais du recours pour excès de pouvoir, largement ouvert et gratuit*».

1067 P. MASSINA, « Plaidoyer pour le fonctionnement de la juridiction administrative au Togo », *Revue des droits des pays d'Afrique*, 1990, pp. 403-421.

1068 Loi n° 2019-015 portant code de l'organisation judiciaire.

1069 *Ibid.*, art. 80 et 81.

préfectoral prévu par la loi n° 2019-006 du 26 juin 2019. En effet, l'article 125 de la Constitution togolaise réserve à la chambre de la Cour de suprême un certain nombre de matières. Il dispose que, « *la chambre administrative de la Cour suprême a compétence pour connaitre : — des recours en cassation formés contre les décisions rendues en matière de contentieux administratif, — des recours pour excès de pouvoir formés contre les actes administratifs des autorités et des administrations nationales — du contentieux des élections locales — des pouvoirs en cassation contre les décisions des organismes et institutions statuant en matière disciplinaire* ». En dehors de cette réserve de compétence, les chambres administratives du TGI peuvent être saisies de tout contentieux administratif, y compris celui de l'annulation des actes des collectivités territoriales. En outre, les chambres administratives des TGI se sont vues confier une mission de conseil notamment sur demande des autorités déconcentrées, les préfets et les gouverneurs de région[1070]. On aurait pu imaginer qu'a la place des autorités déconcentrées que les organes décentralisés soient ceux qui bénéficient de cette fonction consultative. Une telle démarche aurait été plus cohérente, nous semble-t-il, pour renforcer les fonctions d'assistance et de conseil nécessaires aux collectivités tout en les préservant de l'influence de l'État.

459. - Au-delà de l'apparente simplicité de la chose se pose la difficile question de l'effectivité de la justice administrative et de son impact sur la qualité du contrôle de légalité. Gageons qu'un véritable contentieux de l'annulation des actes des collectivités territoriales verra bientôt le jour grâce notamment à l'amélioration de la composition des chambres administratives. La nouvelle loi ouvre la possibilité aux enseignants de droit public des universités de pouvoir siéger à la chambre administrative comme juges en service extraordinaires[1071]. Cette option inédite permettra assurément de renforcer cette jeune juridiction afin de combler le manque de spécialisation des magistrats en droit administratif. Pour l'heure, l'économie générale de la tutelle reste marquée par des arrangements paternalistes avec le représentant de l'État sous la forme d'une assistance-conseil aux collectivités territoriales.

1070 Article 83, Loi n° 2019-015 portant code de l'organisation judiciaire au Togo.

1071 *Ibid.*, art. 73, al. 5.

ii) La transformation du contrôle *a posteriori* en une activité d'assistance-conseil *a priori*

460. - Le législateur consacre dans les pays de la présente étude une double tutelle. Au sens de la loi n° 2019-006 au Togo, de l'ordonnance n° 2010-54 au Niger et de la loi n° 2017-051 au Mali, la tutelle de l'État sur les actes des collectivités territoriales est instituée sous la double forme du « contrôle de légalité » et d'une « assistance-conseil »[1072]. La loi n° 2009-65 modifiant le code général des collectivités territoriales au Burkina Faso dispose également que l'État a envers les collectivités territoriales un devoir d'assistance qui s'exerce notamment sous la forme d'une mise à disposition d'un appui technique et financier[1073]. L'institution de l'assistance-conseil aux collectivités territoriales semble relever d'une part, d'une obligation constitutionnelle dans la mesure où certaines constitutions africaines font peser sur l'État l'obligation de « *veiller au développement harmonieux de toutes les collectivités territoriales sur la base de la solidarité nationale, des potentialités régionales et de l'équilibre inter-régional* »[1074]. Il est possible de déduire de cette disposition la volonté des constituants de consacrer un droit à la solidarité en faveur des collectivités territoriales[1075] qui se traduirait sous la forme de l'assistance-conseil en matière de décentralisation. D'autre part, l'accroissement des domaines de compétences des collectivités territoriales dans la conception et la mise en œuvre des actions de développement local semble être aussi l'une des principales justifications de la mise en place d'une fonction d'assistance et de conseil au profit des collectivités territoriales africaines. Cependant, l'assistance-conseil est source d'ambiguïté en ce qu'elle contribue à la multiplication des arrangements entre l'autorité de tutelle et les autorités décentralisées plutôt que d'aller vers un traitement juridictionnel du contentieux des actes locaux. Le déroulement du contrôle est condamné à demeurer à la phase administrative sans que le juge puisse se prononcer. L'on serait tenté de voir dans cette affirmation un gout prononcé pour le contentieux. Mais, en réalité, il s'agit non seulement de permettre au juge

---

1072 *Cf.* art. 76 Loi n° 2019-006 *préc.*, art. 305 CGCT au Niger ; art. 303 du code des collectivités territoriales au Mali.

1073 Art. 49 CGCT au Burkina Faso.

1074 Art. 142 de la Constitution togolaise de 1992 modifié ; art. 165 de Constitution du Niger de 2010.

1075 A. HAMADOU, « Le principe de libre administration des collectivités territoriales au Niger », *Annales africaines*, 2020, p. 395.

de dire le droit, mais aussi d'assurer une meilleure protection de l'autonomie des collectivités territoriales.

461. - Assurément, la fonction de conseil et d'assistance aux collectivités territoriales répond à des besoins réels. Les expériences actuelles de gestion des collectivités démontrent que ces dernières ne disposent pas de compétences suffisantes pour exercer effectivement et valablement les missions qui leur sont dévolues par la loi. Dans un tel contexte, une dose d'appui-conseil et d'assistance aux entités décentralisées parait même indispensable afin de les accompagner techniquement dans l'exercice de leurs compétences[1076]. Le principe étant que, mieux les fonctions de conseil seront assurées, moins le contrôle aura des motifs de sanctions. La fonction de l'État au regard des collectivités territoriales n'est pas seulement de contrôler et de sanctionner, mais aussi d'accompagner[1077]. Il s'agit pour le législateur d'anticiper les insuffisances qui pourront se révéler à l'occasion de l'application des textes qui sont de plus en plus nombreux avec un enchevêtrement des dispositions législatives et règlementaires. L'encadrement *a priori* de l'action des collectivités a en ce sens une vertu préventive. L'assistance et le conseil offrent ainsi dans l'esprit du législateur le cadre propice pour la concertation et la coordination des actions publiques nécessaires pour donner sens et contenu à la solidarité nationale[1078]. Elle vise également à travers la tutelle technique à répondre à un objectif de sécurité juridique pour les collectivités.

462. - Cependant, malgré l'importance que semble lui porter le législateur, les législations des pays de l'Afrique de l'Ouest francophones restent très laconiques sur le contenu exact de cette fonction qu'il confie au représentant de l'État. Le ministère de la Décentralisation et de la Ville au Mali estimait, dans son guide pratique d'exercice de la tutelle de l'État sur les collectivités territoriales élaboré en 2012, que « *l'assistance-conseil encore appelée appui-conseil est une approche hétéroclite de la tutelle administrative classique, sans contenu* précis *pour le moment ni moyen de censure, contrairement au contrôle de légalité assorti de modalités* préc*ises* »[1079]. Ainsi, à la différence du

---

1076 *Ibid.*, p. 396.

1077 *Cf.* Exposé des motifs du projet d'ordonnance portant code général des collectivités territoriales de la République du Niger.

1078 C. NANAKO, *La libre administration des collectivités territoriales au Bénin et au Niger, op.cit.*, p. 160.

1079 *Guide pratique d'exercice de la tutelle de l'État sur les collectivités territoriales*, 2012, p. 13.

contrôle de légalité dont les modalités sont amplement détaillées, seuls un ou deux articles au maximum sont expressément consacrés à la fonction de conseil et d'assistance dans les textes de loi. C'est d'ailleurs ce qui fait dire à Cossaba Nanako que le législateur lui-même cerne mal l'utilité et l'objet réel de cette fonction[1080]. Il faut toutefois reconnaitre au législateur malien de 2017 le mérite d'avoir donné un contenu à l'appui-conseil. Au sens de l'article 303 de la loi n° 2017-051 du 2 octobre 2017, « *l'appui-conseil consiste notamment aux conseils, avis, suggestions et informations fournies par les représentants de l'État aux collectivités territoriales dans l'exercice de leurs compétences* ». Cet appui-conseil, renchérit l'article 304, « *est donné à la demande de la collectivité territoriale. Il peut être suscité par le représentant de l'État. Les avis, conseils et suggestions donnés par le représentant de l'État à la collectivité territoriale ont un caractère consultatif* ».

À la lecture de ces dispositions, le législateur n'a pas entendu donner force obligatoire aux avis, conseils et suggestions données par le représentant à la collectivité. Toutefois, la libre administration des collectivités n'en sort pas renforcée. Dans un premier temps, l'institutionnalisation d'une assistance à l'élaboration des actes locaux même voulue par la collectivité territoriale contrevient au sens et à la portée du principe de libre administration des collectivités territoriales[1081]. Dans un second temps, la loi prévoit que l'appui-conseil peut être suscité d'office par le représentant de l'État sans que la collectivité concernée ne l'y invite. L'autorité de tutelle peut alors s'immiscer dans le processus de réflexion et de prise de décision de la collectivité territoriale en méconnaissance de l'autonomie décisionnelle dont elle bénéficie. En outre, comme le rappelle le professeur Bertrand Seiller, « *en conseillant l'autorité administrative fait plus que renseigner l'intéressé, elle donne un avis par lequel elle cherche à orienter l'action de celui-ci. Elle prend alors parti sur ce qu'il est possible de faire en lui indiquant ce qui lui semble préférable. Reposant sur une analyse subjective de la situation, le conseil confère un rôle plus actif à la personne publique* »[1082]. Le conseil délivré sous le sceau de l'expertise juridique détient la valeur d'une recommandation impérative. Ce rôle de conseil conforte la tendance au rapprochement de la tutelle administrative à la tutelle en droit civil. Les collectivités territoriales sont des majeurs incapables qu'il faut assister au

---

1080 C. NANAKO, *La libre administration des collectivités territoriales au Bénin et au Niger, op.cit.*, p. 160.

1081 Voir en ce sens, P. CADIEU, « le "résistible déclin" du contrôle de légalité », *RLCT*, mai 2008, p. 5

1082 B. SEILLER, « Les conseils donnés par les personnes publiques », *Dr. adm.*, n° 11, 2005, Étude 18.

quotidien pour les actes de la vie administrative[1083]. En lui ajoutant la fonction d'assistance et de conseil, le représentant de l'État voit son autorité renforcée par la mise à sa disposition d'un pouvoir d'influence considérable sur toutes les décisions locales.

463. - À la recherche des illusions perdues à propos du contrôle de légalité en France, Jean-François Brisson livre une analyse que l'on peut transposer au contexte africain. Pour ce dernier en effet, la conjonction des fonctions de contrôle et de conseil est une source supplémentaire d'ambiguïté aussi bien pour les collectivités que pour l'État. Cette ambiguïté s'explique d'une part, par « *le risque d'une instrumentalisation par les administrations d'État de la fonction de conseil au service de la fonction de contrôle administratif. Dans ces conditions, la tentation des services de l'État, associés par le biais du conseil à la prise de décision, peut être de jouer de leur latitude dans l'interprétation des textes pour exercer en amont une "régulation" plus politique que juridique, largement fondée sur les objectifs et préoccupations propres à l'État, quitte à rogner sur les marges d'appréciation laissées par la loi aux collectivités locales* »[1084]. L'expertise juridique détenue par les services déconcentrés de l'État renforce la tutelle étatique au point de faire de la consécration de l'autonomie des collectivités territoriales une arlésienne. D'autre part, l'activité de conseil peut créer « *une situation de connivence incompatible avec l'idée de contrôle qui suppose pour le moins la "mise à distance" du contrôlé par rapport au contrôleur* »[1085]. En ce sens, la confusion des fonctions peut contribuer à fragiliser la mission du préfet en tant que gardien de l'intérêt général. Dans le cadre de sa fonction de conseil et d'assistance, les services déconcentrés de la préfecture interviennent en amont de la prise de décision et ce faisant on peut légitimement s'interroger sur la qualité de l'exercice *a posteriori* de sa fonction de contrôle. La phase juridictionnelle du contrôle perd aussi de son sens. Le préfet pourrait-il déférer au juge un acte pour lequel il avait préalablement donné un avis favorable ? La réponse n'est pas évidente. Il

---

[1083] Au sens civiliste, la tutelle exprime la prise en main, par une tierce personne, des décisions notamment juridique et patrimoniale qu'un homme ou une femme ne peut pas prendre en raison de son état physique ou mental, ou de sa condition de minorité légale.
[1084] J.-F. BRISSON, « Le contrôle de légalité dans la loi du 13 aout 2004. A la recherche des illusions perdues », *AJDA,* 2005, p. 126 et s.
[1085] *Idem.*

est nécessaire qu'un texte règlementaire intervienne pour préciser les modalités d'exercice de cette fonction d'assistance et de conseil[1086].

464. - *In fine*, on peut nier l'importance d'une fonction d'assistance et de conseil au profit des collectivités territoriales qui sont en manque d'expertise. La peur de voir l'opérationnalisation de la décentralisation conduire davantage à l'engagement de la responsabilité des élus locaux va susciter quasi systématiquement le recours aux services déconcentrés quand bien même cette solution n'est pas véritablement compatible avec le principe de libre administration. Il est impérieux de développer une alternative fiable à l'offre de conseils par les services déconcentrés. L'État devrait davantage investir pour renforcer les capacités d'expertise des collectivités. La mutualisation des services entre plusieurs communes s'avère en ce sens être une approche intéressante. On pourrait aussi imaginer que les associations nationales des communes se dotent d'un service efficace capable de répondre au besoin des communes qui sont en manque de moyens.

465. - La tutelle de l'État sur les actes des collectivités territoriales s'accorde mal avec l'autonomie des collectivités territoriales. On n'est dans une situation où la phase juridictionnelle du contrôle n'est pas privilégiée en raison de l'absence d'une véritable juridiction administrative et de l'institutionnalisation de la fonction d'assistance et de conseil. Cette forme de tutelle prévue par la loi est, outre son manque de précision, particulièrement handicapante pour la liberté de décision et d'action des collectivités territoriales. Le contrôle *a posteriori* considéré comme relativement respectueux de l'autonomie décisionnelle des collectivités est davantage amoindri par la persistance d'une forte tutelle *a priori*. Le représentant de l'État dispose encore de procédés de tutelle qui renforce l'emprise de l'État sur les collectivités territoriales.

2. Les pouvoirs étendus du représentant de l'État

466. - Le représentant de l'État dispose d'une large gamme de procédés qui lui permettent de s'immiscer dans l'élaboration des actes des collectivités territoriales. Le changement opéré dans la nature du contrôle

---

1086 Pour une étude approfondie en droit français sur une procédure proche : le rescrit préfectoral, L. JANICOT, « La demande de prise de position formelle - L'extension du rescrit au profit des collectivités territoriales », *RFDA*, 2020, p. 254.

par les récentes lois précitées sur la décentralisation dans les États ouest-africains a semblé améliorer l'autonomie des collectivités territoriales. L'autorité de tutelle perd en principe son pouvoir d'annulation des actes au profit du juge administratif placé au centre de la procédure de contrôle *a posteriori*. Il faut néanmoins relever que le contrôle *a posteriori* connait quelques exceptions. Non seulement, le pouvoir du juge de dire le droit est devenu accessoire pour les raisons précédemment décrites, mais aussi la révolution n'a pas eu lieu en matière de contrôle. L'autorité de tutelle exerce toujours un contrôle d'opportunité qui lui permet de mettre son veto aux actes des collectivités territoriales (**a**). Aussi, son contrôle peut prendre une forme positive qui lui permet d'obliger la collectivité dans certaines hypothèses à agir et, le cas échéant, de se substituer à elle (**b**).

a) La conservation d'une tutelle a priori sur les actes des collectivités territoriales

467. - Malgré la tendance générale à la consécration d'un contrôle *a posteriori* des actes des collectivités territoriales, certaines formes de la tutelle *a priori* sont maintenues dans le droit de décentralisation des États d'Afrique subsaharienne francophone. Ces formes de tutelles sont celles qui concentrent les privilèges les plus inconvenants du pouvoir central. Elles confortent l'idée d'une amputation de l'autonomie aux collectivités territoriales. En effet, les délibérations et les actes des collectivités territoriales sont encore soumis au pouvoir d'annulation direct du représentant de l'État (**i**) et certains font, en plus, l'objet d'un contrôle d'opportunité avant de pouvoir devenir exécutoires (**ii**).

i) Le maintien d'un pouvoir d'annulation des décisions locales

468. - L'annulation est une arme mise à la disposition du pouvoir central qui lui permet d'atteindre les délibérations de l'organe délibérant de la collectivité[1087]. Le législateur burkinabè de 2004 et le législateur nigérien de 2010 distinguent sous le régime de l'annulation deux réalités différentes : la nullité de plein droit qui peut frapper à tout moment une délibération et la simple annulabilité qui doit intervenir dans un délai limité de quarante-cinq jours au Burkina Faso et de deux mois au Niger. Les hypothèses dans lesquelles une délibération encourt l'annulation ainsi que le régime juridique qui s'y applique correspondent à la situation qui existait en France avant la réforme introduite par la loi du 2 mars 1982.

[1087] G. MELLERAY, *La tutelle de l'État sur les communes, op.cit.*, p. 167.

469. - Ainsi, selon les alinéas 2 et 3 de l'article 318 du CGCT au Niger, « sont nulles de plein droit les délibérations portant sur un objet étranger aux attributions des organes délibérants des collectivités territoriales ou prises en violation de la législation et la règlementation en vigueur. La nullité de droit est constatée par l'autorité de tutelle. Elle peut être prononcée à tout moment d'office ou à la demande des parties intéressées ». Dans le même sens, l'alinéa 1 de l'article 61 du CGCT au Burkina Faso précise que « sont nuls de plein droit toutes décisions, toutes proclamations et adresses, tous avis qui sortent des attributions des autorités locales, ceux qui sont contraires aux lois ou règlements et ceux qui sont pris par des organes illégalement réunis ou constitués. Cette nullité peut être invoquée ou opposée à tout moment par les parties intéressées. La nullité est constatée par l'autorité de tutelle. Notification est faite par l'autorité de tutelle au président du conseil qui en informe le conseil à sa plus prochaine session ». À côté de ces hypothèses de déclaration de la nullité de plein droit, la loi prévoit également les cas de l'annulation simple. Cette dernière peut intervenir au terme de l'article 319 du CGCT du Niger lorsqu'a pris part à la délibération « un conseiller intéressé soit à titre personnel, soit comme mandataire, ou comme conjoint, ascendant ou descendant direct, à l'affaire qui a fait l'objet de la délibération ». Pour le législateur burkinabè, « l'autorité de tutelle peut suspendre ou annuler soit d'office, soit à la demande de toute personne ayant intérêt, les délibérations et décisions présentant un intérêt personnel pour l'une ou l'autre des autorités ayant participé à la décision ou au règlement, en personne, ou par mandataire »[1088].

Il ressort de l'ensemble de ces dispositions que la constatation de la nullité de droit par le représentant de l'État peut résulter à la fois des conditions dans lesquelles est intervenue la délibération que de l'objet même de cette dernière. Ainsi, la nullité de droit peut être constatée soit parce que la délibération a été adoptée dans un domaine étranger aux attributions de l'organe délibérant de la collectivité, soit parce qu'elle est contraire à une loi ou à un règlement en vigueur. Le droit positif précise les objets sur lesquels le conseil de la collectivité territoriale peut délibérer. Comme nous l'avons montré précédemment, le législateur africain ne reconnait pas aux collectivités territoriales une clause de compétence générale. Leurs actions doivent s'inscrire uniquement dans le champ de compétences que la loi leur attribue expressément. Par ailleurs, pour

---

[1088] Art. 63 du CGCT au Burkina Faso.

qu'une réunion du conseil de la collectivité territoriale soit considérée comme légale, le conseil doit être régulièrement convoqué. L'article 62 du CGCT du Burkina Faso précise à cet effet que, « *lorsque le conseil de la collectivité délibère hors d'une session régulièrement convoquée ou lorsqu'il est illégalement constitué, l'autorité de tutelle prend toutes mesures nécessaires pour mettre immédiatement fin à ladite session. Le cas échéant, l'autorité de tutelle constate l'illégalité des actes découlant de cette session* ». La séance du conseil pour être régulière doit être publique, en dehors des cas limitativement énumérés par la loi où il est autorisé à se réunir à huis clos[1089]. La convocation du conseil doit également se faire dans le respect des exigences de forme et de délai. Une fois régulièrement convoqué, le conseil de la collectivité doit pouvoir délibérer effectivement sur toutes les affaires qui lui sont soumises. Cela suppose donc de pouvoir garantir les conditions d'une libre expression des élus pendant le débat et la délibération[1090].

470. - L'annulation simple est quant à elle prononcée lorsqu'un conseiller intéressé a participé à la délibération. Les conditions de cette annulation peuvent l'objet de diverses ap*pré*ciations. On pourrait se demander à partir de quel moment un conseiller peut être considéré comme ayant un intérêt à l'adoption d'une délibération et aussi, dans quelle mesure sa participation entache de nullité la délibération. À défaut de réponse dans les textes et la jurisprudence des pays concernés, on peut supposer qu'il est possible de rechercher l'annulation d'une délibération dans laquelle un élu à un intérêt pécuniaire ou même moral. Pour conduire à l'annulation, la participation de l'élu doit avoir été de nature à influencer les résultats de la délibération[1091]. L'annulation des délibérations de l'organe délibérant de la collectivité permet au représentant de l'État de vérifier qu'elles respectent le champ de compétences qui leur est attribué et que l'intérêt public local n'est pas corrompu par l'octroi d'avantages particuliers aux élus.

471. - La compétence pour l'annulation des délibérations du conseil de la collectivité territoriale appartient au représentant de l'État, plus proche des collectivités. Il peut prononcer la nullité soit d'office soit à la demande de personnes intéressées. Si la nullité de droit peut être déclarée à tout moment, l'annulabilité ne peut intervenir que dans le respect des

---

1089 Art. 11, 162 & 242 du CGCT au Burkina Faso ; art. 15, 186 du CGCT au Niger.
1090 Sur toutes ces questions, voir les développements *infra*, n° 573 et s.
1091 G. MELLERAY, *La tutelle de l'État sur les communes, op.cit.*, p. 167 et s.

conditions de délai fixées par le législateur. L'autorité de tutelle dispose d'un délai de quarante-cinq jours au Burkina Faso[1092] et de deux mois au Niger[1093] pour prononcer l'annulation d'office de la délibération. Dans l'hypothèse où l'annulation est demandée par un tiers, elle « *doit être déposée sous peine de forclusion dans un délai de quarante-cinq jours à partir de la date de publication ou de notification de l'acte. L'autorité de tutelle statue dans le délai d'un mois* »[1094]. Ce délai passe à trente jours suivant la clôture de la session concernée au Niger[1095]. La décision d'annulation du représentant de l'État doit être motivée[1096] et, en cas de contestation, « *le recours gracieux devant l'autorité de tutelle est obligatoire avant l'exercice des recours pour excès de pouvoir devant les juridictions* »[1097].

472. - L'annulation des délibérations par le représentantde l'État confère au pouvoir central un privilège exorbitant, celui de pouvoir prononcer directement la nullité des décisions locales qu'il estime contraire aux dispositions législatives et règlementaires en vigueur. Cette prérogative de l'autorité de tutelle est inadaptée aux exigences de la nouvelle décentralisation. L'annulation continue d'être un pouvoir spécial à la disposition de l'État tout comme la persistance d'un régime d'approbation préalable des délibérations.

ii) La persistance d'un régime d'approbation des décisions locales

473. - La tutelle a priori n'a pas disparu. Le régime de l'entrée en vigueur des actes des collectivités territoriales est encore conditionné par leur approbation expresse ou tacite voire dans certains cas par leur autorisation préalable. Le dispositif de contrôle mis en place dans les nouvelles législations africaines ne procède pas à une suppression radicale du contrôle *a priori* comme cela a été le cas en France avec la loi du 2 mars 1982. Simplement, les législations africaines ont entendu réduire le champ du contrôle *a priori* sans pour autant y mettre fin. La tutelle *a priori* continue ainsi de concerner une catégorie d'actes importants de la vie juridique locale. Ce qui conduit à nuancer l'étendue de la liberté d'action reconnue aux collectivités territoriales. Le pouvoir central nigérien s'en défend dans

---

1092 Art. 65 du CGCT au Burkina Faso.
1093 Art. 319 al. 2 du CGCT au Niger.
1094 Art. 64 du CGCT au Burkina Faso.
1095 Art. 319 al. 2 du CGCT au Niger.
1096 *Idem.*
1097 Art. 61 al. 2 du CGCT au Burkina Faso.

l'exposé des motifs de son projet d'ordonnance portant code général des collectivités territoriales en affirmant que, « *l'objectif* [de la tutelle *a priori*] *n'est nullement de limiter ou d'entraver l'autonomie des collectivités territoriales, mais plutôt de s'inscrire dans une approche pédagogique, notamment dans un contexte de début de réforme et d'apprentissage collectif, et surtout dans un contexte où ni l'infrastructure judiciaire ou encore moins l'expertise juridique n'est encore disponible au niveau local* »[1098]. Cette affirmation confirme à nouveau l'idée que les collectivités territoriales sont considérées comme des mineures et que leur accorder une liberté totale d'action pouvait entrainer des risques tant pour l'État que pour elles-mêmes[1099].

474. - Les dispositions de l'article 316 du CGCT au Niger sont très claires à cet effet : « par dérogation au caractère exécutoire des actes prévus à l'article 312 du présent code, restent soumises à l'approbation préalable du représentant de l'État les délibérations portant sur : les outils d'aménagement du territoire et d'urbanisme ; les opérations domaniales et foncières et toute intervention impliquant la cession de biens ou de ressources de la collectivité ; les conventions, dons, legs et subventions assortis de conditions ; les emprunts et garanties d'emprunts ; les actions de coopération entre collectivités ». Au Mali, ce sont les articles 23 pour les communes, 96 pour les cercles et 164 pour les régions qui précisent que les délibérations sur les matières ci-après ne deviennent exécutoires qu'après l'approbation du représentant de l'État. Il s'agit des budgets et du compte administratif ; de l'aliénation des biens du patrimoine et des emprunts de plus d'un an[1100]. Le pouvoir d'approbation est aussi aménagé au Togo en ce qui concerne les actes à caractère financier[1101] ainsi que pour les conventions de coopération entre les collectivités[1102]. En plus de l'approbation, la loi n° 055-2004/AN du 21 décembre 2004 portant code général des collectivités territoriales au Burkina Faso prévoit un régime d'autorisation préalable. Au sens de l'article 57 de ladite loi, « l'approbation et l'autorisation préalable ne concernent que les actes et les délibérations à caractère financier ou ayant une incidence financière, conformément aux

---

1098 Exposé des motifs du projet d'ordonnance portant CGCT au Niger, p. 24.

1099 Voir l'exposé des motifs précité : « ces actes sont généralement pris dans des domaines qui, en dépit du transfert des compétences, impliquent l'État dans leur exécution ou dans la prise en charge des conséquences de cette exécution ».

1100 *Cf.* code des collectivités territoriales au Mali.

1101 Loi n° 2019-006, art. 80, 354 & 376.

1102 *Ibid.*, art. 43.

dispositions des articles 150 pour ce qui concerne la région, et 229 pour ce qui concerne la commune ».

475. - L'analyse rapprochée des dispositions qui précèdent permet de relever que les domaines dans lesquels l'approbation est maintenue concernent majoritairement sur les actes et les délibérations de nature financière ainsi que les opérations immobilières des collectivités territoriales soit les domaines les plus importants de la vie locale. La véracité de la maxime selon laquelle *« le contrôle a posteriori est la règle et le contrôle a priori l'exception »* devrait, remarque Ismaël Madior Fall être remise en cause[1103]. Pour cet auteur, si la maxime est quantitativement vraie, elle est qualitativement contestable. En effet, si du point de vue quantitatif, de nombreux actes sont exemptés de la tutelle *a priori* et exécutoire de plein droit dans les conditions que nous avons présentées *supra,* du point de vue qualitatif, les actes les plus déterminants pour la libre administration des collectivités territoriales demeurent encore soumis à l'approbation préalable de l'autorité de tutelle[1104].

476. - L'approbation préalable de l'autorité de tutelle doit intervenir dans un certain délai à compter du dépôt de la délibération. Ce délai est de quarante-cinq jours maximum selon l'article 58 du code général des collectivités territoriales au Burkina Faso et de trente jours au Mali[1105] et au Togo[1106]. Les délibérations soumises à approbation ont force exécutoire à l'expiration de ce délai à dater du récépissé délivré par le représentant de l'État ou du dépôt desdites délibérations constatées par tout autre moyen. Si l'autorité saisie à fin d'approbation n'a pas fait connaitre sa décision dans ce délai, l'approbation est considérée comme acquise au bout de trente jours[1107]. Il faut noter que le législateur nigérien est resté muet sur le délai de l'approbation. On peut alors supposer que s'applique en la matière le même délai que celui des actes soumis à transmission soit deux mois. L'autorité de tutelle n'approuve la délibération que s'il considère qu'elle remplit les conditions de validité prévues par la loi. À défaut, il le notifie à

1103 I. M. FALL, « Le contrôle de légalité des actes des collectivités locales au Sénégal », *op.cit.*, p. 86.

1104 *Ibid.*, p. 87.

1105 Art. 294 du code des collectivités territoriales au Mali.

1106 Art. 354 de la loi n° 2019-006 relatif à la décentralisation et aux libertés locales au Togo.

1107 *Idem* ; art. 58, 152 & 231 du CGCT au Burkina-Faso ; art. 294 *préc.*, au Mali.

la collectivité territoriale en lui communiquant ses observations que la collectivité territoriale est tenue de prendre en compte[1108].

477. - Cependant, il se pose la question de savoir quelle attitude devra tenir la collectivité en cas de refus d'approbation ou si elle refuse de suivre les observations de l'autorité de tutelle. Ce sont les codes maliens et burkinabè qui offrent ici clairement une réponse. Ils organisent expressément en effet, la possibilité pour les collectivités territoriales de contester devant la juridiction administrative le refus d'approbation. C'est ce que précise l'alinéa 3 de l'article 294 du code des collectivités territoriales au Mali, « *la collectivité territoriale peut déférer au tribunal administratif la décision de refus d'approbation* ». La saisine du tribunal doit intervenir dans le même délai que le déféré préfectoral, soit deux mois. Au Burkina Faso, la contestation est soumise à un recours préalable devant l'autorité de tutelle. L'article 59 du CGCT prévoit en ce sens que, « *lorsque l'autorité de tutelle refuse l'approbation ou l'autorisation, le conseil de la collectivité peut se pourvoir en justice, à la condition d'avoir saisi l'autorité de tutelle au préalable d'un recours gracieux. Dans ce cas de saisine, l'autorité de tutelle dispose d'un délai de trente jours au maximum pour réagir* ». Ce délai supplémentaire suspend le délai de recours et permet à l'autorité de tutelle de procéder à une conciliation des intérêts.

478. - C'est donc par ce contrôle préventif de l'approbation préalable que l'autorité de tutelle s'immisce dans le fonctionnement des collectivités territoriales en appréciant l'opportunité des délibérations adoptées par ces dernières. Le pouvoir central s'érige par la sorte en censeur de ce qui est bon ou mauvais pour les collectivités territoriales. Cette procédure de contrôle est particulièrement lourde en ce qu'elle retarde l'action des collectivités. Elle apparait manifestement contraire au principe de la libre administration par des conseils élus. L'augmentation du domaine du contrôle *a posteriori* des actes locaux est peut-être conséquente, mais la liberté d'initiative des collectivités territoriales en sort affaiblit. Il s'agit pour le dire plus simplement d'une réforme en trompe-l'œil qui permet de maintenir des formes de tutelle énergique comme le pouvoir de substitution.

[1108] Voir en ce sens l'alinéa 2 de l'art. 294 *préc.*

b) L'exercice du pouvoir de substitution par le représentant de l'État

479. - L'autorité de tutelle ne dispose pas seulement de la faculté d'empêcher les collectivités territoriales d'agir dans des domaines qu'il considère comme ne relevant pas de leurs attributions. En cas de négligence ou d'abstention d'agir des autorités décentralisées, le représentant de l'État dispose de la possibilité d'agir au lieu et place de la collectivité défaillante : c'est le pouvoir de substitution d'action[1109]. Il s'agit d'une prérogative redoutable pour la décentralisation en ce qu'elle permet au représentant de l'État d'agir dans un domaine de compétence attribué à l'autorité locale. Ce procédé ne peut que faire l'objet d'une utilisation parcimonieuse en raison de l'atteinte qu'il porte à l'autonomie des collectivités territoriales[1110]. L'existence du pouvoir de substitution d'action témoigne avec netteté de la présence de la centralisation dans la décentralisation[1111] rendant difficile la division entre pouvoir hiérarchique et pouvoir de tutelle[1112]. René Chapus voyait d'ailleurs dans la substitution est « *un pouvoir remarquable parce qu'il permet à l'autorité de tutelle de s'ingérer de façon particulièrement marquée dans les affaires de l'institution décentralisée, et en même temps de faire beaucoup plus que ce qui est permis au supérieur hiérarchique* »[1113]. La tutelle par substitution d'action est en ce sens, un pouvoir exceptionnel qui ne peut être mis en œuvre que si un texte le prévoit expressément et seulement après une mise en demeure de l'autorité décentralisée[1114]. La tutelle par substitution s'exerçant par défaut, l'autorité de tutelle doit impérativement s'assurer du refus ou de la négligence de l'autorité décentralisée à exercer une compétence personnelle afin de justifier son intervention.

1109 Voir en ces sens, B. PLESSIX, « Le pouvoir de substitution d'action », in P. COMBEAU (dir.), *Le contrôle de l'État sur les collectivités territoriales aujourd'hui*, L'Harmattan, 2007, p. 67 ; Du même auteur, « une prérogative de puissance publique méconnue : le pouvoir de substitution d'action », *RDP*, 2003, p. 579.

1110 C. NANAKO, *La libre administration des collectivités territoriales au Bénin et au Niger, op.cit.*, p. 145 ; G. MELLERAY, *La tutelle de l'État sur les communes, op.cit.*, p. 201.

1111 *Ibid.*, p. 202.

1112 Voir les développements *supra* n°434 et s.

1113 R. CHAPUS, *Droit administratif général, op.cit.*, p. 390.

1114 La doctrine y ajoute aussi deux conditions moins connues mais non moins importante en ce qu'elles apportent une précision supplémentaire : - l'autorité substituée doit être en situation de compétence liée ; - il doit avoir refus d'agir de la part de l'autorité substituée. v. B. PLESSIX, « Le pouvoir de substitution d'action », *op.cit.*, p. 72 et s.

480. - Le principe général selon lequel il n'y a pas tutelle sans texte s'applique avec une rigueur particulière dans le cadre de la substitution d'action. À cet effet, les textes relatifs à la décentralisation dans les États d'Afrique subsaharienne francophone reconnaissent au représentant de l'État le pouvoir de se substituer à l'autorité décentralisée dans l'exercice de ses responsabilités notamment en matière de police administrative et en matière budgétaire. L'article 66 du CGCT au Burkina Faso prévoit que, « *en cas d'inexécution par les autorités de la collectivité territoriale des mesures prescrites par les lois et règlements ou en vertu de ceux-ci, l'autorité de tutelle peut, après une mise en demeure, se substituer à celles-ci et prendre à cette fin toutes mesures utiles* ». Plus spécifiquement, en matière de police municipale, l'article 157 de la loi n° 2019-006 relative à la décentralisation et aux libertés locales au Togo prévoit que, « *les pouvoirs dévolus au maire en vertu des articles 154 et 155 de la présente loi* [ces articles renvoyant respectivement à la police du stationnement et à l'autorisation de bâtir] *ne font pas obstacle au droit du préfet de prendre, pour toutes les communes de son ressort territorial ou pour certaines d'entre elles et, dans le cas où il n'y a pas été pourvu par les autorités municipales, toutes mesures relatives au maintien de l'ordre, à la salubrité, à la sureté et à la tranquillité publique. Ce droit ne peut être exercé à l'égard d'une commune qu'après une mise en demeure adressée au maire, restée sans résultat* ». On retrouve des dispositions identiques au Niger. Le CGCT prévoit à l'article Article 83 que, « *en cas de troubles graves ou lorsque les circonstances l'exigent, le représentant de l'État peut se substituer à une ou plusieurs communes en vue d'assurer le maintien de l'ordre et de la tranquillité publics dans le strict respect de la loi* ». La demande de substitution par le préfet peut également intervenir à l'initiative des autorités décentralisées. C'est ainsi que l'article 84 du CGCT au Niger dispose que « *lorsque l'ordre public est menacé dans une ou plusieurs communes du même département, l'autorité de tutelle, saisie par le ou les maires concernés peut se substituer à ceux-ci pour exercer les pouvoirs de police administrative. Elle a le même pouvoir de substitution en cas de mise en demeure restée sans effet* ». Le législateur malien reste également dans la même dynamique en prévoyant qu'en cas de défaillance en matière de police administrative, « *le représentant de l'État doit, après mise en demeure restée sans effet, se substituer à celle-ci pour prendre les mesures de police nécessaires. Elle peut également intervenir sur la demande expresse de l'autorité décentralisée* »[1115].

481. - L'autorité de tutelle dispose de l'aptitude à se substituer au maire pour l'ensemble des mesures concernant le maintien ou le rétablissement

---

1115 Art. 300 du Code des collectivités territoriales au Mali.

de l'ordre public. On est ici dans les cas d'hypothèses prévus par le code général des collectivités territoriales en France à l'article L. 2215-1[1116]. Toutefois, à la différence de ce texte, la règle de la mise en demeure préalable est formellement énoncée dans les textes africains précités. Le préfet ne peut intervenir en lieu et place de l'autorité décentralisée sans avoir pris le soin de la mettre en demeure. Cette règle de mise en demeure préalable est considérée en droit français comme un principe général inhérent au droit de la tutelle[1117]. C'est ce qui justifie que malgré son absence dans l'énoncé de l'article 2215-1 CGCT, le juge administratif rappelle constamment le respect de cette exigence préalable au préfet[1118]. Si ces exigences sont respectées alors la responsabilité de la commune sera retenue en cas de dommage dû à l'exercice du pouvoir de substitution par le préfet. En revanche, l'abstention par l'État de l'exercice de son pouvoir de tutelle entraine la mise en jeu de sa responsabilité pour faute lourde[1119].

482. - Le pouvoir de substitution du préfet en matière administrative s'applique aussi en matière budgétaire. L'autorité de tutelle peut régler le budget d'une collectivité en cas de carence de l'organe délibérant ou d'incapacité de celui-ci à adopter un budget équilibré. La procédure comporte une navette entre les services du représentant de l'État et la collectivité territoriale afin de permettre à l'organe délibérant de rectifier lui-même son budget. Toutefois, si après une seconde lecture du budget, la nouvelle délibération ne procède pas « *aux corrections demandées ou si le budget n'est pas retourné dans le délai d'un mois à compter de son renvoi à l'ordonnateur, l'autorité de tutelle, après mise en demeure adressée à la collectivité restée*

1116 Art. L. 2215-1 al. 1 et 2 du CGCT en France : « Le représentant de l'État dans le département peut prendre, pour toutes les communes du département ou plusieurs d'entre elles, et dans tous les cas où il n'y aurait pas été pourvu par les autorités municipales, toutes mesures relatives au maintien de la salubrité, de la sureté et de la tranquillité publiques.
Si le maintien de l'ordre est menacé dans deux ou plusieurs communes limitrophes, le représentant de l'État dans le département peut se substituer, par arrêté motivé, aux maires de ces communes pour l'exercice des pouvoirs mentionnés aux 2° et 3° de l'article L. 2212-2 et à l'article L. 2213-23 ».

1117 B. FAURE, *Droit des collectivités territoriales, op.cit.*, n° 886.

1118 CE, 9 février 1995, Epoux Le Brigand, Lebon, p. 78 ; CE, 27 novembre 1974, Min. de l'Intérieur contre Bertranuc et a., Lebon, p. 584.

1119 Voir en ce sens la jurisprudence française, CE, 25 juillet 2007, *Sté France Télécom, Sté Axa Corporate solutions assurances, Lebon*, p. 707 ; *AJDA*, 2007, p. 1557.

infructueuse pendant quinze jours, apporte les modifications nécessaires et arrête le budget dans les quinze jours suivants »[1120].

483. - De la même façon, l'autorité de tutelle peut inscrire d'office les dépenses obligatoires lorsque le conseil de la collectivité territoriale n'a pas alloué les sommes suffisantes ou n'a pas prévu d'allocation correspondante à cet effet. Là encore, la substitution n'est possible qu'après une mise en demeure. En ce sens, l'article 356 de la loi n° 2019-006 au Togo prévoit que, « *lorsque l'autorité de tutelle constate qu'une dépense obligatoire n'a pas été inscrite au budget de la collectivité ou l'a été pour un crédit insuffisant, elle adresse une mise en demeure à la collectivité concernée. Si dans un délai de trente jours, la collectivité ne se conforme pas à la mise en demeure, l'autorité de tutelle inscrit d'office cette dépense au budget de la collectivité et fait injonction au conseil de l'accompagner de ressources nécessaires* ». La tutelle budgétaire du représentant de l'État s'exerce toutes les fois que ce dernier intervient à travers son pouvoir de substitution d'action pour régler le budget d'une collectivité territoriale : soit parce qu'il n'a pas été adopté dans les délais requis, soit parce qu'il n'est pas voté en équilibre, soit encore parce qu'une dépense obligatoire a été omise. Les pouvoirs de l'autorité de tutelle pour sanctionner l'inaction d'une collectivité en matière budgétaire sont ainsi considérablement étendus. La substitution d'action lui permet d'exercer les pouvoirs de la collectivité en matière budgétaire. Or le budget est l'instrument par excellence qui permet de traduire la politique locale en action concrète et réalisable.

484. - Aussi contestable qu'il puisse paraitre, en raison du dessaisissement temporaire des compétences décentralisées, le pouvoir de substitution d'action de l'autorité de tutelle permet d'éviter un déni d'administration. Si, après rappel de son obligation d'agir, la collectivité locale refuse de prendre les mesures nécessaires, c'est elle-même qui ne respecte pas sa propre compétence, et se met en situation illégale. Dès lors, l'autorité de tutelle peut agir pour corriger cette illégalité, avec une compétence liée quant au but à atteindre, moins liée quant aux moyens[1121]. Pour cela, il faut relativiser l'influence du procédé de substitution sur l'autonomie des collectivités territoriales. En refusant d'agir, les collectivités territoriales peuvent ainsi obliger l'autorité de tutelle à agir

---

1120 Art. 221 du CGCT au Niger.

1121 H. K. NOUBISSI, *Décentralisation et centralisation au Cameroun. La répartition des compétences entre l'État et les collectivités locales, op.cit.*, p. 334.

dans les cas où elles souhaitent faire porter la responsabilité politique d'une décision à l'État[1122]. Toutefois, il aurait été préférable de renforcer les garanties de la libre administration des collectivités territoriales en faisant intervenir un tiers : le juge des comptes. Malgré sa présence formelle dans les textes, le juge de la Cour des comptes n'est que très peu, voire jamais, à notre connaissance, sollicité pour participer de manière active à l'action budgétaire des collectivités territoriales. Son action se résume dans les rares cas à la vérification des atteintes aux deniers publics. L'autonomie reconnue aux collectivités territoriales est confrontée sans cesse aux interventions du pouvoir central dans le cadre du contrôle des actes. En outre, ainsi que nous allons voir, l'État dispose d'un pouvoir de sanction sur les élus pris individuellement ou collectivement en tant que membre d'un organe délibérant.

B- Le contrôle exercé par l'État sur les organes des collectivités territoriales

485. - Le pouvoir central dispose des moyens de réprimer les collectivités territoriales. Dans le contexte actuel marqué par une volonté de renforcement de la décentralisation, les mécanismes de sanctions prévues par le législateur à l'encontre des organes décentralisés peuvent avoir quelque chose de malséant. En effet, la tutelle sur les organes encore appelée la tutelle sur les personnes utilise des procédés qui dépassent le cadre du contrôle généralement admis sur les autorités décentralisées. La tutelle de l'État sur les organes des collectivités territoriales implique la compétence du pouvoir central pour l'exercice d'un pouvoir disciplinaire soit sur un agent soit sur un organe en fonction au sein d'une personne morale de droit public distinct de lui[1123]. Ce faisant, il ne correspond plus en réalité à un pouvoir de tutelle, mais à un pouvoir hiérarchique[1124] identique à celui qu'exerce l'autorité supérieure sur ses subordonnées dans la fonction publique. Le pouvoir de sanction sur les organes induit une subordination des autorités décentralisées au pouvoir central. Si la volonté de préserver une bonne administration des services publics locaux permet de justifier ce pouvoir de l'État, il faut néanmoins trouver l'équilibre entre la nécessaire préservation des prérogatives de l'État et la protection de l'autonomie des collectivités territoriales. Parmi les mécanismes de

---

1122 G. MELLERAY, *La tutelle de l'État sur les communes, op.cit.*, p. 215.
1123 G. VEDEL, P. DELVOLVE, *Droit administratif général, op.cit.*, p. 395.
1124 R. CHAPUS, *Droit administratif général, op.cit.*, n° 412 et s.

sanction des collectivités, on distingue traditionnellement celles qui concernent les élus individuellement et celles qui les touchent collectivement. Les développements qui vont suivre s'articuleront donc autour des sanctions individuelles (**1°**) et des sanctions collectives (**2°**).

1. Le pouvoir de sanction individuelle de l'État

486. - Les membres des organes délibérants des collectivités territoriales peuvent, individuellement, faire l'objet d'un contrôle qui sanctionne leur mauvaise volonté dans l'exercice de leur fonction d'élu ou leur négligence. Toutefois, le régime des sanctions individuelles n'est pas uniforme. Il varie en fonction de ceux qui sont frappés par les sanctions. Ainsi, le droit positif distingue deux catégories de sanctions : celle qui concerne l'ensemble des conseillers des collectivités territoriales, la démission d'office (**a**) et celle qui ne touche que les autorités exécutives c'est-à-dire la suspension et la révocation (**b**).

a) La démission d'office des conseillers des collectivités territoriales

487. - De toutes les hypothèses qui peuvent entrainer la démission d'office d'un élu local, il y en a une qui fait l'unanimité absolue dans le droit positif des pays de la présente étude : la démission d'office pour cause d'inéligibilité ou d'incompatibilité. À celle-ci, il faudrait ajouter au Mali les cas de démission d'office dus à « *l'application des décisions de justice, la perte de la capacité électorale, le refus de remplir une des fonctions qui lui sont dévolues par les textes en vigueur* »[1125] et au Niger, « *l'incapacité physique ou mentale dûment constatée, l'absence non motivée à trois sessions successives* » ainsi que « *la perte de la capacité électorale* »[1126]. L'autorité de tutelle intervient dans la procédure de démission d'office. Elle conserve en la matière un rôle important. Toutefois, des voies de recours sont prévues par les textes en cas de contestation.

488. - Doit être déclaré démissionnaire d'office, tout membre d'un conseil élu d'une collectivité territoriale qui, pour une cause quelconque, se trouve dans l'un des cas d'inéligibilités ou d'incompatibilités prévu par la loi[1127]. Cette inéligibilité ou incompatibilité intervenue postérieurement

---

1125 Art. 18, 91 & 159 du code des collectivités territoriales au Mali.
1126 Art. 41 du CGCT au Niger.
1127 Loi n° 2019-006 *préc.*, art. 12.

à l'élection est constatée par le ministre chargé des collectivités territoriales[1128] ou dans certains cas par le représentant de l'État dans la circonscription[1129]. L'autorité de tutelle se trouve dans ce cas dans une situation de compétence liée pour prononcer la démission d'office. Il se borne simplement, sans aucune appréciation personnelle, à tirer les conséquences d'un cas d'inéligibilité prévue par les textes de loi. Cette inéligibilité peut résulter de la perte de la capacité électorale à la suite d'une condamnation, le plus souvent pénale, devenue définitive. Il faut signaler à ce niveau que la loi nigérienne prévoit une révocation de plein droit de tout conseiller ayant fait l'objet d'une condamnation pour crime ou délit[1130]. Cette disposition montre, ainsi que l'écrit Bertrand Faure à propos du droit français, que sous l'expression « démission d'office », c'est un véritable pouvoir de révocation des conseillers qui est reconnue à l'autorité de tutelle[1131]. Le conseiller déclaré démissionnaire d'office peut intenter un recours devant la juridiction compétente[1132]. La même faculté est reconnue aux conseillers pris individuellement et à tout citoyen en cas de refus de l'autorité de tutelle de déclarer la démission d'office[1133].

489. - La démission d'office procède également du refus d'un conseiller d'exercer une des fonctions qui lui sont dévolues par la loi. C'est une hypothèse expressément prévue par le code des collectivités territoriales au Mali. Le problème est alors de déterminer quelles sont les obligations à la charge des conseillers élus qui peuvent être sanctionnées dans ce cas de figure. Les textes de loi font peser peu d'obligations directes à la charge des élus. En ce sens, la question de l'obligation d'assister aux réunions de l'assemblée délibérante fait l'objet d'une appréciation différente selon les pays. Au Burkina Faso, tout conseiller qui s'absente des sessions de l'assemblée délibérante « *de façon continue durant une année est considéré comme démissionnaire, sauf cas d'invalidité temporaire dûment établie. Le constat est effectué par l'autorité de tutelle rapprochée au regard des états de présence du secrétariat des sessions du conseil* ». Il en est de même au Niger où la démission d'office peut être prononcée en cas d'absence non motivé à trois sessions successives.

1128 *Idem* ; art. 41 du CGCT au Niger.

1129 C'est le cas du Burkina Faso et du Mali. Voir les articles 173 & 253 du CGCT au Burkina-Faso et les articles 19, 92 & 160 du code des collectivités territoriales au Mali.

1130 Art. 42 du CGCT au Niger.

1131 B. FAURE, *Droit des collectivités territoriales, op.cit.*, n° 108.

1132 Art. 19, 92 & 160 du code des collectivités territoriales au Mali ; art. 41 du CGCT au Niger ; Loi n° 2019-006 *préc.*, art. 12.

1133 *Idem.*

La loi n° 2019-006 au Togo prévoit en ces articles 112 pour les communes et 259 pour les régions que « *tout membre du conseil* [régional ou municipal] *qui, sans motif reconnu légitime par le conseil, a manqué à trois sessions ordinaires successives, peut, après avoir été admis à fournir ses explications, être suspendu pour trente jours par* [le gouverneur ou le préfet] *sur proposition du président du conseil* ». Les absences répétées et successives d'un conseiller constituent dans les deux premiers pays un cas de démission d'office alors qu'au Togo le conseiller ne peut que faire l'objet d'une suspension. Cette dernière entraine par voie de conséquence la perte de la jouissance des avantages due à la fonction, notamment les indemnités de sessions et de déplacement[1134]. Au Mali, ce cas qui est considéré comme un refus d'un conseiller de remplir une des fonctions qui lui sont dévolues par la loi a été supprimé en 2012 par la loi n° 2012-007 du 7 février 2012 portant code des collectivités territoriales au Mali. Sous le régime de la loi n° 95-034 du 12 avril 1995, une absence non motivée à plus de deux sessions au cours d'une année pouvait entrainer la démission d'office.

490. - L'imprécision de ce cas de figure de la démission d'office renforce l'emprise de l'État sur les collectivités territoriales. Alors que dans les hypothèses précédentes, l'autorité de tutelle est en situation de compétence liée, elle se retrouve ici en situation de déterminer, au regard de l'imprécision de la loi, ce qui constitue une obligation ou non à la charge d'un conseiller. Il faut signaler qu'en droit français, les absences successives aux séances de l'assemblée délibérante ne sont plus sanctionnées, car elles ne sont plus considérées comme un refus du conseiller de remplir une fonction dévolue par la loi. Il s'agit d'un droit qui appartient en propre au conseiller du fait de son élection[1135]. Aussi, lorsqu'un conseiller refuse d'accomplir une fonction qui lui dévolue par la loi, notamment de surveiller un bureau de vote, c'est au tribunal administratif qu'il revient de prononcer la démission d'office[1136]. La juridictionnalisation de la procédure offre une plus grande garantie aux élus en allégeant la tutelle de l'État. Son absence ainsi que l'insuffisance des garanties procédurales dans certains cas renforcent le pouvoir des autorités de tutelle à l'égard des autorités décentralisées.

[1134] Loi n° 2019-006 *préc.*, art. 111 pour les conseillers municipaux et art. 258 pour les conseillers régionaux.

[1135] CE, 6 novembre 1985, *Mairie de Viry-Châtillon*, *Lebon*, p. 520.

[1136] CGCT, art. 2121-5 pour les conseillers municipaux ; art. 3121-4 pour les conseillers départementaux ; art. 4132-2-1 pour les conseillers régionaux ; CE, 20 février 1985, *Behuret*, *Lebon*, p. 311.

b) La suspension et la révocation des autorités exécutives locales

491. - Hormis les hypothèses d'inéligibilités ou d'incompatibilités, de démission, de destitution ou de décès qui peuvent conduire à mettre fin au mandat en cours de l'exécutif local, ce dernier peut faire l'objet d'une suspension ou d'une révocation. La suspension ou la révocation des autorités exécutives locales prend la forme d'une sanction disciplinaire. La nature de ces sanctions renforce ainsi que nous l'avons montré précédemment l'ambiguïté autour de la notion de tutelle[1137]. Elle semble se doubler d'un contrôle hiérarchique exercé par l'autorité de tutelle sur l'exécutif local. L'indépendance de ce dernier en sort fortement limitée. L'existence du contrôle hiérarchique du préfet sur les maires s'explique par le dédoublement fonctionnel du maire qui agit à la fois comme autorité du pouvoir central et aussi en tant qu'autorité décentralisée. Toutefois, il faut relever que la compétence de l'autorité de tutelle va au-delà des seuls cas dans lesquels le maire agit en qualité d'agent de l'État. Sa compétence est étendue à toutes les fautes commises par le maire, peu importe à quel titre ce dernier a agi. Cette solution s'applique à l'ensemble des autorités exécutives, quel que soit le niveau de collectivité. L'autorité de tutelle dans les pays ouest-africains dispose d'une relative liberté dans le choix de la sanction. Cependant, l'exercice de cette liberté même encadrée au niveau procédural n'est que très peu tempéré par le contrôle du juge administratif au regard de la gravité de la sanction.

492. - À la différence du Mali et de la France ; le Niger, le Burkina Faso et le Togo apportent dans la loi des précisions sur les faits qui sont de nature à justifier la révocation des autorités exécutives locales. Seules les fautes d'une certaine gravité peuvent justifier la sanction. Sont considérées comme tels au sens de l'ordonnance n° 2010-54 portant code général des collectivités territoriales au Niger les faits ci-après : « *détournement des biens et/ou des deniers publics dûment constatés par les services compétents ; concussion et/ou corruption ; prêts irréguliers d'argent sur les fonds de la [collectivité] ; faux en écriture publique et usage de faux ; endettement de la* [collectivité] *résultant d'une faute de gestion ; refus de signer et/ou de transmettre à l'autorité de tutelle une délibération du conseil régional ; refus de convoquer et de réunir le conseil* [de la collectivité] *conformément aux textes en vigueur ; absence de plus de six (6) mois pour*

[1137] Voir *supra* n° 433 et s.

des raisons autres que celles relatives à l'intérêt de la [collectivité] *ou de santé ; condamnation pour des faits et actes punis par la loi, à l'exception des contraventions de simple police ou des délits d'imprudence, hormis les cas de délit de fuite concomitant*»[1138]. La loi togolaise ainsi que le code burkinabè prévoient la révocation des organes exécutifs pour les mêmes faits[1139]. L'énumération des cas dans lesquels peuvent intervenir ces mesures exceptionnelles par leur gravité constitue une garantie supplémentaire pour les élus et protège mieux leur indépendance.

493. - En cas de survenance de l'un de ces cas, la révocation du maire est prononcée par décret pris en conseil des ministres sur rapport du ministre chargé de la décentralisation[1140]. Dans un État de droit, un tel décret doit être motivé ainsi que le prévoit expressément la loi malienne[1141] et les autorités décentralisées doivent pouvoir en demander l'annulation devant la juridiction administrative. L'autorité décentralisée doit être avertie de la mesure prise en son encontre et être en mesure de s'expliquer en défense avant l'intervention de la sanction. Les codes maliens, nigériens et burkinabè, prévoient à cet effet que toute révocation d'une autorité décentralisée doit se faire dans le respect des droits de la défense. L'autorité décentralisée est ainsi invitée à fournir des explications soit de manière écrite[1142] ou orale[1143].

Aussi, l'autorité de tutelle peut prononcer la suspension des autorités exécutives. Elle dispose donc d'une liberté dans le choix de la sanction à infliger. La mesure de suspension est prononcée par arrêté du ministre chargé de la décentralisation ou par le représentant de l'État. Au sens des articles 62 et 134 de l'ordonnance n° 2010-054 précitée, la suspension peut intervenir « *sur saisine d'au moins un membre du conseil et après rapport du représentant de l'État par arrêté motivé du ministre en charge de la tutelle des collectivités territoriales pour une durée n'excédant pas un (1) mois* ». Ce délai est de trois mois

---

[1138] Art. 63 pour les communes et art. 136 pour les régions du CGCT au Niger.

[1139] *Cf.* Loi n° 2019-006 *préc.*, art. 135 pour les communes et art. 283 pour les régions ; CGCT Burkina-Faso : art. 184 pour les régions et art. 272 pour les communes.

[1140] Voir not., Loi n° 2019-006 *préc.*, art. 134 pour les communes et art. 282 pour les régions.

[1141] Loi n° 2017-051 du 02 octobre 2017 *préc.*, art. 57 pour les communes, articles 131 pour les cercles et 197 pour les régions.

[1142] *Cf.* art. 63 et 134 respectivement pour les communes et les régions du CGCT au Niger ; Loi n° 2017-051 du 02 octobre 2017 *préc.*, art. 55, 128, 197 respectivement pour les communes, les cercles et les régions.

[1143] Art. 191 & 271 respectivement pour les communes et les régions du CGCT au Burkina-Faso.

au Burkina Faso[1144] et au Mali[1145]. À l'expiration du délai, le maire ou le président du conseil reprend ses fonctions. Les deux procédés de sanction : la suspension et la révocation sont complètement indépendantes. Une mesure de révocation peut être prononcée sans une suspension préalable. En l'état actuel, le droit de la décentralisation togolais ne garantit pas à l'élu mis en cause le respect du principe des droits de la défense. Le législateur togolais ne prévoit pas non plus la possibilité de suspension des autorités décentralisées.

494. - Cependant, à l'instar des autres pays, l'autorité décentralisée dispose d'une voie de recours pour faire annuler la sanction. C'est le sens de l'article 284 qui dispose que, « *toute décision portant* [...] *révocation est susceptible de recours devant la juridiction compétente* ». C'est seulement en cas de contestation que le juge sera amené à se prononcer sur la régularité de la révocation. En l'absence d'un contentieux administratif, on peut toujours s'interroger sur l'étendue du contrôle du juge. L'interrogation est d'autant plus pertinente quand on sait que l'administration publique en Afrique est marquée par une politisation à outrance[1146]. L'absence de garanties formelles couplée à celle d'une pratique administrative dont les règles et les modalités de gestion ne sont pas toujours bien rodées conduit inévitablement vers des sanctions prononcées sur des motifs politiques. En droit français, le juge administratif exerce un contrôle normal sur l'ap*pré*ciation de la faute de nature à justifier la sanction et un contrôle poussé de proportionnalité sur l'adéquation entre la faute et la sanction prononcée[1147]. Ce type de contrôle prend mieux en compte le fait que le maire est une autorité démocratiquement élue, réalité avec laquelle le pouvoir central doit concilier l'exercice de son contrôle[1148].

Dans ces circonstances, le fait pour le Conseil constitutionnel français d'avoir jugé à l'occasion d'une QPC que la possibilité de révoquer un maire est conforme à la Constitution, et notamment au principe de la libre administration parait moins choquant[1149]. L'autorité exécutive dispose de

---

1144 *Idem.*

1145 Loi n° 2017-051 du 02 octobre 2017 *préc.*, art. 57, 130, 196 respectivement pour les communes, les cercles et les régions.

1146 A. KONTCHOU KOUOMEGNI, « Administration publique et politique en Afrique francophone (Nord et Sud Du Sahara) », *Présence Africaine*, n°127/128, 1983, pp. 261 ; M. BAH, *La politisation des administrations publiques en Afrique,* L'Harmattan, 2021.

1147 CE, 22 mars 2010, *Dalongeville*, *AJDA*, 2010, p. 664 ; note C.-A. Dubreuil.

1148 B. FAURE, *Droit des collectivités territoriales, op.cit.*, n° 324.

1149 Déc. n° 2011-210 QPC du 13 janvier 2012, *M. Ahmed S. Lebon*, p. 78.

garanties suffisantes permettant de protéger efficacement son indépendance. On ne peut pas en dire autant des collectivités territoriales africaines. La rareté voire l'absence de contrôle du juge se conjugue avec l'absence de garanties procédurales suffisantes pour accroitre le pouvoir de l'État sur les élus locaux. La mise en œuvre du régime disciplinaire peut très vite devenir une occasion pour un règlement de comptes politiques. De plus, l'autorité de tutelle dispose du pouvoir de dissolution de l'assemblée délibérante.

2. Le pouvoir de sanction collective de l'État

495. - Les assemblées délibérantes peuvent être dissoutes par le pouvoir central. Il s'agit de la sanction la plus grave qui frappe collectivement l'ensemble des élus locaux en mettant fin, avant le terme normal, à leur mandat électif. La dissolution, faisait remarquer Guy Melleray, reste en la forme un procédé de centralisation administrative[1150]. Elle procède d'une décision de l'État. Une décision grave écrit Bertrand Faure au regard du principe d'indépendance des collectivités vis-à-vis de l'État que manifeste l'élection des autorités locales au suffrage universel[1151]. En Afrique, certains États notamment, le Niger, le Mali ont tenté d'améliorer le régime de la dissolution en prévoyant en amont une possibilité de suspension préalable. Toutefois, comme c'est le cas en droit français, la volonté d'« humanisation » de la tutelle sur les collectivités n'a pas remis en cause le pouvoir de dissolution[1152]. Au regard de la gravité de ce procédé, sa mise en œuvre est soumise à un formalisme procédural rigoureux (**a**) et emporte des conséquences sur le fonctionnement des collectivités (**b**).

a) Les conditions d'exercice du pouvoir de dissolution

496. - Les articles 116 et 263 de la loi n° 2019-006 précitée prévoient au Togo que, « en cas de dysfonctionnement, notamment de dissension grave en son sein, le conseil municipal [et le conseil régional] peut être dissous par décret en conseil des ministres sur le rapport du ministre chargé de la décentralisation ». Les codes nigérien et burkinabè prévoient également que, lorsque le fonctionnement d'un conseil municipal ou

---

1150 G. MELLERAY, *La tutelle de l'État sur les communes, op.cit.*, p. 270.
1151 B. FAURE, *Droit des collectivités territoriales, op.cit.*, n° 113.
1152 *Idem.*

régional se révèle impossible, sa dissolution est prononcée par décret pris en conseil des ministres sur rapport du ministre chargé de la tutelle des collectivités territoriales[1153]. L'analyse rapprochée de ces dispositions permet de dégager comme condition standard de l'exercice du pouvoir de dissolution le « fonctionnement impossible » de l'organe délibérant constaté notamment par une « dissension grave en son sein ». Ce standard s'applique donc à la situation du Mali puisque le législateur ne précise pas dans quels cas les conseils municipaux, de cercles et régionaux peuvent être dissous. Les articles 8, 81 et 148 du code des collectivités territoriales au Mali se contentent d'affirmer que les assemblées délibérantes des trois catégories de collectivités peuvent être dissoutes. Si les causes réelles de la dissolution ne sont précisées, la loi malienne prévoit qu'en cas de dissolution, l'organe délibérant « est admis préalablement à fournir des explications écrites, par l'entremise de son président, au représentant de l'État » dans la circonscription. La dissolution ne pouvant intervenir qu'après, par décret motivé pris en conseil des ministres[1154].

497. - En somme, la dissolution de l'organe délibérant d'une collectivité territoriale ne peut être obtenue que par décret pris en conseil des ministres. L'initiative d'une telle sanction est confiée par la loi à l'autorité de tutelle, en l'occurrence le ministre chargé de la décentralisation. La dissolution ne peut être prononcée que, d'une part, dans le cas du Burkina Faso, du Niger et du Togo après avoir entendu le rapport du ministre chargé de la tutelle, d'autre part, dans le cas du Mali après les observations écrites du président du conseil représentant l'organe délibérant. Les lois africaines précitées sont particulièrement claires sur la procédure de dissolution. En revanche, les cas et les motifs de dissolution ne bénéficient pas du même traitement. En effet, la loi évoque l'hypothèse d'un « dysfonctionnement » ou d'un « fonctionnement impossible » dû notamment à de graves dissensions au sein des assemblées locales sans toutefois préciser en quoi consisteraient ces dissensions. Il est possible de considérer que la dissension grave peut être caractérisée par un désaccord profond entre le maire et le conseil municipal mettant en péril la gestion administrative de la commune. Bien que cette analyse puisse donner une grille de compréhension, l'impossibilité de fonctionner normalement et de gérer la commune est aussi incertaine. D'une part, si l'on peut considérer que le fonctionnement normal de l'assemblée locale dépend par exemple

---

1153 Art. 177 du CGCT au Niger ; art 177 et 251 du CGCT au Burkina-Faso.
1154 Art. 10, 82 et 150 du code des collectivités territoriales au Mali.

de la tenue à des intervalles réguliers des sessions de l'assemblée[1155] ; il faudrait d'autre part, déterminer le délai à partir duquel l'autorité de tutelle peut considérer que le fonctionnement d'une assemblée est devenu impossible. Cette impossibilité doit-elle être durable ou s'agit-il de difficultés répétées dans le temps ?

498. - La connaissance des causes exactes de la dissolution n'est pas facile à identifier malgré l'obligation de motivation qui pèse sur une telle décision. Le respect de cette exigence ne peut être contrôlé que par le juge administratif, mais encore faut-il qu'il ait eu un recours contentieux. Ce qui semble très rare. L'étude empirique des cas de dissolution permet de déceler les conditions réelles pour l'exercice du pouvoir de dissolution. Le gouvernement du Niger a dissous le 12 juin 2020 les conseils municipaux des communes rurales de Hamdallaye, de Doungou, de Gamou et de Akoubounou, en indiquant « *des irrégularités et pratiques frauduleuses dans la gestion administrative, financière et comptable desdites communes* », entrainant un « *dysfonctionnement avéré se traduisant par un blocage prolongé de leurs organes délibérants* ». Le 14 novembre 2007, un décret portant dissolution des conseils communaux a dissous les conseils des communes rurales de Baguindadougou, Fanga et Korarou au Mali aux motifs d'une « *violation caractérisée des dispositions de certains textes régissant l'administration et la gestion de ces communes, en plus de graves et persistantes dissensions qui ont fini par entrainer une paralysie complète de leurs conseils communaux* ». Le gouvernement burkinabè a procédé également à la dissolution de 15 conseils municipaux et un conseil régional en mars 2017. Sur les raisons de cette dissolution, le décret explique qu'à la suite des élections municipales du 22 mai 2016, « *le processus de mise en place des exécutifs locaux a été émaillé d'incidents ayant empêché le fonctionnement et l'élection des membres des exécutifs locaux dans 15 communes et un conseil régional* ». *In fine*, la dissolution des assemblées délibérantes ne peut intervenir qu'à une double condition : l'existence d'une violation de textes d'une particulière gravité mettant en péril la gestion administrative de la collectivité et des dissensions graves ayant des répercussions sur le fonctionnement normal des assemblées.

499. - Par ailleurs, le législateur nigérien prévoit à l'article 175 du CGCT que, « en cas d'urgence et lorsque son fonctionnement est bloqué,

---

1155 H. K. NOUBISSI, *Décentralisation et centralisation au Cameroun. La répartition des compétences entre l'État et les collectivités locales, op.cit.*, p. 320.

le conseil municipal ou régional peut être suspendu par arrêté du ministre chargé de la tutelle des collectivités territoriales sur rapport du représentant de l'État aux fins de chercher une solution à la crise. La durée de la suspension ne peut excéder un mois ». Cette procédure de suspension provisoire de l'assemblée délibérante est également prévue dans la loi malienne pour une durée de trois mois sans toutefois que soient précisées les circonstances dans lesquelles l'autorité de tutelle doit y avoir recours. À titre exceptionnel, le CGCT burkinabè prévoit que la dissolution peut être prononcée par voie de mesure générale « en situation de dysfonctionnement grave et général menaçant la paix et la cohésion sociales au sein des collectivités territoriales »[1156]. Renforçant la solennité de la procédure, l'alinéa 3 de l'article 172 prévoit dans le cas des conseils régionaux que le gouvernement doit en informer le parlement dans un délai de trente jours[1157]. En application des dispositions précitées, il a été prononcé le 1er février 2022 la dissolution de tous les conseils de collectivités territoriales au Burkina Faso[1158] par le président du mouvement patriotique pour la sauvegarde et la restauration créé à l'issue du coup d'État du 23 janvier 2022. Une fois la dissolution prononcée, elle emporte un certain nombre de conséquences.

b) Les conséquences de l'exercice du pouvoir de dissolution

500. - L'assemblée locale dissoute, la loi organise des mesures transitoires pour la gestion des affaires courantes. Ainsi, le secrétaire général de la collectivité sous le contrôle du représentant de l'État ou le président du conseil assure l'expédition des affaires courantes en attendant la mise en place d'une délégation spéciale au Niger et au Mali[1159]. Cette tâche revient directement à l'autorité de tutelle, le gouverneur ou le préfet au Burkina Faso[1160]. Le législateur habilite aussi le gouvernement à procéder à l'institution d'une délégation spéciale qui va se substituer à l'organe délibérant dans la gestion de la collectivité. La délégation spéciale est instituée non seulement dans le cas de la dissolution, mais aussi en cas de démission de tous les membres du conseil ou de l'annulation définitive

---

1156 Art. 172 al. 2 & 251 al. 2 du CGCT au Burkina Faso.

1157 Cette procédure en prévue en droit français pour les cas des conseils départementaux et régionaux. Voir CGCT, art. 3121-5 pour les départements et art. 4132-3 pour les régions.

1158 Décret n° 2022-004 portant dissolution des conseils des collectivités territoriales.

1159 Art.177 al. 3 du CGCT au Niger.

1160 Art. 173 et 252 du CGCT au Burkina-Faso.

de leur élection ou encore dans l'hypothèse prévue à l'alinéa 2 des articles 172 et 251 précités.

501. - La délégation spéciale est nommée, dans les quinze jours qui suivent la dissolution, la démission ou l'annulation par décret en conseil des ministres sur proposition du ministre chargé de la décentralisation[1161]. Au Mali, seule la nomination de l'autorité intérimaire au niveau régional se fait par la voie du décret. Pour les communes et les cercles, c'est un arrêté du ministre en charge de la décentralisation qui est prévu par la loi[1162]. L'acte instituant la délégation spéciale désigne le président et le vice-président. La loi burkinabè prévoit quant à elle que la délégation spéciale est présidée par le représentant de l'État[1163]. La composition de la délégation spéciale varie d'un pays à l'autre. La loi togolaise prévoit que la délégation spéciale est composée de cinq à sept membres en fonction d'un seuil de 100 000 habitants pour les communes et de neuf membres pour toutes les régions[1164]. Aucun membre du conseil dissout ne peut faire partie de la délégation spéciale[1165]. La délégation spéciale ainsi nommée fait à la fois office d'organe délibérant et d'organe exécutif. Les fonctions exécutives sont assurées par le président et le vice-président de la délégation spéciale. Elle se substitue aux organes de la collectivité dans la gestion des affaires courantes. La délégation spéciale ne peut en aucun engager les finances locales au-delà des ressources disponibles de l'exercice en cours[1166], ni « *aliéner ou échanger les biens meubles et immeubles des collectivités ; créer des services publics ; contracter des emprunts ou autoriser le recrutement du personnel* »[1167].

Enfin, la délégation spéciale est mise en place jusqu'à la tenue des prochaines élections qui doivent intervenir dans un délai allant de trois à six mois selon les pays à compter de la date de nomination. Des possibilités de prorogation d'une durée de six mois sont prévues en cas d'empêchement dans l'organisation des élections au Niger[1168] et au Mali. Le code des collectivités territoriales de ce dernier pays prévoit en ce sens

---

1161 Loi n° 2019-006 *préc.*, art. 116 et 264.

1162 Art. 12, 85 et 152 du code des collectivités territoriales au Mali.

1163 Art. 173 & 252 *préc.*

1164 Loi n° 2019-006 *préc.*, art. 118 pour les communes et art. 265 pour les régions.

1165 Art. 179 al. 2 du CGCT au Niger ; art. 13 al. 3 du code des collectivités territoriales au Mali.

1166 Loi n° 2019-006 *préc.*, art. 118 pour les communes et art. 265 pour les régions.

1167 Art. 178 al. 2 du CGCT au Niger.

1168 *Ibid.*, art. 180 al. 2.

que, « *lorsque les circonstances ne permettent pas le déroulement normal de la consultation électorale en vue du renouvellement d'un conseil communal dissous* [...] *le ministre en charge des collectivités territoriales, sur proposition du représentant de l'État dans la région et avant l'expiration du délai de six mois prévu à l'alinéa* précédent, *peut proroger par arrêté la durée des pouvoirs de l'autorité intérimaire. Cette prorogation ne peut excéder douze mois*»[1169]. Aucune possibilité de prolongation n'est prévue par le législateur togolais, mais on peut douter de la tenue d'élections dans les délais prévus. Les communes togolaises ont été déjà placées auparavant sous le régime de la délégation spéciale pendant 32 ans avant la tenue des dernières élections locales en juin 2019.

502. - Le contrôle des actes des collectivités territoriales par le juge a une portée relativement limitée. Non seulement ce contrôle est subordonné au bon vouloir du représentant de l'État, mais il existe également une certaine lenteur dans l'opérationnalisation d'un contentieux des actes administratifs locaux.

1169 Art. 14 al. 2 du code des collectivités territoriales au Mali.

Conclusion du chapitre 1

503. - Les relations entre l'État et les collectivités territoriales sont complexes. Créées et reconnues comme des personnes distinctes de l'État, les collectivités territoriales sont intégrées dans l'appareil administratif étatique. Dans une conception unitaire de l'État, les collectivités territoriales ne peuvent disposer d'un pouvoir législatif comparable à celui du parlement et par conséquent déterminer elles-mêmes leur champ d'intervention. Les collectivités territoriales ne doivent leurs compétences qu'à l'État par l'intermédiaire de la loi. En effet, pour garantir l'unité normative de l'État et l'uniformité de l'action publique, le législateur possède à titre exclusif, sur habilitation formelle de la Constitution, de l'aptitude pour définir les compétences des collectivités territoriales.

504. - L'intérêt de la collectivité territoriale, c'est-à-dire la volonté de satisfaire aux besoins spécifiques de sa population, ne coïncide pas toujours avec celui de l'État. Mais, dans l'expression de cet intérêt à agir pour répondre à un besoin spécifique, la collectivité ne peut aller au-delà de ce que la loi lui autorise expressément. L'existence d'intérêt public concurrents entre l'État et les collectivités territoriales explique le lien consubstantiel entre les concepts de tutelle et de décentralisation. Le contrôle de l'État sur les collectivités territoriales permet ainsi de garantir l'unité nationale et l'unité de l'ordre juridique. Le législateur africain se distingue en ce sens par la gamme variée des procédés de tutelle en vigueur. Malgré sa volonté affichée d'alléger la tutelle en passant d'un contrôle *a priori* à un contrôle *a posteriori*, la mutation demeure inachevée. L'intensité des contrôles se révèle même handicapante pour la liberté d'action des collectivités territoriales. *In fine*, dans la définition des compétences locales comme dans leur exercice, les collectivités territoriales subissent l'influence permanente de l'État sans pouvoir recourir utilement au juge administratif. Le processus de décentralisation ne peut s'épanouir véritablement sans une nouvelle approche des rapports entre l'État et les collectivités territoriales.

Chapitre 2 – L'omnipotence de l'État dans l'attribution des moyens

505. - Le désengagement de l'État de la gestion de certains domaines dans le cadre du processus de décentralisation a pour conséquence l'accroissement progressif des compétences des collectivités territoriales. L'effectivité du processus de décentralisation ainsi entamée implique une prise en charge autonome des compétences transférées par les collectivités territoriales. La réalisation de la décentralisation suppose en effet que les collectivités territoriales puissent exécuter librement de façon autonome leurs missions dans les domaines que leur réserve l'État-législateur. Pour cela, elles doivent disposer de connaissance et de savoir-faire, c'est-à-dire d'une administration décentralisée capable d'assumer et de gérer efficacement ces nouvelles compétences. Aussi, la prise en charge concrète de ces nouvelles missions par les collectivités territoriales nécessite qu'elles puissent disposer de moyens financiers suffisants. La libre administration des collectivités territoriales qui se traduit notamment dans leur capacité d'agir de façon autonome doit bénéficier à la fois de la vigueur interne des collectivités et de la complicité bienveillante du pouvoir central[1170]. C'est sans doute le principal point d'achoppement de tout processus de décentralisation. Se posent en effet deux problèmes majeurs, celui de la réticence de l'État à laisser les collectivités se déployer librement et les difficultés notamment, en matière d'hommes et de finances que les autorités décentralisées rencontrent dans la gestion de leur collectivité. Il s'agira précisément ici de savoir où en sont les collectivités territoriales africaines sur ce dernier point.

506. - Si les instrumentistes composant l'orchestre sont identifiés et que chacun d'entre eux s'est vu affecter une partition, « ils ne disposent pas véritablement des instruments leur permettant d'interpréter la symphonie encore inachevée de la décentralisation »[1171]. Cette métaphore que nous empruntons à Jacques Baguenard résume assez bien l'état de la décentralisation dans les États d'Afrique noire francophone. Les collectivités territoriales ne disposent pas de ressources nécessaires pour mener à bien la mission de développement qui leur est confiée. Elles sont dans une situation de dépendance vis-à-vis du pouvoir central, unique

---

1170 J. BAGUENARD, *La décentralisation*, PUF, coll. « Que sais-je ? », 2004, p. 23.
1171 *Ibid.*, p. 26.

pourvoyeur de ressources humaines de qualité (**Section 1**) et de moyens financiers (**Section 2**).

Section 1 : La capacité déficitaire des collectivités territoriales en matière de ressources humaines

507. - La nouvelle organisation administrative de l'État implique pour le fonctionnement des collectivités territoriales un personnel dont la situation juridique est assortie de garanties suffisantes pour assurer l'efficacité de l'action publique locale. Composée d'agents publics et de fonctionnaires, l'administration publique locale doit non seulement assurer la gestion des équipements et des infrastructures dont dispose la collectivité territoriale, mais aussi assurer la satisfaction des besoins des populations par l'octroi de prestations de qualité. À l'instar de la France en 1984[1172], les pays de l'Afrique subsaharienne francophones se sont dotés — tardivement certes — de statuts propres à la fonction publique territoriale. Cependant, malgré l'adoption de ces statuts, les collectivités territoriales africaines ont du mal à pourvoir à leurs besoins en ressources humaines[1173]. Ce déficit en ressources humaines des collectivités territoriales s'apprécie à la fois qualitativement et quantitativement (**§1**).

508. - En outre, la détermination des règles relatives au recrutement et à l'emploi du personnel des collectivités territoriales est un élément important de leur libre administration. La fonction publique territoriale doit servir l'autonomie des collectivités territoriales. Or, il y a comme un paradoxe lorsqu'on observe les règles relatives à la fonction publique territoriale dans les pays africains. En effet, il semble difficile de servir l'autonomie des collectivités territoriales en dotant les agents d'un statut unique calqué sur le modèle de la fonction publique de l'État. La fonction publique territoriale dans les pays de la présente étude est marquée du sceau de l'État. Ce paradoxe est d'autant plus inquiétant dans la mesure où la politique de gestion des ressources humaines dans la sphère publique étatique n'est pas exempte de critiques. Elle apparait très souvent comme

---

[1172] Loi n° 84-53 du 26 janvier 1984 portant dispositions statutaires relevant relatives à la fonction publique territoriale.

[1173] C. NACH MBACK, *Démocratisation et décentralisation. Genèse et dynamiques comparés des processus de décentralisation en Afrique subsaharienne, op.cit.*, p. 434.

peu innovante et dépourvue de dynamisme[1174]. En réalité, la réforme qu'appelle le statut de la fonction publique territoriale constitue une occasion de repenser l'administration étatique sur le modèle de laquelle s'est construite la première (**§2**).

§1.- Une administration publique locale en manque de compétences

509. - Les collectivités territoriales ont pour vocation principale de concourir au développement local, notamment par une offre de services publics capables de répondre aux besoins exprimés par la population. À ce titre, le législateur a entendu mettre à leur disposition des moyens juridiques afin d'exercer au mieux les compétences transférées. Le bilan de la décentralisation en matière de ressources humaines locale est loin d'être reluisant. Les collectivités territoriales ne disposent pas d'un personnel suffisamment qualifié pour la mise en œuvre de ses compétences. Cette incapacité des collectivités à disposer d'un personnel compétent en nombre suffisant pour répondre aux nombreux défis posés par l'exercice des compétences locales sert de justification à une recentralisation de l'exercice de certaines compétences par le biais des agences nationales placées sous la tutelle du pouvoir central (**A**). Aussi, pour combler le déficit en capacité des collectivités, certaines législations privent les autorités décentralisées de leur liberté de recrutement aux emplois supérieurs de la fonction publique territoriale remettant ainsi en cause leur autonomie de gestion (**B**).

A- Une carence en ressources humaines handicapante pour les collectivités territoriales

510. - Dans le cadre de l'exercice des compétences locales, le législateur confère aux autorités décentralisées le pouvoir de recruter et de gérer le personnel requis. L'article 307 de loi n° 2019-006 précitée consacre en des termes généraux cette possibilité offerte aux autorités décentralisées : « *l'exercice des compétences transférées confère aux collectivités territoriales le pouvoir de recruter et de gérer le personnel requis* ». Plus spécifiquement, la loi fait du

---

[1174] K. B. BEKPOLI, *L'approche statutaire de la fonction publique. Approche comparée de la notion de statut général en France et au Togo,* thèse, dactyl., Université de Poitiers, 2020.

président du conseil de la collectivité l'autorité compétente pour procéder au recrutement. L'article 87 du code général des collectivités territoriales au Niger prévoit en son article 87 que « *le maire est le chef hiérarchique du personnel municipal. Il nomme aux différents emplois de la commune conformément aux lois et règlements en vigueur* ». Toute création de postes d'emploi doit être autorisée par le conseil de la collectivité et inscrite au budget. L'article 17 de la loi n° 003-2017 portant statut de la fonction publique territoriale au Burkina Faso impose sous peine de nullité du recrutement, l'expression du besoin ainsi que sa justification « *dans la note explicative jointe au budget au moment de son examen par le conseil de la collectivité* ». Par ailleurs, la loi précise que le transfert de compétences aux collectivités doit s'accompagner du transfert de personnel de l'État affecté aux services transférés. Le personnel des collectivités territoriales est donc composé de deux catégories d'agents publics. D'un côté, les agents titularisés et nommés dans les emplois permanents de la collectivité et de l'autre, les fonctionnaires de l'État en situation de détachement ou de mise à disposition relevant du statut général de la fonction publique de l'État. À ces deux catégories s'ajoutent dans une certaine mesure des agents contractuels de droit public. On peut donc constater que les dispositions textuelles existent et sont relativement claires, mais le problème de la faiblesse des ressources humaines persiste et constitue un frein à l'aboutissement du processus de décentralisation.

511. - La décentralisation se matérialise par un transfert de compétences de l'État vers les collectivités territoriales. Ce transfert de compétences suppose l'existence de compensations c'est-à-dire le transfert concomitant de la part de l'État des ressources nécessaires à l'exercice des nouvelles compétences. Les compensations doivent être non seulement financières, mais aussi humaines. L'État doit transférer aux collectivités les agents publics dont les services ont été transférés. Il s'agit d'un élément consubstantiel à la politique de décentralisation ainsi que nous l'avons montré précédemment. Le transfert des ressources humaines s'effectue à travers deux techniques que sont le détachement et la mise à disposition ainsi qu'il ressort de l'article 74 de la loi n° 2019-006 précitée : « *des agents de l'État peuvent être détachés ou mis à disposition et affectés à l'exécution de tâches locales* ». Ces deux techniques sont garanties aux fonctionnaires par le statut général de la fonction publique. En réalité, seule la dernière technique permet de garantir un transfert de personnel aux collectivités, car elle relève de la seule décision de l'autorité de nomination. C'est un principe commun à toutes les fonctions publiques[1175]. Le détachement d'un fonctionnaire de l'État dans une collectivité territoriale ne peut être prononcé que sur une demande expresse de ce dernier[1176]. Or, vu l'état peu reluisant de la fonction publique territoriale, il y a peu de chance que des fonctionnaires se décident spontanément à rejoindre une collectivité territoriale. Au Burkina Faso, le tableau de bord statistique de 2017 sur la décentralisation réalisée en 2020 montre que seul 0,8 % du personnel des collectivités territoriales est détaché. Ce chiffre est resté relativement constant entre 2012 et 2017 avec une hausse de 1,2 % en 2016, l'effectif total du personnel des collectivités étant de 6162 agents[1177].

512. - Ainsi, la mise à disposition apparait comme le moyen pour une administration ne disposant pas de ressources budgétaires lui permettant de nommer un fonctionnaire ou de l'accueillir dans le cadre d'un détachement[1178]. Elle est définie par l'article 101 de la loi n° 2013-002 du 21 janvier 2013 portant statut général de la fonction publique togolaise comme *« la position du fonctionnaire qui, placé hors de son Administration d'origine, continue à bénéficier dans son corps d'origine de ses droits à l'avancement et à la retraite, ainsi que de sa rémunération »*. En principe, elle est censée permettre aux collectivités décentralisées de bénéficier de l'expertise de cadres en provenance de l'État. Cependant, en pratique, le transfert de services et des agents déconcentrés n'est pas toujours effectif en raison de l'insuffisance des effectifs au niveau des directions régionales. Dans un rapport réalisé en 2017 sur la problématique des transferts des services techniques déconcentrés de l'État aux collectivités territoriales au Mali, il a été relevé notamment dans la région de Sikasso que la direction régionale de la jeunesse compte un (1) seul agent, quatre (4) au niveau de la direction régionale de l'agriculture et cinq (5) au niveau de la direction régionale des eaux et forêts[1179]. L'insuffisance de personnel qualifié au niveau même de l'État pour l'exercice des compétences transférées est criante À titre illustratif, les dernières statistiques de répartition des fonctionnaires

---

1175 Cf. Statut général de la fonction publique, art. 102 al. 2 : « La mise à disposition ne peut être décidée qu'à l'initiative de l'autorité de nomination ».

1176 *Ibid.*, art. 90 al. 2.

1177 Annuaire statistique 2017 de la décentralisation DGESS/MATDC.

1178 P. COMBEAU, « Transferts de compétences et fonctions publiques », in J.-F. BRISSON (dir.), *Les transferts de compétences de l'État aux collectivités*, *op.cit.*, p. 494.

1179 https://www.maliapd.org/plateformegt/wpcontent/uploads/2018/11/RAPPORT-FINAL-DE-L'ETUDE-SUR-LA-PROBLEMATIQUE-DES-TRANSFERTS-DES-SERVICESTECHN.pdf.consulté le 1 juillet 2022.

territoriaux datant de 2016 montrent que les fonctionnaires territoriaux sont au nombre de 2945 pour environ 700 collectivités territoriales[1180]. Bien que les chiffres soient en hausse par rapport à 2007 où on dénombrait seulement 2627 fonctionnaires[1181], le besoin en ressources humaines qualifié demeure une préoccupation majeure.

513. - Au Togo, à l'issue de l'élection municipale du 30 juin 2019, les communes nouvellement constituées ont hérité du personnel des anciennes délégations spéciales supprimées par décret présidentiel en novembre 2019[1182]. On note au sein de ce personnel une catégorie d'agents majoritairement peu qualifiés en charge de missions d'exécution dans les domaines notamment, de l'assainissement et de la voirie. Devrait s'ajouter à ce personnel hérité des délégations spéciales, pour l'exercice des nouvelles compétences, le personnel issu des services déconcentrés de l'État dont les compétences ont été transférées aux nouvelles collectivités. Or, il se trouve, comme dans les autres pays d'Afrique, que l'administration étatique n'est pas elle-même en mesure de pourvoir à l'ensemble de ces besoins en personnel au niveau central et déconcentré. Le ministre de la Fonction publique, du Travail et du Dialogue social lors du lancement du concours de recrutement général au premier trimestre de l'année 2022 affirmait que le recrutement de nouveaux agents est nécessaire pour faire *« face au déficit quasi généralisé en ressources humaines auquel sont confrontés les ministères et institutions de l'État »*. Pour se réaliser, le transfert doit être matériellement possible. L'État a transféré aux collectivités territoriales des domaines de compétences dans lesquels il ne disposait pas lui-même de ressources nécessaires pour leur accomplissement.

1180 Rapport Général du « contrôle physique des fonctionnaires du cadre de l'administration générale et des contractuels des collectivités territoriales » Septembre 2016.

1181 C. VERGNE, « Décentralisation en Afrique francophone : Réflexions sur la réforme de la fonction publique territoriale », Banque Mondiale, Région Afrique, AFTPR, Septembre 2009, p. 7.

1182 Décret n° 2019-153/PR du 11 novembre 2019 portant dissolution des délégations spéciales dans les préfectures et les communes. Cette suppression intervient après deux décennies d'exercice des délégations spéciales. Les délégations spéciales ont été introduites au Togo dans les années 2000 pour mettre fin à la durée exceptionnelle des conseils municipaux élus en 1987 pour un mandat de cinq (5) ans mais dont le renouvellement n'a pu être organisé du fait des troubles socio-politiques qu'a connu le pays au début des années 1990.

514. - Les collectivités territoriales ont pour la plupart procédé au recrutement des agents décisionnaires par la voie du contrat afin de compléter leurs effectifs. Il en résulte une forte tendance de certains présidents de conseil à recruter selon des critères extraprofessionnels en totale inadéquation avec les besoins réels des communes. Les collectivités territoriales sont donc remplies de personnel occupant des postes sans aucun rapport avec leur qualification professionnelle et des agents d'exécution. On y retrouve des chauffeurs, assistants, des techniciens de surface, etc., mais très peu de cadres pour les fonctions d'encadrement et de conception[1183]. Cette pratique est rendue possible en raison notamment de l'absence de textes cohérents organisant les services publics locaux. Les communes ne disposent pas d'un organigramme type opérationnel. L'organisation administrative interne est en panne et accentue le problème du recrutement des ressources humaines compétentes au niveau local. Les services municipaux sont largement sous encadrés. Même les communes de la capitale et des grandes villes ne font pas exception à ce constat. Pour preuve, dans la commune de Kozah 1 situé à Kara considéré comme la capitale économique du Togo, seules deux directions sont opérationnelles à savoir la direction des affaires financières et la direction des affaires techniques. À Lomé, la commune du Golfe 3 ne dispose que d'une direction fonctionnelle, celle des affaires administrative et financière. La direction de la planification du développement est rattachée au secrétariat général dont l'occupant a été désigné par l'autorité de tutelle. Les services sociaux de proximité : santé, culture, sport, etc. ne sont pas organisés.

515. - L'objectif de satisfaction des besoins des populations par des services publics de proximité est loin d'être atteint en raison du déficit qualitatif et quantitatif du personnel local[1184]. L'autonomie de fonctionnement et de gestion des collectivités territoriales est dans ce contexte fortement limitée. Les insuffisances observées dans le fonctionnement des collectivités ont conduit une recentralisation de la maitrise d'ouvrage dans certains pays. Au Togo, le décret n° 2021-105/PR portant répartition du Fonds d'appui aux collectivités territoriales a fait de l'Agence nationale d'appui au développement à la base le maitre d'ouvrage pour l'exécution des dépenses d'investissements des collectivités territoriales. Ainsi, les collectivités doivent dans le cadre de la réalisation

---

1183 M. FINKEN, *Gouvernance communale en Afrique et au Cameroun*, L'Harmattan, 2011, p. 215.

1184 On dénombre dans les villes africaines un ratio d'un fonctionnaire pour 1000 habitants. Voir Rapport CGLUA 2018 *préc.*

de leurs projets de développement en matière d'infrastructures scolaires, sanitaires, sportives, etc. se référer à l'Agence qui détient les ressources financières et l'expertise nécessaire[1185].

516. - Le problème du manque de ressources humaines prend une ampleur somme toute dramatique pour la libre administration lorsque l'on observe aussi le taux d'analphabétisme et le manque de qualification du personnel élu[1186]. Les derniers chiffres recensés au Burkina Faso en 2017 montrent que parmi les élus, 47,3 % n'ont jamais été scolarisés ni alphabétisés. Cette proportion est de 48,3 % dans les communes rurales et 42,9 % dans les communes urbaines. Les élus ayant un niveau secondaire ou supérieur sont représentés dans les communes urbaines avec les taux respectifs de 22,4 % et 10,7 %. Pour compenser ce manque de qualification à la fois du personnel et des élus, l'État encadre le pouvoir de nomination des collectivités aux emplois supérieurs de la fonction publique territoriale. Toutefois, dans certains pays, notamment le Togo, cet encadrement limite fortement la liberté des autorités décentralisées.

B- La centralisation du pouvoir de nomination aux emplois supérieurs de fonction publique territoriale

517. - La présence de l'État est permanente à tous les niveaux de fonctionnement des collectivités territoriales, y compris dans la nomination aux emplois supérieurs de la fonction publique territoriale. Les emplois supérieurs recouvrent ici les fonctions de direction administrative et financière au sein des collectivités, notamment le secrétariat général[1187]. À travers la constitution de cette catégorie particulière d'emplois de

---

1185 Décret n° 2021-105/PR portant répartition du Fonds d'appui aux collectivités territoriales.

1186 La seule condition exigée par la loi pour élu local est de savoir lire et écrire le français. Au Niger, l'élection en tant exécutif est subordonnée à un niveau de qualification (Bac). Pour une analyse détaillée sur ce point, voir les développements *supra.* Voir aussi, C. NANAKO, *La libre administration des* coll*ectivités territoriales au Bénin et au Niger, op.cit.*, pp. 199-213.

1187 En droit français cette notion est plus large et recouvre les fonctions de direction ou de cabinet et les emplois fonctionnels dont la nomination relève du pouvoir discrétionnaire des exécutifs. Sur le régime d'ensemble des emplois fonctionnels, B. PERRIN, « Le cadre juridique des emplois discrétionnaires de la fonction publique territoriale », *AJDA,* 1989, p. 147 ; « La revalorisation des emplois de direction », *Cah. fonct. publ.*, 2001, p. 27 ; P. RENARD, « Les emplois fonctionnels », *Gaz. cnes* 11/1998. p. 38 ; G.-D. Marillia, « Le détachement sur un emploi fonctionnel », *AJFP,* 1997, p. 43.

direction, l'État cherche à constituer une fonction publique territoriale compétente selon des logiques qui lui sont propres, mais qui ne répondent pas toujours aux particularités de l'administration locale.

518. - Malgré la consécration de l'autonomie administrative des collectivités territoriales, la loi n° 2019-006 du 26 juin 2019 relative à la décentralisation et aux libertés locales au Togo prévoit en ces articles 160 et 274 que l'autorité exécutive locale est assistée dans ses fonctions par un secrétaire général « *nommé par arrêté du ministre en charge de la décentralisation* ». Le secrétaire est le principal collaborateur du chef de l'exécutif. En effet, le secrétaire général incarne la dimension stratégique et managériale de la collectivité territoriale. Placé à la tête de l'administration locale, il est au sommet de la hiérarchie du personnel de la collectivité ; anime l'ensemble des services administratifs et techniques. Aux termes de l'article 5 du décret n° 98-274 du 2 octobre 1998 fixant les conditions de nomination, les attributions et les avantages accordés aux secrétaires généraux des collectivités territoriales au Niger, le secrétaire général assure la coordination de l'action des services extérieurs mis à disposition ; l'organisation, l'impulsion et la coordination des services techniques ; le suivi en matière de gestion financière, du matériel, et du personnel ; une mission d'assistance et de conseil ; une mission d'assistance du maire dans la préparation et la présentation du budget, du compte administratif et de tout autre acte de gestion courante. Il assiste aux séances du conseil de la collectivité avec voix consultative[1188]. En clair, le secrétaire général est le pivot de l'administration locale.

519. - Dans ces conditions, il peut paraitre surprenant que le principal collaborateur de l'exécutif soit imposé par le pouvoir central. Telle est néanmoins la situation qui prévoit en droit positif togolais. Par une série d'arrêtés adoptés le 11 octobre 2019, le ministre en charge de la décentralisation a procédé à la nomination de secrétaires généraux de communes. Lorsqu'on parcourt l'ensemble des arrêtés de nomination, on constate que le poste de secrétaire général de la commune est confié par l'autorité de tutelle soit à des fonctionnaires de l'État soit à des agents publics contractuels de l'État. La nomination par le pouvoir central des secrétaires généraux porte atteinte à l'autonomie locale. Les autorités locales sont dépossédées de leur pouvoir en matière de recrutement pour les emplois supérieurs de la fonction publique territoriale. L'État n'entend

---

[1188] Loi n° 2019-006 *préc.*, art. 161 pour les communes, art. 294 pour les régions.

donc pas laisser à l'exécutif la marge de manœuvre nécessaire à l'administration de la collectivité territoriale. La nomination du secrétaire général par le pouvoir central parait anachronique par rapport au nouvel objectif affiché de construction d'un État décentralisé. Cette nomination par les autorités étatiques institue une tutelle supplémentaire sur les autorités locales, car même placée sous l'autorité du chef de l'exécutif, le secrétaire général dépend du ministre de tutelle à qui il rend compte de son action.

520. - La centralisation de la nomination aux emplois supérieurs de la fonction publique territoriale pourrait se justifier par la volonté des autorités gouvernementales de limiter les disparités techniques entre les différentes collectivités afin de s'assurer du bon fonctionnement des services administratifs et d'harmoniser la qualité des services publics locaux. Toutefois, ce pari est loin d'être gagné. Pour l'illustrer, il se suffit d'analyser l'adéquation entre le profil et les attributions des personnes désignées au poste de secrétaire général par l'autorité de tutelle. L'animation des services locaux nécessite un certain nombre de compétences et de connaissances techniques et managériales dans la conduite des affaires publiques. Cependant, parmi les secrétaires généraux, on note un nombre important de profils dont on peut douter légitimement de la capacité à mobiliser de telles compétences. Nombreux sont les secrétaires généraux issue de l'enseignement primaire secondaire ou disposant d'un diplôme en anglais ou en histoire sans aucune spécialisation en matière de management public, de décentralisation ou de gouvernance locale. Cela donne l'impression qu'il s'agit davantage pour les autorités gouvernementales de récompenser des militants politiques que de doter les collectivités d'un personnel hautement qualifié. Si l'argument de la recherche de compétence semble pertinent, la nomination des secrétaires généraux par l'État n'offre pas une garantie absolue en la matière. Par conséquent, cet argument ne saurait prospérer. En outre, le législateur aurait pu atteindre cet objectif en confiant la nomination des secrétaires généraux aux autorités décentralisées en procédant à « *la définition légale ou règlementaire des critères de désignation* »[1189]. L'argument d'efficacité ne résiste pas non plus à l'analyse, car le chef de l'exécutif n'ayant aucun pouvoir direct sur le secrétaire général, on n'est pas à l'abri de conflits pouvant paralyser la bonne administration de la collectivité.

1189 H. K. NOUBISSI, *Décentralisation et centralisation au Cameroun. La répartition des compétences entre l'État et les collectivités locales*, thèse, L'Harmattan, 2013, p. 253.

521. - Le législateur togolais assure une conciliation déséquilibrée entre la nécessité de doter les collectivités d'un personnel hautement qualifié pour les postes les plus élevés de la hiérarchie et la volonté de laisser aux exécutifs locaux une marge d'autonomie compatible avec l'idée de décentralisation. Dans les autres pays de la présente recherche, la loi aménage le pouvoir de nomination des autorités décentralisées en prévoyant deux modes d'accès, par recrutement direct ou par voie de détachement. Ainsi, l'article 202 du CGCT au Burkina Faso prévoit que, « *le secrétaire général est nommé par le président du conseil parmi les cadres supérieurs de l'administration générale. Le secrétaire général peut être un agent recruté par la région ou un agent mis à disposition ou en position de détachement par l'État* ». La loi n° 2017-52 du 2 octobre 2017 déterminant les conditions de la libre administration des collectivités territoriales attribue également au chef de l'exécutif la compétence pour nommer le secrétaire général. C'est le sens de l'article 11 qui dispose que, « *dans chaque collectivité territoriale, un secrétaire général* [est nommé] *par le chef de l'organe exécutif* [...]. *Le secrétaire général doit être de la catégorie A de la fonction publique* ». Il en va de même au Niger où, selon le décret de 1998 précité, le secrétaire général de la commune est recruté parmi les administrateurs civils de la catégorie A de la fonction publique, pour les villes et les communes urbaines ou à défaut de la catégorie B de l'administration générale de la fonction publique pour les communes rurales[1190]. La définition législative ou règlementaire des règles de recrutement aux emplois supérieurs contribue au respect des spécificités de la fonction publique territoriale. Les dispositions précitées traduisent en fait la nécessité de laisser aux autorités locales un important pouvoir de décision pour des emplois situés à la frontière de la sphère politique et administrative. Cela permet, par ailleurs, de réduire les risques d'étatisation de la fonction publique territoriale. Cette option somme toute cohérente permet de concilier les besoins de qualification et la nécessaire discrétion de l'exécutif dans le recrutement des plus proches collaborateurs en qui il place une totale confiance pour la direction de la politique locale[1191]. Le manque de personnel qualifié au sein des

---

[1190] Voir aussi l'article 289 du GCGT au Burkina Faso.

[1191] Sur la promiscuité ou les allégeances politiques dans la haute fonction publique territoriale, J.-F. BRISSON, « Les emplois fonctionnels dans la fonction publique territoriale, un excès de loyauté ? », in S. NIQUEGE (dir.), *Les figures de la loyauté en droit public*, Mare & Martin, 2017, p. 181.

collectivités pose le problème général de l'attractivité de la fonction publique territoriale.

§2.- Le paradoxe de la fonction publique territoriale : un cadre de gestion à reformer

522. - Le processus de décentralisation implique que les collectivités territoriales puissent disposer d'un personnel compétent en nombre suffisant. Mais, il y a sans aucun doute quelque chose de paradoxal dans l'institution d'une fonction publique territoriale. La loi va chercher à satisfaire l'autonomie locale des collectivités tout en dotant ses agents d'un statut identique et unifié applicable à toutes les catégories de collectivités sur l'ensemble du territoire national. L'adoption d'un tel statut sur le modèle de la fonction publique étatique à un intérêt certain pour les agents de ces collectivités en termes de sécurité et de garantie d'emploi (**A**), mais il est discutable, car il fige et tant à inhiber les possibilités d'action des collectivités territoriales (**B**).

A- Un régime unifié aligné sur le statut de la fonction publique d'État

523. - Le renforcement de la fonction publique territoriale est l'une des clés de la réussite du processus de décentralisation. La régularité et l'efficacité de l'action publique locale nécessitent la présence de femmes et d'hommes compétents, en nombre suffisant pour répondre aux besoins des collectivités territoriales. La décentralisation doit aller de pair avec la construction d'une fonction publique des collectivités territoriales. En droit français, le processus de décentralisation entamé au début des années 1980 s'est accompagné avec la création de la fonction publique territoriale pour toutes les catégories de collectivités territoriales[1192]. La décentralisation ne peut se concevoir sans l'existence d'une fonction publique territoriale. À cet effet, dans chacun des pays objet de la présente étude, les législations nationales ont consacré, certes tardivement, l'existence d'une fonction publique territoriale.

[1192] Deux textes importants ont été adoptés : la loi n° 83-634 du 13 juillet 1983 relative aux droits et obligations des fonctionnaires et la loi n° 86-53 du 26 janvier 1984 portant dispositions statutaires relatives à la fonction publique territoriale.

524. - Le Mali fut l'un des premiers pays à avoir doté ces collectivités territoriales d'un statut particulier avec l'adoption de la loi n° 95-022 du 20 mars 1995 portant statut des fonctionnaires des collectivités territoriales. Cette loi a fait l'objet d'une modification en 2004[1193] avant d'être remplacée par la loi n° 2018-035 du 27 juin 2018 portant statut des fonctionnaires des collectivités territoriales. Au Togo, c'est la loi du 11 juin 2008 qui régit la situation des agents des collectivités territoriales. Le Burkina Faso et le Niger n'ont adopté que très récemment des textes consacrant la fonction publique territoriale. Le premier en 2017 avec la loi n° 003-2017/AN du 13 janvier 2017 et le second en 2019 avec l'adoption de la loi n° 2019-26 du 1er juillet 2019 portant statut autonome du personnel des collectivités territoriales.

L'économie générale de l'ensemble de ces textes montre que les modalités d'administration et de gestion des fonctionnaires des collectivités territoriales sont empruntées au statut général des fonctionnaires. Les agents publics territoriaux sont régis par un statut qui les assimile dans leur situation aux agents publics de l'administration étatique. Indistinctement, la notion de fonctionnaire s'applique à ces deux catégories d'agents publics[1194] nommés dans un emploi permanent et titularisé dans un grade de la hiérarchie administrative. C'est le sens des articles premiers des législations nationales qui disposent dans des termes similaires que la « *loi s'applique aux agents nommés à des emplois permanents et titularisés dans des grades de la hiérarchie administrative des collectivités territoriales et de leurs établissements publics. Ces agents relèvent de la fonction publique territoriale* »[1195]. L'alignement de la fonction publique territoriale sur le statut général de la fonction publique d'État limite l'autonomie des collectivités territoriales.

525. - Les législations nationales ont ajusté le droit des fonctionnaires des collectivités territoriales au plus près du régime général applicable aux

---

1193 Loi n° 04-033 du 27 juillet 2004 portant modification de loi n° 95-022 du 20 mars 1995 portant statut des fonctionnaires des collectivités territoriales.

1194 La qualité d'agent public a été reconnu au personnel communal en droit français avec la jurisprudence *Cadot* : CE, 13 décembre 1889, *Cadot, Lebon*, p. 1148, Concl. Jagerschmidt. V. M. WALINE, *Droit administratif*, Sirey, 1963, p. 117 et s.

1195 *Cf.* not., art. 1, Loi n° 2018-035 du 27 juin 2018 portant statut des fonctionnaires des collectivités territoriales au Mali ; art. 1 Loi du 11 juin 2008 portant statut des agents des collectivités au Togo ; art. 3, loi n° 2019-26 du 1er juillet 2019 portant statut autonome du personnel des collectivités territoriales.

fonctionnaires de l'État. Les agents locaux sont, à l'instar, de leurs homologues de l'État placés dans une situation légale et règlementaire[1196]. L'attribution de la qualité de fonctionnaire aux agents des collectivités territoriales crée une fonction publique unique et unifiée dans son organisation[1197]. Cette conception du fonctionnaire entraine certaines conséquences, notamment le recours subsidiaire au contrat pour le recrutement, l'application à l'ensemble des fonctionnaires des mêmes droits et obligations et surtout la garantie d'emploi et de carrière. Ainsi, on relève des dispositions communes avec le statut général notamment en ce qui concerne le recrutement, les droits et obligations, les positions, le système de notation et de l'avancement ainsi que la rémunération. La loi reprend le schéma posé par le statut général de la fonction publique en instituant le principe de la séparation du grade et de l'emploi. Comme dans la fonction publique de l'État, l'attribution d'un grade donne vocation à son titulaire à occuper un emploi correspondant et la titularisation prononcée à l'issue d'un stage probatoire a pour effet d'intégrer l'agent dans la fonction publique territoriale pour toute la durée de sa vie professionnelle[1198]. Le fonctionnaire territorial ne sera amené à quitter ses fonctions qu'en cas d'admission à la retraite, de licenciement ou de révocation[1199]. L'entrée du fonctionnaire dans un corps de la fonction publique lui offre des perspectives de carrière. Le déroulement régulier de cette dernière est garanti par l'avancement en échelon dans le grade selon l'ancienneté et la valeur professionnelle ou de grade à grade conformément aux dispositions applicables aux fonctionnaires de l'État[1200].

1196 art. 20, Loi n° 2018-035 du 27 juin 2018 portant statut des fonctionnaires des collectivités territoriales au Mali ; art 2 Loi du 11 juin 2008 portant statut des agents des collectivités au Togo.

1197 Pour un aperçu sur ce sujet en droit en français : J. BOURDON, « Le statut général de la fonction publique territoriales », *AJDA*, 1984, p. 199 ; O. SCHRAMECK, *La fonction publique territoriale*, Dalloz, coll. « Connaissance du droit », 1995 ; L. FAVOREU, « Les sources du droit de la fonction publique territoriale », *RFDA*, 1985, p. 309.

1198 F. MELLERAY, Droit de la fonction publique, Economica, 2003, p. 255. L'auteur écrit avec une certaine ironie que : « M. de La Palice, dont on se souvient qu'un quart d'heure avant sa mort il était encore en vie, aurait sans doute pu écrire qu'une carrière commence au moment du recrutement, puis se poursuit durant ce que "on nomme classiquement le déroulement. Il aurait eu raison… ». Voir aussi : A. TALLEFAIT, Droit de la fonction publique, Dalloz, coll. « Précis », 2022, p. 730.

1199 Art. 125 et s. de la Loi portant statut des agents des collectivités territoriales.

1200 *Ibid.*, art. 85.

526. - La parité entre les fonctions publiques s'observe à travers l'organisation des deux fonctions publiques en corps identiques. Elle permet d'assurer le respect des principes constitutionnels d'unité et d'égalité. Les fonctionnaires sont dans des situations comparables au niveau des conditions d'entrée, au même niveau de qualification pour l'entrée dans un corps et à l'exercice de fonctions analogues avec une rémunération identique fixée selon la grille indiciaire retenue par l'État pour le régime général. Cette idée ne semble pas convaincre l'ensemble de la doctrine. Certains auteurs[1201] estiment que l'égalité dans la procédure de recrutement implique des corollaires non seulement pour les candidats, mais aussi pour les organisateurs, notamment celui de l'unicité de concours suivant lequel tous les candidats doivent concourir aux mêmes épreuves en même temps. Ensuite, celui de l'unicité de jury, égalité de traitement et égalité de chances obligent. Or, la fonction publique territoriale est constituée d'une multiplicité d'employeurs publics potentiels et les concours sont organisés selon les besoins propres exprimés par chaque collectivité. Il s'ensuit qu'il est matériellement impossible de garantir précisément les implications sus évoquées[1202] sauf à prévoir une date unique pour l'organisation du concours comme c'est le cas au Mali. L'article 11, alinéa 2 de la loi précitée du 27 juin 2018 prévoit que « *la mise en compétition des emplois à pourvoir a lieu à dates périodiques pour l'ensemble des emplois vacants au niveau de toutes les collectivités territoriales* ».

527. - Néanmoins, on ne peut qu'être frappé par la similitude quasi parfaite entre les deux fonctions publiques. Les législations nationales n'ont pas véritablement entendu les séparer. Cet alignement conduit dans certains cas à limiter fortement le pouvoir de décision des autorités décentralisées en matière d'administration et de gestion du personnel. On assiste alors à une recentralisation de la gestion du personnel territorial. La fonction publique territoriale est tiraillée entre le pouvoir central et son ancrage « normal » auprès des autorités décentralisées tel que le consacrent les textes afin de garantir le principe de libre administration. La Charte africaine des valeurs et des principes de la décentralisation, de la gouvernance locale et du développement local stipule en son article 5.2 que « *les gouvernements locaux ou les autorités locales ont, conformément à la législation nationale, le pouvoir de gérer, de manière responsable et transparente, leur*

1201 Cf. C. NANAKO, *La libre administration des collectivités territoriales au Bénin et au Niger*, *op.cit.*, p. 213.

1202 *Idem.*

administration… ». Les différents textes organisant la décentralisation et la fonction publique territoriale reconnaissent le président du conseil de la collectivité territoriale comme chef de l'administration de son entité. Il est selon l'article 17 de la loi précitée du 1er juillet 2019 au Niger « *le chef hiérarchique du personnel et à ce titre, il nomme aux différents emplois de la collectivité territoriale conformément aux lois et règlements en vigueur* »[1203]. Cependant, les textes de loi en Afrique comportent encore des incongruités qui renforcent les prérogatives de l'autorité de tutelle dans l'administration de la fonction publique territoriale.

Ainsi, au Mali, les conditions de recrutement des fonctionnaires territoriaux sont celles prévues aux articles 9 à 19 de la loi précitée du 27 juin 2018. Il ressort de ces dispositions que le recrutement, l'affectation et la nomination des agents territoriaux relèvent de la compétence du ministre de tutelle en charge des collectivités territoriales. L'ouverture du concours ainsi que les conditions de son organisation matérielle sont fixées par arrêté du ministre en charge des collectivités territoriales[1204]. Aux termes l'article 15 du même texte, à l'issue du concours, « *les candidats déclarés admis sont classés par ordre de mérite. Le ministre chargé des collectivités territoriales procède, par arrêté, à l'intégration, dans la fonction publique des collectivités territoriales, des candidats déclarés admis. L'arrêté d'intégration porte la date de naissance du fonctionnaire et seule cette date fait foi* ». Cette emprise de l'autorité de tutelle sur l'administration de la fonction publique territoriale se combine également avec le contrôle auquel est soumis le budget local dans l'ensemble des pays de la présente étude. La soumission des actes budgétaires au contrôle de l'État offre un moyen supplémentaire de régulation des choix opérés par les collectivités territoriales.

528. - En outre, à la différence du Togo, du Niger et du Burkina Faso, les décisions individuelles relatives à la gestion de la carrière des agents relèvent au Mali de l'autorité de tutelle. C'est le cas notamment du pouvoir de sanction. En application de la règle du parallélisme des compétences, le pouvoir de sanction appartient au titulaire du pouvoir de nomination. Ainsi, le pouvoir de sanction disciplinaire du second degré notamment, l'abaissement d'échelon, la rétrogradation appartient au ministre en charge des collectivités territoriales. Il en va de même de la cessation définitive des fonctions pour cause de révocation ou de licenciement qui est

---

[1203] Une disposition similaire existe au Togo : art. 3, Loi du 11 juin 2008 *préc* ; au Burkina Faso : art. 75, Loi du 13 janvier 2017, *préc.*
[1204] Art. 11 & 12 Loi du 27 juin 2018 *préc.*

prononcé par arrêté du ministre chargé des collectivités territoriales[1205]. Dans ces conditions, on peut émettre des réserves quant à la compatibilité entre de telles prescriptions législatives et le principe constitutionnel de libre administration des collectivités territoriales. Les prérogatives de l'autorité territoriale en tant qu'employeur sont largement méconnues. La gestion du personnel territorial échappe aux collectivités territoriales remettant, ainsi en cause l'effectivité de l'autonomie locale.

529. - La parité entre les deux fonctions publiques fait de la fonction publique territoriale l'un des domaines où les règles sont le plus centralisées laissant peu de marges aux autorités décentralisées. Toutefois, la fonction publique territoriale est marquée par une spécificité au niveau institutionnel avec des organes de participation qui lui sont propres. Ces organes de nature consultative participent, «*par leur avis, à l'organisation et au fonctionnement des services publics locaux, à l'élaboration des règles statutaires et à l'examen des décisions individuelles relatives à la carrière des agents*»[1206]. Il s'agit du Conseil consultatif de la fonction publique territoriale[1207] qui, connait de toutes les questions relatives à la fonction publique territoriale, émet des avis sur les réformes du statut des agents décentralisés ; de la Commission administrative paritaire[1208] qui siège en matière disciplinaire et donne son avis sur les actes de titularisation, d'administration et de gestion du personnel et du Conseil de santé et de réforme des malades[1209] institué au niveau de chaque région, il est chargé de se prononcer sur les aptitudes physiques et mentales des agents des collectivités territoriales. Pour chacun de ces organes de participation, la loi renvoie à des décrets ultérieurs pris en conseil des ministres pour fixer les modalités de fonctionnement et d'organisation[1210].

530. - La définition législative du statut des agents des collectivités territoriales est marquée par la volonté d'alignement de cette dernière au

---

1205 *Ibid.*, art. 95 al. 3.

1206 Voir, art. 4 Loi du 11 juin 2008 *préc.* au Togo ; art. 77 Loi du 13 janvier 2017 *préc.* au Burkina-Faso.

1207 *Ibid.*, art. 5 et 77 ; art. 18 Loi du 1er juillet 2019 portant statut autonome du personnel des collectivités territoriales au Niger.

1208 *Ibid.*, respectivement art. 6, 78 & 18 ; art. 100, Loi du 27 juin 2018 portant statut des fonctionnaires des collectivités territoriales au Mali.

1209 *Ibid.*, art. 7.

1210 art. 100, Loi du 27 juin 2018 portant statut des fonctionnaires des collectivités territoriales au Mali.

régime général de la fonction publique[1211]. Même si les législations nationales évoquées sont relativement récentes, il est possible de montrer que la volonté d'unité de la fonction publique, qui implique qu'une organisation commune soit établie pour toutes les collectivités territoriales est discutable si l'on veut adapter la fonction publique locale au développement de l'idée de décentralisation.

B- Le caractère discutable de l'unité statutaire

531. - Le droit de la fonction publique ne peut se penser sans la notion de statut. Tout le récit sur la fonction publique est construit autour de cette notion[1212]. L'idée statutaire implique que la condition du fonctionnaire est régie de façon unilatérale par l'employeur public sur une base légale et règlementaire qui en fait un droit spécial étranger aux considérations du droit du travail. Dans le cadre des réformes décentralisatrices, il a été question en Afrique comme partout ailleurs de la détermination d'un cadre de gestion du personnel administratif des collectivités territoriales. Dans un premier temps, l'application du statut général de la fonction publique a été la règle avant l'adoption de dispositions spéciales relatives à la fonction publique territoriale.

532. - L'orientation de ce statut vers le régime général suscite quelques interrogations. Les débats relatifs au statut des agents publics au sein des collectivités territoriales révèlent une tension entre d'une part le maintien du cadre statutaire et d'autre part la soumission des agents publics décentralisés aux règles de droit privé. Au-delà de cette tension, le processus de décentralisation souffre moins de la logique statutaire que du cadre même de la gestion publique dans les administrations publiques africaines. La fonction publique dans les États d'Afrique noire francophone est une institution en quête permanente de légitimité[1213]. La gestion des administrations publiques construite autour du système de la carrière est marquée notamment par l'insuffisance de la logique

---

1211 Le constat est partagé partout en Afrique francophone. Pour une perspective plus globale : CGLUA, *1er Rapport sur l'état des ressources humaines dans les collectivités territoriales en Afrique*, 2018.

1212 A. TAILLEFAIT, *Droit de la fonction publique, op.cit.*, n° 34.

1213 Voir sur ce point : M. NEMBOT, *Le glas de la fonction publique dans les États d'Afrique noire francophone,* thèse, L'Harmattan, 2000 p. 254 et s.

managériale, l'absence d'analyse en termes de compétences[1214], la routine, et une extrême politisation. Les mêmes causes produisant les mêmes effets, il ne fait aucun doute qu'en soumettant intégralement la fonction publique territoriale au même régime, son efficacité s'en trouverait réduite. Ainsi que le souligne Cossoba Nanako, l'on devrait prendre en compte le fait que la dynamique qui caractérise les compétences locales et donc les métiers au sein des administrations locales nécessite une certaine souplesse en vue d'une adaptation régulière[1215]. La question du droit applicable aux agents de l'État et aux agents des collectivités territoriales ne se pose pas tout à fait dans les mêmes termes en raison de l'étendue et la nature des compétences décentralisées[1216].

533. - Dans l'exercice de leur fonction d'employeur public, les règles statutaires ne sauraient avoir les mêmes implications pour l'État que pour une commune de petite taille en milieu rural et une grande commune en milieu urbain. Les besoins, les moyens humains ainsi que les modalités de fourniture de services ne peuvent être identiques. Ainsi, uniformiser les statuts et les normes aboutirait à une hypocrisie sournoise qui s'exprimerait en pratique par un contournement des règles établies[1217]. Les personnes publiques territoriales sont tous les jours confrontées à des problématiques nouvelles avec des crises permanentes qui appellent de leur part des réponses rapides et efficaces. Leur proximité avec les populations accroit les interpellations. Pour y répondre, il faut un personnel de qualité, motivé, engagé et doté de capacités d'intervention qui inspirent confiance aux citoyens. Dans cette perspective, il est important d'accorder aux collectivités les marges d'action nécessaires pour manœuvrer entre les différentes exigences. Cela passe notamment par la

---

1214 Ce qui ne constitue pas une spécificité africaine, voir, H. MUSCAT, « La décentralisation des personnels en France et le statut général de la fonction publique », in G. GUIHEUX (dir.), *La mise en œuvre de la décentralisation : étude comparée France, Belgique, Canada, Bruxelles*, Bruylant | LGDJ, 2006, p. 237.

1215 C. NANAKO, *La libre administration des collectivités territoriales au Bénin et au Niger, op.cit.*, p. 224 et s.

1216 H. MUSCAT, « La décentralisation des personnels en France et le statut général de la fonction publique », *op.cit.*, p. 247 ; H. PORTELLI, « Pour une autonomie de gestion du personnel par les collectivités territoriales », *La gazette des communes*, *Cahier détaché*, n° 2-2/1916, 2008, p. 230. Malgré la forte tendance à l'assimilation, tout sépare les administrations d'État et les collectivités locales *« quant à leurs fonctions et quant à leur assiette territoriale et humaine »*.

1217 C. NANAKO, *La libre administration des collectivités territoriales au Bénin et au Niger, op.cit.*, p. 224.

généralisation du contrat comme moyen de recrutement et l'ouverture de la fonction publique aux règles managériales[1218].

Le recours au contrat pour le recrutement des agents territoriaux est encore très restreint. En effet, selon les dispositions de l'article 15, alinéa 3 de la loi précitée du 13 janvier 2017 au Burkina Faso, « *il ne peut être dérogé au principe de recrutement par concours que par décret pris en conseil des ministres* ». Les conditions de recours au contrat pour le recrutement des agents des collectivités territoriales sont particulièrement limitées. Au Niger, la dérogation au recrutement par la voie du concours des fonctionnaires territoriaux n'est possible que dans les deux cas suivants : « - *nécessité de recrutement sur titre pour pourvoir à certains emplois particuliers pour lesquels le nombre des candidats est inférieur au nombre de postes à pourvoir ; — au titre des emplois réservés pour les personnes handicapées dans la limite des quotas fixés par les statuts particuliers* »[1219]. Dans les cas où les collectivités territoriales sont obligées de recourir à des contractuels, le législateur togolais précise que le contrat ne peut être que de courte durée. Il ne peut être conclu que « *pour une durée maximale de deux (2) années et renouvelée une seule fois par reconduction expresse pour pourvoir à des emplois permanents* »[1220].

En droit français, il est reconnu aux collectivités territoriales de recourir à des contrats à durée indéterminée. Dans une décision du 1er août 2019, le Conseil constitutionnel a clairement indiqué que le recours aux contractuels ne porte pas atteinte au principe d'égal accès aux emplois publics dès lors que l'autorité compétente assure la publicité du recrutement et fonde sa décision sur les capacités de l'intéressé à exercer les missions pour lesquelles il postule[1221]. Les législations africaines pourraient s'inspirer de cette décision du juge constitutionnel français. En effet, l'élargissement du recours aux contractuels pour une longue durée permettra aux collectivités territoriales de combler leur déficit en ressources humaines compétentes notamment pour les emplois de conception, d'encadrement, d'orientation, de direction, de coordination, d'étude et de conseil. La fonction publique territoriale en Afrique ne peut se résumer en une simple question juridique. L'efficacité, la performance, l'efficience doivent devenir, sinon des règles juridiques, à tout le moins des

---

1218 Sur cette question, J. CHEVALLIER, « La judiciarisation des préceptes managériaux », *PMP*, n°4, 1993, p. 111.

1219 Art. 29 Loi 1er juillet 2019 *préc.*, au Niger. Voir aussi, art. 14 Loi du 27 juin 2018 *préc.*, au Mali.

1220 Art. 13 Loi du 11 juin 2008 *préc.*, au Togo.

1221 Décision n° 2019-790 DC.

méthodes de gestion de personnel administratif[1222]. Le législateur est ainsi appelé à plus de souplesse dans l'édification des règles d'administration et de gestion du personnel territorial.

534. - Cependant, la fonction publique territoriale ne peut se détacher complètement des grands principes du statut général. On peut en faveur de l'unité statutaire affirmer que l'emploi public présente une particularité qui justifie l'exclusion des règles de droit de privé et la soumission à un droit spécial[1223]. Cette particularité de l'emploi public implique l'affirmation et la promotion de valeurs telles que l'équité, la justice, la neutralité, etc. qui permettent de rendre l'action publique conforme à la logique de cohésion sociale. C'est ce qui explique la technique de la répétition de dispositions identiques ou similaires au statut général de la fonction publique dans les lois portant statut de la fonction publique territoriale. La spécificité de l'objet du service public est identique au niveau de l'État et des collectivités territoriales, donc les règles spécifiques qui s'appliquent aux agents de l'État doivent pouvoir s'appliquer *mutatis mutandis* aux agents des collectivités territoriales[1224]. Il faut simplement s'abstenir en la matière de toute systématisation qui se révèlerait finalement contre-productive[1225].

535. - La question des ressources humaines demeure permanente dans les États d'Afrique noire francophone. Dans l'incapacité de procéder au recrutement du personnel nécessaire à l'exercice des compétences locales, les collectivités restent dépendantes du personnel minimum transféré par l'État. On ne peut attendre des collectivités territoriales qu'elles favorisent le développement local et assurent la satisfaction des besoins des populations sans un personnel de qualité et en nombre suffisant. Les collectivités territoriales doivent s'investir davantage dans la gestion des

---

1222 K. B. BEKPOLI, *L'approche statutaire de la fonction publique. Approche comparée de la notion de statut général en France et au Togo*, thèse, dactyl., Université de Poitiers, 2020. Ces préoccupations sont également partagées par la doctrine française, voir en ce sens, A. TAILLEFAIT, « La loi de fonction publique du 6 aout 2019 de déformations des transitions professionnelles », *RDP*, 2021, p. 77.

1223 À ce propos lire utilement, A. SUPIOT, *Servir l'intérêt général*, PUF, coll. « Les voies du droit », 2000.

1224 H. MUSCAT, « La décentralisation des personnels en France et le statut général de la fonction publique », *op.cit.*, p. 247.

1225 C. NANAKO, *La libre administration des collectivités territoriales au Bénin et au Niger, op.cit.* p. 226 ; F. MELLERAY, « Le code général de la fonction publique : une arlésienne », *AJFP*, 2019, p. 309.

ressources humaines en procédant notamment, à l'élaboration d'un répertoire de métiers avec une approche par domaine d'activité, en partant des domaines d'intervention que la loi leur réserve. Une telle approche permettra un maillage métiers-activités-compétences en croisant à la fois l'analyse de l'organisation actuelle et les niveaux d'intervention potentiels des collectivités à la suite des transferts progressifs à venir[1226]. En fin de compte, ce processus aura l'avantage de fixer clairement les conditions de recrutement, d'établir des organigrammes et de pallier le manque de personnel qualifié. Sans ces hommes et femmes qualifiés, le processus de décentralisation demeurera inachevé et la bonne gouvernance locale restera un souhait vain. La gestion des ressources humaines constitue un défi majeur pour les collectivités tout autant que la question des ressources financières qui se pose d'ailleurs avec plus d'acuité.

Section 2 : Les contraintes liées au financement autonome des collectivités territoriales

536. - La complexité des relations financières entre l'État et les collectivités met en brèche la proclamation de l'autonomie financière des collectivités territoriales. Cette dernière est présentée comme la condition même de la réalisation de la décentralisation et du respect du principe constitutionnel de libre administration des collectivités. Les collectivités territoriales ne peuvent agir que dans la limite de leurs moyens dont la mobilisation s'inscrit dans un périmètre défini par l'État. La question de l'autonomie financière des collectivités territoriales a fait l'objet de multiples études doctrinales en France tout comme dans les États d'Afrique noire francophone[1227]. Elle est sans cesse renouvelée au gré des

---

1226 L. NGONO TSIMI, *L'autonomie administrative et financière des collectivités décentralisées, op.cit.*, p. 335.

1227 T. HOLO, « Décentralisation au Bénin : mythe ou réalité ? », *Revue béninoise des sciences juridiques et administratives*, n° 7, 1986, p. 1 ; A. ESSONO EVONO, « L'autonomie financière des collectivités locales en Afrique noire francophone. Le cas du Cameroun, de la Côte-d'Ivoire, du Gabon et du Sénégal », *Revue Afrilex*, n° spécial Finances Publiques, 2009 ; A. ROUX, « L'autonomie financière des collectivités locales en Europe », *Rapport introductif*, *Annuaire internationale de justice constitutionnelle*, 2006, p. 499 ; R. HERTZOG, « L'ambiguë constitutionnalisation des finances locales », *AJDA*, 2003, p.

réformes décentralisatrices et cristallise l'attention des observateurs du droit de la décentralisation. Impossible à définir malgré cette abondante littérature[1228], on s'accorde à dire que l'autonomie financière désigne l'état d'une collectivité territoriale « *disposant d'un pouvoir propre de décision et de gestion de ses recettes et de ses dépenses* »[1229]. De ce fait, le principe de l'autonomie financière peut être pensé de deux façons dont chacune s'oppose fondamentalement à l'autre. Dans une première acception, le principe confère aux collectivités une capacité juridique de décision en matière de recettes et de dépenses se traduisant pour la première dans un véritable pouvoir fiscal, le pouvoir de créer et de lever l'impôt puis, dans la seconde, en la liberté de décider d'affecter les ressources à telle ou telle dépense. Pris dans ce sens, l'autonomie de décision fiscale est quasi inexistante dans les États d'Afrique noire francophone. La seconde acception du principe renvoie à la possibilité pour les collectivités territoriales de gérer librement leurs dépenses par des ressources propres en quantité suffisante[1230]. Ainsi, une collectivité est considérée comme disposant d'un financement autonome seulement dans la mesure où elle dispose de ressources propres qu'elle a la capacité de gérer. Ces ressources doivent être suffisantes, prévisibles et disponibles pour assurer l'efficacité de l'intervention des collectivités. « *Avoir un patrimoine, avoir des agents, avoir des ressources financières*

---

548 ; L. NGONO TSIMI, *L'autonomie administrative et financière des collectivités décentralisées : l'exemple du Cameroun*, thèse, dactly., Université de Paris Est, 2010 ; J. R. KEUDJEU DE KEUDJEU, *Recherche sur l'autonomie des collectivités territoriales décentralisées au Cameroun*, thèse, dactly., Université de Douala, 2012.

1228 V. DUSSART, *L'autonomie des pouvoirs publics constitutionnels*, CNRS, 2000, p. 13. L'autonomie financière « *se mesure plus qu'elle ne se définit* ». R. HERTZOG, « L'ambiguë constitutionnalisation des finances locales », *AJDA*, 2003, p. 548. « *Si l'autonomie financière constitue un objectif politique faisant consensus, elle devient insaisissable et pétrie de contradictions dès qu'on veut l'enfermer dans une définition juridique apte à produire des effets normatifs* ».

1229 A. ESSONO EVONO « L'autonomie financière des collectivités locales en Afrique noire francophone. Le cas du Cameroun, de la Côte-d'Ivoire, du Gabon et du Sénégal », *Revue numérique Afrilex*, n° spécial Finances Publiques, 2009, p. 2.

1230 Pour A. ROUX, « (…) L'autonomie financière revêt une double dimension. En premier lieu, c'est la reconnaissance d'une capacité juridique de décision qui, en matière de recettes, implique un véritable pouvoir fiscal, le pouvoir de créer et de lever l'impôt et, qui en matière de dépenses implique la liberté de décider d'affecter les ressources à telle ou telle dépense. En second lieu, c'est la possibilité pour les collectivités régionales ou locales d'assurer le financement de leurs dépenses par des ressources propres en volume suffisant », in « L'autonomie financière des collectivités locales en Europe », A. ROUX, « L'autonomie financière des collectivités locales en Europe », op.cit., p. 499.

suffisantes, voilà qui est déterminant pour l'autonomie ainsi entendue. La capacité de produire des règles est, au mieux, accessoire »[1231].

537. - L'analyse globale de la situation des collectivités territoriales africaines révèle une insuffisance des ressources de ces dernières et une forte dépendance vis-à-vis de l'État. Finalement, le cordon ombilical entre le pouvoir central et les collectivités territoriales a du mal à se couper. Les ressources des collectivités territoriales africaines sont essentiellement constituées des concours financiers de l'État (§**1**) rendant impossible la construction d'une autonomie réelle. Les collectivités territoriales sont condamnées en l'état à être de simples gestionnaires de budget (§**2**).

§1.- Des collectivités territoriales pauvres et dépendantes de l'État

538. - Le financement de l'exercice des compétences décentralisées est assuré par l'État. Bien que ce dernier dans les États d'Afrique francophone n'ait pas une situation budgétaire enviable, il est le principal pourvoyeur du budget des collectivités territoriales. La loi impose à l'État de mettre à la disposition des collectivités territoriales les ressources nécessaires à l'exercice des leurs compétences. Le concours financier de l'État dans la mise en œuvre des compétences décentralisées traduit l'idée de coopération, de collaboration dans la mise en œuvre de la décentralisation[1232]. Le pouvoir central participe à l'exercice des compétences locales, notamment par la compensation financière du coût du transfert des compétences. La doctrine distingue généralement deux modalités du concours financier de l'État que sont la dotation et la subvention. La subvention est, en principe, « *un transfert monétaire alloué discrétionnairement et dans un but* précis *par une personne publique à un bénéficiaire quelconque* »[1233]. La dotation quant à elle est aussi un transfert monétaire, mais général et récurrent ayant comme finalité le financement des

---

1231 R. HERTZOG, « L'autonomie en droit : trop de sens, trop peu de significations », in *Mélanges Paul AMSELEK*, Bruylant, 2005 p. 445 et s.

1232 *Cf.* M. SOUSSE, « La notion financière de concours », *RFFP*, n° 53, 1996, p. 157.

1233 R. HERTZOG, « Linéament d'une théorie des subventions », *RFFP*, n° 23, 1988, pp. 6-7.

compétences décentralisées aux collectivités. Le droit positif des États ouest-africains francophones combine les deux modalités avec la mise en place d'un mécanisme national de financement des collectivités territoriales (**B**). Le financement par voie de dotations constitue une garantie de l'autonomie des collectivités territoriales. Toutefois, le financement des collectivités reste altéré[1234] si elles ne disposent de ressources propres suffisantes. Il s'agit là d'un principe clé de la libre administration des collectivités territoriales (**A**).

A- Des collectivités aux ressources propres insignifiantes

539. - Les administrations locales africaines font face à d'énormes difficultés dans la mobilisation des ressources propres pour l'exécution de leurs missions aussi bien pour l'entretien des infrastructures existantes que pour de nouveaux investissements. L'existence de ressources propres suffisantes est un élément indispensable de réalisation de la libre administration des collectivités territoriales[1235]. La mise en œuvre progressive des politiques de transfert de compétences entraine systématiquement une augmentation des responsabilités des collectivités territoriales. Les moyens financiers constituent le ressort principal de l'action des collectivités territoriales. En matière de décentralisation, la provenance et la nature des ressources locales emportent des conséquences sur le degré d'autonomie des collectivités territoriales. Pour André Roux, « *la qualité des ressources, leur origine, permet mieux de qualifier le degré d'autonomie financière des collectivités, car le risque existe, si les financements étatiques sont importants, c'est qu'ils soient attribués sous conditions, limitant la liberté des collectivités* »[1236].

540. - Les ressources des collectivités territoriales proviennent au sens de l'article 115 du CGCT au Burkina Faso des recettes ordinaires principalement, « *du produit des recettes fiscales, de l'exploitation du domaine foncier et des services locaux, des subventions, des ristournes accordés par l'État ou d'autres collectivités publiques sur le montant des impôts et taxes recouvrés à leur profit, des recettes temporaires ou accidentelles et de la répartition annuelle des fonds de dotation*

1234 N. GBEOU-KPAYILE, « Essai sur le financement des collectivités locales dans les États d'Afrique noire francophone : Les cas du Sénégal et du Togo », *Revue béninoise des sciences juridiques et administratives*, n° 33, 2014, p. 76.
1235 L. PHILIP, « L'autonomie financière des collectivités territoriales », *op.cit.*, p. 2.
1236 A. ROUX, « L'autonomie financière des collectivités territoriales en Europe », *op.cit.*, p. 505.

des collectivités territoriales». Pour l'essentiel, les collectivités assurent leur fonctionnement grâce aux recettes fiscales, aux aides de l'État et accessoirement aux recettes non fiscales issues du produit de l'exploitation des biens du domaine et des services rendus. Elles peuvent également recourir à des emprunts dans les conditions prévues par la loi. Les sources de financement des collectivités territoriales sont donc diverses. Parmi toutes ces sources, les recettes fiscales occupent une place importante. En effet, les collectivités territoriales doivent pouvoir disposer d'une fiscalité locale propre. De plus, la part des ressources propres doit être déterminante dans le montant global des ressources de la collectivité.

541. - La loi a entendu reconnaitre une autonomie financière aux collectivités territoriales en leur reconnaissant des prérogatives en matière fiscale. Il ne faut cependant pas se méprendre sur la réelle portée de ce « pouvoir fiscal ». Les collectivités territoriales sont constitutionnellement incompétentes pour créer de nouveaux impôts. Cette compétence appartient exclusivement au législateur[1237]. Toutefois, pour la maitrise de leurs ressources, la loi habilite les collectivités territoriales à fixer le taux des impôts locaux. C'est le sens des alinéas 2 et 3 de l'article 332 de la loi n° 2019-006 précitée qui dispose que, « *la création des impôts et taxes relève du domaine de la loi. Le conseil local, par délibération, en fixe le taux dans les conditions déterminées par la loi de finances* ». La loi n° 2017-051 du 2 octobre 2017 portant code des collectivités territoriales au Mali prévoit également que, « *le principe de la légalité de l'impôt implique que la création des impôts et taxes est du domaine de la loi. L'organe délibérant de la collectivité territoriale, par sa délibération, fixe le taux des taxes et impôts locaux dans la limite du plafond déterminé par la loi* »[1238]. Aucune collectivité territoriale ne peut créer d'elle-même un impôt pour les besoins de son fonctionnement. Par conséquent, le pouvoir fiscal dont il est question ici renvoie à la capacité des collectivités territoriales de déterminer chaque année le taux de l'impôt que le législateur leur a attribué dans la limite des maximas fixés par la loi des finances[1239]. Ce pouvoir fiscal bien que réduit laisse une marge de manœuvre aux collectivités territoriales pour porter leurs ressources au niveau « *adaptés à* [leurs] *structure(s)*

1237 Voir not., art. 84 de la Constitution togolaise de 1992 ; art. 99 de la Constitution de 2010 du Niger ; art. 101 de la Constitution du Burkina Faso de 2012 : « *La loi fixe les règles concernant l'assiette, le taux et les modalités de recouvrement des impositions de toute nature* ».

1238 Art. 221 al. 1 du code des collectivités territoriales au Mali.

1239 CGCT Niger, voir les articles 30 pour le conseil municipal et 105 pour le conseil régional.

économique(s) et à [leurs] *objectifs de développement dans le cadre de la loi* »[1240]. Bien qu'il s'agisse d'une liberté encadrée et contrôlée, une telle habilitation constitue un progrès dans l'aménagement du principe de la légalité fiscale[1241]. Loïc Philip écrivait à ce propos que, « *l'autonomie financière n'implique pas nécessairement un véritable pouvoir fiscal local* »[1242].

542. - Cependant, « le pouvoir de voter les taux d'imposition [...] ne garantit pas mécaniquement la possibilité de mobiliser suffisamment de ressources pour offrir un niveau jugé satisfaisant de services publics. Le rendement peut être très insuffisant si par exemple les bases d'imposition sont étroites ou en régression »[1243]. Il faut, en effet, que les collectivités disposent de nombreuses matières imposables pour garantir leur autonomie. Au Togo, dans les collectivités territoriales où s'exercent des activités particulières, susceptibles d'être imposées, le conseil de la collectivité peut, par délibération, créer des taxes non fiscales y afférentes, après approbation de l'autorité de tutelle et du ministre chargé des finances[1244]. Cette possibilité offerte par la nouvelle loi relative à la décentralisation et aux libertés locales permet de réduire le risque d'exclusion de certaines activités de l'assiette fiscale que relevait Nadjombé Gbeou-Kpayile[1245]. L'auteur soulevait en effet que, « l'un des problèmes majeurs de la fiscalité locale en Afrique est que certaines grandes activités économiques relevant du territoire local ne sont pas soumises à l'imposition locale, et par conséquent ne contribuent pas aux flux financiers locaux ».

543. - L'autonomie financière réelle des collectivités locales ne dépend pas simplement de la consécration d'un pouvoir fiscal au profit des collectivités territoriales. Elle s'ap*préc*ie en rapportant le montant des ressources propres au montant des ressources totales, desquelles il convient de soustraire les ressources d'emprunt qui sont des ressources provisoires. L'autonomie financière exige que ces ressources propres

---

1240 Art. 221 al. 2 du code des collectivités territoriales au Mali.

1241 C. NANAKO, *La libre administration des collectivités territoriales au Bénin et au Niger*, *op.cit.*, p.277.

1242 L. PHILIP, « L'autonomie financière des collectivités territoriales », *op.cit.*, p. 2 et s.

1243 A. GUENGANT, J. M. UHALDEBORDE, « Économie et finances locales », *Annuaire des collectivités locales*, 2003, p. 430.

1244 Art. 332 al. 2 Loi n° 2019-006 *préc.*

1245 N. GBEOU-KPAYILE, « Essai sur le financement des collectivités locales dans les États d'Afrique noire francophone : Les cas du Sénégal et du Togo », *op.cit.* p. 76.

constituent une part déterminante des ressources globales des collectivités locales[1246]. La part que représentent les ressources propres constitue une grille d'appréciation du pouvoir d'action des collectivités territoriales. En effet, elles ne peuvent s'administrer librement sans disposer de ressources propres[1247]. Selon le décret n° 2020-004/PR portant régime financier des collectivités territoriales [1248]*« les ressources propres des collectivités territoriales sont constituées du produit des impôts et taxes de toute nature autorisés par la loi, des redevances pour services rendus, des produits du domaine, des participations d'urbanisme, des produits financiers ainsi que des dons et legs »*[1249]. Cette définition des ressources propres se rapproche de celle française figurant à l'article 1114-2 du CGCT qui dispose que, « *au sens de l'article 72-2 de la Constitution, les ressources propres des collectivités territoriales sont constituées du produit des impositions de toutes natures dont la loi les autorise à fixer l'assiette, le taux ou le tarif, ou dont elle détermine, par collectivité, le taux ou une part locale d'assiette, des redevances pour services rendus, des produits du domaine, des participations d'urbanisme, des produits financiers et des dons et legs* ».

Pour être considérées comme des ressources propres, les collectivités territoriales doivent en avoir la maitrise. En clair, les ressources propres peuvent se définir comme celles dont la création, la suppression, les modalités d'assiette, de liquidation et de recouvrement dépendent des décisions des autorités décentralisées[1250]. Ce pouvoir de modulation est très important. Il permet d'exclure des ressources propres des collectivités

---

1246 A. ESSONO EVONO « L'autonomie financière des collectivités locales en Afrique noire francophone. Le cas du Cameroun, de la Côte-d'Ivoire, du Gabon et du Sénégal », *op.cit.*, p. 17.

1247 Loi n° 2019-006 préc., art. 6 : « Les collectivités territoriales s'administrent librement par des conseils élus dans les conditions prévues par la loi. Elles disposent de ressources propres ».

1248 Ce décret a été adopté en application de la directive n° 01/2011/CM/UEMOA portant régime financier des collectivités territoriales au sein de l'Union économique monétaire ouest africain (UEMOA). Le régime financier des collectivités territoriales de l'espace UEMOA dont fait partie les pays de la présente recherche est quasi-identique. Voir not., l'art. 51 du décret n°2016-302 PRN/MISP/D/ACR/MF du 29 juin 2016, portant régime financier des collectivités territoriales en République du Niger.

1249 Art. 44, décret n°2020-004/PR portant régime financier des collectivités territoriales au Togo. Voir aussi, S. BAZIADOLY, « La Charte européenne de l'autonomie locale et l'autonomie financière des collectivités locales françaises », *RGCT*, n° 29, 2003, p. 721, les ressources propres sont celles « *dont le montant dépend d'une décision prise par les collectivités locales : elles peuvent les instituer ou non et moduler leur assiette, elles en fixent le tarif ou le taux* ».

1250 A. ESSONO EVONO « L'autonomie financière des collectivités locales en Afrique noire francophone. Le cas du Cameroun, de la Côte-d'Ivoire, du Gabon et du Sénégal », *op.cit.*, p. 18.

les concours financiers de l'État regroupant notamment la fiscalité locale partagée. Pour Jean-Marie Pontier, « *du point de vue des principes, en effet, on voit mal comment il est possible de considérer un impôt d'État comme une ressource propre, sauf à enlever l'essentiel de son sens à cette expression, tout au moins si l'on se place du point de vue des collectivités territoriales* »[1251]. Par conséquent, en matière de recettes fiscales, sur les douze catégories d'impôts et taxes prévus par la loi togolaise, seule six[1252] appartiennent en propres aux collectivités et leur sont intégralement reversés. Il s'agit de : la Taxe sur la voirie, la Taxe sur la distribution de l'eau, de l'électricité, et du téléphone, la Taxe sur l'exploitation des entreprises locales de communication, la Taxe sur les spectacles et sur les appareils automatiques procurant un jeu, un spectacle, une audition ou un divertissement (TSA), la Taxe de péage au cordon douanier et la Taxe d'habitation (TH). Au Niger, le code général des collectivités territoriales prévoit dix-huit catégories impôts directs propres aux communes notamment, la taxe sur les embarcations ; la taxe sur les artistes ; la taxe sur les charrettes exploitées à des fins lucratives ; la taxe sur les cycles, etc.[1253]

544. - Le problème fondamental des taxes que l'État réserve aux collectivités en Afrique est que ces taxes n'ont pas un rendement très élevé et ne peuvent donc résoudre les difficultés de financement des collectivités. Cette remarque appelle deux observations sur les ressources locales propres. D'abord sur les difficultés de mobilisation de ressources propres fiscales et non fiscales et ensuite sur la détermination du seuil à partir duquel on peut considérer que les ressources propres sont suffisantes et déterminantes.

Les collectivités territoriales africaines ne disposent ni de méthodes efficaces ni de la maitrise des modalités de mobilisation des ressources non fiscales pour l'obtention d'un rendement optimal. En effet, les collectivités territoriales font face à d'énormes difficultés pour la mobilisation des

---

1251 J.-M. PONTIER, « Sur les notions controversées : Ressources propres, Ensemble de ressources, Part déterminante », *Dr. Adm.*, n° 340, 2004, p. 397 : « *L'uniformisation et la perte d'emprise des* coll*ectivités infra étatiques sur ce type de transfert fragilisent sensiblement leur indépendance fiscale et reviennent sur les espérances des commentateurs de la loi organique [sic] du 28 mars 2003* ».

1252 Les six autres sont ; la Taxe foncière sur les propriétés bâties, la Taxe foncière sur les propriétés non bâties, la Patente, la Taxe professionnelle unique et les taxes directes assimilés, la Taxe de protection et d'entretien des infrastructures et le Taxe sur les produits de jeu de hasard.

1253 Art. 232 du CGCT au Niger.

recettes notamment, la redevance d'occupation du domaine public qui constitue la principale source de revenus. Les causes sont multiples allant du faible niveau de qualification des agents de recouvrement au manque de moyens logistiques. La plupart des agents de recouvrement ne disposent pas d'un bon niveau d'instruction et bénéficient non plus de dispositifs de formation continue. Aussi, les collectivités territoriales manquent de moyens roulants — motos ou véhicules — permettant d'assurer le recouvrement des taxes[1254]. Dans un rapport de 2020 sur la mise en œuvre de la décentralisation au Niger, les auteurs relèvent une politisation extrême dans le choix des collecteurs des taxes et redevances ainsi qu'une immixtion flagrante des élus locaux dans la collecte et le suivi du recouvrement[1255]. Les conseillers se « partagent » les collecteurs, les marchés et autres gares routières. Ils exercent des contrôles sur ces infrastructures économiques très lucratives et s'octroient des avantages indus sur cette base[1256]. Le recouvrement des ressources propres non fiscales devient une prébende pour les élus locaux qui s'assurent ainsi des revenus mensuels au détriment de la collectivité.

La mobilisation des ressources locales propres en Afrique noire francophone bute également sur des problèmes structurels. On constate une certaine désorganisation de la chaine d'émission et de recouvrement des impôts et taxes locaux. La faiblesse du système étant dû principalement à la mauvaise maitrise des rôles, l'insuffisante coordination entre les services, l'insuffisance des moyens matériels, humains et financiers au niveau des services d'assiette et de recouvrement, la motivation insuffisante des agents ainsi qu'à l'incivisme fiscal.

545. - Si le recouvrement des ressources non fiscales relève de l'entière responsabilité des collectivités territoriales, il en va autrement des ressources fiscales propres. Le pouvoir central domine toute la chaine de recouvrement des impôts locaux. La détermination de l'assiette, l'émission et le recouvrement des impôts locaux concédés par l'État aux collectivités relève de la responsabilité de l'administration fiscale de l'État. Les collectivités n'ont aucune maitrise de la « chaine fiscale ». Le principal problème que pose cette organisation de la gestion des principaux impôts locaux est que les collectivités décentralisées sont tributaires des

---

1254 P. AGATE, *Réflexion sur les difficultés de mobilisation des ressources propres non fiscales dans la commune de Kozah* I, Mémoire ENA, Cycle III, 2021, p. 9 et s.

1255 DGDCT, Rapport d'évaluation de la mise en œuvre de la décentralisation au Niger, 2020, p. 112 et s.

1256 *Idem.*

performances des services fiscaux de l'État et subisse le dirigisme de l'administration fiscale, ce qui porte atteinte à l'autonomie locale.

Cependant, le choix du législateur de confier le recouvrement à l'administration fiscale d'État permet d'éviter de grever les charges des collectivités territoriales par la mise en place d'une administration fiscale qui leur est propre au regard du montant relativement faible des ressources collectées[1257]. Dans son étude consacrée à la décentralisation et aux finances publiques en Afrique subsaharienne, Elsa Duret a mis en évidence le caractère marginal des ressources locales (propres et concours financiers). L'auteur évalue les ressources des collectivités locales africaines à moins de 1 % du produit intérieur brut et entre 1,4 et 9,5 % des recettes de l'État[1258]. Pour les pays de l'UEMOA[1259], la situation est encore plus difficile, les ressources locales ne représentent en moyenne que 0,7 % du PIB[1260]. En outre, les ressources propres des collectivités sont marquées par leurs fortes instabilités en raison de la faible prévisibilité des recettes. Les budgets locaux ne sont que très faiblement exécutés avec une moyenne en dessous des 50 %[1261].

546. - Quant à la détermination du seuil à partir duquel les ressources propres de la collectivité sont considérées comme « déterminant » pour l'exercice autonome de ces compétences, les législations africaines sont restées muettes. À partir de quel montant, le ratio d'autonomie financière est-il atteint ? Comment évaluer de la « part déterminante » des ressources propres ? En droit français, la loi organique n° 04-758 du 29 juillet 2004 relative à l'autonomie financière des collectivités locales prévoit que la part déterminante « *ne peut être inférieure au niveau constaté au titre de l'année 2003* »[1262]. Au cours de cette année, le ratio d'autonomie financière des collectivités locales françaises représentait environ 60 % pour les

---

1257 C. NANAKO, *La libre administration des collectivités territoriales au Bénin et au Niger, op.cit.*, p.281.

1258 E. DURET, *Réformes institutionnelles et finances publiques. L'exemple de la décentralisation en Afrique subsaharienne*, thèse de doctorat en sciences économiques, dactyl., Université Clermont Ferrand I, 2000, p. 232 et s.

1259 Il s'agit des huit pays de l'Union économique et monétaire ouest africain : Bénin, Burkina Faso, Côte d'Ivoire, Guinée-Bissau, Mali, Niger, Sénégal, Togo.

1260 C. VERGNE, « Décentralisation fiscale en Afrique francophone : note sur les transferts intergouvernementaux », Banque Mondiale, Région Afrique, AFTPR, Septembre 2009, p. 2.

1261 FMI, Rapport d'assistance technique - réussir la décentralisation financière, 2015, p. 19.

1262 Art. 4, L.O n° 04-758 du 29 juillet 2004.

communes, 58 % pour les départements et 41 % pour les régions. Selon Alexis Essono Evono, si l'on peut supposer qu'une telle référence est pertinente, « *elle demeure très éloignée des ratios enregistrés dans les pays de l'Afrique noire francophones, lesquels sont largement inférieurs à ce(s) chiffre(s). En effet, dans ces pays, les ressources propres des collectivités locales sont généralement insignifiantes par rapport à leurs ressources globales* »[1263]. Les collectivités territoriales dans les États d'Afrique noire francophone sont très pauvres. Leur ratio d'autonomie financière varie entre 7 et 12 %[1264]. La non-mise à disposition spontanée des chiffres par les administrations publiques africaines empêche une actualisation de ces chiffres. Toutefois, on note dans le document de programmation budgétaire et économique pluriannuel 2022-2024 au Mali, une augmentation des ressources propres en moyenne de 10,7 % sur la période 2016-2018. Ce qui porte la part des recettes propres dans le budget global des collectivités à 14,4 %[1265]. Cette augmentation est à saluer, mais à l'instar des autres pays, le montant des ressources propres reste encore insuffisant pour enclencher le développement local.

547. - Il urge dans le cadre de la construction d'un État décentralisé d'accroitre les ressources propres des collectivités territoriales. Les ressources propres composées de la fiscalité locale et autres recettes tirées du domaine ou des redevances pour service rendu doivent être optimisées, car elles constituent un enjeu important de la libre administration des collectivités territoriales. Il existe dans l'ensemble des pays objet de cette étude des marges sérieuses d'amélioration de la fiscalité locale. Cela passe notamment par une amélioration de la chaine fiscale et de la prévisibilité des recettes. L'instauration d'une gestion plus rigoureuse et transparente des équipements marchands, des produits du domaine, de l'expédition des actes d'état civil et de légalisation ; l'amélioration des services publics marchands comme les marchés, les abattoirs, les gares routières peut constituer une source de revenus importante pour les collectivités. En attendant la réalisation de ces réformes structurelles, les concours financiers de l'État représentent une part importante des budgets locaux.

1263 Voir les chiffres évoqués *Supra* ; A. ESSONO EVONO « L'autonomie financière des collectivités locales en Afrique noire francophone. Le cas du Cameroun, de la Côte-d'Ivoire, du Gabon et du Sénégal », *op.cit.*, p. 18.

1264 *Ibid.*, p. 19.

1265 Document de programmation budgétaire et économique pluriannuel 2022-2024 au Mali, p. 47. Disponible en ligne https://budget.gouv.ml/sites/default/files/DPBEP%2020222024_VF_22_04_2021_Après_RI.pdf, Consulté le 4 aout 2022.

Les collectivités sont dans une situation de dépendance, d'éternels assistés vis-à-vis de l'État qui devient le « *premier contributeur local* »[1266].

B- L'origine essentiellement étatique du financement des collectivités territoriales

548. - Les mécanismes de financement des collectivités diffèrent selon les objectifs poursuivis par les États. Les concours financiers gouvernementaux ne sont donc pas régis par un système unique dans les États d'Afrique noire francophone. Ces concours financiers de l'État en faveur des collectivités territoriales sont fondés sur deux motivations principales. Pour le pouvoir central, il est question d'une part de résorber le « déséquilibre vertical » de ressources financières entre l'État et les collectivités locales et surtout, d'autre part, de permettre une certaine péréquation visant à réduire les disparités excessives de ressources entre les collectivités territoriales et donc de prévenir, le « déséquilibre horizontal »[1267]. Ainsi que nous l'avons montré précédemment, les collectivités territoriales d'Afrique subsaharienne ont en partage leur extrême pauvreté, mais certaines collectivités situées dans les grandes agglomérations urbaines ont une capacité de mobilisation des ressources propres généralement plus élevée. Toutefois, même optimisées, les ressources propres des collectivités ne peuvent couvrir l'ensemble de leurs besoins. Les concours financiers de l'État permettent alors à la fois d'augmenter et d'équilibrer les revenus des collectivités locales afin d'assurer un exercice minimal des compétences décentralisées par l'ensemble des collectivités.

549. - Les inégalités étant de plus en plus croissantes dans les États d'Afrique subsaharienne en raison notamment de la montée de l'urbanisation, la péréquation présente l'intérêt de réduire les écarts de richesse entre les collectivités territoriales. Ce principe trouve sa justification dans l'idée de solidarité nationale et d'équilibre interrégional afin de permettre à chaque collectivité de disposer des possibilités comparables pour la fourniture de services publics aux habitants. Dans

---

1266 N. GBEOU-KPAYILE, « Essai sur le financement des collectivités locales dans les États d'Afrique noire francophone : Les cas du Sénégal et du Togo », *op.cit.*, p. 80.

1267 C. VERGNE, « Décentralisation fiscale en Afrique francophone : note sur les transferts intergouvernementaux », Banque Mondiale, Région Afrique, AFTPR, Septembre 2009, pp. 2-3.

certains États, notamment le Togo et le Niger, c'est une exigence constitutionnelle. L'État a l'obligation de « *veille (r) au développement harmonieux de toutes les collectivités territoriales sur la base de la solidarité nationale, de la justice sociale, des potentialités régionales et de l'équilibre inter-régional* »[1268]. Au Mali, c'est la loi n° 2017-052 du 2 octobre 2017 déterminant les conditions de la libre administration des collectivités territoriales qui précise en son article 28 que « *l'État garantit et organise le principe de solidarité entre les collectivités territoriales* ».

550. - La mise en œuvre du principe constitutionnel et législatif de solidarité passe par la création d'institutions financières spécialisées et des Fonds d'appui à la politique nationale de décentralisation dans les États africains. Au Togo, la loi n° 2019-006 précitée relative à la décentralisation et aux libertés locales institue un Fonds d'appui aux collectivités territoriales dont les modalités d'organisation et de fonctionnement sont confiées au pouvoir règlementaire. C'est le sens des dispositions de l'article 14 qui dispose qu'« *il est créé un fond d'appui aux collectivités territoriales (FACT). Les modalités d'organisation et de fonctionnement du FACT sont définies par décret en conseil des ministres* ». C'est également le cas au Burkina Faso avec le Fonds permanent pour le développement des collectivités territoriales (FPDCT) créé par l'article 141 de la loi n° 055-2004/AN du 21 décembre 2004 portant CGCT. Dans le cadre du processus d'accompagnement de la politique de décentralisation, l'État du Niger a également procédé par la loi n° 2008-38 du 10 juillet 2008 à la création d'une Agence nationale pour le financement des collectivités territoriales (ANFICT). Cette Agence est chargée de la gestion du Fonds d'appui à la décentralisation et du Fonds de péréquation. Le Mali quant à lui s'est doté d'une Agence nationale d'investissement des collectivités territoriales (ANICT) avec l'adoption de la loi n° 00-042 du 7 juillet 2000. L'ANICT a pour mission principale d'accompagner les collectivités maliennes dans la réalisation des opérations d'investissements et d'assurer la péréquation entre les collectivités.

551. - La péréquation dans les États d'Afrique francophone est essentiellement une péréquation verticale qui s'exerce par l'intermédiaire des dotations et/ou des subventions de l'État. Les concours financiers gouvernementaux s'exercent donc par des modalités diverses. Quelle que

---

[1268] V. art. 142 de la Constitution togolaise de 1992 ; art. 165 de la Constitution nigérienne de 2010.

soit la forme choisie, c'est la détermination exacte du montant à octroyer qui cristallise les débats. La formule de péréquation choisie doit être à la fois juste et transparente. Il peut s'agir d'un pourcentage fixe de l'ensemble des recettes nationales ou d'impôts et taxes spécifiques. Il peut également être constitué soit d'un montant fixe annuellement déterminé, soit d'un montant variable déterminé annuellement par le pouvoir central. Ainsi, certaines modalités s'établissent sur des références objectives pendant que d'autres fonctionnent à la discrétion des autorités gouvernementales[1269]. Cette dernière hypothèse correspond à la situation togolaise et nigérienne. En effet, il n'y a pas dans ces deux États une véritable planification budgétaire pour la réalisation des transferts financiers. Les montants attribués sont aléatoires. Au Togo, par exemple, la dotation 2022 du Fonds d'appui aux collectivités territoriales est passée à 4 milliards de F CFA contre une dotation de 6, 25 milliards en 2021, soit une baisse de 36 %. Au Niger, selon les derniers chiffres à notre disposition, les concours financiers de l'État aux collectivités territoriales s'élèvent en milliers de FCFA à 15. 677. 504 entre 2014 et 2018. Les partenaires techniques et financiers (PTF) ont contribué au financement des collectivités territoriales à travers l'Agence nationale de financement des collectivités territoriales pour un montant total de 26 547 300 732 F CFA pour la période 2016-2019[1270].

Au Mali, à la suite des Accords d'Alger pour la paix et la réconciliation en 2015, l'État s'est engagé à transférer aux collectivités 30 % de ces recettes budgétaires à l'horizon 2018. Ce pourcentage n'est toujours pas atteint à ce jour. Il faut tout même reconnaitre que le montant des transferts aux collectivités maliennes est en constante évolution. Dans la loi de finances 2018, les ressources transférées aux collectivités territoriales s'élevaient à 19,8 % des recettes budgétaires (recettes fiscales et non fiscales hors recettes en capital). Entre 2019 et 2021, les efforts budgétaires se sont poursuivis au profit des collectivités territoriales. À cet effet, les transferts budgétaires en pourcentage des recettes budgétaires se situent respectivement à 21,4 % et plus de 26 % entre 2020-2021. En somme, entre 2018 et 2021, les transferts budgétaires aux collectivités territoriales ont connu un taux d'accroissement annuel moyen de 40,3 %

1269 C. NANAKO, *La libre administration des collectivités territoriales au Bénin et au Niger*, *op.cit.*, p. 304.

1270 DGDCT, Rapport d'évaluation de la mise en œuvre de la décentralisation au Niger, 2020, p. 112 et s.

tandis que les recettes budgétaires de l'État n'ont évolué que de 6,5 % durant la même période[1271].

Au Burkina Faso, le Plan national de développement économique et social (PNDES) avait fixé comme objectif en 2018 un transfert aux collectivités équivalent au moins à 15 % du budget national à partir de 2020. Cet objectif n'a pu être atteint. En 2020, seuls 11,39 % du budget national a été transféré aux collectivités territoriales. On note une hausse considérable par rapport à 2019 où ce taux était de 4,89 %.

552. - De manière générale, on peut considérer que les États d'Afrique noire francophone fournissent des efforts pour une hausse des montants des dotations, mais ces montants restent somme toute encore très faibles. Toutefois, les transferts ou les dotations accordées par les États aux collectivités territoriales constituent une part importante de leurs ressources globales, soit les 2/3 du budget des collectivités territoriales. Au Mali, entre 2016 et 2018, le poste de recettes de dotation et subvention de l'État représente 75,4 % du montant total des ressources locales[1272].

La dotation annuelle de l'État aux collectivités constituée essentiellement des contributions de l'État fixé par la loi de finances sert à couvrir les dépenses courantes des collectivités. Cette dotation est répartie entre les différentes collectivités en tenant compte du déficit entre leurs charges obligatoires de fonctionnement et leur niveau de mobilisation des ressources budgétaires[1273]. Elle contribue selon l'article 7 du décret fixant les modalités de fonctionnement du Fonds d'appui à la décentralisation au Niger « *à la couverture des charges obligatoires de fonctionnement des collectivités territoriales notamment, le fonctionnement des organes dirigeants et des commissions spécialisées, les charges du personnel minimum ainsi que toute autre charge liée aux transferts de compétences* ». Les modalités de cette répartition sont déterminées par arrêté conjoint du ministre chargé des collectivités territoriales et du ministre chargé des finances[1274] sur proposition de la commission de gestion du Fonds d'appui aux collectivités territoriales[1275].

[1271] *Document de programmation budgétaire et économique pluriannuel 2022-2024 au Mali*, p. 46. Disponible en ligne https://budget.gouv.ml/sites/default/files/DPBEP%2020222024_VF_22_04_2021_Après_RI.pdf, Consulté le 4 aout 2022.

[1272] *Ibid.*, p. 47.

[1273] Art. 6 décret n° 2014-136 fixant les modalités de fonctionnement du Fonds d'appui à la décentralisation au Niger.

[1274] *Ibid.*, art. 7.

[1275] *Ibid.*, art. 19; Loi n° 2019-006, art. 312.

553. - Les collectivités territoriales reçoivent également une dotation/subvention annuelle d'investissement. Elle concrétise la vocation des différents Fonds d'être des « *instrument(s) contribuant à la mise en œuvre de la politique du gouvernement en matière de réduction de la pauvreté et de gouvernance locale* »[1276]. La dotation d'investissement sert également au renforcement des capacités des collectivités territoriales en matière de maitrise d'ouvrage. Pour l'attribution des dotations ou subventions d'investissement, les institutions financières spécialisées s'appuient sur une formule de péréquation qui combine *grosso modo* deux séries de critères : la situation de la collectivité et la performance. Le critère de situation prend en compte le nombre d'habitants de la collectivité, l'éloignement de la collectivité par rapport aux grands centres d'approvisionnement commercial, l'indice de pauvreté local et celui de la performance, notamment la bonne administration fiscale des collectivités[1277].

Au Togo, la dotation d'investissement est composée de dotations affectées et non affectées. Les dotations affectées doivent servir à l'exécution par la collectivité des investissements dans les compétences décentralisées en fonction des destinations prédéfinies par secteur[1278]. Les dotations non affectées sont composées de trois éléments : une dotation de base, une dotation de péréquation et une dotation de performance. La dotation de base tient compte du statut de la collectivité et est répartie de façon égalitaire entre les différentes collectivités, ensuite la dotation de péréquation dont les paramètres discriminants portent sur l'effectif de la population, la superficie de chaque collectivité, le niveau de pauvreté et enfin, une dotation de performance dont l'objectif est d'inciter les collectivités à une meilleure gouvernance. Pour l'application de ce dernier critère, sont pris en compte, entre autres, le nombre de sessions ordinaires tenues par le conseil local et par les commissions permanentes du conseil, le respect des délais pour l'adoption du budget primitif et la transmission du compte administratif. Il s'agit de critères de répartition qui sont globalement suivis par les institutions financières venant en appui aux collectivités territoriales. Dans le cas du Togo, la gestion des dotations affectées d'une valeur d'environ 2,5 milliards de FCFA a été confiée à l'Agence nationale d'appui au développement à la base (ANADEB)

1276 Art. 3 décret n° 2007-254 *préc.*

1277 UE, Rapport 2021 sur la performance de la gestion des Finances publiques au Mali sur la période 2018-2020, octobre 2021, p. 48.

1278 Art. 19 décret n° 2019-130/PR *préc.*

chargée d'exécuter des projets d'investissement socio-économiques et éducatifs des collectivités au titre de la maitrise d'ouvrage déléguée. En outre, elle est chargée de recevoir les projets des communes et d'en évaluer la pertinence. Les communes togolaises sont ainsi réduites à un rôle d'inspecteur de travaux. Que reste-t-il de l'autonomie locale dans ces conditions ? Rien ! du moins pas grand-chose.

554. - Il est possible d'émettre des réserves sur les critères de répartition retenus par les institutions financières africaines. En effet, l'opérationnalisation des critères retenus requiert la collecte d'un certain nombre de données statistiques et des indicateurs fiables. Or, les données statistiques sont une denrée rare dans les pays de l'Afrique noire francophone. Ainsi, la prise en compte des données démographiques peut compromettre la fiabilité de la répartition. Il faut saluer l'introduction du critère de superficie qui permet *in fine* de ne pas défavoriser les collectivités peu peuplées et à très grande superficie.

555. - Le constat général qu'on peut dresser à l'issue de cette démonstration est que les ressources des collectivités territoriales bien qu'elles soient constituées en majorité des concours financiers de l'État en Afrique noire francophone demeurent encore insignifiantes. À défaut de ressources propres suffisantes, les collectivités territoriales se retrouvent dans une situation de forte dépendance vis-à-vis de l'État. Elles sont réduites à un rapport de subordination et condamné à être de simples gestionnaires de budgets.

§2.- Les collectivités territoriales, de simples gestionnaires

556. - Les ressources des collectivités territoriales sont insuffisantes, qu'il s'agisse des ressources propres ou des dotations étatiques, pour leur permettre de fournir des services publics de qualité aux populations. Dans ces conditions, l'autonomie financière est une arlésienne dans les États d'Afrique noire francophone. La consécration constitutionnelle de la libre administration des collectivités territoriales n'a pas entrainé un renforcement du pouvoir de décision des collectivités en matière financière. Si les pouvoirs publics ne manquent pas de faire appel aux collectivités territoriales pour participer à l'effort de développement, ces dernières ne disposent pas de moyens d'agir. En réalité, il est théoriquement difficile dans les États d'Afrique francophone dont la forme constitutionnelle est unitaire de concevoir une véritable autonomie financière avec notamment un pouvoir fiscal (**A**). La situation peu reluisante des finances des collectivités décentralisées les contraint à rester des gestionnaires de budgets limités par le contrôle permanent du pouvoir central (**B**).

A- <u>Une conséquence de la forme républicaine de l'État</u>

557. - La question du pouvoir fiscal n'a pas véritablement fait l'objet de débats lors de l'adoption des textes législatifs relatifs à la décentralisation dans les États d'Afrique noire francophone en raison du caractère unitaire de l'État. Les pays africains dotés de systèmes juridiques d'inspiration française rejettent toute idée de fédéralisme. Malgré les tendances fortes à conduire des réformes en matière de décentralisation, le caractère unitaire de l'État ne peut aucunement être remis en cause. Pas plus qu'elle ne possède un pouvoir législatif, les collectivités décentralisées ne peuvent prétendre à une quelconque souveraineté financière. Pour Michel Bouvier, le pouvoir fiscal avant d'être une technique économique ou juridique est un fait politique majeur, car il ne peut y avoir de pouvoir politique autonome sans pouvoir fiscal. Il poursuit son analyse en montrant que tout pouvoir fiscal concurrent à celui de l'État peut provoquer l'éclatement de ce dernier en de multiples féodalités[1279]. La décentralisation n'est pas le fédéralisme, écrivait Robert Hertzog dans un article sur le système financier local en droit français[1280]. Le pouvoir financier des collectivités demeure alors résiduel. Comme la France, les États d'Afrique noire francophone sont marqués par la doctrine de la politique de « l'égalitarisme républicain » qui trouve son aboutissement juridique, en matière de décentralisation, dans le principe d'égalité entre les collectivités territoriales et les mécanismes de péréquations entre collectivités[1281].

558. - Le principe d'égalité des collectivités territoriales et l'institution d'une solidarité nationale constituent autant d'obstacles à la reconnaissance de l'autonomie financière. Ils permettent à l'État d'intervenir directement dans les ressources des collectivités pour éviter

---

1279 M. BOUVIER, « De la réforme du pouvoir fiscal local à la nouvelle gouvernance financière publique », *RFFP*, n° 98, 2007, p. 3.

1280 R. HERTZOG, « Le système financier local en France : la décentralisation n'est pas le fédéralisme », *RIDC*, n° 2, 2002, pp. 613- 638.

1281 R. DECHAUX, « Les garanties constitutionnelles de l'autonomie financière locale à l'épreuve des concours financiers étatiques », *RDP*, n° 2, 2010, p. 349. Voir aussi art. 9 de la Charte européenne de l'autonomie locale sur la nécessité d'une péréquation.

que s'installent des inégalités territoriales. Raphael Dechaux montre à ce titre qu'en théorie, l'autonomie financière locale et le mécanisme de péréquation nationale sont antinomiques. En conséquence, moins l'État est unitaire, plus les différences entre les collectivités le composant peuvent s'avérer importantes[1282]. Au regard de l'antinomie entre les deux principes, autonomie financière locale et État unitaire, l'auteur propose de choisir entre une véritable décentralisation normative de l'État avec un pouvoir législatif local et donc une autonomie financière des collectivités et le maintien du modèle d'État centralisé sans autonomie financière, mais qui tend au maintien d'une certaine uniformité territoriale avec l'exultation du principe d'égalité[1283].

559. - Pour les États d'Afrique noire francophone, le choix semble clair. L'État reste l'unique garant de l'unité nationale et n'entend pas confier la régulation des rapports financiers aux collectivités territoriales. La réalité est que le pouvoir fiscal n'a pas de place dans la détermination de l'autonomie financière des collectivités territoriales. Pour le reste, elle demeure hypothétique. Le pouvoir central à la maitrise de la composition, du montant ainsi que de la destination des dotations, ce qui restreint la capacité de faire des collectivités et relativise par ailleurs la distinction formelle entre subvention et dotation[1284]. Dans les États d'Afrique subsaharienne notamment au Togo, les dotations comme les subventions sont marquées par leur caractère unilatéral et donc peu confortatif pour la libre administration[1285].

560. - Il semble donc qu'il existe des limites à la fois théorique et pratique à la mise en place d'une autonomie financière réelle des

---

1282 *Ibid.*, p. 349. Voir aussi A. ROUX, « L'autonomie financière des collectivités territoriales en Europe », *op.cit.*, p. 493. C'est la raison pour laquelle en Allemagne, il est impossible de décentraliser totalement certains mécanismes de protections sociales qui sont garantis par l'alinéa 11 du Préambule de 1946.

1283 R. DECHAUX, « Les garanties constitutionnelles de l'autonomie financière locale à l'épreuve des concours financiers étatiques », *op.cit.*, p. 349.

1284 Nous avons montré précédemment que les dotations étaient orientées vers la réalisation d'opérations *préc*ises dans les quatre pays objet de la présente étude.

1285 C. NANAKO, La libre administration des collectivités territoriales au Bénin et au Niger, op.cit., p. 308. ESSONO EVONO déplore aussi leur caractère unilatéral puisque « les collectivités locales n'ont aucune prise sur ces concours financiers : ils sont attribués unilatéralement par l'État, selon des critères qui ne sont pas objectifs » in, « L'autonomie financière des collectivités locales en Afrique noire francophone. Le cas du Cameroun, de la Côte-d'Ivoire, du Gabon et du Sénégal », *op.cit.*, p. 7.

collectivités territoriales. Les collectivités territoriales africaines ont perdu en liberté ce qu'elles gagnent marginalement en ressources et sont finalement réduites[1286] à la simple gestion de ces dernières.

B- L'autonomie financière, une simple liberté de gestion des collectivités

561. - Le lien entre l'autonomie financière et le principe constitutionnel de libre administration a été entretenu pendant longtemps par la doctrine. Cependant, la garantie de l'autonomie financière n'a de sens qu'accompagné d'une certaine autonomie normative. Or, dans les États unitaires décentralisés, les collectivités territoriales disposent uniquement d'un pouvoir limité par la loi en matière de finances locales[1287]. Existe-t-il encore une autonomie financière ? Un principe de libre administration des collectivités territoriales africaines ? Michel Troper concluait un article sur l'analyse théorique du concept de libre administration en 1980 ainsi : *« La "libre administration" est donc un terme vague et vide (privé de référence), dont la fonction est de transposer au niveau administratif l'idéologie politique de la démocratie représentative »*. Jean-François Brisson observe pour sa part que le principe de libre administration, bien que constitutionnalisé en France, comme c'est le cas dans les États africains issu de la tradition républicaine française, reste un concept juridique assez mystérieux, car, au fond, le législateur n'a pas de repères précis quant aux minimums d'exigences relatives à l'application de ce principe[1288]. Malgré l'absence de critères pour donner un contenu précis au principe de libre administration, l'autonomie financière suppose l'existence d'une marge de manœuvre suffisante des collectivités territoriales en matière financière. Force est de constater que les collectivités territoriales africaines ne disposent, ainsi que nous l'avons montré, ni de pouvoirs étendus pour le recouvrement de leurs ressources ni d'une liberté dans l'utilisation des ressources budgétaires. Le « pouvoir financier » local est soumis à un encadrement très strict qui rend impossible la conception d'une autonomie financière locale.

1286 Pour emprunter la formule « La collectivité subventionnée a, peu à peu, perdu en liberté ce qu'elle gagnait en ressources » de J. BOULOUIS, Essai sur la politique des subventions administratives, Armand Colin, 1951, p. 9.

1287 L. NGONO TSIMI, L'autonomie administrative et financière des collectivités décentralisées, op.cit., p. 200.

1288 J.-F. BRISSON, « La réforme du système d'administration aéroportuaire : bilan et perspectives », in *La réforme aéroportuaire de la loi du 13 août 2004 à l'épreuve des faits*, L'Harmattan, 2011, p. 1153.

562. - L'analyse du pouvoir dépensier des collectivités territoriales est un révélateur des contraintes juridiques qui pèsent sur elles et limite leur pouvoir de décision. Les collectivités territoriales n'ont pas la maitrise de leurs dépenses. Le législateur affirme que les collectivités peuvent utiliser leurs ressources pour les dépenses de leur choix. Pourtant, la loi dresse une liste de dépenses obligatoires dont le contenu est loin d'être exhaustif dans certains États. La notion de dépense obligatoire fait son entrée en Afrique à partir de 1955 avec l'adoption de la loi n° 55-1489 du 18 novembre 1955 relative à la réorganisation municipale en Afrique-Occidentale Française, en Afrique-Équatoriale Française, au Togo, au Cameroun et à Madagascar. Cette notion est reprise aujourd'hui dans l'ensemble des législations nationales pour désigner toutes les dépenses qui ont pour objet l'exécution d'une loi, l'accomplissement d'une obligation publique ou privée ou qui touchent essentiellement à l'existence même de la collectivité territoriale. Aux termes de l'article 121 du CGCT au Burkina Faso « *sont obligatoires pour les collectivités territoriales les dépenses mises à leur charge par la loi. Les dépenses obligatoires doivent nécessairement figurer au budget. Elles doivent faire l'objet d'ouverture de crédits suffisants avant qu'il ne soit possible à la collectivité d'inscrire des dépenses facultatives* ». Les dépenses obligatoires couvrent les dépenses de fonctionnement des services, le traitement et les indemnités du personnel, les intérêts et l'amortissement des emprunts, ainsi que les charges résultant des condamnations judiciaires, les dépenses pour la salubrité et la qualité de l'environnement, etc.[1289]

563. - En cas d'omission d'inscription d'une dépense obligatoire par une collectivité ou lorsque les crédits affectés sont insuffisants, l'autorité de tutelle peut les inscrire d'office. En effet, l'article 221 de l'ordonnance n° 2010-054 portant CGCT au Niger prévoit qu'en cas de non-inscription ou d'une inscription insuffisante de dépenses obligatoires, de non-inscription ou de minoration de l'autofinancement brut local, le représentant de l'État peut demander une seconde lecture du budget qui ne peut être refusée. Si après cette deuxième lecture, les observations de l'autorité de tutelle ne sont pas prises en compte, il met en demeure la collectivité de le faire. Si une réaction positive n'est pas enregistrée dans les quinze jours suivant ladite mise en demeure, le budget est réglé par l'autorité de tutelle qui en adresse notification à l'ordonnateur et au comptable de la collectivité. De même au Togo, si après une mise en

---

[1289] Art. 337 Loi n° 2019-006 *préc.* ; art. 243 & 244 du CGCT au Niger.

demeure restée sans réponse dans un délai de trente jours la collectivité ne se conforme pas à la mise en demeure, l'autorité de tutelle inscrit d'office cette dépense au budget de la collectivité et fait injonction au conseil de l'accompagner de ressources nécessaires[1290]. Cette inscription d'office d'une dépense obligatoire au budget de la collectivité lie les autorités décentralisées qui sont tenues de l'exécuter. L'article 273 du code des collectivités territoriales au Mali rappelle en ce sens que, « *lorsque le mandatement d'une dépense obligatoire inscrite au budget n'est pas effectué, le représentant de l'État peut mettre en demeure l'ordonnateur de l'effectuer. Si cette mise en demeure n'est pas suivie d'effet dans le délai d'un mois, il procède au mandatement d'office* ».

564. - Le caractère obligatoire de certaines dépenses réduit la marge de manœuvre dont disposent les collectivités territoriales. Le nombre élevé des dépenses de cette catégorie couvrant plus de la moitié des dépenses des collectivités conduit systématiquement à restreindre l'autonomie financière locale. La notion même de dépense obligatoire s'oppose au principe de la libre administration. Que ce soit dans la doctrine française ou africaine, les auteurs sont unanimes pour dire que l'inscription des dépenses obligatoires peut réduire voire annihiler le pouvoir budgétaire des collectivités[1291]. Pour Alexis Essono Evono, « *l'existence des dépenses obligatoires est en soi une atteinte à l'autonomie financière formelle des collectivités locales, car, avec ces dépenses, le pouvoir de décision des autorités locales disparait* »[1292]. À ce stade, il faut reconnaitre que l'origine et la quantité des ressources locales importent peu dès lors que les collectivités territoriales peuvent décider pleinement de leurs destinations. Dans l'éventualité d'une augmentation des ressources locales, la libre administration financière ne sera pas pour autant préservée si le pouvoir central décide de l'usage de ces ressources. L'analyse faite par le doyen Vedel en droit français dans un article resté célébré[1293] résume la situation actuelle des collectivités territoriales dans les pays de la présente étude. Il écrivait à propos de

1290 Art. 356 Loi n° 2019-006 *préc.*, voir les développements *supra* sur le pouvoir de substitution.

1291 L. PHILIP, « Les garanties constitutionnelles du pouvoir financier local », *RFDA*, 1992, p. 460.

1292 A. ESSONO EVONO « L'autonomie financière des collectivités locales en Afrique noire francophone. Le cas du Cameroun, de la Côte-d'Ivoire, du Gabon et du Sénégal », *op.cit.*, p. 11.

1293 G. VEDEL, « Le droit au logement et le principe de libre administration des collectivités locales », *Pouvoirs locaux*, n° 8, 1991, p. 16.

l'autonomie financière locale que, « *la libre administration peut se trouver compromise lorsque le législateur ne se contente pas de dicter à des collectivités des devoirs dans un domaine d'intérêt national, mais leur impose corrélativement des charges financières dont le poids est tel qu'il affecte pratiquement la liberté d'administration qui leur est reconnue dans d'autres matières, et qui faute de ressources disponibles, ne peut plus en fait s'exercer comme les responsables locaux l'auraient voulu* ».

565. - Outre les dépenses obligatoires qui s'imposent aux collectivités territoriales, leur pouvoir dépensier s'exerce dans un cadre légal strict. Les collectivités sont soumises dans l'élaboration de leur budget qui constitue déjà en soi une contrainte au respect des principes budgétaires définis par l'État. L'article 328 de la loi n° 2019-006 précitée dispose que « *le budget des collectivités territoriales obéit aux principaux généraux du droit budgétaire* ». Ces principes qui figurent dans l'ensemble des législations nationales sont : l'annualité, l'unité, l'universalité, l'antériorité, la sincérité, l'équilibre du budget, la légalité de l'impôt et la spécialité des crédits[1294]. Il s'agit des mêmes principes qui s'appliquent au pouvoir central en matière budgétaire. L'article 211 du CGCT au Niger prévoit à cet effet que, « *le budget est élaboré, adopté et géré conformément aux principes budgétaires et règles de comptabilité publique applicables à l'État* ». Toutefois, pour ce qui est des collectivités territoriales, la mise en œuvre de ces principes pose un certain nombre de difficultés[1295].

566. - À titre d'exemple, la pratique montre que les principes de l'annualité budgétaire et de l'équilibre réel du budget sont d'application difficile pour les collectivités territoriales. Dans le premier cas, l'annualité signifie que le budget d'une collectivité doit être voté pour un exercice budgétaire. Cet exercice budgétaire coïncide avec l'année civile et court du 1er janvier au 31 décembre de l'année considérée. Les budgets locaux étant soumis à des contrôles de la part de l'autorité de tutelle, ils doivent être votés suffisamment tôt pour permettre une exécution au 1er janvier. Les lois nigérienne et malienne prévoient à cet effet que « *le budget primitif est établi et voté en équilibre réel avant le 31 octobre de l'exercice* précédent *celui auquel il s'applique* »[1296]. Dans les autres pays, la loi prévoit simplement une adoption

---

1294 Art. 328 Loi n°2019-006 *préc.*, art. 214 du code des collectivités territoriales au Mali ; art. 112 du CGCT au Burkina Faso.

1295 C. NANAKO, *La libre administration des collectivités territoriales au Bénin et au Niger, op.cit.*, p. 330 et s.

1296 Art. 256 du code des collectivités territoriales au Mali ; art. 212 du CGCT au Niger.

du budget avant le 1er janvier. Il faut rappeler ici que l'État à l'obligation d'adopter la loi de finances au plus tard au 31 décembre. Or, les collectivités territoriales sont tributaires dans l'élaboration de leur budget des données issues de l'administration fiscale de l'État. Aussi, les collectivités ne peuvent voter les taux de l'impôt que dans les limites fixées par la loi de finances. L'alignement de l'adoption du budget des collectivités territoriales sur l'année civile comme c'est le cas de l'État débouche en pratique sur une impasse[1297], car elles ne disposent pas au moment de l'adoption du budget primitif des données nécessaires. Les transferts budgétaires portant sur l'exercice suivant ne sont pas communiqués d'une manière systématique.

Contrairement aux autres États où la question du décalage temporel dans l'adoption des budgets est résolue par un échange informel de documents entre les services de l'État et ceux de la collectivité au moment de l'adoption du projet de loi de finances, le législateur togolais prévoit à l'article 350 de la loi n° 2019-006 que « *les informations relevant des services de l'État indispensables à l'établissement du budget et dont la liste est fixée par décret, doivent parvenir à l'exécutif local au plus tard le 30 octobre de chaque année* ». Si cette solution semble correcte, elle rend toutefois le budget primitif provisoire dans la mesure où les arbitrages au niveau de l'État peuvent toujours changer au dernier moment. L'adoption de collectif budgétaire devient systématique pour les collectivités territoriales. Le principe d'annualité mérite d'être interrogé et revu,[1298] car le manque de prévisibilité sur les ressources dont disposent les collectivités nuit à la préparation et à l'exécution des budgets locaux.

567. - Pour ce qui est de l'équilibre réel, il s'applique de façon stricte aux budgets des collectivités à la différence du budget de l'État. Il vise l'équilibre comptable des recettes et des dépenses. L'ensemble des dépenses ne pouvant être supérieur à l'ensemble des recettes. La collectivité ne peut dépenser plus qu'elle ne gagne. Les prévisions budgétaires doivent être estimées de façon sincère[1299]. Cela implique notamment que la section de fonctionnement et la section

---

1297 Voir en droit français, J. TOUSSAINT, « L'assouplissement du principe de l'annualité dans les budgets locaux », *Revue du Trésor*, 1984, p. 307.

1298 C. NANAKO, *La libre administration des collectivités territoriales au Bénin et au Niger*, *op.cit.*, p. 333.

1299 En France le juge administratif a considéré qu'une sous-évaluation du budget de fonctionnement de 3 % ne peut être constitutive d'insincérité. CAA de Douai, 22 juin 2000, *Dosière*, n°96DA00007.

d'investissement soient respectivement votées en équilibre. La première ne pouvant contrairement à la seconde être équilibrée par l'emprunt. La finalité poursuivit par le législateur ici est d'empêcher les collectivités territoriales d'être déficitaire. Le respect de ce principe est assuré par le représentant de l'État dans les circonscriptions administratives[1300]. Lorsque le budget de la collectivité territoriale n'est pas voté en équilibre réel, l'autorité de tutelle dispose d'un délai de trente jours à compter du vote du conseil local pour proposer à la collectivité les mesures nécessaires au rétablissement de l'équilibre budgétaire et demander au conseil une nouvelle délibération qui doit intervenir dans le délai de trente jours à partir de la communication des propositions de l'autorité de tutelle. En cas de réponse négative ou de mesures jugées insuffisantes par l'autorité de tutelle, il règle le budget qui devient exécutoire dans les quinze jours qui suivent l'expiration du délai précédent[1301].

La difficulté dans l'application de ce principe tient notamment en ce que l'appréciation de la sincérité d'un budget est souvent aléatoire avec la pratique des restes à réaliser. La Cour des comptes française dans son rapport de 1994 relevait à ce propos que « *l'insincérité des comptes se manifeste par le recours abusif aux restes à réaliser dans le seul but de présenter des comptes en équilibre fictif* »[1302]. Au regard du manque de moyens humains dans les pays de la présente recherche, on peut douter de la capacité du représentant de l'État à déceler des prévisions budgétaires fantaisistes[1303] et donc de la pertinence réelle de ce principe. Le FMI dans un rapport de 2015 portant sur le Mali relevait le risque d'un endettement des collectivités ignoré. Les services du Trésor et de la dette publique ne signalent pas d'endettement des collectivités, or ; cette possibilité existe puisque les comptes de collectivités dégagent des déficits chroniques, les collectivités peuvent occasionnellement recourir à l'emprunt, accuser des arriérés de paiement ou de salaires (qui sont une forme de dette), voire réaliser des dépenses extrabudgétaires sur des emprunts informels, inconnus des comptables publics[1304].

568. - En outre, en application du principe de l'unité de trésorerie, les collectivités territoriales doivent déposer leurs fonds au Trésor public.

1300 Art. 220 du code des collectivités territoriales au Mali.

1301 Art. 350, Loi n° 2019-006 *préc.*

1302 Rapport Cour des comptes 1994, p. 219.

1303 C. NANAKO, *La libre administration des collectivités territoriales au Bénin et au Niger, op.cit.*, p. 335.

1304 FMI, *Rapport d'assistance technique - réussir la décentralisation financière*, 2015, p. 21.

Malgré les avantages qu'il peut avoir, l'application de ce principe entraine des retards dans les paiements faute de disponibilité. En effet, les fonds des collectivités territoriales logés au Trésor public ne sont pas souvent disponibles en temps voulu et en quantité souhaitée. Aussi, seuls les agents publics assermentés par l'État peuvent manier les derniers. Le comptable de la collectivité territoriale est un agent issu du Trésor public nommé par le ministre en charge des finances. Les auteurs n'ont pas manqué de relever que la comptabilité constitue un moyen d'exercice d'une tutelle supplémentaire sur les organes de la collectivité[1305].

569. - Finalement, il ne serait pas faux d'affirmer que les collectivités territoriales africaines ne disposent pas d'une autonomie financière. La liberté financière des collectivités est confisquée par le pouvoir central que ce soit dans le recouvrement de ressources ou de leur utilisation. L'autonomie financière en Afrique est un mythe ainsi qu'on a pu le dire aussi à propos des collectivités françaises[1306]. L'autonomie financière des collectivités locales africaines apparait comme un trompe-l'œil. Elle ne répond pas encore aux exigences nécessaires à son effectivité[1307]. La présence de l'État est permanente aussi bien dans la mobilisation des ressources que dans les conditions d'exécution des dépenses. La capacité de faire des collectivités est fortement restreinte. La relative liberté de gestion dont elle bénéficie n'est autorisée à s'exprimer que sous le contrôle de l'État. Néanmoins, les possibilités de coopération entre collectivités prévues par les législations nationales[1308] peuvent dans une certaine mesure contribuer à pallier leurs situations de dépendance en gagnant par la mutualisation une certaine indépendance technique et financière.

1305 *Ibid.*, p. 340 ; R. MUZELLEC, « De la pertinence des principes budgétaires et comptables applicables aux collectivités territoriales », *PMP*, n° 6-7, 1988, pp. 3-37.

1306 A. PARIENTE, « Le mythe de l'autonomie financière », *RFFP*, n° 129, 2015, p. 15.

1307 Y. KPEDU, « La question de l'autonomie financière des collectivités locales en Afrique », *Revue numérique Afrilex*, 2016, p. 1 ; A. SAMAKE, « Les contraintes liées à l'autonomie financière des collectivités territoriales dans l'espace UEMOA. Les cas du Bénin, Mali, Niger, Sénégal », in *Les finances publiques entre globalisation et dynamiques locales, Mélanges en l'honneur de Eloi DIARRA et Salifou YONABA*, , L'Harmattan, 2021, p. 431.

1308 Art. 325 du CGCT au Niger ; art. 132 du CGCT au Burkina Faso ; art 25 et s., Loi 2019-006 *préc.* ; art. 311 du code des collectivités territoriales au Mali

Conclusion du chapitre 2

570. - Les collectivités territoriales doivent bénéficier d'une autonomie dans l'exercice de leurs fonctions. Sans cette indépendance fonctionnelle, la libre administration en sortirait fortement réduite. Ainsi, pour garantir aux collectivités territoriales une réelle autonomie, il faut nécessairement une indépendance financière et technique. En l'état actuel des législations nationales et de la pratique du droit de la décentralisation en Afrique, il est permis de conclure l'examen des moyens de la décentralisation en constatant l'ampleur des obstacles à son effectivité. Des progrès sont réalisés ci et là, mais l'œuvre de décentralisation entreprise ne peut aboutir sans un personnel compétent et disponible et des moyens financiers, qui constituent le ressort principal de l'exercice autonome des compétences décentralisées. Le « placage » de la fonction publique territoriale sur le modèle étatique se révèle à l'analyse contre-productif pour l'autonomie des collectivités territoriales. Aussi, le pouvoir de décision des collectivités territoriales africaines est limité, car elles n'ont ni les ressources financières suffisantes ni la maitrise de leurs dépenses. L'omnipotence de l'État dans l'attribution des moyens nécessaires à l'action des collectivités territoriales place les autorités décentralisées dans une situation de subordination à l'égard du pouvoir central. Cela traduit la permanence de la centralisation administrative dans les États unitaires décentralisés en Afrique noire francophone.

Conclusion du titre 1

571. - Indépendants du pouvoir central en tant que personnes morales de droit public, le droit positif aménage un cadre de liberté d'action pour les collectivités territoriales dans la gestion des affaires présentant un intérêt public local. Dans les États unitaires décentralisés, la liberté d'agir des collectivités territoriales est garantie par la Constitution, mais c'est à la loi qu'il revient d'opérer la répartition des compétences et de déterminer ce qui relève ou non de l'intérêt public local. Le domaine de l'action décentralisée est entièrement défini par le législateur. Ainsi, la reconnaissance de l'existence aux côtés de l'État des collectivités territoriales autonomes n'implique pas le retrait systématique de celui-ci. Au contraire, sa présence est considérée comme une condition de réussite du processus de décentralisation et de la préservation de l'unité de l'État.

572. - Toutefois, le rôle que doit jouer le pouvoir central dans la réalisation de la décentralisation implique qu'il se débarrasse de ces « vieux habits » centralisateurs pour s'adapter à la nouvelle « manière d'être de l'État ». Non seulement l'État en Afrique peine à se réformer, l'incapacité des collectivités territoriales à exercer les compétences décentralisées les maintient dans un état de subordination vis-à-vis de l'État qui réduit les espoirs d'une gestion technique et financière autonome. De plus, les autorités étatiques continuent par exercer sur les collectivités une tutelle pesante combinant divers procédés de contrôle *a priori* et *a posteriori*. La combinaison des deux facteurs — l'insuffisance des moyens humains et financiers ainsi que la multiplicité des mécanismes de tutelle — permet de conclure à une absence en l'état de la libre administration des collectivités territoriales. Quelles que soient les extensions des compétences locales, la logique du système centralisé qui consacre l'ascendant des autorités étatiques sur les collectivités territoriales est préservée. L'administration d'État en Afrique doit se faire modeste pour prendre en compte la capacité contributive des populations par le biais du *« service public - participation »*[1309].

[1309] M. NEMBOT, Le glas de la fonction publique dans les États d'Afrique noire francophone, thèse, L'Harmattan, 2000, p. 300 et s.

TITRE 2 — UNE DÉCENTRALISATION SANS DEMOCRATIE LOCALE

> « Contrairement à ce qu'il [Tocqueville] a avancé, et que des cohortes de "suiveurs" politiques ou juristes ne cessent de répéter à qui mieux mieux, de façon bien plus banale ; il n'y a aucun lien nécessaire entre démocratie et décentralisation [...]. Le couple autocratie-démocratie et le couple centralisation-décentralisation se placent sur deux plans essentiellement différents et indépendants l'un de l'autre »[1310].

573. - Décentralisation et démocratie sont deux notions qui irriguent le droit de la décentralisation. Le lien entre ces deux notions est aujourd'hui généralisé et apparait comme une évidence. Ce lien, unanimement partagé, trouve une expression particulière au niveau communautaire à travers la Charte africaine des valeurs et des principes de la décentralisation, de la gouvernance locale et du développement local qui, à l'instar de la Charte

1310 L. SFEZ (dir.), *L'objet local*, Union générale des éditons, 1977, pp. 67-68.

européenne[1311], stipule dans son article 12 que « *la législation nationale garantit les droits des citoyens et définit leurs responsabilités en ce qui concerne la participation à la vie publique au niveau local. La démocratie est le fondement de la gouvernance locale et revêt une forme participative et représentative* ». Le lien systématiquement établi entre décentralisation et démocratie procède de deux conceptions différentes de la décentralisation. La première considère la décentralisation comme une simple technique d'organisation administrative du territoire et la seconde fait de la décentralisation une liberté. Alexis de Tocqueville ne disait-il pas que c'est dans la commune que réside la force des peuples libres[1312] ? Pour cet auteur, les libertés induites par la décentralisation compensent tous les inconvénients liés à ce mode d'organisation des sociétés, car ce ne sont pas les avantages administratifs de la décentralisation qui importent, mais ses effets politiques[1313].

En réalité, le couplage de la décentralisation à la démocratie n'a pas toujours été aussi évident dans l'histoire de la décentralisation. Les revendications décentralisatrices n'ont pas toujours été motivées par des aspirations démocratiques[1314]. Charles Eisenmann écrivait d'ailleurs que « *la décentralisation est un système sans couleur politique déterminée : elle peut être non démocratique aussi bien que démocratique* » ; il suffit que la désignation de l'autorité locale soit « *indépendante de l'autorité centrale* »[1315]. La décentralisation prise en elle-même en tant que politique institutionnelle ne produit donc pas la démocratie locale[1316].

Toutefois, il faut bien reconnaitre que le développement de la décentralisation a contribué à rendre la démocratie plus accessible. Il a

---

1311 Le préambule de la Charte européenne de l'autonomie locale dispose que : « Le droit des citoyens de participer à la gestion des affaires publiques fait partie des principes démocratiques communs à tous les États membres du Conseil de l'Europe ; c'est au niveau local que ce droit peut être exercé le plus directement ».

1312 A. De TOCQUEVILLE, *De la démocratie en Amérique*, Gallimard, coll. « Idées », 1968, p. 59.

1313 Ibid., pp. 93-95. Dans le même sens Maurice HAURIOU affirmait que « la démocratie et la souveraineté nationale vont à la décentralisation et à l'élection locale comme un fleuve suit sa pente » in, Études sur la décentralisation, éd. Paul Dupont, 1892, et Traité de droit constitutionnel, Sirey 1929 rééd. C.N.R.S. 1965, p. 190.

1314 Voir, S. REGOURD, « De la décentralisation dans ses rapports avec la démocratie », *RDP*, 1990, pp. 961-987, J. CHAPUISAT, « Libertés locales », *AJDA*, 1982, p. 382.

1315 C. EISENMANN, *Cours de droit administratif*, *op.cit.*, p. 277.

1316 J. CAILLOSSE, « L'expérience française de la décentralisation et la question de la démocratie locale aujourd'hui », *Cahiers CNFPT*, n° 37, 1992, pp. 70-92.

conduit à en faire une exigence[1317]. Cette association entre décentralisation et démocratie locale a commencé timidement en droit français par la consécration de la libre administration dans la Constitution française de 1946 puis l'affirmation relative de la prise en compte par la décentralisation des aspirations des citoyens par le rapport *Vivre ensemble*. Elle a fini par s'imposer comme une nécessité avec l'adoption de la loi du 2 mars 1982. Du moins, à partir de la décennie 80, elle est présentée comme telle pour les États d'Afrique par les organismes pourvoyeurs de l'Aide publique au développement[1318].

574. - Malgré l'établissement difficile de la preuve du lien entre les deux notions qui semble relever davantage du construit que du donné, « *il est difficile de concevoir aujourd'hui une décentralisation qui ferait l'économie d'une légitimation démocratique ou qui ne ferait pas intervenir le citoyen dans le contrôle des administrateurs locaux* »[1319]. Ainsi, la démocratie locale est essentiellement politique. Elle suppose une élection des dirigeants locaux et une implication citoyenne dans les affaires locales. L'organisation de l'administration territoriale dans les États d'Afrique ne peut échapper à cette conception renouvelée de la décentralisation associant les citoyens à la gestion des affaires publiques. Toutefois, la démocratie locale en Afrique pose d'énormes difficultés aux pouvoirs publics. D'abord, les nouvelles voies de la démocratie locale appellent, au-delà de l'élection, une amélioration des droits des élus locaux et des mécanismes de participation des citoyens au fonctionnement des institutions locales (**chapitre 1**). Ensuite, les autorités traditionnelles et coutumières, expression même du local, dans les sociétés africaines, revendiquent un droit à la parole que le droit de la décentralisation peine à traduire (**Chapitre 2**).

Chapitre 1 – Les impasses de la démocratisation de la vie publique locale

Chapitre 2 – La difficile intégration de la chefferie traditionnelle dans l'administration locale

---

1317 J. CAILLOSSE, « Éléments pour un bilan juridique de la démocratie locale en France » in, *La démocratie locale. Représentation, participation et espace public*, PUF, 1999, p. 75.

1318 Sur les aspects généraux de cette question, voir les développements *supra* partie 1 — Titre 2 — chapitre 1

1319 C. CHABROT, « Démocratie et décentralisation : un couple platonique ? », *Civitas Europa*, n° 11, 2003, p. 3.

Chapitre 1 – Les impasses de la démocratisation de la vie publique locale

575. - Le couple décentralisation et démocratie est paru assez vite comme légitime aux yeux des acteurs locaux africains. Réunis à Yamoussoukro en 1991 dans le cadre des activités de Cités Unies Développement, les maires africains ont reconnu l'existence d'un « *lien fondamental entre la décentralisation en cours et le processus de démocratisation qui est à l'ordre du jour dans la plupart des États africains* »[1320]. Cette tendance générale à lier décentralisation et démocratie trouve un début d'explication dans l'idée qu'elle favoriserait l'implication des habitants dans la gestion des affaires publiques et le développement d'une démocratie à la base. En outre, certains acteurs voient dans cette démocratie à la base une résurrection des formes anciennes de prise de décision, en particulier « la palabre africaine », censée garantir la participation de tous au processus décisionnel[1321]. L'association entre la décentralisation et la démocratie locale répond non seulement au besoin de développement de la participation des citoyens, mais aussi à la nécessité d'offrir à ces derniers des moyens de contrôle de l'administration locale. Cependant, toute recherche sur la démocratisation de la vie publique locale aboutit à un premier constat : celui d'une absence de démocratie directe permettant aux citoyens d'intervenir directement dans la vie de leur collectivité. Tout comme les États, les collectivités infranationales ont en partage une conception représentative de la démocratie[1322]. Elles sont dirigées par des « conseils élus » qui détiennent la légitimité démocratique. Les citoyens n'intervenant dans le processus de la décision publique que de manière indirecte, au moment de l'élection. L'assemblée délibérante constitue ainsi le lieu par excellence d'exercice de la démocratie locale.

1320 Cités Unies Développement, Développement municipal et coopération internationale, Colloque de Yamoussoukro, 1991, Rapport de synthèse, cité par C. NACH MBACK, *Démocratisation et décentralisation. Genèse et dynamiques comparés des processus de décentralisation en Afrique subsaharienne, op.cit.*, p. 344.

1321 N. KABAMDA, « La démocratie locale en Afrique peut-elle être la route de la reconstruction et de la prospérité : le cas du Zaïre », *African administrative studies*, 1996, pp. 23-28 ; C. NACH MBACK, *Démocratisation et décentralisation. Genèse et dynamiques comparés des processus de décentralisation en Afrique subsaharienne, op.cit.*, p. 345 ; P. KONDOMBO, « La Palabre africaine peut-elle renforcer la démocratie en Afrique ? », *Burkina think.*, 2014.

1322 V. sur ce point, voir l'étude de : J.-A. MAZERES, « Les collectivités territoriales et la représentation », *RDP*, 1990, p. 690.

576. - Toutefois, le régime représentatif local est de plus en plus complété par la consécration de techniques de démocratie dite semi-directe qui permettent aux administrés[1323] de participer sous différentes formes à la vie démocratique de leur collectivité. C'est ce qui explique la consécration de l'expression de « démocratie participative ». Elle offre aux citoyens la possibilité de jouer un rôle plus au moins important dans la définition et la mise en œuvre des politiques publiques locales. La participation citoyenne est présentée comme un correctif au régime représentatif. La vie locale se caractérise désormais dans les démocraties modernes par l'existence des assemblées délibérantes et une ouverture en direction des citoyens invités à collaborer ponctuellement à l'action publique locale[1324].

577. - Dans les États africains, le droit de la décentralisation consacre expressément un régime représentatif au niveau local. Le principe de démocratie locale dans sa constante élective est consolidé. Cependant, cette constatation n'épuise pas à elle seule la démocratie locale. Elle doit également pouvoir s'exprimer à l'intérieur même de la sphère du pouvoir local. Cette question relative à la démocratisation interne des assemblées délibérantes locales n'a d'ailleurs que très peu retenu l'attention de la doctrine africaine contrairement à la France où de nombreuses études lui sont consacrées depuis 1992[1325]. La démocratie locale demeure encore imparfaite au sein même des collectivités territoriales africaines, les droits des élus membres des assemblées délibérantes n'y sont pas suffisamment garantis (**Section 1**). Parallèlement, la démocratie locale se réclame davantage d'une ouverture du pouvoir local vers les citoyens. Ainsi, il se

---

1323 Sur le flou conceptuel autour des notions : administré, citoyen, démocratie administrative, V. CHAMPEILS-DESPLATS, « La citoyenneté administrative », in P. GONOD, F. MELLERAY, P. YOLKA (dir.), *Traité de droit administratif*, Dalloz, 2011, t. 2, p. 413. ; B. FAURE, « Les deux conceptions de la démocratie administrative », *RFDA*, n° 4, 2013, p. 709 ; C. MORIO, *L'administré. Essai sur une légende du droit administratif*, thèse, LGDJ, 2021, pp. 271-277.

1324 CCT, *Rapport régional sur la décentralisation et la démocratie locale dans les pays de l'espace UEMOA*, 2011 ; C.-A. DUBRUEIL, « Réflexions sur la démocratie locale en France », Siècles [En ligne], 37 | 2013, mis en ligne le 11 décembre 2013, consulté le 10 octobre 2020. URL : http://journals.openedition.org/siecles/1225.

1325 V. not., M. VERPEAUX, « Les droits des élus ou le droit des assemblées locales », *RFDA*, 1993, p. 20 ; L. JANICOT, *Les droits des élus, membres des assemblées des collectivités territoriales*, thèse, LGDJ, 2004 ; I. MULLER-QUOY, *Le droit des assemblées locales*, LGDJ, 2001 ; J.-F. BRISSON (dir.), *L'assemblée délibérante, avenir de la démocratie locale* ? L'Harmattan, 2016.

développe progressivement des techniques de démocratie semi-directe sans grande portée pratique (**Section 2**).

Section 1 : Une démocratie locale représentative inachevée : la limitation des droits des élus membres des assemblées délibérantes

578. - La démocratisation de la vie locale s'exprime à l'intérieur des institutions décentralisées, notamment dans les rapports entre l'exécutif local et l'assemblée délibérante. Il est difficile de concevoir une démocratie locale sans une protection effective des droits que les élus tiennent du fait de leur élection au suffrage universel[1326]. Cette élection des membres des assemblées locales au suffrage universel renforce l'ambiguïté de la nature des assemblées locales. En effet, les assemblées locales ont, en raison du caractère politique des élections dont ils procèdent, une origine politique. Toutefois, elles ne sauraient revendiquer la qualité d'institutions politiques en ce que la souveraineté nationale appartient au peuple tout entier, dont la représentation est assurée par le Parlement. Les collectivités territoriales et leurs organes délibérants demeurent malgré leur origine politique des institutions administratives[1327].

579. - Le perfectionnement de la décentralisation et le renouvellement de la démocratie locale appellent à une reconnaissance et à une protection des droits des élus locaux face au risque d'une confiscation du pouvoir par l'exécutif[1328]. On entend par droit ici non pas des droits personnels[1329], mais des droits attachés à la fonction de l'élu en tant que membre d'une assemblée locale. Il s'agit donc des prérogatives reconnues aux membres de l'assemblée locale afin de permettre le règlement des affaires de la

---

1326 L. JANICOT, *Les droits des élus, membres des assemblées des collectivités territoriales*, thèse, LGDJ, 2004, p. 1.

1327 I. MULLER-QUOY, *Le droit des assemblées locales*, LGDJ, 2001, p. 7 : « Les assemblées locales forment une catégorie hybride d'institutions administratives mais représentatives ».

1328 B. FAURE, « La démocratie locale doit s'exercer dans l'intérêt général », in *L'intérêt général, Mélanges en l'honneur de Didier TRUCHET*, Dalloz, 2015, p. 204.

1329 L. JANICOT, Les droits des élus, membres des assemblées des collectivités territoriales, op.cit., p.32.

collectivité[1330] et l'exercice d'un contrôle de l'exécutif local[1331]. Les législations africaines reconnaissent la délibération comme le moyen par lequel l'assemblée réalise ces deux fonctions. La délibération est définie par Maurice Hauriou comme « *une résolution collective sur un objet de gouvernement* (…) *prise à la majorité des voix et après discussion publique, par une assemblée formant corps et constituée en une autorité publique* »[1332]. Pour Geneviève Koubi, la délibération est « *un ensemble de discussions et de prises de position avant de devenir, le cas échéant, un acte administratif* »[1333]. La délibération suppose donc l'existence d'un débat et d'un vote. En ce sens, les droits reconnus aux élus aux fins des fonctions de délibérations doivent non seulement leur garantir un droit d'information et un droit d'expression, mais aussi une information générale sur les affaires de la collectivité territoriale (**§1**). En outre, la protection effective des droits des élus s'avère nécessaire (**§2**).

§1.- Des droits insuffisants à l'accomplissement des fonctions des élus membres des assemblées délibérantes

580. - Pour l'accomplissement de leurs fonctions, les élus locaux doivent en principe disposer d'un droit à l'information sur toutes les affaires soumises à délibération. La reconnaissance d'un tel droit est rendue nécessaire pour deux raisons essentielles. D'abord, on ne peut se prononcer avec certitude que sur ce qu'on connait. La garantie de l'information permet aux élus de délibérer en connaissance de cause. L'information leur permet de préparer éventuellement leurs interventions pendant les réunions du conseil de la collectivité territoriale. Ensuite, les élus locaux n'ayant pas un accès direct à l'information, contrairement à l'exécutif, ce dernier doit, en sa qualité de chef de l'administration locale, leur transmettre les informations nécessaires.

1330 Art. 88 : « *Le conseil municipal règle, par ses délibérations, les affaires de la commune* », Loi n°2019-006 relative à la décentralisation et aux libertés locales au Togo.

1331 Loi n° 055-2004 portant CGCT au Burkina-Faso : Art.148 « Le conseil régional contrôle l'action du président du conseil » ; Art. 227 : « Le conseil municipal contrôle l'action du maire ».

1332 M. HAURIOU, *Précis de droit constitutionnel, op.cit.*, p. 349.

1333 G. KOUBI, « La délibération, manifestation de volonté dans le droit des collectivités locales » *LPA*, n° 71, 1992, p. 6.

581. - Le droit à l'information des élus dans le cadre des fonctions délibératives fait l'objet d'une reconnaissance implicite dans les législations africaines. En effet, à la différence du législateur français, qui reconnait[1334] expressément un droit d'information à tous les membres du conseil sur les affaires faisant l'objet de délibération[1335], les législations africaines ne consacrent le droit à l'information des élus que façon implicite, à travers les règles de convocation des séances du conseil de la collectivité (**A**). De plus, l'exercice des fonctions de contrôle nécessite qu'il leur soit reconnus un droit d'information générale sur les affaires de la collectivité en dehors de toute activité délibérative. Là encore, les législations africaines reconnaissent un droit de contrôle de l'assemblée locale sur l'action de l'exécutif, mais elles ne consacrent aucune disposition à l'affirmation d'un droit général d'information sur les affaires de la collectivité (**B**).

A- L'étendue de la garantie du droit d'information et d'expression dans le cadre des fonctions délibératives

582. - L'implication concrète des élus locaux dans le travail des conseils de collectivités territoriales notamment pendant les délibérations, tient à la mise à disposition des informations générales préalablement à toute délibération (**1°**) et à la garantie de parole pendant le déroulement des séances (**2°**).

1. Une obligation d'information préalable des élus temporellement encadrée

583. - Bien qu'aucun texte législatif n'aborde de manière générale la question de l'information des élus dans le droit de la décentralisation des États d'Afrique francophone, la loi organise les modalités de transmission des informations aux élus avant la tenue d'une délibération. La transmission des informations doit se faire dans une temporalité encadrée par la loi, pour permettre aux conseillers de disposer d'un délai suffisant avant l'ouverture de la session et de pouvoir prendre connaissance des documents relatifs à l'ordre jour. C'est à cette seule condition que la

---

1334 Cette reconnaissance est qualifiée de remarquable par la doctrine française dans la mesure où il n'existe pas d'équivalent dans les autres assemblées. L. JANICOT, *Les droits des élus, membres des assemblées des collectivités territoriales*, *op.cit.*, p. 58.

1335 Art. 28 Loi du 6 février 1992 relative à l'administration territoriale de la République codifié à l'art. L. 2121-13 du CGCT. V. aussi, art. L. 3121-18 et L. 4132-18 du CGCT.

participation active des élus aux délibérations pourra être garantie. Les législations africaines organisent un mécanisme d'information des élus qui se traduit par l'envoi d'une convocation écrite indiquant notamment l'ordre du jour la session. Concomitamment à l'envoi de la convocation, les élus reçoivent une note explicative de synthèse sur les affaires soumises à la délibération (**a**). L'envoi de l'ensemble de ces documents est inséré dans un délai contraignant (**b**).

a) La convocation, un mécanisme classique d'information des élus

584. - « La convocation des assemblées est la condition nécessaire de toute démocratie locale. Une assemblée qui ne se réunit pas à fort peu de chances de fonctionner en général et cette situation conduit souvent à la dissolution de l'organe délibérant »[1336]. La convocation à une réunion de l'assemblée délibérante est une procédure particulièrement importante. Le maire dispose dans la commune de la compétence pour convoquer en session ordinaire le conseil municipal une fois par trimestre[1337], pour une durée maximum de quinze (15) jours[1338]. Mais cette compétence n'est pas exclusive, car le conseil peut se réunir en session extraordinaire à la demande de deux tiers des membres du conseil ou à celle du préfet[1339]. La convocation est censée permettre aux élus locaux de préparer la délibération pour se décider en étant suffisamment informés.

585. - Elle doit donc contenir toutes les informations relatives à l'organisation matérielle de la réunion afin de permettre aux élus de se rendre disponibles pour y participer. Elle contient à cet effet la date,

---

1336 M. VERPEAUX, « Les droits des élus ou le droit des assemblées locales », *RFDA*, 1993, p. 20.

1337 Loi n° 2019-006 préc., art. 93 : « le conseil municipal, sur convocation du maire, se réunit à la mairie, en session ordinaire, une fois par trimestre, pour une durée de quinze jours, au maximum » ; Burkina Faso, art. 233 du CGCT « le conseil municipal se réunit en session ordinaire une fois par trimestre sur convocation du maire » ; Niger, art. 170 du CGCT « le conseil municipal ou régional se réunit tous les trois (3) mois en session ordinaire ».

1338 Cette durée varie selon les pays. Au Niger elle ne peut excéder 4 jours sauf circonstances exceptionnelles (art. 170 al. 2 CGCT), elle est de 5 jours au Burkina Faso (art. 235 CGCT).

1339 Loi n° 2019-006 préc., art. 93 al. 3 : « le conseil municipal est convoqué en session extraordinaire par le maire, sur son initiative ou à la demande motivée du tiers de ses membres ou à celle du préfet ».

l'heure et le lieu de la réunion[1340]. En outre, la convocation doit indiquer obligatoirement les questions portées à l'ordre du jour de la réunion. La détermination de cet ordre du jour relève en principe de l'initiative de l'exécutif local qui a obligation de la mentionner dans la convocation ainsi que le précise l'article 93, alinéa 3 de la loi du 26 juin 2019 précitée[1341]. Selon les dispositions de l'article 174 du CGCT au Niger, « *toute question non inscrite à l'ordre du jour des sessions ordinaires et extraordinaires ne peut être mise en discussion sauf dans les cas d'urgence. L'urgence est constatée par deux tiers (2/3) au moins des membres du conseil présents ayant voix délibérative* ». En l'absence d'un contentieux des actes administratifs locaux opérationnel, le recours au droit français de la décentralisation permet de se rendre compte de l'importance du caractère complet et précis de l'ordre du jour sur l'appréciation de l'étendue de l'information dont bénéficient les élus.

586. - En effet, l'obligation légale de l'inscription de l'ordre du jour dans la convocation se double d'une obligation d'exhaustivité et de précision. Seules les questions formellement inscrites à l'ordre du jour peuvent faire l'objet d'une délibération excepté les hypothétiques cas d'urgence[1342]. L'ordre du jour indiqué dans la convocation doit, pour garantir effectivement le droit à l'information des élus, être suffisamment précis et complet. L'absence de mention d'une question dans la convocation soumise à la délibération[1343] ou la modification substantielle de l'ordre du jour pendant la réunion entraine l'annulation des délibérations. Tout changement d'ordre du jour doit donner lieu à une nouvelle convocation dans le respect des formes et délais prévus par la loi[1344]. En plus de l'obligation de complétude, l'ordre du jour doit être suffisamment précis. Si la mention des « questions diverses » n'est pas interdite, le juge administratif français veille à ce que sous cette rubrique ne figurent pas les questions d'une réelle importance[1345].

1340 Même si les réunions se tiennent en principe au siège de la collectivité, il n'est pas exclu qu'il puisse siéger dans un autre endroit.
1341 Loi n° 2019-006 *préc.*, art. 94 al. 3 « *L'ordre du jour doit figurer sur la convocation* ». V. également, art. 238 al. 2 du CGCT au Burkina Faso ; art. 171 al. 3 CGCT du Niger ; art. 29 al. 3 du code des collectivités territoriales au Mali.
1342 V. en ce sens CE, 25 avril 1913, *Le Moign, Lebon,* p. 465.
1343 CAA, Nantes, 17 décembre 1997, *Gicquel, Lebon*, p. 698.
1344 L. JANICOT, *Les droits des élus, membres des assemblées des collectivités territoriales*, *op.cit.*, p. 91.
1345 CE, 29 septembre 1982, *Richet, Lebon,* p. 532 : à propos de l'irrégularité de la mention dans les questions diverses de projet d'occupation des sols communal.

587. - Le contenu de la convocation des conseillers ne se limite pas seulement à l'indication précise et complète dans l'ordre du jour des questions à examiner. La convocation doit également contenir des informations supplémentaires sur chacune des questions soumises à la délibération. Les législateurs burkinabè et nigérien prévoient à cet effet dans leur code général des collectivités territoriales respectif l'obligation d'accompagner la convocation de l'ensemble des documents afférents à la tenue de la session. L'article 236, alinéa 6 du CGCT burkinabè prévoit que, « *à toute convocation, doivent être joints tous les documents afférents à la tenue de la session* »[1346]. Au Togo, c'est le pouvoir règlementaire qui est intervenu par décret pour préciser que la convocation adressée par le chef de l'exécutif aux conseillers doit être accompagnée pour chaque question inscrite à l'ordre du jour d'une note d'explicative de synthèse contenants les éléments essentiels permettant d'apprécier les motifs des décisions à prendre afin d'en évaluer toute la portée[1347]. En droit français, cette obligation qui figure à l'article 2121-12 du CGCT s'applique aux communes de plus 3500 habitants. Le Conseil d'État considère que l'envoi aux conseillers municipaux de la note explicative de synthèse est une obligation dont le non-respect entache d'irrégularité les délibérations prises, « *à moins que le maire n'ait fait parvenir aux membres du conseil municipal, en même temps que la convocation, les documents leur permettant de disposer d'une information adéquate pour exercer utilement leur mandat* »[1348]. Toutefois, en cas d'incomplétude de la note explicative de synthèse, la juridiction administrative ne prononce pas systématiquement l'irrégularité des délibérations. Le Conseil d'État s'attache à vérifier si l'incomplétude de la note explicative de synthèse est susceptible d'exercer une influence sur le sens de la décision prise ou qu'elle a privé les intéressés d'une garantie[1349]. L'envoi de la note explicative de synthèse est donc considéré comme une formalité substantielle garantissant le droit à l'information des élus. La convocation des conseillers doit être accompagnée d'éléments

---

[1346] Art. 171 al. 3 CGCT au Niger : « La convocation écrite et précisant l'ordre du jour doit être accompagnée des documents à examiner ».

[1347] Art. 6 al. 4 du décret n° 2019-087/PR portant modalités de convocation et mission de la première réunion des conseillers municipaux et fixant le cadre général du règlement intérieur du conseil municipal.

[1348] CE, 14 nov. 2012, *Commune de Mandelieu-la-Napoule*, n° 342327.

[1349] CE, 17 juill. 2013, Société française du radiotéléphone et autres, n° 350380.

suffisamment informatifs sur les affaires portées à l'ordre du jour sous peine d'annulation de la délibération[1350].

588. - Dans le contexte africain, le respect du droit à l'information des élus appelle au moins deux observations. D'une part, sur la forme, la convocation doit être écrite et adressée au domicile des conseillers ou par diffusion publique d'un communiqué. Ici se pose à nouveau la question du niveau d'instruction des élus. Un élu sachant vaguement lire et écrire le français est-il à même de comprendre les principaux problèmes juridiques, techniques ou financiers que soulèvent les questions portées à l'ordre du jour de la session ? Il est permis d'en douter. Même si l'information contenue dans la note de synthèse doit être concise et facilement compréhensible, et si des explications peuvent lui être fournies pendant la délibération, il n'est pas certain, dans ces conditions, que l'élu puisse jouer un rôle actif. D'autre part, sur un aspect pratique, peu de personnes, y compris parmi les élus, disposent d'une adresse postale à leur domicile. L'envoi de la convocation à domicile suppose donc de dégager un personnel spécifique qui fera le tour des domiciles des conseillers afin d'y déposer la convocation. Cette obligation d'envoi peut s'avérer *in fine* une charge supplémentaire pour les petites communes.

589. - Malgré ces quelques remarques, l'information préalable des élus est garantie par les textes législatifs et règlementaires non seulement par la convocation, mais aussi par le respect d'une condition de délai.

b) Le caractère contraignant du délai d'envoi de la convocation et ses implications

590. - Le délai de convocation des élus est essentiel pour le fonctionnement démocratique de l'assemblée délibérante. L'information générale dont disposent les élus doit leur parvenir dans un délai raisonnable pour une préparation adéquate de la délibération. Leur participation active en dépend. Comme le souligne Bertrand Faure, derrière l'apparence technique du délai de convocation des conseillers, se cache un véritable enjeu démocratique, car il faut « *redouter l'absentéisme et la*

1350 *Cf.* TA Lyon, 8 mars 1994, *Decombaz c./Cne de Bourg-en-Bresse, Lebon*, p. 823 ; CE, 12 juillet 1995, *Cne de Simiane Collongue, Lebon*, p. 680. Voir aussi, L. JANICOT, « Les limites de l'information que peuvent exiger les conseillers municipaux », *ADJA*, 2003, p. 787.

docilité »[1351] des élus membres des assemblées délibérantes des collectivités territoriales. En effet, l'auteur rappelle que l'élu local n'est pas un professionnel de la politique et doit donc se rendre disponible pour assister aux sessions. Ainsi, le respect d'un délai suffisant emporte des conséquences sur le bon fonctionnement des assemblées locales.

591. - En premier lieu, la complexité des affaires locales combinée à l'exercice volontaire des responsabilités locales, impose que les élus disposent du temps utile pour l'exercice de leur mandat[1352]. La loi doit préserver les élus de tout préjudice d'ordre financier ou professionnel en raison de l'exercice de leur mandat. Le législateur prévoit à cet effet que « *les employeurs sont tenus, au vu de la convocation régulière, de libérer leurs salariés membres du conseil municipal, le temps nécessaire pour participer aux sessions du conseil ou aux réunions des commissions. Le temps passé par les salariés aux différentes sessions ou réunions est payé par l'employeur comme temps de travail, sur présentation d'une attestation de présence ou d'un ordre de réquisition dûment signé par le maire* »[1353]. Les fonctions de conseillers étant gratuites, la loi prévoit également une indemnisation des frais de déplacement et de session aux élus. Le code des collectivités territoriales au Mali dispose en ce sens que « *les fonctions de conseiller communal sont gratuites. Toutefois, un arrêté du ministre chargé des collectivités territoriales détermine les conditions d'octroi et le taux des indemnités de déplacement et de session des conseillers communaux* »[1354]. Le législateur togolais précise aussi que « *les conseillers chargés de missions spéciales par le conseil perçoivent une indemnité. Les membres des commissions consultatives non permanentes perçoivent également une indemnité* »[1355]. Cependant, les législations nationales ne reconnaissent pas expressément un droit à la formation des élus à la différence du législateur français qui le consacre à l'article L. 2123-12 du CGCT[1356]. En second lieu, et c'est la conséquence immédiate de la précédente constatation, la validité des délibérations des assemblées dépend du nombre de conseillers présents. La loi impose en effet qu'un *quorum* soit atteint pour que l'assemblée puisse délibérer valablement. Ainsi, le conseil « *ne peut délibérer valablement que lorsque la majorité des conseillers*

1351 B. FAURE, *Droit des collectivités territoriales, op.cit.*, n° 126.
1352 *Ibid.*, n° 133.
1353 Art. 237 du CGCT au Burkina Faso.
1354 Art. 31 du code des collectivités territoriales au Mali.
1355 Art. 111 al. 3 Loi n° 2019-006 *préc.*
1356 « *Les membres du conseil municipal ont droit à une formation adaptée à leur fonction* ». V. également art. L. 3123-10 et art. L. 4135-10 pour les départements et les régions françaises.

en exercice sont présents ou représentés à la séance. Lorsqu'après une première convocation, le quorum n'est pas atteint, il n'est plus exigé à la seconde pour les questions qui figuraient à l'ordre du jour de la première convocation »[1357]. Au Togo, le conseil ne peut délibérer valablement lors de la deuxième convocation que *« si le tiers au moins des membres est présent »*[1358].

592. - Au regard de l'importance de ses implications, la détermination du délai de convocation relève de la compétence du législateur. Le délai de convocation varie selon les États. La législation burkinabè prévoit que la convocation doit être adressée aux conseillers au moins cinq (5) jours francs avant la date fixée pour l'ouverture de la session ordinaire[1359]. Au Niger, la loi prévoit un délai d'au moins dix (10) jours francs avant la date de la réunion[1360]. Ce délai est de sept (7) jours francs avant la date de la première séance de la session au Mali[1361]. Quant au législateur togolais, il prévoit un délai d'au moins huit (8) jours sans préciser s'il s'agit de jours francs ou calendaires[1362]. C'est le décret fixant le cadre général du règlement du conseil municipal qui apporte cette précision. La convocation écrite doit être adressée aux élus dans un délai de huit (8) jours francs avant la tenue de la réunion[1363]. Ce délai peut toutefois être raccourci lorsque les circonstances le justifient. La loi autorise un raccourcissement du délai de convocation en cas d'urgence. Le chef de l'exécutif peut donc abréger le délai de convocation, mais dans le respect des exigences légales. D'une part, même en cas d'urgence, le droit d'information des élus dans un délai raisonnable est garanti par le législateur. Les élus doivent dans ce cas être informés dans le respect d'un délai minimal supérieur ou égal à trois (3) jours francs[1364] au Togo et au Niger[1365]. Ce délai est deux (2) jours francs au Burkina Faso[1366]. D'autre

---

1357 Art. 32 du code des collectivités territoriales au Mali ; art. 183 du CGCT au Niger.

1358 Art. 97 al. 3, Loi n° 2019-006 *préc.* ; art. 238 du CGCT au Burkina Faso.

1359 Art. 236 du CGCT au Burkina Faso.

1360 Art. 171 du CGCT au Niger.

1361 Art. 29 du code des collectivités territoriales au Mali.

1362 Art. 94 al. 2, Loi n° 2019-006 *préc.*

1363 Art. 6 al. 1 du décret n° 2019-087/PR portant modalités de convocation et mission de la première réunion des conseillers municipaux et fixant le cadre général du règlement intérieur du conseil municipal.

1364 En droit français ce délai est d'un (1) jour franc.

1365 Respectivement aux art. 95 & 171.

1366 Art. 236 du CGCT au Burkina Faso.

part, la loi impose à l'exécutif une obligation de motivation[1367] en plus de l'obligation de soumettre, dès l'ouverture de la séance, sa décision de convocation d'urgence au vote des élus membre de l'assemblée. Ainsi, les élus pourront décider de renvoyer la discussion de l'affaire à une séance ultérieure s'ils ne reconnaissent pas l'urgence. C'est le sens des dispositions de l'article 95 de la loi précitée du 26 juin 2019 qui prévoit que « *dès l'ouverture de la séance, le conseil se prononce définitivement sur l'urgence et peut décider le renvoi de la discussion, pour tout ou partie de l'ordre du jour, à une séance ultérieure* »[1368]. Cependant, les précautions prises par les lois nationales ont une portée limitée. Le contrôle des membres du conseil ne peut en raison de la règle de la majorité être une limite suffisante au pouvoir du maire comme le relève Laetitia Janicot à propos des assemblées locales françaises[1369]. Cette dérogation légale peut finalement conduire le maire s'appuyant sur la majorité qui l'a élu à « *"forcer la main" du conseil en invoquant l'urgence* »[1370].

593. - En somme, le délai de convocation constitue une formalité substantielle pour la garantie du droit à l'information des élus. Le non-respect de ce délai peut constituer une irrégularité absolue pouvant entrainer l'annulation de la délibération, quand bien même il serait prouvé qu'il n'a pas eu d'incidence particulière sur l'issue de la délibération[1371]. Néanmoins, ainsi que le souligne Bertrand Faure au sujet des collectivités françaises, il est possible que le droit devienne plus pragmatique en application notamment de la jurisprudence *Danthony*[1372]. Ainsi, une délibération ne sera considérée comme irrégulière que si le non-respect d'un délai suffisant a été susceptible d'exercer une influence sur le sens de la décision ou a privé les intéressés d'une garantie.

594. - L'exercice de la fonction délibérative ne saurait être effectif si le législateur ne garantissait pas un droit à la parole aux élus. La reconnaissance d'un droit d'expression donne à la délibération sa nature

---

1367 Art. 171 al. 2 du CGCT au Niger : « En cas d'urgence dûment motivée, le délai peut être abrégé à trois (3) jours francs ».

1368 art. 95 al. 2. Loi n° 2019-006 *préc.*

1369 L. JANICOT, Les droits des élus, membres des assemblées des collectivités territoriales, op.cit., p. 106.

1370 P.-L. FRIER, *L'urgence*, LGDJ, 1987, p. 246 cité par L. JANICOT, *op.cit.*, p. 106.

1371 CE, 21 novembre 1969, *Élection maire et adjoint Cauro, Lebon*, p. 522, *Dr. Adm.*, 1969, p. 379.

1372 CE, Ass. 23 décembre 2011, *Danthony, Lebon*, p. 649.

d'une manifestation de volonté collective des membres de l'assemblée locale.

2. La reconnaissance timorée d'un droit d'expression

595. - La délibération traduit l'expression d'une volonté collective[1373]. Elle suppose que chacun des membres de l'organe collégial puisse exprimer sans entraves son avis sur les questions portées à l'ordre du jour de l'assemblée. On ne peut donc délibérer sans avoir au préalable organisé les conditions d'un échange d'opinions entre les différents conseillers. La délibération est aussi le résultat de cette confrontation d'idées, qui se traduit par la manifestation formelle de la voie de la majorité à travers le procédé du vote. La liberté de participation des membres de l'assemblée délibérante doit être garantie pour préserver la nature d'accord collectif de volonté de la délibération. Ce droit à la participation des élus membres des assemblées locales ne fait l'objet d'aucune disposition dans les législations africaines. En outre, il est encore très peu garanti dans certains États où il suffit d'une majorité simple pour que le conseil puisse valablement délibérer. La combinaison du système majoritaire et de la représentation proportionnelle a pour conséquence dans nombre de situations d'assurer à un parti politique plus de cinquante pour cent (50 %) des sièges à l'assemblée. La marginalisation de la minorité composant l'assemblée locale est dans ces conditions quasi inévitable. En tout état de cause, l'effectivité du droit à la participation des élus passe par une garantie de parole sans entraves sous réserve des exigences de l'efficacité de l'action locale (**a**). Aussi, les élus doivent pouvoir proposer des amendements aux projets soumis à la délibération (**b**).

a) La garantie d'un droit à la parole dans le cadre du délibéré

596. - La délibération sépare deux phases à l'intérieur desquelles l'expression des élus membres des assemblées ne peut être refusée : les débats et le vote[1374]. La parole des élus trouve dans les réunions de l'assemblée délibérante son terrain de prédilection. C'est en effet au sein

1373 « Le terme "délibération" (…) est pris en deux sens : il désigne, d'une part, le fait, pour un organisme collégial, de délibérer et de manifester une volonté collective, et, d'autre part, le résultat de cette activité, qui peut être une décision, un avis, un vœu ou toute autre prise de position », in F.-P. BENOIT, J. BENOIT, « Conseil municipal : régime des délibérations », in Encyclopédie des collectivités locales, 2011, folio n° 300.

1374 B. FAURE, *Droit des collectivités territoriales, op.cit.*, n° 243.

même des assemblées que le droit à la parole des élus doit bénéficier d'une protection singulière. La présidence des séances étant assurée par le chef de l'exécutif, il n'est pas exclu que celui-ci puisse céder à la tentation d'accorder la parole aux seuls membres de la majorité. La délibération doit donner lieu à un échange d'opinions sans aucune entrave et de façon équitable. Toutes les sensibilités politiques présentes au sein de l'assemblée doivent pouvoir s'exprimer librement. Le président de l'assemblée est ainsi dans l'obligation d'accorder la parole à tous les élus et de veiller à ce que les conditions puissent permettre un exercice effectif de ce droit[1375].

597. - Le droit de parole des élus leur appartient à titre individuel dans le cadre de l'exercice de leurs fonctions. L'exercice de ce droit ne peut donc dépendre d'aucune autre autorité que l'élu qui en est titulaire ni même être exercé collectivement par l'intermédiaire d'un quelconque délégué représentant un groupe d'élus[1376]. Ce qui signifie que tout élu local doit pouvoir exercer librement, en toute conscience, son droit à la parole. Il n'existe aucune obligation d'exercice du droit d'expression par les élus, simplement, ils doivent pouvoir prendre la parole quand ils le souhaitent, à tout moment[1377]. Il appartient au chef de l'exécutif présidant la réunion du conseil d'instaurer le débat c'est-à-dire une discussion ouverte entre tous les membres de l'assemblée. Par ailleurs, dans le déroulement de la séance, il faut noter que le débat est précédé par un résumé oral sommaire de l'affaire soumise à discussion par le chef de l'exécutif ou le rapporteur par lui désigné à cet effet. Toutefois, on notera également, avec les auteurs Francis-Paul et Jean Bénoit que « *ce qui conditionne la validité de la délibération n'est pas que se soit instauré effectivement un débat, mais seulement qu'il y ait eu possibilité réelle de discussion au cours de la séance en ce qui concerne la question considérée, même si, en fait, aucun conseiller n'a manifesté le désir de présenter son point de vue* »[1378].

598. - Le décret fixant le cadre général du règlement intérieur du conseil municipal au Togo précise en ce sens, en son article 18 aux alinéas 5 et 6 que la présentation orale de l'affaire avant le débat « *peut être précédée ou suivie d'une intervention du maire ou de l'adjoint compétent et être appuyée*

1375 L. JANICOT, *Les droits des élus, membres des assemblées des collectivités territoriales*, *op.cit.*, p. 172.

1376 *Ibid.*, p. 173.

1377 *Ibid.*, p. 174.

1378 F.-P. BENOIT, J. BENOIT, « Conseil municipal : régime des délibérations », in *Encyclopédie des collectivités locales*, 2011, folio n° 300, § 35.

par la remise de documents écrits ou la présentation de documents visuels»[1379]. Le maire assure la distribution de la parole aux membres du conseil municipal qui en font la demande. Afin de fluidifier les débats, le maire peut être amené à interrompre l'élu qui s'exprime et l'inviter à conclure brièvement[1380], car le droit à la parole des élus doit s'inscrire dans le bon déroulement de la délibération. Dans ce cas, il faudra s'assurer que le chef de l'exécutif utilise ce pouvoir avec une certaine parcimonie pour ne pas empiéter excessivement sur le droit d'expression des élus. Le caractère public des réunions[1381] de l'assemblée délibérante facilite la vérification de la garantie du droit à la parole des élus. La publicité des réunions emporte donc un double avantage : d'abord celui de permettre une transparence des délibérations et la possibilité de vérifier la qualité du travail délibératif ainsi que le contrôle de la garantie d'expression des élus[1382]. Ensuite, elle permet une traduction concrète de l'action collective locale en favorisant la rencontre entre une population et ses représentants[1383].

La publicité des réunions n'exclut pas la possibilité de délibérer à huis clos. Cette décision appartient aux membres de l'assemblée locale ainsi que le prévoient les législations nationales. Ainsi, les lois togolaise et nigérienne précitées prévoient que le conseil de la collectivité peut délibérer à huis clos à la demande des deux tiers (2/3) de ces membres. Le code des collectivités territoriales malien subordonne la mise en œuvre de cette exception à la demande des trois quarts (3/4) des conseillers[1384]. Le législateur burkinabè a quant à lui prévu que « *le conseil peut décider à la majorité simple des membres présents, de siéger à huis clos sur tout ou partie de son ordre du jour*»[1385].

599. - La délibération ne sera complète que lorsque les élus après le débat contradictoire auront pu procéder à la manifestation de la volonté

[1379] Décret n° 2019-087/PR portant modalités de convocation et mission de la première réunion des conseillers municipaux et fixant le cadre général du règlement intérieur du conseil municipal.

[1380] *Ibid.*, art 18 al. 7.

[1381] V. not., art. 185 CGCT Niger : « Les séances du conseil municipal ou régional sont publiques à moins que deux tiers (2/3) au moins des membres présents n'en décident autrement » ; art.101 al. 1, Loi 2019-006 préc. : « Les séances du conseil municipal sont publiques ».

[1382] B. FAURE, *Droit des collectivités territoriales, op.cit.,* n° 143 ; Voir aussi, C.-A. DUBREUIL, « La démocratie et la transparence », *RFDA*, 2016, p. 655.

[1383] *Idem.*

[1384] Art. 34 du code des collectivités territoriales au Mali.

[1385] Art. 241 du CGCT au Burkina Faso.

de l'assemblée à travers le vote. Le mode de scrutin, c'est-à-dire les modalités du vote, est régi par la loi. Ainsi, la publicité des séances conduit le législateur à reconnaitre le vote au scrutin public comme modalité principale[1386]. Le scrutin est ainsi organisé sans excès de formalisme. L'article 197 du CGCT au Niger dispose à cet effet que « *le vote à main levée est le mode de votation ordinaire. Toutefois, pour le choix des personnes et pour toute matière à la demande d'un tiers (1/3) au moins des conseillers élus présents, le vote doit se faire par scrutin secret* ». Ainsi, toutes les fois que le vote porte sur une personne ou à la demande d'un tiers (1/3) des élus, il est organisé un scrutin secret[1387]. Le législateur nigérien précise les modalités de déroulement du vote secret en son article 198 : « *en cas de scrutin secret, il est distribué à chaque conseiller une enveloppe et trois sortes de bulletins : vert, orange et blanc. Le conseiller dépose dans l'urne l'enveloppe contenant le bulletin vert s'il est pour l'adoption ; le bulletin orange s'il est contre, le bulletin blanc s'il désire s'abstenir. Après dépouillement, le président du conseil proclame le résultat. En cas d'égalité de voix, celle du président est prépondérante* ».

Il faut rappeler aussi que le droit de vote est un droit personnel de l'élu. Toutefois, celui-ci peut, en cas d'empêchement, le déléguer[1388]. Un conseiller empêché peut donner à un autre conseiller une procuration écrite légalisée pour voter en son nom ainsi que le prévoit le code des collectivités territoriales au Mali[1389]. La délégation de vote est assortie d'un certain nombre de conditions, notamment sur la forme : elle doit être écrite. De plus, un même conseiller ne peut être porteur que d'un seul mandat pour une même session ou séance du conseil[1390]. La délégation de vote n'est valable que pour une seule séance ou une seule session du conseil[1391]. La procuration sert uniquement pour le vote. Elle ne peut servir pour le décompte du *quorum* de la session[1392].

600. - L'encadrement procédural des débats et du vote permet de protéger l'exercice par les élus de leur droit d'expression. L'exercice de ce

---

1386 *Cf.*, art. 100 Loi n° 2019-006 *préc.*, art. 34 du code des collectivités territoriales au Mali ; art. 161 & 240 CGCT Burkina Faso.

1387 Art. 197 al. 2 du CGCT au Niger ; art. 100 al. 2 Loi n° 2019-006 *préc.*

1388 Art. 195 CGCT Niger : « Le droit de vote des conseillers municipaux ou régionaux est personnel. Il est toutefois permis une délégation dans les conditions déterminées à l'article 182 du présent code ».

1389 Art. 33, 105 & 173.

1390 V. not., art. 182 du CGCT au Niger.

1391 Art. 99, Loi n° 2019-006 *préc.*

1392 Art. 159 & 238 du CGCT au Burkina Faso.

droit est toutefois limité pour des raisons d'efficacité de l'action locale. Le président de séance dispose à cet effet du pouvoir de retirer ou de limiter le temps de parole chaque fois que cela semble nécessaire. Le droit d'expression pendant la délibération ne peut être complet sans une possibilité d'amender les projets soumis par l'exécutif aux fins de délibération.

b) Un droit d'amendement préservant la compétence d'initiative de l'exécutif

601. - La délibération ne se résume pas simplement en un vote en bloc du texte proposé par l'exécutif local. Lors des débats, les élus locaux ont la possibilité de proposer des modifications du texte soumis à l'assemblée locale par l'exécutif. Ce droit d'amendement apparait en droit français comme un principe général « *inhérent au pouvoir délibératif des conseils* »[1393] puisqu'il ne fait l'objet d'aucune consécration législative expresse, comme c'est le cas dans les législations nationales africaines. Il s'exerce pendant les débats, d'où son lien avec la fonction délibérative de l'assemblée exercée par les élus. Le droit d'amendement ne peut s'exercer pour concurrencer l'initiative de l'exécutif[1394]. À la différence du droit d'amendement des parlementaires, celui des élus locaux n'est soumis à aucune contrainte[1395] quand bien même il fait l'objet d'une règlementation.

602. - À l'exception du Togo, les textes dans les États africains n'organisent pas clairement les conditions d'exercice du droit d'amendement des élus locaux. Les règlements intérieurs se contentent de reproduire les dispositions du code des collectivités territoriales, qui ne prévoit guère de dispositions en matière d'amendement. Le droit positif togolais règlemente l'exercice du droit d'amendement par décret. En effet, on peut lire à l'article 21 du décret fixant le cadre général du règlement du conseil municipal que « *les amendements ou contre-projets peuvent être proposés sur toute affaire en discussion soumise au conseil municipal. Ils doivent être présentés par écrit. Le conseil municipal décide si des amendements sont mis en délibération, rejetés ou renvoyés à la commission compétente* »[1396]. Les amendements doivent donc être présentés sous la forme écrite et soumis directement à l'assemblée

---

1393 B. FAURE, *Droit des collectivités territoriales, op.cit.*, n° 246 ; CAA Paris, 12 février 1998, *M. Tavernier,* Concl. C. Lambert, *AJDA*, 1998, p. 742.

1394 L. JANICOT, Les droits des élus, membres des assemblées des collectivités territoriales, op.cit., p. 155.

1395 *Ibid.*, p. 156.

1396 Art. 21 décret n° 2021-087/PR *préc.*

délibérante, à qui il revient de décider de leur examen, du renvoi devant la commission compétente ou encore de leur rejet pur et simple. En principe, les élus doivent avoir la possibilité de déposer des projets d'amendement à tout moment de la discussion. Autrement, la séance de délibération se transformerait en « *une simple ratification du projet de l'organe délibérant* »[1397].

Si l'on se réfère encore une fois à la jurisprudence administrative française, le juge interdit formellement de subordonner la recevabilité des amendements à l'examen préalable d'une commission. Un article du règlement intérieur ne saurait subordonner la recevabilité d'un amendement à son dépôt préalable en commission. Une telle disposition est considérée par le juge administratif comme portant une atteinte à l'exercice effectif du droit d'amendement[1398]. La règlementation du dépôt des amendements doit avoir pour seul objectif d'assurer une meilleure qualité des travaux. Elle doit s'abstenir de toute tentative visant à empêcher les élus d'exercer leur droit d'amendement à tous les stades de la délibération. Toutefois, les élus ne peuvent exercer ce droit que dans le respect des règles prévues. Ainsi, le juge administratif admet que le règlement intérieur d'une commune puisse organiser les modalités d'amendement en exigeant le dépôt des amendements, par écrit, 72 heures avant la séance du conseil municipal sans que cela ne constitue un obstacle à la faculté des conseillers de proposer des modifications aux textes examinés[1399].

603. - De façon générale, il y a lieu de craindre, comme le relève Bertrand Faure, un usage abusif de ce droit d'amendement avec les risques de multiplication des dépôts dilatoires, de dénaturation de la cohérence d'un texte, ou que l'exercice de ce droit soit « *l'occasion de manifestations démagogiques, de surenchères au profit de certaines catégories d'administrés dont l'auteur de l'amendement chercherait à s'attirer les faveurs* »[1400]. Cependant, poursuit l'auteur, le droit d'amendement « *ne saurait être limité de façon excessive sans traduire une déviation du régime représentatif, ou bien en affaiblissant le délibératif par rapport à l'exécutif, ou bien en refusant le débat entre majorité et minorité représentés* »[1401]. Il faut donc concilier les deux préoccupations. Le droit

---

[1397] L. JANICOT, *Les droits des élus, membres des assemblées des collectivités territoriales, op.cit.*, p. 156.

[1398] CAA Paris, 12 février 1998, *M. Tavernier*, n° 96PA01170.

[1399] TA Lille, 29 mai 1997, Carton c. Commune de Roubaix, n° 96-532. Voir aussi, L. JANICOT, *Les droits des élus, membres des assemblées des collectivités territoriales, op.cit.*, p. 159.

[1400] B. FAURE, *Droit des collectivités territoriales, op.cit.*, n° 246.

[1401] *Idem.*

d'amendement des élus locaux étant liés par les affaires soumises à la délibération, son objet doit rester en rapport avec le contenu du texte sur lequel il se prononce. Ainsi, il peut faire l'objet de restrictions lorsque son usage vise à concurrencer le droit d'initiative de l'exécutif sur les questions portées à l'examen de l'assemblée délibérante.

604. - Le droit d'information et d'expression des élus membres des assemblées des collectivités territoriales semble donc particulièrement étendu, malgré l'absence d'une consécration explicite, dans les législations africaines étudiées. Il ne faut cependant pas se méprendre sur la portée réelle de cette hyper formalisation des procédures d'organisation matérielle de la délibération. Le corpus de règles encadrant le « rituel » de la délibération n'est pas en soi une garantie d'un débat contradictoire au sein des assemblées locales. À y voir de plus près, la démocratie locale est enserrée dans le fait majoritaire empêchant de soumettre les projets de délibération à une véritable confrontation des idées[1402]. Le déséquilibre entre la majorité et la minorité penche en faveur de la première qui possède à la fois la qualité de chef de l'exécutif, de l'administration communale et de l'assemblée délibérante. En somme, le « verrouillage » de la démocratie locale par la majorité fait perdre tout enjeu au débat précédant le vote de la délibération. Les garanties accordées aux élus de la minorité leur permettent de siéger au sein des assemblées à la condition qu'elles ne dérangent pas, ou ne dérangent pas trop la majorité[1403]. Le constat que fait Stéphanie Manson dans son étude sur les oppositions dans les assemblées locales françaises est particulièrement éclairant sur la situation africaine. On est loin de l'idéal démocratique et des « palabres » délibératives constructives. On peut également s'interroger sur la valeur de l'information ponctuelle dont disposent les élus locaux africains dans le cadre des fonctions délibératives. Une telle information peut-elle conduire à renforcer leur rôle de contrôleur de l'action de l'exécutif local ? À cette question, nous répondrons par la négative.

1402 Voir sur cette question l'analyse générale de, B. MANIN, « Volonté générale ou délibération. Esquisse d'une théorie générale de la délibération politique », *Le Débat*, n° 33, 1985.

1403 S. MANSON, *L'opposition dans les assemblées locales*, LGDJ, coll. « Systèmes », 2012. V. aussi, « Assemblées locales : l'impuissance délibérative. Pour une délibération locale ranimée », *Pouvoirs locaux*, n° 101, 11/2014, p. 89.

B- L'absence d'un droit général d'information sur les affaires de la collectivité territoriale handicapant l'exercice des fonctions de contrôle

605. - L'information des élus est une condition nécessaire à l'exercice de leur fonction délibérative. Elle permet aux élus membres des assemblées locales de se prononcer en toute connaissance de cause sur les affaires soumises à délibération. En plus de cette fonction délibérative, l'information des élus est tout aussi capitale pour l'exercice des fonctions de contrôle. Les lois décentralisatrices confirment toutes cette nécessité de placer l'action de l'exécutif local sous le contrôle de l'organe délibérant. Par exemple, l'article 36 du CGCT au Niger prévoit que « *le conseil municipal contrôle l'action du maire. Il contrôle également l'exécution du plan de développement communal et du budget et en assure l'évaluation périodique* »[1404]. Cependant, les moyens pour assurer ce contrôle ne sont pas garantis. Les législations nationales restent silencieuses sur les possibilités d'information des élus pour le contrôle de l'action de l'exécutif dans les États d'Afrique francophone tout comme en France[1405]. En ce sens, les élus locaux africains, en l'absence de dispositions réservées, disposent d'un droit à l'information identique à celui de tout administré[1406]. Néanmoins, certaines dispositions spécifiques des lois portant décentralisation permettent aux élus de contrôler l'action de l'exécutif. C'est le cas notamment lors de l'adoption du compte administratif du chef de l'exécutif (**1°**) et des questions écrites et orales qui permettent aux élus d'accéder à l'information utile pour le contrôle de l'exécutif (**2°**).

[1404] Voir aussi, art. 148, 177 & 258 du CGCT au Burkina-Faso ; art. 61, 134 & 202 du code des collectivités territoriales au Mali ; art. 142 de la loi n° 2019-006 *préc.*

[1405] *Cf.* L. JANICOT, *Les droits des élus, membres des assemblées des collectivités territoriales, op.cit.*, p. 227 et s. ; B. FAURE, Droit des collectivités territoriales, op.cit., n° 227 et s. ; I. MULLER QUOY, Le droit des assemblées locales, op.cit., p. 88.

[1406] Loi n° 2016-006 du 30 mars 2016 portant liberté d'accès à l'information et la documentation publiques et son décret d'application adopté en mai 2017. Le gouvernement avait acté dans le cadre de cette politique d'ouverture aux documents publiques la création des deux sites internet www.data.gouv.tg et www.servicespublics.gouv.tg qui sont, au moment où nous écrivons ces lignes, toujours inaccessibles.

1. La reddition des comptes comme moyen d'information des élus sur les affaires de la collectivité

606. - Le droit budgétaire local a développé les moyens de contrôle permettant à l'exécutif de s'expliquer sur sa gestion financière de la collectivité. La reddition des comptes constitue ainsi le moment par excellence du contrôle de l'action de l'exécutif par les assemblées locales. L'ordonnance n° 2010-54 du 17 septembre 2010 portant CGCT de la République du Niger prévoit que « *la comptabilité des collectivités territoriales donne lieu à l'élaboration d'un compte administratif par l'ordonnateur et d'un compte de gestion par le receveur* »[1407]. La présentation de ces deux comptes constitue l'occasion pour les élus locaux d'accéder à une information particulière et d'exercer les fonctions de contrôle de l'exécutif local, d'autant plus que certaines législations nationales leur reconnaissent le pouvoir de délibérer sur les comptes administratif[1408] et de gestion[1409].

607. - Lors de la séance portant adoption du compte administratif du chef de l'exécutif, les élus locaux exercent un contrôle de l'exécution du budget de la collectivité territoriale. Bien que ce contrôle des comptes ne soit pas récurrent, puisqu'il n'intervient qu'une seule fois par an, il offre l'occasion aux élus de juger de la qualité de l'exécution des délibérations budgétaires par l'exécutif. Selon les dispositions de l'article 369, alinéa 3 de la loi précitée du 26 juin 2019, le compte administratif de l'ordonnateur dressé « *en fin d'exercice (…), retrace les opérations d'exécution du budget* ». Il décrit les opérations suivantes : « *la constatation des droits acquis contre les débiteurs ; l'engagement, la liquidation et le mandatement des dépenses budgétaires* »[1410]. La présentation du compte administratif de l'ordonnateur répond à des exigences précises. Les comptes administratifs, à l'instar des lois de règlement, sont soumis à des règles de présentation identiques à celles de l'élaboration des budgets initiaux. Pour assurer à l'assemblée délibérante une information complète sur la gestion, le compte administratif établi par section, chapitre et article présente, d'une part, les prévisions ou autorisations se rapportant à chaque chapitre et à chaque article du budget et, d'autre part, les réalisations constituées par le total des émissions des

---

[1407] Art. 270 du CGCT au Niger.

[1408] Art. 88, Loi n° 2019-006 *préc.*

[1409] Art. 147 & 226 du CGCT au Burkina Faso ; art. 22, 95 & 163 du code des collectivités territoriales au Mali.

[1410] Art. 271 du CGCT au Niger.

titres de recettes et des émissions de mandat correspondant à chaque article budgétaire. Ce faisant, la présentation du compte administratif doit remplir l'ensemble des exigences de transparence qui s'applique à la présentation du budget primitif ainsi que le prévoit l'article 272 du CGCT au Niger : « *le compte administratif est établi par l'ordonnateur suivant la nomenclature des comptes en vigueur* ».

Le compte administratif du chef de l'exécutif local est adopté par délibération du conseil de la collectivité territoriale. Dans les séances où le compte administratif de l'ordonnateur est débattu, l'assemblée délibérante élit en son sein un président de séance. Le président de l'organe délibérant, ordonnateur du budget, participe aux débats, mais doit se retirer au moment du vote. Au Togo, le décret fixant le cadre général du règlement intérieur du conseil de la collectivité prévoit en son article 23 que « *dans les séances où le compte administratif est débattu, le maire peut assister à la discussion, mais il doit se retirer au moment du vote et laisser momentanément la présidence de séance* »[1411]. Cette disposition règlementaire semble contraire aux dispositions de l'article 102 al. 2 de la loi n° 2019-006 du 26 juin 2019 qui dispose clairement que « *dans les séances où le compte administratif du maire est débattu, le conseil municipal élit un président de séance* ». En effet, le pouvoir central entend introduire deux présidents pour une même séance d'adoption du compte administratif, l'un, le chef de l'exécutif, pour la phase des débats et l'autre, un membre du conseil municipal, pour la phase du vote.

Lorsque le compte administratif est adopté, le conseil de collectivité donne au chef de l'exécutif quitus de sa gestion. En cas de rejet du compte administratif, le conseil après en avoir délibéré, peut, au Mali, demander à la section des comptes de la Cour Suprême la vérification de l'exécution du budget de la collectivité[1412]. Le déclenchement de cette opération de vérification relève de la compétence de l'autorité de tutelle au Niger tel que le prévoit l'article 274 : « *en cas de rejet du compte administratif par l'organe délibérant, l'autorité de tutelle saisie, si elle le juge nécessaire, peut demander une vérification de l'exécution du budget de la collectivité territoriale par les corps de contrôle habilités* ». Dans tous les cas, l'assemblée délibérante à l'obligation de transmettre à l'autorité de tutelle qui exerce un contrôle de l'approbation le compte administratif de l'ordonnateur[1413]. En plus du contrôle effectué par les élus locaux et du contrôle d'approbation de l'autorité de tutelle,

---

1411 Décret n° 2021-087/PR *préc.*

1412 Art. 35, 107 & 175 du code des collectivités territoriales au Mali.

1413 V. not., art. 376, Loi n° 2019-006 *préc* ; art. 150 & 259 du CGCT au Burkina Faso.

l'exécution du budget de la collectivité est soumise à un contrôle juridictionnel par le juge des comptes[1414].

608. - À l'exception du Togo, les autres législations nationales prévoient expressément l'examen concomitant du compte administratif et du compte de gestion du comptable de la collectivité. Le code des collectivités territoriales au Mali prévoit en ce sens que « *le compte administratif de l'ordonnateur est soumis au vote de l'organe délibérant de la collectivité territoriale en même temps que le compte de gestion du comptable* »[1415]. Le compte de gestion établi par le receveur principal doit faire ressortir selon l'article 275 du CGCT au Niger « *la situation en début de gestion sous la forme de balance d'entrée ; les opérations de débit et de crédit de la balance générale des comptes ; le développement des opérations effectuées au titre du budget ; le résultat de l'exercice* ». L'organe délibérant vérifie la concordance du compte administratif de l'ordonnateur et le compte de gestion du receveur. La communication du compte de gestion permet à l'assemblée locale de disposer d'une information éclairant utilement son vote, puisqu'elle peut comparer les données chiffrées des deux comptes. Le Conseil d'État français s'est prononcé sur la nécessité pour l'assemblée délibérante de disposer à la fois du compte administratif et du compte de gestion dans un arrêt du 3 novembre 1989[1416]. Il a jugé que, « *un conseil municipal ne peut valablement délibérer sur le compte administratif du maire sans disposer de l'état de situation de l'exercice clos dressé par le receveur municipal* ».

609. - Il existe un flou textuel sur cette question dans le droit positif togolais. L'article 168 du décret n° 2020-004 portant régime financier des collectivités territoriales prévoit que « *le compte administratif de l'ordonnateur, accompagné du rapport d'exécution du budget, est soumis à la délibération du conseil de collectivité territoriale en même temps que le compte de gestion du comptable* ». Pourtant la loi n° 2019-006 du 26 juin 2019 relative à la décentralisation et aux libertés locales ne confère pas aux assemblées délibérantes la compétence pour délibérer sur le compte de gestion. Seul le compte administratif de l'ordonnateur relève, au sens des articles 88 et 374, de la compétence du conseil de la collectivité. L'assemblée locale ne délibère donc pas sur le

---

1414 Art. 278 et s. du CGCT au Niger.

1415 Art. 287 du code des collectivités territoriales au Mali.

1416 CE, 3 novembre 1989, Ecorcheville, Concl. T. Tuot, AJDA, 1990, p. 186. Voir aussi, L. JANICOT, *Les droits des élus, membres des assemblées des collectivités territoriales, op. cit.*, p. 240 et s.

compte de gestion du comptable public. Le contrôle de gestion du comptable principal est assuré par l'autorité hiérarchique, qui n'est autre que le directeur général du Trésor et de la comptabilité publique avant sa transmission au juge financier de la Cour des comptes[1417].

610. - Il faut relever que la restitution publique sur la gestion de la collectivité par le chef de l'exécutif peut porter non seulement sur le compte administratif, mais aussi sur « *l'état d'exécution annuel du Plan de développement économique, social et culturel (PDESC)* [et] *l'état de fonctionnement des organes et services propres de la collectivité* »[1418]. Dans le même sens, la loi nigérienne impose au chef de l'exécutif de rendre compte au cours du premier trimestre de chaque année dans un rapport général notamment, de l'état général de la commune, de l'état d'exécution des délibérations, de la situation économique et financière de la collectivité. Toutefois, si la présentation de ce rapport général donne lieu à des débats, elle n'est pas suivie d'un vote[1419]. Le contrôle de la gestion financière de l'exécutif local constitue un moment d'information des élus locaux sur les affaires de la collectivité. Si ce droit à l'information adéquate pour le contrôle de l'action de l'exécutif est garanti par les lois burkinabè, nigérienne et malienne, on ne peut en dire autant du Togo en raison des contradictions dans les textes qui l'organisent. En plus de l'accès à l'information dans le cadre du contrôle de l'exécution budgétaire local tel que décrit ci-dessus, les élus peuvent également accéder à d'autres informations de façon ponctuelle, grâce à la pratique des questions écrites et orales, qui nécessite de leur part un certain dynamisme.

2. Les questions écrites et orales comme moyen d'information ponctuel des élus sur les affaires de la collectivité

611. - Les questions écrites et orales inspirées de la pratique parlementaire permettent aux élus locaux de leur propre initiative d'obtenir des informations et de participer ainsi au contrôle de l'action de l'exécutif. Ces questions s'apparentent au droit d'interpellation des membres du gouvernement exercé par les élus nationaux. Cette pratique n'a pas fait l'objet d'une institutionnalisation dans les législations africaines étudiées. Toutefois, il faut reconnaitre que les débats qui précèdent le vote

---

1417 Art. 373 & 379 de la loi n° 2019-006 *préc.*

1418 Art. 288 du code des collectivités territoriales au Mali.

1419 Art. 80 du CGCT au Niger.

de la délibération constituent des moyens d'acquisition de l'information et d'expression des élus locaux. En cela, la question orale se passerait bien d'une reconnaissance expresse de loi[1420]. De l'information et de la réflexion de chaque membre aux débats des assemblées délibérantes, il ressort la matière première indispensable à la délibération. Le droit de poser des questions écrites et orales à l'exécutif local sur toutes les affaires présentant un intérêt pour la collectivité est un moyen pour les élus locaux de s'informer sur la gestion de cette dernière. Les questions permettent aux élus de s'émanciper du carcan de l'information en matière de délibération pour s'intéresser à tout ce qui a trait aux affaires de la collectivité en général.

612. - Le Togo fait figure d'exception en ce que même en l'absence de dispositions relatives aux questions écrites et orales dans la loi précitée du 26 juin 2019, le pouvoir central national a introduit par décret la procédure des questions orales et écrites. L'exercice de ces procédures appartient à chaque élu et implique systématiquement un droit de réponse de la part de l'exécutif local. Le décret n° 2019-087/PR portant modalités de convocation et mission de la première réunion des conseillers municipaux et fixant le cadre général du règlement intérieur du conseil municipal prévoit à l'article 31, alinéa 1 que « *tout conseiller municipal peut poser au maire des questions écrites ou orales relatives à la gestion ou à la politique municipales dès lors que les thèmes abordés se limitent aux affaires d'intérêt strictement communal* ».

Les questions écrites peuvent être posées à tout moment[1421]. Le chef de l'exécutif local dispose une fois la question à lui adressée d'un délai de dix (10) francs pour y apporter une réponse écrite. En outre, le décret précise que lorsqu'une réponse ne peut être apportée dans ce délai, en raison notamment de la complexité de la question, il sera porté à trente (30) jours. Le chef de l'exécutif veillera dans ce cas à informer le conseiller auteur de la question dans les dix (10) jours suivant la réception de la question de la prolongation du délai de réponse[1422]. Pour les questions écrites qui nécessitent un avis du conseil de la collectivité, l'élu doit s'assurer de faire parvenir la question à l'exécutif dans un délai de sept (7) jours francs avant la séance. Ce délai permettra à l'exécutif de recueillir les

---

1420 B. FAURE, *Droit des collectivités territoriales, op.cit.*, n° 234.
1421 Art. 31 al. 1 du décret n° 2019-087/PR portant modalités de convocation et mission de la première réunion des conseillers municipaux et fixant le cadre général du règlement intérieur du conseil municipal.
1422 *Ibid.*, art. 31 al. 2.

éléments de réponses possibles qu'il présentera pendant la séance du conseil de la collectivité.

Les questions orales peuvent être posées avant ou pendant les questions portées à l'ordre du jour de la séance du conseil de la collectivité par tout conseiller municipal dans les limites posées à l'article 31, c'est-à-dire sur les affaires présentant un intérêt pour la collectivité[1423]. Le décret prévoit à l'article 32 al. 2 que, « *afin de permettre au maire de réunir les éléments de réponse, le thème abordé dans la question orale doit lui être obligatoirement communiqué sept (7) jours francs avant la séance* ». Au cours de la séance, le conseiller auteur de la question est appelé à présenter brièvement, et de façon concise, sa question à l'oral, de sorte à laisser à l'exécutif le temps suffisant pour une réponse explicite[1424]. En aucun cas, une question orale ne peut servir de prétexte pour un débat sur le thème abordé « *ni d'un vote de quelque nature qu'il soit* »[1425]. Les questions ainsi que les réponses apportées doivent figurer intégralement au procès-verbal de la séance. Le maire peut décider de renvoyer à une séance ultérieure la réponse à une question orale, mais dans ce cas, il doit préciser au conseil les motifs de sa décision[1426].

613. - La règlementation des questions orales est assurée de sorte à ne pas perturber le travail délibératif. Il appartiendra au juge administratif de s'assurer que les procédés d'expression des élus ne sont pas empreints d'un excès d'exigences formelles. À cet effet, on peut s'interroger sur le caractère raisonnable du délai de sept jours francs pour le dépôt au secrétariat de la mairie d'une question. En droit français, le juge administratif a estimé que l'obligation du dépôt du thème de la question orale à la mairie dans un délai de 24 heures au moins avant la séance ne portait pas atteinte à la garantie d'expression des élus[1427]. Il est probable que le juge ne soit pas favorable à un délai plus long[1428]. Les moyens dont disposent les élus locaux pour obtenir les informations sur les affaires de la collectivité sont un complément nécessaire au droit d'information en matière délibérative. Des dispositions législatives et règlementaires éparses ont prévu des mécanismes permettant aux élus d'accéder à une information utile et suffisante pour le travail délibératif et pour le contrôle

---

1423 *Ibid.*, art. 32.
1424 *Ibid.*, art. 32 al. 3.
1425 *Ibid.*, art. 32 al. 4.
1426 *Ibid.*, art. 32 al. 5 & 7.
1427 TA Versailles, 8 septembre 1992, *Cne de Courcouronnes,* n° 925961.
1428 TA Lille, 29 mai 1997, Carton c./Cne Roubaix, Lebon, p. 597. Voir aussi, B. FAURE, Droit des collectivités territoriales, op.cit., n° 236.

de la gestion de la collectivité par l'exécutif. Cependant, ils ne suffisent pas à instaurer une dynamique de contrôle de l'exécutif par l'assemblée délibérante. La piste d'un contrôle orienté vers la sanction mérite d'être explorée.

C- Une mise en jeu de la responsabilité politique de l'exécutif local perfectible

614. - Le temps de l'irresponsabilité politique de l'exécutif local est révolu[1429]. Le droit africain de la « nouvelle » décentralisation semble avoir intégré cette réalité. Les législations nationales prennent en compte la forte demande sociale de responsabilisation des décideurs publics. On assiste à un certain rééquilibrage des rapports institutionnels entre l'exécutif local et l'organe délibérant, ce qui constitue une avancée substantielle dans le droit ouest-africain des collectivités territoriales. Cependant, la prudence doit être de mise dans l'analyse de ce changement, car il existe un décalage entre le schéma institutionnel consacré par les textes de loi et la pratique locale.

615. - Ainsi, la loi prévoit une possibilité de mise en jeu de la responsabilité politique de l'exécutif local par le vote d'une motion de défiance au Burkina Faso et au Niger et d'une procédure de destitution au Togo. L'article 273 du CGCT burkinabè dispose qu'il *« peut être introduit à l'encontre du maire une motion de défiance. Pour être recevable, la motion de défiance doit être signée par au moins un tiers des membres du conseil municipal. La motion de défiance est déposée auprès du maire qui doit en donner accusé de réception dans un délai de deux jours ouvrables. Une copie de la motion de défiance est adressée au haut-commissaire dans le même délai. La motion de défiance doit être motivée »*[1430]. L'article 66 de l'ordonnance n° 2010-54 portant CGCT au Niger prévoit également que, « *en cas d'inconduite notoire, de faute ou de négligence grave, il peut être introduit à l'encontre du maire une motion de défiance. La motion de défiance est constatée par le dépôt d'un document écrit daté comportant : le titre "motion de défiance" accompagné de la ou des motivations et de la signature d'au moins un tiers (1/3) des membres du conseil municipal. Pour être recevable, la motion est déposée auprès du secrétariat général de la mairie qui doit en donner accusé de réception dans un délai de deux (2) jours ouvrables. Une copie de la motion de défiance est adressée au représentant*

1429 P. JAN, « La responsabilité politique des exécutifs locaux, nécessité ou absurdité ? », in J.-F. BRISSON (dir.), *L'assemblée délibérante, avenir de la démocratie locale ? op.cit.*, p. 264.
1430 V. art. 185 du CGCT pour le cas des régions burkinabè.

de l'État dans le même délai»[1431]. Quant au Togo, le législateur dispose à l'article 133 que « *la destitution du maire est décidée par le conseil municipal à la majorité des deux tiers de ses membres* ».

616. - Si l'on peut se satisfaire de l'introduction par le législateur d'un tel contrôle politique, il faut relever que les législations ne posent pas les préalables nécessaires. En effet, les textes de loi n'opèrent pas une distinction franche entre le rôle de l'organe exécutif et celui de l'assemblée locale délibérante. La séparation des pouvoirs telle qu'on aurait pu l'espérer dans le cadre d'un régime représentatif n'y est pas[1432]. Le chef de l'exécutif local préside les séances de l'assemblée délibérante, il est le chef de l'administration de la collectivité. Dans ces conditions, l'emprise de l'exécutif sur le fonctionnement de l'organe délibérant est telle que cette dernière peut difficilement se constituer en contre-pouvoir. Cette observation est vraie aussi bien pour les collectivités africaines que les collectivités françaises[1433]. Il y a une unité de direction et d'action qui se traduit dans la confusion permanente entre le « délibératif » *et l'«* exécutif »[1434]. Bertrand Faure l'explique lorsqu'il relève qu'il y a une contradiction entre les formules générales consacrant une large compétence de l'assemblée et « *la réalité, car le rôle institutionnel de l'exécutif est plus important que son pouvoir normatif propre. Finalement, la supériorité de l'assemblée peut paraitre sans valeur dans le régime local puisque la pratique en a dépassé la lettre* »[1435]. La vertu affichée du mécanisme de mise en jeu de la responsabilité qui aurait pu permettre de mettre fin au cumul des fonctions conduisant l'exécutif local à présider lui-même l'organe délibérant n'est pas pris en compte par les législations nationales africaines.

617. - En outre, l'ordonnance n° 2010-54 précitée semble confondre les motifs de la responsabilité politique et celle de la suspension ou de la révocation de l'exécutif. La responsabilité politique se résume dans

---

1431 V. art. 138 du CGCT pour le cas des régions nigériennes.

1432 En droit français, la Corse constitue l'exemple parfait de cette séparation des pouvoirs. Voir aussi sur la Polynésie française, D. REIGNIER, « La responsabilité de l'exécutif devant l'assemblée délibérante, les enseignements du laboratoire polynésien », in J.-F. BRISSON (dir.), *L'assemblée délibérante, avenir de la démocratie locale ?*, L'Harmattan, 2016, p. 229.

1433 P. JAN, « La responsabilité politique des exécutifs locaux, nécessité ou absurdité ? », in J.-F. BRISSON (dir.), *L'assemblée délibérante, avenir de la démocratie locale ? op.cit.*, p. 264.

1434 *Ibid.*, p. 264 et s.

1435 B. FAURE, *Droit des collectivités territoriales, op.cit.*, n° 404.

l'obligation pour l'exécutif de quitter ses fonctions par suite d'une perte de confiance de l'assemblée délibérante qui l'a élu. La responsabilité en droit a plusieurs acceptions, mais celle qui est politique est la moins juridique de toutes. En effet, contrairement à la responsabilité civile ou pénale, la responsabilité politique ne vise pas à sanctionner une faute[1436]. Sa mise en œuvre ne repose sur aucune cause objectivement appréciable contrairement à la suspension ou la révocation par l'État. Elle repose sur un « *désaccord politique substantiel* » et suggère l'idée d'une subordination de l'exécutif à l'assemblée délibérante, à laquelle il doit rendre compte de ses actes[1437]. Il parait alors curieux que le législateur nigérien subordonne l'introduction d'une motion de défiance à un cas « *d'inconduite notoire, de faute ou de négligence grave* ». Or, la loi prévoit la révocation de l'exécutif par des fautes graves à l'article 63 du CGCT qui dispose que « *sans préjudice des sanctions pénales, le maire et le ou les adjoint(s) peuvent, en cas de faute grave, être révoqués par décret pris en conseil des ministres sur rapport du ministre chargé de la tutelle des collectivités territoriales* ». Il y a donc un mélange de genres entre la mise en œuvre du contrôle politique et la sanction disciplinaire.

618. - Sur le plan procédural, le droit de la décentralisation prévoit une procédure organisée afin de garantir la stabilité du conseil. La motion de défiance, pour être recevable, doit être signée par un tiers (1/3) des membres du conseil de collectivité et mentionner expressément les motifs pour lesquels elle est présentée. Le dépôt d'une motion de défiance donne droit « *à la tenue d'une session extraordinaire dans les sept jours qui suivent la date de dépôt de la motion. En cas de refus du maire, la convocation du conseil est assurée par l'autorité de tutelle* »[1438]. Le délai d'organisation de la session extraordinaire est de quinze (15) jours au Niger[1439]. À l'issue d'un débat contradictoire et d'un vote, la motion est adoptée au Burkina Faso à la majorité absolue des deux tiers (2/3) des membres du conseil[1440] et des trois quarts (3/4) au Niger[1441]. Dans les deux pays, la législation prévoit également qu'« *en cas de rejet, une nouvelle motion de défiance ne peut être déposée pour les mêmes motifs avant un délai d'un (1) an* »[1442]. La conséquence de

---

1436 *Ibid.*, n° 416.
1437 *Ibid.*, n° 416.
1438 Art. 186 & 274 du CGCT Burkina Faso respectivement pour les régions et les communes.
1439 Art. 67 & 139 du CGCT Niger respectivement pour les communes et les régions.
1440 V. art. 187 & 275 du CGCT au Burkina Faso.
1441 Art. 68 & 140 du CGCT au Niger.
1442 *Ibid.* ; art. 275 du CGCT au Burkina Faso.

l'adoption de la motion est la démission du chef de l'exécutif, constatée par arrêté du ministre chargé de la tutelle des collectivités territoriales au Niger et par décret pris en conseil des ministres au Burkina Faso. Il est procédé à la désignation d'un nouveau chef de l'exécutif dans les trente (30) jours au Burkina Faso[1443] et quarante-cinq (45) jours au Niger[1444]. On aurait pu imager, pour garantir davantage la stabilité de la collectivité, que le dépôt de la motion soit précédé d'un consensus autour du nouveau chef comme c'est le cas dans la collectivité Corse[1445]. Au Togo, le législateur s'est encore une fois illustré par son manque de précision. Avare de détails, il prévoit simplement que la destitution du chef de l'exécutif est adoptée à la majorité des deux tiers (2/3) des membres du conseil sans aucune autre précision sur les modalités de mise en œuvre de la destitution.

619. - Le contrôle politique des assemblées locales est encore perfectible. Une meilleure visibilité organique permettrait de distinguer le contrôleur du contrôlé, et ainsi mettre en jeu non pas la responsabilité du seul chef de l'exécutif, mais celle de l'ensemble de l'équipe dirigeante. Aussi, le dispositif devrait tendre vers une motion de défiance constructive pour garantir la stabilité de collectivité territoriale. La rénovation des institutions politiques locales doit passer par une telle réforme. Contrairement à l'enthousiasme suscité par l'introduction d'un contrôle politique, les assemblées locales voient leur rôle réduit dans la détermination des règles internes de leur fonctionnement.

1443 Art. 189 & 277 du CGCT au Burkina Faso.

1444 Art. 70 & 142 du CGCT au Niger.

1445 V. not., B. FAURE, *Droit des collectivités territoriales, op.cit.*, n° 420 et s. ; P. JAN, « La responsabilité politique des exécutifs locaux, nécessité ou absurdité ? », in J.-F. BRISSON (dir.), *L'assemblée délibérante, avenir de la démocratie locale ? op.cit.*, pp. 253-254 ; D. REIGNIER, « La responsabilité de l'exécutif devant l'assemblée délibérante, les enseignements du laboratoire polynésien », in J.-F. BRISSON (dir.), *L'assemblée délibérante, avenir de la démocratie locale ? op.cit.*, p. 229.

§2.- La protection règlementaire des droits des élus membres des assemblées délibérantes : une négation du pouvoir d'auto-organisation des assemblées locales

620. - Les sources des droits des élus locaux constituent, selon Laetitia Janicot, les premières garanties de ces droits[1446]. Plus les assemblées délibérantes ont une marge pour les organiser elles-mêmes, mieux ils semblent garantis. En conséquence, l'absence de pouvoir d'auto-organisation limite encore plus en Afrique la protection des droits des élus membres des assemblées délibératives. Les droits reconnus aux élus locaux possèdent la double caractéristique d'être à la fois des droits propres à chaque élu qu'ils exercent individuellement dans le cadre de son mandat et des droits collégiaux qui s'exercent dans l'intérêt global de l'assemblée délibérante. Dans sa formation collégiale, les droits reconnus à l'assemblée locale ne sont mieux protégés que par la reconnaissance d'un pouvoir d'auto-organisation de l'assemblée[1447]. La notion de « pouvoir d'auto-organisation » renvoie ici, non pas au pouvoir de l'État au sens où l'entendait Carré de Malberg[1448], mais au pouvoir d'organisation institutionnelle[1449] dont dispose tout organe collégial[1450]. Certains espaces d'auto-organisation sont en principe ménagés par les textes pour permettre aux assemblées locales de règlementer leurs procédures de travail et leur fonctionnement[1451]. Il s'agira ici de montrer que ce pouvoir d'auto-organisation des assemblées constitue un élément de protection des droits des élus (**A**), mais qu'il est fortement restreint dans les États d'Afrique subsaharienne par l'omniprésence de l'État dans le fonctionnement interne des assemblées locales (**B**).

1446 L. JANICOT, *Les droits des élus, membres des assemblées des collectivités territoriales op.cit.*, p. 335.

1447 *Ibid.*, p. 339.

1448 R. CARRE DE MALBREG, *Contribution générale à la théorie de l'État, op.cit.*, p. 1559.

1449 M. DOUENCE, *Le pouvoir d'organisation du service public*, thèse, dactyl., Université de Pau, 2003, p. 407 et s.

1450 J.-F. BRISSON (dir.), *L'assemblée délibérante, avenir de la démocratie locale ?* L'Harmattan, 2016.

1451 C. MONEMBOU, « Le pouvoir règlementaire des collectivités locales dans les États d'Afrique noire francophone », *Revue CAMES/SJP*, n° 002, 2015, p. 79 ; B. FAURE, *Le pouvoir règlementaire des collectivités territoriales,* thèse, LGDJ, 1998, p. 32.

A- La reconnaissance d'un pouvoir d'auto-organisation des assemblées délibérantes

621. - Les collectivités territoriales disposent en principe d'un pouvoir d'auto-organisation spontané. Le pouvoir d'édicter un règlement intérieur est reconnu à tout organe administratif collégial afin de lui permettre d'arrêter ses procédures de travail pour les besoins de son bon fonctionnement collectif. Il s'agit d'une question de discipline, affirme Bertrand Faure dans son ouvrage sur le droit des collectivités territoriales françaises[1452]. Au-delà de la question de la discipline, l'intérêt de l'adoption d'un règlement intérieur par les assemblées délibérantes se trouve dans la protection qu'elle offre aux élus membres. Laetitia Janicot souligne à cet effet que le fait, pour les assemblées locales de pouvoir édicter les mesures d'organisation et de fonctionnement interne de leurs assemblées «*permet d'assurer une protection renforcée des droits des élus locaux puisque ces derniers sont à la fois les bénéficiaires de ces droits et ceux qui les règlementent*»[1453]. Ainsi, en édictant elles-mêmes leurs règles de fonctionnement interne, les assemblées délibérantes protègent les droits collégiaux des élus locaux.

622. - Le droit d'auto-organisation des assemblées délibérantes à travers l'adoption d'un règlement intérieur existe même en l'absence de texte. Pour Maurice Hauriou, les assemblées délibérantes possèdent un droit de règlementation interne en ce qu'elles sont des assemblées[1454]. Qu'il soit ou non reconnu expressément par un texte, il résulte de la nature même de l'assemblée. Il s'agit d'un « *droit originel qui résulte de la nature des choses*»[1455]. Aujourd'hui, l'adoption des règlements est une obligation qui pèse sur les assemblées délibérantes. Les législations nationales des pays objet de la présente recherche prévoient toutes que les assemblées municipales et régionales doivent établir dès leur installation un règlement intérieur[1456]. En droit français, le législateur a souhaité, en adoptant la loi du 6 février 1992 relative à l'administration territoriale de la République, rendre obligatoire l'adoption du règlement intérieur pour toutes les

---

[1452] B. FAURE, *Droit des collectivités territoriales, op.cit.*, n° 122.

[1453] L. JANICOT, *Les droits des élus, membres des assemblées des collectivités territoriales, op.cit.*, p. 341.

[1454] M. HAURIOU, *L'institution et le droit statutaire,* Recueil de la législation de Toulouse, 1906, p. 168. Cité par L. JANICOT, *op.cit.,* p. 342.

[1455] J. RIVERO, *Les mesures d'ordre intérieur administratives*, thèse, Sirey, 1934, p. 176.

[1456] Art. 27 & 99 du code des collectivités territoriales au Mali. Voir aussi, art. 32 & 107 du CGCT au Niger ; art. 110 de la loi n° 2019-006 *préc.*

collectivités territoriales[1457] afin d'encadrer et de limiter les pouvoirs du maire, protégeant ainsi les droits des élus, notamment ceux de la minorité. Le règlement intérieur est précisément fait pour permettre la protection des droits de la minorité[1458]. En effet, le développement de la pratique des règlements intérieurs a servi avant 1992 dans les assemblées municipales à limiter les possibilités d'intervention de l'opposition, et donc la liberté d'expression des conseillers minoritaires. Le législateur de 1992 souhaitait ainsi mettre fin à la tentation des exécutifs locaux d'abuser de leurs pouvoirs de chef de l'administration locale dans les collectivités territoriales et empêcher l'exercice par les conseillers de leurs droits[1459]. Le rapporteur à la commission des Lois à l'Assemblée nationale voyait à ce propos dans le règlement intérieur des assemblées locales *« un instrument d'expression de la démocratie locale »*[1460]. Selon Jean Rivero, « *le règlement intérieur est la traduction de cette vérité d'évidence que là où les hommes sont réunis pour délibérer, il est impossible qu'une loi ne vienne diriger leurs délibérations ; la loi qui peut difficilement leur être imposée du dehors, et dont l'autorité sera nécessairement accrue si elle est l'œuvre de l'assemblée elle-même* »[1461].

623. - L'adoption d'un règlement intérieur des assemblées délibérantes des collectivités est un acte qui marque l'autonomie des collectivités territoriales. C'est une manifestation du «*pouvoir souverain*» de l'organe délibérant[1462]. Le fait de considérer le règlement comme une manifestation du pouvoir a influencé la nature juridique du règlement intérieur. Le juge administratif français a pendant longtemps considéré le règlement intérieur des assemblées délibérantes des collectivités territoriales comme une simple mesure d'ordre intérieur insusceptible de faire grief. Il ne pouvait donc faire l'objet d'une contestation devant la juridiction administrative[1463]. Le règlement intérieur n'était qu'un simple code de

---

1457 À l'exception des communes de moins 1000 habitants qui restent libres d'adopter ou non un règlement intérieur.

1458 M. VERPEAUX, « Les droits des élus ou le droit des assemblées locales », *RFDA*, 1993, p. 20.

1459 L. JANICOT, *Les droits des élus, membres des assemblées des collectivités territoriales, op.cit.*, p. 343.

1460 Doc. AN, n°1888, 1990-1991, t. 1, p. 198 cité par L. JANICOT, *op.cit.*, p. 343.

1461 J. RIVERO, *Les mesures d'ordre intérieur administratives, op.cit.*, p. 177.

1462 CE, 22 mars 1912, *Le Moign*, S. 1913, 3, 105, note M. Hauriou ; M.-F. VERDIER, « Le règlement intérieur des assemblées délibérantes des collectivités locales : nouveaux développements », *LPA*, n° 83, 1995, p. 33.

1463 CE, Ass. 2 décembre 1983, *Charbonnel*, *Lebon*, p. 474, concl. M. Roux, ; *RDP*, 1985, p. 827.

bonne conduite pour les élus sans véritable valeur. Il s'agit d'un acte à caractère interne « *fait par les élus pour les élus, donc sans répercussion sur la légalité générale* »[1464]. Toutefois, l'évolution en faveur de l'État de droit ainsi que l'institutionnalisation de la protection des minorités au sein des assemblées ont suscité une évolution jurisprudentielle vers un contrôle juridictionnel des règlements intérieurs des assemblées délibérantes des collectivités par le juge administratif. Le Conseil d'État admet qu'un recours puisse être exercé contre un règlement intérieur établissant le droit d'information des élus dans un sens plus restrictif et contraire aux dispositions de la loi organique[1465]. Cependant, l'invocabilité du règlement intérieur à l'encontre d'une délibération prise en violation de ces dispositions n'est recevable que si elle est constitutive d'un vice substantiel. Seules les formalités particulièrement importantes du règlement peuvent être sanctionnées[1466].

624. - En somme, en reconnaissant le droit et l'obligation aux assemblées délibérantes des collectivités territoriales de fixer elles-mêmes leurs règles d'organisation et de fonctionnement, la loi apporte ainsi une garantie de la protection des droits des élus. Les assemblées délibérantes sont libres d'accorder des garanties supplémentaires aux élus en plus de celles prévues par la loi. Cette liberté s'exprime dans le cadre défini par la loi et sous le contrôle du juge, ce qui permet de limiter l'autocratie locale. Aujourd'hui, la portée du règlement intérieur est minorée par le souci de précision du législateur. Au moment où l'assemblée locale est amenée à élaborer, son règlement près de 2/3 de son contenu relatif aux règles de convocation, de procuration, de vote, de publicité, de séances, de quorum, etc. est déjà fixé par le législateur. Dans les États d'Afrique noire francophone, le tiers restant pour l'exercice d'un quelconque pouvoir d'auto-organisation est comblé par le pouvoir règlementaire national.

B- La négation de fait du pouvoir d'auto-organisation des assemblées délibérantes

625. - La question de l'existence d'un pouvoir d'auto-organisation des collectivités territoriales dans les pays africains de succession française est complexe. La difficulté tient en ce que la reconnaissance d'un pouvoir

---

1464 B. FAURE, *Droit des collectivités territoriales, op.cit.*, n° 124.

1465 CE, 10 février 1995, *Commune de Coudekerque-Branche*, n° 147378, *Lebon,* p. 66.

1466 CE, 31 juillet 1996, *Tête, Lebon,* p. 325 ; RFDA, 1996, p. 1042 ; B. FAURE, *Droit des collectivités territoriales, op.cit.,* n° 124. Voir aussi la jurisprudence Danthony *préc.*

d'auto-organisation des assemblées locales peut conduire à une protection des droits des élus différente selon les collectivités territoriales. Les règlements intérieurs des assemblées délibérantes des collectivités territoriales étant établis par elles-mêmes, que le risque d'une protection inégale des droits est important. Or, les législations nationales africaines semblent particulièrement soucieuses d'assurer une égalité et une garantie homogène des droits des élus sur l'ensemble du territoire national. Cela interroge sur la nature même de l'habilitation législative des collectivités territoriales à établir un règlement intérieur. Le législateur a-t-il entendu reconnaitre un pouvoir d'auto-organisation des assemblées locales en imposant l'établissement et l'adoption d'un règlement intérieur ? L'examen du pouvoir d'auto-organisation des assemblées locales invite à une analyse de la question sous l'angle du droit constitutionnel. Le statut organique des collectivités territoriales étant défini par la Constitution, il s'agit alors de savoir si la Loi fondamentale dans les pays d'Afrique noire subsaharienne permet de reconnaitre aux collectivités territoriales une compétence pour fixer elles-mêmes une part de leur statut[1467].

626. - Le pouvoir d'auto-organisation des assemblées délibérantes des collectivités territoriales pourrait trouver sa justification ainsi que nous l'avons montré dans la consécration par les constituants africains du principe selon lequel les collectivités territoriales s'administrent librement par des conseils élus[1468]. Toutefois, le principe de libre administration des collectivités territoriales ne fait sens que dans les limites organisationnelles que le législateur a définies. Ce dernier invite par ailleurs, au besoin, le pouvoir central à utiliser son pouvoir règlementaire, afin de préciser les modalités des principes qu'il a fixé. L'intervention des collectivités territoriales dans la détermination des règles internes de leur fonctionnement est subordonnée à la marge de manœuvre que leur concèdent la loi et le règlement. Les législations africaines portant sur la décentralisation ont fait le choix d'un alignement de tous les règlements intérieurs sur des dispositions communes. Le pouvoir central est donc en permanence impliqué dans le fonctionnement des assemblées délibérantes des collectivités territoriales afin d'assurer une gestion transparente. Ce faisant, les assemblées délibérantes disposent d'une marge de manœuvre

---

1467 La question se pose dans les mêmes termes en droit français, J. BENOIT, « Conseil municipal : règlement intérieur », in *Encyclopédie des collectivités locales,* folio n° 340, § 10.

1468 V. not., art. 141 Constitution togolaise de 1992 ; art. 164 al. 2 Constitution nigérienne de 2010.

quasi nulle en qui concerne le contenu de leurs règlements intérieurs. La compétence pour organiser le travail délibératif des conseils des collectivités territoriales appartient exclusivement au législateur et au pouvoir règlementaire. Cette démarche est commandée « *par l'idée que les règles légales et règlementaires fixant le régime des assemblées locales sont conçues comme des conditions du bon fonctionnement des collectivités locales et comme constituant une garantie de ce bon fonctionnement, posées dans l'intérêt primordial des administrés et* [...] *des élus eux-mêmes* »[1469].

627. - Ainsi, le code général des collectivités territoriales burkinabè invite le pouvoir règlementaire à intervenir dans le fonctionnement interne des collectivités en adoptant un décret portant règlement intérieur-type[1470], qui précise notamment, les modalités d'application des sanctions aux élus[1471], les attributions et le mode de fonctionnement des commissions permanentes[1472]. Au Togo, la loi du 26 juin 2019 précitée prévoit que « *des décrets d'application en conseil des ministres déterminent notamment, en tant que de besoin, les modalités de transfert de compétences et de ressources correspondantes* [..., le] c*adre général de règlement intérieur des conseils des collectivités territoriales* [...] »[1473]. Lors de son conseil des ministres du 17 juin 2019, le gouvernement togolais affirmait que l'adoption du décret portant cadre général unique du règlement intérieur[1474] pour l'ensemble des communes permettait d'éviter que les assemblées locales délibérantes ne se lancent chacune dans la rédaction d'un règlement intérieur, étant donné qu'elles ne disposent pas d'outils nécessaires pour une bonne organisation du travail au sein des conseils[1475]. Le cadre général du règlement intérieur contient des dispositions relatives aux réunions des assemblées et à l'organisation des débats, à la présidence des séances, au déroulement des séances et des débats, à l'adoption du budget et du compte administratif, au fonctionnement des commissions ainsi qu'au droit d'information et d'expression des conseillers élus. Le contenu de ce décret ne laisse aucune

---

1469 J. BENOIT, « Conseil municipal : règlement intérieur », *op.cit.*, folio n° 340.

1470 Décret n° 2006-208/PRES/PM/MFB/MATD portant règlement intérieur-type des conseils de collectivités territoriales.

1471 Art. 160 du CGCT au Burkina Faso.

1472 *Ibid.*, art 144 & 221.

1473 Art. 385 de la loi n° 2019-006 *préc.*

1474 Décret n° 2019-087/PR portant modalités de convocation et mission de la première réunion des conseillers municipaux et fixant le cadre général du règlement intérieur du conseil municipal.

1475 Communiqué sanctionnant le Conseil des ministres du 17 juin 2019.

marge de manœuvre aux assemblées délibérantes. Le conseil de la collectivité ne peut donc pas déterminer librement le contenu de son règlement. L'obligation faite par la loi à l'assemblée locale d'adopter un règlement intérieur dans les deux mois suivant sa première réunion[1476] consiste *in fine* à reprendre les dispositions règlementaires contenues dans le décret. Si le législateur reconnait aux collectivités territoriales un pouvoir règlementaire local, son contenu[1477], dans son aspect d'organisation institutionnelle, ne peut donc s'exprimer que dans les interstices laissés libres par le législateur et le pouvoir règlementaire national[1478].

628. - Le travail délibératif nécessite une « collaboration confiante » entre l'exécutif et l'assemblée. L'information apparait alors comme le premier des droits des élus locaux qui permet d'établir cette confiance. Cependant, les élus ne bénéficient pas d'un droit d'information expressément consacré par la loi dans les États d'Afrique noire francophone pour l'exercice des fonctions délibératives et de contrôle. En outre, la conception unitaire de l'État a conduit à nier aux collectivités territoriales un pouvoir d'auto-organisation considéré par la doctrine, malgré ses insuffisances, comme une source de protection efficace. La démocratie locale à l'intérieur des collectivités territoriales reste encore à parfaire, notamment à travers l'affirmation par le législateur d'un droit général à l'information des élus. Toutefois, la démocratie locale ne serait pas complète sans une ouverture des collectivités territoriales vers l'extérieur. Ainsi, le législateur a prévu un certain nombre d'outils qui limitent, sans toutefois le remettre en cause, le régime représentatif local. Il est prévu une ouverture de la collectivité territoriale à ses habitants à travers des mécanismes de démocratie participative.

1476 Loi n° 2019-066, art. 110 pour les communes ; art. 270 pour les régions (non encore opérationnelles)

1477 V. utilement sur ce sujet, C. MONEMBOU, « Le pouvoir règlementaire des collectivités locales dans les États d'Afrique noire francophone », *Revue CAMES/SJP*, n° 002, 2015, p. 79-111 ; L. FAVOREU, « Libre administration et principes constitutionnels », in J. MOREAU, J. DARCY (dir.), *La libre administration des collectivités locales*, ECONOMICA-PUAM, 1984 pp. 63-71 ; B. FAURE, *Le pouvoir réglementaire des collectivités locales*, thèse, LGDJ, 1998 ; C. BACOYANNIS, *Le principe constitutionnel de libre administration des collectivités territoriales*, thèse, Economica-PUAM, 1993 ; M. JOYAU, *De l'autonomie des collectivités territoriales françaises. Essai sur la liberté du pouvoir normatif*, thèse, LGDJ, 1998.

1478 M. DOUENCE, *Le pouvoir d'organisation du service public*, thèse, dactyl., Université de Pau, 2003, pp.411-413.

Section 2 : La gestation progressive d'une démocratie locale participative : le développement d'une démocratie au profit des citoyens

629. - La participation des citoyens à la gestion des affaires locales est un droit essentiel[1479]. Il fait l'objet d'une consécration au niveau international. La Déclaration universelle des droits de l'Homme reconnait en son article 21 que « *toute personne a le droit de prendre part à la direction des affaires publiques de son pays, soit directement, soit par l'intermédiaire de représentants librement choisis* ». Les Constitutions africaines intègrent dans leur bloc de constitutionnalité les dispositions issues de cette Déclaration[1480]. Paradoxalement, l'examen des dispositions constitutionnelles dans les États d'Afrique noire francophone montre que l'expression directe du peuple est possible au niveau national, mais qu'elle n'est pas garantie au niveau infranational[1481]. Les articles 4 et 6 des Constitutions togolaise et nigérienne disposent que la souveraineté appartient au peuple qui *« l'exerce par ses représentants et par voie de référendum »*[1482]. Inversement, pour ce qui est des collectivités territoriales, la Loi fondamentale se contente de rappeler qu'elles « *s'administrent librement par des conseils élus dans les conditions prévues par*

1479 M. RASERA, *La démocratie locale*, LGDJ, coll. « Systèmes », 2002, p. 81.

1480 Les Constitutions africaines reconnaissent la DUDH dans leurs blocs de constitutionnalité. Au Burkina-Faso le préambule souscrit « *à la déclaration universelle des Droits de l'Homme de 1948 et aux instruments internationaux traitant des problèmes économiques, politiques, sociaux et culturels* ». Au Togo, le préambule affirme l'attachement du peuple togolais aux valeurs des droits de l'Homme : « *Convaincu qu'un tel État ne peut être fondé que sur le pluralisme politique, les principes de la démocratie et de la protection des droits de l'homme tels que définis par la Charte des Nations unies de 1945, la Déclaration universelle des droits de l'homme de 1948 et les pactes internationaux de 1966, la Charte africaine des droits de l'homme et des peuples adoptée en 1981 par l'Organisation de l'unité africaine* ; l'art. 50 de la Constitution *« Les droits et devoirs, énoncés dans la Déclaration universelle des droits de l'homme et dans les instruments internationaux relatifs aux droits de l'homme, ratifiés par le Togo, font partie intégrante de la présente Constitution* ».

1481 Voir sur la question en droit français avant la réforme de 2003, J. BAGUENARD, J.-M. BECET, *La démocratie locale*, PUF, coll. « Que sais-je ? », 1995, p. 97.

1482 Niger, art. 6 Constitution de 2010 : « Le peuple exerce sa souveraineté par ses représentants élus et par voie de referendum ».

la loi »[1483]. On constate qu'au niveau national la démocratie trouve son sens dans l'expression de la souveraineté nationale alors que pour les collectivités territoriales, elle est associée à la libre administration[1484].

630. - Sur le plan constitutionnel, la démocratie locale ne peut être que représentative. « La démocratie électorale-représentative repose sur l'axiome selon lequel la volonté générale s'exprime directement et complètement dans le processus électoral »[1485]. Cette conception de la démocratie est aujourd'hui en crise, on assiste à l'épuisement du régime représentatif[1486]. Dans un régime représentatif, la démocratie est toujours en état de manque[1487]. Les citoyens ne se retrouvent pas toujours dans le corps des représentants[1488]. Les révoltes populaires et l'utilisation de la participation citoyenne pour légitimer les coups d'État que le continent traverse, notamment au Burkina Faso, en Guinée, au Mali, au Tchad, etc. ces dernières années témoignent de l'exactitude de ce propos[1489]. L'information et la participation publique constituent les principes

---

1483 V. not., art. 141 de la Constitution togolaise de 1992 ; art. 164 al. 2 de la Constitution nigérienne de 2010, art. 145 de la Constitution Burkinabé de 1991.

1484 P. COMBEAU, « Décentralisation et démocratie », in J.-F. BRISSON (dir.), *L'assemblée délibérante, avenir de la démocratie locale ?* L'Harmattan, 2016, p. 12 : « L'idée même d'une démocratie administrative ou d'une démocratie locale représente donc, de ce point de vue, une perversion de l'idéal démocratique ».

1485 P. ROSANVALLON, *La légitimité démocratique. Impartialité, réflexivité, proximité*, Le Seuil, 2008, p. 195.

1486 V. not., J. CHEVALLIER, « La crise de la démocratie : mythe ou réalite ? », in *Les mutations contemporaines du droit public, Mélanges en l'honneur de Benoît Jeanneau*, 2002, p. 361 ; *Nouvelles questions sur la démocratie,* Dalloz, 2010 ; J.-B. AUBY, « Remarques préliminaires sur la démocratie administrative », *Revue française d'administration publique,* n° 137-138, 2011, pp. 13-19 ; A. OGIEN, S. LAUGIER, *Le principe démocratie : enquête sur les nouvelles formes du politique*, La Découverte, 2014.

1487 D. ROUSSEAU, « La démocratie continue : fondements constitutionnels et institutions d'une action continuelle des citoyens », *Confluence des droits _La revue* [En ligne], 02 | 2020, mis en ligne le 11 février 2020. URL : https://confluencedesdroits-larevue.com/?p=726 ; « Conceptualiser la compétence normative des citoyens », *Archives de philosophie du droit*, vol. 62, n°1, 2020, pp. 427-438. Voir aussi, D. ROUSSEAU, (dir.), *La démocratie continue*, LGDJ-Bruylant, 1995.

1488 D. ROUSSEAU, *Six* thèses *pour la démocratie continue*, Odile Jacob, 2022, p. 37 et s.

1489 A. K. SAIDOU, « La participation citoyenne dans les politiques publiques de sécurité en Afrique : analyse comparative des exemples du Burkina Faso et du Niger », *Revue internationale de politique de développement* [Online], 11.1 | 2019, URL : http://journals.openedition.org/poldev/3209 consulté le 30 juin 2022.

cardinaux de la démocratie locale participative[1490]. Aussi, il faut noter que les réformes décentralisatrices conçoivent la participation citoyenne comme un enjeu du processus de la décentralisation. Les citoyens doivent être mis en capacité de pouvoir agir. Pour en finir avec « l'atonie de démocratie de participation »[1491], des dispositifs juridiques sont mis en place dans les nouveaux textes de loi sur la décentralisation afin de garantir l'implication citoyenne. La démocratie participative s'exprime au quotidien grâce à l'élargissement de la transparence de l'action administrative locale (**§1**). Cependant, les mécanismes de participation demeurent sans grande portée pratique (**§2**).

§1.- Une démocratie participative dominée par un droit à l'information des citoyens

631. - La démocratie participative n'a de sens que si les citoyens disposent de l'information. Il s'agit de la condition minimale pour que naisse une démocratie participative. Les citoyens doivent être tenus informés des décisions prises par la collectivité territoriale et des débats qui se tiennent dans les instances délibératives[1492]. L'information permet de rapprocher les citoyens des élus et sert à la bonne administration locale. L'objet de cette information précise Jacques Moreau, doit être plus large que les décisions prises par l'organe exécutif et l'organe délibérant, *« elle doit avoir pour centre la vie locale »*[1493] dans son l'ensemble. La transparence de la collectivité constitue un gage d'une meilleure mobilisation des ressources. Le citoyen bien informé et convaincu de la bonne gestion locale rechignera moins au paiement des taxes et impôts[1494]. Les nouvelles

---

[1490] V. à ce propos le dossier « l'information et à la participation du public, principes cardinaux de la démocratie participative », *AJDA*, n° 42, 2006.

[1491] C. NACH MBACK, *Démocratisation et décentralisation. Genèse et dynamiques comparés des processus de décentralisation en Afrique subsaharienne*, thèse, Karthala-PDM, 2003.

[1492] Y. LUCHAIRE, « Le droit à l'information pertinente », *Annuaire des* coll*ectivités locales*, n° 21, 2001, pp. 109-123 ; P. SARDAN, « Démocratie et décentralisation », in *Études offertes à Jean-Marie Auby*, Dalloz, 1992, p. 294.

[1493] J. MOREAU, « Le droit à l'information des élus locaux et des citoyens », in CURAPP/CRAPS, *La démocratie locale. Représentation, participation et espace public*, PUF, 1999, p. 141.

[1494] G. MARCOU, « La démocratie locale en France : aspects juridiques », in CURAPP/CRAPS, *La démocratie locale. Représentation, participation et espace public*, *op.cit.*, p. 41.

législations adoptées dans les pays d'Afrique francophones semblent avoir perçu l'importance de l'information locale. Elles consacrent un droit à l'information des citoyens sur les affaires de la collectivité. L'article 19 de la loi du 26 juin 2019 précitée dispose que « *le droit des citoyens à être informés des décisions et à être consultés sur les affaires qui les concernent, par les élus locaux, contribue à l'épanouissement de la démocratie locale* ». Ce droit à « *l'information des citoyens sur les affaires locales s'exerce dans le respect des dispositions en vigueur notamment en matière de publicité des actes de l'administration et à la liberté d'accès aux documents administratifs* »[1495]. Le droit à l'information des citoyens prend deux formes distinctes, traduisant chacune une certaine attitude de leurs titulaires. Il s'agit d'abord de la prise de connaissance des décisions locales, qui implique une attitude plus ou moins passive du citoyen (**A**) et de l'exercice du droit à la communication des documents administratifs qui nécessite une démarche proactive de la part du citoyen (**B**).

A- Un accès étendu aux décisions locales

632. - Les citoyens disposent d'un droit à l'information sur les actes pris par les autorités locales. Ce droit comporte l'obligation pour les autorités locales de faire connaitre aux citoyens les moments du débat et les décisions qu'ils prennent. L'obligation qui pèse sur les collectivités territoriales se résume donc dans la publicité des débats tenus par le conseil de la collectivité (**1°**) et la publication des décisions locales (**2°**).

1. Le droit à la publicité des séances de l'assemblée délibérante

633. - La publicité des débats des assemblées délibérantes constitue désormais un acquis irrévocable pour l'information des citoyens. Le caractère public des séances du conseil de la collectivité reconnu par la plupart des législations africaines constitue la première source d'information des citoyens. Cela permet à tout citoyen qui le désire d'assister aux séances de l'assemblée délibérante, de suivre les discussions et de situer la position des différents élus intervenants. Les convocations aux réunions du conseil de la collectivité ainsi que l'ordre du jour sont communiquées au public. L'article 20 de la loi de 2019 prévoit que cette « *information est donnée principalement par voie d'affichage ou par tout autre moyen de*

1495 V. art. 15 du CGCT au Niger ; art. 11 du CGCT au Burkina Faso ; art. 41 du code des collectivités territoriales au Mali.

communication »[1496]. Il pèse sur l'exécutif local une obligation d'organisation de la salle de délibération de sorte à pouvoir accueillir le public.

634. - Cependant, la participation des citoyens à ces séances est soumise à un certain nombre d'exigences. La participation du public aux séances de l'assemblée délibérante doit être silencieuse, afin de ne pas vicier la régularité de la délibération. Toute forme de communication avec les conseillers pendant la délibération est proscrite. Le public doit éviter de perturber de quelque manière que ce soit la réunion de l'assemblée. Ainsi, il est interdit au public présent dans la salle des délibérations de fumer et de troubler par des cris, des paroles, des gestes ou de toute autre façon les délibérations. Bien que les séances du conseil puissent être retransmises, en direct ou en différé, par les moyens de communication audiovisuelle de la collectivité territoriale, il est formellement interdit au public de filmer et d'enregistrer les séances de l'assemblée délibérante[1497]. En France, depuis la loi du 6 février 1992, le public peut enregistrer les séances de l'assemblée délibérante. Le tribunal administratif de Besançon avait même admis, déjà en 1974, la possibilité d'utiliser un magnétophone en raison du fait qu'il s'agissait d'une réunion publique[1498].

635. - La règle de la publicité des débats connait au moins deux exceptions qui méritent d'être mentionnées. La première concerne l'exception du huis clos que nous avons présenté précédemment. Cette procédure permet au conseil de la collectivité de se réunir en secret. Elle porte atteinte aux règles libérales, et par conséquent au droit à l'information des citoyens. Le conseil de la collectivité territoriale peut, sans débats, décider du huis clos à la demande du chef de l'exécutif ou des deux tiers membres de l'assemblée délibérante[1499]. C'est le cas, notamment lorsque les séances portent sur des mesures individuelles ainsi que le prévoit l'article 186 du CGCT au Niger : *« les séances ne sont en aucun cas*

1496 V. également, art. 157 du CGCT pour les régions et art. 236 pour les communes au Burkina-Faso ; art. 15 du CGCT au Niger.

1497 V. not. l'art. 11 du décret n° 2019-087/PR portant modalités de convocation et mission de la première réunion des conseillers municipaux et fixant le cadre général du règlement intérieur du conseil municipal.

1498 *Cf.* M. RASERA, *La démocratie locale*, LGDJ, coll. « Systèmes », 2002, p. 86.

1499 Art. 101 pour les communes et art. 252 pour les régions : « Toutefois, à la demande du président ou du tiers des membres présents, le conseil municipal [régional], sans débat, décide du huis clos, particulièrement lorsqu'il s'agit de délibérer sur des mesures individuelles ».

publiques lorsque les délibérations portent sur des questions de personnes. Le président du conseil prononce le huis clos pour la durée des délibérations afférentes à ces questions ». Toutefois, le secret des débats n'empêche pas totalement le citoyen d'accéder à l'information dans la mesure où les conclusions de ces séances doivent être rendues publiques. « *Le procès-verbal du conseil des séances publiques et les conclusions des séances à huis clos doivent être affichés dans les locaux de la collectivité immédiatement après leur adoption* »[1500]. La seconde exception au caractère public des séances est relative aux commissions permanentes des assemblées délibérantes. La législation nigérienne est la plus explicite à ce propos. En effet, l'article 202 du CGCT dispose clairement que les commissions « *n'ont aucun pouvoir de décision et ne peuvent exercer aucune des attributions dévolues au conseil municipal ou régional ainsi qu'à leurs présidents. Leurs séances ne sont pas publiques. Toutefois, elles peuvent associer à leurs travaux toute personne qu'elles jugent utile d'entendre. À ce titre, elles peuvent associer à leurs travaux, à titre consultatif, le personnel en fonction dans les services de la commune ou de la région* »[1501].

636. - Une fois les séances des assemblées délibérantes achevées, les décisions prises doivent être publiées.

2. La publication des décisions locales, une condition de leur opposabilité

637. - Les décisions prises par les assemblées délibérantes doivent faire l'objet d'une publication afin que l'ensemble des habitants de la collectivité territoriale puisse en prendre connaissance. À ce titre, les législations africaines prévoient que les procès-verbaux des délibérations doivent être rendus publics et accessibles à tous. Cette information est donnée généralement par la voie de l'affichage, qui demeure le mode traditionnel par lequel les administrés sont informés des décisions de l'administration locale en Afrique. Ainsi, le code général des collectivités territoriales burkinabé prévoit que « *les délibérations du conseil municipal sont portées à la connaissance du public par voie d'affichage dans les locaux de la mairie et en tout autre lieu approprié* »[1502]. En outre, le législateur togolais prévoit que « *les extraits du*

1500 Art. 191 al. 7 du CGCT au Niger.

1501 Voir aussi, art. 27 Décret n° 2019-087/PR portant modalités de convocation et mission de la première réunion des conseillers municipaux et fixant le cadre général du règlement intérieur du conseil municipal.

1502 Art. 247 du CGCT au Burkina Faso.

compte rendu de chaque séance sont affichés à la mairie, dans les huit jours qui suivent la séance. La certification de l'affichage du compte rendu est faite par le maire et mentionnée au registre des délibérations »[1503].

638. - Il faut noter également que les actes règlementaires pris par l'organe exécutif ne peuvent pas être exécutoires tant qu'ils ne sont pas portés à la connaissance du public. Les législations en Afrique se conforment toutes à cette exigence sans laquelle les actes règlementaires ne sauraient être opposables à l'égard des administrés. Au Togo, l'article 150 de la loi du 26 juin 2019 relative à la décentralisation et aux libertés locales dispose que « *les arrêtés du maire ne sont exécutoires qu'après avoir été portés à la connaissance des intéressés, par voie de publication ou d'affichage, toutes les fois qu'ils contiennent des dispositions générales* […] ». Lorsqu'il s'agit d'actes règlementaires contenant des mesures d'ordre individuelles, la publication procède par voie de notification aux intéressées[1504]. L'obligation de publication des actes règlementaires s'ajoute à celle de l'obligation de transmission au représentant de l'État dans la collectivité. Le législateur nigérien rappelle le caractère cumulatif de ces deux obligations. Il affirme en effet que « *les actes pris par les autorités des collectivités territoriales sont exécutoires de plein droit dès qu'il a été procédé à leur publication ou à leur notification aux intéressés ainsi qu'à leur transmission à l'autorité de tutelle* »[1505]. On observe les mêmes dispositions du côté du législateur burkinabè, notamment lorsqu'il dispose que les actes de l'exécutif local « *ne sont exécutoires et/ou opposables aux tiers qu'après avoir été portés à la connaissance des intéressés, par voie de publication ou d'affichage, toutes les fois qu'ils contiennent des dispositions générales et dans les autres cas par voie de notification individuelle* »[1506].

639. - Le droit des habitants de la collectivité territoriale à l'information sur les débats et les décisions locales est particulièrement étendu. L'affirmation de ce droit dans les législations africaines constitue une innovation fondamentale dans les rapports entre les autorités locales africaines et les citoyens dans la mesure où la gouvernance par l'opacité est largement répandue[1507]. Cependant, la transparence de l'administration

---

1503 Loi n° 2019-006 *préc.*, art. 106 & 253.
1504 Loi n° 2019-006 *préc.*, art. 150.
1505 Art. 312 du CGCT au Niger.
1506 Art. 213 & 300 du CGCT au Burkina Faso.
1507 C. NACH MBACK, *Démocratisation et décentralisation. Genèse et dynamiques comparés des processus de décentralisation en Afrique subsaharienne, op.cit.*, p. 379.

locale ne sera complète qu'avec une garantie d'accès pour les personnes qui en font la demande aux documents essentiels de la gestion locale.

B- Un droit encadré à la communication des documents administratifs

640. - Le droit d'accès à l'information des citoyens ne peut se limiter à l'information collective dont ils bénéficient du fait de l'obligation qui pèse sur les autorités locales d'ouverture des séances et de publicité de leurs décisions. L'information des citoyens implique aussi que les collectivités territoriales fassent droit à toute demande individuelle visant à s'informer sur la gestion des affaires locales. À cet effet, la plupart des législations africaines reconnaissent dans une certaine mesure un droit à la communication des documents administratifs. La discrétion et le secret[1508] qui caractérisaient l'administration publique africaine en général connaissent une mue, certes lente et encore inachevée[1509], vers un principe de transparence[1510], notamment dans la gestion des affaires locales. Le droit à la communication des documents administratifs permet aux citoyens d'accéder aux documents essentiels de la gestion locale (**1°**), mais demeure limité et enserré dans des contraintes qui ne facilitent pas la lecture de l'action publique locale par les citoyens (**2°**).

1. L'étendue du droit à la communication des documents administratifs

641. - Le droit à la communication des documents administratifs semble particulièrement étendu en Afrique. Les législations nationales relatives à la décentralisation reconnaissent aux citoyens la possibilité d'accéder à certains documents portant sur la gestion locale. Cependant, le droit à la communication des documents administratifs ne s'exerce pas de façon générale et indifférenciée. Le législateur ne reconnait pas, en effet, un droit général à la communication de l'ensemble des documents

---

1508 J.-F. MEDARD, « Les paradoxes de la corruption institutionnalisée », Revue internationale de politique comparée, vol. 13, n° 4, 2006, p. 701 : « *L'État a horreur de la transparence, comme la nature a horreur du vide* ».

1509 J.-P. OMGBA, Re-penser l'administration publique en Afrique, L'Harmattan, 2018.

1510 Sur la transparence voir not., J. MARCHAND, « Réflexions sur le principe de transparence », *RDP*, n° 3, 2014, p. 677 ; M. DELMAS-MARTY, S. MANACORDA, « La corruption : un défi pour l'État de droit et la société démocratique », *RSC*, 1997, p. 696.

produits ou disponibles au sein des collectivités territoriales. Les articles 11 et 15 du code général des collectivités territoriales respectivement burkinabè et nigérien prévoit en des termes quasi identiques la possibilité pour les citoyens d'accéder aux budgets et aux comptes des collectivités territoriales et aux arrêtés municipaux. Ce droit à la communication s'étend également à l'ensemble des actes des autorités locales, notamment ceux relatifs à la création des établissements publics locaux, aux emprunts, à la coopération décentralisée, aux accords passés avec l'État ou d'autres partenaires, à l'acceptation des dons et legs et à la prise de participation dans toute société. La loi prévoit que « *les personnes intéressées peuvent obtenir, à leurs frais, copies desdits documents, auprès des présidents des conseils municipal et régional, ou auprès de tout service public habilité de la collectivité* »[1511].

642. - L'analyse des dispositions qui précèdent amène à s'interroger sur les personnes créancières du droit à la communication des documents par les collectivités territoriales. Les législations burkinabè et nigérienne semblent s'adresser à toute personne sans distinction aucune. L'alinéa 1 des articles 11 et 15 précités prévoit que l'exercice du droit à l'information sur la gestion des affaires locales est reconnu aux habitants de la collectivité. L'accès aux documents administratifs locaux est reconnu à « toute personne physique ou morale » à la condition toutefois qu'elle demeure dans la collectivité. L'appartenance à la collectivité instaure d'une certaine manière une réciprocité dans l'accès au droit à la communication des documents administratifs. Il en va de même au Mali où le code des collectivités territoriales garantit l'accès aux documents administratifs de la collectivité territoriale à tout habitant ou contribuable de la collectivité[1512].

Cependant, la loi togolaise semble introduire une distinction entre les personnes concernées par le droit à l'information et celles concernées par le droit à la communication des documents administratifs. Lorsqu'on reprend les articles 107, 357 et 377, le législateur semble réserver le droit à la communication des documents aux seuls citoyens, donc aux personnes physiques. L'article 357, par exemple, dispose que « *le compte administratif approuvé est mis à la disposition du public pour consultation. Tout citoyen dans le ressort territorial de la collectivité peut en demander copie à ses frais dans les conditions*

[1511] Art. 11 al. 2 du CGCT au Burkina Faso ; Art. 15 al. 2 du CGCT au Niger. Voir aussi, Loi n° 2019-006 *préc.*, art. 107, 357 & 377.

[1512] Art. 44, 117 & 184 du code des collectivités territoriales au Mali.

définies par la règlementation en vigueur»[1513]. Cela revient, dans les modalités de communication des documents administratifs à distinguer au moins trois niveaux[1514]. D'abord, l'information est publiée, par la voie de l'affichage, pour les comptes rendus des séances et les actes de nature règlementaire. En second lieu, pour les documents relatifs à la gestion financière de la collectivité, l'information est à mis à la disposition du public. La notion de « public » est largement ouverte[1515]. En règle générale, elle désigne toute personne physique ou morale. Enfin, l'information est fournie par la collectivité à la demande d'un particulier — citoyen — lorsqu'il s'agit de documents administratifs communicables selon la législation en vigueur sur l'accès aux documents publics. Ainsi, au Togo, toute personne physique ou morale peut avoir accès aux deux premiers niveaux d'information. En revanche, l'accès au dernier niveau ne concerne que les citoyens établis dans le ressort territorial de la collectivité.

Les documents communiqués par les collectivités territoriales doivent être des documents administratifs communicables. Sont considérés comme tels par la loi, « *tous dossiers, rapports, études, documents d'orientation ou de politiques publiques, comptes rendus, procès-verbaux, statistiques, directives, instructions, circulaires, notes de service et réponses ministérielles qui comportent une interprétation du droit ou une description des procédures administratives, les avis, les prévisions et les décisions*»[1516]. Si la notion de document administratif communicable est largement définie par la loi, pour être communicable, il doit exister matériellement et être achevé. Cette dernière condition exclut du droit à la communication les notes préparatoires, les notes de service internes. En somme, les documents n'ayant pas acquis la forme définitive ne peuvent être communiqués. Le document inachevé renvoie selon la loi n° 051-2015/CNT au Burkina-Faso à tout « *document imprimé ou existant sur tout autre support de conservation en cours d'élaboration, de finalisation ou de*

1513 V. aussi art. 107 : « Tout citoyen a le droit de demander communication, de prendre copie totale ou partielle, à ses frais, des procès-verbaux du conseil municipal et des arrêtés municipaux » ; art. 357 : « Le budget voté est affiché au siège du conseil et est tenu à la disposition du public pour consultation. Tout citoyen peut en demander copie à ses frais ».

1514 M. RASERA, *La démocratie locale, op.cit.*, p. 93.

1515 C. MORIO, « Quel modèle de participation dans les projets de réforme institutionnelle ? », *Archives de philosophie du droit*, vol. 62, n° 1, 2020, pp. 439-462.

1516 *Cf.* art. 12 de l'Ordonnance n° 2011-22 du 23 février 2011, portant charte d'accès à l'information publique et aux documents administratifs au Niger ; Art. 44 de la loi n° 051-2015/CNT portant droit d'accès à l'information publique et aux documents administratifs au Burkina Faso ; art. L. 300-2 du CRPA en droit français.

validation»[1517]. Ainsi, le budget d'une collectivité territoriale ne peut être communiqué pendant son élaboration. Il devient un document achevé et communicable une fois la délibération terminée[1518].

643. - Malgré les limites constatées, le droit à la communication des documents administratifs sur la gestion des affaires locales aux citoyens semble assez étendu formellement. La communication des documents administratifs garantit au citoyen local un accès important à l'information afin que ce dernier puisse participer activement à la gestion des affaires locales. L'objectif de transparence souhaitée à travers la consécration de ce droit ne sera atteint que si la collectivité facilite la lisibilité des informations communiquées aux personnes intéressées.

2. Les contraintes à l'exercice effectif du droit à la communication : l'illisibilité des informations communiquées

644. - Les droits afférents au statut de citoyen ne favorisent le rééquilibrage de la relation administrative qu'à la condition que l'information soit véritablement transparente. Selon Jacques Chevallier, « *la transparence va au-delà du simple accès au droit pour donner à l'administré prise sur le fonctionnement administratif : elle vise à rendre l'Administration moins opaque, à dissiper le brouillard qui l'entoure, à déchirer le voile qui la recouvre, à briser le mur de silence qui la protège ; elle implique la levée de la règle du secret, qui constitue la garantie ultime de la construction bureaucratique* »[1519]. Si les législations précitées contribuent incontestablement à ouvrir l'administration locale aux citoyens, elles ne suffisent pas à faire d'elle cette « maison de verre »[1520] totalement lisible par le citoyen. Généralement, les informations diffusées spontanément à l'initiative des collectivités territoriales sont perçues avec une certaine méfiance, car souvent biaisées par la volonté de présenter les autorités locales sous leur meilleur jour[1521]. L'accès à l'information par le citoyen lui-même est un premier gage de la qualité de l'information reçue.

1517 Art. 4 de la loi n° 051-2015/CNT portant droit d'accès à l'information publique et aux documents administratifs au Burkina-Faso.

1518 M. RASERA, *La démocratie locale, op.cit.*, p. 94.

1519 J. CHEVALLIER, « La transformation de la relation administrative : mythe ou réalité ? », *D.* 2000, p. 575.

1520 *Idem.*

1521 C. NACH MBACK, *Démocratisation et décentralisation…, op.cit.*, p. 380 ; G. MARCOU, « La démocratie locale en France : aspects juridiques », *op.cit.*, p. 41.

Encore faudrait-il ensuite qu'elle soit suffisante, complète et compréhensible par ce dernier.

645. - D'abord, il faut noter qu'en ce qui concerne les délibérations, les législations africaines précitées prévoient expressément un accès aux procès-verbaux des sessions du conseil de la collectivité. Certes ils sont rendus publics, toute personne intéressée pouvant en demander la communication à ses frais, cependant, pour ce qui est de l'accès au registre des délibérations « *coté et paraphé par le préfet* »[1522], les législations nationales ne prévoient ni leur publication ni leur communication au public. En droit français, le juge administratif admet depuis bien longtemps que la communication des procès-verbaux des délibérations du conseil municipal signifie également la communication du registre des délibérations proprement dit, c'est-à-dire le registre coté et paraphé par le préfet ou le sous-préfet, lequel doit être mis à la disposition du public[1523]. Toute autre forme de communication serait insuffisante, notamment la remise de simples reproductions du registre ou d'un bulletin imprimé[1524]. Ainsi, l'article L. 2121-26 du CGCT en France, qui dispose que toute personne peut demander communication des procès-verbaux du conseil municipal, est interprété par le Conseil d'État comme concernant aussi le registre des délibérations. Il est possible d'espérer que cette jurisprudence puisse inspirer les juridictions administratives africaines, au moment venu, pour préciser la portée des législations sur le droit à la communication des documents administratifs des collectivités territoriales. Ensuite, en matière financière, les législations africaines permettent la mise à disposition des budgets et comptes des collectivités à toute personne qui en fait la demande. Cependant, il ne pèse sur les autorités décentralisées aucune obligation quant à la lisibilité des documents budgétaires transmis. Les documents budgétaires sont transmis à l'état brut aux citoyens. Les collectivités territoriales n'ont aucune obligation de les assortir d'annexes pouvant contenir, notamment des documents synthétiques et comparatifs

---

1522 V. not., loi n° 2019-006, art. 106 al. 3 ; art. 40 du code des collectivités territoriales au Mali.

1523 CE 21 mai 1926, *Mériot*, *Lebon*, p. 532 ; S. 1926. 3. 57, note Hauriou ; CE 29 octobre 1943, *Debelfont*, *Lebon*, p. 347 ; TA Nice, 21 juill. 1981, *Giudicelli*, *Lebon*, p. 540.

1524 J.-M. BECET, « Communication des documents des collectivités territoriales au public : régimes particuliers », in *Encyclopédie des collectivités locales*, folio n°12212, 2021, § 30.

afin de faciliter la lecture et l'exploitation des chiffres bruts par les citoyens[1525].

646. - La transparence de la gestion locale et la participation locale voulue par les différentes législations ne doivent pas se résumer à la quantité des données mise à la disposition des citoyens. Il faut que les citoyens aient les moyens de se saisir véritablement des informations mises à leur disposition. La participation du citoyen ne peut être conditionnée qu'au seul droit de savoir. La réalisation de cette participation renvoie aussi aux capacités de comprendre et d'interpréter les informations communiquées[1526]. La qualité de l'information et les capacités d'analyse de ces données constituent ainsi le premier niveau d'une démocratie participative bien ordonnée[1527].

Il faut bien se rendre compte que la question de la transparence de l'action administrative locale est finalement moins simple à résoudre qu'elle n'y parait en Afrique. Le niveau d'alphabétisation[1528] des citoyens, que ce soit en milieu rural ou urbain, peut laisser dubitatif, même aux plus optimistes d'entre nous, quant aux nombres de personnes susceptibles d'exercer ce droit à la communication[1529]. Par conséquent, seuls quelques citoyens bien instruits, ayant un intérêt personnel, sont capables dans ces conditions d'effectuer un travail de décodage et de mise en perspective des données brutes. Ce qui confirme encore mieux l'analyse faite par Pierre Sardan à propos de la France. L'auteur estimait que *« la demande de transparence n'existe qu'autant qu'un intérêt individuel concret est en jeu. En dehors des préoccupations personnelles, les dispositifs les plus ingénieux ne rencontrent qu'une indifférence polie »*[1530].

1525 H. K. NOUBISSI, *Décentralisation et centralisation au Cameroun. La répartition des compétences entre l'État et les collectivités territoriales, op.cit.*, p. 395.

1526 H. MICHEL, « Promesses et usages des dispositifs de transparence : entre approfondissement et redéfinition de la démocratie », *Revue française d'administration publique*, vol. 165, n° 1, 2018, p. 5.

1527 M. YAZI-ROMAN, « De l'information à la co-construction : la démocratie participative dans tous ses états », *AJCT*, 2018, p. 240.

1528 Le taux national d'alphabétisation en 2020 était de 30,11 % selon l'Institut national de la statistique du Niger. Au Burkina Faso, ce taux était d'environ 40 % en 2018 et de 30, 8 % au Mali. Au Togo, ce taux avoisine les 67 %.

1529 C. NACH MBACK, *Démocratisation et décentralisation…, op.cit.*, p. 382.

1530 P. SARDAN, « Démocratie et décentralisation », in *Études offertes à Jean-Marie Auby*, Dalloz, 1992, p. 295.

647. - Finalement, pour s'informer efficacement, acquérir une bonne compréhension des sujets, souvent complexes, il faut beaucoup de temps, d'énergie et un peu de connaissance pour décrypter les informations. Le mythe du citoyen actif, faisant usage de sa raison pour participer à la vie de la cité, entretenu par la transparence au sens d'accès aux documents[1531], trouve ses limites en Afrique dans les contraintes qui pèsent sur sa mise en œuvre effective. Le constat est aussi peu reluisant en ce qui concerne l'association des citoyens à la prise de décision.

§2.- La limitation de la participation citoyenne dans le processus décisionnel

648. - La démocratie participative au sens où l'entend Loïc Blondiaux, c'est-à-dire lato sensu, renvoie à « tout ce qui, dans la vie politique des démocraties contemporaines, ne relève pas strictement de la logique du gouvernement représentatif »[1532]. La volonté de dépassement de la logique représentative et d'inflexion du modèle bureaucratique fondé sur la distanciation et l'autorité a conduit à l'instauration, dans les démocraties modernes, de plus de transparence et d'implication citoyenne[1533]. La combinaison de ces éléments invite à s'intéresser aux manifestations de ce que Jacques Chevallier analyse en droit français comme une refondation en cours de la relation administrative[1534]. Il s'agit ici de s'interroger sur les axes que prend cette refondation de la relation administrative dans les États d'Afrique noire francophone au niveau des collectivités territoriales. Assiste-t-on à un véritable renforcement des droits sous la forme de l'apparition progressive d'une citoyenneté administrative impliquant l'octroi au public de « possibilités d'influence et d'emprise sur les processus administratifs » ? Toujours selon Jacques Chevallier, le renforcement des droits des citoyens se traduirait notamment à travers un droit à l'information et un droit de participation aux processus administratifs. L'idée que l'administration est tenue d'associer, d'une manière ou d'une

---

1531 H. MICHEL, « Promesses et usages des dispositifs de transparence : entre approfondissement et redéfinition de la démocratie », *op.cit.*, p. 5 et s.

1532 L. BLONDIAUX, *Le nouvel esprit de la démocratie. Actualité de la démocratie participative*, Seuil, 2008, p. 38.

1533 J. CHEVALLIER, *Science administrative*, PUF, coll. « Thémis », 2013, p. 421.

1534 J. CHEVALLIER, « Le droit administratif vu par la science administrative », *AJDA*, 2013, p. 401.

autre, les administrés à la prise des décisions et à la marche des services s'est désormais imposée[1535]. Sur le continent africain, elle se traduit par la consécration d'une multiplicité d'outils juridiques de la participation citoyenne (**A**), mais à la portée réduite, sans grande influence sur la prise de décision publique (**B**).

A- La consécration de multiples instruments d'implication des citoyens dans la vie publique locale

649. - Les mécanismes juridiques de participation des citoyens à la gestion des affaires locales sont quantitativement importants. S'ils sont faciles à repérer, il s'avère difficile d'en établir une typologie. La panoplie des outils de participation des citoyens aux décisions administratives locales qu'offre les législations africaines permet de distinguer d'un côté, des formes classiques de démocratie participative (**1°**) et de l'autre, de nouvelles approches institutionnelles de la démocratie locale (**2°**), que nous nous attellerons à décrire sommairement.

1. Les formes classiques de la démocratie participative locale

650. - En partant du droit positif et de la pratique institutionnelle, la littérature scientifique sur la démocratie participative distingue généralement quatre grandes modalités de participation des citoyens à la vie publique locale[1536]. Il s'agit de la délégation, la décision, la revendication et la consultation. La délégation se manifeste dans l'élection. C'est la forme principale de l'implication des citoyens dans les démocraties représentatives. Concrètement, elle a pour effet « [d'habiliter] *une ou plusieurs personnes à exercer des compétences préalablement définies* ». La décision consiste en l'adoption directe d'une décision par les participants et prend la forme type du référendum. La revendication relève très souvent de l'action de la société civile organisée et consiste à attirer l'attention des pouvoirs publics sur une idée parfois en opposition avec une décision prise par les institutions politiques légitimes. Elle peut prendre plusieurs formes, comme la manifestation ou la pétition[1537]. Quant à la consultation, elle

---

1535 *Ibid.*, p. 401 et s.

1536 V. not. E. BUGE, C. MORIO, « Le Grand débat national : apports et limites pour la participation citoyenne », *RDP*, n° 5, 2019, pp. 1214-1218.

1537 Sur la capacité des citoyens à s'inviter dans le jeu politique, voir, P. ROSANVALLON, *La contre-démocratie. La politique à l'âge de la défiance*, Seuil, 2006, p. 16.

« *recouvre l'ensemble des procédures qui permettent à une autorité publique de prendre des avis préalablement à sa décision, sans que ces derniers aient une valeur contraignante* »[1538].

651. - Les États d'Afrique noire subsaharienne sont en général traversés par chacune de ces formes de participation, même si prédomine la procédure de l'élection politique. Chacun de ces outils juridiques peut se combiner selon les objectifs poursuivis par les différents acteurs. En Afrique, en dehors de la délégation qui leur confère une légitimité démocratique, les autorités administratives locales recourent le plus souvent à la consultation sous la forme de l'enquête publique et de la concertation. L'ensemble de ces dispositifs juridiques de participation ont vocation à remplir une fonction consultative. Ils permettent aux autorités locales de bénéficier d'une légitimité supplémentaire dans un contexte de désaveu de l'autorité, en faisant avaliser des décisions déjà prises. En cela, les procédés de consultation publique ne font pas des habitants de la Cité des co-décideurs. L'objectif, ainsi que le relève Jacques Caillose dans une étude sur les formes de participation en droit français, n'est aucunement d'investir les populations consultées d'un pouvoir dont les conseils élus sont « *constitutionnellement seuls en charge, mais de faire assumer par une collectivité symboliquement placée en situation de débat des choix déjà effectués ou sur le point de l'être* »[1539].

652. - Ainsi, l'enquête publique permet de recueillir les avis des administrés intéressés préalablement à certaines opérations. En Afrique francophone, l'initiative de l'organisation de l'enquête publique relève généralement de l'État. La consécration de cette procédure était d'abord prévue en matière d'expropriation pour cause d'utilité publique. Le code foncier et domanial du Togo prévoit en ce sens que « *l'expropriation d'immeuble en tout ou partie, ou de droits réels immobiliers ne peut être prononcée qu'autant qu'elle aura été précédée d'une déclaration d'utilité publique intervenue à la suite d'une enquête publique* »[1540]. Désormais, elle trouve une application

---

[1538] C. MORIO, « Quel modèle de participation dans les projets de réforme institutionnelle ? », *Archives de philosophie du droit*, vol. 62, n° 1, 2020, p. 439.
[1539] J. CAILLOSSE, « Quelle(s) participation(s) aux affaires communales ? Éléments de réponse du droit français », *Pouvoirs*, n° 95, 2000, p. 122. V. également, B. FAURE, *Droit des collectivités territoriales, op.cit.*, n° 259 et s.
[1540] Art. 362 du code foncier et domanial au Togo.

également en matière environnementale[1541]. Il faut noter que l'administration ne se confronte à la population qu'une fois qu'elle a arrêté ses différents choix. Le caractère obligatoire[1542] de la procédure de l'enquête publique ne prive pas l'administration de la maitrise de la décision définitive. Elle peut aller à l'encontre des avis exprimés[1543].

La concertation en matière d'urbanisme obéit à la même logique. Elle s'inscrit dans le cadre général de la participation des citoyens à la vie politique nationale et locale. Elle est organisée par les autorités politiques qui déterminent les modalités d'intervention des différents acteurs locaux. C'est dans les modalités d'organisation de la concertation que se trouve sa principale limite, notamment au Burkina Faso, où cette compétence appartient au pouvoir central. Le code de l'urbanisme et de la construction du Burkina Faso dispose en son article 101 que « *des cadres de concertation pour la réalisation des opérations d'urbanisme peuvent être constitués par arrêté du ministre en charge de l'urbanisme et de la construction* »[1544]. Toute personne, physique ou morale, détenant un titre de jouissance et/ou de titres fonciers sur des immeubles sis dans une zone urbaine concernée par une opération d'urbanisme peut participer au processus de concertation[1545]. En revanche, au Niger, l'organisation de la concertation et des enquêtes publiques en matière d'urbanisme relève de la seule compétence des autorités communales. C'est ce que prévoit la loi n° 2008-03 du 30 avril 2008 portant loi d'orientation sur l'urbanisme et l'aménagement foncier. Elle dispose à l'article 5 que « *l'élaboration et la révision des plans d'urbanisme, ainsi que les concertations et les enquêtes publiques y relatives sont de la compétence des autorités communales* ». L'autorité de tutelle c'est-à-dire le représentant de

---

1541 V., art. 30-32, Loi n° 2008-005 du 30 mai 2008 portant loi-cadre sur l'environnement au Togo.

1542 Sauf cas d'urgence dans la DUP, V. art. 363 du code foncier et domanial du Togo. Sur l'action du juge en matière de DUP en droit français, S. ROUSSEL, « Le juge et la régularisation d'une déclaration d'utilité publique », *RFDA*, n° 5, 2021, p. 932.

1543 Pour un panorama complet sur ce sujet, IFDD, *La participation publique dans l'évaluation environnementale en Afrique francophone*, 2013. En droit français, V. not., L. FONBAUSTIER, « La participation du public » *AJDA*, n° 9, 2015, p. 517 ; Y. JEGOUZO, « La réforme des enquêtes publiques et la mise en œuvre du principe de participation », *AJDA*, n° 32, 2010, p. 1812 ; « De la "participation du public" à la "démocratie participative" » ? *AJDA*, n° 42, 2006, p. 2314.

1544 Loi n° 017-2006 du 18 mai 2006 portant code de l'urbanisme et de la construction du Burkina Faso.

1545 *Ibid.*, art. 102.

l'État territorialement compétent dispose cependant d'un pouvoir d'approbation des plans d'urbanisme[1546].

653. - Par ailleurs, le référendum local n'est pas prévu par les textes africains. En droit français, la volonté de promouvoir la démocratie locale a conduit à la consécration par la révision constitutionnelle de 2003 d'un référendum local décisionnel. En effet, l'article 72-1 alinéa 2 issu de cette révision dispose que « *dans les conditions prévues par la loi organique, les projets de délibération ou d'acte relevant de la compétence d'une collectivité territoriale peuvent, à son initiative, être soumis, par la voie du référendum, à la décision des électeurs de cette collectivité* ». Le référendum local n'a que peu de chance devant les mécanismes de consultation indirecte prévus par les lois de décentralisation en Afrique. L'immixtion directe du suffrage universel dans l'administration locale semble *a priori* contraire aux dispositions constitutionnelles précitées qui confient l'administration de la collectivité territoriale à des conseils élus. Une interprétation stricte de ces dispositions semble interdire que d'autres organes puissent gérer la collectivité territoriale avec les conseils élus. Cette situation correspond à celle qui prévalait en droit français avant la réforme de 2003. La lecture stricte de l'article 72 interdisait que les électeurs aient une quelconque attribution directe dans la libre administration des collectivités territoriales[1547].

654. - Les consultations locales sont donc étroitement organisées par les textes en vigueur. Ainsi, le conseil de la collectivité peut prendre l'initiative de mettre en place des organes de concertation ou des comités consultatifs sur des questions présentant un intérêt communal[1548]. L'article 29, alinéa 1 du décret fixant le cadre général du règlement intérieur du conseil municipal au Togo prévoit que « *le conseil municipal peut créer des comités consultatifs sur tout problème d'intérêt communal concernant tout ou partie du territoire de la commune* ». Dans des termes presque identiques, les codes burkinabè et nigérien prévoient que « *le conseil de la collectivité territoriale peut créer des organes de concertation sur toute question d'intérêt local. Ces organes de concertation comprennent des personnes qui peuvent ne pas appartenir au conseil, notamment des représentants des associations locales, des notabilités locales, des*

1546 Art. 5 al. 2 de la loi n° 2008-03 du 30 avril 2008 portant loi d'orientation sur l'urbanisme et l'aménagement foncier.

1547 *Cf.* M. VERPEAUX, « Le "référendum local" et la Constitution », *Les cahiers du Conseil constitutionnel*, n°12, mai 2002.

1548 Art. 12 al. 2 du CGCT au Burkina Faso ; art. 16 du CGCT au Niger ; art. 29 al. 2 du décret fixant cadre général du règlement intérieur au Togo *préc.*

personnalités compétentes dans les domaines traités»[1549]. Ces organes ou comités ont un rôle consultatif. Les avis qu'ils émettent ne sauraient en aucun cas lier le conseil de la collectivité territoriale.

655. - L'ensemble de ces outils juridiques de consultation du public qui se mettent en place au sein des collectivités territoriales ne contribuent pas véritablement à la réaffirmation du « récit » consistant à faire de la décentralisation un processus de démocratisation des rapports sociaux. Que ce soit pour les enquêtes publiques ou les comités consultatifs, le public consulté se contente de formuler une « opinion » ou de donner simplement un « avis ». Tout semble fait afin que l'auteur de la décision demeure le conseil. Le public, ici, ne remplit que le rôle dévolu à un simple organe consultatif. Sur le plan institutionnel, la participation des citoyens dans les affaires des collectivités territoriales africaines prend des formes nouvelles.

2. Les nouvelles approches de la démocratie participative locale

656. - Les réformes décentralisatrices vont contribuer à l'institutionnalisation de formes nouvelles de participation citoyenne. Les législations nationales donnent la possibilité aux collectivités territoriales d'organiser leur espace territorial par la création de structures infra-communales qui servent de canaux d'information du public et à la participation active des citoyens à la gestion des affaires locales. On assiste ainsi à la reconnaissance législative et règlementaire des comités de développement dans les villages et quartiers (**a**) à laquelle s'ajoute la promotion de Bureau du citoyen (**b**) en tant qu'outils de participation et de contrôle citoyen de l'action publique.

a) Les comités de développement à la base

657. - La création des conseils ou comités de développement relève d'une logique d'organisation communautaire qui permet à la collectivité territoriale d'avoir des interlocuteurs au niveau infra-communaux. On peut d'ailleurs lire à ce propos dans le plan de développement communal de la commune Kozah 1 au Togo que « *les comités de développement constituent les bras opérationnels de la commune dans les quartiers* »[1550]. Ils doivent mobiliser

---

1549 *Idem.*
1550 Plan de Développement Communal 2019-2023 de Kozah I, p. 44.

la population autour des actions de développement dans leurs quartiers respectifs. Au Burkina Faso, le code général des collectivités territoriales institue les conseils villageois de développement dans les villages des communes rurales et dans les villages rattachés aux communes urbaines[1551]. Ils constituent des formes anciennes d'organisation spontanée des groupes communautaires apparues en marge du droit avant leur consécration dans les textes. Au Togo, par exemple, les comités de développement à la base ont été créés par un décret du gouvernement en 2012[1552].

658. - Les comités de développement à la base sont des cadres de dialogue entre les élus et les habitants. À ce titre, ils jouent un rôle important dans la promotion de la participation citoyenne. À y regarder de près, ces comités de développement entretiennent une certaine affinité avec la politique française de la ville ayant favorisé la création de conseils de quartiers par la loi n° 2002-276 du 27 février 2002 relative à la démocratie de proximité[1553]. Certains auteurs, notamment Calvin Minfegue, voient également dans ces comités une certaine proximité conceptuelle avec le « mouvement communautaire nord-américain » en ce qu'ils préconisent un traitement territorialement localisé des problèmes communautaires en motivant les habitants à s'auto-organiser[1554]. Les conseils villageois de développement sont portés vers le développement local en favorisant sa prise en charge par les habitants et leur participation au processus décisionnel. Cette participation des citoyens dans les comités de développement prend des formes diverses, allant de la consultation à une implication dans la conception et le suivi des programmes de développement communaux. Au Mali, le législateur impose au conseil communal une consultation préalable des conseils de villages, de fractions ou de quartiers constituant la commune avant l'organisation du vote du

---

1551 Art. 222 du CGCT au Burkina Faso.

1552 Décret n° 2012-005/PR du 29 février 2012 relatif aux comités villageois de développement à la base.

1553 Pour une analyse complète de cette loi, v. M. VERPEAUX, « La loi du 27 février 2002 relative à la démocratie de proximité ou la proximité, une solution pour la démocratie locale ? », *RFDA*, n° 2, 2003, p. 261. V. aussi, B. FAURE, « Démocratie municipale et démocratie de quartier », in J.-F. BRISSON (dir.), *L'assemblée délibérante, avenir de la démocratie locale* ? L'Harmattan, 2016, p. 99.

1554 C. MINFEGUE, « La participation citoyenne en contexte local au Cameroun. Entre demande de légitimité, efficacité versatile et tensions politiciennes », *African Sociological Review*, vol. 20, n° 2, 2016, p. 82.

budget communal[1555]. L'avis des conseils de villages, de fractions ou de quartiers est obligatoirement requis avant toute délibération du conseil municipal sur certaines matières notamment, « *la voirie, le cadastre, les plans d'occupation du sol et les opérations d'aménagement de l'espace communal, la création et l'entretien des puits ou points d'eau* »[1556]. Dans le même sens, l'article 223 du CGCT au Burkina Faso dispose que « *le conseil villageois de développement est chargé sous l'autorité du conseil municipal de : 1. contribuer à l'élaboration et à la mise en œuvre des plans communaux de développement ; 2. contribuer à la promotion du développement local dans le village ; 3. participer aux activités des différentes commissions mises en place par le conseil municipal pour la gestion et la promotion du développement local* ». En outre, le code nigérien laisse aux régions la possibilité, en plus des structures infra-communales, d'organiser des structures infra-régionales qui peuvent contribuer à la promotion du développement de la région[1557].

659. - La participation des citoyens dépend dans ce contexte non seulement de leur capacité à se prononcer sur des points de l'action publique locale, de leur capacité à suivre la mise en œuvre des plans communaux de développement, mais aussi de leur aptitude à proposer et à s'investir dans des réalisations ciblant leur territoire. Or, les comités de développement ne bénéficient d'une manière générale d'aucun appui financier de la commune pour la réalisation de leurs activités. Les comités font face à de nombreuses difficultés dues notamment au manque de ressources financières pour la réalisation des actions de développement, que ce soit pour le suivi des programmes ou la prise en charge directe de certaines actions. Il faut cependant noter que certaines communes togolaises bénéficient d'un appui technique. Les membres[1558] composant le bureau exécutif du comité reçoivent une formation sur leurs rôles et responsabilités ainsi que sur la gestion des infrastructures socio-communautaires. Ces formations sont assurées au Togo en collaboration

---

1555 Art. 245 du code des collectivités territoriales : « Le projet de budget est préparé par l'ordonnateur et soumis au vote de l'organe délibérant de la Collectivité territoriale. Le vote du budget est précédé d'un débat public sur le projet de budget. Pour le budget communal, le débat public est précédé de la consultation des Conseils de villages, de fractions ou de quartiers constituant la Commune ».

1556 Art. 26 du code des collectivités territoriales au Mali.

1557 Art. 204 du CGCT au Niger.

1558 Sur la composition et le rôle des chefs traditionnels dans la création et le fonctionnement des CVD, voir les développements *infra*.

avec l'Agence nationale de développement à la base[1559]. En dépit de cela, il est noté une faible capacité des comités à mobiliser et à faire participer les populations aux actions de développement.

660. - Les comités de développement à la base ne sont pas suffisamment outillés pour participer à l'élaboration et au suivi de la mise en œuvre des plans de développement locaux. L'existence d'une logique d'institutionnalisation par le haut, même si la loi vient consacrer, en l'occurrence, sans innover des dynamiques endogènes préexistantes, fait que les comités de développement sont *in fine* dépendants de la stratégie municipale[1560]. Les habitants continuent à subir l'influence du pouvoir municipal et à être dépendants des maigres ressources que ce dernier pourrait leur octroyer à travers notamment l'organisation des appels à projets.

b) Le bureau du citoyen

661. - Dans un chapitre intitulé « De la participation des citoyens à la gestion des affaires locales », le législateur togolais innove par l'introduction d'une nouvelle institution de participation citoyenne. L'article 14 de la loi n° 2019-006 du 26 juin 2019 relative à la décentralisation et aux libertés locales dispose qu'« *il est créé un bureau du citoyen* ». Il s'agit d'une « *institution locale de contrôle de l'action publique locale par les citoyens. Le bureau du citoyen est un centre d'écoute et de recueil des attentes, préoccupations et suggestions des citoyens de la collectivité territoriale* »[1561]. Il est géré par un coordonnateur issu de la société civile nommé par arrêté municipal. La généralisation de cette instance par loi du 26 juin 2019 précitée vise à formaliser un cadre existant mis en place dans certaines communes pilotes du Togo par la coopération allemande dans le cadre du programme de décentralisation et de gouvernance locale (ProDeGol). Cette institutionnalisation s'inscrit également dans le cadre des recommandations sous régionale, notamment la déclaration de Bamako qui recommande en son principe 15 de « *favoriser la participation des citoyens à*

1559 C'est le cas de la commune Kozah 1, voir le Plan de développement communal *préc.*

1560 C. MINFEGUE, « La participation citoyenne en contexte local au Cameroun. Entre demande de légitimité, efficacité versatile et tensions politiciennes », *op.cit.*, p. 33.

1561 Art. 17 al. 3 de la loi n° 2019-006 *préc.*

la vie publique en progressant dans la mise en place d'une démocratie locale, condition essentielle de l'approfondissement de la démocratie »[1562].

662. - Les habitants disposent à travers cette institution de la possibilité de saisir les élus locaux sur toutes les questions présentant un intérêt communal. Le bureau du citoyen s'érige ainsi en intermédiaire entre les citoyens et les élus. Il sert de courroie de transmission par laquelle les citoyens communiquent leurs avis à l'administration communale sur la gestion des affaires locales. Les attributions du bureau citoyen sont essentiellement le recensement des préoccupations exprimées par les citoyens. Agissant en collaboration avec l'administration communale, le bureau du citoyen assure l'accessibilité et la lisibilité des programmes communaux aux citoyens. En outre, le bureau de citoyen sert d'instance de conciliation des parties en matière de contrôle citoyen et de conflits sociaux. Ce rôle du bureau citoyen est particulièrement important quand on sait que le justiciable africain ne brille pas en général par son dynamisme. Il pèse sur les citoyens des contraintes sociologiques qui freinent le développement d'un contentieux administratif. La faiblesse de l'action juridictionnelle est donc compensée par ce procédé non juridictionnel. Le bureau du citoyen constitue en ce sens un « médiateur territorial »[1563].

663. - Malgré sa formalisation dans la loi, le bureau du citoyen ne dispose pas encore d'un cadre juridique pour son organisation et son fonctionnement. La loi du 26 juin 2019 renvoie à un arrêté du ministre chargé de la décentralisation pour les modalités d'organisation et de fonctionnement du bureau du citoyen. En attendant l'élaboration de cette mesure d'application, les communes qui souhaitent mettre en place un bureau du citoyen se basent sur un guide[1564] élaboré par la coopération allemande dans le cadre de son soutien au processus de décentralisation pour définir précisément les rôles et attributions de cette institution. Cette innovation institutionnelle en vogue est encore à la recherche de son cadre.

[1562] La déclaration de Bamako a été adoptée en 2000 par les chefs d'États africains réunis à Bamako lors du Symposium international sur le bilan des pratiques de la démocratie, des droits et des libertés dans l'espace francophone.

[1563] Le droit français de la décentralisation consacre un médiateur territorial. Voir à ce propos, S. MONNIER, « Un cadre juridique a minima pour le médiateur territorial », *JCP A*, 2020, n° 2031 ; B. FAURE, *Droit des collectivités territoriales, op.cit.*, n° 303.

[1564] GIZ, Le bureau citoyen. Guide méthodologique : Retour d'expériences de la mise en place et du fonctionnement dans les communes pilotes, 2020.

Le bilan de son action dans les communes l'ayant instituée est contrasté. Les rapports élaborés par les coordonnateurs des bureaux citoyens ne sont généralement pas pris en compte par les autorités communales. En réalité, il s'agit moins de la mauvaise volonté des autorités communales que du manque de moyens pour prendre en charge les préoccupations des citoyens. Ces préoccupations portant généralement sur des questions nécessitant d'importants investissements tels que l'accès aux soins de santé, à l'eau, à l'électricité, etc.[1565]

664. - À propos du droit français, Jean-Bernard Auby faisait remarquer en 2010 qu'il offre une panoplie de mécanismes de participation à la décision administrative à la fois conséquente et lacunaire[1566]. Si, dix ans plus tard, ce constat demeure pertinent pour le droit français de la participation, il l'est encore plus pour les outils juridiques de participation en Afrique. Sur le papier, il existe de nombreuses formes de participation dont nous avons dressé sommairement l'inventaire. Toutefois, ces mécanismes d'intervention des citoyens présentés demeurent lacunaires en ce qu'elles sont d'une efficacité démocratique variable et sans grande portée sur le processus décisionnel. L'administration locale n'entend pas accorder aux citoyens un pouvoir d'agir et d'influence sur les processus administratifs. Le citoyen est simplement relégué et maintenu en seconde place.

B- La participation secondarisée des citoyens dans la prise de décision locale

665. - La tradition de la démocratie représentative dans les États africains n'est pas très favorable au développement des mécanismes de participation. Les élus locaux se font une conception de la représentation qui inhibe la promotion de véritables outils juridiques de participation. L'accès aux fonctions de conseiller municipal et surtout du chef de l'exécutif conduit à une reproduction des modes de gouvernance « notablière ». Les détenteurs du pouvoir territorial s'attellent généralement à la construction et à l'entretien de réseaux clientélistes auxquels ils s'emploient à rendre des services aussi individualisés que

---

1565 Pour une analyse approfondie du dispositif dans une commune togolaise, S. SOBOZOU, *La participation citoyenne à la gouvernance locale : le cas de la commune de Zio 1*, Mémoire ENA, Cycle III, 2021.

1566 J.-B. AUBY, « Nouvelles de la démocratie administrative », *Dr. Adm.*, n° 4, repère 4, 2010 ; « Droit administratif et démocratie » *Dr. Adm.*, n° 2, 2006, étude 3.

possible[1567]. Dans ces conditions, il n'est guère possible d'ouvrir la prise de décisions à des mécanismes qui interfèrent avec les circuits classiques de la démocratie représentative. Les élus locaux confortés dans la légitimité qui entoure leur désignation par la voie de l'élection au suffrage universel ne sont pas spécialement séduits par l'idée de la recherche d'une légitimité additionnelle par l'ouverture d'espaces de participation[1568].

666. - Néanmoins, les acteurs locaux sont conscients de la crise de la représentation. Ce faisant, la démocratie participative n'est convoquée dans leurs discours que pour combler cette limite, tout en leur permettant de renforcer leur position. La participation citoyenne ne constitue donc qu'un outil de promotion et de « confortation » de la place des élus locaux[1569]. Par conséquent, la participation citoyenne, considérée comme un simple palliatif, ne peut produire qu'une décentralisation sans démocratie locale. Il s'agit d'instruments symboliques faisant perdre à la décentralisation son idéal démocratique, ainsi que le relève Géraldine Chavrier en droit français[1570]. Il faut relever aussi que, malgré la consécration des mécanismes de participation citoyenne, les législations africaines ne se sont pas suffisamment emparées de la question de la participation. Les cadres définis sont vagues. Cette imprécision des textes aurait pu constituer un terreau fertile à la créativité des élus locaux pour développer des processus ambitieux d'association des citoyens à la prise de décision. Cependant, bousculés de toute part, surtout par les pourvoyeurs de l'aide au développement, les élus locaux s'assurent de garder intactes les conditions de la délibération publique en mettant en place des mécanismes qui ne les exposent pas.

L'appel à la participation n'est accepté que tant qu'il ne remet pas en cause le mode de gouvernement clientéliste précédemment décrit. Ainsi, les élus locaux africains sont particulièrement méfiants des pratiques susceptibles de dérégler et de déplacer le centre du pouvoir. Ils ne s'exposent que très peu dans le cadre des formes de participation mises en œuvre. Ce qui n'est d'ailleurs pas une spécificité des élus locaux

---

1567 Le même constat est dressé en droit français, voir, J. CAILLOSSE, « les élus locaux et la démocratie "directe" », *Pouvoirs locaux*, n° 100, I/2014, p. 133.

1568 C. MINFEGUE, « La participation citoyenne en contexte local au Cameroun. Entre demande de légitimité, efficacité versatile et tensions politiciennes », *op.cit.*, p. 68.

1569 Voir en droit français, P. COMBEAU, « Décentralisation et démocratie », in J.-F. BRISSON (dir.), *Les assemblées délibérantes, avenir de la démocratie locale ? op.cit.*, p. 12 et s.

1570 G. CHAVRIER, « Décentralisation et démocratie locale », *AJDA*, 2013, p. 1317.

africains[1571]. Loïc Blondiaux, observateur de la démocratie participative française, relevait en ce sens que « *la plupart des démarches de participation portent sur des questions relativement secondaires* [...]. *Les autorités publiques hésitent à soumettre à la participation des projets ou des choix importants, par peur de la confrontation politique ou de perdre la maitrise de la décision. Et, dans la grande majorité des cas, la participation n'influe pas sur la décision. Elle reste de l'ordre de la consultation plus que de la co-construction de la décision avec les citoyens.* [...] *Lorsque les citoyens consultés ont le dernier mot, c'est pour une part de décision très résiduelle* [...] »[1572]. Il en résulte un manque de confiance des citoyens dans les dispositifs participatifs et une réticence de leur part à s'y impliquer.

667. - Alors que le continent africain est considéré comme possédant « *un substrat culturel propice à la consultation, au débat et à la participation à la construction de la décision collective* »[1573], la démocratie participative, malgré quelques avancées notables, patine encore. Le « tournant délibératif »[1574], qui fonde l'idéal démocratique théorisé par Jürgen Habermas[1575], semble pour l'instant difficile à concevoir. Or, l'amélioration du processus décisionnel est déterminante pour la gouvernance et la démocratie locale. Si la norme adoptée par les autorités locales se fonde « *sur des raisons publiques résultant d'un processus de délibération inclusif et équitable, auquel tous les citoyens peuvent participer et dans lequel ils sont amenés à coopérer librement* »[1576], leur adhésion contribuera à améliorer la gouvernance locale. L'association des citoyens à tous les niveaux de la chaine décisionnelle a une influence sur la mobilisation des recettes locales. Dans les quelques communes africaines ayant expérimenté les budgets participatifs, les résultats montrent une augmentation des ressources fiscales dans les années qui

---

1571 R. CAYROL, « Enjeux de la démocratie participative en France », in P. IGNAZI, D. REYNIE (dir.), *La vie politique. Pour Pascal Perrineau*, Presses de Sciences Po, 2021, pp. 99-108.

1572 École normale supérieure de Lyon, « La démocratie participative, entretien avec Loïc Blondiaux », Ressources en sciences économiques et sociales, 15 janvier 2018.

1573 CGLU, *La décentralisation et la démocratie locale dans le monde*, 2008, p. 47 ; N. KABAMDA, « La démocratie locale en Afrique peut-elle être la route de la reconstruction et de la prospérité : le cas du Zaïre », *African administrative studies*, 1996, pp. 23-28.

1574 L. BLONDIAUX, B. MANIN (dir), *Le tournant délibératif de la démocratie*, Presses de Sciences Po, 2021.

1575 J. HABERMAS, *Droit et démocratie. Entre faits et normes*, Gallimard, 1997.

1576 L. BLONDIAUX, Y. SINTOMER, « L'impératif délibératif », *Rue Descartes*, vol. 63, n° 1, 2009, p. 28 et s.

suivent[1577]. La délibération sous le modèle de la «*palabre africaine*» à l'avantage de déplacer le moment de l'expression démocratique tandis que la participation telle qu'elle pratiquée aujourd'hui s'inscrit dans le cadre d'un processus de décision finalisée[1578]. Elle ne repose pas sur la discussion et la confrontation des avis[1579]. Au final, le résultat est une multiplication de procédés qui ont peu d'impact sur l'objectif d'ouverture des processus décisionnels des collectivités territoriales aux citoyens. Par ailleurs, il est possible d'imaginer que la démocratie locale africaine pourrait trouver dans le développement d'internet les moyens de redonner un réel pouvoir d'influence et d'action aux citoyens à travers l'organisation de cyber-consultations[1580].

1577 Voir les exemples du Benin, Cameroun notamment : Y. SINTOMER C. HERZBERG, G. ALLEGRETT, « Les budgets participatifs dans le monde. Une étude transnationale », *Dialog Global*, n° 25, 2015.

1578 Voir not., P. COMBEAU, « Décentralisation et démocratie », *op.cit.*, p. 20.

1579 J. M. NZOUANKEU, « La Bonne gouvernance vue du Sud », in *La bonne gouvernance : objet et condition du financement*, *op.cit.*, p. 31.

1580 *Cf.* H. OBERDOFF, *La démocratie à l'ère du numérique*, PUG, 2010 ; H. BELRHALI-BERNARD, « la pratique des consultations sur internet par l'administration », *RFAP*, 2011, p. 181 ; C-A. DUBRUEIL, « L'e-démocratie au service de la démocratie locale », *AJDA*, n° 24, 2018, p. 1383.

Conclusion du chapitre 1

668. - Partout sur le continent africain, l'érection de la décentralisation comme mode d'organisation a constitué un temps d'espoirs pour une plus grande association des populations à la gestion des affaires publiques et un renforcement de la légitimité des autorités locales. Force est de constater que beaucoup reste à faire pour une véritable démocratisation de la vie publique locale. La démocratie participative est condamnée à s'exprimer au sein d'un régime représentatif lui-même inachevé. L'organisation et le fonctionnement des assemblées délibérantes induisent un déséquilibre institutionnel au profit de l'organe exécutif. Il ne fait aucun doute que la vie publique locale doit être profondément rénovée. D'abord de l'intérieur, afin de permettre une réanimation de la délibération locale et de la fonction de contrôle des élus locaux. Ensuite, de l'extérieur, la participation citoyenne doit favoriser la formation d'une opinion publique et non servir simplement de pis-aller. Pour reprendre les mots de Jacques Caillose, « *dans l'expression "démocratie locale", l'épithète s'avère bel et bien décisive, car ce sont des facteurs locaux qui, dans la pratique, favoriseront ou compromettront, c'est selon, l'appropriation collective des instruments juridiques de la démocratie* »[1581]. Il semble que la balle est désormais dans le camp des exécutifs locaux qui doivent profiter notamment des imperfections textuelles pour déployer des processus ambitieux de participation citoyenne impliquant notamment les chefferies traditionnelles.

1581 J. CAILLOSSE, « Éléments pour un bilan juridique de la démocratie locale en France », in CURAP/CRAPS, *La démocratie locale. Représentation, participation et espace public*, *op.cit.*, p. 78.

Chapitre 2 – La difficile intégration de la chefferie traditionnelle dans l'administration locale

669. - Il peut paraitre surprenant d'aborder la question de la chefferie traditionnelle dans un titre consacré à la démocratie locale. Et cela non sans raison, car le processus de décentralisation dans les pays d'Afrique subsaharienne, issue du système juridique d'inspiration française, marginalise les formes traditionnelles d'organisation sociopolitique. L'administration locale relève uniquement des collectivités territoriales, auxquelles est réservé le principe de libre administration[1582]. Cependant, partout sur le continent africain, et particulièrement en Afrique noire, où des formes anciennes d'organisation politique ont existé avant la colonisation, les chefferies imprègnent encore les réalités sociales. Elles continuent d'exercer une influence sur les représentations et le système de référence des populations grâce à leur prégnance sociale[1583]. La décentralisation en Afrique trouve sa justification non seulement dans l'idée de proximité, c'est-à-dire de retour du pouvoir auprès des citoyens, mais aussi dans la volonté de faire de ces derniers des acteurs du développement local. Par conséquent, la prise en compte des chefferies traditionnelles par le droit positif constitue une condition de l'effectivité du processus de décentralisation. L'efficacité de l'administration locale suppose une prise en compte des chefferies traditionnelles au niveau local. Elles ont toujours été décriées en raison de leur mode de gouvernance qui parait « a-démocratique » selon les logiques occidentales[1584]. Il n'en demeure pas moins que les chefferies traditionnelles sont des institutions nécessaires pour assurer la participation citoyenne et de ce fait, promouvoir la démocratie locale.

670. - Il faut relever que, ce que nous désignons ici comme chefferie traditionnelle n'est en réalité que le résultat de la transformation par le colonisateur européen des formes précoloniales d'organisation politique que sont les royaumes, les sultanats, etc. Ayant transformé la chefferie en un simple rouage de l'administration coloniale, le colonisateur n'a pas

---

1582 C. NACH MBACK, *Démocratisation et décentralisation. Genèse et dynamiques comparés des processus de décentralisation en Afrique subsaharienne, op.cit.*, p. 305.
1583 P.-L. AGONDO OKAWE, « L'État africain, un État hybride néocolonial », in *L'État africain. Horizon 2000, Mélanges Pierre-François Gonidec,* LGDJ, 1985, p. 56.
1584 Il existe tout de même en Europe des États qui reconnaissent ces formes traditionnelles. Les Kanak et la création du Sénat coutumier en Nouvelle-Calédonie constitue un exemple.

manqué de créer lui-même, *ex nihilo* des chefferies dites administratives sans aucune légitimité coutumière. C'est cette institution hybride[1585], travestie, qui sera transmise aux États africains à l'issue des indépendances. Ainsi que le rappelle Charles Nach Mback, trois options se présentaient aux nouveaux dirigeants des États africains : le retour aux formes originelles d'organisation des sociétés précoloniales, l'intégration des chefferies dans le droit public moderne hérité de la colonisation, ou la suppression pure et simple de cette institution considérée comme dérangeante par la nouvelle élite africaine[1586]. Seules les deux dernières options furent utilisées. Certains pays ont simplement supprimé la chefferie traditionnelle. C'est le cas du Bénin, du Burkina Faso, de la Guinée, du Mali, etc., mais la suppression ne signe pas pour autant la disparition de la chefferie qui continue d'exister dans certaines contrées[1587]. Au Mali par exemple, après l'indépendance, les autorités traditionnelles ont été formellement supprimées au sud du pays par l'État, mais conservées dans le nord, notamment dans les régions de Kidal et de Tombouctou. Ce n'est qu'avec le processus de décentralisation que les chefferies ont été à nouveau prises en compte dans le code des collectivités territoriales. L'article 26 prévoit que le conseil communal doit consulter les conseils de villages, de fractions et/ou de quartiers sur un certain nombre de matières limitativement énumérées[1588].

D'autres États ont choisi dès les indépendances de maintenir la chefferie dans le prolongement de l'usage colonial. Ainsi, le droit positif nigérien et togolais a entendu garantir une place à la chefferie traditionnelle aux côtés de l'administration locale. Tout l'enjeu est désormais de savoir quel est le degré d'implication de la chefferie traditionnelle dans les processus de décentralisation. L'analyse montre que les textes juridiques cherchent à intégrer la chefferie dans les sphères étatiques, mais la place qui leur réservée dans le fonctionnement et l'organisation de

---

1585 A. BOPDA, « Genèse, mutations et problèmes de la chefferie traditionnelle à Yaoundé », in S. JAGLIN, A. DUBRESSON (dir.*), Pouvoirs et cités d'Afrique Noire, Décentralisations en question*, Karthala, 1993, p. 247.

1586 C. NACH MBACK, *Démocratisation et décentralisation. Genèse et dynamiques comparés des processus de décentralisation en Afrique subsaharienne, op.cit.*, pp. 306-307.

1587 Au Burkina Faso, malgré la suppression des lois sur la chefferie en 1958 et en 1983, les chefs traditionnels gardent un rôle symbolique et un certain pouvoir d'influence dans leur milieu. Lire en ce sens l'interview de Titinga Frédéric Pacéré, chef de Manega au journal Le monde https://www.lemonde.fr/afrique/article/2020/01/23/au-burkina-la-chefferie-traditionnelle-peut-restaurer-le-dialogue-et-la-cohesion-sociale_6026979_3212.html. Consulté le 31 aout 2022.

1588 Art. 26 du code des collectivités territoriales au Mali.

l'administration locale est somme toute assez modeste. Dans une démarche classique, nous reviendrons d'abord sur le problème de la nature juridique de la chefferie traditionnelle (**Section 1**) avant d'aborder ensuite la question de son intégration partielle dans la conduite de l'action publique locale (**Section 2**).

Section 1 : La nature équivoque de la chefferie traditionnelle

671. - Les chefferies traditionnelles en Afrique précoloniale étaient des institutions politiques et administratives. Elles remplissaient également des fonctions de nature économique, sociale et culturelle. La colonisation a bouleversé l'organisation des structures précoloniales en démembrant les anciens royaumes et en créant des chefferies artificielles. Ce faisant, la chefferie a perdu son prestige et son auréole de sacralité avec l'administration coloniale. Toutefois, elle a réussi à se maintenir à travers le temps. Après une longue période d'hésitation due à la crainte de l'effondrement de l'unité nationale des jeunes États nouvellement indépendants, les chefferies apparaissent désormais comme des relais efficaces entre l'État et les populations locales[1589]. Aujourd'hui, un certain nombre de Constitutions africaines reconnaissent ou consacrent l'existence des chefferies traditionnelles[1590].

672. - Pour autant, la place des chefferies traditionnelles dans l'administration demeure ambiguë. Les chefferies s'accommodent mal avec les catégories juridiques du droit administratif « moderne ». En raison de leur préexistence, les chefferies traditionnelles constituent des réalités sociologiques par essence extérieures au droit administratif. Le droit positif africain peine à traduire cette réalité. C'est donc ce qui explique la nature juridique incertaine de la chefferie traditionnelle. Les textes législatifs adoptés en application des dispositions constitutionnelles dans les pays qui reconnaissent la chefferie évitent soigneusement d'y donner une définition univoque (**§1**). En revanche, les législateurs sont

---

1589 Dans la continuité de la logique coloniale, les chefferies ne sont que des « courroies de transmission ».

1590 C'est le cas notamment au Ghana (Constitution de 1992, art. 277), au Niger (Constitution de 2010, art. 167, au Tchad (Constitution de 2018, art. 217), au Togo (Constitution de 1992, art. 143).

particulièrement prolixes pour ce qui est du statut du chef traditionnel. L'analyse de la situation du chef traditionnel permet de conclure qu'il s'agit d'un agent public de l'État d'un genre particulier (§2).

§1.- L'impossible définition juridique de la notion de chefferie traditionnelle

673. - La prise en compte des survivances sociologiques que sont les chefferies traditionnelles est devenue une évidence pour un certain nombre d'États d'Afrique noire francophone à partir de la décennie 1990. Cette volonté de traduire dans le droit positif une réalité sociale ancienne n'est pas allée jusqu'au bout, puisque ni la Constitution ni les textes d'application ne définissent clairement les chefferies traditionnelles. Selon Maurice Glélé, « *la chefferie, ce sont à la fois le commandement et tous les attributs attachés à la personne, inhérents à la fonction de chef ; c'est également la circonscription que couvre l'autorité du chef* »[1591]. La notion de chefferie traditionnelle implique donc l'exercice d'une autorité coutumière par un chef traditionnel sur un groupement humain établi sur une partie du territoire d'un pays. Cependant, la nature juridique de cette circonscription que couvre l'autorité du chef est incertaine. Elle semble formellement relever de l'administration déconcentrée (**A**), mais, l'analyse de ces éléments constitutifs laisse penser qu'elle se rattache davantage de l'administration décentralisée (**B**).

A- La tendance à l'étatisation de la chefferie traditionnelle

674. - Au Niger, la loi n° 2015-01 du 13 janvier 2015 portant statut de la chefferie traditionnelle dispose en son article 2 que, « on entend par la chefferie traditionnelle l'institution qui regroupe l'ensemble des chefs traditionnels dépositaires de l'autorité coutumière. Le chef traditionnel est une personne élue ou désignée pour diriger une communauté coutumière et traditionnelle ». Puis il est précisé que « les communautés coutumières et traditionnelles regroupent les populations d'une fraction, d'un quartier, d'une tribu, d'un village, d'un secteur, d'une chefferie particulière, d'un

---

1591 M. A. GLELE, *Le Danxome : du pouvoir aja à la nation fon*, Nubia, 1974, p. 31.

groupement, d'un canton, d'une province ou d'un sultanat»[1592]. Le législateur togolais dans la loi n° 2007-002 du 8 janvier 2007 relative à la chefferie traditionnelle et au statut des chefs traditionnels considère quant à lui la chefferie traditionnelle comme une institution de l'administration territoriale[1593] animée par des chefs traditionnels[1594], lesquels dont désignés à la tête d'une unité administrative de base à savoir le canton, le village ou le quartier. Cette définition se rapproche de celle du législateur malien de 2006 qui considère le village, la fraction et le quartier comme des entités administratives de base[1595] administrées par un chef, lui-même assisté d'un conseil[1596].

Dans les trois pays, malgré l'intitulé de la loi, le législateur n'accorde que très peu d'attention à la définition même de la chefferie traditionnelle. Il s'emploie davantage à désigner, d'une part, le responsable de la chefferie, et à définir, d'autre part, la base territoriale de son action. Le législateur accorde une importance particulière à l'élément territorial dans son approche des chefferies traditionnelles. Il faut relever à ce propos la coïncidence entre les chefferies traditionnelles et les circonscriptions ou unités administratives de base (canton, village, fraction et quartier). Si l'on devait définir à ce stade une chefferie traditionnelle, on dirait qu'elle recouvre un espace territorial donné formant une localité sur laquelle habite une population, avec à son sommet une personne reconnue comme chef, c'est-à-dire responsable de tous ceux qui y vivent et des différentes activités qu'ils y mènent[1597].

675. - L'élément territorial constitue ainsi le critère pertinent d'identification d'une chefferie traditionnelle dans le droit positif moderne des États de l'espace ouest-africain francophone. Ce qui pose, selon la doctrine africaine, un problème de rationalité. Charles Nach Mback explique dans sa thèse précitée que le fait de construire la notion de chefferie traditionnelle sur une base territoriale contribue à donner au

---

1592 Art. 3, Loi n° 2015-01 du 13 janvier 2015 portant statut de la chefferie traditionnelle en République du Niger.
1593 Art. 1, Loi n° 2007-002 du 8 janvier 2007 relative à la chefferie traditionnelle et au statut des chefs traditionnels.
1594 *Ibid.*, art. 3.
1595 Art. 1, Loi n° 06-023 du 28 juin 2006 relative à la création et à l'administration des villages, des fractions et des quartiers.
1596 *Ibid.*, art. 7.
1597 A. T. KOKO, *Les chefferies traditionnelles africaines face à la dynamique des réformes territoriales : contribution à l'étude des processus de décentralisation, op.cit.*, p. 148.

territoire une importance qu'il n'a pas toujours eue dans les sociétés précoloniales africaines[1598]. En effet, le territoire, malgré son importance, ne constituait pas le socle de l'existence des chefferies traditionnelles. Élément fondamental du patrimoine collectif, la perte du territoire, ou son amputation du fait des guerres, n'a jamais remis en cause l'existence du groupe social. Les liens de sang et de vassalité l'ont toujours emporté sur la fixation territoriale dans la construction des sociétés politiques précoloniale en Afrique[1599]. Le législateur colonial l'avait pourtant bien compris. L'article 2 de l'arrêté du 2 décembre 1949 règlementant le commandement autochtone au Togo disposait en ce sens que « *le village est l'unité administrative autochtone* » comprenant « *l'ensemble de la collectivité qui y réside habituellement et dont les membres sont liés entre eux par des attaches familiales ou ethniques, sous l'autorité de son chef traditionnel* »[1600].

L'absence de mention du fait ethnique comme critère, même subsidiaire, de reconnaissance d'une chefferie traditionnelle par l'État postcolonial révèle la volonté des pouvoirs publics d'achever la maitrise du territoire en fixant définitivement les communautés traditionnelles sur des espaces qu'ils ont précisément définis. Le législateur s'efforce, sur la base du critère territorial, de rattacher les chefferies traditionnelles aux cadres territoriaux de l'administration d'État que sont les circonscriptions ou unités administratives[1601]. Les chefferies semblent faire partie intégrante de l'administration territoriale de l'État. C'est du moins ce qui ressort à la lecture des législations nationales qui intègrent les communautés coutumières et traditionnelles dans l'organisation administrative de l'État[1602]. La création, la modification ou la suppression de ces communautés traditionnelles relèvent de la seule autorité de l'État[1603].

Cependant, l'assimilation voulue de la chefferie traditionnelle à la circonscription administrative se heurte à quelques écueils. L'analyse des relations entre les chefferies et les circonscriptions administratives laissent entrevoir des différences entre les deux institutions administratives. Selon la loi nigérienne, les communautés coutumières et traditionnelles

1598 C. NACH MBACK, *Démocratisation et décentralisation. Genèse et dynamiques comparés des processus de décentralisation en Afrique subsaharienne, op.cit.*, p. 311.

1599 R. G. NLEP, *L'administration publique camerounaise. Contribution à l'étude des systèmes africains d'administration publique, op.cit.*, p. 178.

1600 Arrêté n° 951-49/APA du 2 décembre 1949.

1601 B. MOMO, *Droit administratif. Cours de 2ème année de Licence*, Yaoundé, 1986, p. 94.

1602 Art. 4, Loi n° 2015-01 du 13 janvier 2015 *préc.* (Niger) et art 1, Loi n° 2007-002 du 8 janvier 2007, *préc.* (Togo) ; art. 1, Loi n° 06-023 du 28 juin 2006 *préc.* (Mali).

1603 Art. 6, Loi n° 2015-01 du 13 janvier 2015 *préc.*

participent, avec les circonscriptions administratives et les collectivités territoriales, à l'organisation et à l'administration du territoire[1604]. Cette formulation laisse penser que les chefferies traditionnelles sont distinctes des circonscriptions administratives et des collectivités territoriales. Aussi, contrairement à la circonscription qui se confond avec la personnalité juridique de l'État, la chefferie au Niger tout comme dans les deux autres pays dispose d'un ensemble d'intérêts individualisés, ce qui laisse présumer qu'à l'occasion d'un contentieux, les tribunaux pourraient lui reconnaitre une personnalité juridique distincte de celle de l'État. Au Cameroun, la chambre administrative de la Cour suprême a consacré la personnalité juridique de la collectivité traditionnelle en lui reconnaissant le droit d'ester en justice pour son propre compte[1605]. Cette reconnaissance vaut à la chefferie traditionnelle camerounaise d'être non seulement un sujet de droit, mais aussi une personne juridique au sens plein du terme. « *Cette personnalité juridique est une donnée naturelle liée à l'existence même de la chefferie traditionnelle* »[1606]. Dès lors, la chefferie se distingue de la circonscription administrative. Toutefois, même dans l'hypothèse d'une reconnaissance de la personnalité juridique propre à la chefferie traditionnelle, celle-ci ne peut être assimilée à une collectivité territoriale décentralisée.

B- Le refus d'attribution de la qualité de collectivité territoriale

676. - En l'état du droit positif, la chefferie traditionnelle ne peut être assimilée à une collectivité territoriale[1607]. Aucun texte, que l'on se réfère à la lettre ou à l'esprit de ces derniers n'a entendu faire de la chefferie une collectivité territoriale. Les unités traditionnelles de base ont sans aucun doute été sacralisées par le constituant, mais ce dernier ne leur attribue pas la qualité de collectivité territoriale au sens où l'entend le droit de la décentralisation. Les textes relatifs à la décentralisation consacrent uniquement la région, le cercle et la commune en tant que collectivités territoriales. Ni le constituant, ni le législateur, à qui la Loi fondamentale reconnait la compétence de créer tout autre type de collectivité territoriale,

---

1604 *Ibid.*, art. 4.

1605 V. CS/CA, 25 septembre 1980, Collectivité Déido c/État du Cameroun ; CS/CA 30 novembre 1995, Collectivité Tagne Sobgui c/État du Cameroun.

1606 C. NACH MBACK, *Démocratisation et décentralisation. Genèse et dynamiques comparés des processus de décentralisation en Afrique subsaharienne, op.cit.*, p. 314.

1607 H. K. NOUBISSI, *Décentralisation et centralisation au Cameroun. La répartition des compétences entre l'État et les collectivités territoriales, op.cit.*, p. 340 et s.

ne confèrent ce statut aux chefferies traditionnelles, ou du moins, pas encore.

677. - Pourtant, les ambiguïtés demeurent. Que ce soit au Niger ou au Togo, la Constitution traite de la chefferie traditionnelle dans un titre relatif aux collectivités territoriales. Le constituant togolais en titrant son chapitre XII : « *Des collectivités territoriales et de la chefferie traditionnelle* », semble avoir voulu distinguer les deux entités. La loi fondamentale marque la spécificité des chefferies, qui ne doivent pas être confondues avec les collectivités territoriales. Cependant, le texte tel que rédigé par le constituant peut aussi être interprété dans un sens contraire[1608]. En regroupant « collectivités territoriales » et « chefferie traditionnelle » sous un même titre, le constituant n'entendait-il pas laisser la liberté au législateur, dans le cadre de l'exercice de ses compétences, de préciser le statut des chefferies traditionnelles ? Le laconisme du constituant sur la chefferie traditionnelle conforte d'une certaine manière cette idée. Un seul article traite de la chefferie, sans préciser sa nature. Ainsi, la Constitution togolaise de 1992 dispose que « *l'État togolais reconnait la chefferie traditionnelle, gardienne des us et coutumes. La désignation et l'intronisation du chef traditionnel obéissant aux us et coutumes de la localité* »[1609]. Le texte nigérien va un peu plus loin en disposant que « *l'État reconnait la chefferie traditionnelle comme dépositaire de l'autorité coutumière. À ce titre, elle participe à l'administration du territoire de la République dans les conditions déterminées par la loi. La chefferie traditionnelle est tenue à une stricte obligation de neutralité et de réserve. Elle est protégée contre tout abus de pouvoir tendant à la détourner du rôle que lui confère la loi* »[1610].

678. - L'absence de reconnaissance explicite par le droit positif de la chefferie traditionnelle en tant que collectivité territoriale ne traduit pas la réalité de cette institution. Dans les faits, tout semble la rapprocher d'une collectivité territoriale décentralisée. Il s'agit d'un groupe d'individus, un collectif pour reprendre l'étymologie de la collectivité, disposant d'un patrimoine commun et capable de faire communauté. On peut également rapprocher le chef et son conseil de notables avec l'exécutif et l'organe délibérant de la collectivité territoriale tels que définis par les lois de décentralisation. Tout comme le maire, le chef traditionnel est le chef de

---

1608 A. T. KOKO, *Les chefferies traditionnelles africaines face à la dynamique des réformes territoriales : contribution à l'étude des processus de décentralisation, op.cit.*, p. 59.

1609 Art. 143 de la Constitution togolaise de 1992.

1610 Art. 167 de la Constitution nigérienne de 2010.

l'administration de la communauté coutumière et traditionnelle. Il nomme aux différentes fonctions de sa cour conformément à la coutume et au culte de sa communauté coutumière et traditionnelle. La loi nigérienne lui confie d'ailleurs la nomination des imams[1611].

679. - La cristallisation du débat doctrinal autour de la personnalité morale pour définir juridiquement la notion de chefferie traditionnelle nous parait excessive. Le discours administratif sur la chefferie traditionnelle accorde une importance « quasi religieuse » à la personnalité morale, alors même que cette dernière n'apparait en réalité que bien tardivement dans l'histoire des institutions locales. La personnalité apparaissant comme une évidence, elle n'est que très peu interrogée. Des études ont permis de montrer les ambiguïtés théoriques du rapport entre la collectivité territoriale et la personnalité morale[1612]. Florian Mugnier explique à ce propos en droit français que la personnalité des administrations locales a émergé sans que jamais il ne soit question de penser la qualité de collectivité territoriale de ces dernières. La personne morale est un impensé, du moins jusqu'au début du XIXe siècle, qui voit apparaitre dans le discours doctrinal la notion de personne civile pour rendre compte de la situation des institutions possédant un patrimoine et dont découle la capacité de contracter ou d'ester en justice. Ce n'est qu'à la fin du XIXe siècle que la doctrine requalifiera ces personnes en personne morale de droit public[1613]. Faisant fi de ces éléments historiques, la doctrine juridique africaine arrive à des conclusions identiques, désignant la chefferie traditionnelle comme un « *objet juridique non identifié* »[1614] ou une « *catégorie juridique rarissime* »[1615]. Seul Roger Gabriel Nlep appréhende les chefferies traditionnelles comme des collectivités locales préexistantes à la colonisation, à l'opposé des collectivités locales importées par les différents colonisateurs et réceptionnées par l'élite

---

1611 Art. 17, Loi n° 2015-01 du 13 janvier 2015 *préc.*

1612 F. MUGNIER, *La personnalité juridique des collectivités territoriales, genèse et développement d'une personne morale dans l'État*, thèse, Dalloz, 2022, p. 276 et s. Du même auteur, « Les ambiguïtés théoriques du rapport entre la collectivité territoriale et la personne morale », Communication au Colloque AFDCL, *Peut-on encore définir la collectivité territoriale*, 2021, Publication à venir.

1613 F. MUGNIER, *La personnalité juridique des collectivités territoriales*, op.cit., p. 135 et s.

1614 C. NACH MBACK, « La Chefferie traditionnelle au Cameroun : ambiguïtés juridiques et dérives politiques », *Africa Development*, Vol XXV, n° 3 & 4, 2000, p. 77.

1615 B. MOMO, Droit administratif, op.cit., p. 96.

africaine postcoloniale[1616]. Pour cet auteur, la chefferie traditionnelle possède tous les éléments propres à l'existence d'une personne morale : groupement humain, patrimoine distinct, capacité d'expression collective[1617].

680. - L'étude de la notion juridique de chefferie traditionnelle montre bien l'embarras des législateurs africains. D'un côté, la volonté d'accéder à une certaine modernité conduit à l'appropriation des principes élaborés par les anciennes puissances colonisatrices, eux-mêmes plaqués sur des réalités préexistantes. De l'autre, s'ajoute la crainte d'une réviviscence de formes traditionnelles, qui empêche d'aborder sereinement ces unités traditionnelles. La recherche de modernité l'emportant, les législateurs africains superposent les structures de la décentralisation telles qu'organisées dans les ex-puissances colonisatrices occultant ainsi la réalité locale inhérente à l'organisation territoriale propre aux collectivités humaines africaines[1618] et rendant difficile toute tentative de définition de la chefferie traditionnelle. En l'état, le droit positif ne reconnait pas de personnalité juridique à la chefferie, pas plus que la qualité de collectivité territoriale. Il n'en demeure pas moins que les chefferies traditionnelles avancent, certes à pas lent, mais sans discontinuité, vers la personnalité morale[1619] de droit public. L'analyse du statut du chef traditionnel permet de conforter cette idée.

§2.- La spécificité du statut juridique du chef traditionnel

681. - La notion de « chef traditionnel » tout comme celle de « chefferie » est une invention coloniale. Elle entre dans le vocabulaire juridique et administratif togolais avec l'arrêté précité de 1949. Pour ce qui est du Niger, l'arrêté n° 35 du 14 janvier 1936 portant réorganisation de l'administration indigène introduit pour la première fois dans un texte officiel les notions de chef de village, chef de canton, chef de province,

---

1616 R.-G. NLEP, *L'administration publique camerounaise. Contribution à l'étude des systèmes africains d'administration publique, op.cit.*, p. 88.
1617 *Ibid.*, p. 126.
1618 A. T. KOKO, *Les chefferies traditionnelles africaines face à la dynamique des réformes territoriales : contribution à l'étude des processus de décentralisation, op.cit.*, p. 152.
1619 R. DECOTTIGNIES, « La personnalité morale en Afrique noire », *Annales Africaines*, 1958, p. 31 et s.

chef de groupe et chef de tribu[1620]. Ainsi, possède la qualité de chef traditionnel toute personne désignée à la tête d'une unité administrative de base construite afin d'assurer le maillage territorial de l'État. Cela dit, il n'est pas aisé de déterminer la catégorie d'autorité à laquelle appartient le chef traditionnel. La complexité de la détermination de la place du chef traditionnel dans la nomenclature des cadres de l'administration territoriale s'explique par la dépendance totale du chef traditionnel à l'État dans l'exercice de ces fonctions et par l'intervention de la population locale dans sa désignation. Ainsi le chef traditionnel apparait comme un agent déconcentré de l'État dans ses fonctions (**A**), mais sa nomination répond à des conditions particulières qui en font un agent public singulier (**B**).

A- Le chef traditionnel, un agent administratif de l'État

682. - S'il n'y aucun doute sur la qualité d'agent public[1621] du chef traditionnel dans les législations considérées, il faut tout même relever qu'il présente la particularité d'être soumis dans l'exercice de certaines fonctions à l'autorité du maire, mais aussi à l'autorité hiérarchique du représentant de l'État dans sa circonscription. L'étroite subordination du chef traditionnel à l'administration étatique en fait un agent public. Ses attributions (**1°**), son régime de traitement financier et disciplinaire (**2°**) rendent compte de cette qualité.

1. La multiplicité des fonctions du chef traditionnel

683. - Après la période de résistance qui a suivi les premières années d'indépendance, les chefferies traditionnelles ont été considérées par l'État postcolonial comme un moyen de transmission des informations et d'encadrement des populations locales. Les chefs traditionnels se sont vu confier de nouvelles missions allant parfois au-delà de leurs attributions

---

[1620] M. TIDJANI ALOU, « La chefferie et ses transformations au Niger. De la chefferie coloniale à la chefferie postcoloniale », *LASDEL, Études et Travaux*, n° 76, 2009, p. 40.

[1621] V. en ce sens : CE, 27 Février 1953, Gbogan : « considérant que (…) en vertu de ces textes, lesdits conseils sont juges ordinaires du contentieux local ; qu'ils sont compétents, à moins de textes contraires, **pour statuer sur les litiges ordinaires concernant les agents locaux des circonscriptions où ils ont été institués** (…), il appartient au Conseil du contentieux administratif du Togo de connaître en premier ressort de la requête par laquelle le Sieur Gbogan Toudéka demande l'annulation de la décision du Commandant de cercle d'Anécho réintégrant le Sieur Agbossoumondé dans ses fonctions de Chef du village d'Akoumape Assiko précédemment occupé par le requérant ».

statutaires. Les textes fixant le statut du chef traditionnel lui reconnaissent plusieurs fonctions statutaires. Ainsi, au Mali, l'article 10 de la loi du 28 juin 2008 prévoit que « *le chef de village, de fraction ou de quartier est, dans l'exercice de ses fonctions chargé d'une mission de service public. Il veille à l'application des lois et règlements. À ce titre, il doit faire preuve d'équité et d'impartialité dans l'exercice de ses fonctions* ». Toutefois, la loi togolaise précitée de 2007 marque tout de même une régression par rapport au régime précédent de 1949 et 1959. En effet, l'arrêté n° 951-49/APA du 2 décembre 1949 portant statut de la chefferie traditionnelle et le décret n° 59-121 du 3 août 1959 le modifiant faisaient du chef traditionnel une véritable autorité publique locale. Le chef traditionnel, outre les pouvoirs coutumiers, avait d'importantes attributions notamment en matière de police générale, de police rurale, de développement économique, d'hygiène, de justice et de perception d'impôt[1622]. En revanche, la nouvelle loi relative à la chefferie traditionnelle et au statut des chefs traditionnels prive l'autorité traditionnelle de ses nombreuses fonctions. Le constituant et le législateur togolais cantonnent principalement le rôle des chefs traditionnels à la conservation et au développement des richesses culturelles authentiques c'est-à-dire les « us et coutumes »[1623]. Le pouvoir judiciaire du chef est réduit à une opération de conciliation et d'arbitrage en matière coutumière[1624]. Outre son rôle de gardiens des us et coutumes, le chef assure la représentation de la population locale[1625] et est consulté par les autorités administratives sur les questions de développement local[1626].

684. - La situation est toute autre au Niger. La loi reconnait de véritables compétences régaliennes aux chefs traditionnels, notamment en matière judiciaire, soulignant sa nature d'agent public étatique. Le législateur nigérien, comme le relevait déjà Charles Nach Mback avait, avec l'ordonnance de 1993, fait du chef traditionnel une autorité administrative à part entière[1627]. La loi de 2015 consolide le rôle du chef. Dans sa

---

1622 N. A. GOEH-AKUE, « Relations entre autorités traditionnelles et pouvoir public moderne au Togo : repères, atouts et limites au développement local », *Revue CAMES*, série B, vol. 1, 1999, p. 48.
1623 Art. 143 de la constitution togolaise de 1992.
1624 Art. 20, Loi n° 2007-002 du 8 janvier 2007 relative à la chefferie traditionnelle et au statut des chefs traditionnels.
1625 *Ibid.*, art. 21.
1626 *Ibid.*, art. 22.
1627 C. NACH MBACK, *Démocratisation et décentralisation. Genèse et dynamiques comparés des processus de décentralisation en Afrique subsaharienne, op.cit.*, p. 323.

collectivité, le chef traditionnel dispose d'un quasi-pouvoir juridictionnel en tant que conciliateur en matière coutumière, civile et de transactions foncières. Dans l'accomplissement de cette mission, le chef traditionnel peut déférer des convocations aux parties. Les actes dressés par le chef traditionnel font foi devant les autorités de la République. La loi fait obligation au chef de dresser « *les procès-verbaux de conciliation ou de non-conciliation qui doivent être consignés dans un registre ad'hoc dont extrait est adressé à l'autorité administrative de son ressort et à la juridiction compétente* »[1628]. En cas de demande par l'une des parties à la conciliation, « *les procès-verbaux de conciliation signés par les parties peuvent être revêtus de la formule exécutoire par la juridiction compétente* »[1629]. Le chef traditionnel en tant que magistrat de l'ordre administratif, a le devoir de maintenir l'ordre public à l'intérieur de la communauté dont il a la charge et de rendre compte des faits susceptibles de lui porter atteinte ainsi que de toute infraction à la loi pénale à l'autorité administrative de son ressort[1630].

685. - À ce titre, le législateur prévoit la possibilité pour le chef traditionnel de « requérir les moyens et les agents de l'État disponibles dans son entité, en cas de menace à l'ordre public et de calamité naturelle (incendie, inondation, feu de brousse, invasion des criquets, épidémie, etc.) »[1631]. Le statut d'agent public du chef traditionnel est aussi réaffirmé par la double protection dont il dispose. L'article 28 de la loi nigérienne relative à la chefferie traditionnelle garantit au chef traditionnel une protection « contre les menaces, les outrages, les injures ou les diffamations dont il peut être l'objet dans l'exercice ou à l'occasion de l'exercice de ses fonctions »[1632] en plus du bénéfice « de la protection prévue par toutes les dispositions pénales relatives à cette qualité ou à cette fonction »[1633]. Dans le système nigérien, le chef traditionnel dispose de véritables attributions et pouvoirs. On est ainsi loin de la logique coloniale. Toutefois, la tentation de l'instrumentalisation des chefs traditionnels persiste avec une multiplication des fonctions non statutaires.

1628 Art. 18, Loi n° 2015-01 du 13 janvier 2015 portant statut de la chefferie traditionnelle en République du Niger.
1629 *Ibid.*, art. 18 et s.
1630 *Ibid.*, art. 27.
1631 *Ibid.*, art. 19.
1632 *Ibid.*, art. 28.
1633 *Ibid.*, art. 29.

686. - En plus des fonctions qui leur sont confiées par le droit positif, les chefs traditionnels font l'objet de sollicitations diverses en matière politique. Garant de la cohésion sociale du groupe, le chef traditionnel doit faire preuve d'une certaine neutralité politique. Cette obligation de neutralité est rappelée par le constituant nigérien qui dispose à l'alinéa 2 de l'article 167 que *« la chefferie traditionnelle est tenue à une stricte obligation de neutralité et de réserve »*. Cependant, les acteurs politiques ont recours aux chefs traditionnels en période électorale et également pour la résolution des crises. Le pouvoir d'influence des chefs traditionnels sur les populations locales en fait le parfait allié pour les leaders politiques qui sont très peu suivis par les populations, surtout en milieu rural. L'appui de la chefferie traditionnelle est primordial pour l'accession et la conservation du pouvoir. Le système du parti unique qu'a connu le Togo sous le régime Eyadema a fait la démonstration de l'utilisation des chefs traditionnels à des fins politiques[1634]. À la loyauté du chef traditionnel envers le Parti et son fondateur[1635], la loi de 2007 substitue la loyauté envers l'État[1636]. Quand on sait le niveau de personnification du pouvoir dans les États en construction démocratique en Afrique noire francophone[1637], le respect d'une telle disposition équivaut à une loyauté envers la personne du chef de l'État. Cette utilisation du chef traditionnel à des fins politiques est un héritage de l'époque coloniale. « *L'allégeance personnelle des chefs vis-à-vis du pouvoir colonial apparait comme un élément essentiel de leur recrutement* (…) »[1638]. Aussi, il n'est pas rare de voir le pouvoir central faire appel aux chefs traditionnels en période de crises politiques. Ils sont associés à des débats de portée nationale, notamment lors de la mise en place des politiques de réconciliation[1639]. La présence de Togbui Agboli Kossi F. Agokoli IV, chef de canton de Notsè, en tant que membre du bureau de la Commission Vérité Justice et Réconciliation (CVJR) mise en place au Togo en 2009 à

1634 A. T. KOKO, *Les chefferies traditionnelles africaines face à la dynamique des réformes territoriales : contribution à l'étude des processus de décentralisation, op.cit.*, p. 158.

1635 E. A. B. ROUVEROY NIEUWAAL, *L'État en Afrique face à la chefferie, op.cit.*, p. 125.

1636 Art. 24, Loi n° 2007-002 du 8 janvier 2007.

1637 M. GAZIBO, *Introduction à la politique en Afrique, op.cit.*, pp. 89-116 ; J. BOYON, « Pouvoir et autorité en Afrique noire : État des travaux », *RFSP*, vol. 13, n° 4, 1963, p. 993. Voir aussi, A. MABILEAU, « La personnification du pouvoir dans les gouvernements démocratiques », *RFSP*, vol. 10, n° 1, 1960, p. 39.

1638 M. TIDJANI ALOU, « La chefferie et ses transformations au Niger. De la chefferie coloniale à la chefferie postcoloniale », *op.cit.*, p. 10 : « *L'administration coloniale a tôt fait de comprendre la nécessité de s'appuyer sur les chefs acquis à sa cause pour s'établir de façon durable* ».

1639 A. T. KOKO, *Les chefferies traditionnelles africaines face à la dynamique des réformes territoriales : contribution à l'étude des processus de décentralisation, op.cit.*, p. 159.

la suite des violences postélectorales de 2005 en constitue une parfaite illustration[1640].

687. - Le chef traditionnel est un agent public important non seulement dans l'administration du territoire, mais aussi dans la conduite de la politique nationale. L'importance des attributions qui lui sont dévolues — statutaires ou non — justifie son droit à une rémunération et sa soumission à un régime disciplinaire qui le rapproche de celui des fonctionnaires publics.

2. Le régime rémunératoire et disciplinaire du chef traditionnel

688. - Le chef traditionnel perçoit une indemnité annuelle en contrepartie des diverses charges qu'il assume. Le système rémunératoire varie d'un pays à l'autre. Il repose sur une hiérarchisation des chefferies. Au Togo, seuls les chefs de cantons perçoivent une indemnité, tandis qu'au Niger, peuvent en bénéficier les chefs de cantons, les chefs de groupements, les chefs de secteurs, les chefs des chefferies particulières, les chefs de fractions, les chefs de provinces et les sultans. Sont donc exclus les chefs de villages, de quartiers, de tribus. Au Mali, les chefs coutumiers bénéficient également d'une indemnité dans le cadre de l'assistance à l'administration pour le recouvrement des impôts et taxes. De façon générale, il faut noter que le législateur nigérien a prévu un meilleur traitement pour les chefs traditionnels.

689. - La loi du 13 janvier 2015 précitée prévoit qu'en plus de l'allocation annuelle qu'ils perçoivent, les chefs traditionnels bénéficient d'une prime de rendement pour leur contribution au recouvrement des impôts et taxes. En outre, les chefs perçoivent des indemnités forfaitaires à la charge de l'État en couverture de certains frais, notamment les frais de réception, de tenue de secrétariat, de conciliation, de téléphone, d'électricité, d'eau et de roulage[1641]. L'allocation annuelle du chef traditionnel dépend de la catégorie de classement de la chefferie. Elle varie entre 420 500 FCFA et 2 264 628 FCFA[1642] de la première catégorie à la dernière, hors catégorie. Le traitement des chefs traditionnels ne se résume

---

1640 Il faut noter également la présence des chefs traditionnels lors de conférence nationale souveraine de 1990 au Niger.

1641 Art. 32, Loi n° 2015-01 du 13 janvier 2015.

1642 Décret n° 93-85/PMIMI du 15 avril 1993.

pas à ces avantages financiers. Ils ont droit à des avantages sociaux, notamment une prise en charge médicale et des allocations familiales[1643].

Au Togo, l'article 26 de loi de 2007 dispose que « *le chef de canton bénéficie d'une indemnité annuelle dont le montant est fixé par décret en conseil des ministres* ». Contrairement au chef traditionnel nigérien, son homologue togolais est moins bien loti. Mise à part son indemnité annuelle, la loi ne prévoit pas d'autres avantages financiers ou sociaux. Les indemnités annuelles varient entre 264. 600 FCFA et 529 600 FCFA selon le décret fixant les indemnités attribuées aux chefs de canton[1644]. On est loin de l'allocation annuelle prévue pour la chefferie de première catégorie au Niger. Le caractère insignifiant de cette indemnité couplé à sa fréquence annuelle met les chefs traditionnels dans une situation de dépendance vis-à-vis de l'autorité administrative[1645]. Luc Sindjoun parle à ce propos d'une « coopération prébendière » qui s'installe entre les autorités étatiques et les chefs traditionnels passant parfois par « *l'extorsion des ressources publiques* »[1646]. La subordination des chefs traditionnels à l'administration étatique se traduit également par le pouvoir de sanction dont celle-ci dispose.

690. - L'autorité hiérarchique dispose d'un pouvoir disciplinaire à l'encontre des chefs traditionnels. En cas de manquement à ses obligations administratives, l'autorité administrative compétente peut prendre des sanctions contre le chef traditionnel. Il dispose en fonction de la gravité du manquement de la possibilité de prononcer un avertissement qui peut être verbal ou écrit, un blâme ou la destitution[1647]. La procédure disciplinaire se fait dans le respect des droits de la défense. Le chef est tenu informé des charges retenues contre lui et dispose du droit de présenter ses moyens de défense par écrit et de se faire assister ou représenter par un conseil[1648]. Les droits nigérien et togolais prévoient la possibilité de saisir en amont pour avis le Conseil national de la chefferie traditionnelle

---

1643 Art. 33, Loi n° 2015-01 du 13 janvier 2015.

1644 Décret n° 2020-031/PR du 04 mai 2020 fixant les indemnités de fonction attribuées aux Chefs de cantons et assimilés de la République Togolaise pour l'exercice 2020.

1645 H. K. NOUBISSI, *Décentralisation et centralisation au Cameroun. La répartition des compétences entre l'État et les collectivités locales, op.cit.*, p. 360.

1646 L. SINDJOUN, *L'État ailleurs. Entre noyau dur et case vide*, Economica, coll. « La vie du droit en Afrique », 2002, pp. 83-84.

1647 Art. 20, Loi n° 06-023 du 28 juin 2006 *préc.*, au Mali.

1648 Art. 36, Loi n° 2015-01 du 13 janvier 2015 *préc.*, au Niger, art. 37 Loi n° 2007-002 du 8 janvier 2007 *préc.*, au Togo.

ou les commissions mises en place au niveau national, régional et départemental[1649].

691. - Les attributions, la rémunération ainsi que le régime disciplinaire auquel est soumis le chef traditionnel en font sans aucun doute un agent auxiliaire de l'administration étatique disposant selon le pays de véritables pouvoirs d'action. Toutefois, les chefs traditionnels, particulièrement ceux des niveaux inférieurs, sont aussi soumis à l'autorité du conseil municipal. C'est le cas des villages, quartiers, fractions au Niger et au Mali, notamment dans le cadre de leurs missions de collecte des impôts et taxes. Au Togo, la destitution du chef de quartier, c'est-à-dire le retrait de son acte de reconnaissance est prononcé par le maire. Cette situation place le chef traditionnel dans une position singulière vis-à-vis de l'administration territoriale déconcentrée et décentralisée.

B- Le chef traditionnel, un agent administratif singulier

692. - La singularité du chef traditionnel procède des modalités de sa désignation, qui fait intervenir à la fois la population locale pour le respect des règles coutumières et l'autorité administrative étatique pour la reconnaissance officielle du chef. S'il est un acquis du public africain en matière de chefferie traditionnelle, c'est la nécessité de respecter au moins pour la forme les procédures coutumières dans le choix du chef traditionnel. Ce faisant, le pouvoir de nomination des autorités administratives est tempéré par la procédure préalable qui fait intervenir les populations locales afin de replacer le chef dans le contexte anthropologique et de lui assurer une assise communautaire[1650]. Cette superposition de la procédure de désignation selon les règles coutumières (**1°**) et de la reconnaissance du chef élu par l'autorité administrative (**2°**) renforce le caractère particulier du chef traditionnel.

1649 *Ibid.*, art. 34 & art. 2. V. aussi, le décret n° 2016 - 028/PR du 11/03/2016 portant modalités d'application de la loi n° 2007-002 du 8 janvier 2007 relative à la chefferie traditionnelle et au statut des chefs traditionnels au Togo.

1650 C. NACH MBACK, *Démocratisation et décentralisation. Genèse et dynamiques comparés des processus de décentralisation en Afrique subsaharienne, op.cit.*, pp. 328-329.

1. La désignation du chef traditionnel selon les règles coutumières

693. - La désignation et l'intronisation du chef traditionnel se font dans le respect des us et coutumes de la communauté. Toutefois, dans les législations considérées, la loi fixe un certain nombre de conditions pour être désigné chef traditionnel. Ces conditions, issues du droit public moderne, visent à s'assurer des aptitudes du candidat à la chefferie. En effet, « *nul ne peut être candidat à la chefferie d'une communauté coutumière et traditionnelle donnée s'il a fait l'objet d'une condamnation, par décision judiciaire devenue définitive à une peine afflictive ou infamante, s'il ne jouit de ses droits civiques et politiques ou s'il y a incompatibilité avec son statut professionnel, sauf démission de sa part* »[1651]. Le chef doit être une personne intègre, de bonne moralité, majeure et capable de lire et écrire en langue officielle, c'est-à-dire le français[1652]. Aussi, ne peut prétendre à la chefferie d'une communauté traditionnelle qu'une personne possédant la nationalité du pays, appartenant à cette communauté et habileté par la coutume. La simple résidence, contrairement au cas des élections municipales, ne suffit pas pour la désignation du chef traditionnel.

694. - Ces conditions remplies, il peut être procédé à la désignation proprement dite du chef traditionnel. Au Togo, le législateur prévoit deux voies distinctes pour la désignation du chef traditionnel. Elle se fait soit par voie de succession héréditaire, soit par voie de consultation populaire. Le texte de loi reste silencieux sur les justifications d'une telle distinction, mais cela renvoie au partage entre d'un côté les chefferies traditionnelles ayant des origines précoloniales et de l'autre les chefferies administratives dont la naissance date de l'époque coloniale. Pour les premières, la désignation du chef est confiée au conseil coutumier. En cas de désaccord entre les membres du conseil coutumier sur le choix du postulant, la loi togolaise prévoit qu'il soit organisé une séance de tirage au sort entre les différents candidats remplissant les conditions prévues par la coutume et la loi[1653]. Quant aux secondes, la consultation populaire se fait par

---

1651 Art. 7, Loi du 13 janvier 2015 portant statut de la chefferie traditionnelle en République du Niger ; art. 8, Loi n° 06-023 du 28 juin 2006 relative à la création et à l'administration des villages, des fractions et des quartiers au Mali.

1652 Art. 8, Loi du 8 janvier 2007 relative à la chefferie traditionnelle et au statut des chefs traditionnels au Togo.

1653 Art. 11, Loi du 8 janvier 2007 relative à la chefferie traditionnelle et au statut des chefs traditionnels au Togo.

l'alignement des populations derrière le candidat de leur choix. Le candidat ayant obtenu le plus grand nombre de voix est choisi. En cas d'égalité, le candidat le plus âgé est élu[1654]. L'élection est ici ouverte au suffrage universel. Dans cette dernière hypothèse de désignation du chef par la voie d'une consultation populaire, il est possible de faire un rapprochement avec l'élection des conseillers municipaux. Ainsi que le souligne Adiki Tovemin Koko, penser à ce niveau d'analyse « *que le chef est un élu local ne serait pas contraire au droit, puisqu'il y a élection à laquelle participe la population locale* »[1655]. Cependant, le législateur ne fait pas du chef traditionnel issu d'une chefferie administrative un *élu local.*

Au Niger, le chef traditionnel est élu par deux collèges électoraux restreints. Pour acquérir la qualité de chef traditionnel, le candidat remplissant les conditions établies par la coutume doit être élu par un collège électoral composé de « *l'ensemble des chefs de famille de sexe masculin ou féminin des communautés coutumières et traditionnelles considérées, recensés avant la vacance du poste, pour les quartiers, les villages, les tribus, les chefferies particulières et les fractions* [et] *l'ensemble des chefs de quartiers, de villages ou de tribus en fonction avant la vacance du poste, pour les groupements, les cantons et les secteurs* »[1656].

695. - La désignation du chef est encadrée par les règles coutumières, mais la reconnaissance par l'autorité administrative demeure indispensable.

2. La reconnaissance du chef traditionnel par l'autorité administrative

696. - Le chef traditionnel élu ou désigné par les instances coutumières doit faire l'objet d'une reconnaissance par l'autorité administrative compétente. L'entrée en fonction du chef traditionnel est subordonnée à l'adoption d'un acte de reconnaissance par l'autorité administrative qui selon les cas peut être une autorité relevant de l'État ou une autorité décentralisée. Sous le régime de l'arrêté précité de 1949, la reconnaissance des chefs traditionnels se faisait par gradation au Togo. Les chefs de cantons étaient reconnus par décret en conseil des ministres. Quant aux

---

1654 *Ibid.*, art. 12.

1655 A. T. KOKO, *Les chefferies traditionnelles africaines face à la dynamique des réformes territoriales : contribution à l'étude des processus de décentralisation, op.cit.*, p. 162.

1656 Art. 8, Loi du 13 janvier 2015 portant statut de la chefferie traditionnelle en République du Niger.

chefs de villages et quartiers, leur reconnaissance relevait d'un arrêté préfectoral. Le législateur de 2007 maintient le même principe, à la seule exception que les chefs de quartier sont désormais reconnus par un arrêté municipal.

697. - La reconnaissance au Niger suit également le principe de la gradation, mais est relativement différente. La désignation du chef traditionnel doit être entérinée par arrêté du ministre chargé de l'administration territoriale s'il s'agit de canton, de groupement, de province, de sultanat, de secteur, de chefferie particulière ou de fraction. Pour les autres niveaux de chefferie, le pouvoir de reconnaissance est dévolu à l'autorité administrative locale. Ainsi, la désignation est reconnue par décision du gouverneur pour les quartiers dans les agglomérations des communes chefs-lieux de région et les villages y dépendant, et par décision du préfet pour les chefs de villages, de tribus et de quartiers autres que ceux des agglomérations des communes chefs-lieux de région. Dans les dernières hypothèses, la décision de l'autorité compétente doit être précédée d'une demande d'avis adressée au chef traditionnel de la chefferie supérieure. L'introduction de la procédure de demande d'avis du chef traditionnel supérieur par la législation nigérienne marque une différence fondamentale avec les textes coloniaux. En effet, l'autorité coloniale, à travers son procédé de nomination, plaçait les chefs au même niveau. Les chefs de niveau inférieur bénéficiant d'une reconnaissance identique de la part du colon, ils pouvaient se permettre de désobéir aux chefs supérieurs. Le législateur nigérien, en réintroduisant le chef traditionnel hiérarchiquement supérieur dans la reconnaissance des chefs de niveaux inférieurs rétablit l'autorité des premiers sur les seconds. En outre, le chef de canton, de province, de secteur ainsi que le sultan disposent d'un pouvoir de sanction sur les chefs des villages, des tribus et des quartiers[1657].

698. - Toutefois, il reste une question à laquelle la doctrine peine à répondre[1658]. L'autorité administrative est-elle liée par le choix effectué par les instances coutumières ? Quelle est la valeur juridique de la désignation des chefs par la population locale ? La réponse n'est pas évidente. Pour Roger Gabriel Nlep, le droit pour toute collectivité coutumière de désigner

---

1657 Art. 15, Loi du 13 janvier 2015 portant statut de la chefferie traditionnelle en République du Niger.

1658 NACH MBACK, *Démocratisation et décentralisation. Genèse et dynamiques comparés des processus de décentralisation en Afrique subsaharienne, op.cit.*, p. 333.

ses dirigeants selon des règles qui lui sont propres fait partie des droits fondamentaux qu'il qualifie d'« *essence communautariste* »[1659]. Par conséquent, la désignation du chef par les instances coutumières ou par les procédés qui en tiennent lieu constitue un droit dont le résultat de l'exercice lie l'autorité administrative. Le droit positif semble aller en ce sens. La reconnaissance n'est refusée par l'autorité compétente que si la désignation du chef s'est faite dans des conditions contraires aux dispositions prévues par la loi. Dans ce cas, il est procédé à une nouvelle élection dans un délai de trois mois[1660]. Très souvent, les contestations qui naissent dans le cadre de la désignation des autorités coutumières et traditionnelles sont le fait du pouvoir règlementaire qui cherche à passer outre les règles coutumières pour imposer le candidat de son choix.

699. - Fort heureusement, le juge africain s'est érigé en véritable gardien des spécificités traditionnelles. La Cour suprême togolaise s'est illustrée en tant que gardienne des valeurs coutumières dans une décision du 26 juin 2014 relative à un conflit de succession du chef du canton de Baga[1661]. À la suite du décès de Monsieur Agoda Saguintaah, chef du canton de Baga, le conseil coutumier s'est réuni et a désigné Monsieur Magnétina Saguintaah comme successeur au trône, conformément aux us et coutumes de la localité, ainsi que l'exige, alinéa 2 de l'article 143 de la Constitution, qui dispose que « *la désignation et l'intronisation du chef traditionnel obéissent aux us et coutumes de la localité* ». Contre toute attente, le préfet décide d'organiser une consultation populaire qui porte Monsieur Dalakéna Djadja au trône le 3 juillet 2012. Saisie de l'affaire, la chambre administrative de la Cour suprême togolaise, par un arrêt du 26 juin 2014 a annulé la décision de la consultation du préfet. En dépit de l'annulation de cette première consultation, le gouvernement persiste et organise par décret pris en conseil des ministres une nouvelle consultation populaire. Monsieur Dalakéna Djadja, alors candidat unique, a été à nouveau désigné comme chef traditionnel de la localité.

Or, il est établi que dans cette localité, la dévolution du pouvoir se fait par voie coutumière. La Cour suprême a dans sa décision précitée procédé à l'annulation de cette seconde consultation au motif que les droits

1659 R.-G. NLEP, « Le juge de l'administration et les normes internes, constitutionnelles ou infra constitutionnelles en matière de droits de fondamentaux », *Solon,* vol. 1, 1999, pp. 138-139, cité par C. NACH MBACK, *op.cit.,* p. 334.

1660 Art. 15, Loi du 8 janvier 2007 *préc.*, au Togo

1661 Baga est un canton de la préfecture de Doufelgou du nord-ouest du Togo et regroupant six villages.

fondamentaux contenus dans la Constitution togolaise s'imposaient aux pouvoirs publics. Elle a estimé en l'occurrence que la désignation du chef coutumier par voie élective constituait une violation des articles 143 alinéas 2 de la Constitution et 8 de la loi n° 2007-002 du 8 janvier 2007 relative à la chefferie traditionnelle et au statut des chefs traditionnels au Togo. Par conséquent, la Cour a confirmé la désignation par voie coutumière de Monsieur Magnétina Saguintaah comme successeur au trône. Ainsi, le juge tente à travers son office de limiter l'immixtion des autorités administratives dans la désignation des autorités coutumières et traditionnelles[1662].

700. - Cependant, l'intervention de l'administration à travers l'acte de reconnaissance, même liée par les choix coutumiers, des chefs traditionnels, participe davantage à l'affaiblissement de cette institution. Cette situation, loin d'être une fatalité, constitue un choix des pouvoirs publics dans les États d'Afrique noire francophone. Il suffit pour s'en convaincre d'observer le fonctionnement de la chefferie traditionnelle au Ghana. L'État s'abstient de toute intervention dans la procédure de désignation des chefs et ces derniers ne font l'objet d'aucune validation par une quelconque structure de l'État. La *National et Regional House of Chiefs* assure la participation des chefs à la gestion du pouvoir dans les hautes sphères de l'État[1663].

701. - En somme, l'encadrement de la désignation du chef traditionnel par les règles coutumières n'aura pas suffi à le soustraire de l'autorité de l'État. Au contraire, ses attributions, son régime financier et disciplinaire, son existence subordonnée à une reconnaissance officielle par l'autorité compétente, contribuent à faire du chef un agent de l'État. En revanche, son essence locale et traditionnelle le rapproche davantage d'une autorité décentralisée, renforçant ainsi sa singularité. La permanence de l'institution cheffale dans le droit positif est une caractéristique des sociétés plurales[1664]. Cependant, malgré les déclarations de principe,

---

1662 V. not., B. L. TOUERE ELENGA, *Fait ethnique, droit public et gouvernance de l'État en Afrique subsaharienne francophone : Contribution à la théorie de l'État africain*, thèse, L'Harmattan, 2021, p. 250.

1663 P. JACQUENOT, « Chefferies et décentralisation au Ghana », *Afrique contemporaine*, vol. 221, n°1, 2007, pp. 55-74.

1664 R. G. NLEP, « Administration publique et sociétés plurales en Afrique : essai de réflexion sur l'interaction entre État et société, à travers l'exemple du Cameroun », *Revue française d'administration publique*, n° 94, 2000, p. 276.

l'institution cheffale est insuffisamment prise en compte dans les processus de décentralisation, car les autorités administratives craignent une concurrence du pouvoir traditionnel.

Section 2 : L'intégration relative de la chefferie dans l'administration locale

702. - L'intégration des structures sociopolitiques traditionnelles dans l'organisation administrative et territoriale de l'État pose de grandes difficultés. L'ambiguïté de la chefferie traditionnelle, démontrée à travers les chevauchements entre la circonscription administrative et la collectivité territoriale, rapproche de façon inévitable les pouvoirs ancien et moderne dans l'État postcolonial. Ce rapprochement est inévitable, dans la mesure où les pouvoirs publics cherchent à rapprocher toutes les couches de la société, y compris les plus éloignées de la prise de décision et de la gestion des affaires publiques. C'est du moins ce qui ressort du discours sur la décentralisation dans les pays d'Afrique noire francophones. La consécration par le droit positif des chefferies traditionnelles vise donc à faire d'elles, aux côtés des autorités nationales et locales, des agents du développement local.

703. - Dans la plupart des États africains, les revendications sociales des années 1990 ont permis aux autorités traditionnelles de se réinventer en sollicitant la prise en compte par les réformes institutionnelles des intérêts de la chefferie traditionnelle[1665]. Ainsi, la constitutionnalisation de la chefferie traditionnelle a permis au législateur de consacrer l'intégration de la chefferie dans le fonctionnement de l'administration locale. Cette prise en compte de la chefferie traditionnelle au sein des pouvoirs locaux demeure toutefois partielle et non démocratique[1666]. Le droit positif amenuise le rôle d'agent de mobilisation des forces locales qu'aurait pu jouer la chefferie traditionnelle. Si le processus de démocratisation institutionnel de l'Afrique a permis d'aménager une possibilité de

---

1665 A propos du Cameroun, certains de ces chefs affirmeront plus tard avoir reçu l'assurance que la conférence Tripartite de 1991 marquerait la réhabilitation de la chefferie, en échange de leur soutien aux positions du pouvoir en place. Voir, C. NACH MBACK, « La chefferie traditionnelle au Cameroun : Ambiguïtés juridiques et dérives politiques », *op.cit.*, p. 104.

1666 C. NACH MBACK, *Démocratisation et décentralisation. Genèse et dynamiques comparés des processus de décentralisation en Afrique subsaharienne, op.cit.*, p. 340.

participation des chefs traditionnels à l'administration locale (**§1**), le législateur marginalise toujours la place de la chefferie dans la conduite des politiques locales (**§2**).

§1.- Une garantie de participation au fonctionnement de l'administration locale

704. - La commune est définie dans les législations africaines comme la « collectivité de base ». Toutefois, au regard de l'existence des communautés traditionnelles et coutumières sur l'espace communal, la prise en compte de ces communautés fortement enracinées au sein des sociétés africaines s'avère déterminante dans la conduite de l'action publique locale. De plus, les chefs traditionnels ont été les premiers acteurs de la municipalisation en Afrique dans le cadre des communes mixtes de l'époque coloniale qui intégraient les notabilités coutumières dans les conseils municipaux[1667]. Ainsi, dès l'époque coloniale, se manifeste déjà la tendance à faire coïncider les cadres territoriaux de la décentralisation avec la chefferie traditionnelle, afin de bénéficier de l'influence de cette dernière pour la mobilisation des populations. La période postcoloniale étant marquée par une méfiance à l'égard des chefferies traditionnelles, le Niger et le Togo font partie des rares pays ayant légiféré sur la reconnaissance et la prise en compte formelle de l'existence des chefferies traditionnelles au sein de l'administration locale[1668]. Toutefois, si la participation des chefferies traditionnelles est consacrée (**A**), elles ne jouent qu'un rôle passif au sein des conseils de collectivités territoriales (**B**).

A- La participation consacrée des chefs traditionnels aux conseils de collectivités

705. - Le droit africain de la décentralisation intègre les chefs traditionnels au fonctionnement des assemblées délibérantes. Cette intégration répond à une volonté commune de développement de la démocratie participative faisant intervenir les corps constitués coutumiers

---

[1667]V. en ce sens, NOUBISSI, *Décentralisation et centralisation au Cameroun. La répartition des compétences entre l'État et les collectivités territoriales, op.cit.*, p. 365 ; M. TIDJANI ALOU, « La chefferie et ses transformations au Niger. De la chefferie coloniale à la chefferie postcoloniale », *op.cit.*, p. 37 et s.

[1668] On peut également citer : le Ghana et le Cameroun etc.

à la prise de décision. Dans les pays de cette étude, la participation s'effectue à travers des modalités différentes, mais elle demeure partielle, car le législateur entérine la dichotomie au sein de la chefferie entre le chef traditionnel et la communauté humaine dont il a la charge[1669]. En effet, dans l'intégration de la chefferie à l'administration locale, il est fait abstraction de la communauté, seul le chef en tant qu'« autorité administrative » et coutumière est concerné par la réforme.

706. - La loi n° 2019-006 du 26 juin 2019 relative à la décentralisation et aux libertés locales au Togo fait une place aux chefs traditionnels dans le chapitre 3 de son titre 1 relatif à la démocratie locale. L'article 22 de ce texte prévoit que « *le chef de canton peut participer aux travaux du conseil de la collectivité territoriale dont il relève sur sa demande ou à la demande dudit conseil, avec voix consultative* ». L'initiative de la participation appartient donc concurremment au chef et à l'assemblée locale. La loi ne fixe pas une proportion de la représentation cheffale au sein de l'assemblée délibérante puisqu'il s'agit d'une simple faculté. Toutefois, dans la mesure où la commune peut être constituée d'un découpage de cantons au sens de l'article 57[1670], il est possible de s'interroger sur les modalités de la participation du chef de canton. La non-coïncidence des collectivités territoriales avec les contours de la chefferie traditionnelle introduit une différence dans le niveau de représentativité des populations locales. De quelle commune relèvera-t-il ? Ni le législateur de 2007 ni celui de 2019 ne donne une réponse. On pourrait alors supposer que le chef sera réputé relever de la commune dont l'espace recouvre la majorité du territoire cantonal.

707. - Au Niger, les notabilités coutumières participent également au fonctionnement des conseils de collectivités territoriales au niveau communal et régional. En effet, les sultans, les chefs de province, de canton ou de groupement, sont membres de droit des conseils municipaux et régionaux. Le code général des collectivités territoriales prévoit qu'en « *cas d'empêchement, le sultan, le chef de province, de canton ou de groupement peut se faire représenter par un des membres de sa cour* »[1671]. Contrairement au Togo, la

---

[1669] C. NACH MBACK, *Démocratisation et décentralisation. Genèse et dynamiques comparés des processus de décentralisation en Afrique subsaharienne, op.cit.*, p. 339.
[1670] Art. 57 : « La commune est composée d'un ou de plusieurs cantons. Elle peut également être constituée d'un découpage de canton ».
[1671] Art. 26 du CGCT au Niger.

représentation de la chefferie traditionnelle ne se limite pas seulement aux chefferies de catégories supérieures. Dans les communes urbaines et les communes à statut particulier, « *la représentation de la chefferie traditionnelle est étendue aux chefs de villages, de tribus ou de quartiers* »[1672]. La proportion de sièges réservée à la chefferie traditionnelle varie selon le niveau de collectivité. Ainsi, pour les communes, le nombre de chefs au sein du conseil municipal ne peut excéder le dixième (1/10 ème) des conseillers élus[1673]. Au niveau régional, il ne peut excéder le cinquième (1/5) du nombre des conseillers élus[1674].

La désignation des chefs traditionnels au sein des assemblées locales est laissée à la discrétion des chefs eux-mêmes. Les chefs de villages, de tribus ou de quartiers sont désignés par leurs structures en assemblée générale convoquée à cet effet. Les sultans, les chefs de province, de canton ou de groupement concernés sont désignés par la structure régionale de l'Association des chefs traditionnels du Niger. Les assemblées délibérantes des collectivités territoriales sont ainsi constituées des membres de droits désignés par leurs pairs et des membres élus au suffrage universel. Cette composition particulière, bien que non démocratique, prend en compte les réalités sociales et institutionnelles africaines. Les chefs traditionnels sont également représentés au sein des organes de suivi de la décentralisation[1675]. Aussi, le législateur offre la possibilité pour un chef traditionnel d'être élu conseiller ou maire, à la condition toutefois, qu'il démissionne de son poste de chef traditionnel.

Si les chefs peuvent bénéficier du même droit d'information que les membres élus des conseils, leur participation aux travaux n'est pas assurée sur un mode égalitaire.

B- Une participation passive des chefs traditionnels aux conseils de collectivités

708. - L'impact de l'intégration des chefs traditionnels comme acteurs de la politique locale dans les conseils de collectivités territoriales doit être apprécié avec prudence. Elle apparait davantage comme symbolique. Les législateurs togolais et nigérien ont introduit une inégalité dans la

---

1672 *Idem.*

1673 Art. 26 du CGCT au Niger.

1674 Art. 101 du CGCT au Niger.

1675 Il faut relever que la chefferie est faiblement représentée dans ces instances. Au Togo, seulement 5 chefs de cantons sont membres du Conseil national du suivi de la décentralisation. Or, on dénombre 394 cantons sur le territoire national.

participation aux travaux délibératoires entre les membres élus et la chefferie traditionnelle. Ces derniers n'ont qu'une voix consultative. Ils ne disposent pas d'un droit de vote pour l'adoption des délibérations. Ainsi, les chefs traditionnels ne sont invités à siéger dans les conseils locaux que pour exprimer une opinion, un avis qui ne lie pas l'assemblée délibérante.

709. - Même dans les cas où la loi pose une obligation d'association du chef traditionnel, notamment sur toutes les actions de développement touchant sa communauté, son avis demeure consultatif[1676]. C'est le sens des dispositions de l'article 92 de la loi togolaise du 26 juin 2019 précitée qui dispose que « *le conseil municipal est tenu de consulter le chef de canton sur les questions relatives à l'environnement à la santé, au foncier et à l'élaboration du programme de développement communal. À défaut d'une réponse expresse dans le délai de quinze jours qui suivent la demande, le chef est réputé avoir donné son avis. L'avis du chef de canton est un avis consultatif* ». La faculté de participation à la gestion locale reconnue aux chefs traditionnels ne leur permet pas d'endosser véritablement leur qualité d'agent de développement local qui leur est pourtant expressément reconnue par la loi. Le législateur nigérien prévoit que, « *en matière économique, sociale et culturelle, le chef traditionnel est agent, acteur et partenaire de développement* »[1677]. Les règles modernes d'organisation territoriale mettent à rude épreuve les chefferies traditionnelles. Elles sont partout mises en minorité et condamnées dans les États d'Afrique noire francophone à être de simples institutions culturelles. Il est regrettable que le seul véritable rôle qui leur soit reconnu se cantonne à la garde des « us et coutumes », les laissant en marge des grands enjeux de développement[1678]. La coutume et les traditions africaines, du fait de leur incapacité à intégrer les valeurs et les contraintes issues du droit positif moderne, sont en régression permanente[1679].

710. - Le législateur africain francophone souffre d'un manque d'inventivité dans la création du droit de la décentralisation. Les principes de la décentralisation élaborés en droit français sont repris *in extenso*. Il n'est donc pas surprenant de constater que les réformes en cours en

---

[1676] Art. 26 du code des collectivités territoriales au Mali.

[1677] Art. 23, Loi du 13 janvier 2015 *préc.*

[1678] A. T. KOKO, *Les chefferies traditionnelles africaines face à la dynamique des réformes territoriales : contribution à l'étude des processus de décentralisation, op.cit.*, p. 238.

[1679] C. NACH MBACK, *Démocratisation et décentralisation. Genèse et dynamiques comparés des processus de décentralisation en Afrique subsaharienne, op.cit.*, p. 342.

matière de décentralisation n'accordent que peu de place à la chefferie traditionnelle. Pourtant, les pouvoirs publics s'accordent sur leur potentiel de mobilisation des populations, sans la participation desquelles la décentralisation ne saurait être effective. Tout laisse à penser que les réformes sont faites dans un souci de captation de l'aide internationale, sans réelle prise en compte des acteurs locaux. La marginalisation de la chefferie cache en réalité la volonté de la contenir pour éviter d'en faire un concurrent de l'administration locale.

§2.- La marginalisation de l'apport des chefferies traditionnelles dans l'administration locale

711. - La décentralisation a frayé la voie à la réaffirmation de la chefferie traditionnelle. L'intégration des chefferies traditionnelles à l'administration moderne des collectivités territoriales parait nécessaire pour parfaire les objectifs attendus par le mode démocratique de gouvernance institué à travers les principes de la décentralisation. L'influence des chefs traditionnels sur les populations à la base est légitimée par le droit positif en tant qu'elle participe au développement local (**A**). Il est désormais acquis que la chefferie constitue une entité incontournable de la vie locale, au point que certains craignent qu'elle devienne concurrente de la nouvelle administration locale (**B**). De ce fait, le positionnement des chefferies traditionnelles au sein des collectivités territoriales demeure marginal.

A- L'apport des chefs traditionnels au développement local

712. - La chefferie traditionnelle en Afrique a toujours été au cœur des enjeux politiques nationaux et locaux[1680]. L'échec des politiques d'ajustements structurels menées par les institutions financières internationales, notamment la Banque mondiale et le Fonds monétaire international, a conduit à un changement de paradigme en matière de développement. Désormais, les acteurs locaux sont appelés à jouer un rôle important dans la conduite de l'action publique. Les pouvoirs publics prennent conscience que le développement ne peut se faire sans une

---

[1680] M. TIDJANI ALOU « La chefferie au Niger et ses transformations. De la chefferie coloniale à la chefferie postcoloniale », *op.cit.*, p. 2.

mobilisation effective au niveau local. Les États africains, en adoptant la décentralisation comme mode principal de gouvernance, ambitionnent d'associer les différents acteurs locaux, aux premiers desquels figurent les chefs traditionnels, au nouveau projet de développement.

713. - L'institution « cheffale » gardienne des us et coutumes prend une part active dans la vie politique locale, endossant le rôle d'agent de développement local dotée d'une forte capacité de mobilisation des ressources au niveau local. Les chefs traditionnels associés à cette nouvelle dynamique de développement contribuent à la mise en place des comités de développement et au développement économique social[1681]. Au Togo, le gouvernement a adopté en 2012 un décret relatif aux comités villageois de développement à la base[1682] qui a permis de renforcer l'influence de la chefferie traditionnelle sur la vie publique au niveau local. Le décret du 29 février 2012 prévoit que les Comités villageois de développement (CVD) ou les Comités de quartier de développement (CDQ) sont créés à l'initiative de la population d'un quartier ou d'un village. Toutefois, la convocation de l'assemblée qui consacre la création de ces comités relève de la compétence du chef de village ou de quartier. Une fois les comités créés et leur existence juridique reconnue, les chefs restent continuellement présents dans leur fonctionnement. Ils peuvent décider de le réunir en session extraordinaire. Le décret précité place ainsi sous l'autorité des chefs traditionnels l'action de des comités, considérés comme de véritables outils pour assurer le développement des collectivités locales. L'implication des chefs traditionnels dans le fonctionnement et la coordination de ces entités locales permet aux populations des localités reculées de s'identifier dans les actions menées. Elles constituent aussi un lieu d'apprentissage de la démocratie[1683].

714. - Le chef traditionnel citoyen local, bénéficiant d'une légitimité démocratique, participe et fait participer la collectivité villageoise au développement local[1684]. Au Niger, certains chercheurs, à l'instar d'Olivier

---

1681 A. T. KOKO, *Les chefferies traditionnelles africaines face à la dynamique des réformes territoriales. Contribution à l'étude des processus de décentralisation, op.cit.*, p. 216.
1682 Décret n° 2012-005/PR du 29 février 2012 relatif aux comités villageois de développement à la base.
1683 A. T. KOKO, *Les chefferies traditionnelles africaines face à la dynamique des réformes territoriales…, op.cit.*, p. 220.
1684 B. L. TOUERE ELENGA, *Fait ethnique, droit public et gouvernance de l'État en Afrique subsaharienne francophone : Contribution à la théorie de l'État africain, op.cit.*, p. 236.

de Sardan[1685], Souleymane Abba[1686], Kimba Idrissa[1687], ou encore Mahaman Tidjani Alou[1688], montrent comment la chefferie traditionnelle ne cesse de se métamorphoser depuis l'époque coloniale, marquant ainsi par sa permanence et son poids, la vie sociopolitique du pays au niveau des cantons, villages, fractions ou quartiers. Même si les fonctions autrefois symboliques des chefs traditionnels tendent à disparaitre, les attributions des autorités coutumières et traditionnelles ne cessent de se renouveler à travers les politiques de décentralisation.

715. - Dans les pays africains marqués par une économie qualifiée à tort d'informelle[1689], car échappant au cadre fixé par les bureaucraties administratives, la chefferie traditionnelle est un atout indispensable pour le recouvrement des impôts et taxes locaux. Au Togo, dans les années 1975, pour des raisons politiques liées aux idéaux du parti unique, le chef de l'État d'alors, le Général Gnassingbé Eyadema, a décidé de supprimer la compétence de recouvrement de l'impôt local des chefs traditionnels[1690]. Or, beaucoup d'activités assujetties aux impôts et taxes sont entreprises dans les villages sans que le pouvoir communal ne puisse les contrôler. La proximité du chef traditionnel avec sa population lui permet de connaitre les différentes activités économiques entreprises sur le territoire de la chefferie, ainsi que leurs auteurs. C'est ce qui justifie le maintien du rôle d'agent de recouvrement des chefs traditionnels au Niger et au Mali. En effet, dans la loi précitée du 13 janvier 2015 relative à la chefferie traditionnelle, l'alinéa 2 de l'article 14 dispose que « *sous l'autorité du président du conseil municipal, le chef de quartier, le village, la tribu, de chefferie particulière ou de fraction à la charge de la collecte de la taxe municipale frappant les membres de sa communauté* ».

1685 « L'espace public introuvable. Chefs et projets dans les villages nigériens », *Tiers-Monde*, n°157, 1999.

1686 « La chefferie traditionnelle en question », *Politique Africaine*, n° 38,1990, pp. 51-60.

1687 « La dynamique de la gouvernance : administration politique et ethnicité au Niger », in *Chefferie traditionnelle et l'État postcolonial en Afrique de l'Ouest*, Actes du colloque international, 19-22 mars 2000, Niamey (Niger), pp. 20-24.

1688 « La chefferie et ses transformations au Niger. De la chefferie coloniale à la chefferie postcoloniale », *op.cit.,* p. 1.

1689 C. BEAUCHAMP (dir.), *Démocratie, culture et développement en Afrique noire*, L'Harmattan, 1997, p. 80.

1690 Y. MARGUERAT, « L'État et l'organisation territoriale du Togo », *Afrique Contemporaine*, n°145, 1/88, p. 48.

716. - Les chefferies traditionnelles apparaissent comme de véritables canaux politiques par lesquels passe le développement local en Afrique. La coordination par les chefs traditionnels des initiatives des populations à la base constitue un catalyseur du développement. La chefferie traditionnelle mobilise et assure la participation active des populations locales à la vie politique, économique, sociale et culturelle, par le biais, notamment des comités de développement et de leur contribution à l'amélioration des recettes fiscales des collectivités territoriales. Cependant, l'attitude des pouvoirs publics vis-à-vis de l'institution cheffale demeure ambivalente. Confinées dans un rôle de gardien des valeurs traditionnelles, les chefferies traditionnelles occupent tout de même une place importante dans la vie publique locale. Cette position de la chefferie, de plus en plus affirmée au regard de l'organisation administrative territoriale, fait d'elle un acteur politique concurrent.

B- <u>Les chefs traditionnels, des acteurs concurrents de la vie politique</u>

717. - Dans la plupart des sociétés plurales à l'instar de celles qui composent les États africains, la politique de décentralisation présente des risques essentiellement liés aux tensions socioprofessionnelles, tribales ou ethniques, qui peuvent être ravivées dans les jeunes États en construction[1691]. Ainsi, l'analyse du droit africain de la décentralisation ne peut se faire sans une prise en compte du fait ethnique. Ce qui, au demeurant, n'est pas une spécificité africaine lorsqu'on observe la montée des revendications autonomistes et la consécration d'un droit à la différenciation en Occident, et notamment en France[1692]. Certes, le problème ne se pose pas en des termes identiques. La différence ne réside pas dans la nature, mais dans le degré auquel se pose la question dans les États d'Afrique noire francophone, où l'affirmation identitaire peut être une garantie d'une carrière politique et d'accès à une situation sociale envieuse[1693]. Les chefferies traditionnelles sont le lieu de cristallisation par

---

1691 P. LOUM, *La politique de décentralisation et les nouvelles stratégies de développement, op.cit.*, p.16.

1692 Voir not., Loi n° 2022-217 du 21 février 2022 relative à la différenciation, la décentralisation, la déconcentration et portant diverses mesures de simplification de l'action publique locale.

1693 A. T. KOKO, *Les chefferies traditionnelles africaines face à la dynamique des réformes territoriales. Contribution à l'étude des processus de décentralisation, op.cit.*, p. 225.

excellence de l'appartenance identitaire pouvant constituer un obstacle à l'unité nationale[1694].

718. - Les pouvoirs publics craignent alors qu'une implication trop importante des chefs traditionnels dans la vie politique locale ne favorise l'expansion des partis politiques à base ethnique. Il se poserait alors la question de la neutralité politique des chefs traditionnels. La représentation d'un parti politique quelconque les conduirait à choisir un camp politique, risquant ainsi de compromettre la confiance et le respect des sujets-citoyens qui font partie d'un bord politique différent[1695]. Cette situation porte en elle les germes de conflits interethniques, dans la mesure où la chefferie traditionnelle n'est pas constituée que de membres appartenant à un même groupe ethnique. La crainte de la multiplication, au niveau local, des partis politiques constitués sur des considérations ethniques est d'autant plus grande dans les chefferies où la désignation du chef traditionnel est ouverte à la consultation populaire. Elle se traduirait alors par la volonté d'un groupe ethnique d'assurer par tous moyens l'élection d'un membre de son groupe afin de préserver sa domination sur les groupes minoritaires composant la localité. Le risque d'un conflit ouvert serait alors patent.

719. - Le constat dressé au niveau local vaut également au sommet de certains États en Afrique. Koffi Ahadzi Nonou fait d'ailleurs remarquer que le tribalisme ou l'ethnicité est « *un terreau des activités politiques* » dans la mesure où la composition des partis politiques reflète de moins en moins la diversité nationale. « *Leurs bases sociales respectives demeurent fondamentalement tribales et le pouvoir réel appartient au fondateur ou principal dirigeant entouré d'un petit noyau occulte de conseillers ethniquement proches agissant en marge des organes officiels au profil tribal très avenant* »[1696]. L'identité ethnique reste un élément politique important, surtout en matière électorale. « *Les structures favorisant la participation au pouvoir politique sont essentiellement organisées suivant la logique des clivages ethniques* »[1697], transformant ainsi les groupes ethniques en groupes d'intérêts politiques. Le fait ethnique étant un moyen sûr

---

1694 *Ibid.*, p. 226.
1695 *Ibid.*, p. 229.
1696 K. AHADZI-NONOU, « Réflexions sur un tabou du constitutionnalisme négro-africain : le tribalisme », in *Les voyages du droit, Mélanges en l'honneur du professeur Dominique Breillat*, LGDJ, 2011, pp. 19-25.
1697 M. MARTINIELLO, *L'ethnicité dans les sciences sociales contemporaines*, PUF, 1995, pp. 61-62.

d'accéder à une légitimité démocratique[1698], les pouvoirs publics s'en assurent le monopole avec une instrumentalisation des chefferies traditionnelles, tout en amenuisant le rôle de ces dernières dans les arènes locales.

720. - Le rôle des collectivités villageoises dans la vie socio-économique locale n'est plus à démontrer, en revanche, la part qu'elles doivent prendre dans la conduite des politiques locales reste à clarifier. Conscient de son influence sur les populations locales, notamment en milieu rural, le pouvoir central cherche à en faire des intermédiaires sans pour autant faire du chef traditionnel un substitut au maire. Enrôlé par le pouvoir politique en place, les chefferies traditionnelles d'aujourd'hui ne sont guère mieux loties que celles d'hier soumises à l'autorité coloniale.

1698 Cf. S. H. F. KAKAI, Le vote ethnique au bénin. Contribution à une étude sociopolitique de l'élection, thèse, dactyl., Université d'Abomey-Calavi, 2011.

Conclusion du chapitre 2

721. - « *Deux calebasses placées sur une même surface d'eau finissent par se cogner* »[1699]. La chefferie traditionnelle comme élément de l'organisation administrative et territoriale se laisse difficilement saisir par le juriste habitué à partir des catégories préétablies pour appréhender le réel[1700]. Les politiques de décentralisation ont permis de réintégrer ces structures traditionnelles au fonctionnement de l'administration locale. Cependant, leur nature juridique reste encore incertaine. La chefferie traditionnelle a du mal à intégrer la typologie des cadres classiques de l'organisation administrative et territoriale des États en Afrique noire francophone. C'est donc avec un embarras certain que les faiseurs de lois contemporains, à la suite du colon français, ont désigné comme « unité administrative », la chefferie traditionnelle. Le malaise de cette classification forcée se perçoit aisément à travers les « béquilles et artifices » juridiques mis en place pour gérer cette structure juridique[1701]. Il s'agit d'une véritable « camisole »[1702] de force dont s'accommode difficilement l'institution.

722. - La chefferie traditionnelle est aujourd'hui confrontée aux défis de la décentralisation et du développement local. La communalisation intégrale des territoires étatiques sur la base de regroupement de cantons institue dans l'espace territorial jusque-là contrôlé par les chefs traditionnels des pouvoirs officiels. Si, jusqu'ici, les pouvoirs publics ont réussi à ménager les rapports entre la chefferie et les collectivités territoriales, c'est en marginalisant les premières et en inféodant les chefs à l'administration moderne. Pour redonner toute sa place à la chefferie traditionnelle en Afrique francophone dans l'administration du territoire, le législateur devrait lui reconnaitre la qualité de collectivité décentralisée. Il s'agirait d'une catégorie particulière de collectivité qui se verrait confier des compétences principalement en matière culturelle, touristique et

---

1699 Proverbe africain cité par B. L. TOUERE ELENGA, Fait ethnique, droit public et gouvernance de l'État en Afrique subsaharienne francophone. Contribution à la théorie de l'État africain, op.cit., p. 256.

1700C. NACH MBACK, « La chefferie traditionnelle au Cameroun : ambiguïtés Juridiques et dérives politiques » *Afrique et Développement*, vol. 25, n°3/4, 2000, p. 77.

1701 C. TRIMUA, « Le statut juridique de la chefferie traditionnelle au Togo : incertitudes et ambigüités », *Atelier de Sokodé sur La place du chef traditionnel dans le contexte de la décentralisation*, 2007, https://base.afrique-gouvernance.net/docs/trimua1.pdf. Consulté le 31 aout 2022.

1702 C. NACH MBACK, « La chefferie traditionnelle au Cameroun : ambiguïtés Juridiques et dérives politiques » *op.cit.*, p. 78.

économique. La chefferie-collectivité territoriale serait dirigée par un chef désigné dans le respect des coutumes locales et doté d'un organe délibérant composé pour partie des notables coutumiers et de l'autre des membres élus au suffrage universel[1703]. Ce n'est qu'en procédant ainsi que le législateur africain prendrait en compte la réalité de la société africaine. Cette dernière est marquée par une ambivalence : les citoyens vivent le principe de l'égalité républicaine tout en s'accommodant de l'influence persistante de la chefferie traditionnelle[1704].

1703 *Ibid.*, p. 115.

1704 P. KIEMDE, « Le bicaméralisme en Afrique et au Burkina Faso », *Revue Burkinabé de droit*, n° 2, 1992, p. 27.

Conclusion du titre 2

723. - La décentralisation n'induit pas par elle-même la démocratie locale. L'institutionnalisation d'un espace public local suppose la création d'un cadre propice, dans lequel les décisions des autorités locales sont soumises à l'épreuve de la « déconstruction »[1705]. À défaut, la démocratie locale ne saurait être plus qu'un mythe, dont le récit se construit dans le droit de la décentralisation, faisant ainsi miroiter le rêve d'un renouveau démocratique au sein des territoires locaux[1706]. En Afrique, la responsabilisation progressive des populations locales conduit à une mise à l'épreuve de la légitimité des autorités locales. Le suffrage universel n'épuise plus, à lui seul, l'expression des populations, qui cherchent à prendre davantage part à la conduite de l'action publique locale. La représentation doit être complétée par des mécanismes de participation citoyenne.

724. - Pour l'heure, la démocratie du *dedans* comme du *dehors* reste de l'ordre des professions de foi des acteurs politiques. Le slogan « *nous déciderons par nous-mêmes de nos affaires* » entonné au cours des campagnes de sensibilisation et d'information des populations sur le processus de décentralisation reste encore largement de l'ordre des virtualités[1707]. Les organisations sociales traditionnelles, dont le pouvoir d'influence et de mobilisation n'est aucunement remis en question, cherchent toujours leur place au sein des institutions locales. La forte légitimité des chefs traditionnels étant insuffisamment prise en compte par le processus de décentralisation, l'enracinement ainsi que l'appropriation de cette politique institutionnelle à la base sont mis à mal.

1705 J. CAILLOSE, « Ce que la constitution française donne à voir de la démocratie locale », in *Les collectivités locales, Mélanges en l'honneur de Jacques Moreaux*, Economica, 2002, p. 43.

1706 *Ibid.*, p. 56.

1707 C. NACH MBACK, *Démocratisation et décentralisation. Genèse et dynamiques comparés des processus de décentralisation en Afrique subsaharienne, op.cit.*, p. 342 et s.

Conclusion de la partie 2

725. - À l'issue des analyses qui précèdent, nous pouvons dire sans risque de nous tromper que l'État entretient une emprise certaine sur les collectivités territoriales. L'ampleur de cette emprise est telle qu'elle annihile l'autonomie des collectivités territoriales. Les collectivités territoriales sont fortement dépendantes de l'État pour l'exercice des compétences décentralisées, que ce soit pour les ressources financières ou pour l'expertise technique. En plus de l'attribution des moyens, l'État exerce sur les collectivités territoriales une tutelle qui, si, elle peut se justifier juridiquement, entrave la liberté d'action des collectivités. Les collectivités territoriales africaines évoluent dans un cadre inapproprié, marqué par l'étroitesse des marges de manœuvre dont elles disposent. Il faut ajouter à cette situation le manque de prise en compte des réalités sociologiques dans le processus de décentralisation et la faible implication des citoyens à la gouvernance locale. Dans ces conditions, la décentralisation apparait comme un avatar de la centralisation administrative. Il ne peut être question, dans une telle logique, d'une quelconque autonomie locale. Par conséquent, le processus de construction de l'État décentralisé en Afrique est pour l'heure artificiel. Les réformes en cours doivent se débarrasser du poids du passé et s'en remettre aux ressources du vécu et du spontané qui sont précisément à la racine de la décentralisation[1708]. Dès lors, l'essor d'une véritable décentralisation africaine doit passer par « *un effort d'inventivité dans l'appareillage institutionnel (…) tout en ménageant des cadres juridiques favorables aux expérimentations locales, respectant la diversité des communautés socio-ethniques* »[1709]. L'analyse du discours des différents acteurs semble révéler que l'heure n'est pas encore à ce genre de transformation sociale.

[1708] B. FAURE, « "Penser le changement" ou "changer le pansement" ? », *AJDA*, 2014, p. 600.

[1709] J. FIALAIRE, « L'organisation des métropoles dans les États d'Afrique noire francophones », *op.cit.*, p. 220.

CONCLUSION GÉNÉRALE

> « La malédiction la plus commune, c'est d'être la dupe de bonne foi d'hypocrisie collective, habile à mal poser les problèmes pour mieux légitimer les odieuses solutions qu'on leur apporte »[1710].

726. - L'ambition de la présente recherche était de parvenir à décrire les rapports entre l'État et les collectivités territoriales dans les pays de l'espace ouest-africain francophone afin de pouvoir déterminer le degré de réalisation de l'autonomie locale. Notre objet d'étude était donc orienté sur les collectivités territoriales en tant que structure disposant d'une habilitation constitutionnelle à s'administrer librement dans les conditions déterminées par le législateur. D'emblée, une telle démarche était loin d'être aisée. Ainsi que le remarquait Gérard Marcou dans une étude sur l'autonomie communale en Europe, contrairement à ce que l'on pourrait penser, « *le niveau de décentralisation, ou d'autonomie, ne peut pas se mesurer seulement par le volume des compétences exercées* [...]. *Beaucoup plus importante que le volume des compétences est l'étendue des marges de manœuvre. Or, celle-ci est beaucoup plus difficile à apprécier* »[1711]. En outre, cette difficulté se décuple lorsqu'il s'agit d'évaluer l'étendue de la marge de manœuvre des collectivités territoriales dans les États africains de succession française. En effet, dans ces pays, la question de l'effectivité du droit se pose avec une certaine acuité. À notre sens, les réformes engagées par les États africains subsahariens, à partir de 1990, ne donnent qu'à voir une décentralisation apparente. Présentée au départ comme un contre-pied de la centralisation et un catalyseur du développement, la décentralisation s'est révélée être un moyen de redéploiement de la centralisation et de renforcement des inégalités.

1710 A. CÉSAIRE, *Discours sur le colonialisme,* Présence Africaine, 1950.

1711 G. MARCOU, « L'autonomie communale : étude comparative », *op.cit.,* p. 84-85.

727. - Ainsi formulée, cette conclusion appelle une remarque ; celle de l'inexistence d'un contentieux local, qui constitue, sans aucun doute, une limite de la présente recherche[1712]. Sous prétexte d'une certaine collaboration apaisée, les autorités locales traduisent très rarement l'État en justice[1713]. Or, l'accès à la justice, l'interprétation des normes par le juge ainsi que les pouvoirs de sanction dont il dispose constituent une condition d'effectivité et de concrétisation des règles[1714]. La démonstration s'est donc essentiellement appuyée sur l'analyse des textes législatifs et réglementaires, les travaux de doctrine, et même sur le comportement des différents acteurs impliqués dans le processus de la décentralisation. En partant de ces éléments, la vérification de notre hypothèse de départ s'est articulée autour de deux axes : la réception des normes ; et leur mise en œuvre.

728. - Sur le premier axe, l'étude a mis en évidence les caractères communs, à quelques exceptions près, des institutions décentralisées dans les États de l'espace ouest-africain francophone. Ces dernières sont une reproduction du modèle décentralisé de l'ex-métropole. Plusieurs décennies après les indépendances, les réformes menées en matière de décentralisation n'ont pas permis de rompre avec la dépendance à l'égard de modèles territoriaux étrangers qui demeurent peu adaptés aux contextes et aux peuples auxquels ils s'appliquent. Et pour cause, la colonisation a laissé de profonds traumatismes au continent africain. L'ensemble des travaux des explorateurs[1715] du continent africain ont favorisé la dissémination de l'idée selon laquelle les traditions africaines seraient réfractaires au progrès et incapables de s'autoréguler. Or, on a pu observer en Afrique noire francophone l'existence de groupements humains organisés *a minima* autour des chefferies traditionnelles[1716]. Ces dernières étaient de véritables institutions et marqueurs d'identités culturelles, sociologiques et humaines qui auraient pu servir de socle à la création de collectivités territoriales autonomes si leur évolution n'avait

---

1712 Un constat partagé par les différents auteurs qui entreprennent un travail de recherche sur le droit des collectivités en Afrique. Voir not., C. NANAKO, *La libre administration des collectivités territoriales au Bénin et au Niger, op.cit.* p. 354.

1713 *Idem.*

1714 Cf. J. BETAILLE, *Les conditions juridiques de l'effectivité de la norme en droit public interne, op.cit.*, p. 73 et s.

1715 Ethnologues, anthropologues, juristes etc.

1716 L. NGOMO TSIMI, *L'autonomie administrative et financière des collectivités territoriales décentralisées : l'exemple du Cameroun*, thèse, dactyl., Université Paris-Est, 2010, p. 26 et s.

pas été bouleversée par le fait colonial et l'introduction de l'État « moderne » et de ses techniques. L'autonomie originelle des structures de peuplement n'a donc pas résisté à la création de l'État et de la personnalité juridique, qui a permis la naissance au sein de l'État de collectivités territoriales[1717] à l'autonomie réduite.

Au-delà de l'impact de la colonisation, la « dictature du développement » a permis de projeter sur les sociétés africaines les mythes de l'Occident. Les réformes entreprises depuis les indépendances en faveur de la décentralisation à la suite de la remise en cause de l'État centralisé sont essentiellement dictées par les pays pourvoyeurs de l'Aide publique au développement et les institutions internationales financières. La décentralisation s'est érigée en norme indiscutable du progrès et de la lutte contre la pauvreté. Cette forte implication de la communauté internationale faisant de la décentralisation une valeur absolue — une conditionnalité de l'accès aux financements extérieurs — a contribué au renforcement des tentatives mimétiques. À travers le maintien d'une coopération administrative et décentralisée, l'élaboration des guides sur la conduite des politiques de décentralisation, il fut proposé aux États africains de revêtir des formes institutionnelles produites ailleurs, fruit d'une histoire millénaire dont les initiateurs ne pouvaient imaginer les contours actuels que celles-ci prendraient[1718]. Felwine Sarr, dans son essai *Afrotopia,* s'interroge sur le réel bénéficiaire de ce mimétisme à la fois contraint et voulu. En réalité, le mime ne profite qu'aux États exportateurs et à l'élite dirigeante : on ne peut se réaliser pleinement sous le mode de l'imitation permanente, de la greffe, et l'extraversion[1719]. Desservant les missions qu'on lui assigne, compte tenu de l'application des principes inadéquats aux nécessités d'un pays en voie de développement, ce phénomène mimétique en matière de décentralisation profite, au premier chef, aux « *détenteurs des pouvoirs administratifs, qui ne peuvent faute d'autres ressources sociales, légitimer leur pouvoir que par référence à un modèle utopique, servant un intérêt général mythique, une mission civilisatrice hypothétique, mais confirmée par les organisations internationales* »[1720]. De ce fait, la proposition faite aux États africains de reproduire un modèle préfabriqué où les pratiques sociales

---

1717 F. MUGNIER, *La personnalité juridique des collectivités territoriales. Genèse et développement d'une personne morale dans l'État*, thèse, Dalloz, 2022.

1718 F. SARR, *Afrotopia,* Philippe Rey, 2016, pp. 22-23.

1719 *Ibid.*, p. 24.

1720 D. DARBON, « A qui profite le mime ? Le mimétisme institutionnel confronté à ses représentations en Afrique », in Y. MENY (dir.), *Les politiques du mimétisme institutionnel : la greffe et le rejet, op.cit.*, p. 127.

anciennes n'ont aucune place sert une stratégie d'influence internationale des nations développées.

Dans ces conditions, l'ordre juridique dans les États africains de succession française apparait incohérent. Par conséquent, il est peu susceptible d'influer sur les comportements des acteurs. L'importation de modèles institutionnels étrangers, sans une réflexion préalable sur ce qui conviendrait ou non aux populations locales, constitue la première cause de l'échec des États africains à mettre en place des institutions décentralisées autonomes et viables.

729. - Sur le second axe, nos développements ont porté sur les conditions de construction d'un État décentralisé, notamment sur les facteurs de réalisation de la libre administration et de la démocratie locale. Il a pu ainsi être relevé que malgré la consécration constitutionnelle de la libre administration des collectivités territoriales, la décentralisation en Afrique offre bien de commodités au pouvoir central. Sa mise en œuvre n'a pas sonné l'heure de la transformation de l'État jacobin hérité de la période coloniale et de son emprise sur le territoire.

L'État, « grand architecte du local »[1721], conçoit la décentralisation comme une projection dans le local de sa finalité première qui demeure l'unité et l'indivisibilité de la République. Par conséquent, le mouvement décentralisateur s'accompagne, tout comme pendant la colonisation, d'une négation des autonomies. En clair, les collectivités territoriales sont pleinement intégrées à l'État qui définit leur liberté en fonction de ses besoins au niveau national. La décentralisation en Afrique a donc conduit à un renforcement du pouvoir étatique. En l'absence de règles constitutionnelles claires sur la répartition des compétences, les collectivités territoriales sont soumises au bon vouloir de l'État-législateur qui détermine les conditions d'exercice dans les domaines de la vie locale. À la vérité, l'action est dictée aux autorités décentralisées sans aucune marge d'opportunité en rapport avec les données concrètes de leur territoire[1722]. Aussi, dans l'exercice de leurs compétences, les collectivités territoriales sont placées sous la tutelle permanente de l'État. La mutation annoncée du contrôle *a priori* vers un contrôle *a posteriori* n'a pas eu lieu. Par ailleurs, l'absence d'un contentieux administratif local abondant ne

---

1721 J.-F. BRISSON, « La loi du 16 décembre 2010 portant réforme territoriale ou le droit des collectivités territoriales en miettes », *Dr. Adm.*, n° 3, 2011, étude 5.

1722 *Cf.* B. FAURE, « Différenciation », *AJDA*, 2022, p. 1367 ; « Différenciation des normes : jeu de dupes », *RFDA*, 2021, p. 313.

contribue pas à l'épanouissement des libertés locales ni à l'affirmation d'un pouvoir administratif local. En fin de compte, les autorités décentralisées n'ont pas les moyens de leur politique.

La réussite du processus de décentralisation dépend dans une grande mesure de la qualité de la gestion locale et des ressources financières mises à disposition des collectivités territoriales. L'autonomie locale suppose qu'elles puissent disposer de fonctionnaires territoriaux compétents et en nombre suffisant. Le cadre de gestion du personnel local est calqué sur celui de l'État alors même que ce dernier n'est pas exempt de critiques en matière de management des ressources humaines. L'émergence du développement local passe non seulement par une meilleure gestion du personnel, mais aussi par une amélioration du régime rémunératoire afin d'attirer les meilleurs profils que ce soit pour les agents locaux ou pour les élus locaux. Une telle démarche suppose, toutefois, que les collectivités territoriales disposent de ressources financières suffisantes. La décentralisation territoriale appelle nécessairement à une redéfinition de l'autonomie financière. Les collectivités territoriales sont dépendantes des concours financiers du pouvoir central en raison de l'insuffisance de ressources propres. De plus, le caractère obligatoire de certaines dépenses réduit la marge de manœuvre des autorités décentralisées. Avec la multiplication de ces dépenses, les autorités locales perdent tout pouvoir de décision. Les finances locales doivent donner aux collectivités territoriales une base plus saine et plus autonome à leur activité et faire d'elles de véritables partenaires des autorités centrales sans donner cette impression d'être à la solde de l'État[1723]. Pour l'heure, le cordon ombilical a dû mal à se défaire. L'autonomie des collectivités territoriales en Afrique de l'Ouest demeure une arlésienne.

Au-delà des considérations administrative et financière, l'autonomie locale entretient dans le discours politico-administratif un lien étroit avec la démocratisation de la vie publique. La décentralisation en Afrique devait à la fois faire advenir le développement et la démocratie par la proximité qu'elle implique. En ce sens, il a été relevé que la décentralisation repose également sur une dimension politique. Selon Jacques Baguenard, le fondement de tout pouvoir politique consiste à « *contribuer à satisfaire les besoins élémentaires de survie du groupe concerné et des individus qui le composent* »[1724].

1723 C. CHABROT, *La centralisation territoriale. Fondements et continuité en droit public français*, *op.cit.*, p. 514.

1724 J. BAGUENARD, *L'État, une aventure incertaine*, Ellipses, coll. « Mise au point », 1998, p. 25.

Or, telle est la mission confiée aux collectivités territoriales africaines. La décentralisation implique donc l'existence d'un « conseil élu » pour la mise en œuvre de l'action publique locale. Si le principe de l'élection des organes décentralisés est désormais un acquis, ni le fonctionnement interne des assemblées délibérantes ni la consécration des mécanismes de démocratie directe ne garantissent une réelle participation des élus d'opposition et des populations locales. Les autorités traditionnelles combattues depuis la colonisation demeurent résilientes et posent un véritable problème à l'État moderne postcolonial. Véritable actrice du développement local et de mobilisation des populations locales, la chefferie traditionnelle a du mal à intégrer la typologie des cadres classiques de l'organisation administrative et territoriale des États en Afrique noire francophone.

730. - Fondamentalement, cette recherche sur l'autonomie des collectivités territoriales en l'Afrique de l'Ouest est une invitation à l'invention d'un nouveau modèle. L'Afrique a trop souvent été le continent des expérimentations diverses, elle doit désormais se réinventer en s'appuyant sur ses réalités locales. Les dirigeants africains doivent comprendre que l'Afrique « *n'a personne à rattraper. Elle ne doit plus courir sur les sentiers qu'on lui indique, mais marcher prestement sur le chemin qu'elle se sera choisi* »[1725]. La seule urgence du continent est celle d'être à la hauteur de ses potentialités. Cela passe notamment par la franche constatation de l'inadéquation du modèle et des principes de décentralisation hérités de la colonisation. Il faut donc repenser le système en assumant clairement la mise en place d'une co-administration, plutôt que d'une décentralisation territoriale, dans laquelle l'État guide, autorise, finance, ordonne, et contrôle. La décentralisation doit être pensée comme un facteur d'unité qui permet au système juridique de s'adapter à la diversité des situations concrètes. Il nous semble que l'État en Afrique doit, en ce sens, réinterroger la question ethnique et celle des chefferies traditionnelles dans les réformes territoriales. Ce faisant, ces réformes parviendront à aligner les formes institutionnelles sur les cadres mentaux et les systèmes de significations traditionnelles qui demeurent présents dans l'espace local. La cohérence de l'ordre juridique qui en découlera facilitera l'atteinte des objectifs fixés. L'Afrique doit cesser dans son processus de réinvention d'être une pâle copie de l'Occident. Dès les indépendances, en 1961, Frantz Fanon tirait déjà la sonnette d'alarme en ces termes : « *Ne payons pas de tribut à l'Europe en créant des États, des institutions et des sociétés qui s'en inspirent.*

[1725] F. SARR, *Afrotopia,* Philippe Rey, 2016, p. 152.

L'humanité attend autre chose de nous et cette imitation caricaturale est obscène. Si nous voulons transformer l'Afrique en autre Europe, alors confions à des Européens les destinées de nos pays, ils sauront mieux faire que les mieux doués d'entre nous. Mais si nous voulons la porter à un niveau différent de celui où l'Europe l'a manifestée, alors il faut inventer (…) »[1726]. Le diagnostic est désormais établi. L'État en Afrique est appelé à résoudre ses propres contradictions.

Bordeaux, octobre 2022

1726 F. FANON, *Les Damnés de la terre*, La découverte 2002.

BIBLIOGRAPHIE

(Liste des principaux ouvrages mobilisés dans le travail de thèse)

I- DICTIONNAIRES, REPERTOIRES ET ENCYCLOPEDIES

ALLAND D., RIALS S. (dir.), *Dictionnaire de la culture juridique*, PUF, coll. « Quadrige », 2010.

ANDRIANTSIMBAZOVINA J., GAUDIN H., MARGUÉNAUD J.-P., RIALS S., SUDRE F. (dir.), *Dictionnaire des Droits de l'Homme*, PUF, coll. « Quadrige », 2008.

ARNAUD A.-J. (dir.), *Dictionnaire encyclopédique de théorie et de sociologie du droit*, LGDJ, coll. « Anthologie du droit », 2018.

BECET J.-M., Communication des documents des collectivités territoriales au public : régimes particuliers, *Encyclopédie des collectivités locales*, Dalloz, folio n°12212, act. 2021.

BEQUET L. (dir.), *Répertoire du droit administratif*, 1882-1911, Paul Dupont, 28 volumes.

BENOIT F.-P., « Les attributions générales », in Encyclopédie des collectivités locales, Dalloz, folio n° 2012.

BENOIT F.-P., BENOIT J., « Conseil municipal : régime des délibérations », in Encyclopédie des collectivités locales, Dalloz, 2011, folio n° 300.

COMBEAU P., « Action extérieure des collectivités territoriales », in JCP- A, Fasc. 129-40.

CORNU G.(dir.), *Vocabulaire juridique*, PUF, Coll. « Quadrige » 12ème éd., 2018.

DOUENCE J.-C., « Statut constitutionnel des collectivités locales », in Encyclopédie des collectivités locales, Dalloz, folio n° 63.

JANICOT L., « Coopération intercommunale. Règles communes », in JCP- A, Fasc. 129-20.

KADA N., PASQUIER R., COURTECUISSE C., AUBELLE V. (dir.), *Dictionnaire encyclopédique de la décentralisation*, Berger-Levrault, 2017.

KADA N., MATHIEU M. (dir.), *Dictionnaire d'administration publique*, PUG, 2014.

PONTIER J.-M., « Région : collectivité territoriale », in Encyclopédie des collectivités locales, Dalloz, folio n°1710, act. 2022.

VERPEAUX M., « Principes d'organisation administrative », in JCP- C, Fasc. 10.

II- OUVRAGES, TRAITÉS ET MANUELS

ALBERTINI J.-B., La déconcentration. L'administration territoriale dans la réforme de l'État, Economica, 1997.

AUBY J.-B.,

- *La décentralisation et le droit*, LGDJ, coll. « Systèmes Droit », 2006.
- *La globalisation, le droit et l'État*, Montchrestien, coll. « Clefs-Politique », 2003.

AUBY J.-B., AUBY J.-F., NOGUELLOU R., *Droit des collectivités locales*, PUF, coll. « Thémis », 2015.

BAGUENARD J., *La décentralisation*, PUF, Coll. « Que sais-je ? », 2004.

BLONDIAUX L., Le nouvel esprit de la démocratie. Actualité de la démocratie participative, Seuil, 2008.

BOURJOL M., *Les districts urbains*, Berger-Levrault, 1963.

BURON M., *Décentralisation : l'âge de raison*, La documentation française, 1994.

BADIE B., *L'État importé. L'occidentalisation de l'ordre politique*, Fayard, 1992.

BAGUENARD J., *La décentralisation*, Paris, PUF, « Que-sais-je ? », 2004, p. 45.

BAGUENARD J., BECET J.-M., *La démocratie locale*, PUF « Que sais-je ? », 1995, p. 97.

BALANDIER G., *Anthropologie politique*, PUF, coll. « Quadrige », 2013, p.119.

BAUDIN-CUILLERE F., Principe de subsidiarité et administration locale, LGDJ, 1995.

BAYART J.-F., *L'État en Afrique, la politique du ventre*, Fayard, 1989.

BIGOT G.,

- L'administration française. Politique, droit et société, t. I (1789-1870), LexisNexis, coll. « Manuel » 2014.
- L'administration française. Politique, droit et société, t. II (1870-1944), LexisNexis, coll. « Manuel », 2014.

BUCHMANN J., *L'Afrique noire indépendante*, LGDJ, 1962.

BURDEAU G., *Traité de science politique*, t. I, LGDJ, 1980.

BODINEAU P., VERPEAUX M., *Histoire de la décentralisation*, PUF, 1997.

BOURMAUD D., *La politique en Afrique*, Montchrestien « Clefs », 1997.

CAILLOSSE J., Les « mises en scènes » juridiques de la décentralisation. Sur la question du territoire en droit public français, LGDJ, Coll. « Droit et société », 2009.

CARRE DE MALBERG R., *Contribution générale à la théorie de l'État*, Sirey, 1920.

CHAMPEIL-DESPLATS V., *Méthodologies du droit et des sciences du droit*, Dalloz, coll. « Méthodes du droit », 2016.

CHAPUS R., *Droit administratif général*, Montchrestien, coll. « Domat droit public », 2001.

CHAVRIER G., *Le pouvoir normatif local : enjeux et débats*, LGDJ, coll. « Systèmes », 2011.

CHEVALLIER J.,

- *Science administrative*, PUF, coll. « Thémis », 2013.
- *Institutions politiques*, LGDJ, Coll. « Systèmes », 1996.
- *L'État de droit*, LGDJ, coll. « Clefs », 2017.

COMBACAU J. SUR S., *Droit international public*, Montchrestien, 2014.

COPANS J., *Anthropologie, science des sociétés primitives*, Paris, Ed. Denoël, 1972.

DANDA M., *Niger, une décentralisation importée : une comparaison avec la France*, L'Harmattan, 2013.

DAVID R., *Les grands systèmes de droit contemporains*, Dalloz, 1964.

DEBBASCH C., La décentralisation pour la rénovation de l'État, PUF, 1976.

DE HEUSCH L., *Le Pouvoir et le sacré*, Bruxelles, Annales du Centre d'étude des religions, 1962.

DELAVIGNETTE R., *Les vrais chefs de l'Empire,* Gallimard, 1939.

DEMERS V., Le contrôle des fumeurs. Une étude d'effectivité du droit, Montréal, Thémis, 1996.

DE TOCQUEVILLE A., *De la démocratie en Amérique*, Gallimard, coll. « Idées », 1968.

DIOP ANTA Ch., *L'Afrique noire précoloniale*, Présence Africaine, 1960.

DJIRE M., KEITA A., *Régionalisation au Mali. Regards croisés*, Éditions Tombouctou, 2015.

DUGUIT L., *Traité de droit constitutionnel,* t. 3, 3ème éd., De Boccard, 1930.

DUVERGER M., Méthodes de la science politique, PUF, 1959.

EISENMANN C., *Centralisation et décentralisation. Esquisse d'une théorie générale*, LGDJ, 1948.

ESMEIN A., *Éléments de droit constitutionnel français et comparé*, Paris, éd. Panthéon-Assas, 2001.

FAURE B., *Droit des collectivités territoriales*, Dalloz, coll., « Précis », 2021.

FERAL F., Approche dialectique du droit de l'organisation administrative. L'appareil d'État face à la société civile, L'Harmattan, coll. « Logiques juridiques » 2000.

FERSTENBERG J., PRIET F., QUICHINI P., *Droit des collectivités territoriales*, Dalloz, coll. « Hypercours », 2016.

FINKEN M., *Gouvernance communale en Afrique et au Cameroun*, L'Harmattan, 2011.

FOGUI J-P., *L'intégration politique au Cameroun : Analyse centre - périphérique*, LGDJ, 1990.

FORRAY V., PIMONT S., Décrire le droit… et le transformer. Essai sur la décriture du droit, Dalloz, 2017.

FUKUYAMA F., *La fin de l'histoire et le dernier homme*, Flammarion, coll. « champs essais », 2009.

FRIER P.-L., *L'urgence*, LGDJ, 1987.

GALY M., SANNELLA E., *Les défis de l'État en Afrique*, L'Harmattan, 2007.

GAZIAUX E., L'autonomie en morale : au croisement de la philosophie et de la théologie, Presse universitaire de Louvain, 1998.

GAZIBO M., *Introduction à la politique africaine,* Presses de l'Université de Montréal, 2010.

GERAL DE SEREZIN F., *Le protectorat international : le protectorat sauvegarde, le protectorat de droit des gens, le protectorat colonial*, Pédone, 1896.

GLELE M. A.,

- Naissance d'un État noir, LGDJ 1969.
- Le Danxome : du pouvoir aja à la nation fon, Nubia, 1974.

GOHIN O., DEGOFFE M., MAITROT DE LA MOTTE A., DUBREUIL C.-A., *Droit des collectivités territoriales*, Cujas, 2015.

GONIDEC P.-F.,

- Les droits africains. Évolution et sources, LGDJ, 1976.
- L'État africain : évolution, fédéralisme, centralisation et décentralisation, LGDJ, 1985.

GRANDGUILLOT D., L'essentiel des Institutions politiques et administratives de la France, Lextenso, 2020.

GREWE C., RUIZ-FABRI H., *Droits constitutionnels européens*, PUF, 1995.

GRUBER A., La décentralisation et les institutions administratives, Armand colin, 1986.

HABERMAS J., *Droit et démocratie. Entre faits et normes*, Gallimard, 1997.

HAMON F., TROPER M., *Droit constitutionnel*, LGDJ, 30e éd. 2007.

HAURIOU M.,

- *Précis de droit administratif et de droit public*, Dalloz, 2002, réimpression de la 12ème édition de 1933.
- Précis de droit constitutionnel, Sirey, 1929.

HIBOU B., *L'Afrique est-elle protectionniste ? Les chemins buissonniers de la libéralisation extérieure*, Karthala, 1996.

JAN P., *Institutions administratives*, LexisNexis, Coll. « Objectif droit », 2021.

KAMTO M., L'Urgence de la pensée. Réflexion sur une précondition du développement en Afrique, éd. Mandara, 1994.

LE ROY E.,

- *Cours d'histoire des institutions d'Afrique noire*, Séminaire d'Anthropologie juridique, 1976-1977.
- Les Africains et l'institution de la justice : entre mimétismes et métissages, Dalloz, coll. « Regards sur la justice », 2004.

MANSON S., *L'opposition dans les assemblées locales*, LGDJ, Coll. « Systèmes », 2012.

MASPETIOL R., LAROQUE P., La tutelle administrative, le contrôle des administrations départementales, coloniales et des établissements publics, Sirey, 1930.

MAQUET J., *Les civilisations africaines*, Marabout Université, 1962.

MARTINIELLO M., L'ethnicité dans les sciences sociales contemporaines, PUF, 1995.

MAZERES J.-A., Centre, périphérie, territoire, PUF, 1978

MENY Y., Centralisation et décentralisation dans le débat politique français (1945-1969), LGDJ, 1974.

MENY Y., SUREL Y., Politique comparée. Les démocraties : Allemagne, Etats-Unis, France, Grande-Bretagne, Italie, LGDJ, 2009.

MBEMBE A., De la postcolonie. Essai sur l'imagination politique dans l'Afrique contemporaine, La découverte, 2000.

MILLON-DELSOL C.,

- *L'État subsidiaire*, Paris, PUF, 1992.
- Le principe de subsidiarité, P.U.F. 1993.

MOMO B., Droit administratif. Cours de 2ème année de Licence, Yaoundé, 1986.

MOREAU J., VERPEAUX M., Révolution et décentralisation. Le système administratif français et les principes révolutionnaires de 1789, Economica, coll. « Droit public positif », 1992.

MULLER-QUOY I., Le droit des assemblées locales, LGDJ, 2001.

OBERDOFF H., La démocratie à l'ère du numérique, PUG, 2010.

OMGBA J.-P., Re-penser l'administration publique en Afrique, L'Harmattan, 2018.

LAPIERRE J.W., *Vivre sans État ? Essai sur le pouvoir politique et l'innovation sociale*, Esprit-Le Seuil, 1977.

LEGENDRE P.,

- *Histoire de l'administration de 1750 à nos jours*, PUF, coll. « Thémis », 1968.
- *Trésor historique de l'État en France. L'Administration classique*, Fayard, coll. « Les savoirs », 1992.

LOMBARD J., *Autorités traditionnelles et pouvoirs européens en Afrique noire*, Armand Colin, 1967.

OGIEN A., LAUGIER S., Le principe démocratie : enquête sur les nouvelles formes du politique, La Découverte, 2014.

OUATTARA S., Gouvernance et libertés locales pour une renaissance de l'Afrique, Karthala, 2007.

OST F., VAN DE KERVOCHE M., *Jalons pour une théorie critique du droit* Bruxelles, Publications des Facultés universitaire Saint-Louis, 1987.

PETIT-DUTAILLIS Ch., Les communes françaises caractères et évolution, des origines au XVIII° siècle., Albin Michel, 1947.

PLESSIX B., *Droit administratif général*, LexisNexis, 2020.

KI-ZERBO J.,

- *Le Monde noir*, Abidjan, CEDA, 1963, p.47.
- *Histoire de l'Afrique d'hier à demain,* Paris, Hatier, 1972,

RASERA M., *La démocratie locale*, LGDJ, Coll. « Systèmes », 2002.

ROULAND N.,

- *Anthropologie juridique*, PUF, 1988.
- *Introduction historique au droit*, PUF, Coll. « Droit fondamental », 1998.

ROSANVALLON P.,

- *La légitimité démocratique. Impartialité, réflexivité, proximité*, Seuil, 2008.

- La contre-démocratie. La politique à l'âge de la défiance, Seuil, 2006.

ROUSSEAU D., Six thèses pour la démocratie continue, Odile Jacob, 2022.
ROUX A.,
- *Droit constitutionnel local*, Economica, coll. « Poche droit public », 1995.
- La décentralisation. Droit des collectivités territoriales, LGDJ, coll. « Systèmes », 2016.

SACCO R., Le droit africain. Anthropologie et droit positif, Dalloz, 2009.
SAVONNET-GUYOT C., État et sociétés au Burkina-Faso. Essai sur le politique africain, Karthala, 1986.
SAWADOGO A.-R., *L'État africain face à la décentralisation*, Karthala, 2001.
SCHRAMECK O., *La fonction publique territoriale*, Dalloz, coll. « Connaissance du droit », 1995.
SINDJOUN L., *L'État ailleurs. Entre noyau dur et case vide*, Economica, coll. « La vie du droit en Afrique », 2002.
STIRN B., Les sources constitutionnelles du droit administratif, LCDJ, coll. « Systèmes », 2011.
SUPIOT A., *Servir l'intérêt général*, PUF, « Les voies du droit », 2000.
SY D., *Droit administratif*, Dakar, CREDILA -L'Harmattan, 2014.
SY O., Reconstruire l'Afrique. Vers une nouvelle gouvernance fondée sur les dynamiques locales, Ed. Charles Léopold Mayer, 2009.
TALLEFAIT A., *Droit de la fonction publique,* Dalloz, coll. « Précis », 2022.
THIAM S., *Introduction historique au droit en Afrique*, L'harmattan, 2011.
VEDEL G., DELVOLVE P., *Droit administratif*, PUF, 1990.
VERPEAUX M., JANICOT L., *Droit des collectivités territoriales*, LGDJ, 2019.
VERPEAUX M.,
- *La région*, Dalloz, 2005.
- *Les collectivités territoriales,* Dalloz, coll. « Connaissance du droit », 2020.

VIÉ J.-E., *Les sept plaies de la décentralisation*, Economica, 1986.
WALINE M., *Droit administratif*, Sirey, 1963.

III- OUVRAGES COLLECTIFS

AUBY J.-M. (dir.), *Droit public*, Economica, 1985.
BEAUCHAMP C. (dir.), Démocratie, culture et développement en Afrique noire, L'Harmattan, 1997.
BIOY X., MASTOR W., MOUTON S., MAGNON X. (dir.), *Le réflexe constitutionnel : question sur la QPC*, Bruylant, 2013.
BRISSON J.-F. (dir.),
- *Les transferts de compétences de l'Etat aux collectivités locales*, colloque AFDCL, Bordeaux et Agen, 19-20 Sept. 2007, L'Harmattan, 2009.

- *L'assemblée délibérante, avenir de la démocratie locale ?* L'Harmattan, 2016.

BLONDIAUX L., MANIN B. (dir), *Le tournant délibératif de la démocratie*, Presses de Sciences Po, 2021.

BOUTIN C., ROUVILLOIS F. (dir.) *Décentraliser en France. Idéologies, histoire et prospective*, Paris, éd. F.-X. de Guibert, 2003.

COMBEAU P. (dir.), Le contrôle de l'État sur les collectivités territoriales, L'Harmattan, 2007.

CONAC G., (dir.),

- *Les institutions administratives des Etats francophones d'Afrique noire*, coll. « La vie du Droit en Afrique », Economica, 1979.
- *L'Afrique en transition vers le pluralisme politique*, Economica, 1993.
- *Dynamiques et finalités des droits africains*, Economica, 1976.

COLLECTIF, MELANGES,

- Mélanges offerts à Paul Amselek, Bruylant, 2005
- *Mélanges Gérard Marcou*, IRJS Éditions, 2017.
- Le droit administratif. Permanences et convergences, Mélanges en l'honneur de J.-F. Lachaume, Dalloz, 2007.
- Les collectivités locales, Mélanges en l'honneur de J. Moreau, Economica, 2003.
- Mouvement du droit public, Mélanges en l'honneur de F. Moderne, Dalloz, 2004.
- Long cours, Mélanges en l'honneur de P. Bon, Dalloz, 2014.
- L'État africain. Horizon 2000, Mélanges P.-F. Gonidec, LGDJ, 1985.
- Les voyages du droit, Mélanges en l'honneur du professeur D. Breillat, LGDJ, 2011
- Révolution, constitution, décentralisation, Mélanges en l'honneur de Michel Verpeaux, Dalloz, 2020.
- Les mutations contemporaines du droit public, Mélanges en l'honneur de B. Jeanneau, 2002.
- L'État de droit, Mélanges en l'honneur de Guy Braibant, Dalloz, 1996.
- L'intérêt général, Mélanges en l'honneur de D. Truchet, Dalloz, 2015.
- Terres du droit, Mélanges en l'honneur d'Y Jégouzo, Dalloz, 2009.
- Les finances publiques entre globalisation et dynamiques locales, Mélanges en l'honneur de Eloi DIARRA et Salifou YONABA, L'Harmattan, 2021.

- Itinéraire d'un constitutionnaliste, Mélanges en l'honneur de F. Delpérée, Bruylant, 2007
- Regards sur le droit public en Afrique, Mélanges en l'honneur du Doyen J.-M. Bipoun Woum, L'Harmattan, 2016.

COQUERY-VIDROVITCH C., MONIOT H. (dir.), *L'Afrique noire de 1800 à nos jours,* PUF, 2005.
COULON C., MARTIN D.-C., (dir.), *Les Afriques politiques,* La découverte, 1991.
DARBON D., DU BOIS DE GAUDUSSON J., (dir.), *La création du droit en Afrique*, Karthala, coll. « Hommes et sociétés », 1997.
DE LAUBADERE A., VENEZIA J.-C, GAUDEMET Y., *Traité de droit administratif*, LGDJ, 1990.
DIOP C., DIOUF M. (dir.), Les figures du politiques en Afrique. Des pouvoirs hérités aux pouvoirs élus, Codesria-Karthala, 1999.
DOUENCE M. (dir.), L'exercice des compétences locales entre rationalisation et créativité, L'Harmattan, 2019.
EL MECHAT S., (dir.), *Les administrations coloniales, XIXe-XXe siècles : Esquisse d'une histoire comparée.* Presses universitaires de Rennes, 2009.
FATIN-ROUGE STEFANINI M., GAY L., VIDAL-NAQUET A. (dir.), *L'efficacité de la norme juridique*, Bruxelles, Bruylant, 2012.
FAY C., KONÉ Y. F., QUIMINAL C. (dir.), Décentralisation et pouvoirs en Afrique : En contrepoint, modèles territoriaux français [en ligne]. Marseille : IRD Éditions, 2006.
FERREIRA N., JANICOT L. (dir.), *Quel(s) avenir(s) pour les départements ?* LGDJ, Coll. « LEJEP », 2017.
GAZIBO M., THIRIOT C. (dir.), *Le politique en Afrique : état des débats et pistes de recherche*, Karthala, 2009.

GONOD P., MELLERAY F., YOLKA P. (dir.), *Traité de droit administratif*, Dalloz, 2011.
GUELLEC L. (dir.), *Tocqueville et l'esprit de la démocratie*, Presses de Sciences Po, 2005.
GUIHEUX G. (dir.), La mise en œuvre de la décentralisation : étude comparée France, Belgique, Canada, Bruxelles, Bruylant | LGDJ, 2006.
JAGLIN S., DUBRESSON A. (dir.), Pouvoirs et cités d'Afrique Noire, Décentralisations en question, Karthala, 1993.
KADA N. (dir.),
- *De la réforme territoriale à la réforme de l'État*, PUG, 2011.
- *Les tabous de la décentralisation*, Berger Levrault, coll. « Au fil du débat », 2015

LE POURHIET A.-M., *Droit constitutionnel local*, colloque, Martinique, déc. 1997, coll. « Travaux de l'AFDC », Economica-PUAM, 1999.

LOADA A., WHEATLEY J., (dir.), *Transitions démocratiques en Afrique de l'Ouest. Processus constitutionnels, société civile et institutions démocratiques*, L'Harmattan, coll. « Série politique », 2014.

MATHIEU B., VERPEAUX M. (dir.), *Les mutations constitutionnelles des collectivités territoriales*, Dalloz, coll. « Thèmes & commentaires », 2014.

MONJAL P.-Y., AUBELLE V. (dir.), *La France intercommunale*, L'Harmattan, coll. « GRALE », 2013.

MENY Y., (dir.), *Les politiques de mimétisme institutionnel. La greffe et le rejet*, L'Harmattan, coll. « Logiques politiques », 1993.

MORAND C.-A. (dir.), *Le droit saisi par la mondialisation*, Bruylant, coll. « Droit international » 2001.

MOREAU J., D'ARCY G. (dir.), *La libre administration des collectivités locales. Réflexion sur la décentralisation*, Economica-PUAM, 1984.

NIQUEGE S. (dir.), *Les figures de la loyauté en droit public*, Mare & Martin, 2017.

PAULME D. (dir.), *Classes et Associations d'âge en Afrique de l'Ouest*, Plon, 1971

PARIENTE A. (dir.),

- La séparation des pouvoirs. Théorie contestée, pratique renouvelée, Dalloz, 2007.
- Les chiffres en finances publiques, Mare & Martin, 2019.

PONTIER J.-M., RAPP L. (dir.), Constitution et collectivités territoriales. Réflexions sur le 60e anniversaire de la Constitution, LexisNexis, 2019.

PISSALOUX J.-L., (dir.), *La décentralisation dans les pays francophones d'Afrique de l'Ouest*, L'Harmattan, 2019.

REGOURD S., CARLES J., GUIGNARD D. (dir.), *La décentralisation 30 ans après*, Presses de l'Université Toulouse 1 Capitole, 2013.

ROUSSEAU D. (dir.), *La démocratie continue*, LGDJ-Bruylant. 1995.

SFEZ L. (dir.), *L'objet local*, Paris, Union générale des éditons, 1977.

TOTTE M., DAHOU T., BILLAZ R. (dir.)., *La décentralisation en Afrique de l'Ouest. Entre politique et développement*, Karthala, 2003.

TURK P. (dir.), *Pour un droit constitutionnel des collectivités territoriales : évolutions et débats*, L'Harmattan, 2020.

IV- THESES

AGNAMA-EBOUMI P., *La décentralisation territoriale et le développement local au Gabon*, thèse, Université des sciences sociales de Toulouse, 2004.

AMSELEK P., Perspectives critiques d'une réflexion épistémologique sur la théorie du droit (Essai de phénoménologie juridique), thèse, Paris, 1962, p. 334.

APPIA K.., Recherche sur le contenu de l'autonomie des organes décentralisés. Étude appliquée à l'expérience ivoirienne, thèse, dactyl., Université d'Aix-Marseille, 1985.

AUBY J.-B., La notion de personne publique en droit administratif français, thèse, dactyl., Université de Bordeaux, 1979.

ATANGANA AMOUGOU J-L., *L'Etat et les libertés publiques au Cameroun*, thèse, dactyl., Université Jean Moulin, Lyon 3, 1999.

BACOYANNIS C., Le principe constitutionnel de libre administration des collectivités territoriales, thèse, Economica-PUAM, 1993.

BAUBONNE M., *Les instruments de rationalisation de l'organisation territoriale décentralisée*, thèse, PUAM, 2018.

BARTOLUCCI M., *L'acte plurilatéral en droit public*, thèse, Dalloz, 2022.

BENOIT J., Théorie juridique de la décentralisation administrative, thèse, dactyl., Université Paris, 1990.

BELLINA S., *Droit public et institutionnalisation en situation de pluralisme normatif : le cas de l'Etat malien,* thèse de doctorat en administration publique, dactyl., Université Pierre Mendès France, Grenoble II, décembre 2001.

BETAILLE J., Les conditions juridiques de l'effectivité de la norme en droit public interne : illustrations en droit de l'urbanisme et en droit de l'environnement, thèse, dactly., Université de Limoges, 2012.

BLAIRON K., Le régionalisme et le fédéralisme dans les réformes de l'État italien. Contribution à la définition de la forme d'État italienne, thèse dactyl., Université de Nancy 2, 2005.

BOUET J.-B., *L'administration décentralisée du territoire : choix et perspectives ouverts sous la Cinquième République,* thèse, dactyl., 2006.

CHABROT C., *La centralisation territoriale. Fondements et continuité en Droit public français,* thèse, dactyl., Université de Montpellier I, 1997.

CHAPUISAT J., *La notion d'affaires locales en droit administratif français*, thèse dactyl., Université de Paris II, 1972.

COHENDET M.-A., *L'Épreuve de la cohabitation (mars 1986 – mai 1988)*, thèse, dactyl., Université de Lyon, 1991.

DE MONTALIVET P., Les objectifs de valeur constitutionnelle, Dalloz, 2006.

DIALLO I., Le droit des collectivités locales au Sénégal, thèse, L'Harmattan, 2007.

DIARRA N., *La situation de l'individu face à l'administration malienne. Étude administrative et constitutionnelle*, thèse, dactyl., Université Paris X, 1984.

DIAWARA B., *Territorialisation des politiques publiques et reconstruction de l'État au Mali en termes de gouvernance*, thèse, dactyl., Université Paris-Est, 2019.

DOUENCE M., *Le pouvoir d'organisation du service public*, thèse, dactyl., Université de Pau, 2003.

DURANTHON A., *Subsidiarité et collectivités territoriales. Étude sur la subsidiarisation des rapports entre État et collectivités en droit public français*, thèse, Dalloz, 2017.

DURET E., *Réformes institutionnelles et finances publiques. L'exemple de la décentralisation en Afrique subsaharienne*, Thèse de doctorat en sciences économiques, dactyl., Clermont Ferrand I, 2000.

DUSSART V., L'autonomie des pouvoirs publics constitutionnels, thèse, CNRS, 2000.

DOAT M., *Recherche sur la notion de collectivité locale en droit administratif français,* thèse, LGDJ, 2003.

FAU-NOUGARET M., La conditionnalité démocratique. Étude de l'action des organisations internationales, thèse, dactyl., Université de Bordeaux, 2004.

FAURE B., Le pouvoir règlementaire des collectivités locales, thèse, LGDJ, 1998.

FISCHER B., Les relations entre l'administration et les administrés au Mali. Contribution à l'étude du droit administratif des États d'Afrique subsaharienne de tradition juridique française, thèse dactyl., Université de Grenoble, 2011.

FLOGAÏTIS S., La notion de décentralisation en France, en Allemagne et en Italie, thèse, LGDJ, 1979.

FOGUI J-P., *Autorités traditionnelles et intégration politique au Cameroun*, thèse, dactyl., Université de Bordeaux, 1980.

FOURNIE F., *Recherche sur la décentralisation dans l'œuvre de Maurice HAURIOU*, thèse, LGDJ, 2005.

FORTSAKIS T., *Conceptualisme et empirisme en droit administratif français*, thèse, LGDJ, 1987.

GALLETTI F., *les transformations du droit public africain francophone. Entre étatisme et libéralisation*, thèse, Bruylant, 2004.

INGHLITERRA R. M., La réalisation du droit de la non-discrimination, thèse, LGDJ, 2022,

JANICOT L., Les droits des élus, membres des assemblées des collectivités territoriales, thèse, LGDJ, 2004.

JOYAU M., *De l'autonomie des collectivités territoriales françaises. Essai sur la liberté du pouvoir normatif local*, thèse, LGDJ, 1998.

KAKAI S. H. F., Le vote ethnique au bénin. Contribution à une étude sociopolitique de l'élection, thèse, dactyl., Université d'Abomey-Calavi, 2011.

KAMTO M., *Pouvoir et droit en Afrique noire. Essai sur les fondements du constitutionnalisme dans les Etats d'Afrique noire francophone*, thèse, LGDJ, 1987.

KARZAZI M., *La coopération décentralisée franco-marocaine : entre coopération au développement et diplomatie économique d'influence*, thèse, Université de Cergy Pontoise, 2013.

KEUDJEU DE KEUDJEU J. R., Recherche sur l'autonomie des collectivités territoriales décentralisées au Cameroun, thèse, Université de Douala, 2011.

KOKO A. T., *Les chefferies traditionnelles africaines face à la dynamique des réformes territoriales : contribution à l'étude des processus de décentralisation*, thèse, L'Harmattan, 2019.

LEROY Y., L'effectivité du droit au travers d'un questionnement en droit du travail, thèse, LGDJ, 2011.

LINDITCH F., Recherche sur la personnalité morale en droit administratif, thèse, LGDJ, 1997.

LOUM P., La politique de décentralisation et les nouvelles stratégies de développement : le cas du Sénégal, thèse, dactyl., Université de Montpellier, 2013.

MARRE A., *L'action internationale des collectivités territoriales françaises. La construction d'une coopération interterritoriale et interrégionale dans l'Arc latin et le Maghreb*, thèse, dactyl., Université de Montpellier, 2016.

MELLA E., *Essai sur la nature de la délibération locale*, thèse, LGDJ, 2003.

MEUNIER B., *Les règles relatives aux transferts de compétences entre collectivités publiques*, thèse, dactyl., Université de Clermont I, 2006.

MOURGEON J., *La répression administrative*, thèse, LGDJ, 1967.

MORIO C., *L'administré. Essai sur une légende du droit administratif*, LGDJ, 2021.

MUGNIER F., *La personnalité juridique des collectivités territoriales. Genèse et développement d'une personne morale dans l'État*, Thèse, Dalloz, 2022. **NACH MBACK C.,** *Démocratisation et décentralisation. Genèse et dynamiques comparés des processus de décentralisation en Afrique subsaharienne*, thèse Karthala-PDM, 2003.

NANAKO C., *La libre administration des collectivités territoriales au Bénin et au Niger*, thèse dactyl. Université d'Abomey-Calavi (Bénin), 2016.

NEMBOT M., *Le glas de la fonction publique dans les États d'Afrique francophone. Essai sur la signification d'une institution en quête de légitimité*, thèse, L'Harmattan, 2000.

NGOMO TSIMI L., L'autonomie administrative et financière des collectivités territoriales décentralisées : l'exemple du Cameroun, thèse dactyl., Université Paris-Est, 2010.

NLEP R.-G, *L'administration publique camerounaise. Contribution à l'étude des systèmes africains d'administration publique*, thèse, LGDJ, 1986.

NOUBISSI H. K., Décentralisation et centralisation au Cameroun. La répartition des compétences entre l'État et les collectivités locales, thèse, L'Harmattan, 2013.

NTOLO BEKOA M., *Banque mondiale et développement des pays d'Afrique subsaharienne : l'impact des programmes mis en œuvre au Bénin, au Cameroun et au Togo*, thèse dactyl., Lyon, 2014.

PETITEVILLE F., *La coopération décentralisée. Les collectivités locales dans la coopération Nord - Sud*, thèse, L'Harmattan, 1995.

PONTIER J.-M., L'État et les collectivités locales. La répartition des compétences, thèse, LGDJ, 1978.
PROTIERE G., La puissance territoriale. Contribution à l'étude du droit constitutionnel local, thèse dactyl., Université de Lyon 2, 2006.
REGOURD S., L'acte de tutelle en droit administratif français, thèse, LGDJ, 1982.
RICHARD V., Le droit et l'effectivité – contribution à l'étude d'une notion, thèse, Université Paris II, 2003.
ROMBAUTS-CHABROL T., *L'intérêt public local*, thèse, Dalloz, 2016.
ROUSILLON H., *Les structures territoriales des communes*, thèse, LGDJ, 1972.
RIVERO J., Les mesures d'ordre intérieur administratives, thèse, Sirey, 1934.
SADIO A., *Conditionnalité politique de l'aide publique au développement des partenaires occidentaux à l'Afrique : analyse des actions françaises en Afrique subsaharienne*, thèse, dactyl., Université de Normandie, 2019.
SIMON A., *Les compétences régaliennes et les collectivités territoriales*, thèse, dactyl., Université Grenoble Alpes, 2018.
TADE K., *Impact des projets de coopération décentralisée sur la réduction des inégalités d'accès aux services sociaux de base : Cas des coopérations décentralisées franco-burkinabés et franco-béninoises*, thèse, dactyl., Université Paris-Saclay, 2016.
TARTOUR L., *L'autonomie des collectivités territoriales en droit français*, thèse, LGDJ, 2012.
THALINEAU E., Essai sur la centralisation et la décentralisation. Réflexions à partir de la théorie de Charles Eisenmann, thèse, dactyl., Université de Tours, 1994
TOUERE ELENGA B. L., Fait ethnique, droit public et gouvernance de l'État en Afrique subsaharienne francophone : Contribution à la théorie de l'État africain, thèse, L'Harmattan, 2021.
TOUZEIL-DIVINA M., Éléments d'histoire de l'enseignement du droit public : la contribution du doyen Foucart (1799-1860), thèse, LGDJ, 2007.
TUSSEAU G., *Les normes d'habilitation*, thèse, Dalloz, 2006.
VILLANDRE J., *Les Chefferies traditionnelles en Afrique occidentale française*, thèse, Université de Paris, 1950.

V- ARTICLES DE REVUES, CONTRIBUTIONS ET COMMUNICATIONS

AGONDO OKAWE P.-L., « L'État africain, un État hybride néocolonial », in *L'État africain. Horizon 2000, Mélanges Pierre-François Gonidec*, LGDJ, 1985, p. 56.
AHADZI-NONOU K., « Réflexions sur un tabou du constitutionnalisme négro-africain : le tribalisme », in *Les voyages du droit. Mélanges en l'honneur du professeur Dominique Breillat*, LGDJ, 2011, p. 19.
ALLIOT M.,

- « Un droit nouveau est-il en train de naître en Afrique ? », in *Dynamiques et finalités des droits africains*, Economica, 1976, p. 476.

- « La coutume dans les droits originellement africains », Bulletin de liaison du LAJP, n° 7-8, 1985, p. 79.
- « L'État et la société en Afrique noire, greffes et rejets », Revue française d'histoire d'outre-mer, t. 68, n°250-253, 1981, p. 95.

AUBY J.-B.,
- « Nouvelles de la démocratie administrative », Dr. Adm., n° 4, repère 4, 2010.
- « Droit administratif et démocratie » Dr. Adm., n° 2, 2006, étude 3.
- « Remarques préliminaires sur la démocratie administrative », RFAP, n° 137-138, 2011, p. 13.

AUTEXIER C.,
- « L'ancrage constitutionnel des collectivités de la République », RDP, 1981, p. 595.
- « De la coopération décentralisée », RFDA, 1993, p. 411.

ATANGANA AMOUGOU J-L., « La conditionnalité juridique des aides et respect des droits fondamentaux », Revue numérique Afrilex, n°2, 2001, p. 2.

ARDANT P., « Le néo-colonialisme : thème, mythe et réalité », RFSP, 1965, p. 837.

ASSO B., « De la sacralisation du pouvoir : Essai sur l'Afrique noire animiste », Annales Méditerranéennes d'Histoire et d'Ethnologie, 1980, p.90.

BACOT G., « L'apport de Tocqueville aux idées décentralisatrices », in *Tocqueville et l'esprit de la démocratie*, Presses de Sciences Po, 2005, pp. 203-229.

BALANDIER G., « Réflexions sur le fait politique : le cas des sociétés africaines », Cahiers internationaux de sociologie, 1964, p. 23.

BARELLA X., « La coopération décentralisée à la recherche d'une sécurité juridique renforcée », AJDA, 2008, p.1580.

BAYART J.-F., HIBOU B., SAMUEL B., « L'Afrique "cent ans après les indépendances" : vers quel gouvernement politique ? », Politique africaine, n° 119, 2010, p. 129.

BAZIADOLY S., « La Charte européenne de l'autonomie locale et l'autonomie financière des collectivités locales françaises », RGCT, n° 29, 2003, p. 721.

BECET J.-M., « Généralités. Les grands problèmes relatifs à l'organisation administrative et les choix du législateur de 1982-1983 », in *Droit public*, Economica, 1985, p. 235.

BERGAMASCHI I., DIABATE A., PAUL E., « L'agenda de Paris pour l'efficacité de l'aide. Défis de l'"appropriation" et nouvelles modalités de l'aide au Mali », Afrique contemporaine, n°223-224, 2007, p. 219.

BERIDOGO B., « Processus de décentralisation et pluralité de logiques des acteurs au Mali », in *Décentralisation et pouvoirs en Afrique : En contrepoint,* *modèles territoriaux français* [en ligne]. Marseille : IRD Éditions, 2006 (généré le 28 avril 2021).

BERTUCI G., ALBERTI A., « Le programme des Nations Unies sur l'administration publique : un programme en réinvention pour une réinvention de l'administration publique », RISA, n°2, 2005, p. 360.

BELRHALI-BERNARD H., « la pratique des consultations sur internet par l'administration », RFAP, 2011, p. 181.

BLONDIAUX L., SINTOMER Y., « L'impératif délibératif », *Rue Descartes*, vol. 63, n° 1, 2009, p. 28.

BOLLE S., « La conditionnalité démocratique dans la politique africaine de la France », Revue numérique Afrilex, 2001, p. 1

BOURJOL M.,

- « Libre administration et décentralisation », in *Vertus et limites de la décentralisation*, Les cahiers du droit public, éd. Bergeret, p. 61.
- « La réforme de l'administration territoriale », AJDA, 1992, p. 140.

BOPDA A., « Genèse, mutations et problèmes de la chefferie traditionnelle à Yaoundé », in *Pouvoirs et cités d'Afrique Noire, Décentralisations en question*, Karthala, 1993, p. 247.

BUGE E., MORIO C., « Le Grand débat national : apports et limites pour la participation citoyenne », RDP, n° 5, 2019, p. 1214-1218.

BUGNICOURT J., « Le mimétisme administratif en Afrique : obstacle majeur au développement », RFSP, n°6, 1973, p. 1245.

BOUMAKANI B., « La "bonne gouvernance" et l'État en Afrique », RJPIC, n°1, 2002, p. 21.

BOURDON J., « Le statut général de la fonction publique territoriales », AJDA, 1984, p. 199.

BRANQUART C., « Contrôle de légalité : un réel renouveau ? », AJDA, 2011, p. 198.

BRISSON J.-F.,

- « La loi du 16 décembre 2010 portant réforme territoriale ou le droit des collectivités territoriales en miettes », Dr. Adm., n°3, 2011, p. 4.
- « Les compétences des collectivités territoriales », RFDA, n° 2, 2022, p. 291.
- « Les nouvelles clefs constitutionnelles de la répartition matérielle des compétences entre l'État et les collectivités territoriales », AJDA, 2003, p. 529.
- « Les emplois fonctionnels dans la fonction publique territoriale, un excès de loyauté ? » in, *Les figures de la loyauté en droit public*, Mare & Martin, 2017, p. 181.
- « Synthèse et conclusion », in L'exercice des compétences locales entre rationalisation et créativité, L'Harmattan, 2019, p. 242.

- « La réforme du système d'administration aéroportuaire : bilan et perspectives », in *La réforme aéroportuaire de la loi du 13août 2004à l'épreuve des faits*, L'Harmattan, 2011, p. 153
- « La France est une République indivisible… son organisation est décentralisée », RDP, 2003, p. 111.
- « Clarification des compétences et coordination des acteurs », AJDA, 2014, p. 605.
- « Constitution et compétences locales. Entre théorie introuvable et vaine praxis », in *Constitution et collectivités territoriales. Réflexions sur le 60e anniversaire de la Constitution*, LexisNexis, 2019, p. 51.
- « Décentraliser autrement le territoire », in, *Révolution, constitution, décentralisation.* Mélanges en l'honneur de Michel Verpeaux Dalloz, 2020, p. 539.
- « Les compétences des collectivités territoriales », RFDA, 2022, p. 291.
- « Faut-il inscrire l'expression « *libertés locales* » dans la Constitution ? », in *Pour un droit constitutionnel des collectivités territoriales*, L'Harmattan, 2020 p. 223.
- « Le contrôle de légalité dans la loi du 13 aout 2004. A la recherche des illusions perdues », AJDA, 2005, p. 126.

CABANNES A., ROBBES A., « La coopération décentralisée : comment s'extraire des contingences de l'intérêt public local ? », AJDA, 2003, p. 593.

CABANIS A., GUEYE B., « Dire le droit en Afrique francophone », Droit sénégalais, n° 11, 2013, p. 105.

CABANIS A., MARTIN M.-L., « Décentralisation et pouvoirs locaux dans les constitutions d'Afrique francophone », Politeia, n°6, 2004, p. 75.
CABANIS A., « Le dialogue entre civilisation et culture : le nouveau cadre constitutionnel en Afrique francophone », article consulté sur le site http://www.cmiesi.ma, rubrique notes et analyses, p. 236.
CADOUX C., « La conception française de l'État unitaire (comparaisons) », *Revue de science administrative de la Méditerranée occidentale*, 1er et 2e trimestre 1990, p. 26.
CAILLOSSE J.,

- « Les libertés locales : bréviaire de quelle décentralisation ? » in *Les tabous de la décentralisation*, Berger-Levrault, coll. « Au fil du débat », 2015, p. 61.
- « Éléments pour un bilan juridique de la démocratie locale en France » in, La *démocratie locale. Représentation, participation et espace public*, PUF, 1999, p. 78.
- « Ce que la constitution française donne à voir de la démocratie locale » in *Les collectivités locales*, Mélanges en l'honneur de Jacques Moreaux, Economica, 2002, p. 43.
- « Les élus locaux et la démocratie 'directe' », Pouvoirs locaux, n° 100, I/2014, p. 133.
- « Quelle(s) participation(s) aux affaires communales ? Éléments de réponse du droit français », Pouvoirs, n°95, 2000, p. 121.
- « L'expérience française de la décentralisation et la question de la démocratie locale aujourd'hui », Cahiers CNFPT, n° 37, 1992, p. 70.
- « Éléments pour un bilan juridique de la démocratie locale en France » in, *La démocratie locale. Représentation, participation et espace public*, PUF, 1999, p. 75.

CAMPBELL B., « Stratégies de lutte contre la pauvreté et espaces politiques : quelques interrogations », in *Qu'allons-nous faire des pauvres ?* L'Harmattan, 2005, p. 16.
CAPRON J., « Associations d'âge, économie et pouvoir chez les populations bwa Pweysa », in *Classes et Associations d'âge en Afrique de l'Ouest*, Plon, 1971, p. 33.
CHABROT C.,

- « Démocratie et décentralisation : un couple platonique ? », *Civitas Europa*, n° 11, septembre 2003, p. 3.
- « Une république décentralisée », in Constitution et collectivités territoriales. Réflexions sur le 60e anniversaire de la Constitution, LexisNexis, 2019, p. 29.

CHAMBON M., « Les collectivités territoriales, acteurs de la globalisation du droit administratif ? », RFDA, 2019, p. 994.
CHAMPEILS-DESPLATS V., « La citoyenneté administrative » *Traité de droit administratif*, Dalloz, 2011, p. 413.
CHAPUISAT J., « Libertés locales », AJDA, 1982, p. 382.
CHAVRIER G., « Décentralisation et démocratie locale », *AJDA*, n° 23, 2013, p. 1317.
CHEVALLIER J.,

- « Mondialisation du droit ou droit de la mondialisation ? » in *Le droit saisi par la mondialisation*, Bruylant, 2001, p.39.

- « Le droit administratif vu par la science administrative », AJDA, 2013, p. 401.
- « La reconfiguration de l'administration centrale », RFAP, n°116, 2005, p. 715.
- « La judiciarisation des préceptes managériaux », PMP, n°4, 1993, p. 111.
- « La crise de la démocratie : mythe ou réalité ? », in *Les mutations contemporaines du droit public.* Mélanges en l'honneur de Benoît Jeanneau, 2002, p. 361.
- « La transformation de la relation administrative : mythe ou réalité ? », D., 2000, p. 575.

CAYROL R., « Enjeux de la démocratie participative en France », in *La vie politique. Pour Pascal Perrineau*, Presses de Sciences Po, 2021, p. 99.

CHICOT P.-Y.,
- « Autonomie locale : liberté et responsabilité », in *Un kaléidoscope de l'autonomie locale : théorie, pratique institutionnelle et déclinaisons ultramarines,* Rapport d'information du Sénat, n° 452, 2014, p. 17.
- « Les politiques internationales des collectivités territoriales françaises : une compétence locale atypique ? », in *Les transferts de compétences de l'État aux collectivités locales*, Harmattan 2009, p. 435.

COLARD D., « Une structure supra-communale pour les grandes villes : les communautés urbaines », AJDA, 1967, p. 449.

CONAC G.,
- « Le juge et la construction de l'État de droit en Afrique francophone », in *L'État de droit*, Mélanges en l'honneur de Guy Braibant, Dalloz, 1996, p. 109.
- « La vie du droit en Afrique », in *Dynamiques et finalités des droits africains.* Economica, coll. « La vie du Droit en Afrique », 1980, p. XVI.
- « Les politiques juridiques des États d'Afrique francophone aux lendemains des indépendances », in *Mélanges en l'honneur du Professeur Gustave Peiser*, P.U.G., 1995, p. 145.
- « Le développement administratif des États d'Afrique noire », in *Les institutions administratives des États francophones d'Afrique noire*, Economica, 1979, p. VIII.

CONSTANTINESCO V., « Le principe de subsidiarité : un passage obligé vers l'Union européenne », in *Mélanges Boulouis*, Dalloz, 1991, p. 38

COMBEAU P.,
- « Décentralisation et démocratie », in *L'assemblée délibérante, avenir de la démocratie locale* ? L'Harmattan, 2016, p. 12.
- « Transferts de compétences et fonctions publiques », in *Les transferts de compétences de l'État aux collectivités*, L'Harmattan, 2016, p. 494.

COQUERY-VIDROVITCH C.,

- « Les débats actuels en histoire de la colonisation », Revue Tiers Monde, n°112, 1987, p. 784.
- « Les systèmes politiques précoloniaux jusqu'en 1945 », in *L'Afrique noire de 1800 à nos jours*, PUF, 2005, p. 67.
- « Les structures du pouvoir et la communauté rurale précoloniale », Revue française d'histoire d'Outre-mer, 1981, p. 54.

COULIBALY B., « Le juge administratif, rempart de protection des citoyens contre l'administration en Afrique noire francophone ? », Revue numérique Afrilex, 2013, p. 1.

DEBBASCH R., ROUX A., « L'indivisibilité de la République », in B. MATHIEU ET M. VERPEAUX (dir.), *La République en droit français*, Economica, 1996, p. 75.

DARBON D.,
- « L'État prédateur », Politique africaine, n° 39, 1990, p. 45.
- « L'"aventure ambiguë" des administrations africaines de succession française dites francophones », RIPC, n°2, 1994, p. 211.

DARGENT F., « Les échecs du mimétisme constitutionnel en Afrique noire francophone », RDP, n°5, 2017, p. 1347.**DECHAUX R.**, « Les garanties constitutionnelles de l'autonomie financière locale à l'épreuve des concours financiers étatiques », RDP, n° 2, 2010, p. 349.

DECOTTIGNIES R., « La personnalité morale en Afrique noire », Annales Africaines, 1958, p. 31.

DELAVIGNETTE R., « Des commandants français aux préfets africains », RFFPA, n°2, 1966, p. 30.

DELCAMP A., « Principe de subsidiarité et décentralisation », RFDC, 1995, n° 23, p. 616.

DELMAS-MARTY M., MANACORDA S., « La corruption : un défi pour l'État de droit et la société démocratique », RSC, 1997, p. 696.

DELMAS-MARTY M., « Les processus de mondialisation du droit », in MORAND C.-A., *Le droit saisi par la mondialisation*, Bruylant, coll. « Droit international », p. 72.

DELPERE F., « Organisation territoriale et exception française », Cahiers français, n°318, 2004, p. 15.

DE MONTALIVET P., « QPC et droit des collectivités territoriales », AJDA, 2016, p. 586.

DESMOULIN G., « La recherche de la performance des politiques publiques. De l'illusion à la raison ? », AJDA, 2013, p. 894.

DEROSIER J.-P, « La dialectique centralisation/décentralisation. Recherches sur le caractère dynamique du principe de subsidiarité », RIDC, n°1, 2007, p. 137 et s.

DIALLO I., « L'avenir du déféré préfectoral en droit français », AJDA, 2005, p. 2438

DIOUF M., « L'idée municipale en Afrique. Une idée neuve », Politique africaine, n° 74, 1999, p. 13.

DIOP A. M., « Les Politiques de développement élaborées par la Banque Mondiale à l'Intention des Pays de l'Hémisphère Sud de 1980 à nos jours

: Origines, variantes et implications. » Africa Development / Afrique Et Développement, 2016, p. 180.

DHONDT J., « Petit-Dutaillis et les communes françaises », Annales Économies, sociétés, civilisations, n° 3, 1952. p. 378.

DOAT M., « Vers une conception a-centralisée de l'organisation de la France », RDP, 2003, p. 115.

DONIER V., « Les réformes des collectivités territoriales à l'épreuve de la Charte européenne de l'autonomie locale », in *Mélanges Gérard Marcou,* IRJS Éditions, 2017, p. 169.

DOUENCE J.-C., « le fondement constitutionnel du déféré préfectoral », RFDA, 1997, p. 1200.

DOUENCE M.,

- « La confortation du *statu quo* de la coopération intercommunale ou "beaucoup de bruit pour rien" » ?, *RFDA*, 2020, p. 223.
- « Où en est la jurisprudence du conseil constitutionnel sur le principe de libre administration des collectivités territoriales ? », in *Mélanges Gérard Marcou*, IRJS Éditions, 2017, p. 195.
- « La Corse », RFDA, 2016, p. 645.

DOUMBE-BILLE S., « L'élection en droit administratif », RDP, 1992, p. 1065.

DU BOIS DE GAUDUSSON J.,

- « La décentralisation menacée par la (bonne) gouvernance », in *Mélanges Franck Moderne*, Dalloz, 2004, p.995.
- « Le mimétisme postcolonial, et après ?», Pouvoirs, n° 129, 2009, p. 45.
- « L'État français et le contrôle des actions extérieures des collectivités territoriales », in *Le droit appliqué à la coopération interrégionale en Europ*e, LGDJ, 1995, p. 227.

DUBREUIL C.-A.,

- « L'e-démocratie au service de la démocratie locale », AJDA, 2018, p. 1383.
- « La démocratie et la transparence », RFDA, 2016, p. 655.
- « Réflexions sur la démocratie locale en France », Siècles [En ligne], 37 | 2013, mis en ligne le 11 décembre 2013, consulté le 10 octobre 2020. URL : http://journals.openedition.org/siecles/1225.
- « Libre administration des collectivités territoriales et liberté fondamentale », JCP G, 2012, p. 2048.

DURAND P.,

- « La Charte européenne de l'autonomie locale », BJCL, 2006, p. 628.
- « Le 30ème anniversaire de la charte européenne de l'autonomie locale », AJDA, 2016, p. 2313.

DUPUY F., « Pour le retour de la fonction publique dans le droit commun », JCP A, 2003, p. 320.

DRAGO G.,

- « QPC et droit des collectivités territoriales : premier bilan », JCP A, 2011, p. 221.
- « Le principe de subsidiarité comme principe de droit constitutionnel », RIDC, 1994, p. 583.

EISENMANN (C).,

- « Essai d'une classification théorique des formes politiques », Politique, n°41-44, 1968, p. 5.
- « Quelques problèmes de méthodologie des définitions et des classifications en sciences juridiques », Archives de philosophie du droit, vol. XI, 1966, p. 37.
- « Les structures de l'administration », in *Traité de science administrative*, Mouton, 1966, p. 317.

ESCLASSAN M., « Les ambiguïtés de la fiscalité locale : Une fiscalité en trompe-l'œil », in, *La territorialisation du droit financier, Revue française de finances publiques*, n°129, 2015, p. 27.

ETHIER D., « La conditionnalité démocratique des Agences d'aide et de l'Union européenne », Études internationales, 2001, Vol.32 n°3, p. 495.

FALL I. M., « Le contrôle de légalité des actes des collectivités locales au Sénégal », Revue numérique Afrilex, n° 5, p. 82.

FAU NOUGARET M., « Originalité et convergence des phénomènes de décentralisation en Afrique Sub-Saharienne », Revue électronique Afrilex, 2009, p. 1.

FAURE B.,

- « Différenciation », AJDA, 2022, p. 1367.
- « Différenciation des normes : jeu de dupes », RFDA, 2021, p. 313.
- « Démocratie municipale et démocratie de quartier », in *L'assemblée délibérante, avenir de la démocratie locale* ? L'Harmattan, 2016, p. 99.
- « Le leadership régional : nouvelle orientation du droit des collectivités territoriales », AJDA 2015, p. 1898.
- « La démocratie locale doit s'exercer dans l'intérêt général », in *L'intérêt général*, Mélanges en l'honneur de Didier TRUCHET, Dalloz, 2015, p. 204.
- « Règlements locaux et règlements nationaux », Les Nouveaux cahiers du Conseil constitutionnel, n° 42, 2014/1, p. 43.
- « Penser le changement ou changer le pansement ? », AJDA, 2014, p.600.
- « Les deux conceptions de la démocratie administrative », RFDA, 2013, p. 709.

- « La glorieuse trentenaire. A propos du 30e anniversaire de la loi du 2 mars 1982 », AJDA, 2012, p. 738.
- « Réforme constitutionnelle et décentralisation : des slogans font loi », RDP, 2003, n°1, p. 122.
- « Peut-on ne pas affirmer un particularismes du Droit des collectivités locales ? », in *Les collectivités locales et le droit. Les mutations actuelles*, PUF, 2001, p. 13.
- « Les objectifs à valeur constitutionnelle, une nouvelle catégorie juridique ? », RFDC 1995, p. 47.

FAVOREU L., ROUX A., « La libre administration des collectivités territoriales est-elle une liberté fondamentale ? », Les cahiers du conseil constitutionnel, 2002, n° 12, p. 88.

FAVOREU L.,
- « Les bases constitutionnelles du droit des collectivités locales », in *La nouvelle décentralisation*, Sirey, 1983, p. 83.
- « Libre administration et principes constitutionnels », in *La libre administration des collectivités locales*, ECONOMICA, PUAM, 1984, p. 63.
- « Les sources du droit de la fonction publique territoriale », RFDA, 1985, p. 309.

FERSTENBERT J., « Droit et pratique de l'action extérieure des collectivités territoriales », Revue du centre ouest n° 1, 1988, p.16.

FIALAIRE J., « L'organisation des métropoles dans les États d'Afrique noire francophones », *in Mélanges en l'honneur de Gérard Marcou*, IRJS Éditions, 2017, p. 211.

FLORY M., « Souveraineté des États et coopération pour le développement », in Recueil des cours de l'Académie de droit international, 1974, I, pp. 276-302.

FLÜCKIGER A., « L'évaluation législative ou comment mesurer l'efficacité des lois », Revue européenne des sciences sociales, n° 138, p. 86.

FONBAUSTIER L., « La participation du public », AJDA, 2015, p. 517.

FOUCHER V., SMITH E., « Les aventures ambiguës du pouvoir traditionnel dans l'Afrique contemporaine », Revue internationale et stratégique, n° 81, 2011, p. 33.

FRISON-ROCHE M.-A.,
- « L'efficacité des décisions en matière de concurrence : notions, critères, typologie », LPA, n° 259, 2000, p. 4.
- « Le système juridique français constitue-t-il un atout ou un handicap pour nos entreprises et nos territoires ? », in *Réformes et transformations*, PUF, 2018, p. 431.

GAZIBO M., THIRIOT C., « Le politique en Afrique dans la longue durée : historicité et héritages », in *Le politique en Afrique : état des débats et pistes de recherche*, Karthala, 2009, p. 27.

GBEOU-KPAYILE N.,
- « L'idée de la constitution en Afrique francophone », Revue numérique Afrilex, 2014, p. 1.

- « Essai sur le financement des collectivités locales dans les États d'Afrique noire francophone : Les cas du Sénégal et du Togo », Revue béninoise des sciences juridiques et administratives, n°33, 2014, p. 76.

GERARD P., « Décentralisation : le déféré délaissé », Dr. Adm., 1992, n°1, p. 1

GESCHIERE P., « Le poids de l'histoire », in *Les Afriques politiques*, La découverte, 1991, p. 29.

GOEH-AKUE N. A., « Relations entre autorités traditionnelles et pouvoir public moderne au Togo : repères, atouts et limites au développement local », Revue CAMES, série B, vol. 1, 1999, p. 48.

GOHIN O., « Semper abset. La prétendue clause de compétence générale des collectivités territoriales », in *Pour un droit constitutionnel des collectivités territoriales. Évolutions et débats*, L'Harmattan, 2020, p. 214.

GONIDEC P.-F., « Traditionalisme et modernisme en matière d'institutions publiques en Afrique noire », RJPIC, n° 1, 1966, p. 77.

HAGUENAU-MOIZARD C., MONTALIEU T., « L'évolution du partenariat U.E.-A.C.P. de Lomé à Cotonou : de l'exception à la normalisation », Mondes en développement, n°128, 2004/4, p. 65.

HAMADOU A., « Le principe de libre administration des collectivités territoriales au Niger », Annales africaines, 2020, p. 395.

HAURIOU M., « Décentralisation », in *Répertoire du droit administratif*, éd. Paul Dumont, 1891, p. 482.

HERTZOG R.,

- « L'autonomie en droit : trop de sens, trop peu de signification », in *Mélanges offerts à Paul Amselek*, Bruylant, 2005, p. 446.
- « Le système financier local en France : la décentralisation n'est pas le fédéralisme », RIDC, n° 2, 2002, p. 613.
- « Linéament d'une théorie des subventions », *RFFP*, n° 23, 1988, p. 6.
- « L'ambiguë constitutionnalisation des finances locales », AJDA, 2003, p. 548.

HEUSCHLING L., « Effectivité, efficacité, efficience et qualité », in *L'efficacité de la norme juridique*, Bruxelles, Bruylant, 2012, p. 59.

HUGON P., « Les effets des politiques d'ajustement sur les structures politiques africaines », in G. CONAC (dir.), *L'Afrique en transition vers la démocratie*, Economica, 1993, p. 89 et s.

HOLO T., « La décentralisation au Bénin : mythe ou réalité ? », Revue béninoise des sciences juridiques et administratives, n° 7, 1986, p. 1.

JACQUENOT P., « Chefferies et décentralisation au Ghana », Afrique contemporaine, vol. 221, n°1, 2007, p. 55.

JACQUINOT N., « La coopération décentralisée : entre facilité et complexité », *in Études en l'honneur du Professeur J-A. Mazères*, Litec 2009, p. 392

JAN P., « La responsabilité politique des exécutifs locaux, nécessité ou absurdité ? », in J *L'assemblée délibérante, avenir de la démocratie locale ?*, *op.cit.*, p. 264.

JANICOT L.,
- « Réformer sans réviser », AJDA, 2021, p. 1375.
- « La demande de prise de position formelle - L'extension du rescrit au profit des collectivités territoriales », RFDA, 2020, p. 254.
- « Les départements : quel(s) avenir (s) », AJDA, 2019, p. 2417.
- « Le pouvoir normatif des régions », RFDA, 2016, p. 664.
- « La tutelle de l'État sur les collectivités territoriales trente après la loi du 2 mars 1982 », AJDA, 2012, p. 753.
- « Réflexions sur la notion de compétences propres appliquée aux collectivités territoriales en droit français », AJDA, 2004, p.1574.
- « Les limites de l'information que peuvent exiger les conseillers municipaux », ADJA, 2003, p. 787.

JAUME L., SOLEIL S., « Centralisation/Décentralisation. Retour sur quelques certitudes historiques », AJDA, 2005, p. 760.

JEBEILI C., « La coopération décentralisée », in *La décentralisation 30 ans après.* Presses de l'Université Toulouse 1 Capitole, 2013. p. 307.

JEGOUZO Y.,
- « La réforme des enquêtes publiques et la mise en œuvre du principe de participation », AJDA, n° 32, 2010, p. 1812.
- « De la 'participation du public' à la 'démocratie participative' » ? AJDA, n° 42, 2006, p. 2314.

KADA N.,
- « La réforme de l'état territorial », RFAP, n° 141, 2012, p.109.
- « L'intercommunalité sous les feux de la rampe », RDP, n° 5, 2015, p. 1187.
- « Le préfet est mort, vive le préfet », in *La décentralisation 30 ans après,* Presses de l'Université Toulouse 1 Capitole, 2013, p. 95.

KIEMDE P., « Le bicaméralisme en Afrique et au Burkina Faso », Revue Burkinabé de droit, n° 2, 1992, p. 27.

KEUDJEU DE KEUDJEU J.R., « La gestion des grandes cités en Afrique subsaharienne francophone : Le cas des agglomérations urbaines à statut particulier du Burkina Faso, du Cameroun, du Congo, du Mali et du Sénégal », Revue numérique Afrilex, 2017, p. 1.

KEUTCHEU J., « L' "ingérence démocratique" en Afrique comme institution, dispositif et scène », Études internationales, vol. 45, n° 3, 2014, p. 425.

KOUBI G., « La délibération, manifestation de volonté dans le droit des collectivités locales » LPA, n° 71, 1992, p. 6.

KPODAR A., « Prolégomènes à une virée constitutionnelle en Afrique noire francophone : une approche de théorie juridique », in *Les voyages du droit*, Mélanges en l'honneur de Dominique Breillat, LGDJ, p. 331.

LACHAUME J.-F.,
- « Conditions de légalité de la révocation d'un maire », AJDA, 2014, p. 1384.
- « Remarques sur le contrôle a posteriori de la légalité des actes des autorités locales décentralisées », AJDA, p. 536.

LANGROD G., « Genèse et conséquences du mimétisme administratif en Afrique », Revue Internationale des Sciences Administratives vol. 39, n° 21, 1973, p. 121.

LAPORTE G., « L'après Cotonou : de la coopération Nord-Sud traditionnelle à un partenariat d'intérêts partagés ? », Revue de l'UE ,2019, p. 401.

LAQUIEZE A., « La décentralisation chez les libéraux de la première moitie du XIXe siècle », in *Décentraliser en France. Idéologies, histoire et prospective*, Paris, éd. F.-X. de Guibert, 2003, p. 53.

LASCOUMES P., SERVERIN E., « Théories et pratiques de l'effectivité du droit », *Droit et Société*, n° 2, 1986, p. 127.

LASCOUMES P., « effectivité », in A.-J. ARNAUD (dir.), *Dictionnaire encyclopédique de théorie et de sociologie du droit*, LGDJ, Coll. « Anthologie du droit », 2e éd., corrigée et augmentée, 2018, p. 217.

LE CHATELIER G., « Le débat sur la clause générale de compétence est-il vraiment utile ? », AJDA, 2009, p. 186.

LEGIER H.J., « Institutions municipales et politique coloniale : les Communes du Sénégal », Revue française d'histoire d'outre-mer, n°20, 1968, p. 414.

LERIQUE F., « La métropole : nouvel avatar de l'intercommunalité ou institutionnalisation de la ville ? », in *La France intercommunale*, L'Harmattan, coll. « GRALE », 2013, p. 343.

LEROY Y., « La notion d'effectivité en droit », Droit et société, n° 70, 3/2011, p. 730.

LE ROY E.,
- « Gouvernance et décentralisation ou le dilemme de la légitimité de la réforme de l'Etat africain à la fin du XIXe siècle », in *L'Etat en Afrique : indigénisations et modernités*, Cahier du Gemdev, 1996, p. 97.
- « L'expertise internationale en Afrique, le cas de l'expertise sur les questions juridiques sur les questions foncières », in *Bulletin de l'APAD*, vol 2, pp. 16-25.
- « Les usages politiques du droit », in C. COULON ; D-C. MARTIN, *Les Afriques politiques*, La Découverte, Paris 1991, p. 112.

LOCHAK D., « Le droit administratif, rempart contre l'arbitraire ? », Pouvoirs, n°46, 1988, pp. 44-55.

MADIOT Y., « Les techniques de correction de la répartition des compétences entre collectivités territoriales », RFDA, 1996, p. 966.

MAGNON X., « Bref exercice de méthodologie juridique pratique. Que dire du principe de « libre administration » des collectivités territoriales ? », in *Révolution, constitution, décentralisation*, Mélanges en l'honneur de Michel Verpeaux, Dalloz, 2020, p. 636.

MANIN B., « Volonté générale ou délibération. Esquisse d'une théorie générale de la délibération politique », Revue Le Débat, n° 33, 1985.

MANSON S., « Assemblées locales : l'impuissance délibérative. Pour une délibération locale ranimée », Pouvoirs locaux, n°101, 11/2014, p. 89.

MANAÏ D., PERRIN J., « Le droit entre le mythe, la raison et la réalité sociale », L'Année Sociologique 1940/1948, vol. 31, 1981, p. 403.

MARCHAND J., « Réflexions sur le principe de transparence », RDP, n° 3, 2014, p. 677.

MARCOU G.,

- « L'autonomie communale : étude comparative », Pouvoirs, n°95, 2000, p. 69.
- « Les collectivités locales dans les constitutions des États unitaires d'Europe », Les Nouveaux cahiers du Conseil constitutionnel, 2014, n° 74, p. 63.
- « La démocratie locale en France : aspects juridiques », in *La démocratie locale. Représentation, participation et espace public*, PUF, 1999, p. 41.
- « Le paradoxe de la région », AJDA, 2008, p. 1634.

MARGUERAT Y., « L'État et l'organisation territoriale du Togo », Afrique Contemporaine, n°145, p. 47.

MARIE J., IDELMAN E., « La décentralisation en Afrique de l'Ouest : une révolution dans les gouvernances locales ? », *EchoGéo* [Online], 13 | 2010, En ligne, connection on 23 February 2021. URL : http://journals.openedition.org/echogeo/12001

MARILLIA G.-D., « Le détachement sur un emploi fonctionnel », AJFP, 1997, p. 43.

MASSINA P., « Plaidoyer pour le fonctionnement de la juridiction administrative au Togo », Revue des droits des pays d'Afrique, 1990, p. 403.

MAZERES J.-A.,

- « Les collectivités locales et la représentation. Essai de problématique élémentaire », RDP, 1990, p. 607.
- « Débat : le processus de décentralisation », AJDA, 1992, p. 6.
- « Essai d'analyse archéologique de la décentralisation », Les Cahiers du L.E.R.A.S.S. n° 21, 1990, p. 91.

MEDE N., « L'autonomie « retenue » : étude sur le principe de libre administration des collectivités territoriales en Afrique de l'Ouest francophone », Revue juridique et politique des États francophones, vol. 62, n° 2, 2008, p. 188.

MEDARD J.-F., « Les paradoxes de la corruption institutionnalisée », Revue internationale de politique comparée, 2006, p. 701.

MELIN-SOUCRAMANIEN F., « Le principe d'égalité entre collectivités locales », Les cahiers du Conseil constitutionnel. 2002, n° 12, p. 93.

MELLERAY F.,

- « Le code général de la fonction publique : une arlésienne », AJFP, 2019, p. 309.
- « L'imitation des modèles étrangers en droit administratif », AJDA, 2004, p.1224.

MERCIER P., « L'organisation sociale et les formes de groupement. Problèmes de classification », in *Ethnologie régionale*, tome I, Paris, Gallimard, 1972, p. 958.

MERLE I., « De la "légalisation" de la violence en contexte colonial. Le régime de l'indigénat en question », Revue des sciences sociales du politique, 2004, n°66, vol. 17, p. 137.

MEISEL N., OULD AOUDIA J., « L'insaisissable relation entre "bonne gouvernance" et développement », Revue économique, 2008, Vol. 59 n°6, p 1159.

MICHALON T.,

- « Autoritarisme et hypercentralisation : l'État africain contre les libertés », Revue africaine de droit international comparé, 1991, p. 76.
- « Quel État pour l'Afrique ? », Présence africaine, vol. 3, n° 107, 1978, p. 18.
-

MICHEL H., « Promesses et usages des dispositifs de transparence : entre approfondissement et redéfinition de la démocratie », Revue française d'administration publique, vol. 165, n° 1, 2018, p. 5.

MINFEGUE C., « La participation citoyenne en contexte local au Cameroun. Entre demande de légitimité, efficacité versatile et tensions politiciennes », African Sociological Review, vol.20, n°2, 2016, p. 82.

MILLARD E., MATZNER E., « Note des traducteurs », in *Introduction à l'empirisme juridique*, trad., E. Millard et E. Matzner, présenté et annoté par É. Millard, LGDJ, 2004, p. 11.

MILLARD E.,

- « Deux lectures critiques d'Alf Ross », in *Théories réalistes du droit*, Presses universitaires de Strasbourg, 2000, p. 13.
- « Effectivité des droits de l'Homme », in *Dictionnaire des Droits de l'Homme*, PUF, 2008, p. 349.

MUIR-WATT H., « La fonction subversive du droit comparé », *RIDC*, n° 3, 2000, p. 503.

MULLER-QUOY I.,

- « Le renouveau de la notion juridique d'agglomération », AJDA, 20 Mars 2000, p.197.
- « Le département, une collectivité à abattre ? », BJCL, 2013, p. 640.

MUSCAT H., « La décentralisation des personnels en France et le statut général de la fonction publique », in *La mise en œuvre de la décentralisation : étude comparée France, Belgique, Canada, Bruxelles*, Bruylant |LGDJ, 2006, p. 237.

MONNIER S., « Un cadre juridique a minima pour le médiateur territorial », JCP A, 2020, n° n° 5, p. 23.

MONEMBOU C., « Le pouvoir règlementaire des collectivités locales dans les États d'Afrique noire francophone », Revue CAMES/SJP, n°002/2015, p. 79.

MOREAU J., « Le droit à l'information des élus locaux et des citoyens », in *La démocratie locale. Représentation, participation et espace public*, PUF, 1999, p. 141.

MORIO C., « Quel modèle de participation dans les projets de réforme institutionnelle ? », *Archives de philosophie du droit*, vol. 62, n° 1, 2020, p. 439.

MUKA TSHIBENDE L.-D., « Les Gaulois, nos ancêtres ? Sur la circulation et l'influence du modèle juridique français en Afrique noire francophone », Revue de la recherche juridique. Droit prospectif, n°2 006-1, XXXI-112, p. 384.

NACH MBACK C., « La chefferie traditionnelle au Cameroun : ambiguïtés Juridiques et dérives politiques » Afrique et Développement, vol. 25, n° 3-4, 2000, p. 77.

NGAMPIO-OBELE-BELE U., « Brèves réflexions sur la gestion administrative de la crise du Coronavirus en Afrique francophone », RDSS, 2022, p. 73.
NLEP R. G.,

- « Le juge de l'administration et les normes internes, constitutionnelles ou infra constitutionnelles en matière de droits de fondamentaux », Solon, vol. 1, 1999, p. 138.
- « Administration publique et sociétés plurales en Afrique », RFAP, n° 94, 2001, p. 276.

OST F., « Science du droit », in A.-J. ARNAUD (dir.), *Dictionnaire encyclopédique de théorie et de sociologie du droit*, LGDJ, Coll. « Anthologie du droit », 2e éd., corrigée et augmentée, 2018, p. 543 :
PAGNOU S.,

- « La place du contrôle de légalité dans les processus de bonne gouvernance liés à la décentralisation. Le cas du Togo », Revue internationale des sciences administratives, vol. 79, 2013/3, p. 605.
- « Les données financières de l'État togolais », in *Les chiffres en finances publiques*, Mare & martin, coll. « Droit & gestion publiques », p. 77.

PEKASSA NDAM G.-M., « La classification des communes au Cameroun », Revue africaine des sciences juridiques, vol. 6, n°1, 2009, p. 266.
PARIENTE A., « Le mythe de l'autonomie financière », RFPF, n° 129, 2015, p. 15.
PASQUIER R.,

- « La régionalisation française revisitée. Fédéralistes, mouvement régional et élites modernisatrices, RFSP, n° 1, 2003, p. 101.
- « Les régions dans la réforme territoriale : vers un fédéralisme à la française », AJCT, 2016, p. 71.

PERRIN B.,

- « Le cadre juridique des emplois discrétionnaires de la fonction publique territoriale », AJDA, 1989, p. 147.
- « La revalorisation des emplois de direction », Cah. fonct. publ. 2001, p. 27.
- « Le nouveau cadre juridique des emplois fonctionnels », Cah. fonct. publ. 2002. p. 32.

PONTIER J.-M.,

- « Les contrôles et la tutelle », AJDA, 2022, p. 1311.
- « La notion de compétences régaliennes dans la problématique de la répartition des compétences entre les collectivités publiques », RDP, 2003, p. 193.
- « Autonomie locale, libre administration, décentralisation », in *Itinéraire d'un constitutionnaliste*, Mélanges en l'honneur de Francis DELPÉRÉE, Bruylant, 2007, p. 1221.
- « Charte de l'autonomie locale : enfin ? », *AJDA*, 2006, p. 1577.

- « Actualité, continuité et difficultés des transferts de compétences entre l'État et les collectivités territoriales », RFDA 2003, p. 35.
- « Les principes de répartition des compétences », in Les nouvelles compétences locales : de la loi du 7 janvier 1983 à la loi du 25 janvier 1985, Economica, 1985, p. 4.
- « Semper Manet. Sur une clause générale de compétence », RDP, 1984, p. 1443.
- « L'intercommunalité, hier, aujourd'hui et demain », JCP A, 2010, 2240.
- « La décentralisation et le temps », RDP, 1991, p. 1218.
- « Sur les notions controversées : Ressources propres, Ensemble de ressources, Part déterminante », Revue administrative, n° 340, 2004, p. 397.
- « Une approche historique », in *Quel(s) avenir(s) pour les départements ?* LGDJ, Coll. « LEJEP », 2017, p. 14.

PONTHOREAU M-C.,

- « Trois interprétations de la globalisation juridique. Approche critique des mutations du droit public », AJDA, 2006, p. 20.
- « Le droit comparé en question(s). Entre pragmatisme et outil épistémologique », *RIDC* n° 1, 2005, p. 8.

PLESSIX B.,

- « Le pouvoir de substitution d'action » in *Le contrôle de l'État sur les collectivités territoriales aujourd'hui*, L'Harmattan, 2007, p. 67.
- « Une prérogative de puissance publique méconnue : le pouvoir de substitution d'action », RDP, 2003, p. 579.

PRELOT M., « La signification constitutionnelle su second empire », Revue française de science politique, n°1, 1953, p. 33.

PRIET F., « Sur une notion désormais constitutionnalisée : la décentralisation », in *Terres du droit*, Mélanges en l'honneur d'Y Jégouzo, Dalloz, 2009, p. 249.

RAINAUD J.-M., « Réflexions sur le droit administratif du développement », in *Les grands services publics dans les États d'Afrique noire*, Economica,1984, p. 62.

RANGEON F., « Réflexions sur l'effectivité du droit », in *Les usages sociaux du droit*, PUF, 1989, p. 128.

REGOURD (S.),

- « L'autonomie communale : des vicissitudes de la théorie juridique à l'enseignement de l'histoire », Pouvoirs locaux, n° 1, 1989, p. 73.
- « La génétique de la décentralisation. Tabous, métaphores et autres occultations », in *Les tabous de la décentralisation*, Berger-Levrault, coll. « Au fil du débat », 2015, p. 15.
- « De la décentralisation dans ses rapports avec la démocratie. Genèse d'une problématique », RDP, 1990, p. 964.

REIGNIER D., « La responsabilité de l'exécutif devant l'assemblée délibérante, les enseignements du laboratoire polynésien », in *L'assemblée délibérante, avenir de la démocratie locale ?* L'Harmattan, 2016, p. 229.

RENARD P., « Les emplois fonctionnels », Gazette des communes, 1998, p. 38.

RIVERO J., « Les phénomènes d'imitation des modèles étrangers en droit administratif », *in Miscellanea W.J. Ganshof Van Der Meersch*, Bruylant, 1972, tome 3, p. 619.

ROCHER G., « L'effectivité du droit », in *Théorie et émergence du droit : pluralisme, surdétermination et effectivité*, Montréal, Thémis, 2004, p. 135.

ROIG C., « Théorie et réalité de la décentralisation », RFSP, n° 3, 1966, p. 452.

ROUSSEL S., « Le juge et la régularisation d'une déclaration d'utilité publique », *RFDA*, n°5, 2021, p. 932.

ROUSSEAU D.,

- « La démocratie continue : fondements constitutionnels et institutions d'une action continuelle des citoyens », Confluence des droits _La revue [En ligne], 02 | 2020, mis en ligne le 11 février 2020. URL : https://confluencedesdroits-larevue.com/?p=726 .
- « Conceptualiser la compétence normative des citoyens », Archives de philosophie du droit, vol. 62, n°1, 2020, pp. 427.

ROUX A., « Le principe constitutionnel de la libre administration des collectivités territoriales », RFDA, 1992, p. 435.

SAADA E., « Citoyens et sujets de l'Empire français. Les usages du droit en situation coloniale », Genèses, n° 53, 2003, p. 13.

SAIDOU A. K., « La participation citoyenne dans les politiques publiques de sécurité en Afrique : analyse comparative des exemples du Burkina Faso et du Niger », *Revue internationale de politique de développement* [Online], 11.1 | 2019, URL : http://journals.openedition.org/poldev/3209

SAMAKE A., « Les contraintes liées à l'autonomie financière des collectivités territoriales dans l'espace UEMOA. Les cas du Bénin, Mali, Niger, Sénégal », in *Les finances publiques entre globalisation et dynamiques locales,* Mélanges en l'honneur de Eloi DIARRA et Salifou YONABA, L'Harmattan, 2021, p. 431.

SARDAN P., « Démocratie et décentralisation », in *Études offertes à Jean-Marie Auby*, Dalloz, 1992, p. 294.

SAVONNET-GUYOT C., « La communauté villageoise comme système politique : un modèle ouest-africain », RFSP, n° 6, 1975, p. 1114.

SCHAEFFER E., « Les rapports du droit et du développement », Connaissance de l'Afrique, n° 43, 1973, p. 7.

SCHWARTZ, R. « Le statut de la fonction publique territoriale : vers plus d'unité et d'uniformisme ?», La gazette, Cahier détaché n° 2-18/1692, 2003, p. 216.

SEILLER B.,

- « Le pouvoir disciplinaire sur les maires », AJDA, 2004, p.1637.
- « Les conseils donnés par les personnes publiques », Dr. adm., n° 11, 2005, Étude 18.
- « La contestation du refus d'exercer le déféré préfectoral », RDP, 1992, p. 1149.

SEVERINO, J-M., « Construire les capacités institutionnelles dans les pays en développement : le nouveau défi de la coopération administrative dans la mondialisation », RFAP, 2001, p. 679.

SINDJOUN L., « Présentation générale : éléments pour une problématique de la révolution passive », in *La révolution passive*, Dakar, CODESRIA, 1999, p. 3.

SOHOUENOU M. E.,

- « Contribution à l'élaboration d'une échelle de la décentralisation territoriale en Afrique noire francophone », Revue béninoise des sciences juridiques et administratives, n° 22, 2009, p. 148.
- « L'administration territoriale béninoise et le droit supranational de la décentralisation », in *Leçon inaugurale,*
- Colloque international sur l'effectivité de la décentralisation au Bénin, janvier 2015, p. 2.

SOUSSE M., « La notion financière de concours », RFFP, n° 53, 1996, pp. 157 à 158.

SOME M., « État, pouvoir et démocratie en Afrique de l'Ouest contemporaine : les héritages du passé », in *Transitions démocratiques en Afrique de l'Ouest. Processus constitutionnels, société civile et institutions démocratiques*, L'Harmattan, coll. « Série politique », 2014, p. 53.

STANGOS P., « La conditionnalité politique, en termes de protection des Droits de l'homme, de démocratie et d'État de droit, des relations économiques extérieures de la C.E. et de l'U.E. », in *L'effectivité des Organisations internationales : mécanismes de suivi et de contrôle*, A. Pédone, 2000, p. 273.

TAILLEFAIT A., « La loi de fonction publique du 6 aout 2019 de déformations des transitions professionnelles », RDP, 2021, p. 77.

TARTOUR L., « L'intercommunalité, source de marges de manœuvre financière », RFFP, n° 119, 2012, p. 35.

THOENIG J.-C., « La décentralisation, dix ans après », Pouvoirs, n° 60, 1992, p. 5.

THOMAS L.V., « Réflexions à propos des mythes d'Afrique noire », in *L'autre et l'ailleurs. Hommages à Roger Bastide*, coll. « IDERIC », 1976, p .313.

TIDJANI ALOU M., « La chefferie et ses transformations au Niger. De la chefferie coloniale à la chefferie postcoloniale », LASDEL, Études et Travaux n° 76, 2009, p. 40.

TOGOLO O., « La régionalisation : le défi de la souveraineté économique du pays », in *Regards sur le droit public en Afrique*, Mélanges en l'honneur du Doyen Joseph-Marie Bipoun Woum, L'Harmattan, 2016, p. 79.

TOUZEIL-DIVINA M., « Voyage départemental dans la doctrine publiciste et française », Pouvoirs locaux, n°116, 2019, p. 40.

TRAORE B., « De la genèse de la nation et de l'État en Afrique noire », Présence Africaine, n° 127-128, 1983, p. 157.

TROPER M.,

- « Libre administration et théorie générale du droit », in La libre administration des collectivités territoriales. Réflexion sur la décentralisation, Economica-PUAM, 1984, p. 59.
- « Pour une définition stipulative du droit », Droits, n° 10, 1989, p. 102.

VEDEL G., « Les bases constitutionnelles du droit administratif », in *Pages de doctrine*, t. II, LGDJ, 1980 p. 129.

VENARD J.-L., « Bailleurs de fonds et développement local », in *Pouvoirs et cités d'Afrique noire : décentralisations en question*, Karthala, 1993, p. 20.

VERGNE C., « Décentralisation fiscale en Afrique francophone : note sur les transferts intergouvernementaux », Banque Mondiale, Région Afrique, AFTPR, Septembre 2009, p. 2.

VERPEAUX M.,

- « Réformer sans réviser », AJDA, 2021, p. 1375.
- « Défiance jacobine et réalités constitutionnelles », AJDA, 2021, p. 1750.
- « Droit constitutionnel des collectivités territoriales ou droit constitutionnel local ? », in *Pour un droit constitutionnel des collectivités territoriales*, L'Harmattan, 2020, p. 20.
- « Pavane pour une notion défunte. La clause générale de compétence », RFDA, 2014, p. 457.
- « Libre administration des collectivités territoriales et QPC », in *Le réflexe constitutionnel : question sur la QPC*, Bruylant, 2013, p. 73.
- « Question prioritaire de constitutionnalité et libre administration des collectivités territoriales », AJDA, 2010, n° 28, p. 1595.
- « La Charte de l'autonomie locale et la Constitution », AJDA, 2006, p. 1865.
- « La loi du 27 février 2002 relative à la démocratie de proximité ou la proximité, une solution pour la démocratie locale ? », RFDA, 2003, p. 261.
- « Les droits des élus ou le droit des assemblées locales », RFDA, 1993, p. 20.

VIGNON Y. B., « Les aspects structurels de la mal administration au Togo », Annales de l'Université du Bénin, Série Droit, Économie, Tome XVII, 1997-1998, p. 75.

WILLIAMS-RIQUIER P., « La charte européenne de l'autonomie locale : un instrument juridique international pour la décentralisation », RFAP, n° 121-122, 2007, p. 191.

WISEMAN A., « Démocratisation, réforme économique et conditionnalités en Afrique subsaharienne : contradictions et convergences », in *Développer par la démocratie ? Injonctions occidentales et exigences planétaires*, Karthala, 1995, p. 467.

YAZI-ROMAN M., « De l'information à la co-construction : la démocratie participative dans tous ses états », AJCT, 2018, p. 240.

ZAKI M., « Les entraves à l'autonomie financière des collectivités locales au Sénégal », Revue numérique Afrilex, 2014.

ZERARI F., « L'évaluation de l'autonomie des collectivités territoriales dans les systèmes juridiques d'inspiration française », Les Annales de droit, n°12, 2018, p. 211.

VI- RAPPORTS

BANQUE MONDIALE,

- L'Afrique subsaharienne. De la crise à une croissance durable. Étude de prospective à long terme, Washington D.C., 1989.
- Rapport sur le développement dans le monde, « L'État dans un monde en mutation », Washington 1997.
- Rapport sur le développement dans le monde, « Le développement dans le monde au seuil du XXIè siècle », Eds Banque mondiale, Eska, 2000.
- Rapport sur le développement dans le monde, « La pauvreté », Washington 1990.

CGLU AFRIQUE,

- L'environnement institutionnel des collectivités locales en Afrique, 2015.
- 1er Rapport sur l'état des ressources humaines dans les collectivités territoriales en Afrique, 2018.
- La décentralisation et la démocratie locale dans le monde, 2008.

COHEN D., GUILLAUMONT JEANNENEY S., JACQUET P., Rapport public, « La France et l'aide publique au développement », La Documentation française, 2006.

CONSEIL DES COLLECTIVITÉS TERRITORIALES, Rapport régional sur la décentralisation et la démocratie locale dans les pays de l'espace UEMOA, 2011.

CONSEIL D'ÉTAT,

- Rapport public, « Le cadre juridique de l'action extérieure des collectivités locales », La Documentation française, 2006.
- Rapport public, « L'influence internationale du droit français », La documentation française, 2001.
- Rapport public « Décentralisation et ordre juridique, La documentation française, 1993.

COURSIN F., Rapport public, « La contribution de la France au progrès des pays en développement », La documentation française, 2001.

GIZ, « Le bureau citoyen. Guide méthodologique : Retour d'expériences de la mise en place et du fonctionnement dans les communes pilotes », 2020.

IFDD, « La participation publique dans l'évaluation environnementale en Afrique francophone », 2013.

MINISTÈRE DES AFFAIRES ÉTRANGÈRES,

- Diplomatie et territoire, Pour une action extérieure démultipliée, La documentation française, 2017.

- Rapport public, « L'appui de la France aux processus de décentralisation et de gouvernance locale démocratique », 2012.

LAIGNEL A., Rapport public, « l'action extérieure des collectivités territoriales. Nouvelles approches… Nouvelles ambitions », La documentation française, 2013.

OCDE,

- Les nouvelles formes de gouvernance et le développement économique, Éditions OCDE 2005.
- Dynamiques de l'urbanisation africaine 2020 : Africapolis, une nouvelle géographie urbaine, Cahiers de l'Afrique de l'Ouest, Editions OCDE, 2020.

PERROT H., Rapport public, « De nouvelles perspectives pour la coopération décentralisée et l'action extérieure des collectivités locales », La documentation française, 1996.

POUGNAUD P., *Guide juridique de l'action extérieure des collectivités territoriales*, La Documentation française, 2019.

INDEX ALPHABETIQUE

(Les chiffres renvoient aux numéros de paragraphe)

D.

E.

F.

TABLE DES MATIÈRES

Structures éditoriales du groupe L'Harmattan

L'Harmattan Italie
Via degli Artisti, 15
10124 Torino
harmattan.italia@gmail.com

L'Harmattan Hongrie
Kossuth l. u. 14-16.
1053 Budapest
harmattan@harmattan.hu

L'Harmattan Sénégal
10 VDN SICAP AMITIE 3
BP 45034 Dakar-Fann
direction@senharmattan.com

L'Harmattan Cameroun
Tsinga/Derrière polyclinique FECAFOOT
BP 11486 Yaoundé
Jacques.koukam@harmattan.fr

L'Harmattan Burkina Faso
Villa 281, secteur 23 623, rue 14.32
Ouagadougou
zagbie@outlook.fr

L'Harmattan Guinée
Almamya, rue KA 028 OKB Agency
BP 3470 Conakry
harmattanguinee@yahoo.fr

L'Harmattan RDC
185, avenue Nyangwe
Commune de Lingwala – Kinshasa
matangilamusadila@yahoo.fr

L'Harmattan Congo
219, avenue Nelson Mandela
BP 2874 Brazzaville
deunovmikhael@gmail.com

L'Harmattan Mali
Immeuble Mgr Jean Marie Cissé
Aci Hamadallaye
BP 145 Bamako
syllaka@yahoo.fr

L'Harmattan Togo
Djidjole – Maison Amela
face EPP BATOME
02 BP 20680 Lomé
didieramela11@gmail.com

L'Harmattan Côte d'Ivoire
Résidence Karl – Cité des Arts
Abidjan-Cocody
03 BP 1588 Abidjan
harmattan.ivoire@gmail.com

L'Harmattan Bénin
Rue 604, (Immeuble à l'entrée
de la rue UCAO- Cite de Saint-Jean)
Cotonou
stepcorpus@gmail.com

Nos librairies en France

Librairie internationale
16, rue des Écoles
75005 Paris
librairie.internationale@harmattan.fr
01 40 46 79 11
www.librairieharmattan.com

Librairie des savoirs
21, rue des Écoles
75005 Paris
librairie.sh@harmattan.fr
01 46 34 13 71
www.librairieharmattansh.com

www.ingramcontent.com/pod-product-compliance
Lightning Source LLC
LaVergne TN
LVHW011222061026
849535LV00001BA/141
* 9 7 8 2 3 3 6 5 8 0 1 3 5 *